SHOCK COMPRESSION OF CONDENSED MATTER—1997

SHOCK COMPRESSION OF CONDENSED MATTER—1997

Proceedings of the Conference of the American Physical Society
Topical Group on Shock Compression of Condensed Matter
held at Amherst, Massachusetts, July 27–August 1, 1997

Edited by:

S. C. SCHMIDT
Los Alamos National Laboratory
Los Alamos, New Mexico, USA

D. P. DANDEKAR
Army Research Laboratory
Aberdeen Proving Ground, Maryland, USA

J. W. FORBES
Lawrence Livermore National Laboratory
Livermore, California, USA

 **CD-ROM INCLUDED**

American Institute of Physics
AIP Conference Proceedings 429   Woodbury, New York

EDITORS:

Stephen C. Schmidt
Los Alamos National Laboratory
MS C323
Los Alamos, NM 87545
Email: scs@lanl.gov

Dattatraya D. Dandekar
Army Research Laboratory
AMSRL-MA-F
Aberdeen Proving Ground, MD 21005-5069
Email: ddandek@arl.mil

Jerry W. Forbes
Lawrence Livermore National Laboratory
L-282; Box 808
Livermore, CA 94550
Email: forbes1@llnl.gov

L.C. Catalog Card No. 98-70856
ISBN 1-56396-738-3-Set
ISBN 1-56396-833-9-Cloth
ISBN 1-56396-834-7-CD-ROM
DOE CONF- 970707

Printed in the United States of America

CONTENTS

CHAPTER III: PHASE TRANSITIONS

CHAPTER IV: MODELING AND SIMULATION: Nonreactive Materials

CHAPTER V: MODELING AND SIMULATION: Reactive Materials

xii

CHAPTER XIV: HIGH VELOCITY LAUNCHERS AND SHAPED CHARGES

CHAPTER XV: LASER AND PARTICLE BEAM-MATTER INTERACTION

Preface

The Tenth American Physical Society Topical Conference on Shock Compression of Condensed Matter was held in Amherst, Massachusetts, July 27 - August 1, 1997. Previous conferences were held in 1979 at Washington State University, Pullman, Washington; in 1981 at SRI International, Menlo Park, California; in 1983 at Santa Fe, New Mexico; in 1985 at Spokane, Washington; in 1987 at Monterey, California; in 1989 at Albuquerque, New Mexico; in 1991 at Williamsburg, Virginia; jointly in 1993 with the International Association for Research and Advancement of High Pressure Science and Technology at Colorado Springs, Colorado; and in 1995 at Seattle, Washington. The Eleventh Conference is scheduled for June 27 - July 2, 1999 at Snow Bird, Utah.

The purpose of this conference, as with past conferences was to provide a forum where scientists and engineers studying the response of condensed matter to dynamic high pressures and temperatures could exchange ideas and technical information. During the conference, papers describing the mechanical, chemical, electro-magnetic, and optical responses of condensed phase materials to shock stimuli were presented. Theoretical, computational, and experimental results were discussed. The abstracts of these papers were published in the July 1997 issue of the Bulletin of the American Physical Society.

Three hundred and fifty scientists and engineers from seventeen countries registered at the conference. The countries represented included the United States (240), Russia (26), France (18), the United Kingdom (16), Japan (15), Israel (8), Canada (5), Germany (4), Portugal (4), South Korea (4), India (2), the People's Republic of China (2), Sweden (2), and Belgium, Egypt, Mexico, and The Netherlands (1 each). There were more than 260 technical presentations with 236 of the presentations appearing as written contributions in this volume.

Arthur C. Mitchell and William J. Nellis received the biennial award presented by the American Physical Society Topical Group on Shock Compression of Condensed Matter for outstanding contributions to the field of shock wave physics. William J. Nellis as a recipient of this award gave a plenary talk entitled, "Molecular and Planetary Fluids at High Shock Pressures."

Dennis B. Hayes was the Master of Ceremonies for the conference banquet, which was attended by approximately two hundred and sixty conference participants and guests. Alice Warder-Seely, presented a keynote address, "Decorating Your Home with C-2 Plastic Explosives."

The Organizing Committee for the conference comprised the following people:

Conference Chairman - Dattatraya P. Dandekar, Army Research Laboratory
Technical Program - Jerry W. Forbes, Lawrence Livermore National Laboratory
Poster Sessions Chairman - Robert S. Hixson, Los Alamos National Laboratory
Publication Chairman - Stephen C. Schmidt, Los Alamos National Laboratory
Treasurer - J. Michael Boteler, Army Research Laboratory
Coordinator - Katherine L. Anderson, Army Research Laboratory
Coordinator - Alita M. Roach, Los Alamos National Laboratory
Coordinator - William E. Deal, Los Alamos National Laboratory
Coordinator - Peter T. Bartkowski, Army Research Laboratory
Editor for Publication - Shelly L. Cross, Los Alamos National Laboratory

Technical Program Committee -

Jerry W. Forbes (coordinator), Lawrence Livermore National Laboratory
James R. Asay, Sandia National Laboratories
Lalit C. Chhabildas, Sandia National Laboratories
Dattatraya P. Dandekar, Army Research Laboratory
Yogendra M. Gupta, Washington State University
Robert S. Hixson, Los Alamos National Laboratory
Yasuyuki Horie, North Carolina State University
James N. Johnson, Los Alamos National Laboratory
William J. Nellis, Lawrence Livermore National Laboratory
Stephen A Sheffield, Los Alamos National Laboratory
Gerrit T. Sutherland, Naval Surface Warfare Center
Davis L. Tonks, Los Alamos National Laboratory
Carter T. White, Naval Research Laboratory
Diana L. Woody, Naval Air Warfare Center
Choong-Shik Yoo, Lawrence Livermore National Laboratory

American Physical Society Topical Group Officers and Executive Committee Members:

S. J. Bless
L. C. Chhabildas
D. P. Dandekar
J. N. Johnson (chairman)
A. M. Renlund

R. C. Cauble
D. R. Curran (chairman-elect)
N. C. Holmes (vice-chairman)
E. R. Lemar (secretary/treasurer)

The Organizing Committee would like to express its appreciation to the many people who contributed to the success of the conference: the session chairpersons, the authors, the plenary and the invited speakers, and the participants. The editors are especially grateful to the session chairpersons and other persons who reviewed the manuscripts. They are R. R. Alcon, C. E. Anderson, Jr., B. W. Asay, L. M. Barker, P. T. Bartkowski, J. F. Belak, P. M. Bellamy, L. S. Bennett, J. M. Boteler, R. C. Cauble, R. Cheret, L. C. Chhabildas, J. J. Davis, R. D. Dick, R. M. Doherty, R. Feng, L. E. Fried, J. N. Fritz, G. T. Gray III, Y. M. Gupta, R. H. Guirguis, R. L. Gustavsen, J. E. Hammerberg, D. B. Hayes, W. F. Hemsing, L. G. Hill, R. S. Hixson, N. C. Holmes, Y. Horie, J. M. Howard, J. N. Johnson, V. S. Joshi, R. J. Lee, E. R. Lemar, M. E. Kipp, J. L. Maienschein, P. J. Miller, V. F. Nesterenko, Y. Partom, A. M. Rajendran, H. W. Sandusky, T. Sekine. M. S. Shaw, R. L. Simpson, G. T. Sutherland, C. M. Tarver, D. G. Tasker, N. N. Thadhani, W. M. Trott, P. A. Urtiew, W. H. Wilson, C.-S. Yoo, D. A. Young, and F. J. Zerilli.

Special thanks are due to the many people who willingly gave their time and talents to organize and conduct this conference. These include Katherine L. Anderson, Alita M. Roach, William E. Deal, Paul A. Urtiew, Sharon L. Crowder, Leta K. Picklesimer, Judith J. Scully, and Margaret M. Walton. We appreciate the interesting and popular companion program coordinated and hosted by Lynn Furnish. The conference committee thanks Mickey Daniels and Paul M. Bellamy, for their superb assistance in coordinating the collection of the manuscripts. The editors truly the appreciate the help and expertise of Shelly L. Cross for the editing and correction of the manuscripts. The conference logo was provided through the courtesy of the William E. Deal. We especially appreciate the skillful assistance provided by Mary K. Wong of the University of Massachusetts Lincoln Campus Center, Amherst, Massachusetts.

S. C. SCHMIDT
D. P. DANDEKAR
J. W. FORBES

Foreword

The desire to experimentally observe the insulator-to-metal transition in hydrogen has intrigued scientists both in the United States and in other countries during the past several decades. This year's "American Physical Society Shock Compression Science Award," was presented to Arthur C. Mitchell and William J. Nellis for realizing this goal. This award, sponsored by the American Physical Society Topical Group on Shock Compression of Condensed Matter recognizes the achievements of these two notable scientists, both for this accomplishment and for their many other contributions to shock compression science. Several of these achievements are discussed in the plenary lecture, "Molecular and Planetary Fluids at High Shock Pressures," presented by William J. Nellis.

As with previous conferences, the spirit of this year's meeting was to associate shock wave phenomenology and data with fundamental understanding. This volume embodies the most recent thoughts of the shock wave community toward this goal. Included are sections on equations of state, phase transitions, material properties and synthesis, optical, electrical, and laser studies, hypervelocity phenomenology, and impact and penetration mechanics. Considerable attention continues to be focused on the strain and failure behavior, the weak impulse initiation, and the safety aspects of explosives. Developments in measurement techniques, particularly those employing fast optical methods, also continue to show progress. These latest advances, in addition to the many other results and topics discussed by an international contingent of authors, serve to make this volume the latest authoritative reference source for the high pressure community.

As with previous volumes, the diversity of the subject matter treated in this volume makes it difficult to assimilate. However, an extensive index provides ready entry into various technical subjects. All manuscripts submitted to the editors, of presentations made at the conference have been published in this volume. The editors apologize for any errors that may have occurred, technical and editorial, as part of the publication process.

We sadly note the death of Daniel J. Steinberg, a pre-eminent contributor to equation of state and shock wave science.

Arthur C. Mitchell and **William J. Nellis**

Recipients of the Shock Compression Science Award, 1997

"In recognition of their pioneering experimental investigations of molecular and planetary
fluids using shock compression."

CHAPTER I

PLENARY

CP429, *Shock Compression of Condensed Matter – 1997*
edited by Schmidt/Dandekar/Forbes
© 1998 The American Institute of Physics 1-56396-738-3/98/$15.00

HIGH ENERGY-DENSITY SCIENCE ON THE NATIONAL IGNITION FACILITY

E. M. Campbell, R. Cauble, and B. A. Remington

*Lawrence Livermore National Laboratory, University of California,
P. O. Box 808, Livermore, CA 94551*

The National Ignition Facility, as well as its French counterpart, *Le Laser Megajoule*, have been designed to confront one of the most difficult and compelling problem in shock physics – the creation of a hot, compressed DT plasma surrounded and confined by cold, nearly degenerate DT fuel. At the same time, these laser facilities will present the shock physics community with unique tools for the study of high energy density matter at states unreachable by any other laboratory technique. Here we describe how these lasers can contribute to investigations of high energy density matter in the areas of material properties and equations of state, extend present laboratory shock techniques such as high-speed jets to new regimes, and allow study of extreme conditions found in astrophysical phenomena.

I. INTRODUCTION

The goal of inertial confinement fusion (ICF) is to heat a small capsule of DT fuel, compressing it with multiple staged shocks to densities and temperatures capable of sustaining thermonuclear ignition for the purposes of supporting major national activities: maintaining relevant scientific expertise in the absence of nuclear testing and the development of electrical energy production from fusion.[1] The heating will be done with multi-beam, megajoule-class lasers built with advanced but available technology. Lasers will be the driving source for ICF ignition because they can deliver the necessary high power on target in a precisely staged manner, thereby controlling the adiabat of the fuel. Two such lasers, the National Ignition Facility (NIF) in the US[2] and *Le Laser Megajoule* (3) in France[LMJ], will come on-line during the first decade of the 21st century.

A typical ignition target will be a thin spherical low-atomic-number (low-Z) shell surrounding a solid DT shell; DT gas will fill the interior. The conditions necessary for ignition of the target require that the fuel be compressed to very high density (10^3-10^4 times liquid) and that a central spark region with a temperature of ~ 10 keV be formed.[4] In order to efficiently compress the fuel, multiple carefully-timed strong shocks (1-100 Mbar) are used so that at ignition the fuel is at enormous pressures but not at high temperature; in fact the fuel will be degenerate; *i.e.*, the fuel temperature will less than the Fermi temperature Thus in order to prepare a capsule for ignition, one needs a cold, dense fuel shell enclosing a 10 keV hot spot at pressures near 100 Gbar (Fig. 1). This is a most interesting, and possibly the most difficult, extant shock physics problem. The lasers that will used to create these conditions will be highly flexible, unique facilities for producing and diagnosing matter in a range of extreme conditions in the laboratory, high energy-density regimes that have never before been experimentally explored.

These lasers will use either direct or indirect drive to compress the ignition capsule. These two methods are now, and will be, used in high energy-density shock experiments. Direct drive involves irradiating a target directly with one or more laser beams. At an intensity I the pressure produced in a low-Z material is given by $P(\text{Mbar}) \approx$ few times $[I(10^{14} \text{ W/cm}^2)/\lambda(\mu\text{m})]^{2/3}$.[5]

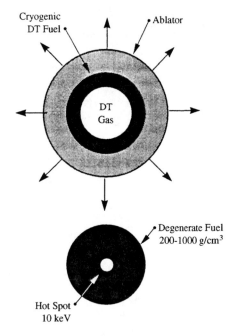

FIGURE 1. Schematic compression of an ICF ignition target showing heating of the ablator and corresponding rocket-like implosion of fuel. The resulting ignition configuration is also depicted. Only precisely staged strong shocks from a carefully shaped drive can lead to ignition.

Therefore shocks of 10's of Mbar are possible even with moderate energy lasers, although the necessary intensity is usually achieved over a small spot and for only a very short time duration. Lasers with high *energy* are needed for most relevant high energy-density experiments. The indirect drive technique uses a high-Z hohlraum into which the laser beams are focused. The resulting x-rays, contained in the hohlraum and characterized by a radiation temperature T_r, are used as the driving source. The pressure generated in the hohlraum is $\approx 10^4 \, T_r(\text{keV})^{3.5}$ Mbar.[6] With indirect drive, more beams can be used and shock pressures of ~100 Mbar are common on present facilities. NIF hohlraum pressures will be higher, near 500 Mbar, and on larger spatial scales. Already indirectly-driven foils employed in the same manner as flyer plate shock experiments have unambiguously produced shock pressures of 750 Mbar.[7]

Beyond the field of ICF, kilojoule-class lasers like Nova[8] at Lawrence Livermore National Laboratory have been used to investigate fundamental aspects of high energy-density science. The following sections discuss a few areas of high energy-density physics that are being explored using Nova. These range from Mbar-pressure equation of state measurements to studies of hydrodynamic processes in supernovas . In each case the advantages of performing similar experiments on NIF or LMJ are described. An important point to emphasize is that not all high energy-density experiments on NIF will be designed and performed by staff members of the national laboratories. A fraction of the shots will be set aside for use by "outside" users, *i.e.*, university researchers. At the time of this writing a similar plan is being proposed for the LMJ.

II. MBAR PRESSURE EQUATIONS OF STATE

The experimental and theoretical investigation of the equations of state (EOS) of materials at high energy density are of interest not only in ICF, but also in astrophysics and other related fields. Strongly shocked matter can be a strongly coupled mixture of molecular, atomic, and ionic species that is extremely difficult to model. Nevertheless EOS tables that include this regime must be constructed not only to design and interpret ICF target data, but also to study the evolution of large planets and stars. This puts a real premium on EOS data in the multi-Mbar regime. Most such data have been generated by nuclear explosions.[9,10]

Lasers are capable of producing Mbar shocks but until recently laser-induced shock waves have proved difficult to use for Mbar EOS measurements.[11] Several challenges must be overcome when using lasers in such experiments. In order to reduce uncertainties in the data it is necessary to have a large, spatially uniform, steady-state shock front driven for as long as possible. These are *energy* considerations, not just intensity considerations. For example, the energy required to maintain a given uncertainty in shock speed at pressure P varies as $P^{5/2}$.[12] Preventing preheat of the sample is a stringent requirement since heating of the sample prior to it being shocked changes the initial sample density in an unknown way. Since laser deposition can produce very high temperatures (~ keV), preheat is a major concern for laser-driven experiments. Finally diagnostics are needed that have few-μm spatial and few-ps temporal resolutions. These are formidable challenges to laser-driven EOS measurements but are not insurmountable.[13]

4

Recent EOS experiments have been performed on Nova to investigate materials that are of intense interest to ICF at relevant Mbar pressures.[14] Hugoniot measurements of polystyrene and beryllium, two candidate materials for the outer shell of NIF ignition capsules, were made using indirect drive. Deuterium, the major constituent of the fuel in fusion capsules was examined using direct drive. All measurements were absolute measurements, that is, they were not dependent on a "standard" material as in the case of the impedance matching technique.

Directly-Driven Absolute Hugoniot of D2 up to 2 Mbar

The first shock in the proposed design for a NIF ignition target will be 900 kbar,[4] which lies in a regime where a transition from a diatomic to a monatomic fluid is expected. This molecular dissociation could significantly affect the EOS of D_2. However, the highest pressure datum on the principal Hugoniot of deuterium was a gas gun result at 200 kbar.[15]

A Nova experiment was designed that utilized a spatially-smoothed laser beam with a 9-ns pulse to drive a strong shock into an aluminum pusher. After transiting the Al the shock propagated into a cryogenic D_2 sample. Employing side-on radiography, the Al/D_2 interface and the shock front were imaged in time on a calibrated streak camera. The motion of the interface provided the particle speed U_p, which is continuous across the interface, and the motion of the shock front revealed the shock speed U_s in the D_2. See Fig. 2a. Shock compression and pressure were determined from the Hugoniot relations.[16] Using Nova, Hugoniot data were thus obtained up to 2 Mbar that showed a significantly more compressible EOS (up to 50%) than had been previously believed.[17] The data, shown in Fig. 2b, include one point at 250 kbar that agrees well with the earlier gas gun data. The Nova data closely follow a model that incorporates molecular dissociation in a new way[18]. However a number of other, often more inclusive theories, including Monte Carlo and molecular dynamics simulations, do not agree with the data (or with each other).

To have confidence in the data, a considerable effort was made to minimize preheat. The use of a

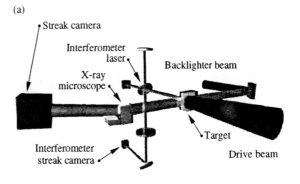

(a)

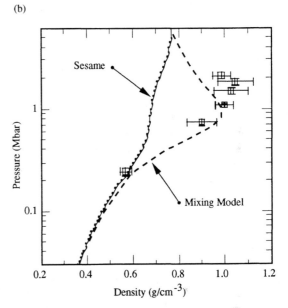

(b)

FIGURE 2. a) Experimental setup for the Nova measurement of the Hugoniot of deuterium by direct laser drive, including interferometry to verify preheat. b) Absolute D_2 Hugoniot data[17] along with gas gun data[15] compared with Sesame[21] and linear mixing model[18] predictions.

low-Z ablator covering the Al pusher reduced high energy x-ray emission in the laser deposition region. One result of x-ray preheat would be expansion of the rear of the Al pusher prior to shock arrival. This was monitored by reflecting a probe laser off the rear surface in a Michelson interferometry arrangement.[19] Results showed no detectable movement before the shock arrived and thus no preheat.

Indirectly-Driven Absolute Hugoniots of Polystyrene and Be

EOS measurements have also be made using indirect drive where higher pressures have been obtained. Because x-rays in the hohlraum undergo multiple reflections, the shock generated in a package – a kind of shock tube – attached to the hohlraum is very uniform. There is the additional advantage of having a much lower temperature x-ray source, reducing the potential for preheating of a sample. The primary disadvantage of using a hohlraum is that the hohlraum fills with plasma and the laser entrance holes close within a few ns; thus shock measurements must be made within this time duration or corrections made for slowing of the shock in the sample.

The Hugoniots of polystyrene (CH) and beryllium were examined using indirect drive as the source for the shock. Eight beams of the Nova laser were focused into a 3-mm-long by 1.6-mm-diameter gold hohlraum; the EOS package was attached to the side of the hohlraum. [20] See Fig. 3a. Additional beams irradiated a metal foil; x-rays from this backlighter were used to radiograph the shock moving through the package. The package consisted of two sections, a pusher of bromine-doped CH, CH(Br), close to the hohlraum and the EOS sample attached to the CH(Br). CH(Br) is opaque to the backlighter x-rays whereas the samples are not. In the same manner as the D_2 experiment, U_p in the sample was obtained by following the interface motion. The shock front was revealed as the boundary between the highly transmissive unshocked sample and the less transmissive (higher density) shocked sample. The shock speed U_s was obtained by following this boundary. Since both U_s and U_p were measured, the Hugoniot data were absolute data. Data on CH was obtained from 10 Mbar to 40 Mbar[14], showing that the EOS for CH from the well-known Sesame EOS library[21] reproduced the data rather well.

Absolute beryllium Hugoniot data were obtained by the same method at 12 and 14 Mbar. Fig. 3b shows these data compared with impedance match data from nuclear-explosion-driven experiments[9,10] and Sesame[21]. The absolute laser data lend confidence to the relative underground results.

These results were produced using a kilojoule-class laser. A megajoule-class laser will allow directly driven

(a)

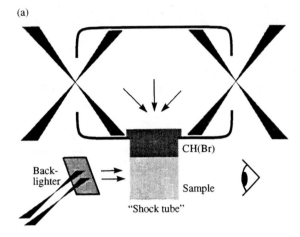

(b)

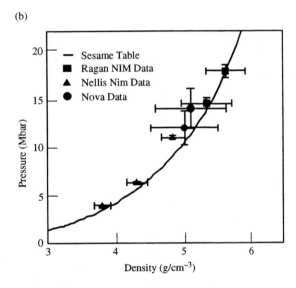

FIGURE 3. a) Experimental setup for the Nova measurement of the Hugoniot of low-Z solids by indirect laser drive employing a gold hohlraum. The experimental package was a two-section shock tube mounted to the side of the hohlraum. The measurements were made with streak radiography using x-rays from a laser-heated backlighter. b) Absolute beryllium Hugoniot data[20] compared with Sesame[21] and nuclear impedance match data[9,10].

shock experiments up to 50 Mbar over longer drive times (up to 20 ns) and larger spot sizes. It will be possible to perform absolute EOS measurements with improved accuracy. Scaling of optimized hohlraums from Nova to NIF are expected to allow 1-2% precision EOS measurements at pressures greater than 1 Gbar in an impedance match arrangement.[22] Further development of laser-driven flyer foils should allow few-percent relative EOS measurements near 10 Gbar.

III. HYDRODYNAMIC MIX AT HIGH ENERGY DENSITY

Matter motion at high energy density can be susceptible to hydrodynamic instabilities. In an extreme example, Nova has been used to create high-Mach-number jets (M=15 to 30) by using a laser-driven "shaped charge."[23] Instabilities are especially problematic in ICF: mixing of cold fuel into the central hot region can reduce the capsule burn performance or entirely quench the burn.[4] The three most common hydrodynamic instabilities are the acceleration-driven Rayleigh-Taylor (RT) instability, its shock analog the Richtmyer-Meshkov (RM) instability, and the shear-induced Kelvin-Helmholtz (KH) instability. High energy-density hydrodynamic instabilities are not only a factor in ICF, but they are also observable in astrophysical objects (see Section IV). The RT instability can occur at an interface between a low density fluid and a high density fluid where the lighter fluid is accelerated into the heavier. In the linear regime, initial perturbations in the interface will grow exponentially in time with a growth rate $\gamma \sim (Ag/\lambda)^{1/2}$, with A being the density-dependent Atwood number, g the acceleration, and λ the perturbation wave length. In the nonlinear asymptotic limit, the interface evolves into bubbles of the lighter fluid rising at their terminal velocity and spikes of the heavier fluid falling through the lighter fluid.

One long-used method to investigate shock-induced mixing is the use of shock tubes.[24] Here, the acceleration is impulsive, the pressures are typically a few bar, the compression is low, and there is no radiation or ionization involved. High explosives generate pressures up to 200-300 kbar[25] and gas guns can generate pressures up to a few Mbar, but with modest compression.[26] Like extraordinary shock tubes, large lasers like Nova and NIF can produce extreme accelerations (10^{13}–10^{14} earth gravities) and pressures of 100's of Mbar. Such lasers can achieve high growth factors, large compressions, and high levels of radiation flow and ionization in arbitrary geometry.

There have been a number of experiments to investigate high energy-density hydrodynamic instabilities using a laser drive.[6,27-29] Because of their flexibility, as evidenced in preparing precisely prescribed ICF pulse shapes, lasers can provide the variety of conditions necessary to probe the evolution of RM and RT instabilities. One excellent example is the recent experimental verification by Marinak et $al.$[30] of the differences between two-dimensional (2D) and three-dimensional (3D) RT-induced perturbation growth. A precisely formed perturbation was pre-imposed on one side of a plastic foil doped with bromine CH(Br). This foil was placed across an opening on a Nova laser hohlraum which was used to create the x-ray drive that ablatively accelerated the foil. The experimental arrangement is similar to that in Fig. 3a, except that the backlighter is positioned above the hohlraum and the detector below so that radiographic images obtained are face-on rather than side-on. Perturbations at the ablation front grow during foil acceleration due to the RT instability. The three perturbations studied all had the same magnitude wavevector, $k=(k_x^2+k_y^2)^{1/2}$, where $k=2\pi/\lambda$, and the same amplitude, differing only in their shape: 3D square $k_x=k_y$, 3D stretched $k_x=3k_y$, and 2D ripple $k=k_x=k_{2D}$. The images shown in Fig. 4a are time-resolved, face-on, in-flight radiographs taken of the accelerated planar foils. Dark regions correspond to spikes, bright regions to bubbles. The growth versus time of the fundamental mode Fourier amplitudes of the perturbations are shown in Fig. 4b. In the linear regime, all three modes grow at the same rate, as expected from linear theory, since they all have the same magnitude wavevector. But in the nonlinear regime, the 3D square mode grows the largest, the 2D ripple grows the least, and the 3D stretched perturbation falls in between. This result can be understood by noting that in the asymptotic limit of terminal bubble velocity, the buoyancy is exactly balanced by the kinematic drag.[31] At the bubble tip, the ratio of drag over buoyancy is smallest for the square mode; this shape has the highest terminal bubble velocity and grows the fastest. The energy of Nova is sufficient to just begin to see these effects. The energy available on NIF will allow 3D experiments to be pushed well into the nonlinear regime. Furthermore we will also be able to perform experiments on NIF with fully multimode 3D foils, not possible on Nova.

(a)

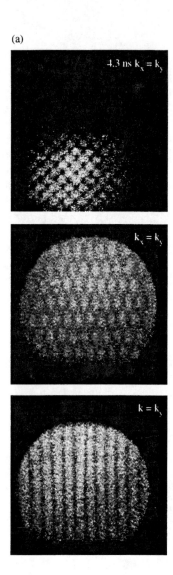

(b)

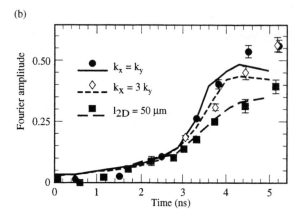

FIGURE 4. a) Face-on radiographs of three different planar Rayleigh-Taylor targets driven indirectly by Nova lasers. The experiment used a hohraum as in Fig. 3a, except that the backlighter was positioned above the hohlraum and the recording device, an x-ray framing camera, was positioned below. The three shapes are 3D square $k_x=k_y$, 3D stretched $k_x=3k_y$, and 2D ripple $k=k_x=k_{2D}$. b) Plot of pertubation Fourier amplitude vs. time for the three shapes showing that the 3D square mode grows fastest

Three-dimensional effects in RT development can also be seen in convergent geometry, the situation encountered in ICF. This is evident in preliminary Nova experiments in which a 3D square ($k_x=k_y$) mode perturbation was inscribed on the outer surface of a doped-plastic hemisphere and, similar to the planar RT experiments described above, the hemisphere was mounted on the wall of a hohlraum, facing inwards.[32] Ablation of the outer surface by the x-ray drive inside the hohlraum causes the hemisphere to radially accelerate. As in the planar case, the perturbations grow as the plastic target is accelerated. But since the hemisphere converges as it is driven, the perturbation wavelength λ decreases as the shell accelerates. Recalling that the RT growth rate $\sim \lambda^{-1/2}$, RT growth was expected to be larger in the linear regime in the convergent case and to saturate sooner. This is indeed what was observed. Although the convergence ratio R_0/R, where R_0 is the initial hemisphere radius and R the final radius, was only about two, RT growth entered the nonlinear regime. Had the foil been planar, the growth would still be in the linear regime at the times observed. Megajoule-class lasers will be required to drive these 3D experiments well into the nonlinear regime especially, as in the case of planar RT, with multimode targets. The ability to perform fully 3D experiments to study high energy-density hydrodynamics finds application not only in ICF but also in astrophysics, as shown in the following section.

Hydrodynamics at high pressure can also be studied in the solid state. Careful tailoring of the drive pulse in a hohlraum can result in Mbar-pressure shock waves in a metallic targets without melting the target. Recent experiments have shown that RT growth in solid Cu is reduced because of strength inherent in the solid.[33] Those experiments also showed that the RT growth reduction could not be explained by extrapolating to Mbar pressures material strength models known to be correct only in the kbar regime. Thus, without these data, Mbar-pressure material strength in Cu would have been underestimated by a factor of 30. As a complement to these integral experiments, high pressure shock processes in solids can be investigated *in situ* by time-resolved x-ray diffraction.[34] As a condensed matter sample deforms under shock loading, the lattice spacing changes and thus the Bragg diffraction angle. Monitoring the diffraction angle as a shock propagates through the sample allows the structure of the shock front to be examined. *In situ* diffraction of a solid shock-compressed to 300 kbar has recently been reported.[35] NIF will be able to drive still larger pressures as well as provide additional and higher flux backlighters for more careful measurements.

IV. HIGH ENERGY LASERS AND SUPERNOVAS

Because of enormous gravitational and nuclear forces, many astrophysical objects are indeed high energy-density phenomena. High energy lasers can contribute to the understanding of astrophysical processes. Exploding stars can create high speed hydrodynamical jets similar to those already described.[36] Supernovas exhibit a range of high energy-density hydrodynamical features that can be scaled to experiments on mega-joule-class lasers. Below we describe two experiments developed to investigate supernova phenomena: (1) hydrodynamic-instability-driven core mixing during the SN1987A explosion and (2) strong shock formation in the SN1987A remnant evolution.[37]

SN1987A Mix Experiments

Gravitational stratification leads to a model of a supernova (SN) progenitor star that is a series of concentric shells with the lightest element, hydrogen, forming the outer envelope, helium just below the H layer, and so on to an iron core. A SN explosion occurs when the Fe core undergoes catastrophic gravitational collapse; a strong radial (rebound) shock is driven outward through the star. At the H/He boundary, the post-shock density, temperature, and pressure are about 2.3 g/cm^3, 6 keV, and 75 Gbar. Much effort has been invested in developing models to understand the

underlying processes of SNs. Most efforts have focused on one-dimensional (1D) stellar evolution models, treating multidimensional hydrodynamics effects with prescriptions from mixing length theory. Observations of SN1987A highlighted two distinct defects with these models.[38]

A primary means of evaluating models is the SN light curve, the total luminosity of the SN versus time. The SN luminosity plummets sharply immediately after the explosion as ejected gas hydrodynamically cools due to expansion. Subsequently the luminosity exhibits a broad peak as the heat wave due to radioactive decay of the core diffuses out of the star. In SN1987A, however, the core became visible much earlier than 1D diffusion models predicted.[39] The reason was unexpected mixing of the core with the stellar atmosphere sitting on top of the core. Looking at just the H and He layers, the boundary between the shells is Richtmyer-Meshkov (RM) unstable as the shock crosses the interface. Further, after shock passage the heavier He decelerates into lighter H and the interface is RT-unstable so that initial density perturbations continue to grow. The observations from SN1987A provided compelling evidence of the need for two-dimensional (2D) stellar models.[40] Multi-dimensional SN simulations indeed showed that RT-induced mix led to earlier emergence of core radiation compared to 1D models.[41]

Given the rapid progress in SN modeling over the last ten years, it is reasonable to ask about the accuracy of these multi-dimensional codes. Can they be benchmarked in a meaningful way? For example, 2D modeling predicts peak core velocities of < 2000 km/s for SN1987A; however, Doppler broadening shows core velocities that are much higher, > 3000 km/s.[42] This is an indication that even 2D modeling is insufficient and 3D codes must be used.

Noting, somewhat remarkably, that we can scale supernova hydrodynamics to the millimeter scales of laser-driven experiments, attempts to simulate SN hydrodynamics have been performed on the Nova laser using a hohlraum x-ray drive with Cu and CH_2 as surrogates for He and H respectively.[37] The indirectly driven shock tube arrangement (as in Fig. 3a) was utilized. The flexibility inherent in laser energy delivery makes a scaled representation of the correct shock-plus-deceleration drive possible. In the experiment, side-on radiographs of the interface show how far and how fast bubbles and spikes form due to RM followed by RT instability. These images are compared to code predictions. Employing a single mode 2D surface (see Fig. 3), gross features of the experiment are reproduced by the simulations.[43] However there is insufficient drive energy available with Nova to allow these experiments to be done in a multimode 3D configuration or experiments in (the correct) exploding geometry. However, NIF will have sufficient energy to drive more realistic experiments.

SN1987A Remnant Collision Experiments

SN1987A is now evolving into the early remnant stage; a 1994 optical image is shown in Fig. 5. The expanding SN ejecta is the central bright spot, surrounded by an assembly of nebular rings, the origin of which is a mystery. Gaseous SN ejecta, moving at $\sim 10^3$-10^4 km/s, will collide with the inner circumstellar ring in 5-10 years in a celestial display that may shed light on the nature of the rings.[40] Using this collision to probe the nature of the rings depends

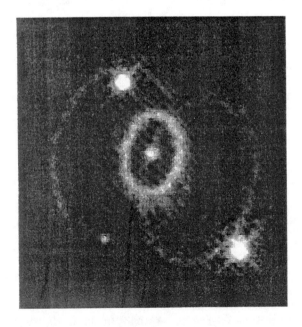

FIGURE 5. An image of SN1987A obtained by the Hubble Telescope in Feb. 1994. The expanding supernova ejecta is the central dot. The rings are planetary nebulae of uncertain origin.

on correctly interpreting observations of a complex, energetic, hydrodynamically unstable collision between two interstellar plasmas. Simulations are being performed now to predict the outcome of the collision.

When the two plasmas collide, shocks will be launched forward into the ring and backward into the SN ejecta. Cooling of the strongly shocked plasma by radiation causes the compressed ejecta to collapse to an even higher density leading to strong RT growth at the contact discontinuity. [44] Qualitatively different mixing evolves depending on the density profile of the ejecta and the initial evolution of the contact discontinuity. It is therefore beneficial to test various models experimentally prior to the collision in order to assess how astrophysics codes handle the complex radiative plasma dynamics.

An experiment has been designed as a proxy to the stellar collision.[37] In the experiment, CH(Br) is used to provide a surrogate for the SN ejecta; SiO_2 aerogel foam represents ambient plasma and a solid CH foil corresponds to the ring plasma. Attached to a Nova hohlraum, x-rays drive a 50 Mbar shock into the CH(Br), ejecting plasma into a gap between the CH(Br) and the foam. The ejecta stagnates in the foam driving shocks forward into the foam and backward into the ejecta. Side-on x-ray backlighting reveals the positions of the shocks and the relative shock densities. Experiments on Nova have successfully produced and imaged both shocks. However, the shocks are not so strong that cooling radiation is a factor in the Nova experiment. Further, longer x-ray drives are needed to evince the expected RT development in the plasma. NIF will provide the drive energy to overcome both of these difficulties.

V. SUMMARY

A primary mission of megajoule-class lasers such as the NIF and LMJ will be to compress, inertially confine, and initiate nuclear fusion in DT capsules. This approach to ICF ignition requires a flexible, high-energy facility capable of producing conditions of matter unattainable by other laboratory methods while simultaneously allowing experimental investigation of shocked, high energy-density matter. We have described several examples of high energy-density experimental investigations that have already been initiated on present-day kilojoule-class lasers and how those studies will be extended on NIF and LMJ.[45] We examined EOS measurements from 1 to 100's of Mbar, production of high-Mach-number jets, and the growth and development of hydrodynamic instabilities in planar and convergent geometries that reveal the limitations of two-dimensional models. Further we described how high energy lasers can be used to investigate shocks in astrophysical phenomena, in particular supernovas. Perhaps most important, experiments at NIF will be performed not only by the staff at US national laboratories, but also in collaboration with and by members of the academic shock physics community, serving to invigorate the field of high energy-density physics.[46]

ACKNOWLEDGMENTS

This work was performed under the auspices of the U.S. Department of Energy by the Lawrence Livermore National Laboratory under contract No. W-7405-ENG-48

References

1. E. M. Campbell, N. C. Holmes, S. B. Libby, R. A. Remington, and E. Teller, to appear in Laser Part. Beams (December, 1997).
2. J. A. Paisner, E.M. Campbell, and W.J. Hogan, Fusion Technology 26, 755 (1994).
3. M. Andre, "Conceptual Design of the French LMJ Laser," First SPIE International Conference on Solid State Lasers for Application to ICF, Monterey, (1995).
4. John Lindl, Phys. Plasmas 2, 3933 (1995).
5. Th. Löwer and R. Sigel, Contrib. Plasma. Phys. 33, 355 (1993).
6. B. A. Remington et al., Phys. Plasmas 2, 241 (1995).
7. R. Cauble et al., Phys. Rev. Lett. 70, 2102 (1993).
8 E.M. Campbell, Laser Part. Beams 9, 209 (1991).
9. C. E. Ragan III, Phys. Rev. A 29, 1391 (1984).
10. W. J. Nellis, J. A. Moriarity, A. C. Mitchell, and N. C. Holmes, J. Appl. Physics 82, 2225 (1997).
11. M. Evans et al., Lasers Part. Beams 14, 113 (1996).
12. N. C. Holmes, private communication.
13. Y. M. Gupta and S. M. Sharma, Science 277, 909 (1997).
14. R. Cauble et al., Phys. Plasmas 4, 1857 (1997).
15. W. J. Nellis et al., J. Chem. Phys. 79, 1480 (1983).
16. Y. B. Zel'dovich and Y. P. Raizer, Physics of Shock Waves and High-Temperature Hydrodynamic Phenomena (Academic Press, New York, 1966).
17. L. B. Da Silva et al., Phys. Rev. Lett. 78, 483 (1997).
18. N. C. Holmes, M. Ross, and W. J. Nellis, Phys. Rev. B 52, 15835 (1995).
19. K. S. Budil et al., "Characterization of Laser-Driven Shock Waves using Interferometry," to appear in Inertial Confinement Fusion Quarterly Report, LLNL report (1997).
20. R. Cauble et al., "Absolute Equation of State Data in the 10-40 Mbar (1-4 Tpa) Regime," submitted to Phys. Rev. Lett. (1997).
21. S. P. Lyon and J. D. Johnson, "SESAME: The Los Alamos National Laboratory Equation of State Database," Los Alamos National Laboratory report LA-UR-92-3407 (1992).
22. N. Landen, private communication.
23. P. L. Miller et al., "Shock-hydrodynamic Experiments on Nova," Proceedings of the 20th International Symposium on Shock Waves," Pasadena (1995).
24. M. Vetter and B. Sturtevant, Shock Waves 4, 247 (1995); R. Bonazza and B. Sturtevant, Phys. Fluids 8, 2496 (1996).
25. R.F. Benjamin and J.N. Fritz, Phys. Fluids 30, 331 (1987).
26. For example, N.C. Holmes et al., Appl. Phys. Lett 45, 626(1984).
27. K. Budil et al., Phys. Rev. Lett. 76, 4536 (1996); G. Dimonte et al., Phys. Rev. Lett. 74, 4855 (1995); T.A. Peyser et al., Phys. Rev. Lett. 75, 2332 (1995).
28. W. Hsing et al., Phys. Rev. Lett. 78, 3876 1997); Phys. Plasmas 4, 1832 (1997).
29. K. Nishihara and H. Sakagama Phys. Rev. Lett. 65, 432 (1990).
30. M. Marinak et al., Phys. Rev. Lett. 75, 3677 (1995).
31. J. Hecht et al., Laser and Part. Beams 13, 423 (1995).
32. S.G. Glendinning et al., in Proceedings of the 6th International Workshop on the Physics of Compressible Turbulent Mixing, Marseille, in press (1997).
33. D.H. Kalantar, in the proceedings of the 6th International Workshop on the Physics of Compressible Turbulent Mixing, Marseille (1997).
34. J. S. Wark et al., J. Appl. Phys. 68 , 4531 (1990).
35. D.H. Kalantar, et al., "Transient X-ray Diffraction Used to Diagnose High Pressure Shocked Solids," Annual Meeting of the American Cyrstallographyic Association, St. Louis (1997).
36. S. Heathcote et al., Astron. J. 112, 1141 (1996).
37. B. A. Remington et al., Phys. Plasmas 4, 1994 (1997).
38. See, e.g., I. Hachisu, T. Matsuda, K. Nomoto, T. Shigeyama, Astron. Astrophys. Suppl. 104, 341 (1994); M. Herant and S.E. Woosley, Astrophys. J. 425, 814 (1994).
39 T. Shigeyama and K. Nomoto, Astrophys. J. 360, 242 (1990).
40. R. McCray, Ann. Rev. Astron. Astrophys. 31, 175 (1993).
41. D. Arnett, Astrophys. J. 427, 932 (1994).
42. B. Fryxell, W. M?ller, and D. Arnett, Astrophys. J. 367, 619 (1991).
43. J. Kane et al., Astrophys. J. 478, L75 (1997).
44. R.A. Chevalier and J.M. Blondin, Astrophys. J. 444, 312 (1995).
45. R. W. Lee, "Science on High Energy Lasers from Today to the NIF," at http://www.llnl.gov/science_on_lasers/ucrl119170.html.
46. See the report "Facility Use Plan for the National Ignition Facility," eds. A. Hauer, R. Kauffman, A. Satsangi, T. Haill, R. Cauble, and T. Saito, available from US Department of Energy, Office of the NIF, 1000 Independence Avenue, S.W. Washington, DC 20585.

CP429, *Shock Compression of Condensed Matter – 1997*
edited by Schmidt/Dandekar/Forbes
© 1998 The American Institute of Physics 1-56396-738-3/98/$15.00

MOLECULAR AND PLANETARY FLUIDS AT HIGH SHOCK PRESSURES

W. J. Nellis and A. C. Mitchell

Lawrence Livermore National Laboratory, Livermore, CA 94550

Shock-compression experiments on liquids using a two-stage gun are described. Results for H_2, He, H_2O, N_2, CO_2, and a mixture of H_2O, NH_3, and C_3H_8O (synthetic Uranus) are discussed and related to explosive reaction products, giant planets, laser-driven fusion, and metallic hydrogen.

INTRODUCTION

Equations of state and electrical conductivities of molecular fluids at high shock pressures and temperatures are important for several reasons. Major products of reacted explosives include H_2O, N_2, and CO_2. The fuel in laser-driven inertial confinement fusion (ICF) is a mixture of the hydrogen isotopes deuterium and tritium (D-T). Interiors of giant planets either contain and/or were formed from massive amounts of molecular species. For example, Jupiter and Saturn contain massive amounts of H_2 and He. Uranus and Neptune were formed from H_2O, NH_3, and CH_4. Giant planets now being discovered close to nearby stars probably contain massive amounts of hydrogen because hydrogen has a cosmological abundance of 90 at.%. The pressure required to metallize highly-condensed hydrogen has been a major scientific question since the early part of the twentieth century.

Conditions of interest span the range 1-100 GPa (0.01-1 Mbar) and temperatures of 1,000-10,000 K. These conditions are readily achieved in liquids using shock waves generated by impact of a 20-g projectile accelerated to a velocity of 1-8 km/s by means of a two-stage light-gas gun. A 20-g projectile traveling at 7 km/s has a kinetic energy of 0.5 MJ, which is equivalent to the total kinetic energy in the proton and antiproton beams at Fermilab. This much kinetic energy produces novel states of condensed matter, analogous to the production of novel states of subnuclear matter in High Energy experiments.

GENERAL EXPERIMENTS

Hugoniot (pressure-volume), temperature, and electrical-conductivity experiments were performed. Single and double-shock states were probed in the Hugoniot and temperature measurements. Electrical conductivities were measured for fluids compressed by both a single shock and by a shock reverberating between stiff sapphire anvils. The experimental configurations are illustrated in Fig. 1. The impactor launched by a two-stage gun is incident from the left. Its velocity is measured by flash x-radiography. Fig. 1a illustrates measurement of single-shock Hugoniot points. Shock velocity is measured in the fluid. The Rankine-Hugoniot equations and the shock-impedance matching principle are used to obtain the pressure, density, and specific internal energy. Figure 1b shows the configuration for measurement of double-shock Hugoniot points, in which shock velocity is measured in an anvil off which the first shock is reflected. Shock temperatures are measured by the technique illustrated in Fig. 1c. Light emitted from the front of the first shock and from the shock reflected off the window is collected by a fiber optic bundle and sent to a fast pyrometer. The spectral intensities give a blackbody or greybody emission temperature. Single-shock electrical conductivities are measured by the method illustrated in Fig. 1d.

In order to study fluids with a two-stage gun, it is necessary to know the equations of state to the highest pressures of solid impactors, basepates, and anvils. For this reason the Hugoniots of solid Al,

13

Cu, Ta, and other materials were measured (1) to twice the pressures achieved previously with plane-wave explosive systems. In addition to qualifying equation-of-state standards for shock-compression experiments, these results also have provided standards for static high-pressure experiments using diamond anvil cells (2).

The key feature of our experiments is the fact that the two-stage gun achieves twice the pressures obtained with plane-wave explosive systems. Intermolecular repulsion is the dominant phenomenon in liquids at shock pressures achieved with planar explosive systems. At pressures just above those obtained with explosives, additional phemomena occur. For example, electronic excitation occurs along the Hugoniot of liquid argon (3) and diatomic N_2 dissociates continuously (4) with increasing shock pressure into a mixture of

N and N_2. The two-stage gun has permitted new physics to be studied in liquids.

EXPLOSIVES

Temporal pressure profiles calculated in a slab of an explosive show a rapid jump to the CJ pressure followed by a typical, relatively slow release. When such a detonation wave reflects off a metal plate, a reshock state is reached (5). This calculation shows that pressures up to 100 GPa, densities up to 3 fold the initial explosive value, and temperatures up to a few 1000 K are reached in one and two shocks. This calculation defines the conditions over which properties of individual molecular products need to be studied. This result also shows an advantage in using a gas gun to

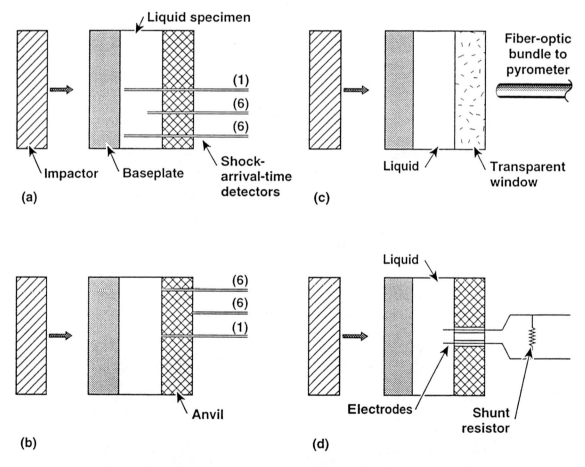

FIGURE 1. Schematics of experimental arrangements for liquid specimens: (a) single-shock Hugoniot; (b) double-shock Hugoniot; (c) shock temperatures under single and double shock; and (d) electrical conductivity of singly-shocked fluid.

14

perform equation of state experiments; namely, the shock wave rises sharply to a constant level behind the shock front. Thus, no material-dependent corrections need be made for decay of the shock-wave behind the shock front.

Chemical reactions occur in individual detonation products. The Hugoniot of water is linear when plotted as shock velocity u_s vs particle velocity u_p for u_p in the range 1.5-7 km/s (6). Despite this simple linear behavior, electrical-conductivity (6) and Raman (7) data show that H_2O ionizes chemically in this regime, becoming a mixture of H^+ and OH^- above 20 GPa. This chemical decomposition needs to be taken into account when theoretically modeling explosive reaction products. Nitrogen dissociates at shock pressures above 40 GPa; double-shock pressures obtained by reshock from single-shock pressures of 20-30 GPa are higher than those at the same density on the Hugoniot; second-shock temperatures are lower than first-shock ones; and the Gruneisen parameter is negative (4,8,9). All these things indicate that nitrogen undergoes a phase transition, which is most probably molecular dissociation. Since the temperatures are quite high in these experiments, dissociation of nitrogen is probably not significant in typical explosives. Liquid CO and CO_2 also decompose on the Hugoniot (4,10). All these experiments are important for obtaining parameters of various interactions within a single fluid. Since reacted explosives are complex mixtures of a large number of chemical species, similar experiments need to be done on two-component mixtures of molecular liquids, such as equimolar concentrations of H_2O and CO_2, to obtain interaction potentials between dissimilar species.

GIANT PLANETS

Because the giant hydrogen planets Jupiter and Saturn contain ~90 at. % H and ~10 at. % He, properties of shock-compressed liquid H_2 and He were measured (11-16). All the Hugoniot and temperature data are consistent simply with intermolecular repulsive interactions between H_2 molecules or He atoms, except for temperature measurements at the highest double-shock points of H_2. At this highest pressure (85 GPa) the measured temperature is lower than expected, assuming that the molecule remains intact (12). This lower temperature is caused by energy absorbed in molecular dissociation, a reasonable assumption given the behavior of the N_2 system.

The electrical conductivity of hydrogen in the mantle of Jupiter was calculated by scaling measured conductivities to values along the hydrogen isentrope calculated from the surface temperature of Jupiter (13,14). Jupiter has no core-mantle boundary separating an insulating molecular mantle from a metallic monatomic core because hydrogen transforms continuously from one to the other, rather than undergoing a sharp first order transition.

The magnetic field is produced by convective dynamo motion of conducting fluid hydrogen. Because hydrogen metallizes in the fluid at half the pressure thought previously for the solid, the conductivity near the surface of Jupiter is larger than previously thought and, thus, the magnetic field is produced much closer to the surface than previously thought. That is, if the Jovian magnetic field is approximated as a dipole, a magnetic dipole decreases as r^{-3}, where r is the distance from the midpoint of the dipole. Since the magnetic field of Jupiter is ~20 times that of the Earth, this might help explain the large magnetic field of Jupiter. That is, because the Jovian magnetic field is produced so close to the surface, it cannot decrease very much from its value at the depth where it is made. In the Earth, however, the magnetic field can only be made relatively deeper in the metallic core and not close to the surface in the electrically insulating mantle. This means the Earth's magnetic field decreases relatively more because of the distance from the metallic core to the Earth's surface. To put the distance scales in perspective, metallic fluid hydrogen is achieved at ~90% of the radius of Jupiter, which corresponds to a depth of 7,000 km, comparable to the radius of the Earth.

Mixtures of water, ammonia, and methane at high pressures and temperatures are thought to be the major components of the giant planets Uranus and Neptune. For this reason equation-of-state, temperature, and electrical-conductivity data were measured for a solution of water, ammonia, and isopropanol (synthetic Uranus) at shock pressures up to 200 GPa (17-19). The chemical composition is similar to that of the fluid mixture thought to be the major constituent of the giant planets Uranus and Neptune. As for hydrogen, the equation of state at very high pressures and temperatures is needed to calculate radial density distributions for comparisons to those derived from gravitational fields measured by spacecraft. Electrical conductivities are needed for calculations of magnetic fields, also measured by spacecraft. Our largest measured pressure corresponds to a depth in Uranus of 10^4 km, about 50% greater than the radius of the Earth.

METALLIC HYDROGEN

Hydrogen has been the prototypical system of the insulator-to-metal transition ever since Wigner and Huntington predicted in 1935 (20) that the insulating molecular solid would transform to a conducting monatomic solid at sufficiently high pressure, P, or density, D, at temperature T=0 K. Substantial pressure is required to do this because solid molecular hydrogen is a wide bandgap insulator (Eg = 15 ev) at ambient. The original theoretical estimate of the required pressure was 25 GPa. Since that time, the estimated pressure has ranged up to 2000 GPa at 0 K. The best recent theoretical estimate is 300 GPa. Extrapolation of recent pressure-volume experimental data at static pressures up to 120 GPa in the hcp diatomic solid phase yields a predicted dissociative transition pressure of 620 GPa. It is also possible that metallization occurs within the diatomic solid, without a transition to the monatomic phase. In this case metallization would be achieved by reduction to zero of the electronic energy gap separating filled valence-band states from empty conduction-band states. Metallization in the solid has not been observed experimentally by optical measurements up to 250 GPa in the diamond anvil cell.

The reason solid hydrogen has not been oberved to metallize at static high pressures is probably caused by phonomena which occur in the ordered solid; namely, structural and molecular orientational phase transitions. Thus, a logical place to look for metallization is in the disordered fluid at temperatures just above melting at high pressures. In this case, metallization is expected when high pressure reduces $E_g \sim k_B T$, where k_B is Boltzmann's constant. When $E_g \sim k_B T$, thermal smearing and fluid disorder fill in the energy gap, a metallic density of states is achieved, and the electronic system has a Fermi surface.

It is extremely difficult to produce a stable hydrogen sample at high temperatures. Hydrogen is so mobile that when it is heated statically, it rapidly diffuses away into the solid walls of the sample holder. Thus, it is essential that hydrogen be heated for a very brief time, say ~100 ns, which is sufficiently fast that hydrogen cannot be lost by diffusion. Shock compression (21-24) is ideal because the sample is both compressed to high pressures and simultaneously heated adiabatically and uniformly. The calculated temperature is ~3000 K, which is larger than the estimated melting temperature of 1500 K at 140 GPa, the observed pressure of metallization. A hydrogen temperature of 3000 K is relatively low because the electronic energygap is 15 ev at ambient pressure and the zero-point vibrational energy of the molecule is 0.3 ev. The time duration of ~100 ns is sufficiently long to obtain thermal equilibrium and an equilibrated configuration of current flow and sufficiently short to avoid the growth of Rayleigh-Taylor instabilities.

As shown below, we made the first observation of metallic fluid hydrogen by using shock compression. Because of the history of this search, this observation has received widespread attention in the scientific and popular press, including the New York Times (March 26, 1996). This publicity demonstrates that shock compression addresses issues of interest to the general scientific community.

Metallization experiments

The experimental configuration is illustrated in Fig. 2. A layer of liquid H_2 or D_2 is compressed dynamically by a high-pressure shock wave reverberating between two stiff, electrically-insulating sapphire (single-crystal Al_2O_3) disks, or anvils. The two sapphire anvils are contained between two Al plates, which are part of a cryostat at 20 K. The compression is initiated by a shock wave generated when a metal plate launched by a two-stage light-gas gun at velocities up to ~7 km/s impacts the Al plate on the left. This shock is amplified when it is transmitted into the first sapphire disk. The first shock pressure in the liquid H_2 is ~30 times lower than the shock incident from the sapphire. The shock then reverberates quasi-isentropically between the two anvils. The P-D states achieved are illustrated in Fig. 3, which shows these states to be relatively close to the 0 K isotherm and at much higher densities than for the Hugoniot.

Electrical resistance of the hydrogen sample was measured versus time by inserting electrodes through the anvil on the right in Fig. 2. Either H_2 or D_2 samples were used, depending on the final

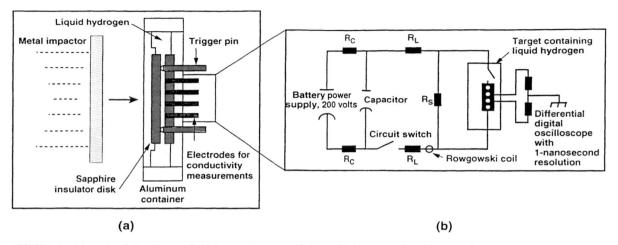

FIGURE 2. Schematic of electrical conductivity experiments on fluid metallic hydrogen. Metal impactor is launched by two-stage gun. Four electrodes in (a) are connected to circuit in (b). Sapphire disks are 25 mm in diameter; liquid hydrogen layer between sapphire disks is 0.5 mm thick. For conductivities lower than metallic, two probes were used. Trigger pins turn on recording system. All cables are coaxial.

density and temperature desired, H_2 giving lower temperatures than D_2. That is, because the initial mass densities of liquid H_2 and D_2 differ by a factor of 2.4, the final shock-compressed densities and temperatures also differ substantially. As shown in Fig. 3, the loading path consists of an inital weak shock followed by a quasi-isentrope. The final temperature is about 1/10 what it would be for a single shock to the same pressure.

At present there is no way to measure D and T because the high-rate deformations caused by the reverberating shock renders the sapphire anvil opaque. Thus, density and temperature were calculated using two reasonable equations of state. The results did not vary significantly. The experimental data are plotted in Fig. 4 as electrical resistivity versus pressure, P_F in Fig. 3a.

RESULTS

In the semiconducting range, 93-135 GPa, the data were fit to the dependence of a thermally activated semiconductor. The result is $E_g(D) = 1.12 - (54.7)(D - 0.30)$, where $E_g(D)$ is the activation energy in ev and D is in moles/cm^3. $E_g(D)$ derived from this fitting procedure and k_BT are equal at a density of 0.32 mol/cm^3 (9-fold initial liquid-H_2 density) and a temperature of ~2600 K (0.22 ev). At 0.32 mol/cm^3 and 2600 K the pressure is 120 GPa, close to 140 GPa at which the slope changes in the electrical resistivity.

At pressures of 140 to 180 GPa the measured hydrogen resistivity is essentially constant at a metallic value of 500 $\mu\Omega$-cm. This value is essentially the same as that of the fluid alkali metals Cs and Rb at 2000 K undergoing the same transition (25). This metallic resistivity of Cs, Rb, and hydrogen is achieved at the same Mott-scaled density of $D_m^{1/3}a^* = 0.30$, where D_m is the density at metallization and a^* is the Bohr radius. Mott's preferred value is 0.25. The value of 500 $\mu\Omega$-cm is bracketed by simple theoretical estimates (24). The free-electron Fermi energy of metallic fluid hydrogen is E_F ~12 ev, as for solid Al. Since T/T_F ~ 0.02, this system is degenerate, highly condensed matter.

Brief History

These metallization experiments require several capabilities: [1] interest in the physics of highly condensed hydrogen (11); [2] a two-stage light-gas gun (26); [3] fast diagnostics (26); [4] liquid-H_2 cryogenics (11); [5] computational simulations of impact experiments (11); [6] computation of the geometrical cell constant relating measured electrical resistance to resistivity (22); [7] experience with single-shock liquid-H_2 conductivities (22). The most important one is an interest in the physics of dense hydrogen because this is the ultimate goal and provided the longterm motivation.

17

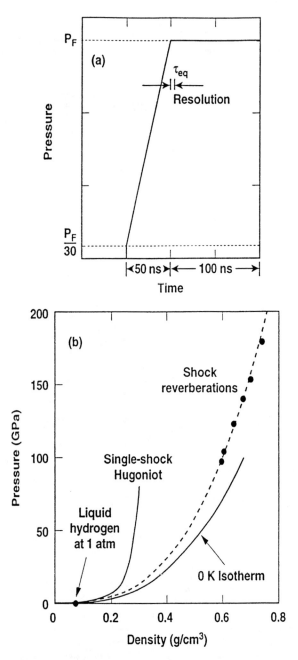

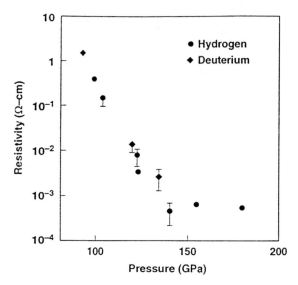

FIGURE 4. Logarithm of electrical resistivity of H_2 and D_2 samples plotted versus pressure. Slope change at 140 GPa is transition from semiconducting to metallic fluid.

FIGURE 3. Effect of rise time on pressure-density states. (a) First pressure in hydrogen is ~ P_f/30, where P_f is incident shock pressure in Al_2O_3. Successive reverberations comprise quasi-isentrope up to pressure P_f. This quasi-isentrope is represented by ramp over ~50 ns from P_f/30 up to P_f. After reverberation is complete, P_f is held for ~100 ns. If P_f were achieved in one jump, this state would be on single-shock Hugoniot. (b) Equation-of-state curves plotted as pressures versus densities: 0 K isotherm, points reached by shock reverberations, and single-shock Hugoniot. Initial point is liquid H_2 at 1 atm.

Two possible approaches were available for supporting this project. One was to convince a technology program that this issue is sufficiently important to support it directly. In this case, substantial support would probably be available for a relatively short period, say 3 years. The other possibility was to treat hydrogen as a long-term science issue and keep the funding level low to minimize visibility. Since the ultimate outcome was unknown, we chose the latter. By doing this we were able to make continual progress over two decades, albeit slowly.

The odyssey to metallic hydrogen took the following path:

The idea	1976
Technology development	1976-1992
Metallization experiments	1992-1994
Physics interpretation	1995
Defense of scientific conclusions	1995-present

One of us (WJN) learned of the hydrogen metallization problem from ACM in 1976. Items [1] to [7] above were developed from 1976 to 1992 (27). The immediate motivation to perform the hydrogen shock-reverberation, or "ringup," experiments came from the fact that we had been simultaneously making single-shock hydrogen conductivity experiments (22) and performing "ringup" experiments to shock compact powders of high-T_c oxides and C fullerenes (28,29). Once we combined these two concepts, it took three years to measure ten hydrogen "ringup" conductivity points, one year to analyze them, and we've been answering questions about this work ever since. Persistence is essential.

ACKNOWLEDGMENTS

We would like to acknowledge people we have worked with from the 1950's to the 1990's. One of us (ACM) would like to thank J. Benveniste, R. Duff, Q. Johnson, N. Keeler, and E. Royce for interactions during the beginning of shock-compression research at LLNL In addition we would both like to acknowledge J. Shaner, N. Holmes, S. Weir, M. van Thiel, H. Radousky, D. Hamilton, J. Trainor, M. Ross, F. Ree, D. Young, C. Henry, R. Silva, and N. Ashcroft for scientific and technical interactions over the past two decades. In particular we want to acknowledge the contributions of S. Weir to the hydrogen metallization experiments and N. Ashcroft for scientific discussions of the hydrogen work. This work was performed under the auspices of the U. S. Department of Energy under Contract No. W-7405-ENG-48.

REFERENCES

1. Mitchell, A. C. and Nellis, W. J., *J. Appl. Phys.* 52, 3363-3374 (1981).
2. Nellis, W. J., Moriarty, J. A., Mitchell, A. C., Ross, M., Dandrea, R. G., Ashcroft, N. W., Holmes, N. C., and Gathers, R. G., *Phys. Rev. Lett.* 60, 1414-1417 (1988).
3. Ross, M., Nellis, W., and Mitchell, A., *Chem. Phys. Lett.* 68, 532-535 (1979).
4. Nellis, W. J., Holmes, N. C., Mitchell, A. C., and van Thiel, M., *Phys. Rev. Lett.* 53, 1661-1664 (1984).
5. Nellis, W. J., *Shock Waves in Condensed Matter-1983*, edited by J. R. Asay, R. A. Graham, and G. K. Straub, Amsterdam: North Holland, 1984, pp. 31-40.
6. Mitchell, A. C. and Nellis, W. J., *J. Chem. Phys.* 76, 6273-6281 (1982).
7. Holmes, N. C., Nellis, W. J., Graham, W. B., and Walrafen, G. E., *Phys. Rev. Lett.* 55, 2433-2436 (1985).
8. Radousky, H. B., Nellis, W. J., Ross, M., Hamilton, D. C., and Mitchell, A. C., *Phys. Rev. Lett.* 57, 2419-2422 (1986).
9. Nellis, W. J., Radousky, H. B., Hamilton, D. C., Mitchell, A. C., Holmes, N. C., Christianson, K. B., and van Thiel, M., *J. Chem. Phys.* 94, 2244-2257 (1991).
10. Nellis, W. J., Mitchell, A. C., Ree, F. H., Ross, M., Holmes, N. C., Trainor, R. J., and Erskine, D. J., *J. Chem. Phys.* 95, 5268-5272 (1991).
11. Nellis, W. J., Mitchell, A. C., van Thiel, M., Devine, G. J., Trainor, R. J., and Brown, N., *J. Chem. Phys.*, 79, 1480-1486 (1983).
12. Holmes, N. C., Nellis, W. J., and Ross, M., *Phys. Rev.* B52, 15,835-15,845 (1995).
13. Nellis, W. J., Weir, S. T., and Mitchell, A. C., *Science* 273, 936-938 (1996).
14. Nellis, W. J., Ross, M., and Holmes, N. C., *Science* 269, 1249-1252 (1995).
15. Nellis, W. J., Weir, S. T., Holmes, N. C., Ross, M., and Mitchell, A. C., *Proceedings of the 1996 US-Japan Seminar on High Pressure-Temperature Research: Properties of Earth and Planetary Materials* (in press).
16. Nellis, W. J., Holmes, N. C., Mitchell, A. C., Trainor, R. J., Governo, G. K., Ross, M., and Young, D. A., *Phys. Rev. Lett.* 53, 1248-1251 (1984).
17. Nellis, W. J., Hamilton, D . C., Holmes, N. C., Radousky, H. B., Ree, F. H., Mitchell, A. C., and Nicol, M., *Science*, 240, 779-781 (1988).
18. Hubbard, W. B., Nellis, W. J., Mitchell, A. C., Holmes, N. C., Limaye, S. S., and McCandless, P. C., *Science* 253, 648-651 (1991).
19. Nellis, W. J., Holmes, N. C., Mitchell, A. C., Hamilton, D. C., and Nicol, M., *J. Chem. Phys.* (in press).
20. Wigner, E. and Huntington, H. B., *J. Chem. Phys.* 3, 764-770 (1935).
21. Weir, S. T., Mitchell, A. C., and Nellis, W. J., *Phys. Rev. Lett.* 76, 1860-1863 (1996).
22. Nellis, W. J., Mitchell, A. C., McCandless, P. C., Erskine, D. J., and Weir, S. T., *Phys. Rev. Lett.* 68, 2937-2940 (1992).
23. Nellis, W. J., Weir, S. T., and Mitchell, A. C., submitted (1997).
24. Nellis, W. J., Weir, S. T., and Mitchell, A. C., *Proceedings of the 1997 AIRAPT Conference* (in press).
25. Hensel, F. and Edwards, P. P., *Phys. World* April, 43-46 (1996).
26. Mitchell, A. C. and Nellis, W. J., *Rev. Sci. Instrum.* 52, 347-359 (1981).
27. Nellis, W. J., Louis, A. A., and Ashcroft, N. W., *Proc. Roy. Soc.* (London) (in press).
28. Weir, S. T., Nellis, W. J., Kramer, M. J., Seaman, C. L., Early, E. A., and Maple, M. B., *Appl. Phys. Lett.* 56, 2042-2044 (1990).
29. Yoo, C. S., Nellis, W. J., Sattler, M. L., and Musket, R. G., *Appl. Phys. Lett.* 61, 273-275 (1992).

19

CHAPTER II

EQUATION OF STATE

CP429, *Shock Compression of Condensed Matter – 1997*
edited by Schmidt/Dandekar/Forbes
© 1998 The American Institute of Physics 1-56396-738-3/98/$15.00

AB FERE INITIO EQUATIONS OF STATE FOR SOLIDS

D. C. Swift

AWE Aldermaston, Reading RG7 4PR, UK / University of Edinburgh

A scheme is under development for predicting the equation of state of solid materials in shock-wave scenarios using a minimum of experimental data. Typically, the only experimental input is a pressure correction to match the ambient density. The equation of state is split as usual into components from the ground state of the electrons in a frozen-ion lattice, the electronic specific heat capacity and the thermal energy of the lattice. The electron ground state is calculated by a quantum-mechanical simulation of a lattice cell. The electron-thermal contribution is calculated from the band structure. The thermal energy of the lattice is found from its phonon modes. These are calculated from the forces acting on an atom displaced from its equilibrium position, e.g. by perturbing the position of atoms in the ground-state calculation. Anharmonic terms may be included. Results are presented for aluminium and silicon.

INTRODUCTION

Theoretical methods are approaching the point where they can be used to deduce the equation of state (EOS) of a substance from first principles. This work is an evaluation of completely *ab initio* techniques, and a minimal normalisation to experiment. The EOS are evaluated in the context of high pressures.

CONSTRUCTION

Completely *ab initio* EOS were generated, and then adjusted slightly to match experimental measurements. The specific internal energy e was split into contributions from the ground state, lattice-thermal and electron-thermal energies. Thermodynamic relations were used to deduce the pressure p.

The EOS were generated as rectangular tables of e and p as functions of density ρ and temperature T. Bilinear or quadratic interpolation was used in finding p and e at states between the ordinates of ρ and T.

Ground state

The variation of ground state energy with density was calculated in the frozen-ion approximation using *ab initio* pseudopotential techniques. A non-local (l-dependent) reciprocal (k)-space pseu-

dopotential was obtained which had been fitted to reproduce the theoretical scattering effects of the core electrons in an isolated atom. Schrödinger's equation was solved to predict the ground state energy e_g of the outer electrons as a function of compression.(1) The electron band structure and Fermi energy were also predicted.

For aluminium, the use of the pseudopotential meant that only three outer electrons from each atom had to be considered. Calculations were made of fcc, hcp and bcc structures. The results agreed with the experimental observation that fcc is the stable phase up to several hundred GPa. For silicon, four outer electrons were considered from each atom. Calculations were made of the diamond structure – other phases are in progress.

These calculations used the local density approximation(2) and generalised gradient correction(3) to model the exchange-correlation effect. The outer electron wavefunctions were expanded in a plane-wave basis set. Plane waves with energies of up to a few hundred eV, were necessary for a converged ground state energy. The wavefunctions were evaluated at a finite number of points in k space, evenly distributed throughout the reciprocal lattice cell. Using 1000 points distributed uniformly throughout the Brillouin zone, the ground state energy had converged to better than 1 GPa. The symmetry of the lattice was used to identify a smaller set of unique k-points, increasing the speed of the ground state calculations.

Lattice-thermal energy

To predict the lattice-thermal energy, the harmonic modes of lattice vibration were determined and a density of phonon states deduced. The phonon modes were populated using Bose-Einstein statistics to obtain the lattice-thermal energy as a function of temperature, at each density.

Lattice modes can be calculated quite efficiently if the atoms can be treated as if they were interacting through an analytic potential. Empirical interatomic potentials were fitted to the frozen-ion cold curve in aluminium, and the resulting density of states compared with experimental data. Radial pair potentials of the inverse power type gave a poor match to the density of states. Morse potentials(4) gave a better match. Multibody potentials of the Finnis-Sinclair type(5) gave the most promising match, but these potentials contained too many degrees of freedom to be fitted only to the cold curve – most of the lattice modes were unstable.

It is not appropriate to represent interatomic interactions in silicon using radial or Finnis-Sinclair potentials. Elements of the dynamical matrix of the lattice were found by perturbing atoms in the ground state calculations, and making use of the relation

$$\frac{\partial^2 \Phi}{\partial [\vec{u}_i]_\alpha \partial [\vec{u}_j]_\beta} \simeq \frac{\partial \Phi([\vec{u}_i]_\alpha)}{\partial [\vec{u}_j]_\beta} \frac{1}{[\vec{u}_i]_\alpha} \qquad (1)$$

where $\vec{u}_i$ is the displacement of atom i from its equilibrium position, Φ is the total potential energy, and square brackets denote components of a vector. $\frac{\partial \Phi([\vec{u}_i]_\alpha)}{\partial [\vec{u}_j]_\beta}$ is the force component $[\vec{f}_j]_\beta$ on atom j when atom i is displaced in the α-direction – a quantity readily obtained from the ground state calculation. Symmetry operations were used to reduce the number of calculations required. For silicon, only a single displacement was necessary of one atom in one direction at each

density. The diamond structure is not centrosymmetric, so an iterative scheme had to be used to enforce the symmetry of the dynamical matrix.(6) Calculations were made with displacements of different magnitudes, in order to estimate the anharmonic effects. These seemed small in silicon. The perturbations were made in a periodic 8-atom diamond cell, so the atomic forces were obtained out to second nearest neighbours.

Similar calculations were made for aluminium, using a periodic array of $2 \times 2 \times 2$ fcc cells. The atomic forces were estimated to third nearest neighbours. The fcc lattice is centrosymmetric, so the symmetry of the dynamical matrix can be enforced by altering the components on the main diagonal.(6) The density of states was closer to experiment than the results obtained using a radial pair potential. The remaining deficiency probably reflects the slower convergence of forces with iterations towards the ground state, compared with the convergence rate of the total energy.

Estimates were made of anharmonic corrections in aluminium using a Morse potential in Monte-Carlo simulations of the lattice. Initial results show some anharmonicity, but more work is needed before the correction can be added reliably to the EOS.

Electron-thermal energy

The density of electron energy levels was estimated from the band structure, and populated using Fermi-Dirac statistics, varying the chemical potential to constrain the total number of electrons.(7) As expected for a simple metal such as aluminium, the resulting electron-thermal energy was very close to estimates made with the free electron model(8) in the range of states considered here. The electron-thermal energy was ignored in silicon.

Completing the equation of state

Given the total specific energy $e(T)$ along each isochore, the specific entropy s was found by integration:

$$s(T) = \int_0^T \frac{dT'}{T'} \frac{\partial e}{\partial T'}. \qquad (2)$$

The specific free energy f was then calculated simply from $f = e - Ts$. The pressure p was obtained from $p = -(\partial f/\partial v)_T$.

Improving the accuracy

It was found that the *ab initio* EOS overpredicted the density at $p = 0$. To retain as much *ab initio* character as possible, a minimal normalisation was made, offsetting the ground state pressure by a constant amount Δp_g. This resulted in a tilt in ground state energy, $\Delta e_g = -v\Delta p_g$. The almost first-principles ('*ab fere initio*') EOS were then calculated as before, but starting from this new ground state energy. A pressure offset of 2 to 5 GPa was required in practice.

EVALUATION

Since the EOS derived from this work are intended for use in shock wave simulations, they were evaluated against measurements for $\rho \geq \rho_0$.

Normal density

Table 1 shows the *ab initio* predictions of normal densities at 293 K compared with experiment. The *ab initio* EOS produced densities which were somewhat too large. In the *ab fere initio* EOS, this discrepancy was corrected to arbitrary accuracy.

TABLE 1. Normal densities (Mg/m^3).

	experiment	calculation
aluminium	2.70	2.76
silicon	2.33	2.42

Ambient isotherm

Diamond anvil measurements have been made of the isotherm of aluminium at room temperature to 220 GPa.(9) The *ab fere initio* EOS matched the data very closely up to about 100 GPa, then fell below it. (Fig. 1.) The pseudopotential model breaks down at high compressions, when the inner electron states change appreciably. This is the likely cause of the discrepancy.

Shock Hugoniot

The principal shock Hugoniot was calculated using the EOS to close the Rankine-Hugoniot form of the hydrodynamic conservation equations.(10) This was compared with data on aluminium alloys 1100 and 6061,(11,12) whose composition and

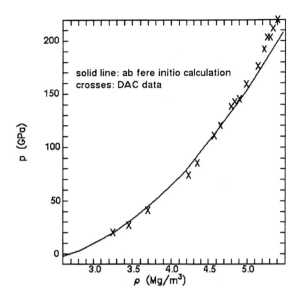

FIGURE 1. Ambient isotherm of aluminium.

equilibrium density are close to that of the element. The effects of non-isotropic stress (e.g. material strength) were ignored in this analysis.

The *ab fere initio* EOS was a reasonable match to the data over its full range. The undulations evident in the density – pressure graph are caused by the use of bilinear interpolation on a relatively coarse density grid. (Figures 2 and 3.)

In silicon, the *ab fere initio* EOS was a good match to the data(13) up to the first phase transition. (Fig. 4.)

CONCLUSIONS

This study demonstrates that completely *ab initio* methods based on non-local pseudopotentials and the quasiharmonic approximation, with minimal fitting to experimental data, can reproduce the compression equation of state of aluminium to over 50 GPa and that of silicon to the first phase transition. The equation of state was adjusted by adding a pressure offset to give the correct normal density.

The scheme described has considerable attractions as a method of generating equations of state. It can be applied to any substance which can be represented by a reasonably small number of atoms in a repeating lattice. The method can be checked against results from other types of exper-

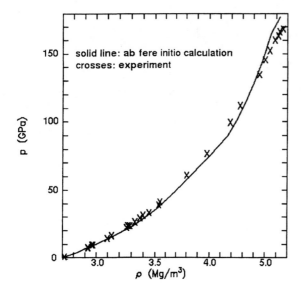

FIGURE 2. Aluminium Hugoniot: pressure and density.

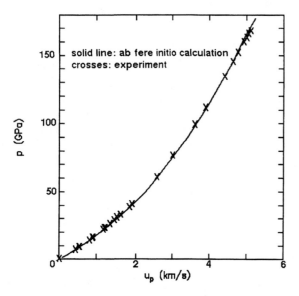

FIGURE 3. Aluminium Hugoniot: pressure and particle speed.

iment, such as the density of phonon states.

ACKNOWLEDGEMENTS

This work would not have been possible without the help of Graeme Ackland, Stewart Clark, Michele Warren (University of Edinburgh) and Mike Payne (University of Cambridge) for theory and algorithms, and Art Ruoff (Cornell University), Bill Nellis and David Young (Lawrence Livermore National Laboratory) for experimental data.

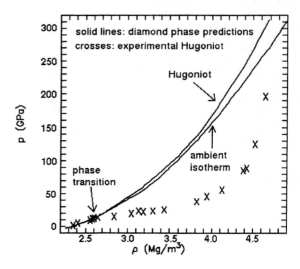

FIGURE 4. Silicon: pressure and density.

REFERENCES

1. Payne, M.C., Teter, M.P., Allan, D.C., Arias, T.A. and Joannopoulos, J.D., Rev Mod Phys **64** 4 (1992).
2. Kohn, W. and Sham, L.J., Phys Rev **140** 4A (1965).
3. Perdew, J., Phys Rev **B46** 6671 (1992).
4. Weins, M.J. in Gehlen, P.C., Beeler, J.R. (Jr) and Jaffee, R.I. (Eds), *Interatomic Potentials and Simulation of Lattice Defects*, Plenum (1972).
5. Finnis, M.W. and Sinclair, J.E., Phil Mag A **50** 1 (1984).
6. Ackland, G.J., Clark, S.J. and Warren, M.C. (University of Edinburgh), submitted for publication (1996).
7. Waldram, J.R., *The Theory of Thermodynamics*, Cambridge (1985).
8. Ashcroft, N.W. and Mermin, N.D., *Solid State Physics*, Holt-Saunders (1976).
9. Greene, R.G., Luo, H. and Ruoff, A.L., Phys Rev Lett **73** 15 (1994).
10. Skidmore, I.C., Applied Materials Research pp 131 – 147 (July 1965).
11. Marsh, S.P. (Ed), *LASL Shock Hugoniot Data*, University of California (1980).
12. Mitchell, A.C. and Nellis, W.J., J Applied Physics **52** 5 (1981).
13. Pavlovskii, M.N., Sov Phys - Solid State **9** 11 (1968).

CP429, *Shock Compression of Condensed Matter – 1997*
edited by Schmidt/Dandekar/Forbes
© 1998 The American Institute of Physics 1-56396-738-3/98/$15.00

THE FEATURES OF THE PRINCIPAL HUGONIOT

J. D. Johnson*

Los Alamos National Laboratory, Los Alamos, NM 87545

The principal Hugoniot jump relations can be cast in a differential form that relates the Grüneisen constant, isentropic bulk modulus, pressure, particle velocity, and shock velocity to each other through an algebraic equation. From this relation I show the general features of the Hugoniot, including a low pressure and a high pressure linear region with a connecting break in slope, limiting forms, and insensitivity to atomic shell structure. The value of the slope for high pressures is explained. The excellent agreement with data is presented.

INTRODUCTION

There has been continuing interest in the behavior of shock Hugoniots for high pressures, in particular for particle velocities greater than $10 \ km/s$. Data has been obtained for such mainly through nuclear shock experiments [1,2]. These data, plotted as shock velocity U_s versus the particle velocity U_p, are remarkably linear and have an almost universal slope different from the well-known low pressure linearity. Modeling shows the same behavior. A general understanding of this high pressure linearity and of the Hugoniot as a whole is needed.

FORMALISM

Rather than go to detailed, complex modeling, we use only the Hugoniot jump conditions and thermodynamics to explain the linear regime and all general features of the principal Hugoniot. We assume that one has the hydrodynamic equation of state $P(\rho, E)$ and the three Hugoniot relations

$$P = \rho_o \, U_s U_p, \qquad \rho / \rho_o = U_s / \left(U_s - U_p \right), \qquad \text{and}$$

$E = U_p^2 / 2$. Here P is pressure, ρ is density, and E is internal energy per gram. We now derive a differential form of the jump conditions. First differentiate the jump relations with respect to U_p

along the Hugoniot, and then express the derivative of P in terms of density and energy derivatives with the use of the partial chain rule. After some algebra and thermodynamic relations, we finally obtain the exact equation

$$B_s / P - x = 3s - 1 + s(2s - 2 - \gamma) / x. \qquad (1)$$

Here, $x = c / U_p$, $\rho\gamma = \partial P / \partial E)_\rho$, and $\rho \partial P / \partial \rho)_E = B_S - P\gamma$, where the isentropic bulk modulus is $B_S = \rho \partial P / \partial \rho)_S$. The lower case s is the slope of the tangent line of the $U_s - U_p$ curve at U_p, and c is the $U_p = 0$ intercept of the tangent. All quantities in Eq. (1) are on the Hugoniot and thus are to be thought of as functions of U_p or another Hugoniot variable.

RESULTS

If one takes the $U_p \to 0$ limit of Eq. (1), we obtain the very initial slope and curvature of the Hugoniot as the Taylor series $U_s / c_o = 1 + s_o U_p / c_o + e \left(U_p / c_o \right)^2 \dots$. After a few thermodynamic manipulations, we obtain $s_o = \left[1 + \partial B_S / \partial P \right)_S \right] / 4$ and $e = \left[0.5 \rho_o \partial^2 B_S / \partial \rho \partial P \right)_S + s_o (2 + \gamma_o - s_o) \right] / 6$.

27

The initial slope is directly given by the pressure derivative of the bulk modulus at constant entropy, and the curvature is given by the higher constant entropy derivative of the B_s with γ_o first entering the expansion at this order. It is common that the Hugoniot is very linear out to $U_p \sim 3-7 \; km/s$. This form for e partially explains why. The two terms for e are dimensionless quantities and in magnitude should lie between one and ten. The sign of the first is negative, the second positive with resulting cancellation. After dividing by six, one expects e to be small, thus a linear $U_s - U_p$. For Nb, we estimate that $e = 0.16$ [3]. The natural variables that follow from the analysis of Eq. (1) for the series are U_s / c_o and U_p / c_o, where c_o is the bulk sound velocity at $U_p = 0$. This implies that new behavior should occur for $U_p \sim c_o$.

One can do large U_p expansions to find the approach to ideal gas. If it is assumed that Debye-Hückel theory [4] describes the very high temperature gas, for large U_p, $\gamma \sim 2/3 - b/U_p^3$, $b > 0$. Then from Eq. (1), $s \sim 4/3 + a/U_p^2 - 2b/U_p^3$, $c \sim -2a/U_p + 3b/U_p^2$ and $P \sim 12 \rho_o a / (\rho/\rho_o - 4)$ with $a > 0$. The parameter a is given by 2/3 the sum of cohesive, ionization, and dissociation energies in going from ambient to $T \to \infty$.

To obtain more details from Eq. (1), we require the qualitative behaviors of γ and B_s / P as functions of U_p which are obtained from the SESAME database [5], where the relevant physics comes from either TFD models or the Inferno model [6,7]. The two models are compatible to the level I need, and one can see that the predicted features of γ and B_s / P are physical. At low U_p, γ is high, say 1.5. For U_p between 3 and 7 km/s, for which the temperature is between 10^4 and 3×10^4 K, γ drops fairly rapidly toward 0.4. Once U_p is greater than 7 km/s or so, γ goes through a very broad minimum with very small variation in γ. Ultimately, at large U_p, γ slowly rises to go to the ideal gas limit of 2/3. For small U_p, B_s / P diverges as $1/U_p$, but, as U_p increases, B_s / P decreases to $\gamma + 1$.

The physics of all this is that for smaller $U_p \lesssim 7 \; km/s$ the equation of state is dominated by the zero temperature isotherm and the phonons. Here, to avoid the left-hand side of Eq. (1) being large, c is constrained to be approximately equal to c_o, thus making the $U_s - U_p$ linear. As U_p increases and goes above $\sim 7 \; km/s$, the electronic thermal excitations start to dominate. The left-hand side of Eq. (1) is now not important, and the $1/x$ term constrains $s \cong 1 + \gamma/2$. Then γ determines s, and the ionization of electrons is pulling γ down below the ideal gas value, but not negative. (Thus it is simple to argue that $\gamma \cong 0.4$ throughout ionization.) All this causes a break in slope. Putting this together in Eq. (1), s is well approximated by $s = 1 + \gamma/2$ for U_p from just above the break at $3 - 7 \; km/s$ to a few hundred km/s. With $\gamma \cong 0.4$, $s \cong 1.2$. Thus the break and high pressure linearity are explained.

EXPERIMENTAL COMPARISON

All of what we have said fits very well with the detailed modeling that goes into the SESAME database, both TFD and Inferno. It also agrees very

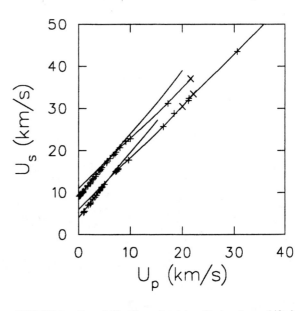

FIGURE 1. Fe and Cu Hugoniot data. Cu has been shifted upwards.

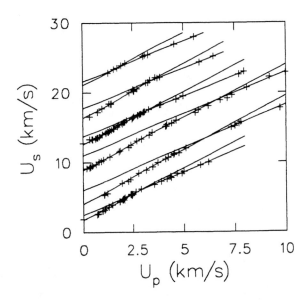

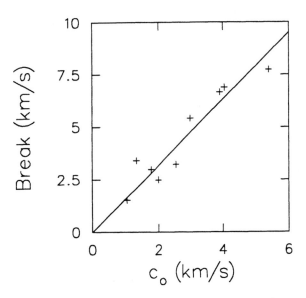

FIGURE 2. Bi, Fe, Cu, Sn, Ar, and Xe Hugoniot data from bottom to top. Shifts have been introduced to avoid overlaps.

FIGURE 3. Breakpoint as a function of c_o.

well with experiment. Figure 1 shows data for iron and two straight line, least-square fits to the upper and lower portions with a well defined break between. The lower data is very linear with a fitted slope of 1.553. The other line, fitted to the upper five crosses, has a slope of 1.213. The uppermost cross, one of the pair next down, and the fourth and sixth crosses down are absolute measurements. The x's are data of Ragan [1] with Mo as a standard, and the remaining two high U_p crosses are measured assuming lead as a standard. The error bars on the uppermost point are $\pm 2\%$ in both U_p and U_s. The high U_p crosses and some of the low U_p come from the Russian literature [8]. The rest of the low data points are from the Los Alamos Shock Compendium [9]. Also, Fig. 1 shows Cu, and Fig. 2 presents Bi, Fe, Cu, Sn, Ar and Xe for lower pressures. All show the break with slopes above it of 1.203, 1.213, 1.170, 1.162, 1.144, and 1.166, respectively [9-11]. The additional data do not go to as high a U_p as that of the iron, but they do strongly support the existence of the break and the linearity with slope $s \sim 1.2$.

We look now to the break and define it by the intersection of the linear fits to the higher and lower portions of the $U_s - U_p$. In Fig. 3, we plot the location of the break as a function of c_o for Al, Fe, Cu, Sn, Bi, Ar, and Xe, going down from the

highest points to the lowest. The solid curve is a fit with a straight line through the origin; the slope is 1.6. Thus the break in slope occurs at $U_p = 1.6c_o$.

SHELL STRUCTURE

Shell structure enters in two ways. One is through the variation in ρ_o in going through the periodic table, but we do not want to focus on this. We look to the shell structure for a given material from the thermal part of the equation of state. Since $s \cong 1 + \gamma/2$, the shell structure comes in only through γ and any γ variations only weakly change s. For Al and elements with higher atomic numbers, from Inferno the maximum variation in γ is ± 0.1. So one finds the variation in s, but what is of interest is ΔU_s which is an integral of Δs. Putting in all the details, which are a little more complicated than outlined, we find that for the ± 0.1 in γ, U_s varies by $\pm 2\%$. These are conservative approximations, and this is a maximum variation at an <u>exact</u> U_p. It will be extremely difficult to see shell structure above the break. If one goes to the $P(\rho)$ Hugoniot and looks in the neighborhood of the maximum density, the small wiggles in U_s are amplified by the presence of the singularity and appear large. But experiments are not known that can measure P and ρ directly, so this is not relevant.

SUMMARY

We have presented a number of results. First is Eq. (1), which relates Hugoniot variables and thermodynamic quantities. From it we have expansions of the Hugoniot for $U_p \to 0$ and $U_p \to \infty$. From simple features of the equation of state, the linear region and slope above the break are understood and the location of the break is estimated. The perturbation of thermal shell structure is quantified. More details of the Hugoniot can be obtained above the high pressure linearity as ideal gas behavior is approached.

We have focused in this paper on elemental metals, and certainly the ideas are valid there. For all other substances, if one is high enough up the Hugoniot that all molecules are dissociated, then all these ideas should be applicable. Further down, the details of the picture will be altered for molecular systems and insulators. Furthermore, below the break where thermal excitations do not dominate, phase transitions with large volume changes introduce structure. Also, for the alkali metals there are shell structure effects for small U_p. But even with these caveats, this is a very powerful overview.

REFERENCES

*Work performed under the auspices of the U.S. Department of Energy

1. C. E. Ragan III, Phys. Rev. A **29** (1984) 1391, and references therein.
2. W. J. Nellis, J. A. Moriarty, A. C. Mitchell, M. Ross, R. G. Dandrea, N. W. Ashcroft, N. C. Holmes and G. R. Gathers, Phys. Rev. Lett. **60** (1988) 1414.
3. I thank J. Wills for furnishing me with an estimate of the double derivative of B_S for Nb.
4. L. D. Landau and E. M. Lifshitz, *Statistical Physics* (Addison-Wesley Publishing Company, Reading, PA, 1969).
5. S. P. Lyon and J. D. Johnson, Sesame: Los Alamos National Laboratory Equation of State Database, Los Alamos report no. LA-UR-92-3407, 1992 (unpublished).
6. R. D. Cowan and J. Ashkin, Phys. Rev. **105** (1957) 144.
7. D. A. Liberman, Phys. Rev. B **20** (1979) 4981.
8. L. V. Al'tshuler, A. A. Bakanova, I. P. Dudoladov, E. A. Dynin, R. F. Trunin and B. S. Chekin, J. Appl. Mech. Techn. Phys. **22** (1981) 145; L. V. Al'tshuler and B. S. Chekin, in: Proceedings of First All-Union Pulsed Pressure Symposium, VNIIFTRI, Moscow, 1974 (unpublished); L. V. Al'tshuler, N. N. Kalitkin, L. V. Kuz'mina and B. S. Chekin, Sov. Phys.-JETP **45** (1977) 167; R. F. Trunin, M. A. Produrets, L. V. Popov, V. N. Zubarev, A. A. Bakanova, V. M. Ktitorov, A. G. Sevast'yanov, G. V. Simakov and I. P. Dudoladov, Sov. Phys.-JETP **75** (1992) 777; R. F. Trunin, M. A. Podurets, B. N. Moiseev, G. V. Simakov and A. G. Sevast'yanov, Sov. Phys.-JETP **76** (1993) 1095; and references therein.
9. S. P. Marsh, Ed., *LASL Shock Hugoniot Data* (University of California Press, Berkeley, 1980).
10. See Ref. 8. Also, S. B. Kormer, A. I. Funtikov, V. D. Ulrin and A. N. Kolesnikova, Sov. Phys.-JETP **15** (1962) 477; B. L. Glushak, A. P. Zharkov, M. V. Zhernokletov, V. Ya. Ternovoi, A. S. Filimonov and V. E. Fortov, Sov. Phys.-JETP **69** (1989) 739; M. V. Zhernokletov, V. N. Zubarev and Yu. N. Sutulov, Zh. Prikl. Mekh. Tekhn. Fiz. **1** (1984) 119; L. P. Volkov, N. P. Voloshin, A. S. Vladimirov, V. N. Nogin and V. A. Simonenko, Sov. Phys.-JETP Lett. **31** (1980) 588; V. A. Simonenko, N. P. Voloshin, A. S. Vladimirov, A. P. Nagibin, V. P. Nogin, V. A. Popov, V. A. Sal'nikov and Yu. A. Shoidin, Sov. Phys.-JETP **61** (1985) 869; and E. N. Avrorin, B. K. Vodolaga, N. P. Voloshin, V. F. Kuropatenko, G. V. Kovalenko, V. A. Simonenko and B. T. Chernovolyuk, Sov. Phys.-JETP Lett. **43** (1986) 308.
11. W. J. Nellis and A. C. Mitchell, J. Chem. Phys. **73** (1980) 6137 and W. J. Nellis, M. van Thiel and A. C. Mitchell, Phys. Rev. Lett. **48** (1982) 816.

CP429, *Shock Compression of Condensed Matter – 1997*
edited by Schmidt/Dandekar/Forbes
© 1998 The American Institute of Physics 1-56396-738-3/98/$15.00

MODIFICATIONS TO OH-PERSSON EQUATION OF STATE

B. E. Fuchs[1], J. Droughton[2], and P.-A. Persson[3]

[1]*U.S. Army ARDEC, Picatinny Arsenal NJ 07806*
[2]*New Jersey Institute of Technology, Newark, New Jersey 07102*
[3]*New Mexico Institute of Mining and Technology, Socorro, New Mexico 87801*

The theory and prior experimental studies for the high pressure shock compaction of porous metals are reviewed. The Mie-Grüneisen and Constant Derivative Mie-Grüneisen equations of state are not accurate for materials of high initial porosity shocked to high pressures. The equation of state proposed by Oh and Persson for these highly porous materials reduces the Grüneisen parameter of high internal energy states. The Oh-Persson equation of state does not predict the larger volumes at higher shock pressures known to exist for material of low initial porosity. We have modified the Oh-Persson equation of state by using an approximate Oh-Persson equation of state that has two asymptotes: the Mie-Grüneisen constant at low energy states and 0.5 at high energy states. The equation of state was applied to porous tungsten and compared with experimental data from the literature and other equations of state. The proposed equation of state shows improved correlation to experimental data for tungsten with high initial porosity, and reflects the larger volumes at high shock pressures for materials with low initial density. This is achieved while maintaining the close correlation to data obtained by most equations of state at higher initial densities.

BACKGROUND

Detonations transmit strong shocks at extreme pressures that are far beyond the mechanical strength of materials. Porous materials have relatively large changes in volume in shock processes. Because of the large changes in volume, shocked initially porous materials have much higher internal energies and temperatures.

MIE-GRÜNEISEN EQUATION OF STATE

The Mie-Grüneisen equation of state was originally developed from statistical mechanics of crystals, Ref. (1). This equation of state assumes:

$$\frac{\gamma}{v} = \left.\frac{\partial P}{\partial e}\right|_v \qquad (1)$$

where: γ = the Grüneisen parameter,

In application, the pressure and energy of an initially solid Hugoniot curve is used to find the pressure of the porous material at the same volume.

For large porosities, increasing shock pressures are predicted by the Mie-Grüneisen equation of state to cause an increase in volume. This conclusion is supported by Krupnikov et. al. Ref. (2). Data from Krupnikov et. al. and other sources, Ref. (2)(3)(4)(5)(6) were examined.

The tungsten data supports the conclusion of Krupnikov et. al. For the lowest initial densities the specific volume on the porous Hugoniot curve increased with increasing shock pressures. This behavior is not as pronounced as the Mie-Grüneisen equation of state predicts.

THE OH-PERSSON EQUATION OF STATE

This equation of state uses the approximate relationship that the change in energy with volume is equal along states of constant pressures and along the Hugoniot curve, Ref. (7)(8),(9).

$$\left.\frac{\partial e}{\partial v}\right|_p \approx \left.\frac{\partial e}{\partial v}\right|_H \qquad (2)$$

Utilizing thermodynamics, and the conservation of mass, momentum, and energy, Oh and Persson developed a new relationship for γ which was later empirically modified to better fit experimental data and to match the low energy states:

$$\gamma(v,e) = v\left.\frac{\partial P}{\partial e}\right|_v = v\frac{c}{\sqrt{2e + \left(\frac{c}{2\gamma_o}\frac{v_o}{v}\right)^2}}\frac{c + 2s\sqrt{2e}}{2c + s\sqrt{2e}}$$

$$(3)$$

The pressure of the initially porous material is given by:

$$P_{por}(V) = P_H(V) + \frac{1}{V}\int_{e_H}^{e_{por}}\gamma(V,e)de \qquad (4)$$

where: $P_H(V) =$ the Pressure of the solid at the same specific volume as the porous material

Oh and Persson's model has been used to compute the Hugoniot curves of several metals, Figure 1.

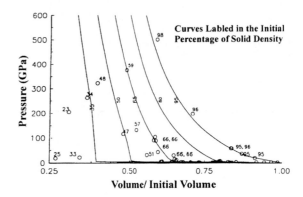

Figure 1. Shock Compression of Porous Tungsten Oh-Persson Equation of State Compared to Data

This plots uses the Oh and Persson's incomplete compaction equation in addition to the Oh-Persson equation of state, Ref. (10). This affects only the lowest pressure portion of the curves. The equation of state performs well, even though the approximation that $\left.\frac{\partial e}{\partial v}\right|_p \approx \left.\frac{\partial e}{\partial v}\right|_H$ and the possible changes in structure (state and crystal structure) are ignored. Oh's gamma parameter at zero energy is equal to γ_o, the value of the Grüneisen parameter at ambient conditions.

Johnson's review, Ref. (9), of the equation of state was highly critical. He determined a loss of accuracy at high internal densities. Oh and Persson's original paper overstated the applicability and range of the equation of state but the usefulness and accuracy can be seen in Figures 1. The equation has introduced an energy term into the relationship which can reasonably be expected.

Behavior of the Oh- Persson Equation of State

At large energies the Oh-Persson gamma, asymptotes to:

$$\gamma(energy \to \infty) = \frac{2c}{\sqrt{2e}} \qquad (5)$$

The value of gamma is then only a function of energy and the speed of sound.

The Oh-Persson gamma predicts infinite pressure only when the solid Hugoniot curve has infinite pressure. This can only happen if the value of $\left.\dfrac{\partial P}{\partial e}\right|_v$ is zero. Al'tshuler Ref. (11) has given the value of γ at high energies as .5, determined from quantum-statistical calculations.

MODIFICATIONS TO THE OH-PERSSON EQUATION OF STATE

Oh-Persson's equation of state requires a numerical integration for solution. This equation of state produces a very slow and inefficient code.

The Oh-Persson equation could be approximated by:

$$P_{por} = P_H(v) + \frac{1}{v}\left(2c\sqrt{2e} - \frac{3c^2}{s}\ln\left(2c\sqrt{2e}\right)\right)\Big|_{E_H}^{E_{por}}$$
(6)

The above equation was further modified to assure a gamma that matches the measured value at low energies. The logarithmic term was also removed and the $\sqrt{2e}$ term was halved.

This equation was further modified to account for the high energy value of the Grüneisen parameter of .5. The Grüneisen parameter is approximated by separating out the high energy value form the low energy:

$$\gamma_{low\ energy} = \gamma_o - .5$$

The equation of state is then:

$$P_{por} = P_H(v) + \frac{c}{v}\left(\sqrt{2e_p + \left(\frac{c}{\gamma_o}\right)^2} - \sqrt{2e_s + \left(\frac{c}{\gamma_o}\right)^2}\right)$$
$$+ \frac{.5}{v_p}\left(e_p - e_s\right)$$
(7)

Comparison between different fully compacted equations of state were made using the Oh-Persson incomplete compaction equation. Figures 1, show the Oh-Persson full-range equation of state in comparison to the data. The equations of state perform extremely well for these materials. Comparison between the proposed equation and the data is shown in Figure 2.

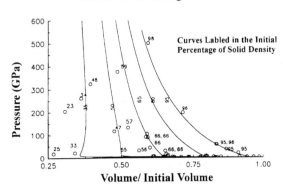

Figure 2. Shock Compression of Porous Tungsten Proposed Equation of State Compared to Shock Data

CONCLUSIONS

The Oh-Persson equation of state, which reduces the Grüneisen parameter for high internal energy states, does not predict the larger volumes at higher shock pressures known to exist for materials of low initial porosity. The Oh-Persson equation of state requires a numerical integration to determine the pressure as a function of volume. This increases the computer time required to model a problem.

A proposed equation of state for shocked porous tungsten was developed. The new equation of state has a minimum value of the Grüneisen parameter at high energies and adds a component based upon the internal energy. The term added, a constant equal to 0.5, is equivalent to the approximate Oh-Persson equation of state with the value of gamma reduced by one-half. This equation of state has the high and low energy asymptotes that match the known values of γ, as well as a functional relationship similar to Oh-Persson's equation of state.

REFERENCES

1. Meyers, M. and N. Thadhani, "Dynamic Deformation of Solids MET 567 Class Notes". New Mexico Institute of Mining and Technology Socorro, NM.

2. Krupnikov, K. K., M. I. Brazhnik, V. P. Krupnikova *Soviet Physics JETP*, Vol. **15**, N0. 3, 470, Sept. 1962.

3. Marsh, Stanley P. LASL Shock Hugoniot Data, University of California Press, Berkeley, Ca. 1980.

4. Dattaraya. P. and R. M. Lamothe, *Journal of Applied Physics*, Vol. **48**, No. 7, 2871, July 1977.

5. Boade, R. R., *Journal of Applied Physics* Vol. **40**, No. 9, August 1969.

6. Gourdin, W. H. and S. L. Weinland, Proceedings of the 4th American Physical Topical Conference on Shock Waves in Condensed Mater, Spokane Wash., July 22-25, 1985.

7. Oh, H. *"Calculation of Shock Hugoniots and Release Isentropes of Porous Materials"* Center for Explosives Technology Research New Mexico Institute of Mining and Technology Socorro, NM, CETR Report A-08-89, July 1989.

8. Oh, H. *"Thermodynamic Properties of Porous Materials under Shock Loading"*. New Mexico Institute of Mining and Technology, Socorro, New Mexico, 1989.

9. Johnson, J. B. *"Analysis of Oh and Persson's (1989) Paper on Equation of State for Extapolation of High-Pressure Shock Hugoniot Data"* Written Communication to P. Persson New Mexico Institute of Technology. 1995.

10. Oh, Ki-Hwan, P. Persson., "Full Range Hugoniot Equation of State for Porous Materials", Presented at American Physical Society Conference on Shock Waves, Albuquerque NM, August 1989.

11. Al'tshuler, L. V., *Usp. Fiz. Nauk* **85**, 197-258, Geb. 1965.

CP429, *Shock Compression of Condensed Matter – 1997*
edited by Schmidt/Dandekar/Forbes
© 1998 The American Institute of Physics 1-56396-738-3/98/$15.00

NEW EQUATION OF STATE MODELS FOR HYDRODYNAMIC APPLICATIONS

David A. Young, Troy W. Barbee III, and Forrest J. Rogers

Physics Department, Lawrence Livermore National Laboratory
Livermore, California 94551 USA

Two new theoretical methods for computing the equation of state of hot, dense matter are discussed. The ab initio phonon theory gives a first-principles calculation of lattice frequencies, which can be used to compare theory and experiment for isothermal and shock compression of solids. The ACTEX dense plasma theory has been improved to allow it to be compared directly with ultrahigh pressure shock data on low-Z materials. The comparisons with experiment are good, suggesting that these models will be useful in generating global EOS tables for hydrodynamic simulations.

INTRODUCTION

An equation of state (EOS) is required for any continuum mechanics simulation, and for processes involving high energy densities, the EOS needs a large range of the density and temperature. There is a longstanding need for more accurate EOS generation models and tabular representations[1]. In this paper we discuss two new models which will be useful in generating accurate EOS tables for hydrodynamic codes.

New data on compressed states of matter are available from the diamond anvil cell and from strong shockwave sources. These data are now numerous enough to stimulate improvements in theory and detailed comparisons between experiments and theory. Two useful new theoretical methods are ab initio phonons and ACTEX.

AB INITIO PHONONS

Ab initio phonon theory is a method for computing lattice phonon frequencies within the electron band structure formalism. For a periodic atomic displacement corresponding to a given lattice wave vector, a self-consistent electron band-structure calculation using the perturbed Hamiltonian yields the harmonic energy and the corresponding lattice mode frequencies[2]. We have carried out this calculation for carbon diamond using the plane wave method with the local density approximation. We have computed a small set of phonons over a range of densities and have averaged them into a density-dependent Einstein frequency. The sum of the vibrational free energy and static lattice energy provides a complete equation of state for solid diamond.

Comparison of theory and experiment is shown in Fig. 1 for the diamond room temperature isotherm[3] and in Fig. 2 for the principal shock Hugoniot[4]. The model predicts a normal density about 1% too large, and this gives a slight offset to the predictions. Remarkably, the model is in good agreement with shock data up to 6 Mbar, where diamond is predicted to melt. Hotter shock states can be obtained from porous diamond with starting densities below the normal density of 3.51 g/cm^3 and the agreement here is also good. Evidently diamond is very nearly a harmonic solid with negligible contributions from anharmonic and electronic terms.

One of the remarkable results from the ab initio phonon theory is the first-principles prediction of the Grüneisen gamma function. This function is frequently used as an empirical parameter, since it is only poorly known from experiments. The ab initio phonon theory has now made it possible to compute the Grüneisen function with much more confidence. The Grüneisen function for diamond is shown in Fig. 3

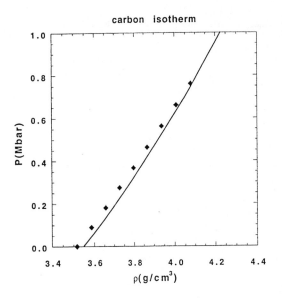

FIGURE 1. Experimental room temperature isotherm for diamond (points) compared with ab initio phonon theory (curve).

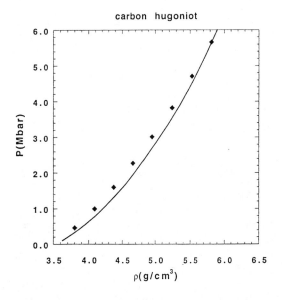

FIGURE 2. Experimental shock Hugoniot for diamond (points) compared with ab initio phonon theory (curve).

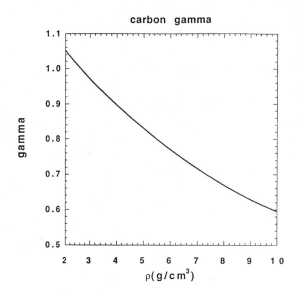

FIGURE 3. Ab initio phonon theory prediction of the density dependence of Gruneisen gamma for diamond.

Ab initio phonon theory can be applied with confidence to low-Z sp-bonded materials where the plane wave technique works well. The use of pseudopotentials allows application of the theory to some higher-Z and d-bonded materials. The theory should be helpful in generating accurate Grüneisen functions for simple metallic, molecular, covalent, and ionic materials, and these can then be used to construct approximate Grüneisen functions for more complex solid materials.

ACTEX

ACTEX (Activity Expansion) is a dense plasma model based on the Abe cluster expansion of the partition function for a Coulomb gas of nuclei and electrons, combined with quantum corrections and a renormalization which includes a representation of atomic species in various states of ionization (5). The ACTEX EOS predictions are exact for weakly coupled plasmas at very high temperature, and become less accurate as the condensed matter regime is approached. ACTEX includes the electron shell ionization region which occurs as a density maximum on the Hugoniot. Recently published ultrahigh pressure Hugoniot data from nuclear explosion and pulsed laser sources overlap the region of validity of ACTEX and a detailed comparison between the two is now possible.

The comparison has been made for D, Be, CH, H_2O, Al, and SiO_2(6). The best overall comparison is with Al, for which nuclear explosion-driven shock experiments have been performed up to 4000 Mbar, where K and L shell ionization is occurring(7). The comparison is shown in Fig. 4. The ACTEX curve shows two large density maxima which correspond to the ionization of the K and L electron shells. The data have large errors and so the test of the theory is not rigorous, but the theory shows the expected form and appears to be semiquantitative in the 10 - 100 Mbar region.

New shock data on liquid deuterium using high energy pulsed lasers have revealed a sharp density maximum in the Hugoniot(8). This feature provides a very strong constraint on any theory of dense, dissociating hydrogen. ACTEX is compared with the data in Fig. 5. The agreement is very good in that ACTEX closely approximates the density and pressure of the experiment near the Hugoniot maximum. Since the ACTEX model does not include the diatomic molecular fluid, it cannot make predictions for the lower part of the Hugoniot curve. Fig. 5 also shows other model predictions for the shock Hugoniot of deuterium. It is clear that there are major discrepancies between theoretical models, largely due to our limited understanding of dissociation in dense fluids.

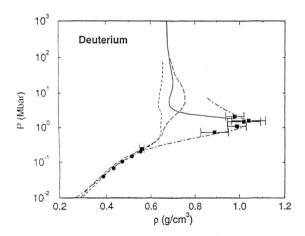

FIGURE 5. Comparison of experimental (points), ACTEX (smooth curve), and other theoretical (dashed curves) shock Hugoniots for deuterium.

Overall, ACTEX is in good agreement with experimental data. It appears that the ACTEX shock pressures fall systematically slightly below the experimental pressures, which may indicate that the ACTEX treatment of the repulsive interactions between ionic cores, taken only to the second virial coefficient level, is not adequate.

DISCUSSION

The ab initio phonon theory is accurate for the solid state and can be extended with corrections into the liquid state, but is valid only when the electronic excitations are negligible, i.e., $T \ll T_F$. ACTEX is a plasma theory valid for conditions where condensed matter interactions are small, i.e., $T \gg T_F$. At present we lack a single theory which covers the entire temperature range from cold solid to hot plasma. A possible candidate for such a theory is the hot band structure model, in which the thermally excited states of the condensed phase are included in the self-consistent LDA calculation. This model should pass smoothly from the $T = 0$ limit to the $T = \infty$ (one-component plasma) limit. Work on this model is underway at Livermore.

The two theories described here increase the precision of theoretical EOS modeling. The application of this work to hydrodynamic code calculations can best be done by incorporating the results into simpler global EOS generators such as QEOS(1) by using adjustable parameters to fit the

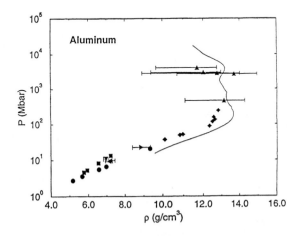

FIGURE 4. Comparison of experimental (points) and ACTEX (curve) ultrahigh pressure shock Hugoniot of aluminum.

theoretical data as if it were experimental. Then smooth and consistent global EOS tables can be generated which incorporate the most accurate physics.

ACKNOWLEDGMENT

This work was performed under the auspices of the U. S. Department of Energy by Lawrence Livermore National Laboratory under Contract No. W-7405-Eng-48.

REFERENCES

1. Young, D. A. and Corey, E. M., *J. Appl. Phys.* **78**, 3748 - 3755 (1995).

2. Giannozzi, P., de Gironcoli, S., Pavone, P., Baroni, S., *Phys. Rev. B* **43**, 7231 - 7242 (1991).

3. Aleksandrov, I. V., Goncharov, A. F., Zisman, A. N., and Stishov, S. M., *Zh. Eksp. Teor. Fiz.* **93**, 680 - 691 (1987) [*Sov. Phys. JETP* **66**, 384 - 390 (1987)].

4. Pavlovskii, M. N., *Fiz. Tverd. Tela* **13**, 893 - 895 (1971) [*Sov. Phys. Solid State* **13**, 741 - 742 (1971)].

5. Rogers, F. J., *Phys. Rev A* **24**, 1531 - 1543 (1981).

6. Rogers, F. J and Young, D. A., preprint.

7. Vladimirov, A. S., Voloshin, V. N., Nogin, V. N., Petrovtsev, A. V., and Simonenko, V. A., *Pis'ma Zh. Eksp. Teor. Fiz.* **39**, 69 - 72 (1984) [*JETP Lett.* **39**, 82 - 85 (1984)].

8. Da Silva, L. B., et al., *Phys. Rev. Lett.* **78**, 483 - 486 (1997).

CP429, *Shock Compression of Condensed Matter – 1997*
edited by Schmidt/Dandekar/Forbes
1998 The American Institute of Physics 1-56396-738-3/98/$15.00

COMMENTS ON THE GRÜNEISEN PARAMETER

Steven B. Segletes and William P. Walters

U.S. Army Research Laboratory, Aberdeen Proving Ground, Maryland 21005-5066

The equation of state developed by Segletes is compared with prior theories. By modifying the functional relationship for the ψ variable and eliminating higher order terms, each of three historical theories for the Grüneisen function can be made to fall directly out of the model—namely, the theories of Slater, Dugdale and MacDonald, and the free-volume theory. One concludes that Segletes' model captures the essence of all of the older theories, each of which being nominally applicable under an appropriate circumstance. The current model differs, however, from these historical models, if the higher order terms are retained.

BACKGROUND

The estimation and measurement of the Grüneisen function, Γ, for metals is an important but difficult task. Simple formulas exist for it, in terms of specific heat and compressibility, or alternately in terms of derivatives of the cold (*i.e.*, 0°) compression curve, or still again in terms of derivatives of the lattice frequency. However, when applied to actual substances, these methods can produce estimates that vary by significant amounts, even at ambient conditions. Some measured variations are due to the approximate nature of the Grüneisen assumption, which has Γ a function of volume only, independent of temperature. Variations in theoretical estimates arise from differences in modeling the propagation velocities associated with lattice vibration. Three widely referenced theoretical models which provide volume-dependent estimates of the Grüneisen function, are those of Slater (1), Dugdale and MacDonald (2), and the free-volume theory (3).

The estimation of the Grüneisen function has been historically complicated by a paucity of cold-compression data. Available high-pressure data were almost exclusively that of shock transition, and so the cold-compression curve, needed by the various models in order to estimate the Grüneisen function, were necessarily back-extrapolated from shock data using a Grüneisen function that was *a priori* unknown. To some extent, this uncertainty in the functional behavior of the Grüneisen function has persisted to the present, as witnessed by hydrocode implementations of the equation of state (4).

In 1984, Rose *et al.* (5) proposed a universal lattice potential, providing a generic function that could be scaled to match theoretical and experimental data for many metals. Based on this work, Segletes (6) proposed a thermal equation of state for metals, which has been shown to capture both the cold and shock behavior of metals into the megabar pressure range. By expressing the binding-energy function in terms of lattice frequency, as opposed to lattice compression, the model links the functions governing cold-compression and thermal effects. Segletes' equation is of the Grüneisen variety and is given by

$$p\,\psi - E = E_b \{\ [(\Theta/\Theta_0)^K - 1] \\ + K(K-1)(\Theta/\Theta_0)^K \ln(\Theta/\Theta_0)\ \}\ , \quad (1)$$

where p and E are the pressure and specific internal energy, E_b is the specific binding energy of the lattice, Θ_0 and $\Theta(V)$ are the reference and current values of the characteristic temperature of the lattice, and ψ is the ratio of specific volume to the Grüneisen parameter, V/Γ, introduced for convenience in manipulating the equations (7, 8). K, assumed constant, is given by $K = C_0/(\Gamma_0 E_b^{1/2})$. Values for K fall in the 2/3 to 4/3 range for metals, near the idealized value of unity, a case that has been studied in greater detail [9]. C_0 and Γ_0 are the reference values of the bulk sound speed and the Grüneisen function, respectively.

In this model, the cold curve, denoted by quantities subscripted 'c', is given by

$$E_c = E_b \{1 - [1 - K \ln(\Theta/\Theta_0)] (\Theta/\Theta_0)^K \} \quad (2)$$

and

$$p_c = (E_b K^2/\psi) (\Theta/\Theta_0)^K \ln(\Theta/\Theta_0) \quad . \quad (3)$$

The Grüneisen function is macroscopically defined by $\Gamma = V/\psi = V(\partial p/\partial E)_V$. Lattice theory relates the characteristic temperature, Θ, to ψ through the relationship that defines the Grüneisen parameter:

$$\Theta'/\Theta = -1/\psi \quad , \quad (4)$$

where the prime denotes ordinary differentiation with respect to specific volume. In the general case, the differentiation in eqn. (4) would be partial at constant temperature. However, for Grüneisen materials, in which $\Gamma = \Gamma(V)$, the derivative in eqn. (4) becomes ordinary. Once ψ is defined, integration of eqn. (4) gives the characteristic temperature, which is the primary equation-of-state variable in eqn. (1). Originally, Segletes [6] incorporated a linear relationship for $\psi(V)$, based on results of his prior work on thermodynamic stability [4, 7, 8], given generally by

$$\psi/\psi_0 = 1 - (\Gamma_0\psi') (1 - V/V_0) \quad . \quad (5)$$

This linear relationship produced an excellent match to both data and first-principles computations over a wide range of specific volumes. However, for very large compressions, this linear relationship can no longer be expected to hold, as the ψ variable must vanish in the high-compression limit ($V \to 0$), whereas the linear form of eqn. (5) will not do so for values of $(\Gamma_0\psi')$ other than unity.

Thus, a power law for the $\psi(V)$ parameter, to replace the linear relationship originally employed, is adopted. A number of positive outcomes result in the context of Segletes' model [6] with this change: the $\psi(V)$ parameter vanishes as $V \to 0$, as is required; binding-energy expansion data for actual metals are better fit with the power law, as compared to the linear model; and most importantly, Segletes' equation of state is shown, by neglecting higher order terms, to reduce to the model of Slater [1], or Dugdale and MacDonald [2], or to the free-volume theory [3], depending on the value of the exponent used in the power law for $\psi(V)$.

A POWER LAW FOR $\psi(V)$

As an alternative to the linear relationship for $\psi(V)$, given by eqn. (5), a power law of the form

$$\psi/\psi_0 = (V/V_0)^x \quad (6)$$

is investigated. The power-law form for $\psi(V)$ retains the mathematical simplicity necessary to manipulate the governing equations conveniently. In terms of compression, $\mu = \rho/\rho_0 - 1$, the power-law ψ translates to a Grüneisen function of the form $\Gamma = \Gamma_0/(1+\mu)^{1-x}$. As far as actual cold-compression data, the power law and the linear assumption for $\psi(V)$ are nearly identical for typical $\Gamma_0\psi'$ values of 0.8 over relative volumes from unity down to 0.5. In this region, the power law provides a slightly more concave-upward curvature to the cold curve, which can be mitigated by choosing a value of exponent x, slightly below the fitted value of $\Gamma_0\psi'$ (e.g., whereas $\Gamma_0\psi' = 0.76$ fits stainless steel, a value of $x = 0.72$ fits comparably well). Over the range of available cold-compression data, the power-law and linear $\psi(V)$ formulations can be made virtually indistinguishable with appropriate selection of x, both resulting in excellent fits to the data.

On the expansion end of the curve, the power-law $\psi(V)$ is seen to provide an improved correlation to data over that given in Segletes [6], especially for lower values of anharmonicity. The power law $\psi(V)$ is able to match the universal potential closely with a single value of exponent x for a given anharmonicity. The value of the exponent x, which provides a match to the universal potential, is given approximately by

$$x = 2/3 + 1/3 \cdot (\eta/10) \quad , \quad (7)$$

40

where η is the value of anharmonicity, as defined by Rose et al. (5). Figure 1 compares the lattice binding energy predicted by the universal potential to that of the current model, when a power-law $\psi(V)$ is employed, the exponent being selected in accordance with eqn. (7). In this graph, the abscissa, a, represents the nondimensional, relative lattice spacing employed by Rose et al. (5). Though not labeled, the solid curve represents the universal potential, while the other five, essentially overlapping, dashed curves are for the current model with anharmonicities of 2, 4, 6, 8, and 10.

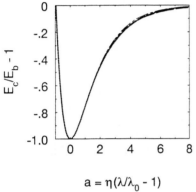

$$a = \eta(\lambda/\lambda_0 - 1)$$

FIGURE 1. Lattice potential vs. nondimensionalized lattice spacing. Solid line is universal potential (5). Dashed lines (essentially coincident) are model of Segletes (6), with power-law $\psi(V)$, for anharmonicity values, η, of 2, 4, 6, 8, and 10.

GRÜNEISEN MODELS

We now intend to show the primary result of this paper—namely, that three historical Grüneisen formulations [those of Slater (1), Dugdale and MacDonald (2), and the free-volume theory (3)] may be recovered through manipulation of the current model, if the material is idealized and higher order terms are neglected. Start with eqn. (3) and multiply by a power of specific volume, V^n, where n is, at this point, still an arbitrary constant:

$$(p_c V^n) = (E_b K^2 V^n/\psi) (\Theta/\Theta_0)^K \ln(\Theta/\Theta_0) \quad . \quad (8)$$

The first and second volume-derivatives of this equation may be obtained in a tedious but straightforward manner. By taking the second derivative multiplied by $(-\psi)$ and dividing by the first derivative, one obtains, upon rearranging,

simplification and selective substitution of V/Γ for ψ [see Segletes and Walters (10) for details]:

$$-(3x-2n)/2 - (V/2)(p_c V^n)''/(p_c V^n)' = \Gamma$$
$$\times \left(K - \frac{1}{2} \frac{[(K+a)^2 + ab]\ln(\Theta/\Theta_0)}{1 + (K+a)\ln(\Theta/\Theta_0)} \right) , \quad (9)$$

where $a = (x-n)(\psi/V)$ and $b = (x-1)(\psi/V)$. The large term in parentheses at the equation's end takes on the value K at ambient conditions, diminishing with compression as the logarithm terms exert influence. For the sake of discussion, consider the large term in parentheses as unity, as an approximation to the situation near ambient volume $[\ln(\Theta/\Theta_0)\approx 0]$ for an idealized material ($K=1$). Eqn. (9) becomes

$$\Gamma = -(3x - 2n)/2 - (V/2)(p_c V^n)''/(p_c V^n)' \quad . \quad (10)$$

In this equation, x, which denotes the power of $\psi(V)$, is a material parameter. If we select the value of parameter n, which has been arbitrary to this point, to be related to x by way of $n = 3x - 4/3$, then x may be eliminated from eqn. (10) to yield

$$\Gamma = -(4 - 3n)/6 - (V/2)(p_c V^n)''/(p_c V^n)' \quad . \quad (11)$$

Eqn. (11) is precisely the generalized result presented by Vashchenko and Zubarev (3) for the Grüneisen function. In this equation, if $n=0$ (i.e., when $x=4/9$), Slater's (1) result is obtained. If $n=2/3$ (i.e., when $x=2/3$), then the result of Dugdale and MacDonald (2) is recovered. Finally, for the case where $n=4/3$ (i.e., when $x=8/9$), the free-volume theory (3) expression follows. What this result tells us is that any one of the three historical Grüneisen models may be nominally applicable, depending on the properties of the material of interest. For comparison, Segletes (6) fit values for $\Gamma_0 \psi'$, corresponding to exponent x at ambient conditions, as low as 0.43 ($\approx 4/9$) for molybdenum and as high as 0.88 ($\approx 8/9$) for rubidium, with the vast majority of fits in the 0.75–0.85 range.

NONIDEAL EFFECTS

Let us consider the practical influence of the large term in parentheses in eqn. (9), idealized as unity in the previous section. The leading K term in the braces affects the ambient value of the Grüneisen parameter, while the remaining term involving logarithms governs the rate at which the Grüneisen function changes with volume. Accepting the model as valid, then the fact that the large term in parentheses diminishes with volume indicates that all three of the historical Grüneisen formulations studied here will tend to overestimate the rate of change of the Grüneisen function with compression when using a given cold-compression curve as the baseline.

To illustrate the model comparison, Fig. 2 depicts the Grüneisen behavior for a hypothetical material with parameter $K=1$ and $\Gamma_0=2$. The figure depicts four curves. There is a curve corresponding to the power-law $\psi(V)$, with exponent x equal to 8/9 [see ref. (10) for complete analysis]. The resulting cold curve from the current model is used to show what the models of Slater, Dugdale-MacDonald, and the free-volume theory would predict for the behavior of Γ, depicted in the figure by curves labeled S, D-M, and FVT, respectively. Per eqn. (9), with K equal to unity, the ambient values for the current thory match the value of Slater when $x=4/9$, that of Dugdale and MacDonald when $x=2/3$, and (as shown in Fig. 2) that of the free-volume theory when $x=8/9$.

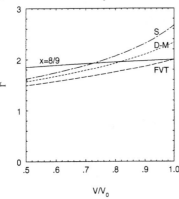

FIGURE 2. Four models of Grüneisen function vs. relative volume for material with $\Gamma_0=2$, $K=1$, and $\psi/\psi_0 = (V/V_0)^{8/9}$.

CONCLUSIONS

The equation of state of Segletes (6) has been examined for compatibility with prior theory. In the original work, an excellent match to available data and first-principles computations was achieved with the use of a linear expression for the $\psi(V)$ function. In the current work, the use of a power law for $\psi(V)$

was examined in lieu of the linear relationship.

Over the range of available compression data, the linear and power-law fits can be made virtually indistinguishable, both providing excellent fits to the data. Also, the binding energy function using a power-law $\psi(V)$ assumption was able to better reproduce the fit of Rose et al. (5) in the area of lattice expansion. In this regard, a simple relation between material anharmonicity and the fitted parameter of the current model has been given, which matches the results of Rose et al. closely.

The primary result of the paper is to show how Segletes' model with a power-law $\psi(V)$, for the case of an idealized material, ignoring higher order terms, is able to reproduce three historical Grüneisen function models, depending on the value of the fitted material parameter in Segletes' model. When the model parameter, which relates to the decrease of the Grüneisen function with compression, takes on a value of 4/9, the result of Slater (1) is obtained. When the parameter is taken as 2/3, the model of Dugdale and MacDonald (2) is recovered. Finally, if the model parameter takes the value of 8/9, the free-volume theory (3) expression follows.

The paper then shows how Segletes' model differs from the historical Grüneisen models, if the higher order terms are retained. It is shown that the historical models of the Grüneisen-function predict a more rapid decrease of the Grüneisen function with compression than does the current model.

REFERENCES

1. Slater, J.C. *Introduction to Chemical Physics.* New York: McGraw Hill (1939).
2. Dugdale, J.S., D.K.C. McDonald. *Phys. Rev.*, **89**, pp. 832–834 (1953).
3. Vashchenko, V.Y., V.N. Zubarev. *Soviet Physics—Solid State*, **5**, pp. 653–655 (1963).
4. Segletes, S.B. *J. Appl. Phys.*, **70**, pp. 2489–2499, (1991). Erratum: **71**, p. 1074.
5. Rose, J.H., J.R. Smith, F. Guinea, J. Ferrante. *Phys. Rev. B*, **29**, pp. 2963–2969 (1984).
6. Segletes, S.B. ARL-TR-1270, U.S. Army Research Lab, Aberdeen Proving Ground, MD, November (1996).
7. Segletes, S.B. *J. Appl. Phys.*, **76**, pp. 4560–4566 (1994).
8. Rajendran, A.M., R.C. Batra (eds.) *Constitutive Laws*, CIMNE: Barcelona, pp. 46–51 (1995).
9. Segletes, S.B. ARL-TR-1357, U.S. Army Research Lab, Aberdeen Proving Ground, MD, May (1997).
10. Segletes, S.B., W.P. Walters. ARL-TR-1303, U.S. Army Research Lab, Aberdeen Proving Ground, MD, March (1997).

CP429, *Shock Compression of Condensed Matter – 1997*
edited by Schmidt/Dandekar/Forbes
© 1998 The American Institute of Physics 1-56396-738-3/98/$15.00

A NEW PROTOTYPE EQUATION OF STATE DATA LIBRARY

Ellen M. Corey and David A. Young

Lawrence Livermore National Laboratory, Livermore California, 94551 USA

Equation of State (EOS) data is a necessary requirement for the simulation of many dynamic processes, including shock wave propagation, high velocity impact, laser-matter interaction, laser medicine, x-ray deposition and planetary and stellar interior evolution. Realistic simulations require high accuracy and smoothness in the EOS. In addition, some processes require independent ionic, electronic and radiation physics. In order to meet these needs, we have implemented a new EOS data library that is platform independent, hierarchically structured and easily extensible for future development. We have just begun our user testing phase and are considering future improvements.

INTRODUCTION

The Equation of State (EOS) is a mathematical relationship between the thermodynamic variables temperature (T), density (ρ), pressure (P) and energy (E). Typically these relationships are expressed as $P=P(\rho,T)$ and $E=E(\rho,T)$. The EOS is a required component for the simulation of many kinds of high energy density dynamic processes, including high velocity impacts, shock wave propagation, laser-matter interaction and laser fusion. A global EOS model is one that is capable of generating data over wide ranges of temperature, density and material composition. These global model tables are becoming the preferred data representation for code simulations because they accurately cover the temperature, density and material composition ranges required.

For many years at Livermore we have successfully used the QEOS(1) model for materials such as mixtures, rocks, plastics and metals. This model has recently been upgraded to provide several desired accuracy improvements, including better representations of shock-Hugoniot curves and diamond anvil isotherms, more accurate liquid-vapor critical points and the addition of molecular degrees of freedom and dissociation(2). The output tables from QEOS typically cover several decades of temperature and density ranges, with at least 10 points per decade, logarithmically spaced.

The efficient use of tables by simulation codes requires a data structure that contains the tables as well as a set of retrieval and interpolation routines to access the data. The structure used at Livermore for many years is a flat, sequential binary file containing pressure, energy and Rosseland mean opacities for several hundred materials. A nine-point biquadratic interpolation method is used.

WHY A NEW DATA LIBRARY

The current EOS data library is not flexible enough to allow the addition of arbitrary functions. It contains a minimal amount of descriptive information about the material or function data. All of the table retrieval and interpolation software is implemented in FORTRAN which limits portability. The nine point biquadratic interpolation method is the method on which the data access routines are based. This method has several inherent problems including: negative bulk moduli at phase transitions, negative opacities and sound speeds which are discontinuous across interpolation box boundaries. These problems limit the overall EOS table accuracy, smoothness and thermodynamic stability and consistency.

The needs of our users require that more detailed, accurate EOS data be available for hydrodynamic simulations. Although our current system is

43

adequate, its limitations have led us to design and implement our new Equation of State library, LEOS.

THE LEOS DATA LIBRARY

The new LEOS data library is a binary file of temperature, density, pressure, energy, entropy, opacity and related data. It was developed using the PDBLib library which is a small set of file management routines useful for storing and retrieving binary data in a portable format(3). It provides a flexible way of managing data files without the user needing to be concerned about the platforms on which the files are created or used. The LEOS library is hierarchically structured, based on a UNIX file system model, portable and easily extensible for future development. The initial set of material properties includes: total, cold, ionic and electronic pressure and energy; entropy; sound speed; melting curve; effective charge; and Rosseland and Planck mean opacities. Each is stored as a function of temperature and density. The library contains data for pure elements, inorganic compounds, explosives, metals, rocks, mixtures and plastics.

THE EOS NUMBERING SYSTEM

Each EOS material in the LEOS library is assigned a unique number for identification. The categories that we have defined are: elements; inorganic compounds, minerals and simple inorganic mixtures; metallic alloys and impure metals; rocks and earth materials; organic compounds and plastics; high explosives; arbitrary mixtures and special tables. When a material is first generated, we typically assign it to the first unused number that is a multiple of ten, within the appropriate category. This allows room for up to ten future versions of the table. These future versions would be generated when there are physics theory upgrades made to the codes or new experimental data becomes available. There is no ordering of materials within each category except for the elements, where the atomic number times ten $(Z*10)$ defines the initial EOS number. It is the responsibility of the person generating the table to select the appropriate category and assign the number. The special tables category is intended for those materials whose use is meant to be temporary or is unusual in some way.

STRUCTURE OF THE LEOS DATA LIBRARY

The LEOS data library is hierarchically organized, based on a UNIX file system model, as shown in Fig. 1. The root contains a master directory and directories for each material in the library. The master directory contains the version number of the library and the date the library was created. The material directories are named by the EOS number preceded by an 'L'. Each material directory has sub-directories containing the actual EOS and opacity data, descriptive information for the material and table data. In general, the (ρ,T) grid for the pressure, energy and entropy functions will be the same. The grid for the opacity data will usually be different from the EOS function grid.

The descriptive information for the material, contained in a separate sub-directory includes: average atomic number, average atomic weight, solid reference density, bulk modulus, cohesive energy, reference temperature, material name, EOS number, chemical formula and material composition. The descriptive information for the table, contained in the same sub-directory as the table data, includes: the name of the person generating the table, generation code name, date of table installation, temperature minimum and maximum values, density minimum and maximum values, number of temperature grid points, number of density grid points, units for temperature, density and function values, comments concerning expected usage for the table and any other relevant comments.

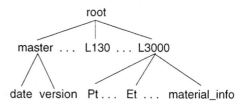

FIGURE 1. LEOS Library Structure

LEOS INTERPOLATION METHODS

The LEOS package offers two types of interpolants, bilinear and bicubic spline, together with their inverse routines(4). The initial step for each method is to divide the (ρ,T) grid into interpolation boxes as shown in Fig. 2.

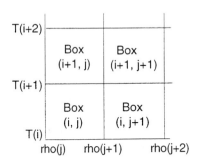

FIGURE 2. Interpolation Boxes

The interpolation coefficients are calculated for each box guaranteeing exact agreement with the EOS function tabular data at the box corner points. The application code inputs the (ρ,T) point at which the EOS function values are wanted. A binary search is done to find the interpolation box containing the data point, the coefficients are retrieved, and the EOS function value is calculated along with the derivatives with respect to temperature and density.

If the temperature or density value is outside the EOS table boundary, the box search will yield the nearest edge box of the table. The coefficients for that box will be used to calculate the EOS function value.

The inverse routines are used when density and energy, (ρ,E), are the known values and temperature is the unknown. This requires the coefficients of the energy table to be explicitly used as a search is done in the energy table for the box containing the known point. The exact temperature value is then calculated by either, in the bilinear case, directly solving the interpolating function or, in the bicubic case, using a bisection method. The temperature value is returned to the user for use in any other (ρ,T) EOS function look-up.

BILINEAR INTERPOLATION ALGORITHM

The bilinear interpolating function is:

$$F(\rho,T) = c_0 + c_1\rho + c_2T + c_3\rho T$$

It is constructed to agree exactly with the EOS data points at the corners of the interpolation box. The coefficients are determined by solving a system of simultaneous linear equations. In general, this is a non-symmetric, indefinite system, solvable by traditional compact methods. A potential problem exists due to loss of numerical significance due to greatly differing magnitudes of temperature and density values. In order to avoid this situation the temperatures are divided by T_i and densities are divided by ρ_j, so that the lower left corner of each interpolation box has coordinates $(1,1)$. This results in a well behaved, less 'ill-conditioned' system. The incoming temperature and density values are scaled before evaluating the interpolating function and unscaled before returning results to the user.

BICUBIC INTERPOLATION ALGORITHM

The bicubic interpolating function is:

$$F(\rho,T) = c_0 + c_1\rho + c_2T + c_3\rho T + c_4\rho^2 + c_5\rho^2 T + c_6\rho^3 + c_7\rho^3 T + c_8T^2 + c_9\rho T^2 + c_{10}T^3 + c_{11}\rho T^3$$

It is constructed by solving a system of 12 linear equations. Rather than directly using the neighboring boxes for the additional points, which results in numerical oscillations due to the coarseness of the (ρ,T) grid, a technique was developed to indirectly use those points. The approach used is to first have the temperature fixed at T_i. A cubic is fitted through the EOS function at ρ_{j-1}, ρ_j, ρ_{j+1} and ρ_{j+2}. This cubic is then used to generate auxiliary points. The process is repeated for temperature fixed at T_{i+1}, density fixed at ρ_j and density fixed at ρ_{j+1}, giving a total of 8 auxiliary points. These points are chosen to fall very close to the actual box corner points so the bicubic interpolant will accurately match the slope of the EOS data. This results in a stable 12 point spline. Once all of the auxiliary points are determined, the bicubic fitting function is computed. The variables are scaled as in the bilinear case. The interpolation boxes along the edge of an EOS table use the bilinear scheme since neighboring boxes do not exist.

TIMING RESULTS

The timing tests for the LEOS package were done with a code that retrieved the requested data table and generated 10000 calls to the interpolation package to lookup and calculate EOS values for a given set of (ρ,T) or (ρ,E) points. In order for the LEOS package to be accepted as the new library by our users, it must be at least as efficient as the current system, and preferably more so. The biquadratic interpolation method was implemented in LEOS to have a direct comparison between the two systems. The initial test was run on the Cray J90 computer on a data table with a grid size of 41 x 51. The LEOS bilinear and bicubic interpolation methods were also timed to gain an understanding of their execution times in relation to the current system. The bicubic is expected to be somewhat less efficient than either the bilinear or biquadratic due to more terms being evaluated for each function call. As shown in Fig. 3, the LEOS methods are at least as efficient as the current biquadratic method.

Initial tests from an actual hydrocode show comparable timings for (ρ,E) calls. We are in the process of improving our methods for (ρ,E) lookups and expect better performance in the near future. Further user testing is required to more fully evaluate the efficiency of the prototype system.

Interpolation Method	(ρ,T) lookup (seconds/point)	(ρ,E) lookup (seconds/point)
Current Biquadratic	1.423e-5	3.523e-5
LEOS - Biquadratic	1.097e-5	N/A
LEOS - Bilinear	1.407e-5	3.430e-5
LEOS - Bicubic	1.814e-5	1.040e-4

FIGURE 3. Interpolation Method Timing Results

FUTURE PLANS

We have plans to improve the execution efficiency and enhance the capabilities of this system in many ways. In the short term, we will be developing a browser to allow users to visualize the EOS data, calculate and visualize data tracks such as Hugoniots and isotherms, compare experimental and table data, compare tables, display data and ASCII information, and print data and ASCII information to an output file. We will be investigating better interpolation methods, including rational polynomials, and robust extrapolation methods that will return physically accurate values of off-table data points.

Our longer term plans include providing some quality metrics with each table that would let users know how well the data meets the requirements of accuracy, smoothness, thermodynamic consistency and stability; and including more material properties such as elastic moduli, electrical resistivity and yield strength.

ACKNOWLEDGMENT

This work performed under the auspices of the U.S. Department of Energy by the Lawrence Livermore National Laboratory under contract number W-7405-ENG-48.

REFERENCES

1. More, R. M., Warren, K. H., Young, D. A., and Zimmerman, G. B., *Phys. Fluids* **31**, 3059-3078, (1988).
2. Young, D. A. and Corey, E. M., *J. Appl. Phys.* **78**, 3748-3755, (1995).
3. Brown, S. A., et. al., PDBLib User's Manual, LLNL Document M-270 Rev. 3, 1995.
4. Aro, C. J., LLNL internal document, 1997.

CP429, *Shock Compression of Condensed Matter – 1997*
edited by Schmidt/Dandekar/Forbes
© 1998 The American Institute of Physics 1-56396-738-3/98/$15.00

SHOCK WAVE DATA BASE

P. R. Levashov, V. E. Fortov, K. V. Khishchenko, I. N. Lomov, and I. V. Lomonosov

High Energy Density Research Center, Russian Academy of Sciences,
Izhorskaya str. 13/19, Moscow, 127412, Russia

The information on thermodynamic properties of matter under conditions of extremely high pressure and temperature is collected and treated. About 10000 experimental points are stored into ASCII files and converted into SQL format. Relational SQL data base is installed on server and graphical interface is elaborated. The information can be accessed via Internet with the help of Netscape Navigator 3.0 browser.

INTRODUCTION

Experimental data on investigations of thermodynamic properties in the waves of shock compression and adiabatic unloading play an important role in modern high energy density physics. Information obtained in such investigations is a result of a number of numerous expensive experiments. Up to now more than 300 sources of information were published, containing more than 10000 experimental registrations. Some of this data has been announced in USA [1-3] and Russia [4]. Nevertheless neither of this editions is complete. For example, Ref. [1-3] contain information up to 1980, and don't contain data on release isentrops. In Ref. [4] there are only data mainly obtained in Arzamas-16, Russia. Such a situation has stimulated the elaboration and creation of shock wave data base.

THE DESCRIPTION OF DATA BASE

The data base contains information on 3 types of experiments: investigations of thermodynamic properties of matter in the waves of shock compression and adiabatic unloading and measurements of sound velocity in shock compressed matter. All available information was recorded into ASCII files

with standard format. For example, consider the file with experimental data for shock compressed Bi:

```
Bismuth, Bi
R0=9.836 g/cc

B. L. Glushak, A. P. Zharkov, M.
V. Zhernokletov, V. Ya. Ternovoy,
A. S. Filimonov, V. E. Fortov,
Experimental investigation of
dense metals plasma at high energy
densities, Zh. Eksp. Teor. Fiz.
96, 1301 (1989) [Sov. Phys. -
JETP, 69, 739 (1989)].
```

m	U	D	P	V0/V
1.00	1.030	3.370	34.030	1.440
	2.060	4.920	99.365	1.720
1.48	1.190	2.660	20.969	1.223
	1.710	3.500	39.646	1.321
	2.240	4.390	65.141	1.380,

where m — the initial porosity of samples, U [km/s] — particle velocity, D [km/s] — shock velocity, P

[GPa] — pressure after shock wave, V_0/V — compression ratio, V [cc/g] — specific volume after shock wave, V_0 and $R_0=1/V_0$ — specific volume and density under normal conditions. Files with sound velocity measurements are the same, but there is sound velocity C_S [km/s] instead of shock velocity D. Files with measurements of release isentrops contain only two columns: pressure P and expansion velocity U.

All substances in shock wave data base are organized into 8 groups:

- Alloys,
- Water solutions,
- Elements,
- Liquid compounds,
- Polymer materials,
- Rocks,
- Solid compounds and minerals,
- Solid organic compounds.

The analysis of available operating systems and available software show that the most simple and inexpensive way of creation of such data base is to use freeware products. Therefore we have chosen Linux operating system (UNIX clone). As an SQL server we have used Postgres95.

The relational SQL data base consists of 8 tables. It contains about 250 substances and 10000 experimental points. The conversion of ASCII files into SQL format was made with the aid of Perl-5.0 scripting language.

SQL-SESSION

One of the ways of accessing to data base is via SQL session. In this case client must be a register user of the data base server. For example, below it can be seen the result of request «Shock compression of aluminum at pressures more than 500 Gpa»:

Request:

```
rusbank=> select a.m, a.U, a.D,
a.P, a.V0V from hugtab a, sub-
stance b where a.subref=b.id and
a.P > 500.0 and b.name='Aluminum';
```

Result of request:

```
m|      U|    D|       P|  V0V|
-+-------+-----+--------+-----+
1|     30|   40|  3254.4|    4|
1|   14.5| 23.4| 920.181|2.629|
1|   15.1| 24.2| 991.019|2.659|
1|  15.67|25.65| 1090.05| 2.57|
1|338.917|  441|  405342| 4.32|
1| 293.64|  379|  301817| 4.44|
1|288.293|  366|  286157| 4.71|
1|283.237|  353|  271153| 5.06|
1|116.691|  147| 46520.5| 4.85|
1|  23.11|32.55| 2038.56|3.448|
1|  12.26|20.05|  666.16|2.574|
1|     11|18.72| 558.049|2.425|
(12 rows)
```

Client can use all standard SQL commands for sorting, searching and treating information.

WWW-INTERFACE

More convenient way of searching information is to use graphical interface. Such interface was elaborated with the help of WWW-browsers. This allow you to access to the data via Internet using every browser which supports java-script and frames (the best is Netscape Navigator 3.0 or higher version). The data base can be accessed via WWW-address http://teos.ficp.ac.ru/rusbank/.

On the first page (Fig. 1) you should choose the type of experiment, the groups of substances and the data fields and click «Send query» at the bottom of the screen. On the second page you will see the list of available substances for the type of experiment you have chosen earlier. Here you should choose the necessary substances. Then click «Send query». On the third page (Fig. 2) you will see the experimental data for substances you have chosen and short reference. To see full reference, click on short reference. Table of contents is generated automatically in every page.

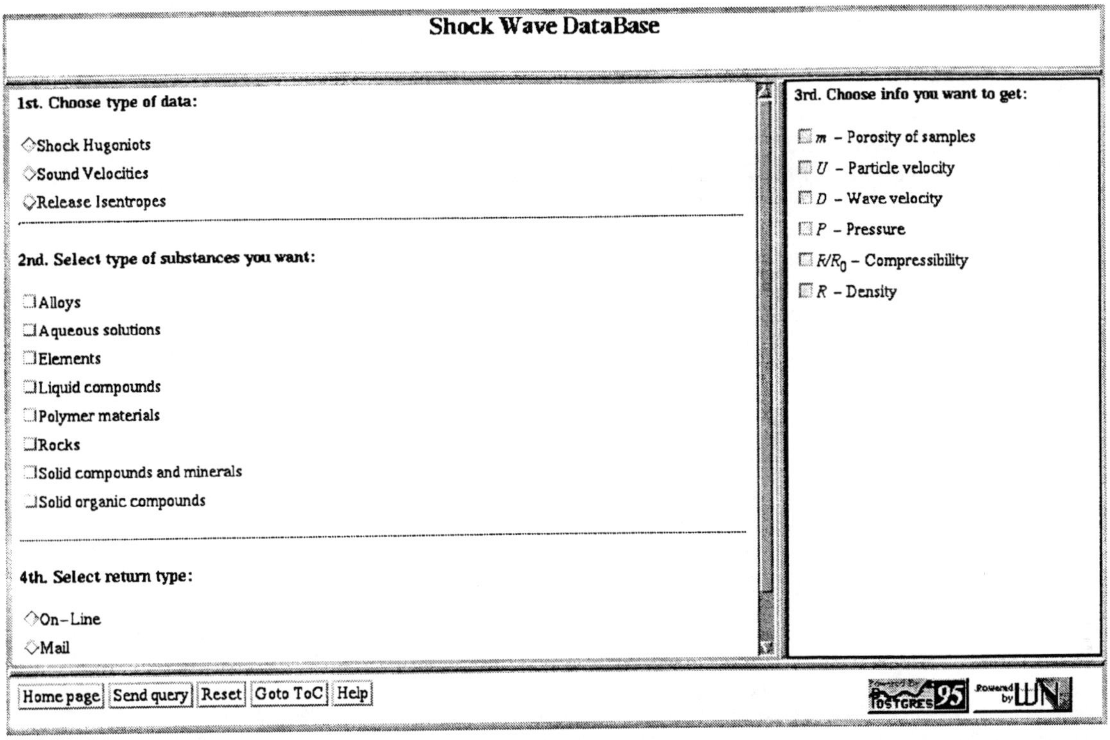

Shock Wave DataBase

1st. Choose type of data:

◇ Shock Hugoniots
◇ Sound Velocities
◇ Release Isentropes

2nd. Select type of substances you want:

☐ Alloys
☐ Aqueous solutions
☐ Elements
☐ Liquid compounds
☐ Polymer materials
☐ Rocks
☐ Solid compounds and minerals
☐ Solid organic compounds

4th. Select return type:

◇ On–Line
◇ Mail

3rd. Choose info you want to get:

☐ m – Porosity of samples
☐ U – Particle velocity
☐ D – Wave velocity
☐ P – Pressure
☐ R/R_0 – Compressibility
☐ R – Density

| Home page | Send query | Reset | Goto ToC | Help |

FIGURE 1. Selection of parameters for request into the data base.

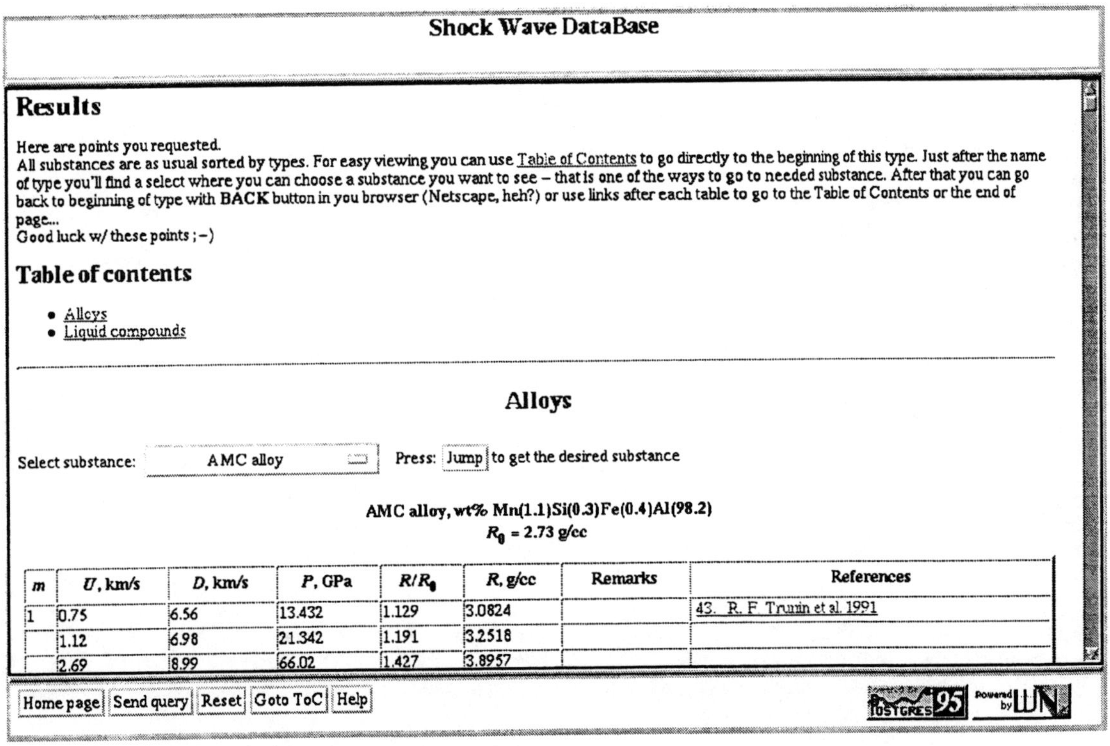

Shock Wave DataBase

Results

Here are points you requested.
All substances are as usual sorted by types. For easy viewing you can use <u>Table of Contents</u> to go directly to the beginning of this type. Just after the name of type you'll find a select where you can choose a substance you want to see – that is one of the ways to go to needed substance. After that you can go back to beginning of type with **BACK** button in you browser (Netscape, heh?) or use links after each table to go to the Table of Contents or the end of page...
Good luck w/ these points ;–)

Table of contents

- <u>Alloys</u>
- <u>Liquid compounds</u>

Alloys

Select substance: [AMC alloy] Press: [Jump] to get the desired substance

AMC alloy, wt% Mn(1.1)Si(0.3)Fe(0.4)Al(98.2)
R_0 = 2.73 g/cc

m	U, km/s	D, km/s	P, GPa	R/R_0	R, g/cc	Remarks	References
1	0.75	6.56	13.432	1.129	3.0824		43. R. F Trunin et al. 1991
	1.12	6.98	21.342	1.191	3.2518		
	2.69	8.99	66.02	1.427	3.8957		

| Home page | Send query | Reset | Goto ToC | Help |

FIGURE 2. Experimental values for the specified request.

CONCLUSION

A great amount of experimental data in the region of high energy density are collected, sorted and treated. All available information are stored into ASCII files. Converters from ASCII files into SQL format is elaborated. The network data base is created and graphical interface for searching information is worked out. The data base is available via Internet by address http://teos.ficp.ac.ru/rusbank/.

ACKNOWLEDGMENTS

The work has been done due to financial support of the Russian Foundation for Basic Research (grant No. 97-07-90370).

REFERENCES

1. McQueen, R.G., Marsh, S.P., Taylor, J.W., Fritz, J.N., Carter, W.J., The equation of state of solids from shock wave studies, in: *High Velocity Impact Phenomena*, New York: Academic Press, 1970, pp. 293-417; appendies on pp. 515-568.
2. Van Thiel, M., *Compendium of Shock Wave Data*, Lawrence Livermore Laboratory report UCRL-50108, **1-3**, Rev. 1 (1977).
3. Marsh, S. P. (Ed.), *LASL Shock Hugoniot Data*, Berkeley: Univ. California Press, 1980.
4. Zhernokletov, M. V., Zubarev, V. N., Trunin, R. F., Fortov, V. E., *Eksperimental'nyye Dannyye po Udarnomu Szhatiyu i Adiabaticheskoi Razgruzke Kondensirovannykh Veshchestv pri Vysokikh Plotnostyakh Energii* (Experimental data on shock compression and adiabatic unloading of condensed substances at high energy density), Chernogolovka: ICPh RAS, 1996.

CP429, *Shock Compression of Condensed Matter – 1997*
edited by Schmidt/Dandekar/Forbes
© 1998 The American Institute of Physics 1-56396-738-3/98/$15.00

DATABASE ON MATERIAL PROPERTIES STUDIED IN EXPERIMENTS USING SHOCK WAVES

M.V.Zhernokletov, R.F.Trunin, L.F.Gudarenko, V.D.Trushchin, O.N.Gushchina

Russian Federal Nuclear Center - VNIIEF, Sarov, Russia

During nearly 50-year period of development of the dynamic methods for studying material properties Russia has accumulated a large amount of experimental data for more that 200 individual materials, compounds, condensed media and gases. It is for systematization of the accumulated experimental data and their visualization that the Database presented was developed. The base is a set of interconnected tables storing: data on shock compressibility of continuous materials and porous materials; on material compressibility by the second shock wave; expansion adiabats of shock-compressed continuous and porous materials; data on velocity of shock-compressed material scattering into air and on sound speed in shock-compressed materials.

INTRODUCTION

The material compressibility data obtained in experiments using shock waves of various intensity currently are the base for construction of equations of state describing thermodynamic properties of these materials at high pressures and temperatures. The equations of state are required at computations of various facilities operating under intensive pulsed loads, for example, for computation of accidental modes of operation of power reactor facilities. During nearly 50-year period of development of the dynamic methods for studying material properties Russia has accumulated a large amount of experimental data for more that 200 individual materials, compounds, condensed media and gases, practically all the experimental data has been obtained at RFNC-VNIIEF.

There are hundreds of publications on various aspects of the areas of research into material properties with dynamic methods (study of shock-wave compressibility, recording of isentropic expansion, double compressibility,

etc.). This data is scattered over numerous publications. Our goal is to develop and publish a general collection containing all shock-wave data available in Russia with their critical analysis and systematization, both that known to experts and that which has not yet been published. This work is being carried out in the context of ISTC Project #373.

GENERAL ARRANGEMENT OF THE EXPERIMENTAL DATA BASE

The experimental data base is developed in the environment of the data base management system (DBMS) Paradox 7 of Windows-95 system using version 5 [1,2]. From the viewpoint of DBMS Paradox the Experimental data base is a set of interconnected tables. Individual tables store the data: on shock compressibility of continuous materials; on shock compressibility of heated materials; on shock compressibility of porous materials; on material compressibility by the second shock wave; on expansion adiabats of shock-compressed porous materials;

data on velocity of shock-compressed material scattering into air; data on sound speed in shock-compressed materials.

In these tables the data of each specific experiment is represented with a separate record which, besides the experimental data, contains an internal code of the material. The general material information is stored in the material table whose each record, in addition to the material name and internal code, contains material density under standard conditions and, if necessary, heated material density and material description. Currently the following material types are recognized: metals; metal hydrides and nitrides; carbides and oxides; solid organic materials; alkali metal halogenides; minerals and rocks; water and saturated water solutions of salts; organic liquids.

For each experiment on shock compressibility of a continuous or heated sample of material under study the Data base stores three experimental values: ρ_{00}, D and U. For heated materials the initial temperature value is also stored. At printing the data table, besides the above values and source code, the following computed values are output for each experiment: ρ, σ, V/V_0, P, E. The values ρ, P, E are computed by the following relations found from the laws of conservation of mass, momentum and energy at the shock wave front:

$$\rho = \frac{D}{D-U} \cdot \rho_0 ; \qquad (1)$$

$$P = \rho_0 \cdot D \cdot U ; \qquad (2)$$

$$E = \frac{1}{2} P \left(V - V_0 \right) = \frac{U^2}{2} \qquad (3)$$

In the table the data appear in the order of increasing U. For data visualization two dependency graphs are constructed for the following values: $D(U)$ and $P(\sigma)$. For close values of ρ_{00} the experimental data is statistically processed in the D-U coordinates. This data is approximated, in the general case, by quadratic dependencies of the form:

$$D(U) = A + BU + CU^2, \qquad (4),$$

where the coefficients A, B and C are calculated with the least square fit using statistical weights w. So, for each element the Data base also stores the w weight value. If at the experimental data

input the value w is not specified, it is taken $w=1$ by default.

If at the computation it appears that $C < 1 \cdot 10^{-4}$, the data is approximated by linear dependencies of the form:

$$D(U) = A + BU , \qquad (5)$$

where the coefficients A and B are calculated in the same manner as for (4).

The dependencies $D(U)$ found in form (4) or (5) are used to construct a solid line in the first graph and the values P and σ calculated by these dependencies are used to construct a solid line in the second graph. In the first graph the quadratic or linear dependency is output which was used to construct the solid line with stating the computed coefficients A, B and, if necessary, C.

If the number of the experiments for a particular material is no more than 3 or the dependency $D(U)$ for a given material has a pronounced break due to a phase transition, the statistical processing of the experimental data in the D-U coordinates is not made and, hence, no solid lines are constructed in the graphs and neither quadratic nor linear dependency is output in the first graph. Given phase transitions leading to a clear break of the D-U dependencies, setting a special flag rules out the statistical processing of the experimental data.

For illustration Figs.1 and 2 present continuous quartz Hugoniots whose $D(U)$ dependency has a pronounced break, as well as continuous copper Hugoniots.

For each studied material porous sample shock compression experiment the Data base stores three experimental values: ρ_{00}, D and U. At printing the data table, besides the above values and source code, the following computed values are output for each experiment: k, ρ, σ, V/V_0, P and E, where P, ρ, E are calculated by (1)-(3), but ρ_{00} is taken instead of ρ_0. The value σ is calculated by the relation

$$\sigma = D / (D-U) \cdot 1 / k$$

In the table the data appears in the order of decreasing ρ_{00} and for identical ρ_{00} in the order of increasing U.

To visualize the porous material shock compression data, the same graphs are constructed as for continuous and heated materials (see example in Fig.3).

Continuous quartz shock compression data

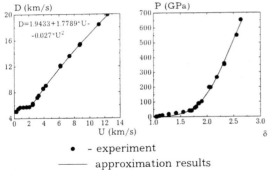

• - experiment

——— approximation results

Fig. 1

Continuous copper shock compression data

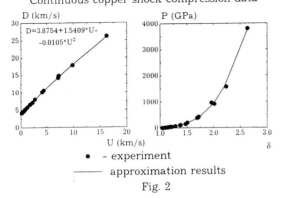

• - experiment

——— approximation results

Fig. 2

For each experiment on material compression by the second shock wave the Data base stores only the values presented in the original. No processing of this data is provided for. The initial material density, as well as parameters characterizing the states behind the front of the first and second shock waves are surely presented. If a reference was used at the experiment, the parameters characterizing its state are presented. In the table the data appear in the order of increasing value U behind the front of the first shock wave and for identical values in the first wave in the order of increasing value U behind the front of the second shock wave.

For each experiment on studying shock-compressed material expansion adiabats (isentropes) the Data base, besides the data source code, stores the following values: initial material density ρ_{00}; the values P and U in the initial state on the Hugoniot; the same parameters on the expansion adiabat; barrier material and barrier shock wave front velocity D.

For the adiabat visualization a graph is

constructed which the available experimental data is put on in the P-U coordinates. The points corresponding to experiments for studying various adiabats (various initial states) are therewith output with different markers. Fig.4 exemplifies the copper data.

For each experiment on studying velocity of shock-compressed material scattering into air the Data base, besides the data source code, stores the following physical values: P and U describing the initial material state on the Hugoniot, as well as W and P describing the material state at discharge into air. The latter value P is therewith input with an accuracy of three significant digits following the point.

For each experiment on studying sound speed in shock-compressed material the Data base, besides the data source code, stores the following physical values: D, P, σ and C. The value σ is therewith input (or output at printing) with an accuracy of three significant digits following the point.

FUTURE WORK

The earlier published shock-wave data compendia are known to be developed by Livermore (edited by Van Thiel M.) [3] and Los Alamos (edited by S.P. Marsh) [4] Laboratories. We are sure that in 1997 Collection developed by VNIIEF will appear. We believe, it would be quite useful to integrate these Collections into a single compendium (apparently, in several volumes) and be able to make use of the experimental data presented both in the printed form and (which is particularly important) in the software product form, i.e. Data base on magnetic media. Moreover, in most cases in the existing Collections the analytical dependencies of the shock wave velocity (D) on the material mass velocity following the front (U) are estimated only by the data presented in the individual Collections. It would be very useful for experts pursuing shock wave physics to have unique D-U dependencies constructed basing on all the data available both in Russia and the USA. In addition, neither Collection contains the data for measured shock-compressed material temperatures which are already fairly numerous. These data are known to be of interest for developers of equation-of-state models.

Porous copper shock compression data

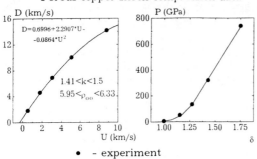

$D=0.6996+2.2907 \cdot U - -0.0864 \cdot U^2$

$1.41 < k < 1.5$
$5.95 < \rho_{00} < 6.33$

• - experiment

——— approximation results

Fig. 3

Shock-compressed continuous copper
expansion adiabats

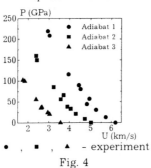

• , ■ , ▲ - experiment

Fig. 4

REFERENCES

1. Diana Tiney. Programming in Paradox for Windows illustrated by examples. Moscow, Binom Publishers. 1995. 732 pp.
2. O.N.Kassandrova, V.V.Lebedev. Processing of observation results. Moscow, Nauka Publishers. 1970, p.79-81.
3. Van Thiel M., Shaner J.W., Salinas E. Compendium of shock wave data. Livermore Lawrence Laboratory Report. OCRL 50-108, 1977. V.1-3.
4. Marsh S.P. LASL Shock Hugoniot data. University California Press, Berkely-LA-London.

54

CP429, *Shock Compression of Condensed Matter – 1997*
edited by Schmidt/Dandekar/Forbes
© 1998 The American Institute of Physics 1-56396-738-3/98/$15.00

EQUATION OF STATE MEASUREMENTS OF D₂ ON NOVA

G. W. Collins, L. B. Da Silva, P. Celliers, K. S. Budil, R. Cauble, D. Gold, M. Foord, R. Stewart, N. C. Holmes, M. Ross, B. A. Hammel, J. D. Kilkenny, R. J. Wallace, A. Ng[1]

Lawrence Livermore National Laboratory, Livermore, CA 94550
[1]University of British Columbia, Vancouver B.C.

INTRODUCTION

Condensed molecular hydrogen at low pressure is an insulator with a 15 eV band gap and 4.5 eV molecular binding energy. Theory predicts that, even at T=0, both the bandgap and molecular dissociation energy will decrease with increasing pressure, ultimately transforming hydrogen into an atomic metal at ~ 3 Mbar. At pressures between ~ .1 and 3 Mbar, thermal dissociation and ionization can occur at temperatures well below the low pressure limit. How hydrogen transforms from a condensed molecular state into a fully ionized dense plasma is of fundamental interest and has a profound impact on the equation of state (EOS) at high density. This EOS is integral to a broad spectrum of disciplines, such as understanding the structure of Jovian planets or designing ignition targets for inertial confinement fusion (ICF).[1,2,3,4] For these reasons, a number of theoretical models of the hydrogen EOS have been proposed.[5-8] The performance of ignition ICF targets on the National Ignition Facility will rely in part on timing the breakout of a sequence of shocks, tuned to minimize the entropy production in the frozen hydrogen (deuterium-tritium) fuel during compression.[9,10] Timing these shocks depends directly on the EOS, again where the molecular fluid transforms to an atomic-partially ionized state.

Hydrogen EOS data at pressures greater than 0.1 Mbar have been obtained by dynamic shock compression and by static compression.[11-15] While both methods can access equilibrium states of

matter, the final-state densities and temperatures obtained by shock compression are directly applicable to the Jovian planets and ICF. In shock compression, a single shock drives the fluid to a point on the principal Hugoniot, which is the locus of all final states of pressure, energy and density that are achieved behind a single shock. With the initial state specified, conservation relations require only two independent parameters be measured to obtain an absolute EOS datum. The shock speed, U_s, particle speed U_p, pressure P, internal energy E, and final density ρ are related by:

$$P - P_0 = \rho_0 U_s U_p \qquad (1)$$

$$\rho / \rho_0 = U_s / \left(U_s - U_p \right) \qquad (2)$$

$$E - E_o = \frac{1}{2}(P + P_o)(\frac{1}{\rho_o} - \frac{1}{\rho}) \qquad (3)$$

where ρ_0 is the initial density, P_0 is the initial pressure, ρ/ρ_0 is the compression, and E_o is the initial internal energy.[16] Equations (1) - (3) are the Hugoniot relations. Notice, Hugoniot measurements do not determine temperature, which shows how internal energy is partitioned among the various modes of the system. Temperature is typically determined by measuring the optical emission intensity from the shock front, and must be measured separately.

While early shock wave hydrogen EOS experiments[13] are well described by an intermolecular pair potential model (RRY)[6], recent reshock temperature measurements of Holmes el al. are significantly lower than the RRY model predicts.[14] These lower temperatures are described by a "dissociation model" based on an ideal mixing of molecular states (using a soft-sphere perturbation theory) and monatomic states (using a one-component plasma model). This model contains one adjustable parameter which is set to agree with all the hydrogen shock data,[13,14] and predicts a significantly higher compressibility in the $P = 0.2 - 5.0$ Mbar regime, than both the RRY model and the Sesame tables.[6,17]

In this paper we describe principle Hugoniot measurements of liquid D_2 up to $P = 2.1$ Mbar. We compressed liquid D_2 with a Nova-laser-driven shock wave launched from an aluminum pusher. The Al/D_2 interface and the shock front in the D_2 are observed with temporally resolved radiography, to determine U_p, U_s, and ρ/ρ_0. The pressure is calculated using Equation (1). These absolute EOS data reveal a compressibility comparable to the dissociation model.

CRYOGENIC TARGET DESIGN

A schematic of the cryogenic target cell is shown in Fig. 1. Liquid D_2 was contained in a 1-mm-diameter, 0.45-mm-long cylindrical cell machined into a copper block. One end of the cell was sealed with an Al disk that served as the shock pusher; the opposite end of the cell was sealed with a 0.5-mm-thick sapphire window. The pusher was 100, 180, or 250 μm thick, depending on the experiment, and had an rms surface roughness of 30 nm. The pusher was coated with 15 to 25 μm of polystyrene (CH) external to the cell, and the polystyrene was overcoated with a 100 nm layer of Al. The thickness of the polystyrene layer was chosen to prevent direct laser ablation of the Al pusher, to minimize x-ray preheat of the pusher. The Al overcoating eliminated direct laser penetration through the plastic at onset of the laser pulse. To accommodate radiography, a 500-μm-diameter window was drilled into each side of the cell and sealed with a 5-μm-thick beryllium foil. D_2 was loaded into the cell at ~ 20 K and then pressurized to a few hundred torr. Temperatures were monitored to within 0.05 K. Initial D_2 densities were determined from the saturation curve[20] to be 0.171 g/cm³. The initial density, ρ_0, for each experiment was known to an accuracy > 99.5%.

Cryogenic Cell Target

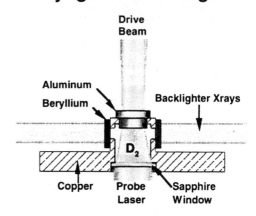

FIGURE 1. Schematic diagram of a cryogenic cell for laser-driven shock compression of liquid deuterium.

EQUATION OF STATE MEASUREMENTS WITH LASERS

It has long been known that lasers are capable of driving very strong shocks into targets.[18] However, laser produced EOS data in the Mbar regiem have been plagued by large errors. There are four issues

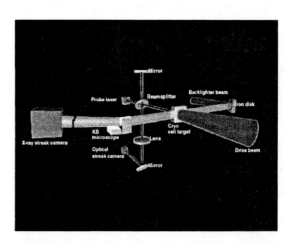

FIGURE 2. Schematic diagram of the experimental setup for simultaneous side-on radiography and end-on interferometry of a cryogenic cell.

typically preventing accurate laser produced EOS data: shock planarity, preheat, shock steadiness, and measurement accuracy. The experimental layout used for our EOS measurements, which addressed each of the above concerns, is shown in Fig. 2 and described below. First, the shock produced must be planar and spatially uniform. This puts constraints on the target planarity and roughness as well as the drive beam uniformity. One beam of the Nova laser (λ = 527 nm) was focused at normal incidence onto the target, ablating the polystyrene layer and driving a shock wave through the Al and into the D_2. A kinoform phase plate[21] was inserted into the Nova beam to smooth and produce a flat top intensity profile. The laser footprint at the target plane, shown in Fig. 3, was elliptical, with major and minor diameters as great as 900 and 600 μm, respectively, depending on focusing. Lineouts taken through the footprint shows speckle-to-speckle variation ~15% with overall smoothness ~10%.

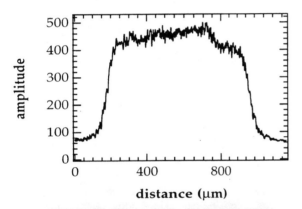

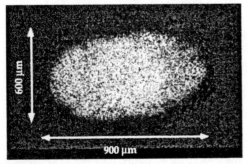

FIGURE 3. Drive laser foot print.

Second, preheat of the sample by penetrating x rays or hot electrons produced in the laser-interaction region must be low. Preheat can cause

an uncertainty in the initial state of the sample which translates directly into an uncertainty in the final state. To determine the optimum ablator/pusher combination a series of Lasnex simulations and thermal expansion measurements were performed. A Michelson interferometer, sketched in Fig. 2, measured the thermal expansion of the pusher as well as shock planarity. The Al pusher (at the Al/D_2 interface) forms one arm of a Michelson interferometer. Thus, before shock arrival, if the aluminum is heated significantly, the sample will begin to expand, causing a shift in the fringe image. Upon shock breakout of the Al, Al unloads into hydrogen, and fringes disappear due to the rapid fringe movement from the large shock velocity. The interferometer-probe beam was a 10-ns-FWHM, 355-nm laser pulse appropriately time-delayed from the Nova drive beam. Results of the calculations and measurements show that a combination of a low Z ablator (~20 μm thick CH) and a thick (100-250 μm depending on drive) Al pusher lowered the preheat of the Al/hydrogen interface to below ~300 C, which is the detection limit of our instrument.

Figure 4 shows the results of 2 different thermal expansion measurements on Al. Figure 4a and 4b show the breakout of aluminum tophat pushers, shown in Fig. 1, during two D_2 EOS experiments. These pushers were 100 μm thick Al with a 1 mm OD and coated with 20 μm of CH and 1000 Å Al. The probe laser beam was reflected off of the rear surface of the Al after passing through a .5 mm thick sapphire window and a .450 mm long reservoir of liquid D_2. The initial cell temperatures were 19.6 K. Figure 4(a) shows the interferogram generated when I ~ 8.5 x 10^{13} W/cm^2. Motion of the D_2/Al pusher interface is clearly observed beginning approximately 4 ns prior to shock breakout. Calculated shock velocities from Lasnex with this ablator/pusher combination scale with intensity, I in W/cm^2, as $Us(\mu m/ns)=24 (I/10^{14})^{0.287}$, so the predicted breakout occurs at ~ 4.4 ns after the start of the drive beam. Thus preheating is occurring early in the drive pulse. The source of the preheat is likely x rays with energies just under the Al k-edge at 1.56 keV. Since the temperature in the laser deposition region is 1-2 keV, there is a significant x-ray flux at this frequency. Finally, Fig. 4a shows the shock is planar over the central 300-400 μm of the target with rarefaction waves moving inward from the edges causing the observed curvature.

distance (μm)

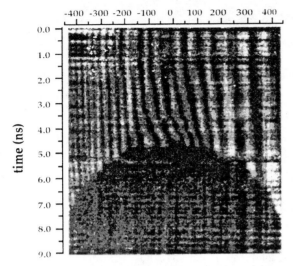

distance (μm)

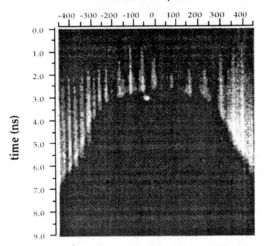

FIGURE 4 streaked fringe images to determine preheat level. Details are discussed in the text

When I is turned down to 1.75 x 10^{13} W/cm^2, no evidence of preheating at the shock front is observed as shown in Figure 4(b). Here the predicted shock breakout time is 6.9 ns after the start of the drive. The region of shock planarity is 200-300 μm. Here, however, a second region of shock curvature is observed. This structure is attributed to the reentrant pusher design. Aluminum plasma moves into the path of the drive beam

during its 8 ns duration, effectively reducing the drive laser intensity at the perimeter.

Interferograms of the thicker (180 and 250 μm) pushers exhibited no rear surface motion for I up to 2.4x10^{14} W/cm^2. For a detection limit of 0.2 fringe, which corresponds to movement of 30 nm at the pusher surface, the maximum pusher surface temperature for these targets prior to shock breakout is estimated to be < 400 K.

Finally, shock steadiness and accurate measurements of Us, Up, and ρ/ρ_o were made with high resolution streaked radiography of the shocked hydrogen. Radiography was performed with ~ 800 eV photons from a plasma x-ray source produced by focusing a second beam of Nova onto an Fe disk (10 ns at 6x10^{13} W/cm^2). The backlighter was placed 12 cm from the target cell to eliminate possible heating of the cell and to produce a near-collimated source. The effective source size in the imaging direction was ~ 150 μm and was set by the width of the laser focal spot. Interferometry shows the x-ray backlighter had no effect on the D$_2$ in the cell. X rays transmitted through the target cell were imaged by a Kirkpatrick-Baez (K-B) microscope onto a streak camera. The K-B microscope's bandpass was 750-840 eV, and the collection half-angle was 2.5 mrad. Two calibrated magnifications were used: 33× and 82×. The resolution of the K-B microscope in this geometry was found to be better than 3 μm over a 300-μm-wide field of view. The microscope imaged a strip 300 μm long by 5 to 30 μm wide, depending on magnification and configuration. The steak camera was calibrated in space and time by a beat mode radiograph of a gold wire grid. The x-ray pulse train exposure and grid shadow were fit to a series of Gaussians to determine the peaks and thus the absolute time or object position vs. position across the film. The temporal resolution was ~ 20 ps over 8 ns.

A streaked radiograph of shock-compressed D$_2$ is shown in Fig. 5. I=10^{14} W/cm^2 for 8 ns. The bright area in the figure is the view through the side windows of the cell. Because the pusher is opaque and the liquid transparent, the Al/D$_2$ interface is the boundary between the light and dark regions. In the figure, the interface is stationary prior to 2 ns. At 2 ns, the laser-driven shock crosses the interface, and the pusher surface accelerates to a steady speed (U_p). The shock front seen moving ahead of the interface is made visible

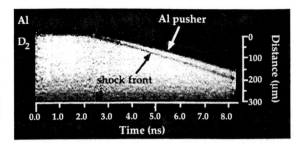

FIGURE 5. A time-resolved, side-on radiograph of laser-shocked D_2. The bright area views the D_2 through beryllium windows bounded by the x-ray-opaque aluminum pusher above. The pusher is seen advancing after breakout at 2 ns; the shock is the dark line in front of and moving faster than the pusher—D_2 interface.

because backlighter x rays grazing the shock-front interface are refracted to greater than 2.5 mrad; out of the angular field of the K-B microscope. Detection of the shock front by refraction is similar to the Schlieren technique for detecting density gradients. The steady propagation of both the shock front and the interface is demonstrated by their linear trajectories until ~ 6 ns, when a stronger shock enters the D_2. First shock U_s and U_p were constant to better than 1%. The second shock is caused by shock reverberations in the pusher. (In this example, no data after 6 ns were used).

U_s and U_p, are evaluated from the slopes, and the single-shock compression can be determined by Eq. (2). Compression can also be measured directly from the film as long as U_s and U_p are constant. At any time t, the compression is equal to the ratio of two lengths: the distance between the shock front $X_2(t)$ and the initial interface position X_o, and the distance between the shock front and the interface $X_1(t)$.

$$\rho / \rho_0 = \left(X_2(t) - X_0\right) / \left(X_2(t) - X_1(t)\right).$$

Because all the measurements are made on one piece of film in the streak camera, uncertainties in ρ/ρ_0 due to magnification and sweep speed are canceled. Experiments show a steady U_s and U_p for 4 to 8 ns and values of ρ/ρ_0 calculated directly or using Eq. (2) gave the same results within experimental error, except with larger experimental uncertainty.

The shock position observed in the radiograph is the leading part of the shock front that emerged from the center of the pusher. In some experiments, the apparent Al/D_2 interface position at $t = 0$ on film was not identical to the actual value of X_o because

the rotation of the cell about the axis perpendicular to both the backlighter path and the shock path could be controlled only to within 3 mrad. This resulted in the center of the pusher being shadowed by an edge of the pusher at very early times. In these cases, X_O was determined by extrapolating the shock and interface position to the point of intersection. This resulted in an increased uncertainty in ρ/ρ_0 from approximately ±3% to approximately ±5%.

The Al/D_2 interface is subject to the Richtmyer-Meshkov hydrodynamic instability (RM). However, using the measured pusher surface finish of 30 nm, we calculated that the largest perturbation expected from RM is less than 0.5 μm during the time of observation.

RESULTS AND CONCLUSION

Figure 6 shows pressure versus final density for our data, the Sesame D_2 EOS table[17], the dissociation model[14], and the D_2 gas gun data[13]. Data above 300Kbar show a larger compression as compared to the low pressure data. Final densities determined from the known initial densities and the measured compressions. As explained previously, the error bars are governed predominantly by accuracy in determining the slopes of the shock and interface trajectories in the radiographs. The figure also plots the D_2 Hugoniots from the dissociation model,[14] the Sesame D_2 EOS table[17] and the D_2 gas-gun data.[13] At the lowest compression, our data are in agreement with the earlier results; at higher compressions where there is no gas gun data, the laser data show a significantly enhanced compressibility as compared to the Sesame prediction but similar to that of the dissociation model. The Dissociation model shown in Fig. 6 differs slightly from that shown in Ref. 11. The previously reported Hugoniot was a preliminary calculation, and a small conceptual improvement in the theory led to the difference.

In conclusion, these experiments demonstrate that laser-driven shocks can effectively be used for EOS studies at pressures beyond those attainable by traditional techniques. Our results suggest the mass distribution in the Jovian planets is different than previously thought. Also, the more compressible EOS of hydrogen offers higher performance and improved margin for NIF ignition capsules.

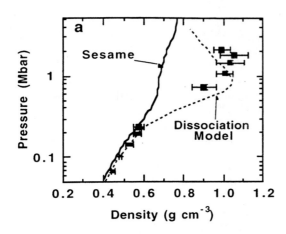

FIGURE 6. D_2 P versus ρ for our laser produced shock data (dark squares) , gas gun data (dots) Sesame model (solid line), and dissociation model (dashed line).

REFERENCES

1. S. Ichimaru, H. Iyetomi, and S. Tanaka, *Phys. Rep.* **149,** 91 (1987).
2. N. W. Ashcroft, *Phys. World* **8,** (7) 43 (1995).
3. R. Smoluchowski, Nature 215, 691 (1967); V. N. Zharkov and V. P. Trubitsyn, *Jupiter,* T. Gehrels, Ed. (University of Arizona Press, Tucson, 1976) pp. 135-175; W. B. Hubbard, *Science* **214,** 145 (1981); W. J. Nellis, M. Ross, and N. C. Holmes, *Science* **269,** 1249 (1995).
4. J. D. Lindl, *Phys. Plasmas* **2,** 3933 (1995).
5. W. B. Hubbard, *Astrophys. J.* **152,** 745 (1968).
6. M. Ross, F. H. Ree, and D. A. Young, *J. Chem. Phys.* **79,** 1487 (1983).
7. D. Saumon, G. Chabrier, and H. M. Van Horn, *Astrophys. J. Supp.* **99,** 713 (1995).
8. W. R. Magro, D. M. Ceperley, C. Pierleoni, and B. Bernu, *Phys. Rev. Lett.* **76,** 1240 (1996).
9. S. W. Haan, et al., *Phys. Plas.* **2,** 2480 (1995);
10. W. J. Krauser, et al., *Phys. Plas.* **3,** 2084 (1996).
11. L. B. Da Silva, et al., *Phys. Rev. Lett.* (1997).
12. H. K. Mao and R. J. Hemley, *Rev. Mod. Phys.* **66,** 671 (1994).
13. W. J. Nellis, et al., *J. Chem. Phys.* **79,** 1480 (1983).
14. N. C. Holmes, et al., *Phys. Rev. B* **52,** 15835 (1995).
15. S. T. Weir, et al., *Phys. Rev. Lett.* **76,** 1860 (1996).
16. Y. B. Zel'dovich and Y. P. Raizer, *Physics of Shock Waves and High-Temperature Hydrodynamic Phenomena* (Academic Press, New York, 1966).
17. G. I. Kerley, *A Theoretical Equation of State for Deuterium,* Los Alamos Scientific Laboratory Report LA-4776 (New Mexico, January 1972).
18. R. J. Trainor, et al., *Phys. Rev. Lett.* **42,** 1154 (1979).
19. R. Cauble, et al., *Inertial Confinement Fusion, 1993 ICF Annual Report,* UCRL-LR-105820-93, pp. 131-136.
20. P. C. Souers, *Hydrogen Properties for Fusion Energy* (University of California Press, Berkeley, 1986).
21. S. N. Dixit, M. D. Feit, M. D. Perry and H. T. Powell, *Opt. Lett.* **21,** 1715 (1996).

CP429, *Shock Compression of Condensed Matter – 1997*
edited by Schmidt/Dandekar/Forbes
1998 The American Institute of Physics 1-56396-738-3/98/$15.00

SOUND VELOCITIES IN SHOCKED LIQUID DEUTERIUM

N. C. Holmes, W. J. Nellis, and M. Ross

Lawrence Livermore National Laboratory, Livermore, CA 94550

Recent measurements of shock temperatures and laser-driven Hugoniot measurements of shocked liquid deuterium strongly indicate that molecular dissociation is important above 20 GPa. Since the effect of dissociation is small on the Hugoniot pressure up to the 30 GPa limit of conventional impact experiments, other methods must be used to test our understanding of the physics of highly compressed deuterium in this regime. We have recently performed experiments to measure the sound velocity of deuterium which test the isentropic compressibility, a derivative quantity. We used the shock overtake method to measure the shock velocity at 28 GPa. These preliminary data provide support for a recently developed molecular dissociation model.

INTRODUCTION

During the past several years, the properties of shocked liquid deuterium and hydrogen have been the subject of many new experimental studies. The equation of state (EOS) along the principal Hugoniot was determined by Nellis, *et al.* (1), and theoretical models were developed subsequently. (2,3) Recently, new shock temperature measurements led to the development of a new model, in which the effective molecular dissociation energy decreases with decreasing volume. (4) Extrapolation of those results to pressures above 100 GPa indicated that deuterium would be substantially more compressible in this pressure region than previously thought. More recently, laser-driven experiments were performed to test that prediction, (5) and those results support the predictions of the model in large part. However, the uncertainties in the latter experiments are large. In other experiments, Weir, *et al.* (6) found evidence for the formation of metallic hydrogen at higher densities and lower temperatures, using quasi-isentropic compression of the liquid.
Measurements of the EOS of the fluid in the pressure range of the most accurate experimental methods (im-

pacts) generally will not provide convincing evidence of small amounts of molecular dissociation, since their effect on the EOS below 30 GPa is small. This suggests the measurement of derivative quantities such as sound velocity or the Grüneisen parameter, both of which should be sensitive to the to the rate of change of compressibility. The sound velocity c of a rarefaction wave is given by

$$c^2 = -\left(\frac{\partial P}{\partial \rho}\right)_S \qquad (1)$$

where P is pressure, ρ density, and the derivative is at constant entropy S. The sound velocity is also related to the Grüneisen parameter through the expression (7)

$$c = V\left[\left(\frac{\partial P}{\partial V}\right)_H \left[(V_0 - V)\frac{\gamma}{2V} - 1\right] + P\frac{\gamma}{2V}\right]^{1/2} \qquad (2)$$

where the Grüneisen parameter given by

$$\gamma = V\left(\frac{\partial P}{\partial E}\right)_V \qquad (3)$$

can be obtained from shock Hugoniot experiments. Measurements of the sound velocity using the over-take method (8) can provide high accuracy, and an unambiguous test of our understanding of dense, hot, fluid deuterium.

EXPERIMENTAL

The shock overtake method exploits the fact that the sound velocity behind a shock is greater than the shock velocity, and thus will overtake it eventually. A target is impacted with a thin impactor; when the impact-generated shock reaches the back surface of the impactor, the pressure drops and a rarefaction wave moves at the sound velocity into the target. If the sound velocities in all layers at high pressure are known, then the time when the overtaking rarefaction reaches the shock provides a direct measure of the sound ve-

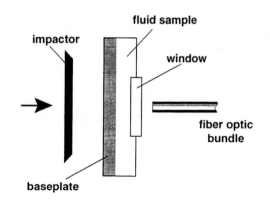

FIGURE 1. Schematic of the optical arrangement for sound velocity experiments. The Ta impactor in embedded in a polycarbonate sabot (not shown), and impacts the Al baseplate, generating a strong shock. Light emitted by the shocked liquid D_2 sample is collected by a seven-fiber bundle which is placed near the Al_2O_3 window. Not shown are trigger pins and a Pt resistance thermometer placed inside the sample cavity.

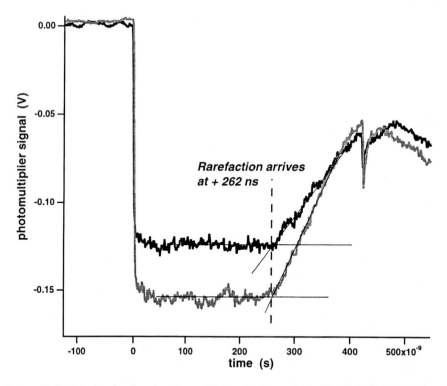

FIGURE 2. Photomultiplier data showing time dependence of light emitted from shocked liquid deuterium at 506 and 700 nm, for the experiments described in the text at a pressure of 28 GPa. The regions of relatively constant signal ($0 < t < 260$ ns) indicate steady shock propagation in the D_2 sample. The overtaking rarefaction wave decreases the pressure and emitted intensity when it arrives at the shock front. To determine the arrival time, we made lineat fits to the data, and took the intersection as the arrival time. The sharp and brief intensity peak at about 420 ns is due to shock arrival at the Al2O3 window.

locity. For shocked fluids at high temperatures such as deuterium, the intensity of the emitted light varies very strongly with shock velocity or pressure. Thus, observation of the light intensity emitted by the shock front provides a sensitive way of determining the overtake time.

We used a two-stage light gas gun to accelerate a thin (0.5 mm) Ta projectile to 7.41 km/s in the experiment described here. The design of the cryogenic targets and sample preparation methods for hydrogen experiments were described elsewhere. (1) The projectile was incident on a 1.5 mm Al baseplate, and the D_2 sample was 6 mm thick, with an initial temperature of 20 K and density of 0.171 g/cm^3. Light emitted from the shocked sample was collected by a 7-fiber bundle placed next to an Al_2O_3 window in contact with the sample. This system has been described previously (9,4) and is depicted schematically in Fig. 1. Measurement of the steady part of the signal provides a measure of temperature and emissivity of the shocked D_2 at the Hugoniot pressure of 28 GPa. Since the rise

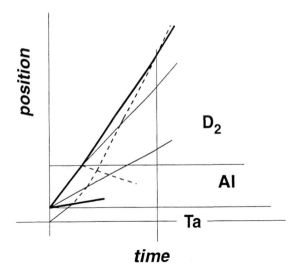

FIGURE 3. Wave interactions and graphical solution for the overtake experiment in liquid D_2. Bold lines indicate shock fronts, thin lines the position of material interfaces, and dashed lines the rarefaction waves. Note that the rarefaction which eventually overtakes the D_2 shock traverses a region of shocked and released Al

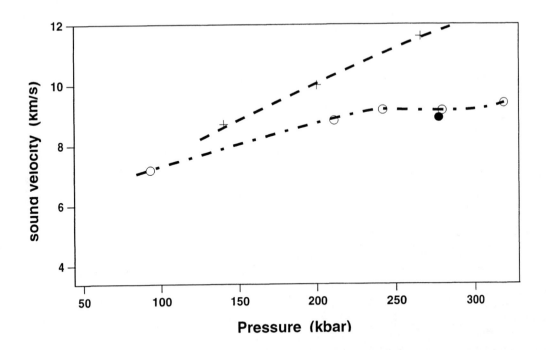

FIGURE 4. Comparison of two calculations of the sound velocity in shocked deuterium vs.shock pressure. The dashed line is calculated from Hugoniot data on liquid H_2 and D_2, the dot-dash line and open circles from the model presented in Ref 4. The experimental point at 28 GPa and a compressed density of 0.6 g/cm^3, with a sound velocity of 8.91 km/s is depicted by the solid circle. Since this datum is preliminary, the experimental uncertainty has yet to be determined, and no error bars are shown.

of the light emission occurs when the shock enters the sample (it becomes optically thick in < 1 ns), the overtake time can be located easily as shown in Fig. 2.

We used a graphical method to determine the overtake velocity, as depicted in Fig. 3. In the figure, the positions of the material interfaces are shown by thin solid lines, the shocks by bold lines, and the leading edge of the rarefaction waves are represented by dashed lines. The sound velocity in shocked Ta at the initial shock pressure of 189 GPa was found by Shaner, *et al.* (10), and we used unpublished measurements of the sound velocity in shocked Al at this pressure by Brown. (11) The experimental geometry we used is complicated by a backward rarefaction wave at the Al/D_2 interface. We used a simple Grüneisen model of the Al to calculate the sound velocity in Al initially shocked to 189 GPa and released to 27.6 GPa.

At the conditions of the experiment, the D_2 sample density and temperature are 0.653 g/cm^3 and 5500 K, respectively. However, we note that these values are based on a modest extrapolation of the measured Hugoniot. The sound velocity was found to be 8.91 km/s. Since the sound velocity in the shocked and released Al is still somewhat uncertain, this value should be viewed as preliminary.

Now let us compare this datum with the predictions of the new model proposed earlier (4), and calculations based on previous experiments. The experiments reported by Nellis, *et al.* (1) in shocked H_2 and D_2 allow us to calculate an approximate Grüneisen parameter, since we can determine Eq. (3), and thus sound velocity (Eq. 2). These are plotted in Fig. 4, and show that the present data support the idea of enhanced dissociation of D_2 at moderate shock pressures. Clearly, more data are needed over a wider pressure range.

ACKNOWLEDGMENTS

We are grateful to many for success of this experiment, and wish to credit J. Crawford, K. Stickel, and W. Brocious for two-stage gun operations, and E. F. See, Jr. for target fabrication and characterization. We thank J. M. Brown for the use of his unpublished data for the sound velocities in shocked Al, and A. C. Mitchell for many helpful discussions. This work was supported by the U. S. Department of Energy under contract W-7405-ENG-48.

REFERENCES

1. W.J. Nellis, A.C. Mitchell, M.van Thiel, G.J. Devine, R.J. Trainor and N. Brown, *J. Chem. Phys* **79**, 1480 (1983).

2. D. Saumon and G. Chabrier, *Phys. Rev.* **A44**, 5122 (1991) ; D. Saumon and G. Chabrier, *Phys. Rev.* **A46**, 2084 (1992).

3. M. Ross, F.H.Ree and D.A.Young, *J. Chem. Phys.* **79**, 1487(1983).

4. N. C. Holmes, W. J. Nellis, and M. Ross, *Phys. Rev.* **B52**, 15835 (1995).

5. L. B. Da Silva, P. Celliers, G. W. Collins, K. S. Budil, N. C. Holmes, T. W. Barbee, Jr., B. A. Hammel, J. D. Kilkenny, R. J. Wallace, M. Ross, R. Cauble, A. Ng, and G. Chiu, *Phys. Rev. Lett.* **78**, 483 (1997).

6. S. T. Weir, A. C. Mitchell, and W. J. Nellis, *Phys. Rev. Lett.* **76**, 1860 (1996).

7. R. G. McQueen, S. P. Marsh, and J. N. Fritz, *J. Geophys. Res.* **72**, 4999 (1967).

8. R. G. McQueen, J. W. Hopson, and J. N. Fritz, *Rev. Sci. Instrum.* **53**, 245 (1982).

9. N. C. Holmes, *Rev. Sci. Instrum.* **66**, 2615 (1995).

10. J. W. Shaner, J. M. Brown, and R. G. McQueen, in *High Pressure in Science and Technology,* Homan, MacCrone and Whalley, eds. North-Holland (New Yor,k, 1984) p. 137 ff.

11. J. M. Brown, private communication

CP429, *Shock Compression of Condensed Matter – 1997*
edited by Schmidt/Dandekar/Forbes
1998 The American Institute of Physics 1-56396-738-3/98/$15.00

ANALYTIC EQUATION OF STATE FOR H6

HERMENZO D. JONES and FRANK J. ZERILLI

Naval Surface Warfare Center, Indian Head Division
Indian Head, MD 20640-5035

The fluid constituents of the reaction products for H6 are described by a perturbation technique based on their intermolecular interactions, which may include angular dependent contributions. Standard solid-state approaches are applied to the solid component of the reaction products. A flexible time dependent model for the reaction of aluminum in explosive mixtures has been implemented in EPIC2, a Langrangian material dynamics computer program. Predicted detonation velocity and cylinder test wall velocities for H6 are in good agreement with experimental data.

INTRODUCTION

Time dependence of the energy release of metallized explosives has been observed and must be accounted for in predicting their performance. A preliminary investigation of the combustion of aluminum indicates that it probably takes place behind the reaction zone. This is an important factor in the partition of energy between the shock wave and the bubble in underwater applications.

For hydrocode calculations of underwater shock-wave phenomena, the Jones-Wilkins-Lee (1) (JWL) equation of state (EOS)

$$p = Ae^{-R_1 V} + Be^{-R_2 V} + \omega E/V \qquad (1)$$

with parameters based on the cylinder test is the most widely used description of the reaction products of explosives. In Eq. (1) p is the pressure, E is the internal energy, and V is the relative volume, while A, B, C, R_1, R_2 and ω are constants. However, the JWL EOS is an empirical formulation which combines a solid-like Gruneisen part and a gas-like contribution. In this approach, it is usually assumed that the Gruneisen parameter is a constant, but calculations for N_2 at pressures and temperatures characteristic of the release isentrope for reaction products show this to be erroneous. The resulting isentropic exponent, $q=-(\partial \ln p/\partial \ln v)_s$, exhibits a double maxima which cannot be explained on physical grounds.

In this work the formalism of Weeks, Chandler and Anderson (2) is employed to describe fluids whose molecules interact via a spherically symmetric, modified Buckingham (exp-6) potential. It is assumed that the repulsive forces provide the dominant contribution to the properties of dense fluids. The intermolecular potential is divided into a reference part that is repulsive in character and a perturbation that is attractive. Division of the potential is carried out so that the resulting free energy is a minimum or essentially constant with respect to the break point as suggested by Ree (3). Intermolecular potential parameters for the appropriate molecular species are chosen to be consistent with Hugoniot data and are given by Jones and Zerilli (4).

The descriptions for the thermodynamic properties of liquid Al and Al_2O_3 as well as their solid counterparts are also given in Ref 4. A semi-empirical EOS suggested by Cowan and Fickett (5) is used for solid carbon. The details of this model for solid carbon is described by Wienbenson, et al. (6).

Chemical equilibrium calculations for the

Chapman-Jouguet (CJ) parameters for several aluminized explosives, including H6 , which consist of a heterogeneous mixture of fluid and solid constituents, are performed with the Jones-Zerilli (JZ) code (4) with the constraint that the Al remains inert. A comparison is made between the measured (7) and predicted detonation velocities. For H6, two tabular EOS surfaces are constructed. The first corresponds to Al remaining inert, while for the second, the Al is allowed to react completely. From this an analytic form for the $p(E,V)$ surface, the pressure as a function of internal energy and volume, is constructed for both reaction regimes. A time dependent function is introduced to join the two surfaces. This approach is used in the EPIC code (8) to model cylinder tests for H6. The calculated wall velocities are compared with experimental data (9).

EOS FOR HYDROCODES

Utilizing the JZ chemical equilibrium code (4), tabular representations of the $p(V,E)$ surfaces were generated for H6 adjacent to the isentropes , shown in Fig. 1, associated with Al remaining inert and complete reaction, respectively. These surfaces were fit to a polynomial representation of the form

$$p(E,V) = \sum A_{mn} E^{m} / V^{n} \qquad (2)$$

and incorporated into the EPIC material dynamics computer program.(8) The time dependence of the aluminum was accounted for with a time dependent pressure given by

$$p(t) = [1 - \lambda(t)]p_1(E,V) + \lambda(t)p_2(E,V) \quad (3)$$

where

$$\lambda(t) = a \frac{1 + \tanh(\frac{t - t_d}{t_r})}{2} \qquad (4)$$

In Eq. (3), p_1 is the equilibrium pressure for the case in which the aluminum is unreacted, and p_2 is the pressure for the case in which the aluminum has totally reacted to form Al_2O_3. The parameters a, t_d and t_r in Eq. (4) are the extent of the reaction, the delay time of the reaction and the rise time of the reaction, respectively. It is expected that they are particle size dependent.

RESULTS

The potential parameters used for all of the reaction product constituents, except solid carbon

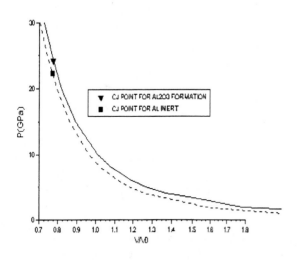

FIGURE 1. Isentropes for H6 with $\rho_0 = 1.75$ Mg/m³.

TABLE 1. CJ Parameters for Aluminized Explosives

HE	$\rho_0(Mg/m^3)$	$D_{exp}(km/s)$	$D_{calc}(km/s)$	$P_{calc}(GPa)$	T(K)
H6	1.75	7.490	7.586	20.9	3211
HBX-1	1.712	7.310	7.395	19..5	3328
HBX-3	1.84	7.120	7.237	18.2	2885
PBXN-109	1.697	7.630	7.532	19.3	2994

are given in Ref. 4. Details of the solid carbon EOS used in these calculations is given in Ref. 6.

The theoretical calculations for the detonation properties of the explosives include the allowance of a fluid-fluid separation associated with the species N_2 and H_2O. Therefore, three separate phases are permitted in the chemical equilibrium calculations:

Fluid 1 - N_2 , H_2O , CO_2 , CO , NH_3 , CH_4 , NO ,

H_2 , O_2 , $Al(l)$, $Al_2O_3(l)$

Fluid 2 - N_2 , H_2O , CO_2 , CO , NH_3 , CH_4 , NO ,

H_2 , O_2 , $Al(l)$, $Al_2O_3(l)$

Solid - $C(s)$, $Al(s)$, $Al_2O_3(s)$.

The detonation properties of several aluminized explosives are now considered with the assumption that the aluminum does not react. In Table 1 it is seen that the experimental results (7) are adequately predicted by the current model, which suggests that the aluminum reacts behind the wave front.

Cylinder test calculations were carried at for H6 with a modified EPIC2 (8) utilizing Eq. (3) for the representation of the constitutive relation for H6. The reaction kinetic parameters employed in Eq. (4) are taken as follows: $a=1$, $t_d = .2\mu s$ and $t_r = 1.5\mu s$. Predictions for the wall velocity for a 1" diameter cylinder test with H6 are compared with experiments(9) in Fig. 2. It is seen that the time

dependence of the energy release is quite important. The theoretical velocity rise is a little below the measured value initially, but matches well at later expansions. The theoretical results are even better when

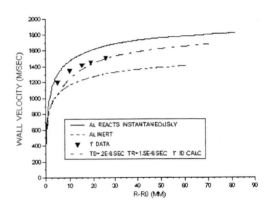

FIGURE 2. Cylinder test with 1"ID for H6 with $\rho_0 = 1.75Mg/m^3$

compared with the 2"diameter results(9) in Fig. 3.

CONCLUSION

The EOS for the reaction products of aluminized explosives, have been calculated using liquid-state perturbation theory for the simple fluid constituents and semiempirical descriptions for the solid components and the aluminized products. Calculations for the

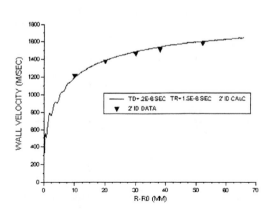

FIGURE 3. Cylinder test with 2" ID for H6 with $\rho_0 = 1.75$ Mg/m³

detonation velocities of several aluminized mixtures with the assumption that the Al reacts behind the shock front were in good agreement with measured values. A constitutive relation was constructed for H6 and interfaced with the EPIC2 Code. Theoretical predictions of the wall velocities from 1" and 2" diameter cylinder tests for H6 described the experimental results very well.

ACKNOWLEDGMENTS

This work was supported by the NSWC TIP Program.

REFERENCES

1. Lee, E. L., Horning, H. C. , and Kury, J. W., Lawrence Livermore Laboratory, Report UCRL-50422 (1968).
2. Weeks, J. D., Chandler, D., and Anderson, H. C., *J. Chem. Phys.* **54**,5237 (1971).
3. Ree, F. H., *J. Chem. Phys*. **84**, 5845 (1986).
4. Jones, H. D., and Zerilli, F. J., *J. Appl Phys*. **69**, 3893 (1991)
5. Cowan, R.D., and Fickett, W., *J.Chem. Phys*. **24**, 932 (1956).
6. Wienbenson, W.E., Zwisler, W.H., Seely, L.B. and Brinkley, Jr., S.R.,, "TIGER Documentation Volume," Stanford Research Institute (prepared for Ballistic Research Laboratory, Aberdeen, Maryland-Contract No. DA- 04-200-AMC-3226(X), Nov. 1968), SRI Publication No. Z106
7. Hall, T.N., and Holden, J.R., "Explosion Effects and Properties: Part III. Properties of Explosives and Explosive Compositions," NSWC MP88-116, Oct. 1988, NSWC, Dahlgren, VA 22448-5000.
8. Johnson, G.R. and Cook, W.H., in *Proceedings of the Seventh International Symposium on Ballistics*, The Hague, The Netherlands, 1983, p. 541.
9. Clairmont, Jr., A.R., "PBX's in Large Navy Munitions Program, Vol. 7, Cylinder Expansion Testing, NSWC TR80-43, November 15, 1988, NSWC, Silver Spring, MD 20903-5000.

CP429, *Shock Compression of Condensed Matter – 1997*
edited by Schmidt/Dandekar/Forbes
© 1998 The American Institute of Physics 1-56396-738-3/98/$15.00

AN EQUATION OF STATE FOR DETONATION PRODUCTS INCORPORATING SMALL CARBON CLUSTERS

M. Sam Shaw

Group T-14 MS B214, Los Alamos National Laboratory, Los Alamos, New Mexico 87545

A theoretical equation of state for detonation products is presented that incorporates the small cluster behavior of the carbon. For small diamond clusters of the size found in recovery experiments, the fraction of carbon atoms on the surface can be as much as 25%. The composition and properties of the clusters are modeled with the dangling bonds capped by various radicals composed of C, H, N, and O from the background molecular fluid mixture. A perturbation theory approach is used for the mixture of molecular fluids that also includes features based on Monte Carlo simulations. For example, the effect of cross potentials on nonideal mixing in chemical equilibrium simulations, is shown to be well approximated by an entropy shift and ideal mixing. Comparison is made of the EOS with individual species Hugoniot data and with detonation velocity data for a variety of explosives. In addition, recent data for PBX-9501 is utilized which characterizes sound speed, overdriven Hugoniot, adiabatic γ, Grüneisen γ, and a precise thermodynamic CJ state(1).

INTRODUCTION

High explosives detonation products form a very complicated system at rather extreme conditions. Despite the high densitiy (up to 3 g/cm^3), the products are primarily a fluid mixture of H_2O, CO_2, CO, N_2, and additional minor molecular species. In addition, there are carbon clusters consisting of a few thousand atoms in graphite-like and diamond-like structures with the added complication of a large fraction of the cluster being on the surface.

EQUATION OF STATE

We describe here three interrelated areas of approach being used to develop a theoretical EOS to accurately describe these complicated systems. First, thermodynamic simulation methods (Monte Carlo and molecular dynamics) are developed and implemented to provide a benchmark of "exact" results. These exact results provide computational experiments for thermodynamics results for a given potential. Second,

a perturbation theory based method is used for the practical implementation. Particular choices of perturbation theory approximations are made which give accurate thermodynamics and chemical equilibrium for a given set of potentials. Third, data analysis methods are used to extract EOS information from new precision experimental data. These data as well as detonation velocities and Hugoniots for individual species are used to constrain the choices of potentials.

The benchmark simulation methods provide the testing ground for the various approximations that are used in the practical implementations. Initially, we used molecular dynamics simulations to characterize the thermodynamics of individual molecular fluids (at pressures and temperatures characteristic of detonation products) with very nonspherical interactions(2) (e.g. N_2 and CO_2). The simulations demonstrated that these nonspherical molecules were not freely rotating under typical conditions in detonation products. A density of states tranformation Monte Carlo method(3) was developed that efficiently simulates a large range of states from a single reference simulation. Most important, we have developed the $N_{atoms}PT$ ensemble Monte Carlo method(4). This method incorporates the chemical equilibrium of a molecular mixture as a natural extension of standard Monte Carlo methods. From an atomic simulation perspective, correlated moves are attempted which interchange atoms between molecules. With a proper accounting of the acceptance probability of these type moves, chemical reactions are allowed in a manner such that the chemical equilibrium composition is determined by an average over states sampled by the simulation. The effect of cross potentials on nonideal mixing was studied by this method. The shift in chemical equilibrium (resulting from a shift in cross potential) had a larger effect on the EOS than the shift in cross potential at fixed composition. A related consequence is that nonideal mixing can be well approximated by ideal mixing plus a constant entropy shift of each constituent that results in shift in chemical equilibrium. In Fig. 1, we illustrate the effect of a small shift in cross potential on the chemical equilibrium composition. In this case, the simulation is for a mixture of N_2, O_2, and NO at 30 GPa and 3000 K. The potentials are those described previously(5). The only difference in the two simulations is that the scaled radius of the $N_2 - O_2$ potential for the set designated X is expanded by about 2% to get the set O. Now, using the perturbation theory outlined below, ideal mixing (lower solid line) goes right through the set X. That is, ideal mixing corresponds to a particular choice of cross potential. This simulation method can even be used to find the CJ state of a molecular fluid mixture. With the increased speed of workstations, the direct construction of a tabular EOS for detonation products may be a practical alternative to the perturbation methods discussed below.

The goal of the practical EOS implementation is to have an accurate, predictive, and physically based method. For the molecular fluid components, we use Ross's perturbation theory method(6) for spherical potentials that is accurate to 1% for single species fluids in this regime. In some cases, we have used an effective spherical

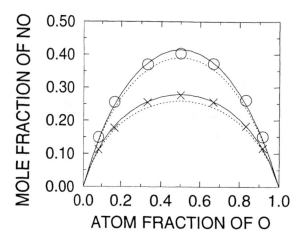

Figure 1. Simulation for two different cross potentials, X and O (see text). Lines are perturbation theory with ideal mixing and entropy shifts.

potential to characterize the very nonspherical interaction. The effective potential is determined through an equally accurate perturbation theory approach that we have developed and tested against simulation benchmarks. The constant entropy shift approximation mentioned above is used to include nonideal mixing effects from the largely uncharacterized cross potentials.

The solid carbon is treated as a cluster rather than a bulk solid. Recovery experiments have found residual diamondlike clusters on the order of 20 Å diameter. With roughly 1000 carbon atoms in a cluster, around 20-30% of the carbon atoms are on the surface in the diamond phase. We allow for the surface dangling bonds to be capped with H, OH, NH_2, NO_2, and the like. A qualitative picture of a diamond cluster of around 20 Å diameter is shown in Fig. 2. In the current model, only the net composition of the surface is prescribed along with an effective density and Debye temperature. Future implementations will allow for the surface equilibrium composition to vary according to free energy contributions of individual capping groups. The detonation products EOS is strongly dependent on the carbon phase and on the surface composition.

Because of the cluster nature of the carbon and the difference in composition between phases both graphite and diamond clusters can exist over a limited range (typically a few GPa). These

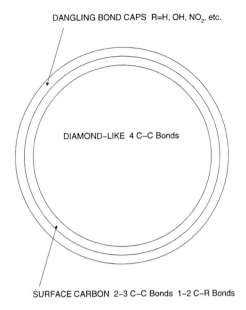

DIAMOND–LIKE 4 C–C Bonds

SURFACE CARBON 2–3 C–C Bonds 1–2 C–R Bonds

Figure 2. A qualitative diagram of the structure of a 20 Å diameter diamond cluster.

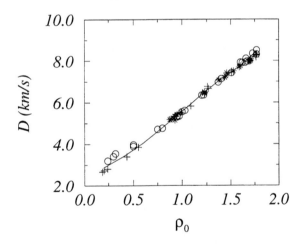

Figure 3. Detonation velocity of PETN versus ρ_0. Comparison of experiment (symbols) and theory (line).

shifts in the phase of carbon are reflected by small, but significant changes in the CJ detonation velocity as a function of initial density, ρ_0, as seen in Fig. 3 for PETN[7]. The current theory has no solid carbon for ρ_0 below 1.0 g/cm^3, a graphite-like phase up to ≈ 1.4, a mixed phase up to ≈ 1.5, and a diamond-like phase at higher values of ρ_0.

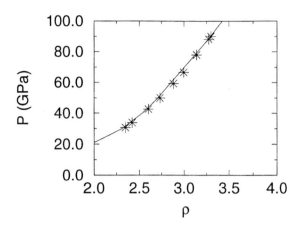

Figure 4. Overdriven Hugoniot data for PBX-9501 (*) compared with theory (line).

The individual potential for a given species is primarily determined from Hugoniot data and where available, spectroscopy measurements leading to a temperature[8]. Constants affecting nonideal mixing and carbon are determined in a coupled manner from detonation velocity data for various compositions and initial densities. In addition, new data from group DX-1, discussed below, provides a more precise constraint[1,9].

The two types of experiments discussed here have in common that the initial state in the detonation products is overdriven. This has the advantage of minimizing the reaction zone effects and any coupling of hydrodynamics to the EOS. The combination of Lagrange sound speed and overdriven Hugoniot data allows for the precise determination of the CJ state. The sonic condition is readily found from the intersection of the two curves. With a combination of a well chosen local EOS form and some statistical analysis, we have determined the CJ state for PBX-9501 to about 1% accuracy. In addition, derivative quantities such as the adiabatic γ and the Grüneisen γ are determined from the data. These provide a much stronger constraint on the detonation products EOS than simple detonation velocities. In Fig. 4, we compare the theory with the overdriven Hugoniot for PBX-9501 with the carbon constrained to be in the diamond phase. In Figures 5 and 6, we see the comparison with the adiabatic γ and the Grüneisen γ, respectively.

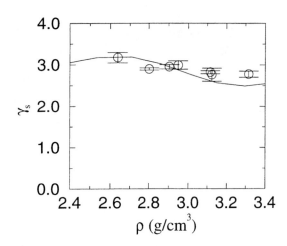

Figure 5. Adiabatic γ data (O) and theory (line) as a function of density.

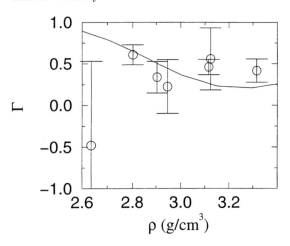

Figure 6. Grüneisen γ data (O) and theory (line) as a function of density.

The release isentrope experiments require the inversion of interface velocimetry data. Here we use a tabular reference isentrope form and hydrodynamic calculations. A computationally intensive nonlinear least squares procedure is used to extract the best EOS match to the velocity data. The high precision of the data leads to a very accurate isentrope that provides strong constraints on the EOS with the advantage of starting from a well characterized initial state. A comparison of the theory with preliminary results for PBX-9501 are shown in Fig. 7.

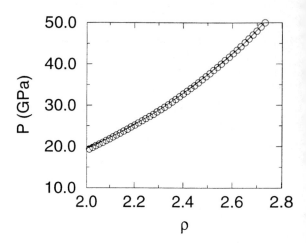

Figure 7. Inverted release isentrope from velocimetry data (lines) and theory (O).

DISCUSSION

The calibration of nonideal mixing terms and the model of the diamond-like carbon clusters leads to an very good representation of the detonation products equation of state. Further refinement of the EOS will be made using a nonlinear least squares procedure to simultaneously fit a variety of data. More precision data of the type described above, will be included as available. Allowance for shifts in the chemical composition of the surface of the diamond-like clusters will probably be needed to improve the global accuracy.

REFERENCES

1. J. N. Fritz, R. S. Hixson, M. S. Shaw, C. E. Morris, and R. G. McQueen, *J. Appl. Phys.* **80**, 6129 (1996).
2. M. S. Shaw, J. D. Johnson, and J. D. Ramshaw, *J. Chem. Phys.* **84**, 3479 (1986).
3. M. S. Shaw, *J. Chem. Phys.* **89**, 2312 (1988).
4. M. S. Shaw, *J. Chem. Phys.* **94**, 7550 (1991).
5. M. S. Shaw, Proceedings of the 10th International Detonation Symposium, pp. 401-408 (1995).
6. M. Ross, *J. Chem. Phys.* **71**, 1567 (1979).
7. H. C. Hornig, E. L. Lee, M. Finger, and J. E. Kurrie, Proceedings of the 5th Symposium (International) on Detonation, pp. 503-510 (1970).
8. S. C. Schmidt, D. S. Moore, and M. S. Shaw, *J. Chem. Phys.* **107**, 325 (1997).
9. P. K. Tang, R. S. Hixson, and J. N. Fritz, this conference.

CP429, *Shock Compression of Condensed Matter – 1997*
edited by Schmidt/Dandekar/Forbes
© 1998 The American Institute of Physics 1-56396-738-3/98/$15.00

SOUND SPEED AND THERMAL PROPERTY MEASUREMENTS OF INERT MATERIALS:
LASER SPECTROSCOPY AND THE DIAMOND-ANVIL CELL

Joseph M. Zaug

University of California, Lawrence Livermore National Laboratory
L-282, P.O. Box 808, Livermore, California 94550

An indispensable companion to dynamical physics experimentation, static high-pressure diamond-anvil cell (DAC) research continues to evolve, with laser diagnostics, as an accurate and versatile experimental technique. Together, static and dynamic high-pressure and temperature physics and geophysical studies of deep planetary properties have bootstrapped each other in a process that has produced even higher pressures; consistently improved calibrations of temperature and pressures under static and dynamic conditions; and unprecedented data and understanding of materials, their elasticity, equations of state (EOS), and transport properties under extreme conditions. A collection of recent pressure and/or temperature dependent acoustic and thermal measurements and deduced mechanical properties and EOS data will be summarized for a wide range of materials including H_2, H_2O, H_2S, D_2S, CO_2, CH_4, N_2O, CH_3OH, SiO_2, synthetic lubricants, PMMA, single crystal silicates and ceramic superconductors. Room P & T sound speed measurements will be presented for the first time on single crystals of β-HMX. New high-pressure and temperature diamond cell designs and pressure calibrant materials will be reviewed.

INTRODUCTION

The condition that we call "room pressure and temperature" (RPT) represents a minuscule fraction of the universe as we know it. In order to understand the physics and chemistry of non terrestrial environments, both natural and man made, requires the marriage of high pressure and temperature experimentation with theoretical and computational efforts. A good measure of this union lies in the physically-based predictive capabilities of computer models. Today, mantle convection, protein dynamics and detonation physics models rely heavily on empirically derived relations.

High pressure and temperature experiments currently fall in one of two categories: dynamic and static.[1] Static high P & T experiments serve to provide virtual snapshots (> 1 second) of dynamic environments and therefore can elucidate unique effects of shock loading experiments.[2] Relative to static experiments, dynamic methods (<ns) intrinsically apply more shear forces thereby accelerating the onset of phase transitions and material deformations. Shock dynamic and static DAC measurements compliment each other insofar as in their ability to completely map out the P-V-T space of candidate materials. DAC experiments are versatile in that one chooses a P-T point for measurements, while

thermodynamically irreversible dynamic experiments can access the highest pressures and corresponding temperatures. The next generation of dynamic experiments will dramatically extend sample volumes to cubic centimeters (ATLAS at LANL) and push accessible pressures beyond one TPa (NIF at LLNL). The current P-T regimes available to DAC's are shown superimposed over the geotherms of terrestrial and Jovian planets Fig. 1.

DAC experiments have been helping to unravel fundamental problems: What are the physical and chemical mechanisms, including the kinetics, responsible for the onset of detonation? How does thermal transport, hot spots and fragmentation affect HE material sensitivity? What are the mechanisms responsible for reversible and irreversible protein denaturation or folding and what are their energetic pathways? Is Earth mantle convection a layered or con

---

[1] For the sake of brevity, all references to static should be thought of as hydrostatic or quasi-hydrostatic.

[2] As an example, consider that the thermodynamic Grüneisen parameter γ decreases, for materials shocked to the fluid state, with increasing pressure in shock-wave experiments due to the increasing temperature along the Hugoniot [1,2].

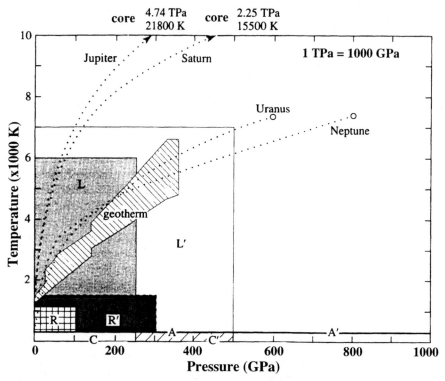

FIGURE 1. Presently accessible P-T ranges for DAC experiments [3]: A, ambient temperature; C, cryogenic; R, external resistance heating; L, laser heating. Corresponding ranges for future development are denoted as A', C', R' and L'. Dotted curves indicate estimated P-T in Jovian planetary interiors [4] and the geotherm covers the temperature range in terrestrial planetary interiors [5].

tinuum process? Aside from these programmatic issues, DAC research remains a vital avenue toward increasing our grasp of solid state physics and opening our eyes to the experimentally wide open fields of liquid and glassy state physics.

This paper brings to light some recent DAC derived results from acoustic and thermal measurements. A focus on laser driven experiments serves to refine this review and should not detract from recent advances in traditional ultrasonic [6] and calorimetric [7] methods.

ACOUSTIC & THERMAL MEASUREMENTS IN THE DAC

A brief description of the techniques used to make acoustic and thermal measurements is given below. Examples of these methods follow in the material studies section.

Acoustic Measurements

The acoustic measurements described here consist of classical, frequency-domain, optical Brillouin scattering (BS); and a time-domain method, impulsive stimulated light scattering (ISLS).[3] A theoretical

comparison between these two methods has been made elsewhere [8] and recently a direct inter-laboratory comparison has been made on San Carlos olivine up to 17 GPa [9]. Thermal DAC measurements employ either external heaters, external coolers, or lasers; or a combination of external and laser controlled systems. There are very few citations that describe simultaneous acoustic and thermal measurements in a DAC (*i.e.*, c(P,T)).

Optical BS measurements represent the natural progression of the original technique [10]. In Brillouin spectroscopy the inelastic Bragg diffraction of incident photons from acoustic phonons, measured as frequency shifts Δv, allows one to determine longitudinal (c_l) and transverse (c_t) sound speeds.

$$\Delta v_x = \pm 2 \frac{n c_x}{\lambda_o} \sin\left(\frac{\theta}{2}\right), \quad \text{(where x = l or t).} \quad (1)$$

In the above expression λ_o is the incident light wave-

---

[3] The name implies that no assumptions are made concerning how light is coupled to the material modes of a given sample.

length, n is the index of refraction and θ is the angle between the DAC axis (normal to the diamond table) and the incoming laser beam. For crystallographically isotropic samples, where n = 1 at θ = 90, running concurrent BS experiments at two different scattering geometries provides a means to determine n(P). Equation (1) is not valid in the presence of dispersion (*i.e.*, where $\Delta v_X / \sin(\theta/2)$ changes with θ at constant P & T). If one studies anisotropic materials, and can determine the density, ρ, using x-ray diffraction (XRD), and can tolerate additional uncertainties in c_X, n can be dropped from (1). Without XRD the density can be obtained by using a high pressure dilatometer [11]. If n(P,T) is known the approximate density can be obtained by Lorentz-Lorenz's formula:

$$\frac{4}{3}\pi N_O \alpha = \frac{M_w}{\rho} \frac{n^2 - 1}{n^2 + 2} \qquad (2)$$

where α is the molecular polarizability, and N_O and M_w are Avogadro's number and the molecular weight, respectively.[4]

Lastly, $\rho(P)$ can be calculated using

$$\rho(P) - \rho_O = \int_{P_O}^{P} \frac{\gamma}{c^2} dP \qquad (3)$$

where ρ_O, P_O, and γ are the RPT density, pressure and heat capacity ratio C_p/C_v. The bulk velocity is given by $c^2 = c_1^2 - (4/3)c_t^2$.

In the most general form of an ISLS experiment, two successive 1064 nm "excitation" pulses, $\sim$80 ps in duration, selected from the output train of a continuously pumped Q-switched and mode-locked Nd:YAG laser, are recombined in the sample volume of the DAC at an angle θ, but otherwise coincident in space and time. When the polarizations of the two excitation pulses are parallel, interference establishes a periodic distribution of intensity in the sample and thus, in the case of an absorbing sample, a (spatially) periodic variation in the temperature and pressure which launches a set of three counterpropagating acoustic waves (one quasi-longitudinal and two quasi-transverse) of wavelength λ_A. The acoustic wavelength, in this case equal to the period of the optical grating, d, may be

expressed in terms of the wavelength of the laser light λ_E

$$d = \lambda_A = \lambda_E / (2 \sin(\theta/2)). \qquad (4)$$

The impulsively excited acoustic waves induce a temporally and spatially periodic variation in the index of refraction of the sample. A third pulse, from the same Q-switched envelope as the excitation pulses is doubled to 532 nm and delayed by time of flight to generate the probe. Monitoring the intensity of the Bragg scattering of the probe by the acoustic grating as a function of probe delay serves to determine the frequency (f_A), and hence the velocity $c = (f_A \cdot \lambda_A)$ of the acoustic waves. In general, the response of any Raman-active vibrational mode of the appropriate wavevector (whether acoustic or optical) to the periodic distribution of electric field generated by the crossed excitation pulses will be, if the period of vibration is long compared to the duration of the excitation pulses, a coherent standing wave which will serve to coherently scatter a delayed probe. For experiments reported here, the pulse duration is such that only acoustic modes are excited. In a DAC, the orientation of the acoustic grating with respect to the crystal axes may be varied by rotating the cell about an axis normal to the sample or diamond faces (See Fig. 2.). The velocity of sound, measured as a function of crystallographic direction (on 2-4 samples) and the density serves to determine the independent elastic constants

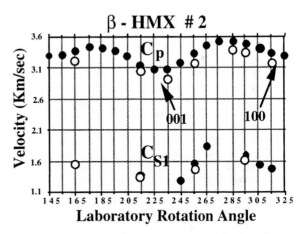

Figure 2. Quasi-longitudinal and quasi-shear sound speeds in single crystal β-HMX as a function of laboratory rotation angle and temperature. Open circles are 380 °K and filled circles are at 297 °K. At the 322.5° rotation angle ultrasonic propagation is most nearly along the crystallographic a axis.

Thermal Measurement

There are two distinct approaches to applying heat to a

---

[4] Affects on n by $\alpha(P)$ are on the order of $1:10^4$ [12]. The polarizability in molecular crystals (N_2O, CO_2) drops by $\sim$0.9% after 1 GPa [13]. It is not clear that assumptions about α hold up say in high P-T salt solution studies.

sample in a DAC: resistive heating (external and or internal) and internal laser heating. External resistive heating of DACs, in evacuated ovens, provides the most stable and accurate means of reaching increasingly better characterized pressures and temperatures up to 1700 °K. More care must be taken with internal (within the sample chamber) applications of heat. Because diamond is an excellent thermal conductor, internal resistive or ohmic DAC heating must be calibrated using known melting points of ionic and metallic solids. The resistivity of internal ohmic heating conductors and the thermal conductivity of deposited insulators will change with pressure and temperature. Unlike externally heated diamonds taken beyond 1400 °K, ohmically heated diamonds may survive beyond one high P-T cycle since the hottest portion of the diamond is within its stability field.

Laser heating can reach the highest static temperatures (> 6000 °K), but not without encountering some rather vexing problems. Apparent sample temperature determinations (made from spectroradiometric interrogation of thin, laser heated, DAC samples) are dependent on the selection of conductivity models or temperature distributions that one inserts into a numerical expression for depth-averaged light intensities [14]. In addition, the magnitude of observed greybody emissivity (ε) hinges on detailed thermal distributions of a sample which can vary with experimental configurations. It would seem that the large temperature gradients intrinsic to laser heating experiments (whether caused by the laser heating process, the measuring process, or a combination of both) is a non trivial issue due to the limited sample area and large variations in sample to pressure medium absorption ratios of the laser light in the DAC. To some degree radial temperature gradients have been reduced by using flat profile multimode lasers while axial temperature gradients have been reduced by simultaneous heating of both sides of the DAC [15]. The use of Raman scattering techniques to determine temperature may prove to be useful on laser heated samples that have Raman active modes [16].

In addition to acoustic studies, ISLS has been used to measure one dimensional thermal diffusivity [17]. As mentioned above, the spatial and temporal overlap of two ~80 ps 1064 nm pulses in a slightly absorbing sample results in a spatially periodic distribution of temperature where the period of the resultant grating, d, is given in (4). When the acoustic disturbance has been fully damped or has propagated beyond the area illuminated by the probe, a spatially periodic variation in the temperature and density remains. The characteristic time, τ_{th}, for the exponential decay of this "thermal" grating in a medium of density (ρ) specific heat at constant pressure (c_p) and thermal conductivity tensor (κ) is given by

$$\tau_{th} = \frac{d^2}{4 \pi^2 D_{th}} \qquad (5)$$

where d is the grating period and the thermal diffusivity is defined by $D_{th} = \kappa / \rho c_p$. The grating spacing is kept sufficiently small so that conduction normal to the plane of the sample is unimportant (i.e., where one dimensional diffusion equation applies).

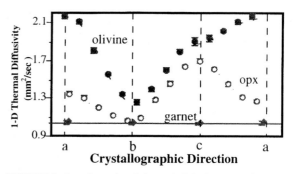

FIGURE 3. One-dimensional thermal diffusivity as a function of crystallographic direction in olivine, orthopyroxene, opx, and garnet at 298 °K and 1 atm [17]. The b-a segment represents a 90° rotation about the c axis from b to a, the b-c segment rotation is about the a axis and the a-c segment is about the b axis. Uncertainties in fits to $D_{th}(t_{th}^{-1}, d^{-2})$ data are also provided.

MATERIAL STUDIES

Room pressure and temperature (gas to solid phase) materials have been investigated for their high pressure acoustic and thermal properties using the methods just described. As a result, determinations of the pressure and/or temperature dependence of structural and super-conducting phase transitions, melting curves, elastic constants, acoustic damping rates, structural relaxation rates and amplitudes, anisotropic thermal diffusivity tensors, EOS's, and molecular polarizabilities have been made.

Planetary physics depends, to some degree, on the direct confrontation of theory and computational models with experimentally determined properties of candidate materials under relevant P-T conditions. The recent availability of elastic moduli on hydrogen (EOS up to 24 GPa) [18,19], methane (up to 5 GPa) [20], nitrous oxide and carbon dioxide (up to 4.5 GPa) [21], water (ice VII) (2.2 - 7.4 GPa) [22], and β-O_2 (6 - 9.5 GPa) [23]; including the pressure dependence of n and α in N_2O and CO_2; will improve our grasp of the compositional stratifications, oscillations and evolutional history of the Jovian planets and their moons.

The acoustic velocities, refractive indices and elastic constants of two sulfides (hydrogen and deuterium) have been determined (up to 5 GPa) [24]. The results indicate that molecular rotation-translation coupling, though present at 1 GPa (stronger in D_2S), diminishes with increasing compression. This is probably due to the strengthening of hydrogen bonds in H_2S & D_2S.

Traditionally the high pressure EOS has been the principle experimental probe in the regime of strong repulsive interactions. An acoustic and thermal study of the equilibrium and structural relaxation in methanol (up to 30.4 GPa) has demonstrated that the relaxation times may well provide more discrimination between model hamiltonians [25]. The relaxation times from this ISLS study were determined by direct measurement of the dispersion $c(f_A)$, where acoustic attenuation is observable, and by fitting the time-domain spectra to the form

$$I = (A_{th} - A_{ac}C \ (t)e^{-\gamma_{ac} t} \cos \omega t + A_{str} \ (1 - e^{-\gamma_m t}))^2 \quad (6)$$

where A_{th} is the amplitude of the thermal grating, A_{ac} the amplitude of the acoustic grating, A_{str} is the amplitude from structural relaxation, $\omega = 2\pi f$ the circular frequency, γ_{ac} the temporal acoustic absorption coefficient, and $\tau_m = 1/\gamma_m$ the characteristic Mountain mode (R. D. Mountain, *J. Res. NBS- A. Phys. & Chem.*, 70, pg. 207, (1966)). $C(t)$ is a geometric runout term where t is the time delay of the probe beam.

Detonation physics is no different than planetary physics in the sense that there is legitimate need for experimental data. Mechanical and thermal transport data on appropriate single crystals of HE and binder materials, as a function of P & T, is scarce. Recently the pressure dependence of the elastic constants (up to 0.30 GPa) and density (up to 0.50 GPa) of poly (methyl methacrylate) were determined [26]. At 0.11 GPa a second-order glass transition occurs resulting in a dramatic divergence in the bulk modulus from Young's modulus. Sound speed measurements in single crystals of β-HMX have been made as a function of temperature [27]. Preliminary one bar measurements indicate the quasi-longitudinal sound speeds (c_p) from 298-380 °K, vary depending on the crystallographic propagation direction, from -3.5 to -5.1% [$(dc_p/dT)_P = -1.4e^{-3}$ to $-2.2e^{-3}$] which corresponds to a 7 to 10% drop in the adiabatic modulus as defined by $K_s = \rho c^2$. Similarly, the variation in the quasi-shear sound speeds (c_s) is -2.2 to -6.7% [$(dc_s/dT)_P = -3.7e^{-4}$ to $-1.3e^{-3}$] which corresponds to a 5 to 13% drop in primarily off diagonal contributions to the overall modulus. At one bar there is little or no hysteresis in quasi-longitudinal or quasi-shear sound speeds when cycling between 298 and 380 °K.

Results from thermal DAC measurements on ammonium dinitramide map out the pressure, and temperature phase diagram, including a new polymorph transition, up to 10 GPa from 198 - 393 °K [28]. A novel DAC technique, high pressure & low temperature matrix isolation, has been developed to trap intermediate (thermally initiated) product species from a thermally initiated propellant HNIW [29]. The method quenches the reaction and allows for spectroscopic determination of the trapped constituents.

Physical properties of lubricants determine the level of elastohydrodynamic lubrication (and hence mechanical wear) of gears, ball bearings and traction devices. Pressure dependent sound speed measurements were made for several paraffinic and naphthenic synthetic lubricants (up to 2.7 GPa) [30,31]. The slope, dc/dP, in some of these lubricants, decreased remarkably at about 0.4 GPa. This change in slope could be an indication of increasing poly-crystal cluster formations which, if true, would have significant industrial ramifications.

The nature of the high pressure phase of silica glass under shock compression has long been a mystery. Shock experiments indicate that a transformation to stishovite should occur, yet there is none to be found in shock-recovered samples. Despite the large temperature variation (< 4500 °K), the bulk velocities determined statically to 57 GPa are broadly consistent with the Hugoniot velocities [32]. In the end there probably are several metastable states produced including poorly crystallized states that can be reached depending on the chosen (P-T)-time route.

Superconductivity in high-T_c superconductors is generally thought to be dependent on j, the number of CuO_2 layers per chemical unit. A cryogenic DAC four-point resistance measurement on j = 3 (up to 21 GPa) and j = 4 (up to 14 GPa) superconductors reveals that it is the type of CuO_2 layer (inner as opposed to outer) that dictates the behavior of $T_c(P)$ [33].

Acoustic and thermal DAC measurements on mineral silicates remains extensive [9, 17, 34, 35, 36]. Conducting ISLS backscattering measurements on gold coated minerals has proven to be a viable means of attaining shear velocities when none are present in the standard transmission measurements.

ADVANCES IN DAC EXPERIMENTATION

DAC technology continues to improve with the addition of new cell designs (hydrothermal studies [37], uniaxial stress [38], XRD [39], IR microspectroscopy [40], IR microspectroscopy featuring *in situ* cryogenic pressure tuning [41], internal ohmic heating [42]) including different gem materials [43], pressure calibration scales (Sm:YAG fluorescence [44], $^{13}C/^{12}C$ Raman scale [45], improvements to the ruby fluorescence scale [46]), gas loading devices [47], near hydrostatic pressure medium fluids [48], and simple microlens attachments that effectively triple optical signals[49]. Significant boosts in pressure have been achieved by increasing the slip friction between diamond-metal gasket interfaces [50].

CONCLUSIONS AND THE FUTURE

The information provided above brings together the most significant acoustic DAC studies reported over the last three and a half years. Some of the static thermal measurement property determinations have been highlighted to give one a feel for where things stand and where they are headed. The next three years will bring more c(P,T) studies conducted on simple fluids and fluid mixtures, mineral silicates, energetic materials and refractory metals. Accessible static pressures will increase by at least 30% and internal heating techniques will be improved upon. The future looks bright!

ACKNOWLEDGMENTS

The author found the library at LANL to be most accommodating to the task demanded by this paper. This work was partially supported by program CU70 at LANL (Phil Howe) and by the auspices of the U.S. Department of Energy at Lawrence Livermore National Laboratory under contract number W-7405-Eng-48.

REFERENCES

1. D. A. Boness, J. M. Brown, *Shock Waves in Condensed Matter*, Presented at the APS Conference on Shock Compression of Matter, Albuquerque, NM., August 14-17, 1989.
2. D. A. Boness, J. M. Brown, and J. W. Shaner, *Rarefaction Velocities in Shocked Lead*, Presented at the APS Conference on shock Waves in Condensed Matter, Monterey, CA., June 20-23, 1987.
3. H.-K. Mao, R. J. Hemley, *Phil. Trans. R. Soc. Lond. A*, **354**, 1315-32 (1996).
4. V. N. Zharov, T. V. Gudkova, *High Pressure Research in Mineral Physics: application to Earth and planetary sciences*, Tokyo: Terra, Geophysics Monograph 67, Mineral Phys., 1992, vol. 3, pp. 393-401.
5. J.-P. Poirier, Introduction to the Physics of the Earth Interior, Cambridge University Press, 1991, p. 264.
6. G. Chen, R. Miletich, R. Mueller, H. A. Spetzler, *P. E. P. I.*, **99**, 273-87, (1997).
7. O. P. Korobeinichev, L.V. Kuibida, A. A. Paletsky, A.G. Shamakov, *Combustion Chemistry of Energetic Materials Studied by Probing Mass Spectroscopy*, Presented at the Materials Research Society Symposium Proceedings, Boston, MA., Nov. 27-30, 1995.
8. Y.-X. Yan, and K. A. Nelson, *J. Chem. Phys.*, **87**, 6257-65 (1987).
9. E. H. Abramson, J. M. Brown, L. Slutsky, and J. Zaug, *J. Geophys. Res.*, Vol. 102 , **B6**, 12,253 (1997).
10. L. Brillouin, *Ann. Phys.*, **17**, 88 (1922).
11. W. Dollhoph, S. Barry, M. J. Strauss, Presented at the Frontiers of High-Pressure Research Proceedings, 1991.
12. P. S. Peercy, G. A. Samara, B. Morosin, *J. Phys. Chem. Solids*, **36**, 1123 (1975).
13. H. Shimizu, H. Sakoh,, S. Sasaki, *J. Phys. Chem.*, **98**, 670 (1994).
14. M. Manga, R. Jeanloz, *Geophys. Research Lett.*, **23**, 1845-1848, (1996).
15. G. Shen, H. K. Mao, R. J. Hemley, *Laser heating diamond-cell technique: double-sided heating with multi-mode Nd:YAG laser*, Presented at ISAM, Tsukuba, Japan, 1996.
16. G. I. Pangilinan, Y. M. Gupta, *J. Appl. Phys.*, **81**, 6662-69, (1997).
17. M. Chai, J. M. Brown, L. Slutsky, *Phys. & Chem. in Minerals*, **23**, 470-75 (1996).
18. C. Zha, T. S. Duffy, H.-K. Mao, R. J. Hemley, *Phys. Rev. B*, **48**, 9246, (1993).
19. T.S. Duffy, W. L. Vos, C. Zha, R. J. Hemley, H.-K. Mao, *Science*, **263**, 1590-93, (1994).
20. S. Sasaki, N. Nakashima, H. Shimizu, *Physica B*, **219 & 220**, 380-82, (1996).
21. H. Shimizu, H. Sakoh, S. Sasaki, *J. Phys. Chem.*, **98**, 670-73, (1994).
22. H. Shimizu, S. Sasaki, *Phys. Rev. Lett.*, **74**, 2820-23, (1995).
23. E. H. Abramson, L. J. Slutsky, J. M. Brown, *J. Chem. Phys.*, **100**, 4518-26, (1994).
24. S. Sasaki, H. Shimizu, *J. Phys. Soc. of Japan*, **64**, 3309-14, (1995).
25. J. M. Zaug, L. J. Slutsky, J. M. Brown, *J. Phys. Chem.*, **98**, 6008-16, (1994).
26. K. Weishaupt, H. Krbecek, M. Pietralla, *Polymer*, **36**, 3267-71, (1995).
27. Work in Progress by the author.
28. T. P. Russell, G. J. Piermarini, S. Block, P. J. Miller, *J. Phys. Chem.*, **100**, 3248-51, (1996).
29. J. K. Rice, T. P. Russell, *Chem. Phys. Lett.*, **234**, 195-202, (1995).
30. Y. Nakamura, I. Fujishiro, T. Tamura, *Jpn. Soc. Mech. Eng.*, **38**, 122-27, (1995).
31. Y. Nakamura, I. Fujishiro, K. Nishibe, H. Kawakami, *J. of Tribology*, **117**, 519-23, (1995).
32. C. Zha, R. J. Hemley, H.-K. Mao, T. S. Duffy, C. Meade, *Physical Rev. B*, **50**, 13105-12, (1994).
33. D. T. Jover, R. J. Wijngaarden, R. Griessen, E. M. Haines, J. L. Tallon, *Phys. Rev. B*, **54**, 10175-85, (1996).
34. C. Zha, T. S. Duffy, R. T. Downs, H.-K. Mao, *J. Geophys. Res.*, **101**, 17535-45, (1996).
35. M. Chai, J. M. Brown, L. J. Slutsky, *Geophys. Res. Lett.*, **24**, 523-26, (1997).
36. M.Chai, J. M. Brown, L. J.Slutsky, *The Elastic Constants of an Aluminous Orthopyroxene to 12.5 GPa*, J. Geophys. Res., in the press, (1997b).
37. W. A. Bassett, A.H. Shen, M. Bucknum, I.-M. Chou, *Rev. Sci. Instrum.*, **64**, 2340-45, (1993).
38. G. Jones, D. J. Dunstan, *Rev. Sci. Instrum.*, **67**, 489-93, (1996).
39. D.R. Allan, R. Miletich, R. J. Angel, *Rev. Sci. Instrum.*, **67**, 840-42, (1996).
40. J. C. Chervin, B. Canny, J. M. Besson, P. Pruzan, *Rev. Sci. Instrum.*, **66**, 2595-98, (1995).
41. R. J. Chen, B. A. Weinstein, *Rev. Sci. Instrum.*, **67**, 2883-89, (1996).
42. S. A. Catledge, Y. K. Vohra, S. T. Weir, J. Akella, *J. Phys. Condens. Matter*, **9**, L67-73, (1997).
43. J.-A. Xu, J. Yen, Y. Wang, E. Huang, *Ultrahigh Pressures in Gem Anvil Cells*, Amsterdam: Gordon and Breach Science Publishers SA, 1996, pp. 127-34.
44. J. Liu, Y. K. Vohra, *Appl. Phys. Lett.*, 64, 3386-88, (1994).
45. D. Schiferl, M. F. Nicol, J. M. Zaug, S. K. Sharma, T. F. Cooney, S.-Y. Wang, T. R. Anthony, and J. F. Fleischer, *The Diamond $^{13}C/^{12}C$ Isotope Raman Pressure Sensor System for High-Temperature/Pressure Diamond-Anvil Cells with Aqueous and Other Chemically Reactive Samples*, (Submitted to, J. Appl. Phys., in the press 1997).
46. M. Chai, J.M. Brown L. J. Slutsky, *Geophys. Res. Lett.*, **23**, 3539-42, (1996).
47. T. Yagi, H. Yusa, M. Yamakata, *Rev. Sci. Instrum.*, **67**, 2981-84, (1996).
48. D. D. Ragan, D. R. Clarke, D. Schiferl, *Rev. Sci. Instrum.*, **67**,494-96, (1996).
49. M. I. Scheerboom, J. A. Schouten, *Rev. Sci. Instrum.*, **67**, 853-54, (1996).
50. (J. Akella, S. Weir, private communication).

CP429, *Shock Compression of Condensed Matter – 1997*
edited by Schmidt/Dandekar/Forbes
© 1998 The American Institute of Physics 1-56396-738-3/98/$15.00

HIGH ACCURACY EOS EXPERIMENTS USING THE AWE HELEN LASER.

S.D. Rothman, A.M. Evans.

AWE plc, Reading, RG7 4PR, U.K.

A knowledge of a material's equation-of-state (EOS) is essential for hydrodynamic calculations. Although laser experiments investigate the pressure range between those attainable by gas guns (<few Mbar) and nuclear underground tests (UGT's) (>10Mbar) where no other data exist, it is still advantageous to obtain high accuracy data to discriminate between EOS models which have been compared with gas gun and UGT data to a few percent in pressure. The AWE HELEN laser is being used to obtain high pressure Hugoniot data by the impedance match method. Indirect drive generates pressures up to 10Mbar in the aluminium reference material. Shock velocities are obtained by observing the visible light emitted on break-out from the surface of the target using optical streak cameras. Experiments have been performed on copper and brominated plastic. Attention to target fabrication and metrology, diagnostic calibration, shock uniformity and attenuation and data analysis have enabled us to measure shock velocities to an accuracy of ~1%.

INTRODUCTION.

Laser driven shocks should be naturally complementary to other techniques for measuring EOS; such as gas-guns and explosives at lower pressures and nuclear driven shocks at higher pressures. The problem with laser experiments has been relatively poor accuracy - 10% or worse - in shock or particle velocities (1,2). This was due to short (~ns), highly peaked laser pulses, small (~100μm) target sizes and bulk heating of the target by X-rays and electrons from the laser-target interaction (preheat).

The technique of indirect drive where laser light is used to heat the inside of a small cavity (hohlraum), generating ~100eV temperatures, gives a longer duration and more spatially uniform X-ray flux for shock generation. This allows the use of larger targets which, coupled with improvements in target characterisation, suggested that 1% accuracy shock velocity measurements might be feasible.

EXPERIMENTAL SET-UP.

This experiment uses the impedance match technique with Al as the standard. A 1mm by 1mm diameter cylindrical hohlraum is heated by two 1ns gaussian, 0.53μm wavelength, nominal 500J laser pulses which drives a shock into the target. This consists of a 5-6μm thick Al ablator, a 3-4μm Au preheat shield and a 20μm Al base layer. On the base layer are a 10μm Al step and an 8 μm Cu or 15μm plastic step. The target sits over a 600μm hole in one flat end face of the hohlraum, the steps are 120-150μm apart and centred on this hole. Two optical streak cameras with slits orthogonal and parallel to the step edges are used to monitor shock breakouts from the steps and base.

Shock velocities are calculated from measured step heights and transit times and the assumed known Al EOS enables the pressure and particle velocity in the unknown material to be found using the impedance match technique.

Indirect Drive Uniformity.

As the shock strength measured at the Al step is being used to calculate that at the unknown it is necessary that the shock be spatially uniform. Analytical and numerical view-factor calculations indicated that the X-ray flux at the target would vary by under 1% over a 163μm radius. This was confirmed by monitoring shock breakout times for flat targets (Al ablator and preheat shield as above plus 25μm Al layer). Shot #6075 showed a breakout time variation of ±4.4ps (RMS) over a 400μm diameter.

Similarly the measured mean shock velocity needs to be the true constant velocity. Hydrocode simulations (AWE 1-D Lagrangian code NYM) showed a pressure of 8.8±0.1Mbar for the 0.6ns after the shock entered the Al step. A constant shock velocity in a wedged target should give a breakout time proportional to position along the wedge and this was seen on shot #5315 where breakout time was best fitted by a linear fit to position with an ~8ps RMS error over 1ns.

Target Fabrication and Metrology.

Targets are produced at AWE by physical vapour deposition (PVD) using appropriate masks and additional machining for wedges by a high precision Cranfield lathe. Oxygen is pulsed in during PVD to disrupt the growth of Al crystallites.

Metrology is done using a Zygo New View 100 white light interferometer with a precision of 1Å. Each individual step target is scanned at several stages during full target assembly and the interferometer is recalibrated against standard steps every 2 hours. By comparison with these steps the accuracy of step measurements is estimated to be 0.5%. Thickness variations and surface roughness combined are ~50nm RMS so the resulting total error on a 10μm step is ~0.7%.

Target density has been measured for other Al PVD samples as $2.70 gcm^{-3} \pm 2\%$ but as density is not strictly a random variable, and has not been measured yet for actual targets, the 2% quoted error is not included in error calculations.

Diagnostics and Data Analysis.

The two streak cameras are Hadland Imacon 675's operating at a wavelength range of 580nm (filter) to 700-800nm (photocathode response fall-off). The streak speed is nominally 80ps/mm giving a time resolution of 14ps, magnification is ~13 and spatial resolution <40μm.

Calibration is performed using a short pulse laser and étalon to illuminate the camera slit with a train of short pulses. From pairs of pulses the streak speeds at points on the camera intensifier are calculated. These are fitted to calculate speed over the whole intensifier and this is integrated to relative time using the streak of one pulse as a time constant. These calibrations are checked by analysing a calibration streak and errors of under 0.7% are estimated.

Streaks of target shots are analysed by finding the 50% points of the shock emission rise. Data are selected away from the step edges where there are effects of rounding of the target steps, pressure release and camera resolution. This typically leaves ~30 points each for the base and step regions which can then be fitted to position along the target. Ideally breakout should be at constant time for each region but there is usually some slope or curvature. Fits to all regions are made, ones with the same higher order coefficients being preferred so that transit times can be found from the differences of the zero order coefficients, but if this is not possible the fits are extrapolated to the step edge positions and time differences found by subtraction. The RMS errors on the fits to breakout time are typically 5-7ps or 1 - 1.5% and are the greatest source of error.

Impedance match analysis is done using calculated second Hugoniots rather than simple reflection of principal Hugoniots.

EXPERIMENTAL RESULTS.

Copper Results.

Three series of copper shots have been fired. In each series the drive uniformity and constant pressure have been verified using flat and wedged targets.

FIGURE 1. Streak of shot 8876, copper step target. Time runs left-right, Cu step is above, Al base middle and Al step lower.

A streak of a copper step target is shown in fig. 1. The calculated principal Hugoniot points are plotted in fig. 2 along with a SESAME 3332 Hugoniot and gas-gun, explosively and nuclear driven data (3-8).

The first series results lie consistently above the SESAME Hugoniot, the second series lie on or below it (9) while the third series data are on the Hugoniot. Laser drive was unchanged between the first two series but a new oscillator led to a shorter pulse length for the third - ~800ps as opposed to ~900ps. However flat and wedged shots did not indicate that drive uniformity had been affected. Similarly, the source of preheat should not have been changed significantly. Targets came from different batches and there were small (few %)

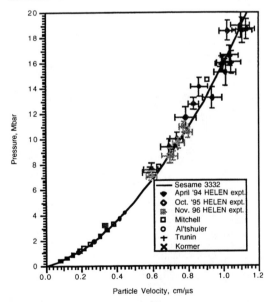

FIGURE 2. Plot of experimental Cu Hugoniot pressure vs particle velocity (with error bars) compared with SESAME 3332 Hugoniot and data from other sources.

differences in step heights and preheat shield thicknesses. The step height should only alter shock decay but this is a second order effect and should not be significant for a few % change. Series III deliberately used 3.2 or 3.8 ±10% preheat shields to investigate the effects of varying this; with no noticeable result. Unless the two sets of targets fortuitously had similar preheat shield thicknesses then this is not the source of disagreement. Series I targets were metrologised using a Zeiss laser scanning microscope and only a few sample step heights were measured leading to an average height with a ±2% error. Later targets were measured individually to higher accuracy with the Zygo interferometer. The cameras were recalibrated for each series and each calibration checked, with series I having the worst (0.7%) error while the later series were ±0.2-0.3%. Finally, small changes were made to the data analysis codes after the first series but the effects were assessed by analysing old data with new codes, and vice versa, and the differences were found to be small: ~1-2% in velocity, comparable to the random errors.

The conclusion from this is that there may be a systematic error in our first series data, probably from the target metrology, leading to higher Cu Hugoniot pressures. Our second and third series data are not inconsistent and both agree with the SESAME EOS.

Brominated Plastic (CHBr) Results.

This material ($CH_{0.946}Br_{0.054}$) is used at LLNL in instability growth and capsule implosion experiments. Its EOS has not been measured and is a source of potential uncertainty in the analysis of some experiments (10). For our experiment target bases and Al steps were produced at AWE and nominal 15µm CHBr steps attached at LLNL by hot pressing of cut foils. Some were flashed with Al, and a thin (1000Å) Parylene layer was coated over the whole target to hold it together.

Metrology was carried out at AWE using the Zygo interferometer. The plastic steps had much greater variations in thickness and surface finish than coated steps - up to several µm - leading to greater

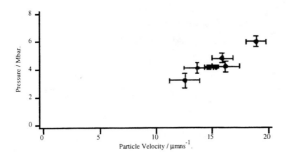

FIGURE 3. Plot of pressure vs particle velocity for brominated plastic (CHBr).

errors unless step measurements were restricted to the regions monitored by the streak cameras.

Two series of shots were fired. The first results for targets with no Al coating showed an optical signal from the CHBr coincident with the laser pulse which was interpreted as due to preheat. The CHBr showed slow (~100ps) shock emission rise as the shock reached the Al base surface and slow (100-200ps) fall when the shock broke out from the CHBr surface, attributed to preheat induced expansion of the CHBr. Shock velocity measurements were of poor accuracy - ~5%. The explanation for the preheat seen in these targets, and not before, is the high opacity of the bromine to gold M-band X-rays from the hohlraum.

The second series used a thicker (4µm) gold preheat shield and these shots have faster shock rise and fall times leading to less noise in breakout times and more accurate - down to 1.2% - shock velocities.

Hugoniot points for CHBr are shown in fig. 3. Although the second series do not show the apparent effects of preheat, there is no inconsistency between the two series' data. This may support the conclusion from the Cu results that preheat is not important.

SUMMARY AND CONCLUSIONS.

We have measured points on the principal Hugoniots of copper and a brominated plastic. Potential sources of error have been investigated - especially with regard to an apparent initial systematic error in our copper data - and minimised.

Our least overall estimated errors are 1.5% in shock velocity leading to ~4% in pressure. If the possible 2% error in target density and a few % in the standard EOS were included these would dominate the error estimate.

Preheat of the targets is also a concern as this leads to expansion of steps and changes in density. Experiments on CHBr indicate that targets with 4µm preheat shields are not significantly preheated while both the Cu and CHBr results suggest that results from targets with both 3 and 4µm preheat shields give consistent Hugoniot data.

Overall we believe we have demonstrated that lasers can make reliable and accurate - ~1% in shock velocity - impedance match EOS measurements.

ACKNOWLEDGEMENTS.

This experiment is very demanding on all aspects of target production and experimental conditions and the efforts of the AWE target fabrication group and laser team are invaluable.

REFERENCES.

1. Romain, J.P., Cottet, F., Hallouin, M., Fabbro, R., Faral, B., and Pépin, H. *Physica* **139** and **140B**, 595-598 (1986).
2. Gu Yuan et al, *Laser and Particle Beams* **10**(3), 611-616 (1993).
3. Mitchell, A.C. and Nellis, W.J., *J. Applied Physics* **52**(5), 3363-3374 (1981).
4. Mitchell, A.C., Nellis, W.J., Moriarty, J.A., Heinle, R.A., Holmes, N.C., Tipton, R.E. and Repp, G.W. *J. Applied Physics* **69**, 2981-2986 (1991).
5. Al'tshuler, L.V., Kormer, S.B. Bakanova, A.A. and Trunin, R.F., *Sov. Physics JETP* **11**(3), 573-579 (1960).
6. Kormer, S.B., Funtikov, A.I., Urlin, V.D. and Kolesnikova, A.N., *Sov. Physics JETP* **15**(3), 477-488 (1962).
7. Al'tshuler, S.B. Bakanova, A.A. and Trunin, R.F., *Sov. Physics JETP* **15**(1), 65-74 (1962).
8. Trunin, R.F., Podurets, M.A., Moiseev, B.N., Simakov, G.V. and Popov, L.V., *Sov. Physics JETP* **29**(4), 630-631 (1969).
9. Evans, A.M., Freeman, N.J., Graham, P., Horsfield, C.J., Rothman, S.D., Thomas, B.R. and Tyrell, A.J., *Laser and Particle Beams* **14**(2), 113-123 (1996).
10. Remington, B.A., Weber, S.V., Marinak, M.M., Haan, S.W., Kilkenny, J.D., Wallace, R.J. and Dimonte, G., *Phys. Plasmas* **2**(1) 241-255 (1995).

CP429, *Shock Compression of Condensed Matter – 1997*
edited by Schmidt/Dandekar/Forbes
© 1998 The American Institute of Physics 1-56396-738-3/98/$15.00

SHOCK COMPRESSION OF HIGHLY POROUS SAMPLES OF COPPER, IRON, NICKEL AND THEIR EQUATION OF STATE

R.F.Trunin, M.V.Zhernokletov, G.V.Simakov, L.F.Gudarenko, O.N.Gushchina

Russian Federal Nuclear Center - VNIIEF, Sarov, Russia

Shock compressibility of copper, iron and nickel samples is studied whose initial density was 5-20 times less than the normal up to~90 GPa pressures. The porous samples were produced from fine-grain powder with sizes of separate grains being several hundreds 0.2-0.3 nm. Explosive generators producing planar and spherical shock waves in the samples were used for the shock loading. The results found are compared with the computations with the semi-empirical equation of state of metals of a variable nuclei and electrons heat capacity.

This paper presents the results of the experiments on generation of non-ideal copper, nickel and iron plasma through compression and irreversible heating of porous targets at the front of powerful shock waves generated by planar or spherically converging shock waves. Recording the shock wave motion velocities in the target and reference samples allows to find the equation of state of shock-compressed plasma using the laws of conservation of mass, momentum and energy and compare it with the computations with the equation of state of non-ideal multi-component plasma.

The parameter range of our concern of severely non-ideal plasma corresponds to reduced (compared to solid-body) values of density ρ_0 and energies higher than the energy of atom and molecule bonds in a solid (~1 ev per particle). To generate such metal states, shock-wave compression of fine (porous) metals was used which enables to markedly enhance the energy dissipation effects at the shock break front and obtain higher plasma heating. Metal powders containing at least 99.5% of the basic element were studied. The samples were pressed cylindrical pellets of diameter D and thickness L. In order to avoid the effect on the shock wave front of the perturbations propagating from the side sample surface, the pellets were made such that to secure the ratio D/L>2.5.

The samples 4<k<8 (where k is the ratio of the metal crystal density ρ_0 to density of the sample being studied ρ_{00}) in porosity were produced from studied metal hydride powder with its subsequent dehydration. The samples of a higher porosity (k>8) were fabricated from fine powder of several (2-3) hundreds of angstrom in sizes of separate grains by the technology based on washing finest particles with a high-speed helium jet from the molten metal surface. The powder oxide impurity content was controlled. In all instances it was no more than a few fractions of percent. The powder particle size was taken accounting the condition of uniform grain warming-up. The technological developments enabled to obtain samples up to k≤ 20.

The dynamic methods of shock-compressed state diagnostics are based on using the general laws of conservation of mass, momentum and energy at the front of a plane stationary shock break.

The wave velocities in the porous samples which were pellets 3-4 mm in thickness were recorded with the electric contact basis method. In doing so, beginning with 10 and more GPa pressures, isolated sensors 0.14 mm in diameter made of PEL-14 wire were used. At lower pressures where operation of these sensors is insufficiently stable the measurements were taken using piezoceramic time markers. Position of the sensors allowed to note the wave velocities within 1-1.5% and, moreover, simultaneously control actuation of the sensors located at the same level.

The tests employed a set of plane-wave and spherical explosive shock wave generators. In all the above devices aluminum was used for the shield the samples under study were placed on as its Hugoniot in the P-U is near the porous metal adiabats and, as it is shown in [1], one may use Hugoniot specular reflection when using the reflection method to estimate the compression parameters in powders under study. This considerably simplifies the estimation of mass velocity following the shock wave front and other shock-wave compression parameters.

Each wave velocity value noted on a particular measuring device is an average value from 3-8 separate individual wave velocity recordings (the maximum number of experiments were made at the highest pressures of shock compression). The newly obtained data is presented in Table. The comparison with the results of [2] in Fig. 1 shows that the copper data obtained on the samples differing in the particle sizes by a factor of 10000 (for k=7.2) coincide with each other. Hence, the shock wave front width which may be related to sizes of individual particles in this case does not affect the experiment results which makes their interpretation more definite. For nickel the new experimental points are logically positioned on the extention of corresponding Hugoniots obtained earlier [3] for lower pressures, see Fig. 2.

TABLE. Experimental data on shock compressibility of nickel, copper and iron porous samples

ρ_{00} g/cc)	k=ρ_0/ρ_0	D км/s	U км/s	P GPa	ρ g/cc
Nickel (ρ_0=8.87 g/cc)					
0,592	15	11,50	9,36	63,7	3,18
0,444	20	11,93	9,74	51,6	2,42
Copper (ρ_0=8.93 g/cc)					
1,240	7.2	5,33	4,79	37,6	5,09
0,893	10	3,40	2,90	8,8	6,01
0,893	10	4,15	3,44	12,7	5,21
0,893	10	5,35	4,29	20,5	4,51
0,893	10	6,58	5,08	29,3	3,92
0,893	10	8,02	6,10	43,7	3,73
0,893	10	11,33	8,69	88,0	3,83
Iron (ρ_0=7.85 g/cc)					
1,570	5	4,16	3,09	2,02	6,10
1,570	5	5,38	3,81	32,2	5,38
0,790	10	4,18	3,50	11,5	4,83
0,790	10	5,35	4,39	18,4	4,37
0,790	10	6,62	5,20	27,0	3,66
0,790	10	8,12	6,23	39,7	3,38
0,790	10	11,68	8,87	81,3	3,27
0,392	20	3,40	3,15	4,20	5,33
0,392	20	5,45	4,79	10,2	3,24
0,392	20	6,75	5,73	15,2	2,59
0,392	20	8,44	6,89	22,8	2,13
0,392	20	12,32	9,87	47,7	1,97

The experimental data description used the equation of state model with a variable heat capacity of nuclei and electrons. In this model the lattice (ion) heat capacity, Grueneisen factor, electron heat capacity are density and temperature functions. Heat capacity of the solid body, molten substance, ideal one-atom gas is described by a unique smooth function of density and temperature.

Melting usually occurs without an abrupt change in thermodynamic values as melting heat is relatively small. These dependencies yield the following limiting transitions:

$C_{vel} \to \beta \cdot T$ at $T << T_f$ degenerate electronic gas,

$C_{vel} \to \dfrac{1}{2} \cdot \beta_0 \cdot T_f$ at $T >> T_f$ free electronic gas.

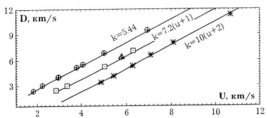

FIGURE 1. Shock wave front velocity D vs mass velocity U following the front in copper.
Experiment: ⊕   k=5.44 [2], ▢ k=7.2 [2];
this paper (Table): ▲ k=7.2, ✳ k=10.
EOS computation ———

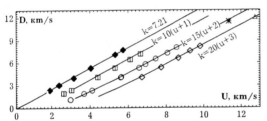

FIGURE 2. Shock wave front velocity D vs mass velocity U following the front in porous nicel.
Experiment: ◆ k=7.21[2], ▢ k=10[3], ○ k=15[3],
◇ k=20[3]; this paper (Table): ✳ k=15, △ k=20.
EOS computation ———

The thermal and caloric forms of the equation of state are as follows:

$$P(\delta,T) = P_x(\delta) + \frac{3 \cdot \psi(\delta) \cdot G_p(\delta) + T}{\psi(\delta) + T} \cdot \rho_0 \cdot T \cdot \frac{C_{vo}}{2} +$$

$$+ \frac{1}{2} \cdot \beta_0 \cdot \gamma_{el}(\delta) \cdot \delta \cdot \rho_0 \cdot \frac{T_f \cdot T^2}{T_f^{f(\delta)} + T}$$

$$E(\delta,T) = E_x(\delta) + \frac{2 \cdot \psi(\delta) + T}{\psi(\delta) + T} \cdot T \cdot \frac{C_{vo}}{2} +$$

$$+ \frac{1}{2} \cdot \frac{\beta_0 \cdot T_f \cdot T^2}{T_f \cdot \delta^{f(\delta)} + T}$$

$$\psi(\delta) = \psi_0 \cdot \delta^{-2/3} \cdot \exp\left[2 \cdot \int_1^\delta \frac{G_p(\tau)}{\tau} d\tau \right]$$

$$G_p(\delta) = a_i \delta^{n_i} + b_i \delta^{m_i} + c_i,$$

where $a_i; b_i; c_i; n_i; m_i$ are constant sets selected for two segments: $0 \le \delta \le 1$ and $\delta \ge 1$

$$\gamma_{el} = f(\delta) + \delta \ln \delta \cdot \frac{df(\delta)}{d\delta},$$

where the function $\psi(\delta)$ is of the same form as $G_p(\delta)$.

$$P_x = a(\delta^n - \delta^m) + b(\delta^l - \delta^m) \text{ if } \delta \le 1$$

$$P_x = \frac{3 \cdot \rho_0 C_0^2}{q - 3 \cdot \mu - 1} \left\{ \delta^{2/3} \exp\left[q\left(1 - \delta^{-1/3}\right) \right] - \delta^{1+\mu} \right\}$$

if $\delta \ge 1$.

The equation of state constructed using this model is completely defined if the parameters $\rho_0, C_{vo}, \psi_0, T_f, \beta_0$ functions $G_p(\delta), \gamma_{el}(\delta), P_x(\delta)$. and values of thermodynamical functions at the two-phase region boundary are known.

These parameters are physically meaningful and their values are known within a certain accuracy. The functions $G_p(\delta), \gamma_{el}(\delta), P_x(\delta)$ are known within some accuracy as well.

The descriptions of the experimental data from studying the shock-wave compressibility of porous Cu, Fe and Ni samples with the equations of state developed using the presented model are given in Figs. 1-3 in the D-U coordinates and in Figs. 4-6 in the P-ρ coordinates.

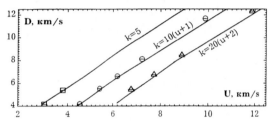

FIGURE 3. Shock wave front velocity D vs mass velocity U following the front in iron.
Experiment: this paper (Table.):
    ⊟ k=5; ⊖ k=10; △ k=20.
EOS computation ———

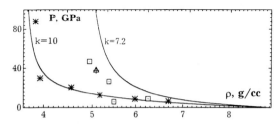

FIGURE 4. Porous copper Hugoniots.
Experiment: ☐ k=7.2 [2];
this paper [Table]: △ k=7.2, ✳ k=10.
EOS computation ———

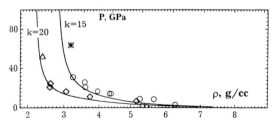

FIGURE 5. Porous nickel Hugoniots.
Experiment: ○ k=15 [3], ◇ k=20 [3];
this paper [Table]: ✳ k=15, △ k=20.
EOS computation ———

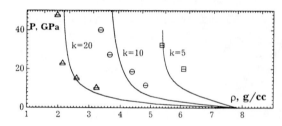

FIGURE 6. Porous iron Hugoniots.
Experiment: this paper [Table]:
    ⊟ k=5, ⊖ k=10, △ k=20.
EOS computation ———

The experimental data presented in the paper provide the information about metal properties in the new area which was not studied earlier.

The work has been done due to financial support of the Russian Foundation for Basic Research (Project № 97-02-16339).

REFERENCES

1. Bugayeva V.A., Yevstigneyev A.A., Trunin R.F. TVT, 1966, V.34, No.5, p.684-690.
2. Trunin R.F., Simakov G.F. et al. ZhETF, 1989, 96(3), 1024-1038.
3. Trunin R.F., Simakov G.F. et al. ZhETF, 1993, 103(6), 2180-2188.

CP429, *Shock Compression of Condensed Matter – 1997*
edited by Schmidt/Dandekar/Forbes
© 1998 The American Institute of Physics 1-56396-738-3/98/$15.00

INVESTIGATION OF TIN THERMODYNAMICS
IN NEAR CRITICAL POINT REGION.

V.Ya.Ternovoi, A.S.Filimonov, V.E.Fortov,
I.V.Lomonosov, D.N.Nikolaev, A.A.Pyalling.

Institute of Chemical Physics in Chernogolovka, Chernogolovka, 142432, Russia.

Near critical point states of tin were generated by the expansion of shocked metal into helium to pressures of several kbars and less. Thermodynamic states on release isentropes originated from shock pressures of 137, 180, and 220 GPa were investigated. Gas dynamic parameters and temperature have been measured by the use of the method of optical pyrometry. The heating of tin on the boundary with shocked gas was described due to the difference between tin and gas temperatures. The problem of the interface heat transfer was solved to obtain the temperature of expanded tin. The point of intersection of experimental and theoretical curves for maximum overheat temperatures was used to estimate parameters of the critical point, which occurred at P_{crit}=0.25 GPa, T_{crit}=7850 K (dT=300K, dP=0.02 GPa). The result is in a good agreement with available evaluations of the critical point. Experimental data obtained have been used for construction of multi-phase wide-range equation of state for tin.

INTRODUCTION

Method of isentropic expansion after intense shock wave loading gives one the opportunity to investigate thermodynamic properties of matter nearby critical point of liquid-vapor phase transition [1,2]. The use of helium as an optically transparent material into which the investigated specimen is expanded makes it possible to measure not only the velocity of the shock wave in the gas, but also the intensity of light emission from specimen-gas boundary. As a result one can obtain information about the pressure and temperature of a specimen at the final state of expansion [3,4]. The density of expanded material can be obtained by calculation of Riman's integral along the isentrope. The release isentrope is traced by variation of initial gas pressure with the same shock wave generator. Gas dynamic material properties in the density range near the critical point are changed from those for condensed matter to gaseous ones. An abrupt in-crease of the expansion velocity was measured with lead on isentropes entering the two-phase region of states from the liquid side of the P-V diagram [3]. Material expansion was described as expansion along the saturation curve below the pressure of this increase. It was proposed [3,5], that this gas dynamic peculiarity is connected with formation of the "boiling wave" [6]. The reason for expansion regime change is the condensation of supercooled metal vapor, evaporated from free surface, and following metal surface instability increase caused by generated compression waves [5]. It was measured also, that the change of expansion regime was realized at final pressure, when expanded supercooled vapor states were close to metal vapor spinodal.

Pulse heating of a metal by helium, shocked to higher temperature, takes place in such experimental conditions. The problem of heat-mass-transfer can be treated in terms of the model of a mixture emitting layer [7]. This layer is formed on specimen free surface due to its disturbance by hydrodynamic instability. To a first approximation, it can be assumed that equal volumes of metal and gas at the final pressure of expansion are mixed. In this procedure of the determination of the final temperature of the emitting layer one needs a value of evaporation heat at the pressures below the critical point and temperatures higher than those of the saturation curve.

In this paper results of determination of tin thermodynamic and gas dynamic properties after unloading from shock states with pressures 137, 180, and 220 GPa down to pressures of one-tenth GPa or less in a manner of [3,4] have been presented. Some results of spectroscopic investigation of the light emission at a final pressure will be discussed to understand the nature of the emitting layer. Finally, a P-T diagram of tin with saturation curve will be presented on the experimental ground and critical point parameters will be evaluated.

EXPERIMENTS

An impact of a stainless steel flyer, accelerated by HE detonation products with experimental assembly [3, 4] has been used to generate plane shock wave in tin foil with thickness of 0.2-0.3 mm (99.99% Sn). Typically the distance of expansion was about ten mm, and the measured time intervals were near one microsecond. The 6-channel optical pyrometer with pin silicon photodiodes and 3 ns time resolution was used to measure the thermal emission intensity at several wavelengths. The interference filters had a band width of 10-15 nm and were centered at 600, 716, 734, 805, 899 and 972 nm, although other wavelengths were also used. To transmit the emission from the experimental assembly to the pyrometer fused quartz optical fibers of 10 meter length were used. The surface area under study had a diameter about 2-4 mm. The signals from photodiodes were amplified and recorded by 500 MHz digital oscilloscopes TDS644 and TDS744. Light emission in spectral range 440-650 nm was registered by OMA [7] with time resolution 20-40 ns in some experiments. Measured voltages were converted to optical intensities by use of a tungsten ribbon lamp at 2700 K and a pulse Xenon lamp as light source with a calibrated spectral radiance.

Measurements with a hemispheric reflector gave the same value of the temperature as that without it, which means that, as usual, emissivity of lead under investigation was near that of a blackbody one.

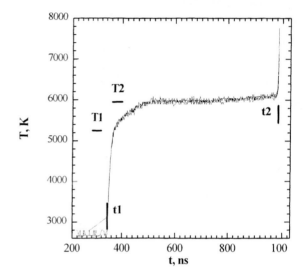

FIGURE 1. T-t diagram of tin expansion from 220 GPa to 37MPa.

The expansion velocity of a sample (U_s) was measured using the optical baselength method. A typical oscilloscope record of the intensity of optical emission corresponding to tin final pressure of expansion above the two-phase region entry is submitted in Fig. 1. Clearly visible is a sharp rise of light intensity in the beginning of free surface expansion (t_1), fast and slow heating periods and nearly steady level with equilibrium condition, the moment of the helium shock wave reflection from the window (t_2) with new increase of measured temperature. Helium properties were calculated using a plasma EOS [8]. The precision of U_s value was about 1-3%. Final pressure of expansion was accepted equal to pressure in helium shock wave, determined from its Hugoniot. The error of pressure determination was due to the accuracy of the U_s measurement and at small pressure also to the accuracy of measurement of helium initial pressure, and appeared to be 3-10%.

EXPERIMENTAL RESULTS AND DISCUSSION

The results of measurements of W_S, P_S, T_S for the tin isentropes investigated are presented in Fig. 2 and 3, where each experimental point corresponded to an average value from 2 - 4 records. Fig. 2 shows the results of W_S measuring on the investigated isentropes in near critical point

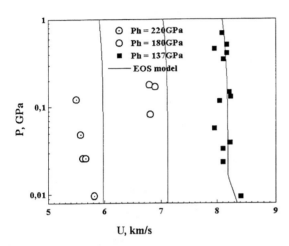

FIGURE 2. P-U diagram of tin expansion.

region vs. pressure of shocked helium. At the low level of shock loading our velocities were smaller then that of EOS model and experimental results of paper [9]. But at highest isentrope agreement is reasonable. In the main part of the investigation only results at last isentrope were used, thus, we not perform additional experiments to eliminate this inconsistency. It should be mentioned that at minimum final pressure the measured velocity also coincides with the EOS model result, and that only the low pressure part of the two-phase region can be investigated with the explosive generators used.

Spectrometric investigation with an OMA confirmed the results of [8], that a thin light emitting layer is on the metal-gas boundary with temperature profile increasing to hot helium. To evaluate the maximum temperature recorded by the pyrometry technique, the process of heat transfer was numerically simulated according to [7]. The temperature in the volume of tin was taken from pyrometry data at

the moment after the first sharp rise (T1 in Fig. 1; 5 in Fig.3). The temperature of shocked helium was calculated using plasma model [8]. Obtained data were described by the model of a turbulent mixing layer on the boundary (4 in Fig.3). As a first approximation it can be assumed that in this region equal volumes of tin and helium are mixing, and then their temperatures become equal adiabatically7]. In the performed calculation the heat of

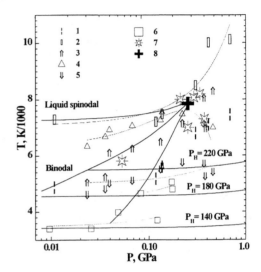

FIGURE 3. P-T diagram of tin near critical point region.
1,2 - T1 and T2 in experiments with glue between
    bottom and sample;
5,3 - T1 and T2 in experiments without glue;
4 - boundary layer temperature due to proposed model;
6 - T1 in experiments with P_H = 137 GPa and 180 GPa;
7 - theoretical estimation of tin critical point parameters
    [10,11] and result of proposed EOS model;
8 - location of critical point on the basis of this work
    experiments following [4];

evaporation was not taken into account. Then, the obtained temperature will coincide with the real one only near the metal critical point for the process of liquid overheating above temperature on saturation curve.

Really, after achieving of saturation curve heating rate was diminished, and this point also can be measured by pyrometer (T2 in Fig. 1; 3 in Fig.3).

Usually in the experiments performed, the maximum temperature was near that on the saturation curve calculated by EOS model. This is connected with the high value of heat of evaporation far from the critical point.

To intensify the process of heating a small intermediate layer of glue between bottom and sample was used (1 and 2 in Fig.3). These experiments gave information about the temperature of tin liquid spinodal. By this method two points of this temperature were measured at pressures 128 and 10.7 MPa. These maximum values were 7450 and 7300 K. These experiments calculated helium temperatures were 12500 and 16100 K. It should be noted here, that tin is an intermediate material in IV group of the Periodic table between Pb and Ge, and the insulator approximation of liquid spinodal [12] can be used. Thus, first step estimation of tin critical point temperature will be $T_{crit} = 8066$ K.

Final estimation of pressure and temperature of tin critical point was made analyzing pressure-temperature diagram shown in Fig. 3. In this diagram EOS model results were shown by solid lines. Fitting lines of heat transfer model results and some fitting lines of measured temperature were shown by thin dashed lines. Following [4] the parameters of the critical point then are $P_{crit} = 0.25$ GPa, $T_{crit} = 7850$ K (dT = 300 K, dP = 0.02 GPa).

In this investigation not only shock wave loading but also pulse heating have been used to realize near critical point states. Shock pressure in tin must be more then 500 GPa to achieve critical point state with unloading process only.

Results of the experiments performed have been used for construction of a tin multiphase semiempirical EOS model, primarily P-T information. Its estimation of tin critical thermodynamic parameters are follows: $P_{crit} = 0.2877$ GPa, $T_{crit} = 8061$ K, $V_{crit} = 0.48$ cm^3/g.

ACKNOWLEDGMENTS

The authors are indebted to V.K.Gryasnov for calculation of plasma parameters presented here. The research described in this publication was made under Russian Fund for Basic Researches Projects N 94-02-03706 and N 97-02-17439.

REFERENCES

1. Zel'dovich Ya.B.,Raizer Yu.P., *Physics of shock waves and high-temperature hidrodynamic phenomena*, Moscow: Science, 1966, 592-605.
2. Fortov V.E.,Yakubov I.T., *Physics of nonideal plasma*, Moscow: Energoatomizdat, 1994, 81-96.
3. Fortov V.E., Bushman A.V., etc., "*Optical properties of dense plasma in shock and rarefaction waves*", in SCCM-1991, 1992, 745-748.
4. Ternovoi V.Ya., Fortov V.E., Kvitov S.V., Nikolaev D.N., "*Experimental study of lead critical point parameters*", in SCCM-1995, 1996, 81-84.
5. Ternovoi V.Ya, Fortov V.E., etc., "*Experimental study of lead thermodynamics at near critical point parameters*", in Physics of Strongly Coupled Plasmas, Singapore-New Jersey: World Scientific Publishing Co Pte Ltd, 1996, 119-124.
6. Labuncov D.A., Avdeev A.A., *Teplofiz.Vis. Temp.(in Russ.)*, **20**, 288-295 (1982).
7. Pyalling A.A., Gryaznov V.K., etc., *Teplofiz.Vis. Temp.(in Russ.)*,, **35**, 288-295 (1996).
8. Ebeling W., Foerster A., Fortov V., Gryaznov V., Polishchuk A., *Thermophysical Properties of Hot Dense Plasmas*, Stuttgart-Leipzig: Teubner, 1991, 142-172.
9. Bakanova A.A., Dudoladov I.P., ets., *Zh. Prikl. Mekh. Techn. Fiz.(in Russ.)*,. **16**, 76-81 (1983).
10. Ohse R.W., von Tippelskirch H., *H.T.-H.P.* **9**, 367-385 (1977).
11. Gathers G.R., *Rep. Prog. Phys.* **49**, 341-396 (1986).
12. Lienhard J.H., *Chemical Engineering Science* **31**, 847-849 (1976)

CP429, *Shock Compression of Condensed Matter – 1997*
edited by Schmidt/Dandekar/Forbes
© 1998 The American Institute of Physics 1-56396-738-3/98/$15.00

FREE ENERGY AND SHOCK COMPRESSION OF DIAMOND

A.M. Molodets, M.A. Molodets, S.S. Nabatov.

Department of High Dynamic Pressures, Institute of Chemical Physics, Chernogolovka, 142432, Russia

The new approach has been developed to calculate the free energy in quasiharmonic approximation for homogeneous condenced matter. Common result has been demonstrated on an example of solid and liquid diamond at high pressures and temperatures of shock compression.

INTRODUCTION

The new Gruneisen factor $\Gamma(V)$ volume V relation [1] and the $\Gamma(V,T)$ temperature T relation [2,3] were received earlier. The equation of state [4] and the free energy [5,6] of solid were constructed on the basis of the $\Gamma(V,T)$. In present work the approach [1-6] is stated and modified for a liquid also.

THEORETICAL PART

The free energy $F=F(V,T)$ in quasiharmonic approximation for Einstein model can be written as

$$F=E_x(V)+E_m+3R(\Theta/2+TLn(1-exp(-\Theta/T)-a_sRT \quad (1),$$

where R-specific gas constant. The key functions E_x, Θ, a_sRT in (1) can be expounded in consent with [6-8]. So, entropy term a_sRT corresponds to the "rough" model of harmonic vibration for a liquid (see, for example, [7]), when $a_s=1$ and $T>>\Theta$. However, similarly to [8], let a_s be a fitting parameter such that $a_s\equiv0$ for solid and $a_s>0$ for liquid.

The potential energy $E_x=E_x(V)$ and Einstein characteristic temperature $\Theta=\Theta(V,T)$ will be considered in identical physical sense for solid and liquid. Followed by [8], let us assume that (1) governs metastable state of liquid at temperature below melting up to $T=0$ K.

Let us assume that the sight of $\Theta(V,T)$, $E_x(V)$ is identical for solid as well as for liquid state. So, $\Theta(V,T)$ can be given as for solid from [4-6]

$$\Theta(V,T) = \Theta_o[(aV_o-V)/(aV_o -V_o)]^2[V_o/V]^{2\,3} \quad (2)$$

where

$$a=a(T)=a_o(1+q_o(T-\Theta_o)/4) \quad at \ T\geq\Theta_o/2$$

and

$$a=a(T)=a_o(1+q_oT^2/\Theta_o) \quad at \ T< \Theta_o/2 \quad (3),$$

$$a_o=1-2/(2/3-bKV_o/C_1), \quad q_o=6bR/(a_oC_P) \quad (4).$$

That is, the individuality of material in (2) is set by initial Einstein characteristic temperature Θ_o, initial volume V_o, adiabatic K (or isothermal K_t) bulk modulus, specific heat at constant volume C_V (or pressure C_P), and volumetric thermal expansion coefficient b. All these properties are calculated at initial temperature T_o and pressure $P_o=P(V_o,T_o)$.

The function for $E_x(V)$ is given by [5,6] also

$$E_x(V)= -a_xV_o(C_1H_x(x)+C_2x)+C_3 \quad (5),$$

where a_x alongside a_s is the second fitting parameter of model. The first approximation for a_x is given (3)

$$a_x=a(0)=a_o \quad (6).$$

In (5) x is $x=V/a_xV_o$, $H_x(x)$ is the polynomial in x

$$H_x(x)=9(x^{-2\,3}/10+2x^{1\,3}+3x^{4\,3}/2-x^{-3}/7+x^{10\,3}/70) \quad (7),$$

where the individual constants of substance C_1, C_2, C_3, are calculated by the account (6) and contain a_x (see [4-6]). The potential energy $E_x(V)$ (5) has a minimum at some point $V=V_{ox}$, where $E_x(V_{ox})=0$. The constant term E_m in (1) serves for a change of the point of reading of potential energy E_x (V).

So, the formulas (1)-(7) are the semiempirical description of thermodynamics properties of homogeneous condenced substance. This description leans upon the five thermodynamics properties Θ_0, V_0, K_t, C_P, b, experimentally determined at the point (T_0, P_0) and two fitting parameters a_x and a_s. The refinement of value of a_x is reached by its fitting under experimental isotherm (see [4,6]). Assuming liquid, the first approximation for a_s is $a_s=1$. The refinement of a_s is reached by selection of such its value, which produce an equality between chemical potential μ of liquid and crystal at any point of melting curve.

We notice, that (1) and many thermodynamics properties have a break at the specific point $V=aV_0$, which corresponds to zero characteristic temperature Θ (2). It means, that the description (1)-(7) will give a large error in some area $V>V_0$ at increase of volume. Therefore at the present work we shall be limited by the area of compression where $V \leq V_0$.

EQUATIONS OF STATE AND SHOCK COMPRESSION

Let us write the pressure P and the energy E at compression, using derivatives of (1) with respect to V and T and Gibbs-Helmholtz equation as usually

$$P=P(V,T)=P_x+3R\Theta\Gamma(0.5+1/(exp(\Theta/T)-1))/V \quad (8),$$

$$E=E(V,T)=E_x(V)+V(P-P_x)/H \quad (9)$$

where potential pressure P_x is given by

$$P_x = P_x(V) = -dE_x/dV=C_1 dH_x/dx + C_2 \quad (10).$$

In (9) $H=H(V,T)$ [5,6] is the extension of the factor of proportionality between thermal pressure $P_t(V,T)=P-P_x$ and density of thermal energy $E_t(V,T)=E-E_x$ as the volume temperature Gruneisen function

$$H=\Gamma/(1+M) \quad (11).$$

In (8), (11) $\Gamma=\Gamma(V,T)$ is the usual Gruneisen factor

$$\Gamma = -(\partial Ln\Theta/\partial LnV)_T = 2/3 - 2/(1-aV_0/V) \quad (12).$$

In (11) $M=M(V,T)$ is a new auxiliary function

$$M = -(\partial Ln\Theta/\partial LnT)_V \quad (13).$$

The description of thermodynamics of the plane one-dimensional shock compression is reduced to an addition the Rankine-Hugonio relation

$$E-E_o=0,5(P+P_o)(V_o-V) \quad (14),$$

that with (1)-(13) permits to compute a temperature and then all thermal properties along shock adiabat.

The computation of strong shock compression of porous substance is executed similarly. The difference implies that the volume V_o of monolithic substance in (14) must be replaced by specific volume of porous sample $V_{oo}=mV_o$ (m-porosity), and the specific surface energy of porous sample must be added to initial specific internal energy E_o before shock front.

Now let us consider the thermodynamics description of the diamond as an example of the general procedure (1)-(13).

FREE ENERGY OF SOLID DIAMOND

Let us consider the thermodynamics description of solid diamond. The data from [9-11] are used for Θ_0, V_o, K_t, C_P, b. The values of a_0, q_0 as (4), a_x as (6), and then C_1, C_2, C_3 was calculated by [4-6]. We accept a graphite as a standard, that gives for a solid diamond $E_m=0.15826$ kJ/g. Let us accept $a_s \equiv 0$ for as in (1). By this means we receive the complete set of parameters of (1) in the context of the author's model for solid diamond (see the =sol= Table 1 column).

Notice, that the first approximation a_x (6) with Table 1 data is sufficient for the wide area of compression of solid diamond. So, the author's computation of isotherm coincides with the experiment [11] up to 40 GPa and the computation of shock adiabat for monolithic diamond coincides with experiment [12] up to 0.5 TPa (see Fig.1). The computation of shock compression of porous samples coinsides with [12,13] (see Fig.2)

This consent suggests that the calculation of properties of solid diamond on (1)-(13) at triple point (T_{tr}, P_{tr}) of carbon is correct also. Some of these properties are $V(T_{tr}, P_{tr})=1/3.4364$ cm^3/g; $K_t(T_{tr}, P_{tr})=546.45$ Gpa; $\mu=-7.32616$ kJ/g. They will take in the next chapter. The values $(T_{tr}=4470$ K, $P_{tr}=13.5$ Gpa) was used from [14].

TABLE 1. Parameters of free energy $F(V,T)$ (1) for solid (=sol=) and liquid (=liq=) diamond

Parameter		=sol=	=liq=
T_o,	K	300.0	4470.0
$1/V_o$,	g/cm^3	3.515	3.2191
Θ_o,	K	1320.0	1214.03
a_o		8.282446	8.513046
q_o,	10-6/K	2.958	3.0201
a_x		8.282446	8.513046
a_s		0.0	2.58895
C_1,	GPa	-27.28853	-19.40608
C_2,	GPa	348.44318	225.52762
C_3,	kJ/g	-761.02331	-614.26260
$1/V_{ox}$,	g/cm^3	3.544392	3.32812
E_m,	kJ/g	0.15826	8.48876

FREE ENERGY OF LIQUID DIAMOND

Let us consider the thermodynamics description of liquid diamond. We accept the initial state for a liquid diamond as (T_{tr}, P_{tr}). Let us estimate V_o, K_b, Θ_o, C_P, b, E_m, a_s, a_x for this phase at point (T_{tr}, P_{tr})

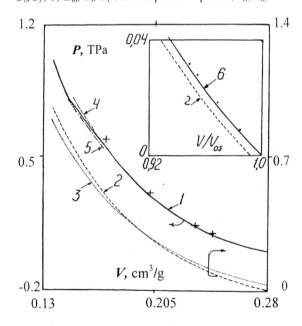

FIGURE 1. Compression of a monolithic diamond. Computation by the help of the Table 1 data: 1 - shock compression; 2 and 3 - $P_x(V)$ (10) respectively for solid and liquid diamond; 4 and 5 - volume respectively of solid and liquid diamond along a melting line; 6-isotherm at 300 K. Experimental data: + - shock data [12]; • - isotherm data [11].

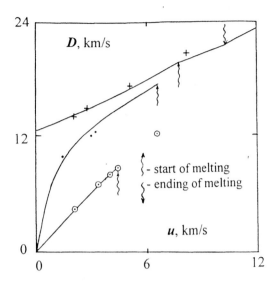

FIGURE 2. Computations D (shock velocity) - u (mass velocity) for a diamond at various porosity m (line from the top down): $m=1.0$; $m=1.098$; $m=1.85$. Experimental data: + - $m=1.0$ [12]; •- $m=1.098$ [13]; ⊙ - $m=1.85$ [12].

using the results of previous chapter, data [14] and ideas of [8,19]. We mark a belonging to liquid and solid state by subscript l and s.

We calculate specific volume of liquid diamond V_{ol} as $V_{ol}=V_l(T_{tr},P_{tr})=V_s(T_{tr},P_{tr})+\Delta V$.

We calculate the point of reading of potential energy for liquid diamond E_{ml} as $E_{ml}\cong E_{ms}+T_{tr}\Delta S-P_{tr}\Delta V$. The values ΔV=0.236 cm^3/mol and ΔS=23.1 J/molK from [14] are used for calculation of V_{ol} and E_{ml}.

We use the rule K_{ll}=0.8 K_{ls} for liquid metals at a melting point from [19], that is we evaluate isothermal bulk modulus of a liquid diamond K_{ll} at point (T_{tr},P_{tr}) as $K_{ll}(T_{tr},P_{tr})$=0.8$K_{ls}(T_{tr},P_{tr})$. Taken into account the idea [8], we assume, that $\Theta_{ol}(V_{ol},T_{tr})=\Theta_s(V_{ol},T_{tr})$. The similar expedient is used for $C_{Pl}(V_{ol},T_{tr})=C_{Ps}(V_{ol},T_{tr})$ and $b_l(V_{ol},Ttr)=b_s(V_{ol},T_{tr})$.

Now we calculate a_{ol}, q_{ol} under common formulas (4) and a_{xl} in the first approximation under common formula (6) and then C_{1l}, C_{2l}, C_{3l}. At last, we make equality of chemical potential μ of solid μ_s (T_{tr},P_{tr}) and liquid $\mu_l(T_{tr},P_{tr})$ at point (T_{tr},P_{tr}) by fitting of a_{sl}. It takes place at a_{sl}=2.58895.

These estimations are listed in =liq= Table 1 column as the parameters of model for liquid diamond.

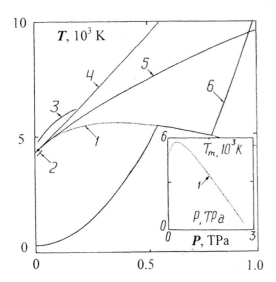

FIGURE 3. Melting curve and shock temperature of diamond. 1 - author's melting curve; 2 - [16]; 3 - [15]; 4 - slope of melting curve 10.2 K/GPa at thriple point (•) [14]; 5 - author's Lindemann melting curve of solid diamond; 6 - author's computation of shock temperature of a monolithic diamond.

MELTING CURVE OF DIAMOND

Let us calculate the melting curve of diamond. For this purpos we solve equation $\mu_s(P,T)=\mu_l(P,T)$ numerically, where formulas (1), (8) with parameters of table 1 are used. This solution is shown in Fig. 3 by lines 1. It has maximum.

As Fig. 3 suggests, the author's melting curve does not countrary to the data of other researchers [14-17] at low pressure, where the slope of the melting line is positive. The negative slope of the melting line at high pressure can be thought reasonable also. Let us show it.

So, the computation of shock compression at high pressure in the area of mixture phases was executed by a support on the author's melting curve. The results of this computation for a monolithic diamond are shown in Figurs 1-3. The computation of change of volume at shock compression of diamond is shown by line 1 in Fig.1. The computation of shock temperature compression of diamond is shown by line 6 in Fig 3. The first bend of line 6 is connected with the beginning melting, gently sloping - a mixture of phases, second bend - ending of melting (these states are marked in D-u plane by wavy pointers in Fig.2).

This computation at high pressure will be agreeed with literature data. Really, the melting of a diamond in a shock wave begins in point (P,GPa;V,cm^3/g) (543;0.1726) and is finished in point (792;0.1477) as viewed in Fig.1. This interval does not countrary to the beginning of shock melting of diamond in point (692;0.1603) from [18].

CONCLUSIONS

Thus, the description of thermodynamics properties of a solid and a liquid on the basis of the author's Gruneisen factor $\Gamma=2/3-2/(1-aV_c/V)$ [1] permits to compute the properties of solid and liquid diamond at compression.

RFERENCES

1. Molodets A.M., JETP **80**, 467-471 (1995).
2. Молодец А.М., Молодец М.А., Химическая Физика №5, 117-121 (1997).
3. Молодец А.М., Молодец М.А., Набатов С.С., "Объемно-температурная зависимость коэффициента Грюнайзена", представлено на 11 Симпозиуме по горению и взрыву, Черноголовка, Ноябрь 18-22, 1996.
4. Молодец А.М., Доклады Академии Наук, **353**, №5, 610- 612 (1997).
5. Молодец А.М., "Изохорно-изотермический потенциал и термодинамика ударного сжатия твердых тел", представлено на 11 Симпозиуме по горению и взрыву, Черноголовка, Ноябрь 18-22, 1996.
6. Молодец А.М., Химическая Физика, №9, 10-18 (1997).
7. Цянь Сюэ-Сень, Физическая механика, Москва:Мир, 1965, глава 9, сс. 332-341.
8. Воробьев В.С., Теплофизика Высоких Температур, **34**, №3, 397-406 (1996).
9. Новиков Н.В. и др., Физические свойства алмаза, Киев: Накова Думка, 1987, с.188.
10. Einstein A., Ann. der Physik **22**, 180-185 (1907).
11. Drickamer H.G. et al, Solid State Physics **19**, 135-228 (1966)
12. Павловский М.Н., ФТТ, **13**, №3, 893-895 (1971).
13. McQueen R.G., Marsh S.P., in Compendium of shock wave data, ed by M.van Thiel, Lawrence Livermor Laboratory,V.1, p.47, 1977.
14. Togaya M., "Thermophysical Properties of Carbon at High Pressure", presented at the 3rd NIRIM ISAM'96, Tsukuba, Japan, March 4-8,1996.
15. Bundy F.P., Physica A 156, 169-178 (1989).
16. Averin A.B. et al, "Equation of state and phase diagram of carbon", presented at the AIP Conference on Seattle, Washington August 13-18, 1995.
17. van Thiel M., Ree F.H., Phys.Rev. B **48**, 3591-3599 (1993).
18. Sekine T., Carbon 31,N1, 227-233 (1993).
19. Webber G.M., Stephens R.W., in Physical Accoustics, V,IV B. ed. by W. Mason: New-York, 1968,

CP429, *Shock Compression of Condensed Matter – 1997*
edited by Schmidt/Dandekar/Forbes
© 1998 The American Institute of Physics 1-56396-738-3/98/$15.00

THE TWOFOLD QUARTZITE SHOCK ADIABAT UNDER PRESSURES OF 55-150 GPA

A.O.Borschevsky, M.M.Gorshkov, A.M.Tarasov

Russian Federal Nuclear Center – Institute of Technical Physics
P. O. Box 245, Snezhinsk, Chelyabinsk region 456770 Russia

By the reflection method with quartzite loading on reference materials (Ti, Sn, Fe, Cu, Ta and W) from initial shock state under $P_1 = 56.4$ GPa up to the pressure near $P_2 = 150$ GPa the points of the twofold quartzite shock adiabat were obtained. Processing of the experiments of the present work and work (1) with taking into account for available porosity of Ta and W samples shows that in these experiments the effect of abnormal compressibility of quartzite in the second shock wave found out in (1) does not take place. The data of the present work and (1) were approximated by the twofold shock adiabat, calculated according to the Mie-Gruneisen equation of state with ordinary values of parameters corresponding to the quartz high pressure phase (stishovite).

INTRODUCTION

Twofold shock compression of quartzite (i.e. loading by two successive shock waves without intermediate unloading) was studied in (1, 2) by measuring the shock waves velocities in the samples of reference materials with known shock adiabats (samples-standards) located behind quartzite samples and subsequent calculation of twofold shock states parameters with using the laws of conservation (this is reflection method presented in (3)). On the ground of the found out break in the quartzite twofold shock adiabat a conclusion about experimental registration of synthesis of a new, more dense than stishovite, phase of silica was made in (1), and this fact then is fixed as a basis of assumptions on constitution of the Earth's interior. In the present work the experiments on quartzite twofold compressibility is carried out by the same method and under the same loading conditions. Obtained results are presented below and their comparison with data (1) is given.

TECHNIQUE OF EXPERIMENT

The set-up of experiments on measuring the shock velocities in the samples-standards is given in Fig. 1.

Flyer plate of aluminum alloy AMtz (with content of Al near 98.3 % and Mn near 1.5% and density of 2.735 g/cm^3) is launched by explosion products of cylindrical charge of cast composition TNT/RDX 50/50 of 400 mm in length and 200 mm in diameter on a way of 42 mm up to velocity close to constant (W=5.46 km/s), therefore a shock wave with a right-angled profile forms in the samples. Electrocontact gauges (pins) made of copper wire were mounted in central part of a sample within the area of 30x30 mm^2, where boundary effects are absent. Voltage of

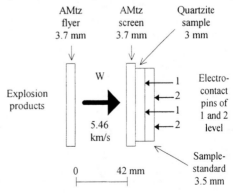

FIGURE 1. Set-up of experiments

400 V was applied to gauges from the registration scheme; an instant of breakdown of the enamel isolation (with thickness of 0.02 mm) of gauges in a shock wave is determined on oscillograms with accuracy of $5 \cdot 10^{-9}$ s. The shock wave velocity in the sample-standard is determined as $D_{ref} = \Delta/(t_1-t_2)$, where Δ - thickness of the sample, t_1 and t_2 - instants of gauges shorting (of 1 and 2 level).

INITIAL STATE OF QUARTZITE SINGLE SHOCK COMPRESSION

The explosive measuring device and the registration scheme described above were used when determining an initial state of quartzite single shock compression, but the gauges were placed on butt-ends of just quartzite sample.

Shock velocity measured in 11 tests was amounted to $D_1 = 6.93$ km/s for quartzite samples of 2.65 g/cm^3 average density with porosity less than 0.5 % and SiO$_2$ contents in the range of 95-98 %. Other parameters of quartzite single shock compression are calculated by reflection method (3) with using D-U relation of reference material AMtz from (4) ($D = 5.231 + 1.485U - 0.0244U^2$): particle velocity $U_1 = 3.073$ km/s, pressure $P_1 = 56.42$ GPa, specific volume $v_1 = 0.2100$ cm^3/g.

Obtained point (point "B" in Fig. 2) is found practically on experimental shock adiabat (5) which is determined by D-U relation $D = 1.739 + 1.7U$.

THE TWOFOLD QUARTZITE SHOCK ADIABAT

Reference materials

Titanium, tin, iron, copper, tantalum and tungsten were used as reference materials. The initial densities of samples are specified in Table 1. The

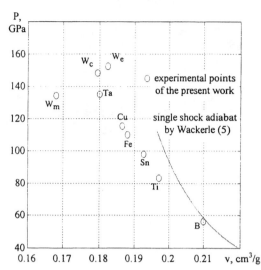

FIGURE 2. Twofold quartzite shock compression. Data of the present work. "B" - initial point of quartzite single shock compression. Point "W" - when processing with using D-U relations for tungsten: W_m - of monolithic W from (6); W_e - obtained by interpolation of data (7, 12); W_c - calculated with EOS (Mie-Gruneisen form, $\gamma=1.3$)

content of the basic component in the samples-standards was no lower than 99 %, and porosity did not exceed 0.1-0.2 % (the tungsten samples had a porosity of ~2.3 %).

Data on twofold quartzite shock compression

In experiments quartzite samples were used with porosity less than 0.5 % and the SiO$_2$ contents of 97-98 %. Series of 3-6 tests was carried out with using each reference material; average values of shock velocities D_{ref} are specified in Table 1. D-U relations of the standards, used during calculations of values of particle velocity U_2, and other parameters of twofold quartzite shock adiabat are pointed there; the obtained points are put in Fig. 2.

TABLE 1. Shock Velocities in the Samples-Standards and Parameters of Twofold Quartzite Shock Compression

Reference material	ρ_{oo}, g/cm^3	D_{ref}, km/s	U_2, km/s	P_2, GPa	D_2, km/s	v_2, cm^3/g	D-U relation of reference material	
Ti	4.500	7.451	2.483	83.3	9.55	0.1970	D=4.918 + 1.02U	(7)[a]
Sn	7.285	6.042	2.226	98.0	10.30	0.1927	D=2.480 + 1.60U	(8)
Fe	7.841	7.060	1.988	110.1	10.38	0.1880	D=3.85 + 1.615U	(9)
Cu	8.922	6.825	1.896	115.5	10.53	0.1865	D=3.94 + 1.55U-0.015U^2	(13)
Ta	16.62	5.282	1.539	135.1	10.77	0.1801	D=3.42 + 1.21U	(11)
W	18.91	5.601	1.267	134.2	9.04	0.1681	D=4.015 + 1.252U	(6)[b]

[a] This relation was obtained on base of data (7) at $1.0 < U < 3.2$ km/s.

[b] D-U relation of monolithic tungsten. The account for porosity is considered below.

Account for porosity of the tungsten samples

• As far as the shock adiabat of tungsten with initial density of ρ_{oo} = 18.91 g/cm³ at U ≈ 1.3 km/s was not investigated, the value of U_2 for the appropriate point of twofold quartzite shock adiabat was determined by calculations.

• Experimental data on shock compression of tungsten of following initial densities are available in literature:
  - 18.91 g/cm³ at U= 1.7-3.2 km/s in work (12),
  - 18.67 g/cm³ at U = 0.3-0.7 km/s in work (7),
  - 19.35 g/cm³: D = 4.015 + 1.252U at U < 3.4 km/s in work (6).

The data (7) were reduced to density of 18.91 g/cm³ by linear interpolation of D(ρ) dependence at given U in a narrow density range of 18.67-19.35 g/cm³, then they were described according to data (12) by dependence D = 3.34 + 1.66U - 0.06U², from which it follows that U_2 = 1.44 km/s for the measured (in Table 1) value of D_{ref} = 5.601 km/s.

• Let us also carry out an estimation of U_2 value with the help of the equation of state in the Mie-Gruneisen form $P-P_{cold} = \gamma\rho (E-E_{cold})$ and the laws of conservation of mass, momentum, energy on shock front for matter with porosity coefficient k = ρ_o /ρ_{oo}, where E - specific internal energy, $P_{cold} = \rho^2 dE_{cold}/d\rho$ and E_{cold} - pressure and specific internal energy at isotherm T = 0 K, γ - Gruneisen coefficient, ρ_o and ρ_{oo} - densities of monolithic and porous matter before loading.

We shall express pressure in porous matter P_{por} (v) through pressure in monolithic (k=1) matter P_{mon} (v) at the same value of v:

$$P_{por}(v) = P_{mon}(v) \cdot \frac{1 - \frac{\gamma}{2}(1 - \frac{v}{v_o})\frac{v_o}{v}}{1 - \frac{\gamma}{2}(1 - \frac{v}{kv_o})\frac{kv_o}{v}} \quad . \quad (1)$$

If dependence D_{mon} (U) = c_o + λU is known, that

$$P_{mon}(v) = \frac{1}{v_o} D_{mon} \cdot U = \frac{c_o^2(1 - v/v_o)}{v_o(1 - \lambda(1 - v/v_o))^2} \quad . \quad (2)$$

That, together with (1), allows to find D and U for porous matter:

$$U_{por}(v) = \sqrt{P_{por}(v) \cdot (kv_o - v)} \quad , \quad (3)$$

$$D_{por}(v) = \frac{U_{por}(v)}{1 - v/(kv_o)} \quad . \quad (4)$$

Calculations by formulas (1-4) give U_2 =1.40 km/s at γ = 1.3 (this is a value of γ at ~200 GPa (12)). Limiting values: U_2 = 1.38 km/s at γ = 2.0 (normal conditions); U_2 = 1.43 km/s at γ = 0.5 (superhigh pressures).

• Thus, U_2 values obtained by interpolation and with the help of the equation of state coordinate between themselves and distinguish more than by 10 % from the value, which follows from D-U relation for monolithic tungsten. The appropriate variants of parameters of a point "W" are given in Table 2, the point locations ("W_e", "W_c" and "W_m") are shown in Fig. 2.

From Fig. 2 it is obviously, that twofold quartzite shock adiabat has a smooth course without any peculiarities at highest achieved pressures (~150 GPa) with the account for porosity of tungsten samples.

COMPARISON OF THE RESULTS OF THE PRESENT WORK AND WORK (1)

In work (1) determination of points of twofold quartzite shock adiabat was carried out with using similar explosive measuring device; the parameters of quartzite single shock compression practically coincide with those (point "B") for the present work: U_1 = 3.02 km/s, P_1 = 55.8 GPa, v_1 = 0.214 cm³/g. In (1) the reference materials, namely, titanium, tin, vanadium, iron, copper and molybdenum were monolithic, tungsten samples had a density of ρ_{oo} = 19.06 g/cm³ being close to the tungsten samples of the present work, but tantalum samples, as opposite to the present work, were porous: k = 1.02. In (1) the values of shock velocities D_{ref}, obtained for the appropriate reference materials, practically coincide with the data of the present work.

Being close to the point "B" of the present work, the corresponding point "P" of work (1) slightly deflects from it by Δv = +0.004 cm³/g (see Fig. 3). It results in the systematic shift of points of twofold quartzite shock adiabat of (1) in relation to those of

TABLE 2. Parameters of a Point of Twofold Quartzite Loading on Tungsten ("W").

Variant of U_2 calculations	U_2, km/s	P_2, GPa	D_2, km/s	v_2, cm³/g
W_e: experiments (7, 12)	1.44	152.5	12.36	0.1823
W_c: equation of state	1.40	148.3	11.53	0.1795
W_m: D-U relation (6)	1.267	134.2	9.04	0.1681

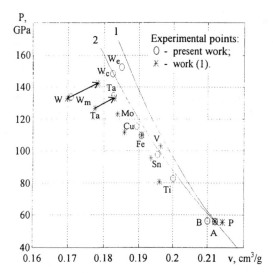

FIGURE 3. Twofold quartzite shock compression. Comparison of the present work data with data (1). Initial points of quartzite single shock compression: "B" - present work, "P" - work (1), "A" - average point; 1, 2 - single shock adiabat and twofold shock adiabat of quartzite with using Mie-Gruneisen EOS.

in Fig. 3). The fact, that adjusted point "Ta" of (1) has coincided with "Ta" of the present work where the tantalum samples were monolithic, testifies to correctness of the approach to the account for porosity.

In Fig. 3 twofold shock adiabat is plotted, which calculated by the equation of state in the Mie-Gruneisen form having constant thermal capacity and constant Gruneisen coefficient (see work (13)), with a set of parameters for quartzite high pressure phase (stishovite): $\rho_o = 4.21$ g/cm^3, $\rho_{oo} = 2.65$ g/cm^3, $c_o = 8.945$ km/s, n = 3 and h = 4.663. That twofold shock adiabat satisfactorily describes the data of the present work and work (1) including as well points "Ta" and "W" calculated with the account of the samples porosity. This fact points out that in the investigated pressures range the effect of abnormal quartzite compressibility in the second shock wave, found out in (1), does not take place.

the present work by $\Delta v \sim +0.007$cm^3/g. However, the experimental shock adiabat of work (5) passes between the initial points of compared works. Thus, it is possible to consider, that experimental uncertainty is the principal reason of that deflection. Therefore, in order to put into agreement the results of compared works we assume, as an initial state of obtained twofold shock adiabats, the average point: $P_1 = 56.1$ GPa, $v_1 = 0.212$ cm^3/g (point "A" in Fig.3). Adjusted with account of the point "A" the data of compared works are plotted in Fig. 3, moreover, data of work (1) were reprocessed with the same D-U relations (from Table 1) as the data of the present work to exclude some distortion of the compared results taking place because of using the particular expressions for shock adiabats of reference materials.

From Fig. 3 it follows that the data of the present work and (1) agree at pressures up to ~120 GPa, but essentially differ at pressures of 130-150 GPa, i.e. for points of quartzite loading on tantalum and tungsten. The reason of that difference is the neglect of samples porosity of these reference materials in (1): at reprocessing by algorithm applied above for tungsten the points "Ta" and "W" of (1) also become in conformity with those of the present work (arrows

REFERENCES

1. Poduretz M.A., Simakov G.V. and Trunin R.F., *Izv., USSR Ac. of Sci., Earth's Phys.* **4**, 30-37 (1990).
2. Poduretz, M.A., Simakov, G.V. and Trunin, R.F., *Izv., USSR Ac. of Sci., Earth's Phys.* **7**, 3-11 (1976).
3. Al'tshuler, L.V., *Rus. J. UFN (Adv. of Phys. Sci.)* **85(2)**, 197-258 (1965).
4. Zhugin Yu.N., Krupnikov, K.K., Ovechkin, N.A., Abakshin, E.V., Gorshkov, M.M., Zaikin, V.T. and Slobodenyukov, V.M., *Rus. J. Earth's Phys.* **10**, 16-22 (1994).
5. Wackerle, J., *J.Appl.Phys.* **33(2)**, 922-937 (1962).
6. Al'tshuler, L.V., Bakanova, A.A., Dudoladov, I.P., Dynin, E.A., Trunin, R.F. and Chekin, B.S., *Rus. J. PMTF (Appl. Mech.&Tech. Phys.)* **2**, 3-34 (1981).
7. *LASL Shock Hugoniot Data.* Ed. Marsh, S.P. Berkeley - Los Angeles - London: Univ. of California Press., 1980, 658 p.
8. McQueen, R.G. and Marsh, S.P., *J.Appl. Phys.* **31(7)**, 1253-1269 (1960).
9. Al'tshuler, L.V., Brazhnik, M.I. and Telegin, G.S., *Rus. J. PMTF* **6**, 159-166 (1971).
10. Argous, J.P. and Aveille, J., "Adiabatique Dynamique de Cuivre a Pression Elvee", in "Comportement Milieux Denses Pressions Dynamique". Paris - New York, 1968, pp. 173-178.
11. Heimke, G. and Leiber, C.O., *Explosivstoffe* **6**, 121-124 (1969).
12. Krupnikov, K.K., Brazhnik, M.I. and Krupnikova, V.P., *Rus. J. Exp.&Techn. Phys.* **42(3)**, 675-685 (1962).
13. Kalashnikov, N.G., Pavlovsky, M.N., Simakov, G.V. and Trunin R.F., *Izv., USSR Ac. of Sci., Earth's Phys.* **2**, 23-29 (1973).

CP429, *Shock Compression of Condensed Matter – 1997*
edited by Schmidt/Dandekar/Forbes
© 1998 The American Institute of Physics 1-56396-738-3/98/$15.00

ISOTHERMAL EQUATIONS OF STATE
FOR NANOMETER AND MICROMETER NICKEL POWDERS

Jin Xiaogang[a] **Zhang Hanzhao**[a] **Che Rongzheng**[b] **Zhou Lei**[b] **Zhao Qing**[b]
Liu Jing[c] **Xiu Lisong**[c]

a). *Laboratory for Shock Wave and Detonation Physics Research, Southwest Institute of Fluid Physics, P.O. Box 523, Chengdu, Sichuan, 610003, China*

b). *Institute of Physics, Academia Sinica, Beijing, 100080, China*

c). *BEPC, Institute of High Energy Physics, Academia Sinica, Beijing, 100039, China*

The volume and pressure for nickel powders with nanometer and micrometer grains were measured using diamond anvil cell (DAC), energy dispersive X–ray diffraction and internal pressure standard of platinum metal in 0–50GPa pressure range. These experiment data were fitted with Birch's and Bridgman's equation of state, respectively, to yield the isothermal bulk modulus and its first derivative. It is found that the nanometer nickel powder is more compressible than the micrometer powder.

INTRODUCTION

Firstly, Gleiter proposed the basic idea of nanocrystalline materials in 1981(1) and made these materials (2). Nanocrystalline materials have stimulated interests in the fields of material science and condensed matter physics, because they possess significant superiorities with respect to the conventional microcrystalline materials and also present an attractive potential for technological applications. A large number of achievements have been summarized in the two reviews (3, 4), but there were few reports with regard to the study of equation of state for nanocrystalline materials.

In present work, in order to provide some information on the equation of state for nanocrystalline materials, the data of the pressure and volume for nickel powder with nanometer grain were measured using a diamond anvil cell apparatus, energy dispersive X–ray diffraction of synchro–radiation and internal pressure standard of platinum metal at room temperature.

For comparison, nickel powder with micrometer grain was also measured. The powders of nickel metal were chosen to study, because their superior oxidation resistance will bring some conveniences on setting operation of test samples in atomspheric conditions.

EXPERIMENT

The DAC apparatus, probe and other supplemental facilities were mounted on a working table in a hut which was shielded with lead plates over its whole surface as indicated in Fig. 1, because the experiments were carried out under the conditions of strong radiation.

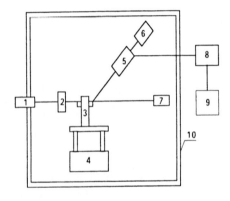

FIGURE 1. Entire distribution of set — up, 1. X — ray, 2. diaphargm, 3. DAC, 4. working table, 5. X — ray probe, 6. laser Ⅰ, 7. laser Ⅱ, 8. multichannel spectrometer, 9. computer, 10. hut.

In our experiments, the diamond anvil cell possesses structurally load system of a lever driven by a worm gear in order to control and adjust automatically the load. The diamond — anvil faces were 0. 36mm in diameter, and the gasket was 0. 2mm in width and made of a work — hardened stainless. The sample chamber was a hole of 0. 12mm diameter drilled into the gasket. The nickel powders were placed

in the sample chamber along with platinum powder as the pressure standard.

As shown in Fig. 1, after the X — ray beam was imported into the hut with a pipe, it passed through a diaphargm and projected on the sample. The facula of $65 \times 65 \mu m^2$ on the sample was required. In order to achieve this requirement, the DAC was precisely aligned in situ using the Laser Ⅰ (see Fig, 1) and direct photography of X — ray with sensitive paper placed on the sample. The diffraction spectrums in $0-50GPa$ pressure range were gathered by a Si(A1)probe, multichannel spectrometer and a computer, and gathered time is about 1000s at each pressure.

Nickel powder with nanometer grain was prepared using an inert gas condensation technique and the average size of this powder grains was 20 nm. The micrometer nickel powder was made with the conventional technique and its grain sizes distributed over the range $\leqslant$ 45μm. The purenesses of the two powders were all 99. 5%

PROCESSING DATA AND EXPERIMENT RESULTS

For nanometer and micrometer nickel powders including platinum standard, the diffraction spectrums under different pressures were showed in Fig. 2 and Fig. 3. These diffraction spestrums followed the Bragg's equation

$$d_{hkl} = hc/2E_{hkl}sin\theta \qquad (1)$$

where d_{hkl} is the interplanar crystal spacing of (hkl) crystal faces, E_{hkl} is, the X — ray energy diffracted (hkl) crystal faces, θ is the diffraction angle, h is Planck constant and c is the light speed. When θ is fixed, the relation among the interplanar crystal spacing and the X—ray energy at ambient and arbitrary pressures may be derived from Eq. (1) as follows

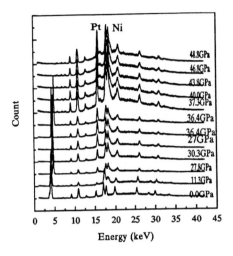

FIGURE 2. Diffraction spectrums of nickel powder with nanometer grains.

$$d_{hkl}/d_{hklo} = E_{hklo}/E_{hkl} \qquad (2)$$

and

$$d_{hkl}/d_{hklo} = (V/V_0)^{1/3} \qquad (3)$$

where v is the specific volume, and the subscript "o" presents the state at the ambient pressure.

The pressure in test sample was estimated with the equation of state of platinum metal.

This equation is

$$P = 3K_{T_0}[(1 - \eta^{1/3})/\eta^{2/3}]e^{a(1-\eta^{1/3})} \qquad (4)$$

where p is pressure, K_{T_0} (bulk modulus) = 278GPa, $\eta = (V/V_0)$, $a = 1.5(K'_T - 1)$, K'_T (first derivative of bulk modulus) = 5.61.

From Fig. 2 and 3, Eqs. (2), (3) and (4), the data of the pressure and the specific volume of nanometer and micrometer nickel powders were obtained, listed in the Table 1 and indicated in Fig 4. The data in the Table 1 were fitted using Birch's equation of state.

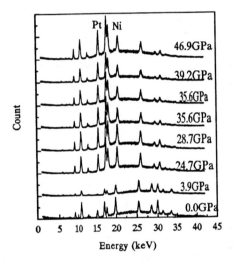

FIGURE 3. Diffraction spectrums of nickel powder with micrometer grains.

Table 1. Experimental Data of Pressure and Volume for Nanometer and Micrometer Nickel Powders

Nanometer	P(GP.)	0.0	11.3	27.8	30.3	32.0	36.4
	V/V₀	1.0000	0.9412	0.8917	0.8838	0.8885	0.8728
Nickel	P(GP.)	36.4	37.3	40.0	43.8	46.8	48.8
	V/V₀	0.8728	0.8667	0.8590	0,8560	0.8530	0.8396
Micrometer	P(GP.)	0.0	3.9	24.7	28.7	35.6	35.6
	V/V₀	1.0000	0.9890	0.9147	0.9033	0.8889	0.8858
Nickel	P(GP.)	39.2	46.9				
	V/VT₀	0.8733	0.8672				

101

$$P = \frac{3K_{T_0}}{2} [(\frac{V_0}{V})^{7/3} - (\frac{V_0}{V})^{5/3}]$$

$$[1 + \frac{3}{4}(K'_T - 1)((\frac{V_0}{V})^{2/3} - 1)] \quad (5)$$

The fitting results were: for nanometer nickel powder, $K_{T_0} = 186 GPa$ and $K'_T = 5.04$; and for micrometer nickel powder, $K_{T_0} = 221 GPa$ and $K'_T = 4.89$. The fitting curves were drawn in Fig. 4.

The data in Table 1 were also fitted with Bridgman's equation of state

$$1 - \frac{V}{V_0} = ap - bp^2 \quad (6)$$

where $a = 1/K_{T0}$, $b = K'_T/2K_{T0}^2$. The fitting results were: $K_{T_0} = 209 GPa$, $K'_T = 2.92$ for nanometer nickel powder, and for micrometer nickel powder, $K_{T_0} = 238 GPa$ $K'_T = 2.96$. The fitted curves were also presented in Fig. 4.

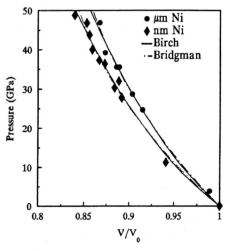

FIGURE 4. Experimental results of pressure and volume for nanometer and micrometer nickel powders.

DISCUSSION AND CONCLUSION

From the distributions of experiment data points in Fig. 4 or K_T and K'_T for two nickel powders with different grains, it is found that the nanometer nickel powder is more compressible than micrometer powder.

Comparing the fitted Birch's equation of state with the Bridgman's, it is seen that for either nanometer or micrometer nickel powder, these two fitting curves in Fig. 4 are basically accordant over a pressure range below 45 GPa, but separate off as increasing pressure. It was appeared that the Birch's equation of state is more suitable than Bridgman's equation to fit those data of nickel powder in Table 1.

ACKNOWLEDGMENT

Experiments in this paper were completed in BEPC of Institute of High Energy Physics, Academia Sinica.

REFERENCES

1. H. Gleiter, in N. Hansen, A. Horsewell, T. Leffers and H. Lilholt (eds), Deformation of polycrystals: Mechanism and Microstructure, Riso National Laboratory, Denmark, 1981, P. 15.

2. H. Gleiter, M. Marquardt, *Metallkude*, 75 (1984) 365.

3. H. Gleiter, *Prog. Mater. Sci.*, 33 (1989) 223.

4. K. Lu, *Mater, Sci. Eng.*, R16 (1996) 161.

CP429, *Shock Compression of Condensed Matter – 1997*
edited by Schmidt/Dandekar/Forbes
© 1998 The American Institute of Physics 1-56396-738-3/98/$15.00

HIGH-TEMPERATURE, HIGH-PRESSURE EQUATION OF STATE FOR POLYMER MATERIALS

K. V. Khishchenko, V. E. Fortov, and I. V. Lomonosov

*High Energy Density Research Center, Russian Academy of Sciences,
Izhorskaya str. 13/19, Moscow, 127412, Russia*

The semiempirical equation-of-state model for polymer materials over wide range of thermodynamic parameters is proposed. Equations of state for polyethylene, polystyrene, and polymethylmethacrylate are developed, and the critical analysis of calculated results describing complicated, existing experimental data is made.

INTRODUCTION

The analysis of the thermodynamic properties of various substances over a wide region of phase diagram is of fundamental as well as practical interest. Structural material thermodynamics under conditions of high temperatures and pressures are a necessary part for carrying out the computer simulation of nonsteady hydrodynamic processes, generated by the influence of intense pulse energy fluxes on condensed media [1].

The base of difficulty confronting a systematic theoretical calculations of the equation of state (EOS) under high energy processes conditions is the need to incorporate correctly the structurally complicated interparticle interaction. The introduction of model simplifications is possible in a limited range of application [2], this possibility being considerably decrease for chemical compounds. Therefore for common description of matter properties over a wide range of thermodynamic parameters on phase diagram it is traditionally to apply semiempirical models in which different experimental data are used to determine the numerical coefficients of general functional dependencies found from theoretical considerations.

In this paper, we propose the semiempirical EOS model for polymer materials over wide range of temperatures and pressures. We present also the results of EOS calculations on the base of developed model for polyethylene (PE), polystyrene (PS), and polymethylmethacrylate (PMMA) in comparison with the set of available at high energy densities experimental data.

EOS MODEL

A thermodynamically complete EOS for polymers is defined by the free energy (F) preassigned as a sum of three components

$$F(V,T) = F_c(V) + F_a(V,T) + F_e(V,T),$$

describing the elastic part of interaction at $T = 0$ K (F_c) and the thermal contribution by atoms (F_a) and electrons (F_e).

The volume dependence of elastic component of energy is expressed as follows

$$F_c(V) = \frac{B_{0c}V_{0c}}{m-n}\left(\sigma_c^m/m - \sigma_c^n/n\right) + E_{coh},$$

where $\sigma_c = V_{0c}/V$, V_{0c} is the specific volume at $P = 0$ and $T = 0$ K, B_{0c} is the bulk modulus

103

$B_c = -V dP_c/dV$ ($P_c = -dF_c/dV$) at $\sigma_c = 1$, E_{coh} is the cohesive energy. A detailed description of the procedure for calculating the coefficients of potential F_c from dynamic measurements can be found in Ref. [3].

The thermal component of free energy is defined by excitation of thermal vibrations of atoms:

$$F_a(V,T) = F_a^{acst}(V,T) + \sum_{\alpha=1}^{3(\nu-1)} F_{a\alpha}^{opt}(V,T),$$

$$F_a^{acst}(V,T) = \frac{3RT}{V}\ln\left(1 - \exp\left(-\sqrt{\theta_{acst}^2 + \sigma^{2/3}TT_a}\Big/T\right)\right) - \frac{RT}{V}D(\theta_{acst}/T),$$

$$F_{a\alpha}^{opt}(V,T) = \frac{RT}{V}\ln\left(1 - \exp\left(-\sqrt{\theta_{opt\alpha}^2 + \sigma^{2/3}TT_a}\Big/T\right)\right),$$

where R is the gas constant, ν is the number of atoms in the repeating cell of polymer chain,

$$D(x) = \frac{3}{x^3}\int_0^x \frac{t^3 dt}{e^t - 1}$$

is Debye function [4], $\sigma = V_0/V$, V_0 is the specific volume at $P = 0.1$ MPa and $T = 298$ K, θ_{acst} and $\theta_{opt\alpha}$ are the characteristic temperatures of acoustic and optical modes of phonon spectrum, T_a is empirical parameter. The volume dependencies of θ_{acst} and $\theta_{opt\alpha}$ are determined by the formula

$$\theta_{acst}(V)/\theta_{0acst} = \theta_{opt\alpha}(V)/\theta_{0opt\alpha} =$$
$$= \exp\left(-\int \gamma_c(V)d\ln V\right),$$

where

$$\gamma_c(V) = 2/3 + (\gamma_0 - 2/3)\frac{\sigma_n^2 + \ln^2\sigma_m}{\sigma_n^2 + \ln^2(\sigma/\sigma_m)},$$

γ_0 is the value of Gruneisen gamma under normal conditions, σ_m and σ_n are free parameters. The values of coefficients θ_{0acst} and $\theta_{0opt\alpha}$ are defined from measured values of isobaric heat capacity C_P

at normal pressure and various temperature [5]. The quality of the proposed form of contribution of thermal vibrations of atoms to the thermodynamic potential is illustrated by Fig. 1.

The electron component of free energy is included in the form:

$$F_e(V,T) = -RZ_e T e^{-T_s/T}\ln\left(1 + AV_0 T^{3/2}\sigma^{-\gamma_e}/RZ_e\right),$$

where

$$A = 4k\left(2\pi m_e k/h^2\right)^{3/2},$$

k and h are the Boltzman and Plank constants, m_e is electron mass,

$$T_s(V) = \Delta_0 \exp((1-\sigma)/\sigma_s)/2k,$$

Δ_0 is the energy gap between the valence band and the conduction band at normal condition, parameter σ_s defines the rate at which the gap is narrowed,

$$\gamma_e(T) = 1 + (\gamma_{e0} - 1)\exp\left(-T/T_g\right)$$

is analogue of the electronic Gruneisen coefficient. Such form of F_e in constructing of EOS for dielec-

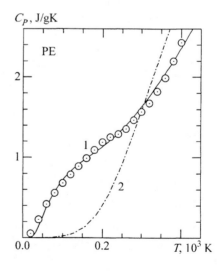

FIGURE 1. Specific heat capacity of PE at normal pressure, 1 — this EOS model, 2 — the Debye model [4]. Experiment [5]

trics takes into account the thermal excitation of electrons into the conduction band which occurs when a substance is heated [6, 7]. Also chosen F_e—T-dependence describes the transition to plasma with average ion charge Z_e at temperature limit $T \to \infty$.

THERMODYNAMIC PROPERTIES OF PE, PS, AND PMMA

The resulting EOS for PE, PS, and PMMA adequately describe the experimental data on the shock compressibility of solid and porous (PS) specimen of these plastics [8-13] over entire range of kinematic and dynamic characteristics realized, as can be seen from Fig. 2-4. A comparison of the calculated temperature values for the shocked PMMA with the results of measurements at the ultrahigh pressure range [12, 14] presented in Fig. 5 shows their good correlation too.

Analysis of the data [3, 8, 9] for PS indicates that there is a physicochemical conversion of the shocked substance. On the principal Hugoniot this conversion begins at $P \approx 20$ GPa. It involves a significant change in the density (by ~20%) and compressibility of the medium. This result is usually attributed to destruction of the polymer caused by the rupture of chemical bonds, resulting in the formation of a slightly compressible mixture of a diamondlike phase of carbon and various low-molecular-weight components [15]. The EOS evaluation for temperature of conversion beginning on the principal Hugoniot is $T \approx 1500$ K.

The phase diagram for PMMA shown in Fig. 4 reveals a region of states investigated in the waves of shock compression and adiabatic unloading. Note that the experimental release isentrope [11] begins with state of highly heated shocked condensed matter and continues up to rarefied-gas states. The isentropic expansion technique [11] has enabled to record the point of boiling of substance. In this case, kink in the calculated curve at the onset of evaporation corresponds to the experimentally observed additional increase of the expansion rate within the two-phase liquid—vapour region. The obtained value of equilibrium evaporation temperature at normal pressure $T_{v0} = 473$ K practically co-

incides with the tabular one for PMMA depolymerisation (methylmethacrylate is gas at such temperature). Calculation of temperature on condensed phase spinodal $(\partial P / \partial V)_T = 0$ at normal pressure gives $T_{sp0} = 790$ K, that is close to experimental value of limiting temperature of attainable overheating $T_1 = 788$ K [7]. The parameters of the critical

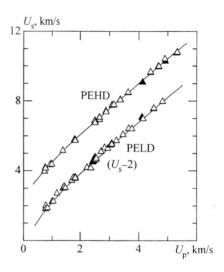

FIGURE 2. Shock Hugoniots of high and low density PE (HD and LD). Experiment [8]

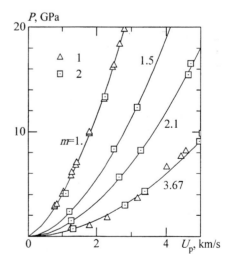

FIGURE 3. Shock Hugoniots of PS, $m = \rho_0 / \rho_{00}$ — initial porosity. Experiment: 1 — [8], 2 — [9]

point for PMMA were evaluated as $P_{cr} = 0.37$ GPa, $T_{cr} = 953$ K, $V_{cr} = 1.64$ cc/g, $S_{cr} = 6.09$ J/gK.

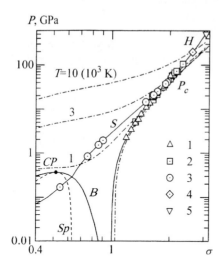

FIGURE 4. Phase diagram for PMMA, H — principal Hugoniot, S — isentrope, P_c — elastic compression curve at $T = 0$ K, T — isotherms, B — condensed phase—vapour equilibrium curve with critical point (CP), Sp — spinodal. Experiment: 1 — [8], 2 — [10], 3 — [11], 4 — [12], 5 — [13]

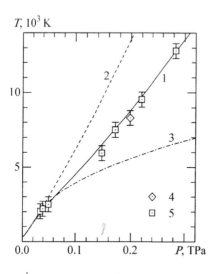

FIGURE 5. Temperature vs pressure for shocked PMMA, 1 and 2 — this EOS model with and without electron component, 3 — with electron component by model [6]. Experiment: 1 — [12], 2 — [14]

CONCLUSION

Calculations by presented EOS model demonstrate that thermodynamic characteristics of plastics are described by analytical formulas which are the same both at the normal conditions and at the highest temperatures and pressures attained in experiments. The resulting wide-range EOS for PE, PS, and PMMA describe consistently all of the available static and dynamic experimental data, and they can be employed effectively in numerical modelling of nonsteady gasdynamic processes at high energy densities.

ACKNOWLEDGMENTS

The work has been done due to financial support of the Russian Foundation for Basic Research (grant No. 97-02-17691).

REFERENCES

1. Bushman, A. V., Fortov, V. E., Kanel', G. I., Ni, A. L., *Intense Dynamic Loading of Condensed Matter*, Washington: Taylor & Francis, 1993.
2. Bushman, A. V., Fortov, V. E., *Sov. Phys. Usp.* **26**, 465 (1983).
3. Bushman, A. V., Lomonosov, I. V., Fortov, V. E., Khishchenko, K. V., Zhernokletov, M. V., Sutulov, Yu. N., *JETP* **82**, 895 (1996).
4. Landau, L. D., Lifshitz, E. M., *Statistical Physics*, Oxford: Pergamon Press, 1980.
5. Wunderlich, B., Baur, H., *Forschritte der Hochpolym. Forshung* **7**, 151 (1970).
6. Ioffe, A. F., *Fizika Poluprovodnikov* (Physics of Semiconductors), Moscow: Izdat. Akad. Nauk SSSR, 1957.
7. Khishchenko, K. V., Lomonosov, I. V., Fortov, V. E., Shlensky, O. F., *Dokl. Akad. Nauk* **349**, 322 (1996).
8. Marsh, S. P. (Ed.), *LASL Shock Hugoniot Data*, Berkeley: Univ. California Press, 1980.
9. Dudoladov, I. M., Rakitin, V. I., Sutulov, Yu. N., Telegin, G. S., *Zh. Prikl. Mekh. Tekh. Fiz.*, No. 4, 148 (1969).
10. Bakanova, A. A., Dudoladov, I. P., Trunin, R. F., *Sov. Phys. Solid State* **7**, 1307 (1965).
11. Bushman, A. V., Zhernokletov, M. V., Lomonosov, I. V., Sutulov, Yu. N., Fortov, V. E., Khishchenko, K. V., *Phys. Dokl.* **38**, 165 (1993).
12. Zel'dovich, Ya. B., Kormer, S. B., Sinitsyn, M. B., Kurekin, A. I., *Sov. Phys. Dokl.* **3**, 938 (1958).
13. Trunin, R. F., *Usp. Fiz. Nauk* **164**, 1215 (1994).
14. Kormer, S. B., *Usp. Fiz. Nauk* **94**, 641 (1968).
15. Ree, F. H., *J. Chem. Phys.* **70**, 974 (1979).

CP429, *Shock Compression of Condensed Matter – 1997*
edited by Schmidt/Dandekar/Forbes
© 1998 The American Institute of Physics 1-56396-738-3/98/$15.00

ANOMALY IN THE TEMPERATURE CALCULATION OF SHOCKED POLYMERS

Kunihito Nagayama and Yasuhito Mori

Department of Applied Physics, Faculty of Engineering, Kyushu University, Fukuoka 812 Japan

Grüneisen parameter in material whose Hugoniot has a sharp kink is discussed based on three theoretical models. Initial values for PMMA are very large, but drop to very a small value at the kink. Shock temperature has been calculated for weak shocks, which seems to show anomalous behavior. These results suggest that shocked states may be in strong non-equilibirium states up to some stress level.

INTRODUCTION

Polymer material like PMMA exhibits non-linear u_s-u_p Hugoniot compression curve in 1GPa stress region.[1,2] The data cannot be extrapolated to the higher pressure Hugoniot function.[3] Curvature of these Hugoniots has not been explained well by the clear physical basis. Stress and particle velocity profiles in the same stress range in polymers have a relaxation structure[4,5] indicating stress relaxation. Relaxation time depends on the material and propagation distance. This phenomena are clear evidence of the dynamical non-equilibrium realized by the shock compression.

Although the profile of these physical variables changes with the shock propagation, the shock velocity was found to be almost constant with propagation. This stationality assures the constant shock impedance depending only on the impact condition. In case of the symmetric impact of a flyer with a target with the same material, the value of the particle velocity can be fixed to the half of the impact velocity irrespective of the relaxation time and propagation distance.

This paper deals with the thermal non-equilibrium nature of the shock waves in PMMA in 1 GPa stress region.[6] Drastic curvature of the u_s-u_p Hugoniot is discussed to be the evidence of this thermal non-equilibrium.

GRÜNEISEN PARAMETER CHANGE AT THE KINK IN HUGONIOT

Non-linear u_s-u_p Hugoniot for PMMA in 1GPa stress region has been plotted in Fig. 1. It show the change in slopes around $u_p = 0.14$ (km/s) and $u_p = 0.5$ (km/s). In this section, this non-linearity has been examined. We will focus our attention to the Hugoniot *kink in the lower stress region*. Although the available Hugoniot data across this region is scattered around the kink, those can be expressed by the piecewise linear segments,

$$u_s = A + B\, u_p, \qquad (1)$$

where the first and second segments can be characterized by the following parameters

$$A_1 = 2.777 \text{ (km/s)}, \qquad 0 < u_p < 0.14 \text{ (km/s)}$$
$$B_1 = 2.324,$$

$$A_2 = 3.017 \text{ (km/s)}, \qquad 0.14 < u_p < 0.5 \text{ (km/s)}$$
$$B_2 = 0.6067,$$

From these two segments, the kink point can be calculated to be

$$u_s = 3.102 \text{ (km/s)},$$
$$u_p = 0.140 \text{ (km/s)}$$

One can see that the initial slope of the Hugoniot function is over 2, while that of the second segment is only 0.6. Physics of this non-linear Hugoniot of polymers has not known yet. Recent study of other

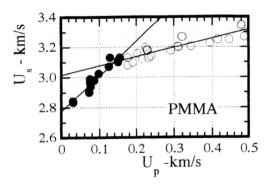

FIG. 1 Non-linear u_s-u_p Hugoniot for PMMA

polymers by our group has found similar properties for various polymers examined.[6] No appreciable evidence has been found for these deflections to correspond to any structural phase transitions.

An abrupt change in the slope in u_s-u_p Hugoniot relation seems to correspond to the change in some physical state variables. It is shown that at least the value of the Grüneisen parameter and the local sound velocity should change by the kink. We will develop the thermodynamic theory of determining the change in these values based on the form of the available u_s-u_p Hugoniot data. By deriving the differential equation for the Grüneisen paramter combined with a state variable instead of the sound velocity, one can obtain an equation to estimate the change in these values across the kink in Hugoniot.

We will assume the following theoretical model for the Grüneisen parameter to estimate the change in gamma across the kink. The model can be expressed as

$$\gamma(v) = \frac{t-2}{3} - \frac{v}{2} \frac{\dfrac{d^2}{dv^2}\left[p_c v^{\frac{2t}{3}}\right]}{\dfrac{d}{dv}\left[p_c v^{\frac{2t}{3}}\right]} \qquad (2)$$

where γ, v, p_c denote the Grüneisen parameter, the sepcific volume and the cold pressure, respectively. Parameter t may have the value, 0, 1, 2 corresponding to the Slater model,[7] the Dugdale-MacDonald model,[8] and the Vaschenko-Zubarev model,[9] respectively. The authors have proposed a new procedure of claculating the Grüneisen parameter by using the available shock Hugoniot data.[10,11] The method is based on the state variable

$$q_S = \left[\frac{u_s-u_p}{u_s}\right]^{\frac{2t}{3}} \frac{(u_s-u_p)^2}{u_s-u_p\dfrac{du_s}{du_p}}\left[u_s+u_p\dfrac{du_s}{du_p}\right.$$

$$\left. -\frac{2}{3}t\frac{u_p\left[u_s-u_p\dfrac{du_s}{du_p}\right]}{u_s-u_p} - \gamma\frac{u_p^2\dfrac{du_s}{du_p}}{u_s-u_p}\right] \qquad (3)$$

which is defined as a derivative along an isentrope and evaluated on the Hugoniot state. Eq.(2) can then be expressed in the form

$$\gamma = \frac{t+1}{3} + \frac{u_s(u_s-u_p)}{2\left[u_s-u_p\dfrac{du_s}{du_p}\right]}\left(\frac{\partial \ln q_S}{\partial u_p}\right)_H \qquad (4)$$

By combining these two equations, one can integrate these equations from the unshocked state up to some Hugoniot point.

Examination of the properties of the kink shows that the kink cannot be a break, corresponding to first order transition in the thermodynamic sense, and furthermore, the kink cannot be an extremely sharp break whose second order derivative has a very large value. The result is the kink should be a smeared change in slope within some velocity interval. The discussion has been reported in detail in a separate paper in the near future.[6]

Consequently, we have an approximate analytical expression for the Grüneisen parameter across the kink region. Calculation has been made for the compression behavior of the Grüneisen parameter for PMMA in this region, and the result is shown in Fig. 2. At the kink, a drastic drop in the parameter γ can be seen in the figure. Attained value of γ is very small consistent with the equilibrium value of γ at the uncompressed state of less than unity. The values depend on the models. For the Slater model, or the Dugdale-MacDonald model, the Grüneisen parameter has a rather large value at the initial state, which decreases slightly up to the kink region, then drops to the value less than unity. Since the value of γ after tha kink is very sensitive to the exact form of Hugoniot, the negative γ for Vaschenko-Zubarev model may be corrected only when we have sufficiently precise form of Hugoniot.

Physical interpretation of the sharp decrease in γ should be given by the following discussion. Dlott et al[12] showed from their ps laser shock experiments

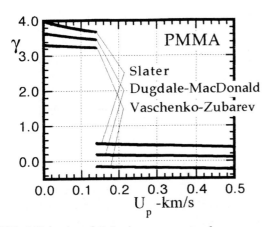

FIG. 2 Behavior of Grüneisen parameter for PMMA in 1 GPa stress region.

that Raman measurement indicates the very fast thermal equilibration after shock front in polymers. On the contrary, very large value of γ estimated in Fig. 2 at the initial value seems to correspond to the non-equilibrium thermal states of polymers at shock compression. These very large values of the Grüneisen parameter correspond to the values of Grüneisen parameter for the inter-molecular phonon mode excitation.

It is very plausible from these discussions that relatively weak shock in polymers will excite exclusively the phonon modes with very large γ values, but not the intra-molecular frequencies with very small Grüneisen parameters. In this sense, we will emphasize that weak shock in polymers is in very strong thermal non-equilibrium as well as in dynamical non-equilibrium. Ultrasonic measurements of polymers under hydrostatic pressure also suggest a very large Grüneisen parameter correspondig to the phonon modes consistent with the present discussion. The situation of the ultrasonic wave propagation and of shock wave seems to be very similar. The present result is considered to be the generalization of the ultrasonic results to large amplitude elastic waves. The present result is in sharp contrast to the result of Dlott et al.[12] It is shown in Fig. 2 that the situation changes after the kink that the Grüneisen parameter of the shocked state of moderate strength shock compression is suggested to be partly equilibrated. Thermodynamic equilibrium Grüneisen parameter for

polymers is estimated to be very small compared with even the value at the kink. The reliability of the absolute value of the Grüneisen parameter after kink cannot be examined, but it seems that the real equilibirium value of the Grüneisen parameter may be reached in higher stress range.

In conclusion of this section, shock wave in polymers in 1 GPa stress region is apparently in dynamical non-equilibrium, but may be almost in thermal equilibrium, only when the shock strength exceeds some stress levels. This conclusion is consistent with the laser shock experiments of exhibiting very fast thermal equilibration for several GPa shock waves.

TEMPERATURE CALCULATION FOR 1GPA SHOCK WAVES IN PMMA

Since the Grüneisen parameter plays an important role in specifying the attained state by shock compression, it may be interesting to discuss how thermal state variables like temperature or entropy for PMMA changes with compression. In this report, we have calculated shock temperature by using the values of Grüneisen parameter obtained in the previous section. The calculation of temperature is based on the following thermodynamic identity

$$\frac{dT}{T} = -\rho \gamma dv + \frac{dS}{C_v} \tag{5}$$

This equation can easily be integrated to give

$$\frac{T}{T_0} = \left(\frac{v}{v_0}\right)^{-\gamma} \exp\left[\frac{S-S_0}{C_v}\right] \tag{6}$$

in case the Grüneisen parameter and specific heat is assumed constant. This equation states that the temperature change have two contributions, temperature change due to isentropic compression, and the one due to the entropy change. In the present case, it is estimated that the entropy change is very small, most of the temperature change is attributed to the isentropic compression. From Fig. 2, the behavior of the Grüneisen parameter has very small changes except at the kink point. We will assume here that the value of it is given by a constant value for each velocity interval as

$\gamma = \gamma_1 = 3.9$ for Slater model

$\quad\ = 3.75$ for Dugdale-MacDonald model

$$= 3.6 \quad \text{for Vaschenko-Zubarev model}$$
$$0 < u_p < 0.14 \quad (km/s) \quad (6)$$

$$\gamma = \gamma_2 = 0.5 \quad \text{for Slater model}$$
$$= 0.2 \quad \text{for Dugdale-MacDonald model}$$
$$= -0.2 \quad \text{for Vaschenko-Zubarev model}$$
$$0.14 < u_p < 0.5 \ (km/s) \quad (7)$$

We further assume that the specific heat at constant volume is almost constant over the stress region considered here. The value of the specific heat in each pressure region is expressed as

$$C_V = \text{constant} \quad 0 < u_p < 0.5 \quad (km/s) \quad (8)$$

In this calculation, the entropy change is estimated by assuming the weak shock approximation. In this situation, we have

$$T_H dS_H = \frac{u_p^2}{u_s} \frac{du_s}{du_p} du_p = \frac{Bu_p^2}{u_s} du_p \quad (9)$$

This relation can be derived from the jump conditions and thermodynamics. Under weak shock approximation, the integration of this equation is straightforward.

Figure 3 shows the calculated shock temperature in PMMA for two model theories for Grüneisen parameter. Very large value for the Grüneisen parameter gives a large temperature rise by shock compression, while for a very small gamma the temperature rise is estimated to be very small. As noted before, the temperature change is almost determined by the value of the Grüneisen parameter, then these results fol-low.

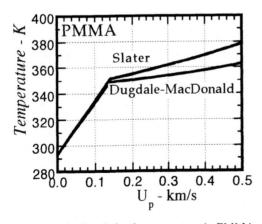

FIG 3. Calculated shock temperature in PMMA

The value of the parameter is very small after the kink but still differs a little. The temperature change apparently depends on the model. We need to know which model is better in this circumstance in order to discuss the temperature rise after the kink. In both models, the estimated shock temperature at the kink point is about 350 K, at which some effect like softening takes place to ease the motion of the conformer much better than at lower temperature.

CONCLUDING REMARKS

This paper delas with the anomalous behavior of the relatively weak shock waves in polymers like PMMA whose shock-compressed state is supposed to be in strong non-equilibrium states. Unlike the discussion of the dynamical non-equilibrium, we have discussed the change in Grüneisen parameter and calculate the shock temperature. It is found that Grüneisen parameter drops to very small value at the kink. We need to proceed to discuss how is the upper kink in Hugoniot, where the scenario assumed here may not be applied.

REFERENCES

1. L. M. Barker, and R.E. Hollenbach, J. Appl. Phys., 41 (1970) 4208.
2. D.N. Schmidt, and M.W. Evans, Nature, 206 (1965) 1348.
3. Los Alamos Shock Lugoniot Data, compiled by S.P. Marsh (University of California, Berkeley, 1981).
4. K.W. Schuler, J. Mech. Phys. Solids, 18 (1970) 277
5. J.W. Nunziato and K.W. Schuler, Trans. Soc. Rheol., 16 (1972) 15-32.
6. K. Nagayama and Y. Mori, submitted to J. Appl. Phys.
7. J.C. Slater, Introduction to Chemical Physics (McGraw-Hill, New York, 1939) Chap. XIII.
8. J.S. Dugdale and D.K.C. MacDonald, Phys. Rev., 89, 832 (1953).
9. V. Ya. Vaschenko and V.N. Zubarev, Sov. Phys. Solid State 5, 653 (1963).
10. K. Nagayama, J. Phys. Chem. Solids, 58 (1997) 271.
11. K. Nagayama and Y. Mori, J. Phys. Soc. Jpn, 63 (1994) 4070.
12. D.E. Hare, J. Franken, and D.D. Dlott, Chem. Phys. Lett., 244 (1995) 224.

CP429, *Shock Compression of Condensed Matter – 1997*
edited by Schmidt/Dandekar/Forbes
© 1998 The American Institute of Physics 1-56396-738-3/98/$15.00

SHOCK COMPRESSION AND THE EQUATION OF STATE OF FULLY DENSE AND POROUS POLYURETHANE

J. R. Maw and N.J.Whitworth

AWE, Aldermaston, Reading, U.K.

The principal shock Hugoniot of fully dense polyurethane is well established experimentally but there is considerable uncertainty in our knowledge of the Gruneisen parameter which is needed if an equation of state is to be developed to describe off-Hugoniot states. Primary and reflected shock Hugoniot data on polyurethane foams have been systematically analysed to determine the shocked densities, pressures and energies. This has enabled the variation with density of the Gruneisen parameter to be more precisely defined at densities above and below the normal density. A Gruneisen form of equation of state is then developed which is shown to give good agreement with results from a number of mutiple shock experiments on polyurethane foams. The equation of state is valid for densities in the range 0.3 - 3.0 g/cc and pressures up to 50 GPa.

INTRODUCTION

In the past a number of investigations have been carried out on the shock compression characteristics of fully dense and porous polyurethane. Marsh [1] has published the most extensive set of data on the principal Hugoniots of both fully dense and foamed material. Data on polyurethane foams obtained at AWE have also been published [2,3].

Mader and Carter [4] considered the double shock behaviour of polyurethane foam at quite high pressures and concluded that chemical decomposition of the foam should be taken into account when formulating a suitable equation of state (EOS).

More recently experiments have been carried out to investigate multiple shock compression characteristics of foam. [5,6].

In this paper we review the data with a view to obtaining an empirically based EOS which can be used to adequately predict the single and multiple shock behaviour of both fully dense and porous polyurethane.

FULLY DENSE POLYURETHANE

Marsh [1] gives principal Hugoniot data on fully dense (ρ_{0s} =1.265 g/cc) polyurethane. At pressures up to ~21 GPa the data can be fitted [7] by a linear shock velocity (U_s) - particle velocity (u_p) relation

$$U_s = C_0 + s u_p$$

with $C_0 = 2.486$ mm/μs and s = 1.577

At higher pressures there is evidence of some form of phase transition and these data can then be represented using additional linear fits

$C_0 = 4.04$ mm/μs, s = 0.97 for $2.6 < u_p < 3.6$

$C_0 = 2.50$ mm/μs, s = 1.40 for $u_p > 3.6$

Figures 1 and 2 compare the data with these fits in the U_s - u_p and pressure - density planes. Although these clearly give an adequate representation of the data we need further information to develop an EOS to describe states off the principal Hugoniot.

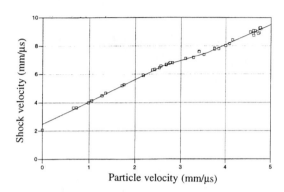

FIGURE 1. Hugoniot of Fully Dense Polyurethane (U_s - u_p)

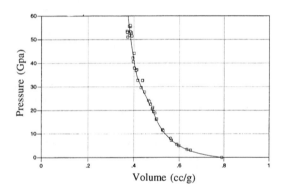

FIGURE 2. Hugoniot of Fully Dense Polyurethane (p - v)

Based on the Hugoniot of the fully dense material a Grüneisen EOS can be constructed in the form :

$$p = p_H(v) + \Gamma(v)[e - e_H(v)]/v \qquad (1)$$

where $p_h(v)$ and $e_H(v) = 1/2p_h(v_{0s} - v)$ are the Hugoniot pressure and energy. The crucial element in this EOS is the magnitude and volume dependence of the Gruneisen parameter $\Gamma(v)$. McQueen and Marsh [7] quote a value $\Gamma_0 = 1.55$ at normal density but the basis for this is not well established. In an effort to more precisely define this parameter we have examined the available data on the principal and double shock response of polyurethane foams.

HUGONIOT DATA ON FOAMS

Figure 3 shows the published principal Hugoniot data on foams [1] in the pressure - volume plane. A notable feature of these data is the fact that the shocked volume is rarely lower than the normal volume of fully dense material. This is a consequence of the high thermal energy generated by the shock compression. There is also considerable scatter in the data, a reflection of the problems encountered in obtaining reliable data on foams using flash gap techniques. In principle it would be possible to use these data to obtain estimates of Γ by considering the expression

$$\Gamma(v) = v [p - p_s(v)]/[e - e_s(v)] \qquad (2)$$

where p and e are the pressure and energy on the foam Hugoniot and p_s, e_s are the corresponding values on the Hugoniot of the fully dense material. However this would only give information on the value of Γ at densities below the normal density and in addition the uncertainty in the porous Hugoniot data leads to large uncertainties in the values of Γ. We therefore looked for more accurate data on both the principal Hugoniot and from double shock experiments where the foam is shocked to higher densities.

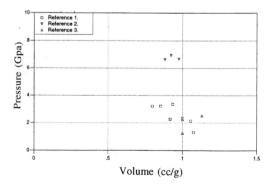

FIGURE 3. Hugoniot data on 0.3 g/cc polyurethane foam

Two gas gun experiments have been carried out [5] which provide limited but more reliable data. In these experiments foams of density ~ 0.3 g/cc backed by an aluminium buffer and LiF window were impacted by an aluminium flyer. A VISAR measurement was made of the velocity of the buffer/window interface. Using standard impedance match techniques it is possible to analyse the data to obtain both the principal Hugoniot state and the state behind the shock reflected from the buffer with greater precision.

Table 1 summarises the results. Also shown in the table is the corresponding value of Γ obtained using equation (2). Although this analysis has determined only two points it is difficult to reconcile these with the value $\Gamma_0 = 1.55$ suggested by McQueen and Marsh. Figure 4 compares the data with a more plausible representation of $\Gamma(v)$ taking $\Gamma_0 = 1.0$ and

$$\Gamma(v)/v = \Gamma_0/v_{0s} \quad \text{for } v < v_{0s}$$

$$\Gamma(v) = \Gamma_0 \quad \text{for } v > v_{0s}$$

Some justification for the reduction in Γ_0 comes from consideration of the basis of the value 1.55 originally proposed by McQueen and Marsh. They quote a value 1.8×10^{-5} Mb-cc/g for the specific heat C_p which with the thermodynamic definition

$$\Gamma_0 = 3\alpha C_0^2 / C_p$$

implies a linear coefficient of thermal expansion, $\alpha = 1.5 \times 10^{-4}$ K^{-1}. However measurements of α made at AWE indicate a value nearer 1×10^{-4} which implies a value $\Gamma_0 \sim 1.0$.

In figure 5 the principal Hugoniot data for foams of density ~ 0.3g/cc are compared with the EOS equation(1)using $\Gamma_0 = 1.55$ and 1.0. The lower value appears to fit the data better although it should be noted that the EOS does not agree with the data at pressures less than ~ 2 GPa. This is probably due to the finite resistance of the foam to compaction which is not modelled with this simple EOS treatment.

TABLE 1. Shock states from analysis of gas gun experiments

	First Shock	Reflected Shock
u (mm/μs)	2.06	0.57
P (GPa)	1.64	9.28
ρ (g/cc)	1.128	1.683
E (kJ/g)	2.13	3.72
Γ	1.03	0.75

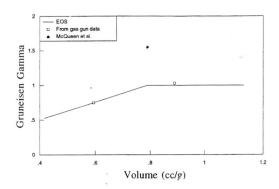

FIGURE 4. Variation of Γ with volume

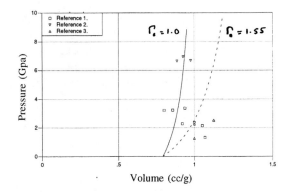

FIGURE 5. Calculated Hugoniot of 0.3 g/cc foam

113

LANL HIGH PRESSURE REFLECTED SHOCK EXPERIMENTS

Mader and Carter [4] reported a series of experiments in which a plane-wave explosive system was used to shock polyurethane foam of density 0.5 g/cc backed by layers of copper, aluminium and magnesium. Measurements were made of the shock velocity in the foam and in the metal backing layers from which the initial and reflected shock states were deduced. The initial shock pressure was ~ 110 kb and the reflected pressures were in the range 260 - 440kb. From their analysis, Mader and Carter concluded that the data could not be explained with a sensible volume dependence of Γ and postulated some degree of dissociation in the polyurethane at these pressures. However, since their analysis was based on an earlier EOS for fully dense polyurethane which is in poor agreement with the Hugoniot data available now, it was thought worthwhile to see how the EOS derived here compares with their data.

The measured shock velocity in the foam (5.7 mm/μs) together with our EOS determines the first shock conditions p = 10.7 GPa, u_p = 3.8 mm/μs, ρ = 1.48 g/cc. (Note that for the 0.5 g/cc foam the shocked density is higher than the normal density of polyurethane.) The p - u diagram in figure 6 shows the measured reflected shock data points which agree very well with the intersection of the calculated reflected shock locus and the Hugoniots of the three metals. This lends further support to the validity of our chosen $\Gamma(v)$.

CONCLUSIONS

The available data on the shock response of fully dense and porous polyurethane have been reviewed. Analysis of the double shock data has enabled the determination of the variation with volume of the Gruneisen parameter and an EOS has been developed which accurately represents single and double shock states in polyurethane foams.

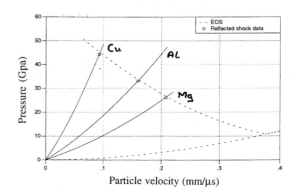

FIGURE 6. P - u diagram corresponding to experiments of Mader and Carter

REFERENCES

1. Marsh, S. P., *LASL Shock Hugoniot Data,* University of California Press, (1980).

2. Skidmore I. C. and Morris E., "Experimental Equation-of-State Data for Uranium and its Interpretation in the Critical Region", *Thermodynamics of Nuclear Materials,* IAEA, Vienna, p173, 1962.

3. James R. M. et al. "The Critical Angle for Mach Bridge Formation in Polyurethane Foam" in *Proceedings of the 6th International Detonation Symposium,* (1976).

4. Mader C. L. and Carter W. J., "An Equation of State for Shocked Polyurethane Foam" *Los Alamos Scientific Laboratory Report , LA 4059,* (1969).

5. Wise J. L. and Cox D. E. SNL internal memorandum (1996).

6. Maw J. R., Whitworth N. J. and Holland R. B., "Multiple Shock Compression of Polyurethane and Syntactic Foams," in *Shock Compression of Condensed Matter -1995,* pp 133-136.

7. McQueen R. G. et al. "The Equation of State of Solids from Shock Wave Studies", in *High Velocity Impact Phenomena* (ed. R Kinslow), Academic Press, (1980).

CP429, *Shock Compression of Condensed Matter – 1997*
edited by Schmidt/Dandekar/Forbes
© 1998 The American Institute of Physics 1-56396-738-3/98/$15.00

SHOCK WAVE EQUATIONS OF STATE OF CHONDRITIC METEORITES

William W. Anderson

Department of Geology and Physics
Georgia Southwestern State University, Americus, Georgia 31709

Thomas J. Ahrens

Lindhurst Laboratory of Experimental Geophysics, Seismological Laboratory
California Institute of Technology, Pasadena, California 91125

We have obtained shock compression data for Murchison and Bruderheim chondritic meteorites. Data for Murchison suggest that the Hugoniot states are described by a smooth curve to $\geq$90 GPa, having ρ_0 = 2.656 Mg/m^3, K_{S0} = 24.2$\pm$.7 GPa, K' = 4.17$\pm$.10, and constant γ = 1.0. The data for Bruderheim suggest more complicated behavior. A mineral mixture model consistent with the Bruderheim data suggests that the Hugoniot state is a low pressure phase below 25 GPa, with ρ_0 = 3.555 Mg/m^3, K_{S0} = 146 GPa, K' = 2.53, and constant $\rho\gamma$ = 7.11 Mg/m^3; and a high pressure phase above 65 GPa, with ρ_0 = 4.40 Mg/m^3, K_{S0} = 225 GPa, K' = 3.25, and constant $\rho\gamma$ = 7.485 Mg/m^3.

INTRODUCTION

Because hypervelocity impacts are a significant process in planetary evolution, accurate knowledge of the shock compression properties of the materials involved is important for models of such events. While most existing shock compression data for relevant materials are for terrestrial rocks, the impactors are extraterrestrial objects with compositions similar to meteorites, chondrites being the most abundant of these. Here, we study the shock compression behavior of two chondritic meteorites and present suggested equations of state.

EXPERIMENTAL PROGRAM

Shock wave equation of state (EOS) experiments were performed on samples of the Murchison carbonaceous chondrite (bulk density = 2.244$\pm$.087 Mg/m^3) and Bruderheim hypersthene

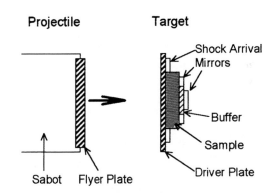

FIGURE 1. Schematic of experimental arrangement.

chondrite (bulk density = 3.337$\pm$.011 Mg/m^3). The samples were cut into rectangular slabs with the surfaces ground flat and parallel and were placed on driver plates, with buffer materials against the rear surfaces of the samples (Fig. 1). These assemblies were impacted by projectiles launched

TABLE 1. Experimental Results for Bruderheim and Murchison Meteorites.

Sample and Shot #	Initial Density (Mg/m³)	Shock State				Release State		
		Particle Velocity (km/s)	Shock Velocity (km/s)	Density (Mg/m³)	Pressure (GPa)	Particle Velocity (km/s)	Pressure (GPa)	Density (Mg/m³)
Bruderheim 900	3.3247 (.0024)	.757[a] (.017)	5.111[a] (.024)	3.902[a] (.016)	12.86[a] (.30)			
		.773[b] (.012)	4.285[b] (.058)	3.920[b] (.016)	13.08[b] (.24)	2.586 (.092)	0	1.98 (.07)
Bruderheim 897	3.3490 (.0025)	1.967 (.009)	6.571 (.072)	4.780 (.028)	43.29 (.43)	3.115 (.407)	24.7 (3.6)	3.57 (.80)
Bruderheim LGG309	3.3293 (.0034)	2.416 (.008)	6.990 (.033)	5.088 (.020)	56.22 (.21))	3.712 (.074)	32.0 (1.1)	3.76 (.16)
Bruderheim LGG306	3.3465 (.0032)	3.913 (.006)	9.433 (.035)	5.719 (.020)	123.53 (.39)			
Murchison 885	2.3914 (.0027)	.398[a] (.016)	2.997[a] (.019)	2.757[a] (.018)	2.85[a] (.12)			
		.736[b] (.008)	2.862[b] (.021)	3.197[b] (.018)	5.15[b] (.12)	3.237 (.719)	0	.65 (.21)
Murchison 1016	2.1761 (.0027)	2.163 (.037)	5.057 (.017)	3.802 (.050)	23.80 (.41)	2.601 (.061)	19.7 (.7)	3.23 (.32)
Murchison LGG255	2.1852 (.0032)	3.046 (.014)	6.386 (.036)	4.178 (.032)	42.51 (.24)	4.129 (.059)	38.5 (.9)	1.88 (.38)
Murchison LGG254	2.2242 (.0049)	4.503 (.008)	8.458 (.092)	4.757 (.067)	84.72 (.80)	5.869 (.074)	72.2 (1.6)	2.79 (.31)

[a]Elastic Precursor
[b]Plastic Wave

from the Caltech 25 mm two-stage light gas gun and 40 mm propellant gun. Shock arrivals at the surfaces of target components were detected on streak camera records by the disappearance of reflections from rear-surface mirrors placed against the target. The shock and release states were determined via the impedance matching method.

Experimental Results

Table 1 presents the experimental results. The U_s-u_p projections of the shock Hugoniot curves (Fig. 2) can be described by straight lines:

$$U_s = C_0 + su_p \qquad (1)$$

with $C_0 = 3.11\pm.06$ km/s and $s = 1.62\pm.02$ for Bruderheim and $C_0 = 1.87\pm.07$ km/s and $s = 1.48\pm.03$ for Murchison. The data for Bruderheim show some evidence that the U_s-u_p Hugoniot might be more complicated than a single straight line. There is no such evidence in the Murchison data.

Both meteorites exhibit two-wave shock behavior at low stresses. In Murchison, the stress of the precursor wave is $2.85\pm.12$ GPa and is consistent with interpretation as an elastic wave. The stress level of the precursor in Bruderheim is $12.86\pm.30$ GPa, which is unexpectedly high for an elastic wave. We suggest that the double-wave structure in this case may indicate a sluggish phase transformation, such as is seen in carbonates (1).

116

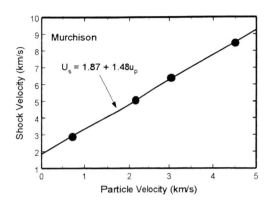

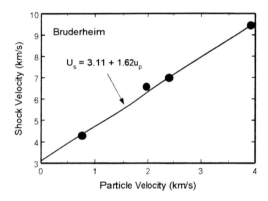

FIGURE 2. Shock velocity-particle velocity projections of the experimental data with linear fits.

EQUATIONS OF STATE

We wish to constrain effective equations of state that can be used in models of planetary impact processes. Both meteorites contain a number of minerals that undergo known shock-induced phase changes. The deviations of the Bruderheim data from linearity suggest such phase changes. The smooth linear trend followed by Murchison data in Fig. 2 can be described by a single "phase," but probably simply does not reflect the complex behavior of the individual components under shock loading. It is important to note that, as these samples are polymineralic aggregates, "phase" is used here to imply an assemblage of phases.

Murchison Equation of State

Since the data suggest that Murchison does not undergo any detectable phase transformations up to at least 90 GPa, we can fit the present data to a single effective equation of state. The zero-pressure nonporous density of Murchison, based on mineral norms calculated from the composition (2), is 2.656 Mg/m^3, indicating that the present samples are ~16% porous. A fit to the shock wave data (Fig. 3a), gives the isentropic bulk modulus and its pressure derivative as $K_{S0} = 24.2 \pm .7$ GPa and $K' = 4.17 \pm .10$. The best fit was obtained under the assumption that the thermodynamic Grüneisen parameter has a constant value of $\gamma = 1.0$.

Composition-Based Estimates

As an alternative to fitting EOS data directly, we can attempt to estimate the shock compression behavior of a material using knowledge of the composition. We use simplified mineral norms, which are based on published composition data (2,3), to estimate the Hugoniot curves of both meteorites. In both cases, we treated the Hugoniot volume of the bulk material as the mass-weighted mean of the Hugoniot volumes of the constituent minerals (4). No attempt was made to correct for the temperature differences between the Hugoniot states of different minerals, since thermal expansion is a second-order effect.

Figure 3a shows the results of the mineral mixing estimate for Murchison. The model consistently overestimates the specific volume of the Hugoniot state. The model also requires both low- and high-pressure phase stability regions, based on the behavior of the constituent minerals, while there is no evidence for multiple phases in the present data. The calculated Hugoniot curves resemble those of serpentine, which is often used as an analog for Murchison and other carbonaceous chondrites. The present results suggest that serpentine, even though a major constituent of Murchison, is a poor analog and that the high content of hydrous phases gives rise to complicated shock compression behavior.

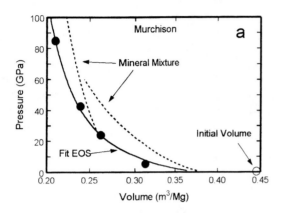

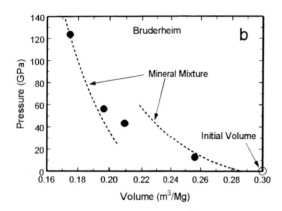

FIGURE 3. (a) Pressure-volume Hugoniot data for Murchison, compared to Hugoniot curves predicted by the fit equation of state and mineral mixture model. (b) Pressure-volume Hugoniot data for Bruderheim, compared with Hugoniot curves predicted by the mineral mixture model.

Figure 3b shows the estimated Hugoniot curves for Bruderheim. In this case, the mineral mixture model gives a good match to the experimental data. Based on these results, we can attempt to obtain estimates of effective EOS parameters for Bruderheim low and high pressure phases. For the low pressure phase, using $\rho_0 = 3.555$ Mg/m^3, based on measured Archimedian densities of the samples, we get $K_{S0} = 146$ GPa, $K' = 2.53$, and constant $\rho\gamma = 7.11$ Mg/m^3. For the high pressure phase, we get $\rho_0 = 4.40$ Mg/m^3, $K_{S0} = 225$ GPa, $K' = 3.25$, and constant $\rho\gamma = 7.49$ Mg/m^3, with an STP transition energy from the low pressure phase of 1.25 MJ/kg. The intervening pressure region, from ~25 GPa to ~ 65 GPa, represents a mixed-phase region.

CONCLUSION

We have obtained shock compression data for Murchison and Bruderheim chondritic meteorites. The data for Murchison suggest that the Hugoniot state is described by a single curve up to at least 90 GPa, having $\rho_0 = 2.656$ Mg/m^3, $K_{S0} = 24.2\pm.7$ GPa, $K' = 4.17\pm.10$, and constant $\gamma = 1.0$. Comparison with a mineral mixture model suggests that such models do not work well for carbonaceous chondrites and that serpentine is a poor analog for carbonaceous chondrites.

Bruderheim data suggest a more complicated Hugoniot, but are insufficient to constrain equations of state. A mineral mixture model consistent with the data suggests that the Hugoniot state consists of a low pressure phase below 25 GPa, with $\rho_0 = 3.555$ Mg/m^3, $K_{S0} = 146$ GPa, $K' = 2.53$, and constant $\rho\gamma = 7.11$ Mg/m^3; and a high pressure phase above 65 GPa, with $\rho_0 = 4.40$ Mg/m^3, $K_{S0} = 225$ GPa, $K' = 3.25$, constant $\rho\gamma = 7.485$ Mg/m^3, and transition energy of 1.25 MJ/kg.

ACKNOWLEDGMENTS

We thank Prof. Dr. Dieter Stöffler of the Museum für Naturkunde, Berlin, for providing the samples and E. Gelle and M. Long for assistance with the experiments. Contribution #6213, Division of Geological and Planetary Sciences, California Institute of Technology. Research supported by NASA.

REFERENCES

1. Ahrens, T. J., and Gregson, V. G., *J. Geophys. Res.* **69**, 4839-4874 (1964).

2. Jarosewich, E., *Meteoritics* **6**, 49-52 (1971).

3. Duke, M., Maynes, D., and Brown, H., *J. Geophys. Res.* **66**, 3557-3563 (1961).

4. Al'tshuler, L. V., and Sharipdzhanov, I. I., *Izv. Earth Physics (Engl. Trans.)* 167-177 (1971).

CP429, *Shock Compression of Condensed Matter – 1997*
edited by Schmidt/Dandekar/Forbes
© 1998 The American Institute of Physics 1-56396-738-3/98/$15.00

SHOCK HUGONIOT AND RELEASE STATES IN CONCRETE MIXTURES WITH DIFFERENT AGGREGATE SIZES FROM 3 TO 23 GPA

C. A. Hall, L. C. Chhabildas, and W. D. Reinhart

Sandia National Laboratories, Albuquerque, NM 87185-1181*

A series of controlled impact experiments has been performed to determine the shock loading and release behavior of two types of concrete, differentiated by aggregate size, but with average densities varying by less than 2 percent. Hugoniot stress and subsequent release data was collected over a range of approximately 3 to 25 GPa using a plate reverberation technique in combination with velocity interferometry. The results of the current data are compared to those obtained in previous studies on concrete with a different aggregate size but similar density. Results indicate that the average loading and release behavior are comparable for the three types of concrete discussed in this paper.

INTRODUCTION

Considerable interest in characterizing the dynamic response of concrete under impact loading exists because it is used extensively as a structural material. Concrete is a heterogeneous composite, typically consisting of quartz aggregate and cement grout. Local variations in the shock and particle velocities due to impedance differences within the material cause fluctuations in the measured particle velocity profiles. A deliberate attempt to average these local variations was made in this study by using the thickest possible copper and tantalum plates which would still allow an adequate number of isentropic decompression states to define a representative curve. The plate reverberation technique has previously been used to determine the shock loading and release states for concrete (1) and quartz (2).

EXPERIMENTAL TECHNIQUE

The experiments were performed on an 89mm diameter, smooth bore powder gun which is capable of generating impacts in the 0.5 km/s - 2.4 km/s range. The tilt between impactor and target plate, exit velocity of the projectile, and particle velocity from the rear surface of the metallic target plate were measured during each experiment. The configuration used for this experimental series is shown in Figure 1. The projectile consisted

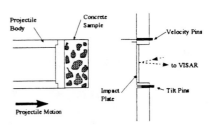

FIGURE 1: Experimental configuration

of a concrete sample attached to the aluminum projectile nose plate and phenolic body. The target was a thin metal plate inserted into an aluminum target holder. The resulting velocity profiles with copper target plates are shown in Figure 2. Release states for the concrete can be inferred by knowledge of the Hugoniot and release adiabat of the metallic plate material. In an isentropic

* Sandia is a multiprogram laboratory operated by Sandia Corporation, a Lockheed Martin Company, for the United States Department of Energy under Contract DE-AC04-94AL85000.

Table 1. Experimental parameters and Hugoniot results

Shot Number	Concrete Density (kg/m^3)	Target Thk / Matl (mm)	Impact Velocity (km/s)	σ_h (GPa)	u_c (km/s)	U_c (km/s)	ε_c
LC-1	2353.9	3.51 / Cu	0.464	2.79	0.387	3.06	0.13
LC-2,b	2356.1	3.52 / Cu	0.797	5.97	0.637 / 0.647	3.98 / 3.66	0.16 / 0.18
LC-3,b	2356.1	3.52 / Cu	1.340	11.43	1.048 / 1.075	4.63 / 4.05	0.23 / 0.27
LC-4,b	2363.4	3.50 / Cu	1.740	15.86	1.368 / 1.360	4.98 / 4.76	0.27 / 0.29
LC-5,b	2356.7	3.50 / Cu	2.150	20.92	1.650 / 1.710	5.38 / 4.55	0.31 / 0.37
LC-7	2354.0	1.85 / Ta	2.140	22.17	1.833	5.12	0.36
SC-1	2340.2	3.50 / Cu	2.143	19.71	1.668	5.05	0.33
SC-2	2347.9	3.50 / Cu	1.748	15.52	1.363	4.85	0.28
SC-3	2321.6	3.50 / Cu	1.330	11.75	1.030	4.91	0.21
SC-4	2340.6	1.86 / Ta	2.175	22.70	1.820	5.33	0.34
SC-5	2327.7	3.52 / Cu	0.830	6.01	0.669	3.86	0.17
SC-6	2327.5	3.52 / Cu	0.451	2.07	0.394	2.26	0.17

release process approximation, an approach using stress and particle velocity decrements can be employed to calculate the release path. This technique will satisfy wave approximations until the first attenuating release wave from the concrete arrives at the impact surface.

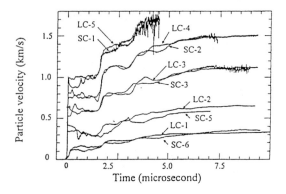

FIGURE 2. Velocity profiles for large and small aggregate concrete experiments which used copper target plates

MATERIAL DESCRIPTION

The concrete used in the present study had two distinct aggregate size distributions. The concrete referred to as large aggregate had an ASTM aggregate size number of 57. This implies that 5% of the material by weight is between 25 mm and 37.5 mm, 40% to 75% is between 19 mm and 25 mm and the balance is 4.75 mm or smaller (3). The concrete referred to as small aggregate had an

ASTM aggregate size number of 7, which means 10% of the material by weight is 12.5 mm, 30% to 60% is 9.5 mm and the balance is 4.75 mm or smaller (3). Cores were taken from large castings in both cases to ensure representative responses. Samples were obtained from each core and measurements made to determine densities. Results are listed in Table 1 with LC and SC referring to large and small aggregate concretes/experiments respectively. The aggregate size for SAC-5 concrete (1) was similar, but slightly finer that of the small aggregate concrete used in this investigation.

HUGONIOT RESULTS

The experimental parameters and Hugoniot results from this investigation are given in Table 1. In typical experiments, measurements of shock velocity and particle velocity are made directly on the sample of interest. For a highly heterogeneous material such as concrete, however, the measurements of these parameters are best made through an averaging medium such as a homogeneous metallic plate. Knowing the equation of state for the target plate, measuring impact velocity, V_i, and free surface particle velocity, u_2, the Hugoniot stress and particle velocity in the concrete can be inferred through stress and particle velocity continuity across the impact interface (1,2). The results, including some lower pressure data on SAC-5 (4), are plotted as stress vs. particle velocity with quadratic curve fits in Figure 3. As can be seen, the curves are tightly grouped. This indicates that the loading response of concrete is somewhat independent of aggregate size at these stresses. Scatter bars representing stress deviations due to local variations in the measured particle velocity at the Hugoniot state for each experiment are included. Greater dispersion of the data can be seen in a stress-strain plot. This is expected since strain varies as the square of the particle velocity. The shock velocity, U_c, versus particle velocity, u_c, Hugoniot data for the concrete has been plotted in Figure 4. Also shown in the figure are the results of previous studies on other concrete both above (1) and within (4) the elastic regime. There appears to be a definite slope change in the concrete behavior above the initial elastic regime. A linear least squares fit to the large aggregate, small aggregate, and corresponding SAC-5 data yields $U_c = 2235 + 1.75u_c$. For comparison, the fit for the lower stress data is given by $U_c = 551 + 4.52u_c$ (4). This behavior can be attributed to both the porosity and heterogeneous nature of the material. The large slope, S, indicated by the lower stress range results suggests relatively large compressions are occurring in this pressure regime. As the stress increases beyond this point, considerably stiffer compaction behavior is indicated by the lower slope value.

RELEASE STATE RESULTS

Once the Hugoniot point is established in the P-u_p plane, subsequent release stress states can be determined within concrete using Δu_c as the change in particle velocity between states of interest. The average wave velocity within the concrete, C_c, can be estimated from $C_c = \Delta\sigma_c / \rho_{0c}(\Delta u_c)$ where $\Delta\sigma_C$ represents the difference in stress. The quadratic curve fits to the release data in the concrete are tightly grouped in the P-u_p plot, as shown in Figure 5. As with the Hugoniot curves, this would indicate that the release response of the concrete does not exhibit a great dependence upon aggregate size at these stress levels. Larger deviations can be seen in a stress-strain plot. This can be partly attributed to the difficulty in determining an average particle velocity and two way transit time within the metallic plate as the particle velocity steps become less discernible late in time. Also, the change in strain between release states is dependent upon the square of the corresponding change in particle velocity.

An indication of residual strain exists in each of the particle velocity records. In the latter part of each race where discrete velocity steps are not

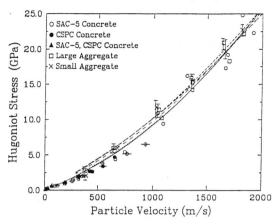

FIGURE 3. Stress vs. particle velocity for large aggregate, small aggregate, SAC-5 and low stress conventional concrete Hugoniot data

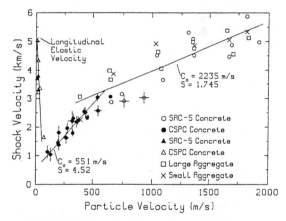

FIGURE 4. U_c-u_c for Large aggregate, Small aggregate, SAC-5 and low stress conventional concrete Hugoniot data

easily distinguishable, the change in particle velocity per two way transit time in the target plate approaches zero. Consequently, $\Delta\varepsilon$ will be quite small. Therefore, it appears that residual strain is present within the concrete samples in the latter stages of the unloading process.

COMPARISON WITH PREVIOUS STUDIES

In order to obtain a measure of the variations in particle velocity that can be expected within a particular type of concrete due to its heterogeneous nature, two VISAR signals were recorded at separate locations on the same experiment where possible. Results indicated the largest deviation in local particle velocity was seen in shot number LC-3, where as much as 29% deviation was seen

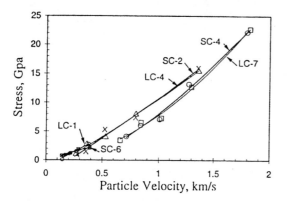

FIGURE 5. Stress versus particle velocity release curves for high, medium and low stress data sets on Large and Small aggregate concrete which used copper target plates.

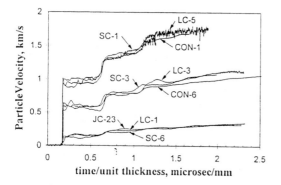

FIGURE 6: Normalized particle velocity traces for Large aggregate, Small aggregate and SAC-5 concrete for high, medium and low stress tests using copper target plates.

between the maximum and minimum values during initial loading, while only a 5% deviation between average values was observed.

Figure 6 shows normalized velocity profiles for large aggregate, small aggregate, and the SAC-5 (1) concrete. The time axis was divided by the corresponding plate thickness in millimeters to obtain transit time per unit thickness for each experiment. The velocity axis was not normalized since variations in impact velocity were small. The average value for particle velocity at

the Hugoniot state, in both the high and medium velocity experimental sets varies by no more than 5%. The average value for the lower velocity set varies by as much as 25%. This indicates that material characteristics such as aggregate size are significant at lower stress levels.

SUMMARY

In summary, Hugoniot and release state data was collected for two types of concrete, differentiated by aggregate size, but with average densities varying by less than 2 percent over a stress range of approximately 3 to 25 GPa using a plate reverberation technique in combination with velocity interferometry. This data set was compared in several ways to data obtained in previous studies on SAC-5 concrete (1), which has a different aggregate size but similar density. Stress versus particle velocity data for both the Hugoniot and subsequent release states were plotted and compared, and particle velocity profiles normalized with respect to plate thickness and overlayed on the same graph. Results indicate that the average loading and release behavior of the three types of concrete discussed in this paper are loosely grouped within scatter bars derived from particle velocity variations due to the heterogeneous nature of the material. Therefore, it appears that the average loading and release response of concrete does not exhibit a strong dependence upon these aggregate sizes in the 3 to 23 GPa stress range.

REFERENCES

1. Grady, D.E., "Dynamic Decompression Properties of Concrete From Hugoniot States 3 to 25 GPa", Sandia National Laboratories Technical Memorandum - TMDG0396, February 1996.

2. Chhabildas, L.C. and Grady, D.E., "Dynamic Material Response of Quartz at High Strain Rates", Material Response Symposium Proceedings, Elsevier 22, pp. 147-150 (1984).

3. American Society for Testing and Materials, C33-93 Standard Specifications for Concrete Aggregates, pp. 3, 1993.

4. Kipp, M.E., Chhabildas, L.C., and Reinhart, W.D., "Elastic Shock Response and Spall Strength of Concrete", Proceedings of the American Physical Society Topical Conference, this volume, 1997.

CHAPTER III

PHASE TRANSITIONS

CP429, *Shock Compression of Condensed Matter – 1997*
edited by Schmidt/Dandekar/Forbes
© 1998 The American Institute of Physics 1-56396-738-3/98/$15.00

QUASI-ISENTROPIC COMPRESSION
OF LIQUID XENON
UP TO THE DENSITY OF 20 g/cm³
UNDER THE PRESSURE OF 700 GPa

V. D. Urlin, M. A. Mochalov, and O. L. Mikhailova

Russian Federal Nuclear Center - VNIIEF, Sarov, 607190

The liquid xenon compressibility has been experimentally investigated up to pressures of ~720 GPa. The material was compressed within an explosively driven cylindrical shell. The density was registered by the gamma-graphic method and the pressure was determined from gasdynamic calculations. Comparison of experimental and calculated results showed that the considered compression process was isentropic to a high degree. Compression of liquid xenon up to a density of ~20 g/cm³ confirms the presence of an anomaly connected with the structural transformation at 8.37 g/cm³.

INTRODUCTION

Inert gases have a filled and symmetric outer electronic shell and crystallize into close-packed structures. This in a certain degree makes easier their theoretical description. Of them xenon has the narrowest energy gap between the valence and conduction band and is a most interesting material. In Reference [1] xenon was investigated at quasi-isentropic compression under dynamic conditions up to a density of 13 g/cm³. In the same work its equation of state (EOS) was found, which describes well the measured pressure, density, and brightness temperature under shock compression.

The American investigators [2-4] measured the optical absorption, performed an X-ray crystal structure analysis and revealed a structural transition from the face-centered cubic (FCC) structure to the hexagonal-close-packed (hcp) structure in statically compressed xenon under pressures of the 70-90 GPa without a density jump. In Ref. [3] investigations of xenon were performed up to a pressure of 200 GPa

and the band gap closure was found to occur at a pressure of 132 GPa and density of 12.3 g/cm³ which corresponds to the metallic state transition, practically without a density jump. In Ref. [4] at a pressure of 150 GPa the authors observed anomalies in the absorption spectrum which they connected with the metallic state transition. On the basis of these works in [1] there was found an equation of state describing the xenon behavior after a structural transition at $\rho = 8.37$ g/cm³.

In our work, being a continuation of work [1], the region of investigations of xenon under quasi-isentropic compression is considerably extended. A new experimental point has been obtained at a density of ~ 20 g/cm³ and pressure of ~ 720 GPa. It has been shown that the obtained equation of state for xenon gives a good description of a newly obtained experimental point if the electron component is taken into account adequately to the metallic state.

125

EXPERIMENTAL RESULTS

In the present work as in Ref. [1] the quasi-isentropic compressibility of liquid xenon is studied in a device in which high pressures are dynamically produced by superposition of direct and reflected shock waves. The material under study is encapsulated in a cylindrical shell and is compressed with the help of a charge generating converging shock waves. After they converge on the shell axis there arises a reflected cylindrical shock wave propagating in the already compressed and heated material. Our estimates show that the entropy increment in the reflected wave is appreciably below the value achieved in the first converging wave and subsequent waves practically do not change the total entropy of matter.

Investigations of the liquid xenon compressibility up to a pressure of ~720 GPa were performed in a specially designed apparatus. As in Ref. [1], the trajectory of the shell with liquid gas was photographically recorded with the help of a powerful "shining through" gamma-graphic facility [5]. The measured value is the inner radius of the shell at fixed moments of time. In a series of successive experiments we determined the trajectory of the shell motion and from it estimates were made of the cavity dimensions at the moment of maximum compression (stopping moment).

Pressure inside the compressed material was estimated from gasdynamic calculations using EOS's of chemical explosives, construction materials, and xenon. In investigations of argon [6] an acceptable boundary contrast was achieved by using shells of heavy metals: copper and tungsten. In the present work the liquid xenon was compressed in a compound shell consisting of an inner aluminum cylinder surrounded by a shell of the tungsten alloy. The use of "light" aluminum ($\rho_0 = 2.71$ g/cm^3), for which the x-ray radiation absorption was found to be appreciably lower than the absorption by a layer of compressed liquid xenon, gave a satisfactory contrast. The experimental and calculated trajectories are given in Figure 1. This figure shows the trajectories of boundary Xe-Al, calculated for Xe in phase I (Xe I) which is in thermodynamic equilibrium at $\rho < 8.37$ g/cm^3 and for phase II (Xe II) which is in thermodynamic equilibrium at $\rho > 8.37$ g/cm^3.

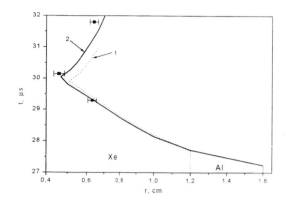

FIGURE 1. Experimental and calculated trajectories of the Xe-Al boundary (the pressure at the moment of stopping is ~720 GPa)

..... calculated for Xe I and
——— for Xe II

EQUATION OF STATE OF XENON IN THE LIQUID AND SOLID PHASES

The xenon equation of state was taken to be analogous to that of Ref. [1]. For the solid and liquid phase the free energy can be written as

$$F_s = E_x(\rho) + 1.125R\Theta + RT[3\ln(1-e^{-\Theta/T}) - D(\Theta/T)] + F_{el}, \quad (1)$$

$$F_L = E_x(\rho) + 3RT\{\ln[\Theta/T\,(1+Z)^{0.5}] - b\} + 3RT_0\,f(\rho) + F_{el}, \quad (2)$$

where T and p are the temperature and density, R is the gas constant, and D(x) is the Debye function. The curve of elastic interaction is approximated by the dependence

$$E_x = \frac{3}{\rho_k}\sum_{i=1}^{} \frac{a_i}{i}\left(\delta^{1/3} - 1\right),$$

where $\delta = \rho/\rho_k$, ρ_k is the density at $P_x = \rho^2 dE_x/d\rho = 0$, a_i are the empirical constants.

The Debye temperature is defined by the dependence $\Theta = \Theta_0 \delta^{1/3}\sqrt{C_x^2 - n\,2P_x/3\rho}$, where $C_x^2 = dP_x/d\rho$, Θ_0 and n are the empirical parameters.

126

TABLE 1. Parameters of the equation of state for XeI in the solid and liquid phases. The values of a_i are in GPa.

ρ_k, g/cm^3	Θ_0	a_1	a_2	a_3	a_4	a_5	a_6
3.805	20.91	-2.251	11.856	-15.849	22.255	-44.382	28.371
T_0, K	ρ_0, g/cm^3	n	η	r	A	B	C	b
161.4	2.985	3	5	2	3.538	-3.429	-2.8105	0.6624

TABLE 2. Parameters of the equation of state for XeII. The values of a_i are in GPa.

ρ_k, g/cm^3	Θ_0	a_1	a_2	a_3	a_4
8.371	41.49	-93.06	523.05	-851.02	460.83

The parameters in the equality (2)

$z = \eta RT / \left(C_x^2 - n\, 2P_x / 3\rho\right)$ and $f = A\delta^r + B \ln \delta + C$ determine the value of deviation of the thermal and elastic properties of liquid from these of the solid phase. The values of η, r, A, B, C are the empirical constants, T_0 is the melting temperature and ρ_0 is the liquid phase density at P = 0. Tables 1 and 2 give the xenon EOS parameters for the first and second phases taken from Ref. [1].

At normal density xenon is a dielectric and its energy gap between the valence and conduction band is equal to 9.3 eV. Beginning with the temperatures on the order of 10^4 K the contribution of thermal excitation of electrons in the equation of state becomes appreciable. This contribution can be taken into account using the free-electron theory, similar to Refs. [1, 6]. For the dielectric this component can be put in the form

$$F_{el} = -\frac{4kT}{\rho_k}\sqrt{n_p n_n}\left(\frac{2\pi m^* kT}{h^2}\right)^{1.5} \exp\left(-\frac{W}{2kT}\right)$$

$m^* = m_0\delta^{-\Gamma}$, where m_0 is the mass of a free electron, n_p and n_n are the orbital degeneracies in the bands. $W = W_0 \ln(\rho_m / \rho)$ is the energy gap between the valence and conduction band. ρ_m is the density at which the gap closes. Γ is the empiric constant which value is found from the condition of the shock adiabat description in a high-temperature region. In calculations the following values have been used: $W_0 = 7.46$ eV, $\Gamma = 0.9$, and $\rho_m = 13.13$ g/cm^3.

As xenon metallization is assumed to occur at $\rho = 13.13$ g/cm^3, then in the region of densities above this value the electron term in the equation of state is estimated using the free-electron theory for metal: $F_{el} = -0.5\beta\, T^2$, where $\beta = \beta_0\delta^{-2/3}$ and $\beta_0 = 1.03\ 10^{-5}$ J/gK.

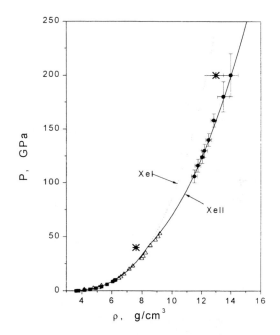

FIGURE 2. Xenon isotherm at T=300 K. Experiment: • from [3], Δ from [8], ♦ from [7] * isentrope from [1]

COMPARISON OF CALCULATED AND EXPERIMENTAL RESULTS.

Fig. 2 shows the experimental data on static compression of solid xenon at T = 300 K up to 200 GPa from Refs. [3, 7, 8]. The curve XeI calculated with the EOS parameters for XeI gives a good description of the experimental data up to ~40 GPa, but above this value a discrepancy is observed. In calculations according to Refs. [2-4] it has been taken that the structural transition without a density jump occurs at $\rho = 8.37$ g/cm^3. The curve

XeII calculated with the parameters of XeII EOS gives a good description of all the experimental data [8, 3].

Fig.1 gives the experimental points on the trajectory of the boundary aluminum shell-liquid xenon at maximally achieved pressure of 720 GPa. The dotted curve 1 corresponds to the gasdynamic calculation with the xenon EOS without a phase transition. The solid curve 2 has been calculated using the xenon equation of state taking into account both structural transformation and metallization. A satisfactory agreement is observed between the experimental trajectory and calculated curve 2.

Fig. 3 compares the calculation with the experiment on the quasi-isentropic compression of liquid xenon performed in Ref. [1] and in the present work. The experimental point located in the region of $\rho < 8.37$ g/cm^3 is well described by the isentrope with the EOS for XeI. The experiment at P~200 GPa is within the region of densities where the transition from fcc to hcp structure has already occurred [3]. The solid curve, which was calculated with this transition taken into account, is in a good agreement with the experiment. For comparison we give the dotted curve calculated without taking into account this transition. The solid squares on this curve denote the Xe state at the calculated moment of stopping of the shell. The experimental point at P~720 GPa is within the region of densities above the phase transition connected with metallization at $\rho = 13.13$ g/cm^3 and is quite well described by the solid curve calculated with the EOS for XeII with metallization taken into account.

CONCLUSION

Thus, in the present work the xenon isentrope has been extended to pressures of ~720 GPa where other experimental data are absent. The new results on quasi-isentropic compression of liquid xenon confirm location of transitions connected according to Refs. [2-4] with the structural transformation and metallization. We succeeded in describing the present experiment and experiments in Refs. [1-4] using the equations of state of the form (1) and (2) with the assumption of transition from the fcc to hcp structure at $\rho = 8.37$ g/cm^3 and metallization at $\rho = 13.13$ g/cm^3.

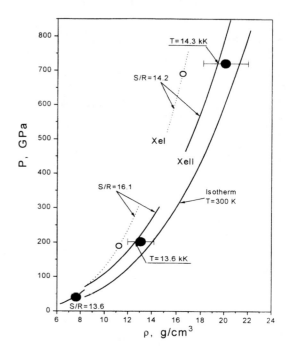

FIGURE 3. Quasi-isentropic compression of liquid xenon.

Experiment: ●

Calculation: — XeII, - - - XeI

○ pressure of Xe at the moment the shell stops calculated using the XeI EOS.

REFERENCES

1. Urlin V.D, Mochalov M.A., and Mikhailova O.L., *High Pressure Research* **8**, 595 (1992).
2. Jephcoat A.P., Mao H-K,. Finger L.W, Cox D.E., Hemley R.J., and Szha C., *Phys.Rev.Lett.* **59**, 2670 (1987).
3. Goettel K.A., Eggert J.H., Silvera I.F., *Phys. Rev. Lett.* **62**, 665 (1989).
4. Reichlin R., Brister K., McMahan A., Ross M., Martin Sne, Vohra Y.K., Ruoff A.L., *Phys. Rev. Lett.* **62**, 6691(1989).
5. Pavlovski A.I., Kuznetsov G.D., Sklizkov G.V. et al., *DAN SSSR* **160**, 68 (1965).
6. Adamskaja I.A, Grigoriev F.V., Mikhailova O.L., Mochalov M.A, Sokolova A.I., and Urlin V.D., *ZhETF* **93**, 647 (1987).
7. Syassen K., Holzapfer W.B., *Phys. Rev.* **18**, 5826 (1978).
8. Jisman A.N., Aleksandrov I.V., Stishov S.M., *Phys. Rev.* **32**, 484 (1985).

CP429, *Shock Compression of Condensed Matter – 1997*
edited by Schmidt/Dandekar/Forbes
© 1998 The American Institute of Physics 1-56396-738-3/98/$15.00

A MODEL FOR THE SHOCK-INDUCED PHASE TRANSITION IN IRON

Jonathan C. Boettger and Duane C. Wallace

Theoretical Division, Los Alamos National Laboratory,Los Alamos, NM 87545

It has long been known that the shock-induced α to ε phase transition in iron exhibits significant metastability in the two-phase region above 13 GPa. We have developed a simple, physically-motivated model that accurately describes the metastable phase transition surface for shock-loaded iron during numerical simulations. It is demonstrated here that the metastability of Fe can have a substantial impact on numerical simulations of time-resolved experiments.

Theoretical interpretation of shock-wave data is frequently complicated by the existence of shock-induced, solid-solid phase transitions in the transmitting medium. For this reason, there has been a long-standing interest in developing techniques for accurately modelling such transitions during simulations of shock processes. One of the more difficult issues to be addressed in this context is the possible metastability of a low-pressure phase beyond its equilibrium phase boundary. One of the best studied examples of such a transition is the $\alpha \to \varepsilon$ transition in iron [1-5], which begins on the principal Hugoniot at about 13 GPa, but is not completed until a shock-stress greater than 20 GPa is achieved. We have recently developed a simple, physically-motivated model for describing the shock-induced $\alpha \to \varepsilon$ transition in iron during numerical simulations [6]. In the present work, we will first review the basic features of our model. We will then use simulations of a time-resolved shock-wave experiment [5] to demonstrate the large impact that metastability can have on such calculations.

Accurate simulations of a solid-solid phase transition require high-quality equations-of-state (EOS) for the two phases involved in the transition. Here we utilize a highly accurate analytical EOS developed by Wallace [7]. The Helmholtz free energy $F(V,T)$ for each crystal structure is written as

$$F = \Phi_0 + F_H + F_A + F_E \qquad (1)$$

where Φ_0 is the static-lattice potential, F_H is the quasiharmonic phonon free energy, F_A is the anharmonic contribution to the lattice free energy, and F_E is the free energy due to thermal excitation of electrons. All other needed thermodynamic functions can then be obtained from the usual thermodynamic relationships involving partial derivatives of the free energy.

The exact expressions used here for the various contributions to $F(V,T)$ and the values of all required parameters, have been provided elsewhere [6] and, hence, are only briefly discussed here. The static-lattice potential is fitted with a modified version [8] of the Vinet-Ferrante-Rose-Smith universal EOS [9]. The quasiharmonic phonon free energy is described by a high-temperature expansion based on moments of the V-dependent phonon frequencies, that should be quite accurate for the temperatures of interest here, room temperature and above. The anharmonic contribution to the free energy is assumed to be negligible for iron. Finally, the thermal electronic contribution is expressed as the sum of two parts, a normal conduction-electron contribution (F_{cond}; approximated with its low-temperature form) and a contribution due to the magnetic ground state of α iron (F_{mag}; obtained from Andrews' fit to the magnetic specific heat of α

iron [3]). The analytical EOSs used here for the α and ε phases provide a good fit to the experimental $\alpha \rightarrow \varepsilon$ phase boundary of iron [10-12] and the 300 K isotherm measured by Mao et al [13-14].

Our analytical EOSs for the α and ε phases of iron have been implemented in the one-dimensional numerical simulation code HYDROX [15]. A dynamic phase mixing scheme described by Boettger et al. [16] is then used to determine the state of the iron in the mixed-phase region. It is assumed during the mixing that: (1) the two phases are in local pressure and temperature equilibrium $\left(P_\alpha = P_\varepsilon \text{ and } T_\alpha = T_\varepsilon\right)$ at all times; (2) for any given P and T, the metastable mass fraction of the ε phase, λ_m, is a function of the Gibbs free energy difference between the phases, $\Delta G(P,T) = G_\alpha - G_\varepsilon$; and (3) all extensive thermodynamic functions can be expressed as sums of α and ε phase contributions; for example, $V = (1-\lambda)V_\alpha + \lambda V_\varepsilon$ where λ is the instantaneous mass fraction of the ε phase.

For the dynamic mixing scheme employed by HYDROX to be useful, λ_m must be simply related to ΔG in the mixed phase region. Ignoring any time dependence of the transition, for the moment, the necessary equation can be developed from a balance of forces. At any instant during the forward transition, the fraction of the material that has not yet been transformed $(1-\lambda)$ will be driven to transform by a thermodynamic force ΔG. Because of the volume change associated with the transformation, the already transformed fraction of the material (λ) will exert a stress that resists any further transformation. The metastable mass fraction (λ_m) will then be the mass fraction at which the driving force and the resisting force just balance;

$$d\lambda_m \propto (1-\lambda_m)d\Delta G. \qquad (2)$$

Integrating this gives

$$\lambda_m = 1 - \exp\left[(A_F - \Delta G)/B_F\right], \qquad (3)$$

where A_F plays the role of an activation energy and B_F determines the range of ΔG over which the transition occurs. An analogous equation can be developed for the reverse transition with parameters A_R and B_R [6].

$$\lambda_m = \exp\left[(\Delta G - A_R)/B_R\right]. \qquad (4)$$

Although Eq. 3 was developed under the assumption that the phase transition is metastable, it can also be used to model an equilibrium transition by setting A_R to zero and using a very small value for B_R. Following this procedure, equilibrium Hugoniot states were obtained for iron by generating compressive waves with constant velocity boundary conditions. (Here, and throughout the remainder of this work, the strength effects in iron were modelled with a simple elastic-plastic model with a constant yield strength of 3 kbar.) In Figure 1, the theoretical equilibrium Hugoniot of iron (dashed line), generated as described, is compared with experimental data [2,5]. In the single-phase regions, the theoretical Hugoniot is in excellent agreement with the data, demonstrating the quality of the analytical EOSs being used here. In the mixed phase region, however, there is a significant difference between the equilibrium Hugoniot and the data, due to metastability.

The metastable mixed-phase region on the principal Hugoniot of iron can be fitted rather well with Eq. 3 by choosing $A_F = 0$ and $B_F = 642$ J mol [6]. The metastable Hugoniot obtained with these parameters is shown in Fig. 1 (solid line). Since $\Delta G = 0$ on the equilibrium phase boundary, the fitted value $A_F = 0$ implies that the shock

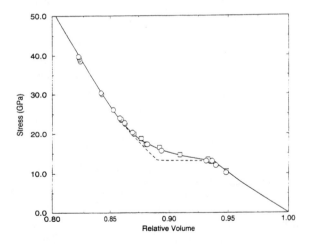

FIGURE 1. Iron Hugoniot. Experimental data are from Bancroft, et al. (Ref. 2; squares) and Barker and Hollenbach (Ref. 5; circles). Dashed line is theoretical equilibrium Hugoniot, and solid line is theoretical metastable Hugoniot.

transition begins at the equilibrium phase boundary, near 13 GPa. The fitted value for B_F ensures that the transition is not completed until after 20 GPa.

Diamond anvil cell (DAC) measurements at room temperature have shown that, even under quasistatic conditions, the $\alpha \to \varepsilon$ transition in iron does not proceed on the equilibrium surface [17]. Indeed, the DAC data for λ vs P is remarkably similar to what we find along the Hugoniot. To allow a direct comparison between the data and the theoretical metastable surface for the forward transition, we used Eq. 3 to determine λ vs P along the 300 K isotherm, with the results shown in Fig. 2. With the exception of the rather small amount of ε phase seen in the DAC data below 13 GPa, the theoretical results provide a good fit to the data.

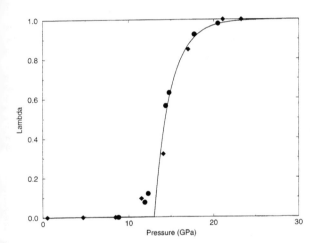

FIGURE 2. λ vs P for the forward transition in iron at 300K.

Between the initial and final states of a shock, the medium passes through a series of non-equilibrium states. To lowest-order, the transition rate will be linear in the distance between the current state and the metastable surface. That is, $\dot{\lambda} = (\lambda_m - \lambda)/\tau$ where τ is the relaxation time for the transition. This relationship is not unique, as written, since the distance to the metastable surface depends on the path along which $\lambda_m - \lambda$ is evaluated. Since the driving force is proportional to $\Delta G(P,T)$, the distance $\lambda_m - \lambda$ should be measured at constant P and T, giving

$$\frac{d\lambda}{dt} = \frac{\lambda_m(P,T) - \lambda(P,T)}{\tau}. \qquad (5)$$

This result differs from previous models, in that Andrews [3] measured $\lambda_m - \lambda$ at constant V and U, Horie and Duvall [18] measured $\lambda_m - \lambda$ at constant V and T, and both used the equilibrium value of λ in place of λ_m.

In our earlier work [6], we tested our phase transition model by simulating four time-resolved impact experiments on iron that were carried out by Barker and Hollenbach [5] with peak stresses ranging between 17 and 30 GPa, roughly the range of the mixed-phase region. We found that Eq. 5 provides an excellent representation of the experimental data, so long as the relaxation time τ was optimized for each experiment. The optimum values of τ ranged from 20 ns to 50 ns for the four experiments considered. Surprisingly, the optimum values of τ obtained for two experiments that were identical other than the sample thickness differed by nearly a factor of two (30 ns vs 50 ns), indicating that the $\alpha \to \varepsilon$ transition in iron exhibits a nonlinear transition rate.

One issue that was not addressed in our earlier work is the question of how important it is to use the metastable surface as the reference for the transition rate equation, Eq. 5, rather than the equilibrium surface used in previous models [3,18]. To answer this important question, we have repeated our calculations for experiment 1 of Barker and Hollenbach [5] with λ_m in Eq. 5 replaced by λ_{eq}. In this experiment, a 6.330 mm thick iron flyer plate was impacted on a 6.317 mm thick iron target at a velocity of 0.9916 km/s, producing a peak stress of 17.3 GPa. The velocity of the free surface of the target was then monitored for 3 μs after the impact. As in our previous work, a relaxation time of 36 ns was used during the simulation, and the parameters for the reverse transition, A_R and B_R, were set equal to the forward transition parameters, A_F and B_F.

In Figure 3, the free surface velocity profile obtained here using λ_{eq} in Eq. 5 is compared with the experimental data of Barker and Hollenbach [5] and the free surface profile calculated using λ_m in Eq. 5. The profile obtained with λ_m is in near perfect agreement with the experimental data. In contrast, use of λ_{eq} in Eq. 5 produces a significant delay in the arrival time of the P2 wave relative to the experimental data. This rather large discrepancy between the current simulation and the data is quite significant, since the arrival time of the midpoint of the P2 wavefront can not be adjusted by simply varying the value of τ, which primarily effects the width of the P2 wave. Thus, it seems clear that the driving force for the time-dependent $\alpha \rightarrow \varepsilon$ phase transition in iron should be determined from the metastable transition surface, not the equilibrium transition surface that has been used in the past.

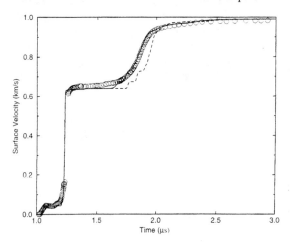

FIGURE 3. Free surface velocity vs time for experiment 1 of Barker and Hollenbach (Ref. 5; circles), compared with calculations using λ_m (solid line) and λ_{eq} (dashed line) in the transition rate equation (Eq. 5).

ACKNOWLEDGMENTS

This work was supported by the U. S. Department of Energy.

REFERENCES

1. Minshall, S., *Phys. Rev.* **98**, 271 (1955).
2. Bancroft, D., Peterson, E. L., and Minshall, S., *J. Appl. Phys.* **27**, 291-298 (1956).
3. Andrews, D. J., *J. Phys. Chem. Solids* **34**, 825-840 (1973).
4. Forbes, J. W., ``Experimental Investigation of the Kinetics of the Shock-Induced Alpha to Epsilon Phase Transition in Armco Iron'' (Washington State University Report WSU-SDL 76-01, Pullman, Washington, 1976).
5. Barker, L. M., and Hollenbach, R. E., *J. Appl. Phys.* **45**, 4872-4887 (1974).
6. Boettger, J. C., and Wallace, D. C., *Phys. Rev. B* **55**, 2840-2849 (1997).
7. Wallace, D. C., *Thermodynamics of Crystals* (Wiley, New York,1972).
8. Boettger, J. C., and Trickey, S. B., *Phys. Rev. B* **53**, 3007-3012 (1996); see also Straub, G. K., and Wills, J. M. (unpublished).
9. Vinet, P., Ferrante, J., Rose, J. H., and Smith, J. R., *J. Phys. Condens. Matter* **1**, 1941-1963 (1989).
10. Johnson, P. C., Stein, B. A., and Davis, R. S., *J. Appl. Phys.* **33**, 557-561 (1962).
11. Kaufman, L., Clougherty, E. V., and Weiss, R. J., *Acta Metall.* **11**, 323-335 (1963).
12. Bundy, F. P., *J. Appl. Phys.* **36**, 616-620 (1965).
13. Mao, H. K., Bassett, W. A., and Takahashi, T., *J. Appl. Phys.* **38**, 272-276 (1967).
14. Mao, H. K., Wu, W. Y., Chen, L. C., and Shu, J. F., *J. Geophys. Res.* **36**, 21737-21742 (1990).
15. Shaw, M. S., and Straub, G. K., ``HYDROX: A One-Dimensional Lagrangian Hydrodynamics Code'' (Los Alamos National Laboratory Report LA-8642-M, Los Alamos, New Mexico, 1981).
16. Boettger, J. C., Furnish, M. D., Dey, T. N., and Grady, D. E., J. *Appl. Phys.* **78**, 5155-5165 (1995).
17. Taylor, R. D., Pasternak, M. P., and Jeanloz, R., *J. Appl. Phys.* **69**, 6126-6128 (1991).
18. Horie, Y., and Duvall, G. E., ``Shock Waves and the Kinetics of Solid-Solid Transitions'' (Washington State University Report WSU-SDL 68-06, Pullman, Washington, 1968).

CP429, *Shock Compression of Condensed Matter – 1997*
edited by Schmidt/Dandekar/Forbes
© 1998 The American Institute of Physics 1-56396-738-3/98/$15.00

SHOCK TEMPERATURES AND THE MELTING POINT OF IRON

Thomas J. Ahrens, Kathleen G. Holland*, and George Q. Chen[†]

Lindhurst Laboratory of Experimental Geophysics, Seismological Laboratory 252-21, California Institute of Technology, Pasadena, CA 91125

New measurements of the ratio of Fe to LiF and Al_2O_3 anvil thermal diffusivities are used to obtain revised shock temperatures for Fe. New results match Brown and McQueen's (1) calculations of the temperatures of 5000 and 5800K at the 200 and 243 GPa transitions in Fe. New sound speed measurements along the Hugoniot of γ-Fe, centered at 1573K, demonstrate that this phase melts at ~70 GPa and ~2800 K and the γ phase does not occur above ~93 GPa. At higher pressures, perhaps over the entire pressure range of the Earth's molten outer core (132 to 330 GPa), the β (dhcp) phase, and not the ε phase, appears to be the solidus phase of pure Fe.

INTRODUCTION

The melting point of iron (Fe) at the pressures of the outer (liquid) core-inner (solid) core (330 GPa) at a depth in the Earth 5150 km was suggested (2) to provide a constraint on the absolute temperature. Initial work on the melting relations in the Fe-Ni-O-S system below 20 GPa (3) indicated that geochemically plausible iron alloys drastically lowered the solidus of Fe from 2200 to 1150K. However, recent measurements (4; 5) indicate a decrease of eutectic melting depression in the Fe-FeO-FeS system at core pressures (>130 GPa).

Brown and McQueen (1) conducted pioneering measurements of the longitudinal wave velocity behind shock waves along the principal Hugoniot (Fig. 1) of Fe and interpreted the 5 and 3.5% decreases at 200 and 234 GPa to the intersection of the Hugoniot with the ε to γ, and γ to liquid phase lines. Assumption dependent temperature calculations gave 4100 to 5300 K and 4900 to 6900 K, for the 200 and 243 GPa transitions, respectively.

SHOCK TEMPERATURE MEASUREMENTS

Urtiew and Grover (6) laid the theoretical basis for shock temperature measurements in metals. In

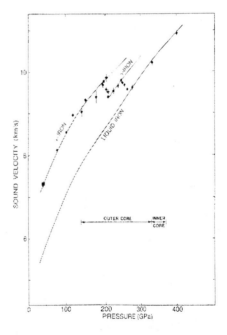

FIGURE 1. Rarefaction velocities for iron along principal Hugoniot as a function of pressure. Circles are present data. The square is from Barker and Hollenbach (7) Brown and McQueen (1) suggested these two phase transitions are related to the temperatures in the Earth as shown in Fig. 2. Copyright 1986, American Geophysical Union.

our experiments a film of metal is deposited upon a transparent anvil material and a shock wave is driven from the metal sample into the transparent anvil.

Lyzenga and Ahrens (8) first reported radiance versus time measurements for such an experimental assembly, for a 51 μm-thick Ag sample sputtered onto a Al_2O_3 anvil shocked to 185 GPa. These experiments demonstrated that the steady interface temperatures predicted by Urtiew and Grover (8) could be obtained via spectral measurements of the grey-body Planck function. Urtiew and Grover (8) showed that the metal Hugoniot temperatures are related to the interface temperatures by

$$T_i = T'_H + (T_a - T'_H)/(1+\alpha) \qquad (1)$$

where T'_H is the Hugoniot temperature of the metal (in the case where the metal and anvil have the same shock impedance), T_a is the Hugoniot temperature of the anvil material and α a correction factor involving the ratio of thermal properties of both media is given by

$$\alpha = \left\{ (\kappa_m \rho_m C_m)/(\kappa_a \rho_a C_a) \right\}^{1/2} \quad (2)$$

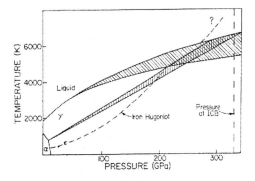

FIGURE 2. Phase diagram for pure iron proposed by Brown and McQueen (9)   Reprinted with permission.

where κ_m and κ_a, ρ_m and ρ_a, and C_m and C_a are the thermal diffusivities, densities, and specific heats for the metal and anvil at the compressed interface state, respectively. If the shock impedance of the anvil is lower than the sample (as in our experiments on Fe using Al_2O_3 or LiF anvils), the value of T'_H is replaced by T_R. The temperature achieved upon wave reflection and partial release at the metal film-anvil interface, T_R, is related to Hugoniot temperature by

$$T_R = T_H \exp \left[\int_{V_R}^{V_H} \frac{\gamma}{V} dV \right] \qquad (3)$$

Similarly, the shock compressed volume is slightly increased upon partial decompression by an amount given by

$$V_R - V_H = \Delta V \cong \frac{\left(u_R - u_H \right)^2}{P_H - P_r} \qquad (4)$$

where V_H and V_R, and u_R and u_H are the Hugoniot and release states specific volume and particle velocity, respectively. Here γ is the metal's Grüneisen ratio. It appears from Eqs. 1, 2, and 3, that thermal parameters of the metal and anvil are required to relate T_i to T_H. However, if the anvil and sample are even approximately matched in shock impedance, then T_a and T'_H are of the same order, and for iron samples, and LiF and Al_2O_3 anvils, since α is ~ 10, the second term of Eq. 1 makes only a 10-15% contribution to T_i. Moreover, adiabatic decompression prescribed by Eq. 3, results in T_R being ~85-90% of T_H. Thus, Eqs. 1-4 allow correcting the measured value of T_i and providing for uncertainties in the EOS parameters for the thermal properties of even ~50%, affects the resulting values of T_H by only some 10%.

SHOCK TEMPERATURES FOR IRON

Measured (10) values of κ_m/κ_a of Eq. 3, are some 12 to 32% greater than that calculated using the Weidemann-Franz law for κ_m (11) and Debye theory for κ_a (12-14). Revised values of T_H for Fe (Fig. 2) allow a smooth curve to be drawn through the data (of Fig. 6 of Bass et al. (15)) with an addi-

tional two data at 178 and 194 GPa (10). The temperature along the principal Hugoniot below 100 GPa are from Table 3 (1). Points at 200 and 243 GPa correspond to T_H = 5000 (4410, 5300) K and T_H = 5800 (5620, 6990) K. The uncertainties plotted in Fig. 3 correspond to γ/V = constant = 20 Mg/ m^3 of Table 4 (cases a and b) (1).

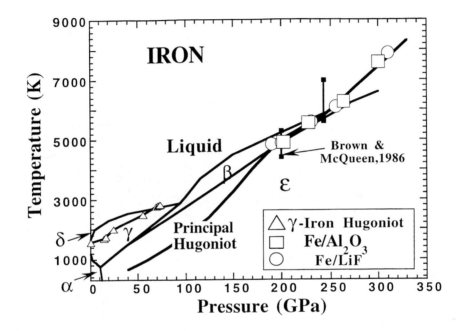

FIGURE 3. Pressure-temperature principal and γ-iron (centered at 1573K) Hugoniot states relative to phase diagram based on Boehler (4) and Saxena et al. (16). Phase transitions of Brown and McQueen (1) now agree closely with the revised shock temperature data for Fe.

PHASE DIAGRAM OF IRON

Fig. 3, also shows the states achieved in our study of preheating γ-Fe (Hugoniot centered at 1573 K) where we measured longitudinal elastic unloading velocities We find a sharp, 19.7% decrease in compressional wave velocity from 7.71 km/sec upon melting of the initial γ phase at 70±2 GPa and 2800 ±30K. This agrees with the phase diagram of Saxena et al. (16) and Boehler (4). Our results are consistent with the γ phase terminating at a γ-ε-liquid triple point at ~2900 K and ~93 GPa (Fig. 3).

CONCLUSIONS

We agree with Boehler (4) that the 200 GPa transition of Brown and McQueen (1) corresponds to the ε to β phase change and the 243 GPa transition represents the onset of melting of the β phase. Thus, the solidus iron phase at pressures of the outer core in the 133 to 243 GPa range is probably the β phase. Finally, as shown in Fig. 3, the extent of the pressure stability regime of the β phase is unknown. This phase's field of stability may extend to the pressures of the outer to inner core boundary at 330 GPa or, even to higher pressures, or there may exist a β-ε-liquid triple point between 243 and 330 GPa.

ACKNOWLEDGMENTS

Research supported by NSF. We appreciate the experimental support of E. Gelle and M. Long. Contribution #8484, Division of Geological and Planetary Sciences, California Institute of Technology, Pasadena, CA.

* Present Address: Sandia National Laboratory, MS 1181, Albuquerque, NM 87185.
† Present Address: The Santa Cruz Operations, Inc., 400 Encinal Street, P. O. Box 1900, Santa Cruz, CA 95061-1900.

REFERENCES

1. Brown, J.M. and McQueen, R.G., *J. Geophys. Res.*, **91**, 7485-7494 (1986).
2. Birch, F., *J. Geophys. Res.*, **57**, 227-286 (1952).
3. Urakawa, S., Kato, M. and Kumazawa, M., in *High Pressure Research in Mineral Physics*, Manghnani, M. H. and Y. Syono (eds.), Terra. Scient. Publ., American Geophys. Union, Wash. DC, 1987, pp. 95-111.
4. Boehler, R., *Nature*, **363**, 534-536 (1993).
5. Boehler, R., *Annu. Rev. Earth Planet. Sci.*, **24**, 15-40 (1996).
6. Urtiew, P.A. and Grover, R., *J. Appl. Phys.*, **45**, 140-145 (1974).
7. Barker, L.M. and Hollenbach, R.E., *J. Appl. Phys.*, **45**, 4872-4887 (1974).
8. Lyzenga, G.A. and Ahrens, T.J., *Rev. Sci. Instrum.*, **50**, 1421-1424 (1979).
9. Brown, J.M. and McQueen, R.G., in *High Pressure Research in Geophysics*, Akimoto, S. and M. H. Manghnani (eds.), Academic Press, New York, 1982, pp. 611-622.
10. Holland, K.G., 1997. Phase Changes and Transport Properties of Geophysical Materials under Shock Loading, Ph.D. thesis, California Institute of Technology, Pasadena, California.
11. Manga, M. and Jeanloz, R., *J. Geophys. Res.*, **102**, 2999-3008 (1997).
12. Tang, W., *Chinese J. of High Pressure Physics*, **8**, 125-132 (1994).
13. Roufosse, M.C. and Jeanloz, R., *J. Geophys. Res.*, **88**, 7399-7409 (1983).
14. Jeanloz, R., in *High-Pressure Research in Geophysics*, Akimoto, S. and M. H. Manghnani (eds.), Center for Academic Publications, Tokyo, Japan, 1982, pp. 479-498.
15. Bass, J.D., Svendsen, B. and Ahrens, T.J., in *High Pressure Research in Mineral Physics*, Manghnani, M. and Y. Syono (eds.), Terra Scientific, Tokyo, 1987, pp. 393-402.
16. Saxena, S.K., Shen, G. and Lazor, P., *Science*, **260**, 1312-1313 (1993).

CP429, *Shock Compression of Condensed Matter – 1997*
edited by Schmidt/Dandekar/Forbes
© 1998 The American Institute of Physics 1-56396-738-3/98/$15.00

AN INVESTIGATION OF THE α–ε PHASE TRANSITION IN SHOCK LOADED EN3 MILD STEEL.

J.C.F. Millett[1], N.K. Bourne, Z. Rosenberg[2]

Shock Physics, Cavendish Laboratory, Madingley Road, Cambridge, CB3 0HE, UK.
[1]Email jcfm100@phy.cam.ac.uk
[2]RAFAEL, P.O. Box 2250, Haifa, Israel.

The α–ε phase transition in a mild steel has been investigated using manganin stress gauges mounted in longitudinal orientation. The phase transformation has been located at 13.3±0.3 GPa. Since the gauges have been mounted within the specimen, it has been possible to directly measure parameters of the phase change such as the transformation stress and the Hugoniot stress without the interference of releases that attend back-surface measurements. Strain gauges have been embedded alongside the stress gauges. In this orientation they are sensitive to lateral strain. Evidence has been collected that suggests the possibility of non-uniaxial strain behind the transformation front.

INTRODUCTION

Dynamic phase transitions have been a subject of great interest ever since Bancroft *et al.* [1] identified the phase change in iron at *ca.* 13 GPa through shock loading experiments. Since then a great deal of work has been carried out on a range of materials, much of it summarised in the review article by Duvall and Graham [2].

A number of techniques can be used to study phase transitions, including free surface velocity determination (VISAR), as used by Barker and Hollenbach [3] in their investigation of iron, and manganin stress gauges, as used in iron [4], bismuth [5] and potassium chloride [6, 7].

In this investigation, we have chosen the latter technique, since it allows the direct measurement of the stresses involved in the phase transformation by placing the gauge inside the sample, as opposed to the free surface techniques such as VISAR, where stresses have to be inferred from reloading signals.

During a plate impact experiment, inertial confinement of the shocked material results in conditions that give rise to a one-dimensional longitudinal strain. However, when loaded by a phase transition wave, strain gauges show signals which we suggest may indicate that conditions of one-dimensional strain are violated, although this is contrary to assumed behaviour. In considering the

changes in structure of iron, Mao *et al.* [8] have measured the volumetric change in iron quasi-statically using a diamond anvil cell, and have shown that there is a volume change of -6.8% from α (BCC) to ε (HCP) . Moreover they suggest that the *c/a* ratio (ratio of principle axis) remains constant with pressure at *ca.* 1.6, hence if one axis is compressed, then the other must also contract to maintain *c/a* at a constant value. Therefore if these conditions also occur during the shock-induced transformation, then the one-dimensional strain state can no longer exist, and thus there should be a measurable lateral strain. The work discussed in this report was carried out to investigate this possibility.

EXPERIMENTAL PROCEDURE

Plate impact experiments on EN3 mild steel were carried out on the single stage, 50 mm gas gun at the University of Cambridge [9]. Manganin stress gauges (LM-SS-125CH-048) were embedded between 6 mm plates, using a low viscosity epoxy. Specimens were impacted with lapped 6 mm copper or tungsten discs. The gauge calibration was taken from the work of Rosenberg *et al.* [10] . Strain gauges (CEA-13-125UW-120) were mounted in a similar fashion, such that the lateral component of strain (should it exist) would be measured.

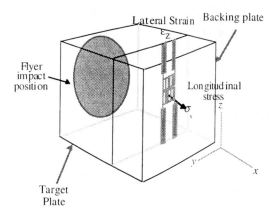

Figure 1. Specimen configurations for the measurement of longitudinal and transverse stresses.

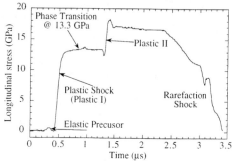

Figure 2. Stress gauge trace of mild steel impacted at 881 m s^{-1}.

The calibration of these gauges follows other work of Rosenberg *et al.* [11]. Gauge mounting positions are presented in fig. 1.

MATERIALS DATA

EN3 mild steel was supplied in the 'bright rolled' condition, *i.e.* the final processing step was a cold rolling pass to increase strength and improve surface finish. The relative materials parameters were: density 7.82 g cm^{-3}, longitudinal sound speed, c_L 5.91 mm μs^{-1}, shear wave speed, c_T 3.25 mm μs^{-1}, Vickers hardness, 254 kg mm^{-2} and grain size 15 μm.

RESULTS AND DISCUSSION

A typical stress gauge trace in mild steel above the phase transition is shown in fig. 2. It clearly shows the presence of the elastic precursor, the plastic shock and the phase transition wave. The phase transition stress has been measured

at 13.3 ±0.3 GPa. The release proceeds in several stages from the Hugoniot stress. Firstly there is a simple release, followed by a rarefaction shock and finally a further simple release region. There are several dips and overshoots associated with the arrival of the elastic wave, phase transition wave and rarefaction shock. These have been shown to be caused by electrical effects due to capacitive linking between the specimen and the gauge. This is more fully discussed elsewhere [12].

The measured Hugoniot of this steel is presented in figure 3. Note the cusp at the phase transition stress of 13.3 GPa. This is a typical behaviour for materials featuring a phase transition. This stress lies in the range 12.9 GPa to 14.1 GPa that Duvall and Graham [2] quote for a range of steels.

Note also in this figure that a point has been included in which a mild steel flyer impacted on a copper target. In this configuration the gauges were placed in copper, where no transitions occur. The significance of this is discussed later.

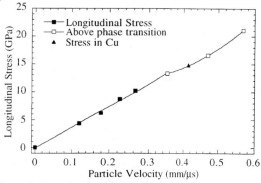

Figure 3. Hugoniot of mild steel. Errors are 2%.

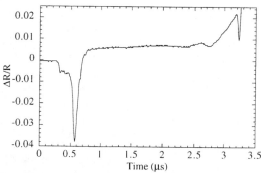

Figure 4. Strain gauge trace in mild steel at 8.6 GPa.

138

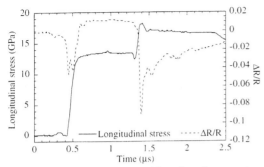

Figure 5. Longitudinal stress and lateral strain gauge traces at 17.0 GPa in steel.

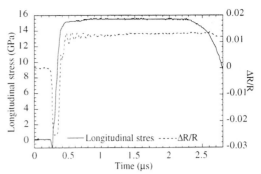

Figure 6. Longitudinal stress and lateral strain gauge traces in copper at 16 GPa.

Figure 4 gives that the relative resistance change ($\Delta R/R$, where R is the initial resistance of the gauge) of a strain gauge embedded in a steel target impacted to 8.6 GPa. The trace shows a large initial dip, before rising to a plateau of 0.007 (0.7%). This value of resistance change is a consequence of the piezoresistive (stress) response of constantan (see [11]), from which our strain gauges are made. For this reason none of the strain gauge results have been converted to strain. The initial dip is the result of capacitive linking, as discussed in the previous gauge trace. It is more noticeable in the strain gauge trace because the stress response of constantan is very small compared to that of manganin. The final rise in the signal occurs very late. This is most likely the result of lateral release arriving at the gauge location.

In fig. 5, a comparison is made between longitudinal stress and lateral strain gauge traces at 17.0 GPa in steel. In the strain gauge trace there is an initial dip, followed by a rise to a plateau of around $\Delta R/R = 0.01$. This corresponds to the phase transition stress of 13.3 GPa, and is similar to the behaviour seen in the previous trace. Note however that there is a rapid dip with the arrival of the plastic II wave itself (as shown in the stress gauge trace), followed by a rise to a near constant value of around -0.02. Besides the very low piezoresistive response, we are assuming that the strain gauge is following the lateral strain (should it exist). Note that the $\Delta R/R$ of the strain gauge, at -0.02 indicates a *compressive* strain, which would be expected if an overall decrease in specific volume accompanies the phase transformation, and if some of this volume change takes place along the lateral direction too.

To confirm that this was not simply a pressure effect, a similar experiment was performed on

copper shocked to a stress of approximately 16.0 GPa. The results are presented in fig. 6. It was observed that the strain gauge trace rises to a flat plateau (as did the trace at 8.6 GPa in steel) without the subsequent dip, even though the stress is well above the phase transition point in steel. From this it is clear that the dips in the strain gauge traces in steel are not due to the high pressures involved. However, there are several issues that need to be discussed.

Firstly, we note that after the arrival of the phase transition, the stress histories do not show the same interference as the strain histories. It is believed that the large piezoresistive response of the stress gauges dominate the (possible) strain response at these high stresses. However, we were concerned to quantify the strain response since it may effect the measured Hugoniot above the phase transition. In fig. 4 a point is shown where a copper target is impacted with a steel flyer. Observe that it does not deviate from the measured Hugoniot using steel targets. A comparison of the $\Delta R/R$ values of the stress and strain gauges in steel above the phase transition show that at 0.41 and -0.02 respectively, the strain contribution to the stress gauge response is negligible. We thus conclude that embedded stress gauges above the transition stress give reliable results.

In addition, the piezoresistive response of the strain gauge in copper at 16 GPa, where no transition occurs, is $\Delta R/R=0.013$. Thus the negative response noted in steel above the phase transition must come from some other source. The fact that it is negative would seem to lend credence that a compressive lateral strain accompanies the arrival of the phase transition.

One final point to consider concerns the electrical behaviour of the gauge itself. Tokheim[14], in his analysis of steel encapsulated flat pack soil stress gauges, showed that the current will depend upon the inductance of the steel encapsulation. This of course will change significantly across the phase transition. Thus there exists the possiblity that the behaviour of the strain gauge in this investigation is dependent upon the magnetic properties of the steel, and its variation below and above the phase transition.

CONCLUSIONS

The dynamic phase transition in EN3 mild steel has been investigated using embedded stress and strain gauges. The stress traces give a phase transition at 13.3 GPa, as expected. Strain gauge traces show anomalous behaviour with the arrival of the phase transition. We suggest that this may be the result of one-dimensional strain conditions being violated at this point, although we do so with great caution. However, the use of strain gauges in shock loading experiments in this work and others [11, 13] provide a useful insight into the behaviour of shock loaded materials. We have also considered the possiblity of the magnetic properties of the steel may also have a significant effect upon the gauge traces. Hopefully this work will provide a basis for further investigation to clarify this issue.

ACKNOWLEDGEMENTS

We acknowledge financial support DERA, Ft. Halstead. We are grateful to Prof. J.E. Field , Dr. B. Goldthorpe and Dr. T. Andrews for suggestions and encouragement, and Mr. D.L.A. Cross for valuable technical support.

REFERENCES

1.   Bancroft, D., Peterson, E.L. and Minshall, S. *J. Appl. Phys* **27** (1956) 291-298.
2.   Duvall, G.E. and Graham, R.A *Rev. Mod. Phys.* **49** (1977) 523-579.
3.   Barker, L.M. and Hollenbach, R.E *J. Appl. Phys.* **45** (1974) 4872-4887.
4.   Rosenberg, Z., Partom, Y. and Yaziv, D. *J. Phys. D: Appl. Phys.* **13** (1980) 1486-1496.
5.   Rosenberg, Z. *J. Appl. Phys.* **56** (1984) 3328-3329.
6.   Al'tshuler, L.V., Pavlovskii, M.N. and Komissarov, V.V. *JETP* **79** (1994) 616-621.
7.   Galbraith, S.D., Bourne, N.K. and Rosenberg, Z., *The determination of the reverse phase transition b2-b1 stress in shock loaded KCl,* in *Shock compression of condensed matter-1995,* S.C. Schmidt and W.C. Tao, Editors. 1996, AIP Press: Seattle. p. 219-222.
8.   Mao, H.-K., Bassett, W.A. and Takahashi, T. *J. Appl. Phys.* **38** (1967) 272-276.
9.   Bourne, N.K., Rosenberg, Z., Johnson, D.J., Field, J.E., Timbs, A.E. and Flaxman, R.P *Meas. Sci. Technol.* **6** (1995) 1462-1470.
10.   Rosenberg, Z., Yaziv, D. and Partom, Y. *J. Appl. Phys.* **51** (1980) 3702-3705.
11.   Rosenberg, Z., Yaziv, D. and Partom, Y *J. Appl. Phys.* **51** (1980) 4790-4798.
12.   Bourne, N.K. and Rosenberg, Z., *Fractoemission and its effect upon noise in gauges placed near ceramic interfaces,* in *Shock compression of condensed matter-1995,* S.C. Schmidt and W.C. Tao, Editors. 1996, AIP Press: Seattle. p. 1053-1056.
13.   Rosenberg, Z., N.K.Bourne and Millett, J.C.F. *J. Appl. Phys.* **79** (1996) 3971-3974.
14.   Tokheim, R.E. *Analysis of electrical noise from shock loading a steel flatpack soil stress gauge* in *Shock Waves in Condensed Matter* Y.M. Gupta, Editor1986, Plenum Press: Spokane p559-564

CP429, *Shock Compression of Condensed Matter – 1997*
edited by Schmidt/Dandekar/Forbes
© 1998 The American Institute of Physics 1-56396-738-3/98/$15.00

SHOCK-INDUCED PHASE TRANSITION OF 6H POLYTYPE SiC AND AN IMPLICATION FOR POST-DIAMOND PHASE

T. Sekine and T. Kobayashi

National Institute for Research in Inorganic Materials, Namiki 1-1, Tsukuba, Ibaraki 305, Japan

The Hugoniot compression curves for alpha and beta SiC were determined to pressures of 160 GPa. A phase transition was detected over 100 GPa with volume reduction of 15±3%, suggesting that the high-pressure phase has sixfold coordination and most probably rocksalt structure. The transitions of the both SiC and the high-pressure behaviors of Si imply that the post-diamond phase may possess a sixfold-coordinated, simple cubic structure as predicted theoretically.

INTRODUCTION

Silicon carbide (SiC) displays unique properties mechanically, chemically, electrically and thermally and more than 100 crystallographic modifications at ambient pressure. The chemical bonds between Si and C are identical and tetrahedrally coordinated, being similar to diamond.

In the system C-Si, the physical and chemical properties of SiC show an intermediate between tetrahedrally coordinated C and Si, that is diamond and diamond-structure Si. With increasing pressure, silicon[1, 2] displays a series of high pressure phases, β-tin at 11 GPa, an orthorhombic phase at 13 GPa, simple hexagonal form at 16 GPa, an intermediate phase at 37 GPa, hexagonal closed-pack structure above 40 GPa, and face-centered cubic form at 78 GPa by static pressures and a high-pressure phase with 22% volume reduction at 13 GPa by shock compression. Diamond itself appears to be stable at least about 600 GPa. It is quite interesting to investigate high-pressure forms of diamond, but it needs to generate ultra high pressures over 1 TPa which is currently very difficult in laboratory.

Several computer simulations[1-3] have been carried out and reveals that diamond transforms to fourfold-coordinated bcc (BC-8) at 1.2 TPa, monoatomic simple cubic (SC1) at 1.9 TPa or metallic phase (SC4) at about 3 TPa. One approach to evaluate possible candidates for post-diamond based on experimental evidence is to look at the high-pressure transformation of SiC because SiC is considered to be an alloy of diamond and diamond-structure Si in terms of chemistry and physics.

There are several high-pressure studies on SiC. For 3C SiC, a phase transition to a NaCl-type structure was observed in DAC about 100 GPa[4]. 6H SiC, however, remains stable at 90 GPa[4] and showed some changes at 95 GPa[5]. 15R polytype did not display any significant difference from 3C at high pressures[6]. Shock compression data on SiC indicated two phase transitions at 24 and 96 GPa[7,8]. Resent Hugoniot data do not indicate the 24 GPa transition to pressures of 49 GPa[9], and the post-shocked SiC from 82.8 GPa did not contain any evidence for presence of a high-pressure phase[10].

The results of pseudopotential calculations[11] suggest that the SiC polytypes such as 3C, 2H, and

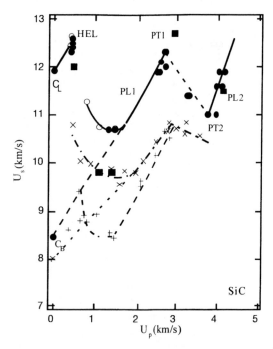

FIGURE 1. Shock velocity (Us) versus particle velocity (Up) of SiC Solid circles are for alpha SiC, solid squares for beta SiC, open circles from [9], crosses from [7] and pluses from [8].

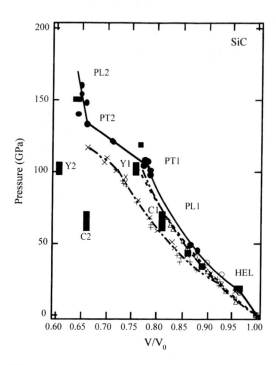

FIGURE 2. Compression curves of SiC. Solid curve with solid circles (alpha SiC) and open circles [9], double dot-broken curve with crosses [7], and pluses [8] are Hugoniots. Broken curve [4] and dot-broken curve with open triangles are static compressions. Bars with Y1 and Y2 and with C1 and C2 indicate a transition from 3C SiC to rocksalt structure [4] and *ab initio* psudopotential calculation results [11].

4H transform into the NaCl phase with six fold coordination around 66 GPa, slightly depending on the hexagonality.

In this paper, we present the shock Hugoniot data on 6H SiC and 3C SiC to pressures up to 160 GPa. We find a shock-induced phase transition with a volume reduction of 15±3%, at a pressure of about 110 GPa.

EXPERIMENTAL

Plates (12 x 10 x 2.530 mm) of alpha SiC and plates (15 x 8 x 2.023 mm) of beta SiC were used. The densities of plates are measured to be 3.226 g/cm^3 and 3.195 g/cm^3, respectively. The alpha SiC is composed mainly of 6H crystals with average grain size of 4 μm and the beta SiC is 3C ploytype with grain size less than 0.1 μm. Each sample plate was mounted on the rear surface of a metal driver disc (1 mm thick, 35 mm in diameter). Two small arrival mirrors on the driver disc, two small arrival

mirrors on the sample, and an inclined mirror on the sample were mounted in order to observe the shock wave transit time using a xenon flash lamp and a streak camera.

The shock conditions were calculated by the free-surface velocity approximation method and by the impedance match method. The sweep rate of the image converter streak camera has been calibrated against a modulated diode laser beam trace. We used a sweep rate of some 55 ns/mm. Shock waves were generated by impacts of flyer plate (22 mm in diameter) glued on the front surface of sabot, which was accelerated by a two-stage light-gas gun (NIRIM 2ST-1)[12]. The projectile velocity was measured with three cw x-ray beams [12].

EXPERIMENTAL RESULTS

Table 1. Summary of shock compression measurements on alpha and beta SiC.

Shot #	Flyer	Impact Velocity (km/s)		U_s (km/s)	U_{fs} (km/s)	U_p (km/s)	Density (g/cm^3)	V/V_0	Pressure (GPa)
alpha-SiC (d_0 = 3.226 g/cm^3)									
T-41	Al	3.830	HEL	12.4	0.94	0.47	3.37	0.961	18.9
			PL1	10.7	2.57	1.29	3.69	0.879	44.8
			[IM[a]			1.43	3.74	0.866	49.6]
T-35	SS	4.286	HEL	12.6		(0.45)[b]	3.35	0.964	18.3
			PL1	11.9	5.0	2.5	4.09	0.790	96
			[IM			2.55	4.12	0.786	98.2]
T-48	W	4.422	HEL	12.4		(0.45)	3.36	0.964	18.1
			PL1	12.0	5.39	2.70	4.17	0.775	104.8
			PL2	11.4	6.61	3.30	4.55	0.711	121.7
			[IM			3.27	4.54	0.713	120.6]
T-49	W	4.980	HEL	12.4		(0.45)	3.36	0.964	18.1
			PL1	12.1	5.15	2.58	4.12	0.787	101.4
			PL2	11.0	7.87	3.93	5.05	0.643	140.2
			[IM			3.74	4.91	0.661	133.2]
T-36	W	5.480	HEL	12.3	0.84	0.42	3.35	0.966	16.7
			PL1	12.3	5.37	2.69	4.14	0.782	107.4
			PL2	11.6	7.90	3.95	4.89	0.659	147.9
			[IM			4.09	4.98	0.650	153.9]
T-50	W	5.600	HEL	12.5		(0.45)	3.35	0.964	18.2
			PL1	12.3	5.5	2.75	4.16	0.776	109
			PL2	11.9	8.0	4.0	4.86	0.664	154
			[IM			4.16	4.98	0.650	160.4]
beta-SiC (d_0 = 3.190 g/cm^3)									
T-72	Al	3.595	HEL	12.0	0.96	0.48	3.33	0.960	18.4
			PL1	9.80	2.18	1.09	3.60	0.889	34.2
			[IM			1.38	3.72	0.860	43.2]
T-74	W	5.495	HEL	(12.0)		(0.50)	3.33	0.967	19.1
			PL1	12.7	5.87	2.93	4.10	0.768	119
			PL2	11.5	8.24	4.12	5.00	0.638	151
			[IM			4.12	4.97	0.642	151]

[a] IM means values calculated by the impedance match method and the others are calculated by the free surface velocity approximation.

[b] These values in parentheses are taken as an average value of T-41 and T-36.

Experimental results are summarized in Table 1 for the alpha and beta SiCs. The relationships between shock velocity (Us) and particle velocity (Up) and between pressure and volume are given in Figs. 1 and 2, together with the previous data by DAC, shock, and calculations. The alpha SiC Hugoniot has been reported elsewhere[13]. Although the beta SiC Hugoniot data are only from two shots, both static and shock compression curves are identical for beta SiC below 50 GPa.

143

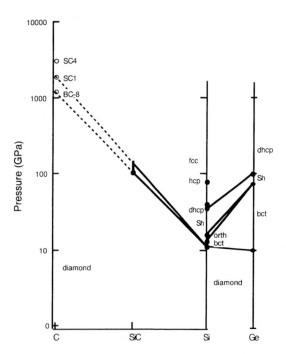

FIGURE 3. A comparison of high-pressure transitions in the system C-Si.

sixfold-coordinated phase such as a simple cubic phase which has been predicted to appear at 1.9 TPa with a volume reduction of 60%, but not BC-8 which has fourfold coordination. A recent energetical and kinetical consideration by computer simulations suggests that diamond may transform into a SC4 metallic, sixfold-coordinated phase about 3 TPa. Figure 3 illustrates an extrapolation of high-pressure transformations in the system Si-C, indicating that the extrapolation of the experimental results on Si and SiC to C gives a range of transition pressure for the diamond to post-diamond.

ACKNOWLEDGMENTS

We thank E. Takazawa, O.V. Fat'yanov, T. Osawa for their help in conducting the shock experiments, and H. Otsuka for preparing the manuscript.

DISCUSSION

In the present experiments, we have determined the Hugoniot states of alpha and beta SiC up to 160 GPa. A phase transition has been detected over 100 GPa and the HEL is around 17-19 GPa. The phase transition is associated with a volume reduction of 15±3%. The high-pressure form is considered to be a NaCl-type based on the DAC results and the *ab initio* density-functional pseudopotential calculations, although the DAC results on 3C SiC indicate 20% volume reduction about 100 GPa and the calculation assuming the hydrostatic pressure condition reveals a volume reduction of 17 to 19% at 66±5 GPa for various polytypes of SiC.

Looking at the system Si-C at high pressures, Si converts initially to beta-tin structure with 6-fold coordination and SiC also converts most probably to a NaCl-type structure with 6-fold coordination. Then these experimental framework strongly implies that the post-diamond phase of C is also a

REFERENCES

1. Young, D.A., *Phase Diagrams of the Elements* (Univ. of California Press, Berkeley, 1991)
2. McMahon, M.I. et al., *Phys. Rev.* **B50** 739 (1984).
3. Scandolo, S. et al., *Phys. Rev.* **B53**, 5051 (1996).
4. Yoshida, M. et at., *Phys. Rev.* **B48** 10587 (1993).
5. Liu, J. and Vohra, Y.K., *Phys. Rev. Lett.* **72** 4105 (1994).
6. Akeksandrov, I.V. et al., *JETP Lett.* **50** 127 (1989).
7. Gust, W.H. et al., T. *Appl. Phys.* **44** 550 (1973).
8. McQueen, R.G. et al., in High Velocity Impact Phenomena, ed. by Kinslow, R. (Academic, New York, 1970) pp. 348, 405, 523,
9. Grady, D.E., *J. Apple, Phys.* **75** 197 (1994).
10. Kovtun, V.I. and Timofeeva, I.I., *Poroshk Mettall* **8** 921 (1988).
11. Karch, K. and Bechstedt, F., *Phys. Rev.* **B53** 13400 (1996) and Chang, K.J. and Cohen, M.L., *Phys. Rev.* **B35** 8196 (1987)
12. Sekine, T. et al., in Shock Compression of Condensed Matter-1995, ed. by Schmidt, S.C. and Tao, W.C. (AIP, Woodbury, 1996) pp. 1201.
13. Sekine, T. and Kobayashi, T., *Phys. Rev.* **B55**, 8034 (1997).

CP429, *Shock Compression of Condensed Matter – 1997*
edited by Schmidt/Dandekar/Forbes
© 1998 The American Institute of Physics 1-56396-738-3/98/$15.00

SHOCK INDUCED AMORPHIZATION OF MATERIALS

S.K. Sikka and Satish C. Gupta

High Pressure Physics Division,
Bhabha Atomic Research Centre, Mumbai, 400 085, India.

The pressure induced crystalline to amorphous (c → a) transition is a topic of considerable current interest (see Sharma and Sikka, Prog. in Mat. Sci. 40 (1996), for a review and references there in). Under static pressures about 50 substances have been amorphized. Some of these have also been vitrified using shock waves. Molecular dynamics simulations have also been done. A comparison shows that the differences in the two loading methods viz. presence of shear, high strain rate, temperature rise and generation of defects in the shocking process produce the usual differences for (c → a) transition as for any other solid - solid transition. Following facts are now firmly established: (1) (c → a) transition under shock loading is a solid - solid transition and not a quenched molten phase, (2) it is a metastable state governed by a three - level free energy diagram, and (3) the density increases observed in some cases in shock recovered samples have structural origins. We will also describe our shock wave results on GeO_2 and $FePO_4$, and molecular dynamics simulations on quartz.

INTRODUCTION

The glasses of minerals, rich in quartz and feldspar, were discovered as early as in 1870 in the meteorite craters and these glasses were realized to be produced by shock compression as a results of meteorite impacts. However, the first evidence of laboratory-production of amorphous phase under compression came from static pressure experiments on gadolinium molybdate. Since then, numerous materials have been amorphized on application of pressure under both static and shock loading conditions (see ref. 1 and references there in).

An amorphous phase is characterized by lack of long range order; thus, a crystal to amorphous transition under pressure can be detected by a number of microscopic techniques, such as, by disapearence of the sharp x-ray diffraction pattern, by the loss of Raman peaks from external modes, and by the discontinuous increase in peak widths in other spectroscopic techniques. Under shock loading, however, the compressed state lasts only for a very short duration (about a few microseconds), which makes it extremely difficult to employ these microscopic probes. Consequently,

the detection of a crystal to amorphous transition could be accomplished using macroscopic measurements: by observing a break in the slope of the shock velocity versus particle velocity plot and a mixed phase region on the Hugoniot and from presence of two wave structure in the stress or particle velocity profiles. A transition, after detection is generally characterized by comparing with static pressure data, or analyzed by performing theoretical calculations, e.g., molecular dynamics simulations. However, if a transition is irreversible, it can be studied by investigating the shock recovered samples by diffraction, Raman and electron microscopy measurements.

Shock compression, being uniaxial and irreversible, is inherently accompanied by shear stresses and rise in temperature. Recent in-situ Raman measurement (2) showed that during compression along Z-axis in quartz, the frequency shift of the 464 cm^{-1} mode, which is related to the Si-O-Si bond angle, is higher than that under hydrostatic pressure. This suggested that the decrease in Si-O-Si angle under compression is enhanced by the presence of non-hydrostatic stresses. Hence, the process of amorphization,

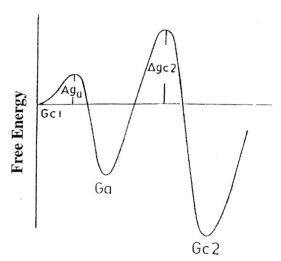

Reaction Coordinate

FIGURE 1. A schematic diagram showing free energy as a function of reaction coordinate for a crystal to amorphous transition.

which is related to the decrease of this angle to about 125° in the tetrahedral network structures, would be aided by shear stresses, thus, lowering the transition pressure. Even under static pressure, the effect of shear is quite large as evidenced by the amorphization of quartz just by mortar grinding (~ 0.4 GPa) as against the quasi hydrostatic pressures of about 30 GPa (3). Apart from this, a c → a transition, reversible under hydrostatic pressures may become irreversible under non- hydrostatic environment. This is indeed found for $AlPO_4$ on shock loading (4).

The high pressure amorphous phase is a metastable phase and results from the structural frustration seen by the parent phase in reaching the high pressure equilibrium phase. The transformation kinetics may be understood from the three level free energy diagram schematically shown in figure 1 in which Gc1, Ga and Gc2 are the free energies of the initial crystalline phase c1 just before the transformation, amorphous phase a, and the high pressure phase c2, respectively; Δg_a, Δg_{c2} and Δg_{c1c2} (not shown in figure) are the barrier heights for c1 to a, a to c2 and c1 to c2 transitions, respectively. The compressed c1 phase energetically prefers to transform to the c2 phase, however the kinetics impedes this and instead a transformation to the amorphous (a) phase occurs. This then may

remain as a metastable phase or slowly transform to the equilibrium phase.

The role of temperature rise in the sample under shock compression on the kinetics of the transition can also be understood in terms of figure 1. Three possibilities could arise depending on the temperature (T) of the specimen and barrier heights Δg_a and Δg_{c2}. (i) Formation of the amorphous phase will occur if kT is comparable with Δg_a, but much smaller than Δg_{c2}. (ii) Both the amorphous and the crystalline C2 phase may result if kT is comparable to both Δg_a, and Δg_{c2}. (iii) If kT is larger than Δg_a, but comparable with Δg_{c2}, the amorphization may be skipped and only the crystalline phase C2 may be formed. Thus, even if amorphization is observed under hydrostatic pressures, it may not be seen on shock compression. In all the three cases, the high pressure phase may or may not be retained on unloading depending on the residual temperature

The strain rates under shock compression are high (10^8 s^{-1}). Deformation at such high strain rates may cause material to undergo heterogeneous heating. The temperature in the locally heated zones can momentarily exceed melting temperature. The material once melted could be quenched and get trapped as vitreous material. This mode of amorphization is interpreted from observations in quartz (5).

The shock compression is accompanied by formation of copious amount of defects. The role of defects on the crystal to amorphous transition is not well understood. However, it is conjectured that the defects may provide nucleation sites for the development of the amorphous phase, so as to increase the transition kinetics.

In this paper we present a few examples of c → a transition from work in our laboratory: experimental work on q-GeO2 and α-FePO4, and molecular dynamical simulations of α-quartz.

AMORPHIZATION OF q-GeO2

The trigonal phase of germanium dioxide (q-GeO2) is isomorphic to the α-quartz structure of silica, which has been observed to undergo a crystalline to amorphous transition under both static and shock pressures in the 15 - 35 GPa pressure range. High resolution electron microscopy measurements on shock recovered samples of quartz revealed two types of disordered materials, one present in the transformation lamellae and the other produced along the microfaults (5). The former has been interpreted as arising due to solid-solid transformation (diaplectic glass) and the latter due to quenching of the molten material (fusion glass). Using electron microscopy, even cristobalite (another crystalline phase of SiO_2) shock amorphized at ~ 28 GPa, has been found to consist of both the diaplectic and fusion glasses (6). The objective here was to examine whether the melt quenched effect could be suppressed in an analogous transition that occurs at a lower pressure. In this regard, q-GeO_2 is the ideal material to study as it has been reported to amorphize in diamond cell experiments at about 10 GPa (7).

The samples of q-GeO_2 were shock loaded to the desired pressures by a reverberating shock wave between two stainless steel plates in a recovery capsule upon the impact of flyer plate accelerated in the gasgun (9). The retrieved samples were characterized using x-ray diffraction and Raman scattering techniques (10). The x-ray diffraction measurements on the shock recovered samples are displayed in figure 2. The pattern of the sample recovered from 5 GPa is identical to the one at ambient conditions. For 6.8 GPa, the XRD pattern contains all the original peaks riding on an amorphous background. The peaks are, however, much broadened indicating that lattice strains are present and/or the particle size is reduced. Besides, the pattern also contains two additional peaks (d spacings 4.04 A° and 3.11 A°) which now have been related to the cristobalite form of GeO_2 (11). The patterns at 7.4 GPa clearly shows that a part of the material is amorphized implying that the amorphization has set in between 6.8 and 7.4 GPa. The amount of amorphous component increases with pressure, e.g., the pattern for the 10 GPa sample shows a broad hump implying the glassy state of the material. The position of the first glass peak is close to the first peak in the structure factor of the vitreous GeO_2 (12), and also of the pressure amorphized GeO_2 under static conditions, suggesting that the short range order of the amorphous phase obtained in static and dynamic loading might be the same. The amorphization of the q-GeO_2 was also confirmed by Raman spectra which contained very broad Raman bands for the 10 GPa sample. The observation that the amorphization occurs at 7 GPa, at a lower pressure than 9-10 GPa under static pressure, supports that

non-hydrostatic stresses aid the process of amorphization. The retention of the amorphous

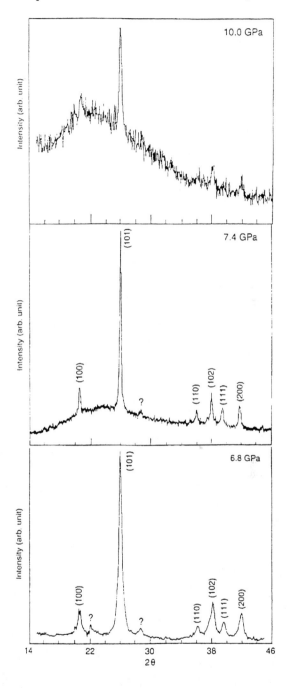

FIGURE 2. X-ray powder diffraction pattern of q-GeO_2 shock recovered from (a) 6.8 GPa (b) 7.4 GPa and (c) 10 GPa.(Ref. 10)

phase on shock unloading is in contrast with the complete reversibility of the transition below 10 GPa in hydrostatic environment (13), and may be related to the excessive distortion of tetrahedra which do not regain their shapes on unloading.

Electron microscopy measurements on shock recovered q-GeO$_2$ could not be done as the retrieved specimens were pulverized. Therefore, we carried out an estimation of the shear band temperatures in GeO$_2$, in order to assess the possibility of the fusion glass in the shock recovered samples.

A parametric study was performed by varying the strain rate values from 10^6 s^{-1} to 10^8 s^{-1} and shear band spacings of 2 μm and 5 μm. In figure 3 the plot of calculated temperatures depicts that the shear band temperatures in GeO$_2$ for shock pressures upto 10 GPa are below the melting temperature of 1388K at ambient (at 10 GPa, the melting temperature is even higher). This suggests that the fused GeO$_2$ will not form up to this shock pressure and implies that the mechanism of shock-induced amorphization in q-GeO$_2$ is a solid-solid one. This inference is supported by the high temperature high pressure work of Yamanaka et al. (14), which reported that the trigonal phase transforms at 850° C to rutile form around 0.1 GPa and remains stable at higher pressures, but at lower temperatures the trigonal phase turns amorphous; this amorphous material does not transform to rutile form even on heating to 3000° C. The absence of the rutile phase in the shock recovered specimen confirms that the amorphization occurs in the solid state.

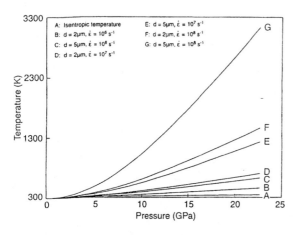

FIGURE 3. Calculated shear band temperatures versus shock pressure in GeO$_2$. (Ref. 10)

SHOCK INDUCED RESPONSE OF FePO$_4$

Recently, Chitra et al (15) conducted static pressure experiments on the berlinite form of iron phosphate α-FePO$_4$, space group P3$_1$21, in which Fe - O(1) - P and Fe - O(2) - P angles are 139.2° and 137.5° respectively. On the basis of Raman measurements in diamond anvil cell they reported a phase transition around 3 GPa, which was confirmed by the X-ray diffraction (XRD) measurements that showed the presence of a broad peak which did not disappear till 10 GPa. The structure of the high pressure phase could not be determined because of the paucity of the diffraction data. In order to compare the static pressure behavior with the shock induced response we carried out experimental investigation on this material

Experiments were conducted on samples with initial density of 1.8g/cm^3 (porosity 40 percent), which were shock loaded in a recovery capsule (16). The sample reached the final pressure after a few reverberations and its peak pressure lasted for around 1-2 μs.

The XRD measurements on the shock recovered samples are displayed in figure 4. The XRD pattern of the sample recovered from 2 GPa is identical to that at ambient conditions. However, for 3.5 GPa sample, the diffraction pattern exhibits new peaks along with the broadening of the peaks of the original phase. The development of these peaks is more clear in XRD pattern of the sample pressurized up to 5.2 GPa. These new diffraction peaks could be associated with an orthorhombic phase having space group Cmcm and cell constants a= 5.227Å , b= 7.770 Å , c= 6.322 Å. Moreover, the peaks are riding over a hump like background indicating the presence of amorphous material. This suggests that the trigonal phase has irreversibly transformed partly to orthorhombic crystalline phase and partly to the amorphous phase around 3.5 GPa. The coexistence of the high pressure crystalline phase and the amorphous phase is quite unusual. These tranformations were confirmed by Raman measurements.

The comparison of the XRD pattern of the shock retrieved samples with that of the statically compressed ones clarified that the new unidentified peaks seen under static pressure belonged to the orthorhombic structure observed in the shocked samples. This was confirmed by the later XRD measurements on the samples pressure quenched from greater than 3 GPa between tungsten carbide anvils cell.

The XRD pattern of the sample shock recovered from 8.5 GPa shows that in comparison with that of the 5.2 GPa pattern the amorphous background has decreased, the peaks belonging to the orthorhombic structure are much weaker and those of the the trigonal phase are sharper. (Two experiments gave almost similar results.) This behaviour of α- FePO$_4$ is abnormal and is in contrast to that of q-GeO$_2$ where the proportion of the amorphous component in the shock retrieved samples increased with the peak loading stress (10).

As the Cmcm phase is the equilibrium phase under pressure for this family of phosphate compounds (17), the above results could be interpreted in terms of three level free energy diagram described in introduction. The fact that both the transitions in FePO$_4$ occur in a small pressure interval suggests that the energy barriers

Δg_a and Δg_{c2} are of similar magnitude (figure 1). It seems that the Cmcm phase is forming from the amorphous phase, however, we can not rule out the direct transformation from berlinite phase to Cmcm phase. In the case of 8.5 GPa experiment, although higher component of orthorhombic and may be amorphous phase could have been formed, these might have not been retained on unloading due to the reverse transformation caused by the higher residual temperature of about 1200 K as the samples have 40% porosity. This leads to the presence of smaller components of the amorphous as well as the higher pressure structure.

MOLECULAR DYNAMICAL SIMULATIONS OF QUARTZ

For the simulations of quartz (18,19), equations of motions have been integrated by the constant-volume method using Verlet's algorithm with a time step of 0.002 ps. The simulations as a function of pressure and temperature use the size and shape of the cell as variables that are determined by equating the internal stress to the external pressure. The starting atomic configuration is the crystallographic structure of α-quartz at ambient conditions. The pair potentials derived by Tsuneyuki el al (20) were used, which reproduce various equilibrium crystal structures of SiO$_2$, determined the structure of the new crystalline phase at 20 GPa prior to amorphization and the amorphous structure factor at 28 and 47 GPa (21) in agreement with experimental data. Simulations were carried out using a periodic macrocell of 162 and 243 atoms with essentially similar results.

The simulations show that on pressure loading around 21 GPa, α-quartz looses the crystalline order in the basal plane. On heating the retrieved amorphous phase to temperatures higher than 1000 K, the order along the c axis is also destroyed and one obtains essentially a three-dimensional isotropic glass. This is in agreement with shock experiments that indicate increase in the percentage of isotropic glass with shock pressure (and therefore temperature). The pressure of amorphization shows a very weak dependence on initial temperatures up to 1000 K, in accord with the experiments done with preheating the samples. The temperature dependence of the pressure characterizing the completeness of transformation, however, seems to be essentially a kinetics effect.

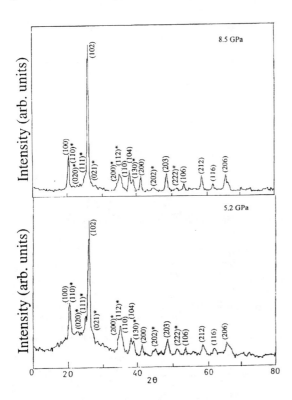

FIGURE 4. XRD patterns of FePO$_4$ shock recovered samples. (hkl) - trigonal structure (hkl)* - orthorhombic structure. (Ref 16)

149

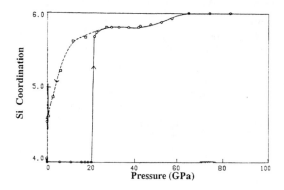

FIGURE 5. Average Si coordination as a function of pressure. (Ref. 19)

The calculated $P - V/V_0$ from these calculations is compared with the experimental data in figure 6. The agreement with the data in the crystalline phase is excellent. At higher pressures, the $P-V/V_0$ behavior of simulated amorphous phase is very close to the shocked α–quartz. The figure also shows that the unloading path in the calculation is much steeper, which is again in conformity with release wave measurements in the shock wave experiments. It suggests that the so called "mixed phase" regime might as well be the mixture of the initial quartz and the amorphous phase rather than the stishovite phase.

The variation of average Si coordination under increasing and decreasing pressure is shown in figure 5. On increase of pressure the coordination changes abruptly and substantially beyond 20 GPa and attains a plateau at 5.8 between 25 GPa and 40 GPa. Full six coordination is approached slowly and is attained fully at ~ 65 GPa. On decreasing pressure the coordination shows a large hysteresis loop and reduces steeply from 12 GPa to reach a final value of ~ 4.6. A fourfold coordination reverts back only on heating this phase to ~1000 K.

REFERENCES

1. Sharma S.M. and Sikka S.K, *Prog. Mater. Sci.* **40**, 1 (1996).
2. Gallivan S.M. and Gupta Y.M., *J. Appl. Phys.* **78**, 1557 (1995).
3. Kingma K.J., Meade C., Hemley R.J, Mao H.K. and Veblen D.R., Science, **259**, 666 (1993)
4. Cordier P., Gratz A.G., Doukhan J. C. and Nellis W.J., *Phys. Chem. Miner.* **21**, 133 (1994)
5. Gratz A.J., Nellis W. J, Christie J.M., Brocious W., Swegle J. and Cordier P., Phys. Chem. Miner. **19**, 267 (1992)
6. Gratz A.J., DeLoach L.D., Clough T. M. and Nellis W. J., Science , **259**, 663 (1993)
7. Wolf G. H., Wang S., Herbert C.A., Durben D.J., Oliver W. F., Kang Z.C. and Halvorson K., High Pressure Physics: Application t o Earth and Planetary Science, Eds. Y. Syono and M.H. Man-ghnani (terra Scientific, Tokyo and American Geophysical Union, Washington DC, 1992) p.503.
8. Madon M., Gillet Ph., Jullien Ch. & Price G.D., Phys. Chem. Miner., **18**, 7 (1991)
9. Gupta S. C., Agarwal R. G., Gyanchandani J. S, Ro y S., Suresh N., Sikka S. K., Kakodkar A. and Chidambaram R. (1992), eds: Schmidt S.C, Dick R.D, Forbes J.W and Tasker D. G *Shock Compression of Condensed Matter- 1991*, Elsevier, Amsterdam, pp. 839.
10. Suresh N., Jyoti G., Gupta S. C., Sikka S. K., Sangeeta, Sabharwal S. C., *J. Appl. Phys.* **76**, 1530 (1994).
11. Suresh N., Joshi K. D., Gupta S.C. and Sikka S.K., to be published
12. Leadbetter A.J. and Wright A.C., Non-Crystalline Solids **7**, 37 (1992)
13. Somayazulu M.S., Garg N., Sharma S. M. and Sikka S. K., Pramana **43**, 1-9 (1994)
14. Yamanaka T., Shibata T., Kawasaki S. & Kume S, in High Pressure Research : Application to Earth & Planetary Sciences, Ed. Y. Syono and M.H. Manghnani, 493-501(1992)
15. Chitra V., Momin S. N., Kulshreshtha S. K , Sharma S. M. and Sikka S. K., accepted for publication in Pramana:
16. Joshi K.D., Suresh N., Jyoti G., Kulshreshtha S.K., Gupta S.C. & Sikka S.K., (under publication)
17. Sharma S. M. and Sikka S. K., *Phys. Rev. Lett* **74**, 3301 (1995).
18. Chaplot S.L. and Sikka S.K., Phys. Rev., **B47** 5710 (1993)
19. Somayazulu M.S., Sharma S.M., Garg N., Chaplot S.L., and Sikka S.K., J. Phys.: Condens. Matter **5**, 6345 (1993)
20. Tsuneyuki S., Tsukada M., Aoki H., and Matsui Yu., Phys. Rev. Lett., **61**, 869 (1988) and Nature, **339**, 209 (1989)
21. Somayazulu M.S., Sharma S.M. and Sikka S.K., Phys. Rev. Lett. **73** 98 (1994)

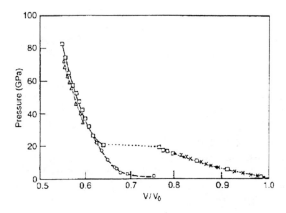

FIGURE 6. Calculated pressure versus V/V_0 for quartz as a function of increasing and decreasing pressures. (x,•) represnt experimental data for crystalline phase under static pressure where as (Δ) represent shock data for x-cut quartz. (Ref. 19)

CP429, *Shock Compression of Condensed Matter – 1997*
edited by Schmidt/Dandekar/Forbes
© 1998 The American Institute of Physics 1-56396-738-3/98/$15.00

SHOCK-INDUCED PHASE TRANSITION OF MnO AND SEVERAL OTHER TRANSITION METAL OXIDES

Y. Syono, Y. Noguchi, K. Fukuoka, K. Kusaba and T. Atou

Institute for Materials Research, Tohoku University, Katahira, Aoba-ku, Sendai 980-77, Japan

Shock compression measurements of MnO have been carried out up to 120 GPa, and a phase transition with a volume decrease, $-\Delta V/V_0$, of about 6 percent is observed at 90 ± 3 GPa. Release adiabat measurements revealed that the volume difference between the high pressure phase and the rocksalt (B1) phase was estimated to be about 23 percent. Although the cesium chloride (B2) structure is a possible candidate for the high pressure phase from the systematics obtained for the B1 - B2 transformation in alkaline earth monoxides, the observed large volume change suggests a drastic change in the electronic configuration of manganese ions at very high pressures. Metallic reflectivity observed in recent in situ observation of the high pressure phase in a diamond anvil cell and a metallic nickel arsenide (B8) structure predicted for the high pressure phase by theoretical calculations lend support for this interpretation. Comaprison with the high pressure behavior of other transition metal oxides is also made.

INTRODUCTION

The electronic state of transition metal ions is known to be sensitively modified under high pressure. The increased ligand field due to contraction of cation-anion distances may induce spin-pairing transition accompanied by reduction of magnetic moments as well as large volume contraction.(1) Band width increase has been pointed out to be more important under high pressure, which may lead to delocalization of valence electron.(2, 3) Magnetic collapse and metallization are key concept for understanding the high pressure behavior of transition metal compounds.

Such pressure-induced modification of electronic states of geophysically important iron compounds like FeO, FeS and $FeSiO_3$ has significant implications for the understanding of elasticity and conductivity of the earth's lower mantle and core. We have already carried out shock compression experiments of FeO (4) and Fe_2O_3 (5), and recently extended to other transition metal monoxides of MnO (6), CoO and NiO (7). In the present article, high pressure behavior of MnO studied by shock compression will be elucidated in comparison with those of other transition metal monoxides.

EXPERIMENTS AND RESULTS

The MnO single crystals used for the shock compression study were grown with the Verneuil method by Nakazumi Crystal Laboratory. The bulk density measured with the Archimedean method was 5.33 (2) g/cm^3, slightly smaller than the X-ray density of 5.365 (4) g/cm^3 which was estimated from the unit cell dimension of 0.4445 (1) nm. This difference would be explained by the presence of a small amount of Mn_3O_4.

The (100) platelets of MnO single crystal with the thickness of 2.1 to 2.6 mm were mounted on the tungsten or 2024 Al driver plate. To measure shock and free surface velocities, the conventional inclined mirror technique with streak photography was adopted. The release path from the shock-induced high pressure phase was measured by means of buffer technique, using fused quartz glass and polymethylmetacrylate (PMMA) for the buffer material. The specimen assembly was illuminated with an intense Xenon flash lamp and the reflected light was introduced to a continuous-access, rotating-mirror type streak camera with a writing speed of 10 mm/μs.

Shock loading experiments were carried out by using a 20-mm bore two-stage light gas gun and a 25-mm bore propellant gun for the velocity range of 1.63 - 3.93 km/s.(8) Hugoniot parameters were determined by symmetrical impact with 1-mm tungsten or 2024 Al flyers. The particle velocity of the final deformational state was obtained by the impedance match solution, which showed generally good agreement with the free surface approximation.

The shock velocity (U_s) versus particle velocity (u_p) relation is summarized in Figure 1. We observed an indication of an elastic-plastic transition in the shot with the lowest impact velocity, although a large uncertainty in determining the Hugoniot elastic limit was inevitable. A linear relation of U_s = 5.32 + 1.18 u_p was obtained for the low pressure B1 phase of MnO. The C_0 value of 5.32 km/s is in good agreement with the bulk sound velocity measured with ultrasonic methods.(9) A break of the linear U_s - u_p relation at U_s = 8.0 km/s was due to a phase transition which was observed as a kink in the inclined mirror image in the streak photograph. Single wave records observed for the runs with the shock velocity above 8.0 km/s lend support for this assignment.

The shocked state was computed from the mass and momentum conservation relations from the measured shock and particle velocities. The pressure - volume relation obtained is shown in Figure 2. The phase

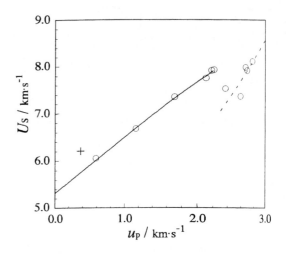

FIGURE 1. Shock velocity, U_s, versus particle velocity, u_p, of MnO (6).

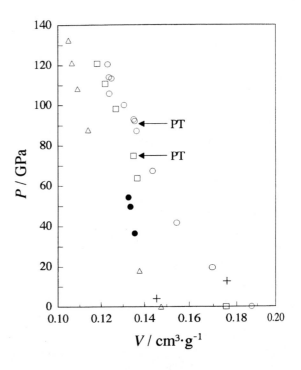

FIGURE 2. Pressure-volume relation of MnO determined by shock compression (open circle) and release adiabat (closed circle) measurements (6), together with shock compression data of $Fe_{1-x}O$ (square) (4) and NiO (triangle) (7). Crosses indicate Hugoniot elastic limit.

transition accompanied by a volume decrease,-$\Delta V/V_0$, of about 6 percent was observed at 90 ± 3 GPa. The measured larger volume change in MnO than in $Fe_{1-x}O$ is consistent with no kinked inclined-mirror records were obtained in the shock compression measurements of $Fe_{1-x}O$. The isotherm of the low pressure B1 phase of MnO was calculated with the aid of the Mie-Grüneisen equation of state, using Grüneisen constant of 1.51 and Debye temperature of 534 K.(10) The temperature at the phase transition was estimated to be about 1000 °C. The bulk modulus, K_0, of the B1 phase MnO was determined to be 142 ± 1 GPa from Murnaghan-Birch fit, assuming $K_0' = 4$. The isotherm determined from the present Hugoniot measurements was slightly more compressible than the previous static compression data,(11) but closer agreement was obtained with recent in situ X-ray work by Kondo et al. using a diamond anvil cell (DAC).(12)

Full description of the high-pressure phase could not be made in this experiment because of the limited capability of our two-stage light gas gun. The decompression behavior determined by the buffer technique is also shown in Figure 2. A simple extrapolation of the release path to zero pressure yields a volume change of about 23 percent.

DISCUSSION

The phase transition pressure of 90 GPa and volume change at the phase transition of 6 % observed in MnO by shock compression measurements are not inconsistent with the B1-B2 transition systematics observed in alkaline earth monoxides.(13) However, release adiabat measurements showed a large volume change at zero pressure, suggesting a remarkable change in the electronic state of Mn ions at high pressure.

The phase transition pressure of MnO above 90 GPa was confirmed by X-ray diffraction using DAC by Kondo et al.,(12) but the observed diffraction pattern of the high pressure phase could not be explained by the B2 structure. Interestingly visual observation of the high pressure phase in DAC under optical microscope revealed metallic lustre, lending support for the pressure-induced electronic transition.

A first principles band structure calculation of transition metal monoxides has been reported by Cohen et al.,(2) who showed a magnetic collapse of the strained B1 phase due to band width broadening at high pressures rather than an increase in crystal field splitting. Fang et al.(3) also carried out similar calculations on MnO and FeO, and predicted that the most stable high pressure phase of MnO will be with a nickel arsenide (B8) type structure with a ferromagnetic spin arrangement and metallic conductivity. They also pointed out that most of the X-ray diffraction peaks observed in the high-pressure phase of MnO (12) can be indexed on the basis of the B8 structure, with a c/a ratio of 2.08, and the pressure-volume relation can also be explained, although the shock compression curve might still be in a mixed phase region between normal and high pressure phases.

It is worth comparing with the high pressure behavior of other transition metal oxides. FeO, among others, has been studied most intensively both experimentally (4, 14-17) and theoretically (2, 3) because of significant implication for the deep interior of the Earth. The phase transition at 70 GPa observed under shock compression (4, 14) could be realized by static compression only at high temperature,(17) but not at room temperature.(15) Metallic conductivity of the high pressure phase was observed under shock compression.(16) X-ray diffraction study using DAC suggested that the high pressure phase was with the B8 structure. Theoretical calculation by Fang et al.,(3) also predicted the inverse, rather than normal, B8 structure for the high pressure phase. Special stability of the inverse B8 structure apparently came from the existence of a band gap inherent for the d^6 of Fe^{2+} configuration, leading to a band insulator in the ordered antiferromagnetic state. The observed metallic conductivity could be explained by carriers produced by well-known non-stoichiometry of FeO or disappearance of the antiferromagnetic order at high temperatures.

No pressure-induced phase transition was observed for NiO by our recent shock compression measurements to 130 GPa,[7] which is in reasonable agreement with the theoretical prediction by Cohen et al.[2]

Worthy to mention here is the high pressure behavior of Fe_2O_3 with isoelectronic d^5 cations in octahedral environments. A phase transition has been observed at 50 GPa both by shock (5, 18) and static (19, 20) compression. Two kinds of Fe ions, non-magnetic and magnetic, were observed above 50 GPa by Mössbauer spectroscopy using DAC.[19] A suggestion for disproportionation to Fe^{2+} and Fe^{4+} together with a perovskite structure of the high pressure phase has been made by in situ X-ray diffaction study,[20, 21] although definite conclusion could not be reached at the present stage.

ACKNOWLEDGMENTS

The authors are indebted to T. Yagi and T. Kondo, ISSP, Univ. of Tokyo, and K. Terakura and Z. Fang, JRCAT, Nat. Inst. Adv. Interdiscipl. Res., for many valuable comments and discussions.

REFERENCES

1. Ohnishi, S., Phys. Earth Planet. Inter. 17, 130-139 (1978).

2. Cohen, R. E., Mazin, I. I., and Isaak, D. G., Science 275, 654-657 (1997).

3. Fang, Z., Terakura K., Sawada, H., Miyazaki, T., and Solovyev, I., Nature, Submitted.

4. Yagi, T., Fukuoka, K., Takei, H., and Syono, Y., Geophys. Res. Lett. 15, 8784-8788 (1988).

5. Goto, T., Sato, J., and Syono, Y., High-Pressure Research in Geophysics, edited by M. H. Manghnani and S. Akimoto, Tokyo: Center. Acad. Publ. Japan, 1982, pp. 595-609.

6. Noguchi, Y., Kusaba, K., Fukuoka, K., and Syono, Y., Geophys. Res. Lett. 23, 1469-1472 (1996).

7. Noguchi, Y., Uchino, M., Hikosaka, K., Kusaba, K., Fukuoka, K., Mashimo, T., and Syono, Y., AIRAPT-16 & HPCJ-38, Kyoto, 1997, to be presented.

8. Goto, T., and Syono, Y., Materials Science of the Earth's Interior, edited by I. Sunagawa, Tokyo: Terrapub, 1984, pp. 605-619.

9. Pacalo, R. E., and Graham, E. K., Phys. Chem. Miner. 18, 69-80 (1991).

10. Anderson, O. L., and Isaak, D. G., Mineral Physics and Crystallography: A Handbook of Physical Constants, AGU Reference Shelf 2, edited by T. J. Ahrens, Washington, D. C.: AGU, 1995, pp. 64-97.

11. Jeanloz, R., and Rudy, A., J. Geophys. Res. 92, 11433-11436 (1987).

12. Kondo, T., Yagi, T., and Syono, Y., AIRAPT-16 & HPCJ-38, Kyoto, 1997, to be presented, and also in this volume.

13. Sato, Y., and Jeanloz, R., J. Geophys. Res. 86, 11773-11778 (1981).

14. Jeanloz, R., and Ahrens, T. J., Geophys. J. R. Astr. Soc. 62, 505-528 (1982).

15. Yagi, T., Suzuki, T., and Akimoto, S., J. Geophys. Res. 90, 8784-8788 (1985).

16. Knittle, E., Jeanloz, R., Mitchell, A. C., Nellis, W. J., Solid State Commun. 59, 513-515 (1986).

17. Fei, Y., and Mao, H. K., Science 266, 1678-1680 (1994).

18. McQueen, R. G., and Marsh, S. P., Handbook of Physical Constants, edited by S. P. Clark, Jr., New York: Geol. Soc. Am., Memoir 97, 1966, pp. 153-159.

19. Syono, Y., Ito, A., Morimoto, S., Suzuki, T., Yagi, T., and Akimoto, S., Solid. State Commun. 50, 97-100 (1984).

20. Suzuki, T., Yagi, T., Akimoto, S., Ito, A., Morimoto, S., and Syono, Y., Solid State Physics under Pressure, edited by S. Minomura, Tokyo/Dordrecht: KTK/Reidel, 1985, pp. 149-154.

21. Olsen, J. S., Cousins, C. S. G., Gerward, L., Jhans, H., and Sheldon, B. J., Physica Scripta 43, 327-330 (1991).

CP429, *Shock Compression of Condensed Matter – 1997*
edited by Schmidt/Dandekar/Forbes
© 1998 The American Institute of Physics 1-56396-738-3/98/$15.00

MELTING OF SHOCK-COMPRESSED METALS IN RELEASE

G.I. Kanel[*], K. Baumung[+], D. Rush[+], J. Singer[+], S.V. Razorenov*, and A.V. Utkin*

[*] *High Energy Density Research Center, Izhorskaya 13/19, Moscow 127412 Russia*
[+]*Forschungszentrum Karlsruhe, P.O. Box 3640, D-76021 Karlsruhe, Germany;*
* *Institute of Chemical Physics, Chernogolovka 142432, Russia*

Parameters of shock waves that cause melting of aluminum, copper, titanium, and molybdenum in release have been measured. A pulsed proton beam was used to launch 20 to 50 μm thick aluminum flier plates of ~8 mm diameter up to velocities of 5 to 10 km/s. As a result of the power density profile of the beam, the radial velocity distribution of the flier was bell-shaped and a range of shock pressures was covered by impact of the flier plate in each shot. Acceleration of the flier and the sample free surface velocities were recorded simultaneously using line-imaging version of the ORVIS laser velocimeter. The sharp loss of the sample surface reflectivity was considered indicative of melting.

INTRODUCTION

Pulsed high-power ion beams are subject of research and development in a number of laboratories all over the world. Since the beam interaction with targets is accompanied with appearance of strong compression waves, modern methods of the shock-wave physics are employed for analysis of the interaction phenomena and for investigations in fields of equation of state (EOS) and other traditional shock-wave problems [1]. However, at a typical beam pulse duration of few tens of nanoseconds pressure, density, and temperature in the energy deposition zone show strong spatial and temporal variations. This circumstance creates problems in investigations of most interesting plasma states of the matter heated by the beam. On the other hand, hypervelocity launching of thin flyer plates and impact opens a way to study the EOS in the 100-1000 GPa pressure range. The small load duration creates new possibilities to study time-dependent processes, such as phase transitions, chemical reactions (dissociation, thermal destruction), and the stress relaxation (visco-elastic response, fracture). The

problem of equilibrium can become very important when we change the time scale by several orders of magnitude.

The most appropriate method for diagnosing beam driven shock waves is the laser Doppler velocimeter [2,3]. It can be used as far as the reflectivity of the surface investigated is maintained.

Shock compression is accompanied by heating of the medium. At a certain peak pressure, the entropy increase in the shock wave results in melting of solids at isentropic unloading. The shock pressure leading to material melting in the release wave is of general interest for the verification of equations of state but also for practical applications, and for the shock-wave experiments themselves.

It is known the reflection of strong shock waves from the surface of a solid body can lead to ejection of material [4,5,6]. The effect is sharply intensified when approaching post-shock melting at release. The development of surface instabilities at melting was also observed with X-ray shadowgraphy of samples after shock compression and release [7,8]. As a result, for optically polished surfaces, when a solid-liquid transition begins on the release isentrope, the reflectivity drops drastically. This phenomenon can be used to detect melting in release of shock-loaded materials [5,6].

In this paper, results of measurements of the threshold pressure leading to melting of aluminum, copper, molybdenum and titanium in the release wave are presented. The method [9] utilizes the particular properties of our experimental capabilities at the **K**arlsruhe **L**ight **I**on **F**acility KALIF [1]. Recently, we have developed an improved line-imaging laser-Doppler velocimeter [10] that can operate both in the VISAR [2] or ORVIS [3] mode. This instrument allows us to measure the radial velocity profile over a ~8-mm line across the ion beam focus.

EXPERIMENTAL

According to the power density profile along the beam cross-section, the radial velocity distribution of the ablatively accelerated flier foil is bell-shaped showing a ~25% decrease over a 4-mm distance from the beam axis. This inspired us with the idea to utilize the nonuniformity of the flier velocity, and the line-imaging velocimeter capabilities to measure the shock melt pressure.

Figure 1 shows the arrangement of the targets used for the shock melt experiments. Shock waves in the targets were created by impact of fliers. In the experiments performed, aluminum foil fliers of 40 to 75 μm initial thickness were accelerated by the ablation pressure. Given the material ablation of 25-30μm up to the impact, the final solid impactor thickness was ~10 to 50 μm, respectively. The upper edge of the target was positioned ~1 mm below the beam axis in order to allow a proper extrapolation of the flier velocity distribution measured in the upper

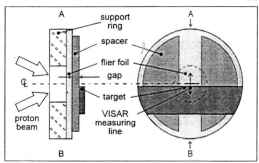

FIGURE 1. Schematic of the target arrangement.

part of the field of view into the region hidden by the target. For the assembly, the flier foil was kept stretched radially while gluing it onto a 40-mm outer and 10 mm inner diameter PMMA support ring to ensure planarity. The impact velocity was adjusted by the initial gap width from 50 to 130 μm between the flier and target and by the flier thickness.

Using line-imaging laser Doppler velocimeter, the velocity history was recorded along a 4-mm distance of both the flier and the target simultaneously. With an appropriate velocity interval chosen the shock melt pressure threshold is contained in the pressure interval observed. The loss of the intensity of the reflected laser light was considered indicative of material spraying that occurs upon the shocked state release to a solid + liquid mixture.

The metals chosen for this study are aluminum, copper, titanium, and molybdenum. All samples were foils 20 to 70 μm thick with the main material content of 99.9%.

RESULTS

Aluminum. Four shots with aluminum samples have been performed covering the impact velocity range 3 to 6 km/s. Figures 2 shows a line-imaging record in the VISAR mode of an experiment where the melting boundary is seen. The upper part of interferogram shows the flier acceleration; lower part is related to the target free surface. In the VISAR mode, the interference fringes represent counts of constant velocity. The velocity-per-fringe constant was 0.773 km/s and the initial jump of flier

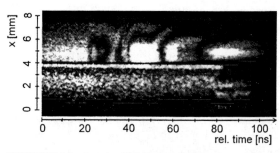

FIGURE 2. Line-imaging VISAR record of impact experiment with melting of the aluminum target. The lower half shows the target. In its upper part (2.5 to 4 mm) the impact causes a loss of reflectivity whereas in the adjacent section (1 to 2.5 mm) two fringes appear indicating a velocity gradient.

velocity corresponds to ~3.5 lost fringes.

Obviously, the target foil was rather deformed at the upper edge when it was cut. As a result, the impact starts ~1.5 mm below the border. Nevertheless, this interferogram shows a clear separating line between the upper dark zone where the reflectivity is lost, and the section below, where the emergence of two fringes indicates a vertical velocity gradient. Taking into account the initial phase of the fringe, the threshold velocity is 5.1±0.2 km/s corresponding to a pressure of 62±3 GPa. In other shot we observed the loss of reflectivity at the impact velocity of 5.3 km/s, so using the Hugoniot of aluminum and presuming the loss of reflectivity to be characteristic of material melting, we find a threshold pressure for melting in the range of 62 to 65 GPa.

Titanium. Four shots with 52 μm thick titanium targets were done using aluminum flyers of 50 μm initial thickness. The line-imaging ORVIS interferogram of one of two experiments where the boundary between melted and non-melted states of the target was covered is presented in Fig. 3.

While the VISAR mode gives directly a clear view of the time dependence of the velocity field, the ORVIS mode provides more precise velocity measurements. For both modes, the velocity is calculated through number of fringes recorded for the given space point to the considered moment. In the line-imaging ORVIS mode, the interference fringes have a slope that is determined both by the acceleration and by the space velocity gradient. If the direction of the fringe deflection coincides with the velocity gradient, the fringe slope increases, if not - it decreases. The fringe slope dy/dx in the interferogram can be described by formula

$$\frac{dy}{dx} = \frac{\Delta y \cdot \left[\dfrac{\partial (v/v_0)}{\partial t} \Big/ \dfrac{dx}{dt} \right]}{1 - \Delta y \cdot \left[\dfrac{\partial (v/v_0)}{\partial r} \Big/ \dfrac{dy}{dr} \right]},$$

where x and y are horizontal and vertical coordinates of some point in the interferogram, v_0 and Δy are the velocity-per-fringe constant and the spacing between fringes respectively, r is the coordinate of the same point on the sample surface. The formula shows that

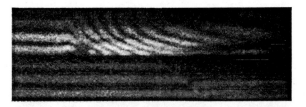

FIGURE 3. ORVIS interferograms of experiments #4144 with titanium. The velocity increase corresponds to the negative fringe slope.

the slope of fringes can even become vertical or change the sign.

The impact velocity in this shot varied between ~5.2 to ~6.3 km/s. At impact velocities of 5.6 km/s and less, both reflectivity and interference contrast are maintained during the whole recording time. This permitted to measure the target free surface velocity at the shock front and by that to recover closed parts of the flyer velocity isolines. Between ~5.6 and 6.0 km/s (82.5 to 90.8 GPa of the shock pressures) the partly maintained reflectivity and interference contrast is continuously decreasing to zero.

Molybdenum. Figure 4 displays interferograms of two experiments using 20 μm thick molybdenum samples and aluminum impactors of 40μm initial thickness. In the range of 7.6 to 8 km/s impact velocity (shot # 4143) a decrease of the reflectivity and interference contrast appear, but fringes remain observable for a relatively long period. In shot #4163, the impact velocity varies from 8.8 km/s near the bottom of the interferogram to 9.8 km/s at upper border of the target foil. In this range, a continuous reduction of the reflectivity and interference contrast is recorded. Nevertheless, some weak fringes can be observed for a short time after shock break-out over the whole range. An intense luminosity of the flyer in the last third of the interferogram is caused by the burn-through of the ablation plasma. In other similar shot a weak fringe was recorded during few nanoseconds at the impact velocity of ~9.4 km/s. Therefore, 9.4±0.4 km/s is a best estimate of the critical impact velocity for molybdenum, corresponding to a pressure of 252±16 GPa.

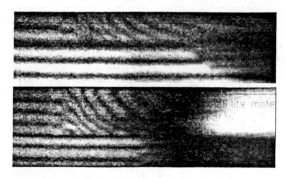

FIGURE 4. Interferograms of experiments #4143 (upper) and #4163 (lower) with molybdenum at the velocity-per-fringe constant of 1261 m/s.

Copper. Two shots were done using 45 μm thick copper foils and aluminum flyers of 50 μm initial thickness as impactors. The interferogram covering the melting boundary is displayed in Figure 5. Weak fringes are maintained clearly visible for >10 ns in the lower part of the interferogram up to impact velocities of 6.6 km/s. Their initially negative slope indicates that there is still some notable tensile strength. The impact velocity of 6.6 km/s corresponds to a shock pressure of 136 GPa.

FIGURE 5. The interferogram of experiment with copper target.

DISCUSSION

The results of experiments performed show a good agreement between the threshold pressure of maintained reflectivity and the calculated [11] shock pressure of melting at release for aluminum (64±2 GPa and 65 GPa respectively), copper (136±5 GPa and 137 GPa), and molybdenum (252±16 GPa and 230 GPa), while there is a large discrepancy in the case of titanium (86.5±4 GPa measured and 120 GPa as follows from the EOS). The good agreement means that the release process of shock-compressed metals is isentropical or nearly isentropical in our conditions. Obviously, the α-ω phase transition in shock-compressed titanium is a reason of the discrepancy with the equation of state data.

The data we obtained are related to melting onset. Since under shock-wave conditions the matter is heated as a result of mechanical work, any imperfection in the material creates conditions for the localization of the energy deposition. The volume increment at melting in such hot spots creates conditions for nucleation of surface instabilities which appear in the observed fast drop of the surface reflectivity.

ACKNOWLEDGMENTS

This work was supported by the Russian-German Cooperation Program, by the NATO Science Programme, linkage grant LG 930 326, and an INTAS grant INTAS-94-3189.

REFERENCES.

1. Baumung, K., Bluhm, H., Goel, B., Hoppé, P. Karow, H.U., Rusch, D., Fortov, V.E., Kanel, G.I., Razorenov, S.V., Utkin, A.V., and Vorobjev, O.Yu., *Laser and Part. Beams* **14**(2), pp. 181-209 (1996).
2. Barker, L. M. and Hollenbach, R.E., *J. Appl. Phys.* **43**, 4669 (1972).
3. Bloomquist, D.D. and Sheffield, S.A., *J. Appl. Phys.* **54**(4), 1717 (1983).
4. Asay, J.R., Mix, L.P., and Perry, F.C., *Appl. Phys. Lett.*, **29**, pp.284-287 (1976).
5. Anrdiot, P., Chapron, P., Lambert, V., Olive, F. *In: Shock Waves in Condensed Matter - 1983*, Eds.: J.R.Asay, B.A.Graham, G.K.Straub. Elsevier Science Publishers B.V., 1984, pp.277-280.
6. Chapron, P., Elias, P., and Laurent, B. *In: Shock Waves in Condensed Matter - 1987*, Eds.: S.C.Schmidt and N.C.Holmes, Elsevier Science Publishers B.V., 1988, pp. 171-173.
7. Belyakov, L.V., Valitski, V.P., Zlatin, N.A., and Motchalov, S.M., *Academy of Sciences of USSR - Doklady*, **170**, p.540 (1966).
8. Werdiger, M., Arad, B., Henis, Z., Horowitz, Y., Moshe, E., Maman, S., *Laser and Part. Beams* **14**(2), pp.133-147 (1996).
9. Baumung, K., Kanel, G.I., Razorenov, S.V., Rusch, D., Singer, J., and Utkin, A.V., *Int. J. Impact Engg.*, **20** (1997)
10. Baumung, K., Singer, J., Razorenov, S.V., and Utkin, A. V., *in: Shock compression of Condensed Matter - 1995"*, (edited by S.C.Schmidt and W.C.Tao), AIP Conference Proceedings 370, Woodbury, New York, 1996, pp. 1015-1018.
11. McQueen, R.G., Marsh, S.P., Taylor, J.W., Fritz, J.N., and Carter, W.J., *in: High-Velocity Impact Phenomena*, R. Kinslow, ed., Academic Press, 1970, Ch. VII, pp. 293-417.

CP429, *Shock Compression of Condensed Matter – 1997*
edited by Schmidt/Dandekar/Forbes
© 1998 The American Institute of Physics 1-56396-738-3/98/$15.00

HIGH PRESSURE *IN SITU* X-RAY DIFFRACTION STUDY OF MnO TO 137 GPa AND COMPARISON WITH SHOCK COMPRESSION EXPERIMENT

T. Yagi[1], T. Kondo[1], and Y. Syono[2]

[1] *Institute for Solid State Physics, University of Tokyo, Roppongi, Minato-ku, Tokyo 106, Japan*
[2] *Institute for Materials Research, Tohoku University, Katahira, Aoba-ku, Sendai 980-77, Japan*

In order to clarify the nature of the phase transformation in MnO observed at around 90 GPa by shock compression experiment, high pressure *in situ* X-ray observations were carried out up to 137 GPa. Powdered sample was directly compressed in Mao-Bell type diamond anvil cell and X-ray experiments were carried out using angle dispersive technique by combining synchrotron radiation and imaging plate detector. Distortion of the B1 structured phase was observed above about 40 GPa, which continues to increase up to 90 GPa. Two discontinuous changes of the diffraction profiles were observed at around 90 GPa and 120 GPa. The nature of the intermediate phase between 90 GPa and 120 GPa is not clear yet. It is neither cesium chloride (B2) nor nickel arsenide (B8) structure. On the other hand, the diffraction profile above 120 GPa can be reasonably well explained by the B8 structure. High pressure phases above 90 GPa have metallic luster and all the transformations are reversible on release of pressure.

INTRODUCTION

High pressure behaviors of monoxides are of great importance for Earth science because both MgO and FeO plays important role for the discussion of the deep interior of the Earth. Magnesiowustite, (Mg, Fe)O, is considered to be one of the major constituent mineral of the lower mantle, together with silicate perovskites. Moreover, FeO is one of the important candidate of the light elements in the core. From these points of view, many studies has been made to clarify the high pressure behavior of alkaline earth metal monoxides (AO: A= Ba, Sr, Ca) and clarified that all these monoxides transforms into B2 structure at high pressure(1). On the other hand, it was found by high pressure and high temperature *in situ* X-ray study that FeO transforms into B8 structure (2), rather than B2 structure. This indicates that the high pressure behavior of transition metal monoxide may different from those of alkaline earth monoxides.

Very recently, an existence of pressure induced phase transformation was reported based on the shock compression experiments to 114 GPa (3). They observed a transition at about 90 GPa which is associated with the volume decrease ($\Delta V/V_0$) of about 6 %. Various hypothesis were made to explain this transition such as B1-B2, B1-B8, and electronic transition (4). Theoretical prediction based on the first-principles density functional calculations was also made, which clarified that the B1 phase of MnO may transform into a metallic B8 structure (5). On the other hand, the only static compression data on MnO available is up to 60 GPa (6) and no transformation was reported in their study. Thus, the nature of the transition of MnO at around 90 GPa remains unclear. The purpose of this study is to make high pressure *in situ* X-ray observation of MnO up to above 100 GPa and clarify the nature of the transition observed in the shock compression study.

EXPERIMENTAL

A Mao-Bell type diamond anvil cell with bevelled anvils (outer culet = 340 μm, inner culet = 120 μm, bevelled angle = 8 °) were used for the present study. A stainless steel gasket of 250 μm thick was precompressed to about 25 GPa and a small hole (60 μm in diameter), which was use as a sample chamber, was opened at the center of the indentation using Q-switched YAG laser. The thickness of the gasket in the indentation was about 40 μm. A single crystal of MnO grown by the Verneuil method (Nakazumi Crystal Laboratory) was crushed and powdered with a pestle and mortar and was placed in the sample chamber. This sample is from the same batch of the crystal used for the shock compression experiments by Noguchi et al.(3). No pressure transmitting medium was used and the powdered sample, together with a small ruby crystal, which works as a pressure marker, were compressed directly by two diamonds.

X-ray measurements were made at the Photon Factory of KEK, Tsukuba. A white X-ray from the multipole wiggler was monochromatized and focused by a double bent monochromater made of silicon. The size of focused X-ray was further reduced by two perpendicular slits of 30 μm and the monochromatized X-ray of 29 keV was irradiated to the sample. Diffracted X-ray was recorded on the imaging plate detector. Exposure time depends on the thickness of the sample and was about 10 to 20 minutes at low pressure while it was about 2 hours above 100 GPa. All the X-ray measurements were made at room temperature. After the pressure was increased to 131 GPa, however, the sample was heated by both YAG and CO_2 lasers, in order to promote the transition. Because MnO became metallic above 90 GPa, as will be mentioned later, heating the sample was difficult when YAG laser was used. Therefore we have irradiated CO_2 laser light to the sample through type-Ia diamond anvil. Heating was made at 60-90 W for 2 hours. No incandescent light was observed during heating, and the temperature was estimated to be below 800 K. The pressure after the heating was increased to 137 GPa, which was probably caused by the flow of the stainless steel gasket.

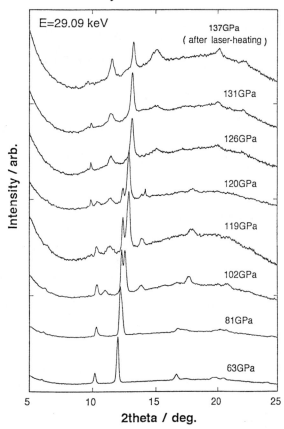

FIGURE. 1. Examples of the X-ray diffraction profiles of MnO above 63 GPa. All the observations were made at room temperature. The last diffraction was obtained after heating the sample to below 800 K by irradiating a CO_2 laser. The pressure after the heating was 137 GPa. Phase transformations were observed at about 90 and 120 GPa.

RESULT

Examples of the X-ray diffraction profiles are shown in Figure1 and all the d-spacings observed in this study are plotted as a function of pressure in Figure 2. With increasing pressure, a splitting of the (220) diffraction line was observed above about 40 GPa. On the other hand, no splitting was observed in other diffraction lines such as (111) and (200) of the B1 phase. This splitting is probably caused by the distortion of the B1-type structure and the detailed analysis of this phase is in progress. Jeanloz and Rudy(6) did not report such transition, although they performed similar X-ray diffraction study up to 60 GPa. This is probably because they used film method and only two strong diffraction lines, (111) and (200) were mainly used for their analysis. They sometimes used (220) diffraction as well, but as is clear from Figure1, the splitted line is much weaker than the original (220) line and it is easily overlooked when the film was examined visually.

At about 90 GPa, new diffraction lines appeared, and the intensity of these new peaks increased with pressure. When the pressure was increased above 120 GPa, the diffraction profile has changed again. At 131 GPa, the strongest line was still the line from the intermediate phase, which appeared above 90 GPa. After heating the sample, diffractions from the high pressure phase above 120 GPa became stronger, while the intensity of the strongest diffraction at $2\theta=13.2°$ decreased considerably, accompanied by the increase of the shoulder in the low angle side of this peak. It is also clear that the line width of the peak at $13.2°$ is quite different from those of other lines. All these observations suggest that there are two successive transitions at around 90 GPa and 120 GPa.

The color of the sample changed considerably with pressure. The single crystal of MnO had dark green color and powdered sample had dark brownish color. It was so dark that no transmitting light was observed when compressed in diamond anvil cell. The reflectivity of the visible light was very low at low pressure but above 90 GPa, the sample had metallic luster and it became indistinguishable from the surrounding stainless steel gasket. In the pressure decreasing cycle, the metallic luster disappeared and

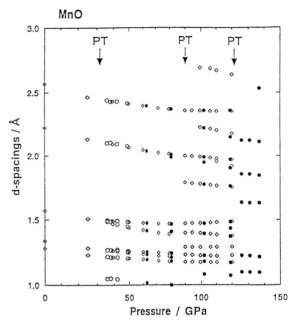

FIGURE. 2. Observed d-spacings of MnO as a function of pressure.

TABLE 1. Observed and calculated d-values of the X-ray diffraction at about 137 GPa. Calculation was made based on the NiAs structure with a hexagonal unit cell a= 2.437(2)Å and c= 5.089(13)Å.

Index			dobs	dcal	dobs/dcal -1
0	0	2	2.5340	2.5446	-0.0042
1	0	0	2.1101	2.1107	-0.0003
1	0	1		1.9497	
	?		1.8444		
1	0	2	1.6279	1.6246	0.0021
1	1	0	1.2184	1.2186	-0.0002
1	1	2	1.0989	1.0991	-0.0002

a=2.437(2)   c=5.089(13)   c/a=2.088
V=26.18(8)   V/Vo=0.5953

the sample looked dark brown again. The changes of the diffraction profiles were also reversible on release of pressure, although considerable hystereses of pressure was observed.

DISCUSSION

In shock compression study (3,4), no transformation was observed at around 40 GPa. Although the structure of the phase above 40 GPa remains unknown, the d-spacings and the intensity of (111) and (200) diffractions varied continuously across this pressure. This fact suggests that the nature of this transition is probably a distortion of the B1 structure and that the transition is in second order. In this case, it is reasonable that no transition was observed by shock compression experiment. Noguchi et al. (3) reported that the volume compression curve determined by the shock compression deviates slightly from that of the static compression data(6). This is quite reasonable since Jeanloz and Rudy(6) calculated the density of MnO assuming B1 structure, which is different from the real structure. Our preliminary analysis indicates that this distorted phase can be explained by a hexagonal unit cell and the volume compression curve obtained by this calculation is in good agreement with the shock compression data(7).

Transition pressure observed at about 90 GPa is in good agreement with the shock compression experiment (3,4) but the structure of the high pressure phase after the transition is not yet clear. The observed diffractions cannot be explained neither by cesium chloride (B2) nor by nickel arsenide (B8) structures. On the other hand, the diffraction observed at 137 GPa is reasonably well explained by the nickel arsenide structure, in accordance with the result of theoretical calculation (5). Although it is difficult to explain the strongest line which has the d-spacing of 1.844 Å $(2\theta=13.2°)$, this line probably belongs to the intermediate phase, as was discussed before. Although (101) reflection of the B8 structure was not observed clearly, the expected position is in harmony with the small shoulder of the 1.844 Å peak, which appeared after the laser heating. Comparison of the observed and calculated d-spacings are shown in Table 1. At around 120 GPa, this B8 structure phase is about 13 % more dense than the extrapolation of the B1 phase and this density is in good agreement with the theoretical calculation(5). On the other hand, the density increase observed by shock compression at 90 GPa is only 6 %(4). The reason of this discrepancy is not yet clear. It could be either because shock compression has observed the density of the intermediate phase or because the sample is still in a mixed phase region in shock experiment.

Shock compression experiment clarified the existence of one high pressure phase at above 90 GPa and the theoretical calculation predicted that the B8-structured phase become stable at very high pressure. Although our X-ray diffraction study is in harmony with these observation and the prediction, the real behavior of MnO under pressure is much more complicated. Further experiments and analysis will be required to fully understand the high pressure behavior of this simple oxide.

ACKNOWLEDGMENTS

Authors are grateful to T. Kikegawa and O. Shimomura of KEK for their help in constructing the X-ray system at BL13 of the Photon Factor. K. Terakura of NIAIR, Y. Noguchi and K. Kusaba of Tohoku University are greatly acknowledged for valuable comments and discussions.

REFERENCES

1. Liu,L.,and Bassett,W.A.,*Elements ,Oxides, Silicates,High- Pressure Phases with Implications for the Earth's Interior*, NewYork: Oxford Univ.Press,1986.
2. Fei,Y., and Mao,H.K.,*Science* **266**,1678-1680 (1994)
3. Noguchi,Y.,Kusaba,K.,Fukuoka,K.,and Syono,Y., *Geophys.Res.Lett.***23**,1496-1472(1996)
4. Syono,Y.,Nogughi,Y.,and Kusaba,K. *Higt Pressure-Temperature Research:Properties of Earth and Planetary Materials* ,AGU,in press.
5. Fang,Z.,Terakura,K.,Sawada,H.,Miyazaki,T.,and Solovyev,I.,*Nature*,submitted.
6. Jeanloz,R.,andRudy,A.,*J.Geophys.Res.***92**,11433-11436(1987)
7. Kondo,T.,Yagi,T.,and Syono,Y., *J. Appl .Phys.*, submitted.

162

CP429, *Shock Compression of Condensed Matter – 1997*
edited by Schmidt/Dandekar/Forbes
© 1998 The American Institute of Physics 1-56396-738-3/98/$15.00

SHOCK WAVE INDUCED PHASE TRANSITIONS FROM THE TRIGONAL PHASE TO THE COEXISTING AMORPHOUS AND ORTHORHOMBIC PHASES IN α-FePO4

K. D. Joshi, N. Suresh, G. Jyoti, S. K. Kulshreshtha*
Satish C. Gupta and S. K. Sikka

High Pressure Physics Division, *Chemistry Division
Bhabha Atomic Research Centre, Mumbai, 400 085, India.

The shock induced response of berlinite form of α-FePO4, which has been studied recently under static pressure, has been investigated in order to examine the effect of shear and high temperature on the process of amorphization in this material. The samples were shock loaded up to 8.5 GPa in a gas gun and after recovery were analyzed using x-ray diffraction (XRD) technique. The sample retrieved from 5.2 GPa revealed an irreversible phase transformation of some of the material to an amorphous and a crystalline orthorhombic structure (space group Cmcm), which are co-existing. The XRD pattern of the 8.5 GPa sample on the other hand displayed the presence of only the orthorhombic phase along with the ambient structure. The absence of the the amorphous phase is attributed to the reverse transformattion due to the high residual temperature in the 8.5 GPa sample. Since the Cmcm phase is equilibrium high pressure phase of such materials, the results could be interpreted on the three level free energy diagram. The comparison of these results with those reported under static pressure is presented.

INTRODUCTION

Many crystals have been observed to undergo crystalline to amorphous phase transitions under static and dynamic compression (1). Among them quartz family of compounds have been studied extensively because of their geophysical importance. Compounds belonging to this family have structures based on corner linked tetrahedral network. This tetrahedral framework structure, which does not favour close packing, has an important role in deciding the various kinds of pressure induced phase transformations in these materials. Earlier studies indicate that the phase transformation and amorphization in these compounds are related to the decrease in non bonded O---O distances due to reduction of T - O - T angle (T= Si, Al, P etc.) of tetrahedral network. These transformations are sensitive to shear component of stress, as is evident from the results of shock wave Raman experiments that demonstrated that the decrease in T - O - T angle under compression is increased by the presence of non-hydrostatic stresses (2). Also, shear stresses may cause irreversibility in a transition that is reversible under hydrostatic conditions. Crystalline to amorphous phase transition in berlinite form of AlPO4 is reversible or irreversible depending on whether the applied stress is hydrostatic or not (3-6). Recently, Chitra et al (7) conducted static pressure experiments on the berlinite form of iron phosphate α-FePO4 , space group P3₁21, in which Fe - O(1) - P and Fe - O(2) - P angles are 139.2° and 137.5° respectively. On the basis of Raman measurements in diamond anvil cell they reported a phase transition around 3 GPa, which was confirmed by the X-ray diffraction (XRD) measurements that showed the presence of a broad peak which did not disappear till 10 GPa. The structure of the high pressure phase could not be determined because of the paucity of the diffraction data.

Shock wave compression of materials is always accompanied by shear, high temperature and defects. All these factors could influence the nature as well as kinetics of a transformation. In the present work, we have examined the behaviour of α-FePO$_4$ under shock loading in order to compare it with that under static pressures.

EXPERIMENTAL

Experiments were carried out using the gas gun with bore size of 63 mm at our laboratory (8). Four experiments have been conducted on α-FePO$_4$, with initial density of 1.8g/cm^3, a nominal diameter of 13 mm and 0.9mm thickness. The samples were emplaced in a recovery fixture (9) and loaded to shock pressures up to 8.5 GPa.

The pressure history in the sample was estimated by performing numerical simulations using a two dimensional hydrodynamic code with the measured projectile velocity and estimated Hugoniot of FePO$_4$ as input. The peak pressures for the four experiments corresponding to the measured projectile velocities of 0.11, 0.19, 0.27 and 0.45 km/sec were estimated to be 2$\pm$ 0.5, 3.5$\pm$ 0.5, 5.2$\pm$ 0.5 and 8.5$\pm$ 0.5 GPa, respectively. The sample reached the final pressure after a few reverberations and its peak pressure lasted for around 1-2 μs. The shock recovered samples were characterized using powder x-ray diffraction.

RESULTS AND DISCUSSION

The XRD measurements on the shock recovered samples are displayed in figure 1. The XRD pattern of the sample recovered from 2 GPa is identical to that at ambient conditions (figure 1a). However, for 3.5 GPa sample, the diffraction pattern exhibits new peaks along with the broadening of the peaks of the original phase (figure 1b). The development of these peaks is more clear in XRD pattern of the sample pressurized up to 5.2 GPa (figure 1c). These new diffraction peaks could be associated with an orthorhombic phase having space group Cmcm and cell constants a= 5.227A^o, b= 7.770A^o, c= 6.322 A^o. Moreover, the peaks are riding over hump like background indicating the presence of amorphous material. This suggests that the trigonal

phase has irreversibly transformed partly to orthorhombic crystalline phase and partly to the amorphous phase around 5 GPa. The coexistence of the high pressure crystalline phase and the amorphous phase is quite unusual.

Raman spectra of shock recovered α-FePO$_4$ under different pressure is shown in figure 2. Around 5.2 GPa, the intensity of all the Raman modes decrease drastically and some new peaks emerge in the region of 100-500 cm^{-1}. This agrees with the XRD results indicating the amorphization of the starting material along with transformation to a new crystalline structure.

It is interesting to compare the above results with those of the static pressure measurements of Chitra et al. (7) performed concurrently in our laboratory. In the in-situ XRD measurements on the sample above 3 GPa, they observed the disappearance of the strong (012) original peak of the trigonal structure with emergence of a weaker peak close to it and appearance of a new broad peak near (110) original peak. This indicated a transition to new crystalline phase. However, because of paucity of the diffraction data, the new structure could not be determined. On comparing our XRD pattern with that of Chitra et al, we found that the new unidentified peaks seen under static pressure belonged to the orthorhombic structure observed in the shocked samples, and suggested that the unidentified high pressure structure could be orthorhombic. This was confirmed by their later XRD measurements on the pressure quenched samples prepared by compressing above 3 GPa between tungsten carbide anvils in a 100 ton press.

The XRD pattern of the sample shock recovered from 8.5 GPa (figure 1d) shows that in comparison with that of the 5.2 GPa pattern the amorphous background has decreased, the peaks belonging to the orthorhombic structure are much weaker and the peaks corresponding to the parent trigonal phase are sharper. Two experiments gave almost similar results. This behaviour of α- FePO$_4$ is abnormal and is in contrast to that of q-GeO$_2$ where the proportion of the amorphous component in the shock retrieved samples increased with the peak loading stress (9).

As the Cmcm phase is the equilibrium phase under pressure for this family of phosphate compounds (10), the above results could be interpreted in terms of a three level free energy diagram. Figure 2 shows a schematic diagram of the free energy versus the reaction coordinate,

164

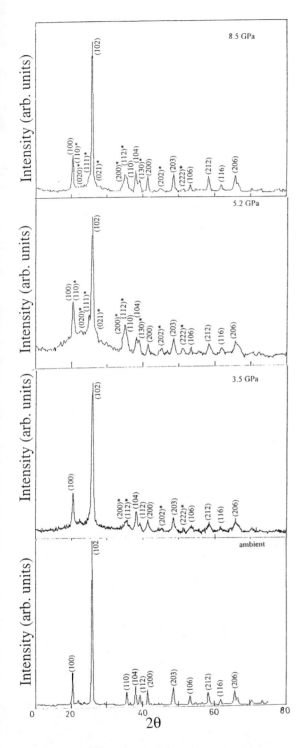

FIGURE 1. XRD patterns of FePO$_4$ samples recovered from different pressures. (hkl) trigonal structure (hkl)* orthorhombic structure.

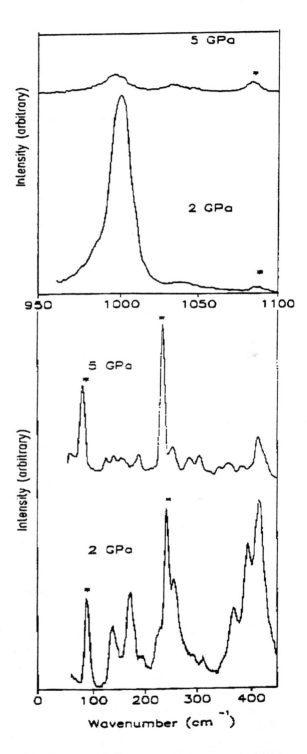

FIGURE 2. Raman Spectra of shock recovered FePO$_4$. * indicate plasma peaks

where G_{c1}, G_a and G_{c2} are the free energies of the initial crystalline phase c1 just before the transformation, amorphous phase a, and the high pressure phase c2, respectively; Δg_a, Δg_{c2} and Δg_{c1c2} (not shown in figure) are the barrier heights for c1 to a, a to c2 and c1 to c2 transitions, respectively. The compressed berlinite phase (c1) energetically prefers to transform to the Cmcm phase (c2) above 3.5 GPa. However the kinetics impedes this and instead a transformation to the amorphous (a) phase occurs. This then transforms to the equilibrium phase. This is similar to the behaviour of quartz SiO_2, where the c→a transition occurs at 21 GPa and the traces of the equilibrium stishovite phase are detected at 70 GPa (11). The fact that both the transitions in $FePO_4$ occur in a small pressure interval suggests that the energy barriers Δg_a and Δg_{c2} are of similar magnitude. However, we can not rule out the direct c1 to c2 transformation which depends on the barrier Δg_{c1c2}. The transformation kinetics under shock compression is much influenced by the peak shock and the residual shock temperatures. Figure 3 shows the estimated shock temperature as a function of peak pressure in α-$FePO_4$. In the case of 8.5 GPa experiment, although higher component of orthorhombic and may be amorphous phase could have been formed, these might have not been retained on unloading due to the reverse transformation caused by the higher residual temperature of about 1200 K as our sample has 41% porosity. This leads to the presence of very small components of the amorphous as well as higher pressure structure.

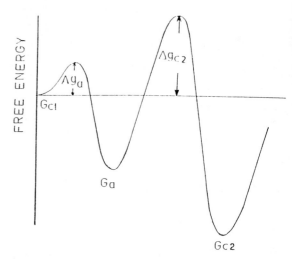

FIGURE 3. A schematic diagram showing free energy as a function of reaction coordinate for crystal to amorphous transition.

ACKNOWLEDGMENTS

We thank M. Chitra for providing the static pressure results prior to publication and for help in the Raman measurements.

SUMMARY

We find the onset of shock induced amorphization in α-$FePO_4$ around 3 GPa along with transition to an orthorhombic phase. However, unlike under static high pressures where the amount of amorphous material continues to increase on increasing the pressure, under shock compression proportion of the amorphous material has substantially decreased at 8.5 GPa, because of the reversion due to higher post shock temperature. Our results support the validity of the three level free energy diagram for interpreting c→a transitions.

REFERENCES

1. Sharma S.M. and Sikka S.K, *Prog. Mater. Sci.* **40**, 1 (1996).
2. Gallivan S.M. and Gupta Y.M., *J. Appl. Phys.* **78**, 1557 (1995).
3. Gillet P., Badro J., Varrel B., Macmillan P. F., *Phys. Rev. B.* **51**, 11262 (1995).
4. Sankaran H., Sharma S.M., Sikka S.K. and Chidambaram R, *Pramana J. Phys.*, **35**, 177 (1990).
5. Krugger M.B and Jeanloz R, *Science* **249**, 647 (1990).
6. Cordier P., Doukhan J. C. and Peyronneau J., *Phys. Chem. Miner.* **20**, 176 (1994)
7. Chitra V., Momin S. N., Kulshreshtha S. K , Sharma S. M. and Sikka S. K., *Proceedings of the International Conference fon Condensed Matter under High Pressures* (India), 1996 (under publication).
8. Gupta S. C., Agarwal R. G., Gyanchandani J. S, Ro y S., Suresh N., Sikka S. K., Kakodkar A. and Chidambaram R. (1992), eds: Schmidt S.C, Dick R.D, Forbes J.W and Tasker D. G *Shock Compression of Condensed Matter- 1991*, Elsevier, Amsterdam, pp. 839.
9. Suresh N., Jyoti G., Gupta S. C., Sikka S. K., Sangeeta, Sabharwal S. C., *J. Appl. Phys.* **76**, 1530 (1994).
10. Sharma S. M. and Sikka S. K., *Phys. Rev. Lett* **74** , 3301 (1995).
11. Hazen R. M. and Finger L.W. *Phase Trans.* **1**, 1 (1979); *Am. Sci.* **72** , 143 (1984); *Sci. Am.* **252**, 110 (1985); Moffat W. G., Pearsall G. W. and Wulff J., *The Structure and Properties of Materials*. Wiely Eastern Limited, New Delhi (1980) vol.1, Chap. 5.

CP429, *Shock Compression of Condensed Matter – 1997*
edited by Schmidt/Dandekar/Forbes
© 1998 The American Institute of Physics 1-56396-738-3/98/$15.00

SHOCK WAVE EQUATION OF STATE AND SHOCK-INDUCED PHASE TRANSITION OF HALLOYSITE

Gong Zizheng, Tan Hua and Jing Fuqian

Laboratory for Shock Waves and Detonation Physics Research,

Southwest Institute of Fluid Physics, P.O.BOX 523-61, Chengdu, Sichuan, P.R.China.

Hugoniot measurements for halloysite with two different initial densities have been performed at the shock pressures up to 100GPa. Three distinct regions appear along the Hugoniot of samples with ρ_{00}=1.375g/cm^3: a low-pressure phase (LPP) exists within the shock pressure up to about 22GPa, a mixed phase region (MP) begins at 22GPa and completes at about 31GPa, and then a high-pressure phase (HPP) occurs at shock pressure between 31GPa and 50GPa. The fitted linear U_s-u_p relation of LPP and HPP can be expressed, respectively as U_s =0.2219+1.916 u_p, and U_s =1.8668+1.2602 u_p, U_s is the shock wave velocity and u_p the particle velocity. But for the sample of ρ_{00}=2.001g/cm^3, the experimental data can only be fitted with a single linear U_s-u_p relation: U_s =1.9563+1.5474 u_p, without observable discontinuity in the U_s-u_p plot or a significant volume change up to at least 100GPa. Through the comparison between the experimental Hugoniots and theoretical ones calculated using additive principle, the phase transition procedure of halloysite was determined.

INTRODUCTION

Hugoniot equations of state of minerals and rocks provide the basis for describing shock wave propagation from intense explosions in the Earth and the effects of meteorite impact on the Earth and planets. Hydrous minerals may play a critical role in controlling the partial pressure of H_2O within the Earth's interior and hence affecting the lower crust and mantle rheology and melting behavior and hence the Earth's evolution, such as regulating the water budget or triggering deep focus earthquakes (1,2). Therefore, the study of hydrous minerals at high-pressure and temperature is crucial for understanding the structure and evolution of the Earth's crust and mantle. Recently much more previous experimental investigations have been performed on many important member of hydrous minerals (1,2,3,4,5,6). However, few hydrous aluminum silicates have been studied before. In this

report, the Hugoniot measurements were conducted at pressures up to 100GPa, for polycrystalline halloysite with two different initial densities of 2.001g/cm^3 and 1.375g/cm^3.

EXPERIMENTAL PROCEDURES

Preparation for Starting Sample

The polycrystalline halloysite used in these experiments were collected from Xuyong county, Sichuan Province, China. Its ideal chemical formula is $Al_4 (H_2O)_4 [Si_4O_{10}](OH)_8$, and the corresponding theoretical density is 2.50g/cm^3. The result of bulk chemical analysis is given in Table 1.

The halloysite rock was ground into powder (<0.05mm in diameter) and then pressed into disk-like samples 18mm in diameter and 2.5mm in thickness. The average bulk density (calculated using sample dimensions and mass) of the starting samples used in the experiments are ρ_{00}= 2.001g/cm^3 and ρ_{00}=1.375g/cm^3.

TABLE 1. Chemical composition of Halloysite sample

Oxides	Wt.%
SiO_2	45.54
TiO_2	0.017
Al_2O_3	37.39
CaO	0.11
MgO	0.15
FeO	0.017
Fe_2O_3	0.009
MnO	<0.0061
Na_2O	0.029
K_2O	0.0024
P_2O_5	0.012
H_2O^+	16.31*
Total	99.58

* It includes structural water (OH) 10.67% and crystal molecular water (H_2O) 5.64%, determined with TGA method.

Hugoniot Measurement

The shock equation of state (EOS) experiments were carried out with the 37mm two-stage light gas gun in our laboratory using metal flyer plate bearing projectiles to impact samples at speeds of up to 6.5km/s. Figure 1 is schematically the cross-section of the experimental set-up. In all experiments, the symmetric impact technique was utilized for the purpose of convenience for data reduction. The impact velocity just prior to impact, V_{imp}, of the projectile or flyer and the shock velocity, Us, in the sample are measured using magnetoflyer method and the PZT(95/5Nb-1) probe technique, respectively. The particle velocity behind the shock front, u_p, and pressure-density

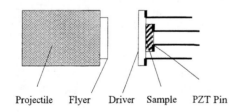

Projectile   Flyer   Driver   Sample   PZT Pin

FIGURE 1 Cross-section of target assembly.

states were calculated through the impedance match method (7). The Hugoniot parameters of the materials made into the flyers and the drivers come from Ref. (8).

RESULTS AND DISCUSSION

Hugoniot EOS

The experimental results on both eight samples of the two different initial density samples are plotted in Fig. 2 and Fig. 3. The data of ρ_{00} =2.001g/cm^3 samples fall close to a straight line, and there is no measurable discontinuity in slope throughout the pressure range of data. This indicates that no phase transition with an appreciable volume change is occurring within the pressure and temperature range of the present experiments, although there is a small amount of scatter in the data. An unweighted linear least squares fit to the data of ρ_{00}=2.001g/cm^3 yields:

$$U_S = 1.9563 + 1.5474u_p \quad (1)$$

However, the data of ρ_{00}=1.375g/cm^3 samples shows a change in slope in U_S-u_p plane (see Fig. 2). Figure 3 indicates that over the pressure range of our experiments 0-50Gpa, the data of ρ_{00}=1.375g

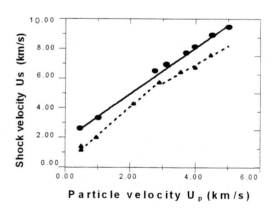

FIGURE. 2 Shock wave velocity-particle velocity (Us-Up) plot of halloysite. ● and solid line represents experimental points and fitted line for the samples of ρ_{00}=2.001g/cm^3, respectively. ▲ and dashed line represents experimental points and fitted line for the samples of ρ_{00}=1.375g/cm^3, respectively.

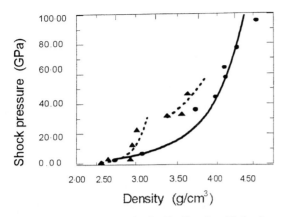

FIGURE 3 Shock pressure-density P_H-ρ_H plot of halloysite The points and the lines represent the same meaning as in Fig. 1.

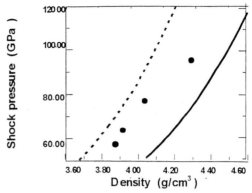

FIGURE 4 Pressure-density Hugoniot of halloysite at high pressure phase ($\rho_{00}=2.001 \text{g/cm}^3$). '●' represent experimental points

/cm³ samples shows three distinctive regions on the Hugoniot: the low-pressure phase region (LPP), the mixed-phase region (MP), and the high-pressure phase region (HPP). The LPP region lies between 0 and about 22GPa , the mixed phase region occurs over a narrow pressure interval between approximately 22-31GPa, the HPP region lies between approximately 31-50GPa. An unweighted linear least squares fit the five points in LPP and the three points in HPP yields:

$$Us = \begin{cases} 0.2219 + 1.916 u_p \\ 1.8668 + 1.2602 u_p \end{cases} \quad (2)$$

Shock-induced Phase Transition of Halloysite

In order to understand the phase transition and minerals' composition of the high-pressure phases, we calculated the theoretical Hugoniots using the additive principle of Hugoniot EOS (7) for possible decomposition products of halloysite. There are two kinds of possible decomposition reactions for halloysite as follows (9):

$$Al_4 (H_2O)_4 [Si_4O_{10}](OH)_8 \rightarrow 2Al_2O_3 + 4SiO_2 + 8H_2O \quad (3)$$

$$3 Al_4 (H_2O)_4 [Si_4O_{10}](OH)_8 \rightarrow 2Al_6Si_2O_{13} + 8SiO_2 + 24H_2O \quad (4)$$

where $Al_6Si_2O_{13}$ is mullite. All the EOS parameters used in calculations come from Ref. (8).

Model Hugoniots are plotted in Fig. 4 and Fig. 5, and compared with experimental data. It is noted that the Hugoniot parameters must be corrected to porosity using the following formula (7):

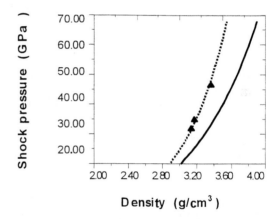

FIGURE 5 Pressure-density Hugoniot of halloysite at high pressure phase ($\rho_{00}=1.375 \text{g/cm}^3$). '▲' represent experimental points

$$P_H' = P_H [\gamma(V_0 - V) - 2V] / [\gamma(V_{00} - V) - 2V] \quad (5)$$

where V_0 and V_{00} are the ambient volume for nonporous and porous samples, respectively. P_H and P_H' are the Hugoniot pressure for nonporous and porous samples at the same volume V, respectively. γ is the Grüneisen parameter, its value at zero pressure is given by $\gamma=(\alpha K_{0S})/(\rho_0 C_p)$. The volume coefficient of thermal expansion, α of halloysite is unknown. We estimate $\alpha \approx 4 \times 10^{-6}$/K based on data for kaolinite (9). C_p, the specific heat at constant pressure, is 1.193J/g·k for halloysite (9). Combining these with ρ_0 (2.5g/cm³) and $K_{0S}=32.2$ GPa (11), we yield a zero-pressure γ_0 of 0.43. The volume dependence of γ was modeled by $\gamma = \gamma_0(\rho_0/\rho)$.

As seen in Figure 5, the measured results in HPP of $\rho_{00}=1.375\text{g/cm}^3$ samples are in very good agreement with the model Hugoniot of the product mixtures of reaction (4), which was dashed line in the Figures. As shown in Fig. 4, the experimental data of $\rho_{00} = 2.001\text{g/cm}^3$ samples lie between the model Hugoniot of the product mixture of reaction (3) (solid line in the Figures) and reaction (4). So we can say that the HPP compositions of $\rho_{00}=1.375\text{g/cm}^3$ are the mixture system of $(Al_6Si_2O_{13}+SiO_2+H_2O)$, the compositions of $\rho_{00}=2.001\text{g/cm}^3$ are the mixture system of $(Al_6Si_2O_{13}+SiO_2+H_2O)$ and $(Al_2O_3+SiO_2+H_2O)$ above 50GPa along its Hugoniot.

The theoretical Hugoniots constructed in the above manner are not corrected for temperature differe nces on the individual Hugoniots .However, it's found that the temperature correction reduces the density of the mixture not more than 0.1g/cm^3 (4), and therefore the inferences of the above are not influenced significantly.

It is well known that the Hugoniot behavior of a porous material differs that of a nonporous one. Being compressed by a shock wave, whether at the same pressure or at the same specific volume, the increment of internal energy in a porous material is more than that of a nonporous one. So the high-pressure phase is much more readily achieved by shock compression of porous rather than nonporous samples. The comparison of the Hugoniot data with the calculated Hugoniots reveals the following important feaures of halloysite phase transition under shock co mpression:

Shock induced chemical decomposition phase transition of halloysite occurs through its intermediate phase assemblage $(Al_2O_3+SiO_2+H_2O)$ (see reaction (3)) to its end phase assemblage $(Al_6Si_2O_{13} +SiO_2+H_2O)$ (See reaction (4)) ; however, the dehydration phase transition of reaction (3) has volume changes that are too small to be detected on the Hugoniot .This analysis of the phase transition procedure for halloysite is consistent with that of heating decomposition analyses (10) and the results of the shock recovery experiments (11).

ACKNOWLEDGMENTS

The research was supported by China Academy of Engineering Physics (CAEP) science grants.

REFERENCES

1. Sekine, T., Rubin, A.M., and Ahrens, T.J., *J.Geophys. Res.*, **vol.96(B12)**, 19,675- 19,680, 1991.

2. Huang, E., Li, A. and Xu, J., *Geophys.Res.Lett.*, **vol.23(22)**, 3083-3086, 1996.

3. Meade, C. and Jeanloz, R., *Geophys.Res.Lett.*, **vol.17(8)**, 1157-1160, 1990.

4. Duffy, T. S. and Ahrens, T. J., *J.Geophys. Res.*, **vol.96(B9)**, 14,319-14,330, 1991.

5. Tyburczu, J. A., Duffy, T. S., Ahrens, T. J. and Lang, M. A., *J.Geophys. Res.*, **vol.96(B11)**, 18,011-18,027, 1991.

6. Duffy, T. S., Meade, C., Fei, Y., Mao, H. K. and Hemley, R. J., *Amer. Mineral.*, **vol.80**, 222-230, 1995.

7. Jing, F. Q., *Introduction to Experimental Equation of State*, Beijing: Academic Press, 1986, Ch. 4 (in Chinese).

8. Ahrens,T.J., (Ed.), *Mineral physics and Crystallograpty:a handbook of physical constants,*, published by AGU,2000Florida Ave.,N.W., Washington, DC20009

9. He, H.P., Hou, C., Guo, J. G.,*et al.*, *Chinese Science Bulletin*, **vol.38 (6)**, 570-572 , 1991.

10. Lin, C. X. , Bai, Z. H. *etal.*, *Thermodynamic Data of Minerals and Compounds*, Beijing: Academic Press, 1985.

11. Gong, Z. Z., Tan, H. and Jing, F. Q., *Science in China*, (in press), 1997.

170

CHAPTER IV

MODELING AND SIMULATION:
Nonreactive Materials

CP429, *Shock Compression of Condensed Matter – 1997*
edited by Schmidt/Dandekar/Forbes
© 1998 The American Institute of Physics 1-56396-738-3/98/$15.00

Shockwave-Induced Plasticity via Large-Scale Nonequilibrium Molecular Dynamics *

Brad Lee Holian

Theoretical Division, Los Alamos National Laboratory Los Alamos, New Mexico 87545 USA

Nonequilibrium molecular-dynamics (MD) simulations of shock waves in single crystals have shown that, above a threshold strength, strongly shocked crystals deform in a very simple way. Rather than experiencing massive deformation, a simple slippage occurs at the shock front, relieving the peak shear stress, and leaving behind a stacking fault. Later calculations quantified the apparent threshold strength, namely the yield strength of the perfect crystal. Subsequently, pulsed x-ray experiments on shocked single crystals showed relative shifts in diffraction peaks, confirming our MD observations of stacking faults produced by shockwave passage. With the advent of massively parallel computers, we have been able to simulate shock waves in 10-million atom crystals with cross-sectional dimensions of 100x100 fcc unit cells (compared to earlier 6x6 systems). We have seen that the increased cross-section allows the system to slip along all of the available {111} slip planes, in different places along the now non-planar shock front. These simulations conclusively eliminate the worry that the kind of slippage we have observed is somehow an artifact of transverse periodic boundary conditions. Thus, future simulations are much more likely to show that weak-shock plasticity is nucleated by pre-existing extended defects embedded in the sample.

INTRODUCTION

Almost 20 years ago, nonequilibrium molecular dynamics (NEMD) simulations, where Newton's equations of motion are solved on the computer for thousands of strongly interacting atoms, were first used to study true shock waves at the atomistic level[1, 2, 3, 4]. True shock waves exhibit steady profiles (density, velocity, stress, or energy), which accompany dissipative, irreversible flow of atoms in the directions transverse to the planar wave propagation. In fluids, Klimenko and Dremin observed that viscous flow led to steady waves[3], and subsequently, Hoover showed that the continuum constitutive model, Navier-Stokes hydrodynamics, could explain the observed NEMD profiles[5]. Further NEMD simulations for even stronger fluid shock waves, where the shock thickness is only 2-3 mean-free paths, showed that Navier-Stokes is still valid[4].

Shockwave propagation in solids is inherently more complex than in fluids, where viscous flow is highly localized, yet ubiquitous. By contrast with fluids, point defects in solids are widely separated, while extended defects, such as dislocations, are themselves characterized by long-range displacement and stress fields. More important, the energies required to make defects in solids are significantly larger than in fluids (comparing defect formation energies to kT, the thermal energy available, where T is temperature, and k is Boltzmann's constant).

In general, plastic deformation in solids is accomplished by the creation and motion of dislocations. Dislocations are caused in a crystal lattice when a shear stress or strain is applied normal to a set of planes of atoms, causing them to break their registry along an intersecting slip plane parallel (or nearly so) with the direction of shear. An edge dislocation can thus be thought of as an extra half plane of atoms moving along a slip plane in the direction of the Burgers vector, which quantifies the lattice mismatch. The

---

*This work supported by the US Department of Energy.

energy penalty for this misregistry slippage mechanism is not large, especially compared with the energy barrier for all atoms to move over each other along the slip plane in unison. (Dislocation motion has been described as similar to getting a rug to move by pulling on it and at the same time making a wave-like pulse propagate, versus simply pulling on it; the latter approach involves overcoming static friction along the whole surface of the rug, while the former makes use of the lower value of propagating, localized, dynamic friction.) Under applied stress, an edge dislocation (extra half plane of atoms) moves toward a free surface, emerging as a step.

In NEMD simulations, there are three principal ways to generate a shock wave: (1) As in planar shockwave experiments, one can hurl a flyer plate toward a stationary target at a velocity of $2u_p$, where u_p is the "piston" or particle velocity. This is equivalent to slamming the two plates together at $\pm u_p$; in this symmetric-impact case, a pair of shock waves move out from the interface at the shock velocity u_s. (2) The symmetric impact can be generated by inhomogeneously shrinking the longitudinal periodic length. This is useful, particularly for fluids, for eliminating entirely any free surfaces. (3) Material can be pushed by an infinitely massive piston moving at velocity u_p; all particles coming in contact with the piston face are specularly reflected (momentum mirror). Equivalently, the piston can be at rest, with the unshocked target material given a velocity of $-u_p$; the shock wave then moves out from the stationary piston at velocity u_s. We have found that, while there are some minor differences between these three shock-generation approaches, the first and third are most suitable for studying both shockwave and release-wave phenomena in solids.

So as to model an infinite plate of material, periodic boundary conditions (pbc's) are used in the transverse directions to minimize edge effects. In the past, a typical simulation might be on the order of 100 lattice planes in length of shock propagation, and approaching 100 atoms in each cross-sectional plane, or 10,000 atoms total in three dimensions (3D). Computational times are then limited by the sound-traversal time in

the shock-propagation direction, on the order of 25 vibrational periods ($\sim$ 5 picoseconds). Shock thicknesses are then observed to be less than 10 nanometers, with rise times of the order of a picosecond.

NEMD calculations done as recently as a decade ago were severely limited in the length of run in the direction of shockwave propagation, or time required to achieve a steady wave, and in the transverse cross-sectional area. Thus, weak shock waves, whose thicknesses are measured in fractions of micrometers, with presumably similar sizes of cross-sectional structures, were well beyond the reach of atomistic simulations.

In spite of these limitations in time and space, it is remarkable that shock waves are nevertheless amenable to NEMD simulations. In the early 1980's, shockwave structure in solids was elucidated at Los Alamos in NEMD simulations by Holian, Straub, and Swanson[1, 2, 6, 7]. These calculations showed that shock waves in single crystals became steady waves by virtue of transverse displacements of atoms – not by viscous flow, as in fluid shock waves[3, 4, 5], but rather, by plastic flow, or slippage of atoms over each other. In the Lennard-Jones solid, represented by the face-centered cubic (fcc) lattice, the slippage incurred by a shock wave traveling in the $\langle 100 \rangle$ direction is accomplished by slippage along one of the four available {111} planes, due to emission of a Shockley partial dislocation, which results in a stacking fault (the usual ABCABC... stacking of triangular-lattice close-packed planes becomes ABABCA...). In a finite system with pbc's transverse to the shock, the result is the appearance of "bands", whose spacing doubles when the cross-sectional dimension L doubles.

In subsequent work, Holian showed that the shear stress τ, which builds up to a maximum at the center of the shock front, is relieved – almost back to zero, as in fluid shock waves – by this transverse plastic flow[6]. Moreover, it was noted that for shock strengths at and above a threshold limit, the ratio of the Hugoniot jump stress P to the maximum shear stress at the shock front τ appeared to be about 10 in 3D[6, 7]. At the threshold, the critical value of $P \approx G$, where G is the shear modulus, so that $\tau \approx G/10$, roughly the ideal-crystal yield strength. For 2D solids, where the cross-sectional area could be taken to be nearly an order of magnitude larger than previous 3D simulations, it was observed that two of the three available slip systems were activated, and the shock front became irregular, rather than perfectly planar as in the early purely elastic phase of propagation[7]. It remained to be seen, however, whether these 2D observations would hold for large 3D systems, where the complexity of dislocation emission is considerably enhanced[8].

Long after the earliest MD results showed that stacking faults were generated in perfect crystals by propagating shock waves, experimental verification was obtained by Russian investigators, using dynamic shockwave x-ray measurements in fcc crystals [9]. Earlier computer simulations by Mogilevsky in the USSR[10], which were contemporary with our dynamic shockwave NEMD simulations, but instead used quasi-static relaxation in uniaxially stressed fcc crystals, showed dislocation-emission events that were similar to those we saw. Recently, Belonoshko and Osiptsov have used NEMD to simulate shock waves in the $\langle 100 \rangle$ direction in bcc iron, where the interactions were modeled by an embedded-atom-method (EAM) many-body potential[11]; there, though the authors did not remark upon it, stacking faults along {110} close-packed planes can be clearly seen, suggesting that bcc materials also produce them. In our early small-scale NEMD simulations, we observed a threshold in shock strength that appeared to be close to the perfect-crystal yield strength (provided that we could someday unambiguously rule out the effect of transverse pbc's). We speculated that pre-existing defects (such as vacancies, dislocations, or grain boundaries) might lower this ideal crystal threshold considerably[6, 7].

In this paper, we will explore the questions of system size in making these evaluations of shockwave-induced plasticity. Recent advances in computer technology, primarily massively parallel machines, have enabled us to simulate the atomistic behavior of solid materials exposed to moderately strong shock waves for much larger systems than were possible just a decade ago.

NEMD RESULTS

We have studied 3D crystalline systems that are considerably wider in cross-sectional area than have ever been considered before. Our preliminary calculations, which were 15x15 fcc unit cells in cross-sectional area, demonstrated a startling new feature. Instead of a single slip system being triggered by the shock wave, slippage occurred along two different {111}-type planes. This behavior is reminiscent of the idealized model of C.S. Smith[12], where the shock front in a perfect lattice creates pairs of dislocations that accommodate the increased density of the shocked material, leaving the crystal orientation virtually unchanged, at the same time that the difference in normal stresses (shear stress τ) is relieved. But the dynamic results are richer than this static picture: the slippage on the two systems "compete" with each other, causing the front to become somewhat unstable, with the leading slip system making the front bulge out ahead by two or three lattice spacings. These slipped regions are clearly stacking faults, as can be seen by looking at the oncoming shock front along the shock-propagation direction. It appears that the interaction potential does not affect this shock-induced slippage so much as the geometry of the fcc crystal structure itself; both Lennard-Jones pair-potential and EAM many-body potential materials (e.g., copper[8, 13]) exhibit this behavior.

To test the hypothesis that pre-existing point or extended defects could trigger plastic deformation (by superposing the rise in stress at the shock front on top of the stress field around a vacancy, for example), we placed a vacancy, and then a di-vacancy, in the path of a shock wave whose strength was insufficient to initiate plastic flow in the small-cross-section perfect crystal. No plastic deformation was triggered by these defects, though there was a certain amount of "jangle" in the front as the wave passed over them. (This is in contrast to shock waves in 2D lattices, where vacancies appear to be more effective in triggering plastic flow[14].)

The critical transition in shock strength between elastic and plastic behavior is essentially independent of the initial temperature T_0 (provided that T_0 is not strictly zero, where purely elastic behavior is "frozen in," regardless of shock strength). We tested this by varying the initial temperature all the way from half the melting temperature T_M down to $0.001 T_M$, with no visible effect on the "cliff" in plasticity versus shock strength.

We can definitively answer the question of whether the existence of this apparent onset of plasticity with critical shock strength is due to transverse periodic boundaries, by expanding the cross-sectional area to 100x100 fcc unit cells in a 10 million-atom simulation. With 4x4, 6x6, 10x10, or 15x15 cross-sections, we can at best observe two stacking faults in the region of the transition ($u_p/c_0 \approx 0.2$). Now, with 100x100, we see that the shock wave propagates about 60 lattice planes, at which time a large number of stacking faults are generated relatively quickly. These are distributed randomly on all four {111} slip systems and they begin propagating back through the shocked, unslipped material, as well as forward with the shock front, at a speed that is comparable to the sound speed in the compressed material. As the shock propagates further, the front becomes pronouncedly non-planar, with slippage appearing to make the front bulge in a "galloping" fashion. When viewed at an arbitrary {100} plane, the intersections emerge as a randomly spaced plaid pattern. This demonstrates conclusively that we are no longer limited by periodic boundaries, but are seeing the true nature of plastic flow in this intermediate regime of shock strength.

When the shock strength u_p/c_0 is reduced to 0.15, no plasticity is observed, just as in the smaller cross-section systems. Thus, the "cliff" in plasticity versus shock strength is clearly due to ideal crystal yielding, rather than to periodic boundaries. We note that when the shock wave reaches the free surface and a rarefaction (relief) wave is produced, the stacking faults that were produced by shock compression are mostly annihilated. This is consistent with the observation of much smaller dislocation densities in recovered shocked materials[15].

CONCLUSIONS

We have demonstrated that the production of partial dislocations, which leave stacking faults in their wake, are a significant mechanism for plasticity in moderate-strength, solid-state shock waves. Large-scale NEMD simulations were required to settle the question of whether transverse periodic boundary conditions or ideal-crystal yielding is the cause of stacking-fault production. We have found that pbc's are not the reason, nor is temperature, nor are point defects the primary nuclei for this mode of plasticity.

The next test is to see whether *extended defects* (beyond vacancies or di-vacancies) can cause plastic deformation at shock strengths below the perfect-crystal yield strength. To that effect, we will introduce extended inhomogeneities (such as grain boundaries or stacking faults) into the unshocked material.

ACKNOWLEDGMENTS

I would like to thank my colleagues Peter Lomdahl, Jim Hammerberg, David Beazley, Shujia Zhou, and Ramon Ravelo for their many contributions to the work reported here. We also benefitted from discussions with Anatoli Belonoshko, Evgeny Zaretsky, Pavel Makarov, Ralf Mikulla, and Robb Thomson.

REFERENCES

1. B.L. Holian and G.K. Straub, Phys. Rev. Letters **43**, 1598 (1979).
2. G.K. Straub, B.L. Holian, and R.E. Swanson, Bull. Am. Phys. Soc. **25**, 549 (1980).
3. V.Y. Klimenko and A.N. Dremin, in *Detonatsiya, Chernogolovka*, edited by O.N. Breusov et al. (Akad. Nauk, Moscow, 1978), p.79.
4. B.L. Holian, W.G. Hoover, B. Moran, and G.K. Straub, Phys. Rev. **A 22**, 2498 (1980).
5. W.G. Hoover, Phys. Rev. Letters **42**, 1531 (1979).
6. B.L. Holian, Phys. Rev. A **37**, 2562 (1988).
7. B.L. Holian, Shock Waves **5**, 149 (1995).
8. S.J. Zhou, D.M. Beazley, P.S. Lomdahl, and B.L. Holian, Phys. Rev. Letters **78**, 479 (1997); S.J. Zhou, D.M. Beazley, P.S. Lomdahl, A.F. Voter, and B.L. Holian, in *Ninth International Conference on Fracture*, edited by B.L. Karihaloo *et al.* (Pergamon, Sydney, 1997), p. 3085.
9. E.B. Zaretsky, G.I. Kanel, P.A. Mogilevsky, and V.E. Fortov, High Temp. Phys. **29**, 1002 (1991).
10. M.A. Mogilevsky, in *Shock Waves and High Strain Rate Phenomena in Metals*, edited by L.E. Murr and M.A. Meyers (Plenum Press, New York, 1981), p. 531.

11. A.B. Belonoshko and A.N. Osiptsov, unpublished (private communication, 1997).

12. C.S. Smith, Trans. Metall. Soc. AIME **212**, 574 (1958); J.W. Taylor, J. Appl. Phys. **36**, 3146 (1965); J.J. Gilman, Appl. Mech. Rev. **31**, 767 (1968).

13. B.L. Holian, A.F. Voter, N.J. Wagner, R.J. Ravelo, S.P. Chen, W.G. Hoover, C.G. Hoover, J.E. Hammerberg, and T.D. Dontje, Phys. Rev. **A 43**, 2655 (1991).

14. Robin Selinger, unpublished (private communication, 1997).

15. E. Zaretsky, Acta metall. mater. **43**, 193 (1995); M.A. Meyers, Scripta metall. **12**, 21 (1978).

CP429, *Shock Compression of Condensed Matter – 1997*
edited by Schmidt/Dandekar/Forbes

A COUPLED VISCOELASTIC-VISCOPLASTIC FINITE STRAIN MODEL FOR THE DYNAMIC BEHAVIOUR OF PARTICULATE COMPOSITES: NUMERICAL ISSUES

A. Fanget*, H. Trumel*, A. Dragon**

*Centre D'Etudes de Gramat, 46500 Gramat, France, **ENSMA, 86960 futuroscope*

The dynamic behaviour of a propellant like material is modelized with viscoelastic-viscoplastic behaviour in the lagrangian finite strain frame work. Employment of logarithmic strain allows elastic-plastic decomposidon. The viscoelastic part is performed in differential form and is integrated by an explicit method. The plastic deformation is split into volumetric and distorsional parts which constitute two distinct state variables for compaction and yielding. This model has been implemented in a finite element 2D code. The algorithm of the implementadon is presented and numerical and experimental results are shown.

INTRODUCTION

Modern propellant media, composed of crystalline grains bound by ten or twenty percent of synthetic rubber, are rather uncommon materials. Their mechanical behaviour is relevant to non linear viscoelasticity, viscoplasticity phenomena. To understand the vulnerability of that kind of material, a model taking into account these properties has been built [1], [2], [3] . Defined in the frame of thermodynamics, the model describes a hyperelastic, viscous, plastic with non associated law, damageable behaviour. The paper deals with the introduction of this model in an explicit finite element code.

MODEL REVIEW

Kinematics.

Based on the multiplicative decomposition of the deformation gradient as postulated by Lee [4], namely $\mathbf{F}=\mathbf{F}^e\mathbf{F}^p$, using the polar decomposition $\mathbf{F} = \mathbf{R}\mathbf{U}$ relative to the lagrangian frame while disregarding the rotations by application of the indifference principle the following equation is obtained :

$$\mathbf{E} = \mathbf{H}^e + \mathbf{E}^p \qquad (1)$$

with $\mathbf{E}=\text{Ln } \mathbf{U}$, $\mathbf{H}^e = \text{Ln } \mathbf{U}^e$ and $\mathbf{H}^p = \text{Ln } \mathbf{U}^p$ where $\mathbf{H}^e$ is the lagrangian elastic strain with respect to the intermediate configuration (contrary to $\mathbf{U}^e$ which is referred to the reference configuration).

Taking into account (1) , the generic kinetic variables used in the model are $I^p = \text{Tr } (\mathbf{E}^p)$ and. $\overline{\mathbf{E}}^p = \mathbf{E}^p - I^p \mathbf{1}$ In the following, the notation will describe the deviatoric part, and bold letters denote tensor mathematical tool.

Viscoplasticity.

From the thermodynamic potential, the elastic stress $\mathbf{S}^R$ and the viscous stress $\mathbf{S}^V$, following the expressions are obtained respectively:

$$\mathbf{S}^R = (K1 + K3\text{ I3} +2\text{ G3 I II })\mathbf{1}+2\text{ (G1 + G3 I2) }\mathbf{He} ,$$
$$\mathbf{S}^V = \eta\Gamma$$

where $\text{I} = \text{Tr } (\mathbf{H}^e)$ and $\text{II} = \text{Tr } (\mathbf{H}^e \mathbf{H}^e)$.

The bulk (K1, K3), shear (G1, G3) and viscous moduli depend on I^p according to :

$$\mathbf{K_i} = K_i^c \text{ q}(I^p), \quad G_i = G_i^c \text{ q}(I^p), \quad \eta = \eta^c \text{ q}(I^p),$$

where, K_i^c, G_i^c, η^c denote the respective moduli of the fully compacted material.

The visco-elastic model is complemented by :

$$\dot{\Gamma}+\frac{1}{\tau}\Gamma=\dot{\overline{H}}^e$$ which for initial conditions implies

$$\text{Tr }\Gamma = 0$$

179

Compaction and plasticity.

The variables of compaction and plasticity are "measured" respectively by the mean of I^P and E^P.

The thermodynamic affinities are respectively S^p and $\overline{SP}$ which are the compaction stress and the plastic stress. The expressions of S^P and are :

$$\overline{SP} = 2\,(G1 + G3\,I2)\,\overline{\mathbf{H}}^e$$

$$S^p = (\,K1\,I + K3\,I2 + 2\,G3\;III)-$$
$$(\tfrac{1}{2}K_1'I^2 + \tfrac{1}{2}K_3'I^2 + (G_1' + G_3'I^2)\,II + \tfrac{1}{2}\eta'\,\Gamma{:}\Gamma)$$

where ( )' means the derivative versus I^P.

The compaction threshold is expressed by :
$Sp = Y$ where Y depends on I^P through the expression $Y=.\;Y_E \exp\!\left(-(I^P+\varsigma)/a_y\right)$ Note since S^P contains viscous terms, it may be considered that the compaction threshold is rate dependent.

The yield locus is given by $f\,(\overline{SP})=K(Sm,I^P,\Lambda,T,)$ with $Sm=(Tr\,S^R)/3$ and the plastic flow is governed by a flow potential $g(S^{vp})$, so :

$$\mathbf{N}^{vp} = \left.\frac{\partial g}{\partial \mathbf{S}^{vp}}:\right]\frac{\partial g}{\partial \mathbf{S}^{vp}}\quad;\quad \dot{E}^p = \theta\,\mathbf{N}^{vp}\;(2)$$

with $S^{vp} = S^p + S^{v.}$

Yielding does not preserve volume: the dilatancy damage effect is accounted for in (2). Pressure prevents excessive dilatancy, this effect is taken into account in the form of g by:

$$g(\mathbf{S}^{vp}) = \frac{\overline{S}^2}{\hat{S}^2}\left\{1.-\exp\!\left(-\left(\frac{S^p}{S}\right)^2\right)\right\} + \overline{\gamma}^2\,\overline{S}^p{:}\overline{S}^p$$

with $\overline{\alpha}, \hat{S}$ constants and $\overline{\gamma}$ being a function of $.\Lambda$ The respective equipotential surfaces undergo a transition between an ellipsoïdal form for low pressure and a Von Mises cylinder for high pressure.

NUMERICAL INTEGRATION
Frame work.

The model has been implemented in the explicite finite element code OURANOS. The space is discretised with quadrangular elements (4 nodes in 2D, 3 nodes in 3D). The space function bases is under integrated. The spurious deformations are dealt with an hourglass algorithm. The equation of motion is discretised with a Wilkins second order scheme.

NUMERICAL IMPLEMENTATION

To solve the kinematic part of the model, it is necessary to work in the spectral representation for both lagrangian and intermediate configurations. A detail of this procedure can be found in [5]. From the computational standpoint, the problem consists in formulating numerical integration procedure for updating the known state variables:
$\left\{\mathbf{F}_n^e,\mathbf{F}_n^p,\mathbf{S}_n^p,\mathbf{S}_n^v,\Gamma_n,\Lambda_n,T_n\right\}$ at the configuration n, to obtain that at the configuration n+1. The numerical procedure can be described as follows:

Kinematic: Box 1

- Update deformation gradient
  $\mathbf{F}_{n+1} = \mathbf{F}_u\,\mathbf{F}_n$; $\mathbf{F}_u = 1.+\mathrm{Grad}u\delta t$.
- Eigenspace of the right Cauchy tensor
  $\mathbf{C}_{n+1} = {}^T\mathbf{F}_{n+1}\mathbf{F}_{n+1}$; $N_a\,a=1,3$
- Evaluations of the transport matrix Q from N_a to e_a spaces and of the right pure deformation tensor
  U^{-1}
- Evaluation of the rotation R
  $R = F\,U^{-1}$
- Evaluation of the deformation tensor
  $\mathbf{E}_{n+1} = Q_{n+1}\,\widetilde{\mathbf{E}}_{n+1}\,{}^T Q_{n+1}$

Elasticity :   Box 2

- elastic precursor deformation
  $\mathbf{H}_{n+1}^e = \mathbf{E}_{n+1} - \mathbf{E}_n^p$
- evaluation of the invariants and elastic moduli
  $IE = \mathbf{H}_{ii}^e$; $IIE = \mathbf{H}_{ij}^e\mathbf{H}_{ji}^e$
  G_i, K_i , $i=1,2$
- evaluation of the elastic stress
  $\mathbf{S}_{n+1}^R$

Viscosity : Box 3

- evaluation of rate deformation of the right Cauchy-Green tensor
  $\dot{\mathbf{C}}_{n+1} = {}^T\dot{\mathbf{F}}_{n+1}\mathbf{F}_{n+1} + {}^T\mathbf{F}_{n+1}\dot{\mathbf{F}}_{n+1}$
- → projection in the $\overline{N}$ basis

→ evaluation of $\dot{\boldsymbol{\varepsilon}}_{n+1}$ in the $\vec{N}$ basis

→ back to the $\vec{e}$ basis

•evaluation of the elastic rate deformation

$$\dot{\mathbf{H}}_{n+1}^{e} = \dot{\boldsymbol{\varepsilon}}_{n+1} - \dot{\boldsymbol{\varepsilon}}_{n}^{p}$$

•evaluation of the viscous deformation

$$\Gamma_{n+1} = \Gamma_{n}\left(1. - \frac{dt}{\tau}\right) + \overline{\mathbf{H}}_{n+1}^{e}$$

•evaluation of the viscous stress

$$\mathbf{S}_{n+1}^{v}$$

Compaction : Box 4

•check for compaction

If $S^{p} - Y \leq 0. \rightarrow$ box 5

else

•evaluation of internal variables

IP, Γ ..with a Newton or Runge Kutta algorithm

Plasticity : Box 5

•Check for plastification

If $\overline{S}^{p} - F(\overline{S}^{p}) \leq 0. \rightarrow$ box 6

else

•evaluation of the normal to the potential surface

$$\mathbf{N}^{vp}$$

•return mapping algorithm

•evaluation of $\mathbf{\varepsilon}_{n+1}^{p}, \Lambda_{n+1}, I^{P}, \Gamma$

Updating cycle : Box 6

•update configuration

• $\sigma = \mathbf{R}\,\mathbf{S}\,{}^{t}\mathbf{R}\,/\det \mathbf{F}$

•evaluation of the dissipation

• $\rho\,c_{v}\,\dot{T} = \mathbf{S}^{vp}:\dot{\mathbf{\varepsilon}}^{P}$

COMMENTS

A lot of numerical difficulties appeared to accompagny this kind of behaviou, in particular in the iterative algorithms used in the return mapping treatment. Newton and Rugge Kutta (fourth order) have been used and each of them showed instable behaviour for some loading. In each case the large deformations are not incremented, but the strong non-linearities of the cofficients incoming in the model, particulary in the plastic part. In the calculation presented later, flags were positioned to

be sure that the convergence criteria were satisfied all the time.

NUMERICAL RESULTS

The configuration consists of the penetration of a bullet of velocity 1000 m/s in an explosive target confined by an aluminum plate at its front. The behaviour of the bullet is elastic-plastic. The behaviour of the explosive is described by the model presented above. The figures 1, 2 and 3 present the isopressure, isodamage and isotemperature curves at 54 µs after impact.

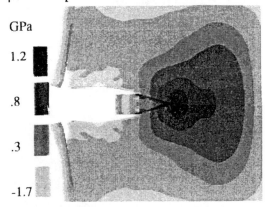

GPa

1.2

.8

.3

-1.7

FIGURE 1. Isopressure at time 54 µs.

It may be considered that at this time the penetration is quasi stationary. The range of values of pressure is [-1.7x10^{8}, 7x10^{8} Pa] and are expected for this kind of loading. The internal variable I^{P} measures the compaction, the decohesion of the grains in the medium. For the reason it has the meaning of a damage variable. During this time of penetration, the compaction is close to 5% in the zone between crater lips and the free surface, whereas the strongest "decohesion" is close to the lip. This decohesion is physically acceptable because once the sample is recovered, it is found that the crater can be filled with small particles and dust. Likewise, the temperature is localized on the crater lips. The reason is that the shear deformation is very important and generates large plastic work. The rise of temperature is 40 degrees for the modeled material. But in fact the grains do not contribute much to the plastic work in shear deformation; they have tendance to rotate, so of coursely, this dissipative work is absorbed by the binder. If we take into account the proportionnality of binder in

181

the sample, its rise temperature will be close to 200°C. For this temperature it is possible that a flame can be initiated.

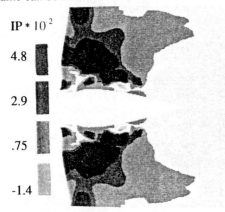

FIGURE 2. Isodamage at time 54 μs.

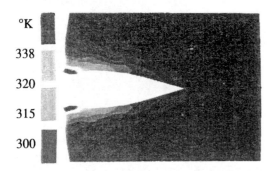

FIGURE 3. Isotemperature at time 54 μs.

DISCUSSION AND CONCLUSION

Although many results have been obtained in 1, 2 and 3D with this model, a lot of problems remain to be solved. From the numerical point of view the iterative algorithms need to be studied deeply for the reason that the variations and the non linearities of the coefficients are strong. Furthermore the mixing between the iterative algorithms and the scheme used to discretize the evolution of the viscous deformation may present in some cases non convergence. It is also possible that loading with a softening process occurs while the convergence algorithm induces instabilities. It may be noticed that this phenomenon is not associated with large distortions. For example, tests performed on a cell which is at the same time strongly compressed and sheared present no numerical problems and remains physically coherent . From the mechanical point of view, important hypotheses have been defined concerning the decomposition of the rotation tensor. These hypotheses are standard and come from the lack of experiments. A mesoscopic approach followed by an homogeneization work will allow sense to be made of the macroscopic hypothesis, or require changes to them.

ACKNOWLEDGEMENTS
This work has been performed with the financial support of the French Ministry of Defence, which is gratefully acknowledged.

REFERENCES
1. H.Trumel, A.Dragon,A.Fanget: a finite strain model for an explosive simulant. Journal de physique III, vol 4, sept. 1994
2. H.Trumel, A.Dragon: a hyperelastic non linear model for a nearly incompressive particulate composite. Int. J. Engng Sci Vol 7, 1994
3. H.Trumel, A.Fanget, A.Dragon: A finite strain elastic-plastic model for the quasi-static behaviour of particulate composites. Int. J. Engng Sci Vol 34, 1996
4. E.H.Lee, J.Appl.Mech 36, 1 1969
5. D.Peric, D.Owen,M.Honnor: a model for finite strain elasto plasticity based on logarithmic strains: computational issues. Comp. Meth. Appl. Mech and Engng 94, 1992

CP429, *Shock Compression of Condensed Matter – 1997*
edited by Schmidt/Dandekar/Forbes
© 1998 The American Institute of Physics 1-56396-738-3/98/$15.00

NUMERICAL STUDY OF SHOCK PROPAGATION IN INHOMOGENEOUS MATERIAL

Alexei D. Kotelnikov and David C. Montgomery

Dept. of Physics & Astronomy, Dartmouth College
Hanover, New Hampshire 03755-3528, U.S.A.

Recently (1), we have developed a kinetic-theory-based numerical method for computing shock-wave behavior in multi-phase materials with macroscopic-sized density inhomogeneities, and have applied it to problems of shocks propagating through an array of "bubbles" of higher-density material embedded in a lower-density background medium (2). Here, we remark upon three additional applications of this technique: (i) shocks are propagated into a random bubble array, to explore possible differences from propagation into the the more computationally convenient "lattice" array used in our earlier work; (ii) following Olim et al (3), we compute reflected shocks which rebound from and are amplified by rigid material walls in the presence of "foam"; and (iii) we launch a high Mach number shock into the turbulent debris left behind by the passage of a previous shock into a quiescent two-phase medium. Applications of the method are far from exhausted.

INTRODUCTION

In a variety of situations, we would like to compute shock passage through strongly inhomogeneous fluids. We have recently developed a kinetic theory based numerical method in a lengthier paper to which reference should be made for background and details (1). In a later paper (2), we have applied the method to strong planar shock propagation into a half-space of compressible fluid in which was embedded a doubly-periodic lattice of two-dimensional parallel "bubbles" of heavier material. The emphasis in (2) was on characterizing the strong and highly-compressible turbulence left behind by the passage of the shock, the mixing of the shocked materials, the reflected shocks, the vorticity, the pressure field, etc. [Refs. (1) and (2) are to be referred to for some terminology.] Here, we wish to remark three other situations to which the code has been applied, all of which involve shocks in two-phase materials. First, we propagate a shock into a half space in which an array of bubbles is embedded that is not periodic in the direction of shock propagation, with an eye to determining the effects of the regularity of the array in the previous (2) computations. Second, we study the effects of finite cross-species relaxation time on the structure of a shock reflected from a wall in the presence of a two-species mixture, following an experimentally motivated paper by Olim et al (3). Finally, we elaborate the work of reference (2) by studying the behavior of a second strong shock which is launched into a non-uniform material after the first, and therefore propagates into the turbulent debris left by the first.

SUMMARY OF PREVIOUS WORK

We have computed a passage of strong shock wave (Mach number $M = 7.3$) through a two-dimensional doubly-periodic array of bubbles of density 4 times background density. Fig. 1 is a contour plot of total mass density at $t = 40ps$, after the shock has passed through the first several bubbles (at $t = 0$ the shock located at 40 $\overset{\circ}{A}$ apart from the first bubble); the broken line at the right boundary of the irregular shaded area is the shock front which becomes irregular but retains its identity and moves to the positive x-direction. The periodicity intervals in the undisturbed region in x and y are 600 $\overset{\circ}{A}$ and 500 $\overset{\circ}{A}$, respectively. The heavy bubbles in the undisturbed region are of mass density $1.0 g/cm^3$ and the background material is of $0.25 g/cm^3$. Both materials are treated as ideal gases, with initial temperatures of $0.025 eV$. Shock transit times are short compared to diffusion time scales for the materials into each other. The various fields behind the shock (density, vorticity, pressure, temperature, kinetic energy, mass fraction) soon become so irregular that spatially averaged values become more useful for characterizing the turbulence left behind than visual graphics do, and the major part of Ref. (2) is the assembly of such spatial averages and their physical interpretation.

A RANDOM ARRAY OF BUBBLES

Fig. 2 shows a passage of the same shock as in the previous example through an array of bubbles of the same higher-density material embedded in the same lower-density material, however, now the array is random in x rather than periodic, and has a longer periodicity length in y (1000 $\overset{\circ}{A}$, now). The mean mass density and bubble number density in the upstream undisturbed region are the same as in Fig. 1. The intent is to assess the effects of the regularity of the array in Fig. 1. In some applications, the bubbles will represent a tangle of fibers with no discernible periodicity, and one may wonder if interference or diffraction effects, for example, might be controlling some of the variables in the purely periodic case. This would be unfortunate, since the periodic array is computationally much more economical, and one would like to be able to draw conclusions on the basis of it. Space limitations preclude a thorough catalogue here of similarities and differences in the two cases. In summary, most of the features and their statistical averages were rather similar: the gross features of the shock propagation such as speed, planarity,

and rate of vorticity generation were the same. The most notable difference showed up in the reflected shocks which were launched backwards, propagating to the left (in the local rest frame of the fluid), from the bubbles upon their encounter with the original shock. These were notably weaker and less numerous in the random bubble case, leading to lower values of the rms divergence of the velocity field. Figs. 3 and 4 show spatially-averaged values of the divergence of the velocity field (a good index of the degree of compressibility of the turbulence) with the periodic array in the undisturbed region shown in Fig. 1 and the random array in the undisturbed region shown in Fig. 2, at a time $t = 100 ps$. The spikes between $x = 4 \cdot 10^3 \overset{\circ}{A}$, which is the trailing edge of the shocked heavy material, and $x = 1.5 \cdot 10^3 \overset{\circ}{A}$, which is the leftmost boundary of the reflected shocks, are noticeably less pronounced for the random-array case.

In summary, we can, within limits, assume that the turbulence left by shock passage through a random array of density inhomogeneities is similar to that left behind by a periodic array. One principal difference lies in the level of reflected-shock turbulence to the left of the region occupied by the heavy material, where some interference and cancelations seem to occur in the aperiodic case that do not occur in the periodic one. Turbulence randomizes itself rather quickly upon generation, and interference and reinforcement effects are probably less to be expected than might occur in the case of linear wave propagation.

WALL-REFLECTED SHOCKS IN FOAM

Olim et al (3) and van Dongen et al (4) have studied shocks reflecting from a rigid material wall in a liquid-foam mixture, motivated by experiments on the effects of foam on explosion-generated shocks. A weak shock in a shock tube may be propagated through air and across an interface into an air-foam mixture, at the end of which stands a rigid wall. Pressure sensors, two in the air, one at the interface and one near the wall, record time histories of pressure as the shock propagates into the quiescent region and

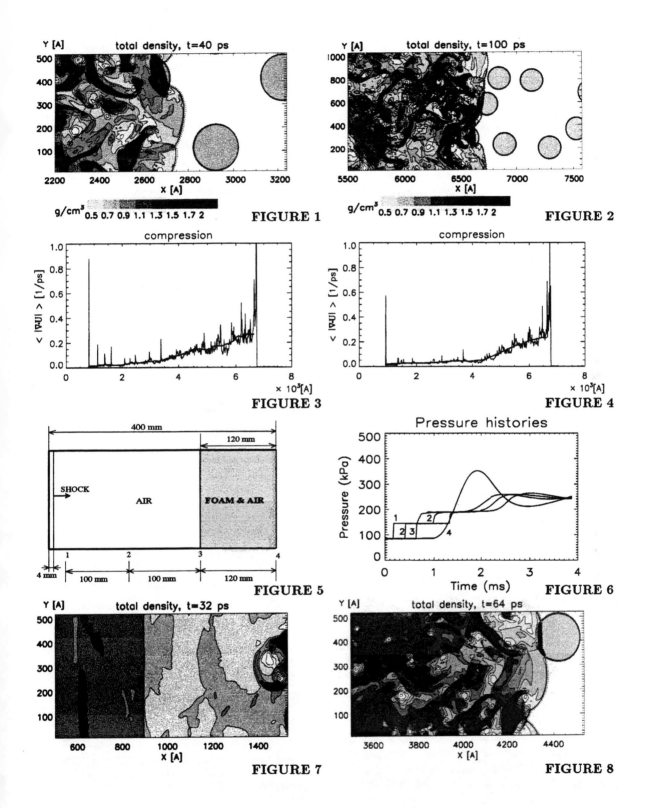

FIGURE 1

FIGURE 2

FIGURE 3

FIGURE 4

FIGURE 5

FIGURE 6

FIGURE 7

FIGURE 8

185

then reflects from the wall. We may compute the passage of the shock two-dimensionally using our code, and attempt to duplicate the pressure histories recorded by the four sensors by plotting the pressure versus time at four selected locations. In this case, we approximate the air-foam region as a uniform mixture of ideal gases, but we choose the adjustable cross-species relaxation time (1) sufficiently long that the two species lag behind each other in their dynamical response to the shock passage; a relaxation time between the air and foam broadens the pressure signal noticeably. In most applications, the code has been run with molecular relaxation times that are so short that we were essentially solving the Euler equations with discontinuities but without finite transport coefficients, away from the discontinuities.

Fig. 5 shows schematically the shock tube arrangement and dimensions into which the shock (Mach number $M = 1.25$) was launched, with the locations of the pressure sensors indicated. The mass density of the air was chosen as $1.29 kg/m^3$, and that of the air-foam mixture, $14.8 kg/m^3$. Representing these as ideal gases requires a choice of mass ratio of the two species as 210.4. The ambient air pressure was taken to be $85 kPa$, and the ratio of specific heats was 5/3 for both species. The computational domain was a modest 400 x 100 cells. Fig. 6 shows the four pressure histories, experimental and computational, in units of kPa vs. ms. The agreement with (3) is quite satisfactory.

SUCCESSIVE SHOCKS LAUNCHED INTO FOAM

The third application we wish to note is the effect of turbulence generated by a previous shock on the passage of a second, stronger one, when the first has been launched into a periodic array of bubbles. Many laser-fusion scenarios (e.g., (5),(6)) propose the successive launch of multiple shock waves in an implosion in order to minimize the generation of entropy during the compression. If the first shock passes through an impurity fiber-laden shell, and generates turbulence at the level indicated by our previous computations (2), it is natural to ask how that will affect the propagation of subsequent shocks. There-

fore, we have launched a second shock of pressure $200 kBar$, following on a first shock of pressure $1 kBar$, into a quiescent two-phase material consisting of background and bubbles, arranged periodically as in the introductory example (2). In Fig. 7, we see the second planar shock at time $t = 32 ps$ after its launch; it has not yet encountered the compressed heavy material that the first shock has distorted and shed vorticity from. A great deal happens as the shock moves through the turbulent region, and its identity can be lost sight of completely in the contour plots. Nevertheless, it gains on and eventually captures the first shock, and Fig. 8 shows the composite shock emerging into the as yet unshocked bubble array. A spatially averaged pressure plot (not shown) reveals considerable widening of the shock front that suggests considerable thickness in the case of many shocks being launched in succession.

SUMMARY

Though there is far to go in being able to simulate accurately the passage of shocks through composite materials, there is now substantial reason to believe that kinetic theory based numerical codes are a useful research tool for this mechanical problem. The turbulence launched in fluids by strong density inhomogeneities is, for strong shocks, quite intense and likely to interfere with scenarios that require high degrees of spatial uniformity.

ACKNOWLEDGMENT

This work was supported in part by the U.S. Naval Research Laboratory under grant N000014-96-1G005.

REFERENCES

1. Kotelnikov, A.D., and Montgomery, D.C., *J. Comp. Phys.* **134**, 364 (1997).
2. Kotelnikov, A.D. and Montgomery, D.C., "Turbulent shocks in composite materials," submitted to *Phys. Fluids* (June, 1997).
3. Olim, M., van Dongen, M.E.H., Kitamura, T., and Takayama, K., *Int. J. Multiphase Flow* **20**, 557 (1994).
4. van Dongen, M.E.H., Smeulders, D.J.M., Kitamura, T., and Takayama, K., *Acustica* **81**, 63 (1995).
5. Rubenchik, A. and Witkowski, S., editors, *THE PHYSICS OF LASER PLASMA* (Amsterdam, North-Holland, 1991).
6. Lindl, J., *Phys. Plasmas* **2**, 3933 (1995).

CP429, *Shock Compression of Condensed Matter – 1997*
edited by Schmidt/Dandekar/Forbes
© 1998 The American Institute of Physics 1-56396-738-3/98/$15.00

ACCURACY OF DIFFERENTIAL SENSITIVITY FOR ONE-DIMENSIONAL SHOCK PROBLEMS

R. J. Henninger, P. J. Maudlin, and M. L. Rightley

Los Alamos National Laboratory, Los Alamos, New Mexico 87545

The technique called Differential Sensitivity has been applied to the system of Eulerian continuum mechanics equations solved by a hydrocode. Differential Sensitivity uses forward and adjoint techniques to obtain output response sensitivity to input parameters. Previous papers have described application of the technique to two-dimensional, multi-component problems. Inaccuracies in the adjoint solutions have prompted us to examine our numerical techniques in more detail. Here we examine one-dimensional, one material shock problems. Solution accuracy is assessed by comparison to sensitivities obtained by automatic differentiation and a code-based adjoint differentiation technique.

INTRODUCTION

Expert use of a hydrocode for design or design-optimization purposes requires information about how some result (or response, R) will change when some code parameter (α) is changed. One can always start by changing parameters one at a time and form a finite difference sensitivity. This method, which we call the Direct Method, requires $n+1$ computer runs to determine sensitivities to n problem parameters.

A sensitivity technique (1-3) used successfully in the early eighties (4-6), which is called Differential Sensitivity Theory (DST) when addressing the differential equations and Differential Sensitivity Analysis (DSA) when addressing the finite-difference equations, is applied in this effort to a system of time-dependent continuum mechanics equations. Differential sensitivity can be used either in the forward or adjoint mode to determine exact sensitivity derivatives, i.e., if a calculational result or response of interest is R_j and α_i is an input parameter, then (dropping the subscripts) $\partial R / \partial \alpha$ is the sensitivity.

DSA and DST both address the differentiated physical equations (i.e., physical equations differentiated with respect to a parameter α) that are subsequently solved numerically by some code.

DSA solves the difference equations (a set of algebraic equations), whereas DST numerically solves the differential equations (a set of partial differential equations, PDEs) and, in principle, is not tied to any particular choice of hydrocode or numerical scheme. In the adjoint mode, DST and DSA obtain the sensitivities to all of the problem parameters in only two computer runs.

Automatic Differentiation (AD) and the Adjoint Differentiation In Code Technique (ADICT) (7), other methods applied to this problem, address the numerical code itself. AD programs take as input the original code and analyze it line by line, providing a code that can produce the needed derivatives. AD programs utilized here include GRESS (GRadient Enhanced Software System) (8) and ADIFOR (Automatic DIfferentiation of FORtran) (9). The ADICT method is similar in concept to AD when utilized in the adjoint mode, but the implementation of the idea varies. Whereas AD operates on a line by line basis, the ADICT method advocates grouping together code in subroutines whenever possible, and each subroutine is analyzed individually. This results in fewer global gradient variables to store and is often more efficient than AD-generated code. Individual derivative expressions produced by AD-generated code, however, are often useful in the implementation of the ADICT method. Of the AD programs, ADIFOR can only operate in the forward mode, while GRESS can operate in both forward and adjoint modes. Here, ADICT is applied only in the adjoint mode.

In previous works (10-12) we have described the derivation and solution of DST adjoint sensitivity equations for the purpose of computing sensitivities for high-rate, 2D, multi-component, high-deformation problems that contain material strength. Here, we begin with a description of the various sensitivity methods available. Three methods: Forward DSA (FDSA), Adjoint DSA (ADSA), and Adjoint DST (ADST) will be briefly described, and then applied to a simple 1D metal flow problem to investigate the equivalence and numerical accuracy of the methods.

PHYSICAL EQUATIONS

The physical system of equations is the set of 1D Eulerian conservation equations for mass, momentum, and internal energy and an equation-of-state (EOS) augmented with an expression for artificial dissipation. As an example, we follow the development of the momentum equation for the various tecniques:

$$\rho\left[\frac{\partial u_z}{\partial t} + u_z \frac{\partial u_z}{\partial z}\right] = -\frac{\partial P'}{\partial z} \qquad (1)$$

In this equation the dependent variables are density ρ, velocity u_z, and the pressure P' (the sum of an EOS pressure (a function of ρ and internal energy i) and a scalar artificial dissipation).

The finite difference form of Eq. 1 used in this effort assumes an explicit solution, staggered spatial grid, and upwind/donor differences for the convective derivative terms:

$$\rho_{j+1/2}^n \, CD(u_z)_{j+1/2}^{n+1} = -GRAD(P')_{j+1/2}^n \quad (2a)$$

where the finite-difference operators CD, GRAD and DIV are defined using the arbitrary scalar s, vector v and scalar/vector q with the indices n and j designating the time and spatial grids, respectively:

$$GRAD(s)_{j+1/2}^n \equiv \frac{s_{j+1}^n - s_j^n}{\Delta z} \qquad (2b)$$

$$CD(q)_j^{n+1} \equiv \frac{q_j^{n+1} - q_j^n}{\Delta t} + u_{z,j}^n \, DONOR(q)_j^n \quad (2c)$$

$$DIV(v)_j^n \equiv \frac{v_{j+1/2}^n - v_{j-1/2}^n}{\Delta z} \qquad (2d)$$

$$DONOR(q)_j^n = \frac{q_j^n - q_{j-1}^n}{\Delta z}, \quad \text{for } u_z > 0 \quad (2e)$$

DSA EQUATIONS

The finite difference form of the physical equation given by Eq. 2a is next differentiated with respect to α giving the DSA forward equation set in terms of differentiated dependent variables,

$$\bar{y} = \left(\left(\frac{\partial \rho}{\partial \alpha}\right)_j^n, \left(\frac{\partial u_z}{\partial \alpha}\right)_{j+\frac{1}{2}}^n, \left(\frac{\partial i}{\partial \alpha}\right)_j^n, \left(\frac{\partial P'}{\partial \alpha}\right)_j^n\right)^T$$

$$\equiv \left(\Psi_j^n, \Phi_{z,j+\frac{1}{2}}^n, I_j^n, \Pi_j^n\right)^T \qquad (3a)$$

to obtain:

$$\Psi_{j+\frac{1}{2}}^n \, CD(u_z)_{j+\frac{1}{2}}^{n+1} + \rho_{j+\frac{1}{2}}^n \times$$

$$\left[CD(\Phi_z)_{j+\frac{1}{2}}^{n+1} + \Phi_{z,j+\frac{1}{2}}^n \frac{u_{z,j+\frac{1}{2}}^n - u_{z,j-\frac{1}{2}}^n}{\Delta z}\right] \quad (3b)$$

$$+GRAD(\Pi)_{j+\frac{1}{2}}^n = 0$$

The system of linear algebraic equations represented by Eq. 3b can be rearranged into matrix form as

$$\bar{y}^{n+1} = \underline{A}\,\bar{y}^n + \bar{s} \qquad (4)$$

where $\bar{y}$ is the dependent variable vector given by Eq. 3a. With this matrix form, the DSA adjoint equation set can be obtained by transposing in space and time to give

$$\bar{y}^{*n} = \underline{A}^T\,\bar{y}^{*n+1} + \bar{s}^* \qquad (5)$$

The forward equations given by Eq. 4 can be easily solved for the sensitivities given by Eq. 3a with an appropriate source $\bar{s}$ specified for a single parameter of interest. Alternatively, the adjoint or transposed equations given by Eq. 5 can be solved for an adjoint solution with an appropriate source $\bar{s}^*$ specified for a single response of interest; this adjoint solution is then combined with the physical solution in various integrals to obtain the sensitivities to all the parameters (10-12).

ADJOINT DST EQUATIONS

Deriving the DST adjoint for Eq. 1 was presented in Refs. 10 and 11 giving the following equation in 3D:

$$
\begin{aligned}
&-\rho\nabla\Psi^* - \rho\left[\frac{\partial\bar{\Phi}^*}{\partial t} + (\bar{u}\bullet\nabla)\bar{\Phi}^*\right] + \rho(\nabla\bar{u})\bullet\bar{\Phi}^* \\
&-\nabla\left(\frac{\partial P'}{\partial i}I^*\right) + \rho\nabla i\, I^* + \nabla\left(\frac{\partial P'}{\partial(\nabla\bullet\bar{u})}\Pi^*\right) = s_\Phi^*
\end{aligned} \quad (6)
$$

where the gradient, divergence and dyad operators appearing in Eqs. 6 are taken to be 1D for the purposes of this effort. The dependent variables appearing in Eq. 6 (with SI units given in parentheses, [R] indicating units of the response) are the adjoint density Ψ^* ([R]/kg), the adjoint velocity Φ_z^* ([R]/N-s), the adjoint energy I^* ([R]/J) and the adjoint pressure Π^* ([R]/Pa-m^3-s). The definition of the adjoint source $\bar{s}^*$ depends on the desired response, and is discussed in more detail in Refs. 10 and 11.

ADJOINT DST DIFFERENCE OPERATORS

In order to numerically solve the DST adjoint equations an appropriate choice for finite difference approximations for the differential operators (divergence, gradient, etc.) must be selected. One choice might be to prescribe ad hoc difference operators, taking care to correctly propagate the adjoint boundary conditions and initial conditions into the solution domain. Previous efforts (10-12) utilized this approach, realizing reasonable but not highly accurate sensitivity results using a donor-cell, staggered-mesh scheme much like the scheme used for solution of the physical equations. A better approach is to use the method-of-support or compatible operators advocated by Shashkov (13)

that utilizes inner product properties for continuous functions. This methodology can be used to derive difference operators for the DST equations that are compatible to the difference operators used in the solution of the physical equation set. Although discussed in detail in Ref. 13 an example is given here that finds the difference operator for $\rho\nabla\Psi^*$ appearing in Eq. 6. This example uses the integral identity (i.e., integration-by-parts) for the gradient operator that when specialized to our purpose of finding the difference operator for away from domain boundaries gives the result

$$
\begin{aligned}
\int_V \rho\,\Psi^*\left(\nabla\bullet\bar{\Phi}\right)dV = &-\int_V \rho\,\bar{\Phi}\bullet\nabla\Psi^*\,dV \\
&-\int_V \Psi^*\,\bar{\Phi}\bullet\nabla\rho\,dV
\end{aligned} \quad (7)
$$

Now identifying as prime operators (13) the original DIV and DONOR difference operators defined by Eqs. 2d and 2e for the physical equations, and substituting these into Eq. 7 gives the finite difference form:

$$
\begin{aligned}
\sum_j \rho_j\,\Psi_j^*\,\mathrm{DIV}(\Phi_z)_j\,\Delta z = &-\sum_j \Phi_{z,j}\,\rho\nabla\Psi^*\Delta z \\
&-\sum_j \Psi_j^*\,\Phi_{z,j}\,\mathrm{DONOR}(\rho)_j\,\Delta z
\end{aligned} \quad (8)
$$

Expanding the summations, substituting in the DIV and DONOR definitions, and solving for the unknown gradient operator $\rho\nabla\Psi^*$ gives the compatible finite difference approximation (for $u_z > 0$):

$$
\begin{aligned}
\left(\rho\nabla\Psi^*\right)_{j+\frac{1}{2}} = &\frac{\rho_{j+1}\Psi_{j+1}^* - \rho_j\Psi_j^*}{\Delta z} \\
&-\frac{1}{2}\left[\Psi_j^*\frac{\rho_j - \rho_{j-1}}{\Delta z} + \Psi_{j+1}^*\frac{\rho_{j+1} - \rho_j}{\Delta z}\right]
\end{aligned} \quad (9)
$$

Note that the spatial grid point for $\rho\nabla\Psi^*$ is located at the cell edge $j + 1/2$, appropriate for use in the cell-edged adjoint momentum equation. This procedure can be repeated using various integral identities to define all the difference operators in

189

Eq. 6, forcing all these operator definitions to be compatible with the physical equation operators given by Eq. 2b-e. *Comparison of these derived DST compatible operators to the implied spatial operators appearing in the adjointed (transposed) DSA equations represented by Eq. 5 shows the two difference operator sets to be identical, and therefore the two methods should produce identical numerical results.*

METAL FLOW TEST RESULTS AND DISCUSSION

Consider the 1D flow of a metal plate which has an initial velocity of 500 m/s, an initial density of 8000 kg/m^3, and a sound speed of 4000 m/s. The plate is 2 mm in thickness and is divided into ten 0.2 mm cells for the numerical computations. At the initial time the right side of the plate instantaneously decelerates to 400 m/s. This produces a left-going shock that compresses the material to a Hugoniot pressure of 21 GPa. Numerical solution was performed using the boundary and initial conditions given above to a final time of 1.0 µs. Sensitivities for representative problem parameters using the six different sensitivity methods discussed above were then generated, and the results are given in Table 1 for a space/time averaged pressure response. The parameters as listed in the table are initial density, initial velocity, and the EOS sound speed. ADIFOR, ADICT and FDSA all show excellent agreement for 5, 6 or more digits. The direct method sensitivities were only converged to 3 or 4 digits. For our purposes the AD results can be considered exact. The two adjoint techniques, i.e., DSA adjoint and DST adjoint, also all show excellent agreement for 5, 6 or more digits. However comparison of the forward results with the adjoint results reveal agreement to only 3 to 4 digits; *in principle all the methods should agree.*

We speculate that this lack of consistency between the forward and adjoint DSA/DST results becomes more severe for stronger shock problems containing additional complexity in terms of constitutive modeling, multi-material discontinuities and increased dimensionality, thus explaining the inaccuracy observed in our previous work.

Table 1. Sensitivity Comparisons for an Average Pressure Response

Method	Sensitivity		
	Initial Density	Initial Velocity	Sound Speed
Direct	1.594500·10^7	146500.0	154080.0
ADIFOR	1.592504·10^7	146255.7	153885.7
ADICT	1.592504·10^7	146258.3	153885.6
FDSA	1.592504·10^7	146254.3	153888.0
ADSA	1.592504·10^7	145760.2	153870.5
ADST	1.592504·10^7	145766.0	153870.4

REFERENCES

1. E. M. Oblow, *Nucl. Sci. Eng.*, **68**, 322 (1978).
2. D. G. Cacuci, C. F. Weber, E. M. Oblow and J. H. Marable, *Nucl. Sci. Eng.*, **75**, 88 (1980).
3. D. G. Cacuci, P. J. Maudlin and C. V. Parks, *Nucl. Sci. Eng.*, **83**, 112 (1983).
4. P. J. Maudlin, C. V. Parks and C. F. Weber, "Thermal-Hydraulic Differential Sensitivity Theory," ASME paper No. 80-WA/HT-56, Proc. ASME Annual Winter Conference (1980).
5. C. V. Parks and P. J. Maudlin, *Nucl. Technol.*, **54**, 38 (1981).
6. C. V. Parks, "Adjoint-Based Sensitivity Analysis for Reactor Applications," ORNL/CSD/TM-231, Oak Ridge National Laboratory (1986).
7. K. M. Hanson and G. S. Cunningham, "The Bayes Inference Engine," *Maximum Entropy and Bayesian Methods*, edited by K. M. Hanson and R. N. Silver, 125-134, Kluwer Academic, Dordrecht (1996).
8. J. E. Horwedel, E. M. Oblow, B. A. Worley, and F. G. Pin, "GRESS 3.0 Gradient Enhanced Software System," Oak Ridge National Laboratory RSIC Peripheral Shielding Routine Collection Report PSR-231 (1994).
9. C. Bischof, A. Carle, P. Khademi, and A. Mauer, "The ADIFOR 2.0 System for the Automatic Differentiation of Fortran 77 Programs," Argonne National Laboratory Report ANL-MCS-P481-1194 (1995).
10. R. J. Henninger, P. J. Maudlin, and E. N. Harstad, "Differential Sensitivity Theory Applied to the MESA Code," Proceedings of the Joint AIRAPT/APS Meeting on High Pressure Science and Technology, 1781, Colorado Springs, CO (June 28–July 2, 1993).
11. P. J. Maudlin, R. J. Henninger, and E. N. Harstad, "Application of Differential Sensitivity Theory to Continuum Mechanics," Proc. ASME Winter Annual Meeting, 93, New Orleans, Louisiana (November 28-December 3, 1993).
12. R. J. Henninger, P. J. Maudlin, and E. N. Harstad, "Differential Sensitivity Theory Applied to the MESA2D Code for Multi-Material Problems," Proceedings of the APS Meeting on Shock Compression of Condensed Matter, 283, Seattle, WA, (August, 1995).
13. M. Shashkov, *Conservative Finite Difference Methods on General Grids*, CRC Press, Boca Raton (1996).

CP429, *Shock Compression of Condensed Matter – 1997*
edited by Schmidt/Dandekar/Forbes
© 1998 The American Institute of Physics 1-56396-738-3/98/$15.00

AN ANALYTIC SOLUTION TO A DRIVEN INTERFACE PROBLEM

J.E.Hammerberg and J.Pepin

Applied Theoretical and Computational Physics Division
Los Alamos National Laboratory, Los Alamos, New Mexico 87545

The frictional properties of sliding metal interfaces at high velocities are not well known either from an experimental or theoretical point of view. The constitutive properties and macroscopic laws of frictional dynamics at high velocities necessary for materials continuum codes have only a qualitative validity and it is of interest to have analytic problems for sliding interfaces to enable separation of model from numerical effects. We present an exact solution for the space and time dependence of the plastic strain near a sliding interface in a planar semi- infinite geometry. This solution is based on a particular form for the strain rate dependence of the flow stress and results in a hyperbolic telegrapher equation for the plastic strain. The form of the solutions and wave structure are discussed.

INTRODUCTION

The numerical treatment of dynamic thermo-plastic flow requires sophisticated numerical algorithms and sophisticated models of the physics of dynamic material flow. For general problems and general geometries, it is at times difficult to separate algorithmic and numerical effects due to physics models. Test problems play an important role in separating numerical from model dependent effects since they provide closed form solutions with which to compare numerical integrations for the relevant continuum equations. For pure fluids a large number of such test problems exist (1). For bulk dynamic plastic flow there is a small number of exactly soluble test problems (2). However, for sliding material interfaces in situations where plastic dissipation is the major irreversible effect, there are very few test problems. We present here one such test problem which is relevant to a driven material interface and rate dependent plastic flow.

A PLANAR SEMI-INFINITE MODEL PROBLEM

The model problem we consider consists of a material whose shear properties are described by a constant shear modulus G_0. We further assume that the constitutive relation for the flow stress, τ, is given by a linear rate dependent hardening law of the form,

$$\tau = G_0(\hat{\tau}_0 + \beta\dot{\psi}). \qquad (1)$$

Here, $G_0\hat{\tau}_0$ is the yield strength of the material, which we take to be a constant and β is another constant, having units of time, describing a linear rate dependent hardening. $\dot{\psi}$ is the equivalent plastic strain rate. This form is a gross simplification of the plastic properties of a real material but does capture, in an average way, some of the rate dependent properties of real materials where $\tau \propto \dot{\psi}^\delta$ at very high strain rates with $0 < \delta < 1$, typically.

We consider a situation in which this material occupies the half space $z \geq 0$ and consider a tangential force $F_x(t)$ which is uniform in the z=0 plane forming the interface. We further assume that the density, ρ, is a constant ρ_0. The latter assumption limits us to shear excitations only. For this symmetric situation, all field quantities depend on (z,t). The equations of mass, momentum and energy conservation are

$$\frac{D}{Dt}\rho + \rho \nabla \cdot v = 0$$

$$\rho \frac{D}{Dt} v_i = \frac{\partial \sigma_{ij}}{\partial x_j}, \qquad (2)$$

$$\rho \frac{D}{Dt} e = \sigma_{ij} \frac{\partial v_i}{\partial x_j} ,$$

where v_i is the material velocity, e is the specific internal energy, σ_{ij} is the Cauchy stress tensor, and $\frac{D}{Dt}$ is the usual material derivative.

We consider the case of purely transverse displacement, i.e. $v_z = 0$, $v_y = 0$. Since ρ is constant, $\frac{\partial v_z}{\partial z} = \frac{\partial v_x}{\partial x} = 0$. Furthermore, steady motion in the x direction and spatial homogeneity in the x and y directions imply $\frac{\partial \sigma_{xx}}{\partial x} = \frac{\partial \sigma_{xz}}{\partial x} = \frac{\partial v_z}{\partial x} = 0$. We look for a solution with $v_z = 0$ and, consistent with this, $\frac{\partial \sigma_{zz}}{\partial z} = 0$. Then,

$$\rho_0 \frac{\partial v_x}{\partial t} = \frac{\partial \sigma_{xz}}{\partial z}, \qquad (3)$$

$$\rho_0 \frac{\partial e}{\partial t} = \sigma_{xz} \frac{\partial v_x}{\partial z}, \qquad (4)$$

Because of the homogeneity and transversality, the components of the strain rate tensor satisfy $\dot{\varepsilon}_{xx} = \dot{\varepsilon}_{zz} = \dot{\varepsilon}_{yy} = \dot{\varepsilon}_{xy} = \dot{\varepsilon}_{yz} = 0$ and $\dot{\varepsilon}_{xz} = \frac{1}{2} \frac{\partial v_x}{\partial z}$. We write $\dot{\varepsilon}_{ij}$ in terms of its elastic and plastic components, $\dot{\varepsilon}_{ij} = \dot{\varepsilon}_{ij}^{(el)} + \dot{\varepsilon}_{ij}^{(pl)}$, and assume that plastic flow is described by the Prandtl-Reuss relation,

$$\dot{\varepsilon}_{ij}^{(pl)} = \frac{3}{2} \frac{S_{ij}}{\tau} \psi , \qquad (5)$$

where S_{ij} is the deviatoric stress tensor. Upon differentiating equation (3) we have,

$$\frac{\partial^2 \varepsilon_{xz}}{\partial t^2} = \frac{1}{2\rho_0} \frac{\partial^2 \sigma_x}{\partial z^2}, \qquad (6)$$

and assuming isotropic elasticity with shear modulus G, we have

$$\frac{\partial}{\partial t} \left[\frac{1}{2G} \frac{\partial S_{xz}}{\partial t} + \frac{3}{2} \frac{S_{xz}}{\tau} \frac{\partial \psi}{\partial t} \right] = \frac{1}{2\rho_0} \frac{\partial^2 S_{xz}}{\partial z^2} . \qquad (7)$$

We consider an isotropic von Mises flow surface, $\frac{3}{2} S_{ij} S_{ij} = \tau^2$. For an initially unstrained sample (assuming ρ constant) we have $S_{xz}^2 = \frac{1}{3} \tau^2$ and S_{xz} satisfies

$$\frac{\partial}{\partial t} \left[\frac{1}{G} \frac{\partial S_{xz}}{\partial t} + \sqrt{3} \frac{\partial \psi}{\partial t} \operatorname{sgn}(S_{xz}) \right] = \frac{1}{\rho_0} \frac{\partial^2 S_{xz}}{\partial z^2} , \qquad (8)$$

or in terms of the flow stress, τ,

$$\frac{\partial}{\partial t} \left[\frac{1}{G} \frac{\partial \tau}{\partial t} + 3 \frac{\partial \psi}{\partial t} \right] = \frac{1}{\rho_0} \frac{\partial^2 \tau}{\partial z^2} . \qquad (9)$$

In deriving equation (9) we have made no assumption about G. If we now take G to be the constant, G_0, and assume the constitutive relationship given in equation (1), then equation (9) reduces to the telegrapher's equation for $\psi(z,t)$:

$$\frac{\partial^2 \psi}{\partial t^2} + \frac{3}{\beta} \frac{\partial \psi}{\partial t} = c^2 \frac{\partial^2 \psi}{\partial z^2} , \qquad (10)$$

with $c^2 \equiv \frac{G_0}{\rho_0}$.

192

SOLUTIONS TO THE MODEL PROBLEM

The solution to the boundary value problem for equation (10) may be obtained in terms of the Bessel function I_1 using the method of Green's functions (3). The general solution may be written as

$$\psi(z, t) = e^{-\lambda z}\psi\left(0, t - \frac{z}{c}\right)$$

$$+ \lambda z \int_0^{t - \frac{z}{c}} e^{-\lambda c(t - t')}\psi(0, t')\frac{I_1[\lambda\sqrt{c^2(t - t')^2 - z^2}]}{\sqrt{c^2(t - t')^2 - z^2}}, \quad (11)$$

where $\lambda = \frac{3}{2}\frac{1}{\beta c}$. For a given $\psi(0,t)$, equation (11) is a closed form solution. One particularly interesting case is that of a constant external force. In that case, $F_x = \frac{A\tau}{\sqrt{3}}$ where A is the surface area. Then $\psi(0, t)$ is a constant, $\psi(0, t)) = \beta^{-1}\hat{\tau}_0(\alpha - 1)$ with $\alpha = \frac{\sqrt{3}F_x}{AG_0\hat{\tau}_0}$. Since we are assuming perfectly plastic behavior, $\alpha > 1$. Figure 1 shows the solutions for ψ when $\alpha=10$ and $\hat{\tau}_0 =0.00021$, $\beta=0.0946$ μs, $\rho_0=8.84$ g cm^{-3}, $G_0=0.9397$ Mbar (parameters roughly corresponding to copper) at times t=1, 2, 4, and 8μs.

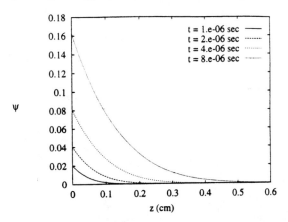

FIGURE 1. Solution to equation (10) for constant force boundary condition. Times 1, 2, 4, and 8 μs.

The broad damped excitation at the interface is diffu-

sive in nature and the width at half-maximum is proportional to the square root of time. Because of the causal nature of equation (10), ψ vanishes beyond z=ct.

The above solutions are continuous solutions to the initial value problem for equation (10). The full set of conservation laws, equations (2), may also admit discontinuous solutions which in the present case correspond to a steady plastic wave. If such a solution is to exist, a nontrivial solution to the jump conditions,

$$s\Delta[u] = \Delta[F], \quad (12)$$

is necessary. In equation (12) we have written the conservation laws in standard form (4),

$$\frac{\partial u}{\partial t} + \frac{\partial F}{\partial z} = 0,$$

$$u = \left[\varepsilon_{xz}, \rho_0 v_x, \rho_0\left(\frac{1}{2}v_x^2 + e\right)\right], \quad (13)$$

$$F = \left[-\frac{1}{2}v_x, -S_{xz}, -S_{xz}v_x\right],$$

where s is the wave speed. It is easy to show that $s = -\frac{S_{xz}}{\rho_0\Delta[v_x]}$ and in order to see whether such a steady plastic wave exists we look for a point (z*,t*) such that the solution given by equation (11) connects continuously to a constant plastic state. Using equation (12) and eliminating s in favor of $\Delta[v_x]$, it is possible to show that this compatibility condition results in the expression

$$\frac{1}{3}\hat{\tau} + \hat{\tau}\psi = \left(\frac{1}{c}\Delta[v_x]\right)^2. \quad (14)$$

where $\hat{\tau} \equiv \frac{\tau}{G}$. The right hand side of the above equation can be expressed in terms of the spatial derivative of ψ and hence we look for the earliest time t* for which a z* exists such that equation (14) is satisfied. This is, in effect, a solution of a Riemann problem for the restricted set of conservation laws. We find, typically, that there is a minimum time, t_{min}, below which there is no solution. For $t > t_{min}$, solutions exist and Fig. 2 shows the progression of solutions using the same parameters as for Fig. 1. The magnitude of ψ is very small relative to the main peak and the wave speed is very nearly the transverse wave speed, c.

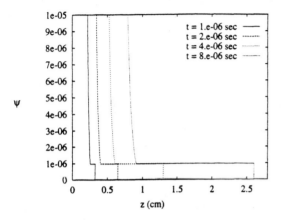

FIGURE 2. Weak solutions of equation (12) at times 1, 2, 4, and 8 μs. Same boundary condition as Fig. 1.

CONCLUSIONS

We have presented a test problem for use in evaluating materials dynamics codes which treat interfacial slip under circumstances where plastic deformation is an important dissipation mechanism. The semi-infinite test problem assumes purely plastic deformation with no elastic transient and accommodates rather general time dependent boundary conditions at the interface. The analytic solution for plastic strain has been given for a particular choice of flow rule which although simple incorporates, in an approximate way, the rate dependence of the flow stress at very large rates of plastic deformation. More general flow rules give rise to a more complicated partial differential equation than the telegrapher's equation and it is more difficult to find closed form solutions. The numerical solution of equation (9) for such flow rules may, however, provide helpful benchmarks for computer codes.

We have limited ourselves to purely transverse excitations. When longitudinal effects are important

as well, one must return to the full set of conservation laws and analytic results for interfacial motion are much more difficult to obtain. We have also neglected thermal conductivity. Thermal conduction introduces a new time scale. For large plastic deformation and high plastic strain rates, it is possible to include such effects via perturbation theory. This is frequently the case for dynamic experiments. When this is not the case, the equation for entropy production must be treated explicitly (5). Nevertheless, closed form solutions such as we have presented can afford some simple guidance for when a full treatment is necessary. We hope that, these restrictions not withstanding, the present test problem will be beneficial for code development where interfacial effects are important.

REFERENCES

1. Coggeshall, S.V., Phys. Fluids A3, 757-769 (1990) and references therein.
2. Verney, D., Symposium H.D.P.F.U.T.A.M., Paris, 293-303. (Gordon&Breach, 1968).
3. Zauderer, Erich, *Partial Differential Equations of Mathematical Physics* (New York, Wiley-Interscience, 1983).
4. Lax, P.D., *Hyperbolic Systems of Conservations Laws and the Mathematical Theory of Shock Waves* (Philadelphia, Society for Industrial and Applied Mathematics, 1973).
5. Glimm, J.G., Plohr, B.J., and Sharp, D.H., Mechanics of Materials 24, 31-41 (1996).

CP429, *Shock Compression of Condensed Matter – 1997*
edited by Schmidt/Dandekar/Forbes
© 1998 The American Institute of Physics 1-56396-738-3/98/$15.00

ON VON NEUMANN REFLECTION OF SHOCK WAVE IN CONDENSED MATTER

S. Itoh[a], Y. Natamitsu[b], Z. Y. Liu[a] and M. Fujita[a]

[a]*Department of Mechanical Engineering, Kumamoto University, Kumamoto 860, Japan*

[b]*Department of Mechanical Engineering, Daiichi College of Industry, Kokubu, Kagoshima, Japan*

In the irregular reflection of shock waves in gaseous media, a kind of reflection pattern, termed as von Neumann reflection(vNR), has been being studied by experimental, numerical and theoretical techniques. In condensed matter, such as liquid , metal and polymer, this kind of reflection pattern has not yet been noticed. This paper will present the studies on von Neumann reflection of shock waves in water and polymethylmethacrylate(PMMA). The basic characteristics of von Neumann reflection would be given.

INTRODUCTION

In the irregular reflection of shock waves, another kind of reflection pattern was discovered in experiment. Differing from the traditionally all-called Mach reflection, this reflection pattern is not of an obvious triple-point and a distinguishable reflected shock wave. When employing the von Neumann's 'three-shock theory' (1) to treat this problem, it would give out unreal physical solution. The difference between the theoretical solution and the physical existence once time was termed as 'von Neumann paradox' (2). Until recently, some investigators uncovered the veil of 'von Neumann paradox' and called this reflection pattern as von Neumann reflection (3, 4, 5). But all of their analyses and experimental observation were based on the circumstance of gaseous media. In condensed matter such concept was rarely put forwarded, although some experimental phenomena reported before already indicated the characteristics of von Neumann reflection (6, 7, 8).

The occurrence of von Neumann reflection is generally accompanied in the interaction or reflection of weak shock waves. In condensed matter there more easily achieves a state of weak shock than

in gaseous media because the condensed matter is usually of a much higher sound speed. In metal forming by underwater explosion, and explosive synthesis and explosive compaction in cylindrical arrangement, the interaction or reflection of shock waves easily lead to the von Neumann reflection. In this paper, we will select water and polymethylmethacrylate (PMMA) as the typical

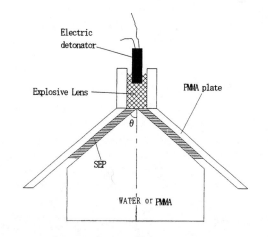

FIGURE 1. A diagram of the apparatus for generating shock wave reflection in water or PMMA.

representatives of condensed matter to study the von Neumann reflection in these two substances.

EXPERIMENTAL AND NUMERICAL METHODS
Experimental Method

To observe the phenomenon of von Neumann reflection of shock wave in water and PMMA, it is firstly necessary to establish an experimental apparatus that could generate shock waves in two media. Figure 1 shows such an experimental arrangement for that purpose. Two explosive sticks were set to form a V-shape configuration with an open angle of 2θ and connected with a common explosive lens and electric detonator. The whole setup was placed in an aquarium full of water. In the case of PMMA, a PMMA block with the same V-shape was added under the two explosive sticks. After the detonator was ignited, the two explosive sticks would be simultaneously detonated. At the same time, two shock waves would symmetrically produced in water or PMMA and spread out to come near to the symmetrical line. As the detonation propagated to some distance in each explosive stick, two shock waves would collide with each other to exhibit a reflection configuration.

The explosive used in the experiment was the so-called SEP, a kind of high-efficient explosive, which consisting of PETN (65wt%) and paraffin (35wt%). The dimensions of each explosive stick are 110 mm long, 50 mm wide and 5 mm thick. Its packing density was 1.31 g/cm^3 and the detonation velocity was 6.97 km/sec.

The usual shadowgraph system was utilized to visualize the reflection configuration of shock waves, as shown in Fig. 2. The system exploited an image-converter-camera of IMACON 790, made by HADLAND PHOTONICS, with a maximum framing speed of 20 million frames/sec and a maximum streak capability of 1.0 nsec/mm as well as an Xenon-flash-light source (HL 20/50 type Flash-light, also made by HADLAND PHOTONICS, with output 500J and flash time $50\,\mu$sec). The scale of the distance on the streak photographs was calibrated by using a block gauge and the time by a known pulse wave.

Table 1. Constants in Mie-Grüneisen equation of state.

Material	ρ_0(kg/m^3)	C_0(m/s)	S	Γ
PMMA	1181.0	2260.0	1.816	0.75
Water	1000.0	1489.0	1.786	1.65

Numerical Method

The numerical technique used was an Arbitrary-Lagrangian-Eulerian method (ALE) (9). In the simulation, we simplify the problem in plane geometry and also assume that all media can be treated as ideal compressible fluid. So combining the common equations of continuity, momentum and energy of two-dimensional inviscid fluid with equations of state of all media, it is possible to make a simulation to the whole shock reflection process. Importantly, here will introduce the equations of state of all media concerned. For water and PMMA, the common Mie-Grüneisen equation of state (10) is used,

$$P = \frac{\rho_0 C_0^2 \eta}{(1-s\eta)}\left(1 - \frac{\Gamma\eta}{2}\right) + \Gamma\rho_0 e \qquad (1)$$

where C_0 is sound velocity, $\eta = 1 - \rho_0/\rho$, ρ is density, Γ is Grüneisen parameter, s is material constant, respectively. These data (10) are listed in Table 1. The equation of state for detonation products is JWL equation of state and expressed as follows:

$$P = A\left(1 - \frac{\omega}{R_1 V}\right)\exp(-R_1 V)$$
$$+ B\left(1 - \frac{\omega}{R_2 V}\right)\exp(-R_2 V) + \frac{\omega\rho_e e}{V} \qquad (2)$$

where A, B, R_1, R_2, ω are JWL parameters to be determined by experiment, P is the pressure and V is the ratio of the initial density of explosive to the density of detonation products, e is the internal energy, ρ_e is the initial density of explosive. The determined JWL parameters (11) are listed in Table 2.

Table 2. Experimentally determined parameters of JWL equation of state.

A(GPa)	B(GPa)	R_1	R_2	ω	e (J/Kg)
376.40	2.15	4.40	1.00	0.28	5.24×10^6

RESULTS AND DISCUSSION

Figures 2 and 3 show a series of framing photographs of the configuration of shock reflection in water and PMMA, respectively. The photographs were taken first at 0μ sec and then at each 4μ sec time step. Three kinds of open angle circumstances were considered here as typical representatives, i.e., $\theta =15^\circ$, 30° and 45°. Because even under the same explosive amount, when the open angle was different, the intensity of shock waves and the angle of collision were also different. Hence, the reflection pattern would, of course, distinguish from one another. Both Figs.3 and 4 have confirmed this point. From the figures, one can clearly see that in the case of $\theta =45^\circ$, the largely curved Mach stem appeared after underwater shock waves interaction at a later time instant and the length of the Mach stem is not

distinguishable, whereas, in the case of $\theta =45^\circ$, the Mach stem is almost straight and it can be seen that the length of the Mach stem is still short in spite of a long time progressing. The former means that the von Neumann reflection takes place, the later, otherwise, shows a typical Mach reflection.

Figure 4 and 5 show the numerically simulated shadowgraphs of shock wave interaction in the cases of $\theta =15^\circ$ and 45°, respectively, at 8 μsec time instant, for water and PMMA experiments. In these figures, S represents the second-order differential value of the dimensionless density ($S= \nabla^2 \rho$), the darker parts represent shock waves, the letters WS stand for the incident shock wave, MS for the Mach stem and RS for the reflected shock wave. The simulated shadowgraphs clearly indicate that there exists a great difference on the reflection pattern between two circumstances. When the open angle is small, the reflection pattern exhibits an obvious reflected shock and a short-length Mach stem. It is belonged to the traditional Mach reflection. When the open angle is large, for instance, $\theta =45^\circ$, there is only a much obscure reflected shock reflected shock. The Mach stem seems to exhibit a continuous

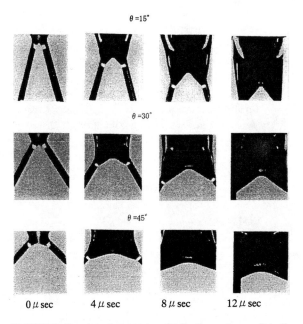

FIGURE 2. Framing photographs of reflection patterns of shock wave in water: (a) corresponding to the case of the open angle of 15°; (b) to the case of the open angle of 30°; (c) to the case of the open angle of 45°.

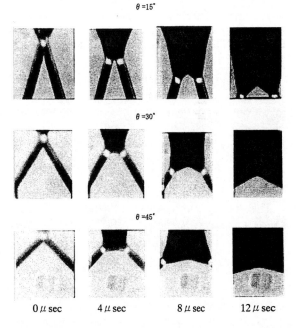

FIGURE 3. Framing photographs of reflection patterns of shock wave in PMMA: (a) corresponding to the case of the open angle of 15°; (b) to the case of the open angle of 30°; (c) to the case of the open angle of 45°.

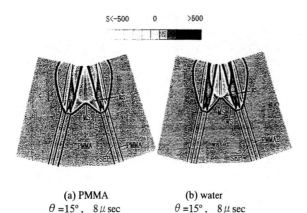

(a) PMMA
$\theta = 15°$, $8\,\mu$ sec

(b) water
$\theta = 15°$, $8\,\mu$ sec

FIGURE 4. Numerically simulated shadowgraphs of the reflection patterns of shock wave in PMMA (a), and in water (b), corresponding to the case of the open angle of 15° at $8\,\mu$ sec instant, respectively.

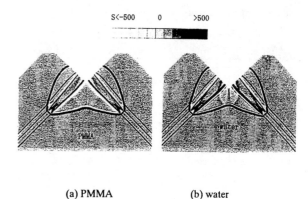

(a) PMMA
$\theta = 45°$, $8\,\mu$ sec

(b) water
$\theta = 45°$, $8\,\mu$ sec

FIGURE 5. Numerically simulated shadowgraphs of the reflection patterns of shock wave in PMMA (a), and in water (b), corresponding to the case of the open angle of 45° at $8\,\mu$ sec instant, respectively.

connection with the incident shock wave. So, this reflection pattern implies that the von Neumann reflection occurred.

CONCLUSIONS

The von Neumann reflection of shock waves in water and PMMA were investigated by experimental and numerical techniques. The reflection pattern differing from the traditional Mach reflection was confirmed. Because PMMA is of a higher sound speed than water, the von Neumann reflection in PMMA displayed more intense than in water under the same explosive amount and open angle shape conditions. At present, there is not an adequate theory to analyze the mechanism of the von Neumann reflection, the numerical means has been proved to be a good effective one.

ACKNOWLEDGEMENTS

The studies were conducted at the High Energy Rate Laboratory, Faculty of Engineering of Kumamoto University. The authors would like to thank Mr. Yukiyasu Ishitani of the Laboratory for his helpful advice. The authors also express their thanks to Nippon Oil&Fat Company for affording some substances needed in experiment.

REFERENCES

1. Neumann, J. von, *Collected Works*, Vol. **6**, Pergamon, 1963, pp. 238-308.
2. Birkhoff, B., *Hydrodynamics: A study in Logic, Fact and Similitude,* Princeton University Press, 1950.
3. Colella, P., and Henderson, L. F., *J. of Fluid Mech.*, **213**, 71-94 (1990).
4. Ben-Dor, G., *Shock Wave Reflection Phenomena*, Springer-Verlag, New York, 1991, ch. 2, pp.70-74.
5. Sasoh, A., and Takayama, K., J. of Fluid Mech., **277**, 331-345 (1994).
6. Fowles, G. R., and Isbell, W. M., *J. of Appl. Phys.*, **36**, 1377-1379 (1965).
7. Krehl, P., Hornemann, U., and Heilig, W., *Shock Tube and Shock Wave Research* (eds. Ahlborn, B. *et al.*), University of Washington Press, 1977, pp. 303-312.
8. Carton, E. P., and Stuivinga, M., and Verbeek, R., *Explomet'95*, 1996, pp. 29-36.
9. Amsden, A. A., Ruppel, H. M., and Hirt, C. W., SALE: A simplified arbitrary Lagragian-Eulerian method. LA-8095, UC-32, 1980.
10. Mader, C. L., et al., *Los Alamos Series on Dynamic Material Properties,* University of California Press, 1980.
11. Itoh, S., Kubota, S., Kira, A., Nagana, S., and Fujita, M., *J. Japan Explosives Soc.*(in Japanese), **55**, 202-208 (1994).

CP429, *Shock Compression of Condensed Matter – 1997*
edited by Schmidt/Dandekar/Forbes
1998 The American Institute of Physics 1-56396-738-3/98/$15.00

THE HUGONIOT ELASTIC LIMIT DECAY LIMIT

J. P. Billingsley

U. S. A. MICOM, Redstone Arsenal, Alabama 35898, U. S. A.

When the Hugoniot Elastic Limit particle velocity, U_{PHEL}, is greater than $V_1 = h/(2md_1)$, it will decay asymptotically with propagation distance, x, to the V_1 level.

INTRODUCTION

Since its discovery, the Hugoniot Elastic Limit (HEL) decay has been the subject of numerous experimental and theoretical studies. This article supplements these efforts by demonstrating via graphical comparisons with experimental U_{PHEL} data, that U_{PHEL} decays asymptotically with travel distance, x, to the V_1 magnitude. These pictorial comparisons are provided for five rather diverse solid materials which are: Iron (Fe), Aluminum Alloy (6061-T6), Nickel Alloy (Mar-M200), Lithium Fluoride (LiF), and Plexiglas (PMMA).

$V_1 = h/(2\,md_1)$ where h = Planck's constant, m is the mass (or average mass) of one atom and d_1 is the closest (or average) distance between atoms. Values of V_1 for the five materials investigated are listed in Table 1.

BACKGROUND

The motivation to investigate a possible relationship between U_{PHEL} magnitudes and V_1 was supplied by Reference [1]. This Reference delineates the importance of the DeBroglie momentum (p) - wave length (λ) relation, $\lambda = h/p = h/(mV)$, with respect to stressed solid material behavior. Reference [1] also strongly emphasizes that $\lambda_1 = 2d_1$ is the smallest particle momentum (PM) wave length possible in an unperturbed lattice of atoms. This leads to V_1 which is the largest possible PM wave particle velocity in a stationary lattice.

From microscopic kinematic considerations, it is shown in Reference [1] that particle velocities, V_i, which are greater than V_1, can only be accommodated if the lattice (or part of it) moves forward with velocity, V_L, toward V_i. This forward lattice motion requires energy expenditure from the lattice itself and will be a dissipative unstable situation. A stable condition ($V_i \rightarrow V_1$) will be sought which requires no lattice energy dissipation ($V_L = 0.0$).

Although derived from a microscopic viewpoint, this so called "V_1 effect" is apparently manifesting itself enough to be observable on a macroscopic scale in the U_{PHEL} decay phonomenon.

ELASTIC WAVE PRESSURE

The elastic wave pressure, Pv_1, corresponding to the wave velocity, C_L, and the particle velocity, V_1 is given by $Pv_1 = \rho_o C_L V_1$ where ρ_o is the material density (grams/cc). Pv_1 can also be expressed in terms of acoustical phonons as follows:

$$Pv_1 = \rho_o C_L\left(\frac{h}{2md_1}\right) = \left(\frac{\rho_o}{m}\right)h\left(\frac{C_L}{2d_1}\right)$$

$$= N_v h\left(\frac{C_L}{\lambda_1}\right) = N_v\,(hf_1).$$

This indicates that Pv_1 is equivalent to each atom in the unit volume having one phonon of energy (hf_1). Perhaps this also contributes to sympathetic stabilization for $U_{PHEL} = V_1$ conditions.

U_{PHEL} AND V_1 COMPARISON

Figures 1 through 7 which compare the experi-

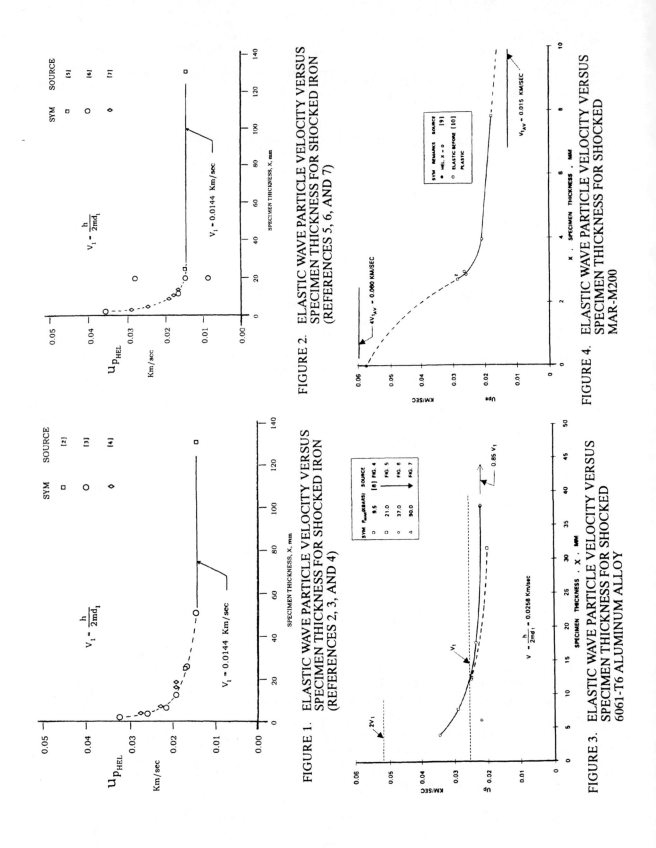

FIGURE 1. ELASTIC WAVE PARTICLE VELOCITY VERSUS SPECIMEN THICKNESS FOR SHOCKED IRON (REFERENCES 2, 3, AND 4)

FIGURE 2. ELASTIC WAVE PARTICLE VELOCITY VERSUS SPECIMEN THICKNESS FOR SHOCKED IRON (REFERENCES 5, 6, AND 7)

FIGURE 3. ELASTIC WAVE PARTICLE VELOCITY VERSUS SPECIMEN THICKNESS FOR SHOCKED 6061-T6 ALUMINUM ALLOY

FIGURE 4. ELASTIC WAVE PARTICLE VELOCITY VERSUS SPECIMEN THICKNESS FOR SHOCKED MAR-M200

200

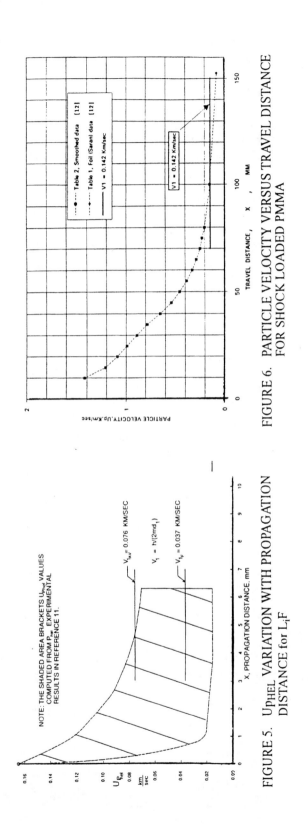

NOTE: THE SHADED AREA BRACKETS U_{Phel} VALUES COMPUTED FROM P_{hel} EXPERIMENTAL RESULTS IN REFERENCE 11.

FIGURE 5. U_{PHEL} VARIATION WITH PROPAGATION DISTANCE for L_iF

FIGURE 6.   PARTICLE VELOCITY VERSUS TRAVEL DISTANCE FOR SHOCK LOADED PMMA

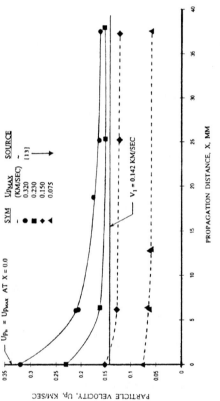

FIGURE 7.   INSTANTANEOUS PARTICLE VELOCITY, U_{PI}, VERSUS PROPAGATION DISTANCE, X, FOR PMMA

201

mental $U_{P_{HEL}}$ (or P_{HEL}) data with V_1 (or Pv_1) are basically self explanatory and all exhibit the same trend which is the theme of this article.

SUMMARY

Because this trend ($U_{P_{HEL}}$ decay to V_1 magnitude) exists for five diverse solid materials, the probability of fortuitous or circumstantial occurrence is greatly diminished. Consequently, this should stimulate both theoretical and experimental research about the "V_1 effect" and its role in stress/strain wave particle motion and shock induced reactions.

REFERENCES

1. E. R. Fitzgerald, Particle Waves and Deformation of Crystalline Solids, Interscience Publishers, a Division of John Wiley and Sons, Inc. New York, (1966).
2. Stanley F. Minshall, Article in *Response of Metals to High Velocity Deformation*, July 11-12, (1960), (Edited by P. G. Shewmon and V. F. Zackay), *Interscience Publishers, Inc.* New York, 249-274, (1961)
3. John W. Taylor, and Melvin H. Rice, *Journal of Applied Physics*, 34 [2], 364-371, February (1963).
4. L. M. Barker, and R. E. Hollenbach, *Journal of Applied Physics*, 45, [11], 4872-4887, November (1974).
5. Denison Bancroft, Eric L. Peterson, and Stanley Minshall, *Journal of Applied Physics*, 27, [3], 291-298, March (1956).
6. D. S. Hughes, L. E. Gourley, and Mary F. Gourley, *Journal of Applied Physics*, 32, [4], 624-629, April (1961).
7. Z. Rosenberg, A. Erez, and Y. Partom, *Journal of Physics E: Scientific Instruments*, 16, 198-200, (1983).
8. J. N. Johnson and L. M. Barker, *Journal of Applied Physics*, 40, [11], 4321-4334, October (1969).
9. D. P. Dandekar, A. G. Martin, and J. V. Kelly, Article in *Proceeding of the Army Symposium of Solid Mechanics, 1980 - Designing for Extremes: Environment, Loading, and Structural Behavior, AMMRC MS* 80-4, September (1980), 317-329.
10. D. P. Dandekar, and A. G. Martin, Chapter 34 in *Shock Waves and High-Strain Rate Phenomena in Metals, Concepts and Applications*, (Edited by M. A. Myers and L. E. Murr), Plenum Press, New York, (1981), 537-587.
11. J. A. Asay, G. R. Fowles, G. E. Duvall, M. H. Miles, and R. F. Tinder, *Journal of Applied Physics*, 43, [5], May (1972), 2132-2145.
12. T. P. Liddiard, Paper in *Fourth Symposium (Int.) on Detonation*, October 12-15, (1965), Published as ACR-126 by the Office of Naval Research, Dept. of the Navy, 214-221.
13. K. W. Schuler, *Journal of the Mechanics and Physics of Solids*, 18, (1970), 277-293.

Table 1. Tabulation of V_1 and Pv_1 for Iron, 6061-T6, MAR-M200, LiF, and PMMA

Fe = Iron atom
Al = Aluminum atom
F = Fluorine atom

MATERIAL ~	ρ_o g/cc	m grams x 10^{-23}	d_1 Å (10^{-8} cm)	V_1 km/sec	C_L km/sec	Pv_1 Kbars
IRON	7.84	9.2700 Fe	2.4800 N	0.01440	6.000	6.77
6061-T6	2.70	4.4800 Al	2.8600 Al	0.02580	6.230	4.34
MAR-M200	8.59	9.8232 Av	2.2530 Av	0.01497	5.780	7.43
LiF	2.64	2.1534 Av	2.0130 Av	0.0764	6.557	13.23
LiF	2.64	3.1546 F	2.8410 F	0.03700	6.557	6.40
PMMA	1.18	1.1083 Av	2.1099 Av	0.14170	3.100	5.18

Av = Average
N = Nearest

CP429, *Shock Compression of Condensed Matter – 1997*
edited by Schmidt/Dandekar/Forbes
© 1998 The American Institute of Physics 1-56396-738-3/98/$15.00

SPH SIMULATION OF HIGH DENSITY HYDROGEN COMPRESSION

R. Ferrel and V. Romero

Energetic Materials Research & Testing Center, New Mexico Tech, Socorro, NM 87801

The density dependence of the electronic energy band gap of the hydrogen has been studied with respect to the insulator-metal (IM) transition. The valence conduction band gap of solid hydrogen is about 15eV at zero pressure, therefore very high pressures are required to close the gap and achieve metallization. We propose to investigate what will be the degree to which one can expect to maintain a shockless compression of hydrogen with a low temperature (close to that of a cold isentrope) and verify if it is possible to achieve metallization. Multistage compression will be driven by energetic materials in a cylindrical implosion system, in which we expect a slow compression rate that will maintain the low temperature in the isentropic compression. It is hoped that pressures on the order of 100Mbars can be achieved while maintaining low temperatures. In order to better understand this multistage compression a smooth particle hydrodynamics (SPH) analysis has been performed. Since the SPH technique does not use a grid structure it is well suited to analyzing spatial deformation processes. This analysis will be used to improve the design of possible multistage compression devices.

INTRODUCTION

High pressures have a fundamental effect on the electron distribution in a material, one of these effects is the transition from insulation to metallic form. The prototype element to understand this transition is hydrogen. Many experiments have confirmed the metallization of hydrogen at high pressures an it is believed to occur by a band-crossing mechanism in the molecular solid[1] above 2.5Mbars. Livermore researchers have experimentally achieved this transition[2] at 1.4 Mbars and temperatures up to 3000° K. This was accomplished by the use of a two-stage gas gun to create enormous shock pressure on a target containing liquid hydrogen cooled to 200° K. Metallic character is most directly established by electrical conductivity measurements which are not yet possible in diamond anvil cells at these pressures.

OBJECTIVE

The objective of this simulation is to find out the possibility to achieve a higher compression for the hydrogen. Using a SPH code name MAGI[5] we intent to model the 1-D cylindrical implosion system based on an axis symmetric multistage compression, and observe the degree to which we can increase the pressure.

SPH THE METHOD

The foundation of Smoothed Particle Hydrodynamics is interpolation theory[7]. The conservation laws of continuum fluid dynamics, in the form of partial differential equations, are transformed into integral equations through the use of an interpolations function that give the "kernel estimate" of field variables at a point. Computationally, information is known only at discrete points, so that the integrals are evaluated as sums over neighboring points. The reason that an underlying grid is not needed is that functions are evaluated using their values at the discrete points (particles) and an interpolation kernel. An integration by parts then moves spatial derivatives from operating on the physical quantities to operating on the interpolation kernel which is analytic, this allows a griddles structure, well suited to analyzing rapid spatial deformation processes like the one expected in our case.

EOS SESAME

For the equation of state of the hydrogen we use a cold isentrope curve of pressure and sound speed obtained from SESAME[4] provide by LANL. The pressure curve used to maintain a low temperature in the compression while trying to avoid the shock heating is shown in Figure 1. We start with cryogenic solid hydrogen (4° K) under the highest pressure that can practically be achieved statically, in the order of 10kbar, and to minimize the shock heating the rate of compression is limited by the sound speed, in Figure 2.

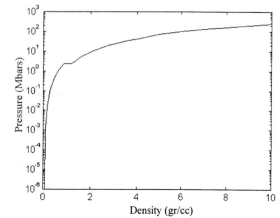

FIGURE 1. Pressure versus density from the SESAME equation of state for Hydrogen.

The approximation used for densities $\rho \geq 4.0$ is

$$c_s = 0.6954\rho + 1.9666 \qquad (1)$$

GEOMETRY OF THE SIMULATION

The simulation was perform using the axis symmetry setup, in a 1-D simulation it represents a mirror effect from the origin. We use an explosive

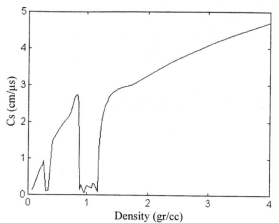

FIGURE 2. Sound speed versus density from the SESAME equation of state for Hydrogen.

driven cylindrical implosion system[3]. The main compression system for reaching the peak pressure consist of a set of four cylindrical shells. The materials utilized for the simulation, their configuration and radial distance are display in Table 1. The mass of each shell decreasing sequentially from the outside. The implosion velocity increases due to convergence and the decreasing mass with each shell collision. The mass of the inner most shell is far greater than the hydrogen mass, and therefore a modest velocity of the inner most shell results in a kinetic energy that is comparable to the internal energy of compressed hydrogen. If the transformation from kinetic energy of the inner most shell to internal energy of both the shell and the hydrogen is accomplished with out shock, then a heavy metal shell (Tantalum, Tungsten, Steel) could reach a pressure of about 200 Mbar at turnaround[6].

The outermost shell is made from steel and it is surrounded by high explosives. The detonation wave drives it radially inward and accelerates the first steel shell. The impact of the steel shell on the second shell is carefully cushioned by a plastic layer (Acrylic) which prevents dissipation of the kinetic energy of the first shell upon impact. The resulting collision is nearly elastic. With a mass ratio between the shells of approximately 2 to 1, the second shell starts off at an implosion speed of about 1.3 times that of the previous shell.

TABLE 1. Materials Description and Configuration.

Material	EOS type	Density (g/cc)	Radial distance (cm)	Thickness (cm)
Hydrogen	SESAME cold isentrope	0.0720	0.223	0.223
Tungsten	mie - gruneisen solid	19.230	0.423	0.200
Acrylic	mie - gruneisen solid	1.4000	0.880	0.457
Tungsten	mie - gruneisen solid	19.230	1.080	0.200
Acrylic	mie - gruneisen solid	1.4000	1.370	0.290
Vacuum			1.650	
Acrylic	mie - gruneisen solid	1.4000	1.850	0.200
Tungsten	mie - gruneisen solid	19.230	2.050	0.200
Acrylic	mie - gruneisen solid	1.4000	2.250	0.200
Vacuum			4.900	
Acrylic	mie - gruneisen solid	1.4000	5.100	0.200
Steel	mie - gruneisen solid	7.8900	5.300	0.200
Acrylic	mie - gruneisen solid	1.4000	5.500	0.200
Vacuum			10.600	
Acrylic	mie - gruneisen solid	1.4000	10.800	0.200
Steel	mie - gruneisen solid	7.8900	11.000	0.200
Explosive	he burn mie - gruneisen for solid, jwl for gas	1.7200	15.000	3.000

RESULTS

The results obtained from the simulation where satisfactory with respect to our expectations. The maximum pressure achieved of 119 Mbars is shown in Figure 3, this pressure surpasses our expectations, because its much higher than the minimum pressure in which metallization is expected to occur (2.5 Mbars).

CONCLUSIONS

We found that SPH can be used to simulate this process. The simulation indicates that the desired condition can be achieved. Also the simulation can be used to evaluate changes in the experimental setup.

The next step will be to conduct a 2-D simulation of this same case and study the effect of an asymmetric detonation in the explosive and what are it's repercussion in the maximum possible pressure which the hydrogen can be compressed.

ACKNOWLEDGMENTS

We thank Ted Carney for his support and patience while working with the MAGI code. Stirling Colgate for the conversations sustained. Naresh Thadhani for his support for us to be here.

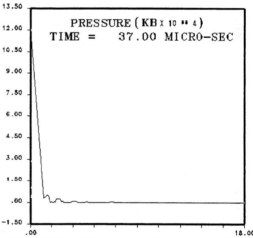

FIGURE 3. Plot of the pressure of the implosion when the hydrogen achieves its maximum pressure of 119 Mbars, the x axis represents the position in cm.

REFERENCES

1. H. K. Mao and R. J. Hemley, *Science* **244**, 1462, (1989).
2. S.T. Weir, A.C. Mitchell, and W.J. Nellis, "Metallization of Fluid molecular Hydrogen at 140 Gpa (1.4 Mbar)" , *Phys. Rev. Lett.* **76**, #11, 1860, (1996).
3. T. Hiroe, H. Matsuo, K. Fujiwara, T. Tanoue, M. Yoshida, S. Fujiwara, "A production of cylindrical imploding shock in solid by exploding wire rows" The Proceedings of the High Shock Pressure Symposium, Boulder, Colorado, 1993,.

4. SESAME, Report on the Los Alamos Equations-of-State Library, Report No. LALP-83-4, T4 Group, LANL, Los Alamos, 1983.

5. A.G. Petschek, L.D. Libersky, "Cylindrical Smoothed Particle Hydrodynamics", *Journal of Computational Physics*, **109**, #1, 76 (1993).

6. S. Colgate, R. White, " The Hydrodynamic behavior of surpernovae explosions", Ap. J., **143**, # 3 , 626, (1996).

7. L.D. Libersky, A.G. Petschek, T.C. Carney, J.R. Hipp, F.A. Allahdadi, "High Strain Lagrangian Hydrodynamics", *Journal of Computational Physics*, **109**, #1, 67, (1993).

CP429, *Shock Compression of Condensed Matter – 1997*
edited by Schmidt/Dandekar/Forbes
© 1998 The American Institute of Physics 1-56396-738-3/98/$15.00

STABILITY OF STRONG SHOCKS IN METALS

I.Rutkevich, E.Zaretsky and M.Mond

Pearlstone Center for Aeronautical Engineering Studies,
Department of Mechanical Engineering,
Ben-Gurion University of the Negev,
Beer Sheva 84105 Israel

The stability of strong shock waves in metals with respect to spontaneous emission of acoustic and entropy-vortex waves is investigated theoretically. The analysis employs the empirical Hugoniot adiabatic (HA) which is commonly represented as a straight line in the plane (U,D) where U is the particle velocity behind the shock and D is the shock velocity. The criterion for spontaneous emission depends on the sound velocity in a shock-compressed medium which is determined from the three-term equation of state. The latter takes into account contributions of atoms and electrons to the total pressure. The atomic Gruneisen parameter and the cold elastic pressure are calculated from a system of coupled differential equations which is based on the empirical HA and the Slater - Landau approach. It has been found that spontaneous emission may occur in metals with relatively low values of the Hugoniot adiabatic slope S=dD/dU such as molybdenum and tantalum.

INTRODUCTION

This paper considers the possibility of spontaneous emission (SE) of sound and entropy-vortex waves (1, 2) from strong shocks in metals. The occurrence of SE means the existence of such two-dimensional sound waves impinging on the shock, for which the reflection coefficient becomes infinite. This results also in spontaneous corrugation of the planar shock front. The problem of corrugation instability of a planar shock was first addressed by Dyakov (1) and Kontorovich (2). A simple thermodynamic consideration of the conditions for SE from strong shocks in solids was given in (3) where it was shown that SE may appear for sufficiently strong shocks.

The purpose of this work is to obtain the equation of state (EOS) which would be consistent with a given linear law D(U) and then to employ this EOS for testing the condition for SE for several metals. An important issue of this modeling is taking account of the contribution of the free electrons to the total pressure and to the total internal energy. That contribution becomes important for sufficiently strong shocks when the temperature of the shock-compressed medium exceeds 10^4K.

THE CRITERION FOR SE

When a planar shock propagates in a uniform cold material with a constant velocity $\mathbf{W} = -D\mathbf{e_x}$, it is convenient to consider the interaction between the shock and small perturbations in the frame of reference K which moves with the same velocity $\mathbf{W}$. In this frame the shock is at rest and its front is specified as x=0. Then the cold medium in the domain x<0, i.e. ahead of the shock, moves with the supersonic velocity $\mathbf{V_1} = D\mathbf{e_x}$, while the compressed medium behind the shock (x>0) moves with the subsonic velocity $\mathbf{V_2} = (D-U)\mathbf{e_x}$. Here U is the particle velocity behind the shock in the laboratory frame of reference.

The perturbations in the supersonic region do not contribute to the spectrum of eigenmodes. Hence, only the perturbations in the subsonic domain x>0 are considered, which, without loss of generality, may be written in the following two-dimensional form

$$\exp[i(k_x x + k_y y - \omega t)] \qquad (1)$$

that describes various types of linear waves. In the framework of an ideal compressible fluid model there are three different types of waves: the upstream acoustic wave, the downstream acoustic wave and the entropy-vortex wave. The latter propagates with the fluid velocity (1, 4). When an incident upstream acoustic wave described by Eq.(1) reaches the shock front, both a downstream acoustic wave and an entropy-vortex wave moving away from the shock are generated. In addition, the initially planar shock front undergoes a sinusoidal corrugation, whose wavelength is defined by the tangential component k_y of the vector $\mathbf{k}$.

The criterion for SE obtained by Kontorovich (2) is:

$$h_c(\overline{\rho}, M_2) < h < 1 + 2M_2 , \quad h = -V_2^2 \left(\frac{d\rho_2}{dp_2}\right)_H \quad (2)$$

Here $\overline{\rho} = \rho_2/\rho_1 > 1$ is the ratio of the densities on both sides of the shock, $V_2 = D-U$ and $M_2 = V_2/c_2 < 1$ are the flow velocity behind the shock in the frame K and the Mach number for this flow, c_2 is the sound velocity in the shock-compressed medium, h is the Dyakov parameter. The derivative $(d\rho_2/dp_2)_H$ is taken along the HA. The critical value h_c of the parameter h determining the threshold of SE is given by

$$h_c = \frac{1 - M_2^2(\overline{\rho} + 1)}{1 + M_2^2(\overline{\rho} - 1)} \quad (3)$$

The Kontorovich criterion for SE (2) cannot be applied until the sound velocity of the shock-compressed material has been found.

Before turning to the full calculation of all needed thermodynamic quantities it is shown here that shock waves with sufficiently small intensity are always stable with respect to SE. For weak shocks $0 < \overline{\rho} - 1 << 1$ we can present the difference $h_c - h$ in the form

$$h_c - h \approx \frac{\rho_1^3 c_1^4}{2} \left(\frac{\partial^2 v}{\partial p^2}\right)_{s=s_1, \rho=\rho_1} (\overline{\rho} - 1) \quad (4)$$

where $v = 1/\rho$ is the specific volume. On the other hand, it is well known (4) that the expansion of the entropy difference $s_2 - s_1$ across the shock in a power series in a small pressure difference $p_2 - p_1$ starts with the third order term

$$s_2 - s_1 \approx \frac{1}{12 T_1}\left[\left(\frac{\partial^2 v}{\partial p^2}\right)_s\right]_{p = p_1} (p_2 - p_1)^3 \quad (5)$$

Therefore, for weak shocks the sign of the difference h_c-h is the same as the sign of the entropy change s_2-s_1 across HA. Thus, the absence of SE from weak shocks is equivalent to the entropy increase behind them. However, the requirement for entropy increase behind the shock does not provide a sufficient condition for the stability of strong shocks, for which Eqs. (4) and (5) cannot be used.

A THERMODYNAMIC MODEL

The problem considered below is formulated as follows: For a given form of the HA find the equation of state (EOS) determining the pressure p as a function ρ and T. In addition, find the specific internal energy ε and the specific entropy s as functions of ρ and T. The solution of this problem should satisfy the condition

$$\varepsilon_2 - \varepsilon_1 \equiv \varepsilon_H(\overline{\rho}) = p_H(\overline{\rho})[1 - \overline{\rho}^{-1}]/2\rho_1 \quad (6)$$

To access a self-consistent thermodynamic description the following three-term EOS is assumed (5)

$$p(\rho, T) = p_c(\rho) + p_T(\rho, T) + p_e(\rho, T) \quad (7)$$

where p_c is the cold (zero Kelvin) elastic pressure, p_T is the thermal pressure corresponding to the thermal motion of the atoms, p_e is the contribution of the free electrons to the total pressure. The terms p_T and p_e have the form

$$p_T = \rho \Gamma(\rho) \varepsilon_T , \quad p_e = \rho \Gamma_e(\rho) \varepsilon_e \quad (8)$$

where ε_T is the thermal part of specific internal energy of the atoms and ε_e is the specific internal energy of the electrons per unit mass:

$$\varepsilon_T = C_V T, \qquad \varepsilon_e = \beta(\rho) T^2/2 \quad (9)$$

where C_V is the specific heat of atoms and $\beta(\rho)T$ is the electronic specific heat. The functions $\Gamma(\rho)$ and $\Gamma_e(\rho)$ are the Grüneisen parameters for the atomic and electronic subsystems, respectively. The lattice specific heat C_V is assumed to be independent on temperature and density. Within the framework of this approach the lattice Grüneisen parameter Γ depends only on the material density ρ. The total internal energy is presented in the form similar to Eq. (7).

$$\varepsilon(\rho, T) = \varepsilon_c(\rho) + \varepsilon_T(\rho, T) + \varepsilon_e(\rho, T) \qquad (10)$$

where ε_c is the elastic potential energy:

$$\varepsilon_c(\rho) = \varepsilon_c(\rho_1) + \int_{\rho_1}^{\rho} \rho^{-2} p_c(\rho) d\rho \qquad (11)$$

Equations (7) - (11) guarantee that the differential form

$$ds = (d\varepsilon - p d\rho / \rho^2) / T \qquad (12)$$

is a total differential of the function $s(T, \rho)$ which is the specific entropy. From Eqs. (7) - (12) one can find the sound velocity $c = \sqrt{(\partial p / \partial \rho)_s}$ as a function of the thermodynamic variables:

$$c(\rho, T) = \sqrt{\begin{array}{l} \dfrac{dp_c}{d\rho} + \dfrac{T(C_V\Gamma + 2\beta T)(C_V\Gamma + \Gamma_e\beta T)}{C_V + \beta T} \\ + C_V T \dfrac{d(\rho\Gamma)}{d\rho} + \dfrac{\beta T^2}{2}\left[\dfrac{d(\rho\Gamma_e)}{d\rho} - \Gamma_e^2\right] \end{array}} \qquad (13)$$

To calculate the sound velocity c_2 as a function of the density along the HA, one has to specify T in Eq. (13) as the temperature $T_H = T(\rho_2, p_2)$ behind the shock. Using the known total pressure $p_2(\rho_2)$ along the HA together with Eqs. (7) - (9) results in a quadratic equation for T_H. To obtain a complete thermodynamic description of the shock-compressed medium one can use the Slater - Landau model (or similar models) giving a connection between the cold pressure $p_c(\rho)$ and the lattice Grüneisen parameter $\Gamma(\rho)$:

$$\Gamma(\rho) = \frac{m-2}{3} - \frac{v}{2}\frac{d^2(p_c v^{2m/3})/dv^2}{d(p_c v^{2m/3})/dv} \quad, \quad v = \frac{1}{\rho} \qquad (14)$$

For particular values of the parameter $m=0,1,2$, Eq.(14) represents the models of Slater - Landau (5,6), Dugdale - MacDonald (7) and Vaschenko - Zubarev (8), respectively.

For the electronic Grüneisen parameter the constant value $\Gamma_e = 2/3$ is assumed. This value is correct for the degenerated electron gas satisfying the Fermi - Dirac statistics. For this value of Γ_e the function $\beta(\rho)$ behaves as $\rho^{2/3}$. When the shock pressure p_H, the atomic specific heat C_V, the electronic specific-heat coefficient $\beta(\rho)$ and the electronic Grüneisen parameter Γ_e are given, all thermodynamic variables can be found after calculating the functions $p_c(\rho)$ and $\Gamma(\rho)$. Since Eq.(14) alone is not enough for determining these two functions, some additional relationship between them should be employed. Such a relationship can be obtained by considering the internal energy along the HA. This quantity is a known function that is provided by Eq. (6). On the other hand, the internal energy along the HA should be obtained from the general model given by Eq.(10), in which the shock temperature T_H is used. Thus, equating those two expressions for ε_H and differentiating the result with respect to ρ provides the desirable relationship, which is:

$$\frac{d\varepsilon_H}{d\rho} = \frac{p_c}{\rho^2} + (C_V + \beta T_H)\frac{dT_H}{d\rho} + \frac{d\beta}{d\rho}\frac{T_H^2}{2} \qquad (15)$$

where $d\varepsilon_H/d\rho$ is a known function of ρ. A specification of the initial data for the system of differential equations (14), (15) in the general case requires the solution of an infinite system of nonlinear algebraic equations for the derivatives of all orders of the cold pressure function in the initial point [9]. For shocks propagating in metals at room temperature, an approximate solution of the above-mentioned algebraic system can be obtained due to the presence of two small parameters

$$\mu = \Gamma_1 C_V T_1 / c_0^2 \ll 1 \quad, \quad v = \beta_1 T_1^2 / 3 c_0^2 \ll 1 \qquad (16)$$

RESULTS AND DISCUSSIONS

The numerical solutions of the system of Eqs.(14), (15) were obtained for the initial temperature $T_1 = 300K$. The calculated shock temperature $T_H(\rho_2)$ and the sound velocity $c_H(\rho_2) = c[\rho_2, T_H(\rho_2)]$ along

HA represent monotonically increasing functions of ρ_2. The curves $\bar{c}_H(\bar{\rho}) = c_H / c_0$ calculated for aluminum with m=0 are shown in Fig.1. The solid curve in this figure corresponds to the three-term EOS [Eq.(7)], while the dashed curve is calculated with the two-term EOS, in which the electronic thermal pressure p_e is neglected. The triangles in Fig.1 present the experimental data for the sound velocity along the HA in aluminum (10). The dot-dashed curves in Fig.1 intersecting the graph of $\bar{c}_H(\bar{\rho})$ show the changes of the sound velocity along various isentropes.

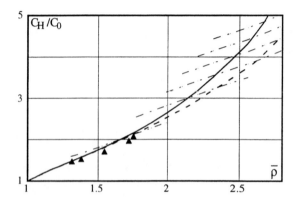

FIGURE 1. Calculated normalized sound velocity $\bar{c}_H(\bar{\rho})$ along the HA for aluminum: the solid curve - three-term EOS, the dashed curve - two-term EOS. Both curves are shown for the Slater - Landau model (m=0). The triangles are the experimental data of Al'tshuler et al. (10). The dot-dashed curves intersecting the graph of $\bar{c}_H(\bar{\rho})$ represent the sound velocity along various isentropes for three - term EOS.

Numerical calculations of the parameters h and h_c defined by Eqs. (2) and (3) have shown that strong shocks may become unstable with respect to SE in those metals, for which the HA has relatively small slope S. Thus, domains of SE were found for molybdenum (S=1.26) and for tantalum (S=1.31). The dependencies $h_c(\bar{\rho})$ for molybdenum calculated on the basis of the two-term and three-term EOS with m=0 and the function $h(\bar{\rho})$ are presented in Fig.2. As is seen from this figure, both EOS result in SE [i.e., a domain in which $h(\bar{\rho}) > h_c(\bar{\rho})$]. Within the framework of the three-term EOS no SE domains were found for aluminum (S=1.35), copper (S=1.5) and lead (S=1.57).

According to obtained domains of SE, the conditions for SE can be achieved in molybdenum under high-velocity impacts created by underground

nuclear explosions (11,12). These conditions may be achieved also in contemporary experiments with laser-driven shock waves (13).

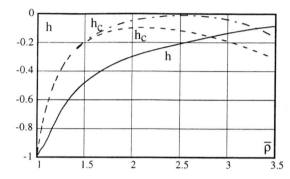

FIGURE 2. The Dyakov parameter h (solid curve) and the critical parameter h_c for molybdenum (S=1.26) calculated along the HA from the three-term (dot-dashed curve) and two-term (dashed curve) EOS for m=0.

REFERENCES

1. Dyakov S.P., *Zh. Eksp. Teor. Fiz.*, **27**, 288 (1954).

2. Kontorovich V.M., *Sov. Phys.- JETP*, **6**, 1179 (1957).

3. Rutkevich I., Zaretsky E. and Mond M., *J. de Physique IV*, **C8**, 728 (1994).

4. Landau L.D. and Lifshitz E.M., *Fluid Mechanics*, Pergamon Press (1987).

5. Zel'dovich Ya.B. and Raizer Yu.P., *Physics of Shock Waves and High-Temperature Hydrodynamic Phenomena*, Academic Press (1984).

6. Slater J.C., *Introduction to Chemical Physics*, McGraw Hill, New York (1939).

7. Dugdale J.S. and MacDonald D.K.C., *Phys. Rev.*, **89,** 832 (1953).

8. Vaschenko V.Y. and Zubarev V.N., *Sov. Phys.- Solid State*, **5**, 653 (1963).

9. Rutkevich I., Zaretsky E. and Mond M., *J.Appl.Phys.*, **81**, 7228 (1997).

10. Al'tshuler L.V., *Sov. Phys. - Uspekhi*, **8**, 52 (1965).

11. Ragan III C.E., *Phys. Rev.*, **A25**, 3360 (1982).

12. Ragan III C.E., *Phys. Rev.*, **A29**, 1391 (1984).

13. Löwer Th., Sigel R., Eidmann K., Földes I.B., Hüller S., Massen J., Tsakiris G.D.,Witkowski S., Preuss W., Nishimura H., Shiraga H., Kato Y., Nakai S. and Endo T., *Phys. Rev. Lett.*, **72**, 3186 (1994).

CP429, *Shock Compression of Condensed Matter – 1997*
edited by Schmidt/Dandekar/Forbes
© 1998 The American Institute of Physics 1-56396-738-3/98/$15.00

Molecular Dynamics Simulation of High Strain-Rate Void Nucleation and Growth in Copper

James Belak

University of California, Lawrence Livermore National Laboratory, Livermore, CA 94550

Isotropic tension is simulated in nanoscale polycrystalline copper with 10nm grain size using molecular dynamics. The nanocrystalline copper is fabricated on the computer by growing randomly oriented grains from seed sites in the simulation cell. Volume strain rates of 10^8 - 10^{10} are considered for systems ranging from 10^5 - 10^6 atoms using an EAM potential for copper. The spacing between voids for room temperature single crystal simulations is found to scale approximately as $l \sim 0.005\ C_s / \gamma$, where C_s is the sound speed and γ is the strain rate. Below strain rates of about 10^9, only one void is observed to nucleate and in the polycrystalline simulation cell.

INTRODUCTION

When tensile shock waves overlap in a ductile metal, extreme states of tension are created and internal failure occurs through the nucleation, growth, and linking of microscopic voids. This process, known as spallation, has been the subject of extensive metallurgical investigation [1-3]. The spallation of ductile metals depends upon the magnitude of the tensile stress, the length of time at stress, and the metallurgical state of the metal (grain size, impurity concentration,...). Experimental measurements of the spall strength of copper range from 1.5 to 4.5GPa depending on the metallurgical state [4]. While mature models for void growth are available [5], there remains several issues regarding the growth of small voids in single crystals, the nature of the nucleation process (homogeneous/heterogeneous) and nucleation sites, and the process of void linking at late times. The purpose of the molecular dynamics study presented in this paper is to examine nucleation and early growth of voids in ductile metals under constant strain rate tensile expansion.

METHODS

A polycrystalline copper metal with nanoscale grains is simulated using the idealized space filling model of Phillpot and coworkers [6]. A thin slice

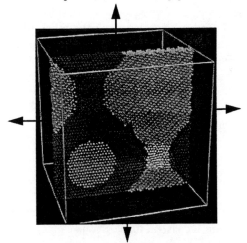

FIGURE 1. A slice through the idealized four grain microstructure used in the void nucleation simulations. The sphere of atoms at the grain center are held fixed during processing and allowed to evolve during failure. The simulation cell is expanded in all directions at a constant rate.

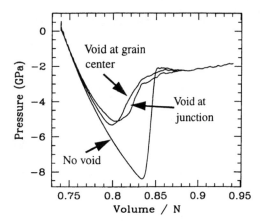

FIGURE 2. Stress as a function of volume per atom (r_e^3, r_e=2.54 Angs) for a strain rate of $7.7 \times 10^8 s^{-1}$ at T=300K.

through the simulation cell is shown in Figure 1. On each of four FCC seed sites, within the cubic simulation cell, a randomly oriented grain is grown to the Voronoi boundary with the neighboring seed sites while maintaining periodic boundary conditions. The center of each grain is held fixed while the surrounding atoms are brought to melting at P=0, then returned to T=300K. Using a 12nm simulation cell, this procedure generated a four grain microstructure with N=141705 atoms.

The embedded-atom method [7,8] is used to model copper. The equations of motion are integrated using a Verlet leap-frog algorithm [9] with a time step of 5fs. We approximate the system at constant temperature (T=300) with a global thermostat [10], and compare to a constant energy (adiabatic) simulation. To simulate a constant uniform expansion strain rate, the position of every atom in the simulation cell is written:

$$\underset{\sim}{x} = H \underset{\sim}{s}$$

where $s \subset [0, 1]$ and $H = \{\underset{\sim}{a}, \underset{\sim}{b}, \underset{\sim}{c}\}$ is a matrix composed of the three vectors of the simulation cell [11]. A constant strain rate is simulated by specifying a constant time derivative of H.

RESULTS AND DISCUSSION

The resulting stress as a function of volume for a simulated strain rate of $7.7 \times 10^8 s^{-1}$ at T=300K is

shown in Figure 2. The system is elastic up to failure, which occurs at 8.4GPa. Also shown in Figure 2 are the results for simulations at the same strain rate with a small void (3.75nm) placed at the center of one grain and at the junction between all four grains.

A thin slice from the simulation with no pre-existing voids is shown at the time of failure in Figure 3a. The system nucleates a single void at the grain junction. This void grows by emitting dislocations into the surrounding grains as can be seen by the mismatch of the lattice planes at the center of the grains and the distortion in grain shape. At a

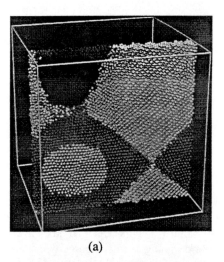

(a)

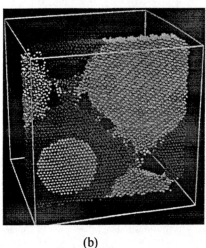

(b)

Figure 3. A slice through the simulation cell at time of failure (a) $\gamma = 7.7 \times 10^8 s^{-1}$, (b) $\gamma = 6.1 \times 10^9 s^{-1}$.

212

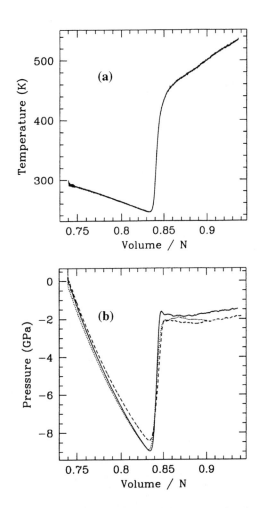

FIGURE 4.(a) Kinetic temperature and (b) stress as a function of volume per atom for an adiabatic simulation at a strain rate of 7.7x108s-1. The stress is compared to isothermal simulations at 300K (dash) and 240K (dot).

times in a viscous manner. The temperature at the yield point is approximately 240K. The rise is temperature upon yielding is well below the melting point (1370K).

The pressure from the adiabatic simulation is compared to isothermal simulations at 300K and at 240K in Figure 4b. The volume at yielding is nearly the same in the three simulations. The 240K simulation yields at a pressure close to the adiabatic simulation and the number of nucleated voids is statistically the same. The results are sensitive to the temperature at yielding though only slightly at the relatively low temperatures studied here. This effect should be more significant at higher temperatures close to melting and during late time growth. It will be quantified in future work.

The affect of system size was studied by simulating failure in two systems: the first was similar to the previous simulation with different grain orientations and N=138490; the second used the same geometric microstructure and grain orientations, but twice the cell length in each direction. The only difference being the number of atoms, N=1141045 in the larger system. The resulting stress curve is shown in Figure 5. The curve for the smaller system is qualitatively the same as before while the larger system yields at a much lower strain. Given the same microstructure and hence the same number of nucleation sites for a slower strain rate, one hypothesis is that the stored elastic energy per nucleation

strain rate of $6.1 \times 10^9 s^{-1}$ (Figure 3b), many more voids nucleate and decorate the grain boundaries with very little distortion of the surrounding grains.

To assess the validity of the isothermal approximation we compare to a constant energy (adiabatic) simulation using the same T=300K initial conditions as in Figure 2 with no pre-existing voids. The result is shown in Figure 4. The kinetic temperature (Figure 4a) initially decreases with increasing volume as we do work against the interatomic potential (Joule-Thompson effect), then rises rapidly at the yield point, and finally rises slowly at late

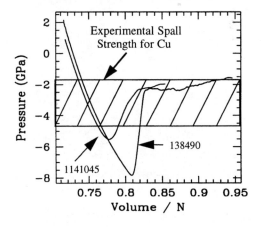

FIGURE 5. Stress curves for two simulations with the same geometric microstructure but differing in the number of atoms per grain. The band is the range of experimental data for copper.

213

site determines the strain at which failure occurs. This energy is greater for a given strain for the larger system and is the energy available to drive the growth of voids.

Figure 2 illustrates the dependence of the number of voids nucleated at failure upon the strain rate. To investigate this further, we consider the failure of a single crystal so that nucleation is homogeneous. A slice through the simulation cell at the time of failure for a $6.1 \times 10^9 s^{-1}$ strain rate is shown in Figure 6. Several voids are visible. By enumerating the number of voids at this and higher/lower strain rates we quantify the dependence of void spacing on strain rate. The result is shown in Figure 7. Several conclusions are evident. The data in the range of this study is nearly linear on this log-log plot and is well approximated by the expression $l \sim 0.005 \, C_s / \gamma$, where $C_s = 3480 m/s$ is the bulk sound speed and γ is the strain rate. At high strain rates $(10^{11} s^{-1})$ the void spacing is approaching the interatomic spacing. The study of slower strain rates, in the experimental range, requires much larger system sizes or a special continuum boundary condition.

The simulations presented here are for a fixed strain rate. Experimentally there will be a range of

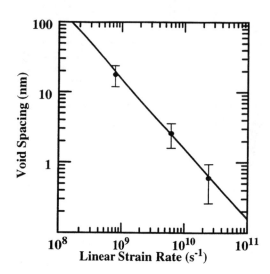

FIGURE 7. The dependence of void spacing on strain rate during simulated isotropic expansion of single crystal copper.

strain rates up to the peak strain rate $(\sim 10^6)$, resulting in a spectrum of void sizes and spacings.

ACKNOWLEDGEMENT

Work performed under the auspices of the U.S. DOE by LLNL under contract No. W-7405-ENG-48.

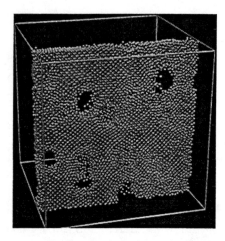

FIGURE 6. A slice through the simulation cell during yielding of single crystal copper at $\gamma = 7.7 \times 10^9 s^{-1}$. Several voids are nucleating.

REFERENCES

1. Barbee, T.W., Jr., L. Seaman, R. Crewdson, and D. Curran, *Journal of Materials*, **7**, 393 (1972).
2. Meyers, M.A. and C.T. Aimone, *Progress in Materials Science*, **28**, 1 (1983).
3. Curran, D.R., L. Seaman, and D.A. Shockey, *Physics Reports*, **147**, 253 (1987).
4. Zurek, A.K. and M.A. Meyers, in **High-Pressure Shock Compression of Solids II: Dynamic Fracture and Fragmentation**, L. Davison, D.E. Grady and M. Shahinpoor eds. (Springer-Verlag, New York, 1996).
5. Gurson, A.L., *J. Eng Materials and Tech.*, **99**, 2 (1977).
6. Phillpot, S.R., J. Wang, D. Wolf, H. Gleiter, *Mat Science and Eng.*, **A204**, 76 (1995).
7. Daw, M.S., and M.I. Baskes, *Phys. Rev B* **29**, 6443 (1984).
8. Oh, D.J. and R.A. Johnson, in **Atomistic Simulation of Materials: Beyond Pair Potentials**, V. Vitek and D.J. Srolovitz eds. (Plenum, New York, 1989).
9. Allen, M.P. and D.J. Tildesley, **Computer Simulation of Liquids** (Clarendon Press, Oxford, 1987).
10. Hoover, W.G., *Phys. Rev. A* **31**, 1695 (1985).
11. Parrinello, M. and A. Rahman, *J. Appl. Phys.* **52**, 7182 (1981).

CP429, *Shock Compression of Condensed Matter – 1997*
edited by Schmidt/Dandekar/Forbes
© 1998 The American Institute of Physics 1-56396-738-3/98/$15.00

DISLOCATION MECHANICS BASED CONSTITUTIVE EQUATION INCORPORATING DYNAMIC RECOVERY AND APPLIED TO THERMOMECHANICAL SHEAR INSTABILITY

Frank J. Zerilli

Energetic Materials Research and Technology Department
Naval Surface Warfare Center Indian Head Division, Indian Head, MD 20604-5035

Ronald W. Armstrong

Department of Mechanical Engineering, University of Maryland, College Park, MD 20742

A closer look into the predicted large strain response and plastic shear instability behavior derived from the so-called Z-A equations, incorporating thermally activated yielding of bcc metals (due to their high Peierls stresses) and thermally activated strain hardening of fcc metals (produced by dislocation intersections), shows the need for including dynamic recovery effects in the strain hardening for both bcc and fcc cases. Recovery effects are observed in the stress/strain behavior of tantalum and the bcc-like Ti-6Al-4V titanium alloy. Critical strains for shear banding are computed for Ti-6Al-4V, copper, and ARMCO iron. In addition, a recent result on ductile fracture is reported.

INTRODUCTION

The stress-strain behaviors of hcp materials and certain alloy steels were described in a single equation incorporating both Peierls stress type interactions (predominant in bcc materials) and intersection-of-forest-dislocations type interactions (predominant in fcc materials) written in the form (1,2)

$$\sigma = \sigma_a + B e^{-\beta T} + B_0 \sqrt{\epsilon}\, e^{-\alpha T} \qquad (1)$$

where

$$\beta = \beta_0 - \beta_1 \ln \dot{\epsilon}, \qquad (2)$$

$$\alpha = \alpha_0 - \alpha_1 \ln \dot{\epsilon}, \qquad (3)$$

and

$$\sigma_a = \sigma_G + k \ell^{-1/2}. \qquad (4)$$

In Eq. (1), σ is the Mises equivalent stress, ϵ is the Mises equivalent strain, $\dot{\epsilon}$ is the strain rate, T is the absolute temperature, ℓ is the average grain diameter, and σ_G is due to the effect of solutes and initial dislocation density. The quantities B, β_0, β_1, B_0, α_0, α_1, k, σ_G are considered to be constant.

A simple dislocation argument given by Taylor (3) leads to the strain hardening being proportional to the square root of the plastic strain. Taylor and Quinney (4), in a pioneering study, observed the saturation in the flow stress of copper compressed to a strain of 4.0.

DYNAMIC RECOVERY

Following work by Bergstrom (5), Klepaczko (6), and Estrin and Mecking (7), it is possible to extend Taylor strain hardening to include dynamic recovery and consequent saturation of the stress-strain curve at large strains (8). A similar approach with somewhat different assumptions was undertaken by Follansbee and Kocks in including dynamic recovery in their description of OFE copper (9).

Because the forces between dislocations are inversely proportional to the distance between them and the areal density of dislocations ρ is inversely proportional to the square of the distance between them, at low temperatures, in the absence of thermal fluctuations, the flow stress is given by

$$\hat{\sigma}_{Th} = \hat{\alpha}\mu b\sqrt{\rho} \qquad (5)$$

where $\hat{\alpha}$ is a constant of order unity which depends on geometry and strength of the dislocation-dislocation interaction, μ is the shear modulus, and b is the Burgers vector.

The dislocation density may be related to the strain by the differential equation

$$\frac{d\rho}{d\varepsilon} = \frac{1}{b\lambda} - \Omega\rho \qquad (6)$$

where λ is the mean free path for immobilization of dislocations and Ω is the probability for mobilizing a stopped dislocation. If $\Omega = 0$, then with constant λ, the solution of this evolution equation results in

$$\varepsilon = \rho b\lambda \qquad (7)$$

which leads to Taylor strain hardening, $\hat{\sigma} \propto \sqrt{\varepsilon}$.

If Ω is constant, a simple extension to Taylor strain hardening is obtained which exhibits a saturation stress. In this case, the cold flow stress becomes

$$\hat{\sigma} = B_0\sqrt{\varepsilon_r(1 - e^{-\varepsilon/\varepsilon_r})} \qquad (8)$$

where $\varepsilon_r = \dfrac{1}{\Omega}$ is a characteristic strain for recovery.

At temperatures above zero, this cold flow stress is reduced by the thermal activation factor, and Eq. (1) becomes

$$\sigma = \sigma_a + Be^{-\beta T} + B_0\sqrt{\varepsilon_r(1 - e^{-\varepsilon/\varepsilon_r})}e^{-\alpha T} \qquad (9)$$

Figure 1 shows a comparison of Eqs. (1) and (9) using constants derived for tantalum from data of Chen, Gray, and Bingert (10). A value of ε_r of 1.0 fits their large strain data well.

THERMOPLASTIC SHEAR INSTABILITY

The criterion for the development of localized large shear deformations is determined by the stability of the

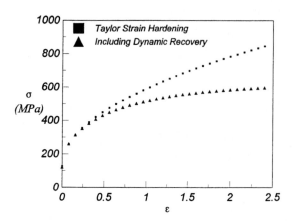

FIGURE 1. Computed stress vs. strain for tantalum at T = 298 K, $\dot{\varepsilon}$ = 10^{-3} s^{-1}. The squares depict the best fit (σ_a = 130 MPa, B_0 = 460 MPa) Taylor ($\sigma_a + B_0\sqrt{\varepsilon}$) strain hardening to the small strain data of Chen, Gray, and Bingert. The triangles represent the relation $\sigma_a + B_0\sqrt{\varepsilon_r(1 - e^{-\varepsilon/\varepsilon_r})}$ where the recovery strain ε_r = 1 and σ_a = 120 MPa, B_0 = 500 MPa.

solutions to the equations of motion when small perturbations are applied. Bai (11) has shown that small perturbations grow exponentially with time if

$$\frac{-\tau_0\left(\dfrac{\partial\tau}{\partial T}\right)_0}{\rho_0 c_v\left(\dfrac{\partial\tau}{\partial\gamma}\right)_0} \geq 1 + \sqrt{\frac{4\kappa\dot{\gamma}_0\left(\dfrac{\partial\tau}{\partial T}\right)_0}{\rho_0 c_v^2\left(\dfrac{\partial\tau}{\partial\gamma}\right)_0}} \qquad (10)$$

where the subscript 0 indicates the quantity refers to the uniform base solution of the simple shear equations and ρ_0 is the initial density (assumed constant), γ is the shear strain, τ is the shear stress, T is the temperature, κ is the thermal conductivity, and c_v is the heat capacity at constant volume (assumed constant). The term under the square root on the right hand side of Eq. (10) vanishes when the thermal conductivity is zero and, in any case, is very small for a typical metal. Thus Eq. (10) amounts to the adiabatic condition proposed by Zener and Hollomon (12) in 1944 -- shear instability develops when the rate of thermal softening overcomes the rate of work hardening.

The critical strain vs. strain rate for the development of shear instability in fcc OFHC copper, calculated from Bai's criterion, is shown in Fig. 2. The constitutive equation is the fcc equation:

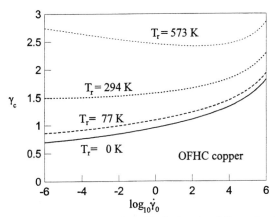

FIGURE 2. Critical strain for adiabatic shear instability to develop in fcc OFHC copper. Curves are plotted vs. strain rate for several initial temperatures, T_r.

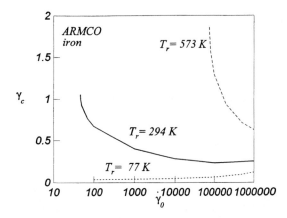

FIGURE 3. Critical strain for adiabatic shear instability to develop in bcc ARMCO iron. Curves are plotted vs. strain rate for several initial temperatures, T_r.

$$\tau = c_0 + B_0 \sqrt{\gamma}\, e^{-\beta T} \qquad (11)$$

where

$$\beta = \beta_0 - \beta_1 \ln \dot{\gamma} \qquad (12)$$

The critical strain vs. strain rate for development of shear instability in bcc ARMCO iron, using Bai's criterion, is shown in Fig. 3. In this case, the constitutive equation is the bcc equation:

$$\tau = c_0 + B_0 e^{-\beta T} + K \gamma^n \qquad (13)$$

The material constants for both figures were given earlier (1). The two figures illustrate the general features of shear banding in fcc and bcc metals which were predicted earlier by analysis of the tensile plastic instability condition and by evaluation of dislocation pile-up stress concentrations (13). The bcc metals are typically more susceptible to shear instability, as shown by the smaller critical strains. The susceptibilty increases with increasing strain rate in the bcc case and decreases with increasing strain rate in the fcc case, although these trends are reversed at the lowest initial temperatures in the bcc case and at the highest initial temperatures in the fcc case. Increasing initial temperature, of course, always tends to suppress shear instability.

The critical strain for shear instability in Ti-6Al-4V is shown in Fig. 4. In this case, a reduction in strain hardening due to dynamic recovery must be included in order to obtain reasonable results. A value of the

recovery strain ε_r of 0.5 is consistent with small strain data for the alloy. The results indicate that this alloy is more susceptible to shear banding than iron, as expected. However, the values of the critical strain are very sensitive to the value chosen for the recovery strain.

DISLOCATION PILE-UPS

Shear band behavior may be described on a more fundamental Hall-Petch dislocation pile-up basis. Armstrong, et al., (14) estimated the upper limiting temperature rise that could be produced by dissipation

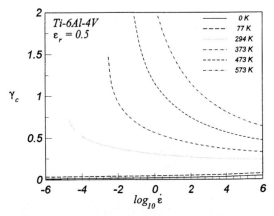

FIGURE 4. Critical strain for adiabatic shear instability to develop in hcp Ti-6Al-4V. Curves are plotted vs. strain rate for several initial temperatures.

of the energy in a dislocation pile-up when catastrophically released during fracturing as

$$\Delta T \leq (k_s \ell^{1/2}/16\pi\kappa)\ln(2\kappa/c_v vb) \qquad (14)$$

where v is the dislocation velocity, b is the Burgers vector, ℓ is a characteristic grain diameter, and k_s is the theoretical shear stress intensity

$$k_s = \pi G b^{1/2}(2-v)/8(1-v) \qquad (15)$$

where G is the shear modulus and v is Poisson's ratio. The dislocation pile-up/avalanche provides the thermal fluctuation to trigger localized shear instability. Thus, the ratio of k_s to κ provides a criterion for susceptibility to shear banding. On this basis, Ti-6Al-4V is very shear band susceptible, followed by iron, with copper being very low in susceptibility, in accordance with experimental observations.

DUCTILE FRACTURE

Ductile fracturing by a hole-joining mechanism eventually occurs for structural metals undergoing extensive plastic straining and, then, the nature and distribution of particles in the material are important.

Figure 5 shows a recent result that is to be discussed in a forthcoming article. The square root of A_0/A_f where A_0 is the initial cross sectional area and A_f is the cross section at fracture in the tensile test is shown to be proportional to the inverse square root of the volume fraction of second phase particles for various copper and steel alloy materials. Note that $(A_0/A_f)^{1/2} = \exp(\varepsilon_f/2)$, where ε_f is the ductile fracture strain. The copper alloy results were reported by Edelson and Baldwin (15) who also quoted the work of Turkalo and Low (16) on carbide steels. Liu and Gurland (17) reported results for a number of spheroidized carbide steels. The result is consistent with a proposal of Petch and Armstrong (18) for fracture to occur at a critical value of particle separation. The slope a of the lines in Fig. 5 is the ratio of particle diameter to critical separation, indicating that the critical separation is proportional to particle diameter. The constant f_B is the volume fraction for which the material becomes brittle and ranges from 0.27 to 0.44 for the materials in Fig. 5.

ACKNOWLEDGMENTS

This work was supported at NSWC by the Independent Research Program and at the University of Maryland by the Office of Naval Research.

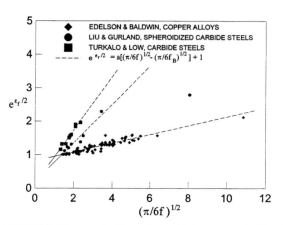

FIGURE 5. Illustrates the linear relationship between $\exp(\varepsilon_f/2)$ and the inverse square root of the volume fraction f of second phase particles in fcc and bcc materials where ε_f ($= 2\ln(A_0/A_f)^{1/2}$) is the ductile fracture strain.

REFERENCES

1. Zerilli, F. J., and Armstrong, R. W., *J. Appl. Phys.* **61**, 1816 (1987).
2. Zerilli, F. J., and Armstrong, R. W., in *High Strain Rate Effects on Polymer, Metal and Ceramic Matrix Composites and Other Advanced Materials*, edited by Y. D. S. Rajapakse and J. R. Vinson, AD-Vol. 48, The American Society of Mechanical Engineers, New York, 1995, pp. 121-126.
3. Taylor, G. I., *Proc. Roy. Soc.* **A145**, 362 (1934).
4. Taylor, G.I., and Quinney, H., *Proc. Roy. Soc.* **A143**, 307 (1934).
5. Bergstrom, Y., *Mater. Sci. Engng.* **5**, 193 (1970).
6. Klepaczko, J., *Mater. Sci. Engng.* **18**, 121-135 (1975).
7. Estrin, Y., and Mecking, H., *Acta metall.* **32**, 57-70 (1984).
8. Zerilli, F. J., and Armstrong, R. W., article in preparation.
9. Follansbee, P. S., and Kocks, U. F., *Acta metall.* **36**, 81 (1988).
10. Chen, S. R., Gray, G. T., III, and Bingert, S. R.,in *Tantalum*, edited by E. Chen, A. Crowson, E. Lavernia, W. Ebihara, and P. Kumar, The Minerals, Metals, and Materials Society, Warrendale, PA, 1996, pp. 173-184.
11. Bai, Y. L., *J. Mech. Phys. Solids* **30**, 195-207 (1982).
12. Zener, C., and Hollomon, J. H., *J. Appl. Phys.* **15**, 22 (1944).
13. Armstrong, R. W., and Zerilli, F. J., *Mech. Mater.* **17**, 319 (1994).
14. Armstrong, R. W., Coffey, C. S., and Elban, W. L., *Acta metall.* **30**, 2111 (1982).
15. Edelson, B. I., and Baldwin, W. M., *Trans. ASM* **55**, 231-250 (1962).
16. Turkalo, A. M., and Low, J. R., Jr., *Trans. Am. Inst. Mining, Met. Pet. Eng.* **212**, 750 (1958).
17. Liu, C. T., and Gurland, J., *Trans. ASM* **61**, 156 (1968); *Trans. TMS-AIME* **242**, 1535 (1968).
18. Petch, N. J., and Armstrong, R. W., *Acta metall. mater.* **38**, 2695 (1990).

CP429, *Shock Compression of Condensed Matter – 1997*
edited by Schmidt/Dandekar/Forbes
© 1998 The American Institute of Physics 1-56396-738-3/98/$15.00

USE OF THE STEINBERG AND CARROLL-HOLT MODEL CONCEPTS IN DUCTILE FRACTURE

L. Seaman

Poulter Laboratory, SRI International, Menlo Park, California 94025

M. Boustie and T. de Resseguier

ENSMA, Futuroscope, France

We have extended the SRI ductile fracture model (DFRACT) for spall behavior of aluminum and copper. The temperature computation procedure, thermal strength reduction function, work hardening, and Bauschinger effects from the Steinberg model were added. The threshold stress for void growth in the DFRACT model was equated to the stress for general yielding in the Carroll-Holt model for porous materials. With these modifications of DFRACT, we simulated a series of earlier impacts in 1145 (commercially pure) aluminum in which partial spall had been reached. The revised model was able to represent the numbers, sizes, and locations of voids through the sample. The use of the Carroll-Holt and Steinberg model features allows the DFRACT model to reach larger void volumes in the simulations and therefore to better represent heavy damage.

INTRODUCTION

Our major interest is in relating the ductile fracture processes under high rate loading – nucleation of voids, void growth, coalescence – to standard properties of ductile materials so that fracture behavior can be predicted without extensive fracture tests. Toward this end we are modifying the model to include several known physical aspects of fracture. Here we describe the nature of our model (called DFRACT from Seaman et al, 1976), review some of the plate impact data on a commercially pure aluminum, outline the current augmentations to the model, and show comparisons with the data.

NATURE OF THE MODEL

Our DFRACT model is a micromechanical model, dealing with void size distributions, not with individual voids. The model is part of the NAG (nucleation and growth) family of SRI fracture models for brittle and ductile fracture and for shear banding (Curran et al, 1987). We have separated the actual fracture processes into a nucleation process (debonding of inclusions, coalescence of vacancies at triple points, etc.) and a growth process (enlargement of the apparent radius of the voids by plastic flow). The model treats the behavior of a void size distribution (a family of voids ranging from zero radius up to the maximum observed) such as those seen in Figure 1a. These data were obtained by counting the voids by sizes on several planes through the thickness of a mm-thick specimen radiated with a laser in ENSMA. The figure shows that the maximum damage (uppermost curve) occurred between 50 and 100 μm from the rear of the plate.

In the model the size distributions are given by the exponential relation for the number per unit volume greater than the radius R (as in Figure 1b):

$$N_g = N_O \exp(-R/R_1) \qquad (1)$$

where N_O is the total number of voids per unit volume of all sizes and R_1 is a characteristic size for the distribution (shown by points on the curves around R = 0.0004 in Figure 1b).

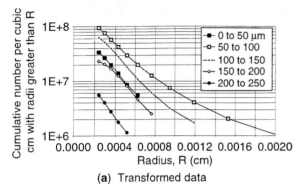

(a) Transformed data

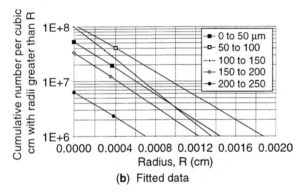

(b) Fitted data

FIGURE 1. Void size distributions for Test 1B1 in aluminum.

DATA FROM IMPACTS

The original void-count data from several planes in five plate impact specimens in 1145 aluminum (Barbee et al, 1972) show considerable scatter, as indicated in Figures 2 and 3 for two of these tests (these data have been transformed from counts on a cross section to volumetric distributions). The variation of the damage on the plane of maximum damage only are shown in Figures 4 and 5 as functions of the impact velocity (and hence, of stress level). Clearly, there is significant scatter from test to test as well as between regions of a single test. As indicated in the table, specimen 939 was half the thickness of the others; therefore, it experienced about half the stress duration.

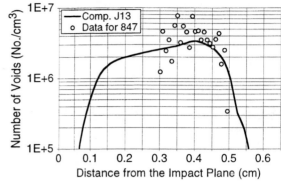

(a) Test 847 at 128.9 m/s (0.9 GPa in tension)

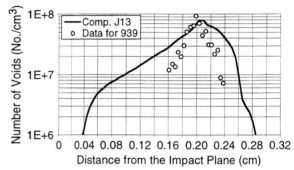

(b) Test 939 at 185.6 m/s (1.3 GPa in tension)

FIGURE 2. Data and computed damage (Set J13) in Al 1145.

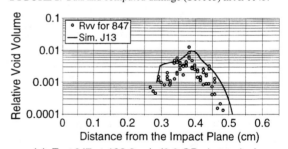

(a) Test 847 at 128.9 m/s (0.9 GPa in tension)

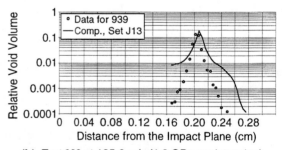

(b) Test 939 at 185.6 m/s (1.3 GPa peak tension)

FIGURE 3. Void volume data and simulation (Set J13) in Al 1145.

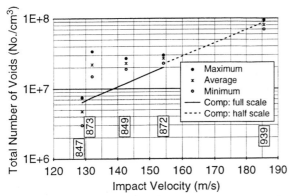

FIGURE 4. Variation of number of voids per unit volume at the plane of maximum damage with impact velocity, 1145 aluminum.

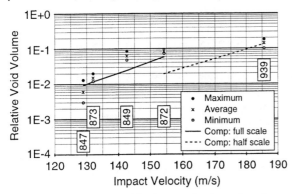

FIGURE 5. Variation of void volume at the plane of maximum damage with impact velocity, 1145 aluminum.

Based on these original data we proposed the following nucleation and growth rate relations to represent the damage processes:

$$dN/dt = T_4[exp((P - T_5)/T_6) - 1] \qquad (2)$$

$$dR/dt = T_1R(P - T_2) \qquad (3)$$

where T_1 through T_6 are material constants and P is the mean stress. N refers to the total number of voids; whereas, R is the radius of an individual void. The usual elastic-plastic stress-strain relations describe the behavior of material around the voids.

CURRENT MODEL

In our current effort, we use the deviator stress model of Steinberg (1980). His model includes work hardening, a Bauschinger effect (both illustrated in Figure 6 for this aluminum), and a thermal strength

reduction effect. In addition we modified the growth law by replacing the constant T_2 by the threshold curve of Carroll and Holt (1972) in Figure 7. This curve (the equation is in the figure) is a function of the current yield strength Y and the relative void volume V_V. The yield value is modified by all the Steinberg features.

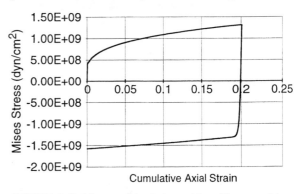

FIGURE 6. Steinberg work-hardening and Bauschinger model.

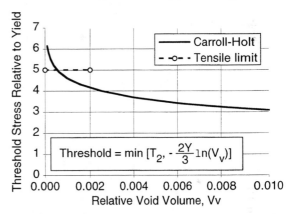

FIGURE 7. Threshold stress for void growth from Carroll and Holt.

The current model was incorporated into our one-dimensional wave propagation code for simulating the impact tests. Simulations were made of the set of five tests and the fracture parameters (T's) were adjusted to optimize the fit. The results are the curves in Figures 2 through 5. An impact of a half-size specimen at 154 m/s was simulated to allow us to draw the dotted lines in Figures 4 and 5. The variation of damage with distance away from the plane of maximum damage in test 847 (with a low level of damage) appears to be represented fairly well by the simulations; however, in test 939 the variation with distance is faster in the data than in the simulations (Figures 2b and 3b).

From the simulations we noted that when the period of tensile stress begins, there is rapid nucleation of voids and a consequent reduction of the tensile stress as the applied strain is taken in void opening rather than in maintaining the stress. Later in this tensile period, void growth dominates and the stress is insufficient to continue nucleation. The declining growth threshold (Figure 7) allows the growth to continue through most of this tensile period. Hence, there is a complex interaction between nucleation and growth. This interaction may be the reason that the dotted curve for a half-size specimen lies above the full-size curve in Figure 4.

SUMMARY

The new fit seems to be an improvement over the earlier fit (Barbee et al, 1972) because now we are able to reach larger levels of void volume. Both the variable threshold stress from Carroll and Holt and the thermal strength reduction effect tend to produce more localization at the plane of maximum damage and thus enhance the void volume there. However, the localization is not yet severe enough to correctly represent the damage appearing in high-damage tests such as 939 (Figures 2b and 3b). By incorporating the Steinberg and Carroll-Holt features into the DFRACT model we have partially reduced the need for special tests to calibrate the model and also improved its representation of the physical processes involved in fracture.

TABLE 1. 1145 Aluminum Plate Impact Conditions

Test No.	Velocity (m/s)	Stress* (GPa)	Thickness (mm) Flyer	Target
847	128.9	0.90	2.36	6.35
873	132.0	0.93	2.36	6.35
849	142.6	1.01	2.36	6.35
872	154.2	1.08	2.36	6.35
939	185.6	1.30	1.14	3.17

* Computed peak tensile stress

ACKNOWLEDGMENTS

The authors acknowledge the Ministère de l'Education Nationale, de l'Enseignement Supérieur et de la Recherche for its grant for the model development and simulations.

REFERENCES

1. Barbee, Jr., T. W., Seaman, L., Crewdson, R. and Curran, D. R., Dynamic Fracture Criteria for Ductile and Brittle Metals, *Journal of Materials, JMLSA*, **Vol. 7, No. 3**, (1972), p 393-401.
2. Curran, D. R., Seaman, L., and Shockey, D. A., Dynamic Failure of Solids, *Physics Reports*, **Vol 147**, Nos. 5 & 6, (1987), pp 254-388.
3. Carroll, M. M. and Holt , A. C., Static and Dynamic Pore-Collapse Relations for Ductile Porous Materials, *J. Appl. Phys.*, **Vol. 43, No. 4**, pp. 1626-1636, (1972).
4. Seaman, L., Curran, D. R., and Shockey, D. A., Computational Models for Ductile and Brittle Fracture, *Journal of Applied Physics*, **Vol. 47, No. 11**, (1976), p. 4814.
5. Steinberg, D. J., Cochran, S. G., and Guinan, M. W., A Constitutive Model For Metals Applicable At High-Strain Rate, *J. Appl. Phys.*, **Vol 51, No. 3**, (1980), p. 1498-1504.

222

CP429, *Shock Compression of Condensed Matter – 1997*
edited by Schmidt/Dandekar/Forbes
© 1998 The American Institute of Physics 1-56396-738-3/98/$15.00

ANALYSIS OF TEMPERATURE INFLUENCE
ON THE DYNAMIC FRACTURE OF METALS

A.A. Bogach

Institute of Chemical Physics in Chernogolovka, Moscow reg., Russia, 142432

It follows, from free surface velocity measurements, that spall strength of metals is not sensitive to the initial temperature and peak pressure of shock-wave pulse until the thermodynamic parameters of damage material are far from the melting point and decreases sharply near it. In assuming of localized melting the expressions for threshold temperature and for dependence of spall strength on temperature were received. According to the solution the threshold temperature and the spall strength depend on stress gradient in shock pulse as power of 1/3. It is provided the calculated and experimental results for Al, Mg, Sn, Pb, Zn and Mo.

INTRODUCTION

There are a number of experimental observations of spall strength of metals at high temperatures. The experiments with preliminary heating [1] and irreversible heating in shock wave [2,3] have shown, that the strength varies little in wide interval of temperatures but sharply falls close to melt. The spall strength of aluminium AD1, magnesium Mg95 and zinc ZHP falls in range 120-140°C, 250-300°C [1] and 70-100°C [4] respectivly. The strength of Mo [5] and Cu [6] does not vary up to temperatures 1400°C and 425°C. In recalculation on residual temperature [7] the dynamic strengths of Sn and Pb fall in ranges 100-150°C [2,3] and 80-100°C [3]. Thus, it is possible to conclude that in conditions of dynamic fracture that metals differing on strength and thermodynamic properties have similar temperature dependence of spall strength. The temperature was entered in a kinetics of nucleation [8] and growth [9] of failure, but it was not received a particular dependence of the spall strength on the temperature. The aim of this work is to get analytical temperature dependence, to describe the experimental results on spall strength of metals at the elevated temperatures.

MODEL OF DYNAMIC FRACTURE

On modern representations the dynamic fracture of ductile solids at pulse tension is caused by growth of voids. The void dynamics was analyzed in detail in work [10,11]. By results of work [12] in high strain rate conditions the viscosity depends on the strain rate as $\eta \sim 1/\sqrt{\varepsilon}$. This dependence gives the pore volume growth rate in form:

$$\dot{V} = \frac{3V_0\Delta P^2 \exp(3t\Delta P^2 / B)}{B}, \qquad (1)$$

where V_0 is the initial specific volume of pores, $\Delta P = P - P_0$, $B = 32/9\eta^2\varepsilon$, P is the tension stress far from the pore, $P_0 = 2Y\ln(b/a)$ is threshold of growth, a is pore radius, b is external radius of material matrix, Y and η are dynamic yield strength and viscosity of the matrix, respectively, t is the time of fracture.

A viscoplastic dissipation of deformation energy results to localized heating in vicinity of pore [10,11]. Early, in the models of compacting both the reduction with growth of temperature [8,13] and independence of Y, η up to melting [14] were used. In present work, the last assumption was accepted as it reflects more authentically the experimental facts on shock loading of metals at increased temperatures [1,4,5,15]. Thus, the main grounds of the model

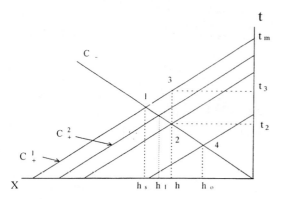

FIGURE 1. X-t diagram of the interaction of shock wave with free surface of sample.

offered are:1) the dynamic yield strength Y and viscosity η in the vicinity of pore do not depend on the temperature up to melting; 2) the effective Y and η decrease after melting, that results to sharp acceleration of pore growth. From the first assumption it follows that the spall strength does not vary up to some initial temperature T_o of material, after which it decreases.

During growth the melting is beginning at pore surface [9,13]. Therefore, in the model T_o is defined as threshold, since which the surface is heated up to melting point T_m. Assuming that there is not thermal loss during this process, the energy balance gives the equation for T_o:

$$T_m - T_o = \frac{t}{c\rho\tau}(2Y + \frac{3}{2}(P - P_o)),\qquad(2)$$

where ρ is density, $\tau = B/(P-P_o)^2$, c is the heat capacity. In this work the fracture time t was accepted as the time of "fracture delay" [16]. The method of definition of this time is explained on Fig.1. In section h, after passage of the last C_--characteristic of reflected wave, the tension stress is equal to P=P(h) in point 2, and $P_{so}=P(h_s)$ for points 1 and $P_o=P(h_o)$ for point 4. If an arrival of minimum W_m on free-surface velocity profile coincides with C_+^1- characteristic, the stress P_s is determined as spall strength of material. As metals have a very low initial porosity ($\sim 10^{-4}$) [17], it is possible to neglect by stress relaxation at the onset of the fracture. For triangular pulse of loading, the stresses increase linearly with time with gradient A. The occurrence of the minimum W_m means, that inside sample the fracture rate is comparable to the strain rate in

unloading part of initial pulse [16]. The fracture rate (1) depends on accumulated failure, therefore the maximum of $\dot{V}$ is reached in point 3. Then, the fracture time is defined as $t=t_3- t_{2,}$, or in acoustic approximation:

$$t = \frac{2}{c_o}(h_s - h),\qquad(3)$$

where c_o is the bulk sound velocity. For linearly increasing stresses it is possible P change to h in (1) through the replacement:

$$P - P_o = A(h - h_o).\qquad(4)$$

Substitution (3) and (4) in (1) permits to define h as:

$$h = \frac{2}{3}h_s + \frac{1}{3}h_o + \delta h,\qquad(5)$$

where δh is the correction value. In variable P, the fracture time is

$$t = \frac{2}{c_o}\left(\frac{P_s - P_o}{3A} - \delta h\right).\qquad(6)$$

Using (2), (4), (5), (6), we receive the following expression for the threshold temperature in the first approximation:

$$T_m - T_o = \frac{(2Y + P_{so} - P_o)(P_{so} - P_o)^3}{12c\rho\eta^2\dot{\varepsilon}Ac_o}.\qquad(7)$$

Just the same, it is possible to define the additional heating of the pore surface in section h_1 (Fig.1) for C_+^2- characteristic reaching the free surface earlier of C_+^1- characteristic:

$$T_m - T = \frac{(2Y + P_s - P_o)(P_s - P_o)^3}{12c\rho\eta^2\dot{\varepsilon}Ac_o},\qquad(7a)$$

where $P_s=P(h_1)$. It is clear from (7) and (7a), that $T_m-T_o > T_m-T$. If the initial temperature $T>T_o$, then the local melting has to begin first in the section h_1. According to the ground 2), the fracture rate increases by a jump in the section h_1 and, thus, reduction of spall strength will be recorded.

The residual heating after release increases temperature at the value δT. In the model, the effect of residual heating is taken into account by subtracting δT from a left-hand part of (7) and (7a). After that, the ratio of (7) and (7a) and the ground 1) give the spall strength dependence on the temperature as follows:

$$P_s = const ,\qquad\qquad T \leq T_o$$

224

$$\frac{(2Y + \Delta P)\Delta P^3}{(2Y + \Delta P_0)\Delta P_0{}^3} = \frac{(T_m - T - \delta T)}{(T_m - T_0 - \delta T)}, \quad T > T_0 \quad (8)$$

where $\Delta P = P_s - P_0$, $\Delta P_0 = P_{0s} - P_0$.

COMPARISON WITH EXPERIMENT

The analysis (8) shows, that a) there is no smooth transition between parts of this dependence, b) the strength in melting point has nonzero meaning P_0. Item a) is a result that we do not take into account that the initial centers of failure is not spherical and the threshold is a statistical value. Item b) is direct consequence of the model and it requires in check. As follows from elastic static decision, the minimum threshold is $P_0 = 2/3Y$, that corresponds to the onset of the formation of a plastic zone.

In present work, it is offered to consider P_0 as variable, instead of it being strictly defined by porosity. It is affirmed, that in the initial period of fracture, the pore grows in elastic-plastic mode. The law of the pore growth in this case is described by (1), where $P = P(b)$, $P_0 = 2Y\ln(b/a)$, b-radius of the plastic zone. It is obvious, that b depends on $P = P(\infty)$, and $P(b) \cong P(\infty)$ for large b. Using Irwin's idea [18,19] about calculation of the size of plastic zone of crack, the expression for threshold of single stationary pore is received in the form:

$$P_0 = 2Y \ln(\sqrt{\frac{3P(\infty)}{Y}} - 0.5). \quad (9)$$

The expression (9) decides the problem of this threshold for materials with a very low initial porosity [10,11], and it corresponds to the results in [17], where it was shown, that porosity of copper samples increases from the free surface without sharp jumps. The definition of the threshold (9) was used in (8) for construction of the dependence of spall strength on the temperature for Al, Mg, Zn (Fig 2,3). In the case of zinc [4], the calculation of P_0 is complicated by significant anisotropy of large grains of the samples. An average yield strength was used in calculation. Table 1 provides calculated and experimental results. Figures 2,3 display a good agreement between the theoretical curves and measured data. The difference is explained by both the experimental error and some indefinite of T_0 and δT. Since the plastic heating in shock wave does not take into account, the values of δT were used a larger of theoretical ones [7]. The best agreement was received, when $\delta T = 40°$ for Mg, $10°$ and $45°$ for Al, that exceeds the theoretical values [7] by factor two. The fit method can provide possibility to make an estimation for δT. From (8) and (9) it follows, that spall strength in melting point is $\sim Y$. The reduction of it below the value is concerned with the hit into the biphase area after release or shock-induced melting, that was not considered. Perhaps, the results for zinc show the existence of the transitional strength.

The spall strength is known to depend on the strain rate in unloading part of the initial pulse [20]. The expression (7) contains pressure gradient A being proportional to the strain rate. It is interesting to define the dependence of threshold T_0 on the pulse this parameter. The condition of minimum on the velocity profile gives the dependence of strength on gradient A in the form:

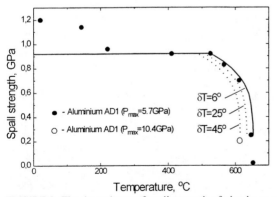

FIGURE.2. The dependence of spall strength of aluminum on the temperature.

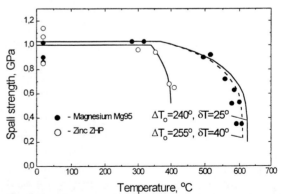

FIGURE 3. The dependence of spall strength of magnesium and zinc on the temperature.

TABLE 1. The parameters of theoretical model and experimental temperature threshold.

Material	η, Pa·c	$\dot{\varepsilon}$, μs^{-1}	Y, GPa	A, GPa/mm	P_s, GPa	δT, $^\circ C$	ΔT_{0exp},* $^\circ C$	ΔT_{0th},** $^\circ C$
AD1	67	3.3	0.2	0.57	0.92	6	120-140	60
Mg95	95	2.2	0.15	0.67	1.03	25	250-300	195
Zinc	100	2.2	~0.4	1.0	1.0	10	70-100	30
Pb	-	-	~0.03	2.5	0.5	0	80-100	30
Sn	-	-	~0.06	1.6	0.7	0	100-150	90
Mo	170	1.6	0.95	3.5	2.3	5	-	50

* $\Delta T_{0exp} = (T_m - T_o - \delta T)$ in (8) from experiments.
** $\Delta T_{0th} = (T_m - T_o - \delta T)$ in (8) from calculations.

$$P_s - P_o = \left(4Ac_o\eta^2\dot{\varepsilon}\ln\frac{8A\eta^2\dot{\varepsilon}}{3\rho c_o V_o \Delta P^2}\right)^{1/3}, \quad (10)$$

where $A=P_s/h_s$, h_s is the thickness of spall plate. If in (10) to neglect weak function $\ln^{1/3}$, we shell receive:

$$P_s - P_o \sim A^{1/3} \quad (11)$$

It follows from (7) and (11), that

$$T_m - T_o \sim const_1 + const_2 A^{1/3} \quad (12)$$

The experimental results for Al and Mg [1] give smaller exponent (~ 0.07) for A. Therefore, it would be expected that the dependence from A with exponent 1/n instead of 1/3. The conclusion received about the decreasing of the threshold T_o with the rise of the gradient A should be checked.

There are not data on the measurements of spall strength of refractory metals near the melting point. The model defines $P_s(T)$ of these metals in wide range of temperatures as (8), where P_{os} and Y can be measured under normal conditions, T_o and P_o are calculated from (7) and (9). The value of viscosity can be determined from measurements of rise time of plastic shock front at normal temperature by the expression:

$$\eta = \frac{\sigma_v}{\dot{\varepsilon}},$$

where σ_v and $\dot{\varepsilon}$ are maximum viscous stress [12] and shear strain rate in shock front, respectively. The results for Mo are presented in Tab 1. The rise times were estimated from [1,4,5]. The data for Pb, Sn are absent and some estimation of η, ε was used in the form of $\eta^2\dot{\varepsilon} \sim 10^{10}$ $Pa^2 \cdot s$.

ACKNOWLEDGMENTS

This work was supported by the Russian Foundation for Basic Research, grant number 96-01-01899a. The authors would also like to thank Prof. G.I.Kanel, and Drs. A.V.Utkin and S.V.Razorenov for a fruitful discussing.

REFERENCES

1. Kanel G.I., Razorenov S.V., Bogatch, A., Utkin A.V., Fortov V.E., *J. Appl. Phys.* 79, 8310-8317 (1996).
2. Grady, D.E., *J. Mech. Phys. Solids* 36, 353-384 (1988).
3. Kanel G. I., Razorenov S.V., Utkin A.V., Grady D.E. The spall strength of metals at elevated temperatures. *Shock Compression of condesed metter*, Seattle, USA, 1995.
4. Razorenov S.V., Bogatch, A.A., Kanel G.I., Utkin A.V., Fortov V. E., Grady D.E. Elastic-Plastic Deformation and Spall Fracture of Metals at High Temperatures. In present proseedings
5. Duffy, T. S., Ahrens, T. J., *J. Appl. Phys.* 76, 835-842 (1994).
6. Bless, S. J., Paisley, D. L., *Shock Waves in Condensed Metter*, ed. Asay J.R., Graham R.A., Straub G.K., Amsterdam: Elsevier, 1984, p. 163.
7. McQueen R.G., Marsh, S. P., *J. Appl. Phys.* 31, 1253 (1960).
8. Curran, D.R., Seaman, L., Shockey, D.A., *Phys.Reports,* 147, 279-289 (1987).
9. Wang Z. P., *J. Appl. Phys.*, 76, 1535-1542 (1994).
10. Carroll M.M., Holt A.S., *J.Appl.Phys.* 43, 1626-1636 (1972).
11. Johnson J. N., *J.Appl.Phys.* 52, 2812-2825 (1981).
12. Swegl J.W., Grady D.E., *J.Appl.Phys.* 58, 692-701 (1985).
13. Carroll M. M., Kim K. T., Nesterenco V.F., *J.Appl.Phys.* 52, 1962-1967 (1986).
14. Dunin S.Z., Surkov, V.V., *J.Prikl.Mech.Tech.Phyz. (USSR),* 1, 131-142 (1982).
15. Asay J.R., *J. Appl. Phys.* 45, 4441-4452 (1974)
16. Utkin A.V., *Prikl. Mech. Tech. Phyz.(Rus),* 4, 140 (1993).
17. Curran D.R., Seaman L., Shockey D.A., *Phys.Reports,* 147, 343 (1987).
18. Irwin G.R., *Proc. 7th Sagamor Ordnance Mater. Res. Conf.* 4, Syracuse: Syracuse University press, 1961, pp. 63-78.
19. Hellan K., ed. Morozov E.M., *Introducthion to Fracture Mechanics*, Moscow: Mir, 1988, ch. 2, pp. 27-30.
20. Kanel G.I., Razorenov S.V., Utkin A.V., Fortov V.E. Shock Wave Phenomena in Condensed Matter, Moscow: Yanus-K, 1996, ch. 5, pp. 149-201.

CP429, *Shock Compression of Condensed Matter – 1997*
edited by Schmidt/Dandekar/Forbes
© 1998 The American Institute of Physics 1-56396-738-3/98/$15.00

PREDICTING OF FRAGMENT NUMBER AND SIZE DISTRIBUTION

Lin Zhang Xiaogang Jin

Laboratory for Shock Wave and Detonation Physics Research, Southwest Institute of Fluid Physics, P.O.Box 523-61, Chengdu Sichuan China 610003

A new statistic model for predicting the fragment number and size distribution in a dynamic fracture process has been proposed. Based on the statistic theory, we consider that there is a minimal size in fragments due to the unloading of release waves arising from the nucleation of fracture, and assume that the positions of fracture are distributed randomly in the sample, we get a set of enclosed equations for predicting the fragment number and size distribution. Expanding cylindrical shell experiments have been carried out to examine such a model, results show that it has a good capability of prediction.

INTRODUCTION

The fragment number and size distribution in dynamic fracture is an interesting subject due to its applications in both military and civil fields. Numerous theoretical and experimental works have been performed in the past decades, see for example, Lienau[1]、 Weibull[2]、 Mott[3]、 Wesenberg[4]、 Grady[5] 、 Grady and Kipp[6,7]. Nevertheless, it seems that a sophisticated model, especially a physical model with capability of prediction, is still needed. In this paper, based on the statistic theory, we have proposed a new model for describing this dynamic fracture behavior, which has shown a good agreement with the experimental observations.

THEORY AND MODEL

For homogeneous material under uniform mechanical state, basically, we assume that: (1) When the mechanic state approaches to a certain criterion, numerous cracks nucleate almost in the same time. (2) Cracks distribute throughout the whole sample completely randomly. Apparently, because of the unloading of the relief waves, vast majority of the nucleated cracks could not run through the sample. In the following, we use the term "superior crack" to denote the crack that could run through the sample. Whether a crack is a "superior crack" or not depends on its nucleating time and position. For any two cracks which are adjacent in the nucleating time, if their distance is less than a certain critical value, the later nucleated one would not run through the sample, thus it is not the superior crack. This also means that for any two cracks that have run through already, they must have a distance greater than the critical value. According to the assumption (1), the time difference in nucleation for all cracks could be neglected, which means the above critical distance would be the same for every pair of the adjacent superior cracks. In this sense, the critical distance is virtually responsible for the extreme minimum fragment size, S_{min}.

From the above discussion, we may see that the fragmentation of sample is determined by both the nucleating time and position of the cracks. We analyzed this problem with statistic method, and obtained the fragment number and size distribtion.

Take the expanding ring as an example, which is an one-dimensional fracture process, suppose the ring's circumference at the fracture moment is S_0, according to the above discussion, only a small number of nucleated cracks can finally run through the sample, and apparently such a number couldn't be more than $n_s = S_0/S_{min}$. Divide the nucleated cracks into two classes, one gets labels showing the cracks with the superior "quality", and the number is n_s; the other gets no labels, which refer to those cracks that can't run through the sample, assume the number is n. Neverthless, it should be noticed that even for the labelled cracks only part of them (assuming the number is N) can surely run through the sample, and the rest of (n_s-N) cracks actually can not achieve their superior advantage, because their nucleating positions or time are not suitable and would lead to unloading in stress. Therefore the probability of finding N fragments is written as:

$$W = C_{n_s}^{N} \cdot C_{n+n_s-N}^{\frac{S_{min}}{S_0}(n+n_s)N} \cdot [\frac{S_{min}}{S_0}(n+n_s)N]! \\ \cdot [n+n_s - \frac{S_{min}}{S_0}(n+n_s)N]! \qquad (1)$$

Let n tend to ∞, and then the most probable value of N is given as:

$$\ln[(\frac{S_0}{NS_{min}} - 1)(1 - \frac{NS_{min}}{S_0})] = \frac{NS_{min}}{S_0 - NS_{min}} \qquad (2)$$

In the above analysis, the effect of size distribution on the fragment number has been naturally taken into account, because the crack nucleating positions have been treated as random.

Apparently, corresponding to a same N, there could be a lot of distribution patterns, among which there must be a way with the most possibility. In the following we will find it out by the statistical method.

The assumption of random fracture implies that the size of any one of the fragments may be any of the value between S_{min} and S_0 with an equal possibility, subject only to the qualification that the total size must be S_0. Given that the number of the fragments with size between S_i and S_i+dS_i is N_i, and

regard each value of S_i as a box with G_i cells, then for the distribution $(N_1, N_2, \cdots, N_i \cdots)$ the number of distributing patterns is:

$$W' = \frac{(N_1 + N_2 + \cdots N_i + \cdots)!}{\prod_i N_i} \prod_i G_i^{N_i} \qquad (3)$$

Then the most probable distribution can be determined by:

$$\frac{N_i}{N} = e^{-\frac{S_i}{S_r}} G_i \qquad (4)$$

where $(1/S_r)$ is the undetermined multiplier. Obviously $G_i \propto dS_i$, so we can write it as $G_i = AdS_i$, where A is a coefficient. Taking $N(>S)$ as the number of fragments whose sizes are greater than S, we get:

$$N(>S) = NAS_r e^{-\frac{S}{S_r}} - NAS_r e^{-\frac{S_0}{S_r}} \qquad (5)$$

Combining with the normalized condition and the limiting condition, and also considering the fact that S_0 is significantly greater than S_r and S_{min}, from Eq.(5) we obtain:

$$N(>S) = NS_r e^{-\frac{S-S_{min}}{S_r}} \qquad (6)$$

$$AS_r e^{-\frac{S_{min}}{S_r}} = 1 \qquad (7)$$

$$N = \frac{S_0}{S_{min} + S_r} \qquad (8)$$

In the above equations, S_{min} could be determined by the dynamic loading conditions and the material properties, generally we can write it as:

$$f(S_{min}, \alpha) = 0 \qquad (9)$$

where α represents all of the necessary loading conditions and material properties. For the stretching rod (one — dimensional stress loading), using the expressions of the stress-released length and the fracture time given by Kipp and Grady[6], we suggest:

$$S_{min} = (\frac{3\gamma}{\rho_0})^{1/3} \dot{\varepsilon}^{-2/3}$$ (9')

Where ρ_0 is the material density, γ is the fracture energy, ε is the strain rate. Now, Eqs.(2) (6) (7) (8) (9) or (9') consist of a group of enclosed equations, with which we can predict the fragment number and size distribution. Generally these equations could also be used to describe the dynamic fragmentation in two or there dimensions

EXPERIMENTAL RESULTS

Expanding cylindrical shell experiments similar to that performed by Winter[8] have been carried out using a one-stage gas gun, the initial size of OFHC copper samples is $\phi(23\text{-}20)$mm$\times$60mm. Figure 1 is a picture for one of the recovered samples, from which we can see the sample has been recovered completely.

Fig 1. A picture for one of the recovered samples

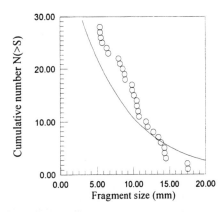

FIGURE 2. Fragment Size distribution. Empty circles " O " are the experimental data, solid line is the theoretical result of this model

Three samples with nearly the same loading condition have been recovered, and their fragment numbers are 9、9 and 10, respectively. In Fig. 2 the " O " is the experimental results of the fragment size distribution. On the other hand, if we consider the fracture characteristics of an expanding ring are mainly determined by circumferential tension, then we can use Eqs. (2) (6) (7) (8) (9') to describe the above experimental results. The strain rate for the present experiments is about $1.5\times10^4 s^{-1}$; The fracture energy γ of OFHC copper is 0.2×10^5J/m$^{2[9]}$; N, S_r and S_{min} are calculated out as 28.98, 6.96mm and 3.10mm respectively. In Fig.2 the solid line is the theoretical fragment size distribution. It can be seen that the present theoretical model has shown a good capability for descripting the experimental results.

CONCLUSION

A new statistic model for predicting the fragment number and size distribution in dynamic fracture process has been proposed, see Eqs.(2) (6) (7) (8) and (9), examination with the experimental results demonstrates that it has a good capability of prediction.

ACKNOWLEDGMENTS

We would like to express our thanks to Dr. Hongliang He for helping us translate this paper into English; This study is supported by the Science Foundation of CAEP.

REFERENCE

1 Lienau,C. C. "Random Fracture of Brittle Solid", J. Franklin .Inst, 1936,221:485

2 Weibull,W. "A statistical Theory of the Strength of Material", Ingretensk .Akad .Handl,1939. 151

3 Mott,N. F. "Fragmentation of Shell Cases", Proc. R. Soc, 1947, 189A:300

4 Wesenberg, D. L. , Sagartz, M. J. , "Dynamic Fracture of 6061-T61 Aluminum Cylinders", J. Appl. Mechanics, 1977 . 643

5 Grady,D. E. , "Local Inertial Effects in Dynamic Fragmentation", J. Appl. Phys, 1982,53:322

6 Kipp, M. E ., Grady, D. E.,"Dynamic Fracture Growth and Interaction in One Dimension", J. Mech. Phys. Solid, 1985 33:399

7 Grady, D. E. , Kipp, M. E., "Geometric statistics and Dynamic Fragmentation ", J. Appl .Phys, 1985, 58(3):1210

8 Winter, R. E. "Measurement of Fracture Stain at High Strain Rate ", In: Mechanical Properties at High Rate of Strain, 1979 . 81

9 Иванов, А. Г. "ОТКОЛ В КВА ЗИАКУСТИЧЕСКОМ ПРИБЛИ —ЖЕНИИ", ФГВ, 1975, 11(3): 475

CP429, *Shock Compression of Condensed Matter – 1997*
edited by Schmidt/Dandekar/Forbes
© 1998 The American Institute of Physics 1-56396-738-3/98/$15.00

MODELLING PRECURSOR DECAY IN AD-99.5 ALUMINA

C. Hari Manoj Simha,[1] S. J. Bless[1] and A. Bedford[2]

[1]*Institute for Advanced Technology, University of Texas at Austin, 4030-2 W. Braker Lane, Austin, TX 78759*
[2]*Aerospace Engineering Department, University of Texas at Austin, Austin, TX 78712*

In this paper we present a simple model to explain the absence of precursor decay in the Coor's AD-99 5 Alumina ceramic, as shown by Grady in his plate impact experiments. The model is incorporated into the Research EPIC 95 finite element code. The simulations compare well with Grady's results.

INTRODUCTION

At the previous topical conference in Seattle (1995), Dr. D. E. Grady presented some results on elastic precursor decay in Coor's AD-99.5[1] alumina (1). He found that there was no appreciable precursor decay in this high purity alumina. We have been investigating the high strain rate properties of the same material (2), so we decided to use Grady's data as a first step in developing a constitutive model. In this work we will present a model that reproduces his results fairly well.

GRADY'S DATA

Grady performed three plate impact experiments on AD-99.5 to study precursor decay. The impact velocity was around 1.5 km/s and he used a VISAR as the diagnostic. The sample thicknesses were nominally 2.5 mm, 5 mm and 10 mm (samples A, B and C), as shown in Fig. 1. He then used self-similar scaling to normalize the velocity-time traces to a thickness of 10 mm and centered the profiles at time t = 0. Self-similar scaling involves multiplying the times of the sample A trace by 4 and the sample B by 2. These are then plotted with the C sample and shown in Fig. 2. He concluded that, though from Fig. 1 it appears as if the precursor is decaying, the self-similar scaling plot shows that there is no precursor

decay. With the exception of Boron Carbide, he obtained similar results in Aluminum Nitride, Silicon Nitride and Titanium Diboride.

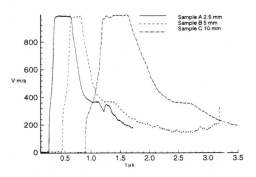

FIGURE 1. Grady's VISAR profiles for AD-99.5 alumina plate impacts, 1.5 km/s.

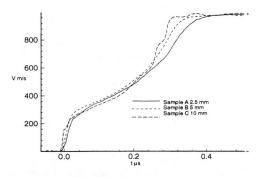

FIGURE 2. Self-similar scaling, normalized to a target thickness of 10 mm.

1. Alumina Dense 99.5% by wt.

MODEL

We assume that the material starts to undergo micro-cracking at the Hugoniot Elastic Limit (HEL). There is some evidence that this is indeed the case. Bourne and Rosenberg (3) have used the phenomenon of fracto-emission during cracking to show that the HEL is the macromanifestation of the onset of microcracking. Using the von-Mises criterion for failure, the reported value of HEL = 6.71 GPa (4), and Poisson's ratio = 0.232 (5), the failure strength at the HEL is found from,

$$Y = \sigma_{HEL}\left(\frac{1-2\upsilon}{1-\upsilon}\right) \qquad (1)$$

to be 4.68 GPa.

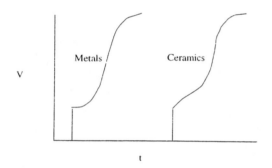

FIGURE 3. V-t profiles in metals and ceramics.

In Fig. 3, the difference between ceramics and metals is the ramp seen between the HEL and the deformation wave (term used by Grady to describe the anelastic portion of the profile).

We assume that the flow stress depends on the deviatoric strain-rate. This leads to a flow rule of the form,

$$Y = Y_{qs} + \frac{3}{2}\frac{\dot{\varepsilon}'^{n}}{\lambda} \qquad (2)$$

where $\dot{\varepsilon}'$ is the deviatoric strain rate, $\frac{2}{3}Y$ is the deviatoric stress, λ is a pressure-dependent parameter, n is a material index and Y_{qs} is the flow stress of the failed material. We assume that Y_{qs} is pressure dependent and is given by,

$$Y_{qs} = min(\alpha P, Y_2) \qquad (3)$$

where Y_2 is the maximum cap, P is the pressure and α is the slope of the dependence (see Figure 4).

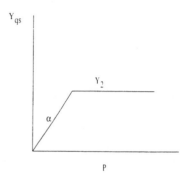

FIGURE 4. Flow stress vs. pressure for the failed material.

Following Partom (11), we assume that λ is pressure dependent as follows,

$$\lambda = \lambda_0 exp(\lambda_1(P - P_H)) \qquad (4)$$

where λ_0, and λ_1 are parameters and P_H is the pressure at the HEL. This is equivalent to assuming that the flow of the failed material is viscous with a pressure-dependent viscosity. This is reasonable since with increasing pressure the friction between the fractured surfaces that flow will increase. We use a Mie-Gruneisen EOS with S = 1.3, and Γ = 1.3 (9).

We have departed from other workers (Rajendran (6), Espinosa (7), and Johnson and Holmquist (8)) who assume that the flow stress depends on plastic strain rate rather than the deviatoric strain rate. The reason is as follows. At the HEL, once the material starts to flow, the strength is the failure strength (see Eq. (1)). At this point, the plastic strain rate is zero and the deviatoric strain rate is non zero. In order to relax to the flow curve for the failed material (see Eq. (3)), we need a non zero strain rate (as indicated by the ramping in the wave profile). The plastic strain rate being zero will not work and would cause the drop in strength to be abrupt and not relax the strength to the lower curve.

SIMULATIONS

We incorporated the model into the Research EPIC 95 finite-element code and determined the best set of parameters to match Grady's experiment at 544 m/s shot AO3 (5). The parameters are shown in Table 1 and the result of the simulation in Fig. 5.

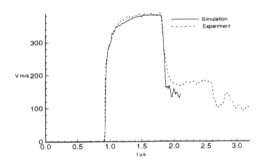

FIGURE 5. Simulation and experiment for SHOT A03.

TABLE 1. Model Parameters

α	1
n	1.2
Y_2	5.9 GPa
P_H	3.15 GPa
λ_0	0.022 1/GPaμs
λ_1	0.6 1/GPa

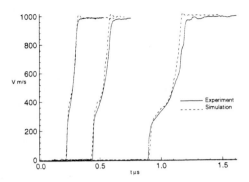

FIGURE 6. Simulation and experiment - precursor decay experiments.

The same set of parameters are then used to simulate the precursor decay experiments. The results are shown in Fig. 6. The simulations agree quite well with the experiments.

DISCUSSION

We have shown that using the deviatoric strain rate instead of the plastic strain rate accounts for the ramp seen in the VISAR profiles. Since we do not see any precursor decay at all we can only conclude that the precursor decays very quickly at thicknesses less than 2.5 mm.

However the results of Cagnoux and Longy (12) and Murray et al. (13) for a different set of aluminas contradict Grady's results. We believe our model can qualitatively reproduce their results. Consider Fig. 8 wherein EPIC calculations with $\lambda_0 = 0.05$ 1/GPaμs are shown. We see that there is indeed pronounced precursor decay. Note that we have reduced the flow stress by increasing the value of λ_0.

We put forward the following hypothesis to explain the above results. Most commercial aluminas have intergranular glassy phases (7). Recently, Sundaram and Clifton (14) have shown that this phase melts during high strain rate loading. With increasing purity the amount of the glassy phase is lower. The amount of glass in AD-99.5 is almost negligible and very hard to detect (16).

233

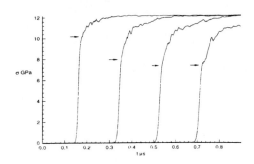

FIGURE 7. Stress profiles showing precursor decay.

At the HEL, once the microcracks develop the material starts to flow. The resistance to flow is mainly due to friction between the fractured surfaces. In the case in which there is melted interfacial glass, the resistance to flow is reduced. From Equations 2 and 4, we see that λ_0 controls the magnitude of the flow stress. For a high value of λ_0, the flow stress is lower and vice versa.

CONCLUSION

Thus one would expect high purity low glass aluminas to display rapid precursor decay, occurring at depths < 2.5 mm, whereas in the relatively low purity high glass aluminas the precursor decay occurs over larger depths in the target. This conclusion is supported by the results of Grady (1) and Murray et al. (13).

We have presented a simple model that can explain the absence/presence of precursor decay in aluminas. The model is not complete at present since we have not accounted for the damage due to microcracking which will affect the elastic moduli of the material. We are working on incorporating this feature into our model.

ACKNOWLEDGMENTS

The authors would like to thank Dr. D. E. Grady for providing his VISAR data. Most of the initial model development was done on WONDY V. We also would like to thank Dr. M. E. Kipp for providing a copy of WONDY V.

This work was performed under U.S. Army Research Laboratory Contract DAAA21-93-C-0101.

REFERENCES

1. Grady, D. E., "Shock-Wave Properties of Brittle Solids," presented at the Shock Compression of Condensed Matter Conference, Seattle, August 13-18, 1995.
2. Simha, C. H. et al., "Dynamic Failure of AD-99.5 Alumina," presented at the Shock-Wave and High Strain-Rate Phenomena EXPLOMET '95 Conference, El Paso, 1995.
3. Bourne, N. K. and Rosenberg, Z., "Fracto-emission and its Effect Upon Noise in Gauges Placed Near Ceramic Interfaces," presented at the Shock Compression of Condensed Matter Conference, Seattle, August 13-18, 1995.
4. Dandekar, D. P. and Barkowski, P., "Shock Response of AD-99.5 Alumina," presented at the High Pressure Science and Technology Conference at Colorado Springs, Colorado, 1993.
5. Grady, D. E. and Moody, R. L., "Shock Compression Profiles in Ceramics," Sandia Report, SAND96-0551, 1996.
6. Rajendran, A. M., "High Strain-Rate Behavior of Metals, Ceramics and Concrete," Wright Laboratory Report, WL-TR-92-4006, 1992.
7. Espinosa, H. D., Brown University, Ph.D. Thesis, 1992.
8. Johnson, G. C. and Holmquist, T. J., Shock-Wave and High Strain-Rate Phenomena EXPLOMET '92 Conference, 1992
9. Munson, J. E., and Lawrence, R. J., *Journal of Applied Physics*, **50**, 6272-6282 (1979).
10. Partom, Y., *Journal of Applied Physics*, **59**, 2716-2727 (1986).
11. Partom, Y., "Calibrating a Material Model for AD-99.5 Alumina from Plate Impact VISAR Profiles," presented at the International Conference on Mechanical and Physical Behavior of Materials Under Dynamic Loading DYMAT 94, Oxford, September, 1994.
12. Cagnoux, J. and Longy, F., "Is the Dynamic Strength of Alumina Rate-Dependent?" presented at the Shock Compression of Condensed Matter Conference, 1987.
13. Murray, N. H. et al., "Precursor Decay in Several Aluminas," presented at the Shock Compression of Condensed Matter Conference, Seattle, August 13-18, 1995.
14. Sundaram, S., and Clifton, R. J., "Pressure-Shear Impact Investigation of the Dynamic Response of Ceramics," in *Advances in Failure Mechanisms in Brittle Materials*, AMD-vol. 29, MD-vol. 75, 1996.
15. Rabenberg, L. K., Department of Material Science, University of Texas at Austin, personal communication, 1997.

CP429, *Shock Compression of Condensed Matter – 1997*
edited by Schmidt/Dandekar/Forbes
© 1998 The American Institute of Physics 1-56396-738-3/98/$15.00

MODELING COMPRESSIVE FLOW BEHAVIOR OF A TUNGSTEN HEAVY ALLOY AT DIFFERENT STRAIN RATES AND TEMPERATURES

Tusit Weerasooriya

Material Division, Army Research Laboratory, Aberdeen Proving Ground MD 21005

Room temperature stress-strain behavior was obtained for a tungsten heavy alloy at 9000, 0.1 and 0.0001/s strain rates. In addition, at the strain rate of 0.1/s, stress-strain data were obtained at 423°K, 573°K and 732°K. Deformation behavior was modeled using standard and modified Johnson-Cook (JC) and Power-Law (PL) models. In the modified models, the temperature terms are replaced by other functions that are proposed in the literature for these models as well as by Arrhenius type exponential functions. The best representation of the data was obtained from modified models with the exponential temperature functions. The model constants were determined using slow rate stress-strain data and the high rate yield stress. This paper presents the modified JC and PL models and the corresponding model constants for the tungsten heavy alloy.

INTRODUCTION

A systematic study of the stress-strain behavior as a function of strain rate and temperature under uniaxial compression has been conducted for a tungsten heavy alloy (WHA) by Weerasooriya[1]. In this paper, the experimental isothermal behavior is represented by two thermo-viscoplastic phenomenological models that are in the literature: the Johnson-Cook (JC) model available in most computer codes and the Power Law (PL) which is frequently used by the Material Scientists. Temperature terms of these models are evaluated against the experimental data and modified to better represent the data.

EXPERIMENTS

In this section, uniaxial compression experiments are summarized. A complete description of these experiments is given elsewhere[1].

Experimental Procedure

A 17% swaged 93W-5Ni-2Fe alloy was used for the experiments. More details of this alloy including chemical composition, microstructure and the processing history are given elsewhere[1].

The specimen used for all the testing was a cylinder of 6.350 mm diameter and 3.175 mm length. On the two loading surfaces of the specimen, concentric grooves were machined to hold lubrication to reduce barreling.

Room temperature slow rate (0.1 and 0.0001 s^{-1}) compression tests were conducted using a servo controlled Instron hydraulic test machine. To obtain the effect of temperature on the deformation, high temperature (423, 573 and 732°K) tests were conducted at a true strain rate of 0.1 s^{-1} using the same test machine. High rate tests at 9000/s were conducted using a compression Hopkinson bar.

Experimental Results

Uniaxial compression behavior at different strain rates is shown in Figure 1. At the lower rates (0.0001 and 0.1/s), this WHA work hardens as the strain is increased. At the 9000/s higher rate, material softens with straining just after yielding.

Yield stress of the material is highly rate sensitive and increases with strain rate.

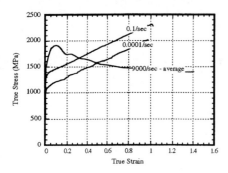

FIGURE 1. Stress-strain behavior as a function of strain rate at 293°K.

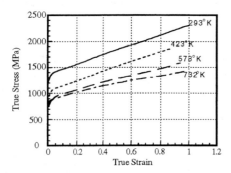

FIGURE 2. Stress-strain behavior as a function of temperature at 0.1 s^{-1} strain rate.

Stress-strain behavior from room and high temperature tests at 0.1 s^{-1} strain rate is given in Figure 2. In general, the flow stress decreases with increasing temperature at a given strain.

MODELING THE DEFORMATION

Linear correlation of yield stress at room temperature with the strain rate from both log-log (Correlation Coefficient, R = 0.99727) and semi-log (R = 0.9892) representations are reasonable. A similar possibility of dual correlation existed for torsion experiments for this alloy(2). This suggests the feasibility of using either a logarithmic law or a power law for modeling the flow stress behavior for rate effects. Researchers have proposed both exponential(3,4) and power(5) law type macroscopic constitutive models to represent plastic flow behavior via slip for many materials.

Johnson-Cook (JC) Model

Johnson and Cook proposed a phenomenological thermo-viscoplastic relationship(4) for flow stress (σ) as a function of plastic strain (ε_p), strain rate ($\dot{\varepsilon}$) and temperature (T°K). This relationship is given by:

$$\sigma = \left(\sigma_{yo} + B \cdot \varepsilon_p^n\right) \cdot \left(1 + C \cdot \ln\left(\frac{\dot{\varepsilon}}{\dot{\varepsilon}_o}\right)\right) \cdot f(T) \tag{1}$$

Here, σ_{yo} (=1186 MPa) is the yield stress at the reference (room) temperature and the reference strain rate ($\dot{\varepsilon}_o$ = 0.1/s). $f(T)$ is defined as

$$f_0(T) = \left[1 - \left(\frac{T - T_R}{T_M - T_R}\right)^\alpha\right]$$ where T_M is the melting

temperature and T_R is the room temperature. B, C, n and α are model constants to be determined from the experimental data. Constants B and n are obtained from the stress-strain curves at the reference strain rate and temperature. Constant C is obtained from yield stresses at 0.0001/s and 9000/s (yield stress at 9000/s is obtained from extrapolation of the data to the initial linear elastic response).

To obtain the temperature effects, experiments can be conducted at temperatures lower than the room temperature. In this case, function $f_o(T)$ is not defined. For the model to be valid at lower temperatures, the following functional forms for $f(T)$ have been also proposed in the literature (see ref. 2 for $f_1(T)$ and ref. 6 for $f_2(T)$:

$$f_1(T) = \left[1 - \left(\frac{T - T_R}{T_M - T_R}\right)\right]^\alpha \tag{2}$$

and $$f_2(T) = \left[1 - \left(\frac{T}{T_R}\right)^\alpha\right] \tag{3}$$

In addition, a function of the Arrhenius form $f_3(T) = \exp\left[\frac{G}{R_g} \cdot \left(\frac{1}{T} - \frac{1}{T_R}\right)\right]$ is considered for $f(T)$ where G is a model constant and R_g is the gas constant (R_g = 8.31 J.mol^{-1}·K^{-1}). This function is proposed because the plastic deformation by dislocation motion is a thermally activated process(7). Here G represents activation energy to overcome the barrier to dislocation glide.

All four functional forms are evaluated using the deformation data at the reference strain rate ($\dot{\varepsilon}_o$) 0.1/s at the four tested temperatures. At the reference strain rate of 0.1/s, Equation 1 reduces to,

$$f(\text{T}) = \frac{\sigma}{\left(\sigma_{yo} + \text{B} \cdot \varepsilon_p^n\right)} \qquad (4)$$

Figures 3(a-d) represent the log-log representation of the Equation 4 with the temperature function indicated in the title of the figure. Y and X axis variables are chosen for the resultant equation to be linear after rearranging its terms. Using these rearranged equations, stress-strain data at the three temperatures are plotted for the strains indicated in the plots. Thus the more linear the data points are represented in the plots, the better the corresponding temperature function. Correlation Coefficients for the plots with each of the four functional forms are: $f_0 \to \text{R} = 0.94906$; $f_1 \to \text{R} = 0.96184$; $f_2 \to \text{R} = 0.90499$ and $f_3 \to \text{R} = 0.99189$. Therefore, from these plots, it can be seen that the exponential function $f_3(\text{T})$ is the best functional form for $f(\text{T})$. The corresponding JC model constants are:

$$\dot{\varepsilon}_o = 0.1/s;\ \text{T}_R = 293°\text{K};\ \text{G} = 1.81\ \text{kJ/mol};$$
$$\sigma_{yo} = 1186\ \text{MPa};\ n = 0..6125;$$
$$\text{B} = 1057\ \text{MPa};\ \text{C} = 0.0227$$

(a). $f_0(\text{T}) = [1 - \{(T\text{-}Tr)/(Tm\text{-}Tr)\}^\alpha]$

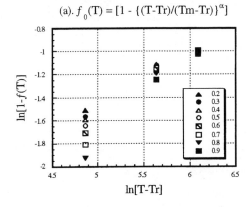

(b). $f_1(\text{T}) = [1 - (T\text{-}Tr)/(Tm\text{-}Tr)]^\alpha$

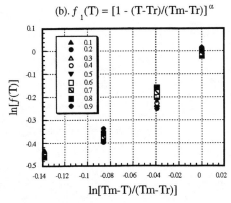

(c). $f_2(\text{T}) = [1 - (T/Tr)^\alpha]$

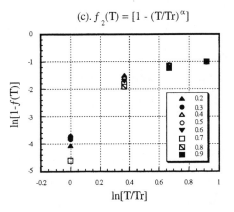

(d). $f_3(\text{T}) = \exp[k.(1/T\text{-}1/Tr)]$

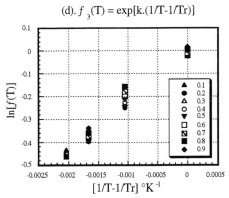

FIGURE 3. Evaluation of different temperature functionals for JC model.

In Figure 4, the comparison between the stress-strain curves generated using this set of constants and the experimental data shows an excellent match.

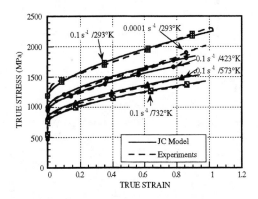

FIGURE 4. Comparison of JC model and experimental (used for model calibration) stress-strain data.

Power Law (PL) Model

Costin et. al.(5) used a power law relationship for the flow stress (σ) as a function of strain (ε), strain rate ($\dot{\varepsilon}$) and temperature (T) of the form:

$$\sigma = \sigma_o \left(\frac{\varepsilon}{\varepsilon_y} \right)^n \cdot \left(\frac{\dot{\varepsilon}}{\dot{\varepsilon}_o} \right)^m \cdot f(T) \qquad (5)$$

where $f(T)$ is given by $f_{0P}(T) = \left(\dfrac{T}{T_R} \right)^\alpha$. Here,

n, m and α are model constants to be determined from the experimental stress-strain data. σ_O (= 1186 MPa) is the yield stress at the reference strain rate ($\dot{\varepsilon}_o$ = 0.1/s) and the reference (room) temperature (T_R = 293°K). The constants ε_y and n are obtained from the room temperature stress-strain data at 0.1/s strain rate. These two parameters change with the strain and are therefore defined in the form a + b Tanh(ε) where a and b are constants. The strain rate hardening exponent m is obtained from the yield stress data at the strain rates 0.0001/s a and 9000/s strain rates.

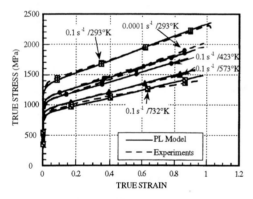

FIGURE 5. Comparison of PL model and experimental stress-strain data.

Since plastic flow is controlled by dislocation motion which is a thermally activated process, an exponential function of Arrhenius type is also chosen to represent the temperature effects. The chosen Arrhenius functional for $f(T)$ is of the form

$$f_{1P}(T) = \exp\left[\frac{G}{R_g} \cdot \left(\frac{1}{T} - \frac{1}{T_R} \right) \right], \text{ where G is the}$$

model constant. Results show that the both $f_{0P}(T)$ and $f_{1P}(T)$ functions can correlate the different temperature data reasonably well. (for $f_{0P}(T)$: α = -

0.4887 with R = 0.9712; for $f_{1P}(T)$: G/R_g = 0.21539 °K with R = 0.9787).

PL model constants that were obtained from the experimental data are:

$$\dot{\varepsilon}_o = 0.1/s; \ T_R = 293°K; \ \sigma_O = 1186 \text{ MPa};$$
$$m = 0.02469; \ G = 1.79 \text{ kJ/mol};$$
$$\text{for } \varepsilon < 0.00349, \ n = 1.0 \text{ and } \varepsilon_y = 0.00349;$$
$$\text{for } \varepsilon \geq 0.00349, \ n = 0.07249 + 0.28874 \text{ Tanh}(\varepsilon)$$
$$\text{and } \varepsilon_y = 0.00408 + 0.12381 \text{ Tanh}(\varepsilon)$$

Figure 5 shows an excellent match between the experimental data and the PL model.

SUMMARY AND CONCLUSIONS

In this paper, an improved expression for the temperature term for both the Johnson-Cook and Power-law strength models has been suggested. Among the various expressions considered, the one that was based on the Arrhenius type relationship reproduced the stress-strain curves at different temperatures very well. Accurate description of the thermal softening using the improved relationship is useful in the computational analysis of projectile penetration into target plates. The material model constants for a tungsten heavy alloy were successfully determined for the JC as well as PL models using experimental stress-strain data at different strain rates and temperatures.

REFERENCES

1. Weerasooriya, T., "Deformation Behavior of 93W-5Ni-2Fe at Different Rates of Compression Loading and Temperatures," Submitted to the Int. J. of Plasticity.

2. Weerasooriya, T., "Modeling Flow Behavior of 93W-5Ni-2Fe Tungsten Heavy Alloy," Proc. of the Joint AIRAPT/APS Conf. on High Pressure Science and Technology, Colorado, July 1993.

3. U. F. Kocks A. S. Argon and M. F. Ashby, Thermodynamics and Kinetics of Slip, Prog. Mat. Sc., v. 19, 1975, Pergamon Press, New York.

4. Johnson, G. R. and Cook, W. H., Proc. 7th Int. Symp. Ballistics, The Hague, The Netherlands, p. 1.

5. Costin, L. S., Crisman, E. E., and Hawley, R. H. and Duffy, J., Mechanical Properties at High Strain Rate, Ed. Harding, J., Proc. 2nd Oxford Conf. Inst. Phys., London, 1980, p.90

6. Gray, G. T., Chen, S. R., Wright, W. and Lopez, M. F., "Constitutive Equations for Annealed Metals under Compression at High Strain Rates and High Temperatures," LA-12669 -MS, Los Alamos National Laboratory, Los Alamos, NM, Jan. 1994

7. Frost, H. J. and Ashby, M. F., "Deformation-Mechanism Maps - The Plasticity and Creep of Metals and Ceramics," Pergamon Press.

CP429, *Shock Compression of Condensed Matter – 1997*
edited by Schmidt/Dandekar/Forbes
© 1998 The American Institute of Physics 1-56396-738-3/98/$15.00

SPALLATION STUDIES ON SHOCK LOADED URANIUM LOS ALAMOS REPORT LAUR-97-3169

D.L. Tonks, R. Hixson, R. L. Gustavsen, J. E. Vorthman, A. Kelly, A. K. Zurek, and W. R. Thissell

Los Alamos National Laboratory, Los Alamos, NM.

Several spallation experiments have been performed on uranium using gas gun driven normal plate impacts with VISAR instrumentation and soft recovery. The shock pressures achieved were 81, 53, and 37 kbar. This paper will focus on modeling the free surface particle velocity trace U with of 300 ppm carbon using the 1 d characteristics code CHARADE. The spallation model involves the growth and coalescence of brittle cracks. Metallographical examination of recovered samples and details of the experimental apparatus are discussed in separate papers.

INTRODUCTION

Several studies of spallation in uranium have been done in the past.(1, 2) They have included VISAR traces and computer modeling but no soft recovery with metallurgical examination. Metallurgical examination should be done where possible, however, as it helps greatly in modeling micromechanical processes.

In recent work, our group has measured VISAR traces and done metallographical examination of recovered samples of pure uranium (30 ppm carbon) and less pure uranium (300 ppm carbon). Shock strengths induced were nominally 81, 52.7, and 37 kbar for the less pure uranium and 53 and 35 kbar for the pure uranium. The details of the gas gun work and of the metallurgical examinations are presented in other papers in this volume. (3, 4)

In this paper, calculations of a simple brittle crack growth model of the less pure uranium VISAR traces are presented. The intent is to show that the brittle crack description has some validity. Preliminary spall strength results for the pure uranium data are also presented using a simple tensile threshold model. More detailed model calculations for the pure uranium will be done in the future.

SPALLATION MODELING

The 1D characteristics code CHARADE(5) was used to model VISAR free surface traces using both ductile and brittle crack damage models. The brittle crack model results were more like the data and these results will be reported on here, with brief comments about the ductile model results. The metallurgical sample examinations of the less pure uranium showed a mixed brittle/ductile mode of fracture with the brittle component predominating.

Before presenting the brittle crack damage model, the equation of state and plasticity modeling will be briefly described. The equation of state (eos) treatment is patterned after the "almost isotropic" approximation of Wallace(6). A "pressure dependent bulk modulus" for use in this treatment was obtained from the Hugoniot relations in the usual way. Strictly speaking, this modulus applies only on the Hugoniot. The elastic moduli were degraded because of damage using the framework for isotropically distributed and oriented brittle cracks of Addessio and Johnson (7). Their Eq. (10) was used to obtain the volumetric "inelastic crack opening strain" as a function of negative pressure. The inelastic strain was then added to elastic volumetric strain to obtain the total strain. The resulting formula leads to the following scaled bulk modulus:

$$B' = B / (1 + D), \qquad (1)$$

where B' and B are the scaled and unscaled bulk moduli, respectively, and D is the effective damage quantity:

$$D = 15B(2 - v)\beta \overline{c}^{3}, \qquad (2)$$

where β is given by:

$$\beta = \frac{64\pi}{15} \frac{(1 - v)}{(2 - v)} \frac{N_o}{G}. \qquad (3)$$

v is Poisson's ratio; G is the "solid" shear modulus, N_0 is the volumetric crack center density; and $\overline{c}$ is the average crack radius. The elastic constants were scaled only under volumetric tension, not for volumetric pressure. A similar treatment was used to obtain the effect of crack growth on the pressure increment for use in the characteristic equations.

In Eq. (1) above, B is the bulk modulus in the "equivalent" solid material. In the expression for B mentioned earlier, B is a function of the compression, so using this equation for B requires an "equivalent solid compression". Such a compression was obtained by dividing B into the "equivalent solid stress", σ_s, given by:

$$\sigma_s = \sigma_{cell}(1 + \phi), \qquad (4)$$

where ϕ is a "porosity" given by:

$$\phi = 1 - \exp(4\pi \overline{c}^{3} N_o / 3), \qquad (5)$$

and σ_{cell} is the longitudinal stress in the computational cell.

The quantity ϕ should approximately account for the unstressed regions around a penny shaped crack: the quantity in the exponent is an effective crack volume and the exponential takes overlaps of such volumes into account.

The deviatoric plasticity, which was not degraded by damage, is the same as used earlier for Ta spall modeling.(8) The "normal" component of the deviatoric plasticity was approximated by writing the plastic strain rate as a power law to the second power in the deviatoric stress. Both a forward and backward yield stress were used. This deviatoric plastic strain rate was supplemented by a simple back stress model (8) in which a release immediately produces reverse plastic flow. These two plasticity models helped to produce a realistic release behavior preceding spall.

The brittle crack damage model is patterned somewhat after that of Grady and Kipp (9) and involves the breakout and growth of a single sized population of cracks of size $\overline{c}$. These cracks have all orientations and uniformly fill a computational cell. The cracks break out when the stress intensity reaches K_{IC} and arrest when it reaches K_{IA}. The applied stress intensity factor K_I is given by:

$$K_I = (2 / \pi)\sigma\sqrt{c(t)}, \qquad (6)$$

where σ is the longitudinal stress and c(t) is the time dependent crack radius. During breakout and arrest, the crack radius grows at the constant rate $v_{crk} = f c_s$, where c_s is a shear sound velocity and f is a reducing factor. Thus, c(t) during the first such breakout is given by:

$$c(t) = c_a + v_{crk}(t - t_a), \qquad (7)$$

where c_a is the initial crack radius and t_a is the time of crack breakout. Before first breakout, the cracks are considered not yet formed and no elastic moduli reduction is performed.

Grady and Kipp's modeling included a time delay in crack loading. In calculations using this delay, the results were almost identical to the ones given here.

The effect of the damage on the equation of state has already been discussed. Spallation is taken to occur when ϕ defined in Eq. (5) reaches 0.30. This rule uses a percolation argument to take approximately into account the regions unloaded by the cracks. When these regions overlap to form a path across the sample, it should be close to total fracture.

TABLE 1. Tensile threshold model spall strengths

Shot #	Shock Strength	Spall Strength
56-96-3	81 kbar	26 kbar
56-96-4	37 kbar	20 kbar
56-96-5	52.7 kbar	25 kbar
56-97-3 (pure)	35 kbar	25 kbar
56-97-5 (pure)	53 kbar	30 kbar

RESULTS

To obtain a rough idea of the spall behavior, a tensile threshold spallation model was used to calculate "spall strengths". In this model, a computational cell is spalled when the normal stress, σ, falls below the "spall strength". (σ is defined positive in compression.) There is no damage and no eos degradation before spall. This model was used to predict the point at which the damage release first occurred in the free surface trace. Table 1 shows these calculated spall strengths for both the impure and pure uranium. Note that these strengths are somewhat larger for the pure uranium (30 ppm carbon) than for comparable impure uranium (300 ppm carbon).

The approximation of one half the pullback velocity times the acoustical impedance gave the following spall strength approximations for the 81, 37, and 52.7 kbar shocks, respectively: 21.9, 17.2, and 22.0 kbar. Cochran and Banner (1) reported a model spall strength of 24 kbar for their experiments of shock strength 40 kbar and less. Grady (2) reported corrected spall strengths of 27.3 kbar and 29.2 kbar for shocks of strength 76.9 kbar and 97.3 kbar.

Figure 1 shows the calculated CHARADE free surface profiles with the experimental data. The experimental times were shifted arbitrarily to obtain correspondence with the calculations. 200 zones were used to model the sample plate in CHARADE. Calculated fits using 500 zones resulted in only

slightly different parameter values but, often, significant changes in profile lineshape resulted.

TABLE 2. Brittle crack model parameter values, Series 56-96

Shock kbar	$V_{f/s}$ mm/μs	K_{IC} $Mbar\sqrt{cm}$	K_{IA} $Mbar\sqrt{cm}$	f
81	0.153	$0.9x10^{-3}$	$0.2x10^{-3}$	0.11
52.7	0.0704	$0.75x10^{-3}$	$0.4x10^{-3}$	0.13
37	0.1078	$0.8x10^{-3}$	$0.5x10^{-3}$	0.092

Table 2 gives the brittle crack model parameters and $V_{f/s}$ values used in the calculations. $V_{f/s}$ is the flyer/sample interface velocity. In many cases, it was adjusted slightly away from the values given by impedance matching and the experimental flyer velocity so that the calculated shock plateau velocity would match the data. This was done to better model the spallation. The value of 3.017 (mm/μs) was used for c_s. The value $2x10^4$ cm^{-3} was used for N_0 in all cases. The initial crack size, C_a, used was 1.5 10^{-3} cm.

Figure 1 shows that the brittle crack model reproduces the general spallation behavior of the data. In particular the extent of rebound after spall is modeled fairly well. The calculated profiles tend to have too many calculated "wiggles". This probably means that the late stage damage modeling needs improvement. The modeling contains no detailed crack coalescence mechanisms beyond the inclusion of overlapping in the "regions of influence" of the cracks. In the modeling it was noticed that spallation occurred over a fairly wide area and that the smallest calculated ring periods were due to spalled cells lying closer than reasonable to the free surface.

Another problem is that Equation 1 for the degradation of elastic moduli gave calculated degradations that were quite extreme. This is because the theory behind Eq. (1) is an approximate linearized theory. A more accurate treatment including crack crack interactions would yield smaller degradations. This behavior also contributed to the smallest ring periods in the calculated profiles.

A ductile void growth model, similar to that of Johnson (10) was also tried. Although the general spallation features could be reproduced, the

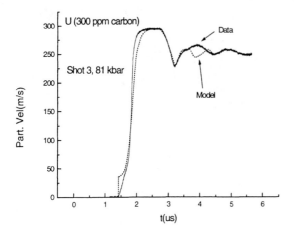

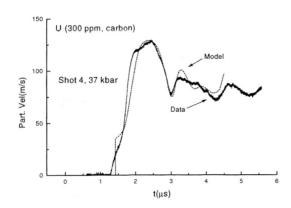

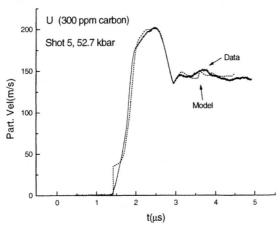

FIGURE 1. Calculated and measured free surface particle velocity profiles.

reproduction of the details was significantly worse. The ductile model did the best job for the 81 kbar

shot, suggesting that the 81 kbar experiment was more "ductile".

The brittle crack fitting parameters are reasonable. The initial crack size of 15 microns and crack growth velocity of about 1/10 the shear sound velocity are plausible. Incidentally, the results are very sensitive to the crack growth velocity. The values of K_{IC} used here are smaller than those typical of low alloy steels(11), e. g. 5×10^{-3} Mbar $\sqrt{cm}$ but larger than the K_{IC} value 0.126×10^{-3} Mbar $\sqrt{cm}$ found by Grady(9) for novaculite, a rock. When quantitative analysis of micrographs is available, the micromechanical realism of the model can be better assessed and improved. In conclusion, it seems that the brittle crack description of spallation in not so pure uranium has some validity. It will be generalized in the future to obtain a mixed brittle/ductile spallation model.

REFERENCES

1. S. Cochran, D. Banner, *J. Appl. Phys.* **48**, 2729 - 2737 (1977).
2. D. E. Grady, in *Metallurgical Applications of Shock-Wave and High-Strain-Rate Phenomena* L. E. Murr, K. P. Staudhammer, M. A. Meyers, Eds. (Marcel Dekker, Inc., New York, 1986) pp. 763-780.
3. A. K. Zurek, et al., Microstructure of Two Grades of Depleted Uranium Under Uniaxial Strain Conditions, APS Topical Conference on Shock Compression of Condensed Matter, Amherst, MA (1997).
4. R. S. Hixson, et al., Spall Wave-Profile and Shock-Recovery Experiments on Depleted Uranium, APS Topical Conference on Shock Compression of Condensed Matter, Amherst, MA (1997).
5. J. N. Johnson, D. L. Tonks, "CHARADE: A Characteristic Code for Calculating Rate-Dependent Shock-Wave Response" *LA-11993-MS* (Los Alamos National Laboratory, 1991).
6. D. C. Wallace, "Thermoelastic-Plastic Flow in Solids" *LA-10119* (Los Alamos National Laboratory, 1985).
7. F. L. Addessio, J. N. Johnson, *J. Appl. Phys.* **67**, 3275-3286 (1990).
8. J. N. Johnson, R. S. Hixson, D. L. Tonks, G. T. Gray, III, in *High Pressure Science and Technology - 1993, AIP Conference Proceedings 309, Part 2* S. C. Schmidt, J. W. Shaner, G. A. Samara, M. Ross, Eds. (American Institute of Physics, New York, 1993) pp. 1095-1098.
9. D. E. Grady, M. E. Kipp, *Int. J. Rock Mech. Min. Sci. & Geomech.* **16**, 293-302 (1979).
10. J. N. Johnson, *J. Appl. Phys.* **52**, 2812-2825 (1981).
11. J. F. Knott, *Fundamentals of Fracture Mechanics* (John Wiley & Sons, New York, 1973).

CP429, *Shock Compression of Condensed Matter – 1997*
edited by Schmidt/Dandekar/Forbes
© 1998 The American Institute of Physics 1-56396-738-3/98/$15.00

MODELLING THE TEMPERATURE AND STRAIN RATE DEPENDENCE OF SPALLATION IN METALS

A. R. Giles, J. R. Maw

AWE, Aldermaston, Reading, U.K.

Further refinements to our implementation of the Johnson void growth model in 2D hydrocodes are described. A consistent treatment of temperature effects is now included and a treatment of failure when voids coalesce has been developed.The temperature dependence of spall strength in the model is tested by comparison with the results of a series of plate impact experiments on aluminium with varying initial temperatures. Observations of spallation in experiments carried out using the AWE HELEN laser also compare favourably with the model predictions. The short pulse durations typical in the latter experiments enable data to be obtained on the strain rate dependence of spall strength which provide a more stringent test of the model's capabilities.

INTRODUCTION

The modelling of spallation in hydrodynamics codes has a long history. Earliest treatments used a simple tensile stress criterion where spallation occured instantaneously when the tensile stress exceeded a particular level. This is inadequate in at least two respects, firstly, particularly at low stresses, the time duration of the spallation process and the effect of partial spallation on the material properties is not accounted for. Secondly experiments have shown that the spall criterion is a function of the strain rate.

More sophisticated models have been developed, notably the SRI Nucleation and Growth (NAG) model [1] which considers the detailed nucleation and growth of voids. Unfortunately this model requires considerable preliminary characterisation of the material properties.

A model which lies between these two extremes is that developed by Johnson [2]. This models growth (but not nucleation) of voids and accounts in a relatively simple manner for the reduction in material strength during spallation.

In a previous paper [3] we discussed the implementation of an extended form of Johnson's model in 2D Lagrangian and Eulerian hydrocodes. In this paper we compare the model predictions with experimental data on the spallation of aluminium at high temperatures [4] and in laser driven experiments carried out using the HELEN laser at AWE. For completeness we begin by briefly reviewing the main features of the extended form of the model.

THE EXTENDED JOHNSON MODEL

The original model

In Johnson's model the material pressure is given by

$$p = f(v/\alpha, e)$$

where v is the specific volume, e is the specific internal energy, $f(v_s, e)$ is the equation of state for the solid material with volume v_s and $\alpha = v/v_s$ is the distention .

Johnson derived an expression for the time rate of change of the distention

$$\dot{\alpha} = -\frac{(\alpha_0 - 1)^{\frac{2}{3}}}{\eta}\alpha(\alpha - 1)^{\frac{1}{3}}(p - p_s)$$

where

$$p_s = -\frac{a_s}{\alpha}\ln\frac{\alpha}{\alpha - 1}$$

is the tensile strength of the material under static conditions which decreases as the distention ratio increases. The parameter a_s controls the spall strength and the parameter η is a viscosity which controls the rate of growth of the voids.

Failure criterion

At some point it may be expected that the individual voids will coalesce and a separated spall plane will form. In our implementation we have simulated this effect by forcing the pressure to zero for values of α greater than a critical value α_f. Denoting values at tme step n by α^n etc. we use the Taylor series expansion

$$p^{n+1} = p^n + \left(\alpha^{n+1} - \alpha^n\right)\frac{\partial p}{\partial \alpha} + \ldots$$

Setting $p^{n+1} = 0$ and noting that

$$\frac{\partial p}{\partial \alpha} = \frac{\partial p}{\partial v_s}\frac{\partial v_s}{\partial \alpha}$$

gives

$$\alpha^{n+1} = \alpha^n + \frac{\bar{p}\rho_s\alpha^n}{\frac{\partial p}{\partial v_s}}$$

Temperature and Melting Effects

We have also included the effects of temperature and melting by making the parameter a_s a function of temperature as follows

$$a_s = a_{s0}[1 - A(T - 300)] \qquad T < T_m$$

$$a_s = 0 \qquad T \geq T_m$$

where the temperature T is calculated in a thermodynamically consistent manner from the EoS and the melt temperature $T_m(v_s)$ is calculated from the Lindemann relation. In the absence of data on the temperature dependence of spall strength we take the magnitude of the parameter A to be the same as that of the parameter used by Steinberg [5] in describing the effect of temperature on yield strength.

APPLICATIONS

Spall at High Temperatures

Kanel et al. [4] have recently published the results of a series of plate impact spall experiments on pre-heated samples of aluminium and magnesium. Spall strength as a function of temperature up to the melting point was obtained from measurements made using VISAR interferometry of the pullback in the free surface velocity. We have chosen to model 3 of the experiments on aluminium, details of which are given in table 1.

TABLE 1. Aluminium Spall Experiments

Expt. No.	Tempera ture (·C)	Flyer thickness (mm)	Target thickness (mm)
1	20	2.21	9.95
2	612	2.07	10.30
3	654	2.05	10.00

The aluminium flyer velocity was 0.70±.03mm/µs. and equation of state and constitutive model parameters for 1100 alloy given by Steinberg [5] were used in the calculations. To correctly simulate the initial temperature in the two higher temperature experiments, an initial internal energy was given to the sample, and the density reduced accordingly to ensure zero pressure. Table 2 gives the spall model parameters used for aluminium and figure 1 shows the calculated free surface velocity profiles.

TABLE 2. Spall Model Parameters

Parameter	
a_s (GPa)	0.18
η (GPa µs)	5.10^{-4}
α_0	1.0003
α_f	1.3
A (K^{-1})	0.001

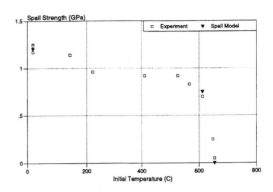

FIGURE 2. Temperature dependence of spall strength

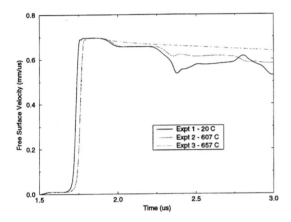

FIGURE 1. Calculated free-surface velocity profiles

Figure 2 compares the measure dependence of spall strength on temperature with that deduced from the calculated pullback in velocity. The agreement is good with only a modest reduction of spall strength for temperatures below the melting point and a drastic reduction as the aluminium melts. It should be noted that the value of the parameter A which determines the temperature dependence of the spall strength was adjusted to give agreement with the data. The value used is of the same order but somewhat higher than the yield strength temperature parameter suggested by Steinberg [5] (6.16×10^{-4}).

Laser Driven Spall

At AWE a number of experiments have been performed to study the spallation of aluminium samples subjected to high intensity short duration laser generated shocks. Initially 100µm thick targets were shocked by pressures generated in a plastic ablator using a 200ps laser pulse. The laser energy and spot size were adjusted to give a range of shock pressures in an effectively 1D geometry. An X-ray backlighting technique was used to image the target to provide evidence of the formation of a detached spall layer. However, although calculated tensions were ~10 GPa no evidence of spall was seen. Calculations of these experiments using the spall model showed that void growth was occurring but the duration of the tensile pulse was not sufficiently long to allow the voids to coalesce and produce an observable detached spall.

An alternative experimental arrangement was therefore designed in which the aluminium target was illuminated directly by the laser pulse and the thickness was increased to 400µm to increase the duration of the tensile phase of the pulse. Figure 3 shows the X-ray backlit image from this experiment. There is clear evidence of a spall plane at about 60µm from the rear surface of the target. The spall model was applied to this experiment with the same parameter values as used in the plate impact experiments. Figure 4 is a plot of the calculated density profile in the target showing clear evidence of spallation with the thickness of the spall layer in agreement with that observed.

It is very encouraging that the model, which had previously been validated against more conventional spall experiments, is capable of good quantitative predictions at the much higher rates of strain encountered in laser experiments.

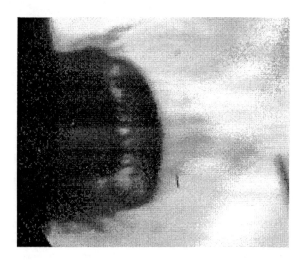

FIGURE 3. Dynamic radiograph of spalled aluminium target

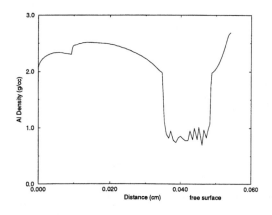

FIGURE 4. Calculated density profile

CONCLUSIONS AND FUTURE WORK

In this paper we have shown that our extended form of Johnson's original void growth model can be used to predict both the temperature and strain rate dependence of spall strength in aluminium.

It is very encouraging that the same model parameters can be used to predict the spall behaviour of aluminium for both plate impact (low tensile stresses, low strain rate) and laser generated spall (very high tensile stresses and corresponding strain rates).

An interferometric velocity measurement technique is currently being developed for use in the laser experiments. We plan to use this to obtain free surface velocity histories in spall experiments to further refine the model.

ACKNOWLEDGEMENTS

We wish to thank Andrew Evans at AWE for making available his results on the laser spall experiments.

REFERENCES

1. Curran, D. R., Seaman L. and Shockey D. A., *Physics Reports* **147** , 1989.

2. Johnson, J. N., *J. Appl. Phys.***52**,1981

3. Maw J. R. and Giles A. R., "Numerical Modelling of Spallation in 2D Hydrodynamics Codes," in *Shock Compression of Condensed Matter - 1995,* pp 295-298.

4. Kanel G. I., Razorenov S. V., Bogatch A., Utkin A. V., Fortov V. E., and Grady D. E., *J. Appl. Phys.* **79**, 8310-8317, 1996

5. Steinberg D. J., *Equation of State and Strength Properties of Selected Materials,* LLNL report UCRL-MA-106439, 1991

CP429, *Shock Compression of Condensed Matter – 1997*
edited by Schmidt/Dandekar/Forbes
© 1998 The American Institute of Physics 1-56396-738-3/98/$15.00

FRACTURE OF BRITTLE MATERIAL WITH INITIAL POROSITY UNDER HIGH ENERGY DENSITY FLOWS

I. N. Lomov, V. I. Kondaurov

High Energy Density Research Center, Moscow, Russia.

The discussion concern mainly with the various types of fragmentation of initially porous brittle materials and their partial evaporation under high energy flows. A model of elastic-viscoplastic damaged medium was build in this work. This model is applicable for description of both continuous and discontinuous flows since the system of equation has completely divergent form. The use of the equations presented in divergent form gives the opportunity to exploit conservative quasi-monotonic method of Godunov's type of second order of approximation. Nonstructural meshes with changed during calculation topology is used for description of strong deformations of the computational region. This technique was applied to solution of problems on action of high energy flows on brittle materials.

INTRODUCTION

Simulation of the materials behavior under intensive energy flows with taking into account real equations of state, phase transitions, fracture and fragmentation, is very important for understanding processes in experiments and real phenomena, for which experiments are unavailable.

In order to describe properties of condensed matter under intensive loading it is necessary to apply correct thermodynamic model. Firstly, we should describe strength and plastic behavior of materials. The ideal elastic-plastic model cannot yields the correct relations on the discontinuites. That's why we chose model with kinematic hardening — the medium of relaxation (Maxwell) type. This model includes the elastic body, ideal elastic-plastic body and nonviscid compressible fluid as limits. Usually used model of elastic-viscoplastic medium has a sufficient drawback: constitutive equations for symmetrical tensor of deformations (stresses) principally cannot be written in divergent (conservation) form in the case of finite strain. This

problem was solved by usage of a nonsymmetrical tensor of the deformation gradient for the description of medium motion [1]. In order to describe fracture of brittle materials under tensile and shear stresses we introduce an equation for damage parameter evolution.

Lagrangian methods are most widely applied for the numerical simulations of the condensed matter because of the absence of the advective smooting and simplicity of the approximation of various rheological models. However Lagrangian meshes is crashed when distortion became sufficiently strong. Implementation of alternative approach — Eulerian methods is connected with difficulties of calculation and interpretation of contact boundaries and increased numerical diffusion. Since the possible solution of this problem is arbitrary Lagrange-Euler (ALE) technique with using of moving grids. The application of non-quadrilateral meshes without fixed topology makes it possible to create near-Lagrangian grids. Hence only shear strains leads to advection between computational cells.

The main purpose of our work is to apply

these thermomechanical model and numerical approach to investigations of fracture and evaporation of matter and to demonstrate solutions of some practical problems.

GOVERNING EQUATIONS

Behavior of continuum is described by the local conservation laws: equations of mass, momentum, energy and compatibility of fields of velocities and deformation:

$$\frac{\partial}{\partial t}\rho + \nabla\cdot(\rho\mathbf{v}) = 0,$$
$$\frac{\partial}{\partial t}(\rho\mathbf{v}) + \nabla\cdot(\rho\mathbf{v}\otimes\mathbf{v} - \sigma) = 0,$$
$$\frac{\partial}{\partial t}(\rho E) + \nabla\cdot(\rho E\mathbf{v} - \sigma\cdot\mathbf{v}) = 0, \quad (1)$$
$$\frac{\partial}{\partial t}(\rho\mathbf{F}^{\mathrm{T}}) + \nabla\cdot(\rho(\mathbf{v}\otimes\mathbf{F}^{\mathrm{T}} - \mathbf{F}\otimes\mathbf{v}) = 0,$$

$\mathbf{F} = \frac{\partial\mathbf{x}}{\partial\mathbf{X}}$ the tensor of displacement gradient, ($\mathbf{x}$ — actual radius-vector, $\mathbf{X}$ — initial one of the particle) $\rho = \rho_0/\det\mathbf{F}$ — density, $\mathbf{v}$ — velocity, σ — Cauchy stress tensor, $E = \epsilon + \frac{1}{2}\mathbf{v}\cdot\mathbf{v}$ — specific total energy, ϵ — specific internal energy, The conservative form of the all equations allows us to describe both continuous and discontinuous flows. These equations yield expressions for jump of flow parameters at the discontinuity. Thus we can construct shocks capturing numerical schemes after providing divergent closing equations.

Equations (1) are closed by constitutive equations: the equation of state (EOS), law of plastic flow and equation of damage evolution. We chose EOS in the form

$$\epsilon = f(\rho, s) + \frac{\mu(\omega)}{\rho_0}(\mathbf{F}_e^{\mathrm{T}}\cdot\mathbf{F}_e - \frac{1}{3}\mathbf{F}_e^{\mathrm{T}}{:}\mathbf{F}_e\mathbf{I})^2{:}\mathbf{I}, \quad (2)$$

where $\mathbf{F}_e$, $\mathbf{F}_p$ are tensors of elastic and plastic displacement gradient in the composition $\mathbf{F} = \mathbf{F}_e\cdot\mathbf{F}_p$, s is entropy, μ is shear modulus, ρ_0 is density of the unloaded state, ω is a scalar damage variable. The first term in (2) is connected with heating and volume compression and is chosen in accordance with semiempirical wide-range equations of state [2]. The second term presents shear energy within second order members. This assumption is acceptable if deviator of elastic strain tensor are small. The stress tensor is obtained from (2) and is equal:

$$\sigma = -p(\rho, \epsilon)\mathbf{I} + \mu(\omega)(\mathbf{F}_e^{\mathrm{T}}\cdot\mathbf{F}_e - \frac{1}{3}\mathbf{F}_e^{\mathrm{T}}{:}\mathbf{F}_e\mathbf{I})$$

, where $p = \rho^2\partial f(\rho)/\partial\rho$. is hydrostatic pressure. Since rotation of unloaded state does not influence on constitutive equation, we may assume that $\mathbf{F}_p$ is symmetrical. The Maxwell type materials obey the equation

$$\frac{DS}{Dt} + \frac{1}{\tau}S = \mu\,dev\left((\nabla\otimes\mathbf{v}^{\mathrm{T}} + \nabla\otimes\mathbf{v})\right),$$

where S is intensity of stress deviator $\tau = \tau(S, \epsilon)$ is relaxation time and D/Dt denotes indifferent derivative. This equation can be rewritten in implicit form with use of plastic gradient:

$$\frac{\partial\rho\mathbf{F}_p}{\partial t} + \nabla\cdot(\rho\mathbf{v}\otimes\mathbf{F}) = \rho\mathbf{\Phi}(\mathbf{F}, \mathbf{F}_p, s). \quad (3)$$

In order to describe fracture of material we introduce a scalar parameter ω, which describes the local degree of substance damage. Equation for description of non-instantaneous kinetic of damage evolution is specified in following form:

$$\frac{\partial\rho\omega}{\partial t} + \nabla\cdot(\rho\omega\mathbf{v}) = \begin{cases} \frac{\rho(f-\omega)}{\tau_\omega(S,\epsilon,\omega)}, & f > \omega, \\ 0, & f < \omega, \end{cases}$$
$$f = \alpha_p(1 - \rho/r_0) + \alpha_S J - \gamma - \omega,$$
$$\tau_\omega = \tau_{\omega 0}\exp\left(-(f - f_0)/f\right),$$
$$J = \mathbf{I}{:}(\mathbf{F}_e\cdot\mathbf{F}_e^{\mathrm{T}} - 1/3(\mathbf{F}_e{:}\mathbf{F}_e^{\mathrm{T}})\mathbf{I}),$$

where $\gamma > 0$ is a coefficient that characterizes the threshold value of stain at which damage starts, $\alpha_p, \alpha_S > 0$ are coefficients defining the effects of dilatation and shear, respectively, on damage evolution.

SCHEME ON UNSTRUCTURED GRIDS

The computational mesh consist of arbitrary polygonal cells which are control volumes of a finite-volume method. These cells are constructed at every time step by searching of nearest neighbors and creating a Delone triangulation which define cells. The topology of this triangulation permits the great flexibility of computational mesh. We do not need additional user-defined rules for grid construction as far as this triangulation is unique excepting very special cases. The mesh flexibility allows to adapt itself to arbitrary interface contour and resolve regions of even highest curvature, if desired. It also permit a rapid and smooth dynamic change in local mesh resolution, which economizes the total number of cells required for a given degree of resolution. Another advantage is the enhancement of mesh isotropy and increased time step due to greater number of neighbor considered. Interior computational cells are defined by vertices that are circumscribed centers of Delone triangles. Cells lying adjacent to the boundaries are partially defined by this technique. The outer vertices of boundary cells are initially specified and then calculated on the base of Riemann solver and Huygen's reconstruction of interface.

Computational quantities are associated with cell points. In order to integrate conservation laws through the computational cell it is necessary to calculate quantities on the cell sides. We apply the Godunov procedure with fast approximate Riemann solver to find values on the cell interfaces. Boundary fluxes also are derived from solution of the Riemann problem, left and right states for which are specified by parameters in boundary cell and boundary condition.

The consideration preceding impose no restriction on the motion of the cell points. With fixed ones calculation will be pure Eulerian. In order to minimize the advection errors it would better to move points with mass velocities. Since a real cell interface velocity is differ from half-sum of point velocities it is not possible to define a purely Lagrangian mesh motion and thus avoid advection errors. Instead we define mesh velocities that are close to Lagrangian velocities, while at the same time minimizing mesh distortion. The near-Lagrangian algorithm thus attempts to preserve Lagrange cell volumes and also to maintain a smooth mesh.

The implementation of this technique removes problems of Lagrangian calculation connected with mesh tangling and crashes. But the desirable properties of unstructured grid are not without price. The data structure is more complicated and the algorithms are more complex.

The time consumption increases more than three times in comparison with calculation on the quadrilateral meshes with the same number of cells, but in the solution of the problems with strong local distortion quadrilateral cells may become scratched and reduction of the time step due to stability limit leads to disappearing of this drawback. Development of this code has been largely experimental and requires further work, but the results achieved show that such an approach is capable of solving nontrivial problems.

COMPUTATIONAL EXAMPLES

1. The behavior of ice plate under influnce of shock wave induced by laser or particle beams. Initial conditions for this calculation was defined as a deposition of specific internal energy of $10 kJ/cm^3$ in a specified region of ice plate, which is equal to heating of this region by particle beam. The diameters of plate and heated region are 20 and 2 mm, thicknesses are 2 mm and 0.3 mm, accordingly. This situations exhibits both large amounts of evaporations, plastic deformations and porosity due to tensile stresses. Contours of hydrostatic pressure levels at time $t = 2mks$ and meshes geometry are shown in fig. 1. The region of material fracture due to tensile stresses are located near back surface of the plate.

2. Simulation the cosmic body motion in planet atmosphere. Second example concerns to investigation of deformation, fragmentation and evaporation of meteoroids in planet atmosphere. Parameters of problems correspond to fall of Tunguska meteoroid on the Earth in 1908. Consistent shear fragmentation of body material leads to filtration of hot gas from layer behind the head shock wave through the cracks and succesively to accelerated heating of substance. For simulation of these processes we specify brake pressure of atmosphere on frontal surface of meteoroid, thermal flux through this surface and heat transfer in the substance of meteoroid with coefficient which is proportional to local damage of material. In fig. 2, which is correspond to $t = 1s$, we can see two different regions of body: almost cold one and completely transform to gas.

This phenomenon can be called as wave of fracture and evaporation.

ACKNOWLEDGMENTS

This work is supported by Russian Foundation for Basic Research under Grant 97-05-65607.

REFERENCES

1. Kondaurov V. I., Nikitin L. V. Theoretical foundations of geomaterials rheology. Moskow: Nauka, 1990.

2. Lomonosov I. V., Bushman A. V., Fortov V. E. Khishchenko K. V. In: High pressure Science and Technology - 1993, N. Y.: AIP Press, 1994, pp. 133-136

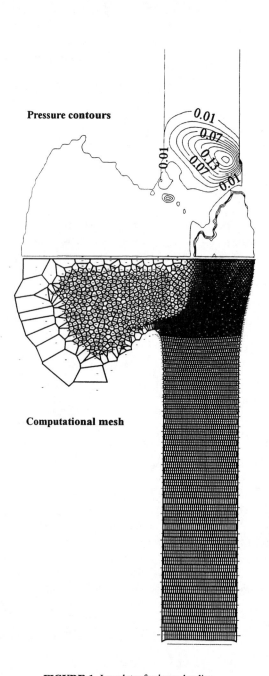

FIGURE 1. Ice plate after beam loading

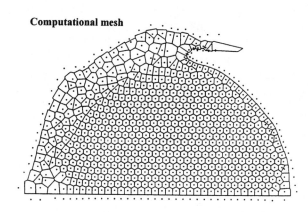

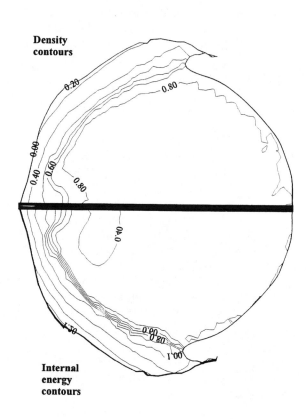

FIGURE 2. Fragmentation and evaporation of Tunguska body

CP429, *Shock Compression of Condensed Matter – 1997*
edited by Schmidt/Dandekar/Forbes
© 1998 The American Institute of Physics 1-56396-738-3/98/$15.00

A PROBABILISTIC MODEL FOR THE DYNAMIC FRAGMENTATION OF BRITTLE SOLIDS

F. HILD[1] and C. DENOUAL[2]

[1] *Laboratoire de Mécanique et Technologie*
E.N.S. de Cachan / C.N.R.S. / Université Paris 6
61, Avenue du Président Wilson, F-94235 Cachan Cedex, France.
[2] *DGA/CREA-Département Matériaux en Conditions Sévères*
16 bis, Avenue Prieur de la Côte d'Or, F-94114 Arcueil Cedex, France.

Impact produces high stress waves leading to the fragmentation of brittle materials such as ceramics. The main mechanism used to explain the size variation of fragments with stress rate is an obscuration phenomenon. When a flaw initiates, the released stresses around the crack prevent other nucleations in an increasing zone. After a presentation of a probabilistic approach, a damage description and an evolution law are derived. Two numerical applications of the model are proposed.

INTRODUCTION

In the bulk of an impacted ceramic, damage in tension is observed when the hoop stress induced by the radial motion is sufficiently large to generate fracture in mode I initiating on micro-defects such as porosities or inclusions. When such a fracture is initiated, the zone affected by fracture is a complex function of time, crack velocity and stress wave celerity. In order to simplify the following development, the shape of the affected (or interaction) zone Z_i is supposed to be constant, *i.e.* all the interaction zones are self-similar and Z_i can be written as

$$Z_i(T - t) = S\left[kC(T - t)\right]^n \qquad (1)$$

where kC is the velocity of a propagating crack, S a shape parameter, C the longitudinal stress wave velocity so that $kC(T-t)$ is a representative length of the relaxation zone at time T around a broken flaw at time t. The parameter $n = 1, 2, 3$ is the space dimension. The shape parameter S is chosen in order to have $d\sigma/dt \leq 0$ in Z_i, *i.e.* no new nucleation can occur in Z_i.

To understand why a crack nucleates, one has to model the interaction of a nucleated defect and other defects that would nucleate. With a constant direction of the maximum principal stress and a small stress gradient, the space dimension can be uncoupled from the tensile stress (or time) dimension and the flaw nucleation can be represented on a space–time graph (Fig. 1). The space location of the defects is represented in a simple abscissa of an x-y graph where the y-axis represents time (or stress) to failure of a given defect. In this graph, a shaded cone represents the expansion of the interaction zone with time due to nucleation and propagation of a crack. A section Z (see Fig. 1) of a cone can be a volume, a surface or a length, depending on the space dimension n. The defects *outside* the shaded cones can nucleate and produce their own increasing interaction zone (*e.g.* defects No. 1 and No. 2). Inside the cones, the defects that should have broken do not nucleate (*e.g.* defects No. 3 and No. 4) since they are obscured.

Because different interaction volumes may overlap (a flaw can be obscured by one or more cracks), its preferable to define the conditions of non-obscuration for a given defect by examining the reverse problem (1). For a given flaw D, a non-interaction zone can be defined so that a de-

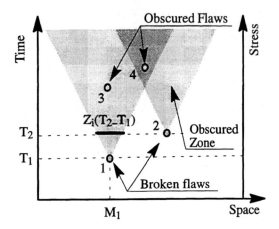

FIGURE 1. Depiction of obscuration phenomena.

fect cannot obscure D (Fig. 2) and the *horizon* of D in which a defect will always obscure D. The total

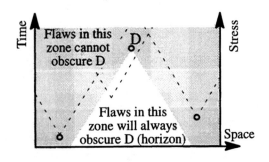

FIGURE 2. Schematic of the obscuration/non–obscuration zones for a defect D.

flaw density $\lambda_t(T)$ can therefore be split into two parts: $\lambda_b(T)$ (the broken flaws) and $\lambda_o(T)$ (the obscured flaws). Furthermore, we assume that the distribution of total flaws in a zone Z is modeled by a Poisson point process of intensity $Z\lambda_t(T)$. New cracks will initiate only if the defect exists in the considered zone and if no defects exist in its horizon so that

$$\frac{d\lambda_b}{dt}(T) = \frac{d\lambda_t}{dt}(T)P_{no}(T) \qquad (2)$$

where P_{no} is the probability of non-obscuration (no defects exist in the horizon). The variable P_{no} can be split into an infinity of events defined by the probability of finding at t a new defect during a time step dt in an interaction zone $Z_i(T-t)$.

This probability increment is written by using a Poisson point process of intensity $d\lambda_t/dt$. Those *independent events* can be used to provide an expression for P_{no}

$$P_{no}(T) = \exp\left[-\int_0^T \frac{d\lambda_t}{dt}(t)Z_i(T-t)dt\right] \qquad (3)$$

where $Z_i(T-t)$ is the measure of the interaction zone at t for a defect that would break at T.

The static description of fracture in brittle materials can be modeled by a two-parameter Weibull law. The density of flaws able to break for a stress less than or equal to σ is then assumed to follow a power law function

$$\lambda_t(T) = \frac{1}{Z_0}\left[\frac{\sigma(T)}{S_0}\right]^m \qquad (4)$$

where m is the Weibull modulus and S_0 the scale parameter relative to the measure of a reference zone Z_0.

DAMAGE DESCRIPTION AND EVOLUTION LAW

One can notice that P_{no} is also the fraction of relaxed zones and can therefore be related to a damage variable D by the relation $1 - P_{no} = D$. This variable, defined in the framework of Continuum Damage Mechanics (CDM), evolves from zero to one when the initial material becomes more and more damaged. It is interesting to notice that the first order approximation of Eqn. (3) leads to the differential equation proposed by Grady and Kipp (2) to describe the evolution of a damage variable. The proposed damage variable is defined with the assumption that many cracks nucleate and propagate due to a tensile stress expressed in the direction of the maximum principal stress. Since the cracks will be strongly oriented, an anisotropic damage description is chosen (3).

This anisotropic description is expressed through a second order damage tensor. The tensor D_{ij} is diagonal in the eigen directions of $\underline{\underline{\sigma}}$ so that only 3 variables (D_1, D_2, D_3) have to be computed. The relationship between the microscopic principal stress σ_i and the macroscopic one (Σ_i) is

$$\Sigma_i = \frac{\sigma_i}{1-D_i} \quad \text{with } i = 1,3 \qquad (5)$$

This approach is useful when multiple crack patterns are superimposed (4).

The evolution of D is expressed in a differential form in order to be implemented in the FE code PamShock (5) by using Eqns. (2), (3) and (4)

$$\frac{d^{n-1}}{dt^{n-1}}\left(\frac{1}{1-D_i}\frac{dD_i}{dt}\right) = \lambda_t(\sigma_i)\,n!\,S\,(kC)^n \quad (6)$$

It is worth noting that D approaches smoothly 1 and does not need any cut-off. According to classical results of CDM, the evolution of D is stopped if $d\sigma/dt < 0$. The eigen directions ($\underline{d_1}$, $\underline{d_2}$, $\underline{d_3}$) associated to D_1, D_2 and D_3 may change at each time step until D_1 reaches a threshold value $D_{th} = 0.01$. Only the direction $\underline{d_1}$ is then locked, the other directions follow the eigen directions of $\underline{\underline{\sigma}}$, with the constraint to be perpendicular to $\underline{d_1}$. When D_2 reaches the threshold value, the whole directions are locked. Equations (2), (3) and (4) are also used to compute the density of broken flaws.

EXPERIMENTS AND COMPUTATIONS

In all the computations, $k = 0.38$, $S = 4\pi/3$ (spherical obscuration zones) and $n = 3$. The two materials used in this section are a natural sintered SiC (SSiC) processed by Céramique & Composites (France) and a SiC–B processed by CERCOM (USA) with the following characteristics

Property	SSiC	SiC–B
Young's mod.	410 GPa	450 GPa
Density	3.15	3.18
Weibull mod.	9.3	18
Mean strength	350 MPa	553 MPa
Effective vol.	1.25 mm^3	1.25 mm^3

The compressive behavior of both ceramics is not modeled in this study because of the very low impact velocities. To test the mesh sensitivity of the above described model, a plane shock wave spall configuration is analyzed. When a compressive stress pulse is generated in a plate, the reflected (tensile) pulse coming back from the free rear surface of the specimen is superimposed on the loading pulse. When the loading pulse duration is properly prescribed, a tensile stress is generated in the bulk of the specimen. The thickness of the damaged zone may strongly vary if the model is not mesh–independent. The shock wave spall simulation is therefore a severe test for damage evolution laws implemented in numerical codes.

A one dimensional code presented in Ref. (3) is used. The plate thickness is 10 mm and the imposed pressure is −1 GPa, for a duration of 10 μs. Four different mesh densities are tested: 20 el./mm, 40 el./mm, 80 el./mm, 160 el./mm. The numerical viscosity parameters are chosen so that they do not modify the loading pulse shape during propagation. The simulation is performed with the SSiC material. The results presented in Fig. (3) show no dependence of the size of the damaged zone on the mesh density. The second

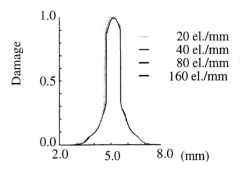

FIGURE 3. Damage location for 4 mesh densities in a plane shock wave spall configuration.

test concerns computational simulations of Edge-On Impacts (EOI) in two different configurations. The principle of EOI is detailed in Ref. (6) and typical results are presented in a companion paper (4). The first simulation is carried out on the configuration designed by the EMI on a SiC–B ceramic with a steel projectile of velocity 185 m/s and a tile of size $100 \times 100 \times 10$ mm^3 (Fig. (4)).

Two differents zones can be depicted in the experimental result. The first one (say the *inner* zone) appears in front of the projectile. This zone widens progressively and finally becomes localized in thinner zones called corridors of varying locations. The second (*outer*) zone is generated at the projectile edges and can roughly be compared to a Herz cone crack. The damaged zone then splits into a thin part that stops at the edge of the tile,

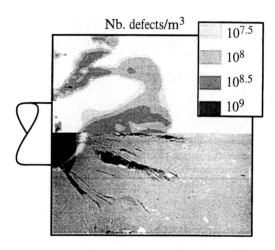

Nb. defects/m³

$10^{7.5}$

10^8

$10^{8.5}$

10^9

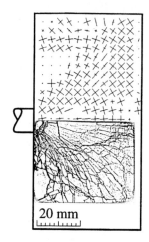

20 mm

FIGURE 4. Simulated (top) and observed (bottom) crack pattern in an EOI test with a projectile velocity of 185 m/s, $T = 7.5$ μs.

FIGURE 5. Simulated (top) and observed (bottom) crack directions in an EOI test with a projectile velocity of 330 m/s, $T = 15$ μs.

and a larger one that propagates through the ceramic at a constant distance from the edge.

The numerical simulation is in good accordance with the general shape of the damaged zone. The inner zone is well reproduced even if the corridors are generated too late and are not sufficiently extended. The simulated outer zone follows closely the observed one except for the bridge observed between the inner and the outer zones.

The second EOI test presented in Fig. (5) is performed with a steel projectile whose velocity is 330 m/s on a SSiC ceramic. The tile size is $100 \times 50 \times 10$ mm³.

The inner/outer zones cannot be clearly defined because the whole specimen is fully broken at $T = 15$ μs. In both experiment and simulation, the ceramic is cracked in front of the projectile in radial and circumferential directions. It is worth mentioning that only an anisotropic damage model (in which the stiffness can be very small in one direction and unaltered in another direction) can provide such a crack pattern.

CONCLUSION

A probabilistic approach has been proposed to describe the stress rate dependence of the fragmentation mechanism. The proposed approach is used to develop a damage evolution law in tensile mode for impact simulations on ceramics. The model can provide an estimation of the number and the directions of activated flaws. Complex damage patterns, in which superposition of cracks of differents directions occurs, can be predicted.

ACKNOWLEDGMENTS

The authors wish to thank E. Straßburger and H. Senf from the EMI for providing the picture of Fig. (4). The authors wish also to thank A. Trameçon from ESI for his valuable help in implementing the model in PamShock. This work was funded by DGA-DRET-STRDT.

REFERENCES

1. Denoual C., Hild F. and Cottenot C.E., *"A Probabilistic Approach for Dynamic Fragmentation of Ceramics under impact Loading,"* Presented at ICF 9, Sydney, Australia, **6**, pp. 2933–2940, 1997.
2. Grady D. E. and Kipp M. E. *Int. J. Rock Min. Sci. & Geomech. Abstr.*, **17**, pp. 147-157, 1980.
3. Denoual C., Cottenot C.E. and Hild F., *"On the identification of Damage During Impact of a Ceramic by a Hard Steel Projectile,"* Presented at the 16th Ballistic Symp., pp. 541–550, 1996.
4. Denoual C., Cottenot C.E. and Hild F., *"Analysis of the Degradation Mechanisms in an Impacted Ceramic,"* Presented at the APS Conference on Shock Compression of Condensed Matter, Amherst (MA), USA, 1997.
5. Pamshock, Users' manual, ESI, 1997.
6. Straßburger E., Senf H., Denoual C., Riou P., and Cottenot C.E., *"An Experimental Approach to Validate Damage Evolution Laws for Brittle Materials,"* To be presented at DYMAT 97, Toledo, Spain, Sept. 1997.

254

CP429, *Shock Compression of Condensed Matter – 1997*
edited by Schmidt/Dandekar/Forbes
© 1998 The American Institute of Physics 1-56396-738-3/98/$15.00

A COMPARISON STUDY BETWEEN SCALAR AND MULTI-PLANE MICROCRACKING CERAMIC DAMAGE MODELS

D.J. Grove and A.M. Rajendran

Weapons and Materials Research Directorate, Army Research Laboratory, Aberdeen Proving Ground, MD 21005

The Rajendran-Grove (RG) ceramic damage model is based on an elastic-plastic-cracking description. A crack density parameter γ ($= N_o^* a^3$) describes the scalar damage. The number of flaws N_o^* is assumed to be constant and the crack size parameter "a" evolves according to a strain energy release based evolution law. Crack orientation is not considered in this model. However, Espinosa's multi-plane (MP) microcracking model considers crack orientations in nine pre-selected directions and computes damage by summing up the crack density contribution from all nine directions. Both models account for crack opening and sliding. These two models have been implemented in the 1995 version of the EPIC code. We simulated plate impact experiments in which a thin alumina flyer plate impacted a thick alumina plate. The experimental data consisted of particle velocities recorded at the back face of the target plate. This paper compares the results from the EPIC code simulations using the RG and MP ceramic models.

INTRODUCTION

During the past decade, several constitutive models that are suitable for hydrocode applications have been developed by various researchers [1-5]. The main objective of this study was to compare the abilities of two of these models [3,4] to reproduce the measured data from plate impact experiments on ceramics. The 1995 version of the EPIC finite element code [6], modified to include the two models, was employed to simulate two plate impact experiments on aluminum oxide ceramics.

In the Rajendran-Grove (RG) ceramic model [3], the permanent strain is due to plastic flow only, the microcracking strains are considered elastic, and the crack damage parameter is a scalar. This model requires two constants to describe plastic flow, five constants to describe microcracking, and two constants to describe the strength of the comminuted ceramic. Espinosa's multi-plane (MP) model [4] describes the behavior of brittle materials through an elastic-microcracking formulation where permanent strains can be achieved only through crack opening and sliding. The MP model considers the effects of crack orientation to compute the contributions to crack damage in nine pre-selected directions.

We simulated two plate impact experiments performed by Grady and Moody [7]: one on a porous aluminum oxide with six percent porosity, and the other on AD995 with only two percent porosity. Computed velocity histories using both the RG and MP models were compared with the recorded laser velocity interferometer (VISAR) data from the experiments.

RAJENDRAN-GROVE MODEL

In the RG model, the total strain is decomposed into elastic strain (ε_{ij}^e) and plastic strain (ε_{ij}^p). The elastic strain consists of the elastic strain of the intact matrix material and the strain due to crack opening/sliding. Plastic flow is assumed to occur in the ceramic only under compressive loading when the applied pressure exceeds the pressure at the Hugoniot elastic limit (HEL). Pore collapse during shock loading is modeled using a pressure dependent yield function, and the strains due to pore collapse are assumed to be viscoplastic. The stress-strain equations for the microcracked material are given by, $\sigma_{ij} = M_{ijkl} \varepsilon_{kl}^e$. The components of the stiffness tensor M are described by Rajendran [5]. The pressure is calculated through the Mie-Gruneisen equation of state.

Microcrack damage is measured in terms of a dimensionless microcrack density γ, defined as $\gamma = N_o^* a^3$, where N_o^* is the average number of microflaws per unit volume and a, the maximum microcrack size, is treated as an internal state variable. The initial values of these two parameters are material model constants. Microcracks are assumed to extend when the stress state satisfies a generalized Griffith criterion. This criterion requires the fracture toughness K_{IC} and a dynamic friction coefficient μ as model constants.

During microcrack extension, the crack density γ increases and stress relaxation occurs. The crack extension (damage evolution) law is derived from a fracture mechanics based relationship for a single crack propagating under dynamic loading conditions: $\dot{a} = n_1^{\pm} C_R [1 - (G_c / G_I)^{\frac{n_2^{\pm}}{2}}]$, where C_R is the Rayleigh wave speed, G_c is the critical strain energy release rate for microcrack growth, G_I is the applied strain energy release rate, and $n_1^{\pm}$ and $n_2^{\pm}$ are the model parameters that are used to limit the microcrack growth rate. The "+" superscript corresponds to microcrack opening under tension (mode I), while the "−" superscript relates to microcrack extension under compression (mode II). The $n_2^{\pm}$ exponents and the n_1^{+} coefficient are all assumed to be equal to "1", but the n_1^{-} coefficient must be calibrated for mode II crack extension. The ceramic material is assumed to pulverize under compression when γ reaches a critical value of 0.75.

Effectively, the RG ceramic model requires only five constants to describe the microcracking of the intact ceramic. The model constants for aluminum oxide AD995 were determined from planar plate impact data. Rajendran and Dandekar [8] reported these constants as: $a_o = 14$ microns, $N_o^* = 5 \times 10^9 / m^3$, $K_{IC} = 4$ MPa $\sqrt{m}$, $\mu = 0.45$, and $n_1^{-} = 0.1$.

MULTI-PLANE MODEL

The MP model assumes that microcracking can occur on a discrete number of orientations. Espinosa [4] selected nine orientations at an interval of 45^0 along three mutually perpendicular planes.

The inelastic strain is entirely due to (penny-shaped) microcrack opening/sliding of the cracks oriented normal to those nine directions. The average inelastic strains are given by,

$$\varepsilon_{ij}^c = \sum_{k=1}^{9} N^{(k)} S^{(k)} \frac{1}{2} \left(\overline{b}_i^{(k)} n_j^{(k)} + n_i^{(k)} \overline{b}_j^{(k)} \right). \quad (1)$$

The subindex k represents the orientation, $N^{(k)}$ is the number of flaws per unit volume, $S^{(k)}$ denotes the surface of the microcrack, $n^{(k)}$ is the corresponding unit normal, and $\overline{b}^{(k)}$ is the average displacement jump vector across the surface $S^{(k)}$. The $\overline{b}^{(k)}$ have been analytically derived for normal tractions under both tension and compression. The corresponding expressions are:

$$\overline{b}_i^{(k)} = \frac{16(1-v^2)}{3E(2-v)} a^{(k)} \left(2\sigma_{ij} n_j^{(k)} - v\sigma_{jl} n_j^{(k)} n_l^{(k)} n_i^{(k)} \right) \quad (2)$$

and

$$\overline{b}_i^{(k)} = \frac{32(1-v^2)}{3E(2-v)} a^{(k)} f_i^{(k)}. \quad (3)$$

E and v are the Young's modulus and Poisson's ratio of the intact ceramic, $a^{(k)}$ is the crack radius of the penny-shaped microcracks on orientation k, and $f^{(k)}$ is the effective shear traction vector on orientation k.

The microcrack growth law is very similar to the one employed in the RG model (see previous section), except that an effective stress intensity factor is used instead of the strain energy release rate: $\dot{a}^k = m^{\pm} C_R [1 - (K_{IC} / K_{eff}^k)^{n^{\pm}}]$, where K_{eff}^k is the effective stress intensity factor for orientation k and $m^{\pm}$ and $n^{\pm}$ are the model parameters that control the microcrack growth rate. As before, the "+" superscript corresponds to microcrack opening under tension (mode I), while the "−" superscript relates to microcrack extension under compression (mode II).

Espinosa et al. [9] reported the following MP model constants for AD995 alumina: Young's modulus = 374 GPa, Poison's ratio = 0.22, density = 3890 kg/m^3, dynamic coefficient of friction = 0.1, fracture toughness = 1.7 MPa$\sqrt{m}$, Rayleigh wave speed = 5000 m/s, initial crack size = 10 microns, the number of flaws per unit volume = $10^{12} / m^3$ for the radial and axial (shock) orientations (planes 1 and 2), and $5 \times 10^{10} / m^3$ for the off-axis orientations

256

(planes 6 and 8). The model constants that appear in the microcrack propagation law are $m^+ = n^+ = 0.3$ and $m^- = n^- = 0.1$. We employed these constants in the simulations described in the next section.

PLATE IMPACT SIMULATIONS

Grady and Moody [7] obtained velocity vs. time history data from a plate impact experiment on a porous aluminum oxide ceramic (similar to Coors AD90) with a porosity of about six percent. In this experiment, the thicknesses of the flyer and target were 4.673 mm and 9.081 mm, respectively, and the impact velocity was 2212 m/s. The target plate was backed by a lithium fluoride window and a VISAR was used to record the velocity history at the target/window interface.

We simulated this experiment using the 1995 version of EPIC, which we modified to include the RG and MP ceramic material models. Computed velocity histories at the target/window interface were obtained for both the RG model (using the elastic-cracking option only) and the MP model. The computed velocity profiles are compared with the experimental data in Fig. 1.

complete absence of the dispersive nature of the wave beyond the HEL point is due to the absence of a strain rate dependent plastic strain description. The real material in the experiment exhibits a strong rate dependent inelastic response above the HEL. It is apparent from the recorded signal that the plastic wave speed is increasing as the stress increases beyond the HEL. In the simulations, both models assumed that strains could result only from elastic and/or microcracking deformation (i.e., no plastic flow was permitted); consequently, the computed velocity profiles exhibited very short rise times.

We re-simulated the plate impact test using the RG model with its elastic-plastic(pore collapse)-microcracking option activated. In this case, the permanent strains associated with plastic flow and pore collapse were computed using a strain rate dependent flow rule derived from a pressure dependent plastic potential. Figure 2 compares the computed velocity histories (with and without pore collapse) with the recorded VISAR data. As the figure indicates, modeling the strain rate dependent pore collapse mechanism significantly improved the model's prediction. These results suggest that pore collapse must be modeled to accurately describe the inelastic response of a porous ceramic.

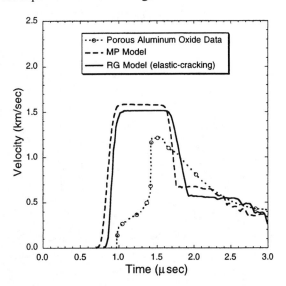

FIGURE 1. Comparison of the RG and MP model results with the porous aluminum oxide plate impact data.

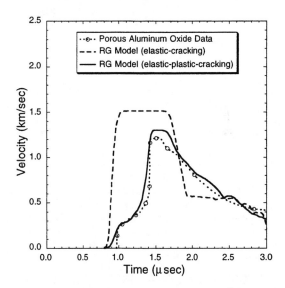

FIGURE 2. Comparison of the RG model results (with and without pore collapse) with the porous aluminum oxide plate impact data.

It is clear from the above figure that both models failed to reproduce the data both qualitatively and quantitatively. In the model predictions, the

To further investigate the strain rate dependent inelastic response of ceramic materials we considered a relatively dense material such as AD995 alumina with a density of 3890 kg/m³. For this purpose, a plate impact experiment on AD995 reported by Grady and Moody [7] was considered. The thicknesses of the flyer and target were 5 mm and 10 mm, respectively, and the impact velocity was 1943 m/s. As in the first experiment, the target plate was backed by a lithium fluoride window and a VISAR was used to record the velocity history at the target/window interface. This experiment was simulated using the RG (elastic-plastic-cracking option) and MP models. Figure 3 compares the model-predicted velocity histories with the recorded VISAR data.

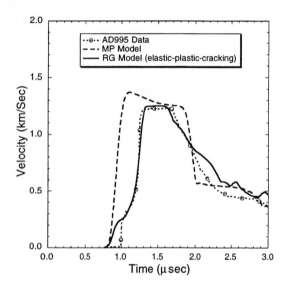

FIGURE 3. Comparison of the RG and MP model results with the AD995 plate impact data.

Since AD995 is a relatively dense material (about two percent porosity), the wave dispersion beyond the HEL is not as pronounced as in the more porous alumina (see Fig. 1). As the above figure indicates, the MP model was once again unable to reproduce the initial portion of the wave profile. However, while the computed velocity profile from the RG model simulation did not perfectly match the spall signal, its overall match with the data is excellent. In the RG model simulation, the model constants were the same as those used in the simulation of the more porous alumina, except for the porosity level (two percent versus six percent). These simulations further demonstrate the need for an accurate description of pore collapse in the ceramic material model.

SUMMARY

The scanning/transmission electron microscopic (SEM/TEM) photographs obtained from both pre-shocked and post-shocked ceramic materials indicate that shock loading of ceramics can induce strains through elastic, microcracking, pore collapse, and dislocation-based plastic deformations. Two high velocity plate impact experiments on aluminum oxide ceramics with different porosities were simulated to compare the abilities of the RG and MP ceramic damage models to describe the measured shock response of ceramic materials. Since the MP model is limited to describing only elastic and/or microcracking deformation, it was unable to reproduce the VISAR data from the two plate impact experiments. However, the RG model could reasonably predict both experimental velocity profiles because of its ability to describe the deformations due to plastic flow and pore collapse in addition to the material's elastic-microcracking behavior. These results clearly demonstrate that an elastic-microcracking assumption to model ceramic behavior is inadequate to describe the shock response (above HEL) of a porous ceramic.

ACKNOWLEDGMENTS

The authors greatly appreciate the funding support of Dr. James Thompson of TARDEC, Warren, MI.

REFERENCES

1. Steinberg, D.J., *Shock Compression of Condensed Matter-1991*, edited by S.C. Schmidt, R.D. Dick, J.W. Forbes, and D.G. Tasker, North-Holland, 1992, pp. 447-450.
2. Johnson, G.R., and Holmquist, T.J., *Shock Compression of Condensed Matter-1993*, edited by Schmidt, et al., AIP Press, NY, 1993, pp. 733-736.
3. Rajendran, A.M., and Grove, D.J., International Journal of Impact Engineering, Vol. 18, No. 6, 1996, pp. 611-631.
4. Espinosa, H.D., Int. J. of Solids and Structures, Vol. 32, No. 21, 1995, pp. 3105-3128.
5. Rajendran, A.M., International Journal of Impact Engineering, Vol. 15, No. 16, 1994, pp. 749-768.
6. Johnson, G.R., Stryk, R.A., Petersen, E.H., Holmquist, T.J., Schonhardt, J.A., and Burns, C.R., Alliant Techsystems Inc., Minnesota ,1994.
7. Grady, D.E., and Moody, R.L., SAND96-0551, Sandia National Laboratory, Albuquerque, NM 87185, March 1996.
8. Rajendran, A.M., and Dandekar, D.P., Int. J. of Impact Engng., Vol. 17, pp. 649-660, 1995.
9. Espinosa, H.D., Zavattieri, P.D., and Emore, G.L., Special Issue of Mechanics of Materials, 1996.

CP429, *Shock Compression of Condensed Matter – 1997*
edited by Schmidt/Dandekar/Forbes
© 1998 The American Institute of Physics 1-56396-738-3/98/$15.00

PARTICLE VELOCITY DISPERSION IN SHOCK COMPRESSION OF SOLID MIXTURES

K. Yano and Y. Horie

North Carolina State University, Raleigh, North Carolina 27695

Two models were considered to investigate the particle velocity dispersion in shock compression of solid mixtures: a continuum mixture model and a discrete element dynamics model. Results for Ni/Al and Ti/Teflon mixtures show (i) a qualitative agreement between the two techniques, (ii) a highly non-equilibrium distribution of particle velocity dispersion in the Ni/Al mixture, and (iii) particle velocity dispersion on the order of 100 m/s for $10 \sim 15$ GPa shock waves. A dispersion of this magnitude is thought to be the mechanism responsible for the initiation of chemical reactions in reactive mixtures.

INTRODUCTION

Heterogeneities of materials are known to cause irregularities in shock waves, which may induce dispersion in particle velocity. In vapor explosions of two-phase liquid systems, the dispersion in particle velocity is thought to be the essential mechanism for high speed mixing (1). The particle velocity dispersion was also observed experimentally in polycrystalline copper (2). Likewise, the existence of particle velocity dispersion during shock compression of heterogeneous solid mixtures is expected as a natural extension of the observations in polycrystalline materials. Also the velocity dispersion is of interest because it has been proposed as one of the major mechanisms for initiating ultra-fast chemical reactions in reactive mixtures (3, 4).

The purpose of this paper is to numerically investigate the particle velocity dispersion in shock compression of heterogeneous solid mixtures using two methods: a continuum mixture theory and a discrete element technique.

CONTINUUM MODEL

For this part of our work we followed the mixture theory of Nigmatulin (5). Assumptions made for the analysis are: one dimensional · hydrodynamic flow, steady state wave propagation, common solid pressure for all constituents, common shock wave speed, no body force, no external heat source, and no chemical reaction or phase transformation.

Integration of the momentum conservation equation, for each constituent, over the shock thickness δ, yields the following equation in the steady-state space coordinate ξ:

$$[\rho_i u_i^2 + p_i] = \int_0^\delta P \frac{d\alpha_i}{d\xi} \, d\xi + \int_0^\delta D(u_j - u_i) \, d\xi \tag{1}$$

where subscripts i and j denote constituents, and $\rho_i, u_i, p_i, P, \alpha_i$, and D are partial density, particle velocity (with respect to shock front), partial pressure, common solid pressure, volume fraction, and drag coefficient, respectively. Bracket "$[f]$" represents the jump in f given by $[f(\xi)] = f(\delta) - f(0)$. Note that all the integrals on the right hand side vanish in the case of a single component material.

For the drag coefficient, D, we used a modified form of the model developed for dispersed two phase flow (6) given by:

$$D = 18 \frac{\bar{\mu}(\alpha_1 \alpha_2)^{\frac{1}{2}}}{d^2} \qquad (2)$$

where $\bar{\mu}$ is the effective viscosity of the mixture, and d is the effective diameter of a particle.

The second integral on the right hand side of eqn. (1) was reduced to the following by assuming a linear change in α_i within the shock front.

$$\int_0^\delta D(u_j - u_i) \, d\xi \approx [\rho_i u_i^2 + p_i] - \eta(\alpha_{i0} - \alpha_{if})P_f \qquad (3)$$

Subscripts "0" and "f" indicate the initial and final states. η is a constant such that $0 < \eta < 1$. A parametric study showed that the second term on the right hand side is negligible in comparison to the first.

The average value of $u_j - u_i$ over the shock thickness can be approximated as follows.

$$\begin{aligned} <u_j - u_i> \; &\approx \; \frac{[\rho_i u_i^2 + p_i] - \eta(\alpha_{i0} - \alpha_{if})P_f}{D\delta} \\ &= \; \frac{[\rho_i u_i^2 + p_i] - \eta(\alpha_{i0} - \alpha_{if})P_f}{18\bar{\mu}(\alpha_{if}\alpha_{jf})^{\frac{1}{2}}U\Delta t/d^2} \quad (4) \end{aligned}$$

U and Δt are shock speed and shock rise time, respectively. The right hand side of eqn. (4) can be calculated using the properties at the final state; i.e., the jump conditions of the mixture and the equation of state of each constituent.

DISCRETE ELEMENT MODEL

The fundamentals of the discrete element model are similar to those of molecular dynamics. Details are given in (7). In the model, materials are represented by a collection of discrete elements interacting with each other according to rigid body dynamics. The interactions are determined based on the chemical bonding status between them. Depending on the status, different types of forces are applied to the elements. Figure 1 shows the types of forces considered in the model. For elements that are linked, a central potential force, a shear resistance force, and a central damping force may be invoked. A repulsive central potential force, a viscous friction force, and a dry friction force may be invoked

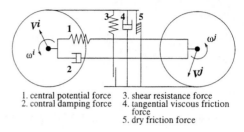

1. central potential force   3. shear resistance force
2. contral damping force   4. tangential viscous friction force
                                       5. dry friction force

Figure 1. Types of forces considered in the discrete element model.

for elements that are in contact. Interaction forces are chosen according to the material property needs.

SIMULATION CONDITIONS

As model systems, Ni/Al and Ti/Teflon mixtures were examined.

For the continuum model, the computation was performed for three different mass fractions (0.3, 0.6, and 0.9) of the denser materials. Parameters used in the computation are summarized in Table 1. Viscosities of metals are known to be on the order of several hundred poise (8). The viscosity of aluminum was that reported in (9). The viscosity of Teflon was assumed to be 20 poise.

Figure 2 schematically shows the boundary conditions for the discrete element simulation. The mixture impacts against a rigid wall at 150, 300, 450, and 600 m/s. Mass fractions of the denser materials were also 0.3, 0.6, and 0.9. In order to simulate viscosity, a central damping force was used between elements of the same material, and a tangential viscous force was used between elements of different materials.

RESULTS AND DISCUSSIONS

Figure 3 shows the spatial particle velocity profile from the discrete element calculation. Since Al is lighter than Ni, Al particles move faster than Ni particles at the shock front. The greatest particle velocity dispersion is seen at the shock front, but there is a recurrence of low

TABLE 1. Parameters used in the continuum mixture model calculation

	μ [dyn·s/cm^2]	r [μm]	Δt [ns]
Ni/Al	200/50	2.0	2.0
Ti/Teflon	200/20	2.0	2.0

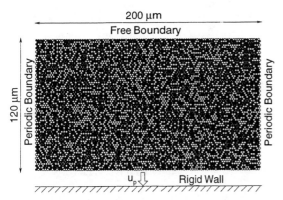

Figure 2. Initial geometry and boundary conditions for discrete element model simulation.

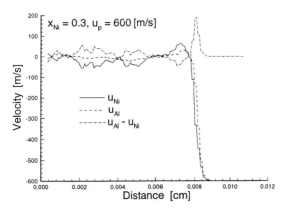

Figure 3. Spatial particle velocity profiles for Ni/Al mixture obtained from discrete element model simulation. x_{Ni} and u_p represent mass fraction of Ni and impact velocity, respectively.

amplitude dispersion. Although results will not be shown in this paper, similar features are observed with the Ti/Teflon mixture. However, it had a wider dispersion at the shock front.

Figure 4 shows the distribution function of the particle velocity dispersion for the Ni/Al mixture, representing the probability of finding ele-

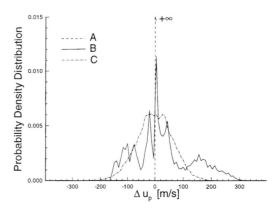

Figure 4. Distribution function of particle velocity dispersion for Ni/Al mixture obtained from discrete element simulation. Curves A, B, and C correspond to the distribution functions at upstream, inside, and downstream of the shock wave front, respectively. $\Delta u_p = u_{Al} - u_{Ni}$

ment pairs having a given value of particle velocity dispersion. The distribution ahead of the shock wave is described by the δ-function. The figure clearly shows the non-equilibrium evolution of the distribution from upstream to downstream of the shock wave. In the case of the Ti/Teflon mixture, it is found that the distribution function is very monotonic in comparison with that of the Ni/Al mixture.

Figures 5(a) and 5(b) show a comparison of the maximum particle velocity dispersion obtained from the discrete element model calculation and the average particle velocity dispersion obtained by the continuum model. There is a semi-quantitative agreement of the trend between the two results. The effect of mass fraction for the Ti/Teflon mixture is more prominent. This is attributed to the large difference in material properties between the constituents.

For shock pressures of 10 ~ 15 GPa, the particle velocity dispersion is on the order of 100 m/s. This velocity corresponds to the local shear strain rate of about 10^7 1/s. This rate is comparable to the estimated shear strain rate in shear bands where chemical reactions were observed (10). Also, it has been proposed (11) that this velocity can be translated into mass mixing on

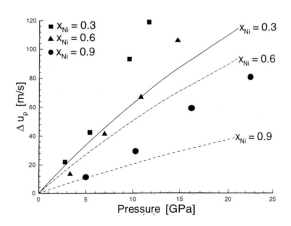

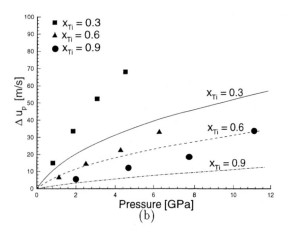

Figure 5(a),(b). Comparison of the maximum particle velocity dispersion obtained from discrete element model and the average particle velocity dispersion obtained from continuum model. Above: Ni/Al mixture. Below:Ti/Teflon mixture.

the spatial scale s given by:

$$s = \frac{36\gamma}{\rho(\Delta u_p)^2}\lambda_s \qquad (5)$$

where γ, ρ, and λ_s are surface free energy, density, and fraction of mass reacted, respectively. Substitution of representative values for metals ($\gamma = 1000$ erg/cm^2, $\rho = 6$ g/cm^3, $\lambda_s = 0.1$, and $\Delta u_p = 100$ m/s) yields $s = 0.06$ μm, indicating that the mixing due to the particle velocity dispersion is feasible on a sub-grain scale.

CONCLUSION

Particle velocity dispersion during shock compression of solid mixtures were numerically investigated using both a continuum model and a discrete element model. Intense particle velocity dispersion was observed at the shock front, reaching 100 m/s in the pressure range of 10 $\sim$ 15 GPa. This particle velocity dispersion corresponds to the local shear strain rate of 10^7 1/s. The localized energy, when used for mass mixing, is sufficient to create a "mixture" on a micron level. At this scale, conventional mechanisms can explain the observed very fast chemical reactions.

ACKNOWLEDGMENTS

This work is supported in part by U.S. Army Research Office (DAAH 04-94-G-0033).

REFERENCES

1. Fletcher, D. F., Anderson, R. P., *Progress Nuclear Energy*, **23**, 137 – 179 (1990).
2. Meshcheryakov, Yu. I., and Astroshenko, S. A., *Izv. Vys. Uch. Zav., Fizika*, **4**, 105 - 123 (1992).
3. Bastsanov, S. S. et. al., *Fizika Gor. i Vzr.*, **22**, 134 - 137 (1986).
4. Horie, Y., "Mass Mixing and Nucleation and Growth of Chemical Reactions in Shock Compression of Powder Mixtures", in *Shock-Wave and High-Strain Phenomena*, eds. Murr, L. E., et al, New York:Elsevier, 1995, pp.603 – 613.
5. Nigmatulin, R. I., *Dynamics of Multiphase Media* Vol.1, New York:Hemisphere Publishing Co., 1991, Chapter 1.
6. Ishii, M., and Zuber, N.,*AIChE Journal*, **25**, 843 – 855 (1979).
7. Tang, Z. P., Horie, Y., and Psakhie, S. G., "Discrete Meso-Element Modeling of Shock Processes in Powders", *High Pressure Shock Compression of Solids IV*, eds. Lee, D., Horie, Y., Shahinpoor, M., New York:Springer, 1997, Chapter 6, pp.143 – 176.
8. Bushman, A. V., et al, *Intense Dynamic Loading of Condensed Matter*, Washington DC:Taylor & Francis, 1993, pp.48 – 49.
9. Sud'enkov, Yu. V., et al, *Sov. Phys. Tech. Phys.*, **26**, 1283 (1981).
10. Nesterenko, V. F., et al, *Metallurgical and Materials Transactions A*, **26A**, 2511 - 2519 (1995).
11. Horie, Y. and Yano, K, "Shock-Wave Initiation of Chemical Reactions in Inorganic Powders" presented at AIRAPT-16, Kyoto, Japan, August 25 – 29, 1997.

CP429, *Shock Compression of Condensed Matter – 1997*
edited by Schmidt/Dandekar/Forbes
© 1998 The American Institute of Physics 1-56396-738-3/98/$15.00

STRESS FLUCTUATION AND ORDER GENERATION IN SHEARING OF GRANULAR MATERIALS

O. J. Schwarz, Y. Horie

Department of Civil Engineering, North Carolina State University, Raleigh, N.C. 27695

A quasi-molecular method is used to simulate the response of a uniformly graded granular material during shear at various strain rates. Samples were sheared to a strain of 66% at rates ranging from $2/3 \times 10^1$ to $2/3 \times 10^4$ (1/s). Simulations were run for both smooth and frictional particles. A stress gage was placed at the sample base which recorded normal stresses transmitted through the sample during shear. Magnitudes of maximum transmitted stresses are on the order of 0.1 to 10 MPa, depending on strain rate. The stress profile recorded generally resemble the results obtained experimentally at Duke University (1). A semi-stable dynamic structure (quasi-laminar flow) was observed to develop during shear. A power law correlation is observed between shear rate and maximum normal stress.

INTRODUCTION

The properties and mechanics of granular material flow have been a subject of study and interest for the past several decades. Previous researchers have focused on the time averaged quantities of shearing stress, normal stress (2,3,4), granular temperature (5), and thermal conductivity (6), as functions of strain rate and solids concentration. They conducted experiments based on the idea that the sheared granular material can be treated much as a liquid or a solid where fluctuations in these quantities are negligible, if not non-existent. Experimental apparatus were designed in an attempt to "damp out" the detection of such fluctuations (7). This way of examining granular flow is similar to the continuum method of system modeling, where material properties are averaged over the body of the system to obtain a picture of the material behavior.

Subsequent computer modeling of sheared granular material flow have utilized discrete element molecular dynamics techniques as opposed to continuum models (4,5). However, results were still obtained based on spatial or temporal averaging of system behavior. Simulations were carried out for relatively low solids concentrations to avoid multiple particle contacts which could not be easily handled computationally (4). For higher solids concentrations, considerable jumps or fluctuations in particle stresses have been seen to develop (4,7).

Recent experimental and computational studies have pointed to the formation of stress chains as the mechanism explaining large fluctuations in particle stresses (1,8,9). Stress chains are collections of contacting particles which carry a large portion of the stress transmitted through the depth of the granular sample. Using a mono-sized dispersion of spherical glass beads undergoing shear in an annular shear cell, researchers at Duke University have recorded the normal stresses at the base of the system as a function of time (1). Peak stresses were noted which exceeded the average normal stress by up to an order of magnitude. These stress peaks were attributed to stress chain formation. Stress chains have been visualized through the use of photoelastic disks or fibers in a two dimensional shear assembly (1,8,9).

The current study makes use of a new two dimensional discrete meso-dynamic method (DM2) to model and study the behavior of granular material undergoing shear at a variety of strain rates ranging from $2/3 \times 10^1$ to $2/3 \times 10^4$ (1/s).

THE MODEL

The DM2 code has the capability of modeling materials as either singular or multi-element particles. The physical and thermo-mechanical states of elements are calculated based on each element's current and past interaction with neighboring elements. Neighboring elements may either be chemically linked and in contact, linked but not in contact, in contact but not linked, or neither linked nor in contact. Depending on each element's current contact status, any of a number of forces based on central pair potential, elasto-plastic shear, viscous friction, tangential viscosity, and dry friction are applied. The DM2 conducts calculations at a user defined time step which must satisfy a discrete element version of the Courant condition. Satisfaction of this condition ensures accurate calculation of the momentum transfer between contacting elements based on admissible physical evidence. A complete description of the theory and equations used in the DM2 code has been previously discussed (10,11). In the current simulation, granular material will be modeled as a collection of discrete, non-cohesive particles which behave elastic-perfectly plastically. Therefore, the only interaction forces which will be considered will be a repulsive central potential force, and a dry friction force.

The model is set up to resemble the annular shear cell geometry of many previous researchers (1,2,3). The sample, consisting of a collection of 1.0 mm diameter elements, has a depth and length of 1.5 cm and 4.0 cm respectively. The bottom boundary of the sample is fixed. The top layer of elements move with a constant velocity in the horizontal (x) direction and are not allowed to translate in the vertical direction. This provides for Couette flow (simple shear) of the granular material. Both top and bottom boundaries are made up of elements consisting of the same physical properties as the bulk material. Elements on the left and right boundaries interact with each other to form, in effect, a cylindrical geometry. Calculations are performed without the consideration of gravity.

The elements within the sample begin from an initial triangular close-packed geometry. Maximum solids concentration for a close packed geometry in two dimensions is roughly 87%. A random distribution of elements are removed from the initial packing to achieve a concentration of 75%. All samples are sheared to a strain of 66%. Breaking of the initial packing is achieved within the first 10% strain and steady state is reached by 33% strain. Steady state is said to be achieved when the behavior of the sample is no longer a function of the original packing, i.e. the average stress per unit time has reached a relatively constant value and the original element packing is no longer evident. The flow geometry of the sample after it has reached steady state is shown in Fig. 1. The strain rate corresponding to Fig. 1 is $1/3 \times 10^3$ 1/s. (Apparent overlapping elements in Fig. 1 are due to a deficiency in the plotting process.) Similar flow geometries were produced at other strain rates. It is evident from Fig. 1, and an examination of the velocity profile, that shearing was achieved throughout the depth of the sample. Uniform shear was observed at strain rates up to $1/3 \times 10^4$ 1/s. At strain rates above this value, localized regions of higher shear rate were observed near the top boundary of the shear flow. However, no "dead" or stagnant zones were observed to occur, as have in previous experimental studies (1,2,3). A linear (or nearly linear) transverse velocity distribution was noted for all strain rates studied, indicating the presence of a laminar flow of particles.

Normal stresses transmitted through the sample are recorded by a gauge consisting of eight consecutive elements located within the stationary base of the sample, in an attempt to capture a stress/time profile similar to that obtained by other researchers (1,7). Average and maximum normal stresses as a function

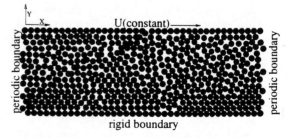

FIGURE 1. Shear flow geometry of sample at steady state.

of strain rate are obtained from the data of the individual stress/time profiles. As calculations were preformed in the absence of gravity and applied normal force, there was no convenient way to non-dimensionalize the data.

RESULTS AND DISCUSSION

Previous researchers have noted a squared dependence of the normal stresses produced in a sheared granular sample on the shear (or strain) rate (2,3,12,13). In these studies, the normal stress was the average pressure applied against a top rotating plate of an annular shear cell apparatus at various strain rates. A linear plot of the average normal stress as a function of strain rate for perfectly smooth particles is shown in Fig. 2. Similar trends of data were seen for frictional particles with an average reduction in stress magnitude of 45% for the strain rates studied. The magnitude of the standard deviation bars in Fig. 2 illustrates the vast range of the distribution of stresses encountered in a sheared sample. For the strain rates simulated, values of standard deviation are typically of the same order of magnitude as the average stress. It would seem, therefore, that the quantity of average transmitted normal stress has little value as a classifying characteristic behavior of a granular shear flow at a given strain rate. However, a dependence of the magnitude of maximum transmitted normal stress on the strain rate is evident.

A logarithmic plot of the maximum normal stress as a function of strain rate is shown in Fig. 3. An approximate linear curve fit of the data yields a relationship where stress is proportional to the strain rate raised to the power of roughly 0.817. (Reductions in maximum normal stress ranging from 0 to 56%, for a given strain rate, were noted when simulations were carried out considering interparticle friction.) This relationship is contrary to the squared dependency. However, deviations from the stress/strain rate relationship, towards a linear relationship, for granular flows having high solids concentrations in experimental studies has been noted (3). Such deviations were attributed to gravitational effects, lasting interparticle contacts, and frictional effects, and occurred at three dimensional solids concentrations of roughly 70% of the theoretical maximum. The current two dimensional simulation was carried out at a solids concentration of 84% of the theoretical maximum.

Normal stress as a function of time at a strain rate of $1/3 \times 10^3$ is shown in Fig 4. Maximum observed stress is on the order of 1×10^7 Pa, while the average stress is on the order of 1×10^6 Pa. Numerical comparison of this data with that of previous researchers is difficult due to the fact that each peak corresponds to a stress felt by an individual gauge element. From examination of the stress/time data files, it was evident that usually only one gauge element "fired" at a given time step. Therefore it is

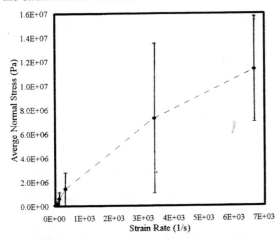

FIGURE 2. Average normal stress as a function of strain rate.

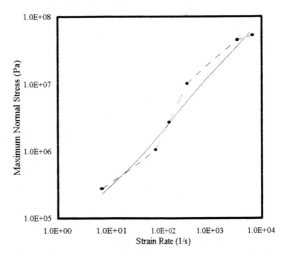

FIGURE 3. Maximum normal stress as a function of strain rate. (Solid line is a linear approximation.)

265

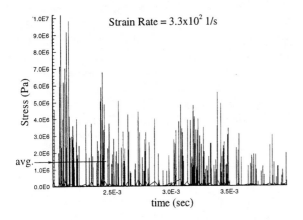

FIGURE 4. Stress versus time profile (strain rate = 333 1/s)

arguable that the magnitude of stress is affected by the total gauge size, as an actual physical gauge would average each stress over its surface area. However, the data compares well qualitatively with the stress fluctuations reported by other researchers (1). The stress/time behavior is also similar to that observed by Miller et al. (1) and simulated by Savage (7).

Strain rate, at a constant solids concentration, does not seem to affect the overall behavior of the sample. Stress/time plots for strain rates ranging from 2/3 $x10^1$ to 2/3 $x10^4$ (1/s) illustrate the same seemingly random distribution of stress peaks as a function of time. The maximum stresses felt are consistently an order of magnitude, or more, greater than the average stress for any given strain rate. Independence of the sample behavior on strain rate was also noted by Miller et al. (1).

CONCLUSIONS

Stress behavior for particles in a granular flow at high strain rates were found to be highly heterogeneous. Typical methods of studying granular materials based on average properties, therefore, neglect features of granular flow behavior which are not negligible. The results obtained from the DM2 model simulation of the rapid flow of a granular material compared well qualitatively with previous experimental results. Individual stress fluctuations were found which exceeded the average normal stress by roughly an order of magnitude. Normal

stress, at a solids concentration of 0.75, was observed to be proportional to strain rate raised to the power of 0.817.

ACKNOWLEDGEMENTS

This work was supported in part by the U.S. Army Research Office (DAAH04-95-1-0269). The authors would like to thank Dr. Clelland and Dr. M. Shearer for their input and comments, and Mr. K. Yano for his help during numerous discussions.

REFERENCES

1. Miller, B., O'Hern, C., and Behringer, R. P., *Physical Review Letters* **77**, 15, 3110-3113, (1996).
2. Hanes, D. M., and Inman, D. L., *Journal of Fluid Mechanics* **150**, 357-380, (1985).
3. Savage, S.B., and Sayad, M., *Journal of Fluid Mechanics* **142**, 391-430, (1984).
4. Lun, C.K.K., and Bent, A.A., *Journal of Fluid Mechanics* **258**, 335-353, (1994).
5. Campbell, C.S., and Brennen, C.E., *Journal of Fluid Mechanics* **151**, 167-188, (1985).
6. Wang, D. G., and Campbell, C. S., *Journal of Fluid Mechanics* **244**, 527-546, (1992).
7 . Savage, S.B., (Bideau, D., and Dodds, J., eds.,) *Physics of Granular Media*, Commack, New York, Nova Science Publishers, 1991, pp. 343-362.
8. Howell, D., and Behringer, R.P., "Fluctuations and dynamics for a two-dimensional sheared granular material", *Proceedings of the third international conference on Powders and Grains,* Durham, N.C., May 18-23, pp 337-340, 1997.
9. Baxter, G.W., "Stress distributions in a two dimensional granular material", *Proceedings of the third international conference on Powders and Grains*, Durham, N.C., May 18-23, pp 345-348, 1997.
10. Tang, Z.P., Horie, Y., and Psakhie, S.G., (Davison, L., Horie, Y., and Shahinpoor, M., eds.), *High Pressure Shock Compression of Solids IV*, Springer, New York, 1997, pp 143-176.
11. Tang, Z.P., Horie, Y., and Psakhie, S.G.,"Discrete Meso-Element Simulation of Shock Response of Reactive Porous Solids", *Shock Compression of Condensed Matter-1995*, 657-660, 1996.
12. Bagnold, R.A., *Proceedings of the Royal Society of London, A.* **225**, 49-63, (1954).
13. Craig, K., Buckholz, R.H., and Domoto G., *Journal of Applied Mechanics* **53**, 935-942, (1986).

CHAPTER V

MODELING AND SIMULATION:
Reactive Materials

CP429, *Shock Compression of Condensed Matter – 1997*
edited by Schmidt/Dandekar/Forbes
© 1998 The American Institute of Physics 1-56396-738-3/98/$15.00

MONTE CARLO CALCULATIONS OF THE PHYSICAL PROPERTIES OF RDX, β-HMX, AND TATB

Thomas D. Sewell

Theoretical Division, Los Alamos National Laboratory, Los Alamos, New Mexico 87545

Atomistic Monte Carlo simulations in the NpT ensemble are used to calculate the physical properties of crystalline RDX, β-HMX, and TATB. Among the issues being considered are the effects of various treatments of the intermolecular potential, inclusion of intramolecular flexibility, and simulation size dependence of the results. Calculations of the density, lattice energy, and lattice parameters are made over a wide domain of pressures; thereby allowing for predictions of the bulk and linear coefficients of isothermal expansion of the crystals. Comparison with experiment is made where possible.

INTRODUCTION

High explosives play an important role in both nuclear and conventional weapons systems, and we must develop a thorough understanding and truly predictive capability of the physics of these complicated materials. As one component in such a capability, it is necessary to have a knowledge of the equilibrium and non-equilibrium thermophysical properties of the explosive for time and distance scales relevant to the controlling hydrodynamic processes. Given the difficulty of conducting experiments which interrogate the wide range of conditions which can arise, there is a need for reliable theoretical and computational tools to allow for predictions of material properties under a diverse set of conditions of pressure, temperature, and strain rate.

The work described here is aimed towards developing a suite of tools and methods for predicting the equilibrium thermophysical properties of high-explosive crystals for temperatures and pressures that range from near-ambient to the extremes which occur in various accident and detonation scenarios. Among the quantities of interest are crystal packing, density and bulk/linear coefficients of isothermal and isobaric expansion, specific heats, and mechanical properties based on the anisotropic elastic coefficient matrix. To date, we have performed calculations for benzene, RDX, β-HMX, and TATB.

THEORETICAL METHODS

Our approach is statistical mechanical, employing the numerical technique of classical Monte Carlo, whereby the thermophysical properties follow from the interaction potential. Specifically, in isothermal-isobaric Monte Carlo (1), the macroscopic property $A(N,p,T)$ of a system of N molecules at temperature T and scalar pressure p is obtained as an average of the microscopic function of configuration $A(q;V)$, the average taken over the states of a Markov chain in the $3N+1$ dimensional configuration space of the system,

$$A\left(N,p,T\right) = \lim_{M \to \infty} \frac{1}{M} \sum_{m=1}^{M} A\left(\mathbf{q}_m\right) \qquad (1)$$

in which the transition matrix between successive states is based on the potential energies $U_N(\mathbf{q}_m)$ and $U_N(\mathbf{q}_{m+1})$ of these states in such a way as to assure detail balance and the equality of $A(N,p,T)$ with the actual ensemble average in the isothermal-isobaric ensemble,

$$\left\langle A_{NpT} \right\rangle = \frac{\displaystyle\int_0^\infty dV A e^{-\beta pV} Q\left(N,V,T\right)}{\displaystyle\int_0^\infty dV e^{-\beta pV} Q\left(N,V,T\right)}, \qquad (2)$$

where V denotes volume, $\beta = 1/\kappa T$, and

$$Q(N, V, T) = \int d\mathbf{q} e^{-\beta U_N(\mathbf{q})}. \tag{3}$$

In practice, the averages are evaluated using a Metropolis algorithm in which trial moves are accepted or rejected according to $P=\min[\exp(-\Delta),1]$, where, for present state $m-1$ and "trial" state m,

$$\Delta = \beta \left\{ \left[U(\mathbf{q}_m) - U(\mathbf{q}_{m-1}) \right] + p(V_m - V_{m-1}) \right\} \\ - N \ln(V_m/V_{m-1}) \tag{4}$$

In the case of rigid molecules, three kinds of trial moves are performed: translations of the molecular centers of mass, rotations of the molecules about their centers of mass, and changes in the size and shape of the simulation cell. The displacements are made in fractional coordinates. If intramolecular flexibility is allowed, then an additional set of moves corresponding to displacements of the internal coordinates is also performed. Maximum displacements were adjusted to yield roughly a 50% acceptance probability for a given kind of move. The battery of analyses described by Hald (2) was used to assess whether a particular realization was under statistical control.

The bulk was simulated by periodic replication in three dimensions of a primary simulation cell containing N molecules. The replication was extended far enough into space to account for all nonbonded interactions between molecules having centers of mass separated by more than 20Å.

The intermolecular potentials are of the form

$$U(\mathbf{R}) = \sum_{A \neq B} \sum_{i \in A} \sum_{j \in B} \left[U_{rep} + U_{disp} + U_{elec} \right], \tag{5}$$

where A and B are molecules, and i and j denote particular atoms. The repulsion and dispersion terms are written as

$$U_{rep} = A_{ij} e^{-B_{ij} R_{ij}} \quad \text{and} \quad U_{disp} = C_{ij}/R_{ij}^6, \tag{6}$$

respectively. The electrostatic contribution U_{elec} to the intermolecular energy is described using an atom centered multipole expansion (ACME). In most cases, only the $l=0$ term was retained, i.e.,

$$U_{elec} = q_i q_j / R_{ij}, \tag{7}$$

in which case potential-derived charges (PDQs) were used. However, in some instances higher order terms were included to investigate the sensitivity of the results to details of the electrostatics. The maximum order of expansion used was octupole-octupole, i.e., $l=3$.

Intramolecular interactions are currently limited to rigid torsional motions of exocyclic nitro (NO_2) and amino (NH_2) groups, and are of the form,

$$U_{intra} = U(\tau) = U^0 \left[1 - \cos^2 \tau \right], \tag{8}$$

where τ is the relevant dihedral angle.

The nonbonded parameters A_{ij}, B_{ij}, and C_{ij} were taken from Williams (3). Potential-derived atomic charges were obtained using the CHELPG method within the Gaussian 92 (4) suite of programs at the Hartree-Fock level using the 6-31g* basis set. Rotational barriers for exocyclic nitro and amino groups were computed using the same level of theory. Higher-order atom-centered multipoles were obtained using the Hirshfeld approach (5), as implemented by Ritchie and co-workers (6).

RESULTS AND DISCUSSION
RDX

The stable form of RDX at room temperature and pressure crystallizes in the orthorhombic space group Pbca, with $Z=8$ molecules per unit cell (7). The measured lattice lengths are $a=13.182$Å, $b=11.574$Å, and $c=10.709$Å, whence $\rho=1.806$ g/cm^3. Using our potential parameters (PDQs), and a primary simulation cell containing eight perfectly rigid molecules (with the measured unit cell as the initial geometry), the calculated density at $T=298$K and $p=1.00$ bar is $\rho=1.740(1)$ g/cm^3. Thus, our calculated density is 3.6% too low.

Olinger et al. (8) reported an x-ray diffraction determination of the lattice parameters of RDX as a function of pressure (0.0 kbar $\leq p \leq$ 39.5 kbar), at a fixed temperature of $T=293$K. (Their measurements actually extended up to $p=91.9$ kbar, but for $p>40.0$ kbar the results corresponded to a high-pressure polymorph denoted as RDX(III).) In Fig. 1 we compare our calculated values for the lattice lengths to those measured by Olinger et al., where it can be seen that the agreement is quite good. The largest discrepancy occurs for the length of a, but the magnitude of the error is only 1.5% at $p=0.0$ kbar and 2.3% at $p=39.5$ kbar. Perhaps more interesting

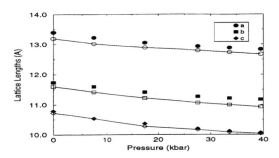

FIGURE 1. The lattice lengths for RDX are shown as a function of pressure, for a fixed temperature of T=298K. Filled symbols: calculated results; open symbols: experiment.

is the fact that the derivatives of the curves are in at least as good agreement with experiment as are the absolute magnitudes. As was observed experimentally, the crystal remains orthorhombic over the entire pressure range studied: the mean values for the lattice angles α, β, and γ never deviated from 90 degrees by more than 0.1 degree.

As would be expected from the results shown in Fig. 1, the bulk compression of RDX is also very well described. A comparison (not shown) of the ratios of the unit cell volumes at pressure p to those at p=0 for the calculated and experimental data indicates a maximum error of 0.7%.

HMX

HMX exists in four different polymorphic forms. The stable structure at ambient temperature is known as β-HMX. It crystallizes in the monoclinic space group P2$_1$/c, with Z=2 molecules per unit cell. The lattice parameters are a=6.54Å, b=11.05Å, c=8.70Å, and β=124.3 deg, yielding a density of ρ=1.894 g/cm^3 (9). In this case, using PDQs and a primary simulation cell containing only two perfectly rigid molecules (and again using the measured unit cell as the initial geometry for our simulation), the calculated density at T=298K and p=1.00 bar is ρ=1.791(2) g/cm^3. Thus, our calculated "baseline" density for β-HMX is 5.4% too low.

Olinger *et al.* (8) performed an x-ray diffraction determination of the lattice parameters of β-HMX as a function of pressure (0.0 kbar $\leq p \leq$ 74.7 kbar), at a fixed temperature of T=293K. We have performed calculations for β-HMX over the same domain of pressures. In general, the results are in good agreement with experiment, although not to the same degree of accuracy as was the case for RDX. The average percent errors compared to experiment

for the lattice lengths a, b, and c are 0.7%, 3.8%, and 1.0%, respectively. The general trends (not shown) with increasing pressure are: (1) decreasing errors for a (2.0% $\to$ 0.0%); (2) increasing errors for b (1.2% $\to$ 5.3%); and (3) decreasing errors for c (2.7% $\to$ -0.4%).

The results for the lattice angle β and the bulk compression of β-HMX are presented in Figs. 2 and 3, respectively. In Fig. 2 one can see that the calculated values of β are in good agreement with experiment, both in magnitude (they never differ by more than one degree) and in general shape. The comparison of V/V^0 in Fig 3 also shows fairly good agreement. By definition, the calculated and measured results are identical at p=0 kbar. At a pressure of 74.7 kbar, the calculated value of V/V^0 is in error by only -1.8% (*i.e.*, slightly too compressible).

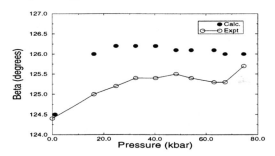

FIGURE 2. The value of the lattice angle β is shown as a function of pressure for β-HMX, for a fixed temperature of T=298K. Filled symbols: calculated results; open symbols: experiment.

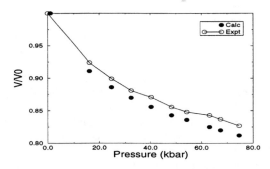

FIGURE 3. The bulk compression of β-HMX as a function of pressure is shown. V/V^0 is the ratio of the unit cell volume at pressure p to that at p=0. Filled symbols: calculated results; open symbols: experiment.

TATB

TATB crystallizes in the triclinic space group P1, with Z=2 molecules per unit cell. The lattice parameters are a=9.010Å, b=9.028Å, c=6.812Å, α=108.59 deg, β=91.82 deg, and γ=119.97 deg, whence ρ=1.937 g/cm^3 (10). Our computed density for the same conditions, using PDQs and a single unit cell of perfectly rigid molecules, is ρ=1.827(3) g/cm^3, a 5.7% error.

In Fig. 4 we present the densities which result from several calculations for TATB as a function of temperature at a constant pressure of p=1.0 bar. The issues of interest are: the effects of varying levels of treatment of the electrostatic potential (PDQ versus ACMEs through order l); the size dependence of the simulation on the results (N=2 versus N=8); and the effect of including exocyclic intramolecular torsions.

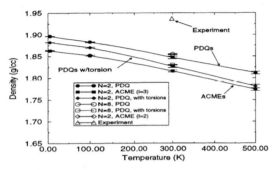

FIGURE 4. Calculated values of the density are plotted as a function of temperature for TATB, for a constant pressure of p=1.0 bar. The objective is to illustrate the effects of several different computational protocol issues. PDQ denotes CHELPG potential-derived charges. ACME denotes atom-centered multipole expansions; l is the order of the expansion. N is the number of molecules in the primary simulation box. "Torsions" indicates simulations for which exocyclic torsions were included.

Very briefly, one can see that, for the limited cases studied, there does not seem to be a strong dependence of the results on the simulation size. The choice of PDQs versus ACMEs leads to a shift in magnitude of the density (ACMEs yielding the lower density) but has little if any effect on the derivative of the curve (essentially the bulk coefficient of volumetric expansion). However, inclusion of exocyclic torsional degrees of freedom does affect the result and, as might be expected, this effect becomes more pronounced at elevated temperatures.

Finally, in Fig. 5 we show a comparison of the calculated and measured lattice lengths of TATB as a function of pressure, for a fixed temperature of T=298K, using PDQs and two perfectly rigid molecules. Only the results for a and c are shown; those for b are qualitatively similar to those for a.

The results for a are at a level of agreement with the experiments of Olinger et al. (11) comparable to what was observed for RDX and β-HMX. However, the results for c are in less satisfactory agreement, with the computed results crossing over the measured values. Also, the results for the lattice angles (not shown) are not under good statistical control - they vary significantly over the course of a given realization. The origins of this are not as yet understood, and constitute an area of ongoing research.

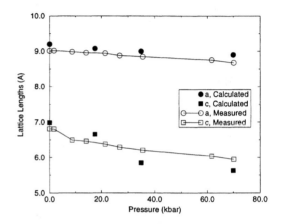

FIGURE 5. As in Fig. 1 but for TATB. Only results for lattice lengths a and c are shown.

REFERENCES

1. Wood, W. W., in *Physics of Simple Fluids*, Temperley, H. N. V., Rowlinson, J. S., and Rushbrooke, G. S., Eds., Amsterdam: North-Holland, 1968, ch. 5, p. 115.
2. Hald, A., Statistical Theory with Engineering Applications, New York: John Wiley & Sons, 1952, ch. 13, p. 338.
3. Williams, D. E. and Cox, S. R., Acta Cryst. **B40**, 404 (1984).
4. Gaussian92/DFT, Revision G.1, Frisch, M. J., *et al.*, Gaussian, Inc., Pittsburgh, PA, 1993.
5. Hirshfeld, F. L., Theor. Chim. Acta **49**, 129 (1977).
6. Ritchie, J. P. and Copenhaver, A. S., J. Comp. Chem. **16**, 777 (1995).
7. Choi, C. S. and Prince, E., Acta Cryst. **B28**, 2857 (1972).
8. Olinger, B., Roof, B., and Cady, H., "The Linear and Volume Compression of β-HMX and RDX to 9 GPa (90 Kilobar)," presented at the Symposium International Sur le Comportement Des Milieux Denses Sous Hautes Pressions Dynamiques, Paris, France, 1978., p. 3.
9. Choi, C. S. and Boutin, H. P., Acta Cryst. **B26**, 1235 (1970).
10. Cady, H. H. and Larson, A. C., Acta Cryst. **18**, 485 (1965).
11. Olinger, B. and Cady, H., "The Hydrostatic Compression of Explosives and Detonation Products to 10 GPa (100 kbars) and Their Calculated Shock Compression: Results for PETN, TATB, CO_2, and H_2O," presented at the Sixth Symposium (International) on Detonation, Coronado, California, August 24-27, 1976, p. 700.

CP429, *Shock Compression of Condensed Matter – 1997*
edited by Schmidt/Dandekar/Forbes
© 1998 The American Institute of Physics 1-56396-738-3/98/$15.00

Modeling, Simulation and Experimental Verification of Constitutive Models for Energetic Materials

K. S. Haberman and J. G. Bennett
ESA-EA, MS P946
Los Alamos National Laboratory, Los Alamos, NM 87544

B.W. Asay, B.F. Henson and D.J. Funk
DX-2, MS C920
Los Alamos National Laboratory, Los Alamos, NM 87544

Simulation of the complete response of components and systems composed of energetic materials, such as PBX-9501 (1) is important in the determination of the safety of various explosive systems. For example, predicting the correct state of stress, rate of deformation and temperature during penetration is essential in the prediction of ignition. Such simulation requires accurate constitutive models. These models must also be computationally efficient to enable analysis of large scale three dimensional problems using explicit lagrangian finite element codes such as DYNA3D (2). However, to be of maximum utility, these predictions must be validated against robust dynamic experiments. In this paper, we report comparisons between experimental and predicted displacement fields in PBX-9501 during dynamic deformation, and describe the modeling approach. The predictions used Visco-SCRAM and the Generalized Method of Cells which have been implemented into DYNA3D. The experimental data were obtained using laser-induced fluorescense speckle photography. Results from this study have lead to more accurate models and have also guided further experimental work.

INTRODUCTION

Finite element simulations are the preferred analysis tool used to predict the response of systems containing explosives. Predicting the correct state of stress, rate of deformation and temperature of a system under going dynamic loading is essential in the prediction of ignition. Thus, the constitutive model used to describe the mechanical behavior of the energetic material (explosive) must be accurate. Classical plasticity based constitutive models have been used, but with limited success due to the fact that the microstructural mechanics producing inelastic behavior and ignition in the energetic material are vastly different from classical plasticity theory. Validating new proposed constitutive models requires robust dynamic experiments where the comparative information is much more than post mortum geometry. In this paper, we report the experimental verification methodology and the comparisons between experimental and predicted displacement fields in PBX-9501 during dynamic deformation.

CONSTITUTIVE MODELS
Visco-SCRAM

This model has been proposed by Johnson (3,4) and combines Maxwell visco elasticity with statistical crack mechanics (5). The result is an isotropic constitutive model that describes the behavior of a visco/brittle material. Statistical crack mechanics is a physically based micromechanical description for the large deformation of brittle materials. During the process of deformation it is assumed that the crack distribution remains random and the size distribution of the cracks is exponential. The Visco-SCRAM approach provides a computational expedient constitutive model that can easily be implemented into a finite element code and exercised on a wide variety of three dimensional problems. Consider an n component Maxwell model, where G^n is the shear modulus and η^n is the time constant for the nth component. For a general visco-elastic solid let the strain rate be defined as,

$$\dot{\varepsilon}_{ij} = \frac{1}{2}\left(\frac{\partial \dot{u}_j}{\partial x_i} + \frac{\partial \dot{u}_i}{\partial x_j}\right) \qquad (1)$$

For an n component Maxwell model, the deviatoric strain rate is equal to the deviatoric strain rate in each component. The deviatoric state of stress is the sum of the deviatoric state of stress in each Maxwell component.

The deviatoric stress rate in each Maxwell component is given by Equation 2,

$$\dot{s}_{ij}^n = 2G^n \dot{e}_{ij}^{ve} - \frac{s_{ij}^n}{\tau^n} \qquad (2)$$

By superposition of strains, the total deviatoric strain rate is the sum of the deviatoric viscoelastic strain rate and the deviatoric cracking strain rate.

$$\dot{e}_{ij} = \dot{e}_{ij}^{ve} + \dot{e}_{ij}^c \qquad (3)$$

From Reference (4) the cracking strains are related to the deviatoric stress using,

$$e_{ij}^c = \beta^e c^3 s_{ij} \qquad (4)$$

where c is the average crack radius. Thus from Reference (4) the relationship between the deviatoric cracking strains and the crack radius is,

$$2G e_{ij}^c = \left(\frac{c}{a}\right)^3 s_{ij} \qquad (5)$$

Or in rate form,

$$2G\dot{e}_{ij}^c = 3\left(\frac{c}{a}\right)^2 \frac{\dot{c}}{a} s_{ij} + \left(\frac{c}{a}\right)^3 \dot{s}_{ij} \qquad (6)$$

Using Equations 6 and 2 an expression for the deviatoric stress rate in each Maxwell element may be obtained.

$$\dot{s}_{ij}^n = 2G^n \dot{e}_{ij} - \frac{s_{ij}^n}{\tau^n} - \frac{G^n}{G}\left[3\left(\frac{c}{a}\right)^2 \frac{\dot{c}}{a} s_{ij} + \left(\frac{c}{a}\right)^3 \dot{s}_{ij}\right]$$

Where,

$$\dot{s}_{ij} = \frac{2G\dot{e}_{ij} - \sum_{n=1}^{n}\frac{s_{ij}^n}{\tau^n} - 3\left(\frac{c}{a}\right)^2 \frac{\dot{c}}{a} s_{ij}}{1 + \left(\frac{c}{a}\right)^3}$$

The crack radius growth rate is determined as a function of K, where

$$K < K' \qquad \dot{c} = v_{max}\left(\frac{K}{K_1}\right)^m \qquad (7)$$

$$K \geq K' \qquad \dot{c} = v_{max}\left[1 - \left(\frac{K_o}{K}\right)^2\right]^m \qquad (8)$$

Where, $K' = K_o\sqrt{1 + \left(\frac{2}{m}\right)}$, and

$$K_1 = K_o\sqrt{1 + \left(\frac{2}{m}\right)}\left[1 + \frac{m}{2}\right]^{\frac{1}{m}}, K = \sqrt{\pi c}\sigma_{eff}$$

The parameters, $a = 0.001\ m$, $v = 0.3$, $m = 10$, $K_o = 5 \times 10^5\ Pa\sqrt{m}$ $c_o = 0.00003\ m$, $v_{max} = 300\ m/s$, have been determined using hopkinson bar data.

Generalized Method of Cells

The Generalized Method of Cells (6) is a physically based unified micromechanical modeling approach used to predict the elastic and inelastic macroscopic behavior of a heterogeneous materials.

The microstructure of the composite material is idealized by a representative volume element (RVE). The constitutive behavior of the constituents is defined along with constitutive behavior of the constituent interfaces. Continuity of displacements and continuity of tractions are imposed across the interfaces in the RVE. The macromechanical behavior is obtained from appropriate averages of the behavior of the constituents and the behavior of the constituent interfaces. A general heterogeneous particulate material may be represented by an idealized RVE shown in Figure 1. The RVE in Figure 1 has 27 subcells, however the number of subcells in any of the three coordinate directions is arbitrary. Each subcell can be filled, by an arbitrary constitutive model,

$$\dot{\sigma}_{ij}^{(\alpha\beta\gamma)} = C_{ijkl}^{(\alpha\beta\gamma)} \dot{\varepsilon}_{kl}^{(\alpha\beta\gamma)} - \dot{\Gamma}_{ij}^{(\alpha\beta\gamma)} \qquad (9)$$

Since the average behavior of the heterogeneous material is sought, it is sufficient to consider a first order theory in which the velocities in each subcell are expanded linearly in terms of the distances from the center of the subcell. Introducing a local coordinate system $\overline{x}$ whose origin is located at the center of the subcell $(\alpha\beta\gamma)$.

$$\dot{u}_{i}^{(\alpha\beta\gamma)} = \dot{w}_{i}^{(\alpha\beta\gamma)}(x) + \overline{x}_{i}\dot{\phi}_{i}^{\prime(\alpha\beta\gamma)} + \overline{x}_{2}\dot{\chi}_{i}^{(\alpha\beta\gamma)} + \overline{x}_{3}\psi_{i}^{(\alpha\beta\gamma)} \qquad (10)$$

With the strain rate in each subcell given by Equation 11,

$$\dot{\varepsilon}_{ij}^{(\alpha\beta\gamma)} = \frac{1}{2}\left(\frac{\partial \dot{u}_{j}^{(\alpha\beta\gamma)}}{\partial \overline{x}_{i}} + \frac{\partial \dot{u}_{i}^{(\alpha\beta\gamma)}}{\partial \overline{x}_{j}} \right) \qquad (11)$$

The strain rate in each subcell may be established as functions of the microvariables.

$$\dot{\overline{\varepsilon}}_{11j}^{(\alpha\beta\gamma)} = \dot{\phi}_{1}^{(\alpha\beta\gamma)}, \quad \dot{\overline{\varepsilon}}_{22j}^{(\alpha\beta\gamma)} = \dot{\chi}_{2}^{(\alpha\beta\gamma)}, \quad \dot{\overline{\varepsilon}}_{33j}^{(\alpha\beta\gamma)} = \dot{\psi}_{3}^{(\alpha\beta\gamma)}$$
$$2\dot{\overline{\varepsilon}}_{23}^{(\alpha\beta\gamma)} = \dot{\chi}_{3}^{(\alpha\beta\gamma)} + \dot{\psi}_{3}^{(\alpha\beta\gamma)}$$
$$2\dot{\overline{\varepsilon}}_{13}^{(\alpha\beta\gamma)} = \dot{\psi}_{1}^{(\alpha\beta\gamma)} + \dot{\phi}_{3}^{(\alpha\beta\gamma)} \qquad (12)$$
$$2\dot{\overline{\varepsilon}}_{12}^{(\alpha\beta\gamma)} = \dot{\chi}_{1}^{(\alpha\beta\gamma)} + \dot{\phi}_{2}^{(\alpha\beta\gamma)}$$

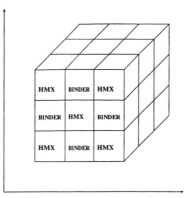

FIGURE 1. Representative Volume Element, RVE

Each subell interface may be described using an interface constitutive law similar to the decohesion laws proposed by Needleman (7). Employing continuity of displacements and continuity of tractions at the interfaces between the subcells of the RVE, and at the interfaces between the neighboring RVE, results in a set of linear algebraic equations which may be solved for the microvariables,

$$\dot{\phi}, \dot{\psi}, \dot{\chi}$$

Once the microvariables have been determined, the strain rate in each subcell may be determined using Equation 12. The stress rate in each subcell is determined using the subcell strain rate and the appropriate subcell constitutive law, Equation 9. The macroscopic state of stress is determined using volume averaging.

EXPERIMENTAL VERIFICATION

The dynamic impact experiments were conducted and reported elsewhere (8) in which a projectile was fired at the nominal velocity of 185 m/s at pushers of various geometry. The in plane displacement field on the surface of the explosive is measured using laser-induced fluorescense speckle photography (8). The measured displacement field can be directly compared with displacement field predicted using the DYNA3D simulation. Constitutive model validation and modification is based on the ability of the constitutive model to reproduce the experimental displacement field. Figure 2 shows the impact test geometry. Figure 3 shows the DYNA3D finite element model that faithfully reproduces the geometry of the impact experiment.

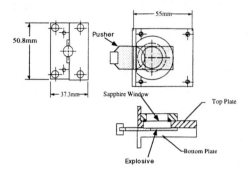

FIGURE 2. Impact test geometry

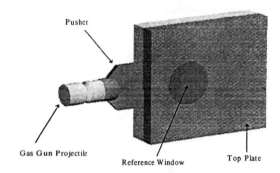

FIGURE 3. DYNA3D Finite element model of the impact test

Figure 4, shows the experimental displacement field in the PBX-9501 speciment located below the reference window, 15 microseconds after the projectile has impacted the Pusher. Figure 5 shows the predicted displacement field using DYNA3D with the Visco-SCRAM constitutive model at the same point in time. The comparison is both qualitatively and quantitatively good. The experimental displacement field is subject to some about of data smothing. Further detailed analysis of Figures 4 and 5 has yielded insight into the necessary parameter modifications that may be imposed on the Visco-SCRAM constitutive model. The Generalized Method of cells has been implemented into DYNA3D in the form of a 2X2X2 and a 3X3X3 RVE. The predicted results are very promising and will appear in a later publication.

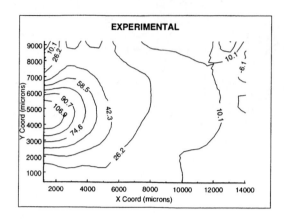

FIGURE 4. Experimental displacement field.

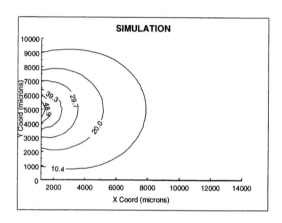

FIGURE 5. Predicted displacement field, obtained using DYNA3D and the VISCO-SCRAM constitutive model.

REFERENCES

1. LLNL Explosives Handbook, UCRL -52997,1982
2. DYNA3D, UCRL-MA-107254, LLNL, 1993
3. J.N. Johnson, LANL, T-1, private communications, 1996
4. Addessio, F.L., J.N. Johnson, Journal of Applied Physics, Vol 67, No. 7, 3275-3286, 1990
5. Dienes, J.K, Mechanics of Materials, Vol 4, 325-335, 1985
6. Aboudi, J., Composites Engineering, Vol 5, No. 7, 839-850, 1995
7. Needleman, A., Ultramicroscopy, Vol 40, 203-214, 1990
8. B.W. Asay, et al., Speckle Photography... To appear, Journal of Applied Physics, July 1997.

276

CP429, *Shock Compression of Condensed Matter – 1997*
edited by Schmidt/Dandekar/Forbes
© 1998 The American Institute of Physics 1-56396-738-3/98/$15.00

MECHANICAL STRENGTH MODEL FOR PLASTIC BONDED GRANULAR MATERIALS AT HIGH STRAIN RATES AND LARGE STRAINS

Richard V. Browning & Richard J. Scammon

Los Alamos National Laboratory, Los Alamos, NM 87545

Modeling impact events on systems containing plastic bonded explosive materials requires accurate models for stress evolution at high strain rates out to large strains. For example, in the Steven test geometry reactions occur after strains of 0.5 or more are reached for PBX-9501. The morphology of this class of materials and properties of the constituents are briefly described. We then review the viscoelastic behavior observed at small strains for this class of material, and evaluate large strain models used for granular materials such as cap models. Dilatation under shearing deformations of the PBX is experimentally observed and is one of the key features modeled in cap style plasticity theories, together with bulk plastic flow at high pressures. We propose a model that combines viscoelastic behavior at small strains but adds intergranular stresses at larger strains. A procedure using numerical simulations and comparisons with results from flyer plate tests and low rate uniaxial stress tests is used to develop a rough set of constants for PBX-9501. Comparisons with the high rate flyer plate tests demonstrate that the observed characteristic behavior is captured by this viscoelastic based model.

INTRODUCTION

Materials that display viscoelastic behavior have a distinctive signature when tested in a standard flyer plate geometry. Tests done recently on a plastic bonded high explosive (PBX) material showed viscoelastic characteristics, but when modeled as purely viscoelastic with constants derived from long term experiments, details of the behavior were not replicated. A PBX is basically a granular material with a polymeric binder. Although modeled successfully at small strains with a viscoelastic model, at the higher strains encountered in the flyer plate experiments some behavior of the granular filler emerged. Granular materials are usually modeled with a cap type plasticity formulation. Our goal here is to construct a model that combines some features of both. This is done by developing micromechanically motivated additions to a viscoelastic model and comparing numerical predictions with the measured velocity histories.

VISCOELASTIC WAVE PROPAGATION

Viscoelastic materials do not necessarily respond with an abrupt change in strain or stress, that is a shock, when loaded with a step velocity input. The waves that are generated are not steady and gradually decay with time or distance traveled, or depending on loading and material properties can get steeper. This behavior was extensively studied (1,2,3) more than a decade ago, particularly the possibility of generating steady acceleration waves. This behavior is quite different from that observed in elastic-plastic materials as described in the detailed summary of Davison and Graham (4).

Recently high quality VISAR experiments on some plastic bonded explosive materials were done by Jerry Dick (5) to characterize the high rate behavior of these materials. While the data can be reduced in standard u_s - u_p form, by ignoring the gradual initial response, the velocity histories are clearly not shock loading waves, but rather a slowly responding wave with viscoelastic characteristics. Figure 1 shows a particular experimental trace, Jerry Dick's G1061, and two computed approximations. In this particular experiment a flyer of Kel-F 800 5.889 mm thick impacts a 10.018 mm thick slab of PBX-9501, backed by a 22.9 mm thick PMMA VISAR window. The initial flyer velocity is 148 m/s. The VISAR records the velocity history at the interface between the PBX-9501 and the PMMA. An elastic-plastic model shows the characteristic sharp shock response of elastic materials. A viscoelastic model result, using a power law in time as the relaxation function, shows the more gradual loading behavior of the experiment, but misses the details of the initial loading time history. Adjustments in the material constants improve the fit in particular regions, but the overall behavior cannot be matched.

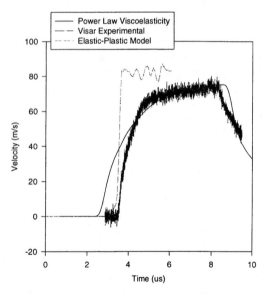

FIGURE 1. Comparison of VISAR measured particle velocity and computed results from two preliminary models.

BEHAVIOR OF GRANULAR MATERIALS

Shock loaded granular materials were studied as porous materials by Herrman (6) and others. The soil mechanics community has studied granular materials extensively under many loading conditions and generated an extensive literature, see (7,8) and their references. The essential characteristics are irreversible bulk behavior, usually modeled by a plasticity model, and dilatency. Dilatency, or the generation of volume under shear deformation is a very important characteristic of granular materials.

MORPHOLOGY OF PBX MATERIALS

Plastic bonded explosives exhibit a mixture of granular and viscoelastic behaviors. In the common manufacturing process, an organic explosive, in crystalline form, is coated with a rubbery polymer used as binder to allow easy fabrication and use. Explosive materials usually only contain 5-10 weight % of binder. At small strain levels modeling with viscoelastic behavior is a good approximation, even over very wide ranges of strain rate. This is the accepted technique for modeling long time behavior and, as we shall see, should be used as well for shock loading conditions if faithful reproduction of experimental results is desired. At larger strain levels we have evidence of granular behavior such as dilatation. For example, cores taken from recovered samples in the Steven Test geometry, where strain rates are 1000/s and there is substantial lateral confinement, still show lower densities than the original material. This is expected based on the observed microstructure. Optical microscopy and particle size analysis done by Skidmore(9) show the wide range of particle sizes, packing of smaller particles between interstitial spaces of the larger particles and binder filling the remaining space. The large number of fines, less than 1 micron in size, are likely acting as reinforcements to the binder but otherwise not involved with the larger particles.

INTERGRANULAR STRESS STIFFENING

A pressing operation at elevated temperatures, 100°C, and high pressures, 100-200 MPa, is used to consolidate the PBX molding powder. Because of the differences in coefficients of thermal expansion and bulk modulus between the crystalline material and the polymer binder, the material is left in a stressed and/or porous state upon cooling back to ambient conditions of 20°C and 0.1 MPa. Estimates of the mismatch indicate residual void volumes of a few percent, assuming no residual stress, in rough agreement with observed densities for pressed parts. This estimate shows the volume change caused by the pressure change to be larger than that from the temperature change, so the binder grows more than the crystal under release of pressure. One mental picture of this situation is to take an array of cubes, remove material from each corner equal to the volume of binder when under pressure. When the cubes are assembled under pressure with the binder, we have small pyramids of binder between the corners of the cubes. As the pressure is released, and temperature lowered, the pucks of binder separate the faces of the cubes, producing some free volume. The actual situation is somewhat more complex because of the distribution of particle sizes and shapes, but the net result is similar. This mental picture leads to behavior that would primarily reflect the binder at low strain levels, but at high pressures the faces of the crystals would again come into contact, generating substantial inter-crystalline forces.

CONSTITUTIVE MODEL

One approach to merging the small strain viscoelastic behavior, binder heating, and intercrystalline forces at high pressures is to add a density dependent set of stresses that represent the intercrystalline forces. Because we are primarily interested in the effects in flyer plate experiments we simplify the intercrystalline force model to generate only pressures, as a function of the current density and a reference density taken as the peak value of the density history. For general strain states shear stresses must be included as well.

The total stress is taken as the sum of a viscoelastic binder stress and interparticle or crystalline stress, $\sigma = \sigma_b + \sigma_c$. The binder stress comes from a standard hereditary integral formulation,

$$\sigma_b(t) = \int_{\tau=0}^{\infty} E(t-\tau)\dot{\varepsilon}(\tau)d\tau , \qquad (1)$$

where the relaxation function is usually taken as

$$E(t) = at^{-n}. \qquad (2)$$

Temperature effects are included by using a WLF shift function, that accelerates the apparent time by a factor w, calculated as

$$\log_{10}(w) = \frac{c_1(T-T_o)}{(c_2+T-T_o)}. \qquad (3)$$

This is implemented numerically using a Prony series approximation to the power law relaxation function. The general form is

$$E(t) = e_0 + \sum_{i=1}^{m} e_i \exp(-b_i t) \qquad (4)$$

The time lags b_i are selected to cover a time interval, usually with a constant ratio from term to term. Enough terms must be used to cover the times encountered in the numerical problem. The coefficients e_i are determined using a least squares fitting procedure.

The intercrystalline force contribution is taken as a pure pressure for these calculations, $\sigma_c = -p_c$, where the pressure and corresponding stiffness are obtained from the Hertz contact solution for two spheres , so

$$p_c = k(\varepsilon_{rel})^{1.5} \qquad (5)$$

if $\varepsilon_{rel} = \varepsilon_v - \varepsilon_o$ is positive, and 0 otherwise.

NUMERICAL RESULTS

This model is implemented in a UMAT routine (10) for use with the structural analysis code ABAQUS. The routine could be adapted to other codes however ABAQUS is well suited for evaluating the viscoelastic part of the model. A detailed 1-D model of the flyer plate experiment was developed using several hundred elements in each layer of material. Figure 2 shows one comparison of the experimental trace and the calculated response with the intercrystalline forces calculated with k of 25000 MPa and ε_o of 0.01. The viscoelastic model constants are a of 526 MPa, for time units of seconds, and n is 0.148. The ABAQUS routine actually has separate Prony series representations for the bulk and shear moduli; for these calculations the two are proportional with values equivalent to a Poisson's ratio of 0.37. The characteristic behavior of the material is captured, although we still need to improve some details of the model in order to better match the initial loading rate and the arrival of the relief wave at late times.

279

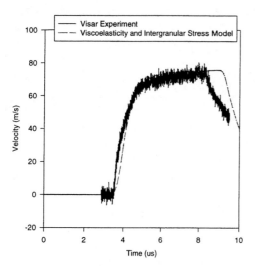

FIGURE 2. Comparison of experimental particle velocity and viscoelastic model result with interparticle stiffening.

CONCLUSIONS

One interesting feature of this study is the extension of a basic viscoelastic model over eight decades of strain-rate with reasonable agreement. Although the polymer physics community might not find this surprising, many in the shock wave community seem to find it odd. Clearly additional work needs to be done before any claims are made about the application range of the model developed. The inclusion of shear strengthening mechanisms in the granular stress is essential, and thermal softening of the binder might be important. The problem is finding appropriate experimental data to support the modeling work. At low rates, bi-axial experiments are used to separate or delineate the bulk and deviatoric behaviors. Under high rate loading conditions this type of experiment is very difficult to arrange. Perhaps the unloading waves seen in long time recordings from flyer plate experiments could be used to obtain some information under non-1D strain loading situations. We also hope to create detailed numerical models at the microstructural level to study the local binder-particle interactions.

ACKNOWLEDGMENTS

We thank Jerry Dick, for the accurate VISAR measurements that provide the foundation for this work, and Phil Howe for providing the funding and encouragement to understand material behavior in polymers.

REFERENCES

1. Warhola, G. T. and Pipkin, A. C., IMA Journal of Applied Mathematics **41**, 47-66 (1988).

2. Nunziato, J. W., Polymer Engineering and Science **18**, No. 14, 1101-1108 (1978).

3. Nunziato, J.W., Walsh, E.K., Schuler, K.W. and Barker, L.M., pp. 1-108 in Encyclopedia of Physics, Vol. VIa/4, Springer-Verlag, 1974.

4. Davison, L. and Graham, R. A., Physics Reports **55**, No. 4, 255-379 (1979).

5. Jerry Dick, Los Alamos National Laboratory, personal communication (1997).

6. Hermann, W., J. Applied Physics, **40**, No. 6, 2490-2499 (1969)

7. Lade, P. V. and Pradel, D., J. of Engineering Mechanics, **116**, No. 11, 2532-2550 (1990).

8. Pradel, D. and Lade, P. V., J. of Engineering Mechanics, **116**, No. 11, 2551-2566 (1990).

9. Skidmore, C.B., Phillips, D.S., Son, S.F., and Asay, B.W., "Characterization of HMX pariticles in PBX-9501", presented this conference.

10. Macek, R., Los Alamos National Laboratory, personal communication (1997).

CP429, *Shock Compression of Condensed Matter – 1997*
edited by Schmidt/Dandekar/Forbes
© 1998 The American Institute of Physics 1-56396-738-3/98/$15.00

VISCOELASTIC MODELS FOR EXPLOSIVE BINDER MATERIALS

S. G. Bardenhagen, E. N. Harstad, P. J. Maudlin, G. T. Gray, J. C. Foster, Jr.[*]

Los Alamos National Laboratory, Los Alamos, NM 87545
[*] *Wright Laboratory, Armament Directorate, Eglin AFB, FL 32542*

An improved model of the mechanical properties of the explosive contained in conventional munitions is needed to accurately simulate performance and accident scenarios in weapons storage facilities. A specific class of explosives can be idealized as a mixture of two components: energetic crystals randomly suspended in a polymeric matrix (binder). Strength characteristics of each component material are important in the macroscopic behavior of the composite (explosive). Of interest here is the determination of an appropriate constitutive law for a polyurethane binder material. This paper is a continuation of previous work in modeling polyurethane at moderately high strain rates and for large deformations. Simulation of a large deformation (strains in excess of 100%) Taylor Anvil experiment revealed numerical difficulties which have been addressed. Additional experimental data have been obtained including improved resolution Taylor Anvil data, and stress relaxation data at various strain rates. A thorough evaluation of the candidate viscoelastic constitutive model is made and possible improvements discussed.

INTRODUCTION

The ability to bridge the gap between the mechanical loading of an explosive and its initiation is a useful tool for assessing munitions performance. It is essential for simulating accident scenarios, where determining initiation is the objective. To bridge this gap requires accurate constitutive modeling of the explosive.

The class of explosives considered, i.e. energetic crystals randomly suspended in a rubber–like polymeric binder, may be idealized as random, two-component composites. The first step in determining the composite's averaged response is accurately characterizing the behavior of the individual constituents. This paper is a continuation of work on the characterization of Adiprene–100, a rubber–like polymeric binder.

In previous work (1) it was found that this material behaves in a strongly rate–dependent fashion, and exhibits creep, stress–relaxation, and recovery. A classical viscoelastic construction formed the basis for a constitutive model, which was calibrated with quasi–static compression tests and used to model a Taylor Anvil experiment. The modeling was reasonably successful. However, modeling shortcomings were apparent, and, due to very large deformations, the Taylor test simulation was hampered by numerical difficulties. This paper discusses an improvement to the constitutive formulation.

EXPERIMENTAL RESULTS

The Taylor Anvil impact test shows clearly how rubber–like the binder material behavior is. A 30 caliber cylinder of polyurethane (diameter

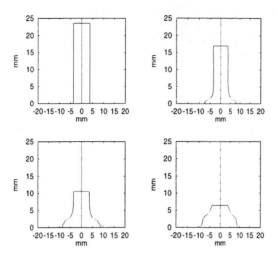

FIGURE 1. Taylor Anvil profiles at various times after impact. The initial configuration profile is shown in the top left frame. The top right frame shows the cylinder profile 17 μs after impact, the bottom left frame 39 μs after impact, the bottom right 55 μs after impact.

7.44 mm, length 22.53 mm) was launched with velocity 303 m/s. A high speed camera photographed the cylinder as it impacted the anvil. Digitization of the photographs gave profiles of the cylinder at various times after impact as shown in Fig. 1. Despite the extreme deformation, when the specimen was retrieved it had returned to its initial configuration.

In an effort to better understand the material response, additional data were gathered. The resolution of the photographic data was improved by performing 50 caliber (diameter 12.37 mm, length 63.22mm) shots at half the velocity (152 m/s), and focusing the camera at the impact interface. Quasi–static uniaxial stress tension and compression tests were performed, as well as stress relaxation tests.

The original data set used to calibrate the viscoelastic model comprised the tension and compression tests. The stress relaxation tests were performed approximately one year later. The material mechanical properties were found to change with age (a stiffening is observed). Aging is beyond the scope of the model to date, so this data could not be used directly, although stress decay rates guided model parameter selection.

CONSTITUTIVE MODEL

The approach taken in the constitutive modeling is described in detail elsewhere (1,2). Here the main features of this model are recapped, and a modification presented. A 1-D spring and dashpot construction exhibits the essential features of the 3-D constitutive formulation, and provides physical insight. The standard linear solid, which consists of a Maxwell element in parallel with a linear spring, provides the basic framework. The spring returns the model to its initial configuration when it is left stress free and unconstrained, a feature observed in the Taylor Anvil experiment. The Maxwell element provides rate dependence, which is made non-Newtonian by selecting for the dashpot viscosity the function

$$\eta(\dot{\epsilon}) = \eta_\infty + \frac{(\eta_0 - \eta_\infty)}{(1 + (\lambda\dot{\epsilon})^2)^{(1-n)/2}}, \qquad (1)$$

which serves to decrease the viscosity η from its initial value η_0 for strain rate $\dot{\epsilon} = 0$ to η_∞ as $\dot{\epsilon} \to \infty$ and consequently provides for shear thinning. Parameters λ and n adjust the rate at which η approaches η_∞ for large $\dot{\epsilon}$. The total stress is the sum of the stress in the Maxwell element, σ^V, and that in the spring, σ^E, i.e. $\sigma = \sigma^V + \sigma^E$.

The 3-D, finite deformation constitutive law is a straight–forward generalization of the 1–D model. The deviatoric and equation of state behaviors are treated separately. It is postulated that the equation of state is hypoelastic,

$$\dot{p} = -3K\dot{\epsilon}_v, \qquad (2)$$

where $p = -\sigma_{ii}/3$ is the hydrostatic pressure, $\dot{\epsilon}_v = D_{ii}/3$ is the volumetric strain rate, and K is the bulk modulus. The Cauchy stress tensor is denoted by σ_{ij} and the rate of deformation tensor by D_{ij}. The deviatoric behavior is postulated to have a viscous character. Summing viscoelastic and hypoelastic stresses as in 1–D, the constitutive law relating the deviatoric stress s_{ij} and the deviatoric deformation rate D'_{ij} may be written

$$\begin{aligned} s_{ij} + \tau \overset{\triangledown}{s}_{ij} &= s_{ij}^E + 2\tau(G^E + G^V)D'_{ij} \\ \overset{\triangledown}{s}{}_{ij}^E &= 2G^E D'_{ij} \end{aligned}, \qquad (3)$$

282

where $s_{ij} = s_{ij}^V + s_{ij}^E$, G^E and G^V are the hypo-elastic and viscoelastic shear moduli respectively, and $\overset{\triangledown}{s}_{ij}$ denotes an objective rate of s_{ij}. The time constant is defined by $\tau(\dot{\epsilon}_e) = \eta(\dot{\epsilon}_e)/2G^V$ and the viscosity is given by Equation (1) where the equivalent strain rate $\dot{\epsilon}_e = \sqrt{2D_{ij}D_{ij}/3}$ is used.

This model gave good results (1). However, it became clear that the formulation lacked an essential ingredient. After the cylinder reached the most deformed state in Fig. 1, it began to (slowly) unload. The hypoelastic deviatoric stress model was too compliant, resulting in continued deformation in the simulation.

To better model stiffening under large compression, a nonlinear, rubber elastic deviatoric constitutive model was incorporated in place of hypoelasticity. The Cauchy stress is given by

$$\sigma_{ij}^R = \frac{2}{\sqrt{I\!I\!I_c}} \left\{ \left(\frac{\partial W}{\partial I_c} + I_c \frac{\partial W}{\partial I\!I_c} \right) B_{ij} \right.$$
$$\left. - \frac{\partial W}{\partial I\!I_c} B_{ik}B_{kj} + I\!I\!I_c \frac{\partial W}{\partial I\!I\!I_c} \delta_{ij} \right\} . \quad (4)$$

where I_c, $I\!I_c$, $I\!I\!I_c$ are the invariants of the right Cauchy–Green tensor (F_{ij} is the deformation gradient) $C_{ij} = F_{ik}^T F_{kj}$, and $B_{ij} = F_{ik}F_{kj}^T$ is the left Cauchy–Green tensor. The elastic deviatoric stresses are then

$$s_{ij}^E = \sigma_{ij}^R - \delta_{ij}\sigma_{kk}^R/3. \quad (5)$$

Specifically, a Blatz–Ko formulation was used. The energy density is given by

$$W = \frac{G^E}{2} \left(\frac{I_c}{I\!I\!I_c} + 2\sqrt{I\!I\!I_c} - 5 \right). \quad (6)$$

Constitutive constants were obtained by fitting compression test data, using stress relaxation and

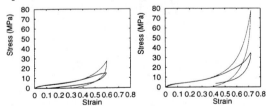

FIGURE 2. Comparison of constitutive model (dashed lines) and experimental data (solid lines). The left graph corresponds to $|\dot{\epsilon}| = .001$, the right to $|\dot{\epsilon}| = .1$.

sound speed data for guidance (3). A comparison of experimental data and constitutive model prediction is given in Fig. 2. The data are for two constant rate loading/unloading tests, at different rates $|\dot{\epsilon}| = .001$, .1. The constitutive model fit is excellent at low to moderate strains. It stiffens somewhat too quickly at higher strains, but gives the correct trend.

COMPARISON AND CONCLUSIONS

The finite deformation, nonlinear viscoelastic constitutive model developed in the previous section was implemented in the explicit, Particle in Cell code FLIP (4,5). The Particle in Cell technique is a mix of Lagrangian and Eulerian computational approaches, Lagrangian particles move through an Eulerian grid, well suited for model-

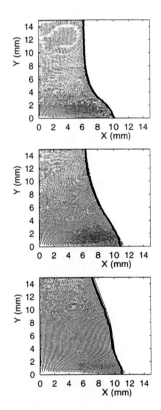

FIGURE 3. Comparison of data (outline) and calculation (points) for 50 caliber Taylor Anvil shots 37 μs (top), 85 μs (middle), and 141 μs (bottom) after impact.

ing large material deformation. The test geometry and isotropy of the material allowed an axially symmetric calculation to be performed.

Comparisons of 50 caliber cylinder profiles are shown in Fig. 3 at 37, 85, and 141 μs after impact. The dark outlines are data from photographs. The shaded region (Lagrangian particles) are the simulation results. Agreement is very good throughout the deformation. Most notable is the successful modeling of the "bulking up" (i.e. the manner in which the deformation proceeds axially) of the cylinder at 141 μs.

Comparison of 30 caliber cylinder profiles are shown in Fig. 4 at 17, 39, and 55 μs after impact. Note the inclusion in the simulation of the plug which follows the cylinder. This plug is used to isolate the cylinder from detonation products. The lift–off of the outer edge of the cylinder at 17 μs is not seen in the data. The simulation is frictionless. Addition of a small amount of friction at the cylinder/Anvil interface may prevent this. Agreement is good at 39 and 55 μs. However, a numerical instability begins to develop at the cylinder axis due to the extreme deformation there (a stagnation point), reducing accuracy and eventually terminating the calculation.

From the data gathered it appears that the incorporation of rubber elasticity provides an essential stiffening mechanism for modeling Adiprene–100. Simulation of the 30 caliber shots continues to pose challenges. The severity of the deformation requires a regularization of the data at some point during the calculation. While the linear equation of state is likely adequate for the 50 caliber shots (6), it's suitability for the 30 caliber shots needs to be determined.

ACKNOWLEDGMENTS

The authors would like to thank J. U. Brackbill, T–3, Los Alamos Nat. Lab., and D. L. Sulsky, Dept. of Mathematics and Statistics, U. of New Mexico, for numerics expertise and the FLIP code. The efforts of L. L. Wilson, Wright Laboratory, Armament Directorate, Eglin AFB, in conducting the Taylor Anvil shots, and M. Lopez, MST–5, Los Alamos, in performing the quasi-static tests, were very much appreciated. This work was performed under the auspices of the United States Department of Energy.

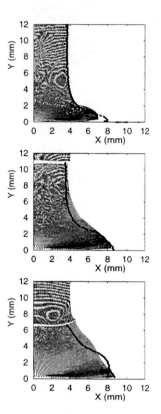

FIGURE 4. Comparison of data (outline) and calculation (points) for 30 caliber Taylor Anvil shots 17 μs (top), 39 μs (middle), and 55 μs (bottom) after impact.

REFERENCES

1. Bardenhagen, S. G., Harstad, E. N., Foster, J. C., and Maudlin, P. J., *Shock Compression of Condensed Matter - 1995*, New York, AIP Press, 1996, pp. 327–330.
2. Bardenhagen, S. G., Stout, M. G., and Gray, G. T., *Mech. Mat.* **25** pp. 235–253 (1997).
3. Johnson, J. N., *Personal Communication.*
4. Brackbill, J. U., and Ruppel, H. M., *J. Comput. Phys.* **65**, pp. 314–343 (1986).
5. Sulsky, D., Chen, Z., and Schreyer, H. L., *Comp. Meth. Appl. Mech. Eng.* **118**, pp. 179–196 (1994).
6. Harstad, E. N., Foster, J. C., Wilson, L. L., Maudlin, P. J., and Schreyer H. L., "Viscoelastic Strain Wave Propagation in Caliber 50 Taylor Anvil Tests", presented at the *Fourth International Conference on Composites Engineering*, Kona, Hawaii, July 7 – 11, 1997.

CP429, *Shock Compression of Condensed Matter – 1997*
edited by Schmidt/Dandekar/Forbes
© 1998 The American Institute of Physics 1-56396-738-3/98/$15.00

SHEAR BAND FORMATION IN PLASTIC BONDED EXPLOSIVE (PBX)

T. N. Dey and J. N. Johnson

Los Alamos National Laboratory, Los Alamos, NM 87545

Adiabatic shear bands can be a source of ignition and lead to detonation. At low to moderate deformation rates, 10-1000 s^{-1}, two other mechanisms can also give rise to shear bands. These mechanisms are: 1) softening caused by micro-cracking and 2) a constitutive response with a non-associated flow rule as is observed in granular material such as soil. Brittle behavior at small strains and the granular nature of HMX suggest that PBX-9501 constitutive behavior may be similar to sand.

A constitutive model for the first of these mechanisms is studied in a series of calculations. This viscoelastic constitutive model for PBX-9501 softens via a statistical crack model. A sand model is used to provide a non-associated flow rule and detailed results will be reported elsewhere. Both models generate shear band formation at 1-2% strain at nominal strain rates at and below 1000 s^{-1}. Shear band formation is suppressed at higher strain rates. Both mechanisms may accelerate the formation of adiabatic shear bands.

INTRODUCTION

Thermal softening as a cause of shear band formation in materials has been extensively studied. Such adiabatic shear bands in explosives may be a source of ignition and lead to detonation (4,5,8,10). At low to moderate deformation rates, two other mechanisms can also give rise to shear bands, and require only a small strain to do so. By concentrating deformation, these mechanisms may accelerate the formation of adiabatic shear bands. The two mechanisms are mechanical softening caused by micro-cracking and a constitutive response characterized by a nonassociated flow rule.

In order to better understand the characteristics of these mechanisms, we have carried out a series of numerical calculations that simulate deformation of a plastic-bonded explosive (PBX) subjected to various loading conditions. In this paper we report results for calculations that use a constitutive model that generates a visco-elastic response together with

mechanical softening due to micro-cracking. This model is intended to mimic the behavior of PBX-9501.

CONSTITUTIVE MODEL

The constitutive model separates the material response into volumetric and deviatoric components. Because of the low pressure obtained in the simulations discussed in this paper, a volumetric response characterized by a constant bulk modulus is sufficient. The deviatoric response is divided into two components acting in series with each other.

One component is modeled as five Maxwell spring and dashpot elements acting in parallel to each other. This component produces a viscoelastic response. A Maxwell spring and dashpot element is the ideal behavior of a linear spring and a linear dashpot acting in series with each other. The spring is characterized by its contribution to the shear modulus of the material while the dashpot has characteristic relaxation time. For deformations occurring more rapidly than the characteristic time, the Maxwell element adds its stiffness to the other elements. For deformations on a time scale much longer than the characteristic time, the Maxwell element adds no stiffness.

In series with the viscoelastic component is a fracture mechanics component. Addessio and Johnson (1) developed this model as a simplification of a statistical crack mechanics model of Dienes(3). The model is characterized by a critical stress intensity factor, a mean crack size, an exponent that determines the crack growth velocity for conditions both below and above the critical stress intensity, and a maximum crack growth velocity. Model values are based on data from room temperature experiments on PBX-9501 by Gray et al. (7).

CALCULATIONS

In order to understand the general characteristics of shear band formation associated with this model, we performed a series of simple plane strain calculations each at a different nominal strain rate. The mesh was 0.01 m wide and 0.02 m high. Roller boundary conditions (no normal displacement but free tangential slip) were applied to the left and bottom edges. The right side was a free surface. The top surface was displaced vertically with no lateral constraint at constant velocity. The average vertical strain rate in each calculation was chosen from the

range of 10-10000 s^{-1}. The meshes had 20 cells in the horizontal direction and 40 in the vertical direction. The time scale associated with the maximum crack growth velocity implies an internal length scale for this model and a finite width for any shear band that may form. The mesh used was found to resolve these shear bands adequately.

One cell at either the lower right or left of the mesh was given a slightly larger initial mean crack length causing it to be slightly weaker. Some form of perturbation like this is required to trigger shear band formation in the absence of loading inhomogeneity.

Figure 1 shows a mesh deformed at a nominal rate of 100 s^{-1}. A shear band has clearly formed from the perturbation at the lower left corner of the sample. The width of the shear band near the perturbation is limited by the small size of the perturbation. Away from the perturbation, the shear band broadens to a width of five or so computational cells.

Figure 2 shows average vertical stress plotted as a function of average vertical strain for a number of different average strain rates. A stress drop is clearly visible in the curves for the lower strain rates. The beginning of this stress drop is coincident with the beginning of localization. The shear band is fully formed by the time the stress drop is complete. For strain rates of 1000 s^{-1} or less, the peak stress is reached at about 1% axial strain.

Figure 3 shows a plot of the mesh for a calculation with a nominal strain rate of 1000 s^{-1}. It is clear from that no significant localization has

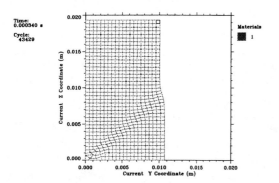

FIGURE 1. Deformed mesh showing shear band formed when average strain rate is 100 s^{-1}.

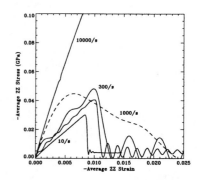

FIGURE 2. Average vertical stress vs. average vertical strain for each calculation.

happened. The stress-strain curve in Figure 2 for this calculation shows a stress drop, but it is apparent that the drop is gradual. Evidently, inertial effects are now sufficiently large and the internal length scale for this material model is sufficiently long that there is not enough time for localization occur. At this and all higher strain rates, this model exhibits no tendency to generate shear bands.

DISCUSSION

A number of studies of shear band formation caused by thermal softening in metals indicate that strains must be substantial, 10-50%, before shear bands form (6,9,11). The numerical studies described here indicate that the mechanical softening in the viscoelastic microcracking model can lead to shear band formation at much smaller strains, only 1% strain in the examples shown here.

The shear band width generated by the model studied here is much greater than observed for the thermal softening mechanism. The numerical results give widths of a few mm, while the thermal mechanism is associated with shear bands that are one to three orders of magnitude narrower. It is unlikely that the mechanism studied here is, by itself, able to generate sufficiently high temperatures to cause ignition of the explosive since the deformation is still being spread over a substantial volume due to the large shear band width.

The mechanism studied here may trigger

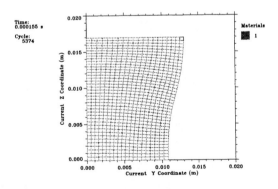

FIGURE 3. Deformed mesh indicating no shear band formation for strain rate of 1000 s^{-1}.

adiabatic shear band formation at much smaller average strains than those occurring if the thermal softening mechanism were acting alone. Even though the viscoelastic microcracking model by itself produces a shear band too wide to generate high temperature, locally it does produce high strain. These locally high strains, which are one to two orders of magnitude greater than the average strain, may activate the thermal softening mechanism. This, in turn, may lead to more localization and to high temperatures while the average strain is still only a few percent. The current viscoelastic microcracking model does not include any thermal softening effects; however, our research plans include examining this issue.

Localization and shear band formation was only observed at average strain rates less than 1000 s^{-1} for the model studied in this work. At sufficiently high strain rate in a metal, above 10^5 s^{-1}, Wright and Walter (11) noted that shear band formation was inhibited due to inertial effects. The shear bands observed to form in this work are about two orders of magnitude larger than those in (11). The ratio of critical strain rates is about the same, indicating results consistent with that previous study.

This low critical strain rate means that the softening mechanism studied here cannot contribute to shear band formation under shock conditions. In many accident scenarios, however, deformation rates are below 1000 s^{-1}. Consequently, the mechanism discussed here may contribute to sensitivity in these cases.

We have performed similar studies on a model with a non-associated flow rule mechanism for generating localization, and found similar results. We have also performed some initial calculations of experiments described by Asay et al. (2) and found good qualitative agreement when either of these models is used, but not when only a thermal softening mechanism is used. These results will be reported as they become available.

ACKNOWLEDGMENTS

This worked was performed for the High Explosives Safety and Performance program of the U. S. Department of Energy. Los Alamos National

Laboratory is operated by the Univ. of California for the U. S. Department of Energy.

REFERENCES

1. Addessio, F. L. and Johnson, J.N., *J. Appl. Phys.*, **67**, 3275-3286, (1990).

2. Asay, B. W., Henson, B. F. and Funk, D. J., "Direct Measurement of Strain Field Evolution During Dynamic Deformation of an Energetic Material," presented at the 1997 Topical Group on Shock Compression Meeting, Amherst, MA, 27 July-1 August 1997.

3. Dienes, J. K., *Mech. Mater.*, **4**, 325 (1985)

4. Field, J.E., Swallowe, G. M. and Heavens, S.N, *Proc. R. Soc. Lond.* **A 382**, 36-42 (1982).

5. Frey, R.B., in *Proceedings of the Seventh Symposium on Detonation*, Office of Naval Research, Arlington, VA, 1981, pp. 36-42.

6. Grady, D. E. and Kipp, M. E., *J. Mech. Phys. Solids*, **35**, 95-118 (1987).

7. Gray III, G.T. ,. Blumenthal, W.R,. Cady, C.M. and Idar, D.J, "Influence of Temperature on the High-Strain-Rate Mechanical Behavior of PBX 9501 and PBXN-9," presented at the 1997 Topical Group on Shock Compression Meeting, Amherst, MA, 27 July-1 August 1997.

8. Kerrisk, J.F., Los Alamos National Laboratory Report LA-13127 (1996).

9. Molinari, A. and Clifton, R. J., *J. Appl. Mech.*, **54**, 806-812 (1987).

10. Winter, R. E. and Field, J.E., *Proc. R. Soc. Lond.* **A 343**, 299-413 (1975).

11. Wright, T. W. and Walter, J. W., *J. Mech. Phys. Solids*, **35**, 701-720 (1987).

CP429, *Shock Compression of Condensed Matter – 1997*
edited by Schmidt/Dandekar/Forbes
© 1998 The American Institute of Physics 1-56396-738-3/98/$15.00

MODELING ENERGY DISSIPATION INDUCED BY QUASI-STATIC COMPACTION OF GRANULAR HMX *

K. A. Gonthier,† R. Menikoff, S. F. Son, and B. W Asay

Los Alamos National Laboratory, Los Alamos, New Mexico 87545

A simple extension of a conventional two-phase continuum model of Deflagration-to-Detonation Transition (DDT) in energetic granular material is given to account for energy dissipation induced by quasi-static compaction. To this end, the conventional model equations are supplemented by a relaxation equation that accounts for irreversible changes in solid volume fraction due to intergranular friction, plastic deformation of granules, and granule fracture. The proposed model, which is consistent with the Second Law of Thermodynamics for a two-phase mixture, is demonstrated by applying it to the quasi-static compaction of granular HMX. The model predicts results commensurate with experimental data including stress relaxation and substantial dissipation; such phenomena have not been previously accounted for by two-phase DDT models.

INTRODUCTION

There has been considerable research during the last thirty years addressing Deflagration-to-Detonation Transition (DDT) in granular energetic materials; much of this work has been motivated by concerns over the accidental detonation of damaged high explosives due to mechanical stimuli. As with DDT experiments, we consider a granular explosive (porosity $\sim$ 30%) as a mock-up for a damaged explosive. It is widely accepted that various dissipative mechanisms induced by compaction of the granulated material give rise to local regions of thermal energy concentration termed hot-spots; such dissipative mechanisms include intergranular friction, plastic deformation of granules, and granule fracture. If the local energy dissipation rate is sufficiently high, chemical reaction is initiated, and transition to detonation is possible. As such, models used to analyze DDT in these systems should accurately account for energy dissipation induced by material compaction.

To this end, we extend the two-phase (inert gas and reactive solid) DDT model of Baer and Nunziato (BN model) (1) to account for energy dissipation induced by quasi-static compaction. Bdzil et al. (2) have recently shown that the BN model predicts no energy dissipation in this limit, contrary to experimental data for granular HMX which indicate substantial dissipation (3); thus, the BN model improperly accounts for compaction energetics. Correctly accounting for dissipation is a necessary step towards the development of an improved burn model based on hot-spots.

In this paper, we first give a brief description of the proposed model, valid in the limit of negligible gas phase effects (i.e., when the gas density is much smaller than the solid density). Though we only consider a granular solid, the model is equally applicable when gas phase effects are included. Next, we show that the model satisfies the Second Law of Thermodynamics for a two-phase mixture. Lastly, we demonstrate the model by applying it to the quasi-static compaction of granular HMX, and give comparisons of model predictions with experimental data.

---

*This research is funded by the Department of Energy under Contract Number W-7405-ENG-36.
†Corresponding author. E-mail: *gonthier@lanl.gov*

MATHEMATICAL MODEL

The model equations, valid in the limit of negligible gas phase effects, are given by

$$\frac{\partial}{\partial t}\left[\rho_s \phi_s\right] + \frac{\partial}{\partial x}\left[\rho_s \phi_s u_s\right] = 0, \qquad (1)$$

$$\frac{\partial}{\partial t}\left[\rho_s \phi_s u_s\right] + \frac{\partial}{\partial x}\left[\rho_s \phi_s u_s^2 + P_s \phi_s\right] = 0, \qquad (2)$$

$$\frac{\partial}{\partial t}\left[\rho_s \phi_s \left(e_s + \frac{u_s^2}{2}\right)\right]$$
$$+ \frac{\partial}{\partial x}\left[\rho_s \phi_s u_s \left(e_s + \frac{u_s^2}{2} + \frac{P_s}{\rho_s}\right)\right] = 0, \qquad (3)$$

$$\frac{\partial \phi_s}{\partial t} + u_s \frac{\partial \phi_s}{\partial x} = \frac{\phi_s(1-\phi_s)}{\mu_c}(P_s - \beta_s), \qquad (4)$$

$$\frac{\partial \tilde{\phi}_s}{\partial t} + u_s \frac{\partial \tilde{\phi}_s}{\partial x} = \begin{cases} \frac{1}{\tilde{\mu}}\left(f - \tilde{\phi}_s\right) & \text{if } \tilde{\phi}_s < f \\ 0 & \text{if } \tilde{\phi}_s \geq f. \end{cases}$$
$$(5)$$

Here, subscript "s" denotes quantities associated with the granular solid. Independent variables are time, t, and position, x. Dependent variables are density, ρ_s; volume fraction, ϕ_s; particle velocity, u_s; pressure, P_s; specific internal energy, e_s; no-load volume fraction, $\tilde{\phi}_s$; equilibrium no-load volume fraction, $f(\phi_s)$; and intergranular stress, $\beta_s(\phi_s, \tilde{\phi}_s)$.

Equations (1-3) are conservation equations for the mass, momentum, and energy of the solid. Equation (4) is a relaxation equation for mechanical stresses, and Eq. (5), not included in the BN model, is a relaxation equation for the no-load volume fraction; the parameters μ_c and $\tilde{\mu}(\dot{\phi}_s)$ determine the relaxation rates, where μ_c is constant, and $\tilde{\mu}(\dot{\phi}_s)$ is compaction rate-dependent ($\dot{\phi}_s$ denotes compaction rate). The dependence of $\tilde{\mu}$ on compaction rate is motivated by experimental data which indicate an increase in the energy dissipation rate (plastic strain rate) with compaction rate (3). The inclusion of $\tilde{\phi}_s$ as an additional internal variable in the theory is not standard, but enables rate-independent dissipation induced by material compaction to be modeled; similar internal variables have been used to model rate-independent plasticity of metals (4). Constitutive relations needed to mathematically close

Eqs. (1-5) are constructed based on both rational thermodynamics and experimental data.

Using a rational thermodynamics development (5), we assume the granular system is 1) in local thermodynamic equilibrium, and 2) can be described in terms of a Helmholtz free energy potential, ψ_s, of the form

$$\psi_s = \psi_s(\rho_s, T_s, \phi_s - \tilde{\phi}_s), \qquad (6)$$

where T_s is the solid temperature. Here, ρ_s, T_s, ϕ_s, and $\tilde{\phi}_s$ are independent thermodynamic variables. Given ψ_s, the dependent thermodynamic variables are defined by

$$\eta_s \equiv -\left.\frac{\partial \psi_s}{\partial T_s}\right|_{\rho_s, \phi_s, \tilde{\phi}_s}, \qquad (7)$$

$$e_s \equiv \psi_s + T_s \eta_s, \qquad (8)$$

$$P_s \equiv \rho_s^2 \left.\frac{\partial \psi_s}{\partial \rho_s}\right|_{T_s, \phi_s, \tilde{\phi}_s}, \qquad (9)$$

$$\beta_s \equiv \rho_s \phi_s \left.\frac{\partial \psi_s}{\partial \phi_s}\right|_{\rho_s, T_s, \tilde{\phi}_s} = -\rho_s \phi_s \left.\frac{\partial \psi_s}{\partial \tilde{\phi}_s}\right|_{\rho_s, T_s, \phi_s}, \qquad (10)$$

where η_s is the specific solid entropy. A functional form of ψ_s which is consistent with Eqs. (6-10), and can reasonably model the material response to both low and high pressure loading, is given by

$$\psi_s = \psi_{sp}(\rho_s, T_s) + B(\rho_s, \phi_s - \tilde{\phi}_s), \qquad (11)$$

where

$$B(\rho_s, \phi_s - \tilde{\phi}_s) = \frac{1}{\rho_s}\int_0^{\phi_s - \tilde{\phi}_s} h(\phi)d\phi \qquad (12)$$

is the reversible compaction energy (discussed below), ψ_{sp} is the Helmholtz free energy of the pure solid, assumed known, and h is an experimentally derived function. Using this definition for ψ_s, Eqs. (8-10) reduce to

$$e_s = e_{sp}(\rho_s, T_s) + B(\rho_s, \phi_s - \tilde{\phi}_s), \qquad (13)$$

$$P_s = P_{sp}(\rho_s, T_s) - \rho_s B(\rho_s, \phi_s - \tilde{\phi}_s), \qquad (14)$$

$$\beta_s(\phi_s, \tilde{\phi}_s) = \phi_s h(\phi). \qquad (15)$$

Based on the quasi-static compaction data of Coyne et al. (3), we take

$$h(\phi) = -\tau\phi\frac{\ln[\kappa - \phi]}{\kappa - \phi}, \quad (16)$$

and

$$f(\phi_s) = \frac{1}{1 - \phi_{so}}\left[(1 - \kappa - \phi_{so})\phi_s + \kappa\phi_{so}\right], \quad (17)$$

where $\tau = 2.3\ MPa$, $\kappa = 0.03$, and ϕ_{so} is the initial solid volume fraction of the granular material. Equations (13), (14), (16), and (17) are sufficient to mathematically close Eqs. (1-5).

Second Law of Thermodynamics

The Second Law of Thermodynamics for a thermally isolated gas-solid mixture reduces to the following mathematical expression in the limit of negligible gas phase effects (2):

$$\dot{\Sigma}_\eta \equiv \frac{\partial}{\partial t}[\rho_s\phi_s\eta_s] + \frac{\partial}{\partial x}[\rho_s\phi_s u_s\eta_s] \geq 0, \quad (18)$$

where $\dot{\Sigma}_\eta$ is the volumetric entropy production rate. Using Eqs. (1), (3), and (11), we obtain the following (we omit details for brevity):

$$\dot{\Sigma}_s = \frac{1}{T_s}\left[(P_s - \beta_s)\frac{d\phi_s}{dt_s} + P_s\frac{d\tilde{\phi}_s}{dt_s}\right], \quad (19)$$

where d/dt_s is the convective time derivative. It is seen from Eqs. (4) and (5) that $\dot{\Sigma}_\eta$ is non-negative provided that $P_s \geq 0$; this condition holds for granular materials as they cannot support tensile stresses. As discussed in the following section, $P_s = \beta_s$ in the quasi-static compaction limit, and Eq. (19) reduces to

$$\dot{\Sigma}_\eta = \frac{\beta_s}{T_s}\frac{d\tilde{\phi}_s}{dt_s}; \quad (20)$$

thus, the model is dissipative in this limit.

QUASI-STATIC COMPACTION

To demonstrate the model, we simulate a quasi-static compaction experiment performed by Coyne et al. (3) on granular HMX. The sample, strongly confined by a movable piston-fixed cylinder apparatus (ID = 2.54 cm), has an initial solid volume fraction of $\phi_{so} = 0.655$ (based on a crystal density of $\rho_s = 1.903\ g/cm^3$), and an initial length of $L_o = 2.36\ cm$. The sample undergoes a (I)loading-(II)unloading-(III)reloading cycle at a constant extension rate of $u_p = 18.8\ cm/min$ during loading, where u_p is the piston velocity.

To this end, we assume 1) $\mu_c u_p/L_o \ll 1$; 2) the solid is incompressible; 3) the kinetic energy of the solid is negligible; and 4) all variables depend only on time. The first assumption implies that the time scale associated with equilibration of mechanical stresses is much smaller than that associated with volume changes due to piston motion; consequently, it can be shown from Eq. (4) that $P_s = \beta_s$ in this limit. With these assumptions, Eq. (1) reduces to a homogeneous ordinary differential equation which can be directly integrated, and the resulting algebraic equation solved for ϕ_s, to get

$$\phi_s(t) = \left(\frac{L(t_o)}{L(t)}\right)\phi_{so}, \quad (21)$$

where

$$L(t) = L(t_o) - u_p(t - t_o). \quad (22)$$

Here, t_o is the time at which the loading or unloading process is initiated. Knowing $\phi_s(t)$, then $f(t)$ is given by Eq. (17). Equation (5) can then be solved directly:

$$\tilde{\phi}_s(t) = \tilde{\phi}_s(t_o)e^{-\frac{t - t_o}{\tilde{\mu}}} + \frac{1}{\tilde{\mu}}\int_{t_o}^t f(t')e^{-\frac{t - t'}{\tilde{\mu}}}dt', \quad (23)$$

where $\tilde{\phi}_s(0) = \phi_{so}$, $\tilde{\mu} = 0.005\ s$ during loading, and $\tilde{\mu} = 0.5\ s$ during stress relaxation. With both $\phi_s(t)$ and $\tilde{\phi}_s(t)$ known, Eq. (16) gives $\beta_s(t)$.

Shown in Fig. 1 are comparisons of the predicted and experimental histories for ϕ_s and β_s. It is noted that only the extrema of t and β_s were reported for the histories shown in Fig. 1 of reference (3). Thus, for purposes of this work, we linearly scaled these variables between their extrema to obtain the experimental histories shown here; as such, these results should be interpreted as semi-quantitative. Nonetheless, the model reasonably predicts experimentally observed trends.

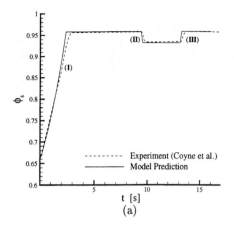

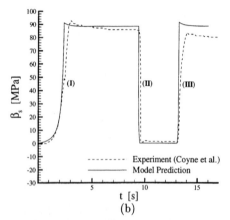

FIGURE 1. Predicted and experimental histories for (a) solid volume fraction and (b) intergranular stress.

Figure 2 gives a plot of the predicted result in the $\beta_s, -\ln\phi_s$ phase plane. The total area under the loading curve is directly proportional to the work per unit mass, w, required to compact the sample from the initial unloaded state to the final loaded state. Using Eqs. (1), (3), and (13), it can be shown that

$$\frac{dw}{dt} = \frac{de_{sp}}{dt} + \frac{dB}{dt} = \frac{\beta_s}{\rho_s\phi_s}\frac{d\phi_s}{dt}.$$

Assuming an incompressible solid, expanding dB/dt in the above expression, and simplifying the result, we obtain

$$\frac{de_{sp}}{dt} = \frac{\beta_s}{\rho_s\phi_s}\frac{d\tilde{\phi}_s}{dt},$$

which, from Eq. (20), is identified as the energy

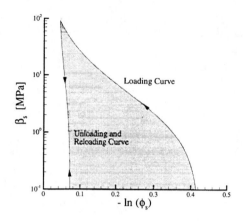

FIGURE 2. Predicted compaction work.

dissipation rate. Thus, the total compaction work consists of reversible and dissipative energy components. The reversible compaction energy (B) is recovered upon unloading of the sample, while the dissipated energy goes into heating the solid. The shaded area in Fig. 2 is proportional to the dissipated energy. Conventional two-phase DDT models do not properly account for this dissipated energy, and thus do not accurately model compaction energetics.

CONCLUSIONS

This paper has outlined the modification of a two-phase DDT model to account for energy dissipation induced by quasi-static compaction. It is important that such models accurately predict compaction-induced dissipation, as it is the localization of this energy which gives rise to the formation of hot-spots in damaged energetic materials. Prior two-phase DDT models do not properly account for this dissipation.

REFERENCES

1 Baer, M. R., and Nunziato, J. W., *Int J of Multi-Phase Flow* **12**, 861-889 (1986).
2 Bdzil, J. B., Menikoff, R., Son, S. F., Kapila, A. K., and Stewart, D. S., *Phys Fluids*, in review.
3 Coyne, P. J. Jr. Elban, W. L., and Chiarito, M. A., *Eighth Int Det Symp*, 645-657 (1989).
4 Kratochvil, J., and Dillon, O. W., Jr., *J Appl Phys* **40**, 3207-3218 (1969).
5 Samohyl, I., *Thermodynamics of Irreversible Processes in Fluid Mixtures*, Leipzig: Teubner-Texte zur Physik, 1987, pp. 28-51.

CP429, *Shock Compression of Condensed Matter – 1997*
edited by Schmidt/Dandekar/Forbes
© 1998 The American Institute of Physics 1-56396-738-3/98/$15.00

MOLECULAR DYNAMICS INVESTIGATION OF THE EFFECTS OF VARIATION IN ENERGY RELEASE ON DETONATION INITIATION

M. L. Elert[a], J. J. C. Barrett[b], D. H. Robertson[c], and C. T. White[d]

[a] *Chemistry Department, U. S. Naval Academy, Annapolis, MD 21402-5026*

[b] *Code 6179, Naval Research Laboratory, Washington, DC 20375-5000*

[c] *Department of Chemistry, Indiana University-Purdue University at Indianapolis, Indianapolis, IN 46202*

[d] *Department of Materials, University of Oxford, Parks Road, Oxford, OX11 3PH, UK*

The amount of energy released in the detonation of an energetic material clearly influences the properties of the detonation, such as peak temperature and detonation front velocity. Using a model diatomic system which has previously been shown to produce realistic detonation properties, we have performed molecular dynamics simulations in which the exothermicity of the chemical reaction supporting the detonation was systematically varied. The minimum energy release necessary to support a chemically sustained shock wave was determined for this model system, as well as the dependence of front velocity, reaction zone temperature, and density on the magnitude of energy release.

INTRODUCTION

In recent years, molecular dynamics simulations have begun to provide useful information regarding nanoscale features of condensed-phase detonations, such as reaction zone width (1) and the effects of voids and defects (2,3). These results nicely complement the available experimental data and hydrodynamic model studies, which generally provide information on a much larger scale ($> 10^{-6}$ m). Another feature of detonation which can profitably be examined via molecular dynamics is the effect of varying the amount of energy released by the exothermic chemical reactions which support the shock front. This question can be approached experimentally by comparing the detonation properties of different energetic materials for which the exothermicity

is known. However, different materials will differ in ways other than just exothermicity; there will typically be variations in density, reaction kinetics, and other factors which influence the detonation properties. Molecular dynamics simulations provide a method for changing the exothermicity in a model system by explicitly changing bond energies of reactants or products. Since other properties of the system are unaffected, this provides an unambiguous measure of the correlation between exothermicity and the properties of the detonation front. It should be noted that exothermicity has also been treated as a variable parameter in some analytical treatments of two-dimensional detonation at the continuum limit, including those of Cowperthwaite (4) and Fickett and Davis (5).

THE MODEL

During the past several years we have developed a two-dimensional condensed-phase model system which is amenable to molecular dynamics simulations and which produces reasonable values for detonation properties such as shock front velocity and temperature (6,7,8). This model has been employed by other groups as well (9,10). In the model, a crystalline array of heteronuclear diatomic molecules "AB" acts as an energetic material, reacting exothermically to produce more stable A_2 and B_2 molecules. Interatomic interactions are governed by a reactive empirical bond order (REBO) potential of the general form first used by Tersoff (11) in his study of silicon. In the REBO formalism, the potential energy $V(r_{ij})$ between two atoms i and j is given by

$$V(r_{ij}) = V_R(r_{ij}) - B_{ij}V_A(r_{ij}) \qquad (1)$$

where V_R and V_A are two-body repulsive and attractive potentials, respectively, and where B_{ij} is a many-body term whose value depends inversely on the number and proximity of other near neighbor atoms. This has the effect of limiting the chemical valence of an atom, and allows for the occurrence of chemical reactions with reasonable transition state behavior.

Details of the AB model have been described elsewhere (8) and will not be repeated here. For present purposes, it suffices to note that the two-body terms V_R and V_A in the potential energy function are given by Morse-like functions of the form:

$$V_R(r) = D_e\left(\frac{1}{S-1}\right)\exp[-\alpha\sqrt{2S}(r - r_e)] \quad (2)$$

and

$$V_A(r) = D_e\left(\frac{S}{S-1}\right)\exp[-\alpha\sqrt{\frac{2}{S}}(r - r_e)] \quad (3)$$

where D_e is the bond energy between a pair of atoms in the absence of other near neighbors ($B_{ij} = 1$). In the original AB model, $D_e(AB) = 2.0$ eV and $D_e(AA) = D_e(BB) = 5.0$ eV, so that the energy released in the reaction $2AB \rightarrow A_2 + B_2$ is 6.0 eV. The exothermicity can be adjusted simply by changing the value of D_e for A_2 and B_2 relative to the value for AB. In the current study, the equality between D_e for A_2 and B_2 is retained, but this parameter is allowed to vary between 3.0 eV and 6.0 eV, giving a range of exothermicity from 2.0 eV to 8.0 eV for the reaction $2AB \rightarrow A_2 + B_2$.

As in studies with the original model, a shock wave is initiated in the AB crystal by impact of a flyer plate of the same material. Periodic boundary conditions are maintained in the direction perpendicular to shock front propagation to simulate an infinitely wide crystal, eliminating edge effects. The equations of motion are integrated using a standard predictor-corrector method (12) with a variable time step.

RESULTS

One detonation property which might be expected to depend on the exothermicity of the reaction is the sensitivity to shock inititation, as measured by the minimum flyer plate impact speed necessary to produce a chemically sustained shock wave of constant velocity. Figure 1 shows a plot of the impact threshold versus

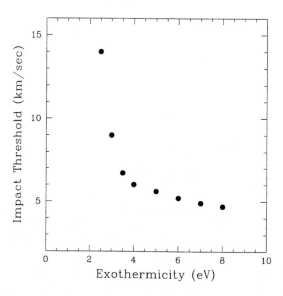

FIGURE 1. Impact threshold versus exothermicity. For exothermicities of 2.0 eV and below, sustained detonation cannot be achieved at any flyer plate impact speed.

exothermicity for a flyer plate three unit cells (six molecular layers) thick. At high exothermicities the impact threshold is low and is relatively insensitive to the amount of energy introduced by the flyer plate. At low exothermicities, the flyer plate energy is critical to inducing sufficient exothermic chemical reactions in the reactant energetic material to produce a self-sustaining shock wave, and so the threshold impact velocity is a strong function of exothermicity. Finally, at exothermicities below 2.5 eV, the shock is not self-sustaining at any flyer plate impact speed. Figure 2 shows the shock front position versus time for an exothermicity of 2.0 eV at a variety of impact speeds. When the energy imparted by the flyer plate has dissipated, there are no longer enough exothermic chemical reactions occurring to maintain the shock wave, and it begins to slow as indicated by the downward curvature of the lines. Reaction zone temperature, pressure, and extent of reaction all begin to decline as well. Higher flyer plate impact speeds cause the shock wave to propagate further before this begins to occur, but even at an impact speed of 20 km/sec the front ceases to be self-sustaining by 20 psec after initiation.

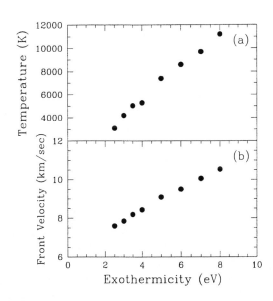

FIGURE 3. (a) Reaction zone temperature and (b) shock front velocity as functions of exothermicity.

Another property which is expected to depend on the magnitude of the exothermicity is the temperature of the reaction zone. In fact, since the shock wave (and the reaction zone behind it) move so rapidly, thermal transport is negligible and the temperature of the reaction zone should be directly proportional to the energy deposited by the chemical reactions occurring in this zone. This is demonstrated in Figure 3(a), which shows the average temperature in the region within 5 nm behind the shock front as a function of exothermicity.

As mentioned above, for a given exothermicity, a chemically sustained shock wave quickly reaches a constant velocity which is independent of the flyer plate impact speed. For the *AB* model, that detonation front velocity is a nearly linear function of exothermicity over the energy range studied, as shown in Figure 3(b).

CONCLUSION

Much of the current effort in applying molecular dynamics to condensed-phase detonation is directed toward extending the method to more complex molecules (2,13), but there is still much

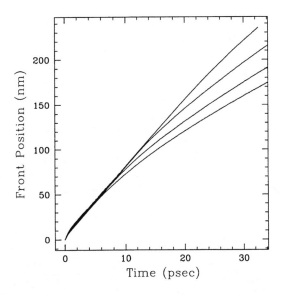

FIGURE 2. Shock front position as a function of time for an exothermicity of 2.0 eV. From bottom to top, the curves represent flyer plate impact speeds of 14, 16, 18, and 20 km/sec.

to be learned from the *AB* model. It provides a simple system in which the effect of changes to various system parameters can be readily observed. This paper presents such an investigation of the role of changing exothermicity on the properties of the detonation. Other factors which could be investigated in a similar way include density, crystal structure, and vibrational frequency.

ACKNOWLEDGEMENTS

This work was supported by the Office of Naval Research (ONR) through the Naval Research Laboratory and directly by ONR through contract # N00014-97-WX-20175. MLE received additional ONR support through the Naval Academy Research Council. CTW thanks the Department of Materials for hospitality during his sabbatical stay at Oxford and for support through the Oxford Materials Modelling Laboratory.

REFERENCES

1. M. L. Elert, D. H. Robertson, J. J. C. Barrett, and C. T. White, *Shock Compression of Condensed Matter – 1995*, S. C. Schmidt and W. C. Tao, eds., (AIP Press, New York, 1996), p. 183.

2. C. T. White, J. J. C. Barrett, J. W. Mintmire, M. L. Elert, and D. H. Robertson, *Mat. Res. Soc. Symp. Proc.* **418** (1996), p. 277; C. T. White, J. J. C. Barrett, J. W. Mintmire, M. L. Elert, and D. H. Robertson, *Shock Compression of Condensed Matter – 1995*, S. C. Schmidt and W. C. Tao, eds., (AIP Press, New York, 1996), p. 187.

3. D. H. Tsai, *Mat. Res. Soc. Symp. Proc.* **418** (1996), p. 281.

4. M. Cowperthwaite, *Thirteenth Symposium (International) on Combustion*, (The Combustion Institute, Pittsburgh, 1971), p. 1111.

5. W. Fickett and W. C. Davis, *Detonation*, (University of California Press, Berkeley, 1979).

6. D. W. Brenner, *Shock Compression of Condensed Matter – 1991*, S. C. Schmidt, R. D. Dick, J. W. Forbes, and D. G. Tasker, eds., (Elsevier, Amsterdam, 1992), p. 115; D. H. Robertson, D. W. Brenner, M. L. Elert, and C. T. White, *ibid.*, p. 123.

7. C. T. White, D. H. Robertson, M. L. Elert, and D. W. Brenner, *Microscopic Simulations of Complex Hydrodynamic Phenomena*, M. Mareschal and B.L. Holian, eds., (Plenum Press, New York, 1992), p. 111.

8. D. W. Brenner, D. H. Robertson, M. L. Elert, and C. T. White, *Phys. Rev. Lett.* **70** (1993) 2174.

9. P. J. Haskins and M. D. Cook, *High-Pressure Science and Technology – 1993*, S. C. Schmidt, J. W. Shaner, G. A. Samara, and M. Ross, eds., (AIP Press, New York, 1994), p. 1341; P. J. Haskins and M. D. Cook, *Shock Compression of Condensed Matter – 1995*, S. C. Schmidt and W. C. Tao, eds., (AIP Press, New York, 1996), p. 195.

10. B. M. Rice, W. Mattson, J. Grosh, and S. F. Trevino, *Phys. Rev. E* **53** (1996) 611; *ibid.*, **53** (1996) 623.

11. J. Tersoff, *Phys. Rev. Lett.* **56** (1986) 632; *Phys. Rev. B* **37** (1988) 6991.

12. C. W. Gear, *Numerical Initial Value Problems in Ordinary Differential Equations* (Prentice-Hall, Englewood Cliffs, 1971), p. 148.

13. L. E. Fried and C. Tarver, *Shock Compression of Condensed Matter – 1995*, S. C. Schmidt and W. C. Tao, eds., (AIP Press, New York, 1996), p. 179.

CP429, *Shock Compression of Condensed Matter – 1997*
edited by Schmidt/Dandekar/Forbes
© 1998 The American Institute of Physics 1-56396-738-3/98/$15.00

SELF-SIMILAR BEHAVIOR FROM MOLECULAR DYNAMICS SIMULATIONS OF DETONATIONS

D. H. Robertson, J. J. C. Barrett[1], M. L. Elert[1] and C. T. White[2]

Chemistry Department, Indiana U. Purdue U. Indianapolis, IN 46202
[1]Chemistry Division, Naval Research Laboratory, Washington, DC 20375-5340
[2]Department of Materials, University of Oxford, Parks Road, Oxford, OX11 3PH, UK

Dissipative processes arising from thermal transport and viscosity together with the shock-induced chemistry introduce a series of length scales into condensed phase detonation profiles. However, far enough behind the detonation front dissipative processes arising from thermal transport and viscosity should vanish on the average and the chemistry should become statistically complete. Hence, in this region the detonation profiles should become self-similar resulting from the loss of all these scales. We have previously introduced a model of a diatomic system which has been shown to support a detonation. Here we report results from a molecular dynamics study of the self-similar characteristics of this model. As the simulation proceeds, we find that the density, particle velocity, and pressure profiles obtained from this piston-supported detonation simulation do become self-similar behind the CJ point when implicitly scaled by displacement over time. These scaled profiles can be used to predict the longer time characteristics of the model behind the CJ point.

INTRODUCTION

Molecular dynamics (MD) simulations have probed the ability to model shock processes at the atomic level. (1–3) Fortuitously, computer-based simulations are ideal for studying shock processes because the physical time and length scales of a shockwave in condensed matter matches those imposed while performing atomistic MD simulations on a computer. These MD simulations, by their very nature, can probe the atomic level details of a shockwave that are very difficult or impossible to investigate by experimental means. But for these simulations to provide new and useful information, the results from atomic simulations must be consistent with results at the continuum level of modeling and with results that are obtained by experimental means. Indeed, early MD simulations of shock waves in Lennard-Jones (LJ) systems have shown that atomic-scale simulations can be of sufficient size to reproduce continuum results. (2,3)

However, to model more complex shockwaves such as detonations a means of energy production through chemical reaction must be incorporated into the model to allow access to the energy necessary to sustain a shockwave propagating through condensed phase matter. Simulations that incorporate chemically-reasonable reactive potentials have shown that MD simulations using these potentials can reproduce a detonation. (4) The resulting detonations using these reactive potentials show the formation of a chemically sustained shock wave with a constant velocity that is intrinsic to the parameterization of the potential and independent of initiating conditions. Additionally, the shock profiles for this model show shapes consistent with the classic Zel'dovich–von Neumann–Doering model for planar detonations. (5–7) A careful study of the Rankine-Hugoniot (6,7) relations across the shock front shows that the simulations rapidly approach near steady-flow conditions over a successively wider region as the simulation progresses. (4) This shows

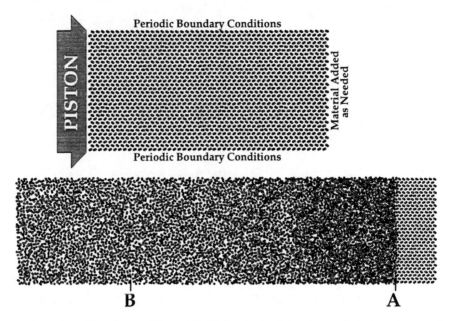

FIGURE 1. *Top:* Initial material and boundary conditions applied during the simulation. *Bottom:* Snapshots of the atomic positions for the detonation at 25 ps. The shock wave is propagating from left to right. Point A shows the position of the shock front and point B denotes where the product region has reached the front of near constant region preceding the piston.

that, on the time scale accessible to MD calculations, these simulations can approach continuum behavior and therefore have the potential to address macroscopic issues.

In this paper, we discuss results based on relatively-large long-time simulations studying the self-similar nature of these shockwaves. We find that the detonation profiles show a self-similar behavior when the timescale is implicitly removed. The only region not self-similar is that from the shock front to the CJ point which should eventually diminish in the time-scaled plots as time increases.

MODEL AND RESULTS

The 2-D simulations presented in this paper are based on an AB model represented using reactive empirical bond order potentials (4) that are based on an Abell-Tersoff formalism (8,9) which incorporates the possibility for chemical reactions into these shock wave studies. This model has the potential to undergo reaction from the energetic AB reactant molecules to the more stable A_2 and B_2 product molecules with an associated energy release of 3 eV per AB reactant.

The boundary conditions used in this simulation are depicted at the top of Figure 1. The black and white atoms are the A and B atoms, respectively. This simulation is performed in two dimensions using periodic boundary conditions perpendicular to the direction of shock propagation with material being added in front of the advancing shock wave as necessary. These boundary conditions, as implemented, allow the calculations to simulate the propagation of the shockwave into a semi-infinite piece of the reactant molecular solid. The width of material used in this simulations as depicted in Figure 1 is 10 nm wide. While most of the previous reported simulations using this model were unsupported detonations, in this simulation a following piston traveling at 2.0 km/s is used both to compress and initiate the starting material as well as to contain the reacted material in the rear of the detonation allowing the reacted material to attain a well-defined final state. After setting the initial and boundary conditions as described above the trajectories of the individual atoms are followed on the potential energy surface by integrating Newton's equation of motion with a high-order predictor-corrector strategy. (10)

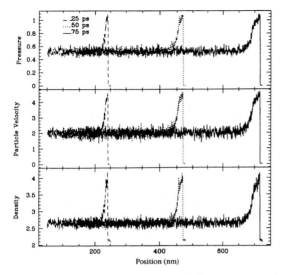

FIGURE 2. Profiles of the pressure (eV/Å^2), density (amu/Å^2), particle flow velocity (km/s) at 25, 50, and 75 ps.

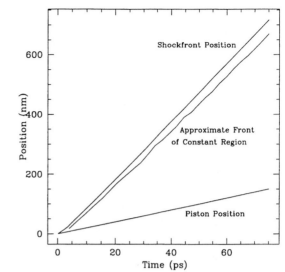

FIGURE 3. Plot of the positions versus time of the shock front, piston position and the front of the constant steady flow material in front of the piston

For this paper the simulation was integrated up to 75 ps allowing the detonation to travel over 700 nm and eventually include over 110,000 atoms. A snapshot of the atomic positions at 25 ps after starting the simulation is given in the bottom of Figure 1. The reactant molecules can be seen to the right, the shock front is clearly visible at point A, and then the products can be seen to the far left.

Profile plots of the pressure, particle velocity, and density are given in Figure 2 at times of 25, 50, and 75 ps during this simulation. It can be seen from this figure that the shockfronts are very sharp on the scale of these plots. Also, a short distance (or time) after the shockfront has passed, the product material comes to a well-defined final state imposed by the piston velocity. Further, it appears that the distance from the wavefront to the point at which the material reaches the final state conditions increases as time progresses. Additionally, although it is difficult to see in this plot, the width of the constant final state region also grows during the simulation. The point at which the products reach the final state is not a well-defined or a sharp point but one which is approached gradually during expansion of the products. However, operationally and numerically, this point can be defined during the simulation as the point at which the density, particle flow velocity and pressure first reach their final state values. This point will be referred to in the text as the front of the constant region (FCR).

The FCR position is shown as B for the 25 ps simulation in Figure 1. Visually there are no distinguishing features at this point compared the region surrounding it, in sharp contrast to the very distinct shockfront at point A. Figure 3 graphs this FCR point together with the shockfront position and the piston position. The piston has a constant velocity and likewise, as seen in Figure 3, the detonation front rapidly stabilizes to a constant velocity. The line corresponding to the position of the FCR in Figure 3 also shows an average constant velocity throughout the simulation but with more variability than the other two lines. This again reflects that the FCR is not a well-defined point or shock-like in the simulation but rather a variable but on average constantly moving point.

The constant but differing velocities of the piston position and the FCR point indicates that the final state region is an expanding region that because it is constant, its profile would be self-similar (simplest case) if scaled appropriately. The region between the FCR and the shock front is more complex because it contains the CJ point. However, the CJ point should remain a fixed distance behind this

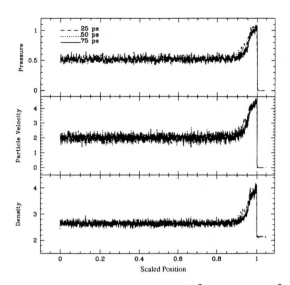

FIGURE 4. Profiles of the pressure (eV/$\mathring{A}^2$), density (amu/$\mathring{A}^2$), particle flow velocity (km/s) at 25, 50, and 75 ps where time has been implicitly removed by scaling the distance between the piston and shockfront to fit in the range from 0.0 to 1.0.

shock front and previous simulations studying the failure diameter and reaction zone length (11,12) have implied that the CJ point in this model should be much less than 4 nm from the shock front. This means that the distance between the shock front and the CJ point is only a very small portion of the region from the FCR to shock front and should become an increasingly smaller portion as time progresses. This implies that as long as the chemistry is statistically complete by the CJ point and dissipative processes arising from thermal transport or viscosity are negligible then the region from the CJ point (practically the shockfront) to the FCR should have a self-similar profile.

To test whether or not dissipative processes are non-negligible after the CJ point using this simulation, time is implicitly removed from these profiles by transforming the domain of the profile is scaled to fit in the range from 0.0 to 1.0. These scaled profiles are shown in Figure 4 for the times of 25, 50, and 75 ps. Clearly from this figure it can be seen that, in fact, the detonation profiles are self-similar. The non-self-similar region from the shock front to the CJ point is a vanishingly small region of these plots and would eventually only be a build-up visible right at the shockfront as time progressed.

CONCLUSIONS

This paper examines the self-similarity of the detonation profiles resulting from a piston-contained simulation using a model energetic material. These profiles from this simulation do show self-similarity when time is implicity removed. The only region in which self-similarity is not expected to exist—in front of the CJ point—is a vanishingly small portion of these profiles especially at the longer time scales examined in this paper. The self-similarity of the profiles for this model shows that, as according to the definition of the CJ point, behind the CJ point the chemistry is statistically complete and any dissipative processes involving thermal transport or viscosity are negligible. Because of this self-similarity, these scaled profiles can be used to predict the longer time characteristics of this model to macroscopic dimension beyond that currently addressable by computer simulations.

ACKNOWLEDGMENTS

This work was supported in part by the ONR through the NRL and ONR Contract # N0001497WX20175. CTW thanks the Department of Materials for hospitality during his sabbatical stay at Oxford and for support through the Oxford Materials Modelling Laboratory.

REFERENCES

1. Tsai, D. H., and Beckett, C. W., J. Geophys. Res. 71 2601 (1960).
2. Holian, B. L., Hoover, W. G., Moran, W., and Straub, G.K., *Phys. Rev. A* **22** 2798 (1980).
3. Dremin, A. N., and Klimenko, V. Yu., *Prog. Astronaut. Aeronaut.* **75** 253 (1981).
4. Brenner, D. W., Robertson, D. H., Elert, M. L., and White, C. T., *Phys. Rev. Lett.* **70** 2174 (1993).
5. Davis, W. C., *Sci. Am.* **256** 106 (1987).
6. Fickett, W., *Introduction to Detonation Theory,* Berkeley; U. Calif. Press, 1985.
7. Zel'dovich, I. B., and Kompaneets, A. S., *Theory of Detonations,* New York; Academic Press, 1960.
8. Abell, G. C., *Phys. Rev. B* **31**, 6184 (1985).
9. Tersoff, J., *Phys. Rev. B* **37**, 6991 (1988).
10. Gear, G. C., *Numerical Initial Value Problems in Ordinary Differential Equations,* Englewood Cliffs; Prentice-Hall, 1971.
11. Robertson, D. H., Brenner, D. W., and White, C. T., *High-Pressure Shock Compression of Solid III,* Davison and Shahnipoor, eds., New York; Springer-Verlag, 1997, Ch. 2.
12. Elert, M. L., Robertson, D. H., Barrett, J. J. C., and White, C. T (1996). *Shock Compression of Condensed Matter - 1995,* Eds. S. C. Schmidt and W. C. Tao; Woodbury NY; AIP Press, 1996, p. 183.

CP429, *Shock Compression of Condensed Matter – 1997*
edited by Schmidt/Dandekar/Forbes
© 1998 The American Institute of Physics 1-56396-738-3/98/$15.00

CHEMICAL REACTION AND EQUILIBRATION MECHANISMS IN DETONATION WAVES

Craig M. Tarver

Lawrence Livermore National Laboratory,
P.O. Box 808, L-282, Livermore, CA 94551

Experimental and theoretical evidence for the nonequilibrium Zeldovich-von Neumann-Doring (NEZND) theory of self-sustaining detonation is presented. High density, high temperature transition state theory is used to calculate unimolecular reaction rate constants for the initial decomposition of gaseous norbornene, liquid nitromethane, and solid, single crystal pentaerythritol tetranitrate as functions of shock temperature. The calculated rate constants are compared to those derived from experimental induction time measurements at various shock and detonation states. Uncertainties in the calculated shock and von Neumann spike temperatures are the main drawbacks to calculating these reaction rates. Nanosecond measurements of the shock temperatures of unreacted explosives are necessary to reduce these uncertainties.

INTRODUCTION

The nonequilibrium Zeldovich-von Neumann-Doring (NEZND) theory of detonation (1-5) was developed as a framework in which to study the major chemical and physical processes that precede and follow exothermic chemical reaction. These nonequilibrium processes determine the time required for the onset of chemical reaction, control the energy release rates, and supply the mechanism by which the chemical energy sustains the leading shock wave front structure. The three-dimensional shock front structure, the nonequilibrium excitation and relaxation processes, and the chemical reaction rates in gaseous detonation waves are fairly well understood (1,2). The high pressures (20 - 40 GPa), densities (2.5 g/cm^3), and temperatures (3000 - 5000 K) generated in less than a microsecond in condensed phase detonation waves traveling at velocities approaching 10 mm/µs create environments that are extremely difficult to study. Calculations of initial unimolecular decomposition rates under shock and detonation conditions are presented in this paper.

THE NEZND THEORY OF DETONATION

Figure 1 shows the four major regions and several

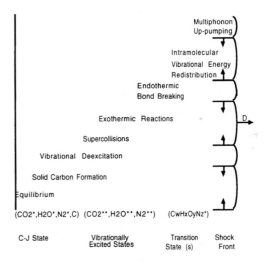

FIGURE 1. Nonequilibrium ZND (NEZND) model of detonation in an organic solid or liquid explosive

chemical processes that occur within the reaction zone of an organic solid or liquid explosive, $C_wH_xO_yN_z$. The first region is the leading shock wave front, which consists of a three-dimensional arrangement of Mach stem interactions. The detonation wave front structure in condensed phase explosives was discussed by Tarver et al. (4).

Following shock front compression, the second region in Fig. 1 is dominated by the flow of the excess phonon energy into the low frequency vibrational modes of the molecule by "multiphonon up-pumping" (6) and the subsequent energy flow from the low frequency "doorway" modes to the high frequency modes by intramolecular vibrational energy redistribution (IVR) (7). Hong et al. (8) measured the up-pumping and IVR rates in liquid nitromethane (NM) and found that complete equilibration required approximately 100 ps. Molecular dynamics studies of these relaxation rates have shown that the equilibration times depend on the number of doorway modes and the shock pressure. Large molecules with several doorway modes, such as octahydro-1,3,5,7-tetranitro-1,3,5,7-tetrazocine (HMX), triaminotrinitrobenzene (TATB), and pentaerythritol tetranitrate (PETN), equilibrate in approximately 10 ps (9). The establishment of vibrational equilibrium is a necessary condition for chemical decomposition, because the initial bond breaking reaction proceeds through a transition state created by high vibrational excitation of one of the highest frequency modes.

The third region of Fig. 1 begins at the transition state (or states), which is followed by the chemical reconstitution process in several small stable reaction product molecules are produced. This region is called the "von Neumann spike" or the "chemical peak" in the condensed phase explosives literature. Hydrodynamic pressure and particle velocity measurements with nanosecond time resolution have yielded considerable information about the average properties of this state (10). Once the exothermic chain reaction process is initiated, highly vibrationally excited products form and interact with the unreacted molecules and each other to greatly increase the rates of decomposition. Large quantities of vibrational energy have been shown to be rapidly transferred by "supercollisions" involving highly vibrationally excited molecules (5).

The fourth region in Fig. 1 is dominated by the expansion and vibrational deexcitation of the stable reaction products plus the diffusion controlled formation of solid products, such as carbon particles in underoxidized explosives. Nanosecond experimental techniques have measured the average pressures, particle velocities, and temperatures as the Chapman-Jouguet (C-J) state of thermal and chemical equilibrium is approached (11). In addition to vibrational-rotational and vibrational-translational

energy transfer, an essential physical process in the attainment of thermal equilibrium and the continued propagation of a self-sustaining detonation wave is the amplification of pressure wavelets by the chemical energy released during transitions from higher to lower vibrational levels during compression by these wavelets of specific frequencies. These process was recently shown to be the mechanism by which the internal chemical energy of the product molecules sustains the leading shock wave front at an overall constant velocity (5).

REACTION RATE CONSTANTS

Calculating the reaction rate constants for the transition states in shocked and detonating materials is the main subject of this paper. Reaction rate constants calculated for the shock-induced decomposition of gaseous norbornene, liquid NM, and solid PETN using high density, high temperature transition state theory are compared to the available experimental induction time data for various shock temperatures. Experimental data for unimolecular gas phase reactions under shock conditions has shown that the reaction rates obey the Arrhenius law:

$$K = A \, e^{-E/RT} \qquad (1)$$

where A is a frequency factor, E is the activation energy, and T is temperature, at low temperatures, but "fall-off" to less rapid rates of increase at high temperatures (12). Nanosecond reaction zone profile measurements for solid explosives overdriven or "supracompressed" to pressures and temperatures exceeding those attained in self-sustaining detonation waves have shown that the reaction rates increase very slowly with shock temperature (13). Eyring (14) attributed this "falloff" in unimolecular rates at the extreme temperature and density states attained in shock and detonation waves to the close proximity of vibrational states, which causes the high frequency mode that becomes the transition state to rapidly equilibrate with the surrounding modes by IVR. Thus these interacting modes form a "pool" of vibrational energy in which the energy required for decomposition is shared. Any large quantity of vibrational energy that a specific mode receives from an excitation process is equilibrated among the modes before decomposition can occur. Conversely, sufficient vibrational energy from the entire pool of oscillators is statistically present in the transition state long enough to cause reaction. When the total energy in the interacting vibrational modes equals the activation energy, the reaction rate constant K is:

$$K = (kT/h) \, e^{-s} \sum_{i=0}^{s-1} (E/RT)^i \, e^{-E/RT}/ i! \qquad (2)$$

where k, h, and R are Boltzmann's, Planck's, and the gas constant, respectively, and s is the number of the vibrational modes interacting with the dissociation mode. The main effect of this rapid IVR among s+1 modes at the high densities and temperatures reached in detonation waves is to decrease the rate constant dependence on temperature. Tarver (3) demonstrated that reasonable reaction rate constants could be calculated for detonating solid and liquid explosives using Eq. (2) with realistic equations of state and values of s. For the high densities but lower temperatures attained in shock initiation of homogeneous liquid and solid explosives, the reaction rate constants from Eq. (2) are greater than those predicted by Eq. (1). To determine whether Eq. (2) is valid under shock conditions, rate constants from Eqs. (1) and (2) are compared to shock initiation induction time results for NM and PETN.

COMPARISON OF RESULTS

The most complete set of gas phase unimolecular rate measurements is for norbornene decomposing to 1,3-cyclopentadiene and ethylene (15). Figure 2 shows that the reaction rates calculated using Eq. (2) with s = 20 and E = 45.39 kcal/m agree extremely well with the measured rates, which are slower than the Arrhenius rate in Fig. 2 that fits several sets of experimental data below 1000K (16). Eyring (14) and Tarver (3) found that twenty is a reasonable number of neighboring modes for many reactions, because only the modes involving 6 or 7 atoms are close enough to rapidly IVR with the reacting mode.

The second comparison is for the shock initiation data for liquid NM, summarized by Sheffield et al. (17) and Yoo and Holmes (18). The induction times for the onset of chemical reaction for transparent liquid explosives are measured at various input shock pressures. The induction time τ_i in thermal ignition theory (19) is related to the reaction rate constant K by:

$$\tau_i^{-1} = QEK/C_vRT^2 \qquad (3)$$

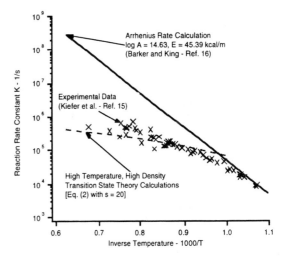

FIGURE 2. Reaction rate constant versus inverse temperature for the unimolecular decomposition of norbornene

where Q is the heat of reaction and C_v is the heat capacity. Thus, to relate the input shock pressure to temperature, an equation of state is required. Two excellent equations of state for NM have been published. Both included the increase in C_v as T increases, but they made different assumptions concerning the dependence of $(dp/dT)_v$ on pressure. Cowperthwaite and Shaw (20) assumed that $(dp/dT)_v$ is a constant, while Lysne and Hardesty (21) assumed that it increases as the density increases. Figure 3 shows the reaction rate constant versus inverse temperature results for NM calculated from the experimental induction times using Eq. (3) with shock temperatures calculated by these two EOS's. Also shown in Fig. 3 are the rates calculated using Eq. (1) with the gas phase values of A = 4.0 x 10^{15}s^{-1} and E = 59 kcal/m (19) and using Eq. (2) with s = 14, since NM has 15 vibrational modes. Despite the uncertainties in shock temperatures, Eq. (2) with s = 14 agrees well with the experimental data. The von Neumann spike rates of approximately 10^8 s^{-1} shown in Fig. 3 agree with the nanosecond reaction times for detonating NM measured by Sheffield (22).

Figure 4 shows the induction time data for single crystal PETN reported by Dick et al. (23). The Jones-Wilkins-Lee (JWL) equation of state for unreacted PETN developed by Tarver et al. (24) was used to calculate the shock temperatures with a constant C_v intermediate between the ambient temperature value and the maximum value. Also shown in Fig. 4 are the straight line for Eq. (1) with

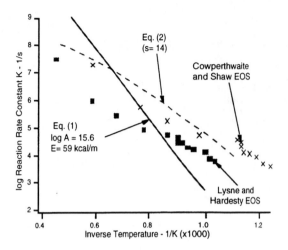

FIGURE 3. Reaction rate constants for nitromethane as functions of shock temperature

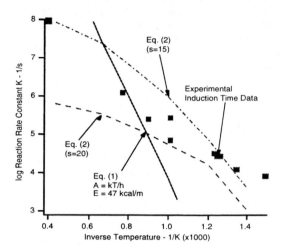

FIGURE 4. Reaction rate constants for single crystal PETN as functions of shock temperature

$E = 47$ kcal/m and $A = kT/h$ and the curves for s = 15 and s = 20 in Eq. (2). As for NM, the high density, high temperature transition state rate constants agree well with the PETN data, even when extrapolated to the von Neumann spike state of 2300 - 2500 K (25). The s = 15 calculation appears to be closer to the experimental data than s = 20, but accurate shock temperature measurements are needed before conclusions can be drawn about the number of vibrations interacting in the PETN transition state.

SUMMARY

High density, high temperature transition state calculations yield good agreement with measured reaction rates in shocked gases, liquids, and solids.

ACKNOWLEDGMENTS

This work was performed under the auspices of the U.S. Department of Energy by Lawrence Livermore National Laboratory (contract no. W-7405-ENG-48).

REFERENCES

1. Tarver, C. M., *Combustion and Flame* **46,** 111-134 (1982).
2. Tarver, C. M., *Combustion and Flame* **46,** 135-156 (1982).
3. Tarver, C. M., *Combustion and Flame* **46,** 157-176 (1982).
4. Tarver, C. M., Fried, L. E., Ruggerio, A. J., and Calef, D. F., *Tenth International Detonation Symposium,* ONR 33395-12, Boston, MA, 1993, pp. 3-10.
5. Tarver, C. M., *J. Phys. Chem. A* **101**, 4845-4851 (1997).
6. Tokmakoff, A., Fayer, M. D., and Dlott, D. D., *J. Phys. Chem.*. **97**, 1901-1911 (1993).
7. Weston, Jr., R. E. and Flynn, G. W., *Ann. Rev. Phys. Chem.* **43**, 559-599 (1993).
8. Hong, X., Chen, S., and Dlott, D. D., *J. Phys. Chem.* **99**, 9102-9108 (1995).
9. Fried, L. E. and Tarver, C. M., *Shock Compression of Condensed Matter-1995*, Schmidt, S. C. and Tao, W. C., eds., AIP, New York, 1996, pp. 179-182.
10. Tarver, C. M., Tao, W. C., and Lee, C. G., *Propellants, Explosives, Pyrotechnics* **21**, 238-246 (1996).
11. Hayes, B. and Tarver, C. M., *Seventh Symposium (International) on Detonation*, NSWC MP86-194, Annapolis, MD, 1981, pp. 1029-1037.
12. Kiefer, J. H. and Shah, J. N. *J. Phys. Chem.* **91**, 3024-3033 (1987).
13. Green, L. G., Tarver, C. M., and Erskine, D. J., *Ninth Symposium (International) on Detonation*, OCNR113291-7, Portland, OR, 1989, pp. 670-678.
14. Eyring, H., *Science* **199**, 740-743 (1978).
15. Kiefer, J. H., Kumaran, S. S., and Sundaram, S. J., *J. Chem. Phys.* **99**, 3531-3541 (1993).
16. Barker, J. R. and King, K. D., *J. Chem. Phys.* **103**, 4953-4966 (1995).
17. Sheffield, S. A., Engelke, R., and Alcon, R. R., *Ninth Symposium (International) on Detonation*, OCNR113291-7, Portland, OR, 1989, pp. 39-49.
18. Yoo, C. S. and Holmes, N. C., High-Pressure Science and Technology-1993, Schmidt, S. C., Shaner, J. W., Samara, G., and Ross, M., eds., AIP, New York, 1994, pp. 1567-1570.
19. Hardesty, D. R., *Combustion and Flame* **27**, 229-251 (1976).
20. Cowperthwaite, M. and Shaw, R. *J. Chem. Phys.* **53**, 555-560 (1970).
21. Lysne, P. C. and Hardesty, D. R. *J. Chem. Phys.* **59**, 6512-6523 (1973).
22. Sheffield, S. A., LANL, private communication, 1997.
23. Dick, J. J., Mulford, R. N., Spencer, W. J., Pettit, D. R., Garcia, E., and Shaw, D. C., *J. Appl. Phys.* **70**, 3572-3587 (1991).
24. Tarver, C. M., Breithaupt, R. D., and Kury, J. W., *J. Appl. Phys.* **81**, 7193-7202 (1997).
25. Yoo, C. S., Holmes, N. C., and Souers, P. C., *Shock Compression of Condensed Matter-1995*, Schmidt, S. C. and Tao, W. C., eds., AIP, New York, 1996, pp. 913-916.

CP429, *Shock Compression of Condensed Matter – 1997*
edited by Schmidt/Dandekar/Forbes
© 1998 The American Institute of Physics 1-56396-738-3/98/$15.00

SHOCK-INDUCED REACTIONS IN ENERGETIC MATERIALS, STUDIED BY MOLECULAR DYNAMICS WITH DIRECTLY EVALUATED QUANTUM MECHANICAL POTENTIALS

P J Haskins and M D Cook

Defence Evaluation & Research Agency, Fort Halstead, Sevenoaks, Kent TN14 7BP, England

In this paper we describe the development of a coupled molecular dynamics and quantum chemistry approach to the study of shock-induced chemical reactions. The method has been applied to study the interaction between small numbers of energetic molecules undergoing collisions at velocities typical of detonations. Preliminary results for nitromethane and nitric oxide are presented, and discussed in the context of mechanisms operating at, or just behind, detonation fronts. The paper concludes by describing our plans to extend the approach to consider larger systems.

INTRODUCTION

In recent years the use of a modified Abell-Tersoff potential, pioneered by Brenner and co-workers (1,2,3), has demonstrated the potential of Molecular Dynamics (MD) to address fundamental issues relating to detonation. In a previous paper (4), we described the use of this type of potential to study both shock and thermal initiation thresholds for a model heteronuclear diatomic system. In this work we showed that reaction under shock loading preceded thermal equilibration, and that the main driving force for chemical reaction was an increase in the local density. However, the use of an empirical potential clearly gives rise to some doubt about the relevance of these findings to real systems. The possibility of non-equilibrium processes playing a large part in the initial reaction steps had been postulated previously (5), and is clearly an important issue which needs to be resolved. To this end we have developed an MD technique employing directly computed quantum mechanical interactions, and have begun to apply this to some small molecular ensembles. In the following sections we give a brief description of the method, and the results of some preliminary simulations.

DESCRIPTION OF THE METHOD

The method we have developed employs a general purpose quantum chemistry code (GAUSSIAN 94 (6)) to directly compute energies and forces at each time step of the MD simulation. The technique allows a wide range of quantum chemical methods to be used. Our work to date has concentrated on the use of Unrestricted Hartree-Fock (UHF) and Density Functional (DF) approaches, but it is also possible to employ post-HF treatments for electron correlation, or to use semi-empirical methods. When using ab-initio techniques (like UHF and DF) the calculations, even for very small numbers of molecules, are quite demanding in terms of computer time. It has been found necessary to use time-steps of 0.1fs to achieve satisfactory energy conservation, and this clearly limits the overall simulation time which can be considered.

The calculations reported here treat the complete simulation ensemble as one "supermolecule" and use no boundary conditions. Other strategies are being developed, and are discussed later.

NITROMETHANE CALCULATIONS

Nitromethane (NM) has been extensively studied, both experimentally and theoretically. As the simplest member of the important nitro compound class of explosives it is an ideal prototype material. We have previously reported (7,8,9) a number of ab-initio molecular orbital investigations of potential energy surfaces for uni- and bi-molecular decomposition reactions of NM. In the calculations reported here we have looked at the interaction between two NM molecules, aligned head to tail along their C-N bonds, such that the methyl group of molecule 1 faced the nitro group of molecule 2. The molecules were given initial velocities towards each other, and the evolution of the system followed using the MD technique described above.

The molecules were set up in their equilibrium geometries with a 3.5A° initial spacing between the carbon atom of molecule 1 and the nitrogen atom of molecule 2. In the simulations reported here the molecules were initially at 0K (although some studies are underway to look at the effects of finite initial temperatures). Calculations were carried out at two levels of theory, namely UHF with a 6-31G** basis set (UHF/6-31G**), and DF employing an unrestricted Becke-style 3 parameter hybrid functional with the Perdew/Wang 1991 non-local correlation expression and the same 6-31G** basis (UB3PW91/6-31G**).

Before running the MD simulations described above, relaxed potential energy surface scans were carried out on the C-N bond of NM at the two levels of theory. Both methods correctly predict dissociation to methyl and nitro radicals, but there is a significant difference in the dissociation energy required. The UHF/6-31G** calculation predicts a bond dissociation energy of 47.51 kcal/mol, whereas the UB3PW91/6-31G** calculation predicts 61.63 kcal/mol, in closer agreement with the experimental value of 60.3 kcal/mol. The improved agreement of the DF result with experiment probably arises

from the inclusion of some correlation effects in the calculation.

MD simulations were carried out using both levels of theory at a range of initial velocities. The simulations were run until reaction occurred, or for sufficient time to decide that reaction would not occur without further collisions (which could only happen if a larger number of molecules, or boundary conditions had been used).

The results of the calculations were similar at both levels of theory, with a C-N bond scission reaction being observed above a threshold velocity. The UHF/6-31G** calculations gave a threshold velocity for reaction of between 7.5 and 8 km/s, whereas the UB3PW91/6-31G** calculations required a velocity between 8.4 and 8.5 km/s. This difference in predicted threshold velocity is understandable in the light of the different bond dissociation energies discussed above. At the threshold velocity it was observed that molecule 1 decomposed into methyl and nitro radicals after 2 vibrations of the C-N bond. Molecule 2 did not decompose. This behaviour is illustrated in Fig. 1, which shows the kinetic energy and C-N bond lengths from the UHF/6-31G** calculation at 8 km/s. At velocities a little above the threshold (e.g. 8.5 km/s for UHF/6-31G**), molecule 1 was observed to decompose immediately upon rebound from the collision, but molecule 2 remained unreacted.

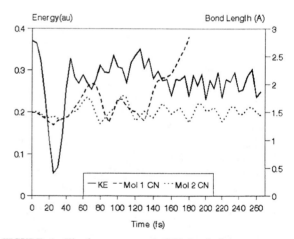

FIGURE 1. Kinetic energy and C-N bond distances as a function of time for a simulation at 8 km/s, using UHF/6-31G** theory.

It is clear from these simulations that strong shocks can result in rapid decomposition of nitromethane, in times far too short for thermal equilibrium to have been established. In the orientation that has been considered here, it is clear that the conversion of translational to vibrational energy is initially concentrated preferentially in the C-N bond. This leads to either immediate decomposition on rebound, or shortly after when other vibrational modes couple to increase the C-N stretch. Naturally other orientations of the molecules may give rise to different results, and it is planned to consider some of these in the near future.

The system studied here corresponds to a gas phase collision, but in condensed phase nitromethane the neighbouring molecules would be expected to play a significant role. Firstly, the presence of a number of near neighbours, after shock compression, could substantially alter the bond strengths and energy transfer processes. Secondly, the presence of neighbours will allow molecules to be excited by multiple collisions. Consequently, molecules which fail to absorb enough energy from the initial collision may still react upon favourable secondary collisions. The inclusion of these effects, in future calculations on larger systems, may be expected to reduce the threshold velocity. The current threshold velocity is considerably in excess of the particle velocity associated with normal detonation, and therefore relates only to highly overdriven states.

NITRIC OXIDE CALCULATIONS

We have also carried out some preliminary studies of shock-induced reactions in nitric oxide. This system was chosen because its simplicity allows a larger number of molecules to be considered, and it may ultimately offer the prospect of a direct comparison with empirical AB models (1,2,3,4). In these calculations we have considered four NO molecules aligned in a plane at $45°$ to the shock axis. The molecules were initially at their equilibrium bond length ($1.15A°$), and were arranged as illustrated in Fig. 2. The left and right hand pairs of molecules were given initial velocities towards each other, and the resulting interactions studied as

$$
\begin{array}{cccc}
N & O & N & O \\
/ \ - 2.5A° - \ / \ & - 5.0A° - \ & / \ - 2.5A° - \ / \\
O & N & O & N
\end{array}
$$

FIGURE 2. Initial arrangement of 4 x NO molecules.

above. The calculations to date have all been at the UB3PW91/6-311G** level of theory.

At impact velocities of 6.0 km/s, and greater, the reaction shown below was observed to occur.

$$4NO \Rightarrow O_2 + 2N_2O \qquad (1)$$

This reaction took place immediately upon rebound from the collision. No reaction was observed at velocities of 5.5 km/s and below. Unlike the essentially uni-molecular decomposition process seen with nitromethane, this is clearly a concerted reaction mechanism. The reaction was observed to be exothermic by 45.22 kcal/mol (cf. 48.06 kcal/mol as determined from published heats of formation). This rapid release of energy is encouraging from the viewpoint of future simulations, as it suggests the possibility of simulating a detonation process with a relatively small ensemble. However, it should be noted that, as for nitromethane, the velocities are still in excess of detonation particle velocities. Clearly, in a larger system, or with boundary conditions, subsequent reactions of N_2O to yield more stable N_2 and O_2 molecules would probably occur, with the release of additional energy.

FUTURE WORK

We plan to continue the preliminary studies reported here by extending them to consider other orientations, the effects of finite initial temperatures, and, most importantly, to study larger ensembles. Unfortunately, the computational time for ab-initio quantum chemistry calculations usually scales with the third power of the system size. This means that systems with sufficient atoms to model a detonation process would be prohibitively slow / expensive with current computers. To help overcome this problem we are developing a methodology which will enable the main features of chemical bond making and breaking processes to be treated

307

quantum mechanically, but interactions between more distant molecules to be treated empirically. The approach is similar in principle to the tight binding method, and should allow a more linear scaling to be achieved.

DISCUSSION AND CONCLUSIONS

The results reported in this paper clearly show that strong shocks are capable of producing chemical reactions prior to thermal equilibrium. These may result from preferential excitation of a vibrational mode, or from a specific bi-(or multi-)molecular reaction. In both cases orientation of the molecules, with respect to each other, and the shock, are likely to be critical. The experimental work on shock initiation by Dick (10), which shows a strong orientational dependence for PETN, lends some support for the importance of such processes.

Although these calculations suggest that such non-equilibrium processes may play a key role in shock initiation and detonation, considerably more work is required before any definite conclusion can be reached on this issue. It will certainly be necessary to consider larger ensembles, where the potential for many collisions exists, as these may provide routes to reaction at lower shock energies than those observed here. It may, of course, prove highly system specific, with some materials exhibiting dependence on non-equilibrium mechanisms, whilst others may always react by slower thermal routes.

It is hoped that our future plans, as described above, will help to answer this important issue. It is also hoped that, in the longer term, such MD approaches will be able to provide kinetic schemes, for implementation in hydrocode-based ignition and growth models.

ACKNOWLEDGEMENTS

The authors would like to thank Mr A. Wood for his assistance with the coding, and Mr J. Fellows for carrying out the potential scans on nitromethane.

REFERENCES

1. Brenner, D.W., Elert, M.L. and White, C.T., "Incorporation of reactive dynamics in simulations of chemically-sustained shock waves", in *Proceedings of the APS Topical Conference on Shock Compression of Condensed Matter*, 1989, pp. 263-266.
2. Brenner, D.W., "Molecular potentials for simulating shock-induced chemistry", in *Proceedings of the APS Topical Conference on Shock Compression of Condensed Matter*, 1991, pp. 115-121.
3. Robertson, D.H., Brenner, D.W., Elert, M.L. and White, C.T., "Simulations of chemically-sustained shock fronts in a model energetic material", in *Proceedings of the APS Topical Conference on Shock Compression of Condensed Matter*, 1991, pp. 123-126.
4. Haskins, P.J., and Cook, M.D., "Molecular dynamics studies of thermal and shock initiation in energetic materials", in *Proceedings of the APS Topical Conference on Shock Compression of Condensed Matter*, 1995, pp. 195-198.
5. Dremin, A.N., "Shock discontinuity zone effect: the main factor in the explosive decomposition detonation process", in *Proceedings of the Royal Society Discussion Meeting on Energetic Materials*, 1991, pp. 355-364.
6. Gaussian 94 (Revision C.2), Frisch, M.J., Trucks, G.W., Schlegel, H.B., Gill, P.M.W., Johnson, B.G., Robb, M.A., Cheeseman, J.R., Keith, T.A., Petersson, G.A., Montgomery, J.A., Raghavachari, K., Al-Laham, M.A., Zakrzewski, V.G., Ortiz, J.V., Foresman, J.B., Peng, C.Y., Ayala, P.A., Wong, M.W., Andres, J.L., Replogle, E.S., Gomperts, R., Martin, R.L., Fox, D.J., Binkley, J.S., Defrees, D.J., Baker, J., Stewart, J.P., Head-Gordon, M., Gonzalez, C., and Pople, J.A., Gaussian Inc., Pittsburgh PA, 1995.
7. Cook, M.D., and Haskins, P.J., "Decomposition mechanisms and chemical sensitisation in nitro, nitramine, and nitrate explosives", in *Ninth Symposium (International) on Detonation*, 1989, pp. 1027-1034.
8. Cook, M.D., and Haskins, P.J., "Chemical sensitisation in C-nitro explosives", in *Tenth Symposium (International) on Detonation*, 1993, pp. 870-875.
9. Cook, M.D., Fellows, J., and Haskins, P.J., "Probing detonation physics and chemistry using molecular dynamics and quantum chemistry techniques", in *MRS meeting on Decomposition, Combustion, and Detonation Chemistry of Energetic Materials*, 1995.
10. Dick, J.J., "Orientation-dependent shock response of explosive crystals", in *Proceedings of the APS Topical Conference on Shock Compression of Condensed Matter*, 1995, pp. 815-818.

CP429, *Shock Compression of Condensed Matter – 1997*
edited by Schmidt/Dandekar/Forbes
© 1998 The American Institute of Physics 1-56396-738-3/98/$15.00

A MODEL TO STUDY THE ELECTRONIC RESPONSE
TO AN IMPACT IN ENERGETIC MATERIALS

D. Mathieu [a], P. Simonetti [a] and P. Martin [b]

[a] CEA - Le Ripault, BP16, 37260 Monts, France
[b] CISI, Aéropôle, Im. Rafale B, 1 rue Charles Linbdbergh, 44340 Nantes-Bouguenais, France

It has been suggested that excited electrons might play a significant role in impact initiation of energetic materials, despite their very small number at thermal equilibrum and the lack of any detailed physical model to support such a view. Nonetheless, as pointed out in this paper, a compression wave may enhance excitations throught non-adiabatic processes. To study such effects, a computational approach is proposed.

INTRODUCTION

Understanding the initiation of chemical decomposition in energetic materials from a microscopic viewpoint is a long-standing problem. Some time ago, it was suggested that electronic excitations might be involved in this initiation process (1). In the following years, this assumption has been used to estimate impact sensitivities (2,3). However, the estimation procedure suffered from a lack of physical ground, and eventually it turned out to be highly doubtful (4). Nonetheless, the role of excited states has been recently revisited by Gilman in a effort to account for experimental facts not consistent with standard models : according to him, sensitivity correlates with the formation of delocalized electrons (5). His work has stimulated recent studies of the electronic structure of energetic materials, one of the aims being to get further insight into the microscopic mechanisms involved in initiation (6-7).

However, the closure of the bandgap assumed by Gilman's model requires very high densities, at which the chemical structure of the material is already altered. Thus, bandgap closure cannot be invoked as the primary step of initiation. However, as the valence and conduction bands come closer, transitions across the gap may be enhanced through non-adiabatic processes. This should result in the appearance of delocalized electrons in the conduction band. Such transitions depend not only on the gap energy but also on other parameters which cannot be derived from the band structure but only through inclusion of non-adiabatic effects. Such parameters will be hereafter referred to as dynamic parameters. Therefore, while much attention is paid to the gap in many attempts to associate impact sensitivities with electronic features (6-7), some dynamic parameters may be even more significant.

A model is presently under development in our laboratory to estimate the magnitude of non-adiabatic effects and their dependence on the features of the material, e.g. electronic structure, vibrational frequencies, temperature, defects, strength of the impact... The goals are to get a better understanding of the interplay between a compression wave and the electronic structure, and to identify relevant parameters which might be valuable descriptors for recent statistical approaches to sensitivity prediction (8). Details of the model and its implementation have been previously described (9-10). In the present paper, the different contributions to non-adiabatic effects are emphasized. Results of simulations are presented to illustrate what can be done with the computer code currently being developped. Atomic units are used unless otherwise mentionned.

309

MODEL CALCULATIONS

Briefly, the present simulations consist in solving the Newton equation for ions as well as the time-dependent Schrödinger equation (TDSE) $id\Psi/dt = H\Psi$ for the electrons. In addition, a set of time-independent equations $H\Psi_k = \omega_k\Psi_k$ are also solved for various instants in the time interval considered. Comparing Ψ to the eigenstates Ψ_k allows us to study the time evolution of the electronic populations in the states Ψ_k. To clarify non-adiabatic effects in the semi-classical context, we first consider a two-level system in motion, carrying two electronic basis functions corresponding to the electronic eigenstates Ψ_0 and Ψ_1 of frozen nuclei. Starting from the ground-state Ψ_0 at time $t=0$, the system is then put into motion and the electrons are described by the wavefunction Ψ, decomposed within the space spanned by the eigenstates :

$$|\Psi\rangle = \left(b_0|\Psi_0\rangle + b_1|\Psi_1\rangle\right)e^{-i\omega t} \qquad (1)$$

where ω is the transition frequency between both states. The time-dependence of the hamiltonian in the TDSE arises via the V_{Ne} potential describing the interaction of the electrons with the ions.

Sudden acceleration

First, let's examine the influence of an acceleration of the whole system on the electronic wavefunction Ψ. At $t=0$ the system initially at rest is suddenly provided with a constant velocity $\mathbf{V}$. Solving the TDSE it is a simple matter to show that the squared amplitude of Ψ on the excited state Ψ_1 varies in time according to :

$$p_1 = |b_1|^2 = \frac{\Omega^2}{\Omega^2 + \omega^2}\sin^2\left(t\sqrt{\Omega^2 + \omega^2}\right) \qquad (2)$$

where $\Omega = |\langle\Psi_0|d/dt|\Psi_1\rangle|$ is a typical frequency of the system, proportional to the magnitude of $\mathbf{V}$, and $\omega = \omega_1 - \omega_0$ is the electronic transition frequency. The population p_1 oscillates with a frequency larger than both Ω and ω, but in any case remain small as long as $\omega \gg \Omega$. For instance, a velocity $V \sim 500$ m/s yields $\Omega \sim 300$ cm^{-1}, assuming $\langle\Psi_0|\nabla|\Psi_1\rangle \sim 1$ bohr^{-1} and $\omega \sim 5$ eV, i.e. 4.10^4 cm^{-1}, and $p_1 \sim 10^{-4}$. These oscillations arise

because the eigenstates are not invariant upon change of galilean frame (11). On the other hand, no relaxation mechanism is present after the sudden acceleration. This very simple model already shows the significance of two key parameters, namely the transition frequency (or gap energy) ω and the magnitude Ω of the non-adiabatic coupling integral $\langle\Psi_0|d/dt|\Psi_1\rangle$.

Forced oscillations

In this section, the model system initially at position $X=0$ along a given axis is supposed to oscillate for $t>0$ according to :

$$X(t) = X_0\left(1 - \cos(\omega_N t)\right) \qquad (3)$$

Introducing the non-adiabatic typical frequency $\Omega = \langle\Psi_0|\nabla|\Psi_1\rangle\omega_N X_0$, the evolution equation for the coefficients of Ψ in equation (1) becomes :

$$\dot{b}_0(t) = -\Omega\sin(\omega_N t)e^{-i\omega t}b_1(t)$$
$$\dot{b}_1(t) = -\Omega\sin(\omega_N t)e^{+i\omega t}b_0(t) \qquad (4)$$

Figure 1 illustrates a typical solution of these equations, for $\Omega \sim 300$ cm^{-1} as before and $\omega/\omega_N = 50$. The fast oscillations of the electronic wavefunction above the ground state are modulated by the forced vibrations of the system at the lower frequency ω_N.

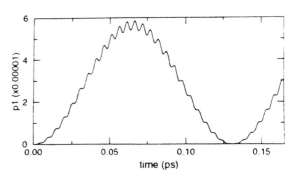

FIGURE 1. Evolution of the excited state population p_1 after the system initially at rest is forced to oscillate from $t=0$.

For long periods of time, the time-averaged value of p_1 depends mostly on the ratio ω/ω_N and, to a lesser extent, on the impact frequency Ω. The dependence of excited populations on these characteristic frequencies are illustrated in Figure 2.

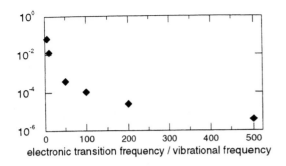

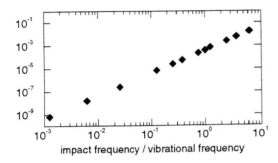

electronic transition frequency / vibrational frequency

impact frequency / vibrational frequency

FIGURE 2. Maximum value reached by p_1 long after the system is put into forced oscillations, as a function of Ω/ω_N and ω/ω_N.

So far, we have studied only the electronic response to molecular motion. The excited state population increases as the ions are accelerated, since the magnitude Ω of the coupling integral $\langle \Psi_0 | d/dt | \Psi_1 \rangle$ depends linearly on ionic velocities. However, in real systems under impact, an additional effect is likely to enhance p_1 values obtained so far. Indeed, as a compression wave propagates through the material, the band structure of the solid is distorted. In high density regions, the localization of the electrons decreases because of the overlap of molecular wavefunctions. In other words, bandwidths increase and the bandgap E_g of the material decreases. If the latter gets significantly lower than the transition energy ω of the isolated molecule, then non-adiabatic transitions across the gap should be dramatically enhanced.

COMPUTATIONAL APPROACH

To simulate the non-adiabatic evolution of the material wavefunction, taking into account both ionic motion and changes of the band structure resulting from lattice distortions, the approach

described above for very simple models has been generalized to atomic clusters. Some complications arise mainly for two reasons : first, we deal with N-electrons systems, and second, suitable eigenstates of the hamiltonian are not known *a priori*. Therefore, they cannot be used directly as a basis set.

Description of the electronic structure
Wavefunctions

Because the non-adiabatic wavefunction Ψ is assumed to differ only slightly from the ground state Ψ_0, there is no need for an accurate description of excited states Ψ_k. Instead, two sets of orbitals $\{\phi_k\}$ and $\{\varphi_k\}$ are used to represent the ground state Ψ_0 and the dynamic state Ψ, respectively. Every orbital is written in term of atomic basis functions $\{\chi_\nu ; \nu=1..K\}$. Thus, instead of using explicit excited states, the empty static orbitals $\phi_{k+1}...\phi_K$ are assumed to span empty bands of the material.

Effective hamiltonian

Using orbital representations for Ψ_0, the equation $H\Psi_0=\omega_0\Psi_0$ yields the Hartree-Fock equations for the orbitals ϕ_k, while the TDSE translates into a set of time-dependent equations for the orbitals φ_k. Both sets involve an effective hamiltonian F. For the time being, a simple EHT hamiltonian is used (12). The latter is obtained empirically without considering the actual orbitals, either static or dynamic : it is not self-consistent.

Electronic populations

Since no excited states are available, electronic populations are defined according to the following expressions, for the occupation of orbital k and mean occupation in the conduction band, respectively :

$$n_k = \sum_\mu \left| \langle \phi_k | \varphi_\mu \rangle \right|^2 \qquad n_C = \frac{1}{N} \sum_{k \in CB} n_k \qquad (5)$$

Description of the cluster dynamics
Interatomic forces

Preliminary simulations were carried out for model atomic clusters. The nuclei being treated classically, their trajectories are obtained from the

Newton equation. The empirical potentials used to describe interatomic forces (in the present case Lennard-Jones potentials) include the force $F_{e \to N}$ of the electrons in the ground state on the ions. The non-adiabatic contribution to $F_{e \to N}$ is, however, neglected. Here is the key approximation of the present model, which allows us to study extended systems over long times (~1 ns) while usual models for non-adiabatic dynamics can only be used in very simple cases, since they are concerned with situations where non-adiabatic effects are crucial to get meaningful trajectories. In contrast, assuming adiabatic trajectories yields good results with regard to the non-adiabatic behaviour of electrons in a material under impact, as illustrated in Figure 3, where non-adiabatic corrections to $F_{e \to N}$ were derived in order to ensure energy conservation of the whole system electrons+nuclei as explained elsewhere (13).

Compression wave

The flyer plate model (13) is used to simulate a dynamic compression. So far, only 1D atomic clusters are considered. The flyer plate is a small cluster launched from the left into the model cluster under study. To minimize edge effects, only atoms located in the bulk of this cluster are provided basis function for explicit description of their electrons (Figure 4).

CONCLUSION

We develop a new approach to gain insight into a possible role of excited states for the initiation of explosives. The present model focusses on how electronic excitations could be enhanced by very intense mechanical perturbations such as shockwaves. Although their absolute values are not expected to be accurate, relative

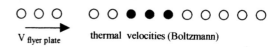

FIGURE 4. Model system (1D) : quantum atoms (those showing basis functions) are drawn in black, others in grey.

populations in excited states obtained from this model may be compared to each other. It will be specially interesting to see whether a significant enhancement of thermal equilibrum populations in excited states is possible under certain circumstances. Moreover, the present model is consistent with the non-radiative character of the transitions. As such, it departs from earlier studies dealing with vertical electronic excitations from the ground state equilibrium geometry (1), notwithstanding the fact that such transitions necessarily involve photons.

REFERENCES

1. Delpuech, A. and Cherville, J., *Propellants, Explos., Pyrotech.* **3**, 169, 1978; **4**, 61, 1979; **4**, 121, 1979
2. Heming, X., Yongfy, L. and Peilei, F., *Proc. Symp. Internat. on Detonation*, 1983.
3. Gamézo, V.N., Odiot, S., Blain, M., Fliszar, S. and Delpuech, A., *Theochem*, **337**, 189, 1995
4. D. Delpeyroux, unpublished results
5. Gilman, J.J., *Chem. Prop. Inf Agency*, **589**, 379, 1992
6. Kunz, A. B., *Phys. Rev. A*, **53**, 9733,1996
7. Brunet, L., Lombard, J.M., Blaise, B. and Morin-Allory, L., *Proc. of the EUROPYR093 conference*. pp. 89-96, 1993
8. Nefati, H., Cense, J-M., Legendre, J-J., *J. Chem. Inf. Comp. Sci.* **36**, 804-810, 1996
9. Mathieu, D., *Proc. Conf "Ab initio Calculation of Complex Processes in Materials"*, pp. 79-80, Schwäbisch Gmünd, Germany, 1996
10. Mathieu, D. and Martin, P. *Comp. Mat. Sci.* accepted
11. Landau, L. and Lifchitz, E., *Mécanique Quantique* Moscow: Mir, 1984
12. Rivail, J-L., *Eléments de Chimie quantique*, Paris: Interéditions, 1989
13. Lescouezec R., Martin P. and Mathieu, D. in preparation.
14. Brenner D.W., Robertson D.H., Elert M.L. and White, C.T., *Phys. Rev Lett.* **70**, 2174, 1993

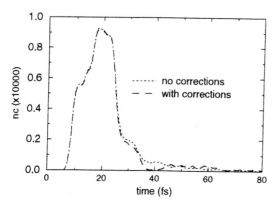

FIGURE 3. Evolution of the fraction of excited electrons, n_C, after an impact is applied at $t=5$ fs, with ionic trajectories obtained either from the adiabatic forces or using non-adiabatic empirical corrections.

CP429, *Shock Compression of Condensed Matter – 1997*
edited by Schmidt/Dandekar/Forbes
© 1998 The American Institute of Physics 1-56396-738-3/98/$15.00

DEFORMATION POTENTIALS AND PLASMON ENERGIES
- MEASURES OF SENSITIVITY

John J. Gilman

Department of Materials Science and Engineering, UCLA, Los Angeles, CA 90095

Covalent bonds in molecules and crystals are stabilized by gaps in the energy spectra of the bonding electrons. Finite deformations (strains), epecially shear deformations, tend to narrow and then close the gaps. The coefficients that describe the amounts of the closures are deformation potentials. Since the closures destabilize the covalent bonds, the deformation potentials provide a measure of the sensitivity towards ignition of these materials. Stability also correlates with plasma frequencies which are easier to measure than deformation potentials.

When a crystal, or molecule, is deformed its electronic structure is changed. This changes its electrical resistvity, and its optical properties. The primary effect of the deformation is to shift the energies of the HOMO and LUMO levels. Shear deformation is particularly effective. For finite deformations the HOMO level usually rises while the LUMO level falls. Thus the energy gap between the two levels diminishes until the gap closes. Since the gap stabilzes the structure, diminishing it causes instability which leads to structural transformation, and metallization. This, in turn leads to fast decomposition of unstable substances. The coefficient that relates the changes in energies of the HOMO and LUMO levels to the deformation (strain) is the deformation potential coefficient. For general deformations, it is a tensor, Ξ_{ij}.

The advantage of relating straininduced destabilization to the deformation potential is that Ξ_{ij} is known to be related to other readily available parameters such as bond lengths, and plasmon frequencies.

The deformation potential is defined for small strains. Here, the definition will be extrapolated to include large strains as well. This will be justified by showing that the extrapolation leads to results consistent with measurements. Shear strains affect the electronic structure differently than do dilatational strains because they change the symmetry of the structure. However, most measurements have been made either by applying uniaxial stresses (yielding a mixture of shear and dilatation), or by applying pressure (yielding only dilatation). A comprehensive review has been presented by Bir and Pikus [1], and a recent review of experiments by Cardona [2].

If the applied deformation (dilatation) is $-(\Delta V/V)$, the change in energy, E_g is $\Delta E_g = - \Xi_d (\Delta V/V)$ where Ξ_d is the dilatational deformation potential. For cubic crystals there is one other independent potential that is usually measured for unixial applied stress. It is designated Ξ_u. For example, Ξ_{ud} has a magnitude of about 8.3 eV. for the deformation of silicon [3].

Unfortunately, there do not appear to be any data for pure applied shear, although some shear coefficients have been obtained by subtracting the effects of pure pressure from the effects of uniaxial stresses. This assumes linear superposition which may not be valid.

When strains of order 20% are applied to them, covalently bonded crystals tend to undergo phase changes, and these structural changes have been associated with closure of the energy band-gap. Thus, the deformation potential can be expected to be a declining function of the strain until the change in energy equals the magnitude of the gap. Since it is symmetric, the cosine of the displacement gradient is appropriate, and we write: $\Delta U = \Xi(1 - \cos \pi \epsilon)$ where ΔU is the change in the energy gap induced by a strain, ϵ. Using silicon as an example: $\Delta U = 1.1$ eV., and $\Xi = 8.3$ ev.; so solving for ϵ yields 17% which is approximately what is observed.

In crystals with more than one kind of atom in the unit cell, internal deformation of the relationships between the two, or more, kinds of atoms may be important. For example, the distance between the two atoms might change differently than the sizes of the energy equals the magnitude of the gap. Since it is symmetric, the cosine of the displacement gradient is appropriate, and we write: $\Delta U = \Xi(1 - \cos \pi\epsilon)$ where ΔU is the change in the energy gap induced by a strain, ϵ. Using silicon as an example: $\Delta U = 1.1$ eV., and $\Xi = 8.3$ ev.; so solving for ϵ yields 17% which is approximately what is observed.

In crystals with more than one kind of atom in the unit cell, internal deformation of the relationships between the two, or more, kinds of atoms may be important. For example, the distance between the two atoms might change differently than the sizes of the axes of the overall unit cell, thereby creating a local electric dipole.

Deformation potentials are usually measured by studying the effects of applied stresses on the optical properties of specimens. Unfortunately, the numerical values for various substances do not exhibit systematic patterns so no useful generalizations regarding the values have been discovered. As a result, the author turned to other potentially useful measures of stability, keeping in mind that a desireable parameter should be easy to measure, and should be insensitive to microstructural defects. Plasma frequencies (energies) are such a parameter.

In molecular crystals the analog of the plasma frequency is the molecular polarization frequency. Thus, in the plasma case the oscillating electrons are bound to the solid as a whole, and the oscilations are collective ones. Whereas, in the molecular polarization case, the oscillations are within the molecules (or atoms), and the electrons are bound to the molecules (or atoms). The analogy can be seen clearly by comparing the approximate expressions for the frequencies.

The plasma frequency, ω_p is given by:
$$\omega_p^2 = 4\pi n e^2/m^* \qquad (1)$$
where n is the electron density, e is the electron charge, and m^* is the effective mass of the electron. For semiconductors, ω_p is related to stability because m^* depends inversely on the minimum band gap. Hence, a large gap gives a small m^* and a large ω_p.

The molecular polarization frequency when the applied frequency is far away from the resonant frequency, and one excitation frequency is dominant is given by:
$$\omega_{hl}^2 = e^2/\alpha m \qquad (2)$$
where $h\omega_{hl}$ is the energy gap between the LUMO and

the HOMO levels, m is the electron mass, and α is the polarizability of the molecule.

Notice that $1/n$ (the volume per electron) plays the same role as α (approx. volume of the molecular valence electron).

Both frequencies are measures of stability. An experimental verification of this is given in Figure 1 which shows bulk moduli plotted versus plasma frequencies [4] for various semiconductors. The data

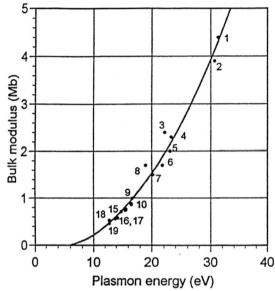

Figure 1 - Relationship between bulk moduli and plasma energies ($h\omega/2\pi$). Key to numbers:

1 - C	6 - BP	11 - AlP	16 - AlSb
2 - BN	7 - BAs	12 - Ge	17 - InAs
3 - GaN	8 - InN	13 - GaAs	18 - Sn
4 - SiC	9 - Si	14 - AlAs	19 - InSb
5 - AlN	10 - GaP	15 - InP	

indicate an approximate parabolic dependence of B on ω_p which is consistent B being proportional to the electron concentration according to Equation (1) if the variation in the effective mass is relatively small.

Another verification comes from the critical pressures at which various semiconductors transform from their normal crystal structures to another structure (typically from diamond/zincblende to β-tin/rocksalt). It is shown in Figure 2 that this tranformation pressure is proportional to the plasmon energies for the pressurized substances. The critical pressures are mostly from reference [5]. Figures 1 and 2, taken together, imply a relationship between B and the critical transformation pressures, of course.

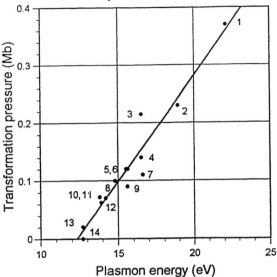

Figure 2 - Correlation between critical phase transformation pressures and plasmon energies ($\hbar\omega_p/2\pi$) for various tetrahedral semiconductors. Key to numbers:

1 - GaN	5 - AlAs	9 - Ge	13 - InSb
2 - InN	6 - GaAs	10 - AlSb	14 - Sn
3 - GaP	7 - Si	11 - InAs	
4 - AlP	8 - InP	12 - GaSb	

Given that the critical transformation pressure is proportional to the bulk modulus, another implication is that there is a critical amount of strain at which transformation occurs; namely, the strain corresponding to the proportionality coefficient.

In accordance with a suggestion made originally by Jamieson [6], it was confirmed previously that the work done during the application of a strain equals (at least approximately) the energy of the minimum band gap (or the LUMO-HOMO gap) when the first phase transformation occurs [7]. This confirms the idea that these gaps stabilize the initial structure. Therefore, it is expected that the gaps will be related to other stability parameters. Figure 3 confirms that this is the case for the gap energy and the plasmon energy.

The gap energies depend very strongly on bond lengths. Moss [8] showed for the Group IV crystals that $E_g(eV) = 6*exp(-3.6*10^{-4} a_o{}^5)$ where a_o is in Angstrom units. It is expected that the plasmon energies will also depend on bond lengths, although not so

strongly as the Moss relationship (Figure 4).

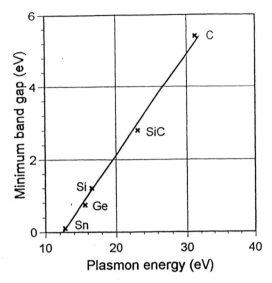

Figure 3 - Dependence of minimum band gap energy on plasmon energy for the Group IV crystals.

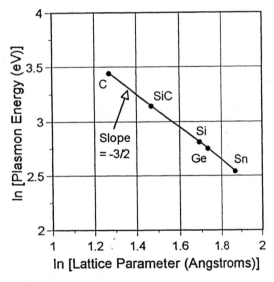

Figure 4 - Relationship between plasmon energies and chemical bond lengths (lattice parameters) in the Group IV crystals.

For molecular crystals, it is the polarizabilities of the bonds that determines their stabilities; the smaller the polarizability, the greater the stability. A simplified expression for the polarizability is as follows [9]:

315

$$\alpha = (N/m)(he/2\pi\Delta)^2$$

where N is the number of electrons in the molecule, m is the electron mass, h is Planck's constant, e is the electron charge, and Δ is the LUMO-HOMO energy gap. Note that, if Δ is small, molecules tend to be colored, along with having high polarizabilities. Since they determine refractivities, and refractive indices, polarizabilities are convenient to measure using standard optical instruments; as are plasma frequencies. Theoretical techniques for calculating them are also available, although these tend to be estimates for complex molecules, or molecules that contain atoms with complex valence states. Thus, as measures of ignition sensitivity these optical (dielectric) parameters may be equally (if not more) effective compared with thermochemical parameters. Since Δ equals twice the chemical hardness, the chemical and optical parameters are closely related, of course.

Plasmon energies for covalently bonded molecules tend to be relatively high, lying in the vacumn ultraviolet region of the spectrum. However, they can be determined by means of ultraviolet, or syncrotron, spectroscopy. Also, they can be determined by measuring the dielectric function at lower energies, and then extrapolating.

Since it is known that the sensitivities of a large number of eplosives correlate with their chemical resonance energies [10], and with their chemical bond energies [11], it is to be expected that they will also correlate with the optical properties being discussed.

The gap that becomes closed as a result of strain, particularly shear strain [12], leads to a drastic change in the electronic structure of a material, and to chemical decomposition in the case of energetic materials [13]. For crystals with dense structures, a simple criterion for this was provided by Herzfeld a long ago [14]. This criterion is that when the molar refractivity, R (which has the dimensions of a volume) is greater than the molar volume, a material becomes metallic (that is, becomes a dense plasma of nearly free electrons and positive ions). But the molar refractivity is given by:

$$R = (4\pi/3)A\alpha_o$$

where A is Avogadro's number and α_o is the long-wavelength polarizability. This indicates how the optical properties of energetic materials are related to chemical stability, and to chemical reactivity.

In a plasma, the barriers (activation energy) to the occurence of chemical reactions between atoms and/or molecules are small or nonexistent. That is why nascent metals are so highly reactive. Metals would be useless for constructing structures if they were not passivated by dense films of oxides, or other nonreactive compounds.

It should be noted that the sensitivity discussed here is the intrinsic chemical sensitivity of a nearly perfect material. Several extrinsic factors play a role in determining the sensitivities of real (practical) materials. However, the intrinsic behavior underlies the extrinsic, so it is important in the overall behavior.

ACKNOWLEDGMENTS

The author is grateful to the Sutter Foundation for financial support; and to to J. Sharma for bringing the paper of F. J. Owens to his attention.

REFERENCES

1. G. L. Bir and G. E. Pikus, *Symmetry and Strain-Induced Effects in Semiconductors*, Halsted Press, (division; John Wiley & Sons), NewYork (1972).
2. M. Cardona, Phys. Stat. Sol. B, **198**, 5 (1996).
3. J. E. Aubrey, W. Gubler, T. Heningsen, and S. H. Koenig, Phys. Rev., **130**, 1667 (1963).
4. J. C. Phillips, Vol. I, Chap. 2, p.47 in *Handbook on Semiconductors*, Revised Edition, Ed. by T. S. Moss, Elsevier Sci. Publ. B. V. (1992).
5. A. L. Ruoff and T. Li, Annu. Rev. Mat. Sci., **25**, 249 (1995).
6. J. C. Jamieson, Science, **139**, 845 (1963).
7. J. J. Gilman, Phil. Mag. B, **67**, 207 (1993).
8. T. S. Moss, *Photoconductivity in the Elements*, p. 61, Butterworths, London (1952).
9. P. W. Atkins, *Molecular QuantumMechanics*, 2nd Edition, p.354, Oxford University Press, Oxford (1983).
10. A. V. Belik, V. A. Potemkin, and N. S. Zefirov, Dokl. Akad. Nauk SSSR, **308**, 882 (1989).
11. F. J. Owens, Jour. Molec. Struc. (Theochem), **370**, 11 (1996).
12. J. J. Gilman, Czech. Jour. Phys., **45**, 913 (1995).
13. J. J. Gilman, Phil. Mag. B, **71**, 1057 (1995).
14. K. F. Herzfeld, Phys. Rev., **29**, 701 (1927).

CP429, *Shock Compression of Condensed Matter – 1997*
edited by Schmidt/Dandekar/Forbes
© 1998 The American Institute of Physics 1-56396-738-3/98/$15.00

MODELING THE DETONATION STRUCTURE OF HETEROGENEOUS EXPLOSIVES

J.P. Dionne, J.H.S. Lee

Dept. of Mechanical Engineering, McGill University, Montréal, Québec, Canada, H3A 2K6

A simplified ZND calculation for the one-dimensional detonation (infinite diameter) of heterogeneous explosives is proposed. The effects of thermal relaxation within the two-phase products is incorporated into a source term in the chemical rate law. The explosive is then approximated as a homogeneous mixture of two phases. The source term is based on the rate of heat transfer from the liquid explosive to the inert heterogeneities. The effect of the properties of the heterogeneities (size, heat capacity, density) on this additional term is discussed.

INTRODUCTION

For homogeneous explosives, the Chapman-Jouguet (CJ) hypothesis leads to an adequate prediction of the detonation properties in general, if the equations of state for the condensed products are known.

However, for heterogeneous explosives, and especially those containing a large amount of inert solid material, Dionne (1) found that the measured detonation properties deviate significantly from the equilibrium CJ predictions.

A model non-ideal heterogeneous explosive composed of chemically sensitized nitromethane (NM) in a packed-bed of inert glass beads (GB) has been studied extensively by J.J. Lee (2). This relatively simple NM+GB system can be characterized by two length scales, namely the chemical length scale of the liquid explosive component, which depends on the degree of sensitization of the NM, and the bead size, which describes the physical characteristic of the inert particles.

The detonation velocities measured for this model explosive, for different bead sizes, were of the order of 4.5 km/s. However, the computed equilibrium CJ detonation velocity for this explosive mixture is only 3.3 km/s, as calculated

with the IDeX thermodynamic equilibrium code (3). This represents a deviation of about 26%.

A time scale analysis reveals that the equilibrium assumption as required by the CJ detonation calculation breaks down for this model explosive of NM+GB. The thermal relaxation time scale associated with the inert particles is found to be orders of magnitude larger than the chemical and mechanical relaxation time scales. As a result, the detonation products are not in thermal equilibrium when the sonic condition is reached. The experimental detonation velocity is thus larger than the CJ predictions, since the effective energy available to the fluid phase to drive the precursor shock is higher.

THE DETONATION STRUCTURE

To account for the different relaxation time scales in heterogeneous explosives, a non-equilibrium model for the detonation structure must be used. The ZND detonation structure describes the time-history of all the variables of interest within the reaction zone of the explosive. It is obtained through the integration of the conservation equations from the shock front to the sonic plane. Appropriate equations of state for the explosive, and a chemical rate law must be used. These equations are written with respect to a fixed shock front.

The relaxation rates of momentum and heat exchange between the phases may be considered in a ZND calculation. These relaxation rates can be modeled as source terms in the conservation equations. The simplified approach proposed in the present work to model the heterogeneous mixture consists of lumping all the heterogeneous effects into the source term for the thermal relaxation. Using this assumption, the conservation equations for a homogeneous mixture are used:

$$\frac{d(\rho u)}{dx} = m \qquad (1)$$

$$\frac{d}{dx}\left(P + \rho u^2\right) = Dm - f \qquad (2)$$

$$\frac{d}{dx}\left[u\left(P + \rho e\right)\right] = \frac{1}{2}D^2 m - Df + q \qquad (3)$$

Equations (1-3) are the mass, momentum and energy conservation equations in differential form respectively. The variables P, ρ, u and e are thus respectively the phase-averaged values for the pressure, the density, the particle velocity and the internal energy. The variable D is the velocity of the shock, and x is the distance from the shock front. The source term in the mass equation, m, can account for the divergence of the flow in a finite charge explosive, the momentum source term f can account for the frictional drag between the two phases and the energy source term q can be used to model thermal non-equilibrium. The terms Dm, $1/2 D^2 m$ and Df are due to the choice of a travelling reference frame.

To study the effect of thermal relaxation, the simple case of an infinite diameter explosive (m=0) assuming negligible internal friction (f=0) is considered. A polytropic equation of state (EOS) will be used, along with a simple single-step Arrhenius rate law.

Since there are no source terms in the mass and momentum equations, the detonation process follows a Rayleigh line within the reaction zone. The pressure and the density of the products are therefore uniquely determined for a given particle velocity. The steady-state ZND structure equations based on the above assumptions can then be written as a set of two coupled ordinary differential equations

(ODE) in terms of the extent of chemical reaction λ, and the particle velocity u:

$$\frac{\partial \lambda}{\partial t} = \dot{\lambda} = \left(1 - \lambda\right) Z \exp\left(\frac{-E_a}{RT_s}\right) \qquad (4)$$

$$\frac{\partial u}{\partial t} = \frac{\left[\dot{\lambda} Q_{max} + \frac{u}{\rho_o D}\, q\right]}{u\left[\frac{\gamma + 1}{\gamma - 1}\right] - \frac{\gamma D}{\gamma - 1}} \qquad (5)$$

$$T_{nm} = T_s + \lambda\left(\frac{Q}{C_v}\right) \qquad (6)$$

where Z and E_a are constants fitted for a given explosive, R is the universal gas constant, D is the detonation velocity, γ is the ratio of the specific heats, T_{nm} is the products temperature, T_s is the shock temperature, C_v is the average heat capacity of the detonation products, λ is the chemical rate law and Q_{max} is the maximum available heat release for the fluid detonation products if no heat was transferred to the inert particles.

The form of the numerator, where the heat release by chemical reactions ($\dot{\lambda} Q_{max}$) is added to the thermal relaxation heat transfer term ($\frac{u}{\rho_o D}\, q$) suggests a generalized form for a chemical rate law encompassing the effect of thermal relaxation. Equations (4) and (5) can be rewritten as :

$$\frac{\partial Q}{\partial t} = \dot{Q} = \dot{Q}_{Chemical} + \dot{Q}_{Thermal\ Relaxation} \qquad (7)$$

$$\frac{\partial u}{\partial t} = \frac{\dot{Q}}{u\left[\frac{\gamma + 1}{\gamma - 1}\right] - \frac{\gamma D}{\gamma - 1}} \qquad (8)$$

where $\dot{Q}$ is the generalized rate law and $\dot{Q}_{Chemical} = \dot{\lambda} Q_{max}$. These two ODE's can then readily be integrated from the shock front to the sonic plane. The boundary conditions (BC) at the shock front are :

$$Q = 0 \qquad u = D\left[\frac{\gamma - 1}{\gamma + 1}\right]$$

The BC for the particle velocity is obtained from the shock wave theory, by considering a strong shock wave. The BC for Q comes from the fact that the shock is reactionless and considered infinitely thin, so that no heat transfer has occurred yet.

At the sonic plane, the denominator of the ODE for the particle velocity vanishes. The singularity can be removed if the numerator also goes to zero at that point. These two conditions are called the generalized Chapman-Jouguet conditions. The numerator being equal to zero implies the following heat balance :

$$\dot{Q}_{\text{Chemical}} = -\dot{Q}_{\text{Thermal}\atop\text{Relaxation}}$$

As a result, $\dot{Q}_{\text{Chemical}}$ is not equal to zero at the sonic plane. The available chemical energy is not completely released, which causes a reduction in detonation velocity. The incorporation of the additional source term in the chemical rate law therefore has an effect on the detonation properties, as calculated at the sonic plane.

THE THERMAL RELAXATION TERM

For the purpose of modeling the heat transfer from the detonation products to the inert particles, the particles are considered as lumped heat capacity systems. It is thus assumed that the temperature throughout the particles is uniform (i.e. negligible conduction time scale). The differential equation for the heat transfer process can be written as follows :

$$\frac{dT}{dt} = \frac{hA}{C\rho V}\left(T_{\text{nm}} - T\right) = k\left(T_{\text{nm}} - T\right) \quad (9)$$

$$\text{with} \quad k = \frac{hA}{C\rho V} \quad (10)$$

where h is the heat transfer coefficient, A, T, V, C, and ρ are respectively the area, the temperature, the volume, the heat capacity and the density of a single inert particle. The parameter k has inverse time units. The heat loss to the inert particles can then be calculated as :

$$Q_{\text{Thermal}\atop\text{Relaxation}} = C_{\text{particles}}\left(T - T_{\text{o}}\right)$$

where $C_{\text{particles}}$ is the heat capacity of the inert particles. Upon differentiation with respect to time, the following expression for the heat transfer rate to the inert particles is obtained:

$$\dot{Q}_{\text{Thermal}\atop\text{Relaxation}} = C_{\text{particles}}\frac{dT}{dt}$$

or

$$\dot{Q}_{\text{Thermal}\atop\text{Relaxation}} = k\,C_{\text{particles}}\left(T_{\text{nm}} - T\right)$$

RESULTS AND DISCUSSION

Equations (7), (8) and (9) have been integrated within the reaction zone for different detonation velocities below the maximum velocity (D_{max}) corresponding to no heat transfer to the inert particles. For each velocity, the numerical code iterated for the corresponding value of the parameter k that leads to the satisfaction of the generalized CJ criterion. Figure 1 shows this k-D relationship. Although the curve is normalized with respect to D_{max}, it is dependent on the relative values used for the parameters such as those in the chemical rate law and the thermal relaxation term.

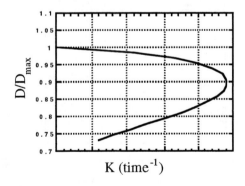

FIGURE 1. Relationship between the iterated value of the heat transfer parameter k and the normalized detonation velocity.

It can be seen from this graph, that the k function is not single valued. As k is proportional to the inverse of the particle size, it is expected that as it is increased (decrease in particle size), the heat transfer to the particles becomes more important, thus causing a reduction in the detonation velocity. This corresponds to the upper part of the k-D curve. On this basis, one may claim that the lower part of the curve is non-physical. It may correspond to unstable solutions.

The parameter k being also inversely proportional to the heat capacity and the density of the particles, it can be concluded that an increase in these properties will cause an increase in the detonation velocity, if we only consider the upper part of the k-D curve.

When k=0 ($D=D_{max}$), the heat release rate is maximum, and the heat release time is minimum. For higher values of k (lower detonation velocities), the peak in heat release is lowered, and the heat release time is increased. Figure 2 shows the effective heat release rate for three different normalized detonation velocities.

When k=0, there is no heat loss to the inert particles. Their temperature thus remains at their initial value just downstream of the shock. As the parameter k is increased, more heat is transferred to the particles. This results in an elevation of their temperature at the sonic plane. This effect is shown on Fig. 3, where the temperature of the particles is non-dimensionalized with respect to the temperature of the fluid products at the sonic plane. This temperature function is single-valued.

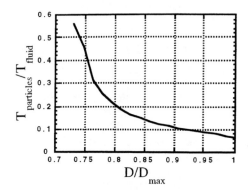

FIGURE 3. The ratio of the temperatures of the inert particles to the temperature of the fluid detonation products at the sonic plane, as a function of the normalized detonation velocity.

CONCLUSION

The present analysis indicates that for a multi-phase explosive, the delay in thermal relaxation of one of the phases can result in a detonation velocity higher than the equilibrium Chapman-Jouguet value. This is in accord with the experimental observation for a heterogeneous mixture of glass beads and nitromethane where the observed detonation velocity is of the order of 4.5 km/s whereas the equilibrium Chapman-Jouguet velocity assumed thermal equilibration of the glass beads and explosion products is only 3.3 km/s.

Quantitative modeling of the thermal equilibration effect would require the knowledge of the appropriate chemical and thermal equilibration rate laws, as well as the equations of state of the shocked phases and the condensed products.

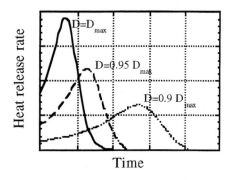

FIGURE 2. The effective heat release rate $\dot{Q}$ as a function of time for different values of the normalized detonation velocity.

REFERENCES

1. Dionne, J.P., "Chapman-Jouguet Properties of Heterogeneous Explosives", Master's Thesis, McGill University, 1996.
2. Lee, J.J., "Detonation Mechanism in a Condensed Phase Porous Explosive", Ph.D. Thesis, Université de Sherbrooke, 1997
3. Freeman, T. L., I. Gladwell, Braithwaite, M., Byers Brown, W., Lynch, P.M., Barker,I.B., *Math. Engng Ind.* 3(2): 97-109, 1990.

CP429, *Shock Compression of Condensed Matter – 1997*
edited by Schmidt/Dandekar/Forbes
© 1998 The American Institute of Physics 1-56396-738-3/98/$15.00

SOLVING CURVED DETONATION RIEMANN PROBLEMS

Bruce Bukiet

Department of Mathematics, Center for Applied Mathematics and Statistics
New Jersey Institute of Technology, Newark, NJ 07102

It is desirable to compute accurate solutions to detonation problems without numerically solving differential equations in the thin reaction zone. For planar detonation waves, algebraic jump conditions can be used to compute the one parameter family of behind states comprising the burned Hugoniot and wave curves. For curved detonations, the state at the end of the reaction zone depends on both the detonation velocity and the curvature of the wave. Recently, curved detonation jump conditions have been derived which contain "source terms" involving front curvature, reaction zone length and integrals of physical quantities in the reaction zone. In this paper, we parameterize the source terms by the detonation velocity and curvature and show how the curved detonation jump conditions can be solved to compute the curved detonation Hugoniot. We demonstrate the method using the Forest Fire rate law with HOM equation of state for the explosive PBX-9404.

INTRODUCTION

In detonation wave computations involving curved detonation fronts, accurate solutions can be obtained by resolving the reaction zone. That is, the partial differential equations for mass, momentum and energy can be solved at each time step on a fine grid. However, using fine grids in the reaction zone requires small time steps and leads to expensive computations. Thus, it is desirable to compute solutions to detonation problems without numerically resolving the reaction zone.

For planar detonation waves, algebraic jump conditions which do not depend on the dynamics within the reaction zone can be used to compute the burned Hugoniot and wave curves. One can then solve the Riemann problem. Solving differential equations through the reaction zone is avoided. There is a one parameter family of physical behind states to which an ahead state may

be connected by a planar detonation. The behind states can be parameterized by the detonation velocity. The Random Choice[1-3] and Front Tracking[4-6] Methods have been applied to detonation problems[7-9] taking advantage of these jump conditions.

For curved detonations, the state at the end of the reaction zone depends on both the detonation velocity and the curvature of the wave. Recently, curved detonation jump conditions have been derived[10]. However, these jump conditions contain two "source terms" which involve front curvature, reaction zone length and integrals of physical quantities in the reaction zone.

We parameterize the source terms by the detonation velocity and curvature. The source terms are linear in curvature, κ, and an exponential fit vs. detonation velocity is used. We show how the curved detonation jump conditions can be solved to compute the curved detonation Hugoniot. By finding the Hugoniot (and the related wave curve), Riemann problems involving curved detonations can be solved. We demonstrate the method using the Forest Fire [11] rate law with HOM equation of state [12] for the explosive PBX-9404.

CURVED DETONATION JUMP CONDITIONS

We start with the Euler equations for conservation of mass, momentum and energy and an equation for reaction progress in curved geometry.

$$
\begin{aligned}
(\rho A)_t + (\rho A u)_x &= 0 \\
(\rho A u)_t + \left(\rho A \left[u^2 + PV\right]\right)_x &= PA_x \\
(\rho A \mathcal{E})_t + (\rho A u \left[\mathcal{E} + PV\right])_x &= -PA_t \\
(\rho A \lambda)_t + (\rho A u \lambda)_x &= \rho A \mathcal{R} \quad (1)
\end{aligned}
$$

where ρ, u, E and λ are the fluid density, particle velocity, specific internal energy and mass fraction of the reaction products ($\lambda = 0$ unburnt, $\lambda = 1$ completely burnt), $V = 1/\rho$ is the specific volume, $\mathcal{E} = u^2/2 + E$ is the total specific energy, $P(V, E, \lambda)$ is the pressure, $\mathcal{R}(V, E, \lambda)$ is the specific reaction rate, and A is the cross-sectional area.

If the flow in the reaction zone is quasi-steady, the PDEs can be reduced to the following ODEs.

$$
\begin{aligned}
\left[\rho v\right]_x &= \rho u \kappa \\
\left[\rho v^2 + P\right]_x &= \rho v u \kappa \\
\left[E + PV + \frac{v^2}{2}\right]_x &= 0 \\
-v \lambda_x &= \mathcal{R} \quad (2)
\end{aligned}
$$

where $v = D - u$, $\kappa = A/A_x$ is the curvature and x is the distance relative to the front. In this form, the influence of the curvature of the wave on the standard planar jump conditions is apparent. Integrating through the reaction zone yields the following set of jump conditions.

$$
\begin{aligned}
\Delta[\rho v] &= \kappa w \langle \rho u \rangle \\
\Delta \left[(\rho v)^2 V + P\right] &= \kappa w \langle \rho v u \rangle \\
\Delta \left[E + PV + \frac{v^2}{2}\right] &= 0 \quad (3)
\end{aligned}
$$

where $\Delta[f] = f(x_a) - f(x_b)$ is the change of variable f across the detonation reaction zone. The subscripts a and b denote ahead of the detonation front and behind the reaction zone, respectively. The reaction zone width is $w = x_a - x_b$ and

$$
\langle f \rangle = \frac{1}{w} \int_{x_1}^{x_0} dx \, f \quad (4)
$$

is the average value of f in the reaction zone.

PARAMETRIZING THE SOURCE TERMS

In order to solve the jump conditions (3) for the Hugoniot curve, we must find explicit approximations for the reaction zone quantities $w \langle \rho u \rangle$ and $w \langle \rho v u \rangle$. Thus, the ODEs in equations (2) must be solved. First the ODEs are put into the equivalent form

$$
-[c^2 - (D-u)^2] \frac{d}{dx} \begin{pmatrix} V \\ D-u \\ \lambda \\ P \end{pmatrix} = \begin{pmatrix} [c^2 \sigma \mathcal{R} - v^2 u \kappa] V/v \\ c^2 (\sigma \mathcal{R} - u \kappa) \\ (c^2 - v^2) \mathcal{R}/v \\ -\rho \, v \, v_x \end{pmatrix} \quad (5)
$$

Here, $c^2 = \partial_\rho P|_{S, \lambda}$ is the square of the frozen sound speed, and $\sigma = (\rho c^2)^{-1} \partial_\lambda P|_{V, E}$ is the thermicity.

The ZND model for detonations is employed. The lead front shocks the material up to a high pressure (von Neumann spike). Planar jump conditions without reaction are used, and then the ODEs (5) are solved through the reaction zone. The state at the end of the reaction zone depends on both the detonation velocity and the front curvature, κ. (The HOM equation of state [12] for PBX-9404 with the Forest Fire reaction rate parameters [11] was used for the computations discussed below. The the Forest Fire rate is derived from distance of run to detonation experiments (Pop plots). It is computed for values of pressure below the Chapman-Jouguet (CJ) point and the rate blows up above CJ. A spline was fit to the log of the rate and extrapolated to pressures above CJ.)

The source terms in equations (3) are linear in κ. This was verified by performing computations with constant detonation velocity while varying κ. In Fig. 1, the other portion of the source terms are plotted. The values of these terms for a given detonation speed were computed for several values of κ (corresponding to both converging and diverging detonations) and averaged. The range of values for D was from slightly below CJ ($D = 0.886$ cm/μsec) up to about 8% above CJ. Detonation velocities below CJ apply only to diverging waves. The values decay exponentially over the region studied. A good fit is obtained using the functions

$$w\langle\rho u\rangle \approx 0.000852 e^{-22.15(D-0.9123)}$$
$$w\langle\rho uv\rangle \approx 0.000492 e^{-22.4(D-0.9123)} \quad (6)$$

SOLVING THE JUMP CONDITIONS

In this section, we describe how to find the state behind a curved detonation wave from the jump conditions (3) using the source term fits (6). Thus, we need to solve

$$\Delta[\rho v] = \kappa F_1$$
$$\Delta\left[(\rho v)^2 V + P\right] = \kappa F_2$$
$$\Delta\left[E + PV + \frac{v^2}{2}\right] = 0 \quad (7)$$

where F_1 and F_2 are the right hand sides of (6).

A range for the density behind the reaction zone for our PBX-9404 example is 2-3 gm/cm^3. Given a detonation speed and curvature, we iterate on ρ_I (I for iterate) as follows. $V_I = 1/\rho_I$. The first and second of equations (7) give

$$v_I = (\rho_a v_a - \kappa F_1)V_I$$
$$P_I = \rho_a v_a^2 + P_a - \rho_I v_I^2 - \kappa F_2 \quad (8)$$

and we compute

$$E(\rho_a, P_a, 0) + P_a V_a + \frac{v_a^2}{2} - E(\rho_I, P_I, 1) - P_I V_I - \frac{v_I^2}{2} \quad (9)$$

We iterate using the secant method until the expression (9) is zero.

A comparison of the results of solving the full ODEs and the jump conditions is presented in

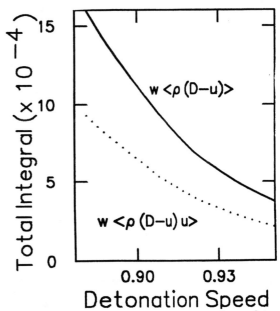

FIGURE 1. A plot of source term integrals vs. detonation velocity. The solid curve represents the source term for the mass equation and the dotted curve represents the source term for the momentum equation. The full source terms are multiplied by the curvature κ.

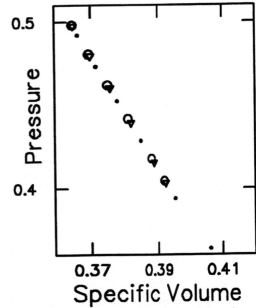

FIGURE 2. Hugoniot points computed by solving the ordinary differential equations through the reaction zone (dots for planar, circles for $\kappa = 3$) and using the jump conditions (triangles for $\kappa = 3$). Using the jumps conditions is more than 1000 times faster.

323

Fig. 2. The dots represent points on the planar Hugoniot curve; the circles represent Hugoniot points for $\kappa = 3$ computed using the full ODEs; the triangles represent Hugoniot points for $\kappa = 3$ computed using the jump conditions. The lowest dot corresponds to the lowest circle and triangle. The second lowest dot corresponds to the second lowest circle and triangle, etc. The $\kappa = 3$ and planar Hugoniots are almost indistinguishable. However, for a given detonation speed, the Hugoniot values are quite different (especially near CJ) and the jump conditions capture this difference accurately. Solving the ODEs for the HOM equation of state takes over 20 CPU seconds for a single Hugoniot point. The jump conditions can compute 100 Hugoniot points in 1 CPU second.

DISCUSSION AND CONCLUSION

We have shown how one can efficiently find the Hugoniot curve for curve detonations once the source terms are parametrized. The secant method solution can solve for a state behind the reaction zone more than 1000 times as fast as solving using the ODEs. Thus, in a computation in which Riemann problems are to be solved at each time step (such as in a Front Tracking framework), the preprocessing involved in order to find the jump correction terms (6) is well worth the effort.

Although we have used a realistic equation of state and rate law, the rate law is probably too high in the ZND reaction zone. The form of the rate law may have given rise to the exponential fit for the source terms. This may not be the case in general. In the future, we plan to use a model for burning on the way up to CJ pressure. The theory is also valid for such models.

ACKNOWLEDGMENTS

The author would like to thank R. Menikoff and K. Lackner for useful discussions. The author also acknowledges support from NSF grant DMS-9508298 for computer equipment.

1. Glimm, J., *Comm. Pure Appl. Math* **18**, 695-715 (1965).

2. Chorin, A. J., *J. Comp. Phys.* **22**, 517-533 (1976).

3. Moler, C., and Smoller, J., *Arch. Rat. Mech. Anal.* **37** 309-322 (1970).

4. Glimm, J., Klingenberg, C., McBryan, O., Plohr, B., Sharp, D. and Yaniv, S., *Adv. Appl. Math.* **6** 259-290 (1985).

5. Chern, I.-L., Glimm, J., McBryan, O., Plohr, B., and Yaniv, S., *J. Comp. Phys.* **62** 83-110 (1986).

6. Glimm, J., Grove, J., Li, X. L., Shyue, K., Zeng, Y., and Zhang, Q., *SIAM J. Sci. Comp.* To appear.

7. Chorin, A. J., *J. Comp. Phys.* **25**, 253-272 (1977).

8. Saito, T. and Glass, I. I., *Prog. Aerospace Sci.* **21**, 201-247 (1977).

9. Bukiet, B. *SIAM J. Sci. Stat. Comp.* **9**, 80-99 (1988).

10. Menikoff, R., Lackner, K. and Bukiet, B., *Comb. and Flame* **104** 219-240 (1996).

11. Johnson, J., Tang, P., and Forest, C., *J. Appl. Phys.* **57** 4323-4334 (1985).

12. Mader, C., *Numerical Modeling of Detonation*, Los Angeles: University of California Press, 1979.

CP429, *Shock Compression of Condensed Matter – 1997*
edited by Schmidt/Dandekar/Forbes
© 1998 The American Institute of Physics 1-56396-738-3/98/$15.00

Size Effect and Detonation Front Curvature

P. C. Souers and Raul Garza

Energetic Materials Center, Lawrence Livermore National Laboratory, Livermore, California 94550 (USA)

Explosive sonic reaction zone lengths are obtained from two sources: the size effect and detonation front curvature, where the edge lag is close to being a direct measure. The curvature comes from a constant energy source plus extra energy released near the walls. The presence of defects can eliminate the central flow of transverse energy to the walls and create a turbulent central section in small reaction zone explosives.

THE SIZE EFFECT

In the size effect in cylinders, the detonation velocity declines with decreasing radius. We assume that all energy is lost out of the side of a cylinder in a skin layer of thickness R_e, and that energy flows in from the axis to make up for it. If energy is lost out the side of the cylinder, we suggest:[1]

$$\left(\frac{<E_o>}{E_o} \right)^{1/2} = \frac{U_s}{D} = 1 - \frac{<x_e>}{\sigma R_o} \quad (1)$$

where $<E_o>$ and U_s; E_o and D are the detonation energies and velocities in cylinders of radius R_o and infinite size. Also, $<E_o>$ is an average value taken across the cylinder. The skin layer is related to the average sonic reaction zone thickness, $<x_e>$, by $R_e = <x_e>/\sigma$. Also, σ is a wall expansion function always larger than one, empirically set for unconfined samples to

$$\sigma = 11 exp \left(-8 \frac{<x_e>}{R_o} \right) + 2 \quad (2)$$

Eq. 2 was created empirically by considering the probable reaction zone lengths of many explosives. For metal-confined samples, the amplitude is set to 22.

The infinite radius D is obtained from an extrapolation of the data versus the inverse radius.

This works for pure and heterogeneous explosives, but is almost certainly wrong for composites, where D is variable.

Table 1 lists the reaction zone lengths obtained from the size effect as well as the measured edge lags, L_o, from detonation front curvature.[2-6] The treatment above suggests the interesting relation

$$< x_e > \approx L_o. \quad (3)$$

DETONATION FRONT CURVATURE

We have found two types of curves, both shown in Figure 1 for LX-04.[6] The smaller radius curves are of the "quadratic" type, which are rounded and fairly smooth everywhere. The larger-radius curve is of the "non-quadratic" type, where the center is flat or turbulent with steep sides.

The more common "quadratic" curve becomes steeper in the last 20% of the radius near the edges. The lag is better described by the equation

325

$$L = AR^2 + BR^6 \qquad (4)$$

where the quadratic term dominates and the fourth power term has a 10-15% effect at the edges.

We now consider the detonation front, using the mathematics for uniform heat flow in a cylinder with an internal heat source, ie. the temperature in a radioactive fuel rod.[7] We replace temperature with the detonation front lag, L, as the cause of an energy flow from the cylinder center to the edge. The "thermal conductivity" becomes a heat flow

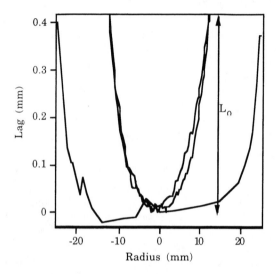

FIGURE 1. Example of quadratic and non-quadratic detonation fronts for the same explosive, LX-04, at two radii of 12.7 and 25.4 mm. The detonation is proceeding downward. The edge lag, L_o, of the 12.7 mm cylinders is shown.

constant, K, with the units MW/mm^2. If R is the radius, we have

$$\nabla^2 L = \frac{1}{R}\frac{\partial}{\partial R}\left(R\frac{\partial L}{\partial R}\right) = \frac{A_o}{K}, \qquad (5)$$

where A_o is the power released per unit volume. If the power were constant, then the calculated lag would be purely quadratic. The power represents energy being sent sideways to sustain the moving edge, so that it will be a fraction of E_o.

The main power source comes from the sideways energy, $E_o - <E_o>$, which is, from Eq. 1

$$E_o - <E_o> \approx \left[1 - \left(1 - \frac{<x_e>}{\sigma R_o} \right)^2 \right] E_o = \frac{2<x_e>E_o}{\sigma R_o} \qquad (6)$$

The energy must be divided by the time to cross the reaction zone ($<x_e>/U_s$) to get power. We add the higher-power term from Eq. 4 to get the average power delivered across the cylinder of

$$\frac{A_o}{K} \approx \frac{2U_sE_o}{K\sigma R_o} + 36BR^4 \qquad (7)$$

E_o is the detonation energy let loose in the reaction zone. This is the thermochemical total only for nearly-homogeneous explosives. For PBXN-111, for example, the aluminum and AP probably do not react within the reaction zone, so that $E_o \approx 2$ kJ/cc whereas the complete reaction produces 13 kJ/cc. From the first term in Eq. 7, the thermal conductivity is

$$K = \frac{U_sE_o}{2A\sigma R_o}. \qquad (8)$$

The best data in Table 1 is for the TATB explosives where $K \approx 80\ MW/mm^2$. This may be the limiting value for fine-grained nearly-homogeneous explosives. PBXN-111, where much of the material is inert in the reaction zone, has a thermal conductivity of about 10 MW/mm^2.

The non-quadratic curves, all of which come from small reaction-zone explosives, show a ragged center section, as seen in Figure 2. We postulate that defects such as inert grains or large voids scatter the front so that only the energy from the outermost regions supports the edges. If the center turbulence

TABLE 1. Summary of reaction zone data taken from detonation front curvature and the size effect.

"QUADRATIC" unconfined Explosive	ρ_o (g/cc)	R_o (mm)	Lag, L_o (mm)	SizeE $<x_e>$ (mm)	U_s (mm/ μs)	D (mm/ μs)	A (mm^{-1})	B (mm^{-5})	K (W/ mm^2)	ref.
PBX-9502	1.890	5.0	0.7	0.9	7.455	7.78	1.7E-02	1.9E-05	77	2
PBX-9502	1.890	5.0	0.8	0.9	7.458	7.78	1.6E-02	2.2E-05	79	2
PBX-9502	1.890	6.0	0.8	1.0	7.495	7.78	1.3E-02	5.7E-06	79	2
PBX-9502	1.890	9.0	1.0	1.4	7.553	7.78	7.5E-03	6.6E-07	87	2
T2	1.855	25	1.9	2.5	7.620	7.65	2.3E-03	2.1E-09	78	3
PBX-9502	1.890	25.0	2.1	2.5	7.677	7.78	2.0E-03	2.8E-09	87	2
PBX-9502	1.890	25.0	2.2	2.5	7.672	7.78	2.2E-03	2.5E-09	80	2
T2	1.855	50	2.9	3.0	7.633	7.65	9.2E-04	4.4E-11	75	3
NM	1.124	9.57	0.9	0.3	6.208	6.24	6.3E-03	3.5E-07	28	4
NM	1.124	13.78	0.8	0.2	6.229	6.24	1.8E-03	6.6E-08	61	4
NM	1.124	18.42	0.9	0.2	6.230	6.24	9.5E-04	1.2E-08	84	4
NM-guar	1.171	5.26	1.0	1.2	5.800	6.55	3.0E-02	7.3E-06	27	4
NM-guar	1.171	6.80	1.0	0.9	6.090	6.55	1.5E-02	2.9E-06	29	4
NM-guar	1.171	9.57	1.0	1.1	6.128	6.55	6.0E-03	5.7E-07	46	4
NM-guar	1.171	18.59	1.1	1.9	6.140	6.55	9.6E-04	1.7E-08	138	4
PBXN-111	1.790	20.45	4.8	6	5.154	5.81	9.6E-03	1.2E-08	9	5
PBXN-111	1.790	20.52	4.3	6	5.158	5.81	8.5E-03	1.2E-08	10	5
PBXN-111	1.790	24.01	4.8	6	5.309	5.81	8.2E-03	2.7E-09	8	5
PBXN-111	1.790	24.06	4.9	6	5.311	5.81	6.5E-03	6.2E-09	10	5
PBXN-111	1.790	34.12	5.7	7	5.572	5.81	3.6E-03	1.0E-09	11	5

NON-QUADRATIC (all confined in copper)

Explosive	ρ_o	R_o	L_o	$<x_e>$	U_s	D	A	B	K	ref.
LX-14	1.825	12.7	0.16		8.778		not used	4.1E-08		6
PETN	1.743	12.7	0.10		8.201		not used	2.4E-08		6
LX-04	1.87	25.4	0.39		8.470		not used	4.5E-08		6
LX-10	1.87	25.4	0.26		8.820		not used	8.9E-10		6

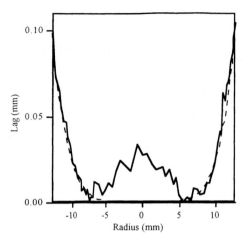

FIGURE 2. Detonation front curvature of 1.76 g/cc PETN taken with a writing speed of 30 mm/μs. The outer lags fit BR^4.

is not extreme, the lag may be described by

$$L = BR^4. \qquad (9)$$

Fig. 2 also shows a real inversion effect with a center lag of 0.03 mm. The reason is unknown, although lack of flatness of the end face of the cylinder or density gradients could be a cause.

D. Roberts has found that LX-14 has not smoothed out the placement of detonators 10 mm in, even though many reaction zone lengths have been covered.[8] This suggests that the turbulent sections might retain a memory of past ignition history whereas quadratic explosives will smooth out their fronts.

However imperfect, the reaction zones form a source of kinetic data that may be used to calibrate upcoming time-dependent codes.[9]

ACKNOWLEDGEMENTS

We wish to thank John Bdzil and Jerry Forbes for their kind assistance in supplying curvature data from their laboratories.. This work was performed under the auspices of the U. S. Department of Energy by the Lawrence Livermore National Laboratory under contract number W-7405-ENG-48.

REFERENCES

1. P. C. Souers, "Size Effect and Detonation Front Curvature," *Propellants, Explosives, Pyrotechnics*, to be published.

2. Los Alamos National Laboratory, Los Alamos, NM, unpublished data, courtesy John Bdzil, private communication, 1996.

3. F. Chaisse' and J. N. Oeconomos, "The Shape Analysis of a Steady Detonation Front in Right Circular Cylinders of High Density Explosive. Some Theoretical and Numerical Aspects," *Proceedings Tenth Symposium (International) on Detonation, Boston, MA, July 12-16, 1993*, pp. 50-57.

4. R. Engleke and J. B. Bdzil, *Phys. Fluids* **26**, 1210 (1983).

5. J. W. Forbes, E. R. Lemar, G. T. Sutherland and R. N. Baker, Detonation Wave Curvature, *Corner Turning and Unreacted Hugoniot of PBXN-111*, Naval Surface Warfare Center Report NSWCDD/TR-92/164, Silver Spring, MD, 1992.

6. LLNL Cylinder Test.

7. H. S. Carslaw and J. C. Jaeger, *Conduction of Heat in Solids*, 2nd ed. (Clarendon Press, Oxford, 1959), pp. 130-132, 191, 232.

8. D. Roberts, Lawrence Livermore National Laboratory, private communication, 1997.

9. L. E. Fried, W. M. Howard and P. C. Souers, "Kinetic Modeling of the Non-Ideal Explosives with CHEETAH," *J. Chem. Phys.*, to be submitted.

CP429, *Shock Compression of Condensed Matter – 1997*
edited by Schmidt/Dandekar/Forbes
© 1998 The American Institute of Physics 1-56396-738-3/98/$15.00

DETONATION HUGONIOT FOR OZONE FROM MOLECULAR DYNAMICS SIMULATIONS

J. J. C. Barrett [a], **D. H. Robertson** [b], **M. L. Elert** [c], and **C. T. White** [d]

[a] *Naval Research Laboratory, Washington, DC 20375–5000*

[b] *Indiana University - Purdue University at Indianapolis, Indianapolis, IN 46202*

[c] *U. S. Naval Academy, Annapolis, MD 21402–5026*

[d] *Department of Materials, University of Oxford, Parks Road, Oxford, OX11 3PH, UK*

We have developed a reactive empirical bond order potential for the ozone system for use in molecular dynamics (MD) studies of shock-induced chemistry. Simulations using this potential have been shown to exhibit the essential characteristics of a chemical detonation. In this paper we examine the detonation Hugoniot for the ozone system directly from a series of MD simulations of supported detonations. This approach avoids having to determine the equation of state of the system from the model potential. The state of the system at the CJ point and detonation velocity are determined from the detonation Hugoniot.

INTRODUCTION

Over the past several years we have established that it is feasible to employ molecular dynamics in the detailed study of chemically sustained detonations using reactive empirical bond order (REBO) potentials (1-4). In our studies of chemical detonations we have developed several model potentials based on a generic AB diatomic molecule (1-2). We have used these models to study a wide variety of shock-induced phenomena including shock-wave splitting, which displays a dissociative phase transition (2) and chemically sustained detonations (2,3). We have also developed a REBO potential for a prototypical energetic material, ozone (4). Molecular dynamics studies with our model ozone potential show that this system exhibits a self-sustained chemical detonation. The versatility of the REBO potential is further revealed

in studies with a REBO potential optimized for hydrocarbon systems. Molecular dynamics studies using this potential explore a variety of chemical systems and mechanical processes at the molecular level including nanotribology and lubrication (5), nanoindentation (6), and fullerene reactive scattering (7).

Molecular dynamics simulations show that when the ozone crystal is shocked above the initiation threshold to detonation a near steady flow detonation wave is quickly established and propagates through the material at a constant detonation velocity of 5.3 km/s. In Figure 1 we show profiles of the local particle flow velocity in the direction of shock front propagation, u_x, local particle density, ρ, and the pressure, P, at several times during the simulation. These plots show that in a short simulation time the properties of the detonation quickly stablize near the front.

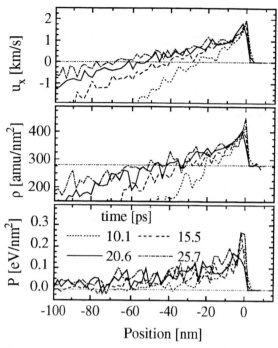

Figure 1. Detonation properties from ozone detonation simulation.

DETONATION HUGONIOT

In shock physics determination of the Hugoniot for a system is a standard method for analyzing material properties. In a reactive system an extent of reaction parameter, λ, ranging from 0 for the unreacted material to 1 for the completely reacted system establishes a family of Hugoniots. The case where $\lambda = 1$ is generally referred to as the detonation Hugoniot.

In our molecular dynamics simulations we determine the detonation Hugoniot for the 2D model ozone material by establishing a rear boundary condition in the detonating solid with a piston (modeled as an infinitely massive, quadratic repulsive wall) moving at a constant velocity. By conducting several simulations at different piston velocities the detonation Hugoniot for the model ozone system can be found directly from the simulation data. More specifically to determine the detonation Hugoniot shown in Fig. 2 for our model ozone system we begin with a simulation in which a steadily propagating detonation has been established in the

system. At a position in the material far behind the detonation front, where the reaction has essentially gone to completion ($\lambda = 1$), we determine the local particle flow. A piston is then inserted at that point, moving at the velocity of the local flow, and the simulation is then allowed to run for several tens of ps to insure equilibrium conditions at the rear boundary. We then calculate the local pressure and specific volume for the completely reacted system immediately in front of the piston. By repeating this process for several different piston velocities we determine the detonation Hugoniot for the material directly from the simulation data without assuming an equation of state for the material.

Once we have the detonation Hugoniot for the model ozone system we can analyze the detonation properties for the material in terms of the ZND model. We determine the Rayleigh line from the initial state of the material to a point tangent to the detonation Hugoniot, the CJ

Detonation Hugoniot for Ozone System

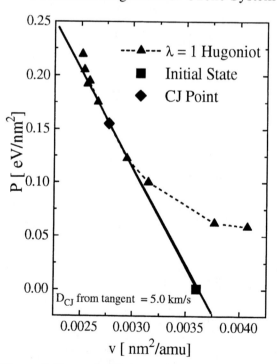

Figure 2. Detonation Hugoniot for ozone model system determined directly from the molecular dynamics simulations.

point. From the slope of this line we determine the detonation velocity for the system to be about 5.0 km/s in good agreement with that determined from the wave front position as a function of time. We can also calculate the condition at the CJ point for the model ozone material as $P = 0.155 \, eV/nm^2$, $v = 2.77 \times 10^{-3} \, nm^2/amu$.

CONCLUSIONS

We have shown how to determine the detonation Hugoniot, Rayleigh line, and CJ point directly from the molecular dynamics simulation data without having to resort to an equation of state for the material. The results of this study show that the classic ZND theory of detonation yields a good approximation to our simulation data for ozone.

ACKNOWLEDGEMENTS

This work was supported by the Office of Naval Research (ONR) through the Naval Research Laboratory and directly by ONR through contract # N00014-97-WX-20175. MLE received additional ONR support through the Naval Academy Research Council. JJCB thanks the National Research Council for a Postdoctoral Associateship. CTW thanks the Department of Materials for hospitality during his sabbatical stay at Oxford and for support through the Oxford Materials Modelling Laboratory. The authors also thank L.E. Fried for fruitful comments.

REFERENCES

1. D. W. Brenner, *Shock Compression of Condensed Matter – 1991*, S. C. Schmidt, R. D. Dick, J. W. Forbes, and D. G. Tasker, eds., (Elsevier, Amsterdam, 1992), p. 115; D. H. Robertson, D. W. Brenner, M. L. Elert, and C. T. White, *ibid.*, p. 123.

2. C. T. White, D. H. Robertson, M. L. Elert, and D. W. Brenner, *Microscopic Simulations of Complex Hydrodynamic Phenomena*, M. Mareschal and B.L. Holian, eds., (Plenum Press, New York, 1992), p. 111.

3. D. W. Brenner, D. H. Robertson, M. L. Elert, and C. T. White, *Phys. Rev. Lett.* **70** (1993) 2174.

4. J. J. C. Barrett, D. H. Robertson, D. W. Brenner, and C. T. White *Phys. Rev. B*, submitted.

5. D. H. Robertson, D. W. Brenner, and C. T. White, *J. Phys. Chem.* **96** (1992) 6133.

6. J. A. Harrison, C. T. White, R. J. Colton, and D. W. Brenner, *Surf. Sci.* **57** (1992) 271.

7. R. C. Mowrey, D. W. Brenner, B. I. Dunlap, J. W. Mintmire, and C. T. White, *J. Phys. Chem.* **95** (1991) 7138.

CP429, *Shock Compression of Condensed Matter – 1997*
edited by Schmidt/Dandekar/Forbes
© 1998 The American Institute of Physics 1-56396-738-3/98/$15.00

AB INITIO DETERMINATION OF THE V OF D OF AN OCTOL FROM THE STATISTICS OF ITS CRYSTAL STRUCTURE

J Brown, J P Curtis and P R Lee

DERA Fort Halstead, Kent TN14 7BP UK & Peter Lee Consulting Co Ltd, 8 Upton Quarry, Langton Green, Tunbridge Wells, Kent TN3 0HA, UK

A One-D technique is described for the estimation of the Velocity of Detonation (VoD) of a composite explosive (Octol), based on the proportions, VoDs, and crystal sizes of its constituents. A large number of possible paths through the composition are simulated by a Monte Carlo method which examines the order in which different types of explosive and various sized crystals are encountered. The overall VoD of the explosive is calculated for each pass, dividing total distance run by the sum of the transit times through the assemblages of individual crystals. This procedure yields a mean VoD close to the accepted reference value, and also interesting statistics about local variability.

INTRODUCTION

As the distance run by a detonation wave through explosive increases, the shape of the detonation front eventually bears less and less relation to that predicted by the use of Huygens wave constructions (see e.g. Lambourn and Swift, 1989[1]). Finite run-up distances to full detonation, reflections, free surface effects, and so on, affect the wavefront; but there are obviously other factors, which are not understood in detail. The present analysis suggests that it may not be possible to regard a detonation front merely as a lamina with a thickness that of the reaction zone in the detonating explosive. Such zones are usually considered to be about 0.1 mm thick in typical high explosives. It appears possible to argue, as will be shown, that there is an intrinsic uncertainty in the position of regions of a detonation front of the order a millimetre or more, even after a run of 100 mm from initiation. This arises from local variations in VoD occasioned by microstructural effects.

This paper aims to show how it is possible to estimate the detonation velocity of a multi-component explosive, from the proportions of the individual constituents and empirical values for their VoDs and crystal sizes.

From the weight by weight (w/w) proportions and physical properties of the constituents, the volume for volume (v/v) composition is obtained, and, with the VoDs, used as the basis for modelling. Analysis of the passage of the detonation through the explosive is undertaken on a 1-D basis, neglecting edge effects. A Monte Carlo method has been used to determine the proportion of track through each explosive species present. The implications for the formation of shaped charge jets are briefly reviewed. The initial objective of this study was to try to see whether some of the lateral velocities observed in such jets might arise in part from unavoidable imprecision in detonation waves (see e.g Brown et al, 1996[2]). Previous analyses of asymmetric jet formation are applied to investigate this question.

PHYSICAL PROPERTIES AND DETONATION CHARACTERISTICS

TABLE 1. Data for Octol and constituents.

Explosive	w/w %	VoD (km/s)	Density (kg/m³)	Comment
HMX Type A	42.1	9.11	1890	Measured[4]
HMX Type B	28.1	9.11	1890	Measured
RDX	4.00	8.78	1800	From UK Density Data[3]
pure TNT	-	7.00	1620	Vacuum cast[4]
R-T Eutectic	24.8	7.06	1626	Calc. from above
Beeswax	1.00	-	0900	
Complete Octol mixture	-	8.33	1760	At density used in UK

We have modelled an Octol which consists of several components: see Table 1. Trace additives are ignored. RDX dissolves in the TNT and forms a eutectic at 3.6% RDX as the mixture solidifies. The HMX also forms a eutectic with TNT, but only at a level of about one part in 300, which we have neglected.

Type A and Type B HMX crystals have the following characteristics:

- Type A HMX consists of 'equant' (i.e. essentially spherical) crystals of between 50 and 1000 micrometres (μmetres) diameter, with a median diameter of 400 μmetres and a standard deviation (s.d.) about that diameter of 130 μmetres.
- Type B HMX consists one third of crystals of median diameter 100 μmetres and an sad of 45 μmetres, and two thirds of crystals distributed randomly between 0 - 45 μmetres diameter.

The RDX is in the form of crystals of mean diameter 210 μmetres and an s.d. of 67 μmetres. The RDX/TNT (R-T) eutectic fills the spaces between the crystals of RDX and HMX and binds the composition together. The beeswax is assumed to be spread uniformly throughout the composition in infinitesimally small particles.

Densities and VoDs of HMX, RDX and TNT are readily accessible from such documents as, 'LASL Explosive Properties Data', edited by Gibbs and Popolato[4] and 'LLNL Explosives Data Handbook', edited by B M Dobratz[5]. Table 1 shows some data available for HMX, RDX and TNT. The properties of the R-T eutectic have not been published. The VoD and density of this material have been calculated by assuming that they are proportional to the mole fractions of the constituents.

It is necessary for us to determine the proportion of any path of the detonation through the Octol occupied by each component of the mixture. The densities and the w/w composition yield the v/v composition of the Octol as: HMX 66.7%, R-T eutectic 28.3%, RDX 3.1%, Beeswax 1.9%. This reveals the relative proportions of each component of the composition encountered along any line, such as AA' or BB' in Fig 1, a diagrammatic cross-section through the explosive.

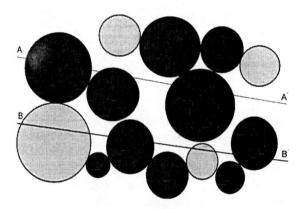

FIGURE 1. Diagrammatic section through Octol, showing differing paths through constituents

Grist determinations for RDX and HMX provide a picture of the frequency of occurrence and sizes of these crystals in the matrix. We do not have a complete picture for the R-T eutectic. Although it is, in effect, the binder, it too can be considered to be divided mostly into discrete volumes between the nitramine crystals. It can be estimated roughly from packing theory that the average diameter of each of the eutectic 'islands' will be about 200 μmetres, with an s.d. of around 70 μmetres.

ANALYSIS

Inevitably, as the wave passes through a crystal, there is an element of lateral spreading. As the detonation travels through an HMX crystal, it will tend to outstrip the wave in either the RDX or the R-T crystals around it, generating new detonation waves at an angle to its main line. Nevertheless we expect local irregularities in the front; corner-turning is not instantaneous, even in HMX. Because of this delay we argue that such spreading takes place *after* the front has passed. Even if not strictly true, this is regarded as a sensible first assumption. If we confine consideration to the central regions of a charge, it is assumed that all crystals detonate at their theoretical maximum velocity. Hence, we can build up a picture of the passage of a detonation wave through the explosive by examining its passage through each individual crystal and summing the effects over a suitable distance. A path represented e.g. by the line AA' in Fig. 1 is different from that represented by BB'. In principle, the average VoD determined from a large number of simulations should be valid. However, locally anomalous values will appear due to the stochastic nature of the composition.

Any line through an Octol charge will meet crystals of HMX, of R-T eutectic and of RDX (ignoring beeswax) with a frequency dependent upon the volume fraction of each component. Selecting a random rectangularly-distributed deviate, (i.e. a random number between 0 and 1) enables in the modelling process the assignment of a chemical identity to any crystal type encountered. If the v/v composition is p_h, p_{rt} and p_r for the three explosive components of the Octol, a random number, p, in the ranges $0 \leq p \leq p_h$, $p_h < p \leq (p_h + p_{rt})$ and $(p_h + p_{rt}) < p \leq (p_h + p_{rt} + p_r) = 1$, will indicate, an HMX, R-T eutectic or an RDX crystal respectively.

Our simulation next assigns the size of the crystal, using statistical data from measurements on the explosive. Some of these crystal sizes are described by Gaussian (Normal) distribution functions, for which we are given mean diameters and standard deviations. We assign a size to a crystal by generating a random Normal deviate and scaling from the empirical data on crystal diameters. Other crystals have diameters randomly distributed between certain limits, so uniformly distributed random deviates are generated and scaled to assign sizes to these. Finally, we allow a statistical variation in the length of the chord of each crystal traversed because we are dealing with a line detonation. The angle at which the detonation line intersects the surface of the sphere is similarly assigned by a uniformly distributed random deviate. The maximum chord of a sphere is D, the diameter, and the mean chord is $2D/\pi$. These processes are repeated for crystal after crystal, to build up a run of any desired overall length or duration. Random number selections, then, assign the species, size and geometrical proportion of each crystal across which the 1-D detonation travels. We have assumed that RDX and HMX crystals are approximately spherical. No doubt the modelling technique could be extended to other crystal shapes.

RESULTS

Mean VoDs and the s.d.s about these means have been compiled for 16,000 replications of 1-D detonations through this Octol, each simulating up to 30 microseconds(μs), or 250 mm run through the explosive. Table 2 shows an example of few raw results from a 30-run sample out to 100 mm distance from initiation. Reducing the recorded values by the factor .99 to allow for the effect of the inert beeswax returns a mean VoD of 8.42 km/s. This is adequately close to the 'book' value of 8.33 km/s[5] for the Monte Carlo method to be adjudged successful.

The extreme spread of arrival times after these 30 runs to 100 mm was found to be 266 nanoseconds. The observed variation in time of arrival at this distance implies a degree of uncertainty in the exact location of any particular microscopic region of the detonation front of around ±1 mm. The whole 16000 runs yielded a nominal detonation detonation velocity of 8.416 km/s (correcting for beeswax).

TABLE 2. Sample Results :VoDs after 100mm run in Octol

Dist.in RDX(mm)	Dist. in R/T(mm)	Dist. in HMX(mm)	Number of steps	Time (µs)	Mean VoD (km/s)
3.987	19.389	76.651	665	11.610	8.613
1.612	24.663	73.728	660	11.768	8.497
1.904	25.809	72.461	628	11.804	8.471
2.832	21.780	75.558	682	11.680	8.561
2.476	28.108	69.959	639	11.876	8.420

The mean detonation velocity was quite stable after even brief runs through the Octol. The mean VoD was sensibly the same for all run times, but the *standard deviation* turns out to be a strong function of the elapsed time or distance run since initiation. The reciprocal of the variance of the VoD against run time is shown in Fig 2.

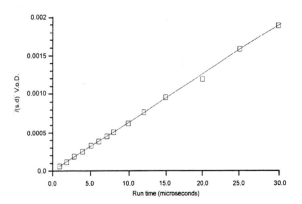

FIGURE 2. Plot of 1/Variance of VoD vs. run time.

The standard deviation about the mean VoD falls from 131 m/s after one µs run, to 23 m/s after 30 µs run. In the limit of a very long charge, one would of course expect the standard deviation to tend to a very small value due to averaging out of the statistical effects.

Table 2 showed that the difference between the maximum and minimum values of the VoD after 12.5 µs run is of the order of 200 m/s, corresponding to time-of-arrival differences of the wave at a distance of 100 mm of around a quarter of a µs. Returning briefly to the shaped charge jet collimation question, analytical jet formation work, e.g. Pack and Curtis[6] and Brown et al.[7], shows that this order of difference could, in the worst case engender off-axis velocities in jet elements on the order of 100 m/s. The effect would, however be much reduced in a truly rotationally symmetric charges as a result of the 'averaging' effect of combination with azimuthal sections where the time differences were much less.

CONCLUSIONS

It is possible to determine *ab initio* the detonation velocity from a statistical analysis of the detonation characteristics of its microcrystalline structure. For the Octol, this results in a possibility of small regions of the detonation wave reaching points 100 mm from a notional initiation site over a 'window' of perhaps a quarter of a µs, conferring on the detonation front localized 'fuzziness' of up to 2 mm. The quantification of the corresponding effect on shaped charge jet formation has been undertaken and will be published later.

REFERENCES

1. B D Lambourn and D C Swift: *Ninth Symposium (International) on Detonation*, Portland, Ore., USA (1989).2.
2. J Brown, P J Edwards and P R Lee: *Propellants, Explosives and Pyrotechnics*, **21**,59-63 1996.
3. Private communication of data.
4. T R Gibbs and A Popolato: *LASL Explosive Property Data*, University of California Press, Berkeley, Cal., USA (1980).
5. B M Dobratz: *LLNL Explosives Handbook*, University of California, Livermore, Ca., USA (1981).
6. D C Pack and J P Curtis: *Journal of Applied Physics*, 67, 6701(1990).
7. J Brown, J P Curtis and DD Cook: *Journal of Applied Physics*, **72**, 2136 (1992).

CP429, *Shock Compression of Condensed Matter – 1997*
edited by Schmidt/Dandekar/Forbes
© 1998 The American Institute of Physics 1-56396-738-3/98/$15.00

THE DEVELOPMENT OF A NEW ARRHENIUS-BASED BURN MODEL FOR BOTH HOMOGENEOUS AND HETEROGENEOUS EXPLOSIVES

M D Cook and P J Haskins

Defence Evaluation and Research Agency, Fort Halstead, Sevenoaks, Kent TN14 7BP, England

This paper describes the methodology behind the development of a new explosive burn model based on Arrhenius kinetics for both homogeneous and heterogeneous explosives types. The model has been developed over the last four to five years and is incorporated into the DYNA2D hydrocode as a new equation of state. The basic homogeneous model has been extended to consider heterogeneous materials by specifying a size and number density of hotspots. The hotspots are gas filled bubbles which are assumed to collapse adiabatically. Using the model it has been possible to match threshold values for shock initiation of a secondary explosive subjected to fragment impact.

INTRODUCTION

Models to describe the initiation and growth of reaction behaviour in energetic materials have traditionally been based on empirical fits to experiment. Whilst these models have certainly played an important role in helping to expand the capabilities of hydrocodes to address some specific problems, they do not address the fundamental physics and chemistry. Consequently, there are serious doubts concerning their applicability to model events far removed from those they were calibrated against. This severely limits their usefulness.

It therefore seemed desirable to attempt to build a new model within the constraints of a hydrocode that was capable of describing, at least in a general way, some of the complex interactions between the physics and chemistry of energetic materials. Consequently, over the last five years we have been engaged upon a programme aimed at developing a realistic ignition and growth model. Initially, the model described only homogeneous explosives, and the results of some of that early work were reported at the last conference in this series (1). More recently, we have implemented a heterogeneous

mechanism based on the notion of gas bubble collapse. This has allowed the model to describe both homogeneous and heterogeneous explosive behaviour. It should be noted that a recent paper by Bonnett et al. (2), closely describes the assumptions within our model, but with some very important differences. In particular, their model could not be directly implemented in a hydrocode without some modification, as noted by the authors themselves.

The model that we describe here has primarily been developed to consider shock initiation. This has more to do with the constraints of constructing models within a hydrocode than a desire to restrict the model to shock initiation problems. This is an important aspect that is often over-looked.

The model described in this paper has been based on Arrhenius chemical kinetics and has been implemented in the DYNA2D hydrocode as a separate equation of state. The development of the model is described in the following sections.

HOMOGENEOUS MODEL

The burn model uses Arrhenius chemical kinetics to evaluate the rate of global chemical reactions. In the present implementation the global chemistry is described by a similar three-step formalism to that employed by McGuire and Tarver (3), and used in the TOPAZ2D heat flow code to describe time to ignition in cook-off events. The Arrhenius scheme determines the reaction rate as a function of the temperature and the concentrations of the reactants. There are no explicit switch-on or off criteria applied to the reactions, they are active at all times during a simulation. However, at room temperature the rate is so small as to be negligible. It is only once the explosive is heated (typically to a few hundred degrees) that the rate increases to a noticeable level.

In a real explosive, there are many possible chemical reaction pathways that enable the initial unreacted material to decompose to the final products. However, for the purposes of this model, this is simplified to a set of either one, two or three schematic steps. These steps are coupled, for example, the intermediate products of the first step may form one of the reactants of the second. There are three unique reaction schemes in the model which, depending how they are parameterised, can span a single exothermic reaction step through to a mixture of endothermic and exothermic steps. The last step is assumed to generate the final detonation products in a gaseous state.

In the homogeneous model, the internal energy is converted into a temperature through the heat capacity at constant volume. The heat capacity can be either a constant or allowed to vary with temperature. There are two equations of state, one for the unreacted explosive (Murnaghan), and one for the products (JWL). At each DYNA timestep the chemistry is allowed to run at its own speed. An iterative pressure equalisation routine allows the pressure of the solid and gas within an element to be calculated for each DYNA time-step. If there is sufficient energy input into the energetic material during a simulation, the energetic material will burn, and this burn may subsequently build-up to detonation with a characteristic velocity and pressure. Alternatively, if insufficient energy is input into the energetic material, the energy losses can out-way the energy released through chemical reaction and the material may 'burn' or else the reaction may die completely.

The initial heating of the material is assumed to be driven by the rapid compression induced by the passing shock-front. Other heating mechanisms are not specifically considered.

The model itself covers the control of the reactions, heating due to compression, the release of chemical energy and the pressure and temperature balance. As the material is homogeneous the entire material is treated in the same manner.

Future developments of the model include the incorporation of more realistic equations of state to describe both the unreacted energetic material and the product gases. It would be particularly desirable to include equations of state derived from theoretical statistical thermodynamics such as the Williamsburg EOS (4).

HETEROGENEOUS MODEL

The heterogeneous model is treated as an extension of the homogeneous model by providing additional local heating. A method based on pore collapse has been developed although it is recognised that this is not the only hotspot mechanism that is important in explosive initiation.

A solid heterogeneous explosive is assumed to contain pores, which are small pockets of gas trapped within the material. These pores are compressed adiabatically under shock loading, causing a temperature rise in the gas. This enables heat to flow out of the pore and into the neighbouring explosive material. The corresponding temperature rise may then ignite the surrounding explosive, possibly leading to a full chemically supported shock-wave, or detonation.

To maintain the simplicity of the model it has been assumed that each given region of the material is defined by a single, uniform temperature. Hence, in order to capture the effect of pore heating causing a temperature rise in the material closest to the pore surface, the explosive material has been modelled as two regions. These regions can be visualised as concentric shells surrounding the pore. The inner shell temperature will tend to rise more rapidly than that in the outer shell and hence, if the appropriate conditions are met, a burn out from the pore surface may result. The temperature in each region is maintained independently, with explicit heat transfer mechanisms applied to allow heat flow between the regions.

338

The heterogeneous part of the model can be explicitly turned on or off through use of an input parameter. Additional parameters needed for the heterogeneous model are: the initial size of the pores, the porosity of the material, the conductivity from the pore gas to the explosive, the conductivity of the explosive, the shell mass fraction, and the initial pore pressure. The model handles the whole process of the compression and heating of the pores, and the heat transfer from the pores to the explosive material.

PERFORMANCE OF THE MODEL

The model has been calibrated and tested against experimental fragment impact data on explosive charges. The results of these experiments have been reported elsewhere (5,6). In these experiments, explosive charges covered by aluminium barrier plates were impacted by flat-nosed mild steel projectiles. Both the velocity of the projectile and the aluminium cover plate thickness was varied and a shock threshold response curve obtained. A theoretically derived response curve for both homogeneous and heterogeneous materials is shown in figure 1. It should be noted that whilst the homogeneous threshold curve is smooth, the heterogeneous response curve (obtained from the inclusion of the pore collapse routine) has a discontinuity. This is also observed in experiment.

The performance of the homogeneous model is very much as expected. The model has been parameterised and tested against experimental data on nitromethane charges. This work has been reported previously (1). As expected, for a given barrier thickness if the initial shock is sufficiently high, the energetic material is bulk heated to a degree that allows build-up of reaction and subsequent growth to detonation. For thin barriers, it is often observed that there is a delay between the initial shock passing and the on-set of significant reaction leading to detonation. This is particularly true near the threshold velocity and becomes significantly less distinguishable as the stimulus is increased above this value. Because the model does not rely on any switches to control the onset and build-up of reaction and therefore heat release, such features as detonation wave curvature should be a

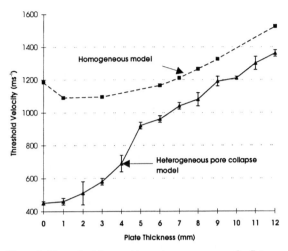

Figure1. Theoretical fragment impact response curves for flat-nosed projectiles impacting an explosive covered by varying thicknesses of aluminium. The same Arrhenius parameterisation was used for both curves. Detonations were obtained above the respective lines.

natural outcome of the model. This is of course assuming that the global chemistry and therefore heat release profile is realistic. It is also assumed that appropriate material models and parameters are used, and that the equations of state for both the unreacted and gaseous products are realistic. A detailed study of the accuracy of predicted curvature of detonation fronts has not been carried out to-date. However, it should be noted that parameters such as detonation velocity and pressure are consistent with experimental observation.

Current work is attempting to parameterise the heterogeneous model for a number of energetic materials. Figure 1 shows the effect of turning on the heterogeneous pore collapse routine. It can be seen from figure 1 that a discontinuous change of slope is observed when the pore collapse routine is added. This is in contrast to homogeneous materials which exhibit a smooth response curve (1). We find that when we include the effects of pore collapse in our model, the general shape of the heterogeneous experimental fragment impact curve is reproduced without extensive re-parameterisation. Whilst the effects of the pore collapse routine are still being investigated, a few points are worthy of note. The pore collapse routine has only six parameters. The initial pressure is a constant, and generally has a value of one atmosphere. The porosity can be estimated from the difference between the theoretical maximum density and the actual density. A value for the pore radius can be estimated from the average value of pores found in the explosive. It is harder to estimate values for the conductivities from gas to solid and solid to solid as they are not well-known especially, as a function of temperature. The final parameter is the shell mass fraction (a ring of energetic material surrounding the pore used to model heat flow into the bulk). Of the six parameters, only the shell mass fraction is a strictly unknown modelling parameter.

A parametric study of the main features of the pore collapse routine has shown that increasing the porosity increases the overall sensitivity of the explosive. It also has the effect of shifting the discontinuity of the fragment impact curve to thicker barriers. The dependence of the sensitivity on porosity is closely linked with the shell mass fraction and the pore radius.

By decreasing the pore radius the effects of the pore collapse model are greatly reduced and the shape of the bullet impact curve is similar to that of the homogeneous model. On increasing the radius, the discontinuity is shifted to thinner barrier thicknesses.

Of the two conductivities used in the model, the gas to solid conductivity is dominant with the solid to solid conductivity having little effect on the response curve. A decrease in gas to solid conductivity results in a steeper discontinuity, whereas increasing the conductivity moves the profile to thinner plates thicknesses.

The shell mass fraction has been shown to cause an increase in sensitivity for decreasing mass fraction.

CONCLUSIONS

Overall, the performance of the model is very encouraging. In particular, the pore collapse functionality has proven to be a major step forward in the path to a physically meaningful model, that will not only provide a predictive capability, but an understanding of the important underlying processes of energetic material behaviour. There is still much to do both on model development and on rigorous testing against experiment. Future enhancements include the development of further hotspot functionality, improvements in the equations of state used to describe both the unreacted and reacted material, and improved material models.

REFERENCES

1. M. D. Cook and P.J.Haskins, "Projectile Impact Initiation of a Homogeneous Explosive", *American Physical Society Topical Group on Shock Compression of Condensed Matter Conference,* held at Seattle, Washington, 14-18 August 1995 p823.
2. D. Bonnett and B. P. Butler, "A Thermochemical Model for Analysis of Hotspot Formation in Energetic Materials," Univ Iowa, UIME PBB95-003, Iowa City, IA 1995.
3. R. R. McGuire and C. M. Tarver, Proc. *7th Symposium (International) on Detonation*, Annapolis, 1981.
4. W. Byers Brown and M Braithwaite, "Analytical Representation of the Adiabatic Equation for Detonation Products based on Statistical Mechanics and Intermolecular Forces", *American Physical Society Topical Group on Shock Compression of Condensed Matter Conference*, held in Williamsburg, Virgina, 17-20 June, 1991, pp 325-328.
5. Cook, M.D., Haskins, P.J., and James, H.R., "Projectile Impact Initiation of Explosive Charges" *in the Proceedings of The Ninth Symposium (International) on Detonation*, 1989, pp 1441-1450.
6. Cook, M.D., Haskins, P.J., and James ,H.R., "An Investigation of Projectile and Barrier Effects on Impact Initiation of a Secondary Explosive", *in the Proceedings of the APS Topical Conference on Shock Compression of Condensed Matter*, 1991, pp. 675-678.

CP429, *Shock Compression of Condensed Matter – 1997*
edited by Schmidt/Dandekar/Forbes
© 1998 The American Institute of Physics 1-56396-738-3/98/$15.00

REACTION PATH OF ENERGETIC MATERIALS USING THOR CODE

L. Durães*, J. Campos** and A. Portugal*

Lab. of Energetics and Detonics
*Chem. Eng. Depart., **Mech. Eng. Depart. - Fac. of Sciences and Technology
University of Coimbra - 3000 Coimbra - PORTUGAL

The method of predicting reaction path, using THOR code, allows for isobar and isochor adiabatic combustion and CJ detonation regimes, the calculation of the composition and thermodynamic properties of reaction products of energetic materials. THOR code assumes the thermodynamic equilibria of all possible products, for the minimum Gibbs free energy, using H_L EoS. The code allows the possibility of estimating various sets of reaction products, obtained successively by the decomposition of the original reacting compound, as a function of the released energy. Two case studies of thermal decomposition procedure were selected, calculated and discussed - pure Ammonium Nitrate and its based explosive ANFO, and Nitromethane - because their equivalence ratio is respectively lower, near and greater than the stoicheiometry. Predictions of reaction path are in good correlation with experimental values, proving the validity of proposed method.

INTRODUCTION

The existing reactions in pyrolisis, combustion or detonation processes, generating intermediary chemical species and compounds, are very hard to follow by experiments, because these processes are very fast and proceed with increasing pressure and temperature. The studies concerning theoretical prediction of the probable pathways of pyrolisis or thermal decomposition of energetic materials are not numerous.

The method of predicting reaction path and final composition of combustion products, as a function of temperature and pressure, uses a thermochemical computer code, named THOR. This predicting code is based on theoretical work of Heuzé et al. (1), (2), later modified by Durães et al. (3), (4). The reaction path is estimated for all the possible compounds, as a function of temperature and pressure, for the minimum Gibbs free energy at thermodynamic equilibrium. In THOR code, several equations of

state may be used, namely Perfect Gas, Boltzmann, BKW, H_9, H_{12} and H_L. The validation of these EoS have been presented in previous works (Durães et al. (3), (4)). H_9 and H_{12} EoS are the natural development of a Boltzmann EoS type, with similar results to the BKW, KHT and JCZ3 EoS (Tanaka (5), Chaiken (6), Heuzé (2) and Campos (7)). H_L EoS (Durães et al. (3), (4)) is supported by a Boltzmann EoS ($PV/RT=\sigma(V,T,X_i)$, being $\sigma=1+x+0.625x^2+0.287x^3-0.093x^4+0.014x^5$ with $x(V,T,X_i)=\Omega/VT^{3/\alpha}$ and $\Omega=\Sigma(X_i\omega_i)$), but based now on physical intermolecular potential of gas components instead of correlations from final experimental results. This EoS takes $\alpha=13.5$ to the exponent of the intermolecular potential and $\theta=1.4$ to the adimensional temperature. Obtained results prove the importance of calculated products composition and the influence of $\Gamma=dH/dU)_S$ value (Brown (8)). The chemical equilibrium equations and validation have also been presented in previous

341

works (Durães et al. (9), (10)). The energetic equation of state is related to the internal energy $E=\Sigma(x_i e_i(T)+\Delta e)$, $e_i(T)$ being calculated from JANAF Thermochemical Tables (11) and polynomial expressions of Gordon and McBride (12).

Two applied examples of thermal decomposition procedure are given for Ammonium Nitrate (AN) and its based explosive (ANFO), and Nitromethane (NM). These examples have been selected to study because they are pure very known energetic substances and their equivalence ratio is respectively lower, near and greater than the stoicheiometry.

ADIABATIC COMBUSTION AND DETONATION CONDITIONS

The combustion regimes of energetic materials are assumed as global adiabatic processes (Durães (9), (10)): - the isobar adiabatic combustion verifies equal initial and final total enthalpy $H_b{}^{Tb}=H_o{}^{To}$; - the isochor adiabatic combustion verifies equal initial and final internal energy $E_b{}^{Tb}=E_o{}^{To}$; - the Chapman-Jouguet detonation condition (mass, momentum and energy balances and $dp/dV]_s=((P-P_0)/(V-V_0))$) is based on the assumption that the detonation velocity D is obtained adding sound velocity a_o with velocity u_p ($D=a_o+u_p$).

The isochor adiabatic combustion regime needs the calculation of the internal energy $E_i{}^T$, for a specified (V,T,X_i), where V represents the volume, T the temperature and X_i the mass fraction. This $E_i{}^T$ can be expressed as a function of the enthalpy and PV for the same conditions, being $PV=\sigma NRT$ obtained from the used EoS, ($E_i{}^T=H_i{}^T-\sigma N_i RT_i$). This expression allows to solve the isochor adiabatic combustion state from the calculation of the corresponding isobar adiabatic combustion state, for P and T conditions. Consequently, the main combustion regime, to the prediction of thermal decomposition is the isobar adiabatic combustion.

The reaction path can now be explained, starting from one initial reactive composition A to the final products composition C. If this reactive initial composition A decomposes in an intermediary composition B, and this composition B decomposes in final products composition C, the reactions can be presented in a simple way like: $A \rightarrow C$, being $A \rightarrow B$ and $B \rightarrow C$. At isobar combustion regime, the total enthalpy released by any reaction is converted to increase the temperature of the products of reactions. Assuming the preceding reaction scheme, from the initial temperature T_0, to the final temperatures T_b, T_b', T_b'', respectively, it can be written, for the total enthalpy: $H_A^{T0} = H_C^{Tb}$; $H_A^{T0} = H_B^{Tb'}$ and $H_B^{T0} = H_C^{Tb''}$. The preceding equations are equivalent to

$$H_A^{T0}-H_C^{T0}=-\Delta H_{A-C}^{T0}=H_C^{Tb}-H_C^{T0},$$

$$H_A^{T0}-H_B^{T0}=-\Delta H_{A-B}^{T0}=H_B^{Tb'}-H_B^{T0} \text{ and}$$

$$H_B^{T0}-H_C^{T0}=-\Delta H_{B-C}^{T0}=H_C^{Tb''}-H_C^{T0}.$$

So, $\Delta H_{A-B}^{T0}+\Delta H_{B-C}^{T0}=H_C^{T0}-H_A^{T0}=\Delta H_{A-C}^{T0}$.

Then, if the reaction is incomplete, the intermediary final products are B, and

$$\Delta H_{A-B}^{T0}=\Delta H_{A-C}^{T0}-\Delta H_{B-C}^{T0}=H_C^{Tb''}-H_C^{Tb}.$$

Obtaining the theoretical results of reactions A-C and B-C, the necessary data to calculate ΔH_{A-B}^{T0} is then obtained. The calculation mechanism for isochor adiabatic combustion is similar, but now related to the internal energy of reaction.

RESULTS AND DISCUSSION

Ammonium Nitrate and Ammonium Nitrate-Fuel Oil Compositions

Applying the preceding scheme and assuming the intermediary components shown in Tables 1 and 2, respectively for pure ammonium nitrate (AN) and ammonium nitrate-fuel oil explosive (ANFO), it can be calculated the three main points of the adiabatic Crussard curves (Fig. 1, where the initial points are also represented). The points are related to the isobar and isochor adiabatic combustion and CJ detonation regimes, from each case of selected products decomposition, proving the importance of

342

the decomposition of pure AN in NH_3+HNO_3 and the small influence of this decomposition in ANFO (all the points are almost in the same position). Experimental results, from decomposition processes of pure ammonium nitrate, show significant influence of (Kolaczkowski (13), Durães et al. (9), (10), (14)) endothermic dissociation above 169 °C ($NH_4NO_3 \rightarrow HNO_3 + NH_3$); exothermic elimination of N_2O on careful heating at 200 °C ($NH_4NO_3 \rightarrow N_2O + 2H_2O$); exothermic elimination of N_2 and NO_2 above 230 °C ($4\ NH_4NO_3 \rightarrow 3N_2 + 2NO_2 + 8H_2O$); and exothermic elimination of nitrogen and oxygen, sometimes accompanied by detonation ($NH_4NO_3 \rightarrow N_2 + 1/2\ O_2 + 2\ H_2O$), in a good agreement with theoretical predictions.

TABLE 1. Enthalpy of Reaction of Ammonium Nitrate and ANFO Decomposition, as a Function of Selected Products (kJ mol^{-1})

Products of Reaction	AN	ANFO
NH_3+HNO_3	184.3	176.1
$N_2+O_2+H_2O$	-97.4	-115.8
N_2O+H_2O	-39.3	-37.5
N_2+NO+H_2O	-30.5	-29.1
$N_2+NO_2+H_2O$	-103.5	-100.9
$N_2+HNO_3+H_2O$	-97.4	-124.3

TABLE 2. Internal Energy of Reaction of Ammonium Nitrate and ANFO Decomposition, as a Function of Selected Products (kJ mol^{-1})

Products of Reaction	AN	ANFO
NH_3+HNO_3	179.5	171.4
$N_2+O_2+H_2O$	-129.3	-123.6
N_2O+H_2O	-46.6	-44.5
N_2+NO+H_2O	-39.1	-37.3
$N_2+NO_2+H_2O$	-112.2	-107.2
$N_2+HNO_3+H_2O$	-133.5	-127.5

Nitromethane

In a similar way results from NM decomposition are presented in Table 3 and Fig. 3. The obtained results prove the main influence of CH_3+NO_2, $2CH_3O+2NO$ and $2CH_2O+2HNO$ as decomposition

products. The thermal decomposition of nitromethane can be expressed (Ornelas (15)), using a calorimetric bomb, by $2CH_3NO_2 \rightarrow H_2O + CO + N_2 + O_2 + CH_4$, with $H_2O + CO \rightarrow H_2 + CO_2$; $CH_4 \rightarrow C(s) + 2H_2$; $2H_2 + O_2 \rightarrow 2H_2O$; $3H_2 + N_2 \rightarrow 2NH_3$, with final main products composition of H_2O, CO, N_2 , H_4, CO_2 , $C(s)$ and NH_3. These mechanisms have been also analised by Bardo (16). Other intermediary products were proposed by Shaw (17), Haskins (18) ($CH_3 + NO_2$) and Agnew (19) ($N_2O + CO_2 + H_2O + CH_4$). These results are in good agreement with predictions.

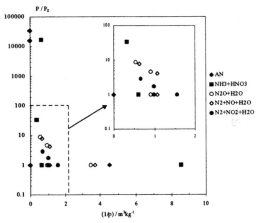

FIGURE 1. Represented States of the Isobar and Isochor Adiabatic Combustion and CJ Detonation of AN, as a Function of Selected Decomposition Products.

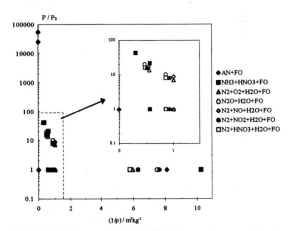

FIGURE 2. Represented States of the Isobar and Isochor Adiabatic Combustion and CJ Detonation of ANFO, as a Function of Selected Decomposition Products.

TABLE 3. Enthalpy of Reaction and Internal Energy of Reaction of Nitromethane Decomposition, as a Function of Selected Products (kJ mol^{-1}) .

Products of Reaction	$\Delta_r H$	$\Delta_r E$
$2CH_3+2NO_2$	291.8	287.0
$2CH_3O+2NO$	161.2	156.3
$2CH_2O+2HNO$	96.7	91.9
$H_2O+CO+N_2+O_2+CH_4$	-101.1	-107.3
$N_2O+CO_2+H_2O+CH_4$	-201.8	-206.7

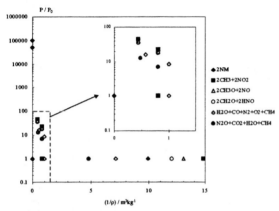

FIGURE 3. Represented States of the Isobar and Isochor Adiabatic Combustion and CJ Detonation of NM, as a Function of Selected Decomposition Products.

CONCLUSIONS

The method of predicting reaction path, using THOR code, allows for isobar and isochor adiabatic combustion and CJ detonation regimes, the calculation of the thermodynamic properties of reaction products. Thermal decomposition of Ammonium Nitrate based explosives and Nitromethane were selected, calculated and discussed. Presented predictions of reaction path are in good correlation with experimental values and prove the validity of proposed method.

REFERENCES

1. Heuzé, O. et al., "The Equations of State of Detonation Products and Their Incorporation into the Quatuor Code", *Proc. of the 8th Symposium (International.) on Detonation*, Albuquerque, New Mexico, pp. 762-769, (1985).
2. Heuzé O., *Cálculo Numérico das Propriedades das Misturas Gasosas em Equilíbrio Termodinâmico*, Universidade de Coimbra, Portugal, (1989).
3. Durães, L. et al., "New Equation of State for the Detonation Products of Explosives." *Proc. of 1995 APS Topical Conference on Shock Compression of Condensed Matter*, Seattle, WA, USA, pp. 385-388, (1995).
4. Durães, L. et al., "Deflagration and Detonation Predictions Using a New Equation of State", *Proc. of the 26th International Annual Conference of ICT*, pp. 67.1-13, (1995).
5. Tanaka, K., "Detonation Properties of Condensed Explosives Computed Using the Kihara-Hikita-Tanaka Equation of State", *Report from National Chemical Laboratory for Industry*, Ibaraki, Japan, (1983).
6. Chaiken, R. et al., "Toxic Fumes From Explosives: Ammonium Nitrate - Fuel Oil Mixtures", *Report of Investigation nº 7867 - Pittsburgh Mining and Safety Research Center*, Pittsburgh, PA, U.S.A., (1974).
7. Campos, J., "Thermodynamic Calculation of Solid and Gas Combustion Pollutants Using Different Equations of State", *Proc. of 1st International Conference on Combustion Technologies for a Clean Environment*, Vilamoura, Algarve, Portugal, pp. 30.4-1-30.4-11, (1991).
8. Brown, W. B., "Sensitivities of Adiabatic and Gruneisen Gammas to Errors in Molecular Properties of Detonation Products, *Proc. of the 9th Symposium (International) on Detonation*, Portland, Oregon, pp. 513-524, (1989).
9. Durães, L. et al., "Thermal Decomposition of Energetic Materials Using THOR Code", *Proc. of the Twenty Second International Pyrotechnics Seminar*, Fort Collins, Colorado, pp. 497-508, (1996).
10. Durães, L. et al., "Combustion and Detonation Modeling Using THOR Code", *Proc. of the 28th International Annual Conference of ICT*, pp. 89.1-89.10, (1997).
11. Janaf, *Thermochemical Tables*, 2nd Edition, National Bureau of Standards, Washington DC., (1971).
12. Gordon, S., McBride, B.J., "Computer Program for Calculation of Complex Chemical Equilibrium Compositions, Rocket Performance Incident and Reflected Shocks and Chapman-Jouguet Detonations", *Report NASA SP 273*, NASA Lewis Research Center, (1971).
13. Kolaczkowski, A., *Samorzutny Rozklad Saletry Amonowej*, Wydawnictwo Politechniki Wroclawskiej, Wroclaw, (1980).
14. Pires, A. et al., "Incineration of Explosives in a Fluidised Bed", *Proc. of the 27th International Annual Conference of ICT*, pp. 80.1-80.1014, (1996).
15. Ornelas, D. L., "Calorimetric Determinations of the Heat and Products of Detonation for Explosives", *LLNL Report*, UCRL - 52821, (1982).
16. Bardo, R., "Calculated Reaction Pathways for Nitromethane and Their Role in the Shock Initiation Process", *Proc. of the 8th Symposium (International) on Detonation*, Albuquerque, New Mexico, pp. 855-863, (1985).
17. Shaw, R., "Discussion on Nitromethane Decomposition Kinetics", *Proc. of the 6th Symposium (International) on Detonation*, Coronado, California, pp. 98-101, (1976).
18. Haskings, P. J. and Cook, M. D., "Quantum Chemical Studies of Energetic Materials", *Proc. of the 8th Symposium (International) on Detonation*, Albuquerque, New Mexico, pp. 827-838, (1985).
19. Agnew, S. F. et al., "Chemistry of Nitromethane at Very High Pressure", *Proc. of the 9th Symposium (International) on Detonation*, Portland, Oregon, pp. 1019-1026, (1989).

CP429, *Shock Compression of Condensed Matter – 1997*
edited by Schmidt/Dandekar/Forbes
© 1998 The American Institute of Physics 1-56396-738-3/98/$15.00

NONEQUILIBRIUM DETONATION OF COMPOSITE EXPLOSIVES

Albert L. Nichols III

Lawrence Livermore National Laboratory
L-282, PO Box 808, Livermore, CA 94550

The effect of nonequilibrium diffusional flow on detonation velocities in composite explosives is examined. Detonation conditions are derived for complete equilibrium, temperature and pressure equilibrium, and two forms of pressure equilibrium. Partial equilibria are associated with systems which have not had sufficient time for transport to smooth out the gradients between spatially separate regions. The nonequilibrium detonation conditions are implemented in the CHEQ equation of state code. We show that the detonation velocity decreases as the non-chemical degrees of freedom of the explosive are allowed to equilibrate. It is only when the chemical degrees of freedom are allowed to equilibrate that the detonation velocity increases.

INTRODUCTION

The detonation properties of uniform materials has a long history in the literature. Less is understood about the properties of mixtures of explosives. In this paper I will examine some of the issues involved with considering the detonation properties of mixed systems within a pseudo Chapmann-Jouget approach.

It is clear from experiments by McGuire et.al. (1) that a mixed explosive need not go to complete chemical equilibration, even after completing the entire detonation process. By examining several mechanisms for non-ideal detonation, it is hoped that we might understand the actual extent of equilibration possible within the detonation front.

There are several mechanisms which can lead to non-ideal behavior in a composite explosive which are not present in an homogeneous explosive. In general, the composite explosives will tend to have different compositions and would have different detonation temperatures if they were

by themselves. They would also have different particle and shock velocities. We define four states that the heterogeneous system can progress through:

1. Pressure equilibrium
2. Pressure and particle velocity equilibrium
3. Thermal but not chemical equilibrium
4. Complete equilibrium

THEORY

Thermal Equilibration

We will not go into the derivation of the standard Chapman-Jouget Detonation theory here. The standard CJ conditions for the detonation of a thermally equilibrated composite explosive are:
Particle Velocity (Conservation of Mass):

$$u = D\left(1 - \frac{\langle v_i \rangle}{\langle v_i^0 \rangle}\right) \qquad (1)$$

Detonation Velocity (Conservation of Momentum):

$$D^2 = \frac{\left\langle v_i^0 \right\rangle^2 \left(P - P^0 \right)}{\left\langle v_i^0 - v_i \right\rangle}, \qquad (2)$$

Hugoniot relation (Conservation of Energy):

$$\left\langle e_i - e_i^0 \right\rangle = \frac{1}{2}\left(P + P^0 \right)\left\langle v_i^0 - v_i \right\rangle. \qquad (3)$$

Here u is the particle velocity, D is the detonation velocity, P is the pressure, v is the volume per unit mass, and e is the energy per unit mass. The CJ state is the state which minimizes the entropy of the entire system. Here the $\langle\rangle$ imply a mass fraction (x_i) weighted sum over the i components.

For complete chemical equilibrium, the product components are homogenized and we can neglect the initial substructure. For thermal equilibrium, we assume that the composition of each explosive does not change and we can use the preceding equations without further modification.

Pressure Equilibration Only

In a composite explosive, the only things which are ensured to be uniform throughout the system are the detonation velocity and pressure. To analyze this situation, it is necessary to use the integral form of the conservation equations. It is assumed that the volume relative to other components of the explosive need not be fixed as it detonates, i.e. one explosive can expand while the other contracts.

If we assume that the volume of integration follows the flow of the material through the detonation, as shown in Fig 1, then the

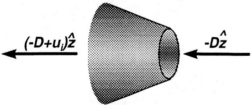

FIGURE 1. An example of a volume element of one of the explosives in the detonation front. The detonation wave is progressing from left to right.

contributions to the conservation equations from the sides will be zero for both the conservation of mass and energy. There will be a contribution due to the conservation of momentum which depends on the rate pressure changes as the system goes through the detonation front. For this work, we will neglect that ill-defined term.

Given these assumptions we find: Particle velocity of each component:

$$u_i = D\left(1 - \frac{\rho^0 x_i^0}{\rho x_i} \right). \qquad (4)$$

Detonation velocity:

$$D^2 = \frac{\left(P - P^0 \right)}{\rho^0 \left[1 - \left\langle \left(\frac{x_i^0 \rho^0}{x_i \rho} \right) \right\rangle^0 \right]} \qquad (5)$$

Mass Fraction:

$$\rho x_i = \frac{\rho^0 x_i^0}{2}\left\{ \frac{\rho_i D^2}{P} + \frac{P^0 \rho_i}{P\rho_i^0}\left[\left(\frac{\rho_i D^2}{P} + \frac{P^0 \rho_i}{P\rho_i^0} \right)^2 - 4\frac{\rho_i D^2}{P} \right]^{\frac{1}{2}} \right\} \qquad (6)$$

Effective Hugoniot Relation for each component:

$$\left(e_i - e_i^0 \right) = \frac{1}{2}\left(v^0 - v\frac{x_i^0}{x_i} \right)\left[P\left(2\frac{v_i x_i}{v x_i^0} - R_i \right) + P^0 R_i \right] \qquad (7)$$

where:

$$R_i = \left(1 - \frac{v x_i^0}{v^0 x_i} \right) \Big/ \left(1 - \left\langle \frac{v x_i^0}{v^0 x_i} \right\rangle^0 \right) \qquad (8)$$

and where $\langle\rangle^0$ is the mass fraction average over the initial mass fraction and ρ is the density. Note that the average in Eq. (5) will result in an increase in the detonation velocity for a given initial composition.

Particle velocity equilibration occurs as the relative velocities between the two explosives are dissipated by viscous effects. A uniform particle velocity requires that the mass fractions not change across the detonation front. This allows us to simplify the effective Hugoniot condition to:

$$\left(e_i - e_i^0\right) = \frac{1}{2}\left(v^0 - v\right)\left[P\left(2\frac{v_i}{v} - 1\right) + P^0\right] \quad (8)$$

IMPLEMENTATION AND RESULTS

All calculations were conducted with the chemical equilibrium-equation of state code CHEQ.(2,3,4,5) CHEQ calculates the equilibrium composition at a given temperature and pressure by minimizing the Gibbs free energy. CHEQ incorporates an effective one-component fluid variational-perturbation treatment of high-temperature, high-pressure, multi-component fluids. This model has been applied to the study of the equation of state of detonation products.(3) CHEQ has equation of state models for three carbon phases (diamond, graphite, and liquid) developed by van Thiel and Ree.(4)

The non-thermally equilibrated states described here require a more complicated method of solution. To calculate the pressure only equilibrium the Hugoniot of the completely equilibrated state at a specified pressure is calculated. This provides an initial values for the function

$$\lambda = \frac{P}{\rho^0 D^2} \quad (9)$$

and the component densities. The mixed detonation velocity and mole fractions are then calculated self-consistently to determine λ. λ is used to determine the detonation velocity which is used in the subsequent solutions of Eq (6). This process is continued until the values of λ between successive iterations converge. This gives the state on the pressure equilibrium only Hugoniot curve at the specified pressure. A series of these calculations are performed until the detonation velocity is minimized.

For the pressure and particle velocity equilibrium, CHEQ uses the same initial state as used in the pressure only equilibrium case. It uses that state to get an initial estimate of the density at the detonation condition. CHEQ then solves Eq (8) for each species. A new estimate for the density is then determined and then Eq(8) is solved again until the densities converge. A succession of these states is calculated until the detonation velocity is minimized.

For purposes of these calculations, two explosives with significantly different chemical composition were chosen. The first is HMX. The density, atomic composition, and heat of formation are 1.89 gm/cm^3, $C_4H_8N_8O_8$, and 75.02 kJ/mol, respectively. The main products from this explosive are nitrogen, water, and either carbon monoxide or carbon dioxide and some form of solid carbon. In general, if it is burned, the products have more of the monoxide, while if it detonates it produces more of the dioxide. The second explosive that was chosen is ADN. The density, atomic composition, and heat of formation are 1.803 gm/cm^3, $H_4N_4O_4$, and -149.787 kJ/ mol, respectively. In contrast to *HMX* which is under oxidized, *ADN* is over oxidized. This implies that the detonation products have a significant amount of molecular oxygen. It also has absolutely no carbon.

The CHEQ calculations described here used the fluid species H_2, O_2, H_2O, CH_4, CO, CO_2, N, N_2, NH_3, NO, NO_2, N_2O, and *CHOOH*, and three phases of carbon: diamond, graphite, and liquid. Two fluid phases were used for the non-carbon fluid species. That is, each fluid species was allowed to have a concentration in two fluid phases. By including two fluid phases, the fluid system can exhibit a super-critical phase separate into nitrogen rich and water rich phases, the existence of which has been postulated by Ree (3).

For those systems where it is assumed that the composition has not equilibrated, a complete set of species are assigned to each explosive component. The composition is allowed to equilibrate within each component's set of species, but the composition is not allowed to migrate between the two component's sets of species.

In Figure 2 we show the effect of the non-equilibrium behavior by comparing the difference of the detonation velocities as a function of composition from a straight baseline connecting the two extremes of detonation velocity.

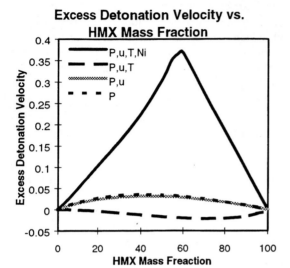

Excess Detonation Velocity vs. HMX Mass Fraction

Legend:
- P,u,T,Ni
- P,u,T
- P,u
- P

FIGURE 2. The difference in the detonation velocity in an HMX/ADN explosive between a straight baseline and various non-equilibrium assumptions, as a function of the weight percent of HMX.

CONCLUSION

In this paper we showed three non-equilibrium cases of detonation and compared them to the standard equilibrium detonation case. The three non-equilibrium states correspond to assumptions regarding the rates of equilibration of kinetic, thermal, and material inhomogeneities, with kinetic equilibration being the fastest process, and material equilibration being the slowest. We note that the detonation velocity need not be a monotonic function of the extent of equilibrium.

In order to verify the conclusions of this work, one must be able to vary the extent of equilibrium in a composite explosive. One way of doing this would be in changing the particle size of the monomolecular explosive components in the composite explosive. As the particles become smaller the mixture of the two components become more intimate, implying that the range of heterogeneity is smaller. This smaller range would allow the viscous and diffusive effects to dissipate more of the non-uniformity in the detonation products. Therefore, we expect that the detonation velocity of a composite explosive will be a non-monotonic function of the particle size.

The pressure only equilibrium case is only possible in a composite explosive where the explosives have been laid out in parallel strands and when the detonation is traveling in a direction parallel to the strands. If the detonation were traveling in any other direction, the components with a higher particle velocity will plow directly into those components with slower velocities, thus rapidly transferring the momentum from one component to the other. When the detonation is traveling parallel to the strands, though, the detonation can set up a standing three dimensional structure where the only momentum which is passed is in the plane of the detonation wave.

There is much evidence that would indicate that even monomolecular explosives fail to attain full equilibrium. We have shown that there are at least three mechanisms in composite explosives beyond those found in the monomolecular explosives which can result in a non-equilibrium detonation. We have also proposed how these mechanisms might be examined with simple detonation tests.

ACKNOWLEDGMENTS

The author would like to thank Craig Tarver, Edward Lee, Mathias van Thiel, and Francis Ree for helpful discussions and encouragement.

This work performed under the auspices of the U.S. Department of Energy by the Lawrence Livermore National Laboratory under contract number W-7405-Eng-48.

BIBLIOGRAPHY

1 McGuire, R.R., Ornellas, D.L., Akst, I.B., *Propellants and Explosives* **4**, 23-29 (1979).

2 Ree, F. H., *J. Chem. Phys.*, **78**, 409, (1983).

3 Ree, F. H., *J. Chem. Phys.*, **81**, 1251, (1984).

4 van Thiel, M., and Ree., F. H., *High Pressure Research*, **10**, 607 (1992).

5 Nichols, A.L. III, and Ree, F.H., *Lawrence Livermore National Laboratory*, UCRL-MA-106754, (1990)

CP429, *Shock Compression of Condensed Matter – 1997*
edited by Schmidt/Dandekar/Forbes
© 1998 The American Institute of Physics 1-56396-738-3/98/$15.00

KINETIC CALCULATIONS OF EXPLOSIVES WITH SLOW-BURNING CONSTITUENTS

W. Michael Howard, P. Clark Souers and Laurence E. Fried

*Energetic Materials Center, Lawrence Livermore National Laboratory
Livermore, CA 94550 USA*

The equilibrium thermochemical code CHEETAH V1.40 has been modified to detonate part of the explosive and binder. An Einstein thermal description of the unreacted constituents is used, and the Einstein temperature may be increased to reduce heat absorption. We study the effect of the reactivity and thermal transport on the detonation velocity. Hydroxy-terminated-polybutadiene binders have low energy and density and would degrade the detonation velocity if they burned. Runs with unburned binder are closer to the measured values. Aluminum and ammonium perchlorate are also largely unburned within the sonic reaction zone that determines the detonation velocity. All three materials appear not to fully absorb heat as well. The normal assumption of total reaction in a thermochemical code is clearly not true for these special cases, where the detonation velocities have widely different values for different combinations of processes.

The detonation velocity of an explosive is usually calculated in a thermochemical code with the assumption of full chemical and thermal equilibrium. This implies that all products are consumed in the detonation wave. This assumption holds in the limit of an infinite-size sample, whereas actual finite cylinders may give different detonation velocities. This arises because of the size effect: some components of the explosive react too late to drive the detonation front. Also, heat may flow too slowly to bring all components into thermal equilibrium.

It is possible to study the results of the size effect with an equilibrium thermochemical code as long as the various possibilities give widely different predictions for the detonation velocity. Here, we use the equilibrium thermochemical code CHEETAH plus the V1.40 BKWC library [1,2], which can make some or all of the starting materials inert. If a particular component reacts, it is converted into gas at the temperature of the overall explosive, whether it has chemical energy to give or not. If it does not react, it takes up volume according to a specified equation-of-state. For the heat capacity we use the Einstein model. By setting the Einstein temperature to 10^5 K, the material has little heat capacity at detonation temperatures (3000 - 5000 K). In effect, we have decoupled it thermally from the hot product gases. Thus, we can have a component that does not react and also absorbs no heat from the product gases.

Also included in this version of CHEETAH is the equation-of-state of the unreacted explosive, where the temperature-independent part of the equation of state is represented by a Murnaghan form [4]. The key is to find initial components such that the degree of reaction and/or thermal transport changes the detonation velocity substantially. The first three samples in Table 1 include large quantities of water which does not react [5]. The results suggest that the water does not transmit the heat of the detonating explosive either.

Finding an illustrative binder is more difficult, but Lawrence has been treating hydroxy-terminated-butadiene (HTPB) as inert in thermochemical codes for some time [6]. It has a low density of 0.907 g/cc, a -0.159 kJ/mol heat of formation and the composition $C_{7.33}H_{11}O_{0.083}$ [7]. Listed densities vary by less than 1% and the heat of formation of HTPB polymers used in explosives varies from -12 to +63 kJ/100 grams [8]. In the explosive, HTPB is cured to a rubber and there is every expectation of uniform distribution. Its unreacted EOS is [9]

$$U_s(mm/\mu s) = 1.63 + 2.24u_p. (1)$$

Six HTPB explosives are listed in Table 1 with combinations of reactivity and thermal transport on and off [10-14]. The results show best agreement with no reactivity of the binder as it affects the detonation velocity, and the spread of detonation velocities is large enough, i.e. greater than about ± 0.2 mm/μs, that we feel confident that the binder is not consumed ahead of the sonic point. The results also suggest that heat transfer may take place in the HTPB. The final samples in Table 1 are aluminum and ammonium perchlorate.[15-16] The aluminum shows no reactivity or heat transfer. The AP shows no reactivity but perhaps some heat transfer.

For effective heat transfer, the heat diffusivity timescale must be much shorter than the sonic reaction zone timescale. The heat diffusivity depends on the thermal conductivity, the particle size, the density and the heat capacity. We estimate that to obtain heat transfer with HTPB binder, we need 0.2 μm particle size, while for AP we require 1.0 μm particle size or less. Aluminum has the highest diffusivity, but the particles are typically large, so that heat transfer does not occur.

Finally, we consider the heat effects of the shock wave as estimated from the Hugoniots. A shock wave of 30 GPa from the explosive will cause compressive heating of: 550 K in Al, 1500 K in kel-F, and 2500 K in AP. Given explosive temperatures of 3000-4000 K, This suggests that shock heating may transfer some of the heat in AP and kel-F but almost none in aluminum.

This work was performed under the auspices of the US Department of Energy by the Lawrence Livermore National Laboratory under contract number W-7405-ENG-48

REFERENCES

1. Fried, L. E. and Souers, P. C., *CHEETAH: A Next Generation Thermochemical Code*, Lawrence Livermore National Laboratory report UCRL-ID-117240 (1994).

2. Fried, L. E. and Souers, P. C., "BKWC: An Empirical BKW Parametrization based on Cylinder Test Data," *Propellants, Explosives, Pyrotechnics* 21: 215-223 (1996).

3. Kittel, C., *Introduction to Solid State Physics*, 2nd ed, New York, John Wiley, 1956, p. 124.

4. Guillermet, A. F., Gustafson, P. and Hillert, M., *J. Phys. Chem. Solids* 46: 1427-1429 (1985).

5. Simpson, R. L., Helm, F. H., Crawford, P. C. and Kury, J. W., "Particle Size Effects in the Initiation of Explosives containing Reactive and Non-Reactive Continuous Phases," *Proceedings Ninth Symposium (International) on Detonation, Portland, OR, August 28-September 1, 1989*, vol. I, pp. 25-35.

6. Lawrence, W., Naval Surface Warfare Center, White Oak, MD, private communication from Jerry Forbes, 1996.

7. Forbes, J. W., Lawrence, W. and Sutherland, G. T., Naval Surface Warfare Center ,White Oak, MD, private communication, 1996.

8. *CPIA/M3 Solid Propellant Ingredients Manual*, Chemical Propulsion Information Agency, John Hopkins University, Columbia, MD, unit 7, pp. 1 and 3.

9. Bernecker, R. and Forbes, J. W., Naval Surface Warfare Center, White Oak, MD, private communication, 1996.

10. F. Bonthoux, F., Deneuville, P. and de Longueville, Y., "Diverging Detonations in RDX and PETN Based Cast-cured PBX," *Proceedings Seventh Symposium (International) on Detonation, Annapolis, MD, June 16-19, 1981*, pp. 408-415.

11. Souletis, J. and Mala, J., "Influence of Test Conditions on the Ballistic Classification of Explosives," *Proceedings Eighth Symposium (International) on Detonation, Albuquerque, NM, July 15-19, 1985*, pp. 625-630.

12. Finger, M., Hornig, H. C., Lee, E. L. and Kury, J. W., "Metal Acceleration by Composite Explosives," *Proceedings Fifth Symposium (International) on Detonation, Pasadena, CA, August 18-21, 1970*, pp. 137-149.

TABLE 1. Calculated and measured detonation velocities for various explosives where some secondary component does not burn within the reaction zone.

Type of Sample	Explosive Binder Density	Secondary Material		Det Velocity (mm/µs)	Composition (wt %)	Conclusion Reactivity /Thermal
		Reactivity	Thermal Transport			
No chemical reaction	HMX water 1.43 g/cc	off Measured off	on off	6.47 **7.06** 7.42	HMX 64 water 36	thermal partly on
	HMX water 1.54-1.55 g/cc	off Measured off	on off	7.45 **7.96** 7.99	HMX 80 water 20	thermal off
	RX-23-AB water 1.356 g/cc	off off Measured	on off	6.51 7.06 **7.48**	hyd nit 69 hyd 5 water 26	thermal off
HTPB binders	B2141 1.63 g/cc	on off Measured off	on on off	7.56 8.16 **8.19** 8.80	RDX 88 HTPB 12	off/on
	P2100B 1.70 g/cc	on off Measured off	on on off	7.80 8.51 **8.57** 9.12	HMX 88 HTPB 12	off/on
	A-589 1.66 g/cc	on Measured off off	on on off	7.55 **8.26** 8.31 9.04	HMX 86 HTPB 14	off/on
	HX-72 1.48 g/cc	on off Measured off	on on off	6.65 7.31 **7.75** 8.41	RDX 80 HTPB 20	off/ mostly on
	IRX-1 1.43 g/cc	on off Measured off	on on off	6.42 6.95 **7.67** 8.50	HMX 70 HTPB 30	off/ partly on
HTPB with some Al	IRX-3A 1.58 g/cc	both on both off Measured both off	both on both on both off	7.08 7.75 **7.87** 9.49	HMX 58.5 HTPB 35.6	off/ on
Al only; no binders	Tritonal 1.695 g/cc	on off off Measured	on on off	5.83 6.20 6.44 **6.52**	TNT 80 Al 20	off/off
	TNM/Al 1.828 g/cc	off on Measured off	on on off	5.54 5.73 **6.01** 6.02	TNM 67.7 Al 32.3	off/off
high Al; other materials	RX-54-AJ 1.811 g/cc	on Al off Al off Measured	on on Al off	6.62 7.27 7.64 **7.65**	HMX 47.4 Al 28.4 TMETN 16.1 NC 8	off/off

TABLE I   Continued

	RDX/Al	on	on	6.38	RDX 62	
	1.92 g/cc	Al off	on	7.33	Al 35.5	off/
		Measured		**7.6**	graphite 0.6	partly
		Al off	Al off	7.79	paraffin 1.9	on
	RX-35-EK	on	on	6.59	HMX 39.5	
	1.814 g/cc	Al off	on	7.08	Al 28	
		Measured		**7.35**	TMETN 24.8	
		Al off	Al off	7.40	PCL 6.7	off/off
AP included	RX-34-AI	off	on	5.61	BTF 47	
	1.824 g/cc	off	off	7.16	AP 53	
		Measured		**7.44**		
		on	on	8.50		off/off
	PBXN-103	both off	both on	4.66	AP 40	
	1.88 g/cc	Measured		**5.85**	Al 27	
		both off	both off	5.97	TMETN 23	
		both on	both on	6.77	NC 6	
		AP only on	both off	7.66	TEGDN 2.5	off/off
	PBXN-111	both off	both on	4.43	AP 43	
	1.78 g/cc	both off	both off	5.38	RDX 20	
		Measured		**5.70**	Al 25	
		both on	both on	6.72	HTPB 5.7	
		AP only on	both off	7.08	IDP 5.7	off/off

13. Volk, F. and Schedlbauer, F., "Detonation Products of less sensitive High Explosives formed under different Pressures of Argon and in Vacuum," *Proceedings Ninth Symposium (International) on Detonation, Portland, OR, August 28- September 1, 1989*, vol. II, pp. 962-969.

14. Sutherland, G. T., Lamar, E. R., Forbes, J. W., Anderson, E., Miller, P., Ashwell, K. D., Baker, R. N. and Liddiard, T. P., "Shock Wave and Detonation Wave Response of Selected HMX Based Research Explosives with HTPB Binder Systems, *"High Pressure Science and Technology-1993, Proceedings American Physical Society Topical Group on Shock Compression, Colorado Springs, CO., June 28-July 2, 1993*, part 2, pp. 1413-1416.

15. LLNL Cylinder Test results.

16. Forbes, J. W., Lemar, E. R., Sutherland, G. T. and Baker, R. N., *Detonation Wave Curvature, Corner Turning, and Unreacted Hugoniot of PBXN-111*, Naval Surface Warfare Center report NSWCDD/TR-92/164 (1992).

CP429, *Shock Compression of Condensed Matter – 1997*
edited by Schmidt/Dandekar/Forbes
© 1998 The American Institute of Physics 1-56396-738-3/98/$15.00

CHARME: A REACTIVE MODEL FOR PRESSED EXPLOSIVES USING PORE AND GRAIN SIZE DISTRIBUTIONS AS PARAMETERS.

G. Demol, J.C. Goutelle, P Mazel

DGA/Centre d'Etudes de Gramat, 46500 GRAMAT, FRANCE

Porosity in pressed explosives is often considered to be responsible for the shock to detonation transition. A model that takes into account the mechanical, thermal and chemical aspects of the pore collapse after a shock was developped by Saurel et al (16). From this model, induction times before the chemical reactions were computed for different pore sizes and shock pressures. These data associated with microscopic observations and the assumption of transition from pore combustion to grain combustion allowed us to build a new reactive model with pore and grain sizes distributions as parameters. All other parameters also have a clear physical meaning. This reaction kinetics was introduced in a lagrangian-eulerian hydrocode (Ouranos) and numerical simulations were conducted and compared to experiments for pressed HMX in 1D and 2D geometries (rod impacts). An important advantage of this new kinetics is that it allows to simulate the effects of variations in pores and grains distributions.

INTRODUCTION

It is now widely believed that shock to detonation transition of explosives is due to "hot spots". Actually, localization of energy is necessary to explain a relatively low threshold of ignition for shocked explosives. Numerous studies have investigated the physical nature of these "hot spots", for instance: Coffey and al. (5), Field and al. (6), Chaudhri (4), Plotard and al. (15). Although it is still controversial, the mechanism of viscoplastic collapse of pores seems to be one of the most important mechanism that can lead to ignition in pressed explosives. Modelling this process of collapse of pores has been tried by a lot of researchers: Caroll et Holt (3), Frey (7), Khasainov and al. (9), Kim and al. (10), Maiden and al.(13), Butler and al. (2), Belmas and al.(1), Saurel and al. (16). Such a model would allow to predict the behaviour of unformulated explosives and would be a tremendous tool for explosives and also for warhead designers. However, the physical phenomenons that occur during the collapse of pores are very complex and the most complete of these models:Saurel et al.'s model (16) is only available in a 1D specific hydrocode designed to support the model. Extensions of this model to 2D and 3D geometries will be very hard and long to do. That is why it was decided to design a simplified model that could be implemented in the Ouranos multi-D hydrocode. This model will use outputs of the complete physical model as parameters. It will also use the pore and grains size distributions to be able to be a design tool for engineers.

BUILDING THE NEW KINETICS

A pressed explosive is made from grains of explosives which are coated with a few percents of polymeric binder and pressed together. The resulting material has a porosity of about 3% with a maximum size of pores of a few microns.

The basic ideas of the model are that, after the shock has passed, there is an induction time followed by the combustion of the pore. After a moment, pores have grown too much for the material to keep on burning this way: a transition happens and the surface of the grains begins to burn.

Inductions times are computed with the complete physical model described by Saurel et al. They depend on the size of the pore and the shock pressure applied. For simplification, the curves induction time versus pore radius have been reduced to two characteristic numbers: one is the minimum pore radius for which ignition can happen (a_{min}) at this shock pressure and the other is the limit of induction time when pore radius increases (τ). Power laws are then fitted to a_{min} and τ. For HMX, we obtained:

$$a_{min} = 352 \cdot 10^{-6} P^{-2.2}$$
$$\tau = 133 \cdot 10^{-6} P^{-2.5}$$
(1)

with P in kbars, a_{min} in meters and τ in seconds.

The pores and grains size distributions are modeled using the same analytical formulae, that is:

$$n(a) = \frac{\alpha+1}{a_{max}}\left(1 - \frac{a}{a_{max}}\right)^{\alpha}$$
(2)

This distribution has the advantage to allow analytical calculation of the reaction rate during both phases: pore and granular combustions. Thus, during the pore phase:

$$\lambda = \frac{N}{V} \int_{a_{min}}^{a_{max}} n(a) \cdot \frac{4}{3} \cdot \pi \cdot [(t - \tau(a)) \cdot V_r]^3 da$$
(3)

and, during the granular phase:

$$\lambda = \frac{N}{V}\left[\int_0^D n(a)\frac{4\pi a^3}{3}da + \int_D^{A_{mx}} n(a)\left(\frac{4\pi a^3}{3} - \frac{4\pi(a-D)^3}{3}\right)da\right]$$
(4)

where V_r is the combustion speed, N/V is the number of pores or grains divided by the volume and D is the burnt depth.

During both phases, combustion happens in a layer by layer mode at a constant speed depending only on the shock pressure. The combustion law is taken from strand burner experiments extrapolated to high pressures. The combustion law had to be accelerated for very high pressures; so, we use an linear law in log-log up to 25kbar (strand burner), then a exponential law in log-log.

The time of the transition between both phases is chosen to have a smooth reaction-rate curve (time derivative of the reaction-rate is continuous). This choice is rather arbitrary but it gives satisfactory results accounting that it is difficult to determine this threshold from a known physical parameter.

RESULTS

All the results that will be given here are related to pressed HMX.

Choice of the parameters

The still unknown parameters are the pores and grains size distributions. We use optical micrographies to determine these distributions. Figure 1 shows such a micrography of pressed HMX.

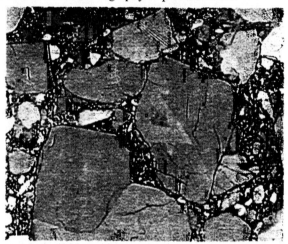

FIGURE 1. Photomicrography of pressed HMX (x90)

The maximum size of pores is observed to be about 20µm (pores of such a size are extremely few). The α coefficient of the distribution is determined for the average size of the distribution to be as observed (between 1and 2µm).

The determination of the grains size distribution lays another problem: the one we need is not the initial one, but the one induced by the shock wave damaging process. According to Loupias et al.(11), grains are severely crushed by the shock. Trying to quantify that using (11), we should made the assumption that the grain sizes are divided by 10 after the shock.

1D calculations

This model called CHARME (french acronym for kinetics inheriting advantageously of results on microstructure of explosive) has been implemented in the lagrangian-eulerian hydrocode Ouranos.

Figure 2 shows a comparison with experiments for pressure gauges at several depths in the explosive after a sustained plane shock at 25kbar. The agreement is good, except for the last gauge which is located just before the transition to detonation and is therefore very sensitive to slight differences in run-to detonation distances.

Computations have been made to determine the Pop-Plot of the explosive. The results are presented in Fig. 3 together with experimental and JTF (Tang and al.(17)) results. JTF and CHARME are able to reproduce fairly well the run-to detonation distance for this explosive.

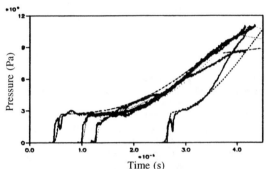

FIGURE 2. Pressure profiles in pressed HMX after a shock at 25kbar (plain line: experiment, dotted line: CHARME).

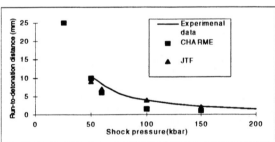

FIGURE 3. Pop-Plot for pressed HMX

As underlined previously, an interesting advantage of this kinetics is to use the pore and size distributions as parameters. This allows to simulate for instance the so-called sensitivity reversal effect: Moulard et al (14), Honodel et al (8). These authors have shown that varying the granulometry of explosive, the sensitivity does not vary monotonically. Starting with a coarse granulometry and diminishing the size of grains, the sensitivity begins to increase, then decreases if the size of the grains is further reduced. A classical interpretation is that pore and grain size distribution are not independant: coarse

grains give large pores while fine grains give smaller pores. When starting from the coarse granulometry, the grain sizes is reduced, the specific surface of the grains is dramatically increased and the sensitivity increases. But, if the granulometry keeps on diminishing, more and more pores become too small to ignite and the sensitivity decreases.

CHARME, because it takes into account both phases of the combustion, inside the pore and granular combustion, is able to reproduce this sensitivity reversal effect. The figure 4 has been drawn for three granulometries with the following maximum grain radii: 40µm, 5µm and 1µm. The pore maximum radii have been chosen varying monotonically with the maximum grain radii. The experiment simulated is a sustained plane shock at 25kbar with a pressure gauge embedded at a given depth. It can be seen that the experimentally observed reversal effect is reproduced. The non-reaction of the finest granulometry is due to the pores that are all too tiny to ignite (maximum radius of pores for this granulometry is 0.1µm).

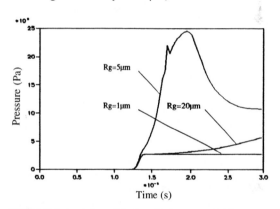

FIGURE 4. simulation of the "sensitivity reversal effect

2D simulations

CHARME was designed and implemented to be able to simulate multi-D geometries. The normal rod impact configuration was chosen because experimental results were available at CEG. Computations were done for several rod diameters ant the threshold for detonation was determined and compared to experiments.

Figure 5 shows a comparison of the detonation threshold observed experimentally and simulated with CHARME and JTF (17). For large diameters of rod, the agreement between experiment, JTF and

CHARME is perfect. However, for more little rods, the two kinetics fail to simulate the experiments. This is due to the amplification of 2D effects. Another explanation for the discrepancy that can be observed between simulations and experiments is that ignition during the experiments may have happened due to another process than shock to detonation transition. Actually, little rods can easily penetrate the explosive and mechanisms based on friction or shear may be relevant.

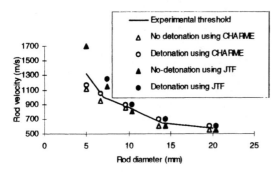

FIGURE 5. Detonation thresholds for pressed HMX after rod impact.

CONCLUSION

Starting from a complete but complex physical model of ignition of pressed explosives, a simpler one, CHARME, has been designed. This model has been successfully compared to experiments in 1D and 2D configurations.

The main advantage of CHARME is that pore and grain size distributions are included in the model as parameters. This allows to study the effect of a variation of these distributions on sensitivity or ignitability. It should be emphasized that taking into account the whole distribution instead of only one "average" size may be important.

ACKNOWLEDGMENTS

We would like to thank Dr. Lambert from Sciences et Applications and Dr. Trumel from our laboratory for their help in obtaining the photomicrographs. This work was conducted under the auspices of DGA/Mission Atome.

REFERENCES

1. Belmas R., Plotard J.P., Bianchi C., "A physical model of shock to detonation transition in heterogeneous explosives", presented at the 10th Symposium (International) on Detonation, Boston, Massachusetts, July 12-16, 1993.
2. Butler P.B., Kang J., Baer M.R. "Hot spot formation in a collapsing void of condensed-phase, energetic material", presented at the 9th Symposium (International) on Detonation, Portland, Oregon, August 28- September 1, 1989.
3. Carroll M. M., Holt A. C., J. Appl. Phys., Vol.43, No.4, April 1972.
4. Chaudhri M.M., "The initiation of fast decomposition in solid explosives by fracture, plastic flow, friction, and collapsing voids", presented at the 9th Symposium (International) on Detonation, Portland, Oregon, August 28-September 1, 1989.
5. Coffey C.S., Frankel M.J., Liddiard T.P., Jacobs S.J., "Experimental investigation of hot spots produced by high rate deformation and shocks", presented at the 7th Symposium (International) on Detonation, Annapolis, Maryland, June 16-19, 1981.
6. Field J.E., Swallowe G.M., Heavens S.N., Proc. R. Soc. Lond., A 382, 231-244, 1982.
7. Frey R.B., "Some aspects of the micromechanics of hot spot formation in energetic materials", Agard Conference Proceedings, May 28-30, 1984.
8. Honodel C.A. et al., "Shock initiation of TATB formulations", presented at the 7th Symposium (International) on Detonation, Annapolis, Maryland, June 16-19, 1981.
9. Khasainov B.A., Borisov A.A., Ermolaev B.S., Korotkov A.I., "Two-phase visco-plastic model of shock initiation of detonation in high density pressed explosives", presented at the 7th Symposium (International) on Detonation, Annapolis, Maryland, June 16-19,1981.
10. Kim K., Sohn C.H., "Modeling of reaction buildup processes in shocked porous explosives", presented at the 8th Symposium (International) on Detonation, Albuquerque, NM, July 15-19, 1985.
11. Loupias C, Fanget A, "Behavior of an unreacted composite explosive on low velocity impact", presented at the 9th symposium on Detonation, Portland, Oregon, August 28-September 1, 1989.
12. Loupias C., Jimenez B., Note Technique CEG T91-35, 1991.
13. Maiden D. E., Nutt G. L., "A hot spot model for calculating the threshold for shock initiation of pyrotechnic mixtures", presented at the 7th International Pyrotechnics Seminar, Vail,Colorado, 1986.
14. Moulard H., "Particular aspect of the explosive particle size effect on shock sensitivity of casr PBX formulations", presented at the 9th Symposium (International) on Detonation, Portland, Oregon, August 28 - September 1, 1989.
15. Plotard J. P., Belmas R., Nicollet M., Leroy M., "Effect of a preshock on the initiation of HMX, TATB and HMX/TATB compositions", presented at the 10th Symposium (International) on Detonation, Boston, Massachusetts, July 12-16, 1993.
16. Saurel R., Massoni J., Baudin G., Demol G., sumitted to Physics of Fluids, 1997.
17. Tang P.K., Johnson J.N., Forest C.A., "Modeling heterogeneous high explosive burn with an explicit hot-spot process", presented at the 8th Symposium (International) on Detonation, Albuquerque, NM, July 15-19, 1985.

CP429, *Shock Compression of Condensed Matter – 1997*
edited by Schmidt/Dandekar/Forbes
© 1998 The American Institute of Physics 1-56396-738-3/98/$15.00

DETONATION ENERGIES OF EXPLOSIVES BY OPTIMIZED JCZ3 PROCEDURES

Leonard I. Stiel[1] and Ernest L. Baker[2]

1) Polytechnic University, Brooklyn, New York 11201
2) U.S. ARMY TACOM-ARDEC, Picatinny Arsenal, New Jersey 07806

Procedures for the detonation properties of explosives have been extended for the calculation of detonation energies at adiabatic expansion conditions. The use of the JCZ3 equation of state with optimized Exp-6 potential parameters leads to lower errors in comparison to JWL detonation energies than for other methods tested.

INTRODUCTION

Previous studies have been conducted on the optimization of EXP-6 potential parameters of the JCZ3 equation of state (1-3). Individual parameters for HCNO explosive product species were established by the analysis of available Hugoniot pressure-density data. Large improvement in the agreement with the Hugoniot data resulted for most of the species considered. An advanced mixing rule for the effective exponent of the EXP-6 potential function has been implemented, and new procedures based on the minimization of the Helmholtz free energy have been developed to enable chemical equilibrium calculations for dissociated species and explosives. C-J detonation velocities and pressures for explosives calculated with the new procedures and optimized JCZ3 parameters demonstrated improved agreement with the experimental values.

ADIABATIC EXPANSION BEHAVIOR OF EXPLOSIVES

In this study procedures for the calculation of detonation properties of explosives with the optimized JCZ3 procedures have been extended to enable the determination of detonation energies at adiabatic expansion conditions. These procedures are based on the determination of chemical equilibrium conditions at each temperature and volume point by the minimization of the Helmholtz free energy of the system by variable metric optimization methods (2-4).

For the principal isentrope, the C-J conditions of the explosive are established by the minimization of the detonation velocity along the Hugoniot curve. The total entropy of the system is calculated as

$$S = \frac{E(T,V,n_i^*) - A(T,V,n_i^*)}{T} \qquad (1)$$

where E and A are the internal energy and Helmholtz free energy of the equilibrium mixture with composition n_i^*. Adiabatic conditions are established by a Newton-Raphson iteration method to find the temperature T and equilibrium composition through the function

$$FX = S - S^{CJ} \qquad (2)$$

where S and S^{CJ} are the total entropies calculated from Equation (1) at the expanded volume and C-J point, respectively. The value of the detonation energy at the expanded state is calculated as

$$EC = .004814 \frac{(e_1 - e_R)}{v_R} \qquad (3)$$

where EC is in J/cc, e_1 and e_R are the specific energies at the expansion and reference (initial) conditions, and v_R is the reference volume. The C-J conditions and entropies are established with optimized JCZ3 parameters and relationships for the heat capacity and molar volume of carbon. The volumetric relationship utilized for carbon diamond is of the Murnaghan form (5):

$$V = \frac{V_O}{(1 + \alpha P)^{\beta}} \qquad (4)$$

For some, but not all, explosives, the diamond form of carbon is indicated by these calculations to be the stable phase at C-J conditions and the graphite form at lower temperatures and pressures. For all calculations, the equilibrium composition is "frozen" at a desired temperature in the range 1800-2200 K by determining the corresponding freeze specific volume by a trial and error procedure.

For a number of explosives, comparisons were conducted between detonation energies calculated with optimized JCZ3 parameters and corresponding values resulting from the JWL relationships presented by Souers and Kury (6). The JWL equation is of the form

$$P_s = A \exp(-R_1 V^*) + B \exp(-R_2 V^*) \\ + C V^{*-(1+\omega)} \qquad (5)$$

where $V^* = V/V_R$.

The following expression for the energy along the adiabat is obtained from Equation (4):

$$E_s = (A/R_1) \exp(-R_1 V^*) + (B/R_2) \exp(-R_2 V^*) \\ + C/\omega V^{*\omega} \qquad (6)$$

The detonation energy at a specified value of V^* is calculated as

$$EC = -[E_s(CJ) - E_s(V^*) - E_c] \qquad (7)$$

where E_c is the energy at the C-J point relative to the reference energy.

Results of the comparisons of detonation energies at $V^* = 6.5$ for a freeze temperature of 2000 K are presented in Table 1 for PETN and HMX at several compositions and for TATB, BTF, TNT, and nitromethane. With the original TIGER JCZ3 parameters, the calculated detonation energies are in general less negative than the JWL values, and the average per cent deviations vary from 4-5% over the range of expansion volumes considered. For method MCO, carbon monoxide parameters $r^*=3.773$, $\varepsilon/k=350$, and $l=14.5$ resulting from Hugoniot data for the disassociating molecule at elevated temperatures were utilized with carbon graphite properties at both C-J and expansion conditions, and the average deviations from the JWL values are lower only at high expansions. For DMCO with the same JCZ3 parameters, the Murnaghan diamond equation of state is used at the C-J point (Equation (4) with coefficients $V_O = 6.563$, $\alpha = 33.92$, and $\beta = 1.22$), and the TIGER carbon graphite relationships at the expansion conditions. For this method the average errors are somewhat lower at low expansions. For most of the range and particularly for high expansions, best results were obtained for method UN which utilizes the alternate parameters for carbon monoxide $r^*=4.4$, $\varepsilon/k = 115$, and $l =11$ and the TIGER graphite equation of state at the C-J point and expansion conditions. These carbon monoxide parameters were established from Hugoniot data for the undissociated molecule at low pressure.

Of the explosives considered, the detonation energies for TATB and TNT are sensitive to the freeze temperature utilized. For UN parameters, the optimum freeze temperature is close to 2100 K, as shown in Table 2 at $V^* = 6.5$. The corresponding errors with the CHEETAH BKW parameters and procedures (BKWC) (7) which utilize a 2145 K freeze temperature are somewhat higher for the whole expansion range, as shown in Table 2 for $V^* = 6.5$.

TABLE 1. Detonation Energies (kJ/cm^3)
V*=6.5 (2000 K Freeze)

	RHO	JWL	TIGER	MCO	DMCO	UN
PETN	1.763	-8.55	-8.09	-8.25	-8.25	-8.29
PETN	1.503	-6.68	-6.31	-6.44	-6.44	-6.51
PETN	1.263	-5.10	-4.80	-4.88	-4.88	-4.97
TNT	1.632	-5.30	-5.07	-5.16	-4.86	-5.24
NM	1.130	-4.04	-3.94	-3.98	-4.01	-4.11
BTF	1.852	-8.96	-8.98	-9.11	-8.90	-9.14
TATB	1.830	-5.63	-5.45	-5.73	-5.58	-5.82
HMX	1.894	-9.47	-9.30	-9.52	-9.45	-9.63
HMX	1.188	-5.11	-4.64	-4.71	-4.71	-4.81
			4.25%	3.04%	3.37%	2.63%

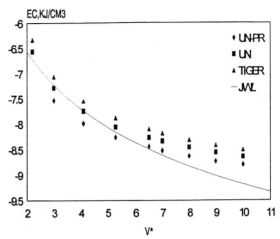

FIGURE 1. Detonation Energies for PETN (Rho=1.763).

At the highest expansions considered, the pressures and temperatures are so low (less than 1000 K and 1 kbar) that the specialized equations of state for explosives such as BKW or JCZ3 are less accurate, and the use of alternate volumetric relationships may be advantageous for high accuracy. Calculations at these conditions have been performed utilizing the Peng-Robinson equation of state which is of a modified Van der Waals form (8):

$$P = \frac{RT}{v-b} - \frac{a(T)}{v(v+b)+b(v-b)} \qquad (8)$$

The constants of the mixture are calculated as

$$a = \sum_i \sum_j x_i x_j a_{ij} \qquad (9)$$

$$b = \sum_i x_i b_i \qquad (10)$$

where a_{ii} and b_i are the parameters of the pure components which are calculated from the critical constants of the fluid, and

$$a_{ij} = \sqrt{a_{ii} a_{jj}}(1 - k_{ij}) \qquad (11)$$

In these calculations the binary interaction constants k_{ij} were taken as zero.

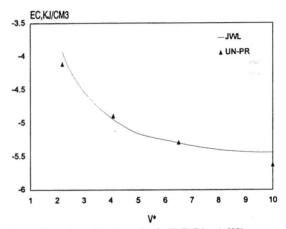

FIGURE 2. Detonation Energies for TNT (Rho=1.632).

TABLE 2. Detonation Energies (kJ/cm^3)
V* = 6.5 (2100 K Freeze)

	RHO	JWL	UN	BKWC	UN-PR
PETN	1.763	-8.55	-8.29	-8.32	-8.44
PETN	1.503	-6.68	-6.51	-6.47	-6.63
PETN	1.263	-5.10	-4.97	-4.97	-5.07
TNT	1.632	-5.30	-5.19	-5.38	-5.30
NM	1.130	-4.04	-4.11	-4.01	-4.20
BTF	1.852	-8.96	-9.13	-9.31	-9.28
TATB	1.830	-5.63	-5.61	-6.23	-5.72
HMX	1.894	-9.47	-9.63	-9.45	-9.77
HMX	1.188	-5.11	-4.81	-4.75	-4.91
			2.41%	3.62%	2.11%

Detonation energies were calculated for the explosives at $V^* = 6.5$ with the Peng-Robinson equation of state, utilizing the JCZ3 UN procedures for the C-J point and for the freeze conditions at 2100 K. It can be seen from Table 2 that improved average errors in EC compared to the JWL detonation energies result by this procedure (UN-PR). In Figure 1 for PETN, it can be seen that at high expansions (V^* greater than 5) the values of EC calculated with the UN-PR procedure are in closer agreement with the JWL curve than those resulting for this range with the corresponding JCZ3 relationships and UN parameters. In Figure 2, detonation energies for TNT calculated in this manner with the UN-PR procedure at high expansions are shown to be in close agreement with the JWL values for the entire range.

Efficient subroutines (designated JAGUAR) have been developed to apply the optimized JCZ3 and variable metric procedures for the calculation of detonation properties of explosives, including at adiabatic expansion conditions (9). The JAGUAR program incorporates the ability to automatically generate JWL and JWLB equation of state parameters.

CONCLUSIONS

The average errors in C-J velocities for substances which form substantial carbon are slightly higher for C-J velocities calculated with alternate carbon monoxide parameters (UN) established for the undissociated molecule at low pressures on the Hugoniot curve (2). However, best results have been presently obtained with these JCZ3 parameters for detonation energies along the adiabatic expansion curve from the C-J point. These results indicate that the experimental error may be higher for the detonation velocity of carbon-forming systems, since less carbon is formed at expansion conditions. There also may be slight inconsistencies for the application of the JCZ3 equation of state at low pressures.

With a freeze temperature of 2100 K, average per cent errors of approximately 2% result with the procedures of this study at high expansions, compared to errors of about 4% with the original TIGER JCZ3 parameters. With C-J parameters and

freeze conditions established at elevated pressures with the JCZ3 equation of state, the use of an alternate equation of state suitable for low pressures leads to improved results for detonation energies at high expansions.

ACKNOWLEDGMENTS

This work was supported by the U.S. Army Armament Research, Development and Engineering Center under the auspices of the U.S. Army Research Office Scientific Services Program administered by Battelle (Contract No. DAAL03-91-C-0034).

REFERENCES

1. Stiel, L.I., Rotondi, P., and Baker, E.L., *Proceedings of 1995 APS Topical Conference of Condensed Matter"*, Seattle, Wa. August, 1995, .389-392..

2. Stiel, L.I. and Rotondi, P., Final Report Contract No. DAAL03-91-C-0034, TCN No. 94062, January, 1995.

3. Stiel, L.I. and Rotondi, P., Final Report Contract No. DAAL03-91-C-0034,TCN No. 95018, October, 1995

4. .Baker, E.L.,"Modeling and Optimization of Shaped Charge Liner Collapse and Jet Formation", Picatinny Arsenal Technical Report ARAED-TR-92017, January, 1993..

5. Nellis, R.J., Ree, F.H., van Thiel, M., and Mitchell, A.C., *J.Chem. Phys.*, **75**, 3055-3063 (1981).

6. Souers, P. C. and Kury J. W., *Pyrotechnics*, **18**, 175-183 (1993).

7. Fried, L.E., CHEETAH *1.39 User's Manual*, March, 1996.

8. Sandler, S.I.,"*Chemical and Engineering Thermodynamics*", Wiley (1989).

9. Stiel, L.I., Final Report Contract No. DAAL03-91-C-0034, TCN No. 96120, October,1996.

CP429, *Shock Compression of Condensed Matter – 1997*
edited by Schmidt/Dandekar/Forbes
© 1998 The American Institute of Physics 1-56396-738-3/98/$15.00

CALIBRATION OF DSD PARAMETERS FOR LX-07 FROM RATE-STICK DATA

Yehuda Partom

RAFAEL, P.O.Box 2250, Haifa 31021, ISRAEL

We calibrate Detonation Shock Dynamics (DSD) parameters for the explosive LX-07 from rate-stick data, taken at different diameters down to the failure diameter. We use a velocity curvature relation proposed by Bdzil, and the model parameters are: the boundary limiting angle θ_c, the failure curvature k_f and the normalized failure detonation velocity D_f. We find that the data is matched when $D_f = 0.939\,\theta_c(\text{rad})$ and $k_f(1/\text{mm}) \cong 0.8\,\theta_c(\text{rad})$.

INTRODUCTION

The Detonation Shock Dynamics (DSD) model was developed by Bdzil & Stewart (1-5) and by Lambourn & Swift (6). It can be used to predict the expansion of diverging detonation waves. The model contains several parameters to be calibrated from experimental data for any specific explosive. Here we calibrate the model parameters for the explosive LX-07 (7) from rate-stick (detonation velocity in a rod) data.

DSD MODEL

The DSD model assumptions are:

- The local detonation velocity D_n of a diverging detonation wave depends only on its local mean curvature.

- The angle θ, on the explosive boundary, between the normal to the detonation front and the boundary, is bounded from below by θ_c.

- A diverging detonation wave fails locally when $k > k_f$ = failure curvature.

For the $D_n(k)$ relation we use:

$$\frac{D_n}{D_{CJ}} = D_f + (1 - D_f)\left(1 - \frac{k}{k_f}\right)^{\frac{1}{2}} \qquad (1)$$

proposed by Bdzil & Fickett (4), where D_{CJ} is the ideal detonation velocity. The parameters we need to calibrate are therefore k_f, D_f and θ_c.

COMPUTATIONS

We compute steady-state detonation front curves in rods. The coordinate system is (r,z) where z is along the rod axis and r is the radial direction. The relevant equations are:

$$\theta = -\arctan z' \qquad (2)$$

$$D_n = D\cos\theta \qquad (3)$$

$$k_T = -\frac{z''}{\left(1 + (z')^2\right)^{\frac{3}{2}}} \qquad (4)$$

361

$$k = \tfrac{1}{2}\left(k_T + \frac{\sin\theta}{r}\right) \quad \text{for } r > 0$$
$$k = k_T \quad \text{for } r = 0 \tag{5}$$

$$D_n = D_n(k) \tag{1}$$

where: θ is the angle from the z direction to the local normal, D is the steady-state detonation velocity in the rod and k_T is the curvature in the r,z plane.

Assuming the value of D we integrate the ODEs numerically along the front starting at the axis. We use the 4th order Runge-Kutta algorithm, and the integration step Δr can be made as small as needed to obtain the desired accuracy. We stop the integration when $\theta = \theta_c$ and register the value of r (denoted by $a = \tfrac{1}{2}d$). Using a range of values for D we obtain pairs (D,d) down to the failure diameter d_f.

To calibrate the model parameters we change them until we get agreement between the computed D(d) curve and the data.

CALIBRATION

From the data for LX-07 we conclude that (see Figs.):

$$D(d_f) = 0.939\,D_{CJ} \tag{6}$$

so that:

$$D_f = 0.939\cos\theta_c \tag{7}$$

Eq. (7) is therefore built into the computations.

To see how each of the parameters k_f and θ_c influences the D(d) curve, we performed runs in which we kept one of them fixed and changed the other. The results are shown in Figs. 1 and 2 (as D/D_{CJ} versus $1/a$). Figure 1 is for $\theta_c = 0.20$ rad and $k_f = 0.15, 0.16, 0.17$/mm. Figure 2 is for $k_f = 0.16$/mm and $\theta_c = 0.17, 0.20, 0.23$ rad.

We conclude that the data can be matched when:

$$k_f\left(mm^{-1}\right) \cong 0.8\,\theta_c\,(rad) \tag{8}$$

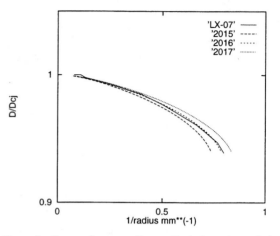

Figure 1. Computed curves of normalized detonation velocity versus inverse rod radius, compared to experimental data. 'LX-07' is the data curve. '2015', '2016' and '2017' are computed curves with $\theta_c = 0.20$ rad and $k_f = 0.15, 0.16, 0.17$/mm respectively.

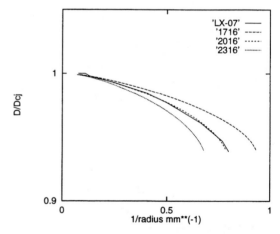

Figure 2. Computed curves of normalized detonation velocity versus inverse rod radius, compared to experimental data. 'LX-07' is the data curve. '1716', '2016' and '2316' are computed curves with $k_f = 0.16$/mm and $\theta_c = 0.17, 0.20$ and 0.23 rad respectively.

To further demonstrate that this is the case, we show in Fig. 3 results of computations with three pairs of values of $\left(k_f, \theta_c\right)$ that obey Eq. (8).

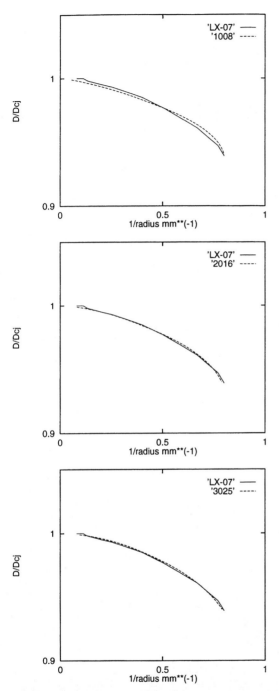

FIGURE 3 confirms that we cannot fully calibrate the DSD model parameters from rate-stick data alone. Additional data is needed. For instance, we can determine θ_c directly from a smear camera picture of the detonation front breakout from the rod end surface.

It may also be interesting to cast our results in terms of the initial slope of the $D_n(k)$ curve. Differentiating Eq. (1) we get:

$$A = \left(\frac{\partial}{\partial k} \frac{D_n}{D_{CJ}} \right)_{k=0} = \frac{1-D_f}{2k_f} = \frac{1-0.939\cos\theta_c}{2 \times 0.8\,\theta_c} \quad (9)$$

and we show $A(\theta_c)$ in Fig. 4.

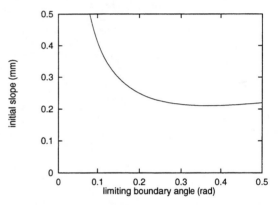

FIGURE 4. Initial slope of the normalized $D_n(k)$ curve as a function of the limiting boundary angle θ_c according to Eq. (9).

We see from Fig. 4 that beyond $\theta_c = 0.2$ rad A is almost constant.

CONCLUSIONS

We look into the possibility of calibrating DSD model parameters from rate-stick data. We use rate-stick data for the explosive LX-07. The DSD model that we use has three parameters: the failure curvature k_f, the failure normalized detonation velocity D_f and the limiting boundary angle θ_c. We show that the data determines a relation between any two of the three parameters.

FIGURE 3. Computational results for three different pairs of k_f, θ_c that obey Eq. (8) compared to the experimental data (LX-07 curves). '1008' is for $\theta_c = 0.10$rad and $k_f = 0.08$/mm, '2016' for 0.20rad and 0.16/mm and '3025' for 0.30rad and 0.25/mm.

For LX-07 we get: $D_f = 0.939\cos\theta_c$ and $k_f(1/mm) \cong 0.8\theta_c(rad)$. Additional data is needed to determine all three parameters.

REFERENCES

1. Bdzil J. B., *J. Fluid Mech.* **10**, 195-226 (1981).
2. Stewart D. S. & Bdzil J. B., *Combustion & Flame* **72**, 311-323 (1988).
3. Bdzil J. b. & Stewart D. S., *Phys. Fluids A* **1**, 1261-1269 (1988).
4. Bdzil J. B. & Fickett W., DSD Technology, A Detonation Reactive Huygens Code, LA-12235-MS (1993).
5. Bdzil J. B. et al., "Level Set Techniques Applied to Unsteady Detonation Propagation," presented at the Mathematical Modeling & Combustion Science Conference, Hawaii (1994).
6. Lambourn B. D. & Swift D. C., "Application of Whithham's Shock Dynamics Theory to the Propagation of Divergent Detonation Waves," presented at the 9th Symposium on Detonation, Portland Oregon, pp. 784-797 (1989).
7. Bobratz B. M. & Crawford P. C., *LLNL Explosives Handbook Properties of Chemical Explosives and Explosive Simulants,* pp. 19/73-19/74 (1985).

CP429, *Shock Compression of Condensed Matter – 1997*
edited by Schmidt/Dandekar/Forbes
© 1998 The American Institute of Physics 1-56396-738-3/98/$15.00

MODELING PBX 9501 OVERDRIVEN RELEASE EXPERIMENTS

P. K. Tang, R. S. Hixson, and J. N. Fritz

Los Alamos National Laboratory, Los Alamos, New Mexico 87545

We show the failure of the standard Jones-Wilkins-Lee (JWL) equation of state (EOS) in modeling the overdriven release experiments of PBX 9501. The deficiency can be tracked back to inability of the same EOS in matching the shock pressure and the sound speed on the Hugoniot in the hydrodynamic regime above the Chapman-Jouguet pressure. After adding correction terms to the principal isentrope of the standard JWL EOS, we are able to remedy this shortcoming and the simulation is successful.

INTRODUCTION

High Explosives (HE) performs work by the expansion of its detonation products. Along with the propagation of the detonation wave, the equation of state (EOS) of the products determines the HE performance in an engineering system. The expansion typically begins at the Chapman-Jouguet (CJ) state and follows the principal isentrope to lower pressure. Since most theoretical EOS works cannot provide an accurate description of the principal isentrope as demanded in some applications, experiments are the only way to extract the EOS information. The cylinder test approximates that condition and is perhaps the most commonly used method for the purpose. But it is doubtful that the EOS so obtained is still adequate in other conditions, particularly when pressure above CJ, known as the overdriven regime, is encountered. In an engineering device, the expansion cannot always be considered isentropic all the way and everywhere. The major factor is the system configuration. For example, the ideal process can be disturbed by wave reflection from various material boundaries causing the expansion to be interrupted by recompression.

Also, the result of detonation waves interaction can lead to an expansion beginning at pressure above CJ. In this paper, we will show how a standard JWL EOS fails to match the Hugoniot data and sound speed in that region and how we can improve the EOS with new corrections. Finally, we show the success of the new EOS to simulate overdriven release experiments in which the detonation products' pressure is maintained and then the expansion begins at pressure above the CJ.

JWL EQUATION OF STATE

The Jones-Wilkins-Lee (JWL) EOS[1] is perhaps the most popular form used in the HE community for a large class of problems but its nonuniqueness is also well recognized. One of the problems is the variability of the CJ state, particularly the CJ pressure. We can attribute the variation to the reaction zone effect,[2] not to the problem associated with the true products EOS. But even with the reaction zone taken care of, we have yet to claim that the right JWL EOS is available for all problems.

The JWL form is essentially an empirical-based EOS with the parameters determined mostly from the cylinder test; its validity is questionable in other hydrodynamic regimes as mentioned. The overdriven Hugoniot pressure is such a useful quality for checking the EOS for the purpose. Experimental overdriven Hugoniot data have been available for some time[3] and the JWL EOS has been shown to always underestimate the result.[4] The comparison of the recently available data for PBX 9501 (95% HMX, 2.5% Estane, 2.5% BDNPA/BDNPF)[5] and the calculation based on a standard JWL EOS[6] in the overdriven regime shows similar trends. The difference is more pronounced farther away from the CJ state.

Another piece of important information is the sound speed, also recently made available experimentally.[5] Again, a substantial difference is observed, and the calculated sound speed is much lower at higher pressure. To compensate for the deficiency of the conventional JWL EOS, more exponential terms can be added[7] but in doing so the original parameter set is perturbed. A new formulation is proposed in this work. It should be noted, however, that no uniqueness can be proven in any of the empirical formulation.

MODIFIED JWL EQUATION OF STATE

We add a correction to the conventional JWL expression to cover the high pressure region only while keeping the low pressure portion unchanged. The dividing line is the CJ state, more specifically, the CJ volume. In doing so we can preserve the utility of the original JWL parameters which are not upset by the new addition. Following the Grüneisen formulation,

$$ p = p_i + \frac{\Gamma}{v} \left(\varepsilon - \varepsilon_i \right). \tag{1} $$

p is the pressure, ε the internal energy, and v the relative volume. Subscript i refers to the quantity on the principal isentrope. A different Grüneisen parameter representation Γ is used here for the reason given later. The new expressions for the pressure and the internal energy on the principal isentrope are:

$$ p_i = \left[1 + F_p(v) \right] A e^{-R_1 v} + B e^{-R_2 v} + C v^{-(1+\omega)} \tag{2} $$

$$ \varepsilon_i = \left[1 + F_\varepsilon(v) \right] \frac{A}{R_1} e^{-R_1 v} + \frac{B}{R_2} e^{-R_2 v} + \frac{C}{\omega} v^{-\omega} \tag{3} $$

A type of compressibility factor is applied to the high pressure exponential term. A polynomial form is chosen for simplicity with the reference point at the CJ volume v_{cj}. We intend to maintain a continuity in pressure as well as in sound speed, and that is why the form is selected. The correction applies only when the volume is less than the CJ volume. The correction term for pressure is:

$$ F_p(v) = A_0 (v_{cj} - v)^2 + B_0 (v_{cj} - v)^3 \tag{4} $$

and for the internal energy,

$$ F_\varepsilon(v) = \left(A_0 - \frac{3B_0}{R_1} \right) \{ \frac{2}{R_1^2} [1 - e^{-R_1 (v_{cj} - v)}] - \frac{2}{R_1} (v_{cj} - v) + (v_{cj} - v)^2 \} + B_0 (v_{cj} - v)^3 . \tag{5} $$

There is no additional parameter introduced here since we have applied the isentropic relationship between the pressure and the internal energy and the new constants appearing in the pressure correction term also show up here. The continuity in the internal energy is thus maintained.

The calibration procedure is as follows. First we choose a Grüneisen parameter. For PBX 9501, we begin with $\Gamma=0.38$, the regular value of ω,[6] and fit the Hugoniot data above the CJ. The constants A_0 and B_0 are thus obtained. However, we notice that any reasonable choice of Γ is also acceptable to fit the Hugoniot data alone; here the difference is in the values of A_0 and B_0, and of course, the resulting isentrope which could be closer or farther from the Hugoniot, depending on the value of Γ. So the second stage is to check whether the selected constants A_0 and B_0 and the chosen Γ can fit the sound speed, a new feature in this work. Not surprising to us, the standard value of 0.38 seems to be a reasonable one. A different approach based solely on the sound speed in the overdriven region leads to a slightly higher value, 0.45, and also a different value of CJ pressure, 355 kbar.[5] So it seems reasonable that Γ should vary from 0.38 to 0.45 as the volume decreases, and that is the reason why we use a different symbol. We will include a volume dependence in Γ for the future work, but for now, a constant of 0.38 is found adequate for PBX 9501. Figures 1 and 2 show both the original JWL EOS and the final modified JWL EOS results for the overdriven Hugoniot pressure and sound speed.

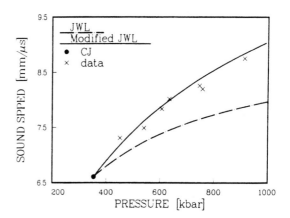

FIGURE 2. Sound Speeds.

OVERDRIVEN RELEASE EXPERIMENT AND MODELING

A brief description of the overdriven release experiment is given here. A piece of sample HE is initiated by a high speed flyer. The HE is not only initiated but also maintained at pressure above the CJ value for a period of time at a given position until the rarefraction wave from the back of the flyer reaches the same location. The information is recorded by measuring the particle velocity between the HE and a transparent window. The velocity-time history shows the constant overdriven state. As the release waves moves in, the expansion begins at a pressure above the CJ state. The experiment we are to simulate uses an aluminum flyer, 4.711-mm thick, impacting on the PBX 9501 at a velocity of 5.414 mm/μs. This provides an overdriven condition at a pressure about 520 kbar. The HE thickness is 13.108 mm and at the end a transparent window of LiF is placed. The experimental result and the two different calculations, one with the standard JWL parameters alone and another with the modified JWL are given in Fig. 3. At this pressure level, both calculations give about the same pressures, but one with the modified JWL indicated a shorter overtaken time, an interval between the arrival of the detonation wave at the HE-window interface and the arrival of

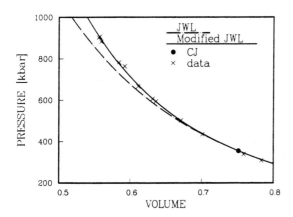

FIGURE 1. Hugoniot pressures.

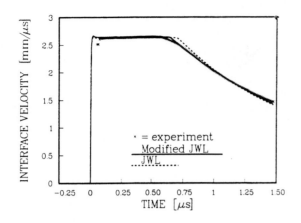

FIGURE 3. Overdriven release experiment and simulations.

the release wave originating from the back of the flyer as a result of a faster calculated sound speed, and therefore it matches the experiment better. Although in this example the improvement is minimal due to a very short HE charge length, we expect for larger systems, that the impact from the sound speed correction can be quite significant.

CONCLUSIONS

A cylinder test alone is not sufficient to cover the overdriven regime for EOS calibration. Nor is the Hugoniot pressure only. More direct methods of measuring the Hugoniot properties, both pressure and sound speed, are available and should be used for the purpose. With the new treatment, we can expand the utility of the conventional JWL EOS to a higher pressure domain using the original set of parameters as a base. The simple modification takes advantage of the fact that the JWL EOS is already available in many hydrocodes and very little programming is required. The only additional information needed is the overdriven data.

ACKNOWLEDGMENTS

The work is supported by the United States Department of Energy under contract W-7405-ENG-36.

REFERENCES

1. E. L. Lee, H. C. Hornig, and J. W. Kury, Lawrence Livermore National Laboratory report UCRL-50422, 1968.

2. P. K. Tang, *J. Prop. Expl. Pyro.*, **22**, 45-50 (1997).

3. J. H. Kineke and C. E. West, "Shock States of Four Overdriven Explosives," presented at the Fifth Symposium (International) on Detonation, Pasendena, CA, Aug. 18-21, 1970.

4. L. Green, E. Lee, A. Mitchel, and C. Tarver, "The Supra-Compression of LX-07, LX-17, PBX-9404, and RX-26-AF and the Equations of State of the Detonation Products," presented at the Eighth Symposium (International) on Detonation, Albuquerque, NM, July 15-19, 1985.

5. J. N. Fritz, R. S. Hixon, M. S. Shaw, C. E Morris, and R. G. McQueen, *J. App. Phys.*, **80**, 6129-6141 (1996).

6. B. M. Dobratz, and P. C. Crawford, Lawrence Livermore National Laboratory report UCRL-52997, Change 2, 1985.

7. E. L. Baker, "An Application of Variable Metric Nonlinear Optimization to the Parameterization of an Extended Thermodynamic Equation of State," presented at the Tenth Symposium (International) on Detonation, Boston, MA, July 12-16, 1993.

CP429, *Shock Compression of Condensed Matter – 1997*
edited by Schmidt/Dandekar/Forbes
© 1998 The American Institute of Physics 1-56396-738-3/98/$15.00

DETONATION IN AN ALUMINIZED EXPLOSIVE AND ITS MODELING

J. Lee, J.H. Kuk, S-y. Song, K.Y. Choi, and J.W. Lee

Agency for Defense Development, Taejon 305-600, Korea

To investigate detonation properties of a heavily aluminized explosive, we calibrated a reaction-rate for an aluminized explosive from two-dimensional steady-state experiments by applying the detonation shock dynamics. To verify short-term behavior of this rate equation, we numerically modeled a detonation-pressure measurement test. Though the calculated peak pressure and duration were slightly higher and longer, respectively, the profile itself was in good agreement with the experimental observation. Using the rate equation, we were able to predict most detonation properties of the aluminized explosive.

INTRODUCTION

A common detonation characteristic of most military explosives is that their reaction rate is fast and strongly dependent on thermodynamic state. As a result, they undergo shock-to-detonation transition (SDT) promptly, and reaction-rate effects on the propagation of the detonation wave can be mostly ignored. The simple rate-independent Chapman-Jouguet (CJ) theory can be used to model these explosives. These explosives are referred to as ideal explosives.

For aluminized explosives, the reaction rate is slow and weakly dependent on state. As a result, aluminized explosives undergo SDT gradually, and consequently, rate effects influence the detonation behavior greatly. The 'ideal' CJ theory cannot be used to describe their detonation behavior, and, thus, these explosives are often referred to as nonideal explosives. Detonation behavior of these explosives can be described only when their reaction rate characteristics are fully understood.

Lee[1] calibrated a reaction rate for an ammonium nitrate-based emulsion explosive, a nonideal explosive, by applying the detonation shock dynamics (DSD) developed by Bdzil[2]. Lee[1,3] re-produced results of wedge tests and an aquarium test by using this rate. This rate calibration method required only 2-D steady-state experimental data which are relatively easy to obtain.

The objective of this study is to calibrate a reaction rate for an aluminized explosive by the method developed by Lee[1] and reproduce results of a detonation experiment through numerical modeling by using the rate.

RATE-CALIBRATION METHOD

In the intrinsic coordinate system employed in the DSD[2], the shock locus is described by the shock angle, ϕ, as a function of arclength, ξ, from the reference point, (z^{l+}, r^{l+}), along the shock, at any instance of time. Then, an arbitrary position in the laboratory-coordinate system, (z_o^l, r_o^l), is given by the distance from the shock in the shock-normal direction, η, and the arclength, ξ, that is, (η, ξ). Figure 1 shows the intrinsic-coordinate system. In this figure, the subscript, o, denotes coordinates on the shock front. The curvature of the shock front, κ, in a cylindrical geometry, is defined by

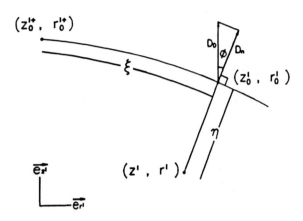

Intrinsic—Coordinate System

FIGURE 1. The intrinsic-coordinate system.

$$\kappa \equiv \phi_{,\xi} + \sin\phi/r_o^l , \qquad (1)$$

where the comma-subscript convention was used to denote a partial derivative; $\phi_{,\xi} \equiv (\partial\phi/\partial\xi)_{l,\eta}$.

Bdzil[2] showed that the detonation velocity in the shock-normal direction, D_n, is a function of the shock-front curvature: $D_n = D_n(\kappa)$, or $\kappa = \kappa(D_n)$. When the detonation is steady, $D_n = D_o\cos\phi$, where D_o is the steady velocity at a given diameter. Equation (1) can be rewritten to yield a shock locus of a cylindrical charge:

$$\phi_{,\xi} + \frac{\sin\phi}{r_o^l} = \kappa(D_o\cos\phi) . \qquad (2)$$

Equation (2) defines the dependence of ϕ on ξ with D_o as a parameter. Once the dependence of ϕ on ξ is known, the laboratory coordinates of the shock locus are obtained from the geometry; $dr = \cos\phi d\xi$ and $dz = \sin\phi d\xi$.

The diameter effect is obtained from integration of Equation (2) by utilizing the relationship that $dr = \cos\phi d\xi$, as follows:

$$r_e^l = \int_0^{\phi_e} \frac{\cos\phi}{\kappa(D_o\cos\phi) - \sin\phi/r_o^l} d\phi , \qquad (3)$$

where r_e^l and ϕ_e are the radial coordinate and the angle at the edge of the shock locus, respectively. Equation (3) defines the dependence of the radius on D_o.

Equations (2) and (3) yield the shock locus and the diameter-effect curve, respectively. Inversely, either the experimental diameter-effect curve or shock locus can serve to define $D_n(\kappa)$.

In this coordinate system, the flow in the reaction zone can be reduced to a quasi 1-D flow in the η-direction for a broadly curved detonation front. For an arbitrary equation of state (EOS), $E(P,\rho,\lambda)$, and a reaction rate, $R(P,\rho,\lambda)$, where P is pressure, ρ is density, λ is reaction extent, the flow in the reaction zone can be described by

$$(D_n - u_\eta)_{,\lambda} = -\frac{D_n - u_\eta}{C^2 - (D_n - u_\eta)^2} \left[\frac{E_{,\lambda}}{\rho E_{,P}} + \frac{\kappa C^2 u_\eta}{R} \right] , \qquad (4)$$

where D_n is detonation velocity normal to the shock front, u_η is particle velocity in the η-direction and C is local sound velocity.

When the flow becomes sonic, $C = D_n - u_\eta$, Equation (4) requires that

$$\frac{E_{,\lambda}}{\rho E_{,P}} R + \kappa C^2 u_\eta = 0 \quad \text{at} \quad C = D_n - u_\eta . \qquad (5)$$

This is called the generalized CJ condition, or the eigenvalue condition.

For known EOS, reaction rate, and κ, Equation (4) can be integrated to yield a flow in the reaction zone. Inversely, if only reaction rate is not known, it can be determined by requiring that integration of Equation (4) satisfy the eigenvalue condition. More details of the rate-calibration method can be found in reference 1.

EXPERIMENTAL TECHNIQUES

The composition of the test explosive, DXD-04, is given by RDX/ammonium perchlorate/aluminum/binder 24/43/22/11 wt%, and its density is 1.78 g/cm^3. In this study, diameter-effect data and a shock-front locus of a 60 mm diameter cylindrical charge were used to obtain $D_n(\kappa)$.

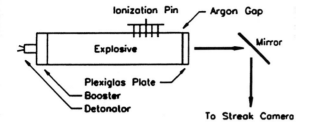

FIGURE 2. Experimental setup for shock-front locus measurement.

Detonation velocity was measured by using ionization pins for charges of diameters from 28.5 to 125 mm. The experimental setup for a shock-front locus measurement test is shown in Fig. 2. An argon flash gap was placed at the end of an explosive charge. The arrival time of the detonation front over the surface was recorded by a streak camera, Cordin 116, at a recording speed of 10 mm/μs through an approximately 100 μm wide slit.

RESULTS AND DISCUSSIONS

In determining the $D_n(\kappa)$ function, a shock-front locus of a 60 mm diameter charge and a diameter effect curve measured in the diameter range of from 27.5 to 125 mm were used. These experimental data were fitted by using the Levenberg-Marquardt algorithm to

$$\frac{D_n(\kappa)}{D_{CJ}} = 1 - 0.2842\kappa^{0.06176} - 0.04418\kappa \quad (6)$$
$$\exp[1.155(D_{CJ} - D_n)] \ ,$$

where $D_{CJ} = 7.693$ mm/μs. Figures 3 and 4 shows experimental data and fitted results.

The reaction rate for DXD-04 was obtained by using the $D_n(\kappa)$ function and the HOM EOS[4]. The EOS parameters for DXD-04 are listed in the companion paper[5]. The result is:

$$\frac{d\lambda}{dt} = 122.6(1 - \lambda)^{7.076} \quad (7)$$
$$\exp(-0.7984\rho^{5.117}/P^{1.202}) \ ,$$

where λ is reaction extent changing from 0 for unreacted explosive to 1 for reaction products,

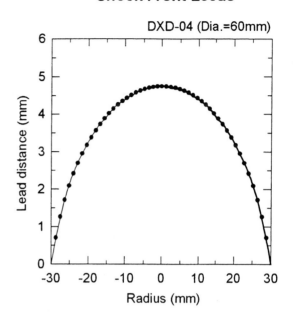

FIGURE 3. The shock-front locus for DXD-04 at 60 mm diameter. Solid line was calculated by using Equation (6).

t is time in μs, ρ is density in g/cm^3, and P is pressure in GPa.

By using this rate, a detonation pressure measurement test was numerically modeled with a 2-D Lagrangian hydrodynamic code. In this test, 25 mm diameter DXD-04 charge was confined in 4.5 mm thick steel tube, and a manganin gauge was placed between the end of the charge and a Teflon block. The detonation wave profiles obtained from the test and the present modeling are shown in Fig. 5.

The calculated peak pressure (22 GPa) was slightly higher than the experimental one (20.7 GPa), and the calculated profile had a slightly longer duration. The calculated velocity (6.2 mm/μs) was also slightly higher than the measured one (5.9 mm/μs). It was suspected that the high velocity and pressure in the calculations were caused by underestimation of energy loss through steel confinement. In general, the calculated profile was in good agreement with the experimental observation, showing that, at least, short-term characteristics of this rate is correct.

Diameter Effect

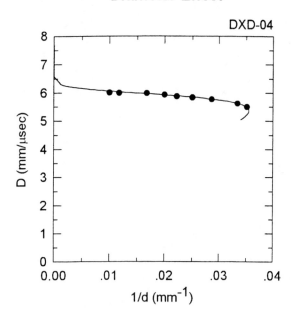

Detonation Wave Profiles

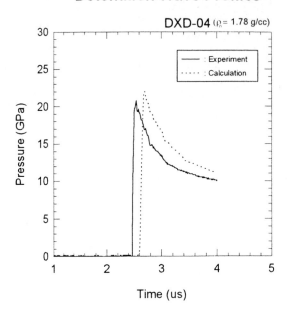

FIGURE 4. The diameter-effect curve for DXD-04. Solid line was calculated by using Equation (6).

FIGURE 5. Results of the detonation pressure measurement test and the present modeling. Arrival times were adjusted arbitrarily for ease of comparison.

For this charge (or at 5.9 mm/μs detonation velocity), the distance from the shock front to the sonic point was calculated to be about 6.1 mm and the pressure at the sonic point to be 13.7 GPa. Within this region, only 50% of explosive energy was released. The rest of the energy was released very slowly after the sonic point.

CONCLUSIONS

In the present study, a reaction rate for an aluminized explosive was obtained from 2-D steady-state experiments based on the DSD theory. This rate reproduced results of a detonation experiment very well. This good agreement demonstrated that this rate behaves correctly, at least, in a short term, suggesting that nonideal detonation behavior can be modeled by using a properly calibrated reaction rate.

REFERENCES

1. Lee, J., *Detonation Shock Dynamics of Composite Energetic Materials*, Ph. D. thesis, New Mexico Tech, New Mexico, Dec. 1990.
2. Bdzil, J.B., and Stewart, D.S., *Phys. Fluids* **A1**, 1261-1267 (1989).
3. Lee, J., Sandstrom, F.W, Kuk, J.H., and Choi, K.Y., "Numerical modeling of an aquarium test for a nonideal explosive", presented at the Tenth Int. Symp. on Detonation, Boston, MA, July 12-16, 1993.
4. Mader, C.L., *Numerical Modeling of Detonation*, Berkeley: Univ. of California Press, appendix A, pp. 316-317.
5. Kuk, J.H., Lee, J., Song, S-y., and Cho, Y.S., "Numerical modeling of an underwater explosion for an aluminized explosive," presented at the APS Symp. on Shock Compression of Condensed Matter, Amherst, MA, Jul. 27 - Aug. 1, 1997.

CP429, *Shock Compression of Condensed Matter – 1997*
edited by Schmidt/Dandekar/Forbes
© 1998 The American Institute of Physics 1-56396-738-3/98/$15.00

SHOCK LOADING AND REACTIVE FLOW MODELING STUDIES OF VOID INDUCED AP/AL/HTPB PROPELLANT

P. J. Miller and A. J. Lindfors

Detonation Sciences Section
Naval Air Warfare Center
China Lake, CA 93555

The unreactive Hugoniot of a class 1.3 propellant has been investigated by shock compression experiments. The results are analyzed in terms of an ignition and growth reactive flow model using the DYNA2D hydrocode. The calculated shock ignition parameters of the model show a linear dependence on measured void volume which appears to reproduce the observed gauge records well. Shock waves were generated by impact in a 75 mm single stage powder gun. Manganin and PVDF pressure gauges provided pressure-time histories to 140 kbar. The propellants were of similar formulation differing only in AP particle size and the addition of a burn rate modifer (Fe_2O_3) from that of previous investigations. Results show neglible effect of AP particle size on shock response in contrast to the addition of Fe_2O_3 which appears to 'stiffen' the unreactive Hugoniot and enhances significantly the reactive rates under shock. The unreactive Hugoniot, within experimental error, compares favorably to the solid AP Hugoniot. Shock experiments were performed on propellant samples strained to induce insitu voids. The material state was quantified by uniaxial tension dialatometry. The experimental records show a direct correlation between void volume (0 to 1.7%) and chemical reactivity behind the shock front. These results are discussed in terms of 'hot spot' ignition resulting from the shock collapse of the voids.

INTRODUCTION

In contrast to 1.1 propellants. 1.3 propellants are widely believed to be non-detonable and essentially unreactive at low pressures. In recent years extensive experimental studies have generated a wealth of information regarding the inherent risk and performance tradeoffs of certain class 1.1 propellant formulations. To date, the more common class 1.3 propellants, have not been afforded such treatment. Recent studies by Boteler et al.(1), Bai and Ding(2) and Huang(3) on the response of several composite propellants to shock loading have indicated that AP/Al/HTPB type propellants begin to decompose behind a 140 kbar shock front. They

observed two distinct peaks in the pressure-time profiles, one at or near the shock front and a second occurring 1-2 microseconds later. They conclude that the observed chemical reactivity behind the shock front is primarily due to the shock decomposition of AP and HTPB, and the second peak is due to the reaction between these products and the Al. The purpose of the present work is to establish the unreacted Hugoniot for an AP/Al/HTPB propellant and investigate the effect of induced damage on the shock response of these same propellant formulations. These results are analyzed in terms of an ignition and growth reactive flow model using the DYNA2D hydrocode. The calculated shock ignition and growth parameters of the model show a dependence on the induced void volume and appears to reproduce the observed reactive growth well.

EXPERIMENTAL

The propellant samples used in this study had a density of 1.836 g/cc, and it contained 20% by weight of 20μ AP, 50%-400μ AP, 20%-30μ Al, and 10%-HTPB and plasticizer. Target samples were cut into 100 mm diameter slabs 6.25 mm thick. Prior to target assembly, the samples were measured for uniformity and density. All samples were obtained from the same formulations batch to minimize non-uniformity.

Shock experiments were performed in a single stage powder gun with a 75 mm bore. The target assembly consisted of a buffer plate (driver) and two or more propellant slabs. Teflon armored gauge packages were placed at each interface . For longitudinal stress less than 75 kbar, piezoresistive manganin gauges were used; otherwise Bauer type PVDF piezoelectric gauges were employed. The stress gauges provided time of arrival for the shock wave in addition to pressure versus time histories. A small layer of urethane adhesive was used to fill any surface voids and bond the slabs together.

Impact configurations were planar and symmetric. Tilt was measured to be 1 miliradian or less. In order to achieve the desired stress range, three impact/buffer materials were used: PMMA, 2024T-4 aluminum, and 304 stainless steel. With these materials, and projectile velocities approaching 1.9 mm/msec, peak input pressures to 140 kbar were attained.

The shock experiments performed on damaged propellant samples used uniaxial tension dilatometry to quantify the state of the material prior to shock loading. This technique was developed by Lepie(4) and Richter(5) and permits simultaneous measurement of stress-strain and void volume-strain properties. For details of the method and diagnostic instrumentation, see the references.

EXPERIMENTAL RESULTS

The experimental Hugoniot for this sample

(obtained with 5 shots with input pressues ranging between 20 and 90 kbar) was determined to be $U_S=2.68+1.58u_P$. It lies within the experimental error of that obtained on 96.4% TMD AP as determined by Sandstrom(6) and is virtually indistinguisable from it. Five shock experiments were performed with static tensile strains of 5, 10, and 15%. The strains were applied along one dimension orthogonal to the shock front resulting in measured void volumes of 0.2, 0.5, and 1.7% respectively. The impact configurations (the table includes flyer/buffer material, % strain applied, % void volume measured, impact velocity, shock and particle velocity, and measured input pressure) are given in Table 1. For a given strain, the void volume (%) was determined from the uniaxial tension dilatometry results. Shock pressure, given in the last column, was measured at the first gauge plane located at the buffer-sample interface. In each case the shock experiment was performed within one hour after the static strain was applied.

The pressure time histories for four of these experiments are illustrated in Figure 1. In this figure the measured shock pressure is given in GPa units and time in microseconds. The time of arrival at the gauge planes and the sample thicknesses are used to calculate the shock velocity. Whereas the pressure growth for 5% strain is questionable, the records for 15% strain show a marked increase in pressure at the second gauge plane. This increase in pressure closely follows the shock front. For an input pressure of 86 kbar, the reactive contribution to the shock front has, clearly, over taken the initial loading.

We have demonstrated that the degree of shock induced chemical reactivity increases with applied static strain and hence with void volume. This result is not surprising since it is well known that damaged materials show greater shock sensivity. However, in a real scenario the true state of a damaged propellant prior to the subsequent arrival of a shock remains uncertain. Relaxation effects must be addressed to better understand real material response.

In the next section, we determine the effect of the applied static strain on the shock ignition and growth reactive rate parameters for this material.

Table 1. Shock Parameters for Gun Shots

SHOT	FLYER/BUFFER MATERIAL	STRAIN %	VOID VOLUME %	IMPACT VELOCITY (mm/ms)	SHOCK VELOCITY (mm/ms)	PARTICLE VELOCITY (mm/ms)	SHOCK PRESSURE (kbar)
3PS2*	2024Al/2024Al	NA	NA	1.05	3.98	0.72	52
1PS2-D	2024Al/2024Al	10.00	0.50	1.10	4.10	0.74	57
2PS2-D	2024Al/2024Al	9.78	0.50	1.11	4.00	0.71	53
3PS2-D	2024Al/2024Al	15.27	1.70	1.22	3.62	0.93	61
4PS2-D	2024Al/2024Al	5.40	0.20	1.18	3.73	0.84	58
5PS2-D	304SS/304SS	14.84	1.70	1.30	4.16	1.15	86

* Baseline study from undamaged Hugoniot

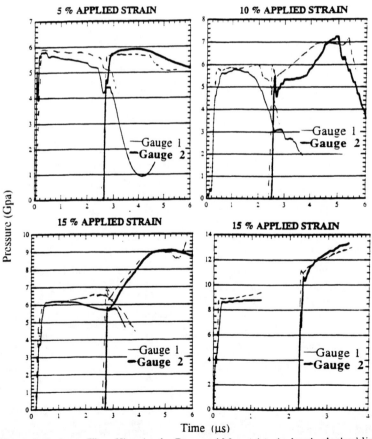

Figure 1. Pressure-Time Histories for Damaged Material (calculated - dashed line)

HYDRODYNAMIC MODELING

An ignition and growth reactive flow model(7) for the ignition and shock induced reactions for this material was developed to investigate the effects of the strain induced voids on the reaction rate constants. This model uses two Jones-Wilkens-Lee (JWL) equations of state, one for the unreacted propellant and the other one for its reaction products. The unreactive Hugoniot was measured in this work and also shown to be essentially equivalent to that of AP. The reactive Hugoniot had to be estimated from from a JWL fit to a TIGER thermochemical code(8) in which the aluminum is assumed to be inert. It is close to other JWL equations of state used for similar propellants(9). These Hugoniots were placed into the DYNA2D hydrocode in the temperature dependent form,

$$P = Ae^{-R1V} + Be^{-R2V} + \omega C_v T/V, \qquad (1)$$

where P is pressure in megabars, V is the relative volume, T is temperature, ω is the Gruneisen coefficient, C_v is the average heat capacity, and A, B, R_1, and R_2 are constants. The reaction rate law used for this propellant for the conversion of reactant into products was

$$dF/dt = I(1-F)^b(\rho/\rho_o-1-a)^x + G_1(1-F)^c F^d P^y, \quad (2)$$

where F is the fraction reacted, t is the time, ρ is the current density, P is pressure, and I, G_1, a, b, c, d, and y are constants. The first term ignites some of the solid as it is compressed by the shock creating 'hot spots' as the voids in the material collapses. The second term represents growth of reaction from the hot spots. This term models the slow deflagration processes of grain burning with pressure exponent close to one (y=1). A third term, not shown here, describes the rapid transition to detonation observed during shock SDT processes. As we are not observing these processes in the present experiments, we have neglected this term.

Using the reactive flow parameters listed in Table 2., the observed pressure-time profiles in Figure 1. were calculated (dashed lines in the figure). The procedure was to fit the high strain rate

pressure profiles and to vary the rate constants I and G linearly with void volume. The idea is that the number density of hot spots that form are proportional to the void volume and that the growth of these hot spots will also be proportional to their number. By choosing values of 9.9 and 1.8 for I and G for the 15% strain sample, and reducing these values proportional with respect to % void volume for the other time history profiles, there appears to be a fairly good dependence for this limited data set.

These results have clearly demonstrated the effects of strain on the sensitivity of propellants and that an ignition and growth reactive rate model can be used to characterize the sensitivity to shock.

Table 2. Reactive Flow Parametrs

Unreacted JWL	Product JWL	Reaction Rates
$\rho_o = 1.836$ g/cc	A=23.35Mbar	I=1.1,2.2,6.6,9.9
A=70.0 Mbar	B =0.551Mbar	a=0
B=-0.0167Mbar	R_1=6.75	b=2/3
R_1=10.0	R_2=2.276	x=4
R_2=1.0	ω=0.242	G_1= .2,.4,1.2,1.8
ω=0.8	E_o=0.70Mbar	y=1
		c=2/3
Shear modulus=0.035 Mbar		d=1/9
Yield strength=0.002 Mbar		Figmax=0.04
C_v(reactant)=1.25e-5 Mbar/K		
C_v(product)=7.73e-5 Mbar/K		

REFERENCES

1. Boteler, J. M. and Lindfors, A. J., "Shock loading Studies of AP/Al/HTPB Propellants" in *Shock Compression of Condensed matterials*, 1995, Seattle, WA, p.767, AIP.
2. Bai, C. and Ding, J., "Response of Composite Propellants to Shock Loading", in *Ninth Symposium (International) on Detonation*, 1989, Portland, Oregon.
3. Huang, F.,Bai, C.and Ding, J.,"Mechanical Response of a composite Propeelant to Dynamic Loading", in *Shock Compression of Condensed Matter*, 1991, Williamsburg, VA
4. Lepie,A. and Adicoff, A., "Advanced Physical Characterization of Solid Propellants", in *Proceedings of the JANNAF Structures and Mechanical Behaviour Group*, CPIA Publication 283, 1977, APL, Johns Hopkins University., Laurel,MD.
5. Richter, H. P., Boyer, L., Lepie, A., "Shock Sensitivity of Damaged Energetic Materials", *in Ninth Symposium (International) on Detonation*, 1989, Portland, OR.
6. Sandstrom, F. W., Persson,P., and Olinger,B., "Shock Compression of AP and AN", in Shock Compression of Codensed Matter, 1993, Colorado Springs, CO, AIP Press.
7. Tarver, C. M. and Hallquist, J. O., *Seventh Symposium (International) on Detonation*, NSWC MP 82-334, 1981, p. 488, Annapolis, MD.
8. Cowperthwaite, M. and Zwisler, W., Stanford Research Institute Report, SRI Publication No. Z106, 1973.
9. Tarver, C. M., Urtiew,P. and Tao, W., *Combustion and Flame*, Vol. 105, p.123-131(1996).

CP429, *Shock Compression of Condensed Matter – 1997*
edited by Schmidt/Dandekar/Forbes
© 1998 The American Institute of Physics 1-56396-738-3/98/$15.00

DISCRETE MESO-ELEMENT SIMULATION OF CHEMICAL REACTIONS IN SHEAR BANDS

S. Tamura and Y. Horie

North Carolina State University, Raleigh, NC 27695-7908

A meso-dynamic simulation technique is used to investigate the chemical reactions in high speed shearing of reactive porous mixtures. The reaction speed is assumed to be a function of temperature, pressure and mixing of materials. To gain a theoretical insight into the experiments reported by Nesterenko et al., a parametric study of material flow and local temperature was carried out using a Nb and Si mixture. In the model calculation, a heterogeneous shear region of 5 μm width, consisting of alternating layers of Nb and Si, was created first in a mixture and then sheared at the rate of $8.0 \times 10^7 s^{-1}$. Results show that the material flow is mostly homogeneous, but contains a local agglomeration and circulatory flow. This behavior accelerates mass mixing and causes a significant temperature increase. To evaluate the mixing of material, average minimum distance of materials separation was calculated. Voids effect were also investigated.

INTRODUCTION

In shock induced chemical reactions, various material are synthesized in very short period in few micro seconds. This high speed reaction is difficult to explain because normal chemical reactions require of the order of seconds to be completed in static case and high pressure decreases diffusion rate. Many people have considered the problem, but there is no clear solution yet. One of the popular explanations is that severe shear flow can initiate the reaction and accelerate the reaction speed. Dremin[1] proposed the roller theory that shear flow creates the nuclei of reaction products. Nesterenko et al. conducted high speed experiments[2] of reactive powder (Nb and Si) where shear deformation was well controlled. They observed that even at a sufficiently low velocity which does not initiate a global reaction, synthesized products could be seen inside intensely sheared region. Their results showed that shear deformation has an important role in initiating chemical reactions.

Reaction speed is increased by temperature, the concentration of reactive materials and melting. In the shear regions, severe deformation

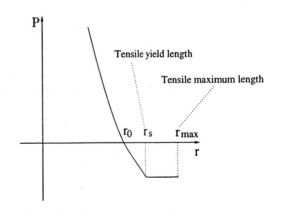

FIGURE 1. A schematic of constitutive relation.

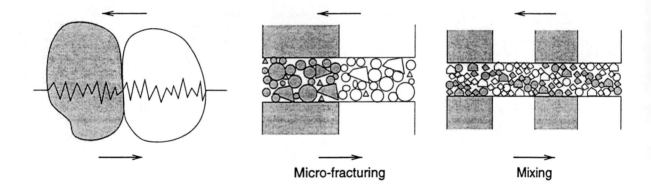

Micro-fracturing   Mixing

FIGURE 2. Formation of shear band in powder material.

causes micro-fracturing of materials, mixing, and circulatory flow of materials. Extremely high temperatures can be attained at the surface of fragmented debris resulting in micro-melting. Reaction heat also raises temperature and accelerates the reaction speed further.

In this research, the meso-dynamic simulation code DM2 is applied to calculate the morphology of material inside a shear region and to compare the results with those of Nesterenko's experiment. Local temperature increase and mixing were calculated to search the parameters that affect the chemical reactions.

NUMERICAL MODELING

DM2 is a code developed at NCSU to simulate dynamic behavior of material in high speed deformation. Materials are discretized into particulate elements. Each element has interactions with neighboring elements and the interactions are time-integrated to trace the dynamic material behavior. For the mechanical interaction, we use Lennard-Jones potential force.

$$P = \frac{-\alpha mn}{r_0(n-m)} \cdot \left\{ \left(\frac{r}{r_0} \right)^{-(n+1)} - \left(\frac{r}{r_0} \right)^{-(m+1)} \right\} \quad (1)$$

Parameters (α, m, n) are determined to represent effective dynamic behavior of the material.

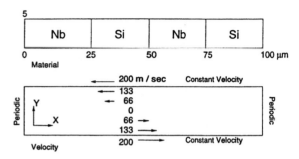

FIGURE 3. Initial structure of the material used in simulation

A schematic constitutive relation of material is shown in Fig.1. Tensile yield strain (r_s) and maximum strain (r_{max}) are also considered. In DM2, heat generation is calculated from the dissipation of mechanical energy. All of the dissipated mechanical energy are converted into heat. The mechanisms of energy dissipation are friction and viscous damping.

MODELING OF SHEAR BAND FORMATION

Nesterenko et al.(2) observed lamellar structure and circulatory flow of material in sheared regions. The width of a shear band is about 10 μm, which is smaller than the particle size (44

TABLE 1. Computational conditions in simulation.

	α/ρ	m	n	r_s	r_{max}	C_p	ρ	η	r_0
Nb	9.65×10^{10}	1	2	1.003	1.006	2.65×10^6	8.59	0.2	1.25×10^{-5}
Si	2.10×10^{11}	1	2	1.003	1.018	7.05×10^6	2.33	0.2	1.25×10^{-5}

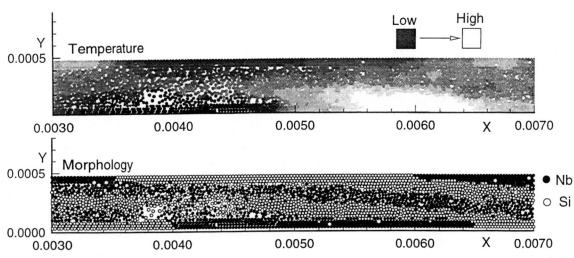

FIGURE 4. Simulated morphology and temperature of material at 2.0×10^{-7} sec.

μm). These micro structures are thought to be formed by the mechanism illustrated in Fig.2. First shear deformation causes micro fracture of powder particles. Then the weakened plane develops into a shear band. Small debris of both materials in the band undergo severe deformation due to intense shear and friction. Severe shear flow also increases temperature and initiates chemical reactions.

In our simulation, we modeled only the sheared region. The width of shear bands is assumed to be 5 μm. Initially two regions of Nb and Si with 25 μm length are created alternatively as illustrated in Fig.3. Left and right boundaries are tied together to form a planar ring. To simulate the initial flow condition, the velocity inside the shear region is assumed to be linear in Y direction. Upper and bottom boundaries move in the opposite direction with constant velocity of 200 m/sec.

SIMULATED RESULTS

Physical parameters of Nb and Si used in simulation are given in Table 1. Tensile yield strain and tensile maximum strain are chosen to represent Si as ductile material and Nb as brittle one. Viscous coefficients are the main factors that control temperature, but we did not have data to normalize them. Thus calculated temperature is qualitative.

Fig.4 shows the morphology and temperature distribution at 2.0×10^{-7} sec for the sample initially having 16 % voids. Total nominal shear strain (γ) for this picture is 16. There is no melting in this calculation. Nb particles were fractured by shear and penetrated into Si particles. Around position X=0.004, there is a concentration of voids and has a lower temperature. On the other hand, around position X=0.006 there is an agglomeration and intense mixing of material. This area has a higher temperature. This is the forward region of a large cluster of Nb. Because Nb is more brittle than Si and has higher den-

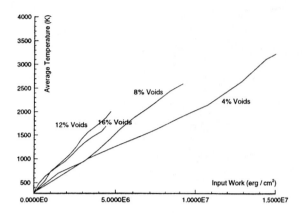

FIGURE 5. Average temperature increase.

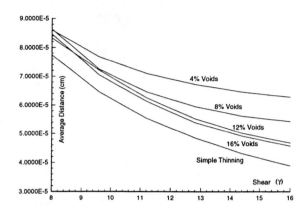

FIGURE 6. Average minimum distance between each materials.

sity, the forward area of the Nb cluster becomes dense area and absorbs much of the deformation energy. This area should become the initiation point of reaction. Fig.5 shows the average temperature of the whole specimen for four initial void ratios at different times. Average temperature increase is described by following equation.

$$\rho c_p \frac{\partial T}{\partial t} \propto \tau \dot{\gamma} \qquad (2)$$

τ is the torque of the upper and bottom boundary, $\dot{\gamma}$ is the shear rate of the specimen. This figure shows that as the void ratio increases, more mechanical energy can be converted into heat. Fig.6 shows the change of the average minimum distance of materials separation as a function of the total nominal shear strain of the specimen. The value in the ideal simple thinning case is also depicted. To calculate this value, the closest element of other material are searched for each element, and the distance between these elements are averaged for all elements. This value is thought to represent the degree of mixing of materials. As seen in the figure, the minimum distance depends on the initial void ratio. This means, voids enhance the fracture of material and accelerate the mixing of materials. It is interesting to see that simple thinning is the best case of mixing.

CONCLUSION

Shear flow of reactive porous mixture of Nb and Si was simulated by a meso-dynamic simulation technique. Temperatures inside the sheared region were calculated qualitatively and the mixing of materials are evaluated quantitatively. Following results are obtained for chemical reaction inside the shear bands. 1) The forward region of a large cluster of brittle material has a higher temperature and could become the initiation points of reaction. 2) Heat generation by input work is greater for higher void ratio. 3) Voids enhance the mixing of materials.

ACKNOWLEDGMENT

This research is supported by the Office of Naval Research under contract N00014-94-1-0240-03 to NCSU (J. Goldwasser).

REFERENCES

1. Dremin, A. N., and Breusov, O. N., Russian Chemical Revires **37** (**5**), 392-402 (1968).
2. Nesterenko, V. F., etc., Appl. Phys. Lett. **65** (**24**), 3069-3071 (1994).

CP429, *Shock Compression of Condensed Matter – 1997*
edited by Schmidt/Dandekar/Forbes
© 1998 The American Institute of Physics 1-56396-738-3/98/$15.00

TWO MECHANISMS OF EXPLOSIVE INITIATION BY THE IMPACT OF A CYLINDRICAL PROJECTILE

V.Yu. Klimenko

High Pressure SIC, Institute of Chemical Physics
Kosygin st, 4, Moscow 117334, Russia

The initiation process was studied numerically by the Multiprocess model of detonation. Numerical simulations show that the process of initiation under critical conditions (i.e., near the initiation threshold) has a different physical nature for different values of the projectile diameter. For a projectile diameter smaller than the failure diameter of the explosive we observe a "fast" initiation mechanism. During the first stage a fast explosive decomposition occurs immediately behind the shock front. The reaction spot (with completely decomposed explosive) is formed which subsequently increases in size. If at the end of the first stage the spot size becomes larger than the failure diameter of the explosive, then during the second stage the spot grows even further and results in a stable detonation. For a projectile diameter larger than the failure diameter we observe a "slow" initiation mechanism. During the first stage a slow explosive decomposition takes place far behind the shock front. If at the end of the first stage the decomposition becomes sufficiently strong (80–100 %), then during the second stage the process turns over into a detonation regime.

I. INTRODUCTION

Initiation of explosive charges by projectile impact is very important for practice, especially the initiation under critical conditions, i.e., near the initiation threshold. Therefore, this process has been studied extensively [1-7]. But, nevertheless, up to this time the process is analyzed solely on the base of the initiation criterion - which is essentially the energetical criterion. There are several versions of this criterion, but the basis for consideration is the same, namely, the quantity of energy received by the explosive from the projectile. Of course, this parameter of the initiation process is very important.

But, we understand that the relaxation processes play an important role in the total pattern of the shock wave processes (in particular, such an important relaxation process as chemical decomposition of the explosive). The characteristic time of this process (and, consequently, of the energy release) is comparable with the time of shock compression of explosive

(and, consequently, of the energy deposition). The side rarefaction reduces the pressure in the compressed explosive and results in suppression of the decomposition reaction. Hydrodynamics affects theexplosive decomposition through temperature and pressure, and the decomposition affects hydrodynamics through the released energy. It is very complex feedback. Therefore, if we analyze the initiation process (and, particularly, under critical conditions), we must take into account this relaxation process (explosive decomposition). The energetical criterion is not sufficient.

It is important for designers of new weapons to understand the physical nature the mechanism of the initiation process under critical conditions in order to know what parameters of the process are essential and what parameters are unessential. However, until now this mechanism has not been investigated thoroughly. It is impossible to study it in detail experimentally, because the process is too complex. Only the numerical simulations by a hydrocode allows one to look inside the initiation process, to understand its intimate nature. In such simulations it is necessary to accurately describe the reaction zone of explosive decomposition. All critical detonation phenomena are very sensitive to the decomposition rate. Hence, one must use a precise numerical model of detonation based on a perfect physical model.

II. CALCULATIONS AND DISCUSSION

To study the mechanism of the initiation process we use the two-dimensional Eulerian RUSS-2DE hydrocode [8,9] including the Multiprocess model of detonation [10,11]. The main calculations have been performed for PBX-9404 explosive. The reaction zone width of the explosive is about 0.1 mm. For good discretization of this zone we use a numerical grid with 0.01 mm cells.

The simulated configuration is shown in Figures 1 and 2. The explosive charge covered by an aluminium plate (0.2 mm) is impacted by a copper cylindrical projectile. The failure diameter of PBX-9404 is 1.2 mm. We have performed calculations for two projectile diameters: $d=0.6$ mm, i.e., half the failure diameter (Fig.2), and $d=2.0$ mm, i.e., about the doubled failure diameter (Fig.1). In simulations we varied the projectile velocity and have found that the critical velocity (threshold) equals 1500 m/s for $d=2.0$ mm (see Figs. 3b and 4b) and 3600 m/s for $d=0.6$ mm (see Figs. 5b and 6b).

In analysing the mechanism of the initiation process under critical conditions we can distinguish two stages. The first stage is the time period before the side rarefaction waves arrive at the region of the compressed explosive. During this stage the impactor plays a decisive role in the process. During the second stage (after arrival of the side rarefaction) chemical decomposition of the explosive controls the development of the process.

Numerical simulations show that the initiation process under critical conditions has a different physical nature for different values of the projectile diameter. If the projectile diameter is larger than the failure diameter then the initiation mechanism is partially similar to the mechanism of the one-dimensional initiation by a flyer plate of a limited width. During the first stage a slow explosive decomposition takes place far behind the shock front. If at the end of the first stage the decomposed fraction is high (80-100 %), then during the second stage the process runs away to a detonation regime (see Fig.7). It is a "slow" initiation mechanism. It develops through the shock-to-detonation process.

If the projectile diameter is smaller than the failure diameter of the explosive, we observe a "fast" initiation mechanism. During the first stage a fast explosive decomposition occurs immediately behind the shock front. The reaction spot with completely decomposed explosive is formed which subsequently increases in size. If to the end of the first stage the spot size becomes larger than the failure diameter of the explosive, then during the second stage the spot grows even further and gives rise to stable detonation (see Fig.8).

III. CONCLUSION

There are two different mechanisms of explosive initiation by the impact of a cylindrical projectile: the "fast" initiation mechanism and the "slow" initiation mechanism. Any attempts to push them in one descriptive formula, i.e., an initiation criterion (for example, V^2d criterion) are incorrect. It is impossible to apply the same initiation criterion for physically different shock initiation processes. It is important to emphasize that both mechanisms belong to a class of shock initiation.

IV. ACKNOWLEDGEMENTS

The author would like to thank Prof. Yves Coer, Dr. J.-P. Choquin (Centre d'Etudes de Vaujours, CEA, France) for useful discussions and Prof. M. de Gliniasty, (Commissariat a l'Energie Atomique / DAM, France) for finantial support. This work has received the computational support from Laboratoire de Modelisation en Mecanique, Universite Pierre et Marie Curie, Paris.

V. REFERENCES

1. Green, L., "Shock Initiation of Explosives by the Impact of Small Diameter Cylindrical Projectiles", *The VII Symposium (Int.) on Detonation*, Annapolis, USA, 1981, pp. 273-277.

2. Bahl, K.L., Vantine, H.C., and Weingart, R.C., "The Shock Initiation of Bare and Covered Explosives by Projectile Impact", *ibid.*, pp. 325-335.

3. Moulard, H., "Critical Conditions for Shock Initiation of Detonation by Small Projectile Impact", *ibid.*, pp.248-257.

4. Barker, M.A., Bassett, J.F., Connor, J., and Hubbard, P.J., "Response of Confined Explosive Charges to Fragment Impact", *The VIII Symposium (Int.) on Detonation*, Albuquerque, USA, 1985, pp. 262-270.

5. Cook, M.D., Haskins, P.J., and James, H.R., "Projectile Impact Initiation of Explosive Charges", *The IX Symposium (Int.) on Detonation*, Portland, USA, 1989, pp. 1441-1450.

6. Cook, M.D., Haskins, P.J., and James, H.R.,"An Investigation of Projectile and Barrier Geometry Effects on Impact Initiation of a Secondary Explosive", *Shock Compression of Condensed Matter-1991*, 1992, pp. 675-678.

7. James, H.R., Grixti, M.A., Cook, M.D., Haskins, P.J., and Stuart Smith, K., "The Dependence of the Response of Heavily-Confined Explosives on the Degree of Projectile Penetration", *The X International Detonation Symposium*, Boston, USA, 1993, pp. 89-93.

8. Klimenko, V.Yu., Kozlov, I.M., Romanov, G.S., and Suvorov, A.E., "Shock Wave Modeling by TVD Method", *International Conference "Shock Waves in Condensed Matter"*, St.Petersburg, Russia, 1994, p. 29.

9. Romanov, G.S., Suvorov, A.E., Kozlov, I.M., and Klimenko, V.Yu., "RUSS-2DE Hydrocode with the TVD Procedure", *International Workshop on New Models and Numerical Codes for Shock Wave Processes in Condensed Media*, Oxford, UK, September 15-19, 1997, p. 5.

10. Klimenko, V.Yu., "Multiprocess Model of Detonation (Version 3)", *Shock Compression of Condensed Matter - 1995*, Eds. Schmidt, S.C., Tao, W.C., 1995, pp.361-364.

11. Klimenko, V.Yu., "Multiprocess Model of Detonation (Version 3)", *Chemical Physics (Russian)*, **17**, N 1, 11-24 (1998).

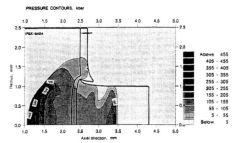

FIGURE 1. A general picture of the numerical experiment. The copper projectile: diameter=2.0 mm, velocity=1700 m/s.

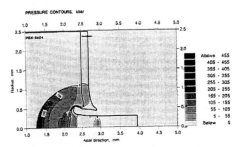

FIGURE 2. A general picture of the numerical experiment. The copper projectile: diameter=0.6 mm, velocity =3800 m/s.

FIGURE 3 (left). Pressure contours (kbar) for simulations of impacts by the copper projectile (diameter =2.0 mm) with velocity: a) 1600 m/s - detonation, b) 1500 m/s - threshold and c) 1400 m/s - failure of detonation.

FIGURE 4 (right). Decomposition contours for the same events as presented at Fig.3. Darkness-comletely decomposed explosive.

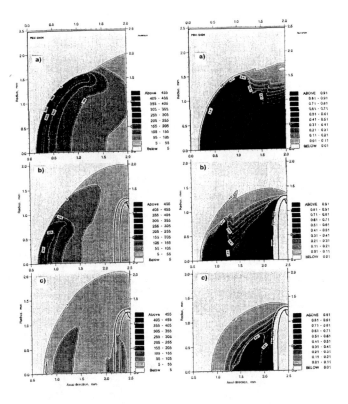

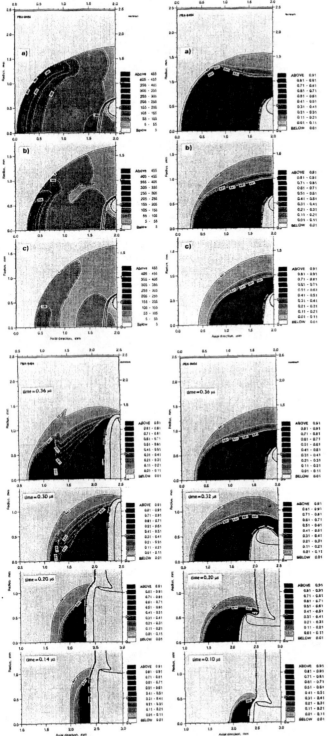

FIGURE 5 (left). Pressure contours (kbar) for simulations of impacts by the copper projectile (diameter = 0.6 mm) with velocity: a) 3800 m/s - detonation, b) 3600 m/s - threshold and c) 3500 m/s - failure of detonation.

FIGURE 6 (right). Decomposition contours for the same events as presented at Fig.5.

FIGURE 7 (left). Decomposition contours. Progress of the explosive decomposition during initiation by the projectile (diameter=2.0 mm, velocity=1500 m/s - initiation threshold). Time=0.30 μs is the critical moment in the initiation development. Darkness - completely decomposed explosive.

FIGURE 8 (right). Decomposition contours. Progress of the explosive decomposition during initiation by the projectile (diameter=0.6 mm, velocity=3600 m/s - initiation threshold). Time=0.20 μs is the critical moment in the initiation developmen.

CP429, *Shock Compression of Condensed Matter – 1997*
edited by Schmidt/Dandekar/Forbes
© 1998 The American Institute of Physics 1-56396-738-3/98/$15.00

OBLIQUE SHOCK WAVE CALCULATIONS FOR DETONATION WAVES IN BRASS CONFINED AND BARE PBXN-111 CYLINDRICAL CHARGES

E. R. Lemar

Naval Surface Warfare Center, Indian Head Division, Indian Head, Maryland 20640

J. W. Forbes

Lawrence Livermore National Laboratory, Livermore, California 94550

M. Cowperthwaite

Enig Associates, Inc., Silver Spring, Maryland 20904

Shock polar theory is used to calculate the angles detonation fronts make with the cylinder wall for brass cased and bare PBXN-111 cylinders. Two extrapolated unreacted PBXN-111 Hugoniot curves are used to calculate these angles. Measured and calculated angles for bare PBXN-111 cylinders are in good agreement for one of the unreacted PBXN-111 Hugoniots. Except for the 100 mm diameter charge, the differences between calculated and measured angles for brass cased charges are beyond experimental error. Limited data suggests that the wave front curvature exhibits a large change right at the brass wall and the resolution in the experiments may not be fine enough to show it clearly.

INTRODUCTION

Oblique shock equations for plane shocks, along with two different unreacted PBXN-111 Hugoniots are used to calculate the angle of the detonation front with respect to the charge axis for both cased and uncased PBXN-111. The angles obtained from these calculations are compared with the angles measured experimentally.

CALCULATIONS

In the following calculations, it is assumed that no reaction occurs in the shock front where the stress jumps from zero to its maximum value. Figure 1 depicts the flow through a stationary oblique plane shock front , δ, where the stress ahead of the

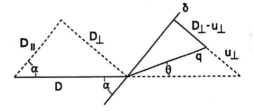

FIGURE 1. Oblique Shock Flow

shock front is zero. Hydrodynamic flow is described by one dimensional shock equations for the flow component normal to the front and conservation of flow across the shock front for the component parallel to the front (1). The equations for this flow are

$$\rho_0 D_\perp = \rho_1 (D_\perp - u_\perp) = \rho_1 \, q \, \sin(\alpha-\theta) \qquad (1)$$

$$P_1 = \rho_0 D_\perp u_\perp = \rho_0 D \sin(\alpha)[D \sin(\alpha) - q \sin(\alpha-\theta)] \quad (2)$$

$$\rho_0 D^2 \sin(\alpha) \cos(\alpha) = \rho_1 q^2 \sin(\alpha-\theta)\cos(\alpha-\theta) \quad (3)$$

where the subscripts o and 1 designate the states ahead of and behind the shock front, respectively. Density is denoted by ρ, stress by P, angle of incoming flow relative to the charge axis by α, angle of flow behind shock front by θ, and flow velocities in and out of the front by D and q, respectively. Note that D is the detonation velocity down the axis of the cylinder.

The determination of the angle the wave makes at the cylindrical brass wall for cased charges requires a boundary condition. In these calculations, we assume that the flow is the same (i.e. P and θ are the same) across the shock front at the explosive/brass boundary (2). Equations 1-3 are used to calculate shock polar (P,θ) curves for both brass and PBXN-111 for a specific value of D (which is a function of the charge diameter). The intersection of these two shock polar curves (see Figure 2) defines the state where the flow is equal through the shock front for the brass and the explosive.

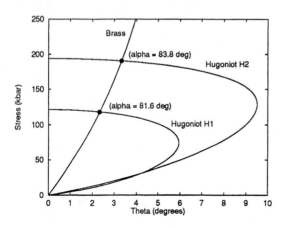

Figure 2. Shock Polars for the 47.0 mm Diameter Cased Charge

For uncased charges, the flow at the charge/air boundary is assumed to be Prandtl-Meyer flow. For this case, the (P,θ) state used is that for which θ is a maximum (see Figure 3). Substitution of these values for P and θ into the oblique shock equations result in a solution for α at the edge of the charge. The calculated values for α are given in Table 1.

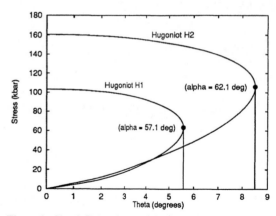

Figure 3. Shock Polars for the 40.9 mm Diameter Bare Charge

EXPERIMENTAL WORK

Figure 4 gives the experimental schematic for measuring the angle at the boundaries. The streak camera technique for making this measurement has been discussed previously (3). The measured angles in this paper were determined by least squares fitting the last 2 to 5 mm of the breakout traces to a straight line. The angles at the tops of the traces, α_T, and at the bottoms of the traces, α_B, are given in Table 1.

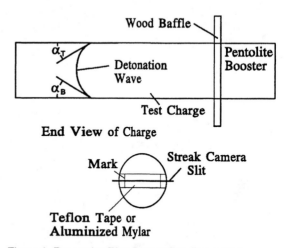

Figure 4. Detonation Wave Breakout Experimental Setup

RESULTS AND DISCUSSION

Calculations of the angles, α, for the cased and bare charges were done using two different extrapo-

TABLE 1. Parameters Obtained From Fitting Wave Breakout Data

Diameter (mm)	Length (cm)	Case Thick (mm)	Meas α_T (deg)	Meas α_B (deg)	Calc α_{H1} (deg)	Calc α_{H2} (deg)
25.2	17.67	5.08	76.2±1.1	75.2±1.1	82.6	84.4
37.9	21.72	5.00	76.0±1.3	72.9±1.1	81.9	84.0
47.0	25.33	5.08	75.6±1.2	73.1±1.3	81.6	83.8
100.0	37.84	5.00	81.7±1.5	81.4±2.0	81.1	83.5
40.9	32.56	0.00	61.4±1.1	61.1±1.3	57.1	62.1
41.05	32.86	0.00	63.9±0.9	63.9±1.0	57.1	62.1
48.02	30.49	0.00	63.8±1.1	62.2±1.1	56.9	61.9
48.12	36.67	0.00	61.9±1.2	63.3±1.1	56.9	61.9
68.25	30.55	0.00	62.1±1.3	63.0±1.9	56.6	61.6
99.9	37.81	0.00	67.3±3.7	61.1±1.4	56.4	61.3

lated Hugoniots for PBXN-111. One unreacted Hugoniot (H1: U_S = 2.118 mm/μs + 2.758 u_p) was obtained from gas gun experiments (4) over a stress range of 7 to 42 kbars. The second Hugoniot (H2: U_S = 2.801 mm/μs + 1.38 u_p) was obtained from wedge tests (5) over a stress range of 39 to 112 kbar. The Hugoniot determined from wedge test data was likely influenced by reaction at the shock front (4). The initial density for PBXN-111 was 1.79 g/cm^3. The Hugoniot for brass (6) is U_S = 3.73 mm/μs + 1.434 u_p for an initial density of 8.45 g/cm^3. Table 1 contains the calculated angles, α, for the brass cased and bare PBXN-111 charges. For the brass cased charges the calculated values range from 81 to 84 degrees. These values show a slight dependence on the particular PBXN-111 Hugoniot used. For the bare charges, the calculated values range from 56 to 62 degrees and and are more dependent on the particular PBXN-111 Hugoniot used.

Figure 2 shows the shock polars for the 47.0 mm diameter brass cased charge, and figure 3 shows the shock polars for the 40.9 mm diameter bare charge. The shock polars for both extrapolated unreacted Hugoniots of PBXN-111 are shown in each figure. These figures show that accurate stress measurements across the charges would help determine the appropriate Hugoniot to use in the shock polar calculations. In addition, very accurate measurements of α at the boundaries would provide unreacted Hugoniot information at high pressures through the oblique shock equations.

Examination of Table 1 reveals that measured

and calculated α's for uncased PBXN-111 are in good agreement if the H2 Hugoniot for PBXN-111 is used. Except for the 100 mm diameter charge, the calculated and measured α's for brass cased PBXN-111 do not agree for either Hugoniot. Both calculated α's for the 100 mm diameter charge are in agreement with the measured α's.

One explanation for the difference between calculated and measured α's for the cased charges is that the wave front has a large change in curvature right at the brass/explosive boundary. The data reported here comes from experiments designed to measure the breakout profile over the entire diameter of the charge, not just at the edge. As a result, the magnification on most of the streak camera films may not be adequate to clearly resolve a change in curvature right at the brass/explosive interface. In the 100 mm diameter cased charge, it was resolved. Evidence of an abrupt change in wave curvature at the brass/explosive interface was also observed on the 47 mm diameter cased experiment. However, this change occurred over a 1.2 mm long region at the edge which does not allow for an accurate determination of α right at the case. These data suggest that experiments with better spatial resolution be performed. Designing experiments to look at the angle right at the brass/explosive interface would give improved spatial resolution and therefore improved accuracy of α values.

A second possible explanation for the disagreement between the measured and calculated α's for the cased charges is that the flow vectors in the brass and the explosive at the interface are not equal. If this is the case, more theoretical work on boundary conditions is needed.

REFERENCES

1. Courant, R, and Friedrichs, K.O., *Supersonic Flow and Shock Waves*, Springer Verlag, NY (1948, reprinted 1976), pp. 297-302.
2. Stewart, D. Scott and Bdzil, John B., *Proc. Tenth Symposium (International) on Detonation*, Boston, MA, July 1993, pp. 781-783.
3. Forbes, J. W., Lemar, E. R., Baker, R. N., *Proc. Ninth Symposium (International) on Detonation*, Portland, Oregon, 1989, pp. 806-815.
4. Forbes, J. W, Lemar, E. R., Sutherland, G. T., and Baker, R. N., NSWCDD/TR 92/164, 19 March 1992 (unpublished).
5. Private communication, J. Dallman, Feb. 17,1987, memo reporting wedge test on PBXW-115.
6. Mitchell, A. C., Van Thiel, M., Coleburn, N. L., Forbes, J. W., *J. Appl. Phys.*, **45**, Sept. 1974.

CP429, *Shock Compression of Condensed Matter – 1997*
edited by Schmidt/Dandekar/Forbes
© 1998 The American Institute of Physics 1-56396-738-3/98/$15.00

MECHANICAL IGNITION OF COMBUSTION IN CONDENSED PHASE HIGH EXPLOSIVES

J. C. Foster, Jr.*, F. R. Christopher*, L. L. Wilson**, J. Osborn[+]

Wright Laboratory/Armament Directorate, Eglin AFB, FL 32542
***Science Applications International Corporation, Eglin AFB, FL 32542*
[+]*Orlando Technology, Inc., Shalimar, FL 32578*

Condensed phased high explosives are metastable materials that release energy when subjected to a range of thermal, mechanical, and electrical stimuli. Recent interest in safety and low order response from these materials has emphasized the importance of the ignition of combustion under low amplitude (<10kbar), long duration (millisec) loads. The major scientific challenge is a physically based understanding of the processes leading to the ignition of combustion which will predict the violence and extent of the reaction based on global characterization of the materials. A model of the process of ignition of combustion has been formulated based on modification to the Frank-Kamenetskii equations for thermal explosions. With the modifications, the model represents a fully coupled thermal/ mechanical/ chemical kinetic global description of the ignition of combustion. The model places emphasis on the low-pressure equation of state parameters, the pressure dependence in the reaction rate kinetics, and the high rate mechanical properties of these materials. The scale of the thermal explosion problem in these materials is discussed and compared to the scale of classical hot spot theory in detonation physics. Literature references to the reaction kinetics resulting from one-dimensional time to explosion experiments and mechanical property data are included in the discussion. New impact experiments are being developed to better understand the relationship between the mechanical properties and the ignition threshold. Current research is focused on tritonal, PBXN109, and other explosives of interest.

INTRODUCTION

Condensed phased high explosives are metastable materials that release energy when subjected to a wide range of thermal, mechanical, and electrical stimuli. The violence of the energy release is dependent on the magnitude of the stimuli. Their distinguishing feature is their ability to support a detonation wave as a very rapid form of energy release, which is the characteristic that warhead designers find desirable for generating blast and fragmentation. The processes leading to detonation are generally characterized as shock to detonation transitions (SDT), deflagration to detonation transitions (DDT), or unknown detonation transitions (XDT). SDT processes are those normally associated with the intentional initiation of the material. DDT and XDT process are generally associated with unintentional initiation that still leads to detonation. The latter results from detonation phenomenology in particulate beds of materials. These processes are well documented in the literature and are the subject of continuing research. However, ignition of deflagration (combustion) without detonation can produce very violent events. These events are a source of safety concerns as well as the cause of failures in penetrating munitions. The present work focuses on this ignition problem. By focusing on the ignition problem, the research will provide information on the entire range of response from low order events to DDT events. The XDT problem involves detonation transitions in particulate beds. The establishment of the fractured bed by mechanical processes requires a fundamental understanding of fracture mechanics that has not been included in the problem definition to date. The engineering problem is associated with undesirable low order responses resulting from low amplitude (<10kbar), long duration (millisec) loads. The major scientific challenge is a physically based understanding of the processes leading to the ignition of combustion which will predict the violence and extent of the reaction based on global characterization of the materials.

THEORY

The problem of ignition of combustion in condensed phase explosives is distinguished from initiation of detonation by scale. The mechanisms associated with the trigger and propagation of a detonation have been characterized as friction, viscous heating, jetting, void collapse (um scale), internal shear, and shock interaction.[1] The scale of the stimuli for the energy release in the detonation process is 10^{-12} m and 10^{-10} sec. An analysis of a fragment impact environment and the problem of survivability of explosives in high velocity penetrating weapons yield a different list of stimuli. We have characterized these as friction (case wall/ charge, fractured bed), PdV heating, void collapse (macro-scale damage), internal shear, and low amplitude long duration shocks. The scale of these processes is 10^{-1} m, 10^{-2} sec. In order to identify the environment producing a sustained reaction, we have characterized the process as a thermal

explosion. Thermal explosions can be modeled by a set of conservation equations addressing the conservation of fuel,

$$V\frac{dc_f}{dt} = -VB_0 fe^{-E/RT} - Sh_f c_f, \qquad (1)$$

and the conservation of energy,[3]

$$V\rho c_v \frac{dT}{dt} = VQ_f B_0 fe^{-E/RT} - Sh_r(T-T_w) \qquad (2)$$

Where:

V = volume element of concern
S = surface area of volume element
c_f = fuel concentration
T = temperature
Q_f = molar heat of reaction
$B_0 fe^{-E/RT}$ = standard Arrhenius expression

The Frank-Kamenetskii equations yield conditions for thermal instability based on the scale of the reactive volume, and the chemical kinetic and thermodynamic properties of the material. The energy equation has been modified to include additional source terms to account for mechanical localization of energy and the Arrhenius expression has been modified to explicitly include pressure dependence on the burn rate and depletion.

Modified Conservation of Fuel: (3)

$$V\frac{dc_f}{dt} = -VB_0(1-\lambda)^n\left(\frac{p}{p_0}\right)^m fe^{-E/RT} - Sh_f c_f$$

Modified Conservation of Energy: (4)

$$V\rho c_v \frac{dT}{dt} = VQ_f B_0(1-\lambda)^n\left(\frac{p}{p_0}\right)^m fe^{-E/RT} + V\frac{d}{dt}\{\int\sigma_{ij}d\varepsilon_{ij}\} - Sh_r(T-T_w)$$

In this form the equations represent a fully coupled thermo-mechanical-chemical kinetic formulation of the thermal explosion problem in a reactive media. Here the mechanical properties can serve as additional source terms for energy localization to enhance the reaction rate, and the reaction rate dependency on pressure is included explicitly. Various researchers have presented reduced forms of the thermal explosion equation to investigate the behavior of one-dimensional time to explosion experiments and shear ignition experiments.[2] The complete form as expressed in equation 4 is being used to guide the materials characterization work. Explicit in the formulation of the problem is a list of mechanical, thermodynamic, and chemical kinetic parameters required to understand the thermal explosive response of energetic materials. These include:

Mechanical:

Mechanical Equation of State- $\sigma(\varepsilon, \dot{\varepsilon}, T, P)$

Thermodynamic:

Thermodynamic Equation of State- $E(P.T)$

Bulk Modulus- $\kappa_s = -\dfrac{1}{V}\left(\dfrac{\partial V}{\partial P}\right)_s$

Heat Capacity - C_v

Thermal Conductivity - h_T

Density- Theoretical Maximum Density (TMD)
　　　　As fabricated density

Chemical Kinetic:

Heat of Combustion - Q_f

Reaction Rate- Arrhenius Constants
　　　　　　　Pressure Dependency

Many of these parameters are part of the traditional thermo- mechanical characterization of energetic materials. The notable exclusions are the high rate parts of the mechanical equation of state together with the pressure dependency on the burn rate. These areas are currently being addressed by the development of experimental tools. The fracture mechanics problem will have to be included in the discussion at a later time.

The relationships between the thermodynamic and mechanical variables are given in terms of the first and second invariant of the stress tensor. The state variables in the thermodynamic equation of state,

$$E = E(P, V), \qquad (5)$$

can be related to the first invariant of the stress and strain tensor, [5]

Stress : $\sigma_{ij} = \sigma_{ji}$

Strain : $\varepsilon_{ij} = \varepsilon_{ji}$

Pressure (1st Invariant): $P = -\dfrac{1}{3}\sigma_{kk}$ (6)

Volumetric Strain : $\varepsilon_v = \dfrac{dv}{v} = \varepsilon_{kk}$

while the mechanical properties are described by the 2nd invariant,

Deviatoric Stress: $S_{ij} = \sigma_{ij} - \dfrac{1}{3}\delta_{ij}\sigma_{kk}$

Deviatoric Strain: $e_{ij} = \varepsilon_{ij} - \dfrac{1}{3}\delta_{ij}\varepsilon_{kk}$ (7)

Experimental designs must incorporate the measurement of mechanical behavior consistent with the methods used in continuum mechanics design codes, [4] e.g.:

Equivalent Stress

$\bar{\sigma} = \bar{\sigma}(\varepsilon, \dot{\varepsilon}, T)$

VonMises Yield (8)

$\bar{\sigma} = \sqrt{\dfrac{3}{2} S_{ij} S_{ij}}$

Equivalent Strain

$\int d\varepsilon^P = \int \sqrt{\dfrac{2}{3} d\varepsilon_{ij}^P d\varepsilon_{ij}^P}$

A variety of mathematical forms of the yield surface have been proposed to describe the mechanical behavior of explosives. These include:

Holmquist, Johnson, and Cook (1993), [6]

$$\sigma^* = \left[A(1-D) + B\varepsilon^n\right]\left(1 + C\ln\dot{\varepsilon}^*\right)\left(1 - T^m\right), \quad (9)$$

Chi Chou, P., et al (1995), [7]

$$\sigma^* = \left(A + BP^{*n}\right)\dot{\varepsilon}^{cT}\left(1 - T^*\right), \qquad (10)$$

and Bardenhagen, Harstad, Foster, and Maudlin (1995), [8]

$$\sigma + \dfrac{\eta(\dot{\varepsilon})}{E}\dot{\sigma} = \bar{E}\varepsilon + \dfrac{\eta(\dot{\varepsilon})}{E}\left(E + \bar{E}\right)\dot{\varepsilon}. \qquad (11)$$

391

The first is a derivative of Swift's equation which addresses damage, work hardening, thermal softening, and strain rate sensitivity and was originally formulated to address the mechanical behavior of concrete. The second addresses pressure hardening and the third is a classical visco-elastic model. These material models range from a brittle material with void volume to a material exhibiting classical visco-elastic/plastic behavior. Costantino and Ornellas [9] investigated initial high-pressure dependency.

EXPERIMENTAL

The experimental characterization of mechanical and thermodynamic equations of state together with the chemical kinetics for the range of state variables identified in the ignition problem requires a range of experimental techniques. The characterization of the low-strain, low-strain rate, low-pressure behavior of many of the materials of interest have been documented.[11] The focus of the current work is the high strain rate mechanical behavior and determination of ignition threshold. Impact techniques are widely used to measure the high rate mechanical properties of metals. Initial techniques yield only a single number, σ_0, which has been characterized as the flow stress of the material[12] Modern instrumentation techniques[13] and improved understanding[14, 15] of the experiment have significantly expanded the amount of information that can be accessed. These techniques have been successfully applied to non-metallic materials and are now being developed for application to energetic materials.

CONCLUSIONS

An approach has been developed based on thermal explosion theory and previously developed experimental techniques to investigate the influence of mechanical properties on the thermal stability of high explosives subjected to long-duration, low-amplitude mechanical stimuli. Experimental techniques are required to measure the pressure dependence in the reaction rate and are under

development. The longer-term goal includes the fracture behavior in order to link up to the technology base associated with DDT/XDT processes.

REFERENCES

1. Davis, W. C., *HIGH EXPLOSVES: The interaction of chemistry and mechanics*, Los Alamos Science, Winter/Spring 1981, Vol. 2 No.1.
2. Boyle, V.; Frey, R.; and Blake, O., "Combined Pressure Shear Ignition of Explosive," presented at the Ninth Symposium (International) on Detonation, Portland Oregon, , pg. 3-17, 28 August 1989.
3. Williams, F. A., *Combustion Theory (2nd ed.),* Addison-Wesley Publishing Company, 1985.
4. McClintock, F. A. and Argog, A. S., *Mechanical Behavior of Materials,* Addison-Wesley Publishing Company, 1965.
5. Malvern, L. E., *Introduction to the Mechanics of a Continuous Medium,* Prentice - Hall, Inc. 1969.
6. Holmquist T., Johnson G., and Cook W., "A Computational Constitutive Model for Concrete Subjected to Large Strains, High Strain Rates, and High Pressure," Presented at the Seventh Int. Symposium on Ballistics, 1993.
7. Chou, P. C., Clark W., and Liang D., "Blunt Cylinder Impact tests for the Determination of the Constitutive Equation of Explosives," presented at the 15th International Symposium on Detonation. 1995.
8. Bardenhagen S., Harstad E., Foster J., and Maudlin P., "Viscoelastic Models for Polymeric Composite Materials," presented at the APS Topical Conference on Shock Compression of Condensed Matter, August, 1995.
9. Costino M. and Ornellas D., "Initial Results for the Failure Strength of a LOVA Gun Propellant at High Pressure and Various Strain Rates," presented at the JANNAF Propulsion Meeting, LLNL, San Diego, CA 1985.
10. Costino M. and Ornellas D., "The Experimental, High Pressure Equation of State of a Very Fast Burning Gun Propellant," presented at the JANNAF Combustion Meeting, Laurel, MD, October 1984.
11. Hall, Thomas N and Holden, J. R., *Navy Explosives Handbook,* NSWC, October 1988, pp. 88-116.
12. Taylor, G. I. Proc. R. Soc. London Ser. A., (1948), pp. 194, 289.
13. Wilson, L. L., House, J. W., and Nixon, M. E., *Time Resolved Deformation from the Cylinder Impact Test*, Air Force Armament Laboratory Report, AFATL-TR-89-76, 1989.
14. Jones, S. E, Gillis, P. P., Foster Jr., J. C., and Wilson, L. L., *Recent Advances in Impact Dynamics of Engineering Structures -1989*, Vol. 17, ASME.
15. Foster Jr., J. C., Maudlin, P. J., and Jones, S. E., "On the Taylor Test: A Continuum Analysis of Plastic Wave Propagation," presented at the APS Topical Conference on Shock Compression of Condensed Matter, August 1995.

CP429, *Shock Compression of Condensed Matter – 1997*
edited by Schmidt/Dandekar/Forbes
© 1998 The American Institute of Physics 1-56396-738-3/98/$15.00

AIR CUSHION EFFECT IN THE SHORT-PULSE INITIATION OF EXPLOSIVES *

J. N. Fritz and J. E. Kennedy

Los Alamos National Laboratory, Los Alamos, New Mexico 87545 USA

When thin flyer plates are used to shock initiate high explosive (HE), any air present ahead of the flyer may cause a significant desensitization of the HE. The effect of the air in cushioning the impact of plastic flyers faced with metal films is analyzed here with MACRAME, a code which calculates wave interactions and traces wave propagation. We find that the second air shock into the HE has sufficient pressure to collapse the HE to crystal density or higher. Precompressed regions of HE do not react rapidly when the main impact pulse does arrive. Define y^* as the depth where the major shock overtakes the precompression wave (for no air $y^* \to 0$). For various flyers and air combinations, we compare pressure profiles at $y = y^* + \epsilon$. The shock pressure profile associated with metal film impact may be greatly attenuated at the depth y^*. Density profiles ($\rho(t)$ at y) show that the shock heating for $y > y^*$ is greater than that for $y < y^*$.

INTRODUCTION

The interplay between shock pressure, P, and shock duration, t, is of interest in defining threshold loading conditions for initiation of detonation of high explosives (HE). An "electric gun"(1) has been used at LLNL for such studies of one-dimensional planar initiation of HE; it uses thin, large-diameter dielectric flyer plates driven electrically. The analysis presented here addresses the use of metal-coated plastic dielectric films for similar studies, and directs attention at the effect of air in the flight path of a flyer plate.

There are two main effects due to air trapped between the flyer plate and the HE. First, the mass and back pressure of the shocked air tends to decelerate the flyer. Secondly, impingement of air shocks onto the HE surface may cause precompression of the HE. If the precompressed region of the HE is significantly compacted, this region can be substantially desensitized and it may not react upon arrival of the main shock due to flyer-plate impact. Setchell(2) showed that granular HNS, precompressed by a ramp-shaped acceleration wave, did not react upon subsequent arrival

of a strong shock, until the ramp wave had steepened into a shock.

The problem we have addressed may be stated as follows. A plastic backing plate is coated with a thinner coating of metal. This coated plate impacts a slab of HE. Air trapped between the flyer and HE mitigates this impact. Our interest is in evaluating the shape of the loading pulse that influences the initiation of the HE. The 1-D geometry of a reference problem is sketched in Fig. 1. We compare variations against the results of this problem. The plastic base layer is $80\,\mu m$ thick, the plate is assigned an initial velocity of $4\,km/s$ and the explosive is TATB at a density of $1.81\,g/cm^3$ in all the examples we address, while all other parameters are varied. Only the mechanical response of the HE is considered in our calculations; we have not introduced a reaction rate law or any energy release.

THE MACRAME CODE AND PHENOMENOLOGY

In analyzing this problem we needed to resolve a rather wide range of interactions, on both distance and pressure scales. This included the states of singly shocked air, multiply shocked air,

---
*This work supported by the US Department of Energy.

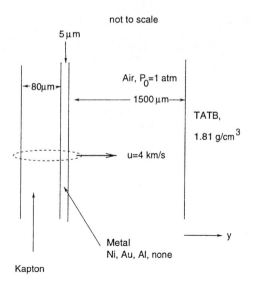

FIGURE 1. Reference One-Dimensional Problem

air-HE and plate-HE interactions.

The code chosen for this analysis was the MACRAME code, which calculates 1-D wave interactions and traces wave propagation. The definition obtained in these calculations is sharp, with no rounding due to artificial viscosity. For MACRAME, coarse resolution in one area and fine resolution in another are compatible. For the EOS of TATB(3) we used a quadratic $u_s(u_p)$, $u_s = 2.75 + 2.35u_p - 0.13u_p^2$ (km/s), appropriate to a compacted density of $1.937\,\mathrm{g/cm^3}$. A snowplow model with a Grüneisen gamma of 2.0 was used to represent the porous explosive. An ideal gas EOS was used to represent the air. This leads to stiffer shocks in the reverberating region; the actual shocks will be softer and more numerous due to extra internal energy channels opening in the air molecules. However, this slightly changed structure will only introduce slight modifications to the overall patterns that we calculate.

Figure 2 illustrates interactions close to the HE surface, showing the propagation paths of the first and second air shock interactions within the HE. The time origin, $t = 0$, was chosen to be that time the flyer would impact the HE if there were no air in the intervening gap. The figure is a section of the overall $x - t$ diagram for the problem. The first air shock interaction drives a pres-

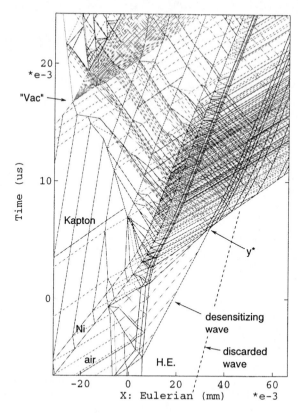

FIGURE 2. Wave Propagation Near the Surface of Granular TATB Acceptor Charge. Shocks and interfaces are shown as solid lines; rarefaction wavelets are dashed. All lines used for calculations are shown: this sometimes makes the strong and important waves difficult to distinguish.

sure of $0.14\,\mathrm{GPa}$ into the HE, which is similar to the static pressure used to construct the sample. Thus it is believed that this transient pressure will not effect any further compaction, and this wave trajectory is designated the "discarded wave" in Fig. 2. Interaction of the second air shock (a reverberation between the HE and the oncoming plate) with the HE drives about $1.8\,\mathrm{GPa}$ into the HE, and this is certainly enough to compact and precompress the HE. The Lagrangian distance where this precompression wave is overtaken by the main shock produced by the flyer plate is designated y^*. It is interesting to note that the value of y^* varied only slightly, 29–35 μm, over the entire set of examples that are described below.

Ordinarily, for a snowplow model, the code

would collapse a material from its porous state to a compacted state for any first wave. This would lead to incorrect behavior for the initial weak air shock. We installed an option to ignore initial weak waves less than some threshold. This permits us to ignore the first air shock in this problem and let the next one compact the porous explosive.

The time for pore closure in the HE was estimated to be 0.4 ns, based upon a free-surface velocity relaxing from a 1.8 GPa shock and a HE particle radius of $0.7 \mu m$, calculated from the surface area of $2.3 m^2/g$. At a shock velocity of 3.5 km/s, this time for pore closure corresponds to shock front travel of $1.4 \mu m$, which is much less than our typical value for y^*. Thus we may conclude that nearly all the material between the impact surface of the HE $(y = 0)$ and y^* has been precompressed and "deadpressed" before arrival of the main shock. The portion of the shock waveform that is useful for initiation of the HE then is the waveform which exists at y^* and deeper into the HE. Density profiles $(\rho(t)$ at $y)$ show that the shock heating for $y > y^*$ is greater than that for $y < y^*$, and this is consistent with the interpretation of deadpressing at $y < y^*$.

Our reference conditions were defined to include 1 atm air and a $5 \mu m$ coating of nickel on the plastic backing sheet. The shock pressure profiles are shown in Fig. 3 at depth intervals of

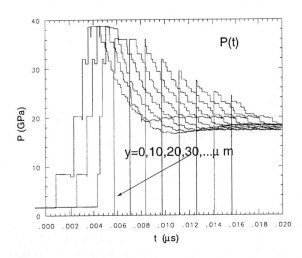

FIGURE 3. Pressure Profiles at Various Depths into HE in Reference Problem.

$10 \mu m$ into the HE. These show "preshock" reverberations that vanish by the time the shock has traveled $30 \mu m$ into the HE. This is a depth sufficiently great that the high-pressure step from impact of the nickel coating has been largely attenuated away.

PARAMETER STUDIES

Parameter studies were done through variations on the reference problem.

We first varied the air pressure ahead of the flyer, at levels of 0, 1 and 2 atm. Figure 4 shows

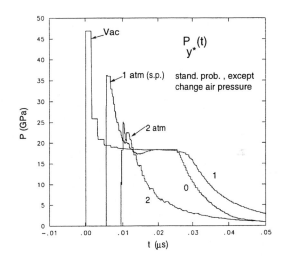

FIGURE 4. Shock History at Depth y^* as a Function of Air Pressure in Flight Path

the shock history at y^*, denoted $P_{y^*}(t)$, for all three cases. Of course, $y^* = 0$ for a vacuum, so there has been no attenuation of the leading pressure step in the waveform. There also is no back pressure to retard the flyer plate in a vacuum, so the magnitude of the leading pressure step is higher. The difference in magnitude of the leading shock step and the impulse delivered to the HE is quite large as the pressure is increased 0-1-2 atm. The 2 atm case may be viewed as an indication of the detrimental effects that any excess pressure ahead of the flyer would have, e.g., due to blowby of the gases that are accelerating the plate.

We next varied the metal material coated onto the plastic backing, holding the metal coating

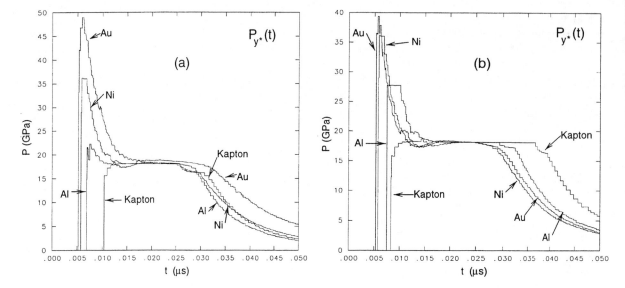

FIGURE 5. Shock History $P_{y^*}(t)$ for Metal-Coated Plastic Flyers: (a) Coating $5\,\mu$m thick, (b) Nickel coating $5\,\mu$m thick, and all other coating thicknesses adjusted to match nickel areal density.

thickness constant at $5\,\mu$m. The coatings were of gold (Au), nickel (Ni), aluminum (Al), and no coating (just Kapton). Because the densities of the coating varied, the initial momentum of the flyer at impact varied, and this is reflected in the $P_{y^*}(t)$ shock waveforms shown in Fig. 5(a). The gold coating delivered the highest shock pressure and the highest impulse into the HE, and these quantities declined with decreasing density and impedance of the coating material.

Perhaps a more equitable comparison of the effects of coating material choice is to hold the momentum constant. This was done by adding coating of a fixed areal density for each material onto the $80\,\mu$m thick Kapton backing plate; the all-Kapton flyer was made thicker to hold the flyer mass constant. The reference problem, with $5\,\mu$m of nickel, was one of this set, and all other coatings were adjusted in thickness to match this mass of nickel. The waveforms shown in Fig. 5(b) show that, with constant impulse in the pulses delivered by the various flyers, there is a trade-off between peak pressure and shock duration. The highest shock pressure is still delivered by the gold coating, but only by a slight margin over the nickel. The shocks from both coatings are clearly being attenuated.

CONCLUSIONS

The main conclusions from this work are:

(1) For thin flyers, air present in the flight path can significantly reduce the impact velocity and momentum of the flyer. This effect is magnified if the pressure is increased, as due to blowby of the driving gas.

(2) Compaction of the acceptor HE occurs due to the compressive loading from interaction of the air shock driven by the flyer, and consequent pre-compression and deadpressing will occur to some depth in granular acceptor HE.

(3) The depth y^* of this deadpressing scales with the initial flyer to HE distance. For the variations we studied it is deep enough to allow attenuation of the loading pulse from thin metal coatings on the face of a plastic flyer plate.

REFERENCES

1. G. Bloom, R. Lee, and W. von Holle, Thin-pulse shock initiation characterization of extrusion- cast explosives, in *Intl. Symp. on Behavior of Media Under High Dynamic Pressure*, La Grande Motte, France, 1989.

2. R. E. Setchell, Initiation of granular explosives by ramp waves, in *Proceedings of the CPIA-JANNAF Combustion Conference*, page 22, Monterey, CA, 1979.

3. G. Kerley, Private communication. This is the EOS used for porous TATB in the CTH codes.

CP429, *Shock Compression of Condensed Matter – 1997*
edited by Schmidt/Dandekar/Forbes
© 1998 The American Institute of Physics 1-56396-738-3/98/$15.00

PRESSURE MEASUREMENTS ON A DEFORMING SURFACE IN RESPONSE TO AN UNDERWATER EXPLOSION

G. Chambers, H. Sandusky, F. Zerilli

Naval Surface Warfare Center, Indian Head Division, Indian Head MD 20640-5035

K. Rye and R. Tussing

Naval Surface Warfare Center, Carderock Division, Carderock MD 20817-5700

Experiments were conducted to benchmark calculations of structural deformation from an underwater explosion with the coupled Eulerian-Lagrangian DYSMAS code. Aluminum (Al) tubes were filled with distilled water except for a small explosive charge in the center that was suspended from hypodermic needle tubing. Tourmaline and carbon resistor gages at the water/wall interface recorded the initial shock loading, cavitation, and reloading as cavitation collapsed. The initial shock loading was consistent with a fine mesh 1-D WONDY calculation and the entire event was mostly simulated by the coarser mesh 2-D DYSMAS calculation.

INTRODUCTION

Computer modeling codes can provide solutions to large or complex events that are not easily tested. Since approximations to physical and chemical processes are required in the codes, they must be validated by precise experiments prior to the intended application. Of interest to this study is the complex interaction between a shock wave in water and a deformable structure, given that the properties of the explosive, water and wall are known. The problem is complex, both experimentally and numerically, because the water momentarily separates (cavitates) during the interaction. In a previous experiment (1), wall deformation and velocity were measured directly using a streak camera and a Doppler laser velocimeter. In the current study, attempts were made to measure pressure at the water/wall interface, free-field pressure, and both inner and outer-wall strain.

The Guirguis Hydro-Bulged Cylinder experiment was conceived for model validation (1). The radially-symmetric arrangement consists of a water-filled tube with a centrally-positioned explosive charge. The symmetrical interior loading removes the complication of bubble collapse driving a water jet towards the structure, which occurs in the usual exterior loading. Interior loading also permits easy access to the deforming structure by various optical techniques for dynamic measurements.

The experimental data have been modeled using the codes WONDY V (2), which is 1-D Lagrangian, and the code DYSMAS/C (3), which is a coupled Eulerian/Lagrangian simulation. Advantages of both Eulerian and Lagrangian scheme are incorporated into DYSMAS which is the code to be validated.

EXPERIMENTAL ARRANGEMENT

The experimental arrangement illustrated in Figure 1 is the same as that previously (1) used, except for a shorter tube to facilitate internal gauge installation. The deformable structure was a 17.8 cm long x 10.2 cm outer diameter Al 5086 tube with a 0.635 cm wall thickness. The runout (roundness) and thickness of the tube wall varied by only .005 cm. The tube was sealed at the bottom by a .254 cm thick polymethyl-methacrylate (PMMA) sheet with a scored circle corresponding to the inner diameter of the tube. With this weakened bottom closure and the tube mostly open on top, there was little resistance to the de-oxygenated water that initially filled the tube from being ejected by the exploding charge. A 2.8 g pentaerythritol tetranitrate (PETN) explosive pellet (1.27 cm diameter x 1.27 cm high) was suspended in the center of the tube. The 3.0 g explosive charges were selected to achieve final wall strains of ~10% based on preliminary DYSMAS [3] calculations. At these strains the tube would deform plastically without rupturing. A 200- mg PETN detonator was positioned above the charge and encased in a 2.7 mm PMMA sleeve with the leads extending up through a hypodermic needle tube. The needle tubing was further filled with epoxy to prevent collapse during shock loading, thereby simplifying the modeling.

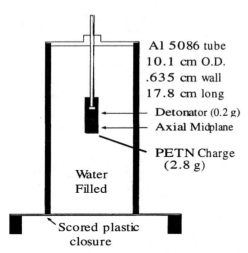

Al 5086 tube
10.1 cm O.D.
.635 cm wall
17.8 cm long

Detonator (0.2 g)
Axial Midplane

PETN Charge
(2.8 g)

Water
Filled

Scored plastic closure

FIGURE 1: Experimental arrangement

Various techniques were attempted to obtain reliable strain and pressure measurements inside the tube while surviving ≥ 100 μs in a complex environment. In addition to the gauges and their mounting, consideration was given to the leads and their connection to the gauges. Protecting the leads by drilling holes in the tube wall was not attempted to avoid perturbing tube deformation; therefore, the leads passed through the water to the top of the tube and were subjected to a similar environment as the gauges.

Strain measurement with a foil gauge on the inner wall was of interest itself and for developing techniques to enhance survivability of similarly constructed pressure gauges. Previously (1), only the onset of deformation (<1% strain) had been measured on the outer wall by circumferentially-oriented constantan foil and optical fiber gauges bonded with cyanoacrylate. In the current experiments, some of the gauges consisted of annealed constantan foils with 20% strain capability and were bonded by an epoxy that could achieve similar strains. It was anticipated that small lead connections to the gauge tabs would avoid premature mechanical failure; either the ends of 32 AWG wires (0.36 mm diameter conductor) were peened flat and soldered to the tabs or the manufacturer had electronically welded even finer magnet wire to the tabs.

Pressure measurements were obtained with both tourmaline and carbon resistor gauges. Each tourmaline gauge had a single 3.2 mm diameter by ~1 mm thick crystal with conductive paint on each end, to which leads were soldered. Completed gauges were bonded to the inner wall of the tube within a bead of 5-Minute epoxy, which also supported the crystal in most gauges. In Shot SV-5, three gauges were positioned around the axial midplane and another four gauges were 38 mm above that plane. Four gauges simply had twin lead or 32 AWG wires soldered to the crystals. For Gauge 5, the crystal was inside a hole in a Mylar strip that had a copper ribbon on each side for leads. The crystal in Gauge 6 was connected to a RG-174 coaxial cable within an oil-filed boot. The crystal in Gauge 7 was connected to RG-174 cable and then coated with polyurethane.

In Shot SV-6, one carbon resistor gauge was located on the inner wall at the axial midplane and another was located 70 mm under the charge in the water. The 470 Ω, 1/8 W resistors were from the same batch used by Wilson (4). A carbon film gauge (Dynasen FC 300-50-EKRTE) was also located on the inner wall at the axial midplane.

RESULTS

All of the strain gauges in Shots SV-5,6 whether located on the inner or outer tube wall, and the carbon film gage in Shot SV-6 failed during the initial shock loading. These thin-film gauges were too fragile to survive at both the water/tube and tube/air interfaces. Most strain gauges were recovered with failures at lead connections.

Of the tourmaline gauges in Shot SV-5, those with wire leads did not provide reasonable records, although the crystals were recovered undamaged. Pressure-time histories, relative to the energizing of the detonator, from two other gauges are shown in Figure 2. Gauge 7, mounted at the midplane on the cylinder wall, recorded an incident shock of 0.33 GPa with a 0.8 μs risetime, a 0.5 μs plateau, and a reflected shock with a 0.8 μs risetime to a peak pressure 0.63 GPa. Separated incident and reflected shocks are observed because the crystal was mounted at least 1 mm from the wall. Gage 5, mounted 38 mm above the midplane, recorded a more diffuse wave beginning at ~30 μs, with a peak pressure of 0.31 GPa.

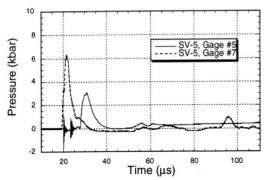

FIGURE 2. Pressure from tourmaline gages at the wall/water interface in Shot SV-5.

Figure 3 shows a comparison at the same location of the pressure-time history from Shot SV-5, with the velocity-time profile from Shot SV-3 (1) taken from a Doppler velocimeter. Shot SV-3 was identical to Shot SV-5 in all respects except that the Al tube was 22.9 cm as opposed to 17.8 cm long. However, differences in tube length should not affect the comparison, since rarefactions

from the end of the tube have insufficient time to interact with the gages during the time frame of interest. As can be seen from Figure 3, the initial pressure spike occurs at about 20 μs, causing the wall to start moving and later separate from the water. The wall velocity increases to a maximum at 35 μs, at which time cavitation has ended the initial shock pulse. The wall then slows down as internal forces in the Al arrest its motion. By 95 μs, the wall has slowed enough for the water to catch up and re-interact with it, as evidenced by the second pressure peak, followed by an immediate increase in wall velocity.

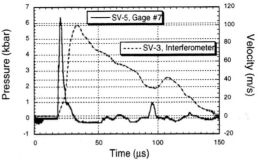

FIGURE 3. Midplane comparison of pressure (Shot SV-5) and wall velocity (Shot SV-3).

In Shot SV-6, the carbon resistor gauge under the charge had a noisy signal, perhaps because of the unshielded loop of resistor leads that passed through the bottom closure. As shown in Figure 4, the gauge on the inner wall at the midplane responded much like tourmaline gauge 7, in Shot SV-5, except for the timing of the second pulse.

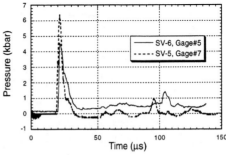

FIGURE 4. Comparison of midplane from tourmaline (Shot SV-5) and carbon resistor (Shot SV-6) gauges.

COMPUTER MODELING

For the 1-D WONDY calculations, a spherical geometry was assumed with zones of 5 and 2.8 μm. This is a good approximation until the shock wave reaches the cylinder wall and is sufficient to calculate the peak pressure achieved near the wall. The 2-D DYSMAS calculations had uniform zones of 1 and 1.5 mm, with a Eulerian treatment of the detonation products in water coupled to a Lagrangian treatment of the tube. Table 1 shows the standard JWL parameters for PETN used in the calculations.

TABLE 1. JWL parameters for PETN

ρ (kg/m^3)	1765	R_1	4.4
D (m/s)	8300	R_2	1.2
A (GPa)	617	ω	0.25
B (GPa)	16.926	E_0 (GJ/m^3)	10.1

The WONDY calculations predicted 0.3 GPa incident and 0.57 GPa reflected shock pressures in the water at the wall versus tourmaline gauge measurements of 0.35 and 0.64 GPa, respectively, thereby verifying consistency of the JWL parameters with the measurements. The midplane measurement from tourmaline gauge 7 in Shot SV-5 is shown in Figure 5, this time compared with DYSMAS calculations at the wall/water interface for the duration of the experiment.

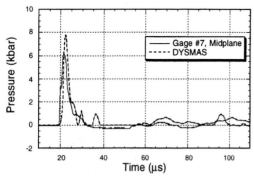

FIGURE 5. Comparison of measured and calculated midplane pressures.

The code calculation for the initial shock had a 3.0 μs risetime, significantly more diffuse than measured, but the peak pressure still exceeded the measurement by 23%. The overprediction is probably caused by the nonconservative method used to reduce numerical diffusion in DYSMAS as contrasted with the underprediction to be expected when artificial viscosity is used to handle shock waves. The code calculation predicted the cavitation but excessively diffused the reloading of the wall at ~90μs following collapse of the cavitation.

SUMMARY AND CONCLUSIONS

Tourmaline and carbon resistor gauges were used to measure pressures at the water/wall interface of a deforming tube in response to an underwater explosion. The gauge measurements of the initial shock loading, cavitation, and reloading as cavitation collapsed corresponded to the timing of previous wall velocity measurements. Fine-mesh calculations with the 1-D WONDY code verified the pressure measurements until just after shock reflection at the wall, when the calculations were no longer valid. The coarser mesh 2-D DYSMAS calculations reasonably predicted the initial shock loading and cavitation, but excessively diffused the subsequent loading.

ACKNOWLEDGMENTS

Discussions with R. Guirguis and funding from the Office of Naval Research. are gratefully acknowledged.

REFERENCES

1. Sandusky, H., Chambers, P., Zerilli, F., Fabini, L., Gottwald, L., 67th Shock and Vibration Symposium, Monterey CA, Nov. 1996
2. Kipp, M., Lawrence, R., Sandia National Labs Report SAND81-0930, Alburquerque, NM, June 1982
3. Dynamic Systems Mechanics Advanced Simulation Coupled Eulerian-Langrangia Hydrocode, IAEG, Ottobrunn Germany. April 1995 version.
4. Wilson, W., *private communication*

CP429, *Shock Compression of Condensed Matter – 1997*
edited by Schmidt/Dandekar/Forbes
© 1998 The American Institute of Physics 1-56396-738-3/98/$15.00

NUMERICAL MODELING OF AN UNDERWATER EXPLOSION FOR AN ALUMINIZED EXPLOSIVE

J.H. Kuk, J. Lee, S-y. Song, and Y.S. Cho

Agency for Defense Development, Taejon 305-600, Korea

Underwater explosion properties of an aluminized explosive were numerically modeled by two burn techniques; a programed-burn and a reaction rate calibrated from two-dimensional steady-state detonation experiments. The programed-burn technique did not reproduce experimental data well; 33% error in peak pressure of shock wave and 9% error in bubble period. The rate reproduced the experimental observations very well; 7% error in peak pressure and 2% error in bubble period. The shock profile also agreed very well with experimental observation. These results demonstrates that the underwater explosion properties for aluminized explosives can be calculated only when the slow energy release is modeled properly.

INTRODUCTION

When an explosive charge is exploded in water, a strong shock wave generated by the explosion propagates through water, and a bubble of gaseous detonation products expands and contracts periodically until it floats to the surface. Most explosive energy is consumed by the shock propagation and the bubble oscillation. Therefore, performance of an underwater explosive charge can be evaluated by properties of the shock wave and the bubble.

Usually, shock-wave and bubble properties for a standard charge of an explosive are measured. For a charge of different weight these properties are estimated by using a scaling law[1]. If necessary, they can be obtained through a series of hydrodynamic calculations, by using a rate-independent programed burn. This type of numerical simulation and the scaling law work well for most military explosives. A common characteristic of these explosives is that their reaction rate is fast and strongly dependent on thermodynamic state.

For aluminized explosives, their reaction rate is slow and weakly dependent on thermodynamic state, in other words, only a fraction of explosive energy is released before the sonic point. Their detonation properties are strongly dependent on size. For these reasons, the scaling law and rate-independent numerical simulation do not produce good results for these explosives.

Lee et al[3] calibrated a reaction rate of DXD-04, an aluminized explosive, from two-dimensional steady-state experimental data by applying the detonation shock dynamics developed by Bdzil[4]. Using the rate, they reproduced a detonation wave profile obtained from a detonation pressure measurement test.

The objective of this study is to investigate long-term reaction characteristics of the above reaction rate by numerically modeling underwater detonation properties of DXD-04.

EXPERIMENTAL TECHNIQUES

The composition of the test explosive, DXD-

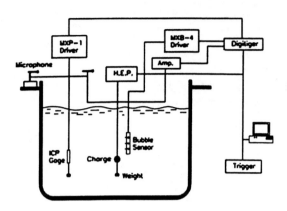

FIGURE 1. Schematic view of experimental configuration.

04, is given by RDX/ammonium perchlorate/aluminum/binder 24/43/22/11 wt%, and its density is 1.78 g/cm^3. To minimize shape effect, this explosive was cast to a sphere, and was coated with a sealant. The test charge with a diameter of 147.5 mm and a weight of 2950 g was initiated by a composition C-4 booster with a diameter of 39.5 mm and a weight of 50 g at the center of the charge.

Figure 1 shows the experimental configuration for this test. The dimension of water tank was 22 m in diameter and 15 m in depth. The test charge was initiated at a location 10 m deep from the surface. The shock pressure in water was measured by using PCB 138 A gauges. located at from 3 to 9 m from the charge in the radial direction at the same depth. Bubble period was measured by using a microphone placed in air and also the pressure gauges in water.

NUMERICAL TECHNIQUES

In this study, the HOM EOS[2] was used to describe the behavior of the mixture of unreacted explosive and reaction products. The HOM EOS assumes that the solid and gas phases are mixed ideally and that the pressure and the temperature of both phases are in equilibrium.

The Mie-Grüneisen EOS for a solid phase is expressed by

$$P_s = P_H + \frac{\Gamma}{V_s}(E_s - E_H) , \qquad (1)$$

where P is pressure, V specific volume, and E specific internal energy. Subscript s refers the solid phase and subscript H does the state on the Hugoniot. Grüneisen parameter, Γ, at an arbitrary state is obtained by assuming $\Gamma/V_s = \Gamma_o/V_o =$ constant[5], where subscript o refers ambient state.

The temperature of a solid phase is obtained by

$$T_s = T_H + \frac{1}{C_{vs}}(E_s - E_H) , \qquad (2)$$

where T is temperature and C_{vs} is specific heat of the solid phase at constant volume. The temperature on the Hugoniot, T_H, is obtained by using the Walsh and Christian method[5].

The shock Hugoniot of DXD-04 was obtained by assuming ideal mixing of the components, and fitted to a linear relationship. That is:

$$D = 2.25 + 1.80\, u , \qquad (3)$$

where D and u are shock and particle velocities in mm/μs. Grüneisen parameter, Γ_o, and heat capacity at constant volume, C_{vs}, were determined to be 1.20, and 0.26 cal/gK, respectively.

The β EOS for a gas phase is expressed by

$$P_g = P_R + \frac{1}{\beta V_g}(E_g - E_R) \qquad (4)$$

$$T_g = T_R + \frac{1}{C_{vg}}(E_g - E_R) , \qquad (5)$$

where subscript g refers the gas phase and subscript R does the state on the isentrope through the CJ point. The parameter, β ($\equiv -(\partial \ln V_g/ \partial \ln T_R)_S$), where S is entropy, is determined on this isentrope calculated by the BKW code[6], and are fitted to fourth-order polynomials:

$$\ln P_R = \sum_{i=0}^{4} b_i (\ln V_g)^i \qquad (6)$$

$$\ln(E_R + e) = \sum_{i=0}^{4} c_i (\ln P_R)^i \qquad (7)$$

$$\ln T_R = \sum_{i=0}^{4} d_i (\ln V_g)^i , \qquad (8)$$

402

TABLE 1. EOS Parameters of DXD-04

Parameter	DXD-04
b_o	1.11233
b_1	- 2.27480
b_2	3.21986×10^{-1}
b_3	- 4.35637×10^{-2}
b_4	2.33193×10^{-3}
c_o	1.67125
c_1	1.51302×10^{-1}
c_2	1.54010×10^{-2}
c_3	2.77603×10^{-3}
c_4	2.02120×10^{-4}
d_o	8.05594
d_1	- 4.54310×10^{-1}
d_2	1.39144×10^{-1}
d_3	- 2.88261×10^{-2}
d_4	2.25685×10^{-3}
C_{vg}	0.5 cal/gK
e	10 GPa·cm^3/g

where e is a constant added to prevent the argument of the logarithm from being negative. The parameters for DXD-04 are shown in Table 1; P_R in GPa, E_R in GPa cm^3/g, and T_R in K.

The underwater explosion was assumed to be one dimensional (1-D) in a spherical geometry, and a 1-D Lagrangian hydrodynamic code was used. The booster charge was divided into 20 cells of 1 mm each and the first cell was used as a hot spot. The booster was burnt by a programed burn, the CJ volume burn[2]. The DXD-04 charge was divided into 60 cells of 0.9 mm each, and burnt by the rate, and the programed burn. The cell size of surrounding water was set to 1.5 mm, and increased with increasing radius. The time step was set to 0.005 μs.

The bubble period at initial hydrostatic pressure from 2000 down to 300 bars was calculated. These period data was fitted to a line as a function of intial pressure in the log (period) - log (hydrostatic pressure). The bubble period at a hydrostatic pressure of 2 bars, then, was obtained by extrapolating the linear relationship. The pressure of shock wave in surrounding water was calculated at the initial pressure of 2 bars.

RESULTS AND DISCUSSIONS

The reaction rate obtained for DXD-04 by Lee

Shock Wave Profiles

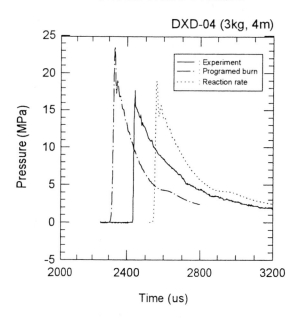

FIGURE 2. Shock wave profiles in water. Arrival times of waves are arbitrary.

et al[3] was given by

$$\frac{d\lambda}{dt} = 122.6(1 - \lambda)^{7.076}$$
$$\exp(-0.7984\rho^{5.117}/P^{1.202}), \tag{9}$$

where λ is reaction extent changing from 0 for unreacted explosive to 1 for reaction products, ρ is density in g/cm^3, and P is pressure in GPa.

Figure 2 compares shock-wave profiles calculated by using the reaction rate and the programed burn with the experimental one, at a location 4 m away from the charge. The peak pressure calculated by using the rate was higher by 7.3% than the experimental data. The duration of the wave was shorter than the experimental observation, but slightly longer than that obtained by using the programed burn. The profile itself, however, agreed well with the experimental one. The peak pressure calculated by using the programed burn was higher by 33% than the experimental measurement.

The bubble oscillation period at the test condition, 2 bars, was obtained by extrapolating pe-

Bubble Period

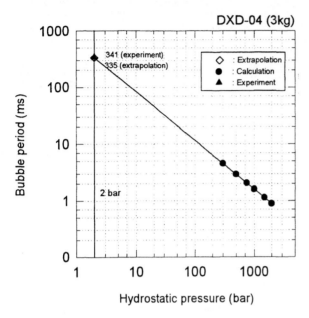

FIGURE 3. Extrapolation of bubble period.

riods at high initial hydrostatic pressure ranging from 2000 down to 300 bars. Figure 3 shows this extrapolation for the bubble period by using the reaction rate. The extrapolated period (335 ms) agreed very well with the experimental result (341 ms). The period obtained by using the programed burn was 310 ms.

Cole[1] showed by using the simple incompressible model that the bubble period is inversely proportional to the hydrostatic pressure to 5/6 (= 0.833) power. Mader[2] obtained the same result only with different value (0.788) of the power for tetryl by using a programed burn. In the present study, the power (0.858) was considerably higher than that in Mader's study. It is because bubble oscillation at low hydrostatic pressure takes longer time and, consequently, more energy is released than that at high hydrostatic pressure for the slowly reacting explosive used in this study. Results of the present modeling supported the above argument: within the first oscillation period, 90.1% explosive energy released at 300 bars while 88.7% was released at 2000 bars.

CONCLUSIONS

In this study, underwater explosion properties of an aluminized explosive were numerically modeled by using a reaction rate calibrated from two-dimensional steady-state experimental data. It was concluded that:

1. Underwater explosion properties of aluminized explosives cannot be modeled by using a programed burn technique;
2. The reaction rate used in this study showed acceptable long-term reaction characteristics and reproduced experimental observations.

REFERENCES

1. Cole, R.H., *Underwater Explosions*, Princeton: Princeton Univ. Press, 1948, ch. 7, pp. 235-245, ch. 7, pp. 332-341.
2. Mader, C.L., *Numerical Modeling of Detonation*, Berkeley: Univ. of California Press, ch. 5, pp. 278-285, appendix A, pp. 316-317.
3. Lee, J., Kuk, J.H., Song, S-y., Choi, K.Y., and Lee, J.W, "Detonation in an aluminized explosive and its modeling," presented at the APS Symp. on Shock Compression of Condensed Matter, Amherst, MA, Jul. 27 - Aug. 1, 1997.
4. Bdzil, J.B., and Stewart, D.S., *Phys. Fluids* **A1**, 1261-1267 (1989).
5. Walsh, J.M., and Christian, R.H., *Phys. Rev.* **97**, 1544-1556 (1955).
6. Mader, C.L., *FORTRAN BKW: A Code for the Detonation Properties of Explosives*, Report LA-3704, Los Alamos Scientific Lab., 1963.

CP429, *Shock Compression of Condensed Matter – 1997*
edited by Schmidt/Dandekar/Forbes
© 1998 The American Institute of Physics 1-56396-738-3/98/$15.00

GAP TEST MODELING TO PREDICT WEDGE TESTS INITIATION OF PBXN-103

Clinton T. Richmond

Naval Surface Warfare Center,Indian Head Division,Indian Head,Maryland 20640

The experimental initiation of PBXN-103 by the standard wedge test has been modeled by using the HVRB initiation and growth model in the CTH code. The P-081 plane wave lens was used as initiator in these experiments. The wedge test was converted to a gap test by replacing the PBXN-103 wedge by a PBXN-103 cylinder. By modeling this gap test, shock initiation in PBXN-103 was calculated. The results of these calculations are in agreement with the data of the wedge test experiments. Comparison of the CTH code calculations with the wedge test data was accomplished by using an auxiliary program called the BCAT code. In particular, it computes the "pop plot" and compares it to the wedge test data. Shock initiation of PBX-9404 was also calculated by the HVRB model and the results compared to the initiation of PBX-9404 using the Lee-Tarver model. The two calculations from both of the models are very compatible.

INTRODUCTION

The Lee-Tarver model is one of the simpler models that has been successful in describing the initiation and growth process in an heterogeneous explosive, such as, PBX-9404 (1). The HVRB is a similar model and it is used in this work to model the initiation of PBXN-103. An analysis of how the HVRB model compares to Lee-Tarver model has been explored by Kerley (2).

Historically, both the wedge test and gap test (3) have been used to investigate the properties of an explosive. Lately, Dallman and Wilson have conducted wedge test experiments to study the shock initiation of PBXN-103 (4). The gap test version of these experiments is used in the present investigation to model the shock initiation of PBXN-103. The results of both studies should be basically the same.

THE GAP TEST SETUP

The setup for the standard gap test consists of three basic sections: the donor explosive, the gap, and the acceptor explosive. The donor explosive is set off by a P-081 plane wave generator (5). Figure 1 illustrates the gap test setup for modeling the initiation of PBXN-103.

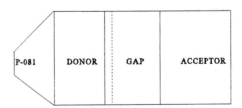

Figure 1. Gap Test Setup

Except for the P-081 initiator, which is a frustum of a cone, each of the other sections is a cylinder of material. The acceptor is always the explosive that is being tested. In this case, the acceptor, or test explosive is PBXN-103. The dotted line indicates that a metal buffer or another explosive may occasionally be inserted between the gap and the donor. The diameter of each cylinder of material is 21.696 cm. The thickness of the P-081 initiator is 8.623. See Table 1 for a listing (4) of other materials and their thicknesses used in each setup.

The gap test setup is initiated from the end of the P-081 lens, in which a plane wave is generated. The simplest method that adequately described the initiation is used in each explosive section. For example, the test setup in ID no. 15-1844, "program burn" is used to initiate P-081, "volume burn" is used to initiate TNT, and the "HVRB" model is used for the initiation of PBXN-103. The character of the shock wave as it proceeds through the various sections of the gap test setup is illustrated in Figure 2.

The pressure contours clearly show the formulation of a plane wave front in the P-081 section(time, t=10usec.). The plane wave character

persists as the wave moves through TNT(time, t=20usec.) and into PMMA(time, t=26usec.). However, curvature of the wave becomes quite evidence as the wave proceeds through PBXN-103(time,t=35usec.).

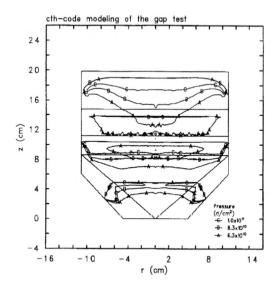

Figure 2. The character of the shock wave in each section of the gap test.

TABLE 1 . VARIOUS GAP TEST SETUPS FOR THE PBXN-103 ACCEPTOR

ID No.	Donor	Thickness (mm)	Buffer	Thickness (mm)	Gap	Thickness (mm)
15-1844	TNT	25.4	None		PMMA	35.9
15-1847	TNT	25.4	BRASS	13.34	PMMA	24.32
15-1849	BARATOL	50.8	None		PMMA	24.4
15-1851	TNT	50.8	BRASS	13.31	PMMA	12.4
15-1854	BARATOL	25.4	TNT	25.4	PMMA	12.4
15-1867	BARATOL	25.4	None		PMMA	33.4
15-1906	TNT	50.8	BRASS	9.17	PMMA	12.4
15-1939	TNT	50.8	BRASS	9	PMMA	12.7

The character of the pressure-time profile of the shock wave as it moves through each section of the explosive is illustrated in Figure 3. The first pressure-time profile

in Figure 3 (reading from left to right) occurs in P-081 at the point (r=6 cm, z=6 cm). The next pressure-time curve(solid line) is the form of the shock wave as it

406

enters TNT(r=1 cm, z=8.7 cm). The next three curves(dotted lines) show a pressure build-up as the TNT is initiated. In PMMA, the shock wave is then attenuated until it reaches the lower peak pressure value of the next pressure-time profile(solid line) in PBXN-103(r=0, z=14.853 cm). The remaining curves(dotted lines) show a pressure build-up as PBXN-103 is initiated.

POP PLOT FOR PBXN-103

The BCAT (6) code is an auxiliary computer program to the CTH code. It is design to do special calculations for the purpose of comparing CTH calculations to experimental data, such as, the "pop plot". This program was used to calculate the pop plot for PBXN-103 and is compared to the experimental curves of Dallman and Wilson (4). This comparison is illustrated in Figure 4 for both "run" distance and "run" time. The solid and dotted lines are the experimental plots from Dallman and Wilson, and the points(circles and squares) are calculations from the BCAT code using the hvrb model for initiation of PBXN-103.

NUMERICAL CALCULATIONS FOR PBXN-103 AND PMMA

All eight cases of the gap test setup in Table 1 was calculated by the CTH code. A tabulation of the numerical results are in Table 2. In this table, the free surface velocity in PMMA, adjacent to PBXN-103 is tabulated first. Then, the initial pressure, the particle velocity, and the shock velocity are tabulated next for the initiation of PBXN-103. Then, these numerical results can be compared to experimental results of the wedge tests (4), which is also in Table 2.

LEE-TARVER VS. HVRB MODEL

A comparison between the Lee-Tarver model and the HVRB model were made by calculating the initiation build-up of pressure over time in PBX9404 at position sites, z=2, 5, 8, 10, and 15 mm.[1] In Figure 5, the solid pressure-time profiles for the six positions were calculated by the HVRB model using the CTH code. The dotted pressure-time profiles for the same corresponding positions were calculated by the Lee-Tarver model using the DYNA2D code. These two

sets of curves seems to tend more toward agreement as the calculations are refined.

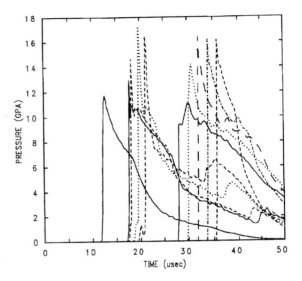

Figure 3. Pressure- time profiles in the P-081 plane wave lens, in the TNT donor, and in the PBXN-103 acceptor of the gap test setup.

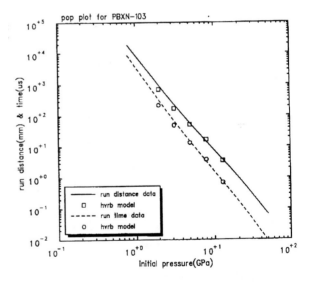

Figure 4. Comparison of wedge test data pop plot(solid and dotted lines) to calculated points (squares and circles) from HVRB model.

TABLE 2. INITIAL SHOCK PARAMETERS FOR PBXN-103 ($\rho_0 = 1.89$ g/cc)

ID No.	PMMA Free-Surface Velocity		Pressure		Particle Velocity		Shock Velocity	
	Experimental	CTH Code	Experimental	CTH Code	Experimental	CTH Code	Experimental	CTH Code
	mm/us	mm/us	GPA	GPA	mm/us	mm/us	mm/us	mm/us
15-1844	3.11	2.34	11.6	11.1	1.29	1.10	4.79	5.34
15-1847	1.68	1.59	4.86	4.1	0.68	0.64	3.76	3.39
15-1849	2.59	2.02	8.84	8.43	1.07	1.02	4.37	4.37
15-1851	2.03	1.96	6.32	6.82	0.83	0.92	4.03	3.92
15-1854	3.71	3.04	15.6	13.9	1.50	1.42	5.50	5.18
15-1867	2.33	1.98	7.9	7.96	0.94	0.97	4.50	4.34
15-1906	2.02	1.90	6.40	6.89	0.80	0.90	4.20	4.05
15-1939	2.07	2.26	6.70	6.42	0.81	0.85	4.35	4.00

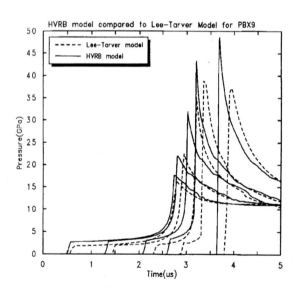

Figure 5. Comparison of Initiation and Growth in PBX9404 Using Lee-Tarver model (dotted curves) and the HVRB model (solid curves).

CONCLUSIONS

These calculations have demonstrated quite convincingly the capability of the CTH code to model correctly a gap test or a wedge test setup. Particularly impressive is the HVRB modeling of initiation in PBXN-103 and PBX-9404. The two-dimensional effects which are included in these calculations were also handled quite well by the CTH code.

REFERENCES

1.Tarver, C.and Hallquist, J., *7th Symposium detona-tion*, 1981,p.488
2.Kerley,G., "CTH Reference Manual: The Equation of State Package",SAND91-0344,UC-410,Sandia National Laboratories,1991
3.Tasker,D. and Baker,R.,NSWCD/TR-92/54, 1992
4.Dallman, J. and Wilson, D.,"Wedge Tests of PBXN-103 Explosive",M-9, Los Alamos National Laboratory,1988
5.Pimbley, G.,Mader,C., and Bowman, A.,"Plane Wave Generator Calculations",LA-9119, Los Alamos National Laboratory
6.Kerley,G.,"BCAT User's Manual and Input Instruct-ions",Sandia National Laboratories,1995

CHAPTER VI

MECHANICAL PROPERTIES:
Nonreactive Materials

CP429, *Shock Compression of Condensed Matter – 1997*
edited by Schmidt/Dandekar/Forbes
© 1998 The American Institute of Physics 1-56396-738-3/98/$15.00

INFLUENCE OF STRAIN RATE AND TEMPERATURE ON THE MECHANICAL BEHAVIOR OF BERYLLIUM

W.R. Blumenthal, S.P. Abeln, D.D. Cannon, G.T. Gray III , and R.W. Carpenter

Los Alamos National Laboratory, Los Alamos, NM 87545

The compressive stress-strain response of three grades of beryllium were studied as a function of strain rate and temperature. Grades S200D, E, and F represent a historical perspective of beryllium processing from the 1960's through 1990's technology. The purpose of this study was to measure the mechanical behavior of beryllium over a range of deformation conditions for constitutive model development and to obtain microstructural evidence for deformation mechanisms. The compressive stress-strain response was found to be independent of grade and strongly dependent on the applied strain rate between 0.001 and 8000 s^{-1}. The strain-hardening response displayed a moderate temperature dependence between 77°K and 873°K. Microstructural examination of SHPB specimens revealed that twinning was extensive at strains above 7%. A SHPB sample deformed to over 20% strain contained both twinning and grain boundary microcracking.

INTRODUCTION

Beryllium metal has many excellent structural properties in addition to its unique radiation characteristics, including: high elastic modulus, low Poisson's ratio, low density, and high melting point. However, it suffers from several major mechanical drawbacks: 1) high anisotropy - due to its hexagonal lattice structure and its susceptibility to crystallographic texturing; 2) susceptibility to impurity-induced fracture - due to grain boundary segregation; and 3) low intrinsic ductility at ambient temperatures thereby limiting fabricability. Commercial beryllium has undergone processing improvements over the last 30 years to control impurity content, minimize crystallographic anisotropy and grain size, and maximize mechanical properties. Grades S200D, E, and F (Brush Wellman, Inc., Elmore, OH) represent a historical perspective of beryllium processing from the 1960's through 1990's technology.

Knowledge of the influence of temperature, strain rate, microstructure, and chemistry on mechanical response is necessary for accurate constitutive model development for current and earlier grades of beryllium which are still in service.

While numerous studies have investigated the low-strain-rate constitutive response of beryllium, the combined influence of high strain rate and temperature on the mechanical behavior and microstructure of beryllium has received limited attention over the last 30 years (1-4). Prior studies have focused on single Be grades (1,4), tensile loading behavior (2), or limited conditions of dynamic strain rate and/or temperature (1-4).

The goal of this study was to measure the compressive stress-strain response of three commercial grades of polycrystalline beryllium over a wide range of temperatures and strain rates and to use microstructural evidence to identify dominant deformation mechanisms. The mechanical response will be considered in terms of the strain-hardening rate, the initial yield behavior, and the rate sensitivity.

EXPERIMENTAL TECHNIQUES

This investigation was performed on three structural grades of vacuum-hot-pressed powder beryllium: S200D, E, and F available from the 1960's to the 1990's. Table 1 shows the impurity content and grain size parameters of the different grades. The total impurity content has improved with each grade (2.037%, 1.158%, and 0.983% for the D, E, and F grades, respectively) along with a reduction in the grain size. Structural grades of beryllium have always been manufactured by powder attritioning methods (5). Improvements in the chemistry and mechanical properties were seen with each new grade introduced. The most significant processing improvements between these grades were: 1) the elimination of recycled machining chips after S200D; 2) a reduction in particle size from -200 mesh to -325 mesh between S200D and S200E; and 3) a change in powder grinding method from Braun type attrition to impact grinding which creates a more blocky powder particle and lower oxide content between grades S200E and S200F. The microstructure of all three grades can be characterized as having equi-axed grains and an extensive grain size distribution with few pre-existing twins.

Uni-axial processes such as die filling, hot-pressing, and hot-rolling are known to preferentially align plate-like grains generated by easy basal plane cleavage in beryllium during grain size reduction. Preferred texture is important because it results in orientation-dependent constitutive properties. The crystallographic texture of the three grades of beryllium was recently measured using neutron diffraction (6). Moderate axisymmetric (fiber) texture was observed in (0001) pole figures oriented along the hot-pressing direction. Texture maxima of 2.0, 1.68, and 1.73 m.r.d. (multiples of random distribution) were reported for the 200D, E, and F grades, respectively.

For this study, samples were machined only for compression along the basal-textured, hot-pressing axis of each plate. A split-Hopkinson pressure bar (SHPB) was used to conduct dynamic testing on 5-mm diameter by 5-mm long specimens as a function of strain rate, 1500-8000 s^{-1}, and temperature, 77°K, 223°K, 293°K, and 473°K to 873°K (in 100 degree increments). Samples were machined to minimize surface damage which is known to reduce tensile ductility, but no additional treatments were used to remove residual machining damage. The dynamic results were compared to quasi-static compression tests conducted on similarly prepared 13-mm diameter by 19-mm long specimens at strain rates of 0.001, 0.01, and 0.1s^{-1} and at temperatures of 423°K to 873°K (5).

SHPB tests were conducted in a hazardous materials containment chamber to minimize potential exposure to beryllium dust in the event of specimen failure. Samples were cooled with liquid nitrogen to achieve temperatures of 77°K and 223°K. A resistance furnace was used to heat samples above room temperature in an argon gas atmosphere.

TABLE 1. Composition (wt%) and Grain Size (μm)

Element/Grade	200D	200E	200F
Beryllium	98.80	99.19	99.28
BeO	1.70	0.860	0.720
Fe	0.13	0.070	0.090
C	0.07	0.080	0.072
Al	N/A	0.036	0.038
Mg	0.03	0.020	N/A
Si	0.04	0.031	0.025
Others combined	0.067	0.061	0.038
Avg. Grain Size	27.0	13.4	11.4
Max. Grain Size	80	56	50

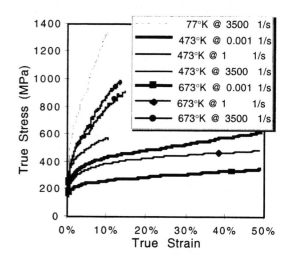

FIGURE 1. Compressive true stress-true strain response of S200F beryllium showing moderate temperature and strain rate dependence.

During high temperature SHPB testing, the lubrication condition of the sample-bar interfaces was observed to substantially affect the flow stress and reproducibility of the stress-strain behavior (although the strain-hardening rate was virtually unaffected). Specifically, specimen barreling was observed and the flow stress of poorly-lubricated specimens was typically 50 to 100 MPa higher than well-lubricated sample values, as previously

412

observed during quasi-static testing (5). A combination of boron nitride (spray-coated) and molybdenum sulfide grease produced reproducible stress-strain results without sample "barreling" up to 773°K.

After SHPB testing, selected specimens were sectioned (parallel to the loading axis) and prepared by conventional polishing for cross-polarized light optical metallography.

RESULTS AND DISCUSSION

The compressive stress-strain behavior of the beryllium exhibits a gradual transition to linear strain-hardening behavior after a few percent strain as shown in Fig. 1 for a wide range of temperatures and strain rates. The strain-hardening behavior was obtained by fitting linear equations to the data over the range of 4% to 9% strain (a range common to both SHPB and quasi-static data sets, but beyond initial yield). The strain-hardening behavior is generally the most important response for modeling constitutive behavior to large strains and is influenced by impurity level, grain size, and the extent of deformation twinning. Quasi-static failure occurred with little ductility at temperatures below 420°K which precluded measurement of the strain hardening behavior in this regime.

All three grades were nearly identical in their compressive high strain rate stress-strain behavior as illustrated in Fig. 2a. The strain-hardening rate (= slope of the linear fit) increases with increasing strain rate or decreasing temperature as shown in the semi-log plot of Fig. 2b. Empirical power-law fits of both dynamic and quasi-static hardening data are offset which suggests that different deformation mechanisms control the strain hardening behavior of Be in these two strain rate regimes.

Because a distinct yield point was not observed under either quasi-static or SHPB conditions, the yield stress-intercept value from the linear strain-hardening fits of the stress-strain curves were instead compared as a function of strain rate and temperature as shown in Fig. 3. The strain-rate sensitivity of the intercept-stress varied from about 0.09 at SHPB strain rates to about 0.04 at quasi-static strain rates which again suggests the operation of different deformation mechanisms.

Microstructural examination of SHPB-deformed specimens revealed extensive deformation twinning. Fig. 4a is a micrograph of a specimen deformed at

room temperature to 7% strain at a strain rate of 1500 s^{-1}. Twins are present in the majority of grains. Twinning was not prevalent in the microstructure of quasi-statically deformed specimens which were tested above 423°K. Hence, the strain-hardening and yield behavior differences between high rate and low strain rates may be due to dislocation- versus twinning-dominated deformation mechanisms.

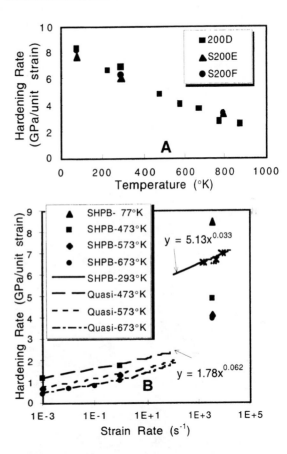

FIGURE 2. a) Strain-hardening behavior is virtually identical for grades S200D, E, and F at 3500 s^{-1} (4-9% strain range). b) Strain-hardening as a function of strain rate and temperature is distinctly offset between high and low strain rates.

Figure 4b shows a sample deformed at room temperature to over 20% strain and a peak stress of 1.48 GPa at 8000 s^{-1}. The microstructure contains both twins and a sizable distribution of microcracks. The microcracks are primarily observed at grain junctions, but are also present intragranularly. The advent of microcracking can be correlated to the end of linear strain-hardening

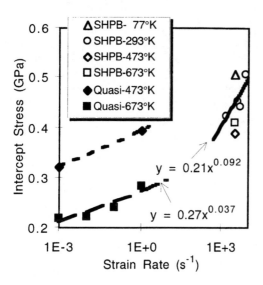

$$y = 0.21x^{0.092}$$

$$y = 0.27x^{0.037}$$

FIGURE 3. Yield stress behavior is analyzed using the strain-hardening intercept versus strain rate and temperature.

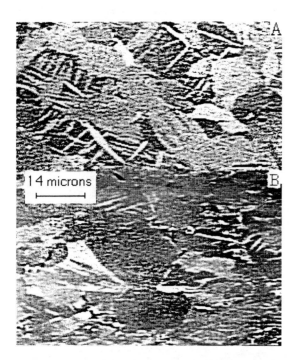

FIGURE 4. a) SHPB specimen deformed to 7% strain reveals twinning as a dominant deformation mode. b) SHPB specimen with 20% strain also shows microcracking at grain junctions.

behavior and a peak in the flow stress. The deviation from linear strain-hardening in this specimen begins at 15% strain with a peak in the flow stress at 18% strain. This response was <u>not</u>

associated with a change in the strain rate or with barreling effects due to interfacial friction. Similar linear hardening response and twin formation under high-strain-rate loading has been observed in other hexagonal metals (7). Twinning is also known to depend on grain size and texture. The grain size and moderate basal texturing of the three grades of beryllium appears to favors twinning for compression along the hot-pressing direction.

SUMMARY AND CONCLUSIONS

A study of the compressive deformation response of three grades of hot-pressed beryllium was conducted as a function of strain rate and temperature. The following were concluded: 1) the high-strain-rate compressive behavior of grades S200D, E, and F are virtually identical; 2) the strain-hardening and the yield intercept stress increase with higher strain rate and lower temperature; and 3) twinning contributes substantially to plastic flow at high strain rates, followed by microcracking at large strains (>15%).

ACKNOWLEDGMENTS

This work was supported under the auspices of the United States Department of Energy. The authors acknowledge the assistance of Dr. Martin Mataya and Carl P. Trujillo.

REFERENCES

1. Green, S.J., and Schierloh, F.L., *General Motors Technical Center Report MSL 68-11*, Warren, MI, 1968.
2. Lindholm, U.S., and Yeakley, L.M., *Air Force Materials Laboratory Report AFML-TR-71-37*, Wright-Patterson Air Force Base, OH, 1971.
3. Breithaupt, D., *Lawrence Livermore National Laboratory Report UCID-19983*, Livermore, CA, 1983.
4. Montoya, D., Naulin, G., and Ansart, J.P., *J. De Physique III*, **1**, 27-34 (1991).
5. Abeln, S.P., Mataya, M.C., and Field, R. , "Elevated Temperature Stress Strain Behavior of Beryllium Powder Product," Proc. 2nd IEA International Workshop on Beryllium for Fusion, Jackson Lake Lodge, WY, Sept 6-8, 1995.
6. Bennett, K., Von Dreele, R.B., and Varma, R., *Los Alamos National Laboratory Report LAUR 97-2942*, Los Alamos, NM, 1997 (submitted to J. Materials Research Society).
7. Gray, G.T., *Twinning in Advanced Materials*, Warrendale, PA, The Minerals, Metals, & Materials Society, 1994, pp. 337-349.

CP429, *Shock Compression of Condensed Matter – 1997*
edited by Schmidt/Dandekar/Forbes
© 1998 The American Institute of Physics 1-56396-738-3/98/$15.00

HIGH STRAIN RATE RESPONSE
OF A TUNGSTEN HEAVY ALLOY

S. N. Chang and J. H. Choi

Agency for Defense Development, Yuseong P.O. Box 35-1, Taejon 305-600, S. KOREA

The effect of thickness change of a tungsten heavy alloy (WHA) on its dynamic behaviors has been studied. Exploding bridgewire (EBW) detonator has been used to drive the alloy plate. The particle velocity at the rear free surface of a specimen was measured by means of VISAR. Simple experimental technique has been introduced herein to obtain the Hugoniot elastic limit and spall strength of materials in the form of small disc plate (diameter of ~7mm) with varying thickness. The peak pressure decay is analyzed as a function of time for traveling of the wave through each specimen. The fracture behavior of WHA caused by the high strain rate herein is similar to that due to the Charpy impact test.

INTRODUCTION

It is generally accepted that the stress pulse traveling through the material is attenuated by an irreversible process. The stress pulse attenuation during traveling through metals immediately takes place when the pulse duration is zero[1,2,3]. Mechanisms of the high strain rate deformation and spallation in tungsten heavy alloys have been developed for their applications. In the present study, a simple technique to measure the dynamic response of a tungsten heavy alloy and an analysis of its Hugoniot stress variation, spall strength and fractography, have been described.

EXPERIMENTS

Time-resolved wave profiles from a disc shaped free surface of WHA were obtained by using a laser interferometer system (VISAR). The chemical composition of the alloy was 90wt%W, 7wt%Ni and 3wt%Fe which was sintered, quenched from 1050℃, and swaged to 24 percent. A cylindrical material with the density of 17.1 g/cm^3 and 7.0mm diameter was cut to have varying thicknesses from 2.15 to 6.07mm. An exploding bridgewire (RP-87 EBW) detonator was contacted and adhered to

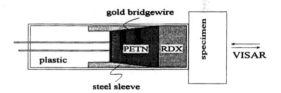

FIGURE 1. Schematic configuration of experimental setup showing an EBW detonator adhered to a tungsten heavy alloy.

a flat-face of the alloy disc as shown in Fig. 1. The detonator consists of initiating explosive of 21mg PETN and output explosive of 47mg RDX, contained in stainless steel cup (4.85mm in diameter and 0.152 mm in thickness).

Optical and scanning electron microscopy were used to study the crack propagation and fracture surfaces of the recovered samples.

RESULTS AND DISCUSSION

Hugoniot and spall data are summarized in Table 1. Figure 2 shows the free surface velocity profiles measured in this series of experiments. As apparent in this figure, the elastic precursor wave shows some variation. In these experimental results, the deviation of Hugoniot elastic limit (HEL) is within 10%.

TABLE 1. Summary of the Hugoniot and spall data

Sample Thickness (mm)	Peak Particle Velocity (km/sec)	Plastic Wave Velocity (km/sec)	Hugoniot Stress (GPa)	Hugoniot Strain Rate ($\times 10^6$/s)	Spall Stress (GPa)	Spall Strain Rate ($\times 10^5$/s)	Hugoniot Elastic Limit (GPa)
2.15	0.580	4.482	23.012	2.39	5.596	3.11	
2.35	0.486	4.415	19.027	1.53	4.653	2.46	
3.20	0.305	4.312	12.171	0.70	4.152	1.73	3.19
3.90	0.242	4.271	9.765	0.44	4.526	1.69	
5.05	0.130	4.195	5.522	0.18	3.741	1.41	
6.07	0.100	4.153	3.959	0.21	3.976	0.81	-

The samples having the thickness between 2.15 and 5.05mm reveal a transition from the elastic precursor rise to the transition ramp region at approximately 60m/sec yielding 2.62GPa from $\rho_o C_l U_p/2$. The elastic wave velocity C_l is 5.114Km/sec measured by ultrasonic device. The HEL can also be estimated to be equal to the particle velocity of 73 m/sec, that is the average precursor amplitude at the center of the strain-hardening ramp. This leads $\sigma_{HEL} = 3.19$ GPa.

The Hugoniot stress provided by Grady[4] is

$$\sigma_h = \sigma_{HEL} + \frac{\rho_o U_s}{2}\left(U_p - U_{HEL} + \frac{C_l - C_o}{C_l C_o} \cdot \frac{\sigma_{HEL}}{\rho_o}\right) \quad (1)$$

and plastic strain rate is given by

$$\dot{\varepsilon} = \frac{\sigma_h - \sigma_{HEL}}{\rho_o U_s^2 \Delta t} \quad (2)$$

where Δt is the rising time of peak plastic wave. We used the mixture theory for the calculation of plastic wave velocity from the constituents of the alloy[5], $U_s = 4.07 + 1.257 U_p$.

The peak particle velocities are shown to decrease very rapidly with increasing the specimen thickness. Using the peak velocities, the Hugoniot stress σ_h is calculated based upon Equation (1). Figure 3 reveals the Hugoniot stress expressed as a function of time, t. The time herein represents the traveling and rising time for a peak stress pulse to travel

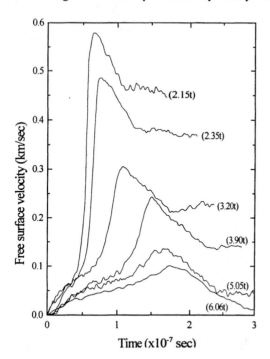

FIGURE 2. Peak-particle velocity measured on the back free surface of tungsten heavy alloy with different thickness.

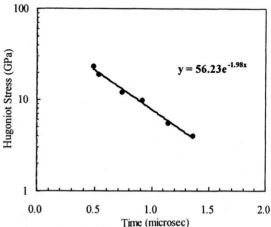

FIGURE 3. Hugoniot stress versus rising time at each thickness.

from detonator interface to the back free surface of a specimen. The relation between Hugoniot stress and traveling time may be modeled by following exponential function,

$$\sigma_p = \sigma_o \exp(\alpha t) \qquad (3)$$

where σ_o is the peak stress at the interface between WHA and the detonator, α is the attenuation coefficient of the peak stress for given materials, and σ_p is the Hugoniot stress at a given time t. In this investigation σ_o and α are shown to be 56.2 GPa and -2.0, respectively. The constant α for the dissipative behavior of the pulse would result from complicated mechanisms such as hydro-dynamics, plastic deformation and material itself.

It is noted that there are two sample groups, showing different strain-hardening ramps prior to arrival of the plastic wave. The separation of the ramp slope may have influence upon the spallation of samples. However, one needs to study the slop of the ramp in detail.

The spall stress and strain rate were calculated by using the methods proposed by Romanschenko and Stepanov[6]. The measured distance from spall line to the back-free face was approximately 0.20mm for all observed samples. Figure 4 shows spall strength depending on Hugoniot strain rate. Spall data for the specimen of 6.07mm thickness should not be included since only few crack initiation

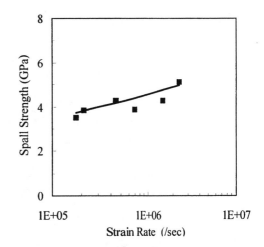

FIGURE 4. Spall strength varying with Hugoniot strain rate.

was observed on the cross-section. The spall stress increases slowly in the period of the related Hugoniot strain rate.

The crack propagation and fracture surface in this liquid-phase sintered alloy are shown in Figs. 5 and 6. Since the W/W-grain boundary is known as the weakest interface, most of the cracks generally initiate in boundaries, followed by plastic deformation of the matrix. As well-known, the tungsten grains are shown to endure until the applied stress exceeds their critical spall strength.

Figures 5(a) and (b) shows the spall cracks occurred in the cross-sectional area of the

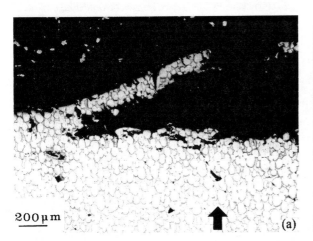

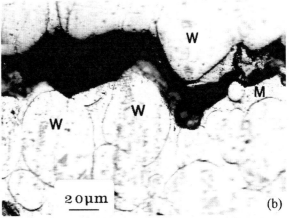

FIGURE 5. (a) Micrograph of spall cracks in the cross-section of WHA (3.2mm thickness); the arrow indicates the moving direction of compressive pulse. (b) Magnified micrograph showing the separation of each W/W particles and transgranular crack.

417

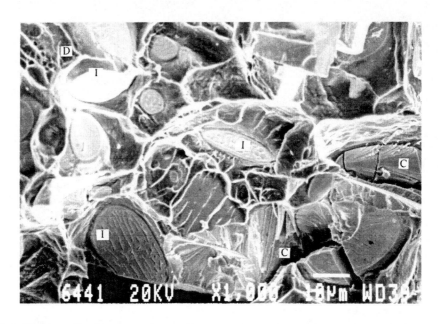

FIGURE 6. Fracture surface showing W/W interface separation(I), ductile matrix dimple(D) and cleavage(C)

specimen. W grains are pulled out from the W/W interface with the elongation of Fe-Ni matrix. Cracks propagate through the extending matrix in zigzag form. When the W grain is large enough to impede the opening of the crack as a barrier, the grain is cleaved in a brittle fracture mode (transgranular crack), indicated in Fig 6.

Thus, W/W-interface is first separated followed by matrix failure with W/matrix separation. Then W-grain would be cleaved. It is observed in some area that internal defects in W-grains formed in sintering process could initiate the failure at the early stage of spalling. These phenomena are similar to those by Charpy impact test[7]. At a low strain-rate[8], W/W-interface separation occurs in the first step, and then the brittle transgranular failure takes place due to the stress concentration in W-grain surrounded by plastically deforming matrix. Therefore, the W-grain cleavages predominate on the fracture surfac.

CONCLUSIONS

The experimental technique described in the present work was shown to be simple in measuring the free surface velocity profiles in a tungsten heavy alloy. The peak pulse attenuation through the alloy is expressed as an exponential function of time. The spall strength increases as the Hugoniot strain rate increases. The fracture behavior of WHA caused by tensile deformation in the strain rate about 1×10^5/sec is similar to that due to the Charpy impact test.

REFERENCE

1. Drummond W.E., *J. Appl. Phys.*, **28**, 1437-1441 (1957).
2. Erkman J.O. and Christensen A.B., *J. Appl. Phys.*, **38**, 5395-5403 (1967).
3. Meyers M.A., *Dynamic Behavior of Materials*, A Wiley-Inter Science Publ., 1994, ch.7.
4. Grady D.E., "Metallurgical Applications of Shock-Wave and High-Strain-Rate Phenomena," ed. L.E. Murr et. al., Marcel Dekker, Inc., 1986, 763-780.
5. McQeen R.G., Marsh S.P., Taylor J.W., Fritz J.N., and Carter W.J., "High Velocity Impact Phenomena," ed. R. Kinslow, Academic Press, 1970, 293-568.
6. Romanchenko V.I. and Stepanov G.V., *J. Appl. Mech. Tech. Phys.*, **21**, 555-561 (1981).
7. Noh J.W., Hong M.H., Kim G.H., Kang S.J., and Yoon D.Y., *Metall. Maters. Trans.*, **25A**, 2828-2831 (1994).
8. Churn K.S. and German R.M., *Metall. Trans.*, **15A**, 331-338 (1984).

CP429, *Shock Compression of Condensed Matter – 1997*
edited by Schmidt/Dandekar/Forbes
© 1998 The American Institute of Physics 1-56396-738-3/98/$15.00

NANOSTRUCTURE FORMATION BY DYNAMIC DENSIFICATION AND RECRYSTALLIZATION OF AMORPHOUS Ti-Si ALLOY

P. J. Counihan, A. Crawford, and N. N. Thadhani

School of Materials Science and Engineering, Georgia Institute of Technology, Atlanta GA 30332-0245

Dynamic densification was used to consolidate mechanically amorphized Ti-Si alloy powders, using a 3-capsule, plate-impact, gas-gun loading system at velocities of 300 and 500 m/s. The recovered amorphous compacts were subsequently annealed above the crystallization temperature. A single-phase nano-structured (50-90 nm) Ti_5Si_3 compound was produced, as revealed by TEM and XRD analysis. In this paper, the influence of dynamic densification on the crystallization behavior of amorphous Ti-Si, and the formation of nano-crystals will be discussed.

INTRODUCTION

Nanostructured materials have unique properties which are derived from the fact that a high percentage of atoms (5-40%) are located at grain boundaries.[1,2] Fabrication of such materials in bulk form is difficult,[3] since inter-particle friction and surface contaminants associated with fine scale powders can inhibit densification during static pressing, and hot-pressing can cause grain growth beyond the nano range. Shock consolidation provides a viable method for densification of amorphous as well as ultra-fine powders,[4-6] without subjecting them to long term thermal excursions, thereby retaining the metastable structure. Shock formed defects can also be introduced, which can enhance the nucleation of precipitates or phases during subsequent thermal treatment.[7] In the present work shock-compression was used to densify mechanically amorphized Ti-Si alloy powders, to investigate its influence on the crystallization behavior and to characterize the structure and properties of the crystallites formed.

EXPERIMENTAL PROCEDURE

Elemental titanium (Ti) and silicon (Si) powders of ~10 μm average particle size, were blended in a 5Ti:3Si molar ratio and mechanically amorphized using a Spex 8000 ball mill for 24 hours. XRD and TEM analysis of the mechanically milled powders (Figure 1 (a,b)), revealed a generally amorphous structure with interdispersed 5-10 nm size crystallites of Ti_5Si_3. Shock densification using a 3-capsule plate-impact fixture was used to produce 10 mm diameter by 3 mm thick compacts.

The two-dimensional radial loading effect, typical with such recovery fixtures, was modeled using AUTODYN-2D to determine the maximum bulk and axial pressures. Compacts made at 300 m/s impact velocity (P = 2.5 GPa bulk and 3.5-4.5 GPa axis) were 82-84% dense, while those made at 500 m/s impact velocity (P = 6-8 GPa bulk and 12-15 axis) were 95-99% dense. Sections taken from bulk regions of the compacts were subsequently annealed above the crystallization temperature.

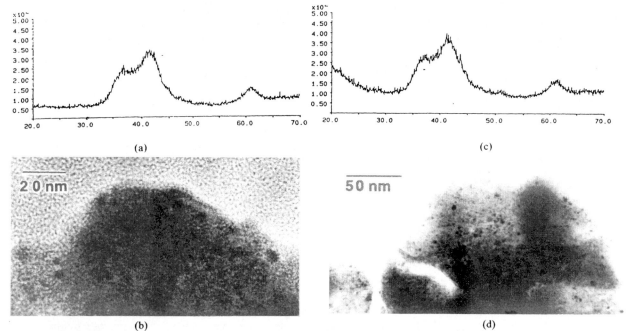

(a)

(c)

(b)

(d)

FIGURE 1. XRD traces and TEM micrographs of (a,b) ball-milled mechanically amorphized powder and (c,d) shock-densified Ti-Si compact showing a generally amorphous structure with interdispersed Ti_5Si_3 crystallites.

RESULTS AND DISCUSSION

XRD characterization of the shock densified compacts revealed that the amorphous state had been retained (Fig. 1(c)). However, with TEM imaging (Fig. 1 (d)), inter-dispersed crystallites of ~10-40 nm size were discerned in the generally amorphous matrix. DTA analysis of the shock-densified amorphous compact showed crystallization at ~600°C, similar to that of the ball-milled powder. Thermal treatments were performed on the mechanically amorphized Ti-Si powders and the shock-densified compacts at 600° to 1200°C for varying times (1, 3, and 12 hours). Figure 2 (a,b) shows XRD traces comparing the typical structure of the crystallized (12 hours at 1000°C) ball-milled powder and the shock-densified compact. It was observed that the ball-milled powder shows primary crystallization to Ti_5Si_3 phase and secondary crystallization to the Si-rich $TiSi_2$ compound (~10wt.% based on intensity ratios of corresponding peaks). In contrast, the shock-densified material showed polymorphous crystallization to a single-phase Ti_5Si_3 compound (Fig. 2(b), with no trace of $TiSi_2$ in any sample.

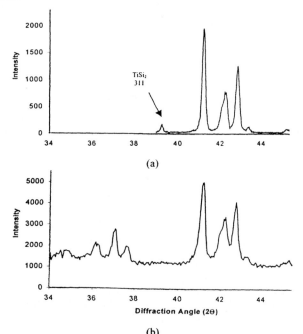

(a)

(b)

FIGURE 2. XRD traces comparing structure of crystallized (a) ball-milled powder and (b) shock-densified compact at 1000°C for 12 hours.

The difference in the crystallization behavior of mechanically amorphized and shock-densified materials may be attributed to the influence of shock pressure in lowering the free energy of the Ti_5Si_3 phase relative to the Si-rich compound. Consequently, the shock-densified compact undergoes a polymorphous glass-to-crystal phase transformation forming single phase Ti_5Si_3.

STRUCTURE-PROPERTY CORRELATION

The average Ti_5Si_3 crystallite size in the annealed shock-densified compacts was obtained from XRD line-broadening analysis. As shown in Figure 3 (a) and (b), the crystallite size range changes from ~38-47nm at 800°C to about 80-90nm at 1000°C. However, with further increase in time (up to 5 hours) and temperature (up to 1200°C), it remains stable at 90nm.

Microhardness measurements made on the crystallized shock densified compacts were observed to be in the range of 1100-1200 kg/mm^2 in samples annealed at up to 800°C and between 1300-1400 kg/mm^2 in samples annealed at 1000-1200°C. The increase in hardness at higher temperatures may be attributed to density increase due to further sintering of the compacts.

The correlation of hardness versus grain size is illustrated in Figure 4, showing (a) Hall-Petch inverse-square root dependence and (b) direct linear dependence. In general, increasing hardness is observed with grain size increase from 40 to 80nm, which is predominantly due to improved density due to sintering at the higher temperatures. Between 80-90nm grain size, the hardness remains fairly constant. Since grain sizes greater than 90nm were not observed, it was not possible to infer if the grain size dependence follows either Hall-Petch, or direct linear behavior.

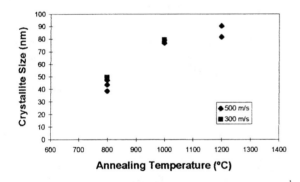

(a)

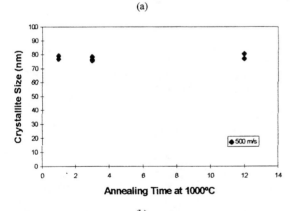

(b)

FIGURE 3. Crystallite size variation with (a) annealing temperature, constant time (1 hour), and (b) annealing time, constant temperature (1000°C).

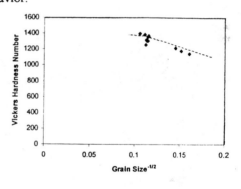

(a)

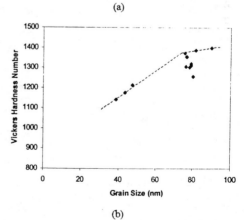

(b)

FIGURE 4. Variation of hardness with crystallite size based on (a) inverse-square root (Hall-Petch) dependence and (b) direct linear dependence.

421

In-situ crystallization of the shock-densified mechanically-amorphized Ti-Si powder compacts was done to observe the evolution of crystallites. Heating of a thin-foil sample was carried out in the TEM, from 400° to 1000°C in steps of 50°C. Bright field images and selected area diffraction patterns were taken from the same location at different temperatures to monitor the crystallization process. Fig. 5 shows (a) image of the region viewed revealing interdispersed crystallites (few to 40 nm) in a generally amorphous matrix, and (b,c) diffraction patterns of this same region at 400° and 900°C.

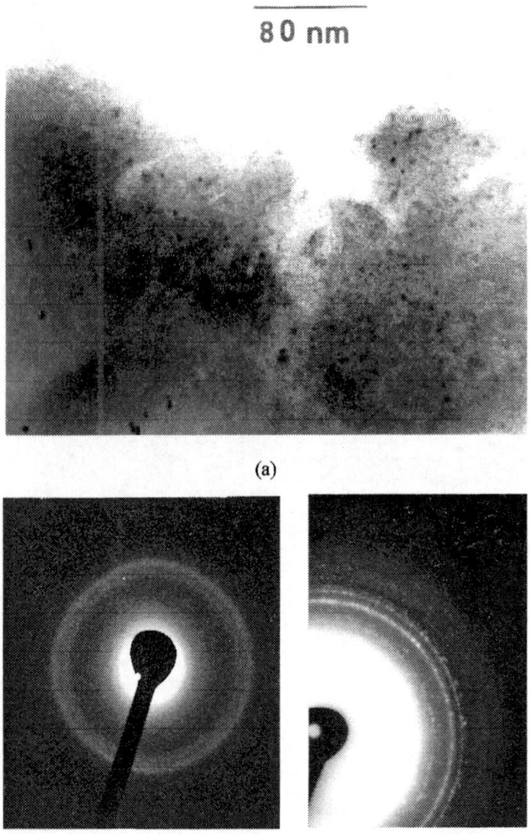

FIGURE 5. (a) TEM image of the region viewed during in-situ crystallization, and (b,c) diffraction patterns of the region at 400° and 900°C.

In general, it was observed that the dispersed Ti_5Si_3 crystallites underwent limited

growth and their size remained practically unchanged. However, the diffraction contrast of the images increased at higher temperatures, and the diffraction patterns showed significantly sharper and spotted rings, indicating increased crystallinity.

This suggests that further crystallization of the shock-densified amorphous Ti-Si compound is occurring mainly via the nucleation of new crystallites rather than the growth of existing ones. It is possible that defects introduced during shock compression, provide the nucleation sites for crystallite formation during subsequent annealing, which may also limit the growth of existing crystallites due to impingement.

SUMMARY AND CONCLUSIONS

Shock densified compacts (97-99% dense) of mechanically amorphized Ti-Si alloy powder were polymorphously crystallized to form a single-phase Ti_5Si_3 compound with 80-90nm grain size and ~1400 kg/mm^2 microhardness. The formation of single-phase structure and limited grain growth are attributed to the effects of shock-compression. The impingement of the extremely large number density of shock-formed nucleation centers results in autocatalytic crystallization during subsequent annealing, thereby inhibiting grain growth.

ACKNOWLEDGEMENTS
Funded by ARO Grant No. DAAH04-93-0062. The help of Prof. Z.L. Wang, and Ms. Yolande Berta with the TEM analysis is much appreciated.

REFERENCES

1. H. Gleiter, *Prog. in Mater. Sci.*, Vol. 33, (4), (1989) pp. 223-315.

2. R. Rosenkranz, G. Frommeyer, and W. Smarsly, *Mater. Sci. and Eng., A152* (1992) 288-294.

3. R.S. Averbach, *Matls. Sci. Eng.*, Vol. A166, (1,2) (1993) pp. 169-177.

4. M. Jain and T. Christman, *Acta Metall.*, Vol. 42, (1994), 1901-1911.

5. W.H. Gourdin, *Prog. in Matls. Sci.*, 30 (1) (1986) 39-80.

6. T. Yamasaki, Y. Ogino, K. Morishita, K. Fukuota, T. Atou, and Y. Syono, *Mater. Sci. and Eng., A179/A180* (1994) 220-223.

7. N.N. Thadhani, A.H. Mutz, and T. Vreeland, *Acta Metall*, Vol. 37 (3), pp. 897-908, 1989.

8. AUTODYN-2D/2.8, "Non-Linear Dynamic Modeling Software, Century Dynamics, 1995.

CP429, *Shock Compression of Condensed Matter – 1997*
edited by Schmidt/Dandekar/Forbes
© 1998 The American Institute of Physics 1-56396-738-3/98/$15.00

MICROSTRUCTURE OF DEPLETED URANIUM
UNDER UNIAXIAL STRAIN CONDITIONS

A. K. Zurek, J. D. Embury, A. Kelly, W. R. Thissell, R. L. Gustavsen,
J. E. Vorthman, and R. S. Hixson

Los Alamos National Laboratory, Los Alamos, NM 87545

Uranium samples of two different purities were used for spall strength measurements. Samples of depleted uranium were taken from very high purity material (38 ppm of carbon) and from material containing 280 ppm carbon. Experimental conditions were chosen to effectively arrest the microstructural damage at two places in the development to full spall separation. Samples were soft recovered and characterized with respect to the microstructure and the form of damage. This allowed determination of the dependence of spall mechanisms on stress level, stress state, and sample purity. This information is used in developing a model to predict the mode of fracture.

INTRODUCTION

Uranium is a very high density material (19.1 g/cm^3) that is relatively strong and easily cast and formed. It is widely used in nuclear and non-nuclear applications as radiation shields or kinetic energy penetrators. The most common form of pure uranium is the U-238 isotope containing some U-235. The most commonly used is low-temperature α orthorhombic phase depleted uranium. This phase is ductile, but its ductility is very dependent on processing and impurity content. The ductile-to-brittle transition of α uranium occurs around 0°C, but decreasing the grain size and hydrogen content can cause it to vary. Impurities have very low solubility in uranium, and they usually form second-phase particles that may decrease macroscopic ductility. Carbon (C) forms carbide inclusions, which may decrease ductility [1] in a manner analogous to carbides in ferritic steels.

Spallation is one of many experimental configurations that can produce controlled dynamic fracture. Spallation is defined as a dynamic uniaxial strain fracture experiment. Fracture occurs during spallation due to tensile stresses generated by the interaction of two release (rarefaction) waves [2].

Spallation is a process of damage accumulation and linkage that differs drastically from fracture damage in the uniaxial tensile test by virtue of the stress state and the rate of extent of damage accumulation. In a tensile test, voids and cracks are subject to a nearly uniaxial stress tensile field; homogeneous plastic strain dominates the flow process for most of the strain history. Due to the uniaxial stress deformation field, the voids or cracks grow to form a fracture surface, and the overall change in porosity in the vicinity of failure is small, on the order of 5% [3]. In contrast, in spallation, voids or cracks are subject to extremely high, nearly isotropic, triaxial, hydrostatic tensile stress fields, and high strain rates, which vary spatially in the sample. Voiding or crack growth dominate all stages of the damage process and produce porosity of up to about 30% at the principal spall plane. The growth rate of voids or cracks is very high, and the distribution of damage is dictated by the large gradient in stress and strain rate generated by the interaction of release waves. Porosity, void or crack formation, growth, and coalescence, therefore, are important variables in descriptions of spallation and the fracture criteria of the material [2,4,5].

In this paper we report on microstructural damage development characterization in depleted α uranium deformed under spall conditions.

MATERIALS AND EXPERIMENTS DESCRIPTIONS

Depleted uranium samples containing 280 ppm C were studied [6]. The as-cast material was wrought to specifications by the process of heat treatment, upset forging, re-heating, and finally hot rolling to 58% of the original thickness in four equal reduction passes. Due to this process the resulting microstructure is not uniform across the plate thickness. Large grains (200 μm) dominate the center of the plate while a substantially reduced grain size (down to 40 μm) exist near the top and bottom plate surfaces. Samples were cut from the center of the plate. Similarly, the high-purity depleted uranium that contained only 38 ppm C was wrought; however, the final microstructure was uniform and consisted of equiaxed 10-μm grains.

Samples were spalled and incipiently spalled using a gas gun under uniaxial strain state conditions. The spall tests were performed under nonsymmetric shock conditions; a z-cut quartz flat plate was impacted against the depleted uranium samples at a shock pressure not exceeding 5.3 GPa for 1-μs pulse duration. VISAR traces of the wave interactions were acquired and are described together with all other shock experimental details in a companion paper authored by R. Hixson *et al.* in this volume. Soft recovered samples were stored immediately in 200-proof dehydrated ethyl alcohol to prevent sample oxidation. Metallographic samples, cut through the center of the spalled sample in the direction parallel to the loading direction, were prepared for quantitative analysis [7].

RESULTS AND DISCUSSION

Plastic-Deformation and High-Strain-Rate-Induced Brittle Fracture

Figures 1a and 1b show typical spalled fracture surfaces in depleted uranium. Predominantly brittle fracture was observed in all the spalled uranium samples, with transgranular cracks for the 280 ppm purity sample, and with brittle intergranular cracks for the 38 ppm purity samples. Some ductility is visible in the 280 ppm purity uranium (Fig. 1a), and in the 38 ppm purity uranium only large grains

showed ductility in the form of ductile dimples (indicated in Fig. 1b).

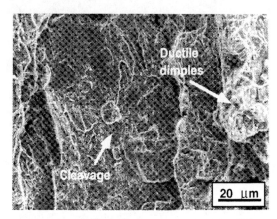

a

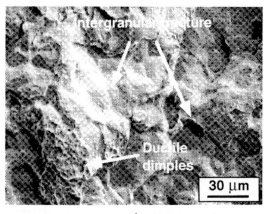

b

FIGURE 1. Fracture surfaces of depleted uranium spalled samples. (a) Depleted uranium with 280 ppm carbon showing predominantly cleavage fracture surface. (b) Depleted uranium with 38 ppm carbon showing predominantly intergranular fracture. Both samples show some areas of ductile dimples. Both samples measured comparable spall strength of -1.9 GPa.

Although α uranium is normally a ductile phase, the tests were performed at room temperature, which is close to the ductile-to-brittle transition temperature (DBTT) in uranium. The nature of the spall test, i.e. deformation at high pressure and high strain rate, contributes to the shift of the DBTT to higher temperatures when the strain rate dependence of the flow strength is taken into account [8]. In addition, the sample is subjected to high strain rate deformation and coincident hardening during the passage of the initial compressive shock wave. This deformation generates a high density of dislocations, and what is more important in the case of uranium, a large number of deformation twins; several twins variants are activated within each grain. Figures 2a and 2b show the highly magnified microstructures of incipiently spalled samples for both purities of uranium.

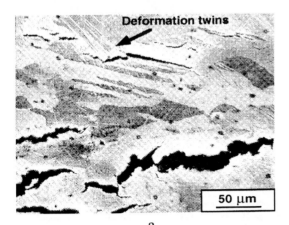

a

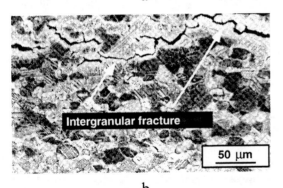

b

FIGURE 2. Cross section of depleted uranium samples incipiently spalled. (a) Depleted uranium with 280 ppm C showing large grains deformed under shock conditions with numerous twins and twin systems. Cracks in this sample are transgranular and frequently run along the twin/matrix interface. (b) Depleted uranium with 38 ppm C showing equiaxed grain structure, deformation twins within the grains, and cracks running along the grain boundaries.

It is evident that the cracks in the 38 ppm purity uranium followed the grain boundaries (Fig. 2b), and the cracks in the 280 ppm uranium are more correlated with the deformation twin systems developed in the sample during the shock.

The hydrostatic tension in the spall test is expected to aid microcrack nucleation. Work-hardening processes are far more rapid in a spall test than in a tensile quasi-static test, which results in a substantial increase in the yield strength and flow stress. This increase makes the accommodation of plastic deformation at a crack tip more difficult and therefore favors intergranular or transgranular brittle fracture. In addition to the great number of deformation twins, twin intersections and twin-matrix interfaces can serve as preferred nucleation sites for sharp brittle cracks.

The ductile-to-brittle transition of α uranium occurs around 0°C as measured under uniaxial stress state conditions in quasi static test [1]. During quasi-static loading, DBT transition is dominated by the temperature dependence of the fracture stress. The hydrostatic pressure in a tensile test is about 1/3 of the flow stress (tensile test $\Rightarrow 1/3 \lesssim -p/2\tau \lesssim 3$), which is very small in comparison to the hydrostatic tension developed under spall conditions (spall test $\Rightarrow 7 \lesssim -p/2\tau \lesssim 30$). In either case the hydrostatic pressure imposed on the sample during quasi-static or dynamic tests influences a shift in the ductile-to-brittle transition temperature. An example of the experimental evidence of this process is shown by Davidson [9] in magnesium tested under a quasi-static tensile stress state. A hydrostatic compression of 0.8 GPa imposed on a Mg sample <u>decreased</u> the DBTT by over 230°C (from 175°C to -55°C) [9].

Because of the nearly isotropic tensile stress state under spall conditions where, very high hydrostatic tension dominates the fracture process, it is reasonable conjecture that the DBTT may <u>increase.</u>

Figures 3a and 3b schematically illustrate this phenomena by plotting the change in material yield strength with respect to temperature, the fracture stress, and the stress state.
The intersection of the fracture stress level with the yield stress curve marks the DBTT. Figure 3a shows an increase in fracture stress for a tensile quasi-static test with an imposed external compressive hydrostatic pressure.
Figure 3b schematically depicts a spall tensile stress state, with its inherent large tensile hydrostatic pressure, that decreases the fracture stress. In a tensile test, the compressive hydrostatic pressure shifts the DBTT to a lower temperature, while in a spall test, large tensile hydrostatic pressure shifts the DBTT to a higher temperature.

This may explain the predominantly brittle fracture exhibited by depleted uranium under spall conditions.

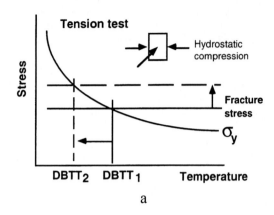

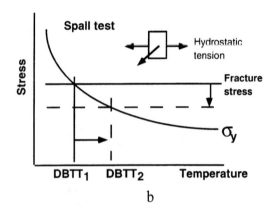

FIGURE 3. A schematic showing the DBT temperature shift resulting from the hydrostatic pressure. In the case of a tension test, compressive hydrostatic pressure imposed on a system shifts the fracture stress to a higher level and thereby the DBTT to a lower temperature (a). In a spall test an inherent to the test— very high tensile hydrostatic pressure—shifts the fracture stress to a lower level and thereby the DBTT to a higher temperature (b).

The change in fracture mode from transgranular fracture for the 280 ppm C uranium (large grain) to intergranular fracture mode for 38 ppm carbon content (small grain) uranium can be attributed to the possibility of hydrogen embrittlement in the latter. The amount of 2 ppm of hydrogen (not outgassed) is sufficient to promote the grain boundary decohesion and therefore intergranular fracture.

SUMMARY

Two purities of depleted uranium samples were tested under spall conditions. The spalled samples showed predominantly brittle fracture: transgranular fracture for 280 ppm C uranium and intergranular fracture for 38 ppm C uranium. Deformation twinning was found to be the dominant form of deformation under spall conditions in the pure uranium samples. Both purities of uranium samples had a comparable spall strength of -1.9 GPa. The high hydrostatic tension and very high strain rates inherent to spall testing are thought to increase the DBTT in pure uranium and thus promote brittle fracture.

ACKNOWLEDGMENTS

The authors would like to thank the Spall and Materials Damage Program Manager, Dean Preston of Los Alamos, for financial support. The work was done under the auspices of the US Department of Energy.

REFERENCES

[1] Eckelmeyer, K. H., in Uranium and Uranium Alloys, ASM International, Materials Park, Ohio, 1991.
[2] Zurek, A. K., Johnson, J. N., and Frantz, C. E., Journal de Physique, 49, (1988), pp. 269-276.
[3] Thomason, P. F., in Ductile Fracture in Metals, Pergamon Press, Oxford, 1990.
[4] Johnson, J. N., J. of Applied Physics, 52, (1981), pp. 2812-2825.
[5] Curran, D. R., Seaman, L., and Shockey, D. A., Physics Reports, 147, (1987), pp. 253-388.
[6] Dunn, P., private communication, 1997.
[7] Kelly, A., private communication, 1997.
[8] Zurek, A. K., Follansbee, P. S., and Hack, J., Metallurgical Transactions A, 21A, (1990), pp. 431-439.
[9] Davidson, T. E., Uy, J. C., and Lee, A. P., Acta Metallurgica, 14, (1966), pp. 937.

CP429, *Shock Compression of Condensed Matter – 1997*
edited by Schmidt/Dandekar/Forbes
© 1998 The American Institute of Physics 1-56396-738-3/98/$15.00

ANALYSIS OF THE DEGRADATION MECHANISMS IN AN IMPACTED CERAMIC

C. DENOUAL,[1] C.E. COTTENOT[1] and F. HILD[2]

[1] *DGA/CREA-Département Matériaux en Conditions Sévères*
16 bis, Avenue Prieur de la Côte d'Or, F-94114 Arcueil Cedex, France.
[2] *Laboratoire de Mécanique et Technologie*
E.N.S. de Cachan / C.N.R.S. / Université Paris 6
61, Avenue du Président Wilson, F-94235 Cachan Cedex, France.

To analyze the degradation mechanisms in a natural sintered SiC (SSiC) ceramic during impact, three edge-on impact configurations are considered. First, the ceramic is confined by aluminum to allow a post–mortem analysis. In the second configuration, a polished surface of the ceramic is observed each micro-second by a high–speed camera to follow the damage generation and evolution. The third configuration uses a high–speed Moiré photography system to measure dynamic 2-D strain fields. Sequences of fringe patterns are analyzed.

INTRODUCTION

In many impact configurations, the stress field generated by a projectile in a ceramic can be assumed to be spherical and to produce damage in both compressive and tension modes in different locations within the ceramic. Damage in compression is generated near the impact surface when shear stresses reach a threshold value which can be dependent on pressure and strain rate. In the bulk of the ceramic, damage in tension is observed when the hoop stress induced by the radial motion of the impacted ceramic is sufficient to generate fracture in mode I initiating on micro–defects such as porosities or inclusions.

Most of the mechanisms initiated during impact (such as nucleation of flaws, propagation and interaction of cracks, stress release) can only be studied during dynamic tests, which have to be chosen to give reliable data on damage evolution.

A quantitative information on the location and evolution of damage can be obtained by an "edge-on" impact (EOI) configuration described in Ref. (1). These configurations are developed by the Ernst-Mach-Institut (EMI) and by the Centre de Recherches et d'Etudes d'Arcueil (CREA). In the latter configuration, a blunt projectile (11 mm in diameter and 20 mm in length) impacts a ceramic tile of size $100 \times 50 \times 10$ mm^3. The velocity of the impactor varies within the range [100 m/s; 330 m/s] and produces in the ceramic a zone of complete fragmentation. It can be shown that the same damage mechanism (*i.e.* damage in tension) is observed in EOI and in real impact configurations (2). The EOI can therefore be used to validate damage evolution laws for numerical simulations of the behavior of light armors.

Three different test configurations described in the following sections are used to understand the damage mechanisms. In all configurations, a nitrogen gas gun launches a steel cylinder which impacts a confined or unconfined SSiC tile with a velocity of 200 m/s or 330 m/s.

1 POST–MORTEM ANALYSIS

The first test is an EOI with an aluminum confinement presented in Fig. 1. A SSiC ceramic tile is impacted at a velocity of 330 m/s. The ceramic characteristics are given in Ref. (3). After impact, the tile is coated in an epoxy resin and polished for macroscopic and microscopic analyses.

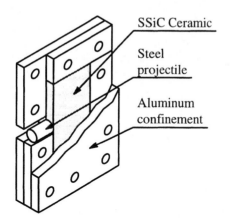

FIGURE 1. Schematic of the ceramic confinement. The tile is confined between two 10 mm thick aluminum plates. The edge confinement is obtained by aluminum tiles screwed on the plates.

The post–mortem crack pattern is presented in Fig. 2. Different zones can be separated with respect to the crack density and orientation. In front of the projectile, a small zone exhibits a randomly oriented crack pattern. In the bulk of the ceramic, one can observe long radial cracks with a second circumferential crack pattern superimposed on the first one.

In this last region, some radial cracks seem to kink in the circumferential direction, kink one more time and propagate in the radial direction. The kinked fractures are made of two small (and hardly visible, see Fig. 2) cracks linked by a small circumferential crack. This complex crack is then opened by the radial motion of the ceramic, widens and becomes a long macroscopic "kinked" fracture. This phenomenon shows that the radial cracks appear prior to the circumferential ones, since the latter do not go through the former. The maximum size of a fracture between two kinks in this zone does not exceed a few millimeters.

Near the tile rear face, a third zone exhibits a high density of cracks. This thin strip remains at a constant distance from the rear surface, like a spalling zone.

A microscopic analysis of the tile section along the projectile axis reveals the presence of some

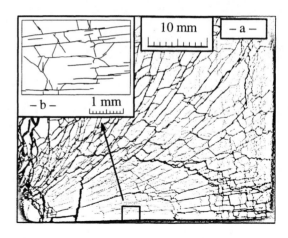

FIGURE 2. a–Half top view of tile impacted at a velocity of 330 m/s. b–The magnification shows many short radial cracks linked by short circumferential ones.

long cracks running through the whole specimen, from the projectile tip to the rear face of the tile. An explanation can be proposed by remembering that the stress field is modified during propagation by the tile edges. The first compressive wave is a uniaxial strain state (longitudinal wave) that is not compatible with the free edges of the tile. Two relaxation waves are therefore generated from both edges and interact with each other in the center of the tile. The stress state becomes tensile and crack nucleation occurs.

2 REAL–TIME VISUALIZATION OF CRACKS

The real–time visualization test is designed to provide an information on the chronology of damage onset. The ceramic target presents one mirror polished face illuminated by a flashlight. A high-speed camera records the reflected light. The pictures shown in Fig. 3 were obtained by Riou (4).

Most of the initiated cracks do not exceed a few millimeters in length (Figs. 3–a and 3–b). This leads to assume that the crack tip cannot follow the crack front, *i.e.* the velocity of the former is less than the velocity of the latter. A crack initiates when the tensile stress reaches a threshold value (depending on the size and shape of the defect) and stops because other defects nucleated in front of it and relaxed the hoop stress. No cir-

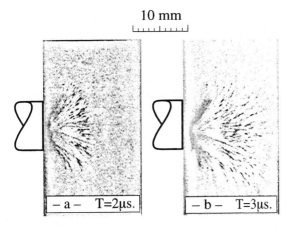

FIGURE 3. An example of pictures obtained by real-time visualization with a SSiC ceramic impacted at a velocity of 203 m/s.

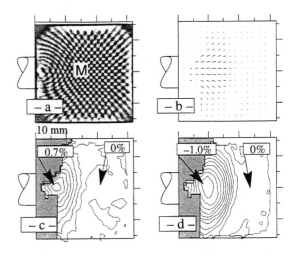

FIGURE 4. a–The initial frame (2 μs after impact). b–Third (negative) eigen direction c–and d–First and third eigen strains plotted with a 0.1% step between each contour.

cumferential cracks are visible in Fig. 3 but are clearly present in the last frame of the same shot (not presented in this paper, see Ref. (4)).

3 HIGH–SPEED PHOTOGRAPHY OF MOIRE FRINGES

Moiré photography is an optical technique that makes use of the geometric interferences occurring when two gratings are superimposed. The main difficulties to observe ceramic impact lie in the time range and the very high wave velocity (around 11,000 m/s). The simple and mobile Moiré optical set-up and the automated fringe pattern analysis are described in Ref. (5).

A typical result of the high–speed Moiré photography is presented in Fig. 4. The velocity of the impactor is equal to 330 m/s. The method used to analyze the fringes cannot give reliable data when they are blurred. To overcome this problem, the artifacts generated during the fringe pattern analysis are automatically reset to zero in a gray colored zone in Figs. 4–c and 4–d.

A set of twelve frames is taken. The first frame is recorded before any physical contact, is free of constrains and constitutes the reference Moiré pattern. The dynamic response of the specimen is recorded on eleven frames from 1 μs to 6.5 μs after impact with a 500 ns interframe time and a 40 ns exposure time.

Figure 4–a is the fringe pattern at $T = 2$ μs (the fourth frame of the whole sequence). Figure 4–b shows the eigen directions of the third (compressive) strain and confirms that the strain wave induced by an EOI has a circular geometry. The first and third eigen strains are given in Figs. 4–c and 4–d, respectively.

To compare the evolution of the radial and hoop strains, a typical result is given in Fig. 5. The strain diagram is plotted for a point M at a distance of 13 mm from the projectile tip (see Fig. 4–a). It can be noticed that the radial strain reaches an important value before any significant evolution of the hoop strain. This is consistent with a cylindrical stress wave in which the tensile strain is induced by the radial motion of the material.

A comparison with a numerical simulation is proposed with the model presented in Ref. (3). The general shape of the curve is very well reproduced by the model, with a tendency to under–estimate the compressive strain. A better modeling of the compressive behavior of the ceramic should improve the numerical/experimental agreement.

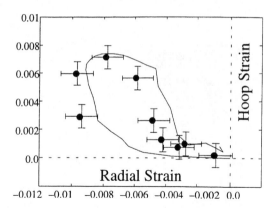

FIGURE 5. Typical example of strain evolution given by a Moiré technique (dots). A comparison (curve) is proposed with the model presented in Ref. (3).

DISCUSSION

A chronology of the damage onset can be proposed by using the results of the previous sections. The post–mortem analysis shows that the radial cracks are first generated and closely followed by the circumferential ones. The former damage is due to the divergent compressive wave that induces a tensile stress state. The latter crack pattern is due to the relaxation waves coming from the broken projectile and to the tensile stress wave emanating from the rear of the tile. When the two relaxation waves cross each other, another zone of dense crack nucleation can be observed and can be compared to a spall region.

The deformation in tension reaches a maximum value of 1% in the frame analyzed in Fig. 4. The ceramic in this zone cannot be undamaged and we can assume that cracks are present even if they are not visible. Moreover, the presence of cracks at $T = 2$ μs is compatible with the proposed chronology. In Fig. 4–a we can observe that the Moiré fringes are not blurred in the zone where cracks are assumed to nucleate. The local deformation induced by the crack nucleation is therefore sufficiently small not to blur the fringes, or, in other words, the nucleation is sufficiently dense so that the fringes location is not modified. This is not compatible with the results presented in Fig. 3 where macroscopic cracks would lead to a visible blur in the Moiré fringes.

The post–mortem analysis shows that many short cracks are nucleated and may be connected (at the end of the interaction) with the neighboring fracture to give macroscopic fractures. The initial short fractures are thus too small to be apparent in both real–time visualization of impact and Moiré photography and the previously estimated crack densities, far less than the real densities, have to be re–estimated. The main mechanism for the degradation of ceramic is therefore a very dense nucleation of cracks, some of them branching to provide macroscopic fracture.

CONCLUSION

Even if some damage mechanisms cannot be directly compared between real armor and EOI configurations, the EOI test can help to understand the ceramic degradation mechanisms during impact. First, the "powder" produced by impact is made of 3 anisotropic crack patterns, superimposing at different times to provide fragments of various shapes. Second, the very high density of nucleated cracks can only be established by using a fine microscopic analysis.

ACKNOWLEDGMENTS

This work was funded by DGA-DRET-STRDT. The authors wish to thank Ms. L. Riolacci for providing the picture of Fig. 2.

REFERENCES

1. Straßburger E., Senf H., Denoual C., Riou P., and Cottenot C.E., *"An Experimental Approach to Validate Damage Evolution Laws for Brittle Materials,"* To be presented at DYMAT 97, Toledo, Spain, Sept. 1997.

2. Denoual C., Cottenot C.E. and Hild F., *"On the identification of Damage During Impact of a Ceramic by a Hard Steel Projectile,"* Presented at the 16th Ballistic Symp., pp. 541–550, 1996.

3. Hild F. and Denoual C., *"A Probabilistic Model for the Dynamic Fragmentation of Brittle Solids,"* Presented at the APS Conference on Shock Compression of Condensed Matter, Amherst (MA), USA, 1997.

4. Riou P., PhD Dissertation, Ecole Nationale Supérieure des Mines de Paris, 1996.

5. Bertin-Mourot T., Denoual C., Deshors G., Louvigné P. F. and Thomas T., *"High Speed Photography of Moiré Fringes: Application to Ceramics under Impact,"* To be presented at DYMAT 97, Toledo, Spain, Sept. 1997.

CP429, *Shock Compression of Condensed Matter – 1997*
edited by Schmidt/Dandekar/Forbes
© 1998 The American Institute of Physics 1-56396-738-3/98/$15.00

DAMAGE QUANTIFICATION IN CONFINED CERAMICS

Yueping Xu and Horacio D. Espinosa

*School of Aeronautics and Astronautics, Purdue University,
West Lafayette, IN 47907*

Impact recovery experiments on confined ceramic rods and multi-layer ceramic targets are performed for failure identification and damage quantification. In-material stress measurements with manganin gauges and velocity histories are recorded with interferometric techniques. Observations on recovered samples are made through Optical Microscopy. Microscopy results show that microcracking is the dominant failure mode in ceramic rods and multi-layer ceramic targets. Macrocrack surface per unit area is estimated on various sections along several orientations. Correlation between dynamic loading and crack density is established. Moreover, *multiple penetrator defeat* is observed in ceramic targets recovered from penetration experiments.

INTRODUCTION

In the analysis of damage mechanisms and the formulation of computational models (1), it is of great relevance to determine the number, size and orientation of cracks in the material upon dynamic impact. In this work, damage mechanisms in confined alumina ceramic bars and target plates penetrated by WHA long rods are examined. Stress histories measured with embedded manganin gauges are reported and analyzed. Crack surface area per unit volume on different orientations is characterized quantitatively by stereographic analyses.

EXPERIMENTAL METHODS

Case Study I : The test geometry for impact of confined ceramic rods is illustrated in Fig. 1. The alumina rod used in this study, AD-94, was manufactured by Coors Porcelain Company, Golden, CO. The rod was produced with a diameter of 0.5". Lateral confinement, on the surface of the cylindrical ceramic specimen, was obtained by shrink fitting a 4340 steel sleeve with a nominal outer diameter of 1". The

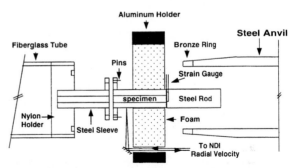

FIGURE 1. Alumina rod on rod recovery experimental configuration.

sleeve was heated such that the inside diameter expanded to a size slightly larger than the ceramic rod diameter. Then, the ceramic rod was slid into the sleeve. The shrinkage of the sleeve during cooling of the assembly provided the required confinement pressure. Dynamic loading of the ceramic rod was produced by launching a steel rod, mounted on a fiber glass tube, in a light gas gun at Purdue University. The surfaces of the rods were lapped and polished flat. The target was aligned to the impactor surface

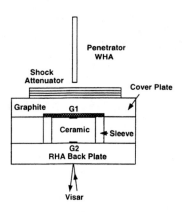

FIGURE 2. Schematic of ballistic experiment.

within 0.5 milliradians by using an optical technique developed by Kumar and Clifton (2). Normal displacement laser interferometry was used to detect the transverse (radial) displacements, at different depths from the impact surface, on the periphery of the sleeve. A distinctive feature of the experiments conducted under this investigation was the recovery of intact specimens. This methodology made it possible to perform extensive microscopic analysis to assess damage. After impact, the sleeve was machined away from the specimen. The rod was sectioned, along the impact axis, grounded, and polished. Chemical etching, by a solution with $HF : HNO_3 : H_2O$ (1:2:5) for 1 minute, was used to reveal the crack pattern.

Case Study II: The test geometry for ballistic impact of confined alumina plates is shown in Fig. 2. Details of the experimental set-up are discussed in (3). After impact, the plates were carefully separated. A two-component epoxy was poured into the cavity, left by the penetrator, to strengthen the fractured ceramic. A diamond saw was used to section the plates on a plane containing the impact direction. To examine the crack pattern, optical microscopy was used. This study provides insight into failure mechanisms and allows the quantification of crack surface area per unit volume.

For estimating the total macrocrack surface area per unit volume, the general relationship, $S_V = 2P_L$ (4), was used. In this formula, S_V is the total crack surface area per unit volume, and P_L is the average value of the number of intersections of a set of test lines of unit length. In principle, the lines can be randomly located with respect to the surfaces of interest. In the case of anisotropic microstructures, the number of intersections of a set of test lines, with the boundaries of macrocracks, depends on the angular orientation of the test lines in the plane. Thus, in order to get a representative average value of the intersection count, it is necessary to perform the measurements on different angular orientations in the different planes. An average of the crack surface per unit volume can be obtained. The dependence of the number of intersections per unit length with the angle of the test array can be used to characterize the degree and type of orientation of a system of lines in a plane. We applied a test array to the system at 30^0, 45^0, 60^0, 90^0, 120^0, 135^0 with respect to the impact direction, and determined P_L separately at each angle θ. From the P_L, we calculated S_V. By plotting S_V versus θ, provides a so-called space rosette which reflected the degree of preferred orientation. For specimens with no preferred orientation, the rosette is a circle. With increasing percentage of preferred crack orientation, the rosette shape is deformed.

EXPERIMENTAL RESULTS AND DISCUSSION

Two rod on rod experiments were performed. The impact velocity for shot 4-1214 was 188 m/sec and for shot 5-0612 was 195 m/sec. For shot 5-0612, the radial velocity was recorded at two points located 12 mm and 15 mm from the impact surface as shown in Fig. 3. The radial velocity at 15 mm has a rising part to a peak velocity of 28 m/sec followed by a decrease and increase resulting from wave release from the sleeve periphery. It should be noted that the radial velocity contains information on the ceramic damage and plastic deformation of the sleeve itself. At approximately 11.5 μsec the trace shows a sudden reduction in particle velocity because of the arrival of an unloading wave from the free end of the impactor rod. Similar features are

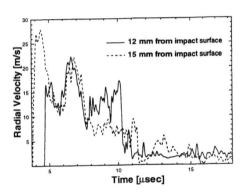

FIGURE 3. Radial velocity history recorded in experiment 5-0612.

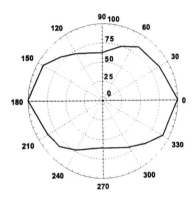

FIGURE 5. Rose of cracks S_V from rod surface sectioned along the impact direction.

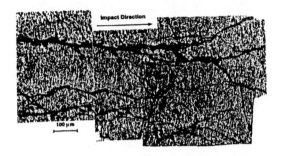

FIGURE 4. Optical micrograph showing macrocrack pattern in recovered rod.

observed in the radial velocity of a point located at 12 mm from the impact surface. A time difference in the arrival of the unloading wave is clearly observed. Further understanding of these traces requires the numerical simulation of the experiments accounting for inelasticity in the ceramic rod and steel sleeve.

Post-test micrographs, of the recovered alumina rod from a region near the impact surface, is shown in Fig. 4. Crack coalescence along the impact direction is observed. The formation of a conical fault, as investigators found in the case of Hopkinson bar experiments (5,6) is not observed. Lack of conical faults can be the result

of low confining pressure or geometric differences in the specimens. The observed macrocracks in the impact direction are evidence of failure by growth and coalescence of microcracks in planes parallel to the compression axis. Our quantification of crack surface, see Fig. 5, shows that the macrocrack pattern does have a preferred orientation. It is evident that macrocrack density decreases with the increase of the angle, and at 0^0, i.e., along the impact direction, the macrocrack surface area reaches a maximum. However, it should be noted that other orientations, e.g., 45^0 also present a significant S_V.

For the ballistic experiment, observations on polished and etched cross-sections along the penetration direction, revealed a well defined two-layer structure of tungsten alloy entrapped by ceramic fragments at different depth along the penetration direction, see Fig. 6. Since the fragments of the fretted penetrator, from the leading surface, have a high initial forwards velocity, they can travel into open cracks and flow laterally, at the penetrator nose. This observation indicates *multiple penetrator defeat* can be achieved in ceramic targets properly confined. Adjacently to the crater left by the penetrator and along the cover plate-ceramic interface, tungsten particles, with greatly elongated grains, are observed. This is the result of the localized shear deformation experienced by the WHA penetrator. Grain distortion decreases with distance

433

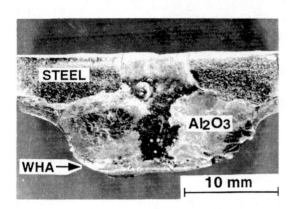

FIGURE 6. Optical micrograph showing penetrator lateral flow at two locations near the ceramic-graphite interface.

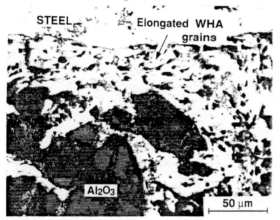

FIGURE 7. Optical micrograph showing penetrator deformation along the steel-ceramic interface.

from the crater as shown in Fig. 7.

Quantification of macrocrack surface area shows that the degree of preferred macrocrack orientation is not as pronounced as in the case of rod on rod impact previously discussed, see Fig. 8. The distribution of macrocracks is close to uniform although the size of fragments is not.

CONCLUSIONS

An attempt to identify failure mechanisms and to quantify crack density in confined alumina rods and plates was made. Fractographic

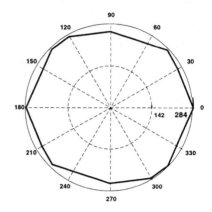

FIGURE 8. Rose of cracks S_V from confined alumina penetration experiment.

observations on penetrated alumina targets show that *multiple penetrator defeat* can be achieved in alumina. The quantification of macrocrack surface area per unit volume, reported in this work, indicates crack patterns with preferred orientations are obtained in confined ceramic rods.

REFERENCES

1. Espinosa, H.D., "On the Dynamic Shear Resistance of Ceramic Composites and its Dependence on Applied Multiaxial Deformation", *Int. J. of Solids and Structures*, Vol. 32, No. 21, 3105, 1995.

2. Kumar, P., Clifton, R.J., 'Optical Alignment of Imapct Faces for Plate Impact Experiments', *J. Appl. Physics*, **48**, 1366, 1977.

3. Brar, S., Espinosa, H.D., Yuan, G., and Zavattieri, P.D., 'Experimental Study of Interface Defeat in Confined Ceramic Targets', presented at the Shock Compression of Condensed Matter Conference on, Amherst, Massachusetts, 27 July - 1 August 1997.

4. Underwood, E.E., 'Quantitative Stereology' Addison-Wesley, Reading, Mass, pp30-33,1970.

5. Arrowood, R., Lankford, J., 'Compressive Fracture Processes in an Alumina-Glass Composite,' *J. Mater. Sci.*, **22**, 3737, 1987.

6. Chen, W.N., 'Dynamic Failure Behavior of Ceramic Under Multiaxial Compression,' Ph.D. Thesis, Caltech, Pasadena, CA, 1995.

CP429, *Shock Compression of Condensed Matter – 1997*
edited by Schmidt/Dandekar/Forbes
© 1998 The American Institute of Physics 1-56396-738-3/98/$15.00

THE INFLUENCE OF TEXTURE AND IMPURITIES ON THE MECHANICAL BEHAVIOR OF ZIRCONIUM

G. C. Kaschner, G.T. Gray III, and S.R. Chen

Los Alamos National Laboratory, Los Alamos, NM 87545

Development of physically-based constitutive models capable of simultaneously describing slip, twinning, and anisotropy requires knowledge of the coincident influence of each on mechanical response. The effects of interstitial impurities and texture on twinning in zirconium (Zr), in addition to variations in strain rate and temperature are utilized to probe substructure evolution and mechanical behavior. The compressive yield response of high-purity crystal-bar and commercial-purity Zr was found to depend on the loading orientation relative to the h.c.p. c-axis, the applied strain rate, varied between 0.001 and 3500/s, and the test temperature, varied between 76 and 298K. The rate of strain hardening in Zr is seen to depend on the controlling defect storage mechanism as a function of texture, strain rate, and temperature. The substructure evolution of high-purity Zr was observed to depend on the applied strain rate and test temperature; the substructure of high-purity Zr was seen to display a greater incidence of deformation twinning when deformed at high strain rate or quasi-statically at 76K.

INTRODUCTION

Plastic deformation of metals may occur mainly by either of two mechanisms: dislocation slip or twinning. Whether slip or twinning is the dominant mechanism depends on which mechanism requires the least stress to initiate plastic deformation.

Knowledge of the influence of temperature, strain rate, microstructure, and chemistry on mechanical response is necessary for developing an accurate constitutive model. Previous studies have demonstrated the influence of temperature and strain rate on the slip and twinning behavior of zirconium (1). This study examines our ability to more accurately model the behavior of such low-symmetry metals by investigating the influence of texture and interstitial impurities on twinning deformation.

The goal of this study was to measure the stress-strain response of zirconium (Zr) over a wide range of temperatures and strain rates and to use microstructural evidence to identify dominant deformation mechanisms. The mechanical response will be considered in terms of the initial yield behavior and strain-hardening rate. The strain-hardening behavior is generally the most important response for modeling constitutive behavior to large strains and is influenced by impurity level, grain size, and the extent of deformation twinning.

EXPERIMENTAL TECHNIQUES

This investigation utilized both high-purity zirconium obtained from Teledyne Wah Chang and commercial-purity (CP) zirconium supplied by the Naval Surface Warfare Center (NSWC). The high-purity material contained (wt. %) approximately 54 ppm Hf and 40 ppm O in contrast to impurity levels of 63 ppm Hf and 510 ppm O in the NSWC stock (referred to hereafter as commercial-purity). The high-purity Zr was clock-rolled at room temperature then annealed at 823K for 1 hour. This produced an equiaxed grain structure with strong in-plane isotropic basal texture in the plate. The commercial-purity Zr was tested in the as-received recrystallized condition. It, too, has in-plane basal texture but exhibits some directionality which reflects that the plate was cross-rolled rather than clock-rolled. Both materials have a mean grain size of approximately 25 μm.

To examine the influence of loading orientation, and thereby texture, on the mechanical response of this strongly basal textured Zr-plate, compression samples were sectioned from both the through-thickness (TT) and in-plane (IP) plate directions. Cylindrical compression samples (5-mm diameter by 5-mm long) were first electro-discharge machined (EDM) normal to the plane of the plate; this orientation, relative to the plate texture, fixed the compression axis of the samples to be nominally 20 degrees to the c-axis orientations of the grains. Samples were also EDM machined in the in-plane plate direction to evaluate texture effects on constitutive behavior and substructure evolution. In-plane samples of high-purity Zr were machined at 0, 45, and 90 degrees orientations to an arbitrary direction in the plate. In-plane samples of commercial-purity Zr were cut at angles of 0 and 90 degrees relative to the rolling direction.

All mechanical tests were performed in compression at temperatures between 76 and 298K. Quasi-static compression tests were conducted at strain rates of 0.001 and 0.1/s using an Instron screw-drive load frame. Results of high-purity IP samples typically exhibited stress-strain curves parallel to one another with small variations in stress levels; +/- 10 MPa about the mean value for a given set of conditions. Dynamic tests at strain rates from 1000 to 3500/s were conducted utilizing a Split-Hopkinson Pressure Bar. The inherent oscillations in the dynamic stress-strain curves and the lack of stress equilibrium in the specimens at low strains make the determination of yield inaccurate at high strain rates; data at strains less than 0.02 have been deleted.

Specimens for optical metallography were sectioned, parallel to the loading axis, from the as-recrystallized and deformed samples. The initial microstructure of both forms may be characterized as having equiaxed grains, a relatively broad grain size distribution, and few residual twinned grains. Specimens for analysis of the deformation microstructures were deformed to a true strain of approximately 20%; selected samples were also deformed to a true strain of 5% to facilitate examination of the microstructure at different stages of deformation. Samples for optical metallographic examination were polished conventionally and etched using an acidic solution (10 ml H_2O, 10 ml HNO_3, 10 ml HCl, and 5 ml HF).

RESULTS AND DISCUSSION

Preferred texture is significant because it results in orientation-dependent constitutive properties Moderate axisymmetric (fiber) texture was observed in (0001) pole figure of clock-rolled high-purity Zr (Fig. 1a). Although the commercial-purity Zr also has a strong texture, the influence of cross-rolling is evident in the distribution of preferred orientations (Fig. 1b). Texture maxima of 5.63 and 3.64 m.r.d. (multiples of random distribution) were reported for the high-purity and commercial-purity, respectively.

The effects of the strong basal texture of the high-purity Zr (Fig. 1a) are demonstrated in the compressive stress-strain data displayed in Figure 2a. Peak flow stresses of quasi-static compression tests performed at room temperature are approximately 2.5 times greater for the TT samples compared to IP samples of the high-purity Zr. When tested at 76K, yield stresses of TT samples were 3 times greater than that of IP samples. The

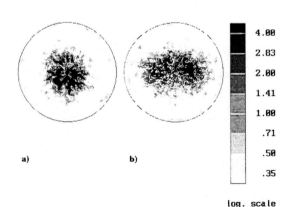

a)   b)

	4.00
	2.83
	2.00
	1.41
	1.00
	.71
	.50
	.35

log. scale

FIGURE 1. (0001) pole figures of (a) clock-rolled, high-purity Zr and (b) cross-rolled, commercial-purity (CP) Zr.

stress-strain response of the TT sample tested under quasi-static conditions at low temperature has an especially sharp transition from elastic loading to nearly linear work hardening. This indicates the onset of twinning with little to no deformation by slip (1, 2).

The stress-strain response at quasi-static strain rates at 298K (Fig. 2b) is consistent with previous observations (1): an increase in strain rate results in a slight shift of the curve but the characteristically continuous curvature is preserved.

Other effects of texture for the commercial-purity Zr are evident in Fig. 3a. The ratio of the yield stresses for TT and IP compression tests is

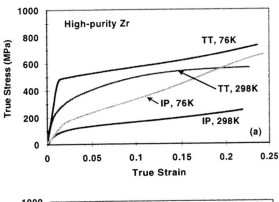

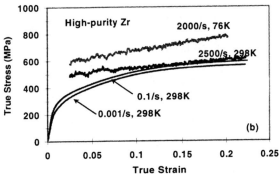

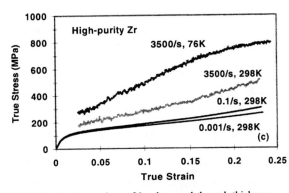

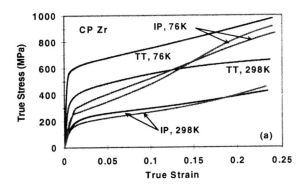

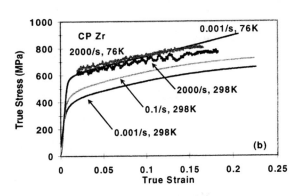

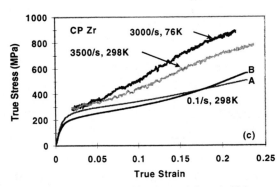

FIGURE 2. a) Comparison of in-plane and through-thickness high-purity Zr tested in compression at 0.001/s. b) Results of through-thickness (TT) tests conducted over a range of rates (0.001 - 2500/s) and temperatures (76 - 298K). c) Results of in-plane (IP) compression tests performed over a range of rates (0.001 - 3500/s) and temperatures (76 - 298K).

FIGURE 3. a) Comparison of in-plane and through-thickness commercial-purity Zr tested in compression at 0.001/s. b) Results of through-thickness tests conducted over a range of rates (0.001 - 2000/s) and temperatures (76 - 298K). c) Results of in-plane compression tests performed over a range of rates (0.001 - 3500/s) and temperatures (76 - 298K).

significantly less than for the case of the more strongly textured high-purity Zr. Data from IP tests (Figs. 3a and 3c) have a distinctive tendency to cross between strain values of 15 to 20 % due to mild texture anisotropy (Fig. 1b).

The contrasts between the IP and TT samples tested under quasi-static conditions (Figs. 2 and 3) are consistent with previously observed results for polycrystalline anisotropic materials (3, 4). The large variation in flow stresses is due to the number of active twinning and slip systems available and the stress levels required to activate them. Compressive loading perpendicular to the basal plane, as in the case of the TT samples, favors deformation twinning over prism, pyramidal, or basal slip. Compressive loading of the IP samples will have more favorable conditions, with respect to Schmidt factors, for activating slip, thus plastic flow is achieved with lower applied forces.

Strain rate sensitivity is observed to vary as a function of loading direction in both the high-purity and commercial-purity Zr (Figs. 2b, 2c, 3b, and 3c). Strain rate sensitivity is greatest in samples tested in the IP orientation. This too, is compatible with slip dominant deformation and the concept of dislocation motion controlled by Peierls forces (1).

Yield stress values are consistently greater for the commercial-purity Zr compared to samples of the high-purity Zr tested under similar conditions. Since the materials used in this study are of similar texture and grain size, the difference in flow stress may be attributed to differences in the interstitial impurities contents (5, 6). Interstitial impurities make twinning less thermodynamically favorable; hence deformation by slip becomes the dominant mechanism (6). In instances where the flow stress of twinning is greater than the flow stress of slip, the tendency to twin is suppressed and the required stress to deform the matrix increases.

Twins are present in the majority of grains after testing. Twins are observed to operate in several different systems in samples deformed at high rates and low temperatures. Hence, the strain-hardening and yield behavior differences between high and low strain rates are thought to be due to dislocation slip versus twinning dominated deformation mechanisms.

SUMMARY AND CONCLUSIONS

The effects of texture and interstitial impurities on the deformation behavior of a low-symmetry metal were demonstrated by comparing high-purity and commercial-purity Zr. The compressive stress-strain response of high-purity crystal-bar and commercial-purity Zr was found to depend on the loading orientation relative to the h.c.p. c-axis, the applied strain rate (varied between 0.001 and 3500 /s) and the test temperature (varied between 76 and 298K). The rate of strain hardening increases with increasing strain rate and decreasing temperature. The rate of strain hardening is also much greater when samples are loaded perpendicular to the c-axis in a slip-dominant orientation. The substructure evolution of high-purity Zr was observed to depend on the applied strain rate and test temperature; a greater incidence of deformation twinning was observed in samples deformed at high strain rate or quasi-statically at 76K. The rate of strain hardening in Zr is seen to depend on the controlling defect storage mechanism as a function of texture, strain rate, and temperature.

ACKNOWLEDGMENTS

This work was supported under the auspices of the United States Department of Energy. The authors acknowledge Mike Lopez for performing the quasi-static low temperature tests and Douglas Cannon for assisting with the metallographic preparations.

REFERENCES

1. Song, S.G. and Gray, G.T. *Metall. Trans. A*, **26A**, 2665-2675 (1995).
2. Song, S.G. and Gray, G.T. *Acta Metall. Mater.* **43**, 2339-2350 (1995).
3. Reed-Hill, R.E. in *Deformation Twinning* (edited by R.E. Reed-Hill, J.P. Hirth, and H.C. Rogers), Proc. TMS, Vol. 25, p. 295-320 (1964).
4. Christian, J.W. and Mahajan, S. *Prog. Mat. Sci.*, **39**, 1-157 (1995).
5. Conrad, H. *Prog. Mat. Sci.*, **26**, 123-403 (1981).
6. Gray, G.T., *Twinning in Advanced Materials*, Warrendale, PA, The Minerals, Metals, & Materials Society, 1994, pp. 337-349.

CP429, *Shock Compression of Condensed Matter – 1997*
edited by Schmidt/Dandekar/Forbes
© 1998 The American Institute of Physics 1-56396-738-3/98/$15.00

MEASUREMENT OF SHOCK-WAVE STRUCTURE OF LiF SINGLE CRYSTAL BY THE VISAR

T. Mashimo, M. Nakamura and M. Uchino

High Energy Rate Laboratory, Faculty of Engineering, Kumamoto University
Kurokami 2-39-1, Kumamoto 860, Japan

Shock-compression behavior of LiF under shock compression in the pressure region between several and several 10s of GPa had not been studied, while the elastoplastic transition in the very low stress region had been well investigated. We had measured the anisotropic Hugoniot-compression curves on LiF single crystal within the region of a few 10s of GPa by the inclined-mirror method. In this study, the shock-wave structures for <100> and <110> axis orientations were measured by the VISAR (Velocity Interferometer System for Any Reflector). Three-waves structure shock waves were observed for both axis orientations, but the cause of the diffused step structure following the elastic precursor remained unknown. The final-wave Hugoniot data also showed a crystallographic anisotropy in the low pressure region.

INTRODUCTION

Lithium fluoride (LiF) is an ionic crystal with B1-type structure, and its equation of state has been well investigated under high pressure. The shock-wave measurements had been performed on the single crystals for <100> and <111> axis orientations in very low pressure region below a few GPa, and the relaxations in elastoplastic transition were well studied (1-6). The Hugoniot data in the high pressure region of up to 100 and 500 GPa had been measured by the flash-gap method and the reflection method, respectively (7,8). On the other hand, this material has been well used as an optical window, and the dislocation behaviors under shock compression have been well investigated. However, the elastoplastic transition in the low pressure region between several and several 10s of GPa have not been studied.

We had performed the inclined-mirror Hugoniot measurements on a LiF single crystal for <100> and <110> axis orientations in the pressure region of up

to 25 GPa (9). The inclined-mirror images indicated diffused three-waves structure of the shock wave. The final-wave Hugoniot data were situated in a higher stress region than the static data, and showed a crystallographic anisotropy in the low pressure region. In this study, the measurement experiments of shock-wave profiles of a LiF single crystal was performed by using a VISAR (Velocity Interferometer System for Any Reflector) combined with a powder gun to clarify the elastoplastic transition and its crystallographic anisotropy.

EXPERIMENTAL PROCEDURE

The VISAR is an absolute integral-type velocity measurement method based on the Doppler phenomenon, and is one of the most powerful measurement methods for the shock-compression research of solids (10). We ourselves constructed a differential-type VISAR (11). An argon-ion laser

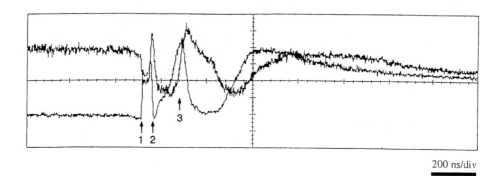

FIGURE 1. Fringe beat signals of S and P components of the free-surface motion by the VISAR of a LiF single crystal shocked to <100> axis direction with an impact velocity of 0.921 km/s (2024 Al impactor).

(wave length: 514.5 nm, total power: 2 W) with etalon was used. For easy alignment of the VISAR apparatus, optical probe and fiber cable were used in leading the coherent laser light to the specimen's surface and reflected light to the system. Two interference fringe signals generated at the surface of the large beam splitter were incorporated by differential amplifiers to reduce the noise (12). The fringe beat signals were recorded by a digital memory (HP54522A of Hewlett Packard : 2 GSa/s, 32 kpts). The data were analyzed by using the Valyn VISAR data reduction program.

The plate-shaped single crystals for <100> and <110> axis orientations of about 4 mm in thickness and about 19x16 mm in width were provided by the Nihon Kessho Koogaku Co., Ltd. The mean bulk density was measured to be 2.643 g/cm^3 by the Archimedean method. The total impurity values were less than 0.04 %. Shock wave experiments were conducted using a keyed powder gun (27 mm in basic bore diameter) (13). The impact velocity was measured by the electromagnetic method with an accuracy better than 0.2 %.

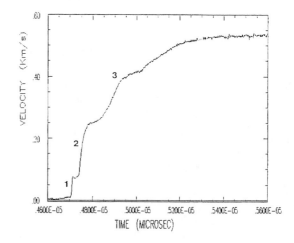

FIGURE 2. Free-surface velocity history along <100> axis of a LiF single crystal, which was derived from the fringe beat signals of Fig.1.

RESULTS

Figure 1 shows the fringe beat signals of S and P components by the VISAR of the free-surface motion shocked to <100> axis direction with an impact velocity of 0.921 km/s. Aluminum alloy (2024 Al) and copper plates were used as impact and driver plates, respectively. In the figure, the acceleration beat signals can be clearly seen at points 1, 2, and 3. The free-surface velocity history along <100> axis derived from the fringe beat signals is shown in Fig. 2. An elastic precursor and a diffused second step structure (three-waves structure shock wave) can be seen at rise. They correspond to the elastic wave [1], the first plastic wave [2], and the second plastic wave [3]. The free-surface velocity history along <110> axis also showed the three-waves structure at rise.

The Hugoniot-elastic limit (HEL) stresses along

<100> and <110> axes were less than 1 GPa, depending on the driving stress. The final-wave Hugoniot data along <100> axis roughly coincided with the static compression curve at below 7 GPa. The final-wave Hugoniot data along <110> axis measured by the VISAR and inclined-mirror methods were situated in a higher stress region than those along <100> axis, and converged to them at up to 15 GPa. The Hugoniot data in the higher pressure region coincided with those obtained by the flash-gap method (7). This indicated that the rigidity for <110> axis orientation under shock compression is larger than the one along <100> one up to about 15 GPa. The present result is consistent with the torsion and shear tests under static condition (14).

We carried out the thermal analysis to discuss the EOS by using the Hugoniot data along <100> axis of up to 50 GPa. The estimated bulk moduli was consistent with the static compression data of the ambient phase (NaCl-type) (15,16). The diffused step structure following the elastic precursor obtained in this study was concluded not to be caused by the phase transition to a CsCl-type phase. It had been understood that the dislocations play an important role in yielding mechanism of LiF. So, the diffused second-step structure following the elastic precursor may be related to the confusing behavior of dislocations or cracks and/or the interaction between shock waves and rarefaction waves.

REFERENCES

1. Asay, J. R., Fowles, G. R., and Gupta, Y. M., *J. Appl. Phys.* **43**, 744 (1972).
2. Asay, J. R., Fowles, G. R., Duvall, G. E., Miles, M. H., and Tinder, R. F., *J. Appl. Phys.* **46**, 2132 (1975).
3. Gupta, Y. M., Duvall, G. E. and Fowles, G. R., *J. Appl. Phys.* **46**, 532 (1975).
4. Asay, J. R., Hicks, D. L., and Holdridge, D. B., *J. Appl. Phys.* **46**, 4316 (1975).
5. Gupta, Y. M., *J. Appl. Phys.* **46**, 3395 (1975).
6. Rosenberg, G. R., and Duvall, G. E., *J. Appl. Phys.* **51**, 319 (1980).
7. Carter, W. J., *High Temperature-High Pressures* **5**, 313 (1973).
8. Kormer, S. B., Sinitsyn, M. V., Funtikov, A. I., Urlin V. D., and Blinov, A. V., *Soviet Phys. JETP* **20**, 811 (1965).
9. Mashimo, T., Kaetsu, M., and Uchino, M., *SPIE* **2869**, 594 (1997)
10. Barker, L. M., and Hollenbach, R. E., *J. Appl. Phys.* **43**, 4669 (1972).
11. Nakamura, M., Uchino, M., and Mashimo, T., *SPIE* **2869**, 1065 (1997)
12. Hemsing, W. H., *Rev. Sci. Instrum.* **50**, 73 (1979).
13. Mashimo, T., Ozaki, S., and Nagayama, K., *Rev. Sci. Instrum.* **55**, 226 (1984).
14. Gilman, J. J., *Acta Met.* **7**, 608 (1959).
15. Yagi, T., *J. Phys. Chem. Solids* **39**, 563 (1978).
16. Boehler, R., and Kennedy, G. C., *J. Phys. Chem. Solids* **41**, 1019 (1980).

CP429, *Shock Compression of Condensed Matter – 1997*
edited by Schmidt/Dandekar/Forbes
© 1998 The American Institute of Physics 1-56396-738-3/98/$15.00

ELASTIC MODULI AND DYNAMIC YIELD STRENGTH
OF METALS NEAR THE MELTING TEMPERATURE

A.V.Utkin, G.I.Kanel, S.V.Razorenov, A.A.Bogach

Institute of Chemical Physics in Chernogolovka, Moscow reg., Russia, 142432

D.E.Grady

Applied Research Associates, San Mateo Blvd. NE Suite A-220 Albuquerque, NM , USA,87110

Under conditions of shock-wave experiments at elevated temperatures, there was observed an anomalous growth of the Hugoniot elastic limit (HEL) in aluminum and magnesium with increasing temperature. In aluminum, the HEL increased by a factor of four as the temperature approached the melting point. For magnesium, the dependence of the HEL on the temperature achieved a maximum at 857K. Both the thermal equation of state and the temperature dependence of the elastic moduli were considered. It was found that the Poisson's ratio of aluminum is practically constant up to 700K, and then sharply grows. The analysis confirmed that the dynamic yield strength of aluminum increases by a factor of two as the temperature approaches the melting point, and for magnesium it achieved a maximum at 857K

INTRODUCTION

Shock wave experiments at different initial temperature provide important information about the elastic-viscoplastic behavior of metals at high strain rates. In our previous paper [1] unexpectedly large HEL of aluminum and magnesium were observed at temperatures near the melting point. Rohde [2] studied the dynamic yield behavior of iron over a temperature range of 76K to 573K. The dynamic yield stress at strain rate of $\sim 10^5$ sec^{-1} was found to be independent of the temperature in contrast to the highly temperature sensitive quasi-static yield stress. According to Asay [3], the elastic precursor amplitude in bismuth is independent of temperature to near 523K, close to the melting temperature T_m of 544.4K. Duffy and Ahrens [4] found that the HEL of molybdenum at 1673K falls by 26-46% relative to its value at room temperature as reported by Furnish and Chhabildas [5]. Tonk [6] analyzed the experimental data [4,5] and found that

peak deviatoric stresses are 1.28 GPa and 2.03 GPa for the hot and cold molybdenum, respectively. This paper deals with analysis of elastic precursors of shock waves in preheated Al and Mg.

EXPERIMENTAL DATA

Details of experiments have been described previously [1]. Metals selected for the testing were aluminum AD1 and magnesium Mg95. AD1 is analogous to the Al 1100 alloy in the AAS specification and Mg95 is "original" cast magnesium of 99.95% purity. Samples were discs 10 mm in thickness (*h*) and 70 mm in diameter. A planar shock wave in the sample was created by aluminum impactor 2 mm in thickness with the velocity of 700±30 m/s. The initial sample temperature was varied from 293K to 927K for aluminum (T_m = 933K) and from 293K to 880K for magnesium (T_m=924K). Free-surface velocity profiles were measured with VISAR.

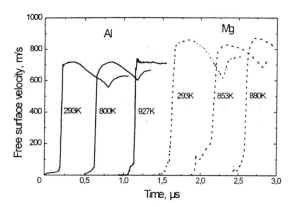

FIGURE 1. Free-surface velocity profiles for aluminum AD1 (solid lines) and magnesium Mg95 (dotted lines) at normal and elevated initial temperatures.

Fig.1 shows the experimental data at different temperatures. The compression shock waves have a two-wave structure in both AD1 and Mg95 and elastic precursors are much more pronounced at temperatures near the melting point. Since the HEL is determined not only by the yield stress but also by the elastic moduli, the result observed does not necessarily mean that the dynamic yield stress increases with temperature. Conclusions will be based on a more in-depth analysis.

ALUMINUM

The normal stress σ_x and volumetric strain $\varepsilon = (V_0-V)/V_0$ (V is the specific volume) are determined from the experimental free-surface velocity. An equation of state (EOS) including unharmonic effects [7] was developed for the material to find the shock wave velocity U_s versus temperature T:

$$F(V,T) =$$
$$= E_c(V) + \frac{RT}{M}\left[3\ln(1-e^{\Theta/T}) - D(\Theta/T)\right] - \frac{A(V)}{\rho_0}T^2 \quad (1)$$

where F is the Helmholts energy, E_c is the cold energy, θ is the Debay temperature, R is the gas constant, M is the molecular weight, ρ_0 is the density at $T=0$, A is the unharmonic term, and D is the Debay function. The cold curve is described by the Birch equation [8] with two parameters, isothermal bulk modulus B_0, and its pressure derivative B_1 at $T=0$. The dependence of the Grüneisen function γ on the volume is given by the Slater equation with a corrective constant δ [7]. It is

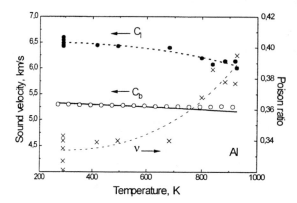

FIGURE 2. Bulk sound velocity, c_b, (open circles from the Ref.[13]) and measured longitudinal sound velocity, c_l (solid circles),and Poisson ratio (crosses) for aluminum.

proposed also that $A(V)=A_0\cdot(\rho_0V)^\gamma$.

The parameters of EOS which were found by fitting the results of calculation to the known experimental data [9-12] are given in the Table 1. Figure 2 shows the comparison between the experimental dependence of the bulk sound velocity c_b on temperature at 1 atm [13] (open circles) and theoretical results from EOS (solid line).

The longitudinal sound velocity, c_l, is determined from the measured free-surface velocity profiles, and the relation, $\Delta t/h = 1/U_s - 1/c_l$, where Δt is the time interval between the elastic and plastic shock wave fronts, and U_s is found from the EOS. Figure 2 shows c_l and the Poisson ratio ν, calculated using c_b and c_l, versus temperature. Starting from 700K, the longitudinal sound velocity decreases and ν increases while temperature increases. The dotted lines are approximations of the experimental data. It is interesting to note that c_l does not approach c_b, and ν does not approach 0.5 as $T \to T_m$. These results agree with the hypothesis [14] that the shear moduli of materials fall continuously through the melting expansion to zero at the completion of melt.

The stress versus strain σ_x-ε, diagram was reconstructed from the measured free-surface velocity profile based on the simple wave approach [15]:

$$d\sigma_x = \frac{c_u}{V}du, \quad dV = -\frac{V}{c_u}du, \quad (2)$$

where c_u is the phase sound velocity corresponding to the propagation of a given velocity level. Assuming that the measured profile is a centered simple wave, c_u is determined by the relationship:

TABLE 1. The parameters of equation of state (1).

Material	ρ_0, g/cm³	B_0, GPa	B_1	Θ_0, K	δ	A_0, GPa
Al	2.74	78.5	4.15	430	0.20	$1.0 \cdot 10^{-7}$
Mg	1.77	36.8	3.90	380	-0.25	$0.5 \cdot 10^{-7}$

$$c_u = c_l \frac{h}{h + c_l(t - t_e)}. \qquad (3)$$

Where t_e is the time at which the elastic precursor front arrives at the free surface. Integrating equations (2) using (3), we obtain $\sigma_x(\varepsilon)$. Taking the pressure $p(\varepsilon)$ from the Hugoniot, we can calculate the deviatoric stress $S_x = \sigma_x - p$.

The results of such treatment are shown in Fig.3 for two temperatures by dotted lines. As it is clear from this figure, the deviatoric stresses decrease with increasing temperature at fixed ε. It is difficult, however, to compare these two stress-strain curves because of different elastic moduli. A plot of deviatoric stress versus plastic strain ε_p is displayed in Fig.3 by solid lines. In these coordinates S_x increases with increasing temperature. It would be natural to propose that the amplitude of elastic precursor is determined by the dependence $S_x(\varepsilon_p)$ at fixed ε_p, but this behavior is not confirmed by experiment. An appropriate description of the elastic precursor amplitude at $T < 700K$ should consider $\varepsilon_p \sim 0.01-0.02\%$, whereas at higher temperatures the plastic strain is closer to ~0.04-0.05%. Fig.4 shows the dynamic yield strength $Y_d = 1.5 \cdot S_x$ as a function of temperature, calculated under these conditions. It is seen to increase by a factor of ~2 as the temperature approaches the melting point.

Consequently, the increase of the HEL at T>700 K observed in aluminum is determined by increases in both the Poisson's ratio and the yield strength as the temperature increases. It must be emphasized, however, that ε_p corresponding to the dynamic yield strength near the melting point is nearly four times greater than that at room temperature. This observation implies that the anomalous growth in Y_d is a result of the plastic flow behavior of Al immediately behind the elastic precursor front.

MAGNESIUM

The EOS of magnesium was considered in the form of equation (1).It accurately reproduces shock-wave experiments [12] and dependencies of density and heat capacity [16] on the temperature at 1 atm. The parameters are presented in the Table 1. Fig.5 shows the bulk sound velocity c_b calculated with the EOS, and longitudinal sound velocity and Poisson ratio found from the free-surface velocity profiles versus temperature at 1 atm. The dotted lines are an approximation of the experimental data. As in the case of aluminum, c_l does not approach c_b, and ν does not approach 0.5 as $T \to T_m$. Significant scatter in the experimental data is observed, especially within the temperature interval from 800K to 900K. One reasons was the large grain size (~3 mm) poly-

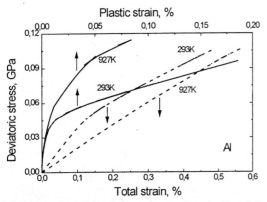

FIGURE 3. Deviatoric stress versus total strain (dotted lines) and plastic strain (solid lines) in elastic precursor of Al.

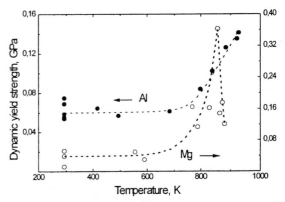

FIGURE 4. Dynamic yield stress of Al (solid circles) and Mg (open circles) versus temperature.

445

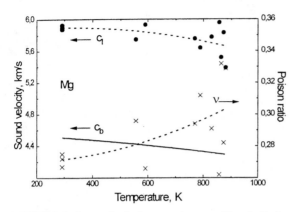

FIGURE 5. Bulk sound velocity, c_b, and measured longitudinal sound velocity, c_l, (solid circles), and Poisson ratio (crosses) for Mg versus temperature.

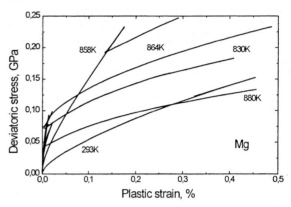

FIGURE 6. Deviator stress versus plastic strain in elastic precursor of Mg at normal and elevated temperatures.

crystalline nature of the material relative to the sample thickness. Experimental differences can be associated with various orientation of magnesium single crystals for experiment to experiment.

Figure 6 shows the S_x-ε_p diagrams for elastic precursors for magnesium at different temperatures. The deviatoric stresses increase with increasing temperature from 293K to 857K. At T=857K the shape of precursor changes with an elastic wave overshoot forming. With further temperature increase the elastic precursor maintains its shape, but deviatoric stresses decrease. Fig.4 presents the dynamic yield stress versus temperature. The value shown are the maximum stress attained in the elastic precursor for temperature above 857K. For T<857K Y_d was calculated at a plastic strain of ~0.04%. The variation of ε_p from 0.02 to 0.05% does not alter the main result, however: Namely, the dynamic yield stress has maximum at 857K.

CONCLUSION

A new phenomenon observed in the present study is a marked increase of the dynamic yield strength of both aluminum and magnesium at temperatures near the melting point. We propose that this unexpected behavior may be explained by the increase of a viscous phonon drag at high temperature. At high strain rates the applied stress exceeds the usual dislocation barriers without aid from thermal fluctuations and the viscous phonon drag becomes dominant.

ACKNOWLEDGMENT

This work was supported by Russian Foundation for Basic Research, grant number 96-01-01899a.

REFERENCES

1. Kanel,G.I.,Razorenov,S.V.,Bogatch,A.A., Utkin,A.V., Fortov, V.E., and Grady, D.E., *J. Appl. Phys*, **79**, 8310-8317 (1996).
2. Rohde, R.W., *Acta Metallurgica*, **17**, 353-363 (1969).
3. Asay, J.R., *J. Appl. Phys.*, **45**, 4441-4457 (1974).
4. Duffy,T.S.,and Ahrens,T.J., *J.Appl.Phys.*, **76**, 835-842 (1994).
5. Furnish, M.D., and Chhabildas, L.C., *in High Strain Rate Behavior of Refractory Metals and Alloys*, edited by R. Asfahani, E.Chen, and A. Crowson (The Minerals, Metals, and Material Society, Warrendale, PA, 1992), pp.229-240.
6. Tonk, D.L., "Shock wave plasticity in Mo at 293K and 1673K", *in Shock Compression of Condensed Matter - 1995*, Seattle, Washington, August 13-18, 1995, pp.507-510.
7. Zharkov, V.N., and Kalinin, V.A., *Equations of State of Solids at high Pressure and Temperature*, Moscow: Nauka, 1968, ch.3, pp.57-62 (in Russian).
8. Birch, F., Phys. Rev., **71**, 809-815 (1947).
9. Gurevich, L.V., Veitz, I.V., et al, *Thermodynamic Properties of individual Matter. Handbook*, 1981, Moscow: Nauka,, v.3, book 2, (in Russian).
10. Syassen, K., and Holzapfel, W.B., *J.Appl.Phys.*, **49**, 4427-4430 (1978).
11. Lityagin, L.M., Malyushitskaya, Z.V., Paskin, T.A., and Kabalkin, S.S., *Phys.Stat.Sol.* (a), **69**, K147-K149 (1982).
12. LASL Shock Hugoniot Data (Ed. Marsh, S.P.), Berkeley: Univ. of California Press, 1980.
13. Tallon, J.L., and Wolfender, A., *J.Phys.Chem.Solids*, **40**, 831-837 (1979).
14. Tallon, J.L., *Philosophical Magazine A*, **39**, 151-161 (1979).
15. Zel'dovich, Ya.B., Raizer, Yu.P., *Shock-Wave Physics and High-Temperature Hydrodynamic Phenomenon*, Moscow: Nauka, 1966 (in Russian).
16. Portnoy,K.I., Lebedev, A.A., Magnesium alloys. Handbook, Moscow: Metallurgy, 1952 (in Russian).

CP429, *Shock Compression of Condensed Matter – 1997*
edited by Schmidt/Dandekar/Forbes
© 1998 The American Institute of Physics 1-56396-738-3/98/$15.00

ELASTIC-PLASTIC DEFORMATION AND SPALL FRACTURE OF METALS AT HIGH TEMPERATURES

S.V.Razorenov, A.A.Bogatch, G.I.Kanel, A.V.Utkin and V.E.Fortov

Institute of Chemical Physics (Chernogolovka), Chernogolovka, Moscow reg., 142432 RUSSIA

D. E. Grady

Applied Research Associates, San Mateo Blvd. NE Suite A-220 Albuquerque, NM 87110 USA.

Measurements of the Hugoniot elastic limits and the spall strengths of aluminum, magnesium, polycrystalline zinc, and zinc single crystals of two orientations have been carried out over a wide range of initial temperatures. The initial temperature of samples was varied from room temperature to near the melting point. The free-surface velocity profiles recorded with VISAR have revealed a marked reduction in the spall strength as temperatures approached the melting point, being most pronounced for polycrystalline materials. The spall strength of zinc single crystals has been found to be unexpectedly large along the weakest base plane and is explained by purely elastic compression of these samples at shock loading at least up to 6 GPa peak stress. The Hugoniot elastic limit for the zinc single crystals was found to be approximately constant over the whole temperature range.

INTRODUCTION

It is well-known that under normal conditions both the yield strength and the tensile strength are strong functions of the temperature. For low rates of mechanical loading, the dislocation motion is aided by thermal fluctuations [1]. A transition to athermal plasticity occurs at high strain rates, when the applied stress is high enough to overcome the usual dislocation barriers without any aid from thermal fluctuations. Since the fracture process includes the plastic flow around growing voids a transition to athermal fracture should be expected at similar strain rates. However, measurements of the yield strengths and, especially, tensile strengths under shock-wave loading at elevated temperatures do not provide unambiguous conclusion concerning the nature of these deformation processes.

In this paper, new results of investigations of the initial temperature effect on the dynamic strength of commercially pure aluminum, magnesium, zinc, and zinc single crystals are presented. The shock-wave measurements were carried out over a range from the ambient temperature up to melting point.

MATERIALS

Experiments were performed with aluminum AD1 (analogous to the Al 1100 alloy in the AAS specification), magnesium Mg95, zinc ZHP (Zinc of High Purity), and zinc single crystals of 99.999% purity at different orientations of the loading direction relatively to the crystal axis. The zinc single crystals were grown by a directed crystallization technique [2]. The magnesium Mg95 and zinc ZHP are "original" cast metals with the grain sizes of 1-3 mm and around 10 mm respectively. Polycrystalline samples were plates of ~10 mm in thickness and 70 mm in diameter. The zinc single crystals were plates with dimensions in plane of 15×10 mm at the thickness varied between 0.5 and 2.2 mm. The samples were subjected to a electrochemical etching in order to remove the surface defects.

EXPERIMENTAL

The details of arrangement of the shock-wave measurements at elevated temperatures have been discussed in a previous paper [3]. The explosive facility launched aluminum impactor plates 0.4 to 2.0 mm in thickness at velocities 500-700 m/s. With a resistive heater, sample temperatures up to 700°C were achieved. The temperature of the sample surface was controlled with a thermocouple to within ±5°C of accuracy. In the experiments, the free-surface velocity profiles, $u_{fs}(t)$ were recorded with a VISAR [4].

RESULTS AND DISCUSSIONS

Figure 1 presents examples of the free-surface velocity profiles recorded in the experiments with polycrystalline zinc samples 10 mm thick. Profile characteristics, including the elastic-plastic compression wave, elastic-plastic unloading features and the spall pullback signal indicative of dynamic tensile failure, were observed over the entire temperature range. These data demonstrated low reproducibility in this coarse-grain material both in their compression and unloading wave profile features. This variability is an obvious consequence of the arbitrarily oriented large anisotropic grains relative to the sample thickness.

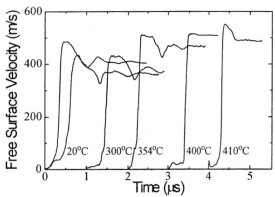

FIGURE 1. Free surface velocity profiles recorded in experiments with polycrystalline coarse-grain zinc of high-purity.

Using the free-surface velocity profiles, the spall strengths, σ^*, were determined from the amplitude of the velocity pullback, u_{fs}, which is the velocity decrement from the peak velocity magnitude of the shock wave emerging at the free surface to its minimum value in front of the spall signal. Spall strengths were calculated from the common relation:

$$\sigma^* = 0.5\rho_o c_b (\Delta u_{fs} + \delta)$$

where ρ_o and c_b are the material density and the bulk sound velocity at zero pressure respectively, and δ is a correction for profile evolution in an elastic-plastic material [5].

Figure 2 shows the spall strength dependence on the initial temperature for aluminum AD1 and magnesium Mg95 measured at a peak compressive shock pressure of 5.8 GPa and 3.7 GPa respectively. For a visual comparison, the data are presented in normalized coordinates. In the figure, the normalized spall strength is σ^*/σ^*_a, where σ^*_a is the average spall strength at ambient temperature that is 1.18 GPa for aluminum and 0.92 GPa for magnesium; normalized temperature is T/T_m, where T_m is the melting temperature in °K. In general, the spall strength maintains almost constant with increasing initial temperature up to 85-90% of the melting temperature. After that, a precipitous drop in the spall strength occurs as temperatures approach the melting point both for aluminum and magnesium.

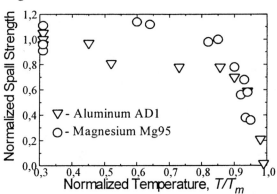

FIGURE 2. Normalized spall strength of aluminum and magnesium as a function of the temperaure devided by the melting temperature.

The spall strength of coarse-grain zinc scatters between 0.8 and 1.2 GPa at the ambient initial temperature, 0.9 to 1.7 GPa at temperatures of 300 to 350°C, and 0.6 to 0.7 GPa at 390 to 410°C. The melting temperature of zinc is 419.5°C. Thus, while there is some reduction in spall strength at temperatures above 350°C, the spall strength of zinc does not reduce dramatically as the melting temperature is approaches.

Zinc has the hexagonal close-packed crystal lattice shown in Figure 3. In the experiments, the collision planes were the basic plane (0001) and the prismatic plane ($10\bar{1}0$) perpendicular to the basic one. The main slip plane for the h.c.p. crystals is the basic plane (0001); the dislocation slip and twinning in the pyramidal ($11\bar{2}2$) plane require much higher shear stress. The slip in zinc is activated also in the ($1\bar{1}01$) plane when the temperature exceeds 225°C [6]. At the orientations chosen, there was no shear stress in the basic plane, so only secondary slip systems were activated in both series of the experiments performed.

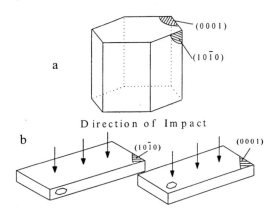

FIGURE 3. The crystal lattice of zinc (a) and the orientations of zinc single crystals samples relative to loading direction (b).

Figure 4 presents the free surface velocity profiles recorded in the experiments with zinc single crystals. At the ambient temperature, the bulk sound velocity in zinc is C_b=3.03 km/s; the longitudinal sound velocity in the direction perpendicular to the hexagonal axis of crystal is C_l=4.73 km/s while along the axis direction it is very close to the bulk sound velocity [7]. In fact for axial loading purely elastic shock waves of ~7 GPa peak stress were recorded in the zinc single crystal. The shock front rise time was close to the resolution limit of 2-4 ns in this case. For the ($10\bar{1}0$) samples the elastic precursors of large amplitude have been revealed.

The pronounced elastic-plastic properties resulted in a fast decay of shock wave in the experiments with impactor-to-sample thickness ratios less than ~0.3. The rise time in plastic shock front of 6 GPa peak stress is around 100 ns.

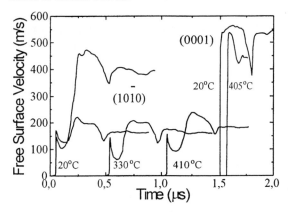

FIGURE 4. Free surface velocity profiles for zinc single crystal samples at ambient and elevated initial temperatures. Impactor and sample thicknesses are 0.85 and 1.7 mm and 0.4 and 1.84 mm at 20°C; 0.4 and 1.72 mm at 330°C; 0.4 and 1.82 mm at 410°C for loading in ($10\bar{1}0$) plane; 0.85 and 1.75 mm at 20°C; 0.4 and 1.7 mm at 405°C for loading in (0001) plane.

It is known that under normal conditions the zinc single crystals can be readily cleaved along the basic plane [2]. The shock-wave profiles shown in Fig.4, however, demonstrate a high spall strength for dynamic fracture on the (0001) basic plane. At 20°C the sharp spall signal front indicates a very rapid fracture process. Over the temperature range of 20°C to 375°C, the spall strength of the (0001) samples is practically constant at 1.86±0.1 GPa. A reduction in spall strength to 1.45 GPa has been recorded at 395°C and to 1.2 GPa at 405°C, which is 0.98 of the melting temperature. As seen in Fig.2, the spall strength reduction of the commercial grade aluminum and magnesium at 0.98 of the melting temperature was much larger approaching ~60-70%.

The experiments with ($10\bar{1}0$) samples at the ratio of impactor-to-sample thickness of 0.5 show a spall strength of 1.4±0.15 GPa at the ambient initial temperature. The other tests were not optimal for precise measurements of the spall strength but comparisons of the low-amplitude free-surface velocity profiles presented in Fig. 4 do not indicate any essential reduction in this value up to 410°C of the initial temperature. Thus, we find the unexpected result that the spall strength is larger for fracture along the weakest plane. Earlier experiments with f.c.c. copper [8] and b.c.c. molybdenum [9] single crystals did not reveal any influence of loading direction on the spall strength. The anisotropy of zinc lattice is much larger, but the main reason of high strength along (0001) plane is that there was no plastic deformation under shock compression so the fracture was initiated in a perfect initial structure. The dynamic tensile strength of single crystals exceeds the strength of polycrystalline material that confirming that a large contribution is due to fracture nucleation on the grain boundaries.

Elastic precursor waves in the ($10\bar{1}0$) zinc samples exhibit a spike-like form. In terms of the dislocation dynamics, this is an evidence of intense multiplication of the mobile dislocations behind the precursor front. Figure 5 presents the free-surface velocity peak and minimum values in the elastic precursor profile.

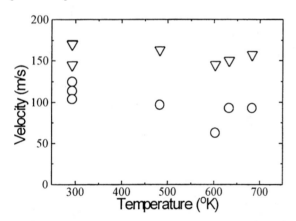

FIGURE 5. Peak (∇) and minimum (O) free surface velocities in the elastic precursor of zinc single crystals .

The stresses in precursors are smallest in both peak and minimum magnitudes at initial temperatures in the vicinity of 600K, but in general we may conclude that the yield strength remains nearly constant to temperatures aproaching the melting point. The peak corresponds to the Hugoniot elastic limit of 2.5 GPa in average. Probably, the observed minimum in the HEL dependence on the temperature is associated with activation of the dislocation slip in the ($10\bar{1}0$) planes. We did not observe significant temperature influence on the rise time of the plastic shock wave. Based on these data, we may conclude that the plastic deformation under shock-wave conditions is athermal in nature. The experiments on aluminum, magnesium, and zinc that were carried out over a wide range of initial temperatures show a significant decrease in spall strength only closely to the melting point. Since we did not observe any signs of significant reduction in the flow stress with temperature, we conclude that the decrease in spall strength is explained by local melting of material as a result of micro-heterogeneities in plastic deformation. Clearly, there are many more reasons for the localization of plastic deformation on grain boundaries and inclusions, and the nucleation of hot spots in polycrystalline materials than in the more homogeneous single crystals.

ACKNOWLEDGEMENTS

This work was supported by the Russian Foundation for Basic Research, grant number 97-02-17701. The authors would also like to thank Drs. Vera Sursaeva and Svetlana Protasova (Institute of Solid State Physics, Chernogolovka) for preparing of samples of zinc single crystal and L.G. Ermolov, and P.V. Skachkov for their great help in experiments.

REFERENCES

1. Kumar A.,Kumble R.G.. *J. Appl. Phys.*, **40**, 3475-3480 (1969).
2. Antonov A.,Kopetskii C.V.,Shvindlerman L.S., Sursaeva V. Sov.Phys.Dokl., **18,** 736-738 (1974).
3. Kanel G.I., Razorenov S.V., Bogach A.A., et.al. *J.Appl.Phys.*, 79, 8310-8317 (1996).
4. Asay J.R. and Barker L.M., *J.Appl.Phys.*, **45**, 2540 (1974).
5. Kanel G.I., Razorenov S.V., Utkin A.V. *High Pressure Shock Compression of Solids - II,* edited by Davison L., Shahinpoor M., Grady D., Springer-Verlag, 1996, ch.1, p.p.1-24.
6. McClintock F.A. and Argon A.S. *Mechanical Behavior of Materials.* Addison-Wesley Publ., 1966.
7. Mason W.P. (ed.), *Physical Acoustics: Principles and Methods, Volume III, Part B: Lattice Dynamics.* Academic Press, New York and London, 1965.
8. Kanel G.I., Rasorenov S.V., and Fortov V.E., *Shock-Wave and High-Strain-Rate Phenomena in Material,* Eds: MeyersM.A., Murr K., and Staudhammer K.. Marcel Dekker, pp.775-782 (1992).
9. Kanel G.I., Razorenov S.V., Utkin A.V., et.al. *J. Appl. Phys.*, 74, 7162-7165 (1993).

CP429, *Shock Compression of Condensed Matter – 1997*
edited by Schmidt/Dandekar/Forbes
© 1998 The American Institute of Physics 1-56396-738-3/98/$15.00

SHOCK RESPONSE OF 5083-0 Aluminum

M. W. Laber and N.S. Brar

Impact Physics Laboratory, University of Dayton Research Institute
Dayton, OH 45469-0182

Z. Rosenberg

RAFAEL, P.O. Box 2250 (24), Haifa, Israel.

Aluminum alloy (5083-0) is used as lightweight armor in armored vehicles. Data on the shock response of this material is useful to simulate ballistic penetration of different nose-shaped penetrators. In this paper we present the dynamic response of 5083-0 aluminum to shock wave loading to 22 GPa. Manganin stress gauges were used to measure the stress wave profiles. Hugoniot elastic limit (HEL) and spall strength were 0.28 GPa and 1.6 GPa, respectively. Shock Hugoniot to stress levels of 10 GPa was determined by embedded in-material gauges and above 10 GPa by measuring shock velocities by embedding manganin gauges at the back surface of stepped targets.

INTRODUCTION

Aluminum alloy 5083-0 is as an armor material used for several light-weight armored vehicles (e.g. Bradley). The dynamic response of this material was determined by measuring the stress histories with commercial manganin gauges. The gauges were embedded in back-surface and in-material configurations. In this paper we present the results on the Hugoniot elastic limit (HEL), spall strength, and Hugoniot of the material.

EXPERIMENTAL METHOD
Material

Physical properties of 5083-0 aluminum specimens are listed in Table 1 (1).

Table 1 Physical properties of 5083-0 aluminum

Density (ρ)	2664 kg/m³
Bulk Modulus	76.5 GPa
Shear Modulus	27.3 GPa
Poisson's ratio	0.34
Long. wave speed (C$_L$)	6510 m/s

Shock wave experiments

Flyer plate impact experiments with embedded manganin gauges (Micro-Measurements) were performed using a 50 mm gas/powder gun at the University of Dayton Research Institute. Targets were assembled by embedding gauges in two configurations:(i) back-surface gauge configuration and (ii) in-material configuration, as shown in Figure 1 (2). In the back-surface gauge configuration, the gauge was sandwiched between 25 μm thick mylar sheet and a 12 mm thick PMMA sheet. This package was glued on the back of the aluminum target plate. In many of the back gauge configuration experiments a stepped target with two thicknesses was used. In the back-surface gauge configuration, a good mechanical impedance match between mylar and PMMA reduces the rise time of the gauge signal to about 20 ns. Such a fast rise time allows the clear separation of elastic precursor from the plastic wave. Furthermore, this configuration is adequate to measure spall strength of the material and shock wave speed at very high shock stresses. In the second configuration, the gauge was sandwiched between two 25 μm thick mylar sheets to electrically insulate it from the aluminum plates. Because of the thick gauge package in this configuration, rise times of the stress profile are relatively longer. Nevertheless, peak shock stresses can be measured with an accuracy of ±3%.

Impactor   Target

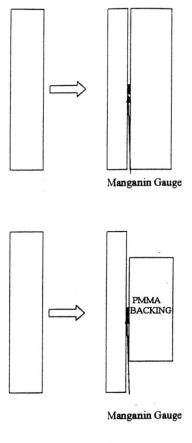

Manganin Gauge

Manganin Gauge

FIGURE 1. Schematic of target configuration; (a) in-material gauge and (b) back-gauge.

RESULTS AND DISCUSSION

A total of 7 experiments were performed and the impact data are summarized in Table 2 (2). The method to use the manganin gauge to measure shock stress and calibration are given in Reference (3).

TABLE 2. Summary of the Hugoniot data

Shot No.	Impactor thick. (mm)/ Velocity (m/s)	Target thick. (mm)	Shock stress (GPa)
(A) Back-gauge configuration			
7-982	(Al) 3/278	6	2.2±.02
7-999	(Al) 3/426	6	3.4±.02
7-1005	(Cu) 2/1093	3.6+7.8	12.3±.02
7-1016	(Al) 2.5/1740	4+8	14.9±.03
7-1024	(Cu) 10/1730	4+8	21.8±.03
(B) In-material gauge configuration			
7-978	(Cu) 5/973	6+12	11.5±.02
7-1006	(Cu) 4/853	6+12	10.3±.02

Manganin gauge profiles from shots 7-982 and 7-999 are shown in Figures 2 and 3. The elastic-plastic wave separation is clearly seen in profile 7-982; the elastic wave amplitude in PMMA is 0.09± .01 GPa. The value of elastic precursor or HEL deduced from stress amplitude in PMMA using impedance matching technique is 0.28±0.05 GPa. The yield strength of the material under uniaxial stress conditions obtained from the value of HEL is 0.15 GPa., which agrees well with the reported value of 0.15 GPa in Reference 5. The profile from shot 7-982 also shows that the target did not spall. The pull back stress in this shot was determined to be 1.3 GPa based on the value of shock stress transmitted in PMMA as 0.4 GPa. This means that the spall strength of the alloy is greater than 1.3 GPa. In the shot 7-999 the profile clearly shows that the target did spall and we observe several stress wave reflections between the target-PMMA interface and the spall plane. The spall strength from this shot was determined following the method outlined in Reference 4. In this method the experimental configuration is numerically simulated using a 1-D shock wave code. The stress history is simulated by assuming different spall strengths, specially the ratio of the first minimum in stress σ_2 to the peak shock stress σ_1. Simulated and measured ratios σ_2/σ_1 match for the optimum value of spall strength. This method yields spall strength for 5083-0 aluminum as 1.6 GPa. This value is indeed greater than 1.3 GPa

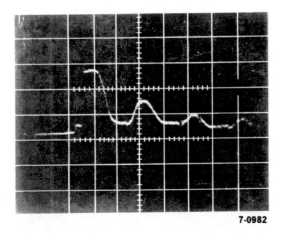

7-0982

FIGURE 2. Gauge profile in shot 7-982; 0.1 V/div. and
1 μs/div.

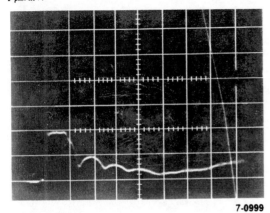

7-0999

FIGURE 3. Gauge profile in shot 7-999; 0.2 V/div. and
1 μs/div.

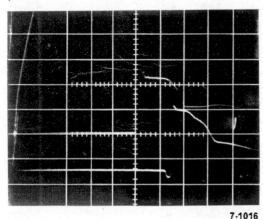

7-1016

FIGURE 4. Gauge profile in shot 7-1016, 1 V/div. and
0.5 μs/div.

pull back tension generated in shot 7-982 and the target did not spall. Peak shock stress in these two low stress shots were estimated assuming elastic impedance for the material.

Hugoniot points at the stress levels of 10 GPa were determined by embedding gauges between two plates of the aluminum alloy. Two shots one at 10.3 GPa (7-980) and the other at 11.5 GPa (7-978) were performed. In both of these shots the stress levels were determined from the flat top portion of the profiles.

Hugoniot points at shock stress levels above 11.5 GPa were determined by measuring shock velocity using stepped targets with two thicknesses. Two gauges, one at each step, in the back-surface gauge configuration, were embedded in the targets. The gauges were backed by PMMA plates. Both gauges were connected to the same two channel scope and triggered by a single source. It is very important that in order to measure shock velocity reliably the two gauge outputs should be triggered by the same source. Profiles from the two gauges embedded at two steps in the target for shot 7-1016 are shown in Figure 4. Measured shock velocities and deduced Hugoniot stresses from Equation (1) are listed in Table 3.

$$\sigma_p = HEL + \rho_1 (U_s - u_e)(u_p - u_e) \quad (1)$$

where ρ_e and u_e are the density and the particle velocity at HEL, respectively; u_p and U_s are the particle and shock wave velocities, respectively, corresponding to Hugoniot stress. Particle velocity u_p at the Hugoniot stress is determined using the

Table 3 Hugoniot data at high shock stress

Shot No.	u_p	U_s	Shock Stress
	(mm/μs)		(GPa)
7-1005	0.74	6.30	12.3
7-1016	0.78	6.57	14.9
7-1024	1.22	7.20	21.8

impedance matching technique and for a symmetric impact it is one half of the impact velocity. Density at HEL is given by equation 2:

453

$$\rho_e = \rho_0 \left(1 + \frac{u_e}{C_L}\right) \qquad (2)$$

where C_L is the longitudinal wave speed in the alloy. Hugoniot stresses for the three shots (7-1005, 7-1016, and 7-1024) were determined from the measured shock velocity and the results are listed in Table 3. The complete Hugoniot in the stress-particle velocity from all the shots is shown in Figure5.

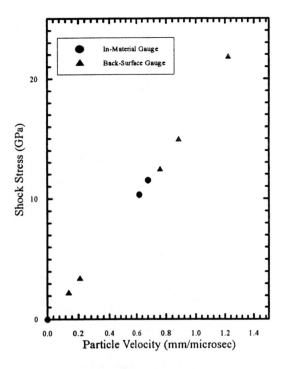

FIGURE 5. Hugoniot of 5083-0 Aluminum

REFERENCES

1. Dandekar, D., Private Communication (1997).
2. Rosenberg, Z., "Determination of the dynamic response of W-2 Tungsten and 5083-0 Aluminum with commercial manganin gauges," The University of Dayton Research Institute Report, UDR-TR-86-122, October 1987.
3. Rosenberg, Z., Yaziv, D., and Partom, Y., "Calibration of foil-like manganin gauges in planner shock wave experiments, " J. Appl. Phys., **51**, 3702-05 (1980).
4. Rosenberg, Z., Luttwak, G., Yeshurun, Y., and Partom, Y., "Spall strength of differently treated 2024 Aluminum specimens," J. Appl. Phys., **54**, 2147-52 (1984).
5. Ryerson, Catalogue for Steel and Aluminum.

CP429, *Shock Compression of Condensed Matter – 1997*
edited by Schmidt/Dandekar/Forbes
© 1998 The American Institute of Physics 1-56396-738-3/98/$15.00

CRACK RESISTANCE OF ALUMINIUM COMPOSITE UNDER SHOCK LOADING AT DIFFERENT TEMPERATURES

S.A.Novikov, V.A.Pushkov, V.A.Sinitsyn, G.T.Gray III

Russian Federal Nuclear Center - VNIIEF
607190, Sarov, Nizhni Novgorod region, Russia

Los Alamos National Laboratory, Los Alamos, NM 87545 USA

The paper presents the technique of study of aluminum composite K_{1d} crack resistance at shock loading at $T=-50 \div 300^{\circ}C$. The technique is based on application of split Hopkinson bars and plane specimen for axial wedging. Heating of specimens was carried out by electric heater, cooling - by vapours of liquid nitrogen. Specimens from aluminum composite A 359 (20 vol.% SiC) of the firm DURAL were tested. K_{1d} values were obtained at mentioned temperatures in the interval of loading rates $\dot{K}_1 = (0.1 - 0.28) \cdot 10^6 \, MPa \cdot m^{1\,2} s^{-1}$.

Level of specified complex of composite properties is designed beforehand. But only after corresponding tests of manufactured composites is it possible to know what these properties are in actuality (1).

Under the framework of collaboration between VNIIEF and LANL there were studied physical-mechanical properties of aluminum composite A 359 (Al + 20% SiC) of the DURAL firm. Part of these efforts was devoted to the study of composite crack resistance under dynamic loading at lowered or elevated temperatures.

For this study the specimens were manufactured from cylindrical blanks with a diameter of 180 mm and a length of 320 mm. Pieces for the specimens were cut out both along and across the blank axis. After manufacturing, the specimens were subjected to the following thermal treatment: heating at $530^{\circ}C$ for 8 hours; cooling in hot water; aging at $154^{\circ}C$ for 5 hours; cooling in the air. Separate measurements showed that composite density was 2.763 g/cm^3, Young's modulus E=102.6 GPa, Poisson's ratio μ=0.308.

The Split-Hopkinson-bar method (2) was used to study dynamic crack resistance of the aluminum composite.

In the experiments heating of specimens was performed by a small-size electric heater, and cooling - by vapors of liquid nitrogen. It was impossible for direct thermal radiation to reach the surface of a specimen at heating due to the existence of a copper screen in the electric heater. To obtain uniform thermal field in the specimens volume, the specimens were subjected to a specified temperature for no less than 5 minutes.

In the experiments we used Hopkinson bars manufactured from hardened steel 30XGSA. We tested compact wedging specimens made like WLCT-specimens (wedge-loaded compact tension specimen) (3). These specimens are used in crack resistance K_{1d} study under plane deformation. The relationship $B \geq 2.5 \cdot (K_{1d}/\sigma_{0.2})^2$ (3) determined the required specimen thickness B. It was obtained by a preliminary calculation indicating B~7.5 mm satisfies this condition. Finally, the B-value was chosen to be 14 mm.

A P-Δ diagram was found for each experiment, where P - wedging stress, Δ - mutual attraction of the rods ends. Values of P and Δ were determined from the formulae:

$$P = E_c \cdot S_c \cdot \varepsilon_T(t)$$

$$\Delta = 2c \int_0^t \left[\varepsilon_I(t) - \varepsilon_T(t) \right] dt \quad,$$

where E_c, c, S_c - elastic modulus, sound velocity, and cross-section area of the support rod, respectively; $\varepsilon_I(t)$ and $\varepsilon_T(t)$ - recorded strains caused correspondingly by stress pulse falling on the specimen and stress pulse arriving (in the support rod) (4).

In Fig.1 one can see typical P-Δ diagrams obtained in experiments with composites at T=-50, +25, 100, 300 °C.

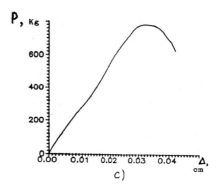

c)

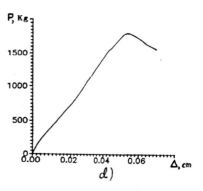

d)

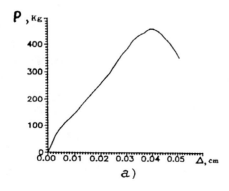

a)

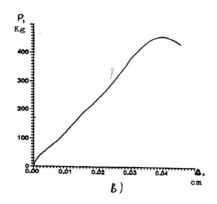

b)

FIGURE 1. Typical P-Δ diagrams of composite at different temperatures and $\dot{K}_1$.

a - T=-50 °C, $\dot{K}_1$ =0.10·10⁶ MPa·m^{1/2}·s^{-1};
b - T =25 °C, $\dot{K}_1$ =0.11·10⁶ MPa·m^{1/2}·s^{-1};
c - T =100 °C, $\dot{K}_1$ =0.23·10⁶ MPa·m^{1/2}·s^{-1};
d - T =300 °C, $\dot{K}_1$ =0.26·10⁶ MPa·m^{1/2}·s^{-1}.

Crack resistance K_{1d} was derived from the diagrams by standard techniques (4) and following the relationship

$$K_{1d} = \left(P_C / S_N \right) \cdot l \cdot Y ,$$

where P_C - critical wedging stress, under which a crack is displaced; S_N - nominal specimen cross-section; l - crack length; Y - a calibration function. The loading velocity $\dot{K}_1 = dK_1 / dt$ was calculated in a linear approximation as K_{1d}/t_c, where t_c is the duration of force increase up to value P_c. A dimensionless calibration function Y was determined in special experiments.

456

TABLE 1. Results of Experiments

N test	T,$^{\circ}$C	$\overset{\bullet}{K}_1, 10^6$, MPa $\cdot$ m$^{1/2} \cdot$ s^{-1}	K_{1d}, MPa $\cdot$ m$^{1/2}$
1		0.10	10.0
2	-50	0.10	9.8
3		0.10	10.5
4		0.12	10.4
5		0.15	10.8
6	25	0.11	9.8
7		0.11	10.5
8		0.18	16.2
9	100	0.20	16.5
10		0.23	17.4
11		0.24	38.4
12	300	0.28	40.5
13		0.26	41.2

The results of the experiments are tabulated in Table 1. Having limited number of samples, we tested only the composite cut from the blank in different directions, for crack resistance at 25°C. The results showed that the crack resistance does not depend on the cut direction.

Specimens fracture character was quasi-brittle. However, the initial signs of viscous destruction were at T=300°C. They were especially more pronounced in layers near the surface on the side planes of the specimens. At E, μ and K_{1d} values, known for 25°C, the intensity of composite destruction energy by Griffiths-Irwin criterion is ~950 J/m^2.

It follows from Table 1 that dynamic crack resistance K_{1d} of the composite has a 4 times increases (with average values) with temperature from -50 to +300°C in the $\overset{\bullet}{K}_1$ interval.

Analysis of the results shows that the aluminum composite (Al+20% SiC) has low crack resistance. For comparison, steel St.3 (5) and aluminum alloy PA6 (3) have K_{1d} values equal to 65.5 and 30 MPa$\cdot$m$^{1/2}$, correspondingly, at 25°C and nearly equal values of $\overset{\bullet}{K}_1$. Probably, the low value of composite K_{1d} can be explained by the discontinuous distribution of SiC particles in the aluminum base and large sizes of these particles up to 0.16 mm. Metallographic analysis of the specimens confirmed this.

ACKNOWLEDGMENTS

The work was supported by Fund of Fundamental Researches of Russian Academy of Science code 97-01-00344.

REFERENCES

1. Song, S.G., Vaidya, R.U., Zurek, A.K., and Gray III, G.T., *Journal Metallurgical and Materials Transactions* A **27A**, 459-465 (1996).
2. Bol'shakov, A.P., Eremenko, A.S., Lupsha, V.A., et al., *Phys.-Chem. Mechanics of Materials (Rus.)* **1**, 79-82 (1981).
3. Klepaczko, J.R., *Trans. ASME J. Eng. Mater. and Technol.* **104**, 29-35 (1982).
4. Eremenko, A.S., Novikov, S.A., Pushkov, V.A., et al., *Appl. Mechanics and Technical Physics (Rus.)* **37**, 149-159 (1996).
5. Novikov, S.A., and Pushkov, V.A.,. "Study of steel crack resistance under dynamic loading and temperature range 20-300 $^{\circ}$C", Proceedings of the Conference of the APS, Seattle, Washington, Aug.13-18, 1995, *American Inst. of Physics*, 511-513 (1995).

CP429, *Shock Compression of Condensed Matter – 1997*
edited by Schmidt/Dandekar/Forbes
© 1998 The American Institute of Physics 1-56396-738-3/98/$15.00

SPALL STRENGTH MEASUREMENTS IN ALUMINUM, COPPER AND METALLIC GLASS AT STRAIN RATES OF $\sim 10^7\,s^{-1}$

B. Arad, E. Moshe, S. Eliezer, E. Dekel, A. Ludmirsky, Z. Henis, I.B. Goldberg
Plasma Physics Department, Soreq NRC, Yavne 81800, Israel

N. Eliaz, D. Eliezer
Department of Materials Engineering, Ben-Gurion University, Beer-Sheva, Israel

Measurements of the dynamic spall strength in Aluminum, Copper and metallic glass shocked by a high power laser to pressures of hundreds of kilobars are reported. An optically recording velocity interferometer system has been developed to measure the free surface velocity time history. The spall strength was calculated from the free surface velocity as a function of the strain rate. The results show a rapid increase in the spall strength with the strain rate at strain rates of about $10^7\,s^{-1}$.

INTRODUCTION

This paper presents an experimental study of the resistance of Aluminum, Copper and metallic glass - Metglas 2605SA1 - (amorphous alloy) to dynamic fracture, based on the detection of the spall phenomenon.

A high power laser pulse induces a shock wave in a solid target. The failure mode occurs upon the reflection of the shock wave from the free surface. When the shock wave reaches the free surface, the compressed material expands freely and accelerates the surface. The expansion leads to material straining and therefore to a deceleration of the surface. The superposition of the two waves, the tail of the initial pressure wave and tensile stress wave, running in opposite directions, yields a residual stress whose amplitude increases with the distance to the free surface. When this stress exceeds the material strength, spalling occurs in a plane parallel to the free surface. The expanded material of the spall region bounces back from the spall surface, generating another pressure wave moving in the direction of the original shock wave and reaccelerates the free surface. This second pressure reaches the free surface and is reflected as a stress wave. The spall pressure or the material strength is determined from the measurement of the free surface velocity time history.

An optically recording velocity interferometer system, ORVIS (1), is used to measure the time evolution of the free surface velocity with an accuracy better than 5%.

The experiments reported here show a rapid increase of the dynamic spall strength with the strain rate, at strain rates $\sim 10^7\,s^{-1}$. This behavior is different, compared to the much smaller spall strength - strain rate dependence (2) in the strain rate range $(2 \cdot 10^3 - 10^6)\,s^{-1}$.

EXPERIMENTS

The experimental set-up for the free surface velocity measurements is described in ref. 1. The laser system that generates the shock waves, operates at a wavelength of 1.06 µm, pulse width at half maximum (FWHM) of 2 ns and 5 ns, energy in the range (10-80) J, and is focused to a spot diameter in the range (200-1000) µm. The laser irradiance was of the order of 10^{13} W/cm². The diagnostic interferometric system produces interference fringes shifts proportional to the Doppler shift of an Argon-ion laser beam focused at the moving free surface. The interference pattern is imaged by a cylindrical lens as bright spots on the entrance slit of a streak camera. The time resolution of the experiments reported here was 70 ps. The interference pattern is analyzed with an image processing system, including a cooled CCD camera, a frame grabber and a PC. The velocity history of the moving free surface is determined from the change in the interference pattern (3-5). The fringe pattern representing the free surface velocity, in an experiment with a 5 ns, 68 J laser pulse focused to a 500 μm diameter spot and a 50 μm thick Aluminum target, is shown in fig.1. Time is increasing from top to bottom. The signature of the wave reverberations in the spall region are seen in the oscillations of the fringe pattern.

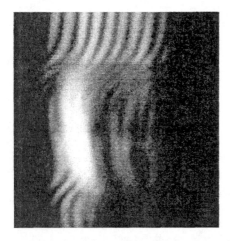

FIGURE 1. The interference pattern obtained with the ORVIS in an experiment with a 50 μm thick Aluminum foil irradiated by a 68 J, 5 ns laser pulse.

In addition to the free surface velocity measurements, the shock wave velocity was estimated from the travel time through the target.

THE SPALL PRESSURE CALCULATION

The spall strength was determined from the measured time dependence of the free surface by two methods: First, in the acoustic approximation (6) the spall strength is:

$$P_{spall} = \frac{1}{2}\rho_0 c(u_{max} - u_{min}) \tag{1}$$

ρ_0 is the initial density of the target, c is the sound velocity, u_{max} is the peak velocity of the free surface and u_{min} is the first minimum in the free surface profile. Secondly, the spall strength was calculated by integrating the Lagrangian equation:

$$\frac{\partial u}{\partial t} + \frac{1}{\rho_0}\frac{\partial P}{\partial h} = 0.$$ h is the Lagrangian coordinate,

$$h = \int_0^x \frac{\rho}{\rho_0}dx,$$ u is the particle velocity, P is the pressure, ρ is the density. A general solution of the wave equation describing the two rarefaction waves running in opposite directions is:

$$P(h,t) = f_1(\frac{h}{c} + t) + f_2(\frac{h}{c} - t) \tag{2}$$

Using the boundary condition P = 0 at the free surface, h = 0, and eq.1-2, we obtain:

$$P(h,t) = -\frac{\rho_0 c}{2}(u_{fs}(\frac{h}{c} + t) - u_{fs}(-\frac{h}{c} + t)) \tag{3}$$

P_{spall} is the absolute minimum of the pressure profile obtained by eq.3.

Both methods for calculating P_{spall} are not independent. Eq.1 can be obtained from eq.3. Solving the Lagrangian equation enabled as to calculate the depth h where the spall actually occurred. Further more, it enabled as also to calculate the pressure history at any distance away from the free surface at any moment up to the time of spallation. Both methods enabled as to use eq.1 to double check the result of eq.3.

The strain rate was calculated by:

$$\dot{\varepsilon} = \frac{du_{fs}}{dt}\bigg|_{t=t_{spall}} \cdot \frac{1}{2c} \tag{4}$$

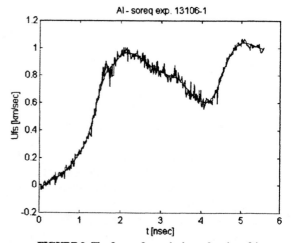

FIGURE 2. The free surface velocity as function of time corresponding to the fringe pattern displayed in fig.1.

Fig.2 shows the free surface velocity corresponding to the fringe pattern from fig.1, as a function of time. The minimum pressure as a function of time, for the time duration of the free velocity measurement (~6 ns), versus the Lagrangian coordinate h, calculated using eq.3, is plotted in fig.3. The coordinate corresponding to the smallest pressure, h = -0.502 µm, is identified with the spall position. The minimum pressure as the function of time at the spall position is defined as the spall pressure and is plotted in fig.4.

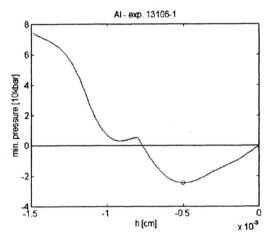

FIGURE 3. The minimum pressure as a function of time versus the Lagrangian coordinate for the experiment shown in fig.1.

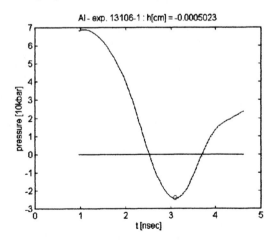

FIGURE 4. The pressure at the spall position as a function of time for the experiment shown in fig.1.

RESULTS AND DISCUSSION

Our spall pressure results for Aluminum, other experimental results (7-9) performed at lower strain rates and the theoretical (10) spall strength of solids are plotted in fig.5. The continuos curve is given by: $P_{spall} = P_0 \cdot \arctan(\dot{\varepsilon}\tau_0 + \varepsilon_0) + P_1$, with the fit parameters : $P_0 = 99$ kbar, $P_1 = 14$ kbar, $\varepsilon_0 = 0.04$, $\tau_0 = 4.7 \cdot 10^{-9}$ s. The measured spall strength in Copper was (71.2±4.9) kbar at a strain rate of $2.1 \cdot 10^7$ s^{-1} and in Metglas (58.5±4.1) kbar at a strain rate of $3.9 \cdot 10^7$ s^{-1} for c=4 km/sec.

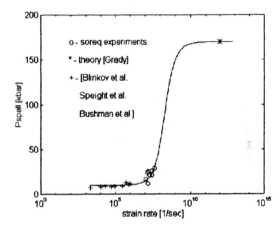

FIGURE 5. The spall strength as a function of the strain rate in Aluminum.

P_{th}, the theoretical spall strength of solids considers interatomic interaction forces only and can be estimated (10) from U_{coh}, the cohesive energy per gram, the normal density ρ_0 and the bulk modulus B_0, $P_{th} = \sqrt{\dfrac{U_{coh}B_0\rho_0}{8}}$. P_{th} is higher by about one order of magnitude than the spall strength measured at strain rates in the range $(10^3 - 10^6)$ s^{-1}. This weakening occurs due to the fact that the mechanical properties of solids are controlled by lattice imperfections and dislocations which weaken the material. It is seen in fig.5 that the spall strength varies slowly with the strain rate in the range $(10^3 - 10^6)$ s^{-1}. In contrast to this behavior, a rapid increase in the spall strength from 7 kbar to 43 kbar in the strain rate range $(9\cdot10^6 - 4\cdot10^7)$ s^{-1} is measured in Al and a similar one in Cu.

The results of the experiments reported here suggest that a critical phenomenon occurs at strain rates of about 10^7 s^{-1}, expressed by a sudden approach to the theoretical value of the spall strength.

When modeling the spallation phenomena in ductile metals at low and medium strain rates (up to 10^6 sec^{-1}) Johnson (11) used the existence of voids which, under the tensile stress, coalesces resulting finally in spallation. Giles and Maw (12) have enlarged that model to also describe the temperature and strain rate dependences of the same phenomena. The phenomenon of the coalescence of already existing voids, takes more time than available at strain rates larger than 10^7 sec^{-1}.

A realistic solid has several kinds of imperfections i.e. points of weakness, e.g. voids, dislocations, interstitials, grain boundaries etc. Each of these needs a different tensile stress in order to turn them into a growing void. The results of the experiments reported here suggests that there exists a critical strain rate of about 10^7 sec^{-1} at which the existing voids no longer suffice and new ones have to be created and added in that short a time.

This critical strain rate is expressed by the lower kink in the arctan curve presented in Fig. 5. From that strain rate and higher, the value of P_{spall} is obtained by the admixture of the density of the initial voids and the various kinds of imperfections involved in creating additional voids. The phenomenological description suggested here is supplementing Johnson's model (11).

ACKNOWLEDGMENTS

We thank S. Maman for his skillful technical assistance and Y. Horowitz, Y. Paiss, R. Shpitalnik and M. Werdiger for helpful discussions.

REFERENCES

1. Moshe, E., Dekel, E., Henis, Z., Eliezer, S., Appl. Phys. Lett. **69**, 1379 (1996).
2. Fortov, V.E., Kostin, V.V., Eliezer, S., J. Appl. Phys. **70**, 4542 (1991).
3. Bloomquist, D.D., Sheffield, A.A., J. Appl. Phys. **54**, 1717 (1983).
4. Baumung, K., Singer, J., Physics of Intense Light Ion Beams and Production of High Energy Density in Matter, Annual Report 1994 (Forschungzentrum Karlsruhe GmBH, Karlsruhe, 1995), p.88.
5. Barker, L.M., Schuler, K.W., J. Appl. Phys. **45**, 3692 (1974).
6. Kanel, G.I., Fortov, V.E., Adv. Mech **10**, 3 (1987).
7. Blinkov, A.V., Keller, D.W., ASTM Spec. Techn. Publ., **336**, 252 (1962).
8. Bushman, A.V., Kanel, G.V., Ni, A.L., Fortov, V.E., Thermophysics and Dynamics of Pulse Action, Chernogolovka, 1988.
9. Speigth, C.S., Taylor, P.F., Wallace, Metallurgical Effects at High Strain Rates, Rhode, R.W., Butcher, Holland, J.R., Carnes. , ed. (Plenum, NY, 1973), p.429.
10. Grady, B.E., J. Mech. Phys. Solids, **3**, 353 (1988).
11. Johnson, J.N. J.Appl.Phys. **52** ,2812 (1981).
12. Giles, A.R. ,Maw, J.R. , in this proceedings.

CP429, *Shock Compression of Condensed Matter – 1997*
edited by Schmidt/Dandekar/Forbes
© 1998 The American Institute of Physics 1-56396-738-3/98/$15.00

HIGH-TEMPERATURE PRESSURE-SHEAR PLATE IMPACT STUDIES ON OFHC COPPER AND PURE WC

K. J. Frutschy*, R. J. Clifton, and M. Mello

Division of Engineering, Brown University, Providence, Rhode Island 02912
*Intel Corporation, 5000 West Chandler Boulevard, Chandler, Arizona, 85226

The pressure-shear plate impact experiment has been modified to test materials at high temperatures. For high strain rate tests, a thin plate of the specimen material is sandwiched between two pure tungsten carbide plates which are heated by an induction heater. To overcome possible misalignment of the impact face of the target due to thermal expansion of the target supports, the alignment of the target assembly is maintained with remote controls and an optical lever in which a laser beam—reflected from the rear surface of the target—is displayed on a distant screen. Photoresist gratings, which normally provide the diffracted beams used in recording the transverse velocity of the target assembly, are replaced by temperature resistant titanium phase gratings produced by SEM lithography. Results are presented for tests on OFHC copper at temperatures ranging from 300 to 700°C and strain rates of 10^5 to $10^6 s^{-1}$. Symmetric impact experiments on pure tungsten carbide provide the compressive and shear responses of the loading plates over the same temperature range.

INTRODUCTION

Measurements of the plastic response of materials at high temperatures and high strain rates are essential for the development of models to describe dynamic shear banding which occurs in applications such as terminal ballistics, high-speed machining, and dynamic fracture. For such so-called *adiabatic* shear bands, the localization of deformation into a narrow band occurs as a thermoplastic instability in which thermal softening allows both strain rate and temperature to increase sharply within the band. For martensitic steels, temperatures within the bands have been measured to be over 600°C[1,2]; strain rates in the range of 10^5 to $10^6 s^{-1}$ are often inferred[1,3]. In order to provide a method for measuring the shearing resistance of materials at these high temperatures and high strain rates, the pressure-shear plate impact

experiment has been modified to enable testing at high temperatures. Copper is the first material to be investigated because of its well documented response under other loading conditions, its low melting temperature, and its applications in armor penetration.

EXPERIMENT

The basic pressure-shear plate impact experiment for high strain rate deformation of a thin foil has been described by Clifton and Klopp[4]. The time/distance diagram for the high temperature, high strain rate experiment is shown in Figure 1. The specimen foil is sandwiched between the center and right tungsten carbide plates which are heated prior to impact. The flyer plate on the left is at room temperature. Upon impact, a pressure wave and a shear wave are generated in both the flyer and the central tungsten carbide plates. The pressure wave arrives at the specimen first and induces a small amount of longitudinal plastic deformation (∼ 2 % according to simulations). More importantly, the pressure wave subjects the specimen/tungsten carbide interfaces to a large normal pressure so that friction is able to prevent slip and allow the shear wave to be transmitted across the specimen. The principal plastic deformation comes from the shear wave which arrives after the specimen is under pressure. The experiment is regarded as completed once the pressure wave—after reflecting from the rear surface of the target sandwich—returns to the specimen. At this point, the normal pressure drops to zero, and shear stress can no longer be transmitted. Clifton and Klopp [4] detail how the stress and strain in the specimen foil are calculated from the measured free surface velocities.

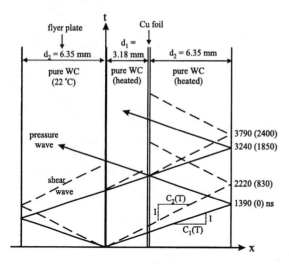

FIGURE 1. t-X diagram for Cu experiments.

The heating for the experiment is provided by an induction heater. Figure 2 shows a schematic of the impact set-up with the induction heating coil wrapped around the circumference of the target sandwich. The target sandwich is supported by a machinable ceramic holder and attached to the frame of the target holder with alumina pins that eventually break away. The inner diameter

of the induction heating coil is made larger than the diameter of the projectile tube to ensure that the coil is untouched throughout the experiment. The target holder (not shown) is shielded from the heat of the target sandwich and the electromagnetic field of the induction heating coil by a water-cooled copper housing that surrounds the target sandwich.

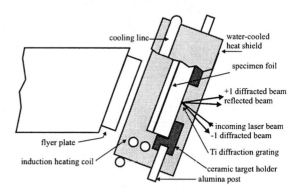

FIGURE 2. Impact schematic.

RESULTS AND DISCUSSION

Figure 3 shows the six tests conducted at nominal shear rates of 2.5×10^5 and $1.2 \times 10^6 \, s^{-1}$. The copper hardens to a peak flow stress within a shear strain of 0.1 for the three lower strain rate tests and 0.5 for the three high strain rate tests. Within each test series, the flow stress decreases with increasing temperature. It is important to note that the curves represent the intrinsic constitutive response of the material only after a strain of $\gamma \approx 0.07$ for the $10^5 \, s^{-1}$ tests and $\gamma \approx 0.3$ for the $10^6 \, s^{-1}$ tests. Until this point, stress gradients exist in the specimen foil making the deformation inhomogeneous. The black circles on this plot correspond to the time when a longitudinal wave arrives at the free surface of the target sandwich (see Figure 1). If any tilt exists at impact, the arrival of the reflected longitudinal wave will induce a small transverse velocity at the free surface, making the interpretation of the record uncertain beyond this point.

From examination of the $10^6 \, s^{-1}$ plots, the strain rate sensitivity of copper at high temperatures is immediately apparent: the

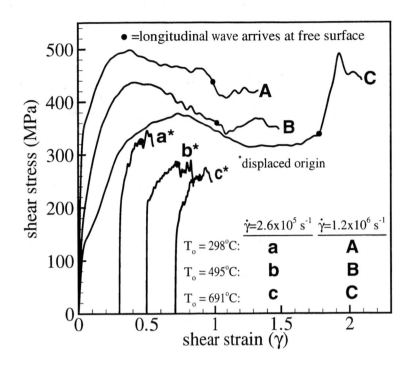

FIGURE 3. OFHC copper experiments.

peak shear stress, for a given temperature, is substantially greater than that for the corresponding test at $\dot{\gamma} \approx 2.5 \times 10^5\ s^{-1}$. The copper hardens, but then softening occurs around a shear strain of $\gamma = 0.5$. This softening continues for significant strains before the flow stress stabilizes – or rehardening begins as shown for the test at 691°C. Figure 4 shows the results from finite difference calculations of one shot using three constitutive models for copper: Follansbee and Kocks [5] (with Johnson and Tonks[6] modification), Zerilli and Armstrong[7], and Johnson and Cook[8]. The models need to be re-addressed because all underpredicted the flow stress and failed to capture its evolution with strain. Frutschy and Clifton[9] give a more detailed analysis of the apparatus and modeling for these copper experiments.

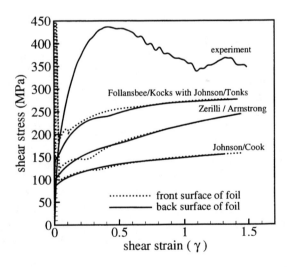

FIGURE 4. Finite-difference computed response for KF9609 (1.2 x 10⁶s⁻¹, 495°C).

465

To investigate the assumption that the pure WC plates remained elastic during the high-strain rate tests, they were tested in pressure-shear symmetric impact over a range of temperatures. In these tests, the target plate was heated prior to impact while the flyer plate was at room temperature. The longitudinal stress is found to be essentially proportional to normal particle velocity up to 7.0 GPa, and the shear stress is found to be essentially proportional to transverse particle velocity up to 0.8 GPa, with an error that increases with increasing stress but remains less than 2% up to these stress values. The impedances for this scaling are given in Table 1. Frutschy and Clifton[10] provide a more detailed analysis of the symmetric impact tests.

TABLE 1. Pure WC impedances

parameter	22°C	495°C	635°C
$\rho c_1 (GPa/\frac{mm}{\mu s})$	105.6	102.8	101.5
$\rho c_2 (GPa/\frac{mm}{\mu s})$	66.2	64.4	63.7

An additional symmetric impact, high-velocity, pressure-shear test was conducted on pure WC. Both the normal and transverse velocities from this experiment are reported in Figure 5. A normal-impact experiment by Dr. Dennis Grady is included in this plot for comparison. The normal velocity shows the same steep ramping initially in all tests. The transverse velocity, however, rises much more slowly in the high velocity test (MM9703) than in the lower velocity test (KF9704). Pure WC is known to have approximately 1.5% porosity which, at high pressures, could result in appreciable damage due to the collapse of cavities. As a result, the shearing response upon arrival of the shear wave could involve dissipative frictional processes as suggested by the ramp-like wavefronts.

ACKNOWLEDGEMENTS

The authors gratefully acknowledge the support of this research by the Army Research Office. They are also grateful to Dr. D. Grady for providing normal-impact data on pure WC.

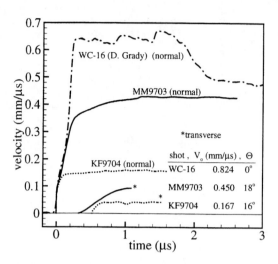

FIGURE 5. Free surface velocities from pure WC symmetric impact tests at 22°C

REFERENCES

1. Marchand, A., and Duffy, J., *J. Mech. Phys. Solids,* **36 (3)**, pp. 251-283 (1988).
2. Zhou, M., Rosakis, A.J., and Ravichandran G., *J. Mech. Phys. Solids,* **44 (6),** pp. 981-1006 (1996).
3. Giovanola, J., in *Proceedings on Impact Loading and Dynamic Behavior of Materials - 1987,* Bremen, Federal Republic of Germany.
4. Clifton, R.J., and Klopp, R.W., *Metals Handbook: Mechanical Testing (nineth ed.)* **8**, pp. 230-239 (1985) .
5. Follansbee P.S., and Kocks, U.F., *Acta Metall.,* **36**, pp. 81-93 (1998).
6. Johnson, J.N., and Tonks, D.L., in *Shock Compression of Condensed Matter - 1991,* Elsevier Science Publishers, New York, pp. 371-378.
7. Zerilli, F.J., and Armstrong, R.W., *J. Appl. Phys.,* **61 (5)**, pp. 1816-1825 (1987).
8. Johnson, G.R., and Cook, W.H., *Engin. Fracture Mech.,* **21 (1)**, pp. 31-48 (1985).
9. Frutschy, K.J., and Clifton, R.J., *submitted to J. Mech. Phys. Solids,* (1997).
10. Frutschy, K.J., and Clifton, R.J., *submitted to Exper. Mech.,* (1997).

CP429, *Shock Compression of Condensed Matter – 1997*
edited by Schmidt/Dandekar/Forbes
© 1998 The American Institute of Physics 1-56396-738-3/98/$15.00

TEMPERATURE DEPENDENCE ON SHOCK RESPONSE OF STAINLESS STEEL

Zhuowei GU Xiaogang JIN

Lab for Shock Wave and Detonation Physics, Southwest Institute of Fluid Physics, CAEP
P.O.Box 523, Chengdu, Sichuan, 610003, P.R.China

Free surface velocity measurements were reported for HR-2(Cr-Ni-Mn-N) stainless steel, initially heated to 300K~1000K and shock-compressed to about 8GPa. The corresponding spall strength σ_f and Hugoniot elastic limit σ_{HEL} were determined from the wave profiles. It is demonstrated that σ_f and σ_{HEL} decrease linearly with increasing temperature T in the range from 300K to 806K, i.e., $\sigma_f = 5.63 - 4.32 \times 10^{-3} T$ and $\sigma_{HEL} = 2.08 - 1.54 \times 10^{-3} T$, and in the range of 806K~980K, σ_{HEL} increases from 0.84GPa at 806K to 0.93GPa at 980K, σ_f has a negligible increase to 2.15GPa from 2.14GPa. Primary TEM test on recovery samples identified the existence of intermatallic compound Ni_3Ti in the sample of 980K.

INTRODUCTION

HR-2 stainless steel is one of the Austenitic stainless steel with nitrogen. Because of its high yield strength and excellent anticorrosion at ambient and low temperature, it has been used widely in areas of petrochemical industry, low temperature engineering and so on. Several studies on the dynamic behavior of this material at room temperature have been reported recently.[1,2] In this study, we reported free surface velocity profiles measurement on HR-2 steel shock-compressed to about 8GPa in the temperature range of 300K~1000K. The results provide insight into the effects of temperature on the elastic and spall properties under shock compression.

EXPERIMENTAL TECHNIQUE

We performed our experiments with a single-stage 100-mm bore gas gun, used a water-cooled copper induction coil powered by a 30-kW radio frequency generator to heat samples, and applied a 15:1 step-down transformer to degrade high voltage of generator. The initial temperature of samples was measured to±0.75% by a Ni90%Cr10%-Ni97%Si3% thermocouple, the motion of the rear

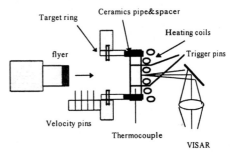

FIGURE 1. The Set-up for high-temperature VISAR experiments.

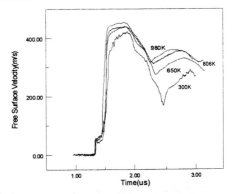

FIGURE 2. Free surface velocity of samples at different temperature(only 3.5us were taken)

TABLE 1. Chemical composition of HR-2 steel (values in weight percent)

C	Mn	Cr	Ni	Si	Ti	Al	N_2	S	P	Fe
0.037	8.95	20.53	7.64	0.41	0.1	0.85	0.29	0.007	0.012	62.1

TABLE 2. Physical constants of HR-2 steel[1]

Materials	Density (g/cm³)	Melting Point (℃)	Logitudial Wave Velocity (km/s)	Shear Wave Velocity (km/s)	Special Heat (Cal/g ℃)	Poisson's Ratio
HR-2	7.806	1410	5.747	3.158	0.115	0.28

free surface of the sample was recorded using a VISAR, of which the time resolution is ~2-3ns, the precision is ±1%.Shock compression of HR-2 steel was carried out using 2.0-mm thick iron flyer plates to impact 6.0-mm thick samples with velocities at about 410m/s.The samples were cut from φ80 bar of HR-2 steel which had never been heat treated before. It's chemical composition and physical parameters were listed in Tables 1 and 2. The experimental set-up was indicated in Figure 1.

EXPERIMENT RESULTS

Six experiments were conducted on HR-2 steel samples at different initial temperature, four of which yielded free surface velocity profiles, shown in Figure 2. In the rest two experiments, data recording failures prevented a complete elastic precursor from being obtained but the whole spall signal were recorded.

To establish timing for the experiments, the toe of the elastic precursor was assumed to propagate at the ambient-pressure compressional wave velocity which were measured in [3]. The time difference between the precursor and the midpoint of the shock established the shock velocity.

The Hugoniot elastic limit, σ_{HEL}, can be obtained from:

$$\sigma_{HEL} = \frac{\rho_{0T} \cdot V_{p0} \cdot u_{fs}}{2}, \quad (1)$$

where V_{p0} is the ambient-pressure, high-temperature compressional sound velocity[3], u_{fs} is free surface velocity. The factor-of-two relationship between the

free surface and particle velocity has been assumed. ρ_{0T} is the initial high-temperature density and in the form of:

$$\rho_{0T} = \rho_0 \cdot (1 + 3\alpha T)^{-1}, \quad (2)$$

ρ_0 is the initial room-temperature density, T is temperature(℃), α is the linear expansion coefficient:[5]

$$\alpha(T) = 1.16422 \times 10^{-5} + 3.69496 \times 10^{-8} T \\ - 3.98813 \times 10^{-11} T^2 \quad (3)$$

The σ_{HEL} data are listed in Table 3. and is plotted as a function of initial temperature in Figure 3. A linear, least-squares fit to the data in the temperature range 300K~806K provides:

$$\sigma_{HEL} = 2.08 - 1.54 \times 10^{-3} T, \quad (4)$$

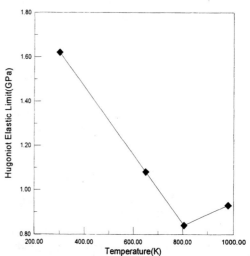

FIGURE 3. Hugoniot Elastic Limit as a function of initial temperature. (Solid line is a least squares fit to the data.)

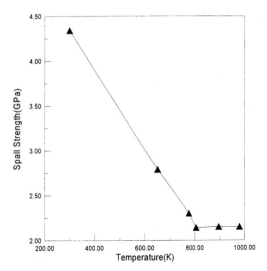

FIGURE 4. Spall strength of HR-2 steel as a function of initial temperature. (Solid line is a least squares fit to the data.)

The spall strength, σ_f, is calculated from:

$$\sigma_f = \frac{1}{2} \cdot \rho_{0T} \cdot D \cdot \Delta u_{pb} , \qquad (5)$$

where D is shock velocity, Δu_{pb} is the pull-back amplitude measured from the peak free surface velocity to the velocity minimum. The spall strength data we get are given in table 3. and is plotted as a function of initial temperature in Figure 4. A linear, least-squares fit to the data in the temperature range 300K to 806K provides:

$$\sigma_f = 5.63 - 4.32 \times 10^{-3} T , \qquad (6)$$

The Hugoniot stress σ_H were calculated in Table 3 according the following expressions[4]:

$$\sigma_H = \sigma_{HEL} + \frac{\rho_{0T} \cdot D}{2}(U_f - U_{fHEL} + \frac{V_{p0} - c_0}{V_{p0} \cdot c_0} \cdot \frac{\sigma_{HEL}}{\rho_{0T}}) \qquad (7)$$

where U_f, the peak free surface velocity ; U_{fHEL}, HEL free surface velocity ; c_0, the bulk velocity.

TEM ANALYSIS ON RECOVERY SAMPLE

In order to explain the results of experiments, primary TEM test on recovery sample were carried out at Materials department of XiNan JiaoTong university. In the recovery sample of 980K, the intermatallic compound Ni_3Ti were found, and it exists along the crystal boundary, shown in figure 5. Because the intermatallic compound has the character of high strength at high temperature. So we can explained the abnormal experiment result at 980K reasonably . Further analysis is still on.

TABLE 3. Experimental results for HR-2 stainless steel at high temperature

Shot No.	Impact Velocity (m/s)	Initial Temperature (K)	Sample Thickness (mm)	Flyer plate Thickness (mm)	Hugoniot Eiastic Limit (GPa)	Spall Strength (GPa)	Shock Stress (GPa)
916	420	300	6.08	2.02	1.62	4.34	8.81
704	409	650	6.08	2.00	1.08	2.79	8.16
1027	426	776	5.90	2.02	\	2.30	\
628	410	806	6.00	2.02	0.84	2.14	7.86
107	400	896	5.80	2.02	\	2.15	\
702	411	980	6.00	2.00	0.93	2.15	7.58

FIGURE 5. The particulate form of Ni$_3$Ti in the sample of 980K. (× 1800)

SUMMARY

Free surface velocities have been measured on HR-2 stainless steel preheated in the temperature range of 300K~1000K in order to investigate its high temperature properties under shock compression. In the temperature range of 300-806K, at 806K to 0.93GPa at 980K, spall strength had a spall strength negligible increase. Primary TEM test on recovery and HEL droped linearly, but when preheated to 980K, Hugoniot elastic limit increased from 0.84GPa samples identified the existence of intermatallic compound Ni$_3$Ti in the sample of 980K, Further analysis result would be published in the following articles.

ACKNOWLEDGMENTS

The authors would like to thank prof. Qingfu Zhang and Yue Sun for their help in Experimental installations. The following colleagues who took part in the experiments are also thanked: Xiang Wang,Wei Wang,Qiuwei Fu,Xiaosong Wang and Hongliang He. This work performed under auspices of Science foundation of China Academy of Engineering Physics.

REFERENCES

1. Shihui Huang et al, " Experimental measurements of 2169 stainless steel under dynamic loading " in Proceedings of the Conference on Shock Waves in condensed Matter, 1994, pp.1083-1086
2. W. Zhang, CAEP report (unpublished).
3. H. .Zhang, CAEP report (unpublished).
4. D.E. Grady, in Metallurgical Applications of Shock-Wave and High Strain Rate Phenomena, New York, 1986, pp. 763-780.
5. S.Lu et al , *Stainless steel*, Beijing: atomic energy Press, 1995) pp.173 (in Chinese).

CP429, *Shock Compression of Condensed Matter – 1997*
edited by Schmidt/Dandekar/Forbes
© 1998 The American Institute of Physics 1-56396-738-3/98/$15.00

SPLIT-HOPKINSON PRESSURE BAR TESTS ON PURE TANTALUM

Richard D. Dick[a], Ronald W. Armstrong[b], and John D. Williams[c]

[a]*Shocks Unlimited, 9737 Academy Rd., NW, Albuquerque, NM 87114*
[b]*Mechanical Engineering, University of Maryland, College Park, MD 20742*
[c]*Talbot Laboratory, University of Illinois, Urbana, IL 61801*

Pure tantalum (Ta) was loaded in compression by a split-Hopkinson pressure bar (SHPB) to strain rates from 450 to 6350 s[-1]. The results are compared with SHPB data for commercial Ta and with predictions from the constitutive model for Ta developed by Zerilli and Armstrong (Z-A). The main conclusions are: (1) the flow stress versus log strain rate agree with the Z-A constitutive model and other reported data , (2) uniform strain exponents computed on a true stress-strain basis for pure Ta are somewhat greater than those determined from SHPB data for commercial Ta, and (3) in both cases the uniform strain exponents versus log strain rate are in good agreement with predictions from the Z-A constitutive model for strain rates above 1500 s[-1] without a clear indication of dislocation generation.

INTRODUCTION

Pure Ta was subjected to high strain rates using a standard SHPB apparatus. Compressive loading produced stresses from 450 to 900 MPa and strain rates from 450 to 6350 s[-1]. The tests provided data to compare to previous experimental data (1-7) and to assess the Z-A constitutive equations for pure bcc Ta (8). The interest in Ta and similar materials is related to the U.S. Army's armor-antiarmor programs. The purpose is to provide material strength information, deformation data, and a constitutive model for use in computer codes to predict high rate loading conditions.

BACKGROUND

Zerilli and Armstrong proposed a set of physically based equations that predict the deformation behavior of face centered cubic (fcc) and body centered cubic (bcc) metals and alloys (9). The bcc equations were applied to experimental data that were available for Ta. The current work is part of a continuing effort to update the constants in the constitutive equations (8).

Figure 1 shows a comparison of predicted and measured true strain at maximum load versus strain rate determined from various studies on relatively pure Ta material (1). The experimental data including data at very high strain rates (1,3,6,7) are shown along with the curve predicted by the Z-A model. The low strain rate data from Hoge and Mukherjee (3) were used to generate the theoretical curve. For strain rates greater than 1000 s[-1], the results of Rajendran et al. (6) and the Regazzoni and Montheillet (7) appear to reach a minimum and then turn upward. This indicates a reversal in ductility, that is, Ta becomes more ductile at very high strain rates. According to the Z-A theory the strain hardening needs to increase with strain rate to explain the increased ductility and additionally, enhanced dislocation generation is necessary with increased strain rate. This phenomena was studied in the recent SHPB tests on pure Ta and the earlier SHPB test data on commercial Ta are included.

EXPERIMENT

The SHPB apparatus consisted of two steel bars,

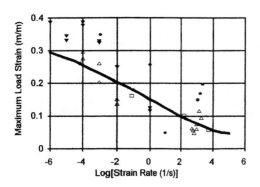

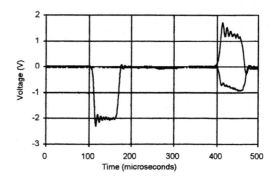

FIGURE 1. Comparison between predicted (solid curve) and experimental (symbols) data. Inverted solid triangles-Ref. 1, solid circles-Ref. 6, open squares-Ref. 3, and open triangles-Ref. 7.

FIGURE 2. Typical voltage-time signals for Test 148-1.

bars, each 1.69 m long by 12.7 mm diameter and a 150 mm long by 12.7 mm diameter striker bar. A gas gun propelled the striker bar. The Ta specimens were sandwiched between the incident and transmitter bars and loaded in compression which resulted from the impact of the striker bar on the incident bar. In the SHPB tests the Ta specimen experienced a 60 μs long stress wave pulse. Strain gage pairs on the two bars detected the three strain signals, i.e., the incident, reflected, and transmitted strain signals. These signals were processed by applying a dispersion correction to the signals and then time shifted to bring the three pulses into coincidence at the sample-bar interfaces. The strain rate, strain, and stress experienced by the Ta samples were computed from the standard bar equations using the corrected strain data.

The pure Ta samples were made from rod 6.35 mm diameter to produce 6.35 mm and 3.18 mm thick samples. The sample ends were machined flat and parallel to provide good contact with the bar ends. A thin grease layer was applied to the Ta sample ends to reduce friction between the bar and the sample during dynamic compression.

EXPERIMENTAL RESULTS

Figure 2 is a typical voltage-time record showing the incident, reflected, and transmitted signals from which the corresponding strains were

used to compute the stress-strain values for the Ta from the bar equations. These are engineering stress-strain values, so as part of the data analysis, the engineering stress-strain was converted to true stress-strain.

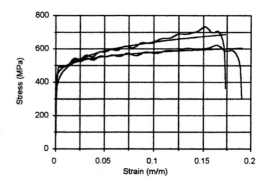

FIGURE 3. Engineering (upper) and true (lower) stress-strain Ta . Fits to the uniform strain equation are shown for each curve

The stress-strain data from previous SHPB tests preformed on Ta samples (2) are also presented in this paper on a true stress-strain basis. Of course the commercial Ta contained more impurities than pure Ta which influenced the deformation behavior. Figure 4 shows the combined stress versus log strain rate results for pure Ta material including the fitted Z-A bcc equation descriptions and the SHPB results for the commercial Ta (1). The flow stress for the pure Ta SHPB tests data from this new study is defined as the stress at two percent strain from the true stress-strain data. Figure 4 also includes recent static compression and tension data (1) and the static and dynamic data obtained by previous researchers (2-7). Considering the strain rate data covers eleven orders of magnitude, all the data for pure Ta in Fig. 4 agree very well with the curve predicted from the Z-A constitutive model. The flow stress data for commercial Ta are greater than those of pure Ta. However, the plot of data for commercial Ta in Fig. 4 does appear to lie on a curve parallel to the predicted Z-A curve..

The coefficients "K" and "n" in the approximate true stress-true strain equation ,

$$\sigma = K\,\epsilon^n ,$$

where σ = stress and ϵ = strain, were determined for each test by fitting the higher strain portions of the true stress-strain curves (see Fig. 3). In this case, n is numerically equal to the maximum uniform strain The exponent, n, is plotted versus log strain rate in Fig. 5 for the Z-A constitutive equations for pure Ta as well as the SHPB data for the pure and commercial Ta samples. For strain rates greater than 1500 s⁻¹, the pure Ta data lie close to the Z-A curve, but for rates less than 1000 s⁻¹, the data are below the predicted curve where the n-values are either small or negative. The commercial Ta data with strain rates greater than 1500 s⁻¹ lie slightly below the predicted Z-A curve for pure Ta. At strain rates under 600 s⁻¹ the n-values for pure and commercial Ta group together. From Fig 5 the two data clusters (<600 s⁻¹ and >1500 s⁻¹) are separated at the 0.014 yield point strain of the Hoge and Mukherjee material (3,8).

DISCUSSION

The curve computed from the Z-A model was calibrated using the Hoge and Mukherjee (3) data for pure Ta (see Fig. 4). Their experiments include compression and tension tests in a static loading frame apparatus and SHPB tests. The commercial Ta SHPB data shown in Fig. 4 have a higher flow stress than pure Ta and appear to lie on a curve parallel to the pure Ta curve, but displaced approximately 100 MPa above. Using this information new constants

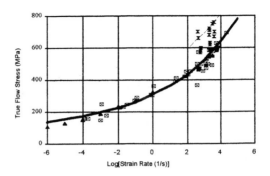

FIGURE 4. Z-A model (heavy solid curve) compared to experimental data. The thin dashed curve marks the commercial Ta. Solid traingle-Ref. 2; open square-Refs. 3,4,5, and 6; solid square-pure Ta; solid hourglass-commercial.

were determined from the Z-A constitutive equations (8,9) to compute a new curve relating the uniform strain n value for impure Ta. This new curve is shown in Fig. 5 as a thin line in comparison to the thick line for the Z-A prediction for pure Ta.

The negative and small n-values obtained from the uniform strain equation shown in Fig. 5 are explained by the yield-point behavior for both pure and commercial Ta materials. Otherwise the n-values agree reasonably well with the Z-A constitutive model for bcc Ta (8) ignoring dislocation generation. The higher commercial Ta flow stress shown in Fig.4 results in the lower Z-A predicted curve in Fig. 5, again without considering dislocation generation. Also, the 0.014 strain value which is the corresponding n-value in Fig. 5 represents an average value for the elongation at the lower yield point in the Z-A constitutive model (8).

The reason for the enhanced uniform strain measurements being obtained by Regazzoni and Montheillet for pure Ta material from sintered powder is puzzling based on the current measurements and continued analysis of the results (4). Figure 5 shows clearly that smaller n-values are obtained in the present study than the uniform strain values shown in Fig. 1. Hence, the complication of the lower yield point behavior and the approximate fit of the single exponential strain term to the total stress behavior may be masking solid evidence of enhanced

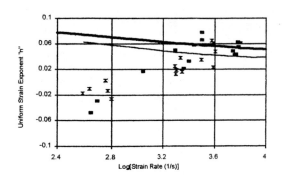

FIGURE 5. Z-A prediction (heavy solid curve), Z-A model for higher flow stress data from Fig. 4, and SHPB data for pure Ta (solid rectangle) and commercial Ta (solid hourglass).

dislocation generation in the Ta at strain rates greater than 1000 s^{-1}.

CONCLUSIONS

There is good correlation among all the experimental data and the Z-A constitutive model for pure Ta. Even though the commercial Ta contains impurities, these SHPB data relate reasonably well with the pure Ta data. As seen in Fig. 4, the flow stress values for commercial Ta are higher than pure Ta, but the data are arranged parallel to the predicted Z-A curve and the experimental data for pure Ta. Purity differences between the Ta materials probably determine how well the commercial Ta data fit relative to the theoretical curve and the pure Ta data. Somewhat smaller n-values were obtained for commercial Ta than pure Ta (see Fig. 5) which seem to complement the observed higher flow stress. At small strain rates the n-values for both Ta materials are negative or small suggesting the deformation observed for these tests have not progressed beyond the yield-point behavior. In addition correlation is distinguished in Fig. 5 between pure and commercial Ta in terms of two major data groupings for n-value versus strain rate. One group is bunched around a strain rate of 500 s^{-1} and another group is observed between 1500 and 5000 s^{-1}. The data group at the higher strain rates agree well with the Z-A theoretical curves, one

for pure Ta, upper curve, and commercial Ta, lower curve (see Fig. 5).

All these results described in this study taken together demonstrate that the constants in the Z-A constitutive equations need slight adjustment to model Ta with different impurity amounts.

ACKNOWLEDGMENTS

This research was supported in part by the U.S. Army Research Office, the Office of Naval Research, and the Department of Mechanical Engineering, University of Maryland, College Park, MD.

REFERENCES

1. Armstrong, R.W., Chen, C.C., Dick, R.D., and Xhang, X.J., "Evaluation and Improvement in Constitutive Equations for Finite Viscoplastic Deformation and Fracturing Behavior Relating to Armor Design," 1997, Final Report, U.S. Army Research Laboratory, Aberdeen, MD.
2. Armstrong, R.W., Zhang, X.J., Feng, C., Williams, J.D. and Zerilli, F.J., *Tantalum*, Minerals, Metals, and Mining Society, Warrendale, PA.
3. Hoge, K.G. and Mukherjee, A.K., J. Matl. Sci.**12**, 1606(1977). 4. Chen, S.R. and Gray III, G.T., "Constitutive Behavior of Tantalum and Tantalum-Tungsten Alloys," Los Alamos National Laboratory Report, LA-UR-94-4345, 1994.
5. Gourdin, W.H., "Constitutive Properties of Copper and Tantalum at High Strain Rates: Expanding Ring Results," Lawrence Livermore National Laboratory Report UCRL-98812, 1988.
6. Rajendran, A.M., Garrett, Jr. R.K., Clark, J.B. and Jungling, T.J., J. Matl. Shap. Tech **9**, 7(1991).
7. Regazonni, G. and Montheillet, F., *Mechanical Properties at High Rates of Strain*, Conf. Series No. 70, J. Harding Ed., London Inst. Phys., 1984.
8. Zerilli, F.J. and Armstrong, R.W., J. Appl. Phys. **68**, 1580 (1990).
9. Zerilli, F.J. and Armstrong, R.W., J. Appl. Phys. **61**, 1816 (1987).

474

CP429, *Shock Compression of Condensed Matter – 1997*
edited by Schmidt/Dandekar/Forbes
© 1998 The American Institute of Physics 1-56396-738-3/98/$15.00

DYNAMIC CONSTITUTIVE RESPONSE OF TANTALUM AT HIGH STRAIN RATES

K. E. Duprey and R.J. Clifton

Division of Engineering, Brown University Providence, RI 02912

Pressure-shear plate impact experiments were conducted on annealed and rolled tantalum foils to determine the dynamic constitutive response of the material at high strain rates between $10^5 s^{-1}$ and $10^6 s^{-1}$ and pressures between 4 and 8 GPa. The experiments were performed to determine the strain rate sensitivity of the flow stress of the metal. The results suggest that both texture and strain rate play an important role in determining the flow stress of tantalum.

INTRODUCTION

The properties of tantalum; high density, high melting point, and excellent ductility through a wide range of temperatures; make it an ideal candidate for impact related applications such as explosively formed projectiles (EFPs). These types of applications can induce strain rates of $10^5 s^{-1}$ and higher and very large strains in the material. Currently, the flow stress of commercially pure tantalum has been studied at strain rates ranging from quasi-static to $10^4 s^{-1}$, temperatures from $-200°C$ to $525°C$, and strains up to 60% by Hoge and Mukherjee (1), Vecchio (2), Meyers *et al.* (3), and Nemat-Nasser and Isaacs (4). However, there is relatively little data on the flow stress of tantalum at higher strain rates and large strains. Hence the motivation for the current study.

Recent polycrystaline modeling work by Kothari and Anand (5) and Wright *et al.* (6) have focused on reproducing the end shape of Taylor cylinder-impact tests conducted on pre-textured tantalum cylinders by Ting (7). The use of rolled foils in the present study allow a more quantitative investigation into the roll of texture in the dynamic constitutive response of tantalum at high strain rates.

EXPERIMENT

Figure 1 shows a schematic diagram of the pressure-shear plate impact experiment described by Clifton and Klopp (8). In this experiment a plate attached to a moving projectile impacts a stationary target plate. Both the flyer and target plates are inclined at an angle relative to the projectile velocity, and the relative alignment of these plates is maintained by means of a keyway machined in the gun barrel which prevents the projectile from rotating. This oblique impact induces both compressive and shear stresses in the target and flyer plates. The configuration used for all but one of the experiments in this study consists of a thick solid flyer plate made of either hardened tool steel or pure tungsten carbide (WC), and a target plate consisting of a tantalum foil sandwiched between two hard plates of steel or WC. The hard plates are chosen for their high elastic limit ensuring that the stress state in these plates remains in

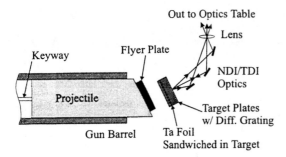

Figure 1: Schematic of Pressure-Shear Plate Impact Experiment

the elastic range for common projectile velocities. Thus the tantalum foil is the only material undergoing any inelastic deformation during the period of interest of the experiment, and elastic wave propagation theory can be used to evaluate the stress evolution in the hard steel or WC plates. This also allows the use of a time-distance (t-x) diagram showing the position versus time of elastic wavefronts in the flyer and target plates.

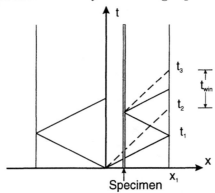

Fig. 2: Time-Distance Diagram for Sandwiched Foil

Figure 2 shows the t-x diagram for this sandwiched foil configuration. On this diagram, the longitudinal wavefronts are shown as solid lines and the shear wavefronts are depicted as dashed lines. From this figure it can be seen that the longitudinal compressive wave reaches the foil interface well before the shear wave. Because of this, the foil is in a state of uniform compression prior to the onset of shearing.

A combined normal and transverse displacement interferometer is utilized to monitor the rear surface motion of the target plate. The shear stress τ and the normal stress σ at the rear surface of the the tantalum foil can be obtained from the measured rear surface velocities using one-dimensional elastic wave propagation theory. The resulting relations are

$$\tau = \frac{1}{2}(\rho c_S)v_{fs}, \quad \sigma = \frac{1}{2}(\rho c_L)u_{fs}, \quad (1)$$

where ρc_S and ρc_L are respectively the shear and longitudinal acoustic impedances of the steel or WC, u_{fs} and v_{fs} are the normal and transverse free surface velocities respectively. Additionally, the nominal shear strain rate is given by

$$\dot{\gamma} = \frac{V_0 \sin \theta - v_{fs}}{h} \quad (2)$$

where V_0 is the projectile velocity, θ is the skew angle of impact, and h is the specimen thickness. Equation (2) can be integrated to yield the nominal shear strain

$$\gamma(t) = \int_0^t \dot{\gamma}(\tau)d\tau. \quad (3)$$

After two or three reverberations of shear waves through the thickness of the specimen, the stress state becomes nominally uniform and the dynamic stress-strain response is obtained by plotting τ from (1) versus γ from (3).

RESULTS AND DISCUSSION

Table 1 summarizes the relevent parameters for this series of experiments. The dynamic shear stress-strain response of the tantalum foils for these experiments are shown in Figures 3, 4 and 5 respectively. In all three curves, the scaling of the vertical Shear Stress

476

Table 1 Pressure-Shear Impact Experiments on Tantalum

Shot #	Material Process	Orient-ation	Specimen Configuration	σ (GPa)	Nominal $\dot{\gamma}$ s^{-1}	Specimen Thickness
KD9401	Annealed	-	Foil on Flyer	3.87	0.23×10^6	0.118 mm
KD9403	Annealed	-	Sandwich	3.72	0.25×10^6	0.119 mm
KD9502	Annealed	-	Sandwich	3.74	0.20×10^6	0.117 mm
KD9601	Rolled	*	Sandwich	3.44	0.42×10^6	0.025 mm
KD9602	Rolled	*	Sandwich	4.25	1.8×10^6	0.010 mm
KD9605	Rolled	*	Sandwich	4.48	0.68×10^6	0.025 mm
KD9612	Rolled	*	Sandwich	3.82	0.84×10^6	0.025 mm
KD9613	Rolled	*	Sandwich	3.78	5.0×10^6	0.010 mm
KD9615	Rolled	*	Sandwich	3.80	1.4×10^6	0.025 mm
KD9702	Rolled	0°	Sandwich	7.96	1.4×10^6	0.025 mm
KD9704	Rolled	90°	Sandwich	8.18	1.6×10^6	0.025 mm

* Angle between shearing direction and rolling direction: 0°, 45°, or 90°

axis is kept constant for comparison purposes.

KD9401 and KD9403 show similar strain hardening slopes.

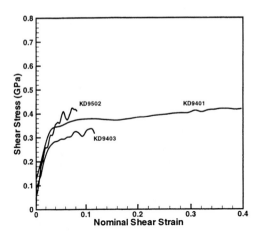

Fig. 3: Dynamic Shear Stress-Strain Curves for Annealed Tantalum Foil

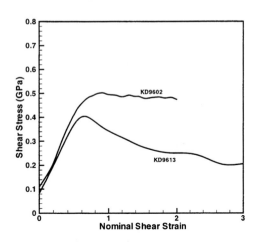

Fig. 4: Dynamic Shear Stress-Strain Curves for $10\mu m$ Tantalum Foil

In Fig. 3, the response of the annealed foils (KD9401, KD9403 and KD9502) are plotted together. Due to the relatively high thicknesses of these samples, only one of the curves shows appreciable plastic strains. All three curves exhibit relatively similar initial flow stresses due to being sheared at similar strain rates. Additionally, experiments

The experiments conducted on the $10\mu m$ rolled foils (KD9602 and KD9613) are shown in Fig. 4. These two response curves show a similar initial response, however during this portion of the curve, the stress has not yet reached a uniform level through the thickness of the foil. Once the stress state becomes uniform, the response of the two samples is very different. Shot KD9602 shows the flow stress saturating, followed by a slight strain softening behavior, whereas KD9613 exhibits substantial softening after the onset of plastic flow, thus leading to a higher nominal strain rate and larger total accumulated plastic strain.

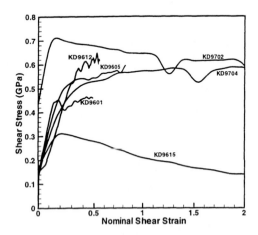

Fig. 5: Dynamic Shear Stress-Strain Curves for $25\mu m$ Tantalum Foil

Finally, Fig. 5 shows the dynamic shear stress-strain responses for the $25\mu m$ rolled foils. Four of these curves (KD9601, KD9605, KD9612, and KD9704) are of a similar nature exhibiting strain-hardening after the onset of plastic flow. These curves are also of a similar nature to the annealed foil experiments shown in Fig. 3. The remaining two curves (KD9615 and KD9702) are different from the other four but are similar to each other in that they exhibit strain-softening behavior after initial flow. The differences between experiments KD9702 and KD9704 can be explained in terms of the known difference in angle between the shearing direction and the rolling direction as indicated in Table 1.

These results suggest that further work is needed to separate the effects of texture and strain-rate sensitivity on the flow stress of polycrystaline tantalum at high strain rates. More experiments combined with polycrystaline modeling could illuminate these differences.

ACKNOWLEDGEMENTS

This work was supported by ARO through a research grant and an AASERT award for the support of a graduate student.

REFERENCES

1. Hoge, K.G., and Mukherjee, A.K., *J. Matls. Sci.* **12**, 1666-1672 (1977).
2. Vecchio, K.S., *Supplement J. de. Physique III* **4**, 301-306 (1994).
3. Meyers, M.A., Chen, Y.J., Marquis, F.D.S. and Kim, D.S., *Metal. and Mater. Trans. A* **26**, 2493-2501 (1995).
4. Nemat-Nasser, S. and Isaacs, J.B., *Internal Report, Center of Excellence for Advanced Materials*, UCSD, San Diego, CA (1996).
5. Kothari, M. and Anand, L., *J. Mech. Phys. Solids* **in print** (1997).
6. Wright, S.I., Beaudoin, A.J. and Gray III, G.T., "Texture Gradient Effects in Tantalum." *Proceedings of the 10th International Conference on Textures of Materials, Part 2*, 1695-1700 (1994).
7. Ting, Chang-Shen, "Constitutive Modeling of Tantalum Dynamic Plasticity based on the Theory of Thermal Activation and the Evolution of Strain Hardening." *High Strain Rate Behavior of Refractory Metals and Alloys*, 249-265 (1992).
8. Klopp, R.W. and Clifton, R.J., *Metals Handbook: Mechanical Testing*, Metals Park, American Society for Metals, 1985, Vol. 8, 9th Edition, pp. 230-239.

478

CP429, *Shock Compression of Condensed Matter – 1997*
edited by Schmidt/Dandekar/Forbes
© 1998 The American Institute of Physics 1-56396-738-3/98/$15.00

SPALL WAVE-PROFILE AND SHOCK-RECOVERY EXPERIMENTS ON DEPLETED URANIUM *

Robert S. Hixson, John E. Vorthman, R. L. Gustavsen, A. K. Zurek, W. R. Thissell, and D. L. Tonks

Los Alamos National Laboratory, Los Alamos, New Mexico 87545 USA

Depleted Uranium of two different purity levels has been studied to determine spall strength under shock wave loading. A high purity material with approximately 30 ppm of carbon impurities was shock compressed to two different stress levels, 37 and 53 kbar. The second material studied was uranium with about 300 ppm of carbon impurities. This material was shock loaded to three different final stress level, 37, 53, and 81 kbar. Two experimental techniques were used in this work. First, time-resolved free surface particle velocity measurements were done using a VISAR velocity interferometer. The second experimental technique used was soft recovery of samples after shock loading. These two experimental techniques will be briefly described here and VISAR results will be shown. Results of the spall recovery experiments and subsequent metallurgical analyses are described in another paper in these proceedings.

INTRODUCTION

Previous shock compression studies of depleted Uranium and alloys have shown a dependence of spall strength on alloy solute concentration as well as strain rate rate (or peak particle velocity).[1] In the case of U-Ti, U-Mo, U-Rh, and U-Nb alloys, a spall strength higher than that of pure U is observed at all realized strain rates. For pure depleted Uranium an increase in spall stress with increasing peak free surface velocity has been observed.[2]

The work described below was undertaken to attempt to obtain a better micro-mechanical understanding of the spall process in both very pure and "dirty" Uranium. Our approach was to do a series of experiments reaching to different final stress states (or free surface velocities). For each material, and at each stress state the free-surface velocity was measured, and a sample recovered in an experiment done as close as possible to the same conditions. In this way we hoped to freeze the spall processes at various stages in its development. Soft recovered samples underwent further

metallurgical analyses.

Spall Recovery Target Assembly

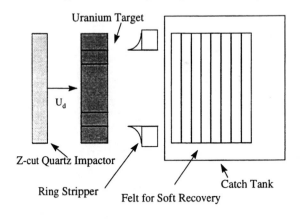

FIGURE 1. Recovery assembly.

EXPERIMENTAL DETAILS

Samples were taken from plates of the two different purity levels of depleted Uranium. The high purity material had carbon impurity levels of approximately 30 ppm. Grain size for this material was measured to be 10 microns. The sec-

*This work supported by the US Department of Energy.

High Carbon Uranium Spall

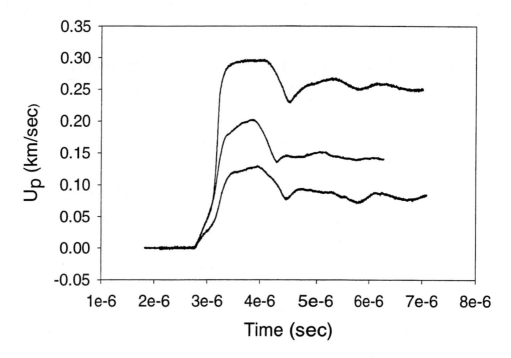

FIGURE 3. Free-Surface Velocities—low-purity uranium.

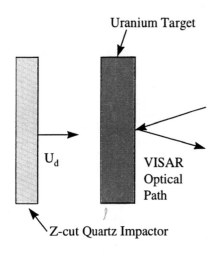

FIGURE 2. VISAR Experimental Geometry.

ond, lower purity material, had about 300 ppm carbon impurity. Grain size was approximately 200 microns. Samples were machined into right circular cylinders for the VISAR (Velocity Interferometer System for Any Reflector(3)) experiments with faces flat and parallel to 0.002 mm. One surface of each sample was polished to a reflecting condition between diffuse and specular. For soft recovery experimental assemblies were made as shown in Fig. 1. The concentric rings act as momentum traps as described by Gray.(4) Experiments were performed on a 50 mm single-stage gas gun. Projectile velocity and tilt were measured using stepped array of six shorting pins, and impacts were measured to be planer to within 1–2 mrad. Z-cut quartz impactors were used for all experiments because their response remains purely elastic over the pressure range of these experiments.

The experimental geometry for the VISAR shots is shown in Fig. 2. A commercially available optical fiber probe was used to illuminate the polished part of the target, and to collect reflected light and return it to the Los Alamos designed

Low Carbon Uranium Spall

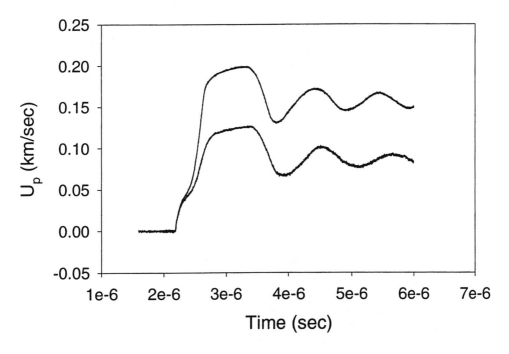

FIGURE 4. Free-Surface Velocities—high-purity uranium.

and built interferometer. For these experiments a high level of sensitivity was required, and so an 8 in. Bk-7 glass etalon was used as a delay leg. The VISAR wave-forms were recorded with a single Tektronix digitizer. This allowed all VISAR data to be collected on a common time base. Data was analyzed with a Los Alamos VISAR data reduction program.

Spall recovery experiments were done in a geometry similar to that shown for the VISAR experiments. These experiments were done separately from the VISAR shots in order to avoid a reshock of the sample due to impact on the probe and holder. Samples were recovered in a felt filled cylinder. In order to minimize oxidation, samples were placed in very pure alcohol immediately after recovery.

RESULTS

Data obtained on the "dirty" material at the three different final free surface velocities are shown in Fig. 3. Well defined elastic precursors

are observed in all experiments as well as a fast rising (30 ns) plastic wave. After the fast rising part of the plastic wave there is a slower rising region leading up to the final particle velocity. For the experiment done at the highest particle velocity we see reverberations of the release wave trapped in the spalled part of the target. We took this to be evidence of complete separation, and this was verified by the spall recovery results. Data obtained for the pure material at two different final particle velocities are shown in Fig. 4. These data indicate that complete spall separation will occur between 35 kbar and 53 kbar. Spall strengths have been calculated using the technique of Romanchenko.(5) A summary of VISAR experiments and calculated spall strength results are given in Table 1. Calculated strain rates shown depend upon many factors including flyer and target thicknesses.

TABLE 1. VISAR Experimental Results.

Material	Flyer u_d (m/s)	Impact Stress (kbar)	Spall Stress (kbar)	Strain Rate (10^5/s)	Target (mm)	Flyer (mm)
high C	285±1	37	15	1.4	5.09	4.07
	399±0.5	53	18	1.8	5.08	4.08
	596±2.0	81	19	1.6	5.09	5.09
low C	268±0.2	35	16	1.4	3.84	4.08
	403±0.2	53	19	2.0	3.84	4.08

CONCLUSION

Spall measurements on the two materials studied reveal very similar strength in tension as determined by the particle velocity change using the Romanchenko correction. The measured free surface particle velocity for the two different purity materials show different behavior in the spall pullback region. For the low purity material, the highest initial stress experiment showed a sharp change in slope in particle velocity. This can be interpreted as indicative of a rapid opening of the spall surface. For the corresponding experiment in the high purity material, the particle velocity change in slope at pullback is more gradual. This can be interpreted as meaning that the spall took longer to occur. The wave profiles from the lower impact stress experiments are not as easy to interpret—the particle velocity pullback seems to be incomplete. Recovery experiments bear this out: for the lower impact stress experiments, damage zones were created but the recovered samples remained in one piece. Calculations of the spall stress for these experiments are therefore less accurate.

ACKNOWLEDGMENTS

The authors would like to thank those who helped us in this project. This includes Dennis Price, Jim Esparza, and Joe Fritz.

REFERENCES

1. J. Buchar, S. Rolc, J. Pechacek, and J. Krejei, Journal de Physique Colloque C3 **1**, 197 (1991).

2. D. E. Grady, Steady-wave risetime and spall measurements on uranium (3-15 gpa), in *Metallurgical Applications of Shock-Wave and High-Strain-Rate Phenomena*, edited by L. E. Murr, K. P. Staudhammer, and M. A. Meyers, pages 763–780, Marcel Dekker, New York, 1986.

3. W. F. Hemsing, Rev. Sci. Instrum. **50**, 73 (1979).

4. G. T. Gray III, Influence of shock-wave deformation on the structure/property behavior of materials, in *High-Pressure Shock Compression of Solids*, edited by J. R. Asay and M. Shahanipoor, chapter 6, pages 187–215, Springer-Verlag, New York, 1993.

5. V. I. Romanchenko and G. V. Stepanov, Zhur. Prik. Mekh. Tekh., Fia **4**, 141 (1980).

CP429, *Shock Compression of Condensed Matter – 1997*
edited by Schmidt/Dandekar/Forbes
© 1998 The American Institute of Physics 1-56396-738-3/98/$15.00

DYNAMIC STRENGTH AND INELASTIC DEFORMATION OF CERAMICS UNDER SHOCK WAVE LOADING

R. Feng, Y. M. Gupta, and G. Yuan

*Shock Dynamics Center and Department of Physics
Washington State University, Pullman, WA 99164-2814*

To gain insight into material strength and inelastic deformation of ceramics under plane shock wave loading, an in-depth study was carried out on polycrystalline silicon carbide (SiC). Two independent methods were used to determine experimentally the material strength in the shocked state: 1) lateral piezoresistance gauge measurements, and 2) compression and shear wave experiments. The two sets of data were in good agreement. The results show that the Poisson's ratio of the SiC increases from 0.162 to 0.194 at the HEL (11.5 GPa). The elastic-inelastic transition is not distinctive. In the shocked state, the material supports a maximum shear stress increasing from 4.5 GPa at the HEL to 7.0 GPa at twice the HEL. This post-HEL strength evolution resembles neither catastrophic failure due to massive cracking nor classical plasticity response. Confining stress, inherent in plane shock wave compression, plays a dominant role in such a behavior. The observed inelastic deformation is interpreted qualitatively using an inhomogeneous mechanism involving both in-grain micro-plasticity and highly confined micro-fissures. Quantitatively, the data are summarized into an empirical pressure-dependent strength model.

INTRODUCTION

A good understanding of shock induced inelastic deformation of ceramics and its relationship to material microstructure is important for impact applications involving these materials. The challenge is how to probe the strength properties and to identify the inelastic deformation mechanisms of ceramics under well characterized, plane shock wave compression. Extensive studies have been conducted to obtain Hugoniot data and longitudinal wave profiles for various ceramics under shock wave loading (1-5). While longitudinal data are useful, they can neither determine uniquely the material strength in the shocked state nor distinguish between the postulated mechanisms for inelastic deformation. Measurements beyond the usual longitudinal data are necessary to resolve these issues.

To determine the material strength in the shocked state (uniaxial strain), we need to know the mean or lateral stress in addition to the longitudinal stress. Efforts to determine the mean stress response of shocked ceramics include: high-pressure hydrostatic data (6-8), hydrodynamic response calculations (4) and experiments (9), and compression and shear *wave* experiments (10). However, these approaches involve either indirect measurements or assumptions which themselves require independent verification. Meanwhile, the use of lateral piezoresistance gauges to directly measure the lateral stresses in shocked ceramics has also been actively pursued (11-13). Although it has been recently shown that the equilibrium response of a lateral gauge can be uniquely related to the sample lateral stress in particular situations (14), determining such a relationship is not straightforward contrary to some earlier claims (see Ref. 15 for more details).

A unique feature of the work presented in this paper is that we used two independent methods (lateral piezoresistance gauge measurements, and compression and shear wave experiments) to determine the material strength of a ceramic under shock wave loading. The results presented here are for a dense polycrystalline silicon carbide (SiC, Cercom type B material). The lateral gauge data were analyzed carefully using a self-consistent approach (14,15). The results of the compression and shear wave experiments on the same material provided an independent verification. Issues related to the inelastic deformation mechanisms of shocked ceramics are also discussed here.

LATERAL GAUGE EXPERIMENTS

The configuration of the lateral, manganin foil gauge experiments used in this work is illustrated schematically in Fig. 1. A manganin foil gauge (nominally 32 μm thick) was glued between two flat SiC sample slabs with a thin layer of Epon 815 epoxy. The target assembly was then subjected to OFHC copper or tantalum plate impact in the direction shown. The gauge resistance change history and impact velocity were measured. Further experimental details can be seen in Ref. 16.

The results of seven lateral gauge experiments are presented in Fig. 2 along with the longitudinal gauge measurements (5) and VISAR data (17) for the same material. The Hugoniot elastic limit (HEL)

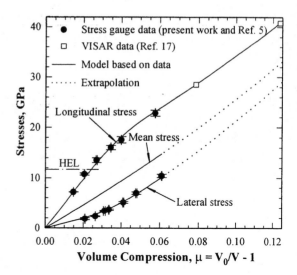

FIGURE 2. Longitudinal, lateral, and mean stresses as a function of volume compression for shocked SiC. The solid and dotted lines represent the pressure-dependent strength model.

of the SiC is about 11.5 GPa (16). The longitudinal and lateral gauge data cover a range of shock stresses from the HEL to twice the HEL, where the evolution of inelastic deformation is expected to be most pronounced. Details of the lateral gauge analysis used can be seen elsewhere (14-16). Briefly, it is a self-consistent approach including a new simplified equilibrium state analysis (15) and rigorous dynamic two-dimensional simulations (16). The combined in-situ longitudinal and lateral stress data fully characterize the stress state in the shocked SiC. It is clear that the material retains significant strength over the stress range examined.

The available longitudinal data (5,17) and the new lateral gauge measurements can be summarized using the following empirical pressure-dependent strength model. The mean stress (σ_m) versus volume compression (μ) relation is

$$\sigma_m = 220.0\mu + 361.3\mu^2 \quad (GPa). \quad (1)$$

Under uniaxial strain, $\sigma_m = (\sigma_x + 2\sigma_y)/3$ where σ_x and σ_y are the longitudinal and lateral stresses, respectively, and $\mu = V_0/V - 1$, where V_0 and V are the initial and final specific volumes, respectively. The compression yield (failure) surface is

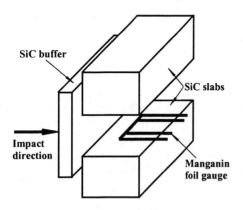

FIGURE 1. Schematic view of lateral, manganin foil gauge experiment showing impact configuration and sample-gauge assembly.

$$Y = 0.639 + 0.8264\sigma_m \quad \text{for } \sigma_m < 5.507, \quad (2a)$$

$$Y = 5.190 + 0.8264\delta - 7.162\times10^{-2}\delta^2$$
$$+ 1.308\times10^{-3}\delta^3 \quad \text{for } 5.507 \leq \sigma_m \leq 13 \ (2b)$$
$$\text{with } \delta = \sigma_m - 5.507,$$

$$\text{and} \quad Y = 7.911 - 2.808\times10^{-2}\Delta - 1.293\times10^{-3}\Delta^2$$
$$- 1.293\times10^{-3} \quad \text{for } \sigma_m > 13 \quad (2c)$$
$$\text{with } \Delta = \sigma_m - 13,$$

where the units of stress are in GPa and $Y \equiv \sqrt{J_2'}$. For uniaxial strain, $\sqrt{J_2'} = (\sigma_x - \sigma_y)/\sqrt{3}$. The results using Eqns. (1) and (2) are plotted in Fig. 2.

COMPRESSION AND SHEAR WAVE EXPERIMENT

To *independently* verify the lateral gauge results, we conducted combined compression and shear wave experiments to determine the mean stress response of the shocked SiC. The experimental configuration is shown schematically in Fig. 3. The parallel inclined plate impact technique (10) was

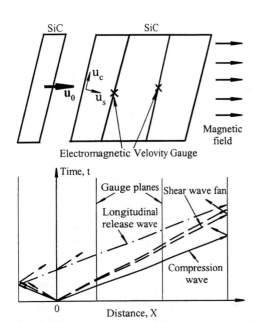

FIGURE 3. Schematic view of compression and shear wave experiment and X-t diagram.

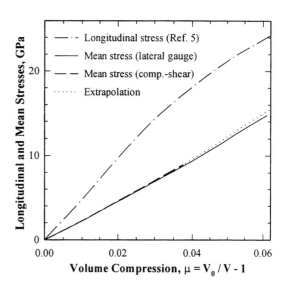

FIGURE 4. Mean stress results from compression and shear wave experiments and piezoresistance gauge experiments.

utilized to impart combined compression and shear loading into the sample. The propagation of the resulting compression, shear and longitudinal release waves of interest is illustrated in the X-t diagram shown also in Fig. 3. The wave propagation velocities were measured using in material, electromagnetic velocity gauges. See Ref. 10 for further technical details.

From the measured propagation velocities of the shear and longitudinal release waves in the sample shocked by the leading compression wave, the shear and longitudinal moduli of the material in the shocked state were determined. Neglecting the anisotropy induced by shock compression, we further determined the bulk modulus and the mean stress response (by numerically integrating the bulk modulus–volume compression relationship) of the shocked SiC. The mean stress results are presented in Fig. 4 as the dashed line in comparison with that of lateral gauge measurements (the solid line). The agreement is very good. This lends confidence to our determination of the material strength in the shocked state since the two results were obtained using completely independent experimental techniques and analytical assumptions.

It was also found that there is a noticeable increase in the longitudinal modulus of the material when shocked elastically to stresses near the HEL.

The shear modulus of the material, on the other hand, remains essentially constant up to a shock stress of 18 GPa (available data). Consequently, the Poisson's ratio of the elastically compressed SiC increases from the ambient value of 0.162 to 0.194 at the HEL. Hence, inferring material strength from the HEL using Poisson's ratio values at ambient conditions may not be appropriate for all materials.

DISCUSSIONS

Post-HEL Strength Evolution

To provide a more detailed examination of the post-HEL strength evolution in the shocked SiC, we present the compression yield (failure) surface in Fig. 5 in terms of the maximum shear stress $[\tau_{max} = (\sigma_x - \sigma_y)/2]$ versus volume compression. Several characteristic features are noteworthy. First, the elastic-inelastic transition at the HEL is nearly indistinctive. Second, the strengthening above the HEL is significant; the maximum shear stress increases from 4.5 GPa at the HEL to 7.0 GPa at a shock stress approximately twice the HEL. Third, the magnitude of the maximum material strength in the shocked state approaches the proximity of theoretical strength [$\tau_{max} = 3.7\%$ of the shear modulus of the material (G), Fig. 5]. Finally, there appears to be a sign of strength reduction when shock stresses approach the highest value examined (24.2 GPa). Though inconclusive, an extrapolation of the available data suggests a gradual softening.

Inelastic Deformation: Cracking vs. Plasticity

As all continuum measurements, the present data cannot provide a definitive determination of the microscopic, inelastic deformation mechanisms. However, given that the post-HEL strength evolution of the SiC has been characterized, some reasonable comments can be made regarding mechanistic issues.

Brittle fracture has been suggested for shocked ceramics (18,19). This mechanism is consistent with the observation of diminishing spall strength in SiC (13,20). However, our data on the shocked SiC show no sign of a catastrophic failure, which may be expected for massive cracking beyond the HEL.

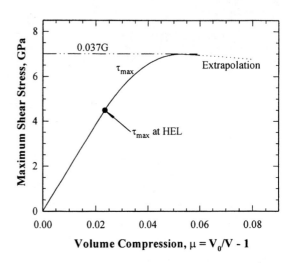

FIGURE 5. Maximum shear stress in the equilibrium state as a function of volume compression for SiC. The shear modulus of the material, G is about 190 GPa.

Hence, if brittle fracture does occur, it is unlikely to proceed via the propagation of pre-existing micro-cracks. Instead, it will necessarily be in the form of homogeneous nucleations of micro-fissures without catastrophic macroscopic failure.

Micro-plasticity has been observed in various shock-recovered ceramics (21,22). This observation coupled with the use of a strain-hardening model to analyze the longitudinal response of SiC (4) raises the possibility of a continuum plasticity mechanism for the measured post-HEL strength. However, the experimental observations discussed in the previous subsection argue against such a mechanism. In addition to the indistinctive elastic-inelastic transition at the HEL and possible softening at stresses beyond twice the HEL, the measured "hardening" from the HEL to twice the HEL (Fig. 5) is too large for the small amount of inelastic deformation (~1.5%) in a single phase material. A strain-hardening plasticity mechanism is also inconsistent with the observation of diminishing spall strength (13,20).

It is clear that the observed strength evolution resembles neither catastrophic failure due to massive crack propagation nor classical plasticity response. We believe that the deformation process is controlled predominantly by the inertial confinement inherently presented in shock wave, uniaxial strain compression. The critical resolved shear stress (CRSS) in a ceramic can vary significantly from loading along one crystal orientation to another as shown in the quasi-static and shock response of sapphire (23,24). If the confinement results in a maximum intergranular friction greater than the lowest CRSS in SiC, it is favorable for some grains to undergo micro-plasticity. As the shock stress increases, micro-plasticity may occur in increasing number of grains and orientations. However, as long as a sufficient amount of grains with higher-strength orientations remain elastic, the macroscopic stress deviators will increase, though at a decreasing rate, with deformation. In this hypothesis, plasticity does not cause strengthening in any individual grain. The macroscopic strength evolution above the HEL is a progressive process to activate various levels of micro-plasticity (in terms of CRSS) permitted by the maximum intergranular friction, which increases with the confinement. This can be viewed as *pressure-dependent strength*. This hypothesis could explain the material response just above the HEL. The macroscopic hardening stops when micro-plasticity sites reach a population for coalescence to occur and result in a global plastic flow. The resulting inelastic deformation can be strongly inhomogeneous. In a highly confined crystal, even micro-plasticity may proceed inhomogeneously through a combination of slipping and twining (25). The HEL might be viewed as a state where micro-plasticity is macroscopically discernible.

This is not to say that micro-plasticity is the only possibility in shocked high-strength ceramics though it was strongly suggested in Ref. 21. Micro-fissures may occur in shocked SiC at places where the material can no longer withstand the microscopic and/or mesoscopic incompatibilities induced by the inhomogeneous deformation. As long as these fine cracks do not propagate to result in catastrophic macroscopic failure, the overall deviatoric stress can still increase with confinement. If the maximum intergranular friction represents the upper bound for the local flow stress as assumed here, even a ceramic completely damaged through percolation of micro-fissures may retain significant strength in the shocked state. It is also very likely that the maximum intergranular friction in shocked SiC can not surpass the CRSS for pyramidal slip, which is necessary for developing global plastic

flow in ceramics with hexagonal or trigonal structures (25,26). In this situation, the combination of micro-plasticity and micro-fissures is a more favorable process for overall flow than micro-plasticity alone. No substantial pyramidal slip has been identified in recovered 6H SiC (22). Hence, an inhomogeneous deformation mechanism involving both in-grain micro-plasticity and highly confined micro-fissures seems to be a more reasonable interpretation for the observed post-HEL strength evolution in the shocked SiC. This mechanism is also consistent with the observation of diminishing spall strength (13,20)

Finally, we emphasize that the hypothesis proposed for inelastic deformation in shocked SiC and the corresponding empirical pressure-dependent strength model though reasonable for uniaxial strain compression (because of the large confining stress) may not be valid for other types of loading conditions. If the ratio of compression to distortion is significantly reduced from that present in uniaxial strain compression, the inelastic deformation of SiC may proceed via a completely different mechanism. A major scientific challenge is how to determine, for arbitrary loading, the conditions for transitions between various deformation mechanisms and to model these mechanisms in a consistent manner. Furthermore, we emphasize that the material strength of a ceramic at the HEL may be irrelevant to that under uniaxial-stress compression, where the material most likely fails through localized crack propagation. It is not meaningful to link the two strengths (19) unless the same failure mechanism is operative for both loading conditions.

CONCLUDING REMARKS

To develop insight into inelastic deformation in shocked ceramics, we determined the material strength of a dense polycrystalline silicon carbide (Cercom type B material) in the shocked state using two independent experimental methods: lateral gauge measurements, and combined compression and shear wave experiments. The shock stresses in the sample ranged from 3 to 24 GPa; the highest value is approximately twice the HEL (11.5 GPa). The two sets of results were in good agreement.

While it was found that the equilibrium response of a lateral gauge can be uniquely (within a reasonable accuracy) related to the sample lateral stress (14), simplified approaches (11-13 and related references) to obtain lateral stresses from lateral gauge data need to be viewed with caution (15). A new approach combining a numerically calculated calibration (15) and rigorous dynamic two-dimensional simulations was used to analyze the lateral gauge data in a self-consistent manner (16). The analysis for the combined compression and shear wave experiments assumes that the anisotropy induced by shock compression is negligibly small.

Our results show that the Poisson's ratio of the SiC increases with elastic shock compression from the ambient value of 0.162 to 0.194 at the HEL. Unlike the typical response of a shocked metal, the elastic-inelastic transition in the shocked SiC is nearly indistinctive. Above the HEL, the material has an extremely high strength; the maximum shear stress increases from 4.5 GPa at the HEL to 7.0 GPa at a stress approximately twice the HEL. The latter value is 3.7% of the shear modulus of the material (~190 GPa). At shock stresses beyond twice the HEL, an extrapolation of our limited data suggests a gradual softening. A definitive answer, however, will require measurements at higher stresses.

The post-HEL strength evolution in the material resembles neither catastrophic failure due to massive crack propagation nor classical plasticity response. Qualitatively, this response is interpreted using a confinement-dependent inhomogeneous deformation mechanism involving both in-grain micro-plasticity and percolation of highly confined micro-fissures. Quantitatively, the available longitudinal and lateral data on the shocked SiC are summarized into an empirical pressure-dependent strength model. For uniaxial strain loading, our interpretation and model presented here are consistent with the available experimental observations. However, for a different type of loading condition (*e.g.*, uniaxial stress loading), inelastic deformation may proceed via a different mechanism and it is not clear that results from experiments that exercise different loading path can be linked in a consistent manner.

ACKNOWLEDGMENT

Dr. D. P. Dandekar is sincerely thanked for providing the samples, sharing his results, and for many valuable discussions. Dr. G. F. Raiser participated in the lateral gauge experiments. D. Savage and K. Zimmerman are thanked for their assistance in the impact experiments. This work was supported by the Army Research Office.

REFERENCES

1. McQueen, R. G, Marsh, S. P., Taylor, J. W., Fritz, J. N., and Cater, W. J., *High Velocity Impact Phenomena*, edited by Kinslow, R., New York: Academic Press, 1970, pp. 293-417.
2. Gust, W. H., Holt, A. C., and Royce, E. B., *J. Appl. Phys.* **44**, 550-561 (1973).
3. Munson, D. E. and Lawrence, R. J., *J. Appl. Phys.* **50**, 6272-6282 (1979).
4. Kipp, M. E., and Grady, D. E., Sandia Report No. SAN89-1461, Sandia National Laboratories, Albuquerque, NM, 1989.
5. Feng, R., Raiser, G. F. and Gupta, Y. M., J. Appl. Phys. **79**, 1378-1387 (1996).
6. Basset, W. A., Weathers, M. S., Wu, T.-C., and Holmquist, T., *J. Appl. Phys.* **74**, 3824-3826 (1993).
7. Dandekar, D. P. and Benfani, D. C., *J. Appl. Phys.* **73**, 673-679 (1993).
8. Dandekar, D. P., Abbate, A., and Frankel J., *J. Appl. Phys.* **76**, 4077-4085 (1994).
9. Grady, D. E., *J. Appl. Phys.* **75**, 197-202 (1994).
10. Gupta, Y. M., *J. Geophys. Res.* **88**, 4304-4312 (1983).
11. Rosenberg, Z., Yaziv, D., Yeshurun, Y., and Bless, S. J., *J. Appl. Phys.* **62**, 1120-1122 (1987).
12. Rosenberg, Z., Brar, N. S., and Bless, S. J., *J. Appl. Phys.* **70**, 167-171 (1991).
13. Bourne, N., Millett, J., and Pickup, I., *J. Appl. Phys.* **81**, 6019-6023 (1997).
14. Feng, R., Gupta, Y. M., and Wong, M. K. W., *J. Appl. Phys.* **82**, 2845-2854 (1997).
15. Feng, R. and Gupta, Y. M., submitted to *J. Appl. Phys.*
16. Feng, R., Raiser, G. F., and Gupta, Y. M., to appear *in J. Appl. Phys.*, 1997.
17. Crawford, D. A. (private communication), 1994.
18. Addessio, F. L. and Johnson, J. N., *J. Appl. Phys.* **67**, 3275-3286 (1990).
19. Rosenberg, Z., *J. Appl. Phys.* **76**, 1543-1546 (1994).
20. Bartkowski, P. and Dandekar, D. P., in *Shock Compression of Condensed Matter-1995*, edited by Schmidt, S. C. and Tao, W. C., New York: AIP Press, 1996, pp. 535-538.
21. Longy, F. and Cagnoux, J., *J. Am. Ceram. Soc.* **72**, 971-979 (1989).
22. Merala, T. B., Chan, H. W., Howitt, D. G., Kelsey, P. V., Korth, G. E., and Williamson, R. L., *Mater. Sci. Eng.* **A105/106**, 293-298 (1988).
23. Snow, J. D. and Heuer, A. H., *J. Am. Ceram. Soc.* **56**, 153-157 (1973).
24. Graham, R. A. and Brooks, W. P., *J. Phys. Chem. Solids*, **32**, 2311-2330 (1971).
25. Castaing, J., Cadoz, J., and Kirby, S. H., *J. Am. Ceram. Soc.* **64**, 504-511 (1981).
26. Hirth, J. P. and Lothe, J., *Theory of Dislocations*, New York: McGraw-Hill, 1968.

CP429, *Shock Compression of Condensed Matter – 1997*
edited by Schmidt/Dandekar/Forbes
© 1998 The American Institute of Physics 1-56396-738-3/98/$15.00

INVESTIGATION OF MECHANICAL PROPERTIES OF CERAMICS USING AXI-SYMMETRIC SHOCK WAVES

G.I. Kanel[*], S.V. Razorenov[**], A.V. Utkin[**], S.N. Dudin[**], V.B. Mintsev[**], S. Bless[+], and C.H.M. Simha[+]

[*]*High Energy Density Research Center of Russian Academy of Sciences;*
[**]*Institute of Chemical Physics in Chernogolovka of Russian Academy of Sciences*
[+]*Institute for Advanced technology, the University of Texas at Austin*

To extend the capabilities of shock-wave experiments to larger deformations, a technique of measurements at cylindrical shock loading of ceramic tube samples has been developed. In experiments, the shock loading of AD998 tubes by cylindrical detonation initiated with electrical explosion of a wire was realized. VISAR measurements of the velocity profiles have been carried out with water windows. Lagrangian 1-D computer code for simulations of the shock-wave processes with axial symmetry has been used for interpretation of experimental data. A phenomenological model of the dynamic response of brittle materials has been developed. The strain range available for analysis has been extended by a factor of 2 or 3 due to the divergent character of flow.

INTRODUCTION

Experiments with plane shock waves provide an ability to study mechanical properties of materials under high pressure and extreme high strain rate. However, the typical strain values are small in the plane shock waves and increase in the strain is usually connected with a inevitable pressure growth. It would be desirable to expand the possibilities of plane-wave experiments to larger deformations. New prospects in this regard can be reached by arrangement of the impact loading with cylindrical or spherical symmetry [1]. An additional transverse deformation will increase the total strain in this case. Behavior of alumina submitted to a divergent spherical stress wave was studied in the ref. [2]. The objective of this work is development of the experimental technique to study behavior of brittle materials at loading by divergent cylindrical shock waves.

EXPERIMENTAL

The experimental technique was based on generation of cylindrical divergent shock waves in the tube-like specimens. The axi-symmetric shock load pulses were created by detonation of an explosive charge placed inside of the specimen (Fig.1). The high explosive was RDX of 1.25 g/cm^3 density. Necessary density and uniformity of the powder-like explosive charge was reached by means of vibration of the whole assembly. The cylindrical detonation wave was initiated by an exploding wire placed along the assembly axes. For this purpose, a standard pulse high-voltage facility was used. As a result of discharge of a 0.1 μF capacitor bank, this facility generates an electrical pulse of 60 kV voltage and ~1 μsec duration. This electrical pulse was applied to the exploding copper wire 100 to 120 mm long and 70 μm in diameter. The explosive driver produces an adequately symmetrical detonation wave over a length of 80 mm at a radial detonation expansion of 15 mm.

As a main diagnostics method, we used a VISAR [3] with the velocity-per-fringe constants of 305 or 80.8 m/sec. We measured the radial velocity history of the interface between the tube sample and the water window. For measurements, aluminum foil 7 μm thick was glued on the sample surface as a reflector for the VISAR laser beam. A rest surface of the tube sample was screened by a black paper in order to shield the VISAR from the detonation luminosity.

The samples for tests were cast AD998 alumina tubes of the Coors Ceramics Company. The tubes were of 36.9±0.3 mm inside diameter, 43.15±0.2 mm outside diameter, and 100 mm long. With a goal to verify the loading conditions, similar measurements were done also with PMMA tubes of 37.0 mm inside diameter and 43 mm outside diameter.

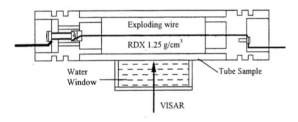

FIGURE 1. Scheme of experiments. The cylindrical divergent shock wave in the tube samples is created by detonation of a RDX charge initiated by the electrical explosion of wire on the charge axis.

The velocity histories of interfaces between the tube samples and the water window are presented in Fig. 2. In the experiment with ceramic tube, the recorded velocity oscillations are the result of multiple wave reverberations inside the alumina wall between the water window and the detonation products. Since the amplitude of oscillations is less than the Hugoniot elastic limit, it is natural to expect that the oscillations period, Δt, should be determined by the longitudinal sound velocity c_l: $\Delta t = 2h/c_l$, but the measured period clearly exceeds this value and rather corresponds to the bulk sound velocity. In the case of PMMA, the Hugoniot of the sample material lies between the detonation products isentrope and

the Hugoniot of water in the pressure-particle velocity plane. Due to that, the wave reverberations inside the PMMA tube wall almost are not visible in the velocity profile.

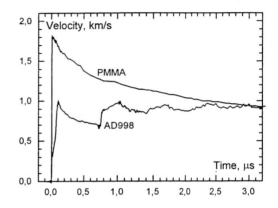

FIGURE 2. The particle velocity profiles measured for the PMMA and ceramic tubes.

Several techniques were tried to launch the tube-like liners. The main problem was insulation of the high-voltage electrical discharge when we used the metal liner. We tested the dielectric liners also, but they are failed during the launching. Figure 3 shows a scheme of a final version of the launching facility. The inside surface of stainless steel liner is additionally insulated by a PMMA tube of 1 mm wall thick. To prevent break-down between the electrical leads and liner, oversize teflon plugs were used. This proved ultimately successful, and we plan to conduct future tests using this improved configuration.

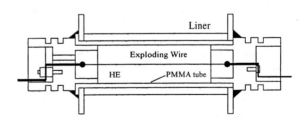

FIGURE 3. Scheme of the explosive launching facility with insulated metal liner.

SIMULATIONS

A simple constitutive model of the brittle material was developed for preliminary analysis of the observed response of ceramic. The model represents each elementary volume of the material as a sum of two parallel elements. One of the elements works as an elastic-plastic body with strain hardening. The second element describes the resistance to deformation of the comminuted component. Both elements have the same elastic modules and their total deformation equals the deformation of elementary volume while the deviatoric stresses in the intact and comminuted components, σ_i and σ_c respectively, are different. The deviatoric stress, S, in the whole elementary volume is determined according to the relationship

$$S = (1-g_c) \cdot S_i + g_c \cdot S_c,$$

where g_c is the comminuted component fraction, $0 < g_c < 1$. The initial fraction of comminuted component is zero. Above the elastic limit, the brittle material is cracking and g_c is growing as a result of the plastic deformation of intact component when the pressure, p, is below the threshold pressure, p_c, according to the relationship

$$\frac{\partial g_c}{\partial \gamma_i} = k_g (p_c - p)(1 - g_c), \ p < p_c,$$

where γ_i is the inelastic strain of the intact component, k_g is a constant parameter. The comminution does not occur when the pressure exceeds the threshold magnitude, p_c. It was supposed the whole material transforms to the comminuted phase when a tensile fracture occurs in the elementary volume considered.

For the high-strain-rate conditions, it is necessary to account for viscosity. Measurements of the shock front rise time [4] show the shear plastic strain rate in the range $\partial \gamma / \partial t > 10^4 \ \text{s}^{-1}$ is related to the shear stress as

$$\frac{\partial \gamma}{\partial t} = \frac{(\tau - Y/2)^2}{A'},$$

where A' is constant. In order to provide such general behavior in the frame of considered model, the plastic strain rates of components were determined as a function of their quasistatic flow stresses [5]. The quasistatic yield stresses of intact and comminuted components, Y_i and Y_c, were described by a strain hardening and friction lows respectively:

$$Y_i = Y_{i0}(1 + k_Y \gamma_i^{0.5}), \ Y_c = f \cdot p \leq k_c Y_i,$$

where γ_i, γ_c are the plastic strains of intact and comminuted components, Y_{i0}, k_Y, f, k_c are constant parameters. The shear modulus dependence on the pressure was calculated assuming a constant Poisons ratio. The tensile strength of the matter was supposed to be negligible. In the ref. [6] we provided a discussion of the motivations for use of a strain-dependent failure criteria.

The constitutive model has been incorporated into a 1-D Lagrangian code with a simple equation of state of the Mie-Gruneisen type. The properties of AD999: the density 3.948g/cm^3, the longitudinal sound velocity 10.85 km/sec, and the Hugoniot in form of $U_s = 7.97 + 1.27 u_p$ were used in the simulations. For the explosion products, a polytropic equation of state $pV^n = const = p_{C-J} V_{CJ}^{\ n}$ was used, where p_{C-J}, V_{CJ} are the pressure and specific volume in the Chapman-Jouguet plane, the exponent $n = U_{det}/u_{CJ} - 1$. The initial distributions of pressure, particle velocity and specific volume in the detonation products roughly corresponded to an approximate solution [7] for the divergent cylindrical detonation wave. According to [7], it was assumed that a central part of the gas cylinder of about half of its total radius is under constant pressure, p_{r0}, and has a zero particle velocity. In the outer part of detonation products, there is a smooth pressure and particle velocity growth to the Chapman-Jouguet state. The initial particle velocity $u_p(r)$ was related to the adjusted pressure distribution, $p(r)$, through the sound velocity:

$$u_p = \frac{2c(p) - U_{det}}{n - 1}.$$

where the sound speed $c = \sqrt{npV}$.

In the Fig. 4, the recorded velocity profiles are compared with the results of computer simulations. In a first step, the experiment with PMMA was simulated to make a reasonable estimation of the initial pressure and particle velocity distributions in the detonation products. Since we used an approximation which did not account for the shock-to-detonation transition range and chemical spike of the stead detonation wave, the solution does not reproduce all features of the initial tube acceleration. Nevertheless, there is quite reasonable agreement for the shot with PMMA. In the next step, a series of simulations was performed for the ceramic tube sample using the same initial state distribution in detonation products, which has been estimated in the first step.

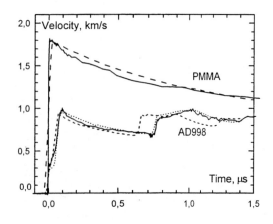

Figure 4. Comparison of the measured and calculated velocity profiles of shock waves created in the PMMA and AD998 tube samples by cylindrical detonation. Measurements with water windows. Solid lines are the experimental data.

Two version of the AD998 alumina response are presented in the Fig. 4. The short-dashed line is a result of simulation with the elastic-plastic model without comminution. The dotted line presents results of simulations using the model of brittle material described above with parameters: p_c=15GPa, k_g=100, Y_{i0}=4GPa, k_Y=5, f=0.1, k_c=0.3. The comparison of recorded and calculated data demonstrates that the simple model of brittle mater provides quite good agreement with experimental data. Moreover, the cylindrical test geometry

provides a good diagnostic tool for the behavior of comminuted material under additional loading.

CONCLUSION

A technique for experiments with cylindrical shock loading of ceramic tube samples has been developed. In experiments, the shock loading of AD998 tubes was realized. The simple model of brittle materials based on the assumption of the comminuted phase formation as a result of plastic deformation of the intact matter provides quite good agreement with the experimental data.

ACKNOWLEDGMENT

This work was supported in part by the Institute for Advanced Technology, under its contract DAAA21-93-C-0101.

REFERENCES.

1. Bauer D P; Bless S J *Strain Rate Effects on Ultimate Strain of Copper.* Scient. tech. Aerospace Rep. 18, (2), 166-67 (1980)
2. Tranchet, J-Y. and Collombet, F. *in: Metallurgical Applications of Shock-Wave and High-Strain-Rate Phenomena*, edited by L.E. Murr, K.P. Staudhammer and M.A. Meyers, Els. Science B.V., pp. 535-542 (1995).
3. Asay, J. R. and Barker, L. M. , *J. Appl. Phys.*, **45**, 2540 (1974).
4. Swegle, J.W and Grady, D.E., *J. Appl. Phys.*, **58**(2), p.692 (1985).
5. Kanel, G.I., *Problems of Strength (USSR)*, 1988, No 9, p.55.
6. Bless, S.J., Satapathy, S., and Simha, H-C, *In: Proc. SUSI96 Conference*, Udine, Italy, July 1996
7. Stanyukovitch, K.P. *Unsteady motions of continuous media (in Russian).* Nauka, Moscow, 1971

CP429, *Shock Compression of Condensed Matter – 1997*
edited by Schmidt/Dandekar/Forbes
© 1998 The American Institute of Physics 1-56396-738-3/98/$15.00

SURFACE FRACTURE ZONES IN SHOCK-LOADED POLYCRYSTALLINE CERAMICS

N.K. BOURNE, Z. ROSENBERG*, J.E. FIELD

Shock Physics, PCS, Cavendish Laboratory, Madingley Road, Cambridge, CB3 0HE, UK.
*RAFAEL, PO Box 2250, Haifa, Israel.

It is now accepted that a range of glasses may fail in uniaxial strain under compression by the propagation of a fracture surface known as a failure wave. It has been further noted that polycrystalline ceramics such as alumina and silicon carbide also may fail by fracture processes although it has been unclear as to how these effects influence the observed stress or particle velocity profiles measured at some distance from the impact face. We have recently demonstrated that polycrystalline materials can show failure near the impact surface which is of the same form as that recorded for the failure wave in glass. The process is, however, localised at the impact face instead of propagating, as is the case for glasses, through the bulk material. In this paper we present further experimental observations to support these interpretations

INTRODUCTION

In recent years there has been some controversy concerning the yielding of polycrystalline ceramics. This debate has concerned the existence of spall strength (above and below the HEL) and observations of precursor decay. Several workers have shown that ceramics such as alumina and boron carbide exhibit a decay in elastic precursor amplitude with thickness. On the other hand others have stated that the effect does not occur (1, 2), although Grady defines his HEL at a different point in the loading history. The effect can also manifest itself as an increase in amplitude of the elastic precursor with increasing driving stress (3-5).

There has also been recent discussion of the phenomenon of delayed failure behind the elastic wave in glass, across a front which has been called a fracture or more lately a failure wave (6). Further work (7) confirmed the existence of these waves by measuring spall and shear strengths ahead of and behind the failure wave, using manganin stress gauges.

In seeking to pursue the failure wave in materials other than glasses, we have conducted a matrix of plate impact experiments on several brittle materials including glasses, aluminas and silicon carbides using a variety of techniques. Some of these results are presented below to contrast the brittle failure behaviour of a range of materials.

We have shown that a wave of fracture passes through glasses at a constant velocity (8). We show below that a similar phenomenon occurs in polycrystalline materials but that the fractured region is localised to a zone near the impact face (9). We believe that this surface fracture causes the break in slope that has been taken as the HEL by many workers. The actual yield point may be at a higher stress as discussed by Grady (10).

EXPERIMENTAL

Plate impact experiments were carried out on the 50 mm bore gun at the University of Cambridge (11). Stress profiles were measured with commercial manganin stress gauges both embedded, or placed on the rear face of the specimens and supported with thick polymethylmethacrylate (PMMA) blocks. These gauges (Micromeasurements type LM-SS-125CH-048) were calibrated by Rosenberg et al. (12). Impact velocity was measured to an accuracy of 0.5% using a sequential pin-shorting method and tilt was fixed to be less than 1 mrad by means of an adjustable specimen mount. Impactor plates were made from lapped tungsten alloy, copper and aluminium discs and were mounted onto a polycarbonate sabot with a relieved front surface in order that the rear of the flyer plate remained unconfined. Targets were flat to within 5 fringes across the surface. Lateral stresses were also measured using manganin stress gauges, this time of type J2M-SS-580SF-025 (resistance 25 Ω). The gauges had an active width of 240 μm and were placed at varying distances from the impact face.

The lateral stress, σ_y was used along with measurements of the longitudinal stress, σ_x to calculate the shear strength τ of the material using the well-known relation

$$\tau = \frac{1}{2}\left(\sigma_x - \sigma_y\right). \qquad [1]$$

This quantity has been shown to be a good indicator of the ballistic performance of the material (13). Our method of determining the shear strength has the advantage over previous calculations of being direct since no computation of the hydrostat is required. In some experiments a VALYN VISAR was used to measured the particle velocity through the rear PMMA window. Relevant materials data are presented in Table 1.

TABLE 1. Material properties for targets cited.

	Al$_2$O$_3$ 97.5%	Al$_2$O$_3$ 99.9%	SiC
ρ (± 0.05 g cm^{-3})	3.80	3.99	3.16
c_L (± 0.01 mm μs^{-1})	10.30	10.82	11.94
c_S (± 0.01 mm μs^{-1})	6.07	6.39	7.57

RESULTS

Figure 1 shows the result of an impact between a copper flyer and a 999 alumina above the threshold normally taken as the HEL. The traces have a simultaneous manganin gauge signal and VISAR velocity history superimposed (for details see ref. 20). In the centre is superposed the output of a photomultiplier unit which sees the light emitted during the experiment. The light is generated at

impact and for a period of 200 ns afterwards and is so intense that it appears on the VISAR trace (dotted) even though a notch filter at the wavelength of the laser is present. Such impact flashes are common in gun systems where there is significant gas in the target chamber at firing. However, in this case the vacuum maintained is such that there is no discernible signal on the photomultiplier (PMT) when metal impacts metal or metal impacts glass. The size of the flash here is indicative of a very high-intensity emission process occurring at the impact face. The shock wave arrives at the rear interface after 400 ns which is exactly as expected for a longitudinal elastic wave. We believe that this emission is associated with fractoemission from the ceramic.

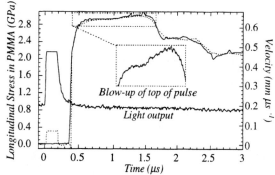

FIGURE 1. VISAR (dotted), gauge (solid) mounted in backsurface configuration. Impact of a copper flyer of thickness 3 mm on a 4 mm 999 alumina target at 778 m s^{-1}. Central trace is light recorded by fast PMT.

The second feature of note on the trace is the reloading signal seen on the top of the pulse, 1 μs after impact which is magnified in the box within the figure. This pulse represents the reflected release from the PMMA window reflecting off a lower impedance boundary within the target. This boundary is believed to separate fractured from intact ceramic.

Figure 2 presents the results of three lateral gauge experiments in the alumina 975 which has a higher glass content and porosity than the 999. The tiles were 25 mm thick and were sectioned to allow gauges to be introduced at 2 mm from the impact face. All were then impacted with 10 mm thick copper flyers travelling at 480 m s^{-1} which induced a stress of 9 GPa in the target. This is incidentally, around the value of the HEL in the material. In a series of experiments the lateral stress histories at 2, 4, 5, 6, and 8 mm from the impact face were recorded. The traces at 2 and 4 mm show an initial plateau followed by a rise to a second higher value. This rise in lateral stress represents a drop in strength and is precisely as noted for the failure wave in glasses. At 5 mm and beyond no failure wave was seen to arrive for the duration of the experiment. These observations indicate that contrary to the glasses, the failure swept into the target from the interface in alumina does not reach beyond the first 5 mm of the target. This distance is approximately that over which the majority of the precursor decay is observed to occur (4) indicating that the precursor decay is governed by the delayed failure swept in from the surface. Interestingly, the initial value of 2τ (calculated from equation 1) which is *ca.* 6 GPa reduces to *ca.* 4 GPa in the surface zone but remains at its initial value in the bulk. The dip in the lateral stress seen before the arrival of the failure front in the 4 mm trace is reproduced in many others. It may indicate a relaxation in the amplitude of the first wave due to the fracture process.

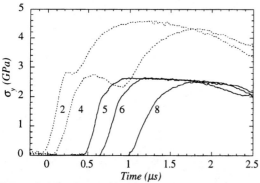

FIGURE 2. Lateral stress histories recorded at various distances from the impact face after impact with copper flyers of thickness 10 mm. Longitudinal stress is 9 GPa.

Finally, we show equivalent experiments for a silicon carbide material further details of which appear elsewhere (14). The measured HEL was 13.5±0.3 GPa as reported in that paper. Experiments were carried out to measure the lateral stress histories at 2 mm from the impact face at five different stress levels; *ca.* 9, 14, 16, 19, and 21 GPa. An additional experiment was carried out in which the gauge was placed 4 mm from the impact face in order to assess the size of the failed surface zone.

Again the impactor was a 10 mm thick copper flyer plate to ensure that longitudinal release did not enter the gauge region until after the compressive failure processes had occurred. The histories are presented in Fig. 3. Similar behaviour was observed in this material to that seen in the aluminas. At the lower stresses, delayed failure was observed whilst at the highest it occurred within the shock front itself. The dotted, 4 mm trace shows that the delay time increases to nearly 1 μs from its previous value at 2 mm of 400 ns. This means that the failure wave is slowing rapidly. It thus appears that, as in the case of aluminas, the failed zone does not propagate more than a few mm into the target. Note again that the lowest stress is below the HEL and shows similar behaviour to histories recorded at much higher stresses.

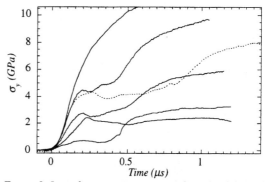

FIGURE 3. Lateral stress measurements at 2 mm in silicon carbide shot with a 10 mm copper flyer plate inducing longitudinal stresses of 9, 14, 16, 19 and 21 GPa. The dotted trace is at 19 GPa but the gauge was placed at 4 mm from the impact face.

DISCUSSION AND CONCLUSIONS

We have presented independent evidence of the formation of a surface zone of fractured material at the impact face in three polycrystalline materials. The light emission processes correlate with the loss of shear strength seen in the surface regions of these aluminas. Further, the VISAR and gauge at the rear surface independently measure the reloading signal seen as the release from the PMMA interface returns to the failed zone, changes phase and comes back as a compression. Such signals have been observed by many workers previously in glasses to indicate the presence of the failure front.

More compelling are the lateral gauge results which show the drop in strength in the surface region as a function of distance. We have investigated the thickness of the failed zone in alumina and SiC at around the accepted HEL in the former case and at 50% above the HEL in the latter. In the alumina case the surface zone was around 5 mm deep at that stress. In the latter case it was found that it was at least 4 mm. Our experiments on SiC suggest that the size of the zone is also stress dependent.

It is seen that the formation of the failed region occurs at stresses below the threshold usually taken as the HEL. At higher stress (*ca.* twice this value) the failure occurs within the shock front. These observations are reminiscent of the results for glasses. In order to illustrate this we take typical results for soda-lime glass. In this material the lower threshold for the appearance of the failure wave is 4 GPa, the HEL has been measured to be 6 GPa and by 9 GPa the failure front is travelling with the shock. In all respects polycrystalline materials show similar behaviour. The only difference is that this behaviour is confined to a surface region and the propagation of the failure front is impeded whereas in the case of glass it can cross the bulk at constant speed until arrested by a longitudinal release returning from a boundary.

It is worth contrasting the behaviours of glasses and polycrystalline materials to seek an explanation for the localisation of the zone. It has been noted in glasses that the presence of an inner interface in a material impedes the passage of a failure front and high-speed photography has suggested that this may be due to a filtering process in which the smaller fractures are unable to cross the interface (9). The polycrystalline material, containing many internal boundaries at a microstructural scale, may show similar behaviour resulting in slowing and eventually arrest of propagating fractures.

Finally, we discuss the measurement of the HEL in polycrystalline materials. It is common to measure a pulse that rises sharply to a cusp and then has a convex rise to a plateau at lower stress (like Fig. 1 for example) and that has an 'S' shaped upper section at higher stresses. It has been conventional to take the limit of elastic behaviour to be the top of the elastic section, and this point has been shown to decay with distance in many cases. Grady on the other hand infers from the lack of hysteresis above

this lower point, that there is an upper threshold which corresponds to yield and should be taken as the HEL (10). This second yield is assumed to be rate-independent. Our work has shown that there are fracture processes occurring at the front face during plate impact. Such processes will be kinetically determined and stress dependent. Such local failure will further act to relieve stresses at the shock front by the transmission of release waves from the rear which travel supersonically with respect to the front. We thus suggest that the threshold that many have been using as the HEL may be determined by the surface fracture we have demonstrated above.

Future investigations will identify the geometrical constraints and the effect of stress upon surface fracture zones for a range of materials.

ACKNOWLEDGEMENTS

NKB acknowledges funding from EPSRC, DERA and Pilkingtons PLC. We gratefully acknowledge results from Drs J.C.F. Millett, N.H. Murray and W.G. Proud used in constructing these analyses. We thank Mr D.L.A. Cross and Mr R.P. Flaxman for technical support.

REFERENCES

1. Grady, D.E., Sandia Natnl. Labs, TMDG0694 (1994).
2. Cagnoux, J. and Longy, F., in *Shock Waves in Condensed Matter 1987,* (North Holland, Amsterdam, 1988), pp. 293-296.
3. Murray, N.H., PhD Thesis, *Univ. of Cambridge* 1997,
4. Murray, N.H., Bourne, N.K. and Rosenberg, Z., in *Shock Compression of Condensed Matter 1995,* (AIP, Woodbury, New York, 1996), pp. 491-494.
5. Bourne, N.K., Rosenberg, Z., Field, J.E. and Crouch, I.G., J. Phys. IV France Colloq. C8, **4**, 269-274 (1994).
6. Rasorenov, S.V., Kanel, G.I., Fortov, V.E. and Abasehov, M.M., High Press. Res., **6**, 225-232 (1991).
7. Brar, N.S., Bless, S.J. and Rosenberg, Z., Appl. Phys. Letts, **59**, 3396-3398 (1991).
8. Bourne, N.K., Rosenberg, Z. and Field, J.E., J. Appl. Phys., **78**, 3736-3739 (1995).
9. Bourne, N.K., Millett, J.C.F., Rosenberg, Z. and Murray, N.H., J. Mech. Phys. Solids, in press (1997).
10. Grady, D.E., in *Shock Compression of Condensed Matter 1995,* (AIP, Woodbury, New York, 1996), pp. 9-20.
11. Bourne, N.K., Rosenberg, Z., Johnson, D.J., Field, J.E., Timbs, A.E. and Flaxman, R.P., Meas. Sci. Technol., **6**, 1462-1470 (1995).
12. Rosenberg, Z., Yaziv, D. and Partom, Y., J. Appl. Phys., **51**, 3702-3705 (1980).
13. Meyer, L.W., Behler, F.J., Frank, K. and Magness, L.S., *Proc. 12th Int. Symp. Ballistics,* 1990), pp. 419-428.
14. Bourne, N.K., Millett, J.C.F. and Pickup, I., J. Appl. Phys., **81**, 6019-6023 (1997).

CP429, *Shock Compression of Condensed Matter – 1997*
edited by Schmidt/Dandekar/Forbes
© 1998 The American Institute of Physics 1-56396-738-3/98/$15.00

EXPERIMENTAL STUDY OF INTERFACE DEFEAT IN CONFINED CERAMIC TARGETS

N.S. Brar[‡], Horacio D. Espinosa[†], G. Yuan[†], and P.D. Zavattieri[†]

[†] *School of Aeronautics and Astronautics, Purdue University, West Lafayette, IN 47907*
[‡] *Impact Physics Laboratory, University of Dayton Research Institute, Dayton, OH 45469-0182*

Recent experimental studies by Hauver et al. reveal that the ballistic performance of ceramic targets depends entirely on how the ceramic is confined in a composite target. If the ceramic confinement is preserved, the penetrator is consumed by lateral flow at the ceramic-cover plate interface; this mechanism is known as interface defeat. A number of variables are important in achieving optimum ballistic performance. The most relevant are: shock attenuation through the use of an attenuator plate, ceramic-cover plate interface, ceramic confinement pre-stress, ceramic-back surface interface properties, and shear localization sensitivity of the cover plate and penetrator materials. In this work several diagnostic tools are used to gain insight of the ballistic performance of ceramic targets. Stress histories produced at the cover plate-ceramic interface and ceramic-back plate interface are recorded with in-material gauges. Velocity measurements, at the back plate free surface, are recorded with velocity interferometry.

INTRODUCTION

There have been a number of investigations to utilize the high compressive strength and low density properties of ceramics to make light weight armor. On the other hand, ceramics fracture upon impact because of their low tensile and spall strength. Thus, in order to design an effective ceramic armor, an optimum confinement is needed to avoid ceramic flow during the penetration process. Bless et al. investigated the effect of confinement on the ballistic performance of 3-inch square 1-inch thick TiB_2 tiles against tungsten heavy alloy (WHA) projectiles shot at 1.5 km/s (1). In these shots the ceramic plate was inserted in a well of the same size as the ceramic machined in 4340 steel block and the confinement on top of ceramic was provided

with a HSLA steel (SAE 4130, BHN 480) cover plate. In this configuration most of the penetrator flowed laterally at the cover plate-ceramic interface and there was only 5 mm penetration in the substrate below the ceramic. This phenomenon was identified as interface defeat (surface magic). Hauver et al. (2-3) investigated this phenomenon further and showed that WHA penetrators, shot at velocities up to 1.6 km/s, can be defeated with improved cover plate configuration. They investigated the performance of a number of round and square, 25 mm thick ceramics (tungsten carbide, titanium diboride, boron carbide, alumina, etc.). They concluded that in targets containing WC/Co or TiB_2, can defeat 93% WHA, L/D =20 penetrators (diameter=0.194") at the 4340 steel cover plate-ceramic interface. Little damage to the ceramics was observed in both the cases. In the case of alumina and boron carbide the targets did not perform as well. In this study no quantitative measurements, such as stress/strain-histories at the ceramic front or back interfaces were made. In this paper we present the results on stress/strain-histories measured at the ceramic-cover plate interface and ceramic-back surface plate with in-material embedded gauges. The ballistic targets were assembled following the configuration used by Hauver et al. (2-3). The first two shots in this series were performed on relatively simpler targets to gain experience in using stress/strain gauges. Measured values of stress/strain and the post shot length of the remaining penetrator were compared with numerically simulated results (4).

EXPERIMENTAL METHOD

Materials Properties

Three types of ceramics were used in the present study. Nominal dimensions, density, and source of the ceramic plates are summarized in Table 1.

Target Configuration and Assembly

Four targets, 7-1795, 7-1796, 7-1797, and 7-1798, were configured and assembled as follows; 7-1795: A manganin/constantan gauge of Type MN/CN-50-EK (Dynasen) was glued on top of 50-mm thick and 150-mm in diameter 4340 steel plate. The gauge was covered with a 25 mm thick mylar sheet. A 13-mm thick steel plate

TABLE 1. Dimensions and density of ceramics.

Ceramic	Shape Dim.(mm)	ρ (g/cc)	Source
Alumina	Square 63x63x13	3.95	Babcock &Wilcox
Ebon-A Alumina	Round ϕ=72; th.=25.7	3.97	Cercom
TiB$_2$	Round ϕ=72; th.=25.7	4.5	Cercom

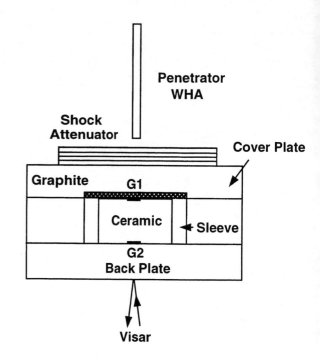

FIGURE 1. Schematic of targets 7-1797 and 7-1798.

was glued on top of the gauge. A shock attenuator block, consisting of 12 alternate layers of 0.8 mm thick 2024 aluminum and plastic, was glued on top of the 12.7 mm thick steel plate. 7-1796: This target was assembled with a 12.7-mm thick alumina plate (Babcock and Wilcox) set in a well machined in 50-mm thick and-150-mm in diameter 4340 steel plate. A Dynasen stress/strain gauge was glued on top of the alumina plate to record the stress/strain histories during the penetration process. On top of the gauge a 13 mm thick steel plate and a shock attenuator of the type in target 7-1795 were glued. 7-1797: This target was designed following Hauver et al. (2-3) and consisted of three plates, as shown schematically in Fig. 1. The top 25.4-mm thick 150-mm in diameter 4340 steel plate had a well machined on the bottom to accept a 2.4-mm thick and 73-mm in diameter graphite disk. The middle disk consisted of Cercom Ebon-A alumina disk shrink fitted in a 5mm wide 17-4 PH steel ring. The steel ring was further shrink fitted into a 25.5-mm thick 4340 steel ring. The bottom plate was 25-mm thick 4340 steel plate. Both the faces of the ceramic containing steel plate, the lower face of the top steel plate, and the upper face of the bottom plate were lapped to ascertain flatness at the interfaces. Two Dynasen stress/strain gauges, one on top of the Ebon-A alumina ceramic and another on top of the bottom 4340 steel plate, were glued to record the stress/strain histories during the penetration process. The three plates were bolted together using 12 grade-8 bolts. A shock attenuator block, consisting of 24 alternate sheets of 0.8-mm thick 2024-aluminum and plastic, was glued on top of the target assembly. A VISAR was set up to measure the free surface motion of the back steel plate. 7-1798: This target was prepared and assembled in the same way as the target 7-1797, replacing Ebon-A alumina disk with the Cercom TiB_2 disk.

Ballistic experiments

Penetrator rods were machined from the Teledyne X21 93% tungsten stock and were launched using a Lexan sabot. Shots 7-1795 and 7-1796 were performed with 6.35-mm diameter L/D=10 penetrators. Shots 7-1797 and 7-1798 were performed with L/D=20 penetrators, (D=4.93-mm), which are similar to those used by Hauver et al. (2-3). Lexan sabots were stripped using a specially designed sabot stripper. The shot data is summarized in Table 2. The outputs of the stress/strain gauges were recorded using a Dynasen pulsed power supply. Strain and stress values were determined following the Dynasen manual (5).

RESULTS AND DISCUSSION

In shots 7-1795 and 7-1796 both stress and strain gauge profiles were obtained. In 7-1795 strain was less than 0.5% until the penetrator

TABLE 2. Summary of the shot data.

Shot No\	Impact Vel. (km/s)	Penet.	Interface Defeat
7-1795	1.3	complete	NA
7-1796	1.41	complete	partial
7-1797	1.7±0.1	complete	partial
7-1798	1.7±0.1	complete	partial

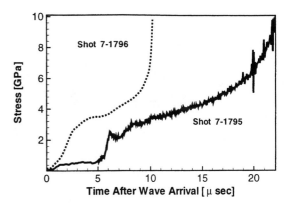

FIGURE 2. Axial stress histories.

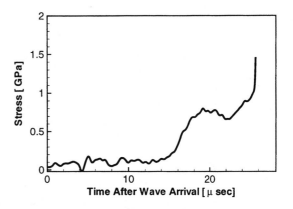

FIGURE 3. In-material stress history in shot 7-1797.

arrives at the gauge location. The stress profile from this shot is shown in Fig. 2. The stress increases to a level of 0.4 GPa and remains at this level for about 5 μs. After this, the stress level jumps first to about 2 GPa and then continuously increases to 10 GPa until gauge failure. The jump in stress may possibly be due to the effect of the approaching penetrator to the gauge location. For numerical predictions of in-material stress histories in ballistic simulations, see (4). The stress history obtained in shot 7-1796 is shown in Fig. 2. Strain was zero until the penetrator arrives at the gauge location. The stress profile shows a slow increase in the

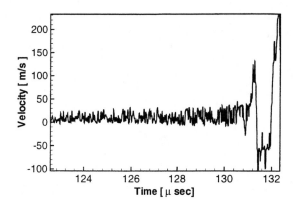

FIGURE 4. Back surface velocity history in shot 7-1797.

form of a ramp to about 4 GPa over a period of 2 μs and continues to increase to about 8 GPa before the gauge fails. The slow increase in the stress level in the form of a ramp is likely due to the penetrator interaction with the attenuator block. The strain and stress gauge failure, at the same time, can be interpreted as due to the arrival of the penetrator at the gauge location. Strain histories recorded at two locations, see Fig. 1, in targets 7-1797 and 7-1798 showed that strain was almost zero before the failure of the gauges on arrival of the penetrator. Stress histories recorded with manganin gauges on top of the ceramics (gauge 1) in both shots, 7-1797 and 7-1798, were not meaningful. The stress history, shown in Fig. 3, is from manganin gauge at location 2, see Fig. 1, in shot 7-1797. There is a slow rise in the stress level to about 1 GPa over a time of 10 μs. Figure 4 shows the back plate free surface velocity. This velocity history represents the early part of the target motion as inferred from numerical simulations (4). An optical micrograph showing *partial interface defeat* is shown in Fig. 5. This phenomenon was observed in both Al_2O_3 and TiB_2.

ACKNOWLEDGMENTS

The research reported in this paper was supported by the Army Research Office through

Purdue MURI grant No. DAAH04-96-1-0331.

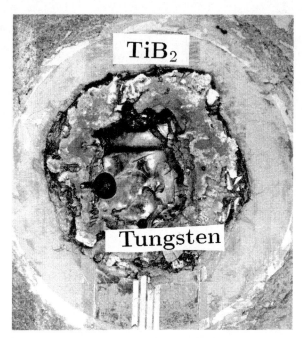

FIGURE 5. Photograph showing penetrator lateral flow at the ceramic-graphite interface.

REFERENCES

1. Bless, S., Ben-yami, M., Apgar, L, and Eylon, D., "Impenetrable targets struck by high velocity tungsten long rods," Proc 2nd Int. Conf. on Structures under Shock and Impact, Portsmouth, UK, 16-18, June, 1992.

2. Hauver, G., Netherwood, P., Benck, R, and Keoskes, L., "Enhanced Ballistic Performance of Ceramics," 19th Army Science Conference, Orlando, FL, 20-24 June, 1994.

3. Rapacki, E., Hauver, G., Netherwood, P., and Benck, R., "Ceramics for Armors - A Material System Perspective," 7th Annual TARDEC Ground Behicle Survivability Symposium, 26-28 March, 1996.

4. Espinosa, H.D., Dwivedi, S., Zavattieri, P.D., and Yuan, G., "Numerical Investigation of Penetration in Multilayer Material/Structure Systems," submitted to *Int. J. Solids and Str.*, 1997.

5. Charest, J. Personal Communication, 1997.

500

CP429, *Shock Compression of Condensed Matter – 1997*
edited by Schmidt/Dandekar/Forbes
© 1998 The American Institute of Physics 1-56396-738-3/98/$15.00

ALUMINA STRENGTH DEGRADATION IN THE ELASTIC REGIME

Michael D. Furnish and Lalit C. Chhabildas

Sandia National Laboratories, Albuquerque NM 87185

Measurements of Kanel et. al. [1991] have suggested that deviatoric stresses in glasses shocked to nearly the Hugoniot Elastic Limit (HEL) relax over a time span of microseconds after initial loading. "Failure" (damage) waves have been inferred on the basis of these measurements using time-resolved manganin normal and transverse stress gauges. Additional experiments on glass by other researchers, using time-resolved gauges, high-speed photography and spall strength determinations have also lead to the same conclusions. In the present study we have conducted transmitted-wave experiments on high-quality Coors AD995 alumina shocked to roughly 5 and 7 GPa (just below or at the HEL). The material is subsequently reshocked to just above its elastic limit. Results of these experiments do show some evidence of strength degradation in the elastic regime.

INTRODUCTION

There is an increasing body of evidence that certain brittle solids, such as glass, undergo strength loss during residence at a Hugoniot state below the Hugoniot elastic limit [HEL], especially for glass. Kanel et al[1] observed a wave arrival corresponding to a reflection from a discrete propagating zone of strength loss in glass, a zone termed a failure wave. They also observed a relaxation in shear stress corresponding primarily to an increase in lateral stress. This is shown schematically in Fig. 1. Other evidence based on spall strength loss[2] and photographic contrast[3] has also been obtained supporting the theory that failure waves propagate in glass.

In alumina, spall strength loss for Hugoniots near (but still below) the HEL suggests that a similar process is occurring[4]. The body of evidence for such phenomena in alumina, however, is not nearly as extensive as for glass.

The objective of the present work is to assess the behavior of alumina initially shocked to below the HEL, and subsequently reshocked to somewhat above the HEL.

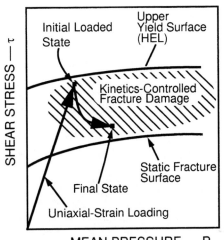

FIGURE 1. Process of sub-HEL material relaxation.

TECHNIQUE

Disks of Coors AD995 ceramic (initial density $\rho_0 = 3.89$ gm/cm^3) were tested in a transmitted-wave geometry as shown in Fig. 2. This configuration provided an initial shock loading of the alumina chosen (via impact velocity) to slightly below the

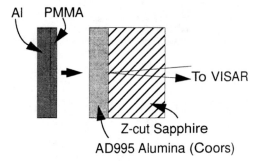

Al PMMA

To VISAR

Z-cut Sapphire
AD995 Alumina (Coors)

FIGURE 2. Configuration used for impact tests.

TABLE 1. Test parameters. Test number corresponds to curve labels in Fig. 3.

Test #	Impact Vel. km/s	Impactor Thick mm	Sample Thick mm	Hugoniot Press. GPa
1	1.11	2.696	5.022	4.9
2	1.10	2.698	9.987	4.8
3	1.54	2.703	4.938	7.3
4	1.51	2.7	10.003	7.2
5	1.98	2.703	9.888	11.

HEL, followed by a reloading above the HEL. A VISAR monitored the motion of the sample/sapphire interface, giving a nearly *in situ* measurement because of the close impedance match between the (unyielded) AD995 alumina and the window.

RESULTS

Observed velocity histories are given in Figure 3., with critical experiment parameters given in Table 1.

The wave profiles have been analyzed through an explicit Lagrangian calculation comparing input and output wave profiles for the sample. Such an analysis includes corrections for sample/window shock impedance mismatch (small for the present samples).

It is interesting to compare the results of this analysis with results of an analogous analysis of Grady's[5] waveforms, which were obtained by introducing a single shock, then release, into the alumina. Lithium fluoride windows were used for Grady's tests. Figure 4 shows an overall agreement between the two sets of results. Results of earlier experiments by Grady[5] and Dandekar[6] suggest that the average Hugoniot Elastic Limit is approximately 6.5 GPa. However, the yield strength of the material at the Hugoniot state may also be estimated as 3/4 of the width of the stress-strain loop, or about 6 GPa.

The behavior of the alumina near the yield point may be seen better by converting the velocity profiles to stress-time profiles via the Lagrangian analysis described above. Results are shown in Fig. 5.

Note that the Fig. 5 curves represent *in situ* stresses rather than stresses which would actually be measured at the sample/window interface. This is

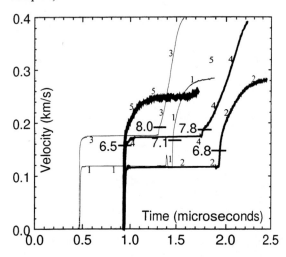

FIGURE 3. Wave profiles from present series of transmitted-wave experiments. Axial stresses at yielding are as shown. Shot parameters corresponding to curves labeled 1 - 5 are in Table 1.

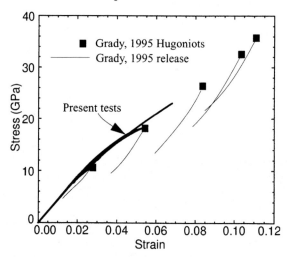

FIGURE 4. Present stress-strain data for alumina, compared with impact data of Grady[5].

important for the Grady results (using LiF windows), although the window used does not appear to have a strong effect on the data plotted in this way. For the tests reported in this study such a distinction is much less important due to the nearly exact impedance match between alumina and sapphire.

Details of the HEL are better seen in a plot of wave speed vs. stress (Fig. 6). The HEL from the Grady single-shock data in fact range from ~6 GPa to 7 GPa. The data show two trends:

1. The axial stress at which the material yields appears to increase after a reshock from slightly below the HEL. Particularly, elastic limit stresses of 7.9 GPa are measured upon reshock, suggesting strength degradation resulting from the first shock loading.

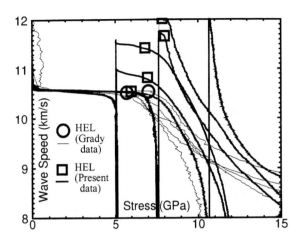

FIGURE 6. Detail of HEL region of present data and Grady[5] data, represented as stress vs. wave speed. Bold lines represent present data (circles show HEL values) and fine lines represent Grady single-shock data (squares show HEL values).

2. The yield stress appears to decrease with increased wave travel distance.

Both of these trends are consistent with a relaxation of shear stress during residence at the Hugoniot state. The evidence is consistent with the strength degradation observed in glass. Further tests to assess this hypothesis should involve adjusting the thickness of the PMMA flyer plate to change the residence time of the material at the Hugoniot state prior to reloading.

FURTHER INTERPRETATION

In the Introduction the problem was stated in terms of shear stress relaxation. It is important to view the present results in those terms.

From the waveforms alone, it is only possible to calculate the axial stress. If elastic behavior persists, the Poisson's ratio (0.23 to 0.24[5]) allows a calculation of shear stress τ and mean pressure $\bar{\sigma}$ via:

$$\tau = 0.5\, \sigma_{axial}\, (1 - 2\nu)/(1-\nu) \qquad \text{(Eq. 1)}$$

$$\bar{\sigma} = \sigma_{axial} - 4/3\, \tau \qquad \text{(Eq. 2)}$$

The initial elastic, uniaxial strain loading is along a path given by (from Equations 1 and 2):

$$\tau = \overline{3\sigma}X/(3-4X) \quad (X \equiv (1 - 2\nu)/(1-\nu)) \qquad \text{(Eq. 3)}$$

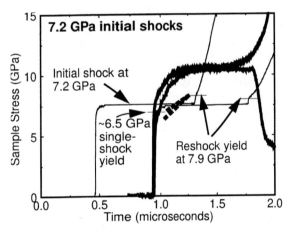

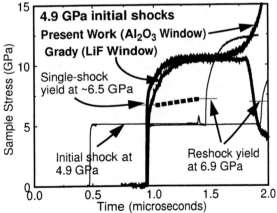

FIGURE 5. Comparison of present results (expressed as stress vs. time) with the lowest-stress single-shock result of Grady[5]. Bold lines represent single shocks to stresses above the HEL.

503

The reload points yield points lie on lines given by (from $\tau = 1/2$ $(\sigma_{axial} - \sigma_{transverse})$ and $\bar{\sigma} = 1/3$ $(\sigma_{axial} + 2\sigma_{transverse})$):

$$\tau = 3/4 \, \sigma_{axial} - 3/4 \, \bar{\sigma} \qquad \text{(Eq. 4)}$$

This, however, assumes that the material behavior remains elastic, i.e. that the Poisson's ratio still reflects material behavior.

These are plotted for the present experiments in Figure 7.

Hence there is a continuum of points in $(\bar{\sigma}, \tau)$ space possibly corresponding to the reshock yield point, and other diagnostics will be required to completely define this material behavior.

ACKNOWLEDGEMENTS

This work was performed at Sandia National Laboratories supported by the U. S. Department of Energy under contract DE-AC04-94AL85000. Sandia is a multiprogram laboratory operated by Sandia Corporation, a Lockheed Martin company, for the USDOE.

REFERENCES

1. Kanel, G. I., S. V. Rasorenov and V. E. Fortov, The failure waves and spallations in homogeneous brittle materials, pp. 451-454 in *Shock Compression of Condensed Matter 1991*, S. C. Schmidt, R. D. Dick, J. W. Forbes and D. G. Tasker (eds.), Elsevier, 1992.

2. Brar, N. S., S. J. Bless and Z. Rosenberg, Impact-induced failure waves in glass bars and plates, *Appl. Phys. Lett., 59*, 3396-3398, 1991.

3. Bourne, N. K., Z. Rosenberg and J. E. Field, High speed photography of compressive failure waves in glasses, *J. Appl. Phys., 78*, 3736-3739, 1995.

4. Rosenberg, Z. and Y. Yeshurun, Determination of the dynamic response of AD-85 alumina with in-material manganin gauges, *J. Appl. Phys., 58*, 3077-3080, 1985.

5. Grady, D. E., Dynamic properties of ceramic materials, Sandia National Laboratories Report SAND94-3266, 1995.

6. Dandekar, D. P. and P. Bartkowski, Shock response of AD995 alumina, pp. 733-736 in *Shock Compression of Condensed Matter 1993*, S. C. Schmidt, J. W. Shaner, G. Samara and M. Ross (eds.), AIP Press, 1994.

7. (e.g.) Bauccio, M. (ed.), *ASM Engineering Materials Reference Book, Second Edition,* ASM International, 1994, pp. 251-254. Note that J. Lankford (Compressive strength and microplasticity in polycrystalline alumina, *J. Mat. Sci, 12*, 791-796, 1977) gives a value of ~2.9 GPa.

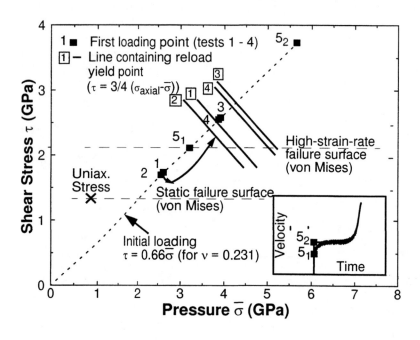

FIGURE 7. Interpretation of the present results in pressure vs. shear stress space. Initial loading is from (0, 0) to the loading points shown for the present experiments. Reloading is to some point along the heavy diagonal lines shown (present data cannot constrain further) Inset shows how alternative values were chosen for the yield point on test 5. Uniaxial stress yield of 2.6 GPa[7] used to calculate static yield point shown.

CP429, *Shock Compression of Condensed Matter – 1997*
edited by Schmidt/Dandekar/Forbes
© 1998 The American Institute of Physics 1-56396-738-3/98/$15.00

IMPACT OF AD995 ALUMINA RODS

L. C. Chhabildas, M. D. Furnish, W. D. Reinhart and D. E. Grady[1]

Sandia National Laboratories, Albuquerque, NM, USA, 87185-1181*
[1]Applied Research Associates, Inc., 4330 San Mateo Blvd., NE, Albuquerque, NM, USA, 871110

Gas guns and velocity interferometric techniques have been used to determine the loading behavior of AD995 alumina rods 19 mm in diameter by 75 mm and 150 mm long, respectively. Graded-density materials were used to impact both bare and sleeved alumina rods while the velocity interferometer was used to monitor the axial-velocity of the free end of the rods. Results of these experiments demonstrate that (1) a time-dependent stress pulse generated during impact allows an efficient transition from the initial uniaxial strain loading to a uniaxial stress state as the stress pulse propagates through the rod, and (2) the intermediate loading rates obtained in this configuration lie between split Hopkinson bar and shock-loading techniques.

INTRODUCTION

There is a need for accurate ceramic material models to facilitate computational and engineering analyses under dynamic loading. Well-controlled impact techniques and high-resolution diagnostics (1) are generally used to determine the baseline material property data, under uniaxial strain conditions. Such a data base forms the foundation for material models that have been developed for engineering analysis in computer codes. Validation and the continued development of ceramic material models appropriate under multiaxial loading conditions will, however, require the existence of a comprehensive material property data base. In this paper, new measurements on alumina under a broader range of loading conditions are reported. Graded-density materials (2) were used to impact both bare and sleeved alumina rods, while the velocity interferometer was used to monitor the axial-velocity of the free end of the rods. Results of these experiments are discussed in this paper.

The aluminum oxide used in this study is generally referred to as Coors AD995 and is the same batch of material used in previous studies on alumina (3-5). Its composition consists of 99.5% alumina and the rest aluminosilicate glass. The density of the material used in this investigation was 3.89 g/cm^3; the average longitudinal and shear wave speeds were determined to be 10.59 km/s and 6.24 km/s, respectively. These yields an estimate of 7.71 km/s, 9.80 km/s, and 0.234 for the bulk wave velocity, bar wave velocity, and Poisson's ratio, respectively.

EXPERIMENTAL TECHNIQUE

These experiments were performed on a 64 mm diameter smooth-bore, single-stage compressed gas gun which is capable of achieving a maximum velocity of about 1.6 km/s. Three electrically shorting pins were used to measure the velocity of the projectile at impact. Four similar pins were mounted flush to the impact plane and used to monitor the planarity of impact. Projectile velocity could be measured with an accuracy of about 0.5% and the deviation from planarity of impact was a few milliradians. The graded-density impactor assembly is fabricated by bonding a series of thin plates in order of increasing shock impedance from the impact surface. The series of layered materials used in these studies were TPX-plastic, aluminum, titanium, and 4340 steel. The thickness of each layer is controlled to tailor the time-dependent input stress pulse into the alumina rod. The exact dimensions of each material assembly is given in Table 1. This layered material assembly is used as a facing on an aluminum projectile and is accelerated on a gas gun to velocities of about 320 m/s, providing

* Sandia is a multiprogram laboratory operated by Sandia Corporation, a Lockheed Martin Company, for the United States Department of Energy under Contract DE-AC04-94AL85000.

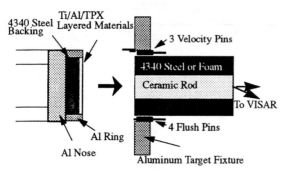

Figure 1. Experimental configuration of a layered/impactor and a ceramic-rod target assembly.

a time-dependent loading to ~ 6.5 GPa. The experimental target assemblies consisted of either a bare or a sleeved alumina rod ~ 19 mm in diameter. The length of the rods in this study were nominally 74 mm or 151 mm. When used, 4340 steel was chosen for the close fitting sleeve material to provide a good shock impedance to the alumina sample. The outer diameter of the sleeve was nominally 39 mm.

When unsleeved, a polyurethane foam was used to decouple the rod from the aluminum target fixture. A 0.055 mm thick tungsten reflector glued onto the free surface of the rod was used to obtain the axial particle velocity measurements using the velocity interferometer, VISAR (6), having a time resolution of ~ 1 ns. These measurements are shown in Figure 2 for the experiments summarized in Table 1.

UNSLEEVED EXPERIMENTS

The experimental result for a single density impact (FW1) is indicated in Figure 2. The wave profile reveals a distinct two-wave structure, i.e., the arrival of an initial elastic compression wave (2.1GPa) at a wave speed of 10.6 km/s followed by a second compression wave traversing at a bar wave speed of 9.8 km/s. This results in loading the alumina to a final stress of 3.4 GPa. However, when a graded density impactor is used to impact the rod (FW5), the leading edge of the initial compression wave traversing at 10.6 km/s loads the material to only 0.2 GPa. A subsequent wave arrives at a bar wave speed of 9.8 km/s and loads the material to a final stress of 3.5 GPa at a strain-rate of ~ 4 x 10^3 /s.

Even though the impact velocity of the experiment FW5 is approximately 6% lower than the single density impact experiment FW1, the peak particle velocity attained in the graded-density impact experiment is slightly higher. In the graded density impact experiment FW6, the rod is ~ 150 mm long, and the

impact velocity is 0.366 km/s, ~ 10% higher than the single density impact experiment. The elastic precompression wave is attenuated to 0.1 GPa, compared to the 0.2 GPa in experiment FW5; the subsequent compression wave traversing at the bar wave velocity loads the material up to 4.2 GPa at a strain rate of 4.5 x 10^3 /s, eventually relaxing to a stress state of ~ 3.6 GPa. The first compression state σ_l is calculated using $\sigma_l = (\rho_o c_1 \delta u_{fs})/2$, where ρ_o is the initial density, c_1 the elastic wave speed, and δu_{fs} the incremental free surface velocity measurement associated with the longitudinal elastic wave

The axial compression state σ_a and the loading strain rates ε associated with the bar wave are calculated using $\sigma_a = (\rho_o c_b \Delta u_{fs})/2$ and $\varepsilon = \Delta u_{fs}/(2c_b t)$, where c_b is the bar wave velocity, and Δu_{fs} the corresponding free-surface velocity measurement, and t the time duration for loading.

SLEEVED EXPERIMENTS

Experimental results for sleeved experiments FW2 (74 mm rod), FW3 & FW4 (151 mm rod) are also shown in Figure 2. Graded density impactors were used in these experiments. The initial elastic compression wave traversing at 10.6 km/s loads the material up to stress states of 0.2 GPa and 0.1 GPa, respectively, for the short and the long rods. The subsequent compression wave traversing at a bar wave speed compresses the material to a final stress of 5.1 GPa and 4.6 GPa, respectively. The corresponding loading rates are approximately 5 x 10^3 /s and 4.5 x 10^3 /s, respectively. The results of these experiments are shown plotted as the failure stress *vs.* strain-rate (7) in Figure 3.

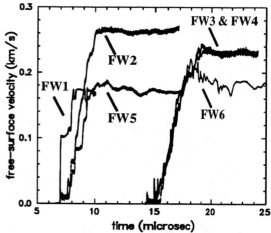

Figure 2. Axial velocity measurements for all experiments.

Table 1: Summary of impact experiments on AD995 alumina rods

Test No.	Rod Diameter/Length (mm)/(mm)	Impactor Materials	Impactor Thickness (mm)	Impactor Velocity (km/s)	Sleeved
FW1	19.164/73.67	Steel	10.59	0.318	no
FW2	19.169/73.67	Steel/Ti/Al/TPX	19.04/1.097/1.199/1.034	0.321	yes
FW3	19.162/150.32	Steel/Ti/Al/TPX	19.05/1.123/1.204/1.024	0.321	yes
FW4	19.172/151.38	Steel/Ti/Al/TPX	19.04/1.102/1.204/1.041	0.322	yes
FW5	19.159/73.668	Steel/Ti/Al/TPX	19.06/1.107/1.204/1.024	0.300	no
FW6	19.192/152.41	Steel/Ti/Al/TPX	19.08/0.998/0.998/0.975	0.366	no

CONCLUSIONS

Previous studies on impact of alumina rods (8-9) have concentrated upon using a single density impactor to evaluate the uniaxial compressive behavior of the ceramics. However, due to the low spall strength of alumina (3,10), the radial stress components will fracture the material (9,11) during the loading phase even though the mean stress of the material indicates compression. The technique proposed herein (i.e., using graded-density impactors to study the uniaxial compressive behavior of the rods) circumvents this problem by reducing the magnitude of tension generated in alumina. A sleeved rod totally prevents the formation of radial tension during the loading process.

It is not surprising that the single-density impact experiment yields a failure stress of 3.4 GPa, the graded-density impact experiments fails at 4.2 GPa, and the sleeved experiments fails at 5.1 GPa. The material that is damaged the most fails at a lower stress. These results are consistent with the hypothesis that for brittle materials the onset of failure depends heavily on the loading rate (shown in Figure 3.) Shock experiments yield higher estimates of strength mainly because rate-dependent kinetics prevent the nucleation and growth of flaws and defects in materials during rapid loading.

CTH-calculational results (11) indicate that the ratio of the lateral stress to the axial stress is ~ 0.23 for the single-density impact of the alumina rod, and ~ 0.1 and ~ 0.08 for the graded-density impact of the unsleeved and sleeved rod, respectively. This apparently indicates that the degree of confinement is least for the sleeved rod. It is consistent with the earlier inference that the stress propagation in the rods transitions to a uniaxial stress motion when it is loaded at finite rates, and is also depicted in Figure 5. There-

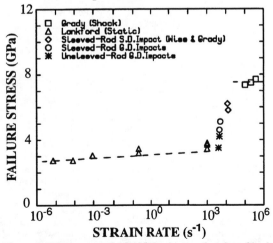

Figure 3. Failure stress of AD995 alumina as a function of loading rate. Quasi-static and shock loading results are also shown.

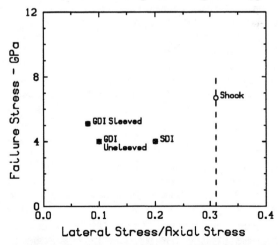

Figure 4. Variation of calculated failure stress as a function of confinement *i.e.*, the ratio of the lateral stress to axial stress.

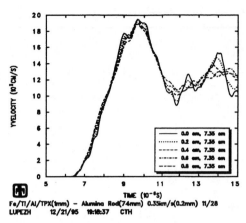

Figure 5. Calculated free-surface axial velocity history at a radius of 0.0, 0.2, 0.4, 0.6, 0.8 cm from the axis of the rod 74 mm in length. Radial velocity dispersion is minimal (11).

fore, the experimental measurements of a higher failure stress (5.1 GPa) for the sleeved rod when compared to the lowest value (3.4 GPa) are not due to the sleeved-confinement of the rod, but are more related to strain-rate sensitivities, as indicated in Figure 3. If the rod were rigidly confined, then one should measure an upper dynamic limit of 6.7 GPa which is the Hugoniot elastic limit (3,10).

The most significant result of this study is that the use of a graded-density impactor allows an efficient transition to the uniaxial stress configuration even though the ratio of the length to diameter of the rods is only around 4 for 74 mm rods when unsleeved, and effectively 2 when it is sleeved. Besides, a finite rate of loading allows a method by which strain-rate effects of the material can be determined. This is obviously not the case for a single density impact (11), as evidenced by a two-wave structure in Figure 2.

The current experiments address strain-rate effects in alumina at strain rates of $\sim 5 \times 10^3$ /s. The strength of alumina is estimated to be ~ 5.1 GPa. The strain-rate loading can be increased by decreasing the thickness of the graded density layers. A factor of four decrease in thickness should load the material at a strain rate of 2×10^4 /s. The technique, therefore, will permit accessibility to intermediate loading rates which are difficult to achieve either using traditional split Hopkinson bar or shock loading techniques. Furthermore, the stress amplitude of the wave propagating at the elastic longitudinal wave speed can be further reduced in these experiments by using a lower impedance material such as foam as the first layer in the series of graded density materials. As indicated in this study, the use of plastic (FW5) as compared to

steel (FW1) reduces the amplitude of the elastic wave by over an order of magnitude from 2 to ~ 0.2 GPa

It appears that loading rates of a few times 10^4/s can be achieved by optimizing the design of the graded density layered materials, the diameter of the bar, and the impact velocity. Concepts are currently being pursued to achieve yet higher loading rates of 10^5/s. One approach under consideration is to use the graded-density materials as an impactor to perform isentropic loading experiments up to its Hugoniot elastic limit. These experiments will, however, characterize the material behavior under uniaxial strain loading.

REFERENCES

[1] Chhabildas, L. C. and Graham, R. A., "Developments in Measurement Techniques for Shock Loaded Solids," *Techniques and Theory of Stress Measurements for Shock Wave Applications*, Editors, R. B. Stout, *et. al.*, **AMD-Vol83**, 1987 pp. 1-18.

[2] Chhabildas, L. C., Kmetyk, L. N., Reinhart, W. D., and Hall, C. A., "Enhanced Hypervelocity Launcher - Capabilities to 16 km/s," *Int. J. Impact Engng.* **17** (1995) pp. 183-194.

[3] Grady, D. E., "Dynamic Properties of Ceramic Materials," Sandia National Laboratories Report, SAND94-3266, February 1995.

[4] Kipp, M. E. and Grady, D. E., "Shock Compression and Release in High-Strength Ceramics," *Shock Compression of Condensed Matter - 1989*, Editors S. C. Schmidt, *et. al.*, North-Holland, Amsterdam, 1990, pp. 377-380.

[5] Wise, J. L., Grady, D. E., "Dynamic, Multiaxial Impact Response of Confined and Unconfined Ceramic Rods," *High Pressure Science and Technology--1993, AIP Conference Proceeding 309*, Edited by S. C. Schmidt *et. al.*, 1994, pp. 733-736.

[6] Barker, L. M. and Hollenbach, R. E., "Laser Interferometer for Measuring High Velocities of Any Reflecting Surface," *Journal of Applied Physics* **43** (1972) pp. 4669-4675.

[7] Grady, D. E., "Shock-Wave Properties of Brittle Solids," *Shock Compression of Condensed Matter - 1995, AIP Conference Proceeding 370*, Edited by S. C. Schmidt *et. al.*, 1996, pp. 9-20.

[8] Cosculluela, A., Cagnoux, J., Collombet, F., "Uniaxial Compression of Alumina, Structure, Microstructure and Strain-Rate," *Journal de Physique IV*, **C3** (1991) pp. 109-116.

[9] Brar, N. S., and Bless, S. J., "Dynamic Fracture and Failure Mechanisms of Ceramic Bars," *Shock-Wave and High-Strain-Rate Phenomena in Materials*, Edited by M. A. Meyers *et. al.*, 1992, pp. 1041-1049.

[10] Dandekar, D. P. and Bartkowski, P., "Shock Response of AD995 Alumina," *High Pressure Science and Technology--1993, AIP Conference Proceeding 309*, Edited by S. C. Schmidt *et. al.*, 1994, pp. 777-780.

[11] Chhabildas, L. C., Furnish, M. D., Grady, D. E, "Impact of Alumina Rods-A Computational and Experimental Study," *Proceedings of the DYMAT International Conference on Mechanical and Physical behavior of Materials*, 1997.

508

CP429, *Shock Compression of Condensed Matter – 1997*
edited by Schmidt/Dandekar/Forbes
© 1998 The American Institute of Physics 1-56396-738-3/98/$15.00

MECHANICAL PROPERTIES OF PRESHOCKED
SAPPHIRE DRIVER

D.N. Nikolaev, A.S. Filimonov, V.E. Fortov,
I.V. Lomonosov, V.Ya. Ternovoi

Institute of Chemical Physics in Chernogolovka, Moscow reg., Chernogolovka, 142432, RUSSIA

Bromoform analyzer method was used to investigate the mechanical properties of sapphire after preshocking up to 130 GPa. The time dependence of shock front emission in $CHBr_3$ was recorded under its impact with the shocked sapphire driver. Expansion distance of the sapphire was varied from 1.45 to 6.05 mm. Previously obtained sound speeds and shock pressure - brightness temperature dependence for bromoform were used to calculate the pressure profile on the driver-bromoform boundary. It was found that the increase in the vacuum gap between sapphire free surface and bromoform pressure gauge causes the smearing of the loading pulse front and the decrease of the maximum detected pressure. This phenomenon has been explain by sapphire fragmentation in the unloading wave.

INTRODUCTION

Optical analyzer technique was used for several years to determine sound speeds in shocked opaque materials. It helped to determine bulk and longitudinal sound speeds and to detect shock melting in several metals and minerals [1,2]. In these experiments a transparent material (analyzer), placed in front of the sample, radiates when shocked. The overtake ratio is determined by varying the sample thickness until rarefaction catch-up at the sample -- indicator interface occurs. The main requirement for analyzer is to have a sharp dependence of the optical emission intensity on the shock pressure, while analyzer EOS is not important. Bromoform is suitable as analyzer because as a liquid it is free from elastic-plastic behavior under shock compression. On the other hand its density (2.87g/cc) is quite high and its compressibility is comparable to that of metals.

If the equation of state, especially dependencies of the emission and sound speed on the shock parameters of bromoform are known, it becomes a real high-pressure analyzer. Registration of its radiation history allows to obtain information about shock and detonation waves [2,3,4] at pressures and with time resolution not available with other common gauge types.

Experiments to be described are investigation of bromoform EOS in 60-200 GPa pressure range. Sound speed and brightness temperature data obtained for shocked bromoform was used to investigate the mechanical properties of sapphire single crystal driver, preshocked up to 130 GPa, which was planed to use as a gas compressor in multiple shock reverberation technique.

BROMOFORM INVESTIGATION

Various explosive launchers were used to generate shock waves in the $CHBr_3$. Stainless steel (1.4-1.6 mm) and aluminum (2 mm) strikers were accelerated by a common one-stage systems. One - stage layered systems [5], accelerating 0.4-0.6 mm steel and copper strikers, were used to increase final velocities. Maximum velocities of molybdenum (0.1-0.2 mm) strikers were reached with two - stage layered systems, where acceleration of the strikers was performed in the evacuated chamber. Accelerated striker was impacted with the bottom of reservoir filled with bromoform. The diaphragmed quartz fiber was mounted on the other side of reservoir. Thin (4-5 μm) mylar films, immersed in the bromoform [2], allowed to measure the shock velocities with good accuracy (1.5-3%). In some shots we used assemblies with thin (5-7 μm) stretched Al foil instead of reservoir bottom. In this case the sound speed calculation becomes very simple, and these shots were used to verify the calculation procedure. These assemblies were also used with thin Mo strikers. Due to the low saturated vapor pressure of $CHBr_3$ it was possible to place the bromoform reservoir, covered only with thin foil, in the vacuum chamber.

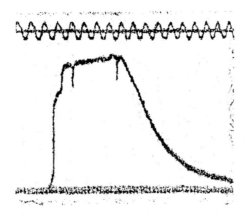

FIGURE 1. Typical oscilloscope snapshot for the sound speed measurement. Impact of the 5.1 km/s 1.5 mm steel striker with 1 mm steel bottom. Frequency of the time marks is 10 MHz.

The registered parameter was the optical emission from the shock wave in the bromoform. We used multichannel optical pyrometer with fiber emission inlet, narrow interference filters with 6-10 nm halfwidth, and silicon photodiodes as emission receivers. The time resolution was about 2-5 ns and was limited mostly by the transient response of preamplifiers and oscilloscopes. Common emission registration snapshot is shown in Fig. 1. It is distinctly seen some lowering of a shock wave intensity at a first time due to large distance of air (25 mm) in which the striker was accelerated by HE detonation products before the impact. Sharp decrease in the radiation intensity indicates the overtake moment. Time marks from the mylar films used for the shock velocity calculation are also well seen.

The pyrometer was calibrated with a tungsten ribbon lamp before every shot. It allowed to calculate the brightness temperatures using Plank relations [6].

The sound speed calculation is very sensitive to the sound speed value in the bottom material, and it was taken into consideration very carefully. Sound velocities in the shocked aluminum, copper and molybdenum were calculated according to [2], using Hugoniots from [7]. The shock and sound velocities in steel were taken from EOS, proposed in [8]. A drop in the sound speed from longitudinal to bulk value under melting was taken into account. It was also taken into account the refraction of the back rarefaction wave under counteraction with the rarefaction wave traveling through the bottom from the boundary with $CHBr_3$. $CHBr_3$ Hugoniot from [2] was used for calculations.

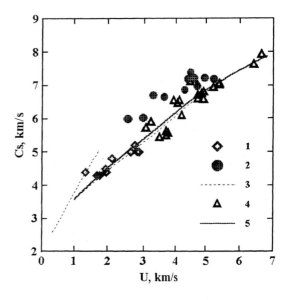

FIGURE 2. Experimental c_s - u dependence. 1 - [9], 2 - [3], 3 - Caloric EOS calculation, 4 - Our data, 5 - Polynomial fit.

Results of the sound speed (c_s) measurements are shown in Fig. 2 vs. particle velocity (u) together with the data from [3,9] and listed in the Table 1. The following polynomial fits have been obtained for the sound speed and temperature: $c_s=2.5+1.1u-0.045u^2$, $T=1.038B+856$, $2<u<7$, where $B=u^2/2$, c_s and u are in km/s, and T is in thousands of K.

510

Gruneisen parameter for bromoform has been also calculated according to [2,10,11]. Caloric - type semiempirical equation of state [12,13] for bromoform describes all gas dynamic parameters quite well. For example, this EOS calculation of the sound speed on the Hugoniot is shown in Fig. 2.

TABLE 1. Shock wave data for bromoform.

P, GPa	Cs, km/s	T, K/1000	γ
49.8	5.7	5.7	0.619
54.2	5.87	6.1	0.639
62.8	5.41	6.73	0.944
69.7	5.47	7.5	0.983
69.7	5.6	7.75	0.944
72.1	5.55	7.97	0.977
78.0	6.52	8.35	0.736
83.3	6.44	9.0	0.807
85.9	6.53	9.7	0.802
88.9	6.08	10.3	0.932
100.0	7.09	12.6	0.749
104.8	7.17	13.05	0.756
109.6	6.57	13.25	0.904
109.6	6.68	13.25	0.882
114.7	6.73	12.8	0.887
117.4	6.56	13.23	0.926
117.2	6.78	13.2	0.892
130.5	6.92	15.1	0.890
137.9	7.01	15.87	0.889
138.8	7.02	14.5	0.887
189.0	7.6	20.7	0.851
201.9	7.9	22.7	0.822

EXPERIMENTS WITH SAPPHIRE.

Analyzer technique described above was used to investigate mechanical properties of the sapphire driver after shock loading up to 130 GPa. Sapphire plate was used in the past as an insulating driver for conductivity measurements in the multiple - shocked hydrogen [14]. The main idea of these experiments was to determine if any fracture takes place in the sapphire. A bromoform reservoir, covered with thin (7µm) Al foil, was used as a pressure gauge as described above. After shock loading by stainless steel striker with 5.1 km/s velocity, sapphire has expanded through the evacuated gap (1.45- 6.05 mm) and then impacted with $CHBr_3$ surface (Fig. 3).

If T-u relation is known, one can calculate the pressure history on the shock front from the

bromoform radiation history. For calculation of pressure history on sapphire - $CHBr_3$ interface procedure, described in [15] was used.

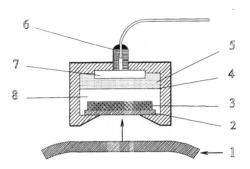

FIGURE 3. Experimental assembly for the sapphire properties investigation. 1. Stainless steel striker (2 mm width). 2. Stainless steel bottom (0.8 mm width). 3. Sapphire disk (1.8 mm width). 4. Al foil. 5. $CHBr_3$. 6. Diaphragmed optical fiber. 7. Glass window. 8. Evacuated gap.

One can see that a pressure on the shock front changes approximately $c_L/(c_L - D)$ (c_L - Lagrangian sound velocity, D - shock velocity) times slower than that on the sapphire - $CHBr_3$ boundary, that allows for 2-4 times resolution enhancement in $CHBr_3$ pressure range up to 150 GPa.

Calculated pressure profiles on sapphire-indicator boundary are shown in Fig. 4. The minimum registered pressure is limited by the sensitivity of the pyrometer and was about 20-25 GPa, so it is impossible to get the time of the beginning of the pressure pulse. In Fig. 4 are shown beginnings of curves, linearly extrapolated to zero pressure and placed on the time axis according to the distances ratios. It is clearly seen that in our case all three curves have a locus and self-modeling process of the smearing of pressure pulse front is realized. The increase in the vacuum gap leads to the increase in the rise time of the pressure pulse front and to the decrease in the maximum pressure of the pulse. The smearing of pressure pulse front was approximately 1/20 of a time need driver to cross vacuum gap.

Maximum impulse pressure has achieved the theoretical value for monolith sapphire impact only for 1.45 mm distance. This phenomenon can be explained by sapphire failure in the shock com-

511

pression process [16] and its failure and fragmentation starting from free surface in the unloading wave. It may be accompanied with mixing process

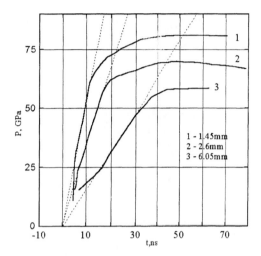

FIGURE 4. Measured (solid lines) and extrapolated to zero point (dashed lines) pressure profiles on driver- CHBr₃ boundary for various vacuum gap widths.

with low density material if this sapphire will used as a driver for its shock compression.

SUMMARY

Hugoniot sound velocity and brightness temperature of the shock front in bromoform were determined in the 50-200 GPa pressure range. Liquid analyzer technique was used to investigate the mechanical properties of shocked sapphire. It was shown that fracture and loss of the initial sapphire mechanical properties take place. So it becomes evident that application of brittle materials such as sapphire as a driver is a very delicate procedure.

ACKNOWLEDGMENTS

The work was supported by RFBR under grants N 94-02-03706-a and N 97-02-17439.

REFERENCES

1. Duffy T.S., Ahrens T.J. Hugoniot sound velocities and finite strain theory. In: SCCM - 1989, ed. by S.C.Schmidt, J.N.Jonson, L.W. Davison. Amsterdam, Elsevier Science Publishers B.V., 1990, 91-94.
2. McQueen R.G., Isaak D.G. Bromoform (CHBr₃) - a very high-pressure shock wave analyzer. ibid., 125-128.
3. Fritz J.N., Morris C.E., Hixson R.S., McQueen R.G. Liquid sound speeds at pressure from the optical analyzer technique. In: High Pressure Science and Technology -1993, ed. by S.C.Schmidt, J.W.Shaner, G.A.Samara, M.Ross. New York: AIP Press, 1994, part 1, 149-152.
4. Voscoboinikov I.M., Gogulya M.F. *Him.Fizika*, 1984, v.**3**, N 7, 1036-1041.
5. Glushak B.L., Zharkov A.P., et.al. *Sov.Phys.JETP*, 1989, v.**69(4),** 739-749.
6. Gogulya M.F. *Temperatures of shock compression of condensed matter*. Moscow: MIFI, 1988, pp. 3-10.
7. Belyakova M.Yu., Zhernokletov M.V., Sutulov Yu.N., Trunin R.F. *Izv. AN USSR, Fizika Zemli*, 1991, v.**1**.
8. Morgan J.A. *High Temperatures-High Pressures*, 1975, v.**7**, 65-70.
9. Voskoboinikov I.M., Dolgoborodov A.Yu. Sound velocities and temperatures on isoentropes of shock-compressed CCl₄ and CHBr₃. In: Detonaciya-1989, ed. by Dremin A.N., Chernogolovka, 1989, 91-95.
10. *Physics of high energy density*. Moscow: Mir, 1974.
11. Morris S.E., Fritz J.N., McQueen R.G. *J.Chem.Phys.*, v. **80**, N **10**, 5203-5218.
12. Khischenko K.V, Lomonosov I.V., Fortov. V.E. Equation of state for organic compounds over wide range of densities and pressures. In: SCCM-1995, ed. by S.C.Schmidt, W.C.Tao, Woodbury, New York, 1996. Part 1, 125-128.
13. Bushman A.V., Zhernokletov M.V., Lomonosov I.V., et.al. *ZETP*, 1996, v.**109**, N **5**, 1662-1666.
14. Weir S.T., Mitchell A.C., Nellis W.J. *Phys.Rev.Lett.*, v. **76**, N **11**, 1860-1863.
15. Kanel G.I., Razorenov S.V., Utkin A.V., Fortov V.E.. *Shock - wave phenomena in condensed matter*. Moscow:, Yanus-K, 1996, pp.72-74.
16. Grady D.E. Shock-wave properties of brittle solids. In: SCCM-1995, ed. by S.C.Schmidt, W.C.Tao, Woodbury, New York, 1996. Part 1, 9-20.

512

CP429, *Shock Compression of Condensed Matter – 1997*
edited by Schmidt/Dandekar/Forbes
© 1998 The American Institute of Physics 1-56396-738-3/98/$15.00

DAMAGE KINETICS IN SILICON CARBIDE

I. M. Pickup and A. K. Barker

Defence Evaluation Research Agency, Chertsey, Surrey, England, KT16OEE.

Three silicon carbides of similar density and grain size but manufactured via different routes (reaction bonded, pressureless sintered and pressure assisted densification) have been investigated. High speed photography in conjunction with Hopkinson pressure bar compression tests has revealed that not only does the manufacturing route confer a significant difference in failure kinetics but also modifies the phenomenology of failure. Plate impact experiments using lateral and longitudinal manganin stress gauges have been used to study shear strength behaviour of damaged material. Failure waves have been observed in all three materials and characteristically different damaged material shear strength relationships with pressure have been observed.

INTRODUCTION

The effectiveness of ceramic armour subject to kinetic energy ballistic penetration is associated with the creation and flow of a damaged (comminuted) region around the tip of the penetrator (1). The relative ballistic protection that different ceramics may offer is due to the relative flow resistance of this dynamically confined transient phase. It is proposed that the flow resistance is strongly dependent on the confining pressure, the deviatoric stress and the comminutia pattern. Shear strengths in excess of 5GPa behind a failure wave in plate impact experiments have been measured for the silicon carbides considered in this paper (2). If similar yield strengths exist in the comminutia ahead of a tungsten alloy penetrator, which has a much

lower yield strength, then, at normal ordinance velocities, the penetrator will erode rather than penetrate. Unfortunately for the armour designers this happy situation will not exist indefinitely; confining pressure will reduce as release waves arrive from free surfaces and the yield properties of the ceramic will decline as fracture processes develop. Clearly it is advantageous for the decline in flow properties to be minimised with respect to time. One of the objectives of this study is to consider *relative* failure kinetics and ultimately relate these to the material microstructure.

MATERIALS

Three silicon carbides are considered, each manufactured from a significantly different method

TABLE 1. The Physical Properties of the Silicon Carbides.

Material	Density (kgm⁻³)	E (GPa)	G (GPa)	ν	Mean grain diameter (μm)	QS Comp. strength (GPa)	SHPB Comp. strength (GPa)
SiC 1	3210	408	174	0.17	1.16	4.48±0.18	6.72±0.27
SiC 2	3163	423	183	0.16	4.48	5.21±0.50	7.47±0.32
SiC 3	3238	456	196	0.16	2.90	5.15±0.35	8.17±0.16

but resulting in similar densities. Table 1 summarises the mechanical and physical properties. SiC 1 is produced from a reaction bonding technique; SiC 2 is pressureless sintered; and SiC 3 is made from a pressure assisted densification method.

EXPERIMENTAL PROCEDURE

Quasi-static (QS) compressive strength was measured at a strain rate of $\sim 10^{-3}s^{-1}$ using waisted, 6mm diameter cylindrical specimens with a parallel gauge length of 5mm.

High strain rate tests ($\sim 10^3 s^{-1}$) were conducted on a 16mm diameter split Hopkinson pressure bar (SHPB) using a compressive stress pulse of approximately 200μs. The specimens were scaled down versions of the QS specimens with a gauge diameter of 3mm. Photographs were taken during testing using an Imco, Ultranac high speed framing camera.

Shear strength measurements of damaged material behind a failure front were measured at various pressures in plate impact experiments (2).

RESULTS

Considering the differences in manufacturing route, the differences in quasi-static strengths of the three materials is small, Fig. 1. However, there is a significant difference in the rate-dependent increase in strength. SiC 1 and SiC 2 have an approximately 2.2GPa increase between the QS and SHPB

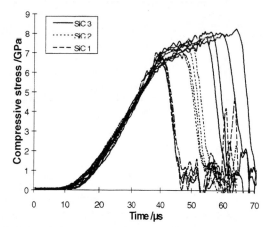

FIGURE 2. SHPB stress histories exhibiting prolonged periods of inelastic behaviour for SiC 2 and SiC 3.

compressive strength compared to 3GPa for SiC 3. This is a good indicator of the relative damage kinetics (3,4) where SiC 3 is the most resistant to damage.

It is useful to look in detail at the high strain rate stress histories of the ceramics. Figure 2 shows several histories for each of the three silicon carbides. SiC 1 fails in an apparently ideally brittle behaviour, typical of that seen in alumina, but SiC 2 and SiC 3 demonstrate a significant period of apparently *inelastic flow.*

After a significant change in slope at around 6GPa, a monotonically increasing load is supported for ~ 10-12μs for SiC 2 and up to 35μs for SiC 3. This is a remarkably long period for what is essentially a brittle material to survive after exceeding the QS failure limit. It is notable that the termination point of this inelastic behaviour is rather more variable for SiC 3 than it is for SiC 2.

In an attempt to determine the operative mechanisms during these periods of *inelastic flow,* 6 specimens each of SiC 2 and SiC 3 were tested in the Hopkinson bar in conjunction with high speed photography. Figure 3 shows a sequence of four frames from a SiC 2 and a SiC 3 test, with corresponding stress histories and fiducial markers.

Significantly different and reproducible failure phenomenology was found for the two materials. Failure initiation for SiC 2 is clearly visible in frame 2, or even frame 1, by the right shoulder of the specimen. By reference to the fiducial markers on

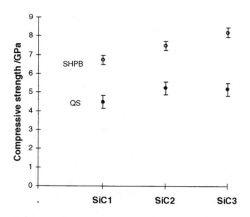

FIGURE 1. Compressive strengths at QS strain rates ($10^{-3}s^{-1}$) and SHPB rates ($10^3 s^{-1}$).

the material's loading trace, initiation of ejection of failed particles occurs well before the maximum load. In fact for SiC 2 initiation for all 6 specimens was around the QS failure strength. Ejecta is produced along the gauge length up to the maximum load, some 2GPa above the initiation stress.

For SiC 3 there is no indication of ejecta until the maximum load, e.g. frame 2, more than 3GPa above the QS initiation stress. However from the loading trace it is apparent that the specimen is becoming softer at a much lower stress indicating damage formation.

The velocity of the ejecta particles has been estimated by measurement of the rate of ejecta plume dilation as approximately 100ms[-1] for SiC 2 and 250ms[-1] for SiC 3.

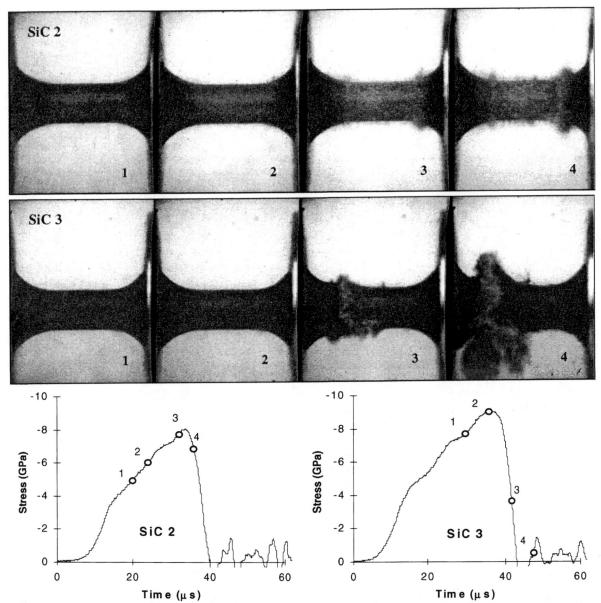

FIGURE 3. Selected frames of the gauge length of SHPB test specimens of SiC 2 and SiC 3 showing the failure process during high rate testing. The respective stress histories have numbered fiducial markers corresponding to the relevant frame number.

DISCUSSION

In general ceramics have slight strain rate-dependent compressive strengths up to strain rates of around 10^2-$10^3 s^{-1}$ due to thermally activated crack kinetics. At higher strain rates inertial effects dominate. Kipp *et al* (3) have proposed an analysis based on inertia dominated fracture kinetics which allows reasonable prediction of the onset of this highly rate-dependent regime. Janach (5) has proposed a rate-dependency model based on the inertia of ejecting comminuted particles. Flaws nucleate and coalesce and failed particles are subsequently ejected at the surface of a specimen. He postulated the existence of an ejecta wave moving towards the axis of the specimen which on arrival caused complete failure. The velocity of this ejecta wave, which is inversely proportional to the volumetric strain associated with comminutia development, dictates the rate sensitivity.

Clearly the three silicon carbides in this study have significantly different rate sensitivities and characteristic failure times, i.e. the period from attainment of the QS failure strength to catastrophic failure under specific high strain rate conditions. For the experimental programme yielding the data in Fig. 2, mean characteristic failure times of 10µs, 18µs and 30µs may be attributed to SiC 1, SiC 2 and SiC 3 respectively. The production route also affected the mode of failure significantly. The pressureless sintered material SiC 2, comminuted very quickly and ejecta was observed at around the QS compressive strength in SHPB tests. This has the appearance of a Janach type behaviour, i.e. fast coalescence; slow ejecta wave. SiC 3 clearly did not behave in the same way. There was no evidence of any ejecta being formed from exceeding the QS compressive strength up to the attainment of the highest stress. The specimen did however show signs of becoming less stiff (reduction in slope of the load-time trace) indicating that flaws were being activated but were reluctant to coalesce. Catastrophic failure was much more energetic than in SiC 2 as unfailed ligaments were subject to much higher elastic strain energy. The relatively wide scatter in failure times for SiC 3 compared to the other materials is probably a reflection of the statistical nature of coalescence.

It is possible that SiC 3 resists flaw activation by microplastic accommodation. In addition this material has fewer large scale intergranular inclusions than the other two materials and consequently has a lower number of inherent flaw sites.

The ranking of relative resistance to damage, presented above, was demonstrated by Bourne *et al* (2) on the same materials. In a series of plate impact experiments, the shear strength was measured behind a failure front, over a range of pressures. The shear strength of damaged SiC 3 pressure hardened over an applied longitudinal stress range of ~1 to 1.4 times the Hugoniot elastic limit, whereas the shear strength of SiC 1 and SiC 2 softened over the same stress range.

Significantly, the ranking of damage tolerance, discussed above, matches the ranking of ballistic performance of the SiC ceramics (6).

CONCLUSIONS

Significant differences have been found in the kinetics of damage during high strain rate compression testing of 3 silicon carbides manufactured from different routes. The ranking of the damage kinetics from 1-D stress experiments matches that of the pressure-yield characteristics of damaged material measured in plate impact experiments and also the ballistic performance. It is believed that microplasticity significantly contributes to reduced damage kinetics.

REFERENCES

1. Curran, D. R., Seaman, L., Cooper, T. and Shockey, *J.Impact Engng.*, **13**, No. 1, 53-83, (1993).

2. Bourne, N., Millett, J. and Pickup, I. M., *J. Appl. Phys.*, **81**, 6019-6023, (1997).

3. Kipp, M. E., Grady, D. E. and Chen, E. P., *J. Int. Fract.*,**16**, 471, (1980)

4. Lankford, J., *Fracture Mechanics of Ceramics Volume- 5*, New York-London, Plennum Press, 1980, pp. 625-637.

5. Janach, W., *Int. J. Rock Mech. Min. Sci. & Geomech. Abstr.*, **13**, 177-186, (1976).

6. James. B. J., "Factors affecting ballistic efficiency tests and the performance of modern armour systems", presented at the European Fighting Vehicle Symposium, Shrivenham,UK., May 1996.

CP429, *Shock Compression of Condensed Matter – 1997*
edited by Schmidt/Dandekar/Forbes
© 1998 The American Institute of Physics 1-56396-738-3/98/$15.00

FLOW BEHAVIOR OF SODA-LIME GLASS
AT HIGH PRESSURES AND HIGH SHEAR RATES

S. Sundaram and R. J. Clifton

Division of Engineering, Brown University, Providence, Rhode Island 02912

Results are reported for a plate impact investigation of the flow behavior of soda-lime glass. A $5\mu m$ thick layer of soda-lime glass is sandwiched between two hard, elastic Hampden steel plates and impacted by another steel plate. The impact is skewed with respect to the direction of travel of the flyer plate in order to generate both compression and shear waves in the target. The dynamic response of the glass to this combined loading is measured using laser interferometry. The measured wave profiles are used to develop a micromechanism-based deformation model for the glass. It is also hoped that the investigation will contribute to an understanding of so-called *failure waves* in glasses through improved understanding of the dynamic shearing resistance of glasses.

INTRODUCTION

Normal impact experiments on soda-lime and similar glasses have shown an increase in the *transverse* compressive stress without a change in the normal compressive stress when a postulated 'failure wave' passes the observation point, thus implying a fall in shear stress at that point[1]. Experimental measurements indicate that the glass has no spall strength behind the failure wave front while retaining substantial spall strength ahead of the failure wave front. The experiments described in this paper were designed specifically to study the influence of shear loading on these oxide glasses.

EXPERIMENTAL TECHNIQUE

The pressure-shear plate impact technique involves the impact of a *flyer plate* travelling at a speed of 100-800 m/s against a target (Fig. 1). The impact is skewed at an angle θ relative to the direction of travel so that both longitudinal (compressive) and transverse (shear) waves are produced upon impact. Details of the theory and techniques of plate impact can be found in other

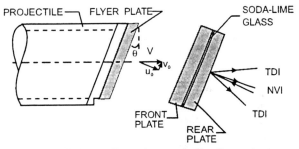

FIGURE 1. Experimental Configuration

publications[4,6]. In the experiments described here, a thin $(5\mu m)$ layer of vapor-deposited soda-lime silica glass $(\rho = 2530 kg/m^3, c_1 = 5.74 mm/\mu s, c_2 = 3.40 mm/\mu s)$ is sandwiched between two hard Hampden steel plates $(\rho = 7861 kg/m^3, c_1 = 5.98 mm/\mu s, c_2 = 3.26 mm/\mu s)$ and is impacted by another Hampden steel plate. The steel plates remain elastic during the loading. The time-distance diagram for the experiment is shown in Fig. 2. The compressive wave generated at impact reaches the specimen first and establishes a uniform state

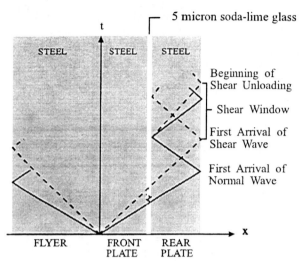

5 micron soda-lime glass

FIGURE 2. Time-Distance Diagram

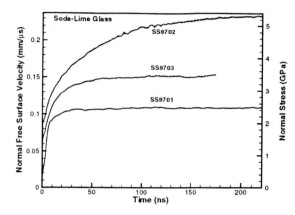

FIGURE 3. Normal-Velocity Profiles

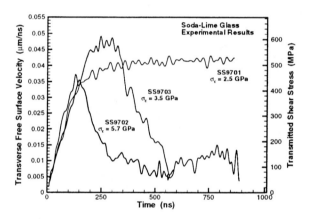

FIGURE 4. Transverse-Velocity Profiles

of compressives stress in the specimen before the shear wave arrives. We are interested in the shear flow strength of the soda-lime glass under the imposed compressive stress. Both, the compressive and shear waves, continue on to the traction-free rear surface of the target where they are monitored using laser intererometry.

EXPERIMENTAL RESULTS

Three plate impact experiments were performed on the soda-lime glass. In all three shots, a skew angle of 22° was used. For the highest velocity shot, SS9702, tungsten carbide ($\rho = 14600 kg/m^3, c_1 = 6.63 mm/\mu s, c_2 = 4.03 mm/\mu s$) impacting plates were used with a Hampden steel rear plate. The impact velocities for shots SS9701,2 and 3 were 118, 198 and 169 m/s respectively.

The initial part of the normal velocity profiles measured at the rear (free) surface are shown in Fig. 3 and the corresponding transverse velocity profiles are shown in Fig. 4. In both figures, the right ordinate shows the corresponding stresses in the soda-lime glass specimen. From Fig. 3 it can be seen that the normal stresses attain a steady value within 100 nanoseconds of impact — much before the shear wave arrives at the specimen. The compressive stresses on the specimen, in order of increasing magnitude, are 2.5, 3.5 and 5.7 GPa. In Fig. 4, the compressive

stress on the specimen corresponding to the shear stress profiles is indicated alongside the curves. One can immediately see the dramatic weakness in shear that soda-lime glass exhibits at the higher impact velocities (coresponding to higher shear strain rates for any given stress level and higher shear strains at any given time). A similar weakness in shear of soda-lime glass has also been observed by other investigators[2] who performed pressure-shear plate impact experiments on thick soda-lime glass specimens as part of a study of the so-called *failure waves* in these glasses. A fairly straightforward analysis[6] of the profiles shown in Fig. 4, in conjunction with the known parameters of the experiment, indicates that the average shear strain in the sample just before the drop in shear stress occurs is 2.05 in shot SS9703 and 1.97 in shot SS9702. In shot SS9701 where no loss of shear strength is observed, the average shear strain at the end of the measurement time (900 ns) is calculated to be approximately 0.5. It appears, therefore, that soda-lime glass exhibits a dramatic loss in shear strength at a shear strain of approximately 2.0.

CONSTITUTIVE MODELING

The conclusion reached in the previous section can now be examined in light of a mechanism of inelastic deformation in oxide glasses suggested by Myuller[5]. This mechanism is shown schematically in Fig. 5 and involves a switching of adjacent covalent Si–O bonds at large shear strains. The combined influence of short-range and directional nature of the covalent bonding in oxide glasses makes such a switching mechanism possible only at very large shear strains. It is also apparent that such a large-scale bond-switching occuring at some large shear strain will cause a strong relaxation of stress in the material. Furthermore, due to the amorphous nature of the network, the bond switching is not likely to be completed at all locations, leading to a large loss in shear strength. It is clear that many of the important features of the experimental results can be explained by the physical mechanism described above. The experimental results suggest a critical shear strain of 2.0 at which this deformation mechanism is activated A detailed discussion of this mechanism can be found in Sundaram[6].

Based on the above considerations, some aspects of a constitutive model for the inelastic deformation of soda-lime glass can now be formulated. Let f_i and f_c represent the fraction of atomic sites where the bonding is, respectively, ionic and covalent. Initially the plastic flow will be confined to the ionic sites of the network where the modifier ions are present. Of these ions only those that have sufficient adjacent free volume can participate in the flow[3]. Following the rate expressions for metallic glasses[3], the

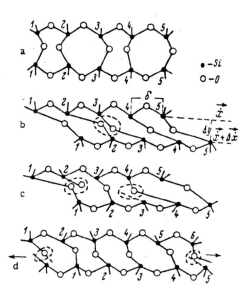

Fig. 5 Bond Switching Mechanism in Oxide Glasses (From Myuller (1960))

inelastic shear strain rate can be written as

$$\dot{\gamma}^p = \dot{\gamma}_i^p = f_i \exp(-\frac{v^*}{\overline{v}_f})\nu \exp(-\frac{\Delta G_i^0}{kT})2\sinh\frac{\tau\Omega_i}{kT} \quad (1)$$

where ΔG_i^0 represents the activation barrier for the ionic bond-switching process, Ω_i represents the shear activation volume for the process, and ν is the attempt frequency. The quantities v^* and $\overline{v}_f$ represent the volume of an ion and the average free volume respectively. The term $\exp(-\frac{v^*}{\overline{v}_f})$ represents the fraction of modifier ions that are adjacent to a site with free volume greater than the ionic volume. It is only these ions that can participate in the flow process. During this initial ionic flow, the covalent network is progressively distorted. The distortion of the network decreases, locally, the free volume v_f in the glass, making it harder for ionic flow to occur. This shear-assisted reduction in free volume will, in place of a more detailed derivation, be simply represented by the following rate equation for the evolution of the average free volume $\overline{v}_f$:

$$\dot{\overline{v}}_f = -A\dot{\gamma}^p. \qquad (2)$$

When the total plastic strain reaches a critical value γ_{cr} (approximately equal to 2.0), a bond-switching mechanism is taken to be activated in the covalent network. Using similar notation as above, with subscripts c to represent the covalent network, we write

$$\dot{\gamma}_c^p = f_c \cdot \nu \exp(-\frac{\Delta G_c^0}{kT}) 2\sinh\frac{\tau\Omega_c}{kT} \qquad (3)$$

for the plastic strain rate in the covalent network. The total plastic strain rate in the glass is now given by

$$\dot{\gamma}^p = \dot{\gamma}_c^p + \dot{\gamma}_i^p \qquad (4)$$

In the above equation, it is expected that the contribution of the ionic flow process to the total strain rate is small compared to the contribution of the covalent relaxation process. The activation barrier height ΔG_c^0 is taken to be equal to the bond dissociation energy for Si–O bonds (approximately 1 eV). The shear activation volume Ω_c can be expected to be at least a few atomic volumes because of the covalent network structure.

The plate impact experiments have been computationally simulated using a second-order accurate finite difference code to solve the wave propagation problem in the specimen. Details of the simulation technique are described in other publications[4,6]. The parameter values are listed in the Appendix and the computed transverse velocity profiles are shown in Fig. 6. The results show fairly good qualitative and quantitative agreement with the experimental profiles. Further discussion of these results is given in Sundaram[6].

ACKNOWLEDGEMENTS

The authors would like to gratefully acknowledge the support of this research by the Army Research Office through the URI at Brown University on the Dynamic Behavior of Brittle Materials.

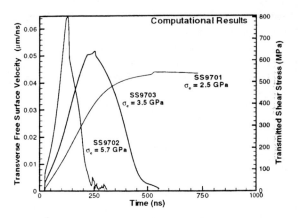

FIGURE 6. Computational Shear Profiles

REFERENCES

1. Brar, N. S. and Bless, S. J., *High Pressure Research*, Vol. 10, pp. 773–784, 1992.
2. Clifton, R. J., Mello, M. M. and Brar, N.S., *Shock Compression of Condensed Matter - 1997*, S. C. Schmidt *et al.*, eds., 1997.
3. Spaepen, F., *Acta Metallurgica*, Vol. 25, pp. 407–415, 1977.
4. Sundaram, S., and Clifton, R. J., "Pressure-Shear Impact Investigations of the Dynamic Response of Ceramics," ASME AMD-Vol. 219, *Advances in Failure Mechanisms in Brittle Materials*, R. J. Clifton and H. D. Espinosa, ed., pp. 59-80, 1996.
5. Myuller, R. L., *The Structure of Glass, Proceedings of the Third All-Union Conference on the Glassy State, Leningrad (1959)*, Vol. 2, pp. 50–57, Consultants Bureau, N.Y., 1960.
6. Sundaram, S., Ph.D. Thesis, Brown University, 1998.

Appendix Values used in Simulations

$f_i = 0.4$

$f_c = 0.6$

$\nu = 10^{13}~\text{s}^{-1}$

$v^* = 33~(\text{A}^o)^3$

$\overline{v}_f^0 = 5~(\text{A}^o)^3$

$A = 1.5~(\text{A}^o)^3$

$\Omega_i = 45.0~(\text{A}^o)^3$

$\Omega_c = 100.0~(\text{A}^o)^3$

$\Delta G_i^0 = 0.2~\text{eV}$

$\Delta G_c^0 = 1.0~\text{eV}$

$\gamma_{cr} = 2.0$

CP429, *Shock Compression of Condensed Matter – 1997*
edited by Schmidt/Dandekar/Forbes
© 1998 The American Institute of Physics 1-56396-738-3/98/$15.00

EFFECT OF SHEAR ON FAILURE WAVES IN SODA LIME GLASS

R. J. Clifton, M. Mello, and N. S. Brar[*]

Division of Engineering, Brown University, Providence, Rhode Island 02912
* University of Dayton Research Institute, Dayton, Ohio 45469-0182

By means of in-material stress gauges, failure waves in shock-compressed soda lime glass have been shown to be distinguished by a marked reduction in shear stress. To explore further the relation between failure waves and shearing resistance, a series of pressure-shear impact experiments have been performed involving the impact of a glass plate by a steel flyer plate and vice versa. The latter configuration is designed to allow direct measurements of the shearing resistance of the failed material. In both configurations, the normal and transverse motion of the free surface of the target is monitored using laser interferometry. The transverse velocity-time profiles show a pronounced loss in shearing resistance of the glass at impact velocities above the threshold for failure waves to occur.

INTRODUCTION

For impact-induced stresses above a certain threshold, plate impact experiments on soda lime glass have led to a reduction of shearing strength and the loss of tensile strength across a "failure wave" that propagates at speeds less than the elastic wave speeds for the glass[1]. Similar behavior has been observed by other investigators for a variety of glasses[2-4]. Further evidence of a pronounced change in the mechanical properties of glass across the failure wave is the occurrence of a so-called "recompression wave" which results when the tensile unloading wave from the traction-free rear surface of the target meets the oncoming failure wave[2,3].

Such failure wave phenomena have not been fully explained although several attempts have been made. Raiser et al.[3] examined the possibility that reported failure wave phenomena did not represent bulk material response, but instead was an artifact of surface preparation. However, their experiments showed that surface roughness had no measurable effect on failure wave phenomena over the full range of surface roughnesses from 10 *nm* to 530 *nm*. It can be argued that these experiments do not rule out the possibility that failure waves are primarily due to surface effects as even the smoothest impact surfaces may still have been rough enough to cause the failure waves. Bourne[4] made a further investigation of surface effects by comparing the propagation of failure waves through homogeneous glass plates to their propagation through stacked plates with the same total thickness. He concluded that the failure waves arrive somewhat earlier when propagating through the stacked plates – suggesting that at least some reinitiation of failure waves occurs as the elastic precursor propagates through the interfaces between stacked plates.

From the initial results of Brar et al.[1] on the reduction in shearing resistance across a failure wave, and from numerous reports of the plastic flow of glasses under large shear stresses, it appears that further examination of the role of shear stresses in failure wave phenomena is warranted. To this end, the

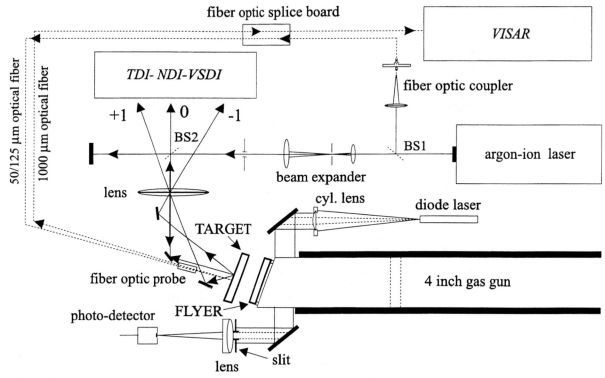

FIGURE 1: Pressure-Shear Impact Configuration

current series of experiments has been conducted under pressure-shear impact conditions for which the shear stresses can be changed independently by changing the angle of inclination of the two plates relative to the direction of approach.

EXPERIMENTAL APPROACH

The experimental configuration is shown in Figure 1. Each experiment involved the impact of a Hampden steel plate ($\rho = 7861 kg/m^3, c_1 = 5.98 mm/\mu s, c_2 = 3.26 mm/\mu s$, thickness $\approx 4.4 mm$)with a soda lime glass plate ($\rho = 2530 kg/m^3, c_1 = 5.74 mm/\mu s, c_2 = 3.40 mm/\mu s$, thickness $\approx 5.6 mm$, except thickness $\approx 12.7 mm$ for shots MM9702 and MM9705). Shots are denoted as steel/glass impact or glass/steel impact with the first named plate being the flyer and the second named plate being the target. The projectile velocity is measured by recording the time for the projectile to cut a sheet of light of known width. Impact occurs in vacuum. The motion of the rear surface of the target plate is recorded by means of a variety of interferometers: the normal velocity is measured with a fiber-optic velocity interferometer for any reflector (VISAR); the normal displacement is also monitored with a normal displacement interferometer (NDI); the transverse motion is obtained by using beams diffracted from a diffraction grating (600 $lines/mm$) deposited on the rear surface of the target – this motion is obtained either from a transverse displacement interferometer (TDI) using the $+1$ and -1 order diffracted beams or a variable sensitivity displacement interferometer (VSDI) using combinations of the $+1$, -1, and 0 order diffracted beams. The t-X diagram for the experiments is shown in Figure 2. Longitudinal wavefronts are shown as solid lines; shear wave fronts are shown as heavy dashed lines. The failure wave shown as a dashed line is suggestive of the failure wave propagation reported in previous studies. The impact velocities are designed to be sufficiently low that the response of the steel plates can be taken to be linearly elastic.

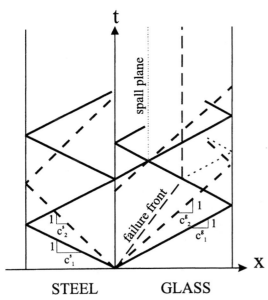

FIGURE 2: t-X Diagram

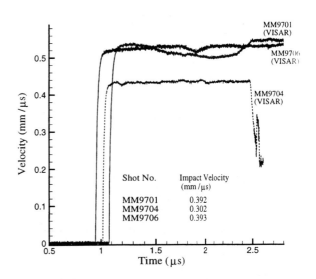

FIGURE 3: Free-Surface Normal Velocity for Steel/Glass Impact ($\theta = 18°$).

RESULTS AND DISCUSSION

Normal and transverse velocity-time profiles for three steel/glass impact experiments are shown in Figures 3 and 4, respectively. These experiments are designed for spall to occur behind the failure wave. The impact velocity of $0.302 mm/\mu s$ of Shot MM9704 is clearly below the threshold for failure waves as the normal velocity record is flat until the spall signal arrives; then, the fall in particle velocity at approximately $2.5\mu s$ indicates substantial spall strength. In contrast, the normal velocity records for the shots at impact velocities of 0.392 and $0.393 mm/\mu s$ are below the level predicted by elastic response, exhibit appreciable structure, and show no spall strength at the time when the spall signal should arrive. Furthermore, both records show an upturn at a time representative of the time at which the "recompression wave" would be expected to arrive. The corresponding transverse velocity-time profiles in Figure 4 show that the transverse velocity rises quickly to a level that is equal (MM9701) or nearly so (MM9706) to the level predicted by elastic response and then fall off continuously. Thus, the shots that exhibit customary failure wave phenomena also exhibit a progressive loss in capacity to transmit shear.

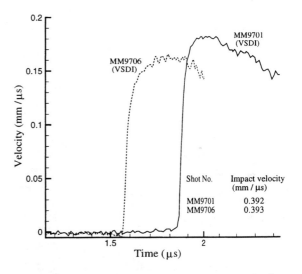

FIGURE 4: Free-Surface Transverse Velocity for Steel/Glass Impact ($\theta = 18°$).

Normal and transverse velocity-time profiles for three glass/steel impact experiments are shown in Figures 5 and 6, respectively. These experiments are designed to probe the stress state in the glass, in the region behind the failure wave (i.e. adjacent to the steel plate). Because

523

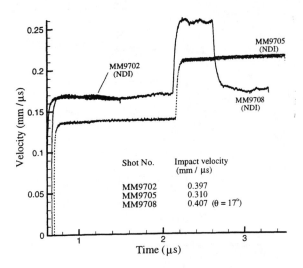

FIGURE 5: Free-Surface Normal Velocity for Glass/Steel Impact ($\theta = 18°$).

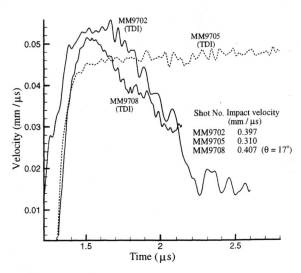

FIGURE 6: Free-Surface Transverse Velocity for Glass/Steel Impact ($\theta = 18°$).

the steel remains elastic under the conditions of impact, the velocity-time profiles at the free surface of the steel plates are proportional to the corresponding stresses at the impact plane (until multiply reflected waves arrive). Again, at the lowest impact velocity of $0.310 mm/\mu s$ the response agrees with elastic predictions, although the last 10% of transverse velocity is transmitted over a rather large risetime. At the higher impact velocities the normal velocity plateaus are below the levels expected from elastic considerations. However, the reflection coefficient inferred from the step at $\approx 2.1\mu s$ due to the reflection of the longitudinal unloading wave originating at the free surface of the steel is essentially the same as the value inferred from the analogous step for the elastic case, shot MM9705. The transverse velocity records show a quick rise to peak values that are below elastic predictions, followed by a substantial fall before stabilizing at a lower plateau. The corresponding fall in shear stress is interpreted as a clear indication of the loss of shearing strength in the glass adjacent to the impact face.

From the experiments reported here it appears that failure wave phenomena is closely related to loss in shearing resistance of the glass. Figure 6 suggests that the shearing response of the glass becomes unstable at sufficiently high shear stresses or shear strains. Figure 5 suggests that, in spite of the loss in shearing resistance, the unloading response of the glass in uniaxial strain is essentially the same as the elastic response of the undeformed glass.

ACKNOWLEDGEMENTS

The authors gratefully acknowledge the support of this research by the Army Research Office through the URI at Brown University on the Dynamic Behavior of Brittle Materials.

REFERENCES

1. Brar, N.S., Bless, S.J., and Rosenberg, Z., *Appl. Phys. Lett.* **59**, 3396-3398 (1991).
2. Kanel, G.I., Rasorenov, S.V., and Fortov, V.E., in *Shock Compression of Condensed Matter - 1991*, edited by S.C. Schmidt, R.D. Dick, J.W. Forbes, and D.G. Tasker (North-Holland, Amsterdam, 1992), pp. 451-454.
3. Raiser, B.F., Wise, J.L., Clifton, R.J., Grady, D.E., and Cox, D.E., *J. Appl. Phys.* **75**, 3862-3869 (1994).
4. Bourne, N. Millett, J. Rosenberg, Z., and Murray, N., *J. Mechs. Phys. Solids*, to appear.

CP429, *Shock Compression of Condensed Matter – 1997*
edited by Schmidt/Dandekar/Forbes
1998 The American Institute of Physics 1-56396-738-3/98/$15.00

SHOCK, RELEASE, AND TENSION RESPONSE OF SODA LIME GLASS

Dattatraya P. Dandekar

Army Research Laboratory, Weapons and Materials Research Directorate
Aberdeen Proving Ground, Maryland 21005

This work describes the result of shock wave experiments on soda lime glass in which the shock wave profiles were recorded simultaneously at or near the impact surface and the free surface of the glass specimen by means of multi-beam VISAR. Since earlier work[1] indicated that the glass under shock compression does not follow the Gladstone-Dale model, these profiles provide accurate and self consistent values of transit times for shock, release, and tensile waves propagating in soda lime glass.

INTRODUCTION

Earlier work on soda lime glass by Dandekar [1] showed that the change in refractive index of soda lime glass with shock compression does not follow the Gladstone-Dale model. This model implies that relative change in the density (ρ) of a material ($d\rho/\rho$) equals $[dn/ (n-1)]$, where n is the refractive index of the material. The non-compliance of soda lime glass with this model suggests that a shock wave experiment could be designed to obtain information pertaining to magnitude of shock, release, and tensile wave velocities from a single shock wave experiment at a given shock stress. Thus in soda lime glass the arrival times of various waves can be measured accurately without depending on the precision with which the time of impact is determined in a given shock wave experiment. The transparency of soda lime glass coupled with the observation that its change in the refractive index under shock compression does not follow Gladstone-Dale model permits the measurement of tensile wave velocity, and thereby the value of its tensile impedance. Primary motivation for initiating the present work was to measure tensile wave velocity in a material and compare the value of tensile impedance obtained from the product of density and velocity with the value of tensile impedance obtained directly from the experiment described by Dandekar [2]. In that work [2] it was shown that through simultaneous measurements of free surface velocity in a conventional symmetric impact shock wave spall experiment and rear surface particle velocity in a companion experiment where the sample is backed by a well characterized low impedance material one can obtain the value of effective tensile impedance of a material. It thus provides a possibility of obtaining two independent estimates of the tensile impedance of a transparent material.

DESIGN OF EXPERIMENT

The experiment consists of symmetric impact of two flat plates of a transparent material.

A general configuration of the experiment is shown in Figure 1 (a). The experiment consists of impacting a stationary plate i.e., a target with a thinner plate of the same material, i.e., an impactor, at a given impact velocity. The particle velocity is recorded both at or near the impact surface and at the free surface of the target. Figure 1 (b) shows the corresponding x-t diagram. For simplicity, a single shock is assumed to be propagating in the plates at the given impact velocity, the rarefaction fan is assumed to have negligible dispersion, and the tensile wave is also assumed to be non-dispersive.

Impact generated shock waves begin to propagate both in the target and the impactor away from the impact surface at the time of impact t_0 [Fig. 1 (b)]. Upon reaching the respective free surfaces of the impactor at time t_1 and the target at time t_2, release

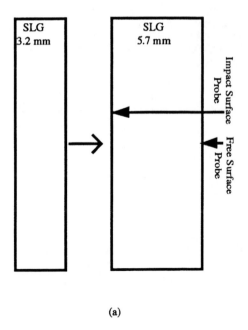

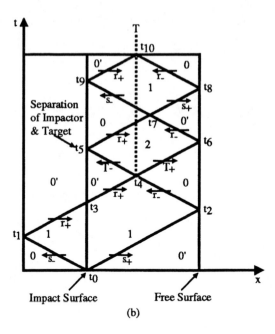

(a)   (b)

Figure 1. (a) Configuration of shock wave experiment in soda lime glass (SLG), (b) x-t diagram of the experiment.

waves begin to propagate towards the impact surface. The release wave originating at the free surface of the impactor reaches the impact surface at time t_3. As a result of the propagation of this release wave the stress in the impactor and target is reduced to zero. The corresponding particle velocity is different from zero only if the material undergoes an irreversible deformation due to the shock wave propagation. Otherwise it is zero. The leftward going release wave, originating at the free surface of the target, meets with the rightward going release wave originating at the free surface of the impactor in the target at time t_4. This generates tensile waves propagating in the directions of the free surface and impact surface of the target. If the target material sustains this tension and preserve its integrity, the leftward going tensile wave reaches the impact surface at t_5 and the rightward going tensile wave reaches the free surface at t_6. These waves interact with the free surface and shock up to zero stress state. The particle velocity of the leftward going shock wave as a result of the interaction with the free surface of the target is zero or differs from zero in case of irreversible deformation. Similarly the particle

velocity of the rightward moving shock corresponding to zero stress state originating at the impact surface of the target due to leftward moving tensile wave is accelerated to a magnitude equal to the impact velocity unless the deformation suffered by the material is irreversible. These two shock waves meet in the interior of the target plate and material is re-shocked to a high stress. All further interactions take place in the target material only because as a result of the interaction of the tensile wave going towards the impact surface the target and the impactor plates separate. Ideally, if the deformation of the material is elastic, reverberations of the propagating waves in the target plate will generate shock and tension as long as one dimensional strain condition remains in effect. The time of arrival of the waves at the impact surface and the free surface of the target were monitored by means of multi-beam VISAR. Aluminum was vapor deposited on the impact surface and the free surface of the glass target plate to facilitate recording of wave profiles at these two surfaces of the target. It should be noted that the free surface velocity profile is not affected by refractive index changes in the glass due to shock, release, and tensile wave propagations.

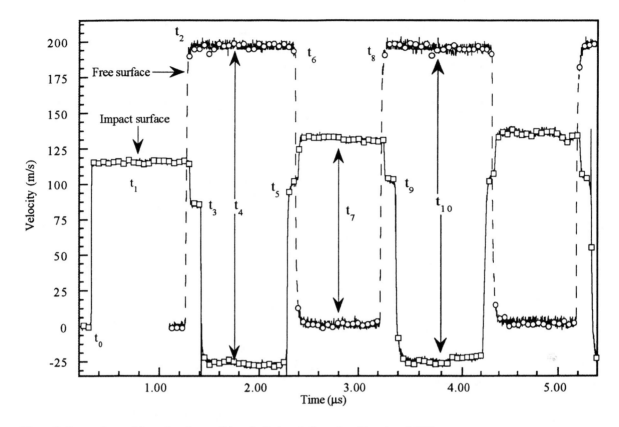

Figure 2. Free surface and Impact surface particle velocity in soda lime glass (Experiment 610).

MATERIAL

Soda lime glass used in the present work was provided by Dr. Brar, University of Dayton, Ohio. The density of the glass is 2.491 ± 0.006 Mg/m^3. The measured values of longitudinal and shear wave velocities measured by a pulse echo-overlap technique are 5.82 ± 0.01, and 3.45 ± 0.03 km/s, respectively. The glass specimens were square disks with linear dimension of at least 70 mm. The thickness of the glass impactors used in the experiments were 3.17 ± 0.01. The thickness of the target specimens varied between and 5.64 -5.67 mm. In experiment #606, the target consisted of two glass plates of thicknesses 1.49 and 5.67 mm bonded together.

RESULTS

The results of shock wave experiments performed on soda lime glass are summarized in Table 1. In all but experiment #606 particle velocity at the impact surface and free surface velocity of the targets were monitored simultaneously. In experiment #606 particle velocity was instead monitored at a distance of 1.49 mm from the impact surface.

The values of shock, release, and tensile wave velocities and measured free surface velocities and pull-back particle velocities obtained in these experiments from the simultaneous monitoring mentioned above are given in Table 1.

The velocity profiles obtained in experiment #610 show several interesting features (Figure 2).

(i) The respective arrivals of shock at the free surface of the target at time t_2, and of release from the free surface of the impactor at time t_3 are clearly indicated by the velocity profile monitored at the impact surface due to the effect of change in the refractive index of glass under shock and release

(ii) The cumulative effect of the changes in the refractive index is observed to make the impact surface particle velocity after time t_3 negative. Discounting for the optical effect it would be zero

TABLE 1. Summary of shock wave experiments performed on soda lime glass.

Experiment	Impact velocity (km/s)	Compression		Release		Tension	
		Wave velocity (km/s)	Impact stress (GPa)	Wave velocity (km/s)	Measured free surface velocity (km/s)	Wave Velocity (km/s)	Pull-back particle velocity (km/s)
606	0.128 ± 0.004	5.83 ± 0.03	0.93	5.83 ± 0.06	0.125	5.82 ± 0.06	0.125
608	0.300 ± 0.003	5.83 ± 0.03	2.17	5.79 ± 0.06	0.305	5.79 ± 0.05	0.305
610	0.197 ± 0.001	5.83 ± 0.03	1.43	5.86 ± 0.06	0.199	5.78 ± 0.06	0.200
623	0.594 ±0.003	5.72 ± 0.04	4.23	5.82 ± 0.04	0.585	5.72 ± 0.06	0.485

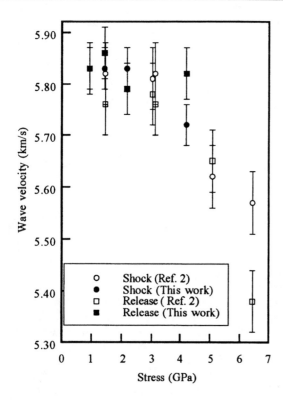

Figure 3. Stress versus shock and release wave velocities in soda lime glass.

in this experiment.

(iii) The free surface particle velocity indicates that glass did not spall.

(iv) The continued reverberation of shock and tension in the glass target is clearly seen in these wave profiles.

The trend in the values of shock and release wave velocities given in Table 1 is consistent with the reported decreasing trend in their values in soda lime glass with an increases in the stress between 1.45 and 6.43 GPa [1]. Figure 3 shows the variation in the measured values of shock and release wave velocities in soda lime glass. It shows that the values of shock wave velocity remain constant at 5.82 ± 0.06 km/s to 3.1 GPa and then decline to a value of 5.57 ± 0.06 at 6.4 GPa. A similar trend is also seen in the variation of release wave velocity.

The tensile wave velocities also show a similar trend with an increase in impact stress(Table 1).

Finally, based on the magnitudes of pull back particle velocity, tensile wave velocity (Table 1), and its density in the release state, i.e., 2.50 Mg/m^3, tensile strength of the glass at 4.2 GPa is calculated to be 3.5 GPa.

FUTURE WORK

Optical aspects of the present work will be described in a future publication. In addition, shock wave experiments described in Ref. 2 will be performed on soda lime glass to determine its tensile impedance directly for comparison with the results of this work.

ACKNOWLEDGMENT

Author thanks S. Spletzer for performing the shock wave experiments on glass.

REFERENCES

1. Dandekar, D. P., *Structures Under Shock and Impact IV,* edited by N. Jones, C. A. Brebbia, and A. J. Watson, Computational Mechanics Publications, Boston, 1996, pp. 439-448.
2. Dandekar, D. P., *Shock Compression of Condensed Matter-1995,* edited by. S. C. Schmidt and W. C. Tao, AIP Press, New York, 1995, pp. 947-950.

CP429, *Shock Compression of Condensed Matter – 1997*
edited by Schmidt/Dandekar/Forbes
© 1998 The American Institute of Physics 1-56396-738-3/98/$15.00

USING REVERBERATION TECHNIQUES TO STUDY THE PROPERTIES OF SHOCK LOADED SODA-LIME GLASS

A. Ginzburg and Z. Rosenberg

RAFAEL, P.O. Box 2250, Haifa, Israel

A series of plate impact experiments on soda-lime glass specimens was performed in order to probe the properties of the glass behind the failure wave front. Shock stresses in the range of 2-6 GPa were induced in the glass specimens by impacting them with thick brass impactors. Stress histories, at the back of the specimens, were recorded using commercial manganin gauges, which were backed by thick plexiglass disks. The gauges followed the shock and release reverberations in the glass resulting from the mismatch of brass and plexiglass on each side of the glass specimen. By analyzing the height and duration of these reverberations we can determine several important features of the glass behind the fracture wave front.

INTRODUCTION

Ever since the first publication on the fracture wave front in shock loaded glass (1), a large number of papers have been devoted to this subject in an attempt to understand the phenomenon of fracture under dynamic uniaxial strain conditions (see (2)-(6), for example). It seems that everyone agrees on the basic experimental findings which include the following facts: (1) Failure wave fronts are observed in glasses when shocked between about 0.5HEL-HEL (Hugoniot Elastic Limit); (2) The front velocity is in the range of 1.5 to 2.5 km/s; (3) Spall strength of the glass behind the front is zero, while lateral stress in the glass increases, denoting a decrease in shear strength behind the front. Several other features are still controversial as is the general explanation of the phenomenon. This was the main motivation for our work in which we tried to use a somewhat different configuration in order to obtain more information about the material state behind the failure front.

EXPERIMENTAL

The experimental technique we used is shown schematically in Figure 1. A relatively thick brass disk (5 mm) impacted planarily at an instrumented target, in our 64 mm gas gun. Impact velocity was measured prior to impact, with four pairs of shorting pins, to within ±0.5%. The commercial manganin gauges (Micro-Measurements) were calibrated for both loading (7) and unloading (8) conditions so their resistive hysteresis is taken into account. The gauges were embedded at the back of the specimens (soda-lime glass 2 mm thick) with a thick plexiglass disk supporting them. Thus, the measured stress signals are those transmitted to the plexiglass. In order to convert them to the stresses in the glass, a factor of $(z_1+z_2)/2z_2$ has to be used where z_1 and z_2 are mechanical impedances of glass and plexiglass, respectively. Since the shock levels in the glass remained below its Hugoniot elastic limit, this procedure yields relatively accurate stress values.

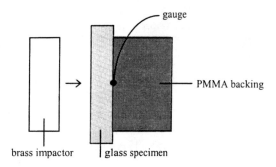

FIGURE 1. Schematic Description of Experimental Arrangement.

Figure 2 shows the X-t diagram corresponding to the impact situation of Figure 1. Ignoring reflections from the free surface of the brass impactor, a shock wave S_1 is induced in the glass upon impact, reflecting from the plexiglass interface as a release wave R_1.

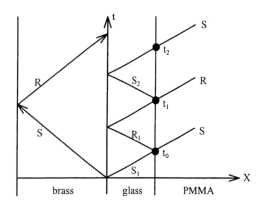

FIGURE 2. Expected X-t Diagrams for a Low Velocity Impact.

The gauge at the interface records the transmitted shock σ_1 (at t_0) which lasts until the release R_1 is reflected from the brass impactor (at t_1). By this time the release is reflected as a compression wave S_2 back into the glass which, after a double transit, comes back as a recompression at time t_2 to the glass-plexiglass interface.

This state of affairs is anticipated for stresses below about 4 GPa, which is considered to be the threshold for the onset of fracture waves in soda-lime glass. For higher stresses (in-between 4 to 6.4 GPa) the fracture front should interact with the various waves, as shown in Figure 3. The first interaction of R_1 with this front results in the well known small recompression at the gauge location at

t'_0. Since the front is known to slow down (or even to arrest) as a result of the interaction with R_1, we expect a second feature at t'_1 resulting from the interaction between S_2 and the fracture front. Also, since the material behind the front is supposedly of lower impedance (totally or mostly comminuted), we expect the time duration of the various stress levels to be different for the higher impact experiments. These expected changes should result in either a different sound speed or changes in density upon fracture (dilation), which are responsible for the lower impedance of the glass behind the failure front.

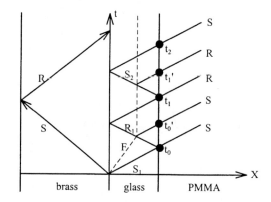

FIGURE 3. Expected X-t Diagram for a High Velocity Impact.

RESULTS AND DISCUSSION

Three shots at impact velocities of 212, 478 and 669 m/s were performed. These resulted in stress levels of (roughly) 2, 4.5 and 6 GPa in the glass. Figure 4 shows the gauge records for the 4.5 and 6 GPa shots (the lowest stress record looks very much like the intermediate one). One can clearly distinguish the three main waves arriving at the gauge location, namely the impact shock S_1 which is relieved (at t_1) by the reflection of R_1 and the recompression (at t_2) from the reflection of S_2. The small recompression is clearly seen in the profile of the high pressure shot, as expected, and its appearance (at t'_0) corresponds to a fracture front velocity of 2.5 km/s. However, except for this recompression pulse the two records are practically identical. Thus, the second reflection (anticipated at t'_1) is absent from the high stress record and the arrival times of the different waves are the same for the two shots indicating no change in sound velocity across the comminuted material. These findings clearly demonstrate the complexity of the fracture-front phenomenon, which seems to "disappear" after interaction with the R_1 release wave. Clearly one should take these features into account in any theoretical framework which is intended to explain the fracture wave in glass under uniaxial strain loading.

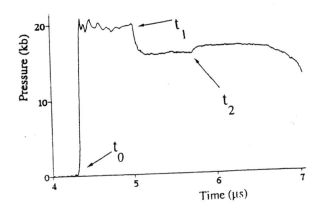

(a)

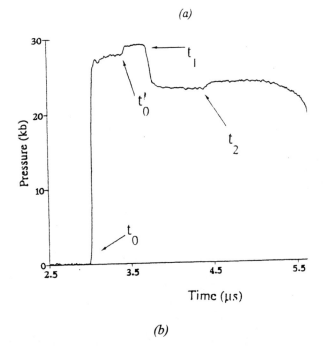

(b)

FIGURE 4. Gauge Records for : (a) 4.5 GPa ; and (b) 6 GPa Experiments.

SUMMARY

Planar impact experiments were performed in order to probe the state of soda-lime glass behind the fracture wave front. While the well known recompression, from the interaction of the release wave with this front, was obtained for the high stress experiment, other features were absent. In particular we did not find any indication for different sound speeds for the waves (shocks or release) going through the supposedly comminuted material. One could argue that the reason for this finding is that no changes in density (dilation), behind the fracture front, occur along the X-direction (shock propagation). Thus, lateral dilation could also explain the fact that lateral stresses increase behind the fracture front. If every material element undergoes a bulking process, in the lateral directions, one can anticipate a certain increase in lateral stress with no changes in stress along the X-direction. The other point to note is the absence of a second recompression wave, resulting from interaction between S_2 and the fracture front. This feature implies that the impedance of the fractured material is the same as that of the intact part ahead of the front. We do not understand this feature and we plan to perform more experiments in the near future in order to explore this point.

REFERENCES

1. Rasorenov, S.V., Kanel, G.I., Fortor, V.E., and Abasehov, M.M., *High Press. Res.*, 6, 225 (1991).
2. Kanel, G.I., Rasorenov, S.V., and Fortor, V.E., in "Shock compression of condensed matter - 1991", S.C. Schmidt, R.D. Dick, J.W. Forbes and D.G. Tasker (eds.), Elsevier Science Publications, 1992, pp. 451-454.
3. Bless, S.J., Brar, N.S., Kanel, G.I., and Rosenberg, Z., *J. Amer. Cer. Soc.*, 75, 1002 (1992).
4. Raiser, G.F., Wise, J.L., Clifton, R.J., Grady, D.E., and Cox, D.E., *J. App. Phys.*, 75, 3862 (1994).
5. Bourne, N.K., and Rosenberg, Z., in "Shock compression in condensed matter - 1995", S.C. Schmidt, J.W. Shaner, G.A. Samara and M. Ross (eds.), North Holland Publ., 1996, p. 491.
6. Bourne, N.K., Field, J.E., and Rosenberg, Z., *J. App. Phys.*, 78, (1995).
7. Rosenberg, Z., Yaziv, D., and Partom, Y., *J. App. Phys.*, 51, 4790 (1980).
8. Yaziv, D., Rosenberg, Z., and Partom, Y., *J. App. Phys.*, 51, 6055 (1980).

531

CP429, *Shock Compression of Condensed Matter – 1997*
edited by Schmidt/Dandekar/Forbes
© 1998 The American Institute of Physics 1-56396-738-3/98/$15.00

SYMMETRICAL TAYLOR IMPACT OF GLASS BARS

N.H. MURRAY[1], N.K. BOURNE[1], J.E. FIELD[1], Z. ROSENBERG[2]

[1]*Shock Physics, PCS, Cavendish Laboratory, Madingley Road, Cambridge, CB3 0HE, UK.*
[2]*RAFAEL, P.O. Box 2250, Haifa 31021, Israel.*

Brar and Bless pioneered the use of plate impact upon bars as a technique for investigating the 1D stress loading of glass but limited their studies to relatively modest stresses (1). We wish to extend this technique by applying VISAR and embedded stress gauge measurements to a symmetrical version of the test in which two rods impact one upon the other. Previous work in the laboratory has characterised the glass types (soda-lime and borosilicate)(2). These experiments identify the failure mechanisms from high-speed photography and the stress and particle velocity histories are interpreted in the light of these results. The differences in response of the glasses and the relation of the fracture to the failure wave in uniaxial strain are discussed.

INTRODUCTION

The Taylor impact test was first devised in 1948 to monitor yield strengths in metal rods (3). It provides a large strain, high strain-rate test of material behaviour. In the original investigation, a rod was impacted against a solid anvil and analysis consisted of post-mortem measurements of the final length of the deformed cylinder. As more sophisticated constitutive models have been produced, it has been found that the ability to predict the final lateral and longitudinal deformation profiles can prove a valuble test of model validity. The technique was improved in the early 1980s by replacing the anvil with a rod identical to the projectile (4). In this way problems due to friction and deformation of the anvil are eliminated and the impact is perfectly rigid in the centre of mass frame. The method has also been improved by using more sophisticated instrumentation such as high speed photography, piezoresistive gauges and velocity interferometry to enable intermediate deformation states to be monitored. Until recently the vast majority of experimental and theoretical investigations using this geometry have been performed on metals. In the last decade several works have also been carried out on rods constructed from brittle materials (1, 5-10). The work of Bless *et al.* has thoroughly investigated the lower velocity regime (6, 7). Nevertheless, their work

has been limited to the region below the threshold at which the compressive failure wave has been noted in plate impact. We emphasise that the confinement there is entirely different to this situation and the fracture observed in the rod cannot be correctly called a failure wave. However, it is of interest to investigate the higher stress regime near 4 GPa which has been observed to be the threshold for the failure wave in uniaxial strain experiments. The experiments below represent the first steps in this program.

EXPERIMENTAL

Targets consisted of borosilicate (BS) and soda-lime (SL) glass rods of diameter 10 mm and length 100 mm (*L/D* 10). The impact configuration is shown in Fig. 1.

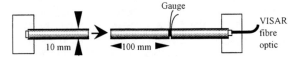

FIGURE 1. Symmetrical Taylor impact set-up.

Stress levels were measured using a commercial manganin piezoresistive stress gauge (Micro-Measurements type C-880113-B) embedded between the two target rods and VISAR was used to monitor the particle velocity of the free rear surface of the second rod. Rosenberg *et al.* have shown that the one-dimensional strain calibration curve can also be used for dynamic uniaxial stress experiments(11, 12). A fully programmable high speed Ultranac FS501 camera was used in framing mode to view the first target rod. The sequences were backlit using a high intensity flash. The lighting was diffuse so that no visualisation of the elastic wave was possible. Impact velocities were measured to an accuracy of 0.5% using a sequential pin shorting method.

Relevant material properties for the glasses are listed in Table 1. Ultrasonic methods were used to determine the longitudinal, c_L, and shear, c_S, wave velocities. HELs were reported in an earlier paper (2).

TABLE 1. Materials Parameters for the Glasses Studied.

	ρ (± 0.05 g cm^{-3})	E (GPa)	μ (GPa)	ν	c_L (± 0.01 mm μs^{-1})	c_S (± 0.01 mm μs^{-1})	HEL (± 0.5 GPa)
BS	2.23	73.1	30.4	0.20	6.05	3.69	8.0
SL	2.49	73.3	29.8	0.23	5.84	3.46	6.0

We will present the results of tests carried out at a nominal impact velocity of 500 m s^{-1} for the two materials.

RESULTS

Figure 2 shows a typical sequence of a soda-lime rod impacted at 500 m s^{-1}. The pin, visible at the bottom of each frame, is mounted on the rod holder and provides a means of aligning each image with the others in the sequence. The times at which each picture was taken relative to the impact time, are marked on each image. The impact has occurred by frame 2 in which a dark front is seen travelling down the rod at a constant velocity of 4.5$\pm$0.4 mm μs^{-1} which appears characteristic for this material at this stress. The elastic wave is travelling down the rod at 5.4 mm μs^{-1} as determined from the modulus. This calculation is independently confirmed by the measured time delay between the gauge and VISAR signals. A feature of the propagation of the elastic wave is that it appears to nucleate damage behind itself visible as the dark spots seen in frames 3 and 4, and the irregular appearance of the dark front in frame 5. The impact site shows radial expansion of a cloud of material in defined jets. It will be seen from frames 3 onwards that these jets assume a periodic appearance, and their separation corresponds approximately to the diameter of the rod. In the last frame a second region of expansion is observed corresponding to a front returning from the gauge location.

The shot of Fig.1 was repeated several times with very reproducible results. In one of these cases the surface of the rod was roughened from its initial polished state. The high-speed camera sequence was very similar to that shown and the velocity of the dark front was identical within errors. It was thus concluded that the density of surface flaws did not play a significant role in the fracture process.

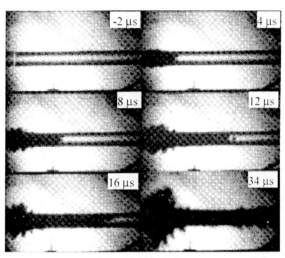

FIGURE 2. Typical symmetrical impact of soda-lime glass. The impact velocity is 500 m s^{-1}. The first frame is taken 2 μs before impact which has occurred by frame 2. The fracture has fully occurred by the last frame.

Figure 3 shows a typical impact on borosilicate glass. Again frame 1 is 2 μs before impact. Note that in this case the dark front down the rod does not display the ragged appearance seen above. Neither do the fractured glass fragments in the region of the impact site show the jetting seen previously. The radial expansion of the rod occurs with a smooth profile which is reminiscent of that seen for metals. The expansion is uniform behind the front and the impact site is largely hidden. This contrasts with the concentration of the expansion in the region around the impact site in the case of soda-lime and with little or no radial expansion evident far from the impact region. It was noted in the case of previous plate impact work (13) that the borosilicate showed a failure wave front that was composed of small uniform fractures whereas the soda-lime showed a much rougher, bifurcating crack front. This may be mirrored in the fracture process here. The dark front travels at uniform speed again at velocity of 4.9$\pm$0.4 mm μs^{-1}. This compares with the elastic wavespeed of 5.7 mm µs⁻¹ as calculated from the moduli.

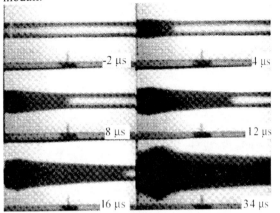

FIGURE 3. Typical symmetrical impact of borosilicate glass. The impact velocity is 512 m s⁻¹. The first frame is taken 2 µs before impact which has occurred by frame 2. The fracture has fully occurred by the last frame.

Figure 4 shows typical signals seen by the gauges placed 10 diameters from the impact site. The traces are displaced on a relative time axis for ease of comparison. The soda-lime gauge signal is superposed on the VISAR trace recorded at the free end of the second rod (20 diameters from the impact plane). The transit times of waves between the gauge and VISAR are as expected from calculations of the elastic wavespeeds.

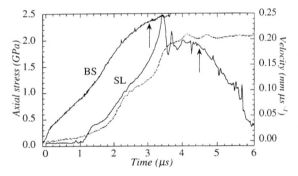

FIGURE 4. Typical gauge (solid) and VISAR (dotted) traces recorded in the experiments. SL indicates soda-lime and BS borosilicate glasses.

The borosilicate gauge sees a maximum axial stress of *ca.* 2.5 GPa whilst in the soda-lime case it reaches 2.0 GPa. It has been shown previously that the stress decays at the interface from its uniaxial strain value asymptoting to its uniaxial stress value at *ca.* 5 diameters from the impact face (6). In this work the stresses induced at the impact face in the uniaxial

strain loading are 3.0 GPa for the soda-lime and 2.6 GPa for the borosilicate. We can apply the Griffith's criterion (14) to calculate the yield stress, *Y*, from the measured HELs using

$$Y = \sigma_x - \sigma_y = \frac{(1-2v)^2}{1-v}\sigma_x. \qquad (1)$$

This calculation gives *Y* values of 2.3 GPa for soda-lime which is *ca.* the stress measured at the gauge location.

The borosilicate trace is smooth and concave following the sharper rise to *ca.* 0.5 GPa. The arrow at 3 µs indicates the time at which the dark front is calculated to arrive at the gauge location. It appears that the pulse flattens at this time and within a microsecond the gauge is destroyed. The soda-lime trace again rises in an initial sharper jump over *ca.* 200 ns to 0.4 GPa. After this it rises in two further concave sections over a microsecond to 1.2 GPa and then over a further microsecond to 2.5 GPa. After this point, there are two rings on the trace which are a consistent feature of each of the repeats done. The general form of the pulse is reproduced at the VISAR location (including the two rings which have displaced to longer times by dispersion). The particle velocity of 0.2 mm µs⁻¹ is less than expected from an elastic analysis. This may be due to processes involved in restarting the fracture at the cut introduced to mount the gauge. It is at first puzzling that the stepped structure is reproduced at 20 diameters since dispersion should have displaced these features by then. This may be a further indication that the fracture restarts at the gauge location (as can be seen in the sequences) which means that the two traces are recorded the same distance from each nucleation interface. The stress starts to drop with the arrival of the dark front and after 6 µs the gauge is destroyed.

DISCUSSION AND CONCLUSIONS

Figure 5 shows the plotted waves for our experiments as compared with the lower velocity work of Bless *et al.* (6, 7). For clarity we have plotted only data for soda-lime glass but the borosilicate data looks qualitatively the same. The impact stress is 2.5 GPa for our experiment (which is shown as a solid line) and 1.4 GPa for theirs (where the line is dotted). Our impactor was L/D 10 whilst theirs was L/D 4 which had the effect of interacting with their fracture front, *F1*, and halting it within the sample.

This was not the case in our experiment so that the fracture *F2* reached the gauge *G2*.

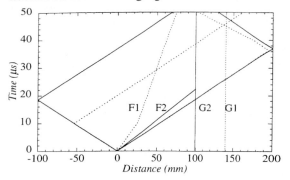

FIGURE 5. Symmetrical impact of soda-lime glass rods. The solid line represents impact at a velocity of 500 m s^{-1}. The dotted line represents an experiment from the work of Bless and Brar.

We can add our data to that of Bless and Brar (7) and Bless *et al.*(6) to determine the asymptotic behaviour of the fracture front velocity as the impact stress is raised. Table 2 summarises this data. Our results are in bold. The stresses quoted are the uniaxial strain values calculated from known Hugoniots of the materials. It will be seen that in the case of the borosilicate, the front speed has stabilised at a velocity of *ca.* 5 mm μs^{-1} whilst it is approaching this value in soda-lime. It thus appears that there is an upper bound for the front velocity.

TABLE 2. Fracture Front Velocities as a Function of Impact Stress.

Borosilicate		Soda-lime	
Stress (±.1 GPa)	Front speed (± 0.4 mm μs^{-1})	Stress (±.1 GPa)	Front speed (± 0.4 mm μs^{-1})
2.5	**4.9**	**3.0**	**4.5**
2.0	5.2	2.0	3.9
1.4	3.6	1.4	1.5
0.8	2.3	-	-

The front appears to be nucleated at the impact face by fracture which accompanies lateral release from the sides of the rod. It is clear that the morphology of the rod expansion depends critically on this fracture process and it appears that the borosilicate forms finer comminutia and has a higher density of fractures. There is also the possibility that this material can densify near the impact region.

The soda-lime shows much more irregular fracture. The jetting phenomena observed may relate to the crossed release structure that must exist behind the elastic front. The periodic 1 μs steps on the stress and VISAR histories are too short to represent a full elastic wave transit of the rod diameter but are of the same order as the time taken for a release to reach the central axis of the rod from the surface.

Further experiments will aim to extend the stress range into the region above the failure wave threshold in uniaxial strain (*ca.* 4 GPa) and to investigate other glass microstructures.

ACKNOWLEDGEMENTS

We would like to thank Dr. W.G. Proud for his help with VISAR operation and Mr D. Cross for providing technical support.

REFERENCES

1. Brar, N.S., Bless, S.J. and Rosenberg, Z., *Appl. Phys. Letts,* **59**, p. 3396-3398, (1991).

2. Bourne, N.K., Rosenberg, Z. and Millett, J.C.F., *The plate impact response of three glasses,* in *Structures under Shock and Impact IV,* Computational Mechanics Publications: Southampton. (1996), p. 553-562.

3. Taylor, G.I., *Proc. R. Soc. Lond. A,* **194**, p. 289-299, (1948).

4. Erlich, D.C., Shockey, D.A. and Seaman, L., *Symmetric rod impact technique for dynamic yield determination,* in *Shock Waves in Condensed Matter – 1981,* American Institute of Physics: New York.(1982), p. 402-406.

5. Bless, S.J., Brar, N.S., Kanel, G. and Rosenberg, Z., *J. Am. Ceram. Soc.,* **75**, p. 1002-1004,

6. Bless, S.J., Brar, N.S. and Rosenberg, Z., *Failure of ceramic and glass rods under dynamic compression,* in *Shock Compression of Condensed Matter - 1989,* Elsevier: Amsterdam. (1990), p. 939-942.

7. Bless, S.J. and Brar, N.S., *Impact induced fracture of glass bars,* in *High-pressure science and technology,* American Institute of Physics: New York. (1994), p. 1813-1816.

8. Cosculluela, A., Cagnoux, J. and Collombet, F., *Two types of experiment for studying uniaxial dynamic compression of alumina,* in *Shock Compression of Condensed Matter - 1991,* Elsevier: Amsterdam. (1992), p. 951-954.

9. Glenn, L.A. and Janach, W., *Trans. ASME: J. Engng Mater. Technol.,* **100**, p. 287-293, (1978).

10. Grove, D.J., Rajendran, A.M., Bar-On, E. and Brar, N.S., *Damage evolution in a ceramic rod,* in *Shock compression of condensed matter - 1991,* Elsevier: Amsterdam. (1992), p. 971-974.

11. Rosenberg, Z., Yaziv, D. and Partom, Y., *J. Appl. Phys.,* **51**, p. 3702-3705, (1980).

12. Rosenberg, Z., Mayseless, M. and Partom, Y., *Trans. ASME: J. Appl. Mech.,* **51**, p. 202-204, (1984).

13. Bourne, N.K., Rosenberg, Z. and Field, J.E., *J. Appl. Phys.,* **78**, p. 3736-3739, (1995).

14. Rosenberg, Z. *J. Appl. Phys.,* **74**, p. 752-753, (1993).

CP429, *Shock Compression of Condensed Matter – 1997*
edited by Schmidt/Dandekar/Forbes
1998 The American Institute of Physics 1-56396-738-3/98/$15.00

COMPRESSION-SHEAR STUDY OF GLASS REINFORCED POLYESTER

J. Michael Boteler

Army Research Laboratory - Weapons and Materials Research Directorate
MS: AMSRL-WM-MF
Aberdeen Proving Grounds, MD 21005

Fiber reinforced organic matrix composites, due to their lower density, are being considered for armor applications where weight is an important factor. In this work, shock wave experiments were performed on Glass Reinforced Polyester (GRP) composite to investigate delamination due to normal and off-axis impact. This study extends the prior work of Dandekar and Beaulieu which examined both the compressive and tensile strengths of GRP under normal loading conditions. Symmetric impact shock wave experiments were performed on a 102 mm slotted barrel single-stage light gas gun over a range of obliquity to 26 degrees. Oblique impact geometry was chosen to investigate the shear strain dependence of the delamination process. Particle velocity history was measured by VISAR and the "pull-back" signal typifying delamination provided a measure of the delamination strength. Delamination was detected from measurements of the particle velocity history using VISAR. Delamination values as low as 0.007 GPa were recorded for shock stress of 0.103 GPa and obliquity of 26 degrees. Overall the delamination threshold in GRP was observed to decrease with increasing obliquity, suggesting shear strain dependence. These results will be discussed and the experimental details describing the compression-shear experimental arrangement will be provided.

INTRODUCTION

Glass Reinforced Polyester (GRP) is a lightweight, composite armor material which has found widespread usage due to its performance, low cost, and availability. Recent investigations into the effect of low-velocity impact on composite laminates suggest that the delamination threshold may be approached by very low input stress [1, 2]. Previous studies performed at ARL have investigated the dynamic response [3] and the compressive and tensile strengths [4] of GRP under shock wave propagation. The latter study by Dandekar and Beaulieu found the delamination threshold to be approximately 0.05 GPa when shock loaded to a stress level of 0.23 GPa for 1.2 μs and 3.5 μs durations. They also found the delamination strength to be negligibly small when shock loaded to a stress level of 0.37 GPa for only 1.2 μs. These results are in agreement with the earlier work of

Letian et al [5, 6] on fiber reinforced composite material.

The overall objective of the prior studies was to develop a predictive methodology for the dynamic response of GRP to impact loading conditions. The present work extends the work of Dandekar and Bealieu to include shear and the role it may play in the delamination process. Pressure-shear impact experiments were performed for several oblique geometries, the experiments and results are described in the following sections.

MATERIAL

The GRP laminates used in this study were composed of S-2 glass woven roving in a polyester resin matrix with a resin content of 32 % by weight. Glass fabric was provided by Owens Corning in a balanced construction with weight of 814 g/m^2. The resin coating of the fabric was performed by

American Cyanamide using Cycom 4102 polyester resin. All samples were fabricated in-house by Elias J. Rigas of the Composites Development Branch at ARL. Surface flatness was measured to be within 25 μm. The laminates were 0.68 mm thick and laid-up in a ± 90° geometry. The mean density was measured to be 1.949 ± .030 g/cc. The samples used in the shock loading experiments were machined to 55 mm squares with mean thickness of 3 mm, 7 mm, and 13 mm.

The elastic constants for GRP were obtained from the phase velocity measurements of ultrasonic waves. An image superposition technique [7] similar to the pulse-echo overlap method was used for the velocity measurements. Table 1 summarizes these results.

TABLE 1. Values of the elastic constants C_{ij} and elastic compliances S_{ij} for GRP composite [3].

Elastic Stiffness GPa	Elastic Compliance GPa^{-1}
$C_{11} = 31.55 \pm 3.8$	$S_{11} = 0.045039 \pm .012$
$C_{33} = 20.12 \pm 0.40$	$S_{33} = 0.062074 \pm .0128$
$C_{44} = 4.63 \pm 1.22$	$S_{44} = 0.2160 \pm .05$
$C_{66} = 4.94 \pm 1.31$	$S_{66} = 0.2024 \pm .06$
$C_{12} = 15.86 \pm 4.53$	$S_{12} = -0.01869 \pm .0082$
$C_{13} = 9.75 \pm 3.83$	$S_{13} = -0.01276 \pm .0077$

EXPERIMENTS

Impact experiments were performed on a 100 mm single-stage light gas gun. In a companion paper we report on a simple modification to the gas gun which greatly extends the lower velocity range [8]. The wrap-around breach is replaced with an aperture plate which is open to atmosphere on one side. A throttling process regulates the air flow through the aperture thus controlling the projectile velocity. Velocities as low as 0.0153 mm/μs have been achieved with this gun modification.

During the experiment the impactor and target material are oriented parallel to each other at an angle Θ to the bore axis. A VISAR probe is located normal to the impact surface. A keyway in the barrel prevents the projectile from rotating during acceleration. At impact longitudinal and transverse stress waves propagate into the sample. The VISAR probe monitors the free surface velocity at the rear of the target and provides a velocity-time history of the impact event. All shock loading experiments reported here were symmetric impact with a pulse duration of ≈ 2.4 μs which was kept constant. For additional details of oblique-plate impact experiments, the reader is referred to the papers of Abou-Sayed [9] and Klopp [10].

For comparison purposes finite strain analysis [11] is used to estimate the transverse wave magnitude to first order in strain. The principal shearing stress is then determined from a knowledge of the longitudinal and transverse stress. The following relation holds [12],

$$\sigma_{shear} = \frac{1}{2} \left| \sigma_{longitudinal} - \sigma_{transverse} \right| \quad (1)$$

Numerous examples of finite strain calculations are available in the literature [13, 14], hence only a brief outline will be given here.

The change in internal energy with respect to strain at constant entropy defines the thermodynamic (Piola-Kirchoff) stresses t_{ij}. To first order in strain they are given by,

$$t_{ij} = C_{ijkl} \, \eta_{kl} \quad (2)$$

where C_{ijkl} are the stiffness constants and η_{kl} is the applied strain. The Piola-Kirchoff stress tensor is not in general a symmetric tensor. It is common practice to work with the Cauchy stress tensor σ_{ij} which is symmetric and defined by,

$$\sigma_{km} = \frac{1}{J} \left(\frac{\partial x_k}{\partial a_j} \frac{\partial x_m}{\partial a_i} \right) t_{ij} \quad (3)$$

where J is the Jacobian for the transformation between the two coordinate systems. For uniaxial strain along the [001] Z-direction, the strain tensor is given by,

$$\underline{\underline{\eta}} = \begin{bmatrix} 0 & 0 & 0 \\ 0 & 0 & 0 \\ 0 & 0 & \eta_{33} \end{bmatrix} \quad (4)$$

and J assumes the simple form $\left(\rho_0 / \rho \right)$. Once the elastic constants are known, it is straightforward to determine the principal stresses for normal impact geometry ($\theta=0°$). However for oblique impact, care must be exercised in transforming the second rank tensors to reflect the new axes. In the primed coordinate system, the expressions for the Cauchy and thermodynamic stresses are,

$$\sigma'_{kl} = \frac{1}{J'}\left(\frac{\partial x'_k}{\partial a'_j}\frac{\partial x'_l}{\partial a'_i}\right)t'_{ij}, \text{ and } t'_{ij} = C'_{ijmn}\eta'_{mn} \quad (5)$$

The values of the primed Cauchy stresses must be determined in terms of the unprimed constants which are only known relative to the principal axes. For arbitrary theta, these expressions are found to be

$$\left. \begin{aligned}
\sigma'_{11} &= \frac{\rho}{4\rho_0}\left\{\left(C_{11}+C_{33}-4C_{44}\right)\sin^2 2\theta + 4C_{13}\right\}\eta' \\
\sigma'_{22} &= \frac{\rho}{\rho_0}\left(C_{12}\sin^2\theta + C_{13}\cos^2\theta\right)\eta' \\
\sigma'_{33} &= \frac{\rho_0}{\rho}\left\{C_{11}\sin^4\theta + C_{33}\cos^4\theta + \left(\frac{1}{2}C_{13}+C_{44}\right)\sin^2 2\theta\right\}\eta' \\
\sigma'_{23} &= \left\{\left[C_{11}\sin^2\theta - C_{33}\cos^2\theta + (C_{13}+C_{44})\cos 2\theta + C_{44}\right]\sin 2\theta\right\}\eta'
\end{aligned} \right\} \quad (6)$$

Transforming these results back to the original basis set we find,

$$\left. \begin{aligned}
\sigma_{11} &= \sigma'_{11}\cos^2\theta - 2\sigma'_{23}\sin\theta\cos\theta + \sigma'_{33}\sin^2\theta \\
\sigma_{22} &= \sigma'_{22} \\
\sigma_{33} &= \sigma'_{11}\sin^2\theta - 2\sigma'_{23}\sin\theta\cos\theta + \sigma'_{33}\cos^2\theta
\end{aligned} \right\} \quad (7)$$

Note that for $\theta = 0°$, equations (6) and (7) reduce to the expressions for the principal stresses in the original frame. In the following section the experimental results will be compared to the finite strain predictions given by Eqtns (6)-(7).

RESULTS

A total of six compression-shear shock experiments have been performed thusfar on GRP. Experimental data is summarized in Table 2. The shot number, projectile velocity and impact angle are given in the first three columns. The remaining four columns list the peak input stress, particle-velocity jump, delamination value, and density compression, respectively. The second experiment did not have adequate recording time to capture the pull-back signal if one occurred. Projectile velocity was measured by shorting pins and recorded on an oscilloscope with an uncertainty of less than 1%.

Table 2. Summary of measurements for pressure-shear experiments.

SHOT #	V_P (m/s)	Θ (degrees)	σ_{IMPACT} (GPa)	Δu_P (m/s)	σ_T (GPa)	μ
1	83.0	0°0'	0.257	9.8	0.060(5)	0.013
2	62.8	8°0'	0.192	*	*	0.0098
3	49.6	8°11'	0.151	6.1	0.040(5)	0.0097
4	53.2	11°54'	0.160	5.4	0.034(5)	0.0081
5	37.9	26°22'	0.103	1.2	0.007(5)	0.0053
6	23.4	0°0'	0.071	**	no spall	0.0073

* Inadequate recording time. ** no delamination signal observed

The experimental results are plotted in Figure 1 as a function impact angle and density compression, respectively. In this figure the dashed and solid lines represent the shear values estimated from the finite strain analysis (Eqtn 7). The curves in the left portion of the figure are for constant density compression while the curves in the right portion are for constant theta. The measured values for delamination are represented by the solid symbols with error bars.

In this figure the delamination values decrease with increasing theta. indicating that shear does

539

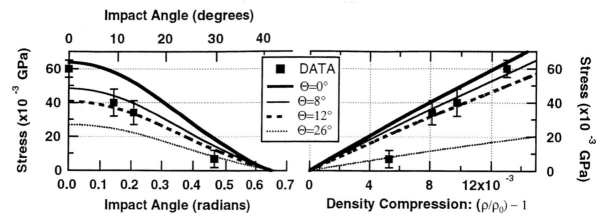

Figure 1. Plot of delamination values vs. impact angle and density compression

contribute to the delamination process. The reasonable agreement between the finite strain curves and the measured stress delamination values also suggest that damage in the GRP was minimal or nonexistent prior to delaminating. The only experiment to date which did not result in delamination was experiment # 6 which had a peak stress of only 0.071 Gpa..

Although the data is insufficient at this time to define a delamination threshold or failure scenario, several observations may nonetheless be made:

(1). Delamination may be initiated in GRP at input stress as low as 0.21 GPa at normal impact for a pulse duration of 2.4 μs.

(2). Delamination in GRP composite is sensitive to the impact angle, suggesting that shear strain plays a role in the delamination process. The most oblique impact of 26°22' resulted in a delamination value of only 0.007 GPa.

(3). Projectile and/or fragment impact on GRP may result in delamination for even very low velocities.

REFERENCES

1. Martin, V., *et al. Proposal for a Experimental Device to Study the Behavior of Laminate Composites Under Low Velocity Impact Loading.* in *Recent Advances in Experimental Mechanics: 10th International Conference.* 1994, Lisbon, Portugal,J.F.e.a. Silva Gomes, A. A. Balkema/Rotterdam/Brookfield.1031-1036.
2. Collombet, F., J. Bonini, and J.I. Lataillade, *A Three-Dimensional Modeling of Low Velocity Impact Damage In Composite Laminates.* Int. J. for Numerical Methods in Eng., **39**, p. 1491-1516,1996.
3. Chou, S.-C. and E. DeLuca, *Dynamic Response of S-2 Glass Reinforced Plastic Structural Armor* Army Research Laboratory, 1993.
4. Dandekar, D.P. and P.A. Beaulieu. *Compressive and Tensile Strengths of Glass Reinforced Polyester Under Shock Wave Propagation.* in *High Strain-Rate Effects on Polymer, Metal, and Ceramic Matrix Composites and Other Advanced Materials.* 1995, Y.D.S. Rajapakse and J.R. Vinson, ASME New York.63-70.
5. Letian, S., *et al., An Experimental Study of Spallation Criterion In Phenolic-Resin Based Woven Roving Glass Fiber Reinforced Composite Material.* ACTA Mechanica Sinica, **1**(1), p. 81-93,1985.
6. Letian, S., Z. Shida, and B. Yilong, *Threshold and Some Microscopic Observations of Spallation of Phenolic-Resin Based Woven Roving Glass Fiber Reinforced Composite Material.* Int. J. Impact Engng., **2**(2), p. 169-178,1984.
7. Martin, A.G., *Phase Velocity Measurements in Dispersive Materials by Narrow Band Burst Phase Comparison* Army Materials and Mechanics Research Center, Watertown, Ma, 1976.
8. Boteler, J.M., P. Bellamy, and P. Bartkowski, *Gas Gun Modification For Achieving Ultralow Velocities.* Manuscript in preparation, ,1996.
9. Abou-Sayed, A.S., R.J. Clifton, and L. Hermann, *The Oblique-plate Impact Experiment.* Experimental Mechanics, **16**, p. 127-132,1975.
10. Klopp, R.W., R.J. Clifton, and T.G. Shawki, *Pressure-Shear Impact and the Dynamic Viscoplastic Response of Metals.* Mechanics of Materials, **4**, p. 375-385,1985.
11. Thurston, R.N., *Wave Propagation in Fluids and Normal Solids,* in *Physical Acoustics,* W. Mason, Editor. 1964, Academic Press:
12. Prager, W., *MECHANICS OF CONTINUA.* Introductions to Higher Mathematics, eds. G. Birkhoff, M. Kac, andJ.G. Kemeny. 1961, Ginn & Co. 230.
13. Horn, P.D. and Y.M. Gupta, *Wavelength Shift of the Ruby Luminescence R Lines Under Shock Compression.* Appl. Phys. Lett., **49**(14), p. 856,1986.
14. Gustavsen, R. and Y.M. Gupta, *Time Resolved Raman Measurements in α-Quartz Shocked to 60 kbar.* J. Appl. Phys., **75**(6), p. 2837-2844,1993.

CP429, *Shock Compression of Condensed Matter – 1997*
edited by Schmidt/Dandekar/Forbes
© 1998 The American Institute of Physics 1-56396-738-3/98/$15.00

POLYURETHANE IN PLANE IMPACT
WITH VELOCITIES FROM 10 TO 400 M/SEC

D. Tsukinovsky, E. Zaretsky and I. Rutkevich

Pearlstone Center for Aeronautical Engineering Studies,
Department of Mechanical Engineering, Ben-Gurion University of the Negev
P.O. Box 653, Beer-Sheva 84105, Israel

Shock response of the rubber-like polyurethane (PU) samples in the plane impact experiments with the impactors made of the same PU has been studied within the impact velocity range 10 - 400 m/sec. The experiments were performed with 25-mm pneumatic gun. The free surface velocity of the samples was measured by VISAR. The shock response of PU is not perfectly elastic and this can be connected with a complicated rheology of the material. The non-elastic behavior is well observed with the lowering of the velocity of impactor.

INTRODUCTION

Quasi-static mechanical properties of elastomers (rubbers) differ strongly from that of other solids. Elastomers are used as a load-bearing capacity in different structures, which, in particular, should work in a shock environment. However, the information about the shock response of elastomers is poor. At moderate shock pressures only elastomer-based composites was studied to characterize the shock response of solid propellants [1]. The major of the data available was obtained in the experiments with strong shocks when the pressure pulse amplitude in the rubber sample was, at least, about 2 GPa, e.g. [2] and [3]. At this pressure no distinction between the shock response of rubber and other solids was found [2]. This agrees with the results of the study of the rubber under static compression [4]; at pressure 2 GPa the rubber should be in the "glassy" state characterized by high shear strength.

The purpose of the present work is to study the shock response of elastomeric material at the impact of low and moderate strength when the difference in the behavior of the elastomer and "normal" solid was proposed to be distinct.

EXPERIMENTAL

The rubber-like elastomer polyurethane (PU) was studied in symmetrical (PU-PU) plane impact experiments with impact of different strengths. The polyurethane sheets of 5.2-mm thickness (in several experiments the samples were made of the thinner sheets) were received from Alexandrovich Plastics Company, Tel-Aviv. The density of the polyurethane was 1200 kg/m^3, the sound velocity measured at the frequency 5 MHz was $C_0 = 1800$ m/sec. The Young modulus of the studied polyurethane was found E = 49±0.5 MPa. The samples and the impactors, the discs of about 25-mm diameter, were machined from the sheets. The impactors were accelerated by the 25-mm, 3 m length, pneumatic gun up to velocities 12.5 - 406 m/sec. During the impact the free surface velocity of the polyurethane sample was continuously monitored by the VISAR [5], having the interferometer constant $U_0 = 97$ m/sec. The experimental setup is shown schematically in Fig. 1a.

The guard ring together with the sample and the embedded electrical pins was lapped before assembling of the velocity pins. The impactor-sample misalignment did not exceed 0.2 mrad. To prevent the impactor bending due to the acceleration in the barrel all impactors were backed with 1-mm discs of Al 6061-T6. The x-t diagram of the experiments together with the outline of the sample free surface velocity profile are shown in Fig. 1b: the time intervals t_1 and t_2, were used for calculation of the velocity D of the shock front propagation in the undisturbed material and of the velocity C of the propagation of the reloading pulse through the compressed impactor and unloaded sample, respectively.

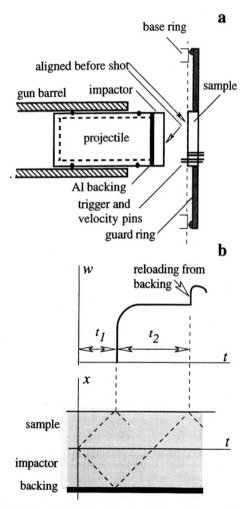

FIGURE 1. a) Experimental setup. 10-μ Al foils were glued on the front surface of the impactor, to short the electrical pins, and on the sample back surface, to enhance the reflectivity. b) Distance - time and free surface velocity - time diagrams of the symmetrical PU-PU impact. The collision takes place at the instant $t = 0$.

RESULTS AND DISCUSSION

Several free surface velocity profiles of polyurethane samples are shown in Fig. 2a,b. For more convenience the profiles obtained after the strong and the weak impacts are shown separately.

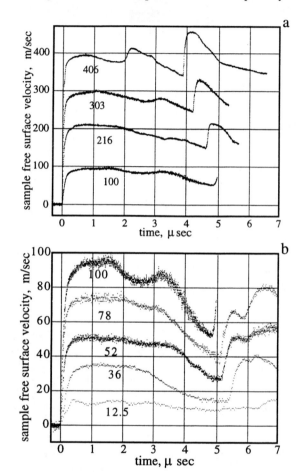

FIGURE 2. Free surface velocity profiles of polyurethane samples after impact with high (a) and low (b) impact velocities (the corresponding velocity values are shown under profiles).

As is seen from Fig. 2, within the time interval t_2 (see Fig. 1b) the arrival of some unloading wave followed by the reloading one takes place. The appearance of these features should be related to the high sound velocity in a shock-compressed rubber-like material; part of the sample lateral surface is free while another part is glued to the guard ring. The VISAR incident beam was focused in the center of the sample free surface with accuracy of about 1 mm. Assuming that the free surface velocity starts to decrease with the arrival of the head unloading characteristic from the sample lateral surface, the longitudinal sound velocity C_l may be estimated. To appear at the free surface the head characteristic of the unloading wave should pass the distance about 13 mm from the intersection of the sample front plane with its lateral surface. The result of C_l estimation based on the time interval between the impactor-sample collision and the instant of the appearance of the unloading feature at the free surface velocity profile is shown in Fig. 3.

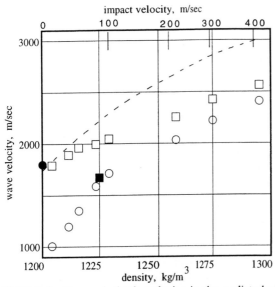

FIGURE 3. Measured shock velocity in the undisturbed polyurethane (circles). Estimation of C_b (squares) and C_l (dashed line). Filled circle ($C_0 = 1800$ m/sec) is the result of ultrasonic measurements of the sound velocity in polyurethane. Filled square is the result of the determination of the velocity from four shots with two-layer samples (see the text).

A striking result of the shock velocity D measurements presented in Fig. 3 is the fact that for the impact velocities less than 100 m/sec, the apparent shock velocity appears to be less than the longitudinal velocity C_0 obtained in ultrasonic measurements. The sample-impactor misalignment may have a strong influence on the measured values of the time interval t_1 (see Fig. 1b). In particular, the time interval, corresponding to 40 - 50-μ spatial separation between the instant of the impact and the shortening of the embedded pins may result in such behavior of D. To verify this assumption an additional set of experiments was performed. The PU disks of different thicknesses, 5.10±.01, 3.50±.01, 1.74±.01 and 0.50±0.03 mm were glued to the

thick, 0.980±.005-mm, disks of aluminum alloy 6061-T6 and these, two-layer, samples were impacted by 6.14-mm impactor made of the same aluminum alloy. In order to provide rectangular shape of the initial pressure pulse the impact velocities were about 40 m/sec, below the HEL of the alloy, 56 - 60 m/sec. The front parts of the velocity profiles obtained from these shots together with corresponding "x - t diagram" are shown in Fig. 4a,b. The velocity $D = 1670$ m/sec obtained from the time-sample thickness dependence of Fig. 4b is in the excellent agreement with the velocities D obtained in the symmetrical, PU-PU, impact, see. Fig. 4a.

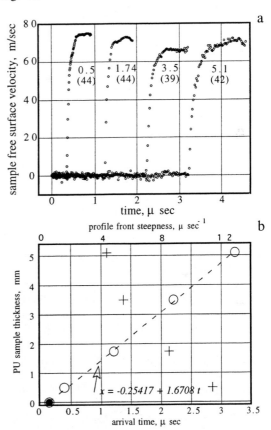

FIGURE 4. a) Free surface velocity profiles of two-layer polyurethane samples. b) Wave arrival time (circles) and profile front steepness (crosses) vs. thickness of PU layer. Filled circle is the arrival time for Al substrate having $C_l = 6380$ m/sec.

Clearly, the discontinuity resulted from the impact never can propagate with the velocity D less than C_0 since C_0 is the maximal characteristic velocity corresponding to the unperturbed material. It seems to be reasonable to assume that at low impact velocities the propagation of the initial discontinuity is accompanied by its strong decay related, to the relaxation of the impact-generated shear stress. This assumption is in agreement with the results of ultrasonic measurements of acoustic velocities in PMMA [6]. It was found in [6] that the attenuation of the longitudinal acoustic signal in PMMA increases from 5 dB/cm to 20 dB/cm (10 times) with the increase of the signal frequency form 6 MHz to 30 MHz. Within the same frequency range the attenuation of the shear signal was found essentially stronger, from 14 dB/cm to 60 dB/cm (100 times). In PU this high-frequency attenuation is expected to be even stronger. Our attempt to record 5-MHz shear signal at the rear surface of the 5-mm sample failed. The decay increases with the increase of the frequency of the signal, i.e. the high-frequency components of the step-like signal cannot arrive at the sample free surface. The value of the free surface velocity corresponding to the discontinuity-generated signal is too small to be detected by VISAR. A confirmation of this statement may be found in the evolution of the shock front rise-time with the sample thickness. The steepnesses of the velocity fronts estimated from the initial slope of the profiles shown in Fig. 4a are: 11.5, 8.5, 5.5 and 4.4 ± 0.5 μsec^{-1} for 0.5, 1.74, 3.50 and 5.10-mm samples respectively, see Fig. 4b. The profile front steepness seems to be stabilized between the values 2.5 - 3 μsec^{-1}., i.e. the shock front of this steepness should not experience strong decay, while higher frequency components of the signal are decayed. It is interesting that the velocity front comparatively smeared during its motion through the unperturbed sample propagates in the sample with the velocity higher than 1800 m/sec. The latter is apparent from the density dependence of the reloading wave velocity C in the polyurethane shown in Fig. 3. The initial steepness of the reloading wave front was determined for several weak-impact experiments, the profiles 36, 52 and 78 m/sec shown in Fig. 2b. The steepness value was found equal 2.7 ± 0.5 μsec^{-1}. This confirms that the low-frequency acoustic signals can travel in the polyurethane without essential attenuation.

Assuming that the impact-induced shear stress relaxed completely during the smearing of the pulse one may consider the velocity C as an estimation of the bulk sound velocity C_b in the compressed material. The Poisson's ratio ν of the shock-compressed polyurethane may be calculated from the

relation $(C_l/C_b)^2 = 3(1 - \nu)/(1 + \nu)$. The calculated values of the Poisson's ratio may be described by the following dependence on the compression degree $\sigma = (\rho-\rho_0)/\rho_0$: $\nu = 0.5 - 5.053\sigma + 77.79\sigma^2 - 398\sigma^3$, $\sigma < 0.1$. The value $\nu = 0.5$ was accepted for Poissson's ratio of uncompressed polyurethane. The Poisson's ratio of polyurethane decreases with the increase of the shock compression degree and achieves, under 7 - 8%-compression, the value $\nu = 0.37$, which is close to that for normal solids. This behavior of the Poisson's ratio agrees with that was found for rubber under shock compression [7] and with the results of static compression experiments [4] with rubber and other elastomers, in particular, the polyurethane, [8]. For these materials the transition from the rubber-like to the glassy state was found at room temperature at the pressure range 0.4 -0.6 GPa.

ACKNOWLEDGMENT

The work is supported by the Israeli Ministry of Defense under grant No 493/7836/7.

REFERENCES

1. Werrick L.J., "Characterization of Booster-Rocket Propellants and their Simulators", The 9th International Symposium on Detonation, Portland, USA, 1989, V.1, pp. 462-470

2. Kolmykov Yu.B., Kanel G.I., Parkhomenko I.P. et al., *Zh. Prikl. Mech. Techn. Fiz.*(1990), N1, 126

3. Danker G.R., Newlander C.D. and Colella N.J., Shock Compression of Condensed Matter, S. Schmidt Ed.(Elsevier Science Publishers, NY, 1990) pp. 213-216

4. Weaver C.W. and Paterson M.S., *Journ. Polym. Sci.*, **7**, (1969) 587

5. Barker L.M. and Hollenbach R.E., *Journ.Appl.Phys.*, **43**, (1972) 4669

6. Asay J.R., Lamberson D.L. and Guenther A.H., *Journ.Appl.Phys.*, **40**, (1969) 1768

7. Kanel G.I., Utkin A.V., Tolstikova Z.G., Shock Compression of Condensed Matter, S. Schmidt Ed.(Elsevier Science Publishers, NY, 1994) pp. 903-906

8. Patterson M.S., *Journ.Appl.Phys.*, **35**, (1964) 176

9. Kolsky H. Stress Waves in Solids, (NY, Dover Publications, Inc., 1963) pp. 158-162

10. Chen P.J. Selected Topics in Wave Propagation, (Noordhoff Internationa Publishing, Leyden, 1976), pp.86-87

CP429, *Shock Compression of Condensed Matter – 1997*
edited by Schmidt/Dandekar/Forbes
© 1998 The American Institute of Physics 1-56396-738-3/98/$15.00

SHOCK RESPONSE OF A UNIDIRECTIONAL COMPOSITE AT VARIOUS ORIENTATION OF FIBERS

S. A. Bordzilovsky, S. M. Karakhanov, and L. A. Merzhievsky

Lavrentyev Institute of Hydrodynamics, Novosibirsk, 630090 Russia

The shock response of a unidirectional laminated composite was studied using the manganin gauge technique for peak stresses of 3.8 to 5.4 GPa and angles of 5 to 90° between the normal to the shock surface and the fiber direction. It was found that the stress-time history of the composite strongly depends on the fiber orientation. Comparison of the data with the results of a static compression test showed that the dynamic yield strength of the composite is dominated either by the yield strength of the epoxy matrix or the interlaminar shear stress depending upon the fiber orientation.

INTRODUCTION

The problem of description of the dynamic mechanical properties of plastic-matrix composites has received considerable attention because of their use under extreme shock loading conditions. In a previous paper (1), a model was formulated which used Maxwell representations of the viscoelastic behavior of the matrix and reinforcing fibers. The characteristics of state were calculated by averaging over an elementary volume. The predictions of the model agree well with our results (2), where the attenuation of shock waves in laminated aramid composites was experimentally investigated.

In all the experiments referred, the direction of loading was perpendicular to that of reinforcing fibers, but otherwise the important feature of composites is their mechanical anisotropy especially large for unidirectional laminated composites. The numerical simulations (3) showed the complex wave structure in the case the shock wave moved along the fibers. So, the objectives of this work are: first, to record the wave profiles in the unidirectional aramid fiber/epoxy matrix composite when the shock wave moves along the reinforcing fibers or at an angle θ to the fibers;

and, second, to measure the shock velocity versus the angle.

EXPERIMENTS

A plane wave lens and a 75 mm diameter, 65 mm thick booster charge of TNT generated the shock wave in a 20 mm thick paraffin pad. The shock then passed through a 10 mm thick copper plate into a 50 mm diameter, 7.5 mm thick specimen backed with a 10 mm thick aramid composite plate. The geometry chosen provided the measuring time 5 μs before the arrival of the second shock reflected from the TNT-paraffin interface. The charge and the set of plates of mismatched impedance produced the shocks of stress ranged from 3.8 to 5.4 GPa in the specimens tested.

The composite tested was made of epoxy resin reinforced with unidirectional aramid 15 μm thick fibers. The volume fraction of aramid fibers was 30%. The initial density of the samples was 1.27 g/cc. The angle θ was measured between the shock wave propagation direction and the fiber direction and equaled 5, 15, 45, 90° for the specimens tested.

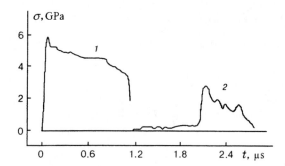

FIGURE 1. Stress-time histories at the impact face (*1*) and at 7.62 mm depth (*2*); $\theta = 5°$.
[6] Reprinted with permission.

The longitudinal stresses normal to the plane wave fronts versus time $\sigma_i(t)$ (*i* is the number of gauges) were recorded by manganin gauges (2). The gauges recorded two stress-time profiles during the experiment: first, at the near interface when the shock entered the specimen, and, second, at the far interface between the specimen and the back plate.

EXPERIMENTAL RESULTS

The stress-time record at the first interface shows the peak stress with the stress decay behind the shock (Fig. 1, curve *1*). The rise time of the signal continues for ≈ 50 ns, which equals the resolution time of the recording system.

The peak stress at the leading edge of some stress-time histories is from 5 to 20% greater than the stress level immediately following the peak. This "blip" is due to shock reverberation in lavsan layers which insulated the manganin strip because lavsan has an impedance greater than that of the composite. The duration of the "blip" corresponds to the shock reverberation time in the insulation layer. To assess this distortion, the maximum stress in the specimen was measured by extrapolating the slope of the decaying part of the profile to the moment the shock entered the specimen.

The stress-time record $\sigma_2(t)$ at the interface between the specimen and the back plate showes one of the following three cases depending upon the angle θ:

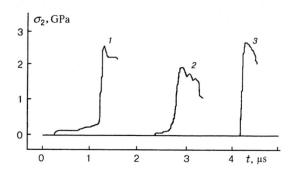

FIGURE 2. Stress-time histories at the specimen-backing plate interface; $\theta = 15°$ (*1*), $\theta = 45°$ (*2*) $\theta = 90°$ (*3*). [6] Reprinted with permission.

1) at $\theta = 5$ and $15°$ one can observe a distinct elastic precursor followed by the main stress rise of short duration;

2) at $\theta = 45°$ the elastic precursor gradually transformes into a "smeared" plastic wave;

3) at $\theta = 90°$ one can observe a single shock of short rise time of about 50 ns.

The distinct precursor is observed at $\theta = 5$ and $15°$. Its velocity ≈ 6 mm/μs is about 1.8 times greater than that of the main shock jump. So, the precursor travels a distance of about 3 mm ahead of the main shock for the specimen thickness of about 7.5 mm. The elastic precursor manifastes itself as a complex structure with a first rise followed by a constant level for ≈ 0.55 μs which then increased during ≈ 0.45 μs by as much as twice of the first rise. On this profile the amplitude of the Hugoniot elastic limit (HEL) and the corresponding yield point can easily be chosen (Fig. 2, curve *1*).

At $\theta = 45°$ the precursor consistes of the first rise with the amplitude three times smaller than in the previous case and the part immediately following which could be characterized as the plastic wave portion with a rise time of about 160 ns. Both the velocities of the first rise and toe of the elastic wave diminish (Fig. 2, curve *2*).

At $\theta = 90°$ no elastic precursor is observed and the second profile has a rise time equals the resolution time (Fig. 2, curve *3*).

The experimental conditions and results are summarized in Table 1, where σ_{HEL} is the Hugoniot elastic limit; $\sigma_{max,1}$, $\sigma_{max,2}$ are the maximum stresses on the first and second profiles, respectively; c is the precursor velocity; u is the velocity of the main shock or the toe of the plastic wave. The relation between the velocities mentioned above and the angle θ are presented in Fig. 3. The plot showes clearly the splitting in the wave trajectories for angles of 5 to 45°.

DISCUSSION

In general, the composite used for the investigation is described as a transverse isotropic material characterized by five independent elastic constants (1). In order to analyze the data at small angles $\theta \approx 5, 15°$, we assume the composite to be in a state of uniaxial strain when loading is parallel to the fibers. In this case, we can use the model of conditionally isotropic material, i.e. the material is described by mechanical properties averaged over volume and the transverse elastic modulus and strength do not depend on the direction in the orthogonal to the fiber plane. According to (4), under these conditions the longitudinal wave velocity is given by

$$c^2 = \frac{E(1 - \mu)}{\rho(1 + \mu)(1 - 2\mu)}, \qquad (1)$$

where E is the equivalent longitudinal elastic modulus of the composite; μ is the equivalent Poisson's ratio characterized transverse strain when the specimen is subjected to uniaxial loading in the fiber direction, ρ is the averaged density of the composite.

Under condition Poisson's ratio of fibers is equal to that of the matrix, the equivalent longitudinal elastic modulus can be estimated using

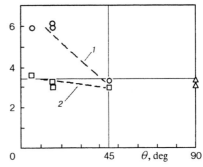

FIGURE 3. Velocities of the elastic precursor (1) and the compression wave (2) vs angle.
[6] Reprinted with permission.

the mixture rule

$$E = vE_f + (1 - v)E_m \simeq vE_f, \qquad (2)$$

where E_f and E_m are the elastic moduli of the fiber and matrix, respectively, v is the volume fraction of fiber.

Because of the lack of static measurements for the aramid composite tested, we compare the dynamic data with the available static results of (5). In Table 2, the elastic modulus of composite calculated from Eqs. (1), (2) and the precursor velocity measured at $\theta = 5°$ is presented.

Correction to the condition of the higher volume fraction in the static experiments gives E ranging from 50 to 82 GPa, which agrees well with the results of (5). This shows that the elastic precursor velocity is determined by the fiber properties, i.e., by the high elastic modulus and

TABLE 1. Experimental Results

$\theta°$	σ_{HEL} GPa	$\sigma_{max,1}$ GPa	$\sigma_{max,2}$ GPa	c, mm/μs	u, mm/μs
5	0.2	5.4	2.8	5.9	3.6
15	0.15	4.8	2.4	6.0	3.1
15	0.15	4.4	2.3	6.1	3.3
45	0.5	3.8	1.3	3.2	3.0
45	0.5	4.0	2.0	3.4	2.9
90	–	5	3	–	3.2
90	–	5	3	–	3.5

[6] Reprinted with permission.

TABLE 2. Comparison Between the Static and Dynamic Material Properties

Load	Sample	E, GPa	Y, GPa	τ, MPa
Static	Fiber	110–160	2.4–4.2	–
$v =$				
0.45–0.75	Epoxy	2	0.13–0.16	–
	Composite	78–95	0.28–0.31	20–44
Shock	Composite	33	0.08–0.11	15
$v = 0.30$				

[6] Reprinted with permission.

its volume content. Under the assumption mentioned above, the Hugoniot elastic limit is related to the longitudinal yield strength of the composite by

$$Y = \frac{(1 - 2\mu)}{(1 - \mu)}\sigma_{\text{HEL}} \qquad (3)$$

This relation and the values measured for σ_{HEL} give the dynamic longitudinal yield strength of the composite tested, which is proved to be smaller by a factor of 1.5 than the static yield strength of epoxy in the compression test (see Table 2). At $\theta = 15°$ the elastic precursor exhibits a more complicated profile that looks like a multiple wave structure: the stress level rises from 150 to 300 MPa at half-time between the arrivals of the first stress rise and the main shock. This behavior can be related to a some form of gradual yielding due to statistical variability of the mechanical properties. Note the value of the stress level just before the main shock agrees well with the static yield strength for the epoxy matrix in the compression test.

It should also be noted that at the small angles the ratio of elastic to main shock velocities is $c/u \simeq 1.8$, and this value is correlated with the ratio of longitudinal to transverse wave velocities in isotropic materials $c_{el}/c_{tr} = \sqrt{3}$ (4). This observation indicates that the simplest mechanism of plastic flow of the composite in the main shock should be complicated by the effect of structure heterogeneity due to the different mechanical properties of constituents. According to (5), the strength of aramid fibers is about 15 or more times greater than the yield strength of epoxy. At stresses greater than the yield strength of epoxy the fiber continues to be in an elastic state up to 2.4 – 4.2 GPa. This stress level characterizes the main shock. Therefore, it may be expected that yielding of the epoxy matrix causes the transverse wave to propagate along the elastic fibers.

At the $\theta = 45°$ fiber orientation the elastic precursor amplitude decreases three times that in the previous case. In Table 2, the maximum shear stress in the precursor (τ) on the laminae plane is compared with the interlaminar shear strength of the aramid composite in statics (5). It is clear by comparing the results that the fail-

ure of the unidirectional composite at the 45° fiber orientation is dominated by the value of interlaminar shear strength.

CONCLUSIONS

The results show that the stress-time history of the unidirectional aramid composite strongly depends on the loading orientation. When loading is nearly paralell to the fibers, the profile exhibits a distinct elastic precursor followed by a shock wave. In the case of loading at the 45° to the fibers, the precursor amplitude decreases and the main shock transformes into a "smeared" compression wave.

The dynamic yield strength of the composite is determined by the smallest of the strength characteristics of the constituents. Thus, in the case of parallel loading, the dynamic strength is dominated by the yield strength of the epoxy matrix; while in the case of loading at the 45° to the fibers, the strength of the composite is dominated by the interlaminar shear strength.

ACKNOWLEDGMENTS

We would like to acknowledge the support of the International Science Foundation (Grant RCC 300).

REFERENCES

1. Merzhievskii L. A., Resnyanskii A. D., and Romenskii E. I., *Dokl. Ross. Akad. Nauk (in Russian)* **327(1)**, 48–54 (1992).
2. Bordzilovskii S. A., Karakhanov S. M., Merzhievskii L. A., and Resnyanskii A. D., *Combustion, Explosion, and Shock Waves* **31(2)**, 236–240 (1995).
3. Merzhievskii L. A., Resnyanskii A. D., and Romenskii E. I., *Combustion, Explosion, and Shock Waves* **29(5)**, 620–624 (1993).
4. Timoshenko S. and Goodier J. N., *Theory of Elasticity*, McGraw-Hill Book Company, Inc., 1951, ch.15, p. 454.
5. Lubin G. ed., *Handbook of Composites*, New York: Van Nostrand Reinhold Company, 1982, ch. 12.
6. Bordzilovskii, S. A., Karakhanov, S. M., and Merzhievskii, L. A., *Fizika Goreniya i vzryva*, **33 (3)**, 132-138, 1997.

CP429, *Shock Compression of Condensed Matter – 1997*
edited by Schmidt/Dandekar/Forbes
© 1998 The American Institute of Physics 1-56396-738-3/98/$15.00

CRACK PREVENTION IN SHOCK COMPACTION OF POWDERS

E. P. Carton[*,**], M. Stuivinga[*], and H. J. Verbeek[*]

[*] *TNO Prins Maurits Laboratory, P.O. Box 45, 2280 AA, Rijswijk, The Netherlands*
[**] *Delft University of Technology, Laboratory for Applied Inorganic Chemistry,*
P.O. Box 5045, 2600 GA, Delft, The Netherlands

The occurrence of macro-cracks in compacts fabricated by shock compaction of powders is a severe problem preventing this consolidation technique from commercial applications. In this paper the sources of important failure types that typically occur in the cylindrical configuration i.e. radial, transverse, and spiral cracks and the Mach stem are described. Subsequently, solutions for their prevention are given supported by experimental results and/or computer simulations. Some conflicting requirements for obtaining bonded crack-free compacts are discussed.

INTRODUCTION

Elimination of macro-cracks in final bodies is one of the most challenging problems to overcome in the shock compaction of powders. Most microstructures of shock compacted ceramic powders reported in the open literature show macro cracks (1, 2, 3). The tendency for crack formation and intensity of cracking increases as the compact approaches theoretical maximal density, TMD (4).

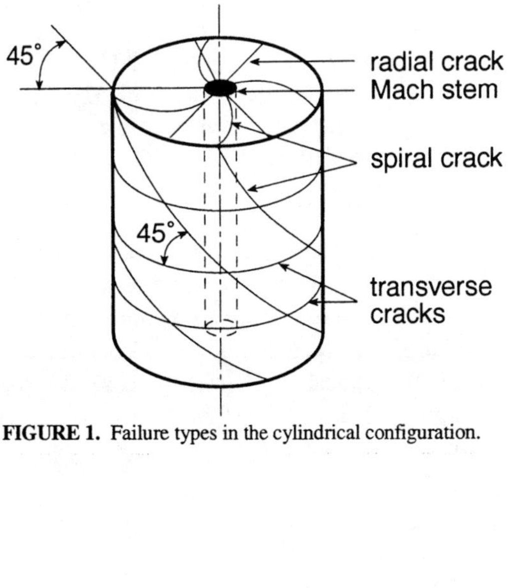

FIGURE 1. Failure types in the cylindrical configuration.

Subject of this article are the failure types typically found in bodies created in the cylindrical configuration (spiral, radial, and transverse cracks, as well as a hole in the compacts center as a result of a Mach stem), as shown in Fig. 1. Benefits of the cylindrical configuration are its simplicity and scale-up possibilities (5), but drawbacks are the macro-cracks and radial inhomogeneity in density that frequently occur in the compacts. The purpose of this article is to show that when the mechanism is known, effective counter-measures for the prevention of those failures are possible.

CYLINDRICAL CONFIGURATION

Shock compaction of powders in the cylind-rical configuration involves the initiation, propa-gation and multiple reflection of shock and rare-faction waves. In Fig. 2, a schematic representa-tion of these processes is shown for two arrange-ments of explosives, a standard or one-layer arrangement (left), and a two-layer arrangement (6) (right), respectively. Compaction takes mainly place in the shock front of the initial shock wave. The strength of the shock wave is a consequence of two competing processes, the absorption and the convergence of the

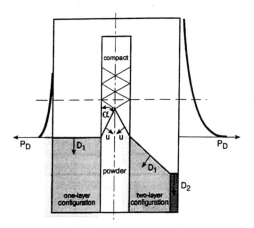

FIGURE 2. Shock, rarefaction, and detonation waves in one-layer (left) and two-layer (right) arrangements of the cylindrical configuration. P_D is the pressure in the detonation products near the container wall.

shock wave energy. At the cylinder axis a self-reflection of the initial shock wave occurs. The reflected shock wave diverges and, upon arrival at the free surface of the cylinder, reduces according to the pressure in the detonation products, P_D. Here the shock reflects as a rarefaction (or release) wave, which converges towards the cylinder axis, where again a reflection occurs. The process of converging and diverging shock and release waves repeats itself many times until the pressure is equalized to atmospheric condition.

CRACK GENERATION

Macro cracks in the compacts, however, occur within the first shock and release waves. Here the intensity of the pressure fluctuations and the amount of strain are maximal. The first danger for cracking is the amount of strain within the initial shock wave. The outer portion of the compact densifies first and must continue to move inward to accommodate the densification of the inner portion of the compact. When the strain is larger then the materials maximal strain, *spiral (or helicoidal) cracks* will occur along planes of maximal shear stress, i.e. 45° with respect to the radius and to the cylinder axis. The mechanism that leads to spiral cracks has been described by Meyers and Wang (7).

Upon arrival of the initial shock wave at the cylinder axis, a regular reflection (see Fig. 2) will

occur, unless the angle of the shock wave front with the cylinder axis (α) exceeds a critical value and an irregular reflection, a so called *Mach stem* occurs. With a Mach stem reflection occurs off axis and at the point of reflection a third shock wave is generated, which travels with the detonation velocity D in axial direction. The Mach stem usually creates a hole in the center of the compact.

At the moment the reflected (or second) shock arrives at the interface between the cylinder and detonation products, the pressure of the two media will equalize. The higher the pressure in the the detonation products will be, the lower will be the pressure reduction, and hence, the intensity of the first rarefaction wave. *Radial cracks* may occur upon the self-reflection of the rarefaction wave at the cylinder axis. Then the compact may fail in tension, with cracks radiating from the cylinder axis. The magnitude of the tensile stress is determined by the amount of convergence of the first rarefaction wave and its initial intensity. In order to decrease this initial intensity, explosives with long pulse duration are used.

The inhomogeneity in starting density of the powder is thought to be the cause of *transverse cracks*. Inhomogeneous compaction occurs mostly with powders that show a large densification during static (pre)compaction. Discontinuity in density leads to local compaction and expansion differences in the powder. This may cause transverse cracks between areas with high and low starting density.

CRACK PREVENTION

Spiral cracking can be avoided by using a high starting density of the powder. This is illustrated by the two B_4C compacts shown in Fig. 3. The left one was obtained with a starting density of 51% TMD and shows numerous spiral cracks, while the right one, with a starting density of 68% TMD obtained by using a trimodal powder mixture, is free of macro cracks.

Also the *Mach stem* can be prevented by useing a higher starting density. In this case the shock wave angle (α) will be larger, enlarging the travel distance of the initial shock wave. The angle will nevertheless stay below the critical angle (α_{cr}), because also the total energy dissipation will be larger. This is due to the longer distance the shock wave travels through

550

the powder until its self-reflection at the cylinder axis. This increased energy dissipation now sufficiently compensates the energy increase due to convergence of the shock wave, avoiding a strong pressure increase towards the cylinder axis.

The avoidance of *radial cracks* requires a long pulse duration of the detonation wave. The shock pressure required for consolidation has been shown to be dependent on the hardness of the powder material (10). To consolidate harder materials explosives with higher detonation velocity, providing higher shock wave pressures, are required. But these have a short pulse duration, which increases the risk of radial cracks. To circumvent this drawback, use can be made of a cylindrical configuration with two explosive layers (6). The outer layer has a high detonation velocity and initiates an oblique detonation wave in the second, slower detonating, explosive surrounding the powder container, as shown in Fig. 2 (left). Due to the oblique detonation angle and the convergence of the slower detonation, high pressures with a longer pulse duration can be attained, see Fig. 2. If the pulse duration obtained this way is insufficient, use could be made of another effect. In Fig. 4 the result of a computer simulation with the hydrocode Autodyn (9) is shown for a configuration in which the layer of faster detonating explosive is situated directly against the metal cylinder. This layer is surrounded by a second explosive layer with a lower detonation velocity. The simulation shows that a reflection of the oblique diverging detonation wave at the wall of the confining cylinder, increases the pressure within the detonation products again after a certain time. After the convergence of this reflected shock wave, it reaches the metal cylinder, temporarily increasing the

pressure at the cylinder wall again. By a careful adjustment of the detonation angle and the dimensions of both the metal cylinder and the explosives arrangement, the moment this pressure increase occurs can be made to coincide with the moment the reflected shock wave in the powder reaches the cylinder wall. This is shown in Figure 5 for two high explosives, PETN with D=8.3 km/s and TNT with D=6.93 km/s.

Since the occurrence of *transverse cracks* is related to the quality of the precompaction of the powder, their prevention can also be obtained there. The particle size distribution and particle morphology should be adjusted in order to obtain a more homogeneous density distribution in the starting powder.

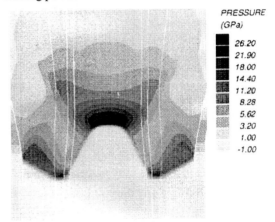

FIGURE 4. Two-layer configuration with reflection of oblique detonation wave at the confining cylinder wall.

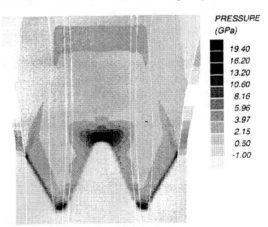

FIGURE 5. Optimized two-layer configuration for the reduction of the intensity of the initial rarefaction wave.

FIGURE 3. Cross-sections of B₄C compacts (both 82 %TMD) which had a starting density of 51 (left) and 68 %TMD (right).

DISCUSSION

Spiral cracks and a Mach reflection can be prevented by using a high starting density of the powder. The prevention of radial cracks requires a long pulse duration of the detonation wave or a precise timing of its reflection. Finally, transverse cracks can be avoided by a careful precompaction of the powder in order to obtain a homogeneous density distribution.

In the examples given, full crystal density (100% TMD) has not been obtained. For a hard and brittle ceramic, like B_4C, the final density of a compact does not exceed 90% TMD, even when a high starting density of 68% TMD is used. For higher final densities much higher pressures should be used. But the tendency for crack formation and intensity of cracking increases as the compact approaches theoretical maximal density (4). The task to obtain dense compacts is further complicated when also particle bonding (consolidation) is required. This introduces further restrictions for the parameters of the process.

Some models for particle bonding require a minimum fraction of melt to be attained during shock compaction (10). For this a minimum amount of energy must be dissipated during the compaction process. As obtained from the shock wave jump-relations, the increase in internal energy can be estimated by $E = 1/2\ P\Delta V$. The pressure P necessary for the compaction of the powder is determined by the hardness of the material and the starting density of the powder (8). This means that the change in volume within the initial shock wave (ΔV) must have a minimum value. To obtain this volume change, the starting density of the powder can not be at its maximal level, as is required for the prevention of most failure types. In order to circumvent these conflic-ting conditions, more parameters should be involved. This could be the starting temperature of the powder or the use of chemical reactions (syn-thesis) during shock consolidation.

CONCLUSIONS

It has been shown that when the failure mechanism is known, in principle it is possible to prevent macro failures during shock compaction.

Generally, a high starting density of the powder reduces the number of spiral cracks in the compact, since its compaction involves a lower amount of strain. Also the Mach stem can be prevented by using a high starting density. Radial cracks can be prevented by making use of one of the two-layer explosive arrangements, which can reduce the intensity of rarefaction waves, while giving a high pressure. Transverse cracks can be avoided by using a homogeneous starting density of the powder.

The avoidance of cracking remains very difficult when full density is to be approached. The counter-measures for crack prevention are not in conflict with each other. However, for particle bonding by the solidification of a melted powder fraction, conflicting requirements can occur.

REFERENCES

1. Adair, J.H., Wills, R.R., and Linse, V.D., *Mat. Sci. Res.*, Vol. 17, R.F. Davis, H. Palmour III, and R.L. Porter (editors), New York, Plenum Press, 1984, pp. 639-655.
2. Shang, S.S., Benson, D.J., and Meyers, M.A., *J. de Phys. IV*, Colloque C8, supplement of Journal de Physique III, **4**, 1239-1242 (1994).
3. Ferreira, A., Meyers, M.A., Thadhani, N.N., Shang, S.N., and Kough, J.R., *Metall. Trans. A*, **22A**, 685-695 (1991).
4. Linse, V.D., *Dynamic Compaction of Metal and Ceramic Powders*, NMAB-394, Natn. Acad. Sci., (1983).
5. Coker, H.L., Meyers, M.A., and Wessels, J.F., *J. Mater. Sci.* **26**, 1277-1286 (1991).
6. Carton, E.P., Verbeek, H.J., Stuivinga, M., and Schoonman, J., *J. Appl. Phys.* **81** (7), 3038-3045 (1997).
7. Meyers, M.A., and Wang, S.L., *Acta Metall.* **36** (4), 925-936 (1988).
8. Ferreira, A., and Meyers, M.A.: *Shock-Wave and High-Strain-Rate Phenomena in Materials*, M.A. Meyers, L. E. Murr, and K.P. Staudhammer (eds.), Marcel Dekker, New York, 1992, pp. 361-370.
9. Birnbaum, N.K., Cowler, M.S.: Proc. Int. Conf. on *Impact Loading and Dynamic Behaviour of Materials*, IMPACT 87", Bremen, Germany, 1987.
10. Schwarz, R.B., Kasiraj, P., Vreeland jr., T., and Ahrens, T.J., Acta Metall. **32** (8), 1243-1252 (1984).

CP429, *Shock Compression of Condensed Matter – 1997*
edited by Schmidt/Dandekar/Forbes
© 1998 The American Institute of Physics 1-56396-738-3/98/$15.00

SHOCK COMPRESSION OF Al+Fe₂O₃ POWDER MIXTURES OF DIFFERENT VOLUMETRIC DISTRIBUTIONS

N.N. Thadhani, K.S. Vandersall, and R.T. Russell,
School of Materials Science and Engineering, Georgia Institute of Technology, Atlanta GA 30332-0245
R.A. Graham,
The Tome Group, 383 La Entrada Road, Los Lunas, NM 87031
G.T. Holman and M.U. Anderson
Advanced Matls. Physics Div., Org. 1152, Sandia National Labs, Albuquerque, NM 87185-0345

The shock compression response of Al and Fe₂O₃ powders has been studied with time-resolved pressure measurements using the PVDF stress-rate gauges, extending the early work of Holman et al.[1] Experiments were performed on Al and Fe₂O₃ powders, mixed in different volumetric distributions, corresponding to 50:50, 40:60, and 25:75 volumetric ratios. The shock-compression response demonstrates a complex effect of volumetric distribution on the densification behavior. The propagated stress wave-forms reveal a change in slope in the rise to peak pressure, indicating the influence of the differences in reactant properties. Differences in the crush strength in powder mixtures of different volumetric distributions are also observed, with the equivolumetric powder mixture showing crush-up to full density at lower pressures.

INTRODUCTION

It is well established that the fundamental mechanisms which control the compression of powders, leading to shock-induced chemical changes or simply microstructural modifications, are dominated by unique processes not observed in other deformation methods.[2,3] These processes occur in time scales of a few to more than a hundred nanosecond rise time of the stress-pulse and microsecond duration of the peak pressure state, and are not clearly understood.[4] Materials variables such as powder particle size, morphology, size distribution, and packing density, as well as the distribution of reactants and differences in their intrinsic properties, significantly influence the shock compression process.[5] By systematically investigating the influence of materials variables, it

may be possible to develop a generalized understanding of the mechanisms of processes occurring during the shock state and leading to the formation of compounds or modified microstructures. The work described in this paper, is part of an effort with this objective in mind and builds upon prior efforts.[1,4]

Prior time-resolved pressure measurements on 2Al+Fe₂O₃ (thermite) powder mixtures[1] have shown that the shock response reveals a complex behavior, with the crush-up to solid density being separately dominated by individual constituents. In the present work, the time-resolved pressure measurements were extended to different volumetric mixtures of aluminum (Al) and hematite (Fe₂O₃) powders, to investigate their shock-compression response.

EXPERIMENTS PROCEDURE

Aluminum and hematite powders of ~15 μm average particle size, were blended in molar ratios of 1:1, 2:1, and 3:1, corresponding to volumetric distributions of 25/75, 40/60 and 50/50 respectively. The mixtures were pressed into capsules, as ~4 mm thick powder layers at ~52-53% packing density. Two different configurations, consisting of (a) copper (front) and Kel-F (back) drivers or (b) PMMA front and back drivers, shown in Fig. 1, were employed to better impedance match the powder with the driver.

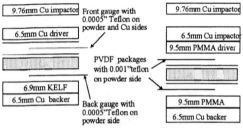

FIGURE 1. Schematic of loading configurations employed for PVDF stress wave measurements.

PVDF gauge packages were placed on either side of the powder sample to monitor the input stress (front gauge), propagated stress (back gauge), and wave speed (knowing transit time between gauges) through the powders. The 25 μm gauges were insulated using an ~12 μm thick teflon insulation. A 0.2 μm layer of Al or Au was vapor deposited on the powder side of the gauge package. Impact experiments were conducted using the controlled symmetric impact-loading facility at Sandia National Laboratories. One experiment was performed using the 80 mm diameter impact facility at Georgia Tech. In both cases Al projectiles with a 10 mm thick copper flyer plate were used to impact the 50 mm diameter powder container. All experiments were designed so that a planar shock wave propagates through the powder thickness without attenuation from the loading or peripheral surfaces. Velocity measurements were performed using charged pins. The piezoelectric current was recorded with two amplifier sensitivities connected to a current viewing resistor

at the PVDF gauge. Details of the PVDF stress gauge measurements and data interpretation have been described in previous publications.[1,4,6]

RESULTS AND DISCUSSIONS

The experimentally measured and calculated parameters from all of the time-resolved PVDF gauge experiments performed on the three types of Al and hematite powder mixtures are listed in Table I. Data in the input stress column corresponds to the stress in the powder measured by the input-shock gauge. The wave speed is the wave velocity through the powder, obtained by measuring transit time between the two gauges placed in direct contact with opposite surfaces of powder compact (less propagation times for the insulation). The propagated wave stress is measured by the gauge placed between the powder and polymer (Kel-F or PMMA) backing. The rise time of the propagated stress wave is also measured based on toe-to-toe and 10-90% peak positions.

The propagated wave-forms of the powder mixtures of the three different volumetric distributions are shown in Fig. 2 (a) and (b) revealing the behavior for input stress <1 GPa and >1 GPa, respectively. In general, the propagated wave-forms show a dispersive behavior, with the equivolumetric $Al+Fe_2O_3$ mixture revealing most significant effects of wave dispersion, both, at low and high pressures. It is also interesting to note that the rise of the stress waves reveals a change in slope at ~0.45 GPa. This change in slope is most dominant in the equivolumetric distribution sample and in particular in the lower pressure experiment.

The relative volume shown in Table I is calculated from the known initial density, measured input stress, and wave velocity, and applying the shock jump conditions for conservation of mass and momentum. Since the stress pulses propagating through the 4 mm thick powder mixtures have a structure characteristic of wave-dispersion effects, calculation of the relative volume based on jump conditions applied to a steady-state shock wave, may not be appropriate. Given the very large compression achieved with such porous materials, one can use the calculated relative volume along with measured input stress to obtain first-order effects of shock-compression of powder mixtures.

TABLE I. Experimental parameters and results from shock compression experiments on Al and Fe$_2$O$_3$ powder mixtures of three different volumetric ratios [Nominal powder thickness is 4mm and packing density is ~53%].

Exp. No.[a]	% Vol. Ratio (Molar)		Density gm/cm^3	Proj. Velocity mm/µs	Measured Input-Stress GPa	Wave Speed mm/us	Propagated Stress GPa	Rise-time (ns)	Relative Volume
	Al	**Fe$_2$O$_3$**				toe-toe/ half-max		toe-toe/ 10-90%	toe-toe/ half-max
2627-B	25(1)	75(3)	2.418	0.234	0.4	0.78/0.76	0.408	330/180	1.38/1.35
2628-B	25(1)	75(3)	2.409	0.592	1.73	1.43/1.41	1.7	90/40	1.27/1.26
2485-A	40(2)	60(1)	2.219	0.314	0.67	1.05/1.04	0.97	310/160	1.39/1.37
2484-A	40(2)	60(1)	2.250	0.635	1.94	1.8/1.77	2.82	140/100	1.38/1.36
9713-B[b]	50(3)	50(1)	2.104	0.310	0.44	0.80/0.78	0.8	700/460	1.12/1.07
2629-B	50(3)	50(1)	2.065	0.546	1.41	1.24/1.16	1.61	560/240	0.95/0.81

[a]A and B refer to experimental configurations in Figure 1; [b]Experiment performed using 80 mm GaTech impact loading facility.

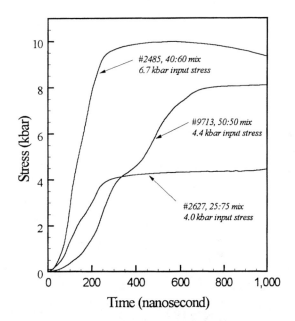

FIGURE 2(a). Propagated stress wave-forms of powder mixtures of 40:60, 25:75, and 50:50 volumetric distributions showing shock compression behavior at input stress <1 GPa.

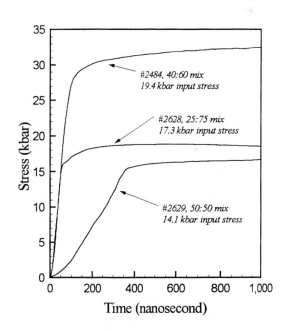

FIGURE 2(b). Propagated stress wave-forms of powder mixtures of 40:60, 25:75, and 50:50 volumetric distributions showing shock compression behavior at input stress >1 GPa.

The plot of measured input stress as a function of calculated relative volume for the three different powder mixtures shown in Fig. 3, reveals significantly different trends in their compression, while showing a similar dual stiffness response. It can be seen that points for 50:50 powder mixture show crush-up to full density at input stress of about 1.4 GPa. In contrast, points corresponding to the 40:60 vol% show crush up at a stress of ~4.5 GPa. The 25:75 vol% mixtures also follow a trend similar to the 40-60 vol% mixture. It is significant to note that the data for both Al-deficient mixtures, indicates crush-up to full density at stress levels higher than that for the equivolumetric mixture. The observed different crush strengths are overt indications of the strong influence of volumetric distribution on the deformation process during shock-compression of powders and their mixtures.

The significance of the results observed in the present work is two-fold. First, the change in slope of rise to peak pressure, observed in the wave profiles (Fig. 2), is indicative of the influence of the differences in properties of powder mixture constituents. The trends provide valuable information useful for process modeling. Second, the observed differences in the crush strength may be expected to influence the threshold conditions for shock-induced reaction initiation. This has, in fact, been observed in mixtures of Ti-Si powders of different morphology.[4] Medium morphology powders of Ti and Si were observed to reveal crush-up to full density at ~1 GPa, followed by reaction initiation at ~1.5 GPa. In contrast, coarse and fine morphology powders had higher crush strengths and showed no reaction at pressures up to 2.5 GPa.

CONCLUSIONS

Propagated wave-forms in Al+Fe$_2$O$_3$ powder mixtures of 25:75, 40:60 and 50:50 volumetric distribution, measured using PVDF gauges show a dispersive behavior. The equivolumetric mixtures reveal most significant effects and also show crush-up to solid density at significantly lower pressures in contrast to the Al-deficient mixtures.

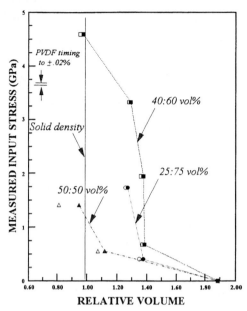

FIGURE 3. Plot of input stress versus relative volume showing compression of Al+Fe$_2$O$_3$ powders of 50:50, 40:60, and 25:75 volumetric distribution. Solid/open symbols refer to volumes calculated using toe-to-toe and half-max values of wave speed.

ACKNOWLEDGEMENTS

Work supported by USDOE Contract No. DE-AC04-76DP00789 at Sandia, and ONR Contract No. N00014-94-1-0169 and ARO Grant No. DAAH04-93-G-0062 at Georgia Tech.

REFERENCES

1. G.T. Holman et al., in *High-Pressure Shock-Compression Science and Technology - 1993*, eds. S.C. Schmidt et al., AIP, 1994, 1119-1122.

2. R.A. Graham, *Solids Under High Pressure Shock Compression: Mechanics, Physics, and Chemistry*, Springer-Verlag, 1993.

3. A. N. Dremin and O.N. Breusov, *Russian Chemical Reviews*, 37 (5), 392-402 (1968).

4. N. N. Thadhani, et al. to be published in *J. Appl. Phys.*, Aug 1997.

5. N.N. Thadhani, *J. Appl. Phys.*, 76, 2125-38 (1994) .

6. M.U. Anderson, R.A. Graham, and G.T. Holman, in High-Pressure Science and Technology - 1993, AIP Conf. Proc. 309, Part 2, eds. S.C. Schmidt, J.W. Shaner, G.A. Samara, and M. Ross, AIP Press, New York, 1994, pp. 1111-1114.

CP429, *Shock Compression of Condensed Matter – 1997*
edited by Schmidt/Dandekar/Forbes
© 1998 The American Institute of Physics 1-56396-738-3/98/$15.00

ELASTIC SHOCK RESPONSE AND SPALL STRENGTH OF CONCRETE

Marlin E. Kipp, Lalit C. Chhabildas and William D. Reinhart

Sandia National Laboratories, Albuquerque, NM, USA, 87185-0820*

Impact experiments have been performed to obtain shock compression, release response, and spall strength of two scaled concrete formulations. Wave profiles from a suite of ten experiments, with shock amplitudes of 0.08 to 0.55 GPa, focus primarily on the elastic regime. Despite considerable wave structure that develops as the shock transits these heterogeneous targets, consistent pullback signals were identified in the release profiles, indicating a spall strength of about 30 MPa. Explicit modeling of the concrete aggregate structure in numerical simulations provides insight into the particle velocity records.

INTRODUCTION

Concrete is a material for which it is desirable to have a comprehensive shock response database in order to assess both local and structural response to projectile impact and explosive loading. The wide range of shock amplitudes that accompany the divergence of a shock from its source requires the acquisition of data over a full spectrum of shock and fracture behavior. The presence of interfaces in the vicinity of shock sources can induce internal fracture (spall) of the material; in addition, shear failure may also occur during propagation of the compressive shock wave.

In this study of two concrete materials with small scale aggregate, gas gun impact experiments are used to obtain Hugoniot data at pressures in the elastic regime. These data are consistent with existing impact data at higher pressure amplitudes. Spall measurements obtained from these low pressure experiments, under the confinement of uniaxial strain, provide insight into one aspect of the fracture response of these two concrete formulations.

In addition, numerical simulations of the impact experiments are reported in which the heterogeneous structure of the concrete is explicitly modeled.

CONCRETE DESCRIPTION

The two concrete formulations considered in the present study - 'SAC-5' and 'CSPC' - differ primarily in the nature of the aggregate (1,2). SAC-5 has a pea gravel and CSPC has an angular gravel, with maximum dimensions of about 10 mm in both cases, constituting about a 40-45% volume fraction of the concrete, and the rest grout. The SAC-5 concrete has a nominal density of 2260 kg/m^3 and an ultrasonic longitudinal velocity of 5060 m/s; the CSPC concrete has a density of 2290 kg/m^3 and an ultrasonic longitudinal velocity of 5200 m/s. These properties are similar to those of a full scale aggregate concrete (3).

EXPERIMENTAL CONFIGURATION

The present plate impact configuration consisted of a projectile launched in a 64 mm light gas gun. The projectile was faced with carbon foam ($\rho = 0.2$ g/cm^3) and a flat polymethylmethacrylate (PMMA) impactor. The PMMA impactor plates had thicknesses of approximately 4.5 mm and 9.5 mm, and a diameter of 57 mm. The PMMA impacts directly onto the concrete sample, 12.7 mm thick. Velocity interferometric techniques, VISAR, were used to monitor the velocity of the concrete rear surface. A thin (10 μm) aluminum foil on the rear surface of the concrete ensured that local surface roughness did not impair the veloc-

---

* Sandia is a multiprogram laboratory operated by Sandia Corporation, a Lockheed Martin Company, for the United States Department of Energy under Contract DE-AC04-94AL85000.

ity measurement. The dispersive nature of heterogeneous materials, however, gives rise to non-unique velocity records whose individual character depends upon the location of the monitoring position. Nevertheless, there will be an average shock response that prevails for the bulk behavior of the material.

For the majority of these experiments, the rear surface of the concrete remained free; in two of the experiments, the concrete was backed with a PMMA window to alter the magnitude of the pullback signal relative to the main shock amplitude, providing a complementary measure of the spall strength.

DATA SUMMARY

The PMMA plate impactor velocities in this series were in the range of 30 to 220 m/s, resulting in compressive stresses of ~ 80 to 500 MPa in the concrete (Table 1). The wave profiles for all ten experiments are shown in Fig. 1a (CSPC) and 1b (SAC-5). Shock rise time rates are nearly uniform across all the experiments; larger amplitudes have correspondingly longer rise times.

Stress and particle velocity must be continuous at the impact interface. Consequently, estimates for the stress, the particle velocity and the shock velocity can be made utilizing the PMMA impact velocity, free-surface particle-velocity measurements and the PMMA equation of state (4) (Table 1). These shock velocity vs. particle velocity data are plotted in Fig. 2, labelled "Elastic", and are seen to be consistent with existing data for this concrete taken at higher pressure levels (1,2). The corresponding Hugoniot stress vs.

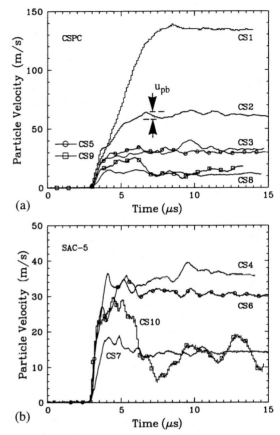

FIGURE 1. Particle velocity profiles for CSPC (a) and SAC-5 (b) concrete experiments.

particle velocity are plotted in Fig. 3, and are seen to be consistent with the elastic modulus expected based on the concrete longitudinal velocity (dashed line).

TABLE 1. Summary and Results of Impact Conditions.

Shot	Concrete Type	Target Thickness (mm)	Impactor Thickness (mm)	Impact Velocity (m/s)	Shock Velocity (m/s)	Hugoniot Stress (MPa)	Strain (u_p/U_S) (cm/cm)	Spall Stress (MPa)
CS1	CSPC	12.743	9.469	220	3561	553	0.0191	23
CS2	CSPC	12.750	9.437	107	3352	260	0.0101	29
CS3	CSPC	12.746	9.535	62	3201	148	0.0063	41
CS4	SAC-5	12.753	9.528	62	3299	149	0.0060	15
CS5	CSPC	12.753	4.427	66	4334	173	0.0040	29
CS6	SAC-5	12.761	4.404	62	3776	155	0.0048	35
CS7	SAC-5	12.730	4.402	32	3690	80	0.0026	29
CS8	CSPC	12.753	4.553	32	5021	87	0.0015	32
CS9	CSPC/Win	12.743	4.630	62	4334	173	0.0040	20
CS10	SAC-5/Win	12.741	4.460	62	3776	155	0.0048	32

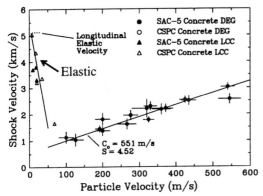

FIGURE 2. Shock velocity vs. particle velocity for these experiments ("Elastic", labelled LCC) and existing data at higher pressures (Grady, (1,2) labelled DEG).

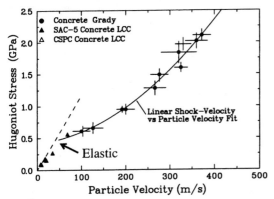

FIGURE 3. Hugoniot stress vs. particle velocity for these experiments ("Elastic") and existing data at higher pressures (1,2).

The results for both SAC-5 and CSPC concretes are included in these figures.

Tensile states form within this heterogeneous material when the release waves from the rear projectile surface and target rear surface interact. When the material tensile limit is exceeded, the signal is transmitted to the monitoring surface, where the change in particle velocity ("pullback") indicates the amplitude of the stress level at which fracture occurred (5,6):

$$\sigma_{spall} = \frac{1}{2}\rho c u_{pb} \ ,$$

where ρc is the material impedance, and u_{pb} is the pullback velocity for free surface conditions (cf, Fig. 1a). Although the most accurate means of determining the fracture amplitude is to iterate with a one-dimensional shock wave code, so that decay of the signal from the fractured region to the diagnostic surface can be included in the analysis, the current heterogeneity invests the wave profile with sufficient noise as to make such an approach impractical.

Impedance differences between the aggregate and the matrix (grout) result in the development of considerable wave structure as the shock transits the target. Despite the large scale heterogeneous composition of the concrete, consistent pullback signals were identified in the release profiles. It was also determined that attenuation effects in the release waves that define the pullback signal were minimal: when the PMMA impactor thickness was reduced from 9.5 mm to 4.5 mm, thereby changing the location of the fracture region in the sample, no alterations appeared in the particle velocity pullback record.

The spall strength of the concrete for each experiment is listed in Table 1 and plotted as a function of impact Hugoniot stress in Fig. 4. Within the elastic re-

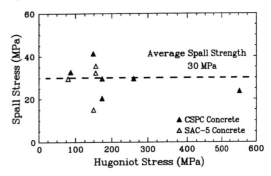

FIGURE 4. Spall stress as a function of Hugoniot stress for these experiments.

gime, there is some scatter at the lower amplitudes of impact stress, but in general the spall strength appears to be rather constant. The low magnitudes of the fracture stress suggest that the failure is being dominated by the grout or interface bonds, since quartz has a much larger failure stress.

NUMERICAL ANALYSIS

The high pressure Hugoniot data included in Figs. 2 and 3 have been utilized in a continuum concrete model developed by Silling (7). The present elastic impact results have motivated numerical simulations in which the target is modeled with explicit aggregate in a grout matrix. An estimated size distribution is used to characterize the dimensions of the aggregate. The position and orientation of each aggregate particle (ellipsoid) is located randomly within a cylindrical envelope that defines the target sample. Similar techniques have been used to define explicit granular structures to explore hot spot formation in explosives (8), and have also been discussed by Amieur, et al (9).

The Eulerian shock-wave propagation code, CTH (10), was used for the simulations. A uniform resolution of 0.2 mm in all three dimensions allowed for about 60 cells through the target thickness. The constituent materials were modeled with a Mie-Gruneisen equation of state and elastic perfectly plastic deviatoric behavior (Table 2).

TABLE 2. Material Constants for CTH Simulations.

Property / Material	Quartz	Grout
Density, ρ_0 (kg/m^3)	2650	2000
Bulk Sound Speed, C_0 (m/s)	3760	2320
Slope of U_s - U_p Hugoniot, s	1.83	1.68
Gruneisen Coefficient, γ_0	1.0	1.0
Specific Heat, C_v (J/kg-K)	86	86
Yield Stress, Y_0 (GPa)	3.1	0.5
Poisson Ratio, ν	0.18	0.22
Fracture Stress, σ_f (MPa)	500	30

Velocity histories at several spatial locations on the rear surface for the CS6 geometry, obtained from a simulation, are plotted in Fig. 5, and compared with the data. The dispersion and scatter created by the heterogeneous structure of the concrete target is apparent in these records.

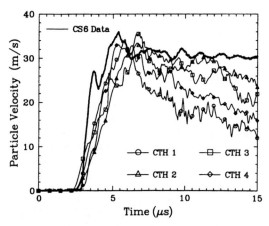

FIGURE 5. Numerical simulations of rear surface particle velocity histories compared with experimental data (CS6).

CONCLUSIONS

The Hugoniot and spall stress have been measured in the elastic regime for two scaled concrete materials. Despite considerable wave structure attributed to the heterogeneous nature of the concrete, the spall amplitudes are quite reproducible, and of nearly constant magnitude over this pressure range. These low pressure shock data are consistent with existing data on these concrete formulations at higher amplitudes. In order to compare with full scale concrete, much larger samples will be required. The current studies were of fully supported shock waves introduced into the concrete. Pulse attenuation studies would also be appropriate to pursue for these materials, to determine how the heterogeneous nature of the concrete influences the decay of the shock. Such data would be important in the simulation of diverging waves in concrete due to projectile impact or explosive loading. It is anticipated that these records will provide additional basis for continued development of damage models for concrete failure.

REFERENCES

[1] Grady, D. E., "Shock and Release Data for SAC-5 Concrete to 25 GPa", Sandia National Laboratories Technical Memorandum - TMDG0595, October 1995.

[2] Grady, D. E., "Dynamic Decompression Properties of Concrete From Hugoniot States - 3 to 25 GPa", Sandia National Laboratories Technical Memorandum - TMDG0396, February 1996.

[3] Read, H. E. and Maiden, C. J., "The dynamic Behavior of Concrete", Systems, Science and Software Topical Report 3SR-707, August 1971.

[4] Barker, L. M. and Hollenbach, R. E., *J. Appl. Phys.*, **41**, 4208-4226 (1970).

[5] Chhabildas, L. C., Barker, L. M., Asay, J. R., and Trucano, T. G., "Spall Strength Measurements on Shock-Loaded Refractory Metals", Shock Compression of Condensed Matter - 1989, Ed. S. C. Schmidt, J. N. Johnson, and L. W. Davison, North-Holland, Amsterdam, 429-432 (1990).

[6] Romanchenko, V. I. and Stepanov, G. V., *Zh. Prikl. Mekh. Tekh. Fiz.*, **4**, 142-147 (1980).

[7] Silling, S. A., "Brittle Failure Kinetics Model for Concrete", in Proceedings of the 1997 ASME Pressure Vessels and Piping Conference, July 27-31, 1997, Orlando, FL.

[8] Baer, M. R., Personal communication, July 1997..

[9] Amieur, M., Hazanov, S., and Huet, C., in "Micromechanics of Concrete and Cementitious Composites", Ed. C. Huet, 181-202, 1993 Presses Polytechniques et Universitaires Romandes, Lausanne.

[10] McGlaun, J. M., Thompson, S. L., and Elrick, M. G., *Int. J. Impact Engng.*, **10**, 351-360 (1990).

CHAPTER VII

MECHANICAL PROPERTIES:
Reactive Materials

CP429, *Shock Compression of Condensed Matter – 1997*
edited by Schmidt/Dandekar/Forbes
© 1998 The American Institute of Physics 1-56396-738-3/98/$15.00

STRUCTURE OF CRYSTAL DEFECTS IN DAMAGED RDX AS REVEALED BY AN AFM

J. Sharma and S.M. Hoover
Carderock Division Naval Surface Warfare Center
Bethesda, Maryland 20817
C.S. Coffey, A.S. Tompa and H.W. Sandusky
Indian Head Division Naval Surface Warfare Center
Indian Head, Maryland 20640
R.W. Armstrong
University of Maryland, College Park, Maryland 20742
W.L. Elban
Loyola College, Baltimore, Maryland 21210

An atomic force microscope (AFM) was employed to reveal the structure of defects produced in single crystals of cyclotrimethylenetrinitramine (RDX), damaged either by indentation, heat or underwater shock. In general, all of these stimuli produced dislocation pits, cracks, fissures and mosaics, however, the details were different. Indentation generated a large number of triangular dislocation pits, which in their turn produced fissures, cracks and holes by coalescing. Heat produced fine parallel cracks. Slivers as thin as sixty molecules across were observed. Shock caused the crystal to become a three-dimensional mosaic structure, 100-500 nm in size, produced by intensive cleavage and delamination. In all cases very fine particles, 20-500 nm in size, were ejected onto the surface as debris from the formation of defects. The AFM has revealed for the first time un-etched dislocation pits in their pristine condition, so that their internal structure could be investigated. A dislocation density of 10^6 cm^{-2} has been observed. RDX is found to behave like a very fragile crystal in which numerous imperfections show up at a level of the stimuli, far below that necessary for the start of chemical reaction.

INTRODUCTION

Sub-micron size crystal defects existing in explosives are believed to play an important role in the formation of hot spots and reaction sites. However, their nature and internal structure is not known because of their small size. The advent of the Atomic Force Microscope (AFM) has removed this limitation and has made it possible to study the structure down to the nanometer or molecular size. In this paper, the structures of defects produced in RDX crystals by stimuli e.g., mechanical deformation, heat or underwater shock are reported on, as revealed by an AM. As expected, the primary effect of mechanical deformation was the production of dislocation pits. The surprising observation of smaller, tens of nm side length pits, that have been associated with local regions of molecular disorder, will be described in a later communication. Historically, the shapes and distribution of dislocation pits were studied only after enlarging them a few hundred times by using a chemical etchant so that they could be seen in a light microscope. The etchants, however, altered the internal structure of the pits. In contrast, the AFM has provided information about the unaltered pits. It has revealed also the structure of the area where the dislocation emerges at the bottom of the pit during glide. For the first time the existence of ejecta, which are nanometer size particles catapulted out during the formation of the pits, has been revealed. The dislocation pits lead to the formation of other defects by coalescing. In the case of the heating, in addition to the kind of defects produced by indentation, a large amount of delamination is shown. Closely packed platelets were produced. The dominant effect of shock was mosaic formation (100-500 nm in size) from intensive three-dimensional cleavage. The present study shows that RDX is a very fragile solid and that its crystal integrity is lost at a very low level of any stimulus.

EXPERIMENTAL

A Digital Instruments Nanoscope II, Scanning Probe Microscope was used. The images were obtained in repulsive mode with a rather high force, 10^{-7} N, to avoid confusion from surface contamination. For indentation, a Knoop indentor with diamond tip was pressed in the (102) growth plane with a force of 10 gf for 5 seconds only. The area surrounding the indentation was investigated for defects caused by the force applied at low rate. To study the effects of heat, a T.A. Instruments DSC machine, Model 2910, was used for heating the samples to desired temperatures at $5°$ C per minute. The shocked samples of RDX studied, were recovered from underwater shock experiments, as described by Sandusky et al. (1). The pressures used were 61 and 129 kbars. The AFM observations were carried out at room temperature, in air.

RESULTS AND DISCUSSION

Figure 1 shows the general effects of mechanical deformation caused by indentation of laboratory-grown RDX crystal at low AFM magnification. Triangular pits, straight crystallographic cracks and curved fissures with ragged banks are observed. The white specks are the particles of RDX which have been ejected onto the crystal surface during the formation of the above mentioned defects. Sometimes the deposits of ejecta reflect the shape of the cracks. Figure 1 shows that the ejecta have been thrown 2-3 microns away from the fissures. Elastic stretching

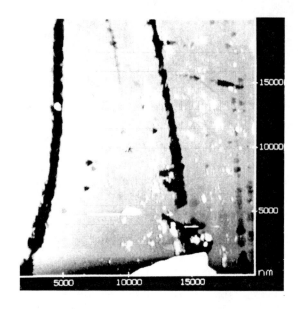

Figure 1. An AFM micrograph of indented RDX showing defects produced: (a) micron-wide curved fissures with ragged banks, (b) straight crystallographic cracks and (c) triangular dislocation pits. The white specks represent particles ejected out during the formation of the defects.

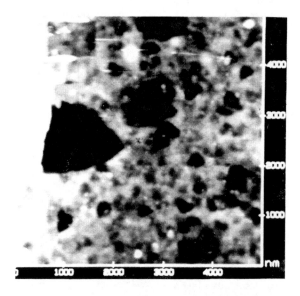

Figure 2. A colony of pits produced from indentation , in an area of large stress.

may have caused this catapulting action. The banks of the fissures and cracks are ragged because they are produced by the coalescing of triangular pits.

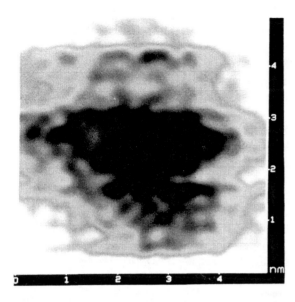

Figure 3. Molecular disorder and worm-hole at the very bottom of a pit, where the dislocation which produced the pit is supposed to emerge.

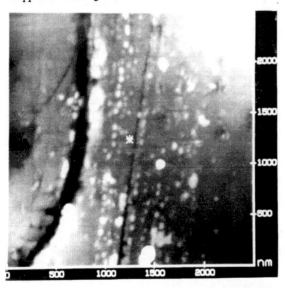

Figure 4. The AFM image of an RDX single crystal, heated to 190° C, showing curved fissures, straight cracks, pits and ejecta, as seen in the case of the indented sample as well.

In some areas, where the stress gets concentrated, a preponderance of the pits is exhibited, as shown in Fig. 2. Most of the pits were triangular, much smaller but not different in

overall appearance from the microscopic-sized (330 microns) etched pits reported by Connick and May (2) for the (111) surface of RDX. If the surface is smooth they seem to be self oriented. Most of the pits were 200-400 nm in size, although the full range is from 20 to 1700 nm. The depth was approximately one tenth of the triangle's side. The ejecta range in size from 20 to 500 nm and have rough surfaces. Large ejecta show dislocation pits on them along with loosened material. It appears that the formation of dislocation pits and the elastic ejection of material is the primary effect. The larger defects such as holes, straight crystallographic cracks and curved fissures are subsequently produced from them. Sometimes the dislocation pits join together and form larger pits or irregularly shaped craters. Very often pits inside pits are seen. Some pits show stairwells which are geometric, but quite often they become irregular. Images taken deep inside the pit show that as the bottom is approached its banks converge, so that the shallow ones are at first eliminated and then the final two banks meet in a ditch. Figure 3 shows molecular arrays at the bottom of a ditch. Perhaps this is the point where the dislocation has emerged in the making of the it. The lattice is twisted and disordered. It could be the remnant of the dislocation pipe. According to Frank (3) and also Gilman and Johnston (4), a stress-driven dislocation would produce a pit, when it intersects the surface during glide. The crystal in that region would be completely disordered and might even show liquid structure.

An RDX crystal which had been heated to 190° C showed defects similar to those produced by indentation. Figure 4 shows wide curved fissures, crystallographic straight cracks and pits which were produced by heating. The dominant effect of heat was the production of parallel straight cracks very close to each other. Some of the ridges are only 30 nm or sixty molecules across. Figure 5 shows a colony of such cracks running over a distance of microns, perfectly parallel to each other. The ejecta in the case of heated RDX were like pebbles and did not show crystal facets. Molecularly resolved images of the heated crystal and of the ejecta showed large disorder and also some molecular vacancies on the surface.

The similarity between the defects produced by mechanical deformation and heat

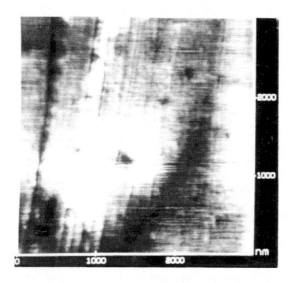

Figure 5. Heated RDX showing a set of parallel cracks, breaking the crystal into thin slivers in the bc plane, some only 60 molecules across.

Figure 6. The surface of an RDX crystal after being subjected to an underwater shock of 129 kbars. The crystal became essentially a lump of three-dimensional mosaics full of cracks and fissures.

indicates that in the latter case the anisotropic expansion may be the main effect and that it

translates into mechanical stress. The calculations carried out by Sorescu et al. (5) have shown that the thermal expansion coefficient along the b and c directions is approximately 47×10^{-6}/K while along the a direction, it is 34.8×10^{-6}/K. It is not surprising then that the material would break into slices in the b-c plane, as it has done.

Underwater shock reduced the crystal into a lump of mosaics, which were 100-500 nm in size as seen in Fig. 6. The straight crystal cracks and curved fissures all wiggled around the mosaic structures, indicating that the mosaics must have formed earlier. The mosaics retain semblance of crystal shapes. In the case of 129 kbar shock, evidence of melting was also exhibited.

At the molecular scale, the surfaces of all of the crystals RDX studied were rough. The free surface never ends in a molecularly flat crystal face. The arrangement of surface molecules shows disorder and reorientation.

In summary, it can be said that RDX is a fragile solid, and that it loses integrity at stresses far below those required for any chemical reaction. The AFM work has drawn attention to the creation of fine particles ranging in size from 20 - 200, nm produced by ejection. These particles have a large surface to volume ratio. The potential involvement of these fine particles in the matter of sensitivity, will need further investigation.

The work reported here was supported by the Office of Naval Research (ONR), # N00014-97-WX-20375.

REFERENCES

1. Sandusky, H. W., Beard, B. C., Glancy, B. C., Elban, W. L. and Armstrong, R. W., *Mat. Res. Soc. Proc.* **296**, 93-98 (1993).
2. Connick, W., and May, F.G. J., *J. Cryst. Growth*, **5**, 65-69 (1969).
3. Frank, F. C., *Acta. Cryst.*, **4**, 497-501 (1951).
4. Gilman, J. J. and Johnston, W.G., *J. Appl. Phys.* **27**, 1018-1022 (1956).
5. Sorescu, D. C., Rice, B. C. and Thompson, D. L., *J. Phys. Chem.* B, **101**, 798-808 (1997).

CP429, *Shock Compression of Condensed Matter – 1997*
edited by Schmidt/Dandekar/Forbes
© 1998 The American Institute of Physics 1-56396-738-3/98/$15.00

DIRECT MEASUREMENT OF STRAIN FIELD EVOLUTION DURING DYNAMIC DEFORMATION OF AN ENERGETIC MATERIAL

B. W ASAY, B. F. HENSON, P. M. DICKSON, C. S. FUGARD, and D. J. FUNK*

Los Alamos National Laboratory, Los Alamos, New Mexico 87545 USA

We previously reported results showing displacement fields (at a single instant in time) on the unconfined surface of an explosive during deformation using white light speckle photography. We have now successfully obtained similar data in confined samples showing the evolution in time of the strain field using laser-induced fluorescence speckle photography. A modified data analysis technique using methods borrowed from particle image velocimetry was used in conjunction with an eight frame electronic CCD camera. For these tests, projectiles of varying shape were fired into an explosive sample. Localization of strain was observed in all cases and was found to be a strong function of the projectile shape, with ignition occurring in those cases where shear appears to play a dominant role. Results from this and continuing studies provide experimental evidence for strain localization, and for the first time allow the direct comparison to computer model predictions. The data are also being used in the design of more realistic and reliable constitutive models.

INTRODUCTION

Optical methods have been used for many years to measure both in-plane and out-of-plane motion of materials during quasi-static and dynamic loading. These techniques include laser speckle interferometry, coherent gradient sensing, laser speckle photography, white light speckle photography, high resolution moiré photography, and digital speckle pattern interferometry, among others. Each method has its particular strengths and weaknesses, and the choice of which technique to use rests upon careful analysis of a host of issues. Among these is the nature of the material to be studied (e.g., viscoelastic or brittle), rate of deformation, magnitude of three dimensional effects, resolution required, cameras and optics available, availability of lasers and other illumination sources,

computational capabilities, and analysis response time requirements.

The many applications of laser speckle photography have been well documented (1). These include the visualization of stress concentration, thermal stress development, and fracture mechanics. It is a useful technique for noncontact strain measurements and has been used extensively in metrology. However, many of these applications are used in quasistatic environments and are not suitable for dynamic measurements.

We have developed a novel technique to perform speckle photography of explosives during dynamic deformation. We are interested in such measurements because we wish to identify energy localization mechanisms which cause ignition at low- to moderate rates of deformation. These measurements are particularly difficult because the

* This work sponsored by the U.S. Department of Energy under contract number W-7405-ENG-36.

material has a low yield strength (~8-80 MPa) and because significant out-of-plane motion and surface disruption occurs during fracture, usually early during the deformation process. We have performed conventional white light speckle photography using coherent illumination and unconfined explosives (2) and more recently developed a technique wherein we use a coherent illumination source and the laser-induced fluorescence from a dye molecule dissolved in a portion of the surface to create the speckle pattern on confined surfaces (3). We here report a variation of that method which increases contrast and thus improves the accuracy of the data. We have also been able to implement the technique using an eight frame CCD array which permits observation of the evolution of the deformation field.

EXPERIMENTS

A full explanation of the method is provided elsewhere (3). A general description will be given here along with the modifications. A rectangular (10 x 25 x 5 mm) PBX 9501 sample is dipped into a solution of Rhodamine 6G dye for 10 s and then rinsed in dichloroethane. The dye is preferentially absorbed into the binder. The explosive is then placed into a steel fixture which permits observation through a 25.4 mm sapphire window on the front and illumination from the rear through the sample (see Fig. 1). We found that by back-illuminating rather than front illuminating as before, we obtained improved resolution and contrast. This configuration maintains a two dimensional confinement for the times of interest. Plungers of different radii are then placed in contact with the explosive. For these experiments we used two plungers, one of 10 mm radius and the other having a 19 mm radius.

The assembly is placed at the end of a gas gun and the plungers are impacted with brass projectiles having nominal velocities of 190 m/s. This introduces a reliable and reproducible strain field into the explosive. Eight (15 mJ) Nd:YAG lasers are focused into a fiberoptic taper and conducted onto the rear of the target, each at a preselected time. The resulting white light speckle pattern is

then imaged on the opposite face with an Imacon 468, eight channel CCD camera (22μ pixels, 586 x 385) at a magnification of 0.875. We found that the 532 blocking filter used in previous front-lit tests was not needed in the back-lit mode. Images were acquired at 3 μs intervals with exposure times determined by the laser pulse width, which is approximately 10 ns.

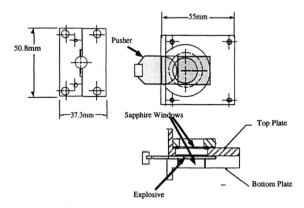

FIGURE 1. Confinement assembly showing explosive containment and optical access.

We previously used a modified CASI data reduction technique (3, 4). Subsequently we have been using a proprietary software package called Visiflow (AEA Technology) originally developed for use in particle image velocimetry. This software has many more capabilities than were originally available to us, and performs a wide range of analyses.

Because the CCD arrays are slightly misaligned with respect to one another, static images were analyzed along with the dynamic images, and apparent motion was removed by subtracting the one from the other. Cross correlation using 32x32 blocks with 50% overlap was used to reduce the data presented here.

RESULTS AND DISCUSSION

We here report a sample of the results obtained with the current method. Figure 2 shows x-component displacement vectors and contours from the experiment at three different times, 3 μs apart, using the plunger of 10 mm radius. Impact occurs at the center of the left edge. The vectors are as-

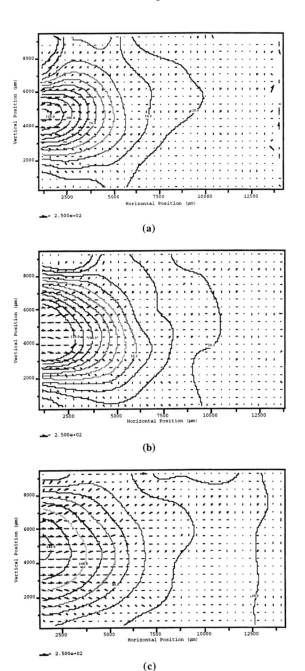

FIGURE 2. Vector field with contour overlay showing displacement after impact with projectile of 10 mm radius at three successive times, 3 μs apart.

measured while the contours were taken from data which were subjected to a kernel smoothing algorithm. The evolution of the displacement field is clearly demonstrated by these data. Note the change in the contour levels in Fig. 2c.

Figure 3 shows another representation of the data found in Fig. 2c. The large amount of curvature in the displacement front can be more easily seen when the two figures are compared.

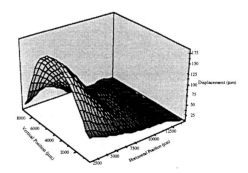

FIGURE 3. Surface plot showing displacement as a function of position. These data are the same as represented in Fig. 2c.

Strain localization is more easily observed by taking selected vertical slices through the displacement data. Figure 4 shows $\varepsilon_x|_y$ taken 2.005

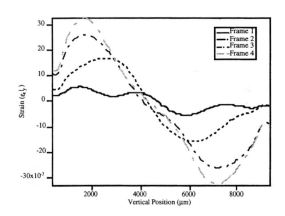

FIGURE 4. Strain ($\varepsilon_x|_y$) computed along a vertical slice of the displacement field 2005 μ from left edge of target after impact by plunger with 19 mm radius. Frames are 3 μs apart.

mm from the left side of the target at four different times, 3 μs apart, after impact by the 19 mm plunger. The development of the strain with time shows a steepening of the curves which represents a localization. Such localizations can lead to shear bands and other structures which can result in high temperatures and ignition.

Figure 5 compares the strain field at the same location as Figure 4, but compares the results from fields resulting from impacts by the two different plungers. The plunger with the smaller radius provides a much higher level of strain in the same volume. Decreasing the radius further is expected to lead to ignition of the explosive.

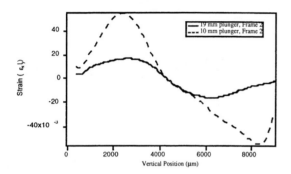

FIGURE 5. Comparison of strain $(\varepsilon_x|_y)$ computed along a vertical slice of the displacement field 2005 μ from left edge of target after impact by plungers with 19- and 10 mm radius.

CONCLUSIONS

Understanding the ignition of explosives by events which do not produce a strong planar shock is very important if we are to adequately address the safety of systems. This understanding cannot come without accurate and reliable models which have been fully tested against resolved dynamic data. We believe that measurements of displacement fields such as those reported in this paper, measurements of temperature fields (5), and similar studies, coupled with three dimensional computer modeling (6) will yield such an understanding. Future work in our laboratory will provide detailed measurements of shear band formation and ignition.

ACKNOWLEDGMENTS

The authors acknowledge the support of Phillip Howe and Joseph Repa throughout this effort.

REFERENCES

1. Khetan, R.P. and F.P. Chiang, *Strain Analysis by One-Beam Laser Speckle Interferometry. 1: Single Aperture Method.* Applied Optics, 1976. **15**(9): p. 2205-2215.
2. Asay, B.W., D.J. Funk, G.W. Laabs, and P.J. Howe. *Measurement of Strain and Temperature Fields During Dynamic Shear of Explosives.* in *Shock Compression of Condensed Matter, American Physical Society Topical Conference.* 1995. Seattle, WA.: AIP Press.
3. Asay, B.W., B.F. Henson, and D.J. Funk, *Speckle Photography During Dynamic Impact of an Energetic Material Using Laser Induced Fluorescence.* Journal of Applied Physics, 1997(July).
4. Chiang, F.P., J. Adachi, R. Anastasi, and J. Beatty, *Subjective Laser Speckle Method and Its Application to Solid Mechanics Problems.* Optical Engineering, 1982. **21**(3): p. 379-390.
5. Henson, B.F., D.J. Funk, G. Laabs, and B.W. Asay. *Surface Temperature Measurements of Heterogeneous Explosives by IR Emission.* in *Shock Compression of Condensed Matter.* 1997. Amherst MA: AIP Press.
6. Haberman, K.S., B.W. Asay, B.F. Henson, and D.J. Funk. *Modeling, Simulation, and Experimental Verification of Constitutive Models for Energetic Materials.* in *Shock Compression of Condensed Matter.* 1997. Amherst MA: AIP Press.

CP429, *Shock Compression of Condensed Matter – 1997*
edited by Schmidt/Dandekar/Forbes
© 1998 The American Institute of Physics 1-56396-738-3/98/$15.00

PARTICLE CHARACTERIZATION
OF PRESSED GRANULAR HMX[*]

N. J. Burnside, S. F. Son, B. W. Asay, C. B. Skidmore

Los Alamos National Laboratory, Los Alamos, New Mexico 87545 USA

It is widely accepted that particle size and morphology in granular beds of HE plays a large role in combustion and detonation events. This work reports the characteristics of coarse granular HMX (Class A) at a range of densities from stock density to 95% TMD. We report measurements of the particle size distribution of original granular HMX, as well as the size distribution of pressed (higher density) samples. Scanning electron microscope (SEM) pictures are presented and are found to be useful in interpreting the size distribution measurements of the granular HMX, as well as in helping to more fully characterize the state of the particles. We find that the particle size distribution changes significantly with pressing. Particles are observed to be highly fractured and damaged, especially at higher pressed densities. Also, we have found that sample preparation can significantly affect size distribution measurements. In particular, even short duration ultra-sonic or "sonication" treatment can have a significant effect on the measured size distributions of pressed HMX samples. Surface area measured by gas absorption is found to be much larger than inferred from light scattering.

INTRODUCTION

It is widely known that damaged explosives can be more sensitive to initiation than undamaged materials. Granular explosives have often been used as a simulant of damaged explosives because it is far easier to characterize the materials than actual damaged explosives and the "damage" is essentially uniform. Still, little material characterization is generally reported in studies that use granular explosives, such as deflagration-to-detonation transition (DDT) experiments. This lack of characterization makes modeling and interpretation of the experiments difficult. Further, very little is known about how particle size changes with compaction processes, even for quasi-static pressing.

Works by Elban *et al.* (1) and Coyne *et al.* (2) have focused on the compaction process of very coarse (~900 μm) granular HMX, and have found fracture at very low pressures. Hardman *et al.* (3) also observed fracturing of other granular material at low pressures. Because of sample consolidation at high densities, however, many past studies have very little particle characterization of high density samples (4) and less characterization of less coarse granular HMX has been performed. Further, measurements of surface area have not generally been made.

As a part of the explosives safety program at LANL our group has worked to develop models to describe the DDT of granular HMX explosives. A main goal of this effort is to develop truly

---

[*] This work sponsored by the U.S. Department of Energy under contract number W-7405-ENG-36

predictive models. DDT experiments by McAfee *et al.* (5) and Burnside *et al.* (6) have extensively used the same batch of Class A HMX, however particle characterization is limited for this material. In this work we try to address this void.

EXPERIMENTAL SETUP

To examine the effect of pressing, eight samples of Class A HMX were prepared starting with poured density, about 64% TMD, and increasing by increments of 5% TMD from 65% TMD to 95% TMD (100% TMD=1.903 g/cc). To reduce density gradients, the samples were pressed at 3 mm increments in a 0.25 in diameter die, and samples were removed from the die after every three increments. Samples that formed pellets (above 80% TMD) due to high density pressing were carefully deconsolidated to powder by hand. The pressures needed to deconsolidate the pellets were very small compared to pressures experienced in the pressing procedure, and were therefore considered to have little effect on particle characteristics. Butler *et al.* (4) made this assumption for low density sugar samples, but did not test higher density samples (>72.2% TMD).

Particle size analysis was done using a Coulter LS 230 Particle size analyzer which uses light scattering of particles to measure size distributions. Samples were taken from solutions of about 0.1 g HMX in a bath of 10 ml of distilled water. Because of quick settling of the larger particles, a magnetic stir bar was used to obtain samples representative of the entire distribution. Samples of approximately 1 ml were quickly transported from the solution to the particle analyzer using a dropper.

Two sets of experiments were performed with the particle size analyzer. In the first set, the 8 samples of HMX which had been pressed then deconsolidated were analyzed for particle size distributions. In the second set of experiments, HMX from the same batch of samples was first put into a low power ultrasonic cleaning bath, 0.28 w/cm^2 at 48kHz, (7) for one minute, then introduced into the analyzer. A magnetic stir bar

was once again used to ensure uniformity in the samples.

Surface area analysis was performed on the samples using a Quantachrome AUTOSORB-1 Surface area analyzer, which measures quantities of gas adsorbed and desorbed on a solid surface. This instrument performs a multipoint Brunauer-Emmett-Teller (BET) analysis using nitrogen as the adsorbate.

Finally, structure of the HMX was qualitatively analyzed using a Scanning Electron Microscope (SEM). Samples were observed and photographed extensively. These images are available at: http://sonhp.lanl.gov/sem_jpg.

RESULTS AND DISCUSSIONS

In this section we present characterization of granular HMX from the 8 pressing densities considered, beginning with the original material.

Unpressed HMX

The original unpressed HMX (a typical particle is shown in Fig. 1) shows well formed crystal structures with identifiable facets, and few cracks or flaws. Particle size analysis at this density shows that the unpressed HMX has a mean particle diameter of 193 μm (see Fig. 2). The largest volume percent of the sample is grouped around 178 μm with a relatively low volume of small diameter particles. A fairly good comparison of this measured size distribution with a sieve analysis was achieved.

FIGURE 1. Unpressed HMX (bar=10 μm).

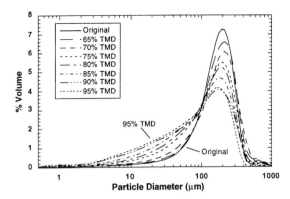

FIGURE 2. Particle size distributions of pressed HMX (No sonication).

Pressed HMX

Figure 3 shows a typical HMX crystal after having been pressed to 70% TMD. Large cracks run throughout the structure of the crystal, however, the bulk of the original particle clings together. Particles at 80% TMD, see Fig. 4 for example are heavily damaged with increased evidence of fracturing and shearing. Finally, particles at 90% TMD, Fig. 5, are crushed to fine pieces which cling together in larger agglomerates of about 100 μm.

As seen in Fig. 2, the mean particle diameter decreases with increasing density from 192 μm to 131 μm at 95% TMD. With increased pressures due to high-density pressing, many of the particles are cracked and sheared, leaving a much larger volume percent of particles in the 20 μm to 40 μm region. It does appear, however, that although highly fractured (see Fig. 3), a large volume of the particles cling together, leaving the distributions of even the high density samples in the 100 μm to 180 μm range.

Sonication Effects

One minute of sonication showed little effect on the mean particle diameter of the original HMX (Fig. 6). Because of the low power of the ultrasonic bath, and the unfractured state of the particles, the distribution was almost unchanged. (Compare "original" distributions from Fig. 2 and Fig. 6). As density increases, however, and the state of the particles becomes increasingly

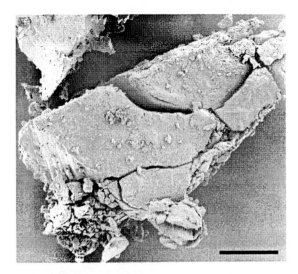

Figure 3. HMX 70% TMD (bar=10 μm).

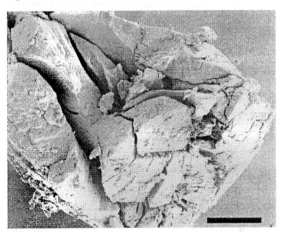

FIGURE 4. HMX 80 % TMD (bar=10 μm).

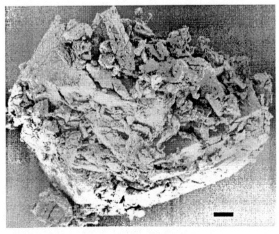

FIGURE 5. HMX 90% TMD (bar=10 μm).

573

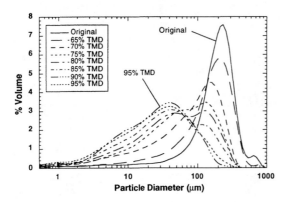

FIGURE 6. Particle size distributions of pressed HMX after one minute of sonication.

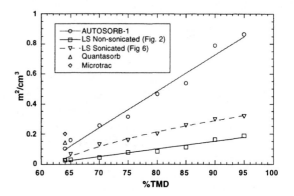

FIGURE 7. Surface area data taken with AUTOSORB-1 gas adsorption unit. Coulter LS data converted from particle size to surface area assuming spherical particles. Earlier data taken with Quantasorb and Microtrac systems.

fractured, the effects of sonication become apparent. Distributions of samples from 65% and 70% TMD show increasing volumes of particles in the <100 μm range. At 75% TMD many of the large, but fractured particles are deconsolidated by the mild stimulus and we begin to see a transition to a bimodal distribution between 40 μm and 180 μm. This new 40 μm mode becomes more prominent with increasing densities until the 180 μm distribution completely disappears, and the remaining distribution tends toward 40 μm.

BET Surface Area Analysis

Increasing density due to pressing results in highly fractured particles with an increased surface area. Surface area analysis, using the AUTOSORB-1, shows that surface area per volume increases approximately linearly with density.

By assuming spherical particles, the size distributions obtained with the light scattering particle analyzer (Fig. 2) were converted to total surface area per unit volume, and plotted along with the AUTOSORB-1 data. Because of the mechanism used by the size analyzer, however, it is incapable of detecting fine cracks in fractured particles (see Figs. 3-5), and thus underestimates the overall surface area. The data, however, also displays a linear increase with surface area. A similar conversion using the size distributions from the sonication samples (Fig. 6) is also plotted. As seen, the

deconsolidation of fractured particles by sonication shifts the conversion closer to the actual surface area analysis. It is also nearly linear with TMD, except at low TMD.

CONCLUSIONS

Surface area analysis of class A HMX shows a nearly linear relationship between density and surface area per volume. Converted particle size analysis, however lacks the ability to account for fractures which leads to an under-prediction of the surface area. The increase of nearly an order of magnitude in the AUTOSORB-1 data corresponds to widespread fracturing and breaking of HMX particles by pressing. These results have significant implications on the modeling used to describe the burning and transition to detonation of granular HMX.

REFERENCES

1. Elban, W.L.,*et al., Powder Technology* **46**, 181-193(1986)
2. Coyne, Jr., *et al., 8th Symp. (Int.) on Det.*, Albuquerque, NM, July 15-19, 1985
3. Hardman, J.S., *et al., Proc. R. Soc. Lond* **333**, 183-199 (1973)
4. Butler, P.B,.*et al. Powder Technology* **62**, 171-181 (1990)
5. McAfee, J.M., *et al., 9th Symp. (Int.) on Det.*, Portland, OR, Aug. 28-Sept. 1, 1989
6. Burnside N.J., *et al.,* JANNAF PSHS Meeting, Naval Postgraduate School, Monterey, CA, Nov. 4-8, 1996
7. Skidmore, C.B., "Effects of Ultrasonic Bath Treatment on HMX Crystals," *Los Alamos National Laboratory Unclassified Report* LA-UR-96-3522, 1996

CP429, *Shock Compression of Condensed Matter – 1997*
edited by Schmidt/Dandekar/Forbes
© 1998 The American Institute of Physics 1-56396-738-3/98/$15.00

POROUS HMX INITIATION STUDIES – SUGAR AS AN INERT SIMULANT [†]

S. A. Sheffield, R. L. Gustavsen, and R. R. Alcon

Los Alamos National Laboratory, Los Alamos, NM 87545

For several years we have been using magnetic particle velocity gauges to study the shock loading of porous HMX (65 and 73% TMD) of different particle sizes to determine their compaction and initiation characteristics. Because it has been difficult to separate the effects of compaction and reaction, an inert simulant was needed with properties similar to HMX. Sugar was selected as the simulant for several reasons: 1) the particle size distribution of C & H granulated sugar is similar to the coarse HMX we have been using (120 μm average size), 2) the particle size of C & H confectioners (powdered) sugar is similar to the fine HMX in the studies (10 μm average size), 3) it is an organic material, and 4) sugar was readily available. Because the densities of HMX and sugar are somewhat different, we chose to do the experiments on sugar compacts at 65 and 73% TMD. As expected, no reaction was observed in the sugar experiments. Compaction wave profiles were similar to those measured earlier for the HMX, i.e., the compaction waves in the coarse sugar were quite disperse while those in the fine sugar were much sharper. This indicates that the compaction wave profiles are controlled by particle size and not reaction. Also, the coarse sugar gauge signals exhibited a great deal of noise, thought to be the result of fracto-emission.

INTRODUCTION

We have been studying porous cyclotetramethylene tetranitramine, HMX, at two densities (65 and 73% theoretical maximum density [TMD]) to determine the low level shock response (less than 1 GPa) of this material to help calibrate computational models being developed in the deflagration-to-detonation transition (DDT) area. Experiments have been carried out in which particle velocity and stress on both sides of an HMX compact were measured so that the stress-particle velocity mapping could be made directly(1,2). From this diagram, it was possible to make estimates of the reaction rate at various input levels. An equation of state was developed for porous HMX that allowed calculation of the Hugoniot at various densities.

Later work involved studies on two different batches of HMX with different particle sizes,(3) one which was called "coarse" HMX, with a mean particle size of about 120 μm, and the other called "fine" HMX, with a mean particle size of about 10 μm. The behavior of these two batches of HMX was quite different, with the coarse material having more disperse transmitted waves and more reaction at lower shock input levels than the fine material. Because reaction and wave dispersion were both occurring at the same levels of input, it was important to find an inert substitute for the HMX so the effects of these two things could be independently determined.

While considering several possible materials, it was determined that granulated sugar had a particle size distribution similar to the coarse HMX and confectioners (powdered) sugar was quite similar to

---

[†] This work supported by the United States Department of Energy.

the fine HMX. Based on this, sugar was chosen as the inert simulant without really worrying about the crystal strength and breakage characteristics.

This paper reports measurements of the shock response of porous sugar compacts (at 65 and 73% TMD) to low level shock inputs (less than 1 GPa).

EXPERIMENTAL DESIGN

C & H brand sugar was selected as the HMX simulant for several reasons: 1) the particle size distribution of C & H granulated sugar was found to be similar to the coarse HMX we used in previous studies (120 μm average particle size) (see Fig. 1), 2) the particle size of C & H confectioners (powdered) sugar was found to be similar to the fine HMX in the studies (10 μm average particle size), 3) sugar is an organic material, and 4) C & H brand sugars were available and some particle size data were available from the company. The crystal density of sugar (1.59 g/cm^3) is less than that of HMX (1.90 g/cm^3) so the material was loaded at the same percentage of TMD as the HMX, namely 65% (1.03 g/cm^3) and 73% (1.15 g/cm^3) TMD. Although pictures of the fine materials are not shown, they were also similar. Information relating to the crystal strength of the two materials was not compared. However, as will be discussed later, some of the crushing properties appear to be similar.

The experimental design for this study was the same as in the earlier studies(1-3). Magnetic "stirrup" particle velocity gauges were attached to the plastic cell front and cell back in contact with the sugar compact so both the input particle velocity and the wave interacting with the cell back were measured. Using this technique, the input to the sugar was accurately determined and the tendency of the wave to spread out was also measured.

It should be noted that confectioners sugar has 3% corn starch added to it to keep the sugar particles from agglomerating. The effect of this on these experiments is unknown at this time.

Gas-gun-driven experiments were conducted; i.e., a projectile faced with polychlorotrifluoroethylene (Kel-F) impacted a Kel-F cell containing the sugar. This is shown schematically in Fig. 2.

Sugar powder was pressed and confined between the Kel-F front face and a polymethylmethacrylate (PMMA) cylindrical plug back. Nylon

FIGURE 1. Pictures of the HMX (left side) and granulated sugar (right side) showing that the particle size of the two batches are similar. The HMX particles are typically diamond shaped while those of the granulated sugar particles are typically cubic.

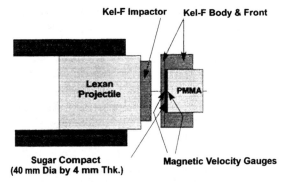

FIGURE 2. Schematic of the sugar experiments. Kel-F cell was impacted by a Kel-F faced projectile. Stirrup magnetic gauges were located on both sides of the sugar compact. Stirrup gauges are single element gauges with a 10-mm long active end that is situated perpendicular to the magnetic field lines.

screws were used to attach the cell front to a Kel-F confining cylinder body (O.D. 68.6 mm and I.D. 40.6 mm). The back PMMA plug was pressed into the Kel-F confining cylinder and held in place with an interference fit. The pressed sugar compact was ≈ 4-mm thick. The magnetic "stirrup" gauges on both sides of the sugar compact were each composed of a 5-μm-thick aluminum stirrup-shaped gauge on a 12-μm-thick FEP Teflon sheet which was glued to the front and back cell pieces.

Magnetic gauging work was started at Los Alamos by Vorthman and Wackerle in about 1980 (4), setting the stage for the magnetic gauge work done at Los Alamos since then.

RESULTS AND DISCUSSION

Eleven experiments were completed in this study; six on coarse sugar and five on fine sugar.

The nominal densities were either 1.03 g/cm³ (65% TMD) or 1.14 g/cm³ (73% TMD). Data obtained from these experiments were surprisingly similar to that obtained for HMX in the earlier studies; the only difference was that no evidence of reaction was observed. In the case of the granulated (coarse particle) sugar, the transmitted waves were disperse in the same way that the coarse HMX waveforms were disperse. This is shown in Fig. 3 for the lower density (65% TMD) HMX and sugar experiments.

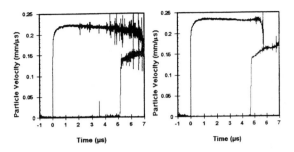

FIGURE 4. Particle velocity waveforms for fine HMX (left side) and C & H confectioners (fine) sugar (right side). The shock transmission times through the compacts are quite close with the sugar being slightly faster, the same as in the coarse materials. The transmitted waves have risetimes of ≈ 100 ns. Impact velocities were 0.279 mm/μs (HMX Shot 982) and 0.299 mm/μs (sugar Shot 1020).

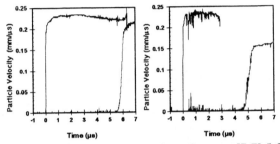

FIGURE 3. Particle velocity waveforms for coarse HMX (left side) and C & H granulated (coarse) sugar (right side). The shock transmission times through the compacts are quite close with the sugar being slightly faster. The transmitted waves are disperse, each with risetimes > 0.5 μs. Impact velocities were 0.288 mm/μs (HMX Shot 912) and 0.295 mm/μs (sugar Shot 1015).

The fine particle (confectioners) sugar experimental waveforms were similar to those obtained in the fine particle HMX experiments. The transmitted wave had a risetime of ≈ 100 ns, indicating that similar phenomena occur both in HMX and sugar. Comparable HMX and sugar waveforms are shown in Fig. 4 for the lower density (64% TMD).

Since the waves move through the sugar faster than through the HMX for both particle sizes, the sound speed in the sugar may be higher than in the HMX. This remains to be experimentally verified.

The sugar waveforms at low input stresses for the two different particle sizes are quite similar to the corresponding HMX waveforms. This indicates that sugar is a good inert simulant for the HMX in that similar processes are obviously occurring in other materials. At the higher input stresses (above about 0.6 GPa) there is reaction in the HMX. One of the input conditions that resulted in considerable reaction in coarse HMX at the lower density (65% TMD) is shown on the left side in Fig. 5. The HMX waveforms are affected a great deal by reaction. Shown on the right side are

corresponding coarse sugar waveforms showing no reaction at all, i.e., this what the HMX waveforms would have looked like with no reaction present. It is now easy to see that the reaction in the HMX is causing the input particle velocity waveform to decrease as a function of time (due to reaction products pushing back on the cell front) and the transmitted wave has grown a great deal as a result of reaction occurring at or near the wave front. Input to the HMX in this experiment was ≈0.8 GPa.

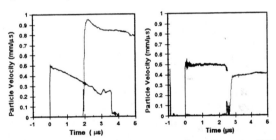

FIGURE 5. Particle velocity waveforms for coarse HMX (left side) and C & H granulated (coarse) sugar (right side). The shock transmission time through the HMX is faster because of the reaction and the wave front buildup. Obviously the HMX waveforms result from reaction occurring in the HMX compact. Impact velocities were 0.696 mm/μs (HMX Shot 913) and 0.700 mm/μs (sugar Shot 1017).

The disperse nature of the transmitted waves measured in the sugar experiments was similar to that previously measured in HMX. This is best illustrated by plotting risetime data as shown in Fig. 6. The risetime of the transmitted wave in each experiment is plotted vs. the projectile impact velocity for that experiment. Risetimes for the sugar are slightly longer than those of HMX and they appear to decrease at slightly higher input velocities. This is probably suggesting there are differences in the crystal strength or crushing characteristics of the sugar and HMX crystals. However, the fact that the behavior was very nearly the same was encouraging. It indicates that in both the sugar and HMX, the particle size controls the wave dispersion, presumably due to the crystal crushing/breakage processes involved in the transmitted wave.

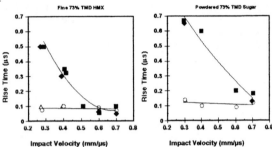

FIGURE 6. Transmitted wave risetime data for the HMX (left side) and sugar (right side) experiments. Data from experiments at both densities ant both particle sizes are shown on each side. Shaded squares are 65% TMD "coarse" materials, shaded diamonds are 73% TMD "coarse" materials, open circles are 65% TMD "fine" materials, and open triangles are 73% TMD "fine" materials.

It is clear that the risetime is not closely associated with density but is directly associated with particle size for both materials.

A difference between the two materials that was substantial was the amount of electrical noise associated with the crystal crushing/breakage process occurring as the wave moved through the compact. Noisy records were not a problem in the HMX experiments. However, all the coarse sugar wave profiles had to be smoothed in order to obtain the average particle velocity waveform. This is shown in Fig. 7 for an experiment on coarse sugar (65% TMD) in which the input to the sugar was ≈ 0.28 GPa. The original data are shown (noisy), along with the smoothed waveform.

This noise indicates that the crushing/breakage process in sugar includes "fracto-emission" to a greater extent than does HMX. This phenomena has been studied in sugar and explosive crystals by Dickinson and coworkers at Washington State Univ. during the 1980s (5-7). They believe the noise is related to photons, electrons, positive ions, etc.,

being emitted as part of the crystal fracture process. We observed this noise in the coarse sugar but not the fine sugar. This could be because the electrical signals were too small to measure or that compaction in the fine material proceeds by a mechanism other than crystal fracture.

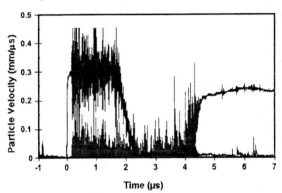

FIGURE 7. Particle velocity waveforms from coarse sugar Shot 1014 with an impact velocity of 0.4 mm/μs. Both the original record and the smoothed record are shown. Notice that the noise stops as soon as all the crushing is complete, i.e., the transmitted wave reaches near the maximum particle velocity in the back gauge.

REFERENCES

1. Sheffield, S. A., Gustavsen, R. L., Alcon, R. R., Graham, R. A., and Anderson, M. U., "Shock Initiation Studies of Low Density HMX Using Electromagnetic Particle Velocity and PVDF Stress Gauges," in the *Proceedings of the Tenth International Symposium on Detonation*, Office of Naval Research Report No. ONR33395-12 (1995), pp. 166-174.

2. Sheffield, S. A., Gustavsen, R. L., Alcon, R. R., Graham, R. A., and Anderson, M. U., "Particle Velocity and Stress Measurements in Low Density HMX," in the *High Pressure Science and Technology – 1993*, Edited by S. C. Schmidt, J. W. Shaner, G. A. Samara, and M. Ross, American Institute of Physics (AIP) Conference Proceedings 309 (1994), pp. 1377-1380.

3. Gustavsen, R. L., Sheffield, S. A., and Alcon, R. R., "Low Pressure Shock Initiation of Porous HMX for Two Grain Size Distributions and Two Densities," in *Shock Compression of Condensed Matter – 1995*, American Institute of Physics (AIP) Conference Proceedings 370 (1995), pp. 851-854.

4. Vorthman, J. E., "Facilities for the Study of Shock Induced Decomposition of High Explosives," in Shock Waves in Condensed Matter -- 1981, Eds. W. J. Nellis, L. Seaman, and R. A. Graham, AIP Conference Proceedings No. 78, American Institute of Physics, New York, 1982, p. 680.

5. Dickinson, J. T., Brix, L. B., and Jensen L. C., J. Phys. Chem. 88, 1698 (1984).

6. Dickinson, J. T., Miles, M. H., Elban, W. L., and Rosemeier, R. G., J. Appl. Phys. 55, 3994, (1984).

7. Dickinson, J. T., Jensen, L. C., Miles, M. H., and Yee, R., J. Appl. Phys. 62, 2965 (1987).

CP429, *Shock Compression of Condensed Matter – 1997*
edited by Schmidt/Dandekar/Forbes
© 1998 The American Institute of Physics 1-56396-738-3/98/$15.00

CHARACTERIZATION OF HMX PARTICLES IN PBX 9501

C. B. Skidmore, D. S. Phillips, S. F. Son and B. W. Asay [*]

Los Alamos National Laboratory, Los Alamos, New Mexico 87545 USA

The particle size distribution and morphology of HMX (cyclotetramethylene-tetranitramine) in the plastic-bonded explosive, PBX 9501 (95% HMX and 5% polymeric binder, by weight), are important to understanding the micromechanical behavior of this material. This paper shows that the size distribution of the "as-received" HMX powder, as measured by light scattering, is not preserved through the processing operations of formulation into molding powder and subsequent consolidation through hydrostatic pressing. Morphological features such as cracking and twinning are examined using reflected light microscopy. This technique helps confirm and interpret the results of the particle size analysis. These results suggest that use of the particle size distribution of the "as-received" powder could potentially yield significant errors in detailed simulations of formulated materials.

INTRODUCTION

HMX is the principal constituent in many energetic materials. Some HMX studies have included characterization data for the particles. Dick (1) reports sieve data and Gustavsen, et al. (2) further describe the powders used in their shock experiments. Elban and Chiarito (3), and Burnside, et al. (4) describe the effects of quasi-static compaction on HMX particles. However, most applications of HMX-based components require the addition of a polymeric binder to improve mechanical integrity, and there is a paucity of characterization data on HMX after it has been processed with a binder.

This work examines HMX particles that have been recovered from PBX 9501 after various stages of processing. The HMX particle size distribution and morphology are important parameters in understanding and modeling the microstructural response of this material to external stimuli. The distribution of the HMX "as-received" powder is controlled by specification (MIL-H-45444B). The HMX (three parts coarse to one part fine) is slurried in water, and binder solution is added to formulate PBX 9501 molding powder. The molding powder is then consolidated into a serviceable article through heating and pressure. This paper reports the particle size distributions of HMX in "as-received powder", after formulation, and after consolidation to a standard density. We confirm these findings with direct observation of PBX 9501 in reflected light microscopy.

EXPERIMENTS

Two principal analytical tools were used in this study. First, all particle size distributions were determined using a Coulter LS 230 Laser Diffraction Analyzer with water as the suspension fluid. Second, a Zeiss Axiophot was used for microscopic inspection. Digital images were acquired with the analyzer oriented parallel to the polarizer.

[*] This work sponsored by the U.S. Department of Energy under contract number W-7405-ENG-36.

All HMX powders were taken from "as-received," commercially produced lots HOL83L030-050 for the coarse portion and HOL82C000E094 for the fine. The specification designates the coarse as Class 1 (previously Class A) and the fine as Class 2 (previously Class B). Both coarse and fine classes are Grade B (previously Grade II) which means there is a minimum RDX content.

Initial samples of material were prepared by the following three processes:
1. A dry blend of the specified proportions (3:1) was prepared by manually mixing the powders.
2. A 3 kg batch of molding powder was produced locally (batch 7249) using the same lots as in the dry blend.
3. A portion of the above molding powder was consolidated at 90 °C and 15000 psi (3 each 3-min. intensifications) to a geometric or calculated density of 1.836 g/cc (99% theoretical maximum density).

Next, it was necessary to remove the binder from samples 2 and 3. Earlier work (5) had shown that ultrasonic bath treatment had a detrimental effect on HMX particles, so an approach with minimum mechanical agitation was chosen. The binder was extracted by soaking samples (14 g each) in dichloroethane at room temperature for 7 days. The solubility of HMX in dichloroethane (6) is reported to be 0.007 g/100 ml of solution at 30 °C. The following sequence was used four times during the week's dissolution:
1. Fresh solvent was added to at least the 50 ml mark on the graduated vial.
2. The soaking material was vigorously shaken by hand to re-suspend particles several times.
3. The material was allowed to settle overnight or forced to settle using a centrifuge.
4. The supernatant was aspirated with vacuum.

The final aspirated samples were dried using a vacuum rotary evaporator to minimize the formation of agglomerates. Dry weights of the samples were obtained before and after extraction to grossly determine that all the binder had been removed. The dried HMX was re-suspended in water prior to particle size analysis.

A portion of sample 1 (dry blend) was also processed with dichloroethane for one day to simulate the effect of the binder extraction and drying process on the HMX itself.

Portions of the molding powder and pressed piece (samples 2 and 3) were also prepared for microscopic examination by mounting in low viscosity epoxy under pressure. After hardening, the mounted sample was polished using methods described in (7).

RESULTS AND DISCUSSION

Figure 1. shows the particle size distributions for "as-received" coarse and fine classes of HMX. The apparent, peak particle diameters are 234 microns for the coarse fraction and 5 microns for the fine one. This may be compared with the peaks from particle size distributions of the processed samples shown in Fig. 2. Here maxima are observed at apparent diameters of 213 microns for the dry blend, 177 microns for the molding powder, and 147 microns for the pressed piece. There is a definite shift toward smaller diameter particles during processing as a result of cumulative particle comminution. A secondary peak is present in the curve for the dry blend at 5 microns diameter which corresponds to that of the "as-received" fine class HMX. The volume percentage is diminished more than expected for a 25% mix. Possible reasons for this are discussed later. An additional, new peak develops for the both molding powder and pressed piece at a diameter of 17 microns. The pressed piece shows a higher fraction of this new peak than the molding

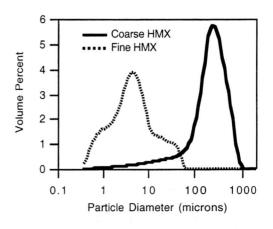

Figure 1. Particle Size Distributions of "As- Received" Coarse (100%) and Fine (100%) Classes of HMX

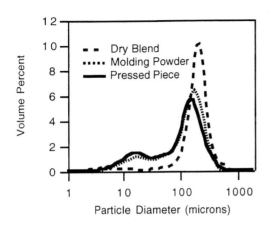

Figure 2. Particle Size Distributions of HMX after Processing.

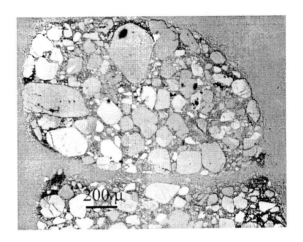

Figure 3. Microstructure of PBX 9501 Molding Powder.

powder. This new peak evidently grows from the breakup of coarse particles during processing.

It is surprising to note that *the HMX size distribution from the molding powder more closely resembles that of the pressed piece than that of the dry blend.* This indicates that the initial processing steps can significantly affect the ultimate particle size distribution. We had previously supposed that the HMX was little affected by the formulation process and that our study erred in using a small mixer rather than the larger equipment of commercial operations. However, there are studies by others that corroborate this unexpected result. Sieve data on HMX from ten different commercially produced lots of PBX 9501 molding powder (8) also show a shift toward a greater fraction of fines than allowed by specification on the original material. The binder was removed in that work by a process involving 24 hours in a wrist-shaker with isobutyl acetate as the solvent. The HMX was sieved wet with the solvent so no drying was necessary. Demol and Alexandre (9) have done binder removal experiments, as well, which demonstrate that damage to the HMX crystals may be minimal or extensive depending on the processing conditions during formulation.

Several micrographs of the mounted PBX samples were taken at magnifications of 50 and 200x. Figure 3. is a 50x image of the molding powder sample. We observe that there are a few cracked crystals which tend to be the larger ones. Twinned crystals are not very common. Visually it is easy to confirm that

the distribution of particle sizes is bimodal. The particles are spaced apart from each other, and the epoxy mounting medium infiltrates large pores within and among the granules. Figure 4. is a micrograph of the pressed piece, also at 50x. Here we note that there are more cracked crystals than in the molding powder. Twinned crystals are abundant. Visually, it is still plausible that the size distribution is bimodal although the coarse particles are notably broken up to form a "new" peak in the particle size distribution. The particles are much more tightly packed than in the molding powder. Quantitative analyses of these and other images are in progress.

A diminished volume fraction of fines in the dry

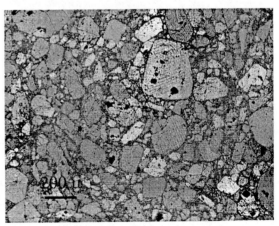

Figure 4. Microstructure of PBX 9501 Pressed Piece

blend was noted previously. Close inspection of Fig. 2. also indicates that there is a *smaller* volume fraction of fines in the molding powder and pressed piece than in the dry blend. Two processes in these experiments may have contributed to a loss of some fines in all three samples and to a greater extent in the molding powder and pressed piece. The first is the aspiration of each sample. This was done only once for the dry blend and four times for the other samples. The supernatant that was aspirated from each settled sample was not perfectly clear, though some turbidity was expected from the dissolved binder. It is possible that fines were lost in this operation. The second is associated with soaking the samples in solvent. The dry blend was soaked in solvent for one day whereas the others were soaked for seven days. There is a mechanism called Ostwald ripening whereby very small crystals are consumed and redeposited onto larger crystals in a saturated solution. The maximum amount of material that could have been lost through normal dissolution is .007 g to all four solvent aliquots. These considerations suggest that the binder removal method may not be optimal but do not invalidate the other results derived here.

CONCLUSIONS

The particle size distributions of commercially produced lots of coarse and fine classes of HMX were determined by a laser diffraction method. Using the same lots of material, PBX 9501 molding powder was produced and a portion was hydrostatically consolidated (pressed). The binder was then removed from these materials and the HMX particle size distributions were determined. The peak volume percentage was observed to shift toward smaller diameter particles indicating some level of comminution in preparation of both the molding powder and pressed piece. Further, the bimodal nature of the distribution is accentuated by the comminution. The distribution of the molding powder more closely resembles that of the pressed piece than that of a dry blend of the "as-received" powders. This demonstrates that formulation can have a significant effect on HMX crystals, whereas

this was previously considered a relatively benign process.

The morphological character of the molding powder and consolidated material was qualitatively investigated by direct observation using polarized light microscopy on polished, mounted specimens. The pressed piece showed a somewhat higher incidence of cracked HMX crystals, a significantly higher frequency of twinned crystals, and a tighter packing of crystals. The bimodal nature of the size distributions can be observed visually in both materials. The microscopy work validates the particle size analysis conclusions.

This work has demonstrated that the particle size distributions of HMX in "as-received" powders do not closely resemble the distributions in PBX 9501 molding powder and pressed pieces

ACKNOWLEDGMENTS

The authors wish to acknowledge support for this work from Los Alamos National Laboratory under the expert guidance of Phil Howe. The assistance of these others is gratefully acknowledged: Rose Gallegos in operating the Coulter instrument, Arnie Duncan for sharing data and helpful discussions, and Nathan Crane for his analysis of the sieve data and investigation into obtaining particle size information through image analysis.

REFERENCES

1. Dick, J.J., *Combustion and Flame* **54**, 121-129 (1983).
2. Gustavsen, R.L., Sheffield, S.A., Alcon, R.R., "Low Pressure Shock Initiation of Porous HMX for Two Grain Size Distributions and Two Densities," presented at the APS conference on Shock Compression of Condensed Matter, Seattle, Washington, August 13-18, 1995.
3. Elban, W.L., Chiarito, M.A., *Powder Technology* **46**, 181-193 (1986).
4. Burnside, N.J., et al., *in these proc.*, (1997).
5. Skidmore, C.B., "Effects of Ultrasonic Bath Treatment on HMX Crystals," *Los Alamos National Laboratory Unclassified Report LA-UR-96-3522*, 1996.
6. Baytos, J.F., unpublished experimental results, Los Alamos National Laboratory, 1954.
7. Skidmore, C.B., Phillips, D.S., Crane, N.B., "Microscopical Examination of Plastic-Bonded Explosives," presented at Inter/Micro 97, Chicago, Illinois, July 21-24, 1997.
8. Duncan, A.A., Mason & Hanger Silas Mason Co., Pantex Plant, Amarillo, Texas, personal communication, June 1997.
9. Demol, G., Alexandre, L., Centre d'Etudes de Gramat, Gramat, France, personal communication, June 1997.

CP429, *Shock Compression of Condensed Matter – 1997*
edited by Schmidt/Dandekar/Forbes
© 1998 The American Institute of Physics 1-56396-738-3/98/$15.00

INFLUENCE OF TEMPERATURE ON THE HIGH-STRAIN-RATE MECHANICAL BEHAVIOR OF PBX 9501

G.T. Gray III, W.R. Blumenthal, D.J. Idar, and C.M. Cady

Los Alamos National Laboratory, Los Alamos, NM 87545

High-strain-rate (2000 s^{-1}) compression measurements utilizing a specially-designed Split-Hopkinson-Pressure Bar have been obtained as a function of temperature from -55 to +50°C for the plastic-bonded explosive PBX 9501. The PBX 9501 high-strain-rate data was found to exhibit similarities to other energetic, propellant, and polymer-composite materials as a function of strain rate and temperature. The high-rate response of the energetic was found to exhibit increased ultimate compressive fracture strength and elastic loading modulus with decreasing temperature. PBX 9501 exhibited nearly invariant fracture strains of ~1.5 percent as a function of temperature at high-strain rate. The maximum compressive strength of PBX 9501 was measured to increase from ~55 MPa at 50°C to 150 MPa at -55°C. Scanning electron microscopic observations of the fracture mode of PBX 9501 deformed at high-strain revealed predominantly transgranular cleavage fracture of the HMX crystals.

INTRODUCTION

The high-strain-rate stress-strain response of energetic materials has received increased interest in recent years related to: 1) the need for predictive constitutive model descriptions for use in large-scale finite-element simulations of collateral damage and energetic systems safety, and 2) focused emphasis on understanding the dynamics of localization phenomena and mechanical failure of polymeric composites. The establishment of more physically-based constitutive models to describe complex loading processes and energetic materials requires a detailed knowledge of the separate and synergistic effects of temperature and strain rate on the mechanical response of Plastic-Bonded eXplosives (PBX's).

A significant number of previous studies have probed the constitutive response of a wide variety of plastic-bonded explosives(1-8). Beginning with the high-rate Hopkinson split-bar studies of Hoge(1) on a range of PBX's and continuing with the high-

strain rate work of Field(3,5), Palmer(6), and Walley it has been found that: a) the effective elastic modulus of PBX's are strongly influenced by strain rate and temperature, b) PBX's during high-rate loading continue straining after the maximum flow stress has been achieved, i.e. viscoelastic-plastic behavior is indicated, c) sample-size and lubrication effects are critical due to the very slow stress wave propagation through PBX's and their susceptibility to shear failure. Low-strain-rate studies on PBX's by Peeters(2), Wiegand(4,7) and Funk(8) have similarly shown that the compressive strength (maximum stress) and the loading modulus increase with decreasing temperature and increasing strain rate. The work of Wiegand(4,7) has further shown a linear correlation between compressive strength and modulus independent of loading rate in compression or temperature, for a range of PBX's. This observation suggests a critical tensile strain criterion to initiate brittle fracture similar to ceramic materials.

The objective of this paper is to present results illustrating the effect of systematic variations of strain rate and temperature at high-strain rate on the constitutive response of PBX 9501.

EXPERIMENTAL TECHNIQUES

This investigation was performed on the plastic-bonded explosive PBX 9501. PBX 9501 is a formulation composed of respectively, 95 / 2.5 / 2.5 / 0.1 weight percent of HMX / Estane / a eutectic mixture of bis(2,2 dinitropropyl)acetal and bis(2,2-dinitropropyl)formal [abbreviated BDNPA-F] / and Irganox (a free radical inhibitor). A molding powder is prepared by the slurry method. The dried powder is then preheated and pressed into cylindrical steel die.

Cylindrical compression samples 6.35-mm in diameter by 6.35-mm in length were machined from the starting billet of PBX 9501. Quasi-static compression tests were conducted at strain rates of 0.001 and 0.1 s^{-1} at 298K in laboratory air exhibiting a relative humidity of ~15%. Dynamic tests were conducted as a function of strain rate, 1500-7000 s^{-1}, and temperature, -55 to 50°C, utilizing a split-Hopkinson pressure bar. The split-Hopkinson bar used for this study was equipped with 9.4-mm diameter Ti-6Al-4V bars that improve the signal-to-noise level needed to test extremely low strength materials as compared to the maraging steel bars traditionally utilized for Hopkinson-Bar studies on metallic materials.

The inherent oscillations in the dynamic stress-strain curves and the lack of stress equilibrium in the specimens at low strains make the determination of yield strength inaccurate at high strain rates. Temperature variations between -55 and 50°C on a split-Hopkinson bar have been achieved utilizing a specially-designed gas manifold system developed at the Los Alamos National Laboratory (LANL) where samples were cooled and heated using helium (He) gas within a 304-stainless steel containment chamber held at a partial vacuum. The He gas is cooled below ambient temperature by first passing the He through a copper coil positioned within a liquid nitrogen dewar, while elevated temperatures are achieved by remotely heating the He in a similar coil within a glycerin-filled beaker warmed to ~100°C by a heating plate. Samples were lubricated using either a thin layer of molybdenum disulfide grease or molybdenum disulfide spray lubricant.

RESULTS AND DISCUSSION

The compressive true-stress versus true-strain response of PBX 9501 was found to depend on the applied strain rate, varied between 0.001 and 2000 s^{-1}, and the test temperature, varied between -55°C and 50°C at a strain rate of 3000 s^{-1}. The yield strength of PBX 9501 at 25°C is shown in Figure 1 to increase from ~8 MPa at 0.0011 s^{-1} to 10 MPa at 0.11 s^{-1} to ~50 MPa at a strain rate of 2000 s^{-1} accompanied by an ~7-fold increase in apparent loading modulus. These results are consistent with previous strain rate studies on PBX 9501(2,7). Due to the documented dispersive nature of wave propagation in ductile polymers and plastic-bonded energetics and the potential influence of sample size on attaining a uniform stress state, the high-rate constitutive response of PBX 9501 was carefully probed to obtain well-posed and accurate data.

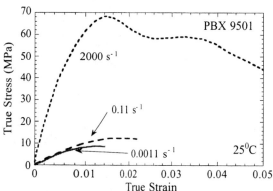

FIGURE 1: Stress-Strain response of PBX 9501 as a function of strain rate at 25°C.

To assure valid high-rate measurements on PBX 9501, it is instructive to examine the different analyses(9) used to calculate sample stress from the Hopkinson bar strain as shown in Figure 2a. In the 1-wave analysis the sample stress is directly proportional to the bar strain measured from the transmitted bar. The 1-wave stress analysis reflects the conditions at the sample-transmitted bar interface and is often referred to as the sample "back stress". This analysis results in more accurate and smoother stress-strain curves, especially near the yield point. Alternatively in a 2-wave analysis, the sum of the synchronized incident and reflected bar waveforms (which are opposite in sign) is proportional to the sample "front stress" and reflects the conditions at the incident/reflected bar-sample interface.

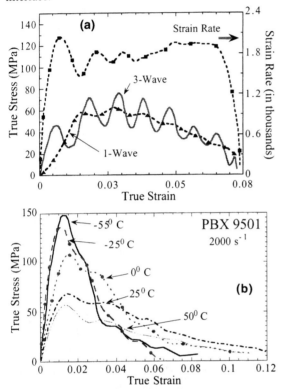

FIGURE 2: Stress-strain response of PBX 9501, a) showing 1- and 3-wave stress curves in addition to the strain rate; and b) as a function of temperature at high strain rate.

Finally a third stress-calculation variation that considers the complete set of three measured bar waveforms, the 3-wave analysis, is simply the average of the 2-wave "front" and the 1-wave "back" stress. A valid, uniaxial Hopkinson bar test requires that the stress state throughout the sample achieve equilibrium during the test and this condition can be checked readily by comparing the 1-wave and 3-wave (or 2-wave) stress-strain response. When the stress state is uniform throughout the sample, then the 3-wave stress oscillates about the 1-wave stress, as seen in Figure

2a. For the current study on PBX 9501 only tests meeting this criterion were deemed acceptable. Previous Hopkinson bar studies of ceramic materials using this 1-wave versus 3-wave comparison have shown quite dramatically that a sample is not in stress equilibrium when divergence is observed(10). In ceramic and cermet materials this divergence correlates very well with the onset of non-uniform plastic flow and/or fracture events before stress equilibrium is achieved.

At high strain rate, the yield strength of PBX 9501 was found to be strongly dependent on temperature as seen in Figure 2b, decreasing from 150 MPa at -55°C to 60 MPa at 50°C. These data on PBX 9501 are consistent with the pronounced influence of strain rate and temperature on the mechanical behavior of energetics(1) as well as ductile polymers(11). Coincident with this flow stress increase upon decreasing the temperature is an ~9-fold increase in the apparent loading modulus with decreasing temperature at high strain rate. In addition, the failure strain (indicated by a peak in the stress independent of temperature at ~ 1.5% strain) is observed to be virtually invariant similar to the findings of Wiegand(7) on various PBX's. This observation suggests a critical tensile initiation criterion in PBX 9501 may be dominant. Attempts to achieve well-posed higher strain rate (> 5000 s⁻¹) data at 298K proved unsuccessful for our sample configuration. At a strain rate of 7000 s⁻¹ the 1-wave and 3-wave signals were found to be divergent for the entire test (invalidating the stress analysis as discussed previously). Accordingly, all high-rate tests were conducted at 2000 s⁻¹.

The PBX 9501 samples loaded at high-rate behaved essentially elastically to the peak stress level and then suffered catastrophic brittle fracture into fragments. Fractographic analysis using a Scanning Electron Microscope(SEM), revealed that at high-rate PBX 9501 fails via predominantly transgranular cleavage through the HMX crystals, as seen in Figure 3a. In several of the larger HMX crystals twins were visible following cleavage fracture (Figure 3b). Further study is required to definitively ascertain the source of these twins. The observation of predominantly transgranular fracture in PBX 9501 is similar to that seen following quasi-static loading(2,7). The findings of this study illustrate that advanced material constitutive models for PBX 9501 will need to incorporate both strain rate and temperature effects on the mechanical behavior.

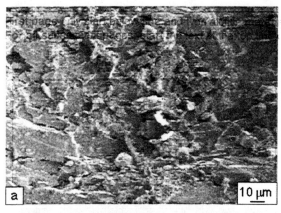

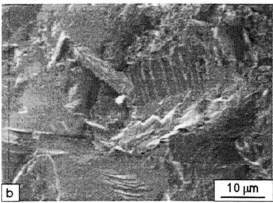

FIGURE 3: Scanning electron micrograph of PBX 9501 following Hopkinson bar testing at 298K and a strain rate of 2000 s^{-1} showing: a) transgranular fracture across the HMX crystals, and b) twins underlying the cleavage fracture in an HMX crystal.

SUMMARY AND CONCLUSIONS

Based upon this study of strain rate and temperature at high strain rate on the constitutive response of PBX 9501, the following conclusions can be drawn: 1) the compressive stress-strain response of PBX 9501 was found to depend on both the applied strain rate; 0.001 to ~2000 s^{-1} and the test temperature; -55 to 50°C at high-rate, 2) decreasing temperature at 2000 s^{-1} was found to increase the maximum flow stress in PBX 9501 from 60 to ~150 MPa at a critical strain of ~1.5%,

and 3) PBX 9501 failed at high-strain rate via predominantly transgranular cleavage fracture of the HMX crystals.

ACKNOWLEDGMENTS

This work was supported under the auspices of the Joint DoD/DOE Office of Munitions. The authors acknowledge Tom Zocco for conducting the SEM work.

REFERENCES

1. Hoge, K. G., "The behavior of plastic-bonded explosives under dynamic compressive loads," Appl. Polym. Symp. **5**, 19-40 (1967).
2. Peeters, R. L., "Characterization of plastic bonded explosives," J. Reinf. Plast. Compos. **1**, 131-140 (1982).
3. Field, J. E., Palmer, S. J. P., Pope, P. H. *et al.*, "Mechanical properties of PBX's and their behaviour during drop-weight impact," in *Proc. Eighth Symposium (Int.) on Detonation*, edited by J.M. Short (Naval Surface Weapons Center, White Oak, Maryland, USA, 1985), pp. 635-644.
4. Wiegand, D. A., Pinto, J., and Nicolaides, S., "The mechanical response of TNT and a composite, Composition B, of TNT and RDX to compressive stress. I: Uniaxial stress and fracture," J. Energ. Mater. **9**, 19-80 (1991).
5. Field, J. E., Bourne, N. K., Palmer, S. J. P. *et al.*, "Hot-spot ignition mechanisms for explosives and propellants," Phil. Trans. R. Soc. Lond. A **339**, 269-283 (1992).
6. Palmer, S. J. P., Field, J. E., and Huntley, J. M., "Deformation, strengths and strains to failure of polymer bonded explosives," Proc. R. Soc. Lond. A **440**, 399-419 (1993).
7. Wiegand, D. A., "Constant Critical Strain for Failure of Highly Filled Polymer Composites," presented at the 3rd Int. Conf. Def. & Frac. of Composites, Guildford, U.K., 1995.
8. Funk, D. J., Laabs, G. W., Peterson, P. D. *et al.*, "Measurement of the stress/strain response of energetic materials as a function of strain rate and temperature: PBX 9501 and mock 9501," in *Shock Compression of Condensed Matter 1995*, edited by S.C. Schmidt and W.C. Tao (American Institute of Physics, Woodbury, New York, 1996), pp. 145-148.
9. Follansbee, P. S. and Frantz, C., "Wave propagation in the SHPB," Trans. ASME: J. Engng Mater. Technol. **105**, 61-66 (1983).
10. Blumenthal, W. R. and Gray III, G. T., "Structure-property characterization of shock-loaded B$_4$C-Al," Inst. Phys. Conf. Ser. **102**, 363-370 (1989).
11. Walley, S. M. and Field, J. E., "Strain rate sensitivity of polymers in compression from low to high strain rates," DYMAT Journal **1**, 211-228 (1994).

CP429, *Shock Compression of Condensed Matter – 1997*
edited by Schmidt/Dandekar/Forbes
© 1998 The American Institute of Physics 1-56396-738-3/98/$15.00

LOW STRAIN RATE COMPRESSION MEASUREMENTS OF PBXN-9, PBX 9501, AND MOCK 9501

D. J. Idar[†], P. D. Peterson[††], P. D. Scott[†], and D. J. Funk[†]

[†]*Los Alamos National Laboratory, MS C920, Los Alamos, NM 87545*
[††]*University of South Carolina, Department of Mechanical Engineering*
300 Main Street, Columbia, SC 29208

Low strain rate (10^{-3} to 10^{-1} s^{-1}) compression measurements have been obtained on three different composite materials: PBXN-9, PBX 9501, and a 9501 sugar mock. These measurements expand on earlier efforts to identify the behavior of PBX 9501 and sugar mocks at different rates, sample aspect ratios (L/d) and temperatures. PBX 9501 samples at three different L/d's were strained at the same strain rate to evaluate L/d effects on the stress-strain parameters. PBXN-9 data were obtained at two different L/d's, two different temperatures, and at three different rates. The PBXN-9 data exhibit similar trends to other energetic materials data, i.e. 1)increased ultimate compressive strength and modulus of elasticity with either an increase in strain rate, or decrease in temperature, and 2)small increases in the strain at maximum stress with decreases in temperature or strain rate. A comparison of the PBXN-9 data to the PBX 9501 data shows that both begin to fail at comparable strains, however the PBXN-9 data is considerably weaker in terms of the ultimate compressive strength.

INTRODUCTION

Quasi-static compression data at room and cold (-55 °C) temperatures, were obtained on PBXN-9 samples with two different sample aspect (L/d) ratios, and at three different rates each. Measurements were also obtained on PBX 9501 samples with three different aspect ratios strained at the same rate to identify possible end effects on the stress-strain measurements. New 9501 sugar mock samples, formulated with a bimodal distribution of coarse and fine grain sugar similar to the HMX grain distribution in PBX 9501, were tested to identify stress-strain parameter differences with the coarse grain sugar mock. Sheffield, et. al.(1), have previously shown with particle size distribution and light microscope measurements that coarse and fine grain sugars have similar particle size distributions to measurements made on coarse(2) and fine particle lots of HMX respectively.

EXPERIMENTAL

Low strain rate compression data were obtained using an Instron 1123 Materials Testing Workstation for PBXN-9 (92 / 2 / 6 wt % HMX / HyTemp 4454 / dioctyl adipate), PBX 9501 (95 / 2.5 / 2.5 / 0.1 wt% HMX / Estane / BDNPA-F / Irganox), and 9501 sugar mock (94 / 3.0 / 3.0 wt% C&H sugar/ Estane / BDNPA-F) samples. Compression samples were machined from pressed billets of the individual composite materials. Load and strain measurements were obtained with an Instron load cell and extensometer. The nominal sample diameters (d), lengths (L), temperatures, and strain rates of the different composite tests are given in Table 1.

DATA ANALYSIS AND RESULTS

An offset accounts for discrepancies in the machine/sample compliance that occur just before

TABLE 1: Test parameters for the new PBXN-9, PBX 9501, and 9501 sugar mock tests.

Sample	Nominal Diameter ≡ d (inches)	Nominal Length ≡ L (inches)	Average Test Temperature	Nominal Strain Rate (s⁻¹)
PBXN-9	0.375 ± 0.003	0.5625 ± 0.003	30 ± 2 °C, and -55 ± 2 °C	0.0015, 0.030, and 0.15
PBXN-9	0.375 ± 0.003	0.750 ± 0.003	30 ± 2 °C, and -55 ± 2 °C	0.0011, 0.022, and 0.11
PBX 9501	0.375 ± 0.002	0.375 ± 0.001	23.4 ± 2 °C	0.0022
PBX 9501	0.375 ± 0.002	0.750 ± 0.001	26.5 ± 2 °C	0.0022
PBX 9501	0.375 ± 0.002	1.500 ± 0.003	27.7 ± 2 °C	0.0022
9501 Sugar Mock	0.375 ± 0.005	0.750 ± 0.001	15.2 ± 2 °C	0.0011

the surfaces completely mate and uniaxial compression is achieved. The adjusted stress-strain curve is then evaluated for the elastic modulus, E, which is defined as the slope of the linear regression fit, the maximum stress or the ultimate compressive strength, σ_m, and strain at maximum stress, ε_m.

Figure 1 is the average, true stress-strain data obtained on PBXN-9 samples at 30±2 °C (room) and -55±2 °C (cold) for three different strain rates.

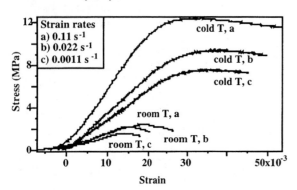

FIGURE 1. True stress-strain data for PBXN-9 samples for 3 different strain rates. Sample diameters were nominally 0.375-in. with $L/d = 2$.

An examination of the data shows: 1)all three stress-strain parameters, E, σ_m, and ε_m, increase with a temperature decrease, 2)both σ_m and E show a greater dependence on the strain rate than ε_m at the lower temperatures, whereas 3)ε_m shows more dependence on the strain rate than the other two values at room·temperatures.

The PBXN-9 compressive mechanical property trends are similar to previous compression results on composite materials(3-7). A comparison of the PBXN-9 true stress-strain data with four of these materials, a)mock 900-21, b)LX-14 coarse grain

sugar mock, c)9501 coarse grain sugar mock, and d)PBX 9501, obtained at a strain rate of 0.0011 s⁻¹ at room temperatures with samples of the same diameter and aspect ratio are shown in Fig. 2.

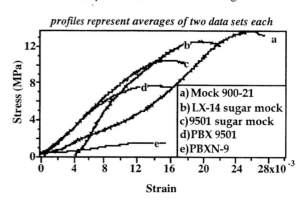

FIGURE 2. Average, true stress-strain data for a)mock 900-21, b)LX-14 sugar mock, c)9501 sugar mock, d)PBX 9501, and e)PBXN-9. Sample diameters were nominally 0.375-in. with $L/d = 2$.

An evaluation of these data show: 1)PBX 9501 displays a higher ultimate compressive strength than PBXN-9 for a comparable maximum strain, 2)9501 coarse grain sugar mock elastic modulus and maximum strain is comparable to PBX 9501, but the sugar mock shows higher compressive strength, and 3)both the LX-14 sugar mock and mock 900-21 exhibit higher compressive strengths and considerably different elastic moduli and maximum strains than the other three materials.

The differences between the PBX 9501 and PBXN-9 compressive strengths were expected based on tactile comparison of the pristine samples, and prior knowledge of the compositions. The PBXN-9 has a pliable, putty-like feel as compared to the PBX 9501. Also, in comparison to

PBX 9501 the PBXN-9 has 1)a lower weight fraction of the hard phase, HMX, and a higher weight percentage of the soft binder phase mixture, 2)a different elastomer, HyTemp 4454, with a polybutyl acrylate backbone, and 3)a higher weight percentage of plasticizer, dioctyl adipate. In contrast to PBXN-9, the 2.5/2.5 wt% Estane/BDNPA-F in the PBX 9501 binder system exhibits higher strength properties due to block co-polymerization properties of the Estane, and the presence of less plasticizer.

The response of the PBXN-9 material to compression at -55±2 °C, was somewhat unexpected. The glass transition temperature of PBXN-9 has been reported as -43 °C(8), similar to -40±5 °C for PBX 9501. Two feasibility tests were first performed with 9501 sugar mock at -55±2 °C, and at the lowest strain rate. Lower temperature testing produced brittle shear failure at 45° to the cylindrical axis in the sugar samples. On the contrary, the PBXN-9 tested -55±2 °C exhibited more ductile behavior.

Additional compression data were obtained on three different aspect ratios of PBX 9501, L/d = 1, 2, and 4, with nominal diameters of 0.375 ± 0.002-in (see Table 1), at the same strain rate and room temperatures to evaluate possible L/d effects on E, σ_m, and ε_m, and the failure mode. It is known(9) from previous work on various materials ranging from solids, e.g. metals, to composites, e.g. concrete, that mechanical properties measurements can be affected by L/d effects. Large L/d's display failure dominated by buckling effects due to unparallel ends and machine misalignment. Small L/d's show failure dominated by frictional end effects, but these can be mitigated or reduced by lubrication. Lubrication techniques and the possible effects on the stress-strain parameters were recently investigated by Liu, and Stout using 9501 sugar mock samples(10). Their analysis indicates that differences in lubrication type only modifies the toe region of the true stress-strain curve, and that the true stress-strain parameters remain unaffected by the lubrication type. Therefore lubrication was used for the compression measurements on PBX 9501. Figure 3 shows the true stress-strain data for six PBX 9501 tests. A comparison of the data shows that all six measurements have nominally the same σ_m, ε_m, and E values within 18.3%, with the highest precision on the ultimate compressive strength. The most

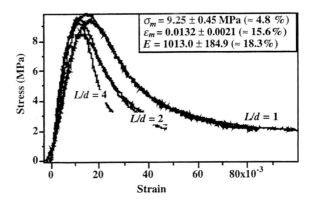

FIGURE 3. True stress-strain data for PBX 9501 samples at three different aspect ratios, L/d =1, 2, and 4, with a nominal diameter of 0.375-in. Measurements were obtained at temperatures ranging from 23.2 to 28.2 °C with a strain rate of 0.0022 s⁻¹.

notably difference occurs in the stress-strain data after the ultimate compressive strength has been reached. A visual analysis of the PBX 9501 samples provides some evidence for these differences. The largest L/d samples exhibited buckling at one of the two sample ends, whereas the smallest L/d samples only exhibited barreling near center and some macrocracking through the length of the sample. Based on these measurements, the stress-strain curve behavior, after the ultimate compressive strength is reached, is indicative of the failure mode behavior for that particular aspect ratio.

The new bimodal, 9501 sugar mock sample data were obtained to evaluate possible mechanical property differences between the bimodal and coarse grain samples. These are shown in Fig. 4.

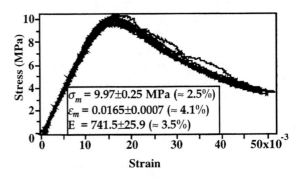

FIGURE 4. True stress-strain data for bimodal, 9501 sugar mock samples with a 3:1 coarse to fine sugar grain distribution. Data were obtained at a rate of 0.0011 s⁻¹, without lubrication, and at room temperatures, using samples with a nominal diameter of 0.375-in. and L/d =2.

With the exception of one measurement, all of the data shows good reproducibility and precision in the σ_m, ε_m, and E values. These data were averaged and compared with the coarse grain sugar mock sample data obtained previously. These data are given in Table 2. A comparison of their σ_m, ε_m, and E values, shows that all three stress-strain parameters are essentially the same within 10%, regardless of the difference in sugar grain distribution.

TABLE 2: Average true stress-strain data for both 9501 sugar mock samples. All values are given with a ± 10% margin.

Sample	σ_m (MPa)	ε_m	E (MPa)
9501 coarse grain sugar mock	10.5 ± 1.1	0.016 ± 0.0016	783 ± 78
9501 bimodal grain sugar mock	9.97 ± 1.0	0.016 ± 0.0016	741 ± 74

SUMMARY

Low strain rate compression measurements on PBXN-9 at 30±2 and -55±2 °C display similar composite material trends seen with other energetic and mock materials. These trends include the dependence of the stress-strain parameters, σ_m, ε_m, and E, on strain rate and test temperature. A comparison of the recent PBXN-9 measurements with data previously obtained on PBX 9501 displays lower overall ultimate compressive strengths and elastic moduli. This behavior was expected based on formulation differences. The PBXN-9 measurements coupled with the PBX 9501 measurements clearly indicate the dependence of the stress-strain parameters on the formulation properties of the composite.

A comparison of the stress-strain parameters obtained from different aspect ratios of PBX 9501 shows that aspect ratios must be considered with compressing testing methods. The failure behavior of the stress-strain data are indicative of L/d failure mode and cracking/fracture behavior differences in the samples.

A comparison of the stress-strain parameters for both the bimodal and coarse grain 9501 sugar mock samples shows comparable values for all 3 evaluated stress-strain parameters within 10%.

ACKNOWLEDGMENTS

This work was supported under the auspices of the Joint DoD/DOE Office of Munitions. We gratefully acknowledge the support and helpful suggestions of the members of the Nonshock Initiation of Energetic Materials project team in this research and George T "Rusty" Gray III for the manuscript review.

REFERENCES

1. Sheffield, S. A., Gustavsen, R. L.. Alcon, R. R., and Archuleta, J. G., LANL DX-1 and DX-2, private communication, December 1996.
2. Dick, J., *Combustion and Flame*, **54**, 121 (1983).
3. Funk, D. J., Laabs, G. W., Peterson, P. D., and Asay, B. W, "Measurement of the Stress/Strain Response of Energetic Materials as a Function of Strain Rate and Temperature: PBX 9501 and Mock 9501," Proceedings of the APS Topical Conference on Shock Compression of Condensed Matter, Seattle, WA, p. 145, August 13-18, 1995.
4. Wiegand, D. A., "Constant Critical Strain for Failure of Highly Filled Polymer Composites," Proceedings of the 3rd International Conference on Deformation and Fracture of Composites, University of Surrey, Guildford, U.K., 558, 1995.
5. Wiegand, D. A., and Pinto, J. J., "Fracture Instabilities, Dynamics, Scaling, and Ductile/Brittle Behavior," Material Research Society Symposium Proceedings, **Vol. 909**, 281 (1996).
6. Wiegand, D. A., "Constant Critical Strain for Mechanical Failure of Several Particulate Polymer Composites and Other Materials," Proceedings of the Army Science Conference, Norfolk, VA, June 1996.
7. Wiegand, D. A., "Influence of Added Graphite on the Mechanical Strength of Pressed Plastic Bonded Explosives," U.S. Army Armament Research, Development and Engineering Center Technical Report ARAED-TR-95018, January 1996.
8. Baudler, B., Hutcheson, R., Gallant, M., and Leahy, J. "Characterization of PBXW-9," NSWC Technical Report TR 86-334, August 1989.
9. Dowling, N. E., *Mechanical Behavior of Materials: Engineering Methods for Deformation, Fracture, and Fatigue*, Upper Saddle River, NJ, Prentice-Hall Inc., 1993, p. 174.
10. Cheng Liu, and Stout, M., LANL MST-5, private communication, October and November 1996.

CP429, *Shock Compression of Condensed Matter – 1997*
edited by Schmidt/Dandekar/Forbes
© 1998 The American Institute of Physics 1-56396-738-3/98/$15.00

MECHANICAL PROPERTIES OF EXPLOSIVES UNDER HIGH DEFORMATION LOADING CONDITIONS

D.G. Tasker,[*] R.D. Dick,[†] W.H. Wilson[*]

[*]*Naval Surface Warfare Center, 101 Strauss Avenue, Indian Head, Maryland 20640*
[†]*Shocks Unlimited, 9737 Academy Road NW, Albuquerque, New Mexico 87114*

The mechanical properties of Navy explosive PBXW-128 were measured using a split-Hopkinson pressure bar (SHPB) apparatus at strain rates up to 2.9×10^4 s^{-1} at room temperature. The true stress-true strain data can be divided into two regions with different moduli, separated by a transition point. Both moduli show significant stiffening with strain rate; above the transition point the modulus increases as the strain rate squared.

INTRODUCTION

When an explosive is subjected to non-uniform, non-planar shocks, the resulting triaxial stresses and high rates of deformation may act synergistically, leading to high-order reactions at stimulus levels significantly below those expected, i.e., from a simple sum of the responses to planar shock and pure shear. A series of experiments, not described here, illustrated this effect. The Navy explosive PBXW-128 was initiated under the combined effects of shock and rapid deformation. PBXW-128 is a soft, rubbery explosive, comprising ~60% by volume solids in a polymer binder, designed to safely undergo large deformations. To model the explosive deformation at the high strain rates typical of those experiments, the explosive stress-strain behavior is required. The mechanical properties were measured, at room temperature, using an SHPB apparatus for strain rates between 2.5×10^3 s^{-1} and 2.9×10^4 s^{-1}; the SHPB study is described here.

EXPERIMENTAL TECHNIQUE

Figure 1 shows the SHPB used (1). The classical compressive SHPB technique involves two elastic bars (usually steel) called the incident and transmit-

ter bars and a gas gun to launch a striker bar made from the same material as the two bars. To improve impedance matching to the sample, either polymethyl-methacrylate (PMMA) or 6061-T6 aluminum (Al) bars were used in this study. The striker impacts the incident bar at a known velocity, producing a uniaxial stress pulse whose duration is twice the travel time of the wave speed in the striker bar. The cylindrical sample is positioned between the incident and transmission bars, which are 25.4 mm in diameter and 1.22 m long. To attain high strain rates, sample diameters smaller than that of the bar were used. The PMMA projectile was 203 mm long, and the Al projectile was 152 mm long; both were 25.4 mm in diameter. The sample faces and the incident and transmitter bar ends were lightly lubricated to reduce friction as the specimen was deformed.

The PBXW-128 samples were cut to the desired nominal length and diameter and then weighed. The actual sample thickness was then determined by measuring the distance between fixtures rigidly mounted on the bars, with the sample in place. The initial sample cross-sectional area, A_0, was calculated from the in-the-bar thickness measurement, the sample weight, and the known explosive density,

591

1514 kg/m³. Strain gages mounted on both bars measured the incident, reflected, and transmitted strain signals - ε_I, ε_R and ε_T - from which the PBXW-128 explosive sample stress, strain, and strain rate - σ, ε, and $\dot{\varepsilon}$ -were determined. Assuming the sample achieved uniform stress after multiple reverberations in the sample, i.e., $\varepsilon_I + \varepsilon_R = \varepsilon_T$, and that the bars behave like one dimensional pistons, the following equations were used to calculate the sample stresses and strains:

$$\dot{\varepsilon}(t) = \left(\frac{2c_{bar}}{l(t)}\right)\varepsilon_R(t)$$

$$\varepsilon(t) = \int\dot{\varepsilon}(t).dt \qquad (1)$$

$$\sigma(t) = E_{bar} \cdot \frac{A_{bar}}{A(t)} \cdot \varepsilon_T(t)$$

A_{bar}, c_{bar}, and E_{bar} are the bar area, wave speed and modulus; A and l are the sample area and length; and (t) denotes a function of time. For the PMMA bars, $c_{bar} = 2168$ m/s and $E_{bar} = 5.58$ GPa; $c_{bar} = 5041$ m/s and $E_{bar} = 68.34$ GPa for the Al bars. The PBXW-128 samples experience large strains, so the sample area, A(t), increases during the test. The true sample stress is obtained from the stress given by Equation (1), modified by the change of area with time, and assuming a constant volume with a Poisson's ratio of 0.5:

$$\sigma_{true}(t) = \sigma_{(t)} \cdot \frac{A_0}{A(t)}$$

$$\left(\frac{A_0}{A(t)}\right) = 1 + \varepsilon(t) \qquad (2)$$

$$\sigma_{true}(t) = \frac{E_{bar}A_{bar}}{A_0} \cdot (1 + \varepsilon(t)) \cdot \varepsilon_T(t)$$

Note that the sample strain, $\varepsilon(t)$, is negative (with a range of 0 to -1), so the true stress is less than $\sigma(t)$ as the sample is squashed. The sample diameter will eventually grow to equal the bar diameter (and beyond). At that point the area corrections are terminated, i.e., frozen at the ratio of areas of unity, where $A(t) = A_{bar}$ and $\sigma_{true}(t) = E_{bar} \cdot \varepsilon_T(t)$. The true strain is:

$$\varepsilon_{true}(t) = \ln(1 + \varepsilon(t)) \qquad (3)$$

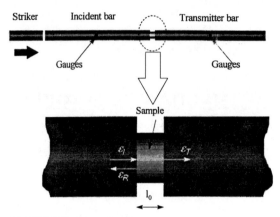

FIGURE 1 Hopkinson bar apparatus.

EXPERIMENTAL RESULTS

The strain data recorded at the center of each of the two bars were translated by the wave propagation times between the gauges and cell interfaces. Corrections for the wave dispersion effects on the three signals - ε_I, ε_R and ε_T - were then made, to find the signals that would be seen at the sample interfaces. (Dispersion corrections were not applied to the PMMA bar strain data because of PMMA's viscoelastic behavior). Compressive strain shown as a positive quantity in all the following tables and figures for simplicity.

Figure 2 shows incident, reflected and transmitted strain data, for test W128-4, obtained by this procedure on PMMA bars. These are the strains computed at the sample interfaces shown in Fig. 1.

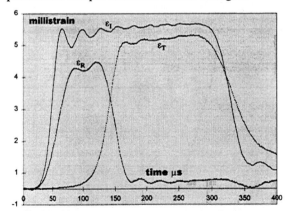

FIGURE 2 Incident - ε_I, reflected - ε_R, and transmitted - ε_T - strain data for shot W128-4.

The true stress-strain data were obtained using Eqs. (2) and (3); these data are shown in Fig. 3. Clearly the explosive has been subjected to large deformations; the strain rate for these data was 9000 s⁻¹. The true stress-true strain data can be divided into two regions with different moduli, E_1 below and E_2 above a transition point, i.e., at a critical stress σ_T and strain ε_T. The values of σ_T and ε_T are subjective, because the slope change is gradual, making the transition point difficult to identify. Linear fits to these two regions are shown in Fig. 3.

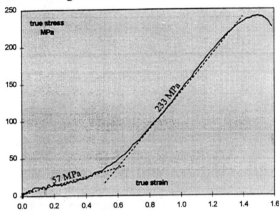

FIGURE 3. True stress versus true strain, shot W128-4.

The results for test #W128-4 are: $\sigma_T = 31$ MPa, $E_1 = 57$ MPa, $E_2 = 233$ MPa. Data for all the successful tests are shown in Table 1.

TABLE 1. Summary of results.

#*	rate ks⁻¹	d_0 mm	l_0 mm	E_1 MPa	E_2 MPa	σ_T MPa	ε_T	Bar
1	3.6	13.3	5.33	29.3	9.0	5.4	0.2	PMMA
2	5.5	5.1	3.94	15.0	35.0	15.0	0.1	PMMA
3	2.5	22.5	8.50	23.1	8.7	3.4	0.2	PMMA
4	9.0	3.9	2.75	57.0	233	31.0	0.5	PMMA
5	3.3	21.8	6.10	14.0	2.2	2.8	0.2	PMMA
6	4.8	13.1	4.00	11.2	288	4.0	0.3	PMMA
10	7.5	19.9	3.40	536	50.0	-	-	Al
11	14.0	13.4	2.60	38.1	61.4	45.0	1.1	Al
12	7.5	13.1	4.60	19.0	19.0	-	-	Al
14	20.0	5.7	1.66	77.5	859	94.0	1.2	Al
15	30.0	4.3	1.50	78.9	1101	95.0	1.0	Al
16	30.0	3.8	1.65	48.4	1290	86.0	1.2	Al
17	5.0	24.2	5.48	360	41.5	-	-	Al
18	3.0	22.5	5.30	292	25.0	-	-	Al

*Key: # - shot W128-#; rate - strain rate; d_0, l_0 - initial sample diameter and length; E_1, E_2 - pre- and post-transition moduli; σ_T, ε_T - transition true stress and transition true strain.

Data Analysis

Two Group Behavior

The data of Table 1 show the explosive often strain-hardened, while in other cases it yielded or softened, above a transition point, i.e., above a critical stress and strain. (For shot W128-12 no transition was observed.) The trend is for the data to fall into two groups. In Group 1, E_1 is greater than E_2; conversely in Group 2, E_1 is less than E_2. The transition from Group 1 to 2 occurs at 3600 s⁻¹ for the PMMA bars and 7600 s⁻¹ for the Al bars.

Material Properties vs. Strain Rate

The results of Table 1 are summarized in Figs. 4 and 5 which show the moduli, E_1 and E_2, and the transition stress, σ_T, as functions of strain rate. (Outlier data for shots W128-10, 17, and 18 are shown in Fig. 7). Even ignoring the outliers, clear trends are evident. E_1 and σ_T show modest increases with strain rate, $\dot{\varepsilon}$. The best fits to the data are:

$$E_1 = 1.89\,\dot{\varepsilon} + 15 \text{ MPa}$$
$$\sigma_T = 3.5\,\dot{\varepsilon} - 4.6 \text{ MPa} \tag{4}$$

where the correlation coefficients are $R^2 = 0.62$ and 0.92 respectively. By contrast, E_2 shows a strong stiffening with $\dot{\varepsilon}$, see Fig. 5. Despite the scatter, the PMMA and Al bar results appear to belong to the same data set. These combined data seem best fitted by $E_2 \propto \dot{\varepsilon}^2$ (R^2 for the log-log plot is 0.71), i.e.:

$$\log_{10} E_2 \text{ MPa} = 2\log_{10}\dot{\varepsilon} - 5.9 \tag{5}$$

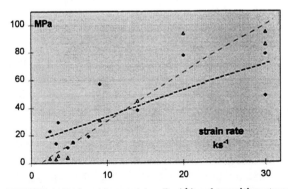

FIGURE 4. Pretransition modulus, E_1, (◆) and transition stress (Δ) versus strain rate, for all shots

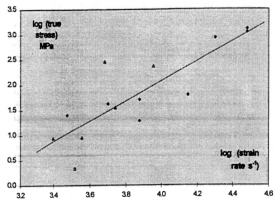

FIGURE 5. Post-transition modulus, E_2, versus strain rate (log-log) obtained on aluminum bars (◆) and PMMA bars (△).

Anomalous structure

Shots W128-10, -17, and -18, all of large diameter on aluminum bars, displayed spikes at the beginning of their transmitted strain records. Figure 6 shows one resulting stress-strain record; the moduli E_1 for all three shots are plotted in Fig. 7.

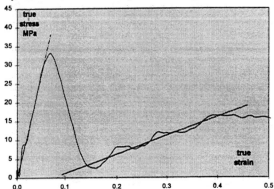

FIGURE 6. True stress vs. true strain, showing anomalous spike, shot W128-10.

DISCUSSION AND CONCLUSIONS

High strain rate data for explosive PBXW-128 were obtained using PMMA and Al. These experiments were difficult to perform and interpret because the explosive is flexible and easily deforms under compression. Viscoelastic effects of the PMMA bars (and explosive) probably contributed to data scatter. Large strains and strain rates were observed, resulting in material flow to the edge of the bars and beyond. Consequently it was necessary to correct the data for the area changes. Despite the

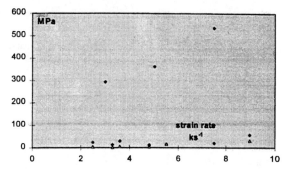

FIGURE 7. E_1 (◆) and σ_T (△) vs. strain rate, showing W128-10, 17 and 18.

scatter in results, the data obtained on the PMMA and Al bars were in general agreement.

The data showed either strain-softening (Group 1) or hardening (Group 2) with increasing strain rate; the cause of the two behaviors is unknown. It may be that large strains cause extrusion of the rubbery binder from between the explosive crystals, leading to crystal-to-crystal contact; similar effects are seen in porous media (2). Perhaps it is due to distortion of the bar faces, because the high strain rates in Group 2 were obtained in small diameter samples. Then the stresses at the bar center are very high, approaching the bar yield strength. So, the bars may not be behaving as one dimensional plane pistons. The distortion may be sufficient to pinch the explosive under these conditions. Overall, the data show the material stiffening at high strains, with the modulus stiffening as the strain rate squared.

The cause of anomalous spikes is yet to be found, and may be initiation of low order chemical reaction, bar friction, radial inertia, or some other effect.

ACKNOWLEDGMENTS

The authors thank William Walton for his experimental skill and resourcefulness.

REFERENCES

1. Tasker, D.G., Dick, R.D., Wilson, W.H., Lee, R.J., and Deiter, J.S., *Response of Wet and Dry Riverbed Sand to High Strain rate Loading Using the Split-Hopkinson Pressure Bar*, IHTR 1861, Feb 1995, Naval Surface Warfare Center, Indian Head, MD 20640.
2. Tasker, D.G., Dick, R.D., Wilson, W.H., Lee, R.J., and Gustavson, P.K., in *APS Proceedings of Shock Compression of Condensed Matter - 1995*, eds. Schmidt, S.C. and Tao, W.C., American Institute of Physics, 585-588, (1996).

CP429, *Shock Compression of Condensed Matter – 1997*
edited by Schmidt/Dandekar/Forbes
© 1998 The American Institute of Physics 1-56396-738-3/98/$15.00

MECHANICAL BEHAVIOR OF ENERGETIC MATERIALS DURING HIGH ACCELERATION IN AN ULTRACENTRIFUGE

Y. Lanzerotti

U. S. Army ARDEC, Picatinny Arsenal, New Jersey 07806 5000

J. Sharma

Naval Surface Warfare Center, Carderock Divison, West Bethesda, MD 20817

The mechanical behavior of explosives subject to high acceleration (high *g*) has been studied in an ultracentrifuge. Melt-cast TNAZ and pressed TNAZ, LX-14, Composition A3 type II, PAX-2A, and PAX-3 have been studied. Failure occurs when the shear or tensile strength of the explosive is exceeded. The fracture strength of melt-cast TNAZ is greater than the fracture strength of pressed TNAZ. The fracture strength of LX-14 is greater than the fracture strength of all pressed explosives studied to date. The fracture strength of Composition A3 type II is smaller than the fracture strength of all pressed explosives studied to date. The fracture strength of PAX-2A is greater than the fracture strength of PAX-3.

INTRODUCTION

We have introduced several new fields of research to study the mechanical behavior of energetic materials during high acceleration by using an ultracentrifuge (1-3). Energetic materials are of significant interest for scientific and practical reasons in the extraction (mining) industry, structure demolition, space propulsion, and ordnance (4). In these applications the materials can be subjected to high, fluctuating, and/or sustained acceleration. The nature of the fracture process of such materials under high acceleration is of particular interest, especially in ordnance and propulsion applications. For example, explosives in projectiles are subjected to setback forces as high as 50,000 *g* during the gun launch process. These high setback forces can cause fracture and premature ignition of explosives.

An energetic material will experience a pressure gradient during acceleration in the gun and under *g*-loading in the ultracentrifuge. The pressure gradient experienced by the explosive during acceleration in the gun and under *g*-loading in the ultracentrifuge is unique and will produce different kinds of behavior and failure than under other material test conditions. Fundamental understanding of the behavior of energetic materials subjected to high acceleration is a key to better practical ordnance designs that solve the problems of abnormal propellant burning and premature ignition of explosives during gun launch. This work is particularly relevant to the future development of insensitive energetic materials to be used in devices with higher acceleration.

We have studied the fracture behavior of TNT (trinitrotoluene) and four types of Octol [70% HMX (cyclotetramethylenetetranitramine, 30% TNT; 75% HMX, 25% TNT; <75% HMX, 25% TNT, <1% HNS (hexanitrostilbene); and ≈ 83% HMX, ≈ 17% TNT) using an ultracentrifuge previously (2). In this work the fracture behavior of cast TNAZ (1,3,3

- trinitroazetidine) and pressed TNAZ have been studied during high acceleration in an ultracentrifuge. The fracture behavior of the plastic bonded explosives LX-14 (95%HMX, 5% Estane), Composition A3 type II [91% RDX (cyclotrimethylene-trinitramine), 9% polyethylene], PAX-2A [85% HMX. 9% BDNPF (bis-dinitropropyl acetal formal, 6% CAB (cellulose acetate butyrate)], and PAX-3 (85% HMX, 9% BDNPF, 6% CAB/25% Aluminum) have also been studied during high acceleration in an ultracentrifuge..

TECHNIQUE

A Beckman ultracentrifuge model L8-80 with a swinging bucket rotor model SW 60 Ti is used to rotate the sample under study up to 60,000 rpm ($\approx$ 500,000 g). The distance of the specimen from the axis of rotation can be chosen as a variable between 6 and 12 cm.

The samples are machined into the shape of the frustrum of a cone. The large diameter is typically 11 mm, the small diameter is 9 mm. The angle between the base and the side is 80°. These samples are fitted into 5-mm long, 11-mm o.d. aluminum cylinders. At one end the i.d. is 11 mm; at the other end the i.d. is 9 mm. The angle between the inner and outer side of the sleeve is 10°. The 9-mm diameter top of the sample faces away from the axis of rotation.

The sample experiences a time rate of change of the acceleration up to a maximum acceleration. The sample then remains at this maximum acceleration for an interval such that for each run there is a combined total elapsed time of five minutes. The sample then decelerates smoothly to zero acceleration. The initial maximum acceleration is less than the fracture acceleration for the material. The maximum acceleration for the sample is then increased systematically in each successive five-minute run. The sample fractures when the shear or tensile strength of the material is exceeded. At this time particles break loose from the surface exposed to the acceleration and transfer to the closed-end tube. A hemispherical fracture surface is formed on the sample.

RESULTS

Melt-cast TNAZ when subjected to high acceleration, experiences brittle fracture at grain boundaries. This occurs when the shear or tensile strength of the TNAZ is exceeded. The fracture acceleration of pressed TNAZ and the plastic bonded explosives LX-14, PAX-2A, PAX-3, and Composition A3 Type II also experience brittle fracture when the shear or tensile strength is exceeded.

TNAZ

Polycrystalline melt-cast TNAZ has been found to fracture at grain boundaries at $\approx$ 105 Kg at 25°C. The cast was made with 70% liquid TNAZ and 30% 40 micron maximum size TNAZ particles. This melt-cast TNAZ was 97.4% of its theoretical maximum density (TMD) of 1.84 g/cc. Pressed TNAZ has been found to fracture at $\approx$ 62 Kg at 25°C. This pressed TNAZ was 97.0% of its TMD of 1.84 g/cc.

Plastic Bonded Explosives

The fracture acceleration of melt-cast TNAZ and pressed TNAZ is compared with the fracture acceleration of LX-14, PAX-2A, PAX-3 and Composition A3 Type II in Table 1. LX-14 has been found to fracture at $\approx$ 84 Kg at 25°C. This LX-14 was pressed at 96.9% of its TMD of 1.849 g/cc. PAX-2A has been found to fracture at $\approx$ 81 Kg at 25°C. This PAX-2A was pressed at $\approx$ 99.3% of its TMD of 1.79 g/cc. PAX-3 has been found to fracture at $\approx$ 43 Kg at 25°C. This PAX-3 was pressed at $\approx$ 97.9% of its TMD of 1.955 g/cc.

DISCUSSION

The fracture acceleration of cast TNAZ is greater than the fracture acceleration of pressed TNAZ. The fracture acceleration of cast explosives is inversely related to the grain size (2). The high fracture acceleration of this cast TNAZ may be due to the

TABLE 1. Fracture Acceleration of Pressed Explosives in an Ultracentrifuge at 25°C

Explosive	Fracture Acceleration, Kg	%TMD
TNAZ (melt-cast)	105	97.4
LX-14	84	96.9
PAX-2A	81	99.3
TNAZ	62	97.0
PAX-3	48	97.9
COMP A3 Type II	33	96.8

very small grain size (≤ 0.1 mm) of the cast. The fracture acceleration of PAX-2A is greater than the fracture acceleration of PAX-3. The fracture acceleration of Composition A3 type II is less than the fracture acceleration of PAX-3. The fracture acceleration of LX-14 is greater than the fracture acceleration of PAX-2A. The fracture acceleration of LX-14 is greater than the fracture acceleration of all pressed explosives studied to date. The fracture acceleration of Composition A3 type II is smaller than the fracture acceleration of all pressed explosives studied to date. The relationship of fracture acceleration to sensitivity needs to be investigated.

REFERENCES

1. Lanzerotti, Y. D. and Sharma, J., App. Phys. Lett. **39**, 455 457 (1981).
2. Lanzerotti, Y. D. and Sharma, J., in Armstrong, R. W., Baker, T. N., Grant, N. J., Ishizaki, K., Otooni, M. A., eds. *Grain-Size and Mechanical Properties - Fundamentals and Applications*, Pittsburgh: Materials Research Society, 1995, **362**, pp. 131-134.
3. Lanzerotti, Y. D., Meisel, L. V., Johnson, M. A., Wolfe, A., and Thomson, D. J. in Hamers, R. and Smith, D., eds. *Atomic Resolution Microscopy of Surfaces and Interfaces*, Pittsburgh: Materials Research Society, 1997, **466**, in press.
4. Borman, S., Chemical and Engineering News **72**, 18-22 (1994).

CP429, *Shock Compression of Condensed Matter – 1997*
edited by Schmidt/Dandekar/Forbes
© 1998 The American Institute of Physics 1-56396-738-3/98/$15.00

MECHANICAL FAILURE PROPERTIES OF COMPOSITE PLASTIC BONDED EXPLOSIVES

D.A.Wiegand

Picatinny Arsenal, NJ. 07806-5000

The initial part of the uniaxial stress versus strain response in compression can be described in terms of a modulus, E, a peak stress, σ_m (a failure stress) and a strain at the peak stress, ε_m (a failure strain). σ_m increases in proportion to E for smaller values of E and in proportion to $E^{1/2}$ for larger values of E with changes in temperature and strain rate. ε_m is constant for the smaller values of E and decreases with increasing E for larger E values. These results indicate two separate failure modes, a constant strain failure mode for smaller E, and a constant strain energy or a critical crack length failure mode for larger E.

INTRODUCTION

This work was initiated to develop understanding of the mechanical properties and in particular the failure properties of a group of particulate polymer composite explosive formulations (1) (2). The composites are made up of polymer binders (with plastizer in most cases) and 80% to 95% organic polycrystalline nonpolymer explosive particulates (see the table). The general approach was to vary properties by controlled changes in temperature and strain rate and to deduce failure mechanisms from observed relationships between properties. The condition of the samples after deformation was also noted, i.e., whether there was evidence of plastic deformation, cracking and/or fracture.

EXPERIMENTAL

Stress versus strain data in uniaxial compression were obtained using an MTS servo-hydraulic system operated at constant strain rates of 0.001 to 10/Sec. (3). Samples were in the form of right circular cylinders one quarter inch to one inch in length and one quarter inch to three quarters inch in diameter and the end faces of the samples were coated with a lubricant, e.g. graphite, to minimize frictional effects between the sample and the loading platens. Samples were conditioned at temperatures between -60 and 75 C for at least two hours before measurements and were compressed along the cylinder axis. Engineering stress and strain were obtained and one to five samples were measured at each temperature and strain rate.

Samples of the composite were prepared either by pressing to size or by pressing into large billets and machining to size (1) (2) (4). Precautions were taken to insure that the cylinder end faces were adequately flat and parallel. The particulate particle sizes are in the micron range for all composites. The densities of all samples were measured and results are presented only for samples having densities in a narrow range close to the maximum theoretical (zero porosity) density.

In Table I the composition of the composites under consideration are given. Glass transition temperatures, Tg, are given where known.

TABLE 1. Composition of Particulate Polymer Explosive Composites

Name	Particulate	Binder		Tg
		Polymer	Plastizer	C
I Pax 2	HMX 80%	CAB 8%	BDNPA/F 12%	-37[a]
II Pax 2A	HMX 85%	CAB 6%	BDNPA/F 9%	-37[a]
III 9404	HMX 94%	NC 3%	CEF 3%	-34[b]
IV 9501	HMX 95%	ESTANE 2.5%	BDNPA/F 2.5%	-41 (B)[c]
V 9502	TATB 95%	KEL F 800 5%		30 (B)[b]
VI LX-14	HMX 95.5%	ESTANE 5702-F1 4.5%		-31 (B)[b]
VII Comp A3 Type II	RDX 91%	Polyethelene 9%		

Nomenclature: HMX - Cyclotetramethylene tetranitramine. TATB - 1,3,5-triamino-2,4,6-trinitrobenzene. RDX - Cyclotrimethylene trinitramine. NC - Nitrocellulose. CAB - Cellulose Acetate Butyrate. BDNPA/F -Bis(2,2-Dintropropyl)Acetal/Formal. CEF - Tris(Beta Chloroethyl) Phosphate. Estane - Polyurethane. KEL F 800 - Chlorotrifluoroethylene/vinylidine flouride copolymer. B - Property of the binder.

a Personal Communication, J. Harris, Picatinny Arsenal, N.J.
b Reference (5)
c Reference (6)

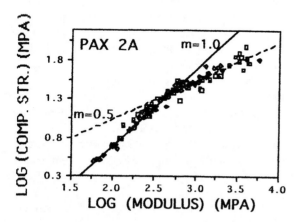

FIGURE 1. Log(Compressive Strength) versus log(Modulus) for PAX 2A. Temperature and strain rate are parameters.

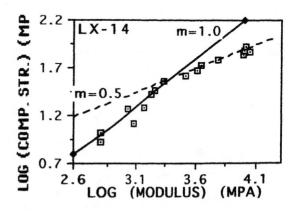

FIGURE 2. Log(Compressive Strength) versus log(Modulus) for LX-14. Temperature and strain rate are parameters.

RESULTS

For uniaxial compression of the materials of the table the stress initially increases linearly with increasing strain, then curves over and passes through a maximum stress with further increases in strain. The stress either decreases continuously for additional increases in strain beyond the maximum stress at higher temperatures or decreases abruptly to near zero at or just beyond the maximum stress at lower temperatures. Three quantities taken from the stress versus strain curves are of interest, the initial slope which is taken as a measure of Young's modulus (E), the maximum compressive stress (the compressive strength), σ_m, and the strain at the maximum stress, ε_m.

In Fig. 1 the log of the compressive strength (σ_m) is given versus the log of the modulus (E) for PAX 2A. For the smaller values of E σ_m increases linearly with E while for the larger values of the E σ_m increases as the square route of E. For a given value of E it appears that the failure branch (linear or square route dependence of σ_m on E) is determined by the branch which requires the lower value of failure stress, i.e. the lower σ_m. Therefore, for values of log E less than about 2.65 the linear branch requires the lower stress while for values of log E greater than about 2.65 the square route branch requires the lower stress. The linear branch was observed for all of the composites of the table, but the square route branch was observed only for PAX 2, PAX 2A, LX-14 (see Fig 2) and to a very limited extent for 9404 (7) and Comp A3. All of the available data for 9501 lies in the linear range while the available data for 9502 for the larger modulus range does not conform to either a linear or a square root relationship. Several other energetic materials also exhibit the linear relationship (7).

From the stress versus strain curve a geometrical relationship between the three quantities σ_m, E, and ε_m can be shown to be

$$\sigma_m = E \varepsilon_m / (1 + a) \qquad (1)$$

where $\sigma_m(1 + a)$ is the stress at which a straight line through the initial portion of the stress versus strain curve, the slope of which defines E, intersects a constant strain line at ε_m (2). a is a measure of the shape of the stress versus strain curve between the point where it deviates from a straight line and the point of maximum stress. In all cases where a linear relationship between σ_m and E is observed, ε_m is found to be approximately constant (1) (2) (7). While the parameter a does change somewhat with temperature and strain rate, the magnitude of the changes in a are such that equation (1) is satisfied within the precision of the data.

Substituting $\sigma_m = k E^{1/2}$ into equation (1) for the square route branch of Figure 1 and rearranging gives

$$\varepsilon_m = k (1 + a)/E^{1/2} \qquad (2)$$

This equation is satisfied within the precision of the data for the appropriate values of E and for those composites which exhibit the square route branch.

DISCUSSION

Consider that failure initiates (or can be first detected) at the point where the stress versus strain curve deviates from linearity in the initial portion of the stress versus strain curve and that at this point the stress and strain are σ_f and ε_f. σ_m and ε_m are then taken as measures of σ_f and ε_f. This corresponds to approximately a 1% offset condition for most of the composites.

The linear branch of the σ_m versus E curve and the attendant constant ε_m are discussed elsewhere in terms of either a constant strain as the criterion for failure or in terms of a mechanism requiring σ_m to be proportional to E (7). In either case equation (1) is satisfied. It should be possible to distinguish between these two types of failure criteria by comparing the results of biaxial, triaxial or torsion tests with the uniaxial data if the failure type, i.e. cracking and fracture versus yield, does not change (8). The linear branch is not discussed further here.

There are several possible reasons for the square root relationship between σ_m and E. Criteria for failure which require a constant total strain energy, a constant strain energy of distortion, and a constant octahedral shearing stress, all have failure stresses which vary as the square route of the modulus (8). However, the latter two failure criteria are equivalent (8). In addition, the Griffith condition for rapid crack growth requires the same dependence of the threshold stress (failure stress) on the modulus (9) and additional measurements are necessary to distinguish between these different criteria. It should also be possible to distinguish between a constant total strain energy criteria and a constant strain energy of distortion criteria by comparing the results of torsion tests with the uniaxial results (8). The Griffith condition, in contrast, is related to the decrease of the total strain energy due to crack growth and occurs when this decrease is equal to the increase in the effective surface energy also due to crack growth. It may be stated as

$$\sigma = [\gamma \, E \, K/c \,]^{1/2} \qquad (3)$$

where σ is the threshold stress, c is the crack length, γ is the effective surface energy per unit area and K is a constant. The slope of the square route branch, Fig. 1, is then a function of the crack length c for this mechanism of failure. This dependence on crack length may be part of the reason for the scatter of the experimental points of Fig. 1 for this branch because of variations of crack length from sample to sample. Measurements of failure strength as a function of crack length are desirable to determine if the square route branch is due to the Griffith condition and so rapid crack growth. While fracture as characterized by an abrupt decrease of the stress to near zero with increasing strain is observed at low temperatures (large E) for most of the composites exhibiting the square root branch, it is not observed for many of the points (Fig. 1) lying on this branch. Therefore, if the square route branch is due to rapid crack growth, it must be concluded that crack arrest is also important because of the lack of fracture in many cases.

SUMMARY

The results indicate two failure modes in the temperature and strain rate ranges studied, a lower modulus mode in which the failure strength is proportional to the modulus and the failure strain is constant, and a higher modulus mode in which the failure strength is proportional to the square root of the modulus. These results suggest that the criterion for failure for the lower modulus mode is a constant strain, and that the criterion for failure for the higher modulus mode is either a constant strain energy or a critical crack length. Additional work is indicated.

An immediate practical result of this work for the low modulus range lies in the simplification to modeling failure in these materials by being able to use a strain criterion (for failure) which is constant over a wide range of temperatures and strain rates rather than a stress criterion which changes significantly with these two variables.

REFERENCES

1. Wiegand, D., Hu, C., Rupel, A., and Pinto, J., *9th International Conference on Deformation, Yield and Fracture of Polymers*, London, The Institute of Materials, 1994, pp. 64/1-64/4 .
2. Wiegand, D., *3rd International Conference on Deformation and Fracture of Composites*, University of Surrey, Guildford, UK, 1995, pp. 558-567.
3. Wiegand, D., Pinto, J., and Nicolaides, S., J. Energetic Materials, 9, pp. 19-80 (1991).
4. Wiegand, D., in eds. Murarka, S.P., Rose, K., Ohmi, T., and Seidel, T., *Interface Control of Electrical, Chemical and Mechanical Properties*, Pittsburgh, Materials Research Society Symposium Proceedings, 1994, Vol 318 pp. 387-392.
5. Dobratz, B. M. and Crawford, P. C., "LLNL Explosives Handbook, Properties of Chemical Explosives and Explosive Simulants", Lawrence Livermore National Laboratory Report UCRL-52997 Change 2, p 6-6 and p6-8 1985.
6. Personal Communication, G.L. Flowers
7. Wiegand, D. A., *20th Army Science Conference*, 1996, Proceedings Vol I pp 63-68, and Picatinny Arsenal Technical Report (In Press).
8. Seely, F. B., and Smith, J. O., *Advanced Mechanics of Materials*, New York, John Wiley and Sons 1952, pp 76-91.
9. Ewalds, H. L. and Wanhill, R. J. H., *Fracture Mechanics*, Baltimore, MD, Edward Arnold 1984, pp 16-19.

602

CP429, *Shock Compression of Condensed Matter – 1997*
edited by Schmidt/Dandekar/Forbes
1998 The American Institute of Physics 1-56396-738-3/98/$15.00

High Strain Rate Testing of AP/Al/HTPB Solid Propellants

Henry J. John Jr., Frank E. Hudson III, and Rodney Robbs

Research & Technology Group, Naval Air Warfare Center, China Lake CA 93555

Ammonium perchlorate (AP), aluminum (Al), and hydroxy-terminater polybutadiene (HTPB) solid propellant samples were subjected to strain rates up to 700 sec[-1] using the Split-Hopkinson Pressure Bar apparatus. The issues are the effects of temperature on the mechanical behavior of these propellants at high strain rates. Strain rates were between 100 sec[-1] and 700 sec[-1] over wide range of temperature regions. Included in this paper are the strain rates, strain, and stress curves and material response properties for Aluminum, AP, and HTPB based propellants formulation.

INTRODUCTION

This program is a joint DOD-DOE effort involving Lawrence Livermore National Laboratory (LLNL), the Air Force Phillips Laboratory (AFPL) and the Navy Laboratory at China Lake. The principal goal is to develop the necessary data base to define and predict the response of deployed 1.3 propellants to various hazard scenarios. Unlike certain 1.1 Class rocket motors which are known to be detonable (via SDT, DDT, XDT), little is known about the shock response of most ammonium perchlorate (AP), aluminum (Al), and hydroxy-terminated polybutadiene (HTPB) binder propellants. This report investigates the mechanical behavior of AP/Al/HTPB propellant by high strain rate compression experiments using the Split Hopkinson Pressure Bar (SHPB) technique. The testing and analysis were conducted at the Naval Air Warfare Center Weapons Division at China Lake, Calif. The experimental results and analysis for the mechanical properties of this propellant are discussed in this paper.

EXPERIMENTAL SET-UP

Experiments were performed on AP/Al/HTPB samples to determine the mechanical response to high strain rate loads. A SHPB apparatus was used to collect the data. The sample, with thickness l_0, is sandwiched between two steel bars, 5/8 inch (15.875 mm) in diameter, labeled in the figure as the incident and transmitter bars. These bars are fabricated from 4150 steel with a measured density of 7.828 g/cc and sound velocity of 5.07×10^5 cm/s. Two pairs of strain gauges are located on the incident and trans-

mitter bars equidistant from the sample. The strain gauge pairs are bonded to the bar opposite to each other to effectively cancel out any bending wave contribution. Each gauge pair is connected to a balanced bridge circuit and the output is displayed on a Nicolet digitizing oscilloscope as shown in Fig. 1.

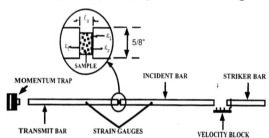

FIGURE 1. Test Set-up of Split Hopkinson Pressure Bar.

For the high and low temperature measurements, the propellant sample was positioned as shown in Fig. 1. Thermocouples were positioned at the end of the incident bar and front of the transmitter bar. For the high temperature measurements, an oven was then positioned along the SHPB apparatus and centered over the sample and the temperature was controlled by an Omega temperature controller. For the low temperature measurements, a device design to allow flow of liquid Nitrogen was positioned along the SHPB apparatus and centered over the sample. Temperature measurements were made on both sides of the propellant sample before each test.

For each experiment the striker bar is driven by a compressed gas gun and impacts the incident bar with a velocity measured by a time interval counter. A one dimensional compression wave propagates down the incident bar with amplitude ε_i and a pulse length that is large compared to the sample thickness l_0. Upon reaching the test sample, part of the compressive pulse is transmitted through the sample with amplitude ε_t and part is reflected with amplitude ε_r due to the impedance mismatch between sample and bars. The stress within the sample is considered to be uniformly distributed. It is easily shown that three equations may be derived to describe the stress, strain rate and resultant strain within the sample. These are given by:

$$\sigma(t) = E_o \left(\frac{A_0}{A_S} \right) \varepsilon_t(t) \qquad (1)$$

$$\dot{\varepsilon}(t) = -\left(\frac{2C_0}{\ell_0} \right) \varepsilon_r \qquad (2)$$

$$\varepsilon(t) = -\left(\frac{2C_0}{\ell_0} \right) \int_0^t \varepsilon_t \, dt \qquad (3)$$

where $\sigma(t)$ is the stress in the sample at time 't', E_0 is Young's modulus, A_0 and A_S are the cross-sectional areas of the bar and sample, respectively, and C_0 is the sound speed (1). The strain is determined by integrating Equation (2) over the duration of the compressive pulse, which is approximately 280 µs for the NAWC Hopkinson bar.

The propellant samples used in this experimental study were cut in the shape of flat disks from uniform slabs of propellant which had been pre-cut to the desired thickness. The aspect ratio was determined by using the results of Davies and Hunter (2) who derived an expression for the uniform diameter-to-thickness ratio from basic energy considerations and stress nonuniformities induced by inertia. They found that an acceptable design criterion for the sample size aspect ratio is given by:

$$\frac{\text{diameter}}{\text{thickness}} = \frac{2}{\sqrt{3\upsilon}} \equiv 2.31 \qquad (4)$$

where υ is Poisson's ratio and is taken to be 0.5 for the purposes of this study (3-4). Although we attempted to maintain a constant aspect ratio, the flexibility of the material resulted in a 10-15% scatter around the aspect ratio.

ANALYSIS OF RESULTS

Typical results for strain rate, strain, and stress-strain curves for AP/Al/HTPB propellant at +70° are shown in Figs. 2-4. In each case the samples show

a brief elastic response before a yield point and a more obvious fail point immediately preceding a negative slope. Each figure displays several curves which correspond to the strain rates as indicated on the strain rate versus time plots.

The least square fit for the strain rate data for AP/Al/HTPB propellant is indicated on the strain rate versus time plots, Fig. 2 by the dotted lines for the temperature tested at +70°C. Figure 3 present the strain versus time data. The stress versus strain plots for AP/Al/HTPB propellant temperature studies are shown in Fig. 4. The results of the high strain testing at various temperatures are summarized in Table 1.

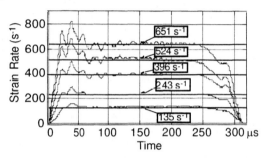

FIGURE 2. AP/Al/HTPB Propellant Strain Rate versus Time Plot at +70°C.

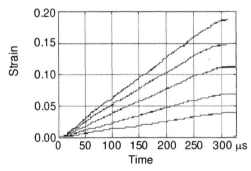

FIGURE 3. AP/Al/HTPB Propellant Strain versus Time Plot at +70°C.

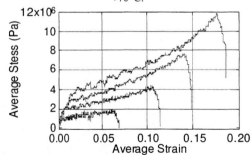

FIGURE 4. AP/Al/HTPB Propellant Stress versus Strain Plot at +70°C.

TABLE 1. AP/Al/HTPB Propellant Summary of Stress versus Strain Results.

Temp.	Strain Rate (S^{-1})	Yield Stress (MPa)	Modulus (GPa)	Failure Stress (MPa)	Slope (MPa)
+200°C	359	0.72	0.04	1.12	4.45
	622	1.08	0.04	1.89	7.06
+150°C	287	1.12	0.12	1.49	9.25
	507	1.45	0.14	1.76	3.87
+100°C	321	1.52	0.16	2.69	15.3
	604	2.04	0.20	3.90	10.5
+70°C	136	0.93	0.21	1.51	23.4
	243	0.97	0.14	2.74	14.1
	396	1.14	0.19	4.44	25.7
	524	2.87	0.21	7.65	37.3
	651	4.19	0.19	11.9	45.2
0°C	105	1.46	0.45	7.40	225.8
	175	3.40	0.73	14.2	205.5
	336	6.74	0.49	19.2	110.9
	454	10.0	0.57	26.	139.9
	583	11.3	0.57	28.1	76.9
-20°C	179	2.51	0.38	16.0	305.4
	304	3.10	0.40	23.8	297.6
	544	3.11	0.38	42.6	280.5
	784	32.5	0.34	54.7	182.7
-40°C	199	3.81	0.47	33.9	369.5
	256	8.16	0.58	40.6	552.9
	453	49.8	1.13	77.1	340.1
	730	56.2	1.09	70.4	164.5
-65°C	174	8.79	0.44	24.4	762.5
	332	12.7	0.64	67.8	1257.8
	606	64.1	1.67	99.1	334.4
	841	88.3	1.94	100.1	195.4
-95°C	104	6.51	5.37	38.1	2279.2
	169	10.0	5.71	57.1	2646.9
	293	46.1	5.86	80.9	832.0
	412	61.3	5.00	84.2	140.9
	554	74.5	3.71	82.6	577.9

The elastic modulus was calculated from the slope of the curve. The fluctuations in the elastic modulus for AP/Al/HTPB propellant at each temperature are attributed to the strain rate not truly being constant at the start of the compressive pulse. The slope seems to decrease with an increase in strain rate and the slope increases with decrease in temperature.

In all the samples tested, as the temperature increases the yield stress point decreases for the stress versus strain curves and the samples become rubberlike at the very high temperature. The raise in temperature typically effects the stress-strain behavior of these materials as would be expected from a polymer or rubber-like material. Each sample has 10 percent binder and plasticizer materials in the formulation. Plots of yield stress versus tem-

perature are shown in. The yield stress temperature results are plotted at approximately the same strain rate 600 s^{-1}, as shown in Fig. 5. The failure stress results are presented in Fig. 6 for the AP/Al/HTPB propellant material at the strain rate of $600s^{-1}$.

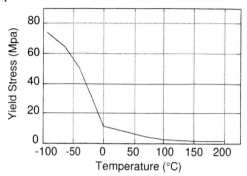

FIGURE 5 Yield Stress versus Temperature Plot for AP/Al/HTPB Propellant..

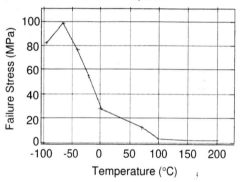

FIGURE 6. Failure Stress versus Temperature Plot for AP/Al/HTPB Propellant..

CONCLUSIONS

The split Hopkinson Pressure Bar apparatus has been used to evaluate the mechanical response of AP/Al/HTPB propellant materials. In general, there is not much data available concerning the high strain rate behavior of solid propellants over a wide range of temperatures. The SHPB technique is a well established method for determining the dynamic stress-strain-strain rate behavior. The strain rates ranged from $100s^{-1}$ to $700s^{-1}$ at nine different temperature regions. All of the samples tested showed higher yield stresses and failure stresses with increasing strain rate. In addition, yield and failure stresses increased with a decrease in temperature. The elastic modulus increased by a factor of 5 or more at the very low temperature. Figure 7 indicates the large increase in the elastic modulus for the propellant samples tested. The large increase in the elastic modulus at the very low temperature indicates that we may have been below the glass transition temperature (T_g) of AP/Al/HTPB propellant materials.

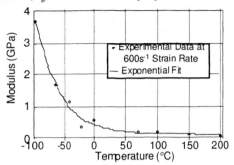

Figure 7. Elastic Modulus Plotted as a Function of Temperature.

During this experimental study, attempts were made to determine the glass transition temperature for the binder material. A differential scanning calorimetry (DSC) method was used to observe the variations of thermodynamic properties with temperature for AP/Al/HTPB propellant. With polymeric materials, DSC tests are commonly used to study temperature transitions (5-6). AP/Al/HTPB propellant samples were tested using the DSC process. By measuring the heat flow to and from the sample, endothermic and exothermic reactions can be studied. The DSC results are presented in Fig. 8. For the sample tested the slope increases from the melt temperature T_M (as shown on the graph) to the exothermic reaction as indicated at about -15.0°C AP/Al/HTPB propellant.

We feel that at the very low temperature, -65°C and below, the characteristics of the polymer material is the driving factor of the stress-strain results. Once the polymer material is very cold, the molecular chains of the material cann't move and the polymer is frozen into what is called a glassy state. This temperature is known as the glass transition temperature T_g. As temperature decreases, the specific volume of the polymer material decreases which implies that the density increases. Therefore, from the DSC it is very difficult to define T_g accurately in terms of mechanical properties due to the volume changes within the material. Usually the transition of a material into the glass transition state occurs over a temperature interval when measured under no stress loading conditions. The T_g is a property of the polymer. Near the T_g temperature, the molecules of a polymer material under prolonged stress have time to rearrange their chains and uncoiling occurs. But under short term stress loadings (such as impact loadings from SHPB), the molecular chains do not have time to uncoil and the material becomes very stiff. This is a physical process of the material which stiffens the polymer and causes cracking of the material under brief impact loads (5-6). At the very low temperatures and higher strain rates, each sample tested exhibited this behavior. At the highest strain rates, the material cracked and the plots of the stress-strain curves were effected in the low temperature plots.

SHPB data collected and analyzed under this experimental study showed typical results that would be expected from elastomers. As the temperature increased, the failure and yield stress decreased. The elastic modulus was relatively constant at each varying strain rate and constant temperature, while the elastic modulus changed slightly with temperature between +70°C and 0°C. The large increase in the elastic modulus at the very low temperature indicates that we may have been below the T_g of these materials. More data and details of this experimental study can be obtain in Reference 7.

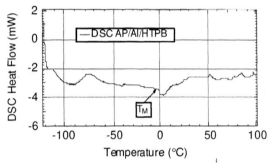

Figure 8. DSC Curve for AP/Al/HTPB Propellant.

REFERENCES

1. Zukas, J. A., Nicholas, T., Swift, H. F., Greszczuk, L. B., and Curran, D. R., *Impact Dynamics*, John Wiley & Sons, 1982, pp. 287-307.

2. Davies, E. D., and Hunter, S. C., J. Mech. Phys. Solids, Vol. 11, 1963, p. 155.

3. Oberth, A. E., Principles of Solid Propellant Development, 1987, pp. 10-11.

4. Zel'dovich, Y. B., and Raizer, Y. P., *Physics of Shock Waves and High Temperature Hydrodynamic Phenomena*, Academic Press, 1967, Vol. II, p. 734.

5. Powell, P. C., *Engineering with Polymers*, Chapman and Hall Ltd., 1983.

6. Andrews, E. H., *Fracture in Polymers*, American Elsevier Publishing Company, 1968.

7. John, H. J., Hudson, F. E., and Robbs, R. L., High Strain-Rate Testing of Solid Propellant, NAWCWPNS TP 8343, June 1997.

CHAPTER VIII

SHOCK-INDUCED MODIFICATION AND MATERIAL SYNTHESIS

CP429, *Shock Compression of Condensed Matter – 1997*
edited by Schmidt/Dandekar/Forbes
© 1998 The American Institute of Physics 1-56396-738-3/98/$15.00

CONTROLLED HIGH-RATE-STRAIN SHEAR BANDS IN INERT AND REACTANT POROUS MATERIALS

V.F. Nesterenko

Department of AMES, University of California, San Diego, CA 92093-0411

Shear localization is considered as one of the main reasons for initiation of chemical reaction in energetic materials under dynamic loading. However despite of widely spread recognition of the importance of rapid shear flow the shear bands in porous heterogeneous materials did not become an object of research. The primary reason for this was a lack of appropriate experimental method. The "Thick-Walled Cylinder" method, which allows to reproduce shear bands in strain controlled conditions, was initially proposed by Nesterenko et al., 1989 for solid inert materials and then modified by Nesterenko, Meyers et al., 1994 to fit porous inert and energetic materials. The method allows to reproduce the array of shear bands with shear strains 10 - 100 and strain rate 10^7 s^{-1}. Experimental results are presented for inert materials (granular, fractured ceramics) and for reactant porous mixtures (Nb-Si, Ti-Si, Ti-graphite and Ti-ultrafine diamond).

INTRODUCTION

The high strain, high strain rate shear flow of fractured ceramics during ballistic impact of ceramic armor plays the important role in governing the depth of penetration as it was shown in model calculations by Curran et al. [1]. The initiation of chemical reactions and phase transformations under quasistatic and dynamic pressure+shear conditions was experimentally demonstrated by Bridgman [2], Dremin and Breusov [3], Graham [4], Batsanov [5], Enikolopyan [6] and others. Winter and Field [7], Frey [8], Kipp [9] used shear localization concept as a feasible mechanism of explosive initiation. Frey [8] emphasized the importance of pressure and shear strain for viscous and pressure dependent materials. Kipp [9] developed a shear band concept based on the detailed analysis of shear band formation by Grady and Kipp [10].

At the same time there is a lack of experimental information on the details of shear band formation in heterogeneous, porous materials and on material behavior within shear bands. Nesterenko, Meyers et al. [11-13] proposed a method providing strain controlled, reproducible high-strain-rate shear bands in heterogeneous porous mixtures. It allows to investigate the array of the shear bands with different measurable strains in single experiment. Material structure inside shear band as function of the shear strain can also be clarified depending on the precisely tuned overall strain. It allows to look at the different stages of chemical reactions and material comminution in high-strain-rate shear.

In this paper the results of the application of the "Thick-Walled Cylinder" method is reviewed for the shear localization process in inert ceramics (comminuted and granular armor ceramics Al_2O_3, SiC with different particle sizes and porosities), and for shear induced reactions in porous Nb+Si, Ti+Si, Ti+C mixtures.

EXPERIMENTAL SET-UP

The shear localization was obtained with the help of the "Thick-Walled Cylinder" method which is

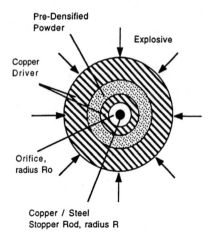

Pre-Densified Powder

Explosive

Copper Driver

Orifice, radius Ro

Copper / Steel Stopper Rod, radius R

FIGURE 1. Initial geometry of "Thick-Walled Cylinder" method.

described in [11-13]. The main idea of this method is based on the strain controlled collapse of the sample into the central orifice which is provided by explosive loading from outside. Initial geometry (Figure 1) and loading conditions ensuring the "smooth" collapse process (finishing close to the center as result of the dissipation without spall, jetting, and fracture in the backward motion and without a hydrodynamic instability) allow to calculate strain in the investigated material. As the predensified powders were usually used, their density at the first approximation was constant during the collapse process. The effective strains can be found according to the equation:

$$\varepsilon_{eff} = \frac{1}{3} \ln \left[1 + \frac{R_0^2 - R^2}{r_f^2} \right],$$

(1)

where the final radii r_f and R (the final radius R of the inner copper tube after collapse is equal to the radius of the central rod) and initial radius R_0 (Figure 1) are experimentally measured.

The state of stresses in these plane strain experiments are determined by the material strength and its dependence on temperature, strain, and strain rate. They can be found based on the modeling of the material behavior or from separate experiments.

The following features of this method should be mentioned. The weak shock waves propagating in materials in the first stage of densification and

during the collapse process have no noticeable influence on the chemical reactions; their amplitudes are less than 1 GPa. This statement also was checked in the experiments with arrested pore collapse but with a corresponding shock loading. The superimposed "hydrostatic" pressure inside collapsing incompressible cylinder due to its acceleration toward the center, is less than 0.1 GPa [12]. Thus, pressure effects can be neglected to the first approximation for consideration of the chemical processes which are in this case mainly strain controlled. However the pressure effects are still very important for the geometry of shear bands in ceramics [14] and for pressure sensitive materials.

The "hydrodynamic" instability is typical for thin shells accelerated toward the center with relatively high velocities [15]. It is not active here due to the thick wall of the cylinder and relatively low velocity. The material instability instead breaks symmetry of the collapse first and results in shear localization. The picture is different if explosives with higher detonation speeds and densities are used to collapse relatively thin shells [16,17].

SHEAR INSTABILITY IN COMMINUTED AND GRANULAR CERAMICS

High-strain, high-strain rate flow of comminuted ceramics represents essential part of ceramic armor flow during the long rod penetration into confined ceramics as was clearly demonstrated by Curran, Seaman, Cooper and Shockey, 1993. At the same time the results on the behavior of ceramics under these conditions are very restrictive. In [13, 18-20] the "Thick Walled Cylinder" method was used to investigate the high-strain, high-strain rate (ε_{eff} ~ 0.35, $\dot{\varepsilon}_{eff}$ ~ 10^4 s^{-1}) respond of comminuted and granular Al_2O_3 and SiC of different particle sizes varying from 0.4 μm to 50 μm. Two steps were used for ceramics. For initially solid material the explosive loading with small strains was initially applied to prepare prefractured (comminuted) material which was subjected to a large strains during collapse process on the second stage.

For the granular material the first step was used to densify initially loose granular material up to the

610

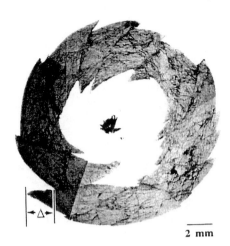

2 mm

FIGURE 2. Sample of SiC after prefracturing and subsequent collapse processes.

85-90% of solid density before starting the collapse process. The overall view of prefractured and deformed SiC is presented in Figure 2.

The main results can be summarized as follows.

Prefractured ceramics [18-21]:

The microstructural differences in initially solid SiC have little effect on the dynamic response of fragmented material;

Shear localization is an important mechanism of deformation of damaged ceramic, approximately 50 to 60% of the total tangential strain is accommodated by shear localization in prefractured SiC and alumina;

Intense comminution inside shear bands was observed with relatively intact material outside shear bands except the areas experienced bending;

Inside shear band a bimodal particle distribution was observed;

The thickness of shear bands increased with strain.

Granular ceramics [13,14,18,19,21]:

Shear localization is also important mechanism of deformation of granular ceramic. Up to 18% of the total tangential strain is accommodated by shear localization in granular SiC with particle size 3 μm and up to 46% in granular alumina with particle size 0.4 μm.

The degree of localized strain strongly depends on initial particle size, for example for granular SiC

with particle size 50 μm the existing shear bands carry negligible amount of total strain.

The thickness of shear bands did not correlate with initial particle size suggesting that this is not essential material parameter for shear band structure under conditions of confined ceramic;

The structure of the shear bands as well as their patterning depend on initial particle size;

The extent of comminution inside shear bands is determined by particle size;

If shear strain inside shear bands is about 10, the material with initial particle size less than 3 μm can be heated up enough to result in plastic flow of SiC and in subsequent bonding process between particles.

One aspect of ceramic flow under high strain rate condition is connected with the particle fracture which depends on the initial particle size [19,20]. The variation in initial particle size results in essential change of overall behavior of granular material [13,14,18,19]. One of the important features can be the dependence of the energy converted into the heat during the flow of granular material if particle fracture is involved. Relatively small particles will not break and whole energy of plastic deformation will be converted into the heat whereas the large particles are able to convert essential part of plastic flow into creation of new surfaces with less heat generation. To evaluate the possibility of this qualitative dependence of granular material on particle size let us find the ratio K_{ph} of thermal energy to the work of plastic deformation W_p. Their difference is connected with creation of new particle surfaces and also with internal damage of newly created particles W_f:

$$K_{ph} = \frac{W_p - W_f}{W_p} = 1 - \frac{\gamma_s \pi (n-1) S(d)}{\tau \gamma d}, \quad (2)$$

where γ_s is the specific energy associated with the creation of new surface area of fragments, τ is the global shear strength of granular material, γ is shear strain inside shear band, d is the initial particle size, n is the ratio of newly created fragment diameter to the initial particle diameter d and $S(d)$ is the factor responsible for the difference between surface area of newly created fragments and the total surface created during the fragmentation. The coefficient S can be of the order of magnitude 10^3 that reflects the fact that the fragment size does not represent the real amount of damage and total created surface as emphasized by Meyers [22]. For numerical evaluations the following parameters for SiC were selected: γ_s = 20 J/m² [23], τ = 2 GPa and n =10 [14], d = 50 μm. At these material parameters and shear strain γ = 0.1 the value of $S(d)$ = 100 will result in K_{ph} = 0.4. For shear strain γ = 10 the same K_{ph} will be obtained with $S(d)$ = 10^4. So it is possible to conclude that particle fragmentation can be an important mechanism of energy dissipation inside shear bands. It is interesting to mention the nonmonotonous dependence of K_{ph} on the initial particle size. If the particle size is getting smaller with the constant n and $S(d)$, it will result in less heat generation for smaller particle sizes. In the contrary for sufficiently small particles with d less than 0.3 μm it is evident that $n \rightarrow 1$ as well as $S(d) \rightarrow 1$ because the fracture process is not developing [14]. For this limit whole energy of plastic work will be converted into heat and K_{ph} = 1.

SHEAR BAND IN REACTANT MATERIALS

It was observed that shear localization of relatively high amplitude can result in chemical reactions in condensed materials[11,12, 21, 24-29]. The different views of the shear localization corresponding to the various strains are presented in Figure 3. A slight increase of overall strain from ε_{eff} = 0.33 (Fig. 3 a,b) to ε_{eff} = 0.38 (Fig. 3 e,f) results in complete reaction in the bulk being initiated within shear band. Small unreacted parts (shown by arrows in Fig.3e) were used to evaluate the reaction rate in a dense, plastically deformed mixture outside shear bands[27].

On the basis of the observations, a mechanism of *shear-assisted* chemical reaction can be proposed [11,12]. The main stages are:

(1) Nb, Ti particles are split into foils with thickness of the order of magnitude of 0.1-1 μm by localized mesoshear. They are heated as a result of localized shear deformation which precedes their formation and have "fresh" surfaces. The spacing between this small scale shear bands in metal particles can be evaluated based on the models developed by Grady and Kipp [10], Wright and

Ockendon [30] and Molinari [31] which provide the numbers of the same order of magnitude. For example the spacing between shear bands in Wright-Ockendon (L_{WO}) model is equal:

$$L_{WO} = 2\pi \left(\frac{kCm^3\dot{\gamma}_o^m}{\gamma^{3+m}a^2\tau_o} \right)^{1/4}, \qquad (3)$$

where $\dot{\gamma}_o$ and $\dot{\gamma}$ are a reference and current strain rates, m is the strain-rate sensitivity, τ_o is the flow stress at the reference temperature T_o and strain rate $\dot{\gamma}_o$, k is thermal conductivity, C is heat capacity, and a is thermal softening term. Nesterenko, Meyers and Wright [32] demonstrated that at strain rate $3.5 \cdot 10^4$ s⁻¹, Wright-Ockendon model provides very good agreement with experiments on Ti. Strain rate inside shear bands can be evaluated from the thickness of shear band δ, its displacement Δ and time of collapse t_c:

$$\gamma \approx \frac{\Delta}{\delta t_c}. \qquad (4)$$

Magnitude of the averaged (over time of experiment and thickness of shear band) strain rates $\dot{\gamma}$ can reach the values up to 10^7 sec⁻¹. Due to the averaging procedure the evaluation of shear spacing will provide only the upper boundary. For Ti it will result in spacing L_{WO} = 7 μm. This value is larger than the thickness of the Ti particles created due to the mesoshear which are about 1 μm and less (Figure 3b). Taking into account the average evaluation of $\dot{\gamma}$ over total time of collapse and thickness of shear band it is possible to conclude that the shear localization is a reasonable mechanism of particle fracture inside a shear band.

(2) Reaction begins due to the extensive relative flow of Nb (Ti) particles and Si inside the shear bands, which is accompanied by melting of Si. Also the relative motion inside shear bands will promote the fracture of the reaction product between metals and silicon accelerating the reaction.

(3) Vorticity inside shear band originating from the instability of gradient flow facilitate the mixing of materials (Fig.3b).

(4) Reaction continues at places where temperatures are sufficiently high and is then quenched (Nb-Si, Ti-Si, Ti-C) or can proceed through surrounding material as it was observed in some experiments with Ti-Si (Fig. 3e,f).

612

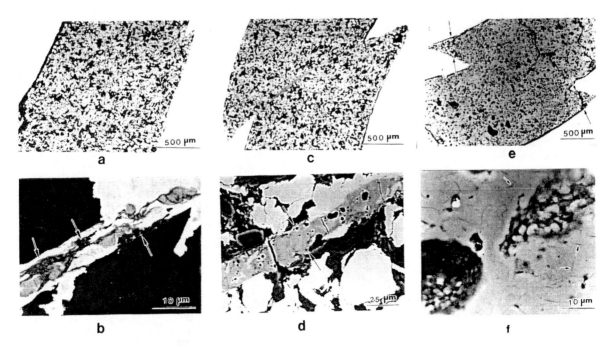

FIGURE 3. Shear bands at various global strain: a,b - $\varepsilon_{eff} = 0.33$; c,d - $\varepsilon_{eff} = 0.35$; e,f - $\varepsilon_{eff} = 0.38$.

The propagation of reaction into bulk in Ti-Si mixture and its extinction inside shear bands in Nb-Si was explained [28] on the basis of thermochemical analysis. The difference is due to the less heat release from the reaction in Nb-Si and higher thermal conductivity of this mixture.

The shear localization in Ti-graphite mixture was suppressed [29] presumably due to the lubricant properties of graphite and reaction initiation was observed in separate localized areas. The replacement of the graphite with ultrafine diamond drastically changed the mechanical behavior: intense shear localization was observed in latter case [29].

Effective speed of reaction can be found from the following arguments. If bulk of reactant material is simultaneously saturated by shear bands with typical spacing L and reaction front propagates with speed V_r, then the typical time for reaction completion does not depend on the sample size and is equal to $t_r = L/V_r$. This time corresponds to the apparent speed of reaction V_{ap} of the whole sample with size L_s by $V_{ap} = L_s/t_r = V_r L_s/L$. Typical size of L is 0.5 - 1mm, maximum reported speed V_r for nanoscale, multilayer material Al-Ni is 10 m/s [33]. From this data the maximum calculated V_{ap} for the sample with

size about 10 cm can be as high as 10^3 m/s. It is reasonable to expect the value of V_{ap} to be at least 100 m/s. It is important that shear band spacing L did not depend on the size of the sample [21,28].

COMPARISON BETWEEN DIFFERENT TYPES OF IMPULSE LOADING

Shock Waves and Shear Bands

Energy released per unit mass under shock loading of porous material with initial density equal half of solid can be evaluated as

$$\varepsilon_{SW} \approx \frac{P}{4\rho_s},\tag{5}$$

where P is shock pressure and ρ_s solid density. The specific energy for material within shear band is:

$$\varepsilon_{SB} \approx \frac{\tau\gamma}{\rho_s},\tag{6}$$

where τ is shear strength averaged over shear deformation and material inside SB is supposed to be densified. These two expressions for specific energies are equal at shock pressure P_c:

$$P_c \approx 4\tau\gamma.\tag{7}$$

From the Eq. 7 it is easy to see that shear band with $\gamma \sim 100$ and $\tau = 50$ MPa is energetically equivalent to the shock at the pressure 20 GPa. The main differences between shear and shock assisted chemical reactions are due to the essential differences in the shear strain resulting in different amount of displacements of neighboring particles. Under shock the relative displacements of particles are comparable or less than the particle sizes ($\gamma < 1$) and cannot induce essential shear localization and fracture inside particles, as it is observed inside shear band. The flow instabilities resulting in vorticities also enhance the mixing and subsequent reactivity of mixtures under shear band conditions.

Laser Heating and Shear Bands

The laser power is enough to ignite mixtures like Ti + 2B, Ti + 2Si (Korotkevich et al., [34]). It is interesting to compare the magnitude of power w_{SB} supplied by plastic work to material inside shear band with laser power. The w_{SB} per unit surface of shear band is

$$w_{SB} \approx \frac{\tau \gamma \delta}{t} \tag{8}$$

For typical shear band with $\gamma \sim 100$, $t = 8 \cdot 10^{-6}$ s, and $\tau = 50$ MPa w_{SB} is $10^5 - 10^6$ W/cm^2. This value is comparable with the laser power to initiate chemical reactions in similar solid materials[34].

ACKNOWLEDGMENTS

This research was performed in collaboration with M. A. Meyers, J.C. LaSalvia, H-C. Chen and C.J. Shih. The support provided by the ONR, N00014-94-1-1040 (Program Officer Dr. J. Goldwasser) and by the ARO, MURI DAAH 04-96-1-0376, is highly appreciated. The author is thankful for the help with experiments to M.P. Bondar, Y.L. Lukyanov, A.A. Stertser, S.M. Usherenko and to S. Indrakantri for paper preparation.

REFERENCES

1. Curran, D.R., Seaman L., Cooper, T., and Shockey D.A., *Int. J.Impact Engng.* 13, 53-83 (1993).
2. Bridgman, P.W., *Phys. Rev.*, **48**, 825-847 (1935).
3. Dremin, A.N., and Breusov, O.N., *Russ. Chem. Rev.*, **37**, 392-402 (1968).
4. Graham, R.A., *Solids Under High Pressure Shock Compression: Mechanics, Physics and Chemistry*, New York: Springer-Verlag, 1993.
5. Batsanov, S.S., *Effects of Explosions on Materials*, New York: Springer-Verlag, 1993.
6. Enikolopyan, N.S., *Doklady Akademii Nauk SSSR*, 302, 630-634 (1988).
7. Winter, R.E., Field, J.E.., *Proc.R. Soc. London, A*, 343, 399-413 (1975)
8. Frey, R.B. *Proc. of the Seventh Sympos. on Detonation*, Albuquerque, NM, 35-41(1985).
9. Kipp, M.E.., *Proc. of the Eigth Sympos. on Detonation*, Annapolis, MD, 36-42 (1981).
10.. Grady, D.E., and Kipp M.E., *J. Mech. Phys. Solids* 35, 95-120 (1987).
11. Nesterenko, V.F., Meyers, M.A., Chen, H.C., and LaSalvia, J.C., *Applied Physics Letters*, **65**, 3069-3071 (1994).
12. Nesterenko, V.F., Meyers, M.A., Chen, H.C., and LaSalvia, J.C., *Metall. and Mater. Trans.*, 26A, 2511-2519(*1995*)
13. Nesterenko, V.F., Meyers, M.A., and Chen, H.C., *Acta mater..*, **44** 2017-2026(1996).
14. Shih, C.J., Meyers, M.A., and Nesterenko, V.F., *Acta Mater.*, submitted.
15. Serikov, S.V., *J. Appl. Mech. and Tech. Physics*, 25, 142-153 (1984).
16. Lindberg, H.E.., *Trans. ASME, E, J. Appl. Mech*, 31, 267-273 (1964).
17. Ivanov, A.G., Ogorodnikov V.A., and Tyun'kin, E.S., *J. Appl. Mech. and Tech. Physics* 33, 871-874 (1992).
18. Chen, H.C. Meyers, M.A. and Nesterenko, V.F. *Proceedings of the Conference of the American Physical Society Topical Group on Shock Compression of Condensed Matter*, Seattle, August 13-18, 1995, AIP Press, 1996, p.607-610.
19. Shih, C.J., Nesterenko, V.F., and Meyers, M.A., *Proceedings of International Conference on Mechanical and Physical Behavior of Materials under Dynamic Loading*, Toledo, September 22-26, 1997.
20. Shih, C.J., Nesterenko, V.F., and Meyers, M.A., *Acta Mater.*, submitted.
21. Chen, H.C., PhD. Thesis, University of California, San Diego, 1997
22. Meyers, M.A., *Dynamic Behavior of Materials*, New York: John Wiley & Sons Inc., 1994, ch. 16, p. 555.
23. Chiang, Y-M, Birnie, D., and Kingery, W.D., *Physical Ceramics*,: John Wiley & Sons, Inc., 1997, ch.5, p.359.
24. Nesterenko, V.F., Meyers, M.A., Chen, Y.J. and LaSalvia, J.C., *Proc. of the Conf. of the American Physical Society Topical Group on Shock Compression of Condensed Matter*, Seattle, August 13-18, 1995, AIP Press, 1996, p.713-716
25. Chen, H.C., Meyers, M.A., and Nesterenko, V.F., in *Proceedings of the 1995 International Conference EXPLOMET-95*, El Paso, August 6-10, 1995, pp. 723-729.
26. Chen, H.C., Nesterenko, V.F., and Meyers, M.A., *Proc. of Int. Conf. on Mechanical and Physical Behavior of Materials uder Dynamic Loading*, Toledo, 1997.
27. Chen, H.C., LaSalvia J.C., Nesterenko, V.F., and Meyers, M.A., *Acta Mater.*, submitted.
28. Chen, H.C., Nesterenko, V.F., and Meyers, M.A., *J. Appl. Phys.*, submitted.
29. Chen, H.C., Nesterenko, V.F., and Meyers, M.A., *J.of Materials Science* , submitted.
30. Wright, T.W., and Ockendon, H., *Int. J. Plasticity* , 12, 927-937(1996).
31. Molinari, A., *Mech. Phys. Sol.* (to be published).
32. Nesterenko, V.F., Meyers, M.A. and Wright, T.W., *Acta Mater.*, (to be published).
33. Weihs, T., JOM, 11, 9 (1996).
34. Korotkevich, I.I., et al., FGV, 17, (1982)

CP429, *Shock Compression of Condensed Matter – 1997*
edited by Schmidt/Dandekar/Forbes
© 1998 The American Institute of Physics 1-56396-738-3/98/$15.00

A NEW APPARATUS FOR DIRECT TRANSFORMATION FROM hBN to cBN

Y. Kuroyama[a], K. Itoh[b], Z. Y. Liu[c], M. Fujita[c] and S. Itoh[c]

[a]*Hokkaido NOF Corporation, Mibai, Hokkaido, Japan*

[b]*NOF Corporation, Aichi Works, Taketoyo Plant, Aichi, Japan*

[c]*Department of Mechanical Engineering, Kumamoto University, Kumamoto 860, Japan*

A new apparatus was devised for direct phase transformation from hBN to cBN by using the cylindrical explosion. The apparatus consisted of multiple thin-metal foils in a metal tube surrounded by a high efficiency explosive and the spaces between the multiple thin-metal foils were filled with hBN containing a small amount of copper powder. This apparatus was expected to produce higher pressure and temperature than the ordinary cylindrical method. These high pressure and temperature has been proved to be satisfactory to result in the direct phase transformation from hBN into cBN.

INTRODUCTION

Cubic boron nitride (cBN) is popularly employed in the lapping and cutting fields of industry because of its super-high hardness. At present, the source of cBN is primarily come from the method of the synthesis from rhombohedral boron nitride (rBN) or hexagonal boron nitride (hBN) under extra-high static pressure via catalyst. For instance, Bundy and Wentorf (1) once realized the direct phase transformation of cBN by a press machine capable of generating the high pressure over 10Gpa. The phase diagram of boron nitride was also given by Bundy and Wentrof. On the other hand, the possibility of the direct phase transformation of boron nitride was also showed by explosive shock synthesis or light gas gun's impact in a very short time, e.g., the order of microseconds (2). These kinds of dynamic methods, especially the explosive shock synthesis, are exhibiting a promising future in the synthesis of cBN, whether in efficient aspect or in economic aspect. Nevertheless, the transformation rate from hBN to cBN is very low because the fact that even at the high pressure over 10Gpa, if not

being enough high temperature, the hBN seems to transform more easily into wurtzite boron nitride (wBN) than into cBN. The main effort that should be devoted to, so far, is to overcome the deficiency of low transformation rate. The promotion of the transformation rate may be accomplished by causing a relative high temperature environment during synthesis processing. Therefore, in explosive shock method, a new cylindrical apparatus, called 'rolled multi-layer method' was devised to achieve this aim. The experimental results showed that the method is of a good effectiveness on the direct phase transformation form hBN to cBN.

APPROACH AND PROCEDURES

The newly developed apparatus is schematically illustrated in Fig. 1. It is primarily composed of high explosive, metal tube, rolled multi-layer metal foil and hBN, and a metal mandrel. Although it is one of the cylindrical implosion methods, its main difference is due to that the hBN is not directly packed in the inside of a metal tube, but trapped between the multi-layers of the rolled metal foil. The

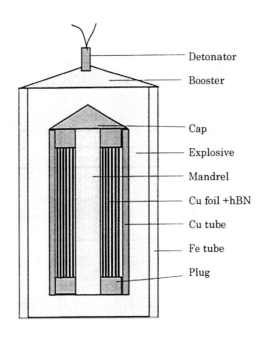

FIGURE 1. A newly developed apparatus for the synthesis of cBN.

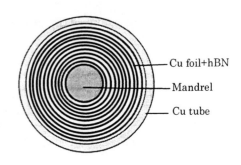

FIGURE 2. A cross-sectional diagram of the main part in the apparatus.

cross-section of this part is shown in Fig. 2.

To test the effectiveness of the apparatus, the practical synthesis experiment was carried out. The explosive was a kind of dynamite, with the detonation velocity of 6500 m/s. Copper tube of 20mm inner diameter and 2 mm thick was used as the metal tube. The metal foil was made of copper with the dimensions of 730mm long, 200mm wide and 0.1mm thick. The hBN was painted on the copper foil. Then, the copper foil with the hBN was rolled around an iron mandrel into multi-layer shape and inserted into a copper tube. After explosion, let the specimen fall into a water pool to have it rapidly cooled, and at the same time, to make a convenience of recovery. The experiment was conducted at the specially designed noise-elimination chamber at Taketoyo Plant, Aichi Works of NOF Corporation.

RESULTS AND DISCUSSION

The original hBN was analyzed by X-ray diffraction method. It is found that the hBN was composed of the part of good crystalline with the complete 3-dimensional order and the one with the disordering atom distribution. Figures 3 shows the X-ray diffraction pattern of the sample obtained from the experiment with the newly developed apparatus. The symbols $\circ$, $\triangle$, $\square$ represent the hBN, cBN and wBN, respectively. It can be seen that the cBN and wBN phases are indeed converted. The peak values of w(100), w(101), w(102) for wBN (symbol $\square$) seem to be a fairly good agreement with the theoretical ones. The peak at 2.095 Å shows the mixed peaks of w(002) and c(111) because that these two peaks are so near that the distinguish of them is become difficult. This peak indicates a fact of the co-existence of wBN and cBN. The same situation was also pointed out by Sekine and Sato (3). Comparatively, in our previous experiment with the ordinary cylindrical apparatus, the X-ray diffraction of the sample showed that only wBN was converted from hBN, no conversion of cBN, as shown in Fig. 4.

CONCLUSIONS

From our study, the following conclusions could be obtained:

1. It is confirmed that the direct phase transformation from hBN to cBN was induced in our developed 'rolled multi-layer method'.

2. The method can bring out a higher transformation rate of cBN than other cylindrical method and plane method.

616

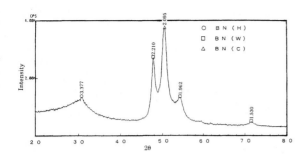

FIGURE 3. The X-ray diffraction pattern of the sample recovered from the 'rolled multi-layer method'.

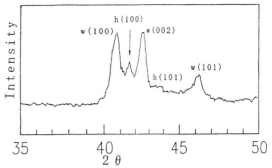

FIGURE 4. The X-ray diffraction pattern of the sample recovered from the ordinary cylindrical method.

ACKNOWLEDGEMENTS

The experiment was conducted at Taketoyo Plant, Aichi Works of NOF Corporation. The NOF Corporation provided the financial aid for the study.

REFERENCES

1. Bundy F. P., and Wentorf R. H., *J. Chemical Phsy.* **38**, 1144-1149 (1963).
2. Adadurov, G. A., et al., *Sov. Phys. Dolk.*, **12**, 173 (1967).
3. Sekine, T., and Sato, T., *J. Appl. Phys.*, **74**, 15 (1993).

CP429, *Shock Compression of Condensed Matter – 1997*
edited by Schmidt/Dandekar/Forbes
© 1998 The American Institute of Physics 1-56396-738-3/98/$15.00

SHOCK WAVE COMPACTED, MELT INFILTRATED CERAMICS

M. Stuivinga[*] and E.P. Carton[*,**]

[*]*TNO Prins Maurits Laboratory, P.O.Box 45, 2280 AA Rijswijk*, The Netherlands
[**]*Delft University of Technology, Laboratory for Applied Inorganic Chemistry,
P.O. Box 5045, 2600 GA Delft, The Netherlands*

Using shock wave compaction followed by melt infiltration with aluminum, B_4C-Al and TiB_2-Al composites have been fabricated. The composites are fully dense and crack-free. They have a high (80-85 vol. %) ceramic content, which gives them good mechanical properties. Due to the infiltration with aluminum, they also have rather good conductive properties. This makes it possible to machine them using spark erosion, in order to obtain complex articles such as nozzles and dies. They are lightweight, an advantage for application in armor and fast turning spindles. In the present article, scanning electron micrographs of the fracture surfaces will be shown and some material properties will be presented.

INTRODUCTION

Shock compaction of ceramic composites to its theoretical density is in principle, a suitable technique for the fabrication of composite materials, because of its limited heating effects. But this will only be true when one find ways to overcome the appearance of cracks (1), especially when approaching full density. One of the solutions to obtain a crack-free ceramic-metal composite is presented here: shock-compaction of a hard and brittle ceramic up to 80-85 % theoretical maximal density (TMD) followed by melt infiltration (2). In this way, dense cermets can be fabricated.

FABRICATION

B_4C-Al (BORCAL) and TiB_2-Al (TIBAL) cermets were fabricated as follows: in order to avoid spiral cracking during shock compaction (3), a high starting density was obtained using a trimodal powder mixture. The mixtures were made with normally distributed powders with particle sizes of < 5 μm, between 16-49 μm and 106-150 μm (ESK, Germany) for B_4C and 10 μm, 10-45 μm and 45-100 μm for TiB_2 (id.) and mass ratios of 12:25:63. The mixture was homogenized with a turbulator. Next, the mixture was

filled batchwise in an Al container (AA 6063) with an inner diameter of 20 or 30 mm. Each batch was mechanically tapped and statically pressed at 20 MPa. In this way a starting density of about 70% TMD was obtained. Shock-wave compaction was done in the cylindrical configuration using non-ideal explosives with detonation velocities of between 2-4 km/s. In this way, homogeneous compacts were fabricated with a final density of about 80-85 % TMD. For the melt infiltration, the sample is put into an oven in a N_2 or Ar atmosphere. In about 5 hours a temperature of 1020 °C is reached. The total infiltration time at this temperature amounts to about 5 hours.

MATERIAL PROPERTIES

Microstructure

In Fig. 1, a back-scattered electron image of a lapped surface of a B_4C-Al sample is shown. One can clearly see that the individual particles are broken, partly into grit, and have formed a network of thin channels into which the liquid Al has penetrated. A fully dense and crack-free composite is the result. Due to the shock-wave, probably all

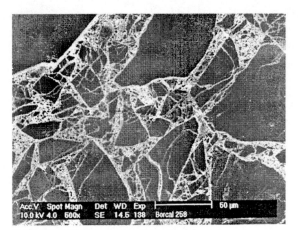

FIGURE 1. SEM image of B_4C-Al composite.

closed porosity - a possible source of flaws - has been removed. Using X-ray analysis it has been shown that Al penetration is non-reactive for TiB_2, but reactive for B_4C. In the latter case, a small part of the Al is used to form ceramic phases such as Al_4BC and AlB_2. This gives a possibility to optimize the properties of BORCAL by a second heat treatment (4). Some material properties are shown in Table 1.

Mechanical properties

Mechanical properties such as the Ball-on-ring strength (5), the fracture toughness and the hardness were measured at the Centre for Technical Ceramics (Eindhoven, Neth.). Ball-on-ring strength is a parameter which can conveniently be measured on slabs of a few mm thickness.

In general, its value is 1.5 to 2 times higher than that obtained from a 4-point bending strength test. The low spread in strength values for BORCAL is an indication that the material is reliable. The surfaces of the TIBAL slabs were not lapped, therefore their flatness was not within 20 μm as is required for that test, in order to minimize the influence of surface defects. Thus real strength values of TIBAL are expected to be higher and to show less scatter.

The value for the strength of BORCAL is comparable with the values for B_4C-Al cermets fabricated by Pyzik and Beaman (4), although their samples had a post-heat treatment in order to improve their mechanical properties.

For TiB_2, no properties of Al-infiltrated (sintered or pressed) compacts were found in the open literature. We found from X-ray photoelectron spectroscopy that shock-compaction modifies the particle surfaces (6): the small concentration of BN present on the particle surfaces decreases and wetting by Al becomes more likely (BN is not wetted by Al). Probably the BN was formed during the production process of the TiB_2 powder.

The fracture toughness of TIBAL, determined in a 3-point bending test with a chevron notch, is very similar to that of pure TiB_2. Fracture appears purely brittle and the force-displacement curve has rather linear behaviour. Such stress-strain behaviour is similar to that obtained by Muscat et al. (7) for melt infiltrated, sintered TiC-Al composites, when the Al content was less than 67 vol. %.

TABLE 1. Material properties; measured values are for 80-85 vol. % Al, unless otherwise stated.

	TIBAL	BORCAL
Materials	TiB_2 - Al	B_4C - Al
Vol % Al	9-18	11-17
Density (g/cm^3)	4.2-4.3	2.5-2.6
Ball-on-ring strength (MPa)	526 ± 51	453 ± 14
Hardness (HV 306 N, GPa)	7.2	-
Fracture toughness (MPa.m$^{1/2}$)	6.1 ± 0.9	-
Young's modulus (GPa)	376	350
Thermal expansion coefficient (10^{-6}/K)	7.8	5.5
Thermal conductivity (W/m.K)	108[a]	20
Electrical resistivity (10^{-6} Ω.m)	0.13	4.7

[a] Value for 28 vol. % Al.

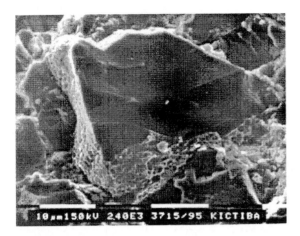

FIGURE 2. Fracture surface of TiB$_2$-Al composite.

However, on a micro-scale there are indications of plastic deformation of the Al. Some Al dimples can be seen on fracture surfaces (see Fig. 2). Such zones of plastic deformation are common for metal-ceramic composites when the metal constitutes the continuous phase (8).

From the strength at fracture, using the chevron notching technique, the fracture toughness of BORCAL could not be determined, because there were indications for some plastic behaviour: the fracture surfaces were not flat but rough and fracture did not occur instantaneously. Probably, the Al$_4$BC and AlB$_2$ phases, formed in the reaction with Al, have some crack-deflecting effect caused by microscopical anisotropy (9). This will further be investigated by measurements of the fracture energy using the force-displacement curve.

Also, the hardness could not be accurately determined with the Vickers method, because the edges of the indentation were not sharp but rounded. Therefore the size of the indentation could not well be established.

In the case of BORCAL, there are also clear indications of zones with plastic deformation, see the fracture surface of BORCAL, shown in Fig. 3 and enlarged in Fig. 4.

Conductive properties

The thermal conductivity of both types of compact was measured by the laser flash method at ECN, Petten (Neth.). The measured value for TIBAL 108 W/m.K, is the same as can be calculated from a parallel conductivity model (10).

The TIBAL material has been tested as a nozzle for rocket motor engines (11), where it has functioned well. Here the combination of the high thermal conductivity and the resistance of TiB$_2$ towards attack by molten Al, has probably played a large role.

The electrical resistivity of BORCAL measured with a 4-point method has a value only 60 times higher than Al. As a result, the material (like TIBAL) can be machined very well by spark erosion. In Fig. 5 a cylinder and a pencil-shaped nozzle are shown, that were machined form the BORCAL compact. The nozzle has an inner diameter of 0.6 mm while its outside was further polished using diamond. This nozzle illustrates the easy machinability of this light weight, BORCAL material; the light weight is an advantage when considering applications such as armor and fast turning spindles.

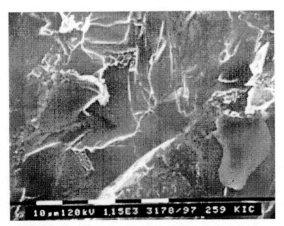

FIGURE 3. Fracture surface of B$_4$C-Al composite.

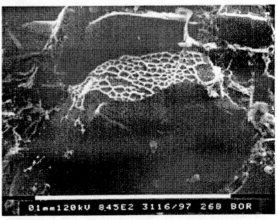

FIGURE 4. Fracture surface of B$_4$C-Al showing regions with Al dimples.

FIGURE 5. B$_4$C-Al compact and articles machined by spark erosion.

ACKNOWLEDGEMENTS

W. Duvalois of the TNO Prins Maurits Laboratory is thanked for his assistance with the scanning electron microscope; A.H.M. van Adrichem of the Faculty of Mechanical Engineering of Delft University of Technology for his skilful electric discharge machining; and L. Dortmans of the Centre for Technical Ceramics for helpful discussions.

REFERENCES

1. Carton, E.P., Stuivinga M., and Verbeek, H.J., "Crack prevention in shock compaction of powders" presented at this conference.

2. Stuivinga, M., Maas, A.M., and Carton, E.P., Int. Patent application No. PCT/NL96/00102 (6 March 1996).

3. Meyers, M.A., and Wang, S.L., *Acta Metall.* **36**, pp. 925-36 (1988).

4. Pyzik, A.J., and Beaman, D.R., *J. Am. Ceram. Soc.* **78**, pp. 305-312 (1995).

5. De With, G. and Wagemans H.M., *J. Am. Ceram. Soc.* **72**, pp. 1538-41 (1989).

6. Carton, E.P., Stuivinga, M., and Schoonman J., "Melt infiltration of shock compacted ceramics" in preparation.

7. Muscat, D., Shanker, K., and R.A.L. Drew, *Mat. Sci. Techn* **8**, pp. 971-976 (1992).

8. Bechner, P.F., and Rose, L. R., *Materials Science and Technology*, Vol. 11, Cahn R.W., Haasen P., Kramer E.J. and Swain, M.V. (eds.), Weinheim (Germany): VCH, 1994, pp. 409-61.

9 R. Telle, ibid. pp 173-266.

10. Van Vlack, L., H., *Elements of Materials Science and Engineering*, Reading (CA): Addison-Wesley Publ. Comp., 1985, pp. 493-95.

11. Carton, E.P., Stuivinga, M., Keizers, H.L.J., Miedema, J.R., and Van der Put, P.J., "Shock wave fabricated ceramic-metal composites", *Proc. of the 5th European Conf. on Advanced Materials, Processes and Applications, EUROMAT '97*, Maastricht, The Netherlands, April 21-23, **Vol. 2** pp. 355-358.

CP429, *Shock Compression of Condensed Matter – 1997*
edited by Schmidt/Dandekar/Forbes
© 1998 The American Institute of Physics 1-56396-738-3/98/$15.00

STRUCTURAL DEFORMATIONS ON FLUOROPHLOGOPITE CRYSTALS OF A PRE-HEATED AND EXPERIMENTALLY SHOCKED MICA GLASS-CERAMIC

M. Hiltl and U. Hornemann

Fraunhofer-Institut für Kurzzeitdynamik, Ernst-Mach-Institut (EMI), Eckerstrasse 4, 79104 Freiburg, Germany

Shock experiments with the reflection method were carried out at room (T_{room})- and pre-shock (T_{pre}) temperatures of 300 and 600°C at pressures ranging from 30 to 75 GPa to investigate the structural deformation on fluorophlogopite ($KMg_3(Si_3AlO_{10})F_2$) crystals of the Macor glass-ceramic. The recovered samples were examined by means of X-ray diffraction, SEM and TEM with respect to their microstructure. The examination shows that with increasing shock pressures and temperatures the crystals loose their morphology and undergo a transition into an amorphous state.

INTRODUCTION

Macor is a trademark of a machinable glass-ceramic produced by the Corning Glasworks N.Y. (1). The excellent mechanical and thermal properties of the material is used in a wide range of technical applications (2). The glass-ceramic consists of a boro-alumino-silicate glass matrix embedded with 50 Vol. % fluorophlogopite crystallites with a size of approximately 2 μm thickness and 5-20 μm in diameter (3). The present paper treats structural deformations and damage behavior of these crystallites under extreme pressure and temperature conditions.

EXPERIMENTAL SETUP

The shock recovery experiments were carried out on 0,5 mm thick discs of Macor with a diameter of 13 mm at peak pressures of 30, 45, 60 and 75 GPa at T_{room} and T_{pre} of 300° and 600°C respectively. The specimen were placed in a Fe-container (Fig. 1) which was embedded in a C60 steel-block surrounded by steel slabs ("momentum traps"). The arrangement was impacted by a flyer plate using a high explosive charge with a plane wave generator

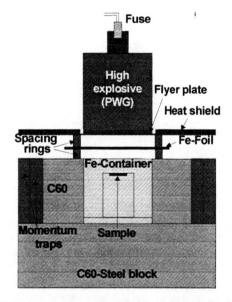

FIGURE 1. Schematic experimental setup for the shock recovery experiments at different pressures and temperatures.

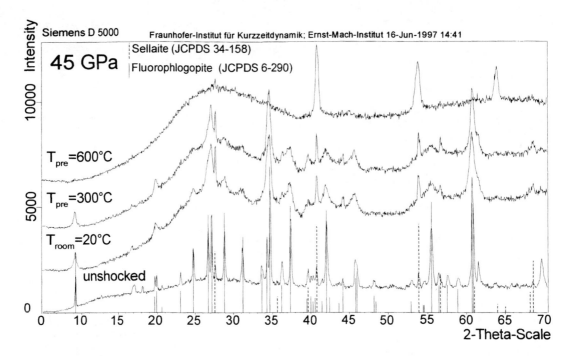

FIGURE 2. Variation of X-ray diffraction patterns for the experimentally shocked Macor-samples at 45 GPa with temperatures of 20, 300 and 600°C. Ni-filtered CuKα radiation.

(PWG). The transmitted shock wave traveled through the upper container wall into the specimen getting the full peak pressure by multiple reflections inside the Macor discs (reflection method). For experiments with pre-heated samples the steel block and the container with the specimen was externally heated in a muffle furnace. Detailed descriptions of the experimental setup are given in (4, 5, 6).

EXPERIMENTAL RESULTS

X-ray diffraction

The X-ray powder diffraction of the unshocked Macor material indicates fluorophlogopite as the main and sellaite (MgF_2) as the second crystalline phase. The small amorphous underground is caused by the glass matrix. The specimen shocked at 30 GPa shows clearly all fluorophlogopite peaks, however by increasing pre-shock temperatures the peak intensities are reduced. This can also be observed on the samples shock-loaded at 45 GPa.

Furthermore, some smaller fluorophlogopite peaks disappear, the reflexes are broadened and the amorphous part rises (Fig. 2). These effects can also be observed on shocked quartz (4), biotite (5) and olivine (6). At T_{pre} of 600°C nearly all fluorophlogopite peaks have vanished and the pattern of sellaite is dominant. This indicates that the shock wave completely transforms the fluorophlogopite crystallites into amorphous state, whereas the sellaite is stable under these shock loading conditions. The higher intensity of the peaks show that this phase has grown. The 60 GPa shocked samples show the same features but the intensities of all peaks were diminished. At 75 GPa the material is almost transformed in an amorphous state and only a few small peaks of sellaite can be observed.

SEM-characteristics

Scanning electron microscopy (SEM) has been used for microstructural study. The 30 and 45 GPa at T_{room} shocked samples indicate that the original

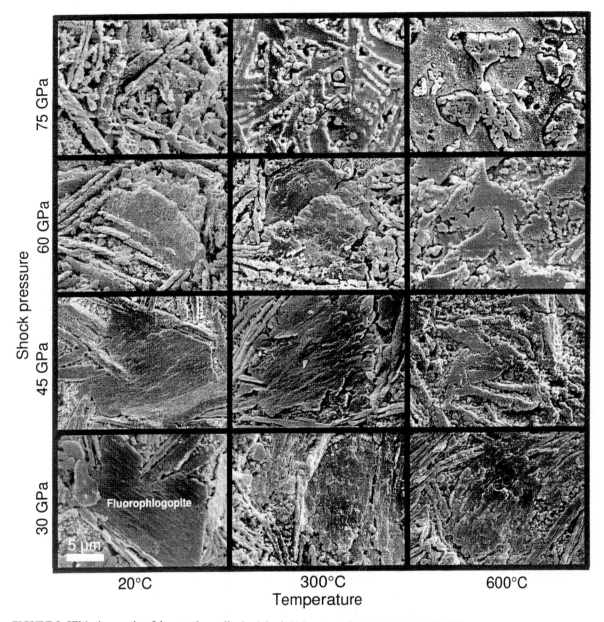

FIGURE 3. SEM micrographs of the experimentally shock-loaded Macor specimen (etched 30 s in 2% HF).

morphology of the crystallites is almost preserved except a kind of lamellar structure on the surface (Fig. 3). This structure seems to be kink bands which are the most characteristic shock effects in sheet silicates (7, 8). With further increasing peak pressures, however, the morphology of the crystals gradually disintegrates and finally the crystals are totally wrecked. The samples shocked at 300°C generally show a higher damage than the samples shocked at T_{room}. In addition, tiny melting spheres formed on the fluorophlogopite surface and at 75 GPa recrystallization effects can be observed. This indicates that the material was molten. At T_{pre} of 600°C the original morphology could be only observed in the 30 GPa shocked sample. With increasing pressure the crystallites gradually trans-

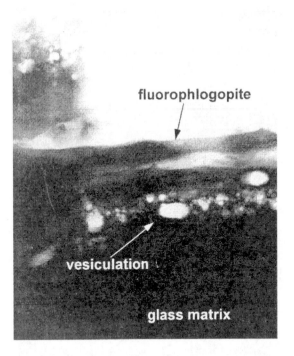

fluorophlogopite

vesiculation

glass matrix

FIGURE 4. Bright field TEM micrograph of Macor shocked at 30GPa and T_{room} (11500x).

formed into a glassy state. Finally the whole morphology is lost.

TEM-observations

Double-sided polished thin sections of the shocked 30 GPa and 45 GPa specimen have been studied with transmission electron microscopy (TEM). Because micas can easily be damaged by electron irradiation (9) the samples had been examined at low magnification and electron doses. The 30 GPa shocked specimen show vesicular parts between the interface of the fluorophlogopite crystals and the glass matrix (Fig. 4). At higher pressure and pre-shock temperature the vesiculation increases and can be found mainly in the glass matrix. Only in the 45 GPa and 600°C pre-heated sample tiny recrystallized sellaite grains could be identified and the electron diffraction pattern of the fluorophlogopite relic shows diffuse scattering due to the transformation in an amorphous state. In the investigated 30 and 45 GPa pre-heated samples no so-called planar deformation features (PDF) could be observed. PDFs (9) usually appear as fine lamellea with a spacing of about 2-10 µm in shocked minerals.

CONCLUSIONS

The shock recovery experiments indicate that with increasing pressures and pre-shock temperatures the structure of the fluorophlogopite is gradually deformed and finally the initial morphology is totally lost. In contrast to this the structure of sellaite under the same shock loading conditions is more stable. But at 75 GPa and T_{pre} of 600°C a condition is achieved where the two phases identified by X-ray to be crystalline components of the Macor glass-ceramic are transformed into an amorphous state. This state is reached because of microstructural deformations in combination with melting processes caused by shock wave and pre-shock temperature. The X-ray investigations also show that at T_{pre} of 300°C and more significantly at 600°C the intensity of the sellaite peaks increases mainly in the 45 and 60 GPa samples. The SEM and TEM observations illustrate that the fluorophlogopite was molten and so a fluorine-rich melt was produced. It is assumed that new sellaite is crystallized from this melt (10).

REFERENCES

1. Grossman, D. G.; Taylor, D. L., *NBS Special Pub.*, 562, (1979) 221-229.
2. Boyd, D. C.; Danielson, P. S.; Thompson, D. A. in *Encyclopedia of Chemical Technology,* (ed.) Howe-Grant, M., **12**, John Wiley & Sons, New York (1994) 627-644.
3. Harrington, S. N.; Brewer, J. A.; Pederson, D. O., *J. Appl. Phys.*, **51** (1980) 2043-2046.
4. Langenhorst, F.; Deutsch, A., *Earth and Planetary Letters*, **125** (1994) 407-420.
5. Schneider, H.; Hornemann, U., *N. Jb. Miner.* Mh H.3/4 (1974) 149-162.
6. Müller, W. F.; Hornemann, U., *Earth and Planetary Letters*, **7** (1969) 251-264.
7. Hörz, F.; Ahrens, T. J., *Amer. J. Sci.*, **267** (1969) 1213-1229.
8. Stöffler, D. *Fortschr. Min.* **49**, 50-113.
9. McLaren, A. C., *Transmissions Electron Microscopy of Minerals and Rocks*, Cambridge Univ. Press, Cambridge (1991).
10. Dalal, K.; Davies, R. F., *Ceram. Bull.* **56** (1977) 991-997.

CP429, *Shock Compression of Condensed Matter – 1997*
edited by Schmidt/Dandekar/Forbes
© 1998 The American Institute of Physics 1-56396-738-3/98/$15.00

SHOCK COMPRESSION RECOVERY EXPERIMENTS ON SOME DIOXIDES

D. Vrel*, X. S. Huang and T. Mashimo

Faculty of Engineering, Kumamoto University, 2-39-1 Kurokami, Kumamoto 860, JAPAN
*Present address : CNRS-LIMHP, Avenue J.-B. Clément, 93430 Villetaneuse, FRANCE

The AX_2 materials are known to undergo phase transitions yielding products with a high coordination structure, a high bulk modulus, and possibly a high hardness. These products may be quenchable at ambient pressure and may then be interesting in material sciences as well as in earth sciences. Shock-compression recovery experiments were performed on some dioxides (ruthenium, hafnium, lead, etc.) using a powder gun to synthesize their high pressure phases. High pressures phases have been successfully produced for several products.

INTRODUCTION

Synthesis of ultra hard materials generally involves the use of small atoms, such as B, C, N, O, with the hope of synthesizing a product with a high density of strong, covalent bonds. The hardest known compounds are in this category, with diamond, cubic boron nitride, boron oxide, and the still un-synthesized C_3N_4. Recently, a special attention has been focused on AX_2 compounds, which can undergo phase transitions under high pressure, yielding products with high coordination numbers and a very dense structure. Interesting results have been obtained on silicon, zirconium, ruthenium and hafnium dioxides, which have proven to possess high pressure phases whose hardnesses are or should be close to the one of diamond. Two of them, zirconium and hafnium dioxides have high pressure phases whose bulk modulus are even higher than the one of diamond (2), but unfortunately, their hardnesses have not been yet measured.

All these previous results have been obtained by static compression, by means of diamond anvil cells or multi anvil cells. The aim we had when we started this research was to find a way to produce these phases in an amount that could make them interesting for an industrial use. Therefore, dynamic compression was seen to be a promising way, as the volume recovered can be several orders of magnitude greater with such techniques.

This paper is a brief presentation of some intermediary results of shock compression recovery experiments concerning hafnium dioxide (HfO_2), ruthenium dioxide (RuO_2), lead dioxide (PbO_2), etc., which were previously studied by means of diamond anvil cells by J.-M. Léger et al. (see Ref. (2) for HfO_2, Ref. (3,4) for RuO_2 and Ref. (5) for PbO_2).

EXPERIMENTAL

Shock compression recovery experiments were performed using a powder gun (1). The samples were prepared from commercial powders of RuO_2 (99.9%, Aldrich Chem. Co.), HfO_2 (99.95%, Rare Metallic Co.), PbO_2 (99.9%, Rare Metallic Co.). These powders were pre-compacted by uni-axial pressing to a 1 mm thick and 12 mm in diameter disk, then placed into an iron capsule, and finally impacted by a 1.5 to 2 mm thick tungsten flyer plate, at a velocity from 1.0 to 1.8 km/s. These samples were made either of the pure dioxide powder or of a mixture with copper (99.99%, Rare Metallic Co.), using 2 volumes of copper for one of the oxide. The use of copper allowed to reduce the initial porosity, to increase the samples impedance and because copper possesses a high heat capacity and a good thermal conductivity, to decrease the extent and the duration of the heat and thus to potentially increase the stability of the high pressure phases.

After impact, these samples were peeled out from the iron capsules using a lathe and then analyzed by

X-ray powder diffraction, using a Cu-Kα_1 radiation.

RESULTS AND DISCUSSION

Hafnium dioxide

When pure hafnium dioxide was used, a distinctive peak from an orthorhombic phase, named phase-II in Ref. (2), was observed at a distance of 2.820 Å with a relative intensity of 47, compared to the 100 attributed to the most intense peak of the pattern (Fig. 1). Some of the other peaks can also be observed but are not as significant, as they are always mixed with the numerous peaks of the monoclinic phase.

When the pressure was increased, the intensity of this peak decreased sharply. However, some faint traces of the cotunnite-type phase (named phase-III in (2)) can be observed. This result is unfortunately not very clear, as the diffraction patterns are quite noisy. Experiments are currently under study in order to confirm this result.

Using a mixture with copper, no decisive proof of a phase transition could be obtained on a diffraction pattern. The fact that the conversion ratio was really observable only with pure hafnium dioxide is a good indication that the temperature helps the transition, either in a thermodynamic way (such as a negative P-T slope) or in a kinetic way.

Ruthenium dioxide

When pure ruthenium dioxide was used, no new peaks could be observed on the diffraction pattern. Shock compression recovery experiments never showed any change on the X-ray pattern, beside a dehydration, even on heat treated powder (500°C, 15h, under vacuum, to dehydrate the powder prior to the experiments, as the as-sold powder present extremely wide peaks), irrespectively of what the shock velocity was (between 1.0 and 1.8 km/s).

The results were very different when a mixture with copper was used, as the CaCl$_2$-type phase was observed for all experimental conditions. This transformation was complete, as no peaks were remaining from the low pressure phase of ruthenium dioxide.

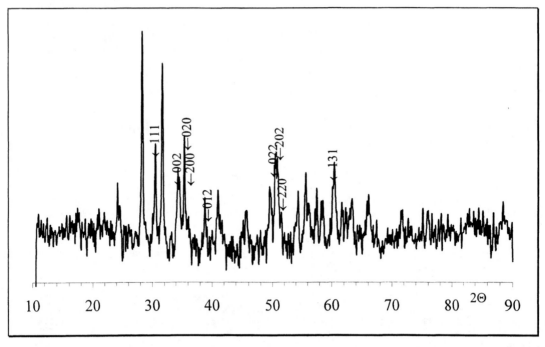

FIGURE 1: diffraction pattern of shock recovered hafnium dioxide (1.382 km/s, W impactor, 1.5mm thick). Arrows represent the expected positions of the peaks of the orthorhombic phase. 111 is the only peak not mixed with a monoclinic one.

We tried, by increasing the pressure, to produce the distorted fluorite structure, but we never found a proof of its presence within our samples. It seems that during static compression experiments a very long time was used, and an important overloading was used to increase the phase transition kinetics. Therefore, the kinetics of the phase transition is probably too slow for the phase to be produced by shock compression.

The complete lack of any phase transition for pure RuO_2 is more difficult to account for. Knowing the different effects of copper, it is possible that the pressure-temperature conditions were quite different for the two types of samples and pure ruthenium dioxide may have been submitted in all cases to conditions where the $CaCl_2$ phase was not stable.

Lead dioxide

The samples were prepared from a PbO_2 powder which was initially made of ~90% of PbO_2-β and ~10% of PbO_2-α, as roughly estimated from the X-ray diffraction patterns. After shock compression, new peaks had appeared on the pattern, with intensities that could reach 20% of the most intense peak of the pattern. These intensities increased when copper was not used during the preparation of the sample, the new phase presenting then the most intense peaks of the pattern (Fig. 2). Unfortunately, these peaks cannot be correlated to any of the six previously determined structures of PbO_2 (5), nor to a combination of them. As the pressures obtained during these dynamic experiments was not higher than the maximum of 47 GPa reached in ref. (5), we cannot explain these results by a new post-cotunnite structure. Moreover, contrarily to what has been obtained previously by Kusaba *et al.* (7), also using shock compression, these peaks do not match any of the other lead oxides found in the JCPDS cards (PbO-litharge, PbO-massicot, Pb_3O_4, $PbO_{1.37}$, $PbO_{1.44}$, Pb_2O_3, $PbO_{1.55}$, $PbO_{1.57}$ and $Pb_{12}O_{19}$)

Contrarily to what had been obtained with static compression, where PbO_2-α was the only low pressure phase to be present after decompression, the shock recovered samples show only the high pressure phase previously mentioned mixed with PbO_2-β.

Because of the low bulk modulus of the different high pressure phases, ranging from 141 to 223 GPa, lead dioxide is not by itself an interesting material for mechanical applications. However, because it is the dioxide with a rutile structure at ambient pressure with the largest cation, it is expected to be a model for the other dioxides of the same family, showing the same phase transitions at lower pressures (5).

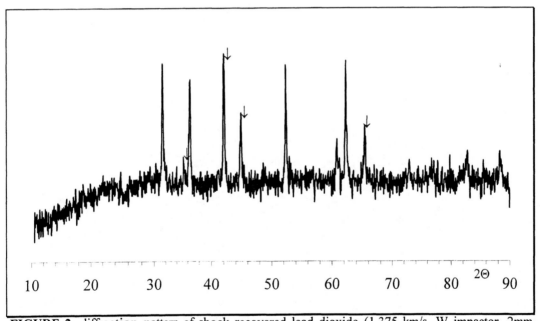

FIGURE 2: diffraction pattern of shock recovered lead dioxide (1.375 km/s, W impactor, 2mm thick). Arrows point to the unidentified peaks.

CONCLUSIONS

It must be pointed out that, with the exception of ruthenium dioxide, the effect of copper was not favorable to the converted ratio of the high pressure phase. It seems that high temperatures are, in most cases, required for either going deeper in the domain of stability of the high pressure phases or to increase the kinetics of the phase transition.

Most of these materials are still under study and more complete results will be published elsewhere, with the results on some other dioxides that we recently started to investigate.

ACKNOWLEDGMENTS

This work was performed at the High Energy Rate Laboratory while D.V. was a post-doctorate researcher, supported by the French Foreign Affairs Ministry (Lavoisier Fellowship).

The authors would like to thank J. Haines and J.-M. Léger for their helpful discussion and E. Ito for his help concerning the static experiments.

REFERENCES

1. Mashimo, T., Ozaki, S., and Nagayama, K., *Revue of Scientific Instruments*, **55**, 226-230 (1984).
2. Léger, J.-M., Atouf, A., Tomaszewski, P. E., and Pereira, A. S., *Physical Review B*, **48**, 93-98 (1993).
3. Haines, J., and Léger, J.-M., *Phys. Rev. B*, **48**, 13344 (1993).
4. Haines, J., Léger, J.-M., and Schulte, O., *Science*, **271**, 629-631 (1996).
5. Haines, J., Léger, J.-M., and Schulte, O., *J. Phys. : Condens. Matter*, **8**, 1631-1646 (1996)
6. Vrel, D., Petitet, J.-P., Huang, X., and Mashimo, T., *Physica B*, **239** (1997). To be published.
7. Kusaba, K., Fukuoka, K., and Syono, Y., *J. Phys. Chem. Solids*, **52**, 7, 845-851 (1991)

CP429, *Shock Compression of Condensed Matter – 1997*
edited by Schmidt/Dandekar/Forbes
© 1998 The American Institute of Physics 1-56396-738-3/98/$15.00

PREPARATION OF METASTABLE ALLOY BULK MATERIAL IN Fe-Cu SYSTEM BY MECHANICAL ALLOYING AND SHOCK COMPRESSION

X. S. HUANG, M. ONO and T. MASHIMO

High Energy Rate Laboratory, Faculty of Engineering, Kumamoto University
Kurokami 2-39-1, Kumamoto 860, Japan

Metastable alloy bulk body including solid solution in iron (Fe)-copper (Cu) system was prepared by mechanical alloying (MA) and shock compression. The MA-treated Fe-Cu system (50:50 in mol%) powder showed an X-ray diffraction pattern of a single phase of face-centered cubic (FCC) structure. The lattice parameter of FCC structure of the MA-treated powder was larger than the one of pure copper. The X-ray diffraction pattern of the shock-consolidated bulk body did not much change from the one of the MA-treated powder. Instrumental chemical analyses revealed only slight changes in compositions by the MA-treatment and shock-compression.

INTRODUCTION

Mechanical alloying (MA) has recently been used for material processing methods: preparation of amorphous phases, nonequilibrium solid solution phases, nanocrystalline phases, high pressure phases, and other phases (1-4). The iron (Fe)-copper (Cu) system does not form intermetallic compounds, and has negligible mutual solid solubility in equilibrium at temperatures below 700℃ (5). The preparation of metastable solid solutions in Fe-Cu system by vapor deposition (6-8) and by ion beam mixing method (9) have been reported. Several studies by MA have been carried out in this system and the preparation of metastable solid solutions have been reported (10-12). The bulk body of metastable solid solution in Fe-Cu system has not been yet prepared.

On the other hand, shock compression has different features from static compression: pulsed short duration, shear stress, heterogeneous state, etc, and can be used as a consolidation method of nonequilibrium materials without recrystallization or

decomposition (13-15). It is important to consolidate nonequilibrium powders for valuations of physical properties and for industrial applications. In this study, the MA treatment and shock compression recovery experiments were performed to prepare metastable solid solution bulk bodies in Fe-Cu system.

EXPERIMENT

Starting powders were provided by Rare Metallic Co., Ltd. These iron and copper powders consisted of irregular particles of 4-7 μm and 325 mesh (<44 μm) in diameter with a total impurity value of less than 0.5 and 0.01 wt%, respectively. The starting powder mixture was prepared by mixing iron and copper elemental powders (50:50 in mol%).

MA experiment was carried out by using the planetary micro ball mill (P-7 of Fritsch Co., Ltd.) in an argon atmosphere (4). A mill capsule with an inner-diameter of 41 mm and a depth of 38 mm and balls with a diameter of 5 mm were used, which were made of silicon nitride (Si_3N_4) and zirconia (ZrO_2), respectively. The powder specimen with a weight of about 20 grams and 200 ZrO_2 balls were contained into the capsule with a ball-powder weight ratio of 4:1. The rotation speed of the ball mill was 2840 rpm. The resultant acceleration was estimated to be about 12 g (gravity). The milling duration was 21 hours, and small amounts of powder were recovered for several different durations of the MA treatment for X-ray analyses.

Shock-compression recovery experiment was conducted using a propellant gun (16). The MA treated powder samples were enclosed in a brass capsule (Cu:Zn=70:30 in wt%) with an inside diameter of about 12 mm and with an inside height of about 3.7 mm. The porosity of powder was about 55 %. Shock loading was carried out by impacting the capsule with an aluminum flat flyer plate whose thickness was 3 mm.

The MA-treated and shock-consolidated specimens were investigated by powder X-ray diffraction (XRD), instrumental chemical analysis and Electron Probe Micro Analysis (EPMA). Powder XRD analyses were carried out using monochromatized Fe-Kα radiation with a Rigaku Goniometer. Calibration of the goniometer was performed by measuring diffraction peaks of pure silicon powder mixed in specimen. The lattice parameter was calculated by least squares method. Instrumental chemical analyses of zirconium (Zr), silicon (Si), carbon (C), oxygen (O) and nitrogen (N) contents were done using the SPS-1200 of Seiko Electric Co., Ltd., the WR-112 and the TC-436 of LECO Co., Ltd. The analyses of the Fe and Cu content distribution were carried out for the shock-consolidated body using the EPMA apparatus, JXA-8900 of JEOL LTD.

RESULTS AND DISCUSSION

Figure 1 shows the powder XRD patterns of the starting powder, the MA treated powder for 21 hours and the shock-consolidated bulk body with an impact velocity of 1.26 km/s. The driving pressure in the capsule was calculated to be 14.9 GPa. For the MA treated sample, the XRD peaks of iron (body-centered

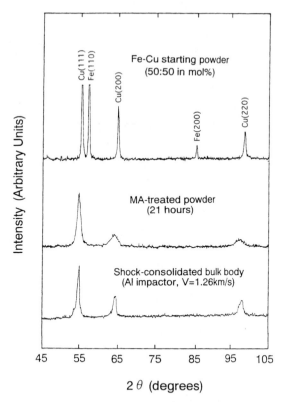

FIGURE 1. Powder XRD patterns of the starting powder (Fe:Cu=50:50 in mol%), the MA-treated powder for 21 hours and the shock-consolidated bulk body.
[17] © 1997 SPIE. Reprinted with permission.

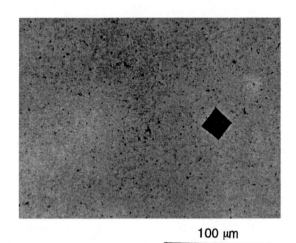

FIGURE 2. Photograph of the polished surface of the shock-consolidated bulk body. Rectangular-shaped hole is a trace of the indentation made for Vickers hardness measurement.

cubic (BCC) structure) disappeared, while the peaks of copper (face-centered cubic (FCC) structure) became broader. In addition, the peaks of the FCC structure shifted to smaller angles. The lattice parameter of the FCC structure was determined to be 0.36400 nm, which is larger than that of pure copper of 0.36153 nm. The nearly equal metallic radii of iron (0.127 nm) and copper (0.128 nm) indicate that the expansion of the FCC structure lattice with an amount of iron atoms disolving in is attributable to magnetovolume effects (10). The XRD pattern at the surface of the shock-consolidated bulk body did not much change from that before shock compression.

Figure 2 shows the photograph of the shock-consolidated bulk body. No large crack can be observed. An almost mono-morphology single phase can be observed over the surface in the figure. The Vickers hardness (300 grams in weight, 15 s in duration time) of the bulk body was measured to be 5-6 GPa, which is much higher than the hardnesses of pure iron and copper polycrystals.

The chemical analysis results on Zr, Si, C, O and N of the MA-treated powder after 21 hours of milling were 1.45, 0.44, 0.20, 1.85 and 1.98 (in wt%), respectively. The compositions of the MA-treated powder changed only slight due to wear debris from the milling tools (Si_3N_4 mill capsule and ZrO_2 balls). The chemical analysis results of C, O and N of the shock-consolidated bulk body were 0.45, 1.84 and 1.93 (in wt%), respectively, which did not much change from those of the MA-treated powder.

The Fe and Cu content distributions were investigated by the EPMA at a polished surface of the bulk body. It was confirmed that Fe and Cu dispersed well within submicron level in the bulk body.

As a conclusion, the metastable alloy powder and bulk body including solid solution in Fe-Cu system (50:50 in mol%) was prepared by mechanical alloying and shock compression, respectively. This nonequilibrium bulk material is expected to show advanced physical properties.

ACKNOWLEDGMENTS

The authors would like to acknowledge Japan New Metals Co., Ltd. for the supports in the instrumental chemical analysis. This work is supported by a Grant in Aid from the Amada Metal Forming Mechanical Technology Foundation.

REFERENCES

1. Mizutani, U., and Lee, C. H., *J. Mater. Sci.* **25**, 399 (1990).
2. Kimura, H., and Takada, F., *Mat. Sci. Eng.* **97**, 53 (1988.)
3. Ma, E., and Atzmon, M., *Mat. Sci. Forum*. **V88-90**, 467 (Trans. Tech. Publication, Switzerland, 1992).
4. Mashimo, T. and Tashiro, S., *J. Mat. Sci. Lett.* **13**, 174 (1994).
5. Hansen, D. P. M., *Constitution of Binary Alloys*., published by Genium Publishing Corporation, P.580 (1985).
6. Kneller, E. F., *J. Appl. Phys.* **35**, 2210 (1964).
7. Sumiyama, K., and Nakamura, Y., in Rapidly Quenched Metals, edited by Steeb, S., and Warlimont, H., (Elsevier, Netherland, 1985), P. 859; *Acta Metall*. **33**, 1791 (1985).
8. Chien, C. L., Liou, S. H., Kofalt, D., Yu, W., Egami, T., and McGuire, T.R., *Phys. Rev. B*. **33**, 3247 (1986).
9. Huang, L. J., and Liu, B. X., *Appl. Phys. Lett.* **57**, 1401 (1990).
10. Yavari, A. R., Desre, P. J., and Benameur, T., *Phys. Rev. Lett.* **68**, 2235 (1992)
11. J. Eckert, J. C. Holzer, C. E. Krill III, and W. L. Johnson, *J. Appl. Phys.* **73**, 2794 (1993).
12. Ma, E., Atzmon, M., and Pinkerton, F. E., *J. Appl. Phys.* **74**, 955 (1993).
13. Yamasaki, T., Ogino, Y., Morishita, K., Fukuoka, K., Atoh, T., and Syono, Y., *Mat. Sci. Eng.* **A179/180**, 220 (1994).
14. Mashimo, T., Tashiro, S., Hirosawa, S., and Ikegami, T., *Trans. Mat. Res. Soc. Jpn.* **14B**, 1079 (1994).
15. Mashimo, T., Huang, X. S., and Tashiro, S., *J. Mater. Sci. Lett.* in press.
16. Mashimo, T., Ozaki, S., and Nagayama, K., *Rev. Sci. Instr.* **55**, 226 (1984).
17. M. Nakamura, M. Uchino, and T. Mashimo, " Measurement of Shock Waves in Nonmetals with a VISAR System," **SPIE - 2869**, 22nd International Congress on High-Speed Photography and Photonics (1997).

CP429, *Shock Compression of Condensed Matter – 1997*
edited by Schmidt/Dandekar/Forbes
© 1998 The American Institute of Physics 1-56396-738-3/98/$15.00

On the Possible Gas Detonation Explosion of BaO$_2$/Zr Powder Mixture

E.A. Dobler, A.N. Gryadunov, and A.S. Shteinberg

Institute of Structural Macrokinetics,
Russian Academy of Sciences, Chernogolovka, 142432, Russia

Possible detonation modes arising due to the formation of an intermediate gaseous phase are considered for the case of the BaO$_2$/Zr system. It appears that explosive destruction of a steel matrix observed at compaction of the BaO$_2$/Zr composition cannot be explained by the system combustion in a closed volume. The mechanism of shock-wave destruction at detonation of a capsule containing the BaO$_2$/Zr mixture is proposed for interpretation of the phenomenon observed.

INTRODUCTION

It has been known since the 1940's, that under conditions of high shear deformations [1] it is possible to initiate in solids chemical processes, which cannot be obtained otherwise, and that these processes are extraordinary fast. Thus, the thermite of Fe$_2$O$_3$ and Al powders placed in Bridgeman's anvil undergoes detonation under static pressure of 2Gpa [1]. To provide it the mass transport mechanisms must be 5-10 order of magnitude faster than the common diffusion mass transport mechanisms, which define powder combustion process at normal conditions [2]. The shock loading of condensed systems creates high shear deformation states directly in the shock front. This implies that transport mechanisms of chemical processes must also be extremely accelerated in this case. This statement has been confirmed in recent Real-time studies of shock-induced chemical reactions in solids [3-5], that showed that some amount of the starting mixture could react within an extremely short period of several hundred nano-seconds. This value is as small or even smaller than the reaction time of common explosives. If the thermodynamical criterion of detonation is satisfied, and the total reaction time is not much greater than the values noted above, then the critical diameter at which unsupported detonation

for such systems is possible will slightly differ from these values for common explosives (some mm). It means that some powder mixtures, in which the fuel and oxidant agents are enclosed in different particles, can provide detonation similar to detonation of common explosives, in which the fuel and oxidant are enclosed in the same molecule. This type of detonation has been established for some perchlorate/aluminum mixtures [6,7].

Thus, two rather different detonation effects in powder mixtures have been obtained under the conditions of high shear deformation:

a) explosions in the Bridgeman anvils (systems can be gas-producing as well as non-gas-producing);

b) explosion of gas-producing aluminum/ perchlorate mixtures under normal conditions.

Unsupported detonation in non-gas-producing mixtures under normal conditions has not been obtained yet, but it is now of great interest and some reports devoted to this problem were published recently [8-11].

The current report presents experiments, which advance a new detonation-capable powder mixture. The detonation process in this system is possible under the conditions, which are someway intermediate between those described in items (a) and (b) above. That is a BaO_2/Zr powder mixture. This system is a nongasproducing and would not detonate under normal condition. However, under conditions of pressing these systems undergoes of explosion, which, as will be shown below, must be of detonation nature. In this case, the pressure constitutes significantly smaller loading than that developed in Bridgeman's anvil. Also, as it will be shown that this detonation must be a gas detonation. Consequently, we have a slightly pressure loaded nongasproducing system, which undergoes a gas detonation explosion under these conditions.

EXPERIMENT

Compaction of the barium peroxide and zirconium powder mixture was performed in hard steel press-forms with 20mm diameter cavity for the substance. The tests showed that the press-forms were capable to withstand a short-time loading up to $2 \cdot 10^4$ kg at compaction of solid powders, which corresponds to 600Mpa. The working inner pressure was defined by the properties of steel and can be estimated not less than 300MPa. The quantity of powder being compressed was 20-50g. The mixture was pressed under pressures not greater than 100MPa up to porosity of approximately 48-50%. Instant ignitions of the mixture were accompanied by explosion, whereupon pressforms were disrupted into shards. The effect is very like the blast of common explosives with the use of somewhat smaller amount of explosive substance. It needs to utilize the energy provided by chemical process of major part of the sample in the form of mechanical work. Also, we rule out the possibility that the observed explosions were due to the build-up of heat tensions in the pressform material and consequential expansion of hot products of combustion. Indeed, a steel wall, which is thick enough to evolve significant tensions in the pressforms, takes at least several hundred milliseconds to heat. Since the reaction time for such systems is about 10^{-2}-10^{-3}s [12] and the velocities of combustion front propagation under pressure can be some centimeters per second [13],

depth of combustion front would be not greater than 1mm. The system will have enough time for whole completing behind the combustion wave before the matrix rupture, while the mixture ahead of combustion wave remains inert. Thus, in a combustion process at any moment the explosions would be accomplished by expansion of barium vapors in reacted part of the sample and by the very insignificant amount of revealed from combustion front free oxygen. Because of the feeble working capacity of these agents (the pressure will be not greater than 5Mpa for this case), this mechanism can not provide the effects observed. The pressform destruction due to the formation of free oxygen at some stage of the chemical process throughout whole sample seems to be a much more relevant explanation. Whereat developed pressures will be at least tens times higher. Also, the work of explosion will grow respectively. Finally, the process cam close to that we should have in case of habitual explosives and this has been in fact observed in the experiments. The possible way to achieve that is the almost simultaneous ignition of the whole sample in a shock wave. It means that it were a detonation explosions.

ON THERMODYNAMIC CRITERIA OF DETONATION

The possibility of unsupported detonation (below referred as 'detonation') systems in the most general form is determined by a simple equation [14]:

$$Q_{pv} > 0, \qquad (1)$$

where Q_{pv} is the heat of the reaction at constant initial volume and pressure of the reactants. This condition is commonly treated as the requirement for the volume growth of the adiabatically reacting substance at the starting pressure. Generally, this statement is implemented for starting and final states of substance. Yet, for a system having a complicated dependence of volume on conversation depth, this criterion must be applied to all states realized in the chemical process. If eq.(1) is valid for some intermediate states, the detonation is possible in the system and thermodynamic properties of these states must be considered in order to determine the possible detonation modes.

Equation (1) can be contented by creating appropriate conditions, for example by applying static pressure. So, for Fe_2O_3/Al thermite, equation (1) is not valid at normal conditions [8], but it becomes valid, when the system is under pressure in the Bridgeman anvils. Some other nongasproducing systems can provide enough mechanical energy to satisfy eq.(1) at normal conditions. Since the greater portion of the mechanical work produced is performed by thermal expansion of condensed products, detonation in such systems can be defined as 'gasless' [9] or 'heat' detonation [8]. Of course, most of these mixtures are extremely high-energy systems such as pyrotechnic metal - metal oxide systems [10, 15]. But, for the extremely energetic reactions of oxygen exchange it is plausible to have free oxygen as an intermediate product. It means that the gas detonation mode is possible in these systems and the properties of gas containing intermediate states should be considered as they can ensure actual detonation.

THERMODYNAMIC EVALUATION OF ISENTROPES AND SHOCK ADIABATES

The calculated isentropes (dashed lines) and shock adiabates (solid lines) of some intermediate states of BaO_2/Zr are shown in Figure 1. All calculations are made on the assumption that the density of all condensed products is the same at all pressures. The error introduced by this assumption is insignificant under pressures of some hundred MPa. The pressures in Figure 1 are equilibrium pressures calculated for the given values of the specific volume and conversion depths (0.34, 0.37 and 0.4). The share of zirconium involved into the chemical process is referred to as the 'depth of conversion'. Hence, labels 0.34, 0.37, and 0.4 (Fig. 1) indicate that the curves were plotted assuming that 0.66, 0.63 and 0.6 moles of zirconium were inert throughout the process and all other constituents reacted at given volumes until they achieved their equilibrium states. For the shock adiabates calculation the inner energy of the system was increased by the energy contributed by the

shock wave and calculated using the Rankee-Hugoniot equation:

$$E_h = (P + P_0)(v_0 - v)/2, \qquad (2)$$

where E_h is an energy increase in the shock wave, $P_0 = 10^5 Pa$; P is the equilibrium pressure; v_0 and v are the specific starting and postshock volumes respectively.

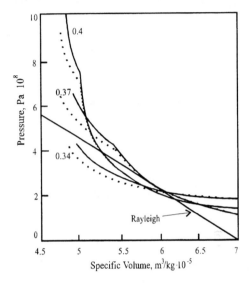

Fig. 1. Isentropes and shock adiabates of a partially reacted BaO_2/Zr mixture.

The volume v_0 was calculated for 40% porosity BaO_2/Zr mixture. As it seen from Figure 1, at the pressures achieved the difference between isentropes and shock adiabates is small. The thermodynamic functions for the components used in the calculations were found in [16]. It was supposed that free energies of the condensed specimens were pressure-independent. The highest and lowest values of the specific volume (Fig. 1) correspond to the 40% and zero porosity, respectively. The inflection points are determined by a change in the mechanism, which leads to the establishment of equilibrium pressure. At larger specific volumes, all non-reacted BaO_2 is decomposed to BaO and O_2, while in the opposite case, oxygen enters the reaction with BaO.

From the data presented in Figure 1, it can be concluded that the pressures achieved in the BaO_2/Zr mixture through the chemical process are not sufficiently high to provide destruction of the hard steel matrices. The detonation mechanism of the process seems to be much more suitable for the effect interpretation. In this case, the detonation parameters can be calculated from the properties of the intermediate state, which has the highest position of the shock adiabates in the pressure-specific volume (P-v) diagram of all states formed in the chemical process (i.e. the detonation velocity calculated for this detonation adiabate has the largest value). For the BaO_2/Zr system of 40% porosity such a detonation mode can proceed according to the Rayleigh line drawned on Fig.1.

CONCLUSION

Gas detonation according to the Rayleigh line tangent to some intermediate state shock adiabates can be expected in some high energetic metal peroxide-metal systems. It is much probable that such mode was obtained in the BaO_2/Zr mixture under static pressure of 100MPa. It does not imply that these systems can provide unsupported detonation at normal condition. Possibly, specific conditions of shear deformation in the closed chamber of steel pressform result in satisfying of the thermodynamical and kinetic criterion of detonation. Nevertheless, such type of detonation can be predicted for a high energetic reaction of oxygen exchange in powders. It must be taken into account when such systems are under treating for their shock or detonation properties.

ACKNOWLEDGMENT

The present study was supported by the Russian Foundation of Fundamental Research (Grant №96-03-32703a).

References

1. Bridgeman P.W., J.Chem. Phys. 1947, Vol.15, pp. 311-313.
2. Enikolopyan N.S, Dokl.Ak.Nauk, 1985, vol..283, №4, pp. 887-899.
3. Boslough M. B., Shock Compression of Condensed Matter, 1991, p. 617-620.
4. Gryadunov A.N., Shteinberg A.S., and Dobler E.A., Dokl. Akad. Nauk, 1991, v. 321 (5), p. 1009-1013.
5. Gogulya M.F., Voskoboinikiov I.M and oth.,Chimicheskaya Fizika, 1991, vol.10, №3, pp.420-422.
6. Belyaev A.F., Nalbandyan A.B., Dokl.Akad.Nauk, 1945, v.46, pp. 113-116
7. Koldunov S.A. Proc. of 11'th Symposium on Combustion and Explosion, Chernogolovka, Russia, 1996, v.2, pp.61-62.
8. Boslough M.B., J. Chem. Phys., 1990, v. 92 (3), p. 1839-1848.
9. Shteinberg A.S., Knyazik V.A. and Fortov V.E. Dokl. Akad. Nauk, 1994, v. 336 (1), p. 71.
10. Dobler, A.N. Gryadunov, A.V. Utkin, and V.E. Fortov., Proc. Intl. Conf. The Current State and Future of High Pressure Physics, Troitsk, Russia, Sept. 7-9, 1995, p. 22.
11. Gordopolov Yu.A., Trofimov V.S. and Merzhanov A.G. Dokl. Akad. Nauk, 1995, v .341 (3), pp. 327-329.
12. Shteinberg A.S., Knyazik V.A. Pure & Appl. Chem., 1992, v.64, №7, pp.965-976.
13. Ponomarev M.A., Shcherbakov V.A., Shteinberg A.S., Dokl.Acad.Nauk., v.340, .№5, pp.642-645.
14. Dremin A.N, Savrov S.D., Trofimov V.S., and Shvedov K.K., «Detonation Waves in a Condensed Matter», Moscow, 1970, 164 p.
15. Boslough M.B. Shock-Wave and High-Strain-Rate Phenomena in Materials (Eds. M. A. Meyers, L.E. Murr, and K.P. Staundhammer), N.Y., 1992, p. 253.
16. Glushko V.P., Gurvich L.V. and oth., «Thermodynamic Properties of Individual Substances», Moscow, 1982.

CP429, *Shock Compression of Condensed Matter – 1997*
edited by Schmidt/Dandekar/Forbes
© 1998 The American Institute of Physics 1-56396-738-3/98/$15.00

NEW EVIDENCE CONCERNING THE SHOCK-INDUCED CHEMICAL REACTION MECHANISM IN A NI/AL MIXTURE

Y. Yang, R. D. Gould, Y. Horie

North Carolina State University, Raleigh, North Carolina 27695

K. R. Iyer

U.S. Army Research Office, Research Triangle Park, North Carolina 27709

Experimental tests were conducted to study the effects of shock wave intensity, particle size and shock loading conditions on the exothermic chemical reaction in a Ni/Al powder mixture. A new 50 mm powder gun was used to reproduce and to extend the real-time observations of ultra-fast exothermic reactions in the Ni/Al powder mixture. Shock pressure-time profiles measured by a manganin gauge show that the threshold pressure for reaction is about 14 GPa for the Ni/Al mixture used here and that the initiation criteria based either on shock energy or melting are not supported by these experimental measurements.

INTRODUCTION

During the past few decades, shock-induced chemical reactions in powder and compound mixtures have been recognized as a potential physical means to synthesize new compounds and compounds with unique microstructures [1]. The high pressure, rapid material motion, intense plastic deformation (even in conventionally brittle materials), high temperature, and short duration of the shock event force material through a unique process. Various shock-compression experiments have shown that the energy from exothermic inorganic chemical reactions can be released in time scales shorter than the duration of the shock state [2]. At the present time, there is no generally accepted theory of such phenomena [3]. However, two predominant theories, the thermochemical theory and the mechanchemical theory, attempt to explain this phenomena. In the former, a sharp energy threshold condition is proposed as the criteria for the reaction [4]. In the latter, the model is based on mechanical concepts, such as plastic deformation, mass mixing, and relative velocity difference [2,5,6]. At present, there are no real-time measurements that can be used to evaluate these theories.

The goals of this research were to obtain some of the first pressure-time measurements with steady-state plateaus and to investigate the effects of powder particle size and shock wave loading conditions on the shock induced chemical reaction in a Ni/Al powder mixture. The time-resolved experiments were conducted using the 50 mm bore powder gun located in the Applied Energy Research Laboratory at North Carolina State University and were specifically designed to test these two theories and to further reveal the mechanisms of chemical reaction initiation.

FACILITY AND EXPERIMENTAL SETUP

Experiments were conducted using a 50 mm powder gun. The target arrangement and corresponding P-Up diagrams are shown in Fig. 1. A flyer plate impacts the front plate with a known velocity and sends a plane shock wave into the target. Two target configurations were used in an effort to determine the effect of shock loading condition on chemical reaction initiation. In both target configurations, all the plates are made of 304 stainless steel. The flyer, front and backing plates are 8.1 mm, 3.0 mm and 10.8 mm in thickness, respectively.

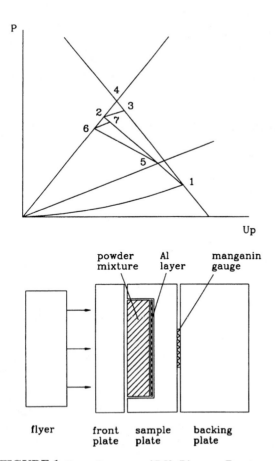

FIGURE 1. Target Structure and P-Up Diagrams. Target a has no Al layer; target (b) has an Al layer.

The sample plate for target (a) is 6.6 mm thick with a 3.1 mm deep pocket. Target (a) is used to reproduce the "excess pressure" results of Bennett [7], but with a larger target and thus a longer observation time. The difference between the computationally predicted shock pressure for the inert system (using mixture theory and assuming complete pore collapse during the incident shock), and the peak measured shock pressure is defined as "excess pressure". This measured "excess pressure", used in conjunction with simultaneously conducted recovery tests, which show products of reaction, indicate that rapid exothermic reactions occur during the shock process.

For target (b), the sample plate has a thickness of 6.6 mm with a pocket 3.35 mm in depth and a thin layer (0.254 mm thick) of Al is placed between the sample mixture and the pocket surface [8]. Target (b) is used to study the effect of the shock loading condition (i.e. pressure-time history) on the chemical

reactions. The thin Al layer has an impedance lower than that of the compressed inert mixture and higher than that of powder mixture. Thus, this Al layer can change a strong shock wave into a multistep shock wave, but has little effect on the shock energy deposited into the powder. If the energy threshold theory is correct, then the Al layer should have no effect on the initiation of the chemical reaction.

In the present research two Ni/Al powder size distributions were used in an effort to determine the effect of particle size on the initiation of chemical reaction. In the first set of tests, 3-7 μm spherical Ni powder with 99.9% purity and 20 μm spherical Al powder with 99% purity were manually mixed and compressed to 55% of the theoretical density inside the sample plate pocket. In the second set of tests, 300 mesh (85 μm) Ni particles and the same 20 μm spherical Al powder were mixed and compressed to 55% of the theoretical density inside the sample plate pocket. A Ni/Al molar ratio of 2.604 was used for both particle distributions. No special treatments were used to control powder surface oxidation. A manganin gauge was mounted to the 304 stainless steel backing plate. Model LM-SS-110FB-048 gauges, made by Micro-Measurement Group, Inc. were used. The P-U_p diagrams (Fig. 1) show that under shock loading, the pressure inside the powder mixture changes as state 1-2-3-4 for the cases with target (a) and as state 1-5-6-7-2-3-4 for the cases with target (b). The manganin gauge can sense pressure 2-4 for the cases with target (a) and pressure 6-2-4 for the cases with target (b).

The 3.1 mm powder sample thickness was chosen to give effective measurement times of ~2 μs for a 846 m/s flyer velocity and ~1 μs for a 1332m/s flyer velocity. The reactions are believed to occur on a 100 ns time scale, and thus the time window provided by this structure is sufficient to capture excess pressure time histories. The 50 mm powder gun was calibrated by conducting symmetric impact tests using OFHC copper as impacting plates. These tests indicate a maximum bias in the pressure measurement system of ±0.6 GPa.

RESULTS AND DISCUSSION

Experimental results from target (a) with 3-7 μm Ni particles are shown in Fig. 2 and were found to be similar to the experimental observations of Bennett [7]. "Excess pressures" result when the flyer impact velocity is greater 1075 m/s, which corresponds to ~21 GPa (the calculated inert reflected shock pressure) in the powder mixture. *Because the 50 mm powder gun can provide longer observation times than Bennett's facility, the measured pressure-time profiles show 100-300 ns duration "steady" plateaus with pressure levels in agreement with the values at state 2 and 4* (see Fig. 1). The present experimental results confirm and extend the previously determined threshold condition in Ni/Al powder systems. It is important to note that the gage fails before the release wave reaches the gage for cases where P > 20 GPa due to the target structure and gage mounting technique. Thus only the early portion (.7 - .9 μs) of the profile is captured and shown. However, this time interval is large enough to determine if a reaction has occurred before gage failure.

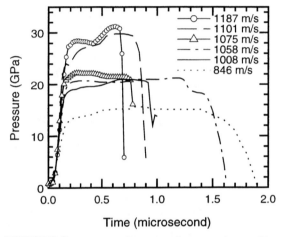

FIGURE 2. Experimentally measured pressure-time profiles for 3-7 μm Ni and 20 μm Al powder mixture.

Figure 3 shows the measured pressure level at state 2 for a range of total input energy deposited into the powder mixture by the shock wave. The solid line shows the predicted maximum pressure vs input energy for an inert Ni-Al mixture using the target (a) configuration. The hollow circles represent the pressures obtained from experiments using target (a) and 3-7 μm Ni particles as shown in Fig. 2 while the solid squares represent the measured pressures when using target (b) and 3-7 μm Ni particles. The triangles represent the pressures obtained from experiments using target (a) and a 300 mesh Ni, 20 μm Al powder mixture.

The results show that for the experiments using target (a), when the total input energy is larger than 483 kJ/kg for the 3-7 μm Ni particle mixture and larger than something less than 473 kJ/kg for the 300

mesh Ni particle mixture, an excess pressure is observed, and that the excess pressure is on the order of 5-7 GPa. For the experiments using target (b), however, even when the total input energy is much larger than the energy input in the reacted cases using target (a), there is no excess pressure observed. It is interesting to note that *the threshold energy for initiation of chemical reaction appears to be slightly less for the larger Ni particle mixture.*

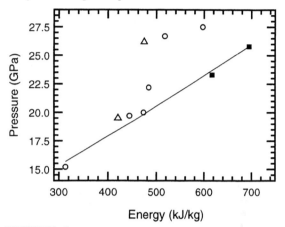

FIGURE 3. Measured pressure vs the total input energy deposited by a shock wave. Solid line is inert mixture prediction; hollow circles represent measured pressures with target (a) and 3-7 μm Ni particles; solid squares represent measured pressures with target (b) and 3-7 μm Ni particles; triangles represent measured pressures with target (a) and 300 mesh Ni particles.

If the chemical reaction is initiated by the total input energy, then these tests indicate that the energy threshold is ~ 483 kJ/kg for the 3-7 μm Ni particle mixture. However, the tests using target (b) (which have an Al layer added), show no excess pressure even when the total input energy is 694 kJ/kg for the 1332 m/s case. It is important to note that the Al layer does not affect the input energy. It does, however, change the mechanical process inside the powder mixture. *Thus, these time-resolved measurements show that the threshold energy theory can not explain the ultrafast chemical reaction in the Ni/Al mixture.* This evidence suggests then, that the initiation of the reaction is influenced by mechanical processes inside the powder mixture and not the energy deposited.

Based on a similar argument, the melting model can not explain the sudden change in the pressure profiles between target (a) and (b) cases. If the excess pressure is caused by material melting and expansion, then the experiments using target (b), where the input energy is much larger than in experiments using target (a), should produce more molten material and one should observe this process as an excess pressure. However, excess pressures were not recorded for any of the target (b) experiments. Thus, the melting model can not explain the ultrafast reaction.

For the experiments with target (b), the shock wave reflected at the interface of powder mixture and the Al layer becomes a multi-wave structure. The resulting pressure-time profiles measured by the manganin gauge for two target (b) tests (1216 and 1332 m/s) are shown in Fig. 4. Also, shown in this figure is the predicted pressure-time profile for the 1332 m/s case using an inert powder mixture Hugoniot model [8] where the constituents are shocked to the same pressure and volume is mass averaged. This profile continues out to ~2.8 - 3 μs and falls off. At the highest impact velocity of 1332 m/s, the powder mixture experiences a series of pressure loading conditions from an incident wave of 9.4 to 15.8 to 24.9 to 25.8 and finally to 27.1 GPa, corresponding to states 1-5-6-2-4 (Fig. 4 shows states 6-2-4). These measurements correspond to four pressure jumps of 6.4, 9.1, 0.9 and 1.3 GPa, and do not show an "excess pressure" indicating that no reaction occurs.

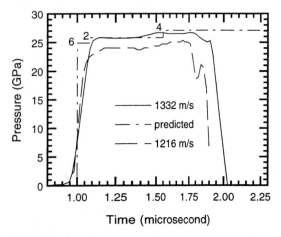

FIGURE 4. The pressure-time profiles for target (b) with 3-7 μm Ni and 20 μm Al powder mixture. Solid line is the experimental measurement for an impact velocity of 1332 m/s; dashed-dot line is the predicted profile for the 1332 m/s case where the numbers correspond to the states given in Fig. 1; dashed line is the experimental measurement for an impact velocity of 1216 m/s.

For the case without the Al layer (target (a)), at an impact velocity of 1101 m/s, Yang [8] reports that the mixture experiences a pressure jump of 14.1 GPa from the incident shock pressure of 6.9 GPa to

the first reflected shock pressure of 21.0 GPa, corresponding to state 1 and 2 shown in Fig. 1, and then the reaction is initiated. This behavior is shown in Fig. 2. These series of target (a) and (b) tests show that the reaction is not initiated by the incident wave, but is initiated by the reflected wave if the pressure jump is large enough. Furthermore, they show that the multistep shock (Al layer target (b) cases) can not initiate the reaction, while the reaction is initiated by a single strong reflected shock in target (a) cases. Under these experimental conditions, the incident wave compresses the powder mixture thus preparing it for initiation, then a strong reflected wave (from the back plate) triggers the reaction.

CONCLUSIONS

A threshold value exists for the initiation of the ultrafast chemical reaction, at a pressure *jump* of about 14 GPa for the Ni/Al mixture used in the present research. The existence of this threshold condition implies that the initiation of chemical reaction is influenced by mechanical factors such as plastic deformation, mass mixing and fluid like behavior of particles, and not energy deposition or melting and expansion.

ACKNOWLEDGMENTS

This research was supported by the U.S. Army Research Office under Contract No. DAAH04-93-D-0003-2 to North Carolina State University.

REFERENCES

1. Graham, R. A., B. Morosin, E. L. Venturini, and M. J. Carr, "Materials Modification and Synthesis Under High Pressure Shock Compression", *Ann. Rev. Mater. Sci.*, **16**, 315, 1986
2. N. N. Thadhani, *J. Appl. Phys.* **76**(4), 2129 (1994).
3. Y. Horie, and A. B. Sawaoka, *"Shock Compression Chemistry of Materials"*, KTK Scientific Publishers, Tokyo, 1993.
4. B. R. Krueger and T. Vreeland, Jr., *Shock Wave and High-Strain-Rate Phenomena in Materials,* Marcel Dekker Inc., New York, 245 (1992).
5. Y. Horie and M. E. Kipp, *J. Appl. Phys.* **63**, 5718 (1987).
6. R. A. Graham, *Solids Under High Pressure Shock Compression: Mechanics, Physics, and Chemistry*, Springer Verleg, New York, 1993
7. L. S. Bennett, F. Y. Sorrell, I. K. Simonsen, Y. Horie and K. R. Iyer, *Appl. Phys. Lett.* 61 (5), 520 (1992).
8. Y. Yang, *Ph.D. dissertation*, North Carolina State University, May (1997).

CP429, *Shock Compression of Condensed Matter – 1997*
edited by Schmidt/Dandekar/Forbes
© 1998 The American Institute of Physics 1-56396-738-3/98/$15.00

SHOCK COMPACTION OF MOLYBDENUM NITRIDE POWDER

S. Roberson*, R. F. Davis*, V. S. Joshi** and D. Fienello***

*North Carolina State University, Raleigh, NC 27695, USA
** Energetic Materials Research and Testing Center, New Mexico Tech, Socorro, NM 87801, USA
*** Eglin Air Force Base, FL 32542, USA

Molybdenum nitride has a potential application in multi layer capacitors. Since this material is not readily available in bulk form, molybdenum nitride powder, consisting of a mixture of the nitrides Mo_2N and MoN has been compacted to 12 mm circular, 1-2 mm thick discs utilizing shock-compression technique. Powders were packed to 55-67 percent of the crystal density and shock compacted using a plate impact shock recovery system at 1.35 to 1.81 km/s impact velocity. The recovered compacts were characterized by scanning electron microscope, x-ray diffraction and cyclic voltammetry to evaluate its electrochemical stability in sulphuric acid. This paper presents the optimization technique used for compaction and the characteristics of the recovered compacts.

INTRODUCTION

Molybdenum nitride is an interesting material. Polycrystalline molybdenum nitride films synthesized either by ion beam assisted deposition [1] or by reactive magnetron sputtering [2] have high densities. These can be used as coatings for wear resistance or as a diffusion barrier material in microelectronic devices [3]. When the material exists in porous form, it can be used as catalyst [4-5] or as high surface area electrode [6]. Currently this material is being investigated as a potential high density capacitor material. Several compounds of molybdenum have been in use for various high temperature and electrical applications, but molybdenum nitride has not yet emerged as a commercial material. This, to some extent is due to its unavailability as a bulk material, although it is now available in powder form.

Commercial molybdenum nitride exists as a mixture of its nitrides, Mo_2N and MoN. The former nitride exists in two different forms as γ-Mo_2N (nitrogen content of 5.1-7%) at high temperature (>850°C) and as β-Mo_2N (nitrogen content of 5.6-7%) at room temperature [7]. MoN contains 12.7% nitrogen, and the mixture invariably consists of traces of elemental molybdenum.

Primary requirement for usability of a material is based on the properties of the material, which in turn relies on the its availability in suitable form. To facilitate measurement of electrochemical and other properties, the powder must be consolidated with minimum loss of properties and its original characteristics. Sintering of powders is generally accompanied by grain growth, which is undesirable. Sintering also requires some binder, which may be detrimental to the properties of the bulk material. Hence these methods are not suitable for obtaining samples for property evaluation.

Shock compression technique is ideally suited for compaction of this material [8] since the process is capable of achieving higher densities in compacts than those obtained by sintering. The process prevents grain growth and also avoids binder additives. Current efforts are focussed in obtaining a high density compact with minimum powder degradation.

EXPERIMENTAL

Dynamic compaction of powder using explosively driven flyer plate impact method is particularly suitable for this material since pressure variations within the compact can be minimized and accurately controlled by controlling the impact velocity.

High purity (99.5%) molybdenum nitride powder, -325 mesh size from CERAC was used in the present work. The powder was tumbled to remove agglomerates, weighed, pressed into stainless steel capsules at the desired packing densities and sealed. Since particle size distribution of powder was not known, range of densities from 55% to 67% of the solid density were achieved by adjusting the hydraulic pressure. A 12-capsule CETR-Sawaoka recovery fixture [9] and modified 4-capsule version of the original recovery fixture were used. The recovery fixture is shown in Figure 1.

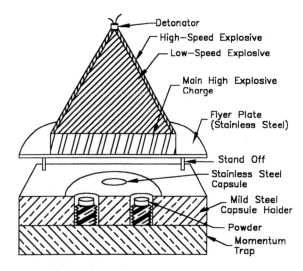

FIGURE 1. Schematic arrangement of the recovery fixture showing various features.

Shock pressures were generated by impacting a standard flyer plate (0.187" thick), which was explosively accelerated to obtain a velocity between 1.35 and 1.65 km/s.

RESULTS

The compacts were recovered in the form of 12 mm diameter and 2 mm thickness single piece discs. The compacts from the first experiments had radial cracks, but had strong mechanical integrity. The capsules from the second set of experiments were blown possibly due to moisture. Powder for subsequent experiment was baked under vacuum using a tube furnace. Compacts recovered from the third experiments appeared to be better than the ones obtained from the first experiment. The recovered compacts were ground to remove the material from the steel capsule and cleaned in acetone prior to analysis.

The samples from the first experiment were observed under scanning electron microscope. The SEM micrograph, as shown in Figure 2 reveals large fraction of material melted, indicating that the pressure was higher than the optimum pressure.

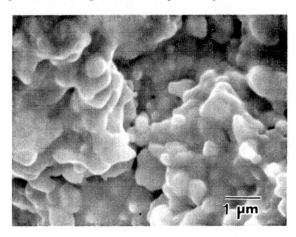

FIGURE 2. SEM micrograph of the ground surface of the compact at 8000 X, shows a molten and resolidified material.

X-ray Diffraction Results

The crystallography of the compacted Mo_xN powders were affected by the compaction process, as interpreted from the results of X-ray diffraction (XRD), which are not shown here. X-ray diffraction pattern of the pre-compacted powder showed the presence of γ-Mo_2N, β-Mo_2N with traces of elemental Mo. However, all post-compacted samples from first experiment were amorphous, confirming the observation of a fused material. One post-compacted polycrystalline sample from the third experiment consisted of γ-Mo_2N and Mo. The loss of β-$Mo_{16}N_7$ in this sample was most likely the result of its lower stability at higher temperatures than γ-Mo_2N.

The formation of the amorphous phase in the other shock compacted samples was most likely due to the presence of moisture in the powder, or adsorption of oxygen during the synthesis of the powder, which led to higher melt fraction and subsequent recrystallization during compaction. The formation of an amorphous phase has also been reported in electrochemical reactions between Mo_xN and 4.4M H_2SO_4 electrolyte above 0.7 V [10]. Although, the reactions in that sample were different than those in the shock compacted samples, it does indicate that under extreme environments Mo_xN can become amorphous.

Cyclic Voltammetry Results

Cyclic voltammetry (CV) of the shock compacted samples were performed in 4.4M H_2SO_4 to compare their voltage stability to the existing data for Mo_xN films made by other processes. The shock compacted samples possessed a lower voltage stability than Mo_xN films, as indicated by the current spikes at 0.5 and -0.5 V, as shown in Figure 3, in which hysterisis curve for different samples have been marked. Furthermore, the unsymmetrical shape of compacted samples with respect to the current axis indicates that an electrochemical reaction occurred during evaluation. In contrast, the Mo_xN film was symmetrical about the current axis and was more stable in this voltage range than the shock compressed samples.

DISCUSSION

The stability of the shock compressed samples were most likely affected by the presence of elemental Mo, which is known to be less stable in H_2SO_4 than Mo_xN, thus the CV's of the shock compressed samples were dominated by the reaction of the less stable Mo and not the Mo_xN. The differences in current between the compacted samples was attributed to the porosity of the compact. As the porosity of a sample increases, more of the total surface area and volume can be accessed, thus increasing the current response of the sample to the applied voltage. The current is much lower in the Mo_xN film sample due to a smaller sample size and a less porous film.

Since the dynamic properties of the material are unknown, one can only speculate its compressive behavior based on the static material property [2]. Successful compaction can only be achieved by consideration of several factors: temperature rise in the sample, magnitude of pressure applied and the duration of pressure.

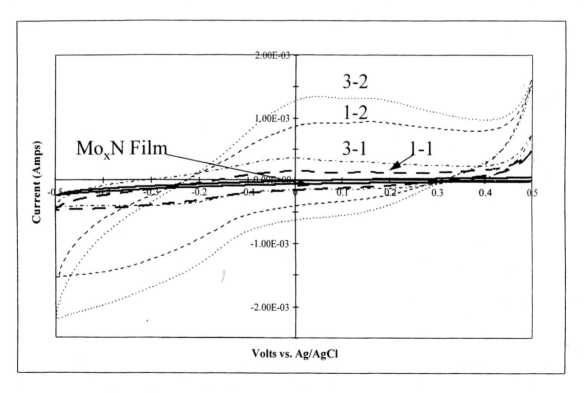

FIGURE 3. Plots for cyclic voltammetry of varoius samples

Since hardness of molybdenum nitride film is about 25 GPa, which is approximately 30% of that of diamond, the yield strength of the material could also be in that proportion to diamond, and the estimated pressure required for compaction would be about 30 GPa, as obtained from the graph given by Ferrerira [11]. This condition would assume substantial melting, which could be detrimental to the properties of the material. Nevertheless, actual compaction results were needed to make any better prediction. The powder contained some free molybdenum, and as a result, actual pressure requirement were reduced. This resulted in larger fraction of melt, producing amorphous material.

CONCLUSION

Molybdenum nitride powders were consolidated to obtain 12 mm diameter and 2 mm thick compacts with good mechanical integrity. Since starting powder contained moisture and elemental molybdenum, CV data could not indicate the true change in the stability due to compaction. Presence of interconnected porosity indicates that compacts were not fully densified, and further improvements could be achieved by using graded powder, free from traces of elemental molybdenum and moisture.

ACKNOWLEDGMENTS

This work was funded by Eglin Air Force Base, FL, through Naval Research Laboratory contract to NOVA Research, with subcontract # NOVA-95-SCI-2009 at New Mexico Tech. One of the authors (VSJ) wishes to thank Dr. J. McDonald of NRL and R. Jeffries of NOVA Research for the contract.

REFERENCES

[1] Mudholkar, M. S., and Thompson, L. T., *J. Appl. Phys.* **77**, pp. 5138-5143 (1995).

[2] Anitha, V. P., Vitta, S., and Major, S., *Thin Solid Films*, **245**, pp 1-3 (1994).

[3] Anitha, V. P., Bhattacharya, A., Patil, N. G. and Major, *Thin Solid Films*, **236**, pp 1-3 (1993).

[4] Colling, C. W., Ph.D. Dissertation, Univ. Of Michigan, Ann Arbor, MI, USA, 1996.

[5] Colling, C. W., Choi, J., Thompson, L. T., *J. Catalysis*, **160**, pp. 35-42 (1996).

[6] Anitha, V. P., Major, S., and Bhatnagar, M., *Surface and Coatings Technology*, **79**, pp. 50-54 (1996).

[7] Jehn, H., and Ettmayer, P., *J. Less-Common Met.*, **58**, pp. 85-98 (1978).

[8] Thadhani, N. N., *Adv. Mater. and Manuf. Proc.*, **3(4)**, pp. 493-550 (1988).

[9] Thadhani, N. N., Holman, G. T., Romero, B., and Graham, R. A., "The CETR/Sawaoka 12-Capsule Plate Impact Shock Recovery Fixture: Design and Experimentation," CETR Report A-01-91 (1991).

[10] Roberson, S. L., Finello D., and Davis, R. F., Mat. Res. Soc. Symp. Proc., **451**, (1996)in press.

[11] Ferreira, A. and. Meyers, M. A, in *Shock-Wave and High-Strain-Rate Phenomena in Materials*, Meyers, M. A., Murr, L. E., and Staudhammer, K. P,. eds., Marcel Dekker Inc. N.Y., 1992, pp. 361-70.

CP429, *Shock Compression of Condensed Matter – 1997*
edited by Schmidt/Dandekar/Forbes

DYNAMIC DENSIFICATION OF MO-SI POWDER COMPACTS FOR REACTIVE SOLID STATE PROCESSING

K. S. Vandersall and N. N. Thadhani

Department of Materials Science and Engineering, Georgia Institute of Technology, Atlanta, GA 30332-0245

Dynamic densification employing explosive techniques were utilized to form un-reacted compacts (~75-95% dense) of Mo and Si powder mixtures and portions with partial to complete reaction. The objective of this work was to study the behavior of the various powder mixture compacts to investigate the formation of intermetallic compounds, and intermetallic ceramic composites, through post-shock reactive solid state processing. In this paper we will describe the densification characteristics of the recovered dense compacts, as analyzed using x-ray diffraction (XRD) and optical and scanning electron microscopy (SEM). The observations shed light on the characteristics of shock-induced and shock-assisted reaction processes.

INTRODUCTION

Molybdenum disilicide provides a combination of ceramic properties like brittleness at low temperatures and oxidation resistance with the similar metallic characteristics of high-temperature plasticity and electrical conductivity (1). Presence of other undesired Mo-Si phases are difficult to avoid by conventional processes. Non-equilibrium processing methods previously used to synthesize $MoSi_2$ include self-propagating high-temperature synthesis (SHS) (2), simultaneous combustion (3), and shock synthesis (4,5,10). All of these methods generally result in a porous final product which in turn must be further processed to achieve a bulk state. A Mo-Si phase diagram is given in Figure 1 for reference (6). Recent developments suggest the use of shock assisted reaction synthesis, wherein a highly defect state is created which increases the solid-state chemical reactivity of the powder mixture (7). These defects sites create a shock modified material, (increasing the internal energy of the system) where transport and/or reaction characteristics will be changed. Thus, during subsequent thermal processing formation of the desired compound occurs in the solid state, thereby limiting the porosity and creation of non-desired phases.

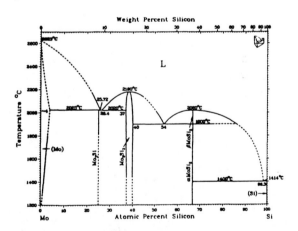

FIGURE 1. Phase Diagram of Mo-Si System

EXPERIMENTAL PROCEDURE

In the current study, commercially available (<44µm or -325 mesh) molybdenum and amorphous silicon powders (obtained from Cerac Inc.) were mixed to a stoichiometric ratio to form $MoSi_2$ and dynamically densified using the double cylinder implosion geometry (8) and the Sandia Bear fixtures (9). The powders were pressed at a density of ~55% and ~60% of theoretical maximum density (TMD) using 35.59-88.96 kN (8,000-20,000 pounds) load. Two different explosive types (ANFO and ANFOIL) were utilized (using the implosion geometry) with each of the packing densities to achieve different shock conditions (temperature and pressure) within each cylinder. The Sandia Momma Bear A (Composition B explosive) and Momma Bear (Baratol explosive) fixtures were used for higher pressure experiments.

X-ray diffraction (XRD), optical microscopy, and scanning electron microscopy (SEM) were used to characterize the starting powders and shocked samples.

RESULTS/DISCUSSION

The primary objective of this work was to fabricate dynamically densified compacts of Mo + Si powders for post-shock-reactive solid state processing. Table I lists the experimental parameters, and the results of general characteristics of the recovered compacts.

General characteristics of the shock densified compacts showed evidence of partial or complete reaction in some of the compacts while the remaining were totally un-reacted. Comparison of the microstructure of the three reacted samples showed interesting differences. Hence, it was decided to investigate the formation mechanics of the reaction

products in the shock densified compacts in more detail.

A comparison of the un-reacted regions in all compacts revealed much similarity. A coalescence of powders with extensive deformation is observed, as shown in Figure 2.

Regions of partial reaction (Fig. 3) were also observed which showed evidence of localized melting, and formation of $MoSi_2$ spherules as observed in other studies (4,5,10). In the case of the fully reacted samples, the most significant differences in reaction product characteristics were observed in sample numbers 3 and 5.

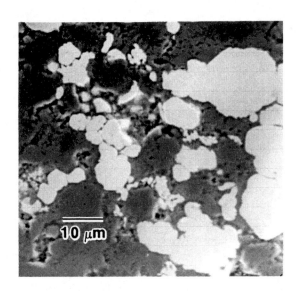

FIGURE 2. SEM image of typical un-reacted region showing Mo (lighter gray) and Si (darker gray).

Table 1. Summary of results.

Sample	Packing Density	Shock Condition	Fixture	Recovered State	Density Achieved
1	55% TMD	ANFOIL	Implosion	Mo + Si, No Reaction Present	86% TMD
2	60% TMD	ANFOIL	Implosion	Mo + Si, No Reaction Present	86% TMD
3	55% TMD	ANFO	Implosion	Mo, Si, $MoSi_2$, 70-95% Reaction	95% TMD
4	60% TMD	ANFO	Implosion	Mo, Si, $MoSi_2$, 40-95% Reaction	95% TMD
5	52% TMD	Comp B	Momma Bear A	Mo, Si, $MoSi_2$, 85-95% Reaction	Porous
6	52% TMD	Baratol	Momma Bear	Mo + Si, No Reaction Present	85% TMD

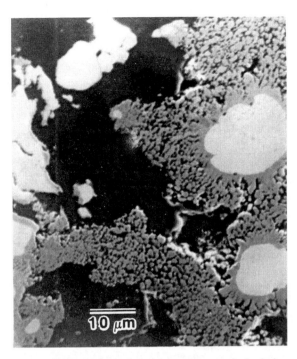

FIGURE 3. SEM image of partial reaction zone in implosion configuration showing reaction zone

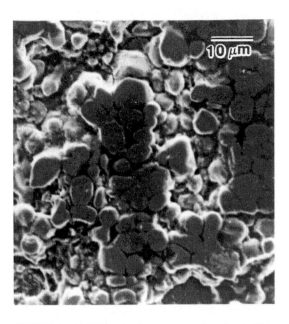

FIGURE 4. SEM image of porous reacted region in Bear Fixture.

A visible difference in morphologies of the two reacted regions can be seen in the SEM micrographs (Figures 4 and 5). Figure 4 shows an SEM image of porous reacted region found in the Bear Fixture (sample 5). This can be contrasted with the implosion geometry sample (No. 3) as shown in Figure 5, which displays dense $MoSi_2$ with larger intergranular dispersed pores. A Mo rich phase is also observed at grain boundaries. Vecchio et. al. (4,5) have made similar observations and identified the grain boundry phase as regions of Mo_5Si_3. It should be noted that Mo_5Si_3 is a very brittle phase and ideally would not be recommended as a grain boundary material due to its brittle nature.

The dramatic difference in morphology suggests two different reaction mechanisms. The reaction product in the cylindrical implosion sample containing a mixture of $MoSi_2$ and Mo_5Si_3 (intergranular) appears to have been formed via a diffusion mechanism over a long time scale. This is consistent with a shock-assisted reaction mechanism occurring subsequent to loading from the peak pressure state. In contrast, the product found in the

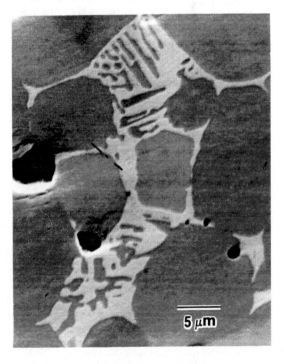

FIGURE 5. SEM image of fully reacted region in implosion geometry sample showing $MoSi_2$ (darker portion) and Mo rich phase (lighter portion) at grain boundaries.

649

Bear sample (No. 5), shows a single phase compound formed via a mechanism consistent with shock induced reactions. The present work shows that microstructural characteristics of reaction products can be useful in identifying possible reaction mechanisms.

CONCLUSIONS

A difference in morphology of reacted regions using two different processing configurations was observed. This difference suggests a different mechanism of reaction. The mechanism in the higher pressure Bear configuration can be described as shock induced characterized by a process occurring in time scales of the shock duration, and that in the implosion configuration can be described as shock assisted occurs involving diffusion on a longer time scale.

ACKNOWLEDGEMENTS

Funding for this work was provided by ARO/AASERT Grant No. DAAH04-95-1-0235 and ARO Grant No. DDAAH04-93-0062. We would also like to thank New Mexico Institute of Mining and Technology for performing the double cylinder implosion and Bear Fixture experiments.

REFERENCES

1. Petrovic, J. J., *MRS Bulletin* July, 35-40 (1993).
2. Deevi, S.C., Material Science and Engineering, A149 (1992) 241-251.
3. Munir, Z. A., *Metallurgical Transactions A*, **23A**, 7-13 (1992).
4. Vecchio, K. S., Yu Li-Hsing, and Meyers, M. A., *Acta metall. mater.* **42**, 701-714 (1994).
5. Vecchio, K. S., Yu Li-Hsing, and Meyers, M. A., *Acta metall. mater.* **42**, 715-729 (1994).
6. Massalski, T. B. Editor in Chief, *Binary Alloy Phase Diagrams*, ASM (1986)1632.
7. Thadhani, N. N., *J. Appl. Phys.* **76**, 2129-2138 (1994).
8. Meyers, M. A., Wang, S. L., *Acta metall.* **36**, 925-936 (1988).
9. Graham, R. A., "Explosive Processing Methods" in *High-Pressure Explosive Processing of Ceramics*, eds., R. A. Graham and A. B. Sawaoka, Transtech Publication, pp. 29-64.
10. Batsanov, S. S., Marquis, F. D. S., and Meyers, M. A., "Shock induced synthesis of silicides", *Metallurgical and Materials Applications of Sock-Wave and High-Strain Rate Phenomena*, 715-722 (1995).

CP429, *Shock Compression of Condensed Matter – 1997*
edited by Schmidt/Dandekar/Forbes
© 1998 The American Institute of Physics 1-56396-738-3/98/$15.00

SHOCK-INDUCED REACTION MECHANISM

TO SYNTHESIZE REFRACTORY METAL SILICIDES

Tatsuhiko Aizawa,   B.K. Yen

Department of Metallurgy, University of Tokyo 7-3-1 Hongo, Bunkyo-ku, Toyo 113

Yasuhiko Syono

Institute of Metal Research, Tohoku University

The pretreatment effect by mechanical alloying on the shock reactive synthesis from elemental powder mixture to refractory metal silicides is studied to describe the essential difference between the shock assisted reaction and the shock induced reaction. Through microstructure observation, non-equilibrium phase formation is precisely investigated to characterize the route of shock induced reaction to yield silicides from the elemental powder mixture.

INTRODUCTION

In the shock assisted reaction, the presence of liquid phase plays an important role in the reaction mechanism. As reported by Batsanov, et al.. (1), in Mo - Si system, the silicon melt reacted with solid molybdenum, resulting in the formation of $MoSi_2$ in liquid phase, which solidified into small particles. Hence, reactions into $MoSi_2$ was partially terminated in the sample, and $MoSi_2$ particles were finely distributed in the unreacted Mo and Si matricies. This shock assisted reaction mechanism is often far from the essential route of shock chemistry taking place during shock loading due to large temperature transients. Author (2,3,4) has been proposing that use of pretreated matters by mechanical alloying enables us to experimentally understand the true mechanism of shock induced reactions. In the present paper, Mo-Si system is employed as a targeting material to investigate the shock induced reaction mechanism. SPEX-8000 is used to yield the pretreated elemental powders with various premixed levels; refined elemental powder mixture with the grain size of Si nearly equal to 20 to 30 nm is used as the

standard starting materials. With reference, elemental powder particle mixture is also shot for various shock pressure; in this case without any pretreatment, almost all reactions were terminated in part with residual Mo and Si just as seen in the shock assisted reactions. On the other hand, the pretreated powders were fully reacted into $MoSi_2$ with 100 % conversion. In addition, this full reaction is indifferent to the applied shock pressure; when the shock pressure exceeds over the critical value, any pretreated powders can be fully reacted into dense silicide compound. Through comparison of microstructure, several features of this shock induced reactions are described for Mo-Si system.

EXPRIMENTALS

Molybdenum powder had a particle size distribution from 4 to 8 μm, while pure Si powder had a size of -200 mesh. These two powders were 99. 9 % or more in purity. The elemental mixes with the chemical composition of Mo33:Si67 were mechanically alloyed by the ball-milling type mechanical alloying. SPEX-8000 was used to control the premixed level by the milling time. The pretreated powders were uniaxially pressed into a green compact billet with the diameter of 10 mm, the thickness of 5 mm, and the relative density of about 80 %. Each billet sample was mounted into a stainless steel capsule with the momentum trap. An one-stage fire-gun was utilized for the shock reactive synthesis. The flyer velocity was measured on-line to calculate the shock pressure in capsules. The recovered specimens were extracted from capsules. The as-synthesized materials were evaluated by XRD, EPMA and SEM.

EXPRIMENTAL RESULTS

In order to demonstrate the refining process takes place during the mechanical alloying, X-ray analysis of several samples with different premixed compositions needs to be examined. Figure 1 depicts the change of XRD profiles for the MA powder with the milling time. The peak intensity of Si decreases monotonically with time, but no other peaks appear in XRD profile. Consideration of the phase diagram, reveals that, no solid solution nor compounds except for $MoSi_2$ are expected to be synthesized even by the mechanical alloying. An abrupt change of XRD profiles in15 min just after fourteen hours of milling is caused by the mechanically-induced self-propagating reaction (MSR). As precisely discussed in Refs. (5,6,7), the formation of the bulk α-$MoSi_2$ during mechanical alloying occurs in less than 15 min once the reaction commences just after 14 h 15 min of milling.

The mechanically alloyed mixture for 10 h was selected as the starting materials for shock reactive synthesis. Using the Scherrer equation, the crystalline sizes for Mo and Si were estimated to be about 30 nm and 20 nm, respectively. In order to understand the pretreatment effects on the shock induced reactions, a compact of elemental powder mixture was also shot by higher flyer velocities.

Figure 2 shows the XRD profiles of recovered samples after shock reactive synthesis from elemental powder mixture with the flyer velocities from 1. 2 to 1. 3 km/s. Although α-$MoSi_2$ was synthesized and its peak intensity became higher with increasing the flyer velocity, residual

Mo and Si were left in the sample. That is, the shock reactions from Mo and Si into $MoSi_2$ might be terminated in part. As discussed later, this reaction can be identified as a typical shock assisted reaction, which is thermally ignited. On the other hand, typical XRD profile of a recovered sample by the shock reactive synthesis from MA-pretreated powders for V = 1. 0 km/s is shown in Fig. 3. The starting materials were completely reacted into α-$MoSi_2$ without any residuals. Of note here is that, the yield of α-$MoSi_2$ in this shock reactive synthesis is 100 % while the residual Mo and Si were still seen in Fig. 1 and the yield was limited to 90 % in MSR. This full reactivity must be attributed to the pretreatment effects on the shock induced reactions.

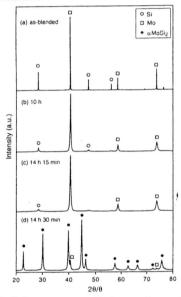

Fig. 1 Refining process of elemental powder mixture during mechanical alloying for Mo33:Si67 system.

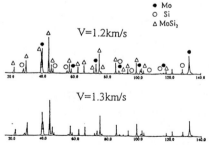

Fig. 2 Shock reactive synthesis from elemental powder mixture.

652

As discussed later, local mass mixing between two elements on the shock front should lead to completion of reactions during shock loading. Next, the flyer velocity was also varied to investigate the effect of shock pressure on the above full reactivity. As shown in Fig. 4, XRD profiles for the whole recovered samples were nearly the same as seen in Fig. 3. The relatively low shock pressure is only needed to ignite this full-reaction and to consolidate the whole sample.

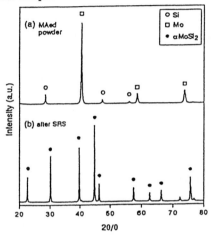

Fig. 3 Shock reactive synthesis from the pretreated powder compact when V = 1. 0 km/s.

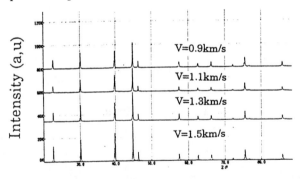

Fig. 4 Effect of flyer velocity or shock pressure on the shock reactivity when starting from the pretreated powder compact by the mechanical alloying.

DISCUSSION

As have been reported in Refs. (8, 9, 10), microstructure of the synthesized materials by the shock reactive synthesis is often quite different from hot-pressed or hipped materials even when starting from the pretreated powders by mechanical alloying. Figure 5 shows microstructure of the synthesized materials both at the center and the end parts of the recovered sample. Although pores were still left in matrix at the end of a sample, the pretreated Mo33:Si67 powder compact was fully reacted into α-MoSi$_2$. On the other hand, dense, Si-rich α-MoSi$_2$ was synthesized at the center part. This difference in the mass density originates from the two dimensional shock wave structure; the center part of a sample often experiences relatively high pressure, while densification is retarded at the vicinity of a capsule. An interesting point to be noted here is that, thin layers with the composition nearly equal to Mo50:Si50 are located in the α-MoSi$_2$ with Si enrichment. Since no regular intermetallic compounds exit except for MoSi2 in the phase diagram for the enrichment side of Si, this Si-diluted layer might be a non-equilibrium phase synthesized by the shock induced reactions.

Through the systematic studies (2,3,4) on the shock induced reactions in 1Ti:1Al and 3Ni:1Al systems, it was found that 1) Mass mixing taking place at the shock front results in the formation of intermediate phase such as solid-solution and amorphous phase, and 2) Application of shock pressure exceeding the critical value to /the above intermediate phase materials should be necessary to make continuation of the shock induced reaction into regular compounds. To be noted is that, this regularization reaction is governed by the formation energy from the above intermediate phase to regular compound and by the structural relaxation time of compound. In the Mo33:Si67 system, no solid solution nor amorphous state materials can be synthesized even by the mechanical alloying (5, 6, 7): other intermediate phase than solid solution or amorphous phases must be synthesized at the hock front. Only intermediate phase to be synthesized by mass mixing is thought to be a nano-particulate mixture of molybdenum and silicon: Mo- and Si-nanocrystals are mixed with various chemical compositions. Due to the significant difference of diffusion constants (11) between Mo into Si and Si into Mo, the solid-phase regularization reaction from the above intermediate materials is ignited by the accelerated diffusion of Si into Mo nanocrystal during shock loading. Since this solid diffusion oriented regularization reaction is intensified with increasing the shock pressure, the amount of residual intermediate phase

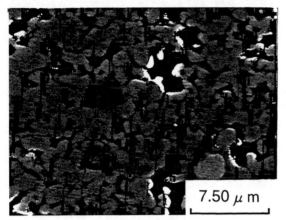

(a) Microstructure of synthesized product in the side-end of a recovered sample.

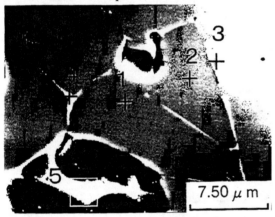

(b) Microstructure of synthesized product in the center part of a recovered sample.

No. Point in (b)	[Si] in at%	[Mo] in at %
1	67. 16	32. 84
2	67. 41	32. 59
3	49. 78	50. 22
4	50. 11	49. 89
5	68. 31	31. 69

Fig. 5   Microstructure analysis of synthesized materials by SEM and EDX.

must be smaller with increasing the flyer velocity.

Microstructure observation reveals that residual amount of nanocrystal mixture phase dramatically decreases with increasing the flyer velocity.

CONCLUSION

Use of the MA-pretreated powders or powder compacts enables us to synthesize various transition metal silicides with 100 % conversion ratio by application of relatively low shock pressure. Different from mechanically-induced reactions in milling, this full reactivity to intermetallic compounds seems to be indifferent to Si concentration. Formation of non-equilibrium phase material must be an intrinsic feature to this shock reactive synthesis from the pretreated powder compact. As suggested in (12), precise control of shock loading pulse enables us to make further understanding of shock induced reaction.

ACKNOWLEDGMENTS

This study is financially supported in part by the Grand-to-Aid from the Ministry of Education, Science and Culture with the contract of # 07555216.

REFERENCES

1) S.S. Batsanov, et al.: Metall. Mater. Appl. Shock Wave and High Strain Rate Pheno. (1995) 715.

2) T. Aizawa, et al.: Annales de Chimie 20 (1995) 181.

3) T. Aizawa, et al: Shock compression of condensed matter (1995) 705.

4) T. Aizawa: Proc. NIRIM Int. Symp. Advanced Materials (1996) 45.

5) T. Aizawa, et al.: J. Graduate School and Faculty of Engineering, Univ. Tokyo. XLIII (4) (1996) 501.

6) B.K. Yen, et al.: J. AmCer. 79 (8) (1996) 2221.

7) B.K. Yen, et al.: Mat. Sci. Eng. A220 (1996) 8.

8) T. Aizawa: Proc. 3rd Int. Conf. Advanced Materials Processing and Synthesis. (1997) (in press).

9) T. Aizawa, et al.: AmCer (1997) (in press).

10) T. Aizawa, et al.: TMS (1997) (in press).

11) P. Gass, et al: Handbook of transition metal silicides. (1995) ttc.

12) N.N. Thdhani, et al.: Shock compression of condensed matter. (1995) 709.

CP429, *Shock Compression of Condensed Matter – 1997*
edited by Schmidt/Dandekar/Forbes
© 1998 The American Institute of Physics 1-56396-738-3/98/$15.00

SHOCK WAVE INDUCED CHEMICAL REACTION IN Mn + S MIXTURE

J. Jiang, S. Goroshin and J. H. S. Lee

Department of Mechanical Engineering, McGill University, Montreal, P. Q., Canada H3A 2K6

Attempts to realize a gasless detonation in Mn+S mixture are reported in this paper. The 25 mm diameter Mn+S charges are typically 100 mm long and encased in a steel tube 25 mm ID and 30 mm OD. A small amount of sensitized nitromethane (~30ml) is used as the initiating charge. Coupling between the reaction front and the shock wave (as in a detonation process) is observed when the shock from the initiation charge is sufficiently strong. Re-acceleration of the reaction front after an initial decay indicates that the chemical energy can be fed back to sustain the shock even in this system when significant products expansion is absent. Results suggest that gasless detonation may be possible when stronger initiation and larger Mn+S charge diameter are used.

INTRODUCTION

SHS (Self-propagating High-temperature Synthesis) or "gasless flame" has been realized since the middle 60's (1). In an SHS process, solid reactants are converted into solid products in a highly exothermic chemical reaction without undergoing any phase change. SHS fronts are deflagrations propagating at typical velocities of the order of centimeters per second. Shock induced reactions in compacted mixtures of powder reactants have also been demonstrated. However self-sustained "gasless detonations" have not been reported to date. The present paper describes some recent efforts to achieve "gasless detonations" in metal-sulfur mixtures.

Theoretical studies of "gasless detonation" had been carried out by Bennett and Horie (2) in Ni-Al mixture through a Hugoniot analysis. Ultra-fast shock induced reactions have been observed in Sn+S by Batsanov (3) and in Ti+C by Gryadunov et al (4). Recently Merzhanov et al. (5) also reported the re-acceleration of the initiating shock wave after an initial decay in a Ti+C mixture suggesting that the energy released by chemical reactions in SHS can be coupled to the shock front. This is the most

encouraging evidence that "gasless detonation" might be possible. However, no conclusive experimental demonstration of self-sustained "gasless detonation" has been reported thus far.

ENERGETIC CONSIDERATION

The general requirements for a reacting mixture capable of sustaining a detonation wave is that energy release must be sufficiently high. To select a suitable mixture, we carried out a thermodynamic calculation of the adiabatic flame temperature of a large number of SHS reactant systems. The adiabatic flame temperature provides a direct measure of the exothermicity of the reactants and its determination is relatively simple as compared to the calculation of the Chapmen-Jouguet detonation state since the equation of state is not required. Only the thermodynamic data and the phase diagram of the products are involved. The thermodynamic software package "Thermo" (6) developed by Dr. A. Shiryaev at the Institute of Structural Macrokinetics (Moscow) was employed in our calculation.

Using the code "Thermo" the adiabatic flame temperatures for some typical SHS systems are

TABLE 1. Adiabatic flame temperature of some gasless reacting mixtures.

Reaction	Pressure, atm	Temperature, K	Products
Ti + C	1	3289	TiC, liquid:solid
Si + C	1	1874	SiC, solid
Ti + 2B	1	3349	TiB, solid
Mo + B	1	2309	MoB, solid
Cr + S	1	2168	CrS, liquid
In + S	1	1980	InS, liquid
Mn + S	4.1	3352	MnS, liquid

TABLE 2. Detonation temperature of several explosives (7).

Explosive	Density (g/ cm^3)	Measured T, K
PETN	1.67	3400
NG	1.59	3470
C(NO$_2$)$_4$	1.64	2800
NM	1.128	3380
TNT (liquid)	1.447	3030

shown in Table 1. Also shown is the phase of the products. We see that the products are all in the condensed phase at their adiabatic flame temperature even at normal atmospheric pressures. As a comparison, the detonation temperature for some common high explosives are given in Table 2 (7). Note that some of the SHS reactants in Table 1 have comparable energetics as the common explosives since these reactants produce a temperature in excess of 3000 ^{o}K which is a typical detonation temperature of common explosives. For the present study we chose Mn+S both for its high energetics and the lower melting points of sulfur so that a solid reactant of theoretical density can be prepared without hydraulic compaction.

EXPERIMENTAL SETUP

Manganese and sulfur of stoichiometric proportions are first thoroughly mixed in a tumbler. The mixture is then heated in a thermal bath at 150 ^{o}C to melt the sulfur. The molten mixture is then poured into the steel confinement. Typical charge diameter used is 25 mm and the outer diameter of the steel casing is 30 mm. The charge length used in all the experiments is 100 mm. In an attempt to

sensitize the charge, glass microballoons (GMB) are also added to the Mn+S mixture. Typical concentrations of glass microballoons used range from 0 - 1% by weight. Ion probes and contact gauges are buried inside the charge and spaced at given intervals to provide measurements for the time of arrival of the reaction front and the shock at the probe locations. When reactions are not initiated by the shock, the ion probes do not register. For initiation, a small charge of nitromethane (NM) sensitized by DETA is placed on top of the Mn+S charge. A schematic diagram of the charge and diagnostics is shown in Fig. 1.

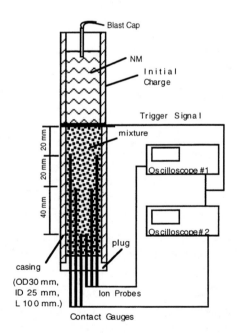

FIGURE 1. The Schematic of Experimental Setup.

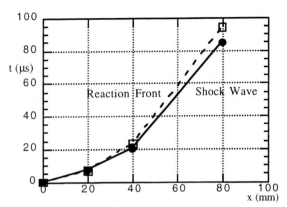

FIGURE 2. Relation between reaction front and shock wave.

EXPERIMENTAL RESULTS

Typical x-t diagram of shock and reaction front trajectories are shown in Fig. 2. In this test, 30 ml of NM is used as the initiation charge and 1% of GMB is used for sensitizing the Mn+S charge. We can observe that the reaction front and the shock wave are coupled initially and then decoupled after propagating a distance of about 20 mm. This indicates that when the shock is strong the reaction rate is sufficiently fast to permit coupling with the shock front, as in a detonation process. However the energy release is insufficient to sustain the shock and prevent its decay.

The effect of using larger initiation charges of NM is illustrated in Fig. 3. To avoid confusion, only the reaction front trajectories are shown. We note that for progressively larger initiation charges (30 ml, 40 ml, 50 ml) the decay rate of reaction front is slower. For the largest initiation charge of 50 ml of NM, the reaction front appears to decay initially and then re-accelerates. The reaction front velocities for the case of using 50 ml of initiation charge of NM is given in Table 3. We note that the averaged velocities for the first 20 mm of propagation is 2.8 mm/μs. It then decayed to an averaged speed of 1.8 mm/μs for the next 20 mm of travel. However it re-accelerates to 2.2 mm/μs for the next 40 mm of propagation. This re-acceleration illustrates the role of the chemical

energy release is sustaining the shock and preventing its decay. A similar observation was reported by Merzhanov et al. (5) for the Ti+C system. A longer test charge would be required to see if the re-acceleration process continues leading to DDT (Deflagration - Detonation Transition).

The sensitization role of GMB is shown in Fig. 4 where shock and reaction fronts for 0% of GMB and 1% GMB are shown. It is clear that glass microballoons increase the velocity of reaction front and delay its decoupling from the initial shock wave.

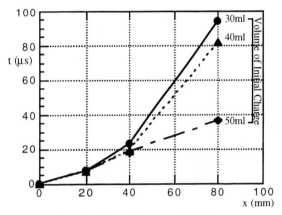

FIGURE 3. Reaction front trajectory vs. initial charge volume.

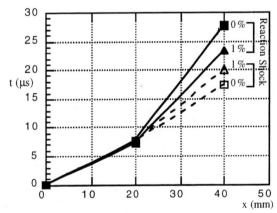

FIGURE 4. Trajectories of reaction and shock fronts in mixtures with 0% and 1%(by weight) glass microballoons.

TABLE 3. Velocity of chemical reaction front obtained in experiment with initial charge of 50 ml explosive.

Distance inside the mixture (mm)	Chemical Reaction Front Velocity (mm/ μs)
0 -- 20	2.8
20 -- 40	1.8
40 -- 80	2.2

CONCLUSIONS

Based on the experiments carried out we have demonstrated that shock induced reactions in Mn+S system can be realized and that the reaction front can be coupled to the initiation shock wave for sufficiently strong shocks. Re-acceleration of the reaction front after an initial decay is observed. This suggests that the chemical energy release can be fed back to sustain the shock even in this system where the products are in the condensed phase and hence significant products expansion (as in conventional explosive) is absent. It appears that a stronger initiation shock and (or) a larger Mn+S charge diameter compared to those used in the present study may lead to successful realization of gasless detonation in the Mn+S system shocked.

ACKNOWLEDGMENTS

This work is supported by DRES (under contract W7702-6-R605/001/EDM) and monitored by Dr. Irene Hooten.

REFERENCES

1. Merzhanov, A. G., "Self-Propagating High- Temperature Synthesis: Twenty Years of Search and Findings", presented at the International Symposium on Combustion and Plasma Synthesis of High-Temperature Materials", San Francisco, California, October 23-26, 1988.
2. Bennett, L. S. and Horie, Y., Shock Waves, 4,127-136, 1994.
3. Batsanov, S. S., Combustion, Explosion, and Shock waves, 32, 102-113, 1996.
4. Gryadunov, A. N., Shteinberg, A. S., Dobler, E. A., Gorel'skii, V. A. and Zelepugin, S., International Journal of Self-Propagating High-Temperature Synthesis, Vol. 3, No. 3, pp. 253-260, 1994.
5. Merzhanov, A. G., Gordopolov, Y. A. and Trofimov, V. S., Shock waves, 6, 157-159, 1996.
6. Merzhanov, A. G., Int. J. SHS 2: 133-156, 1993.
7. Eighth Department of Beijing Institute of Technology, Explosion and Its Application, Beijing. Defense Industries Press, 1979, ch.4, pp. 122-124.

CP429, *Shock Compression of Condensed Matter – 1997*
edited by Schmidt/Dandekar/Forbes
© 1998 The American Institute of Physics 1-56396-738-3/98/$15.00

REACTION SYNTHESIS OF SHOCK DENSIFIED TITANIUM-SILICON POWDER MIXTURES

S. A. Namjoshi, N.N. Thadhani

School of Materials Science and Engineering, Georgia Tech, Atlanta GA 30332-0245

The reaction behavior of shock-densified intermetallic-forming Ti-Si powder mixture compacts was studied to investigate fabrication of net-shaped materials with fine-grained microstructures. Compacts of similar density were also made by uniaxial pressing at 50,000 psi (~0.78 GPa) to compare their reaction behavior with those made by shock compression. Upon reaction, the uniaxially-pressed samples showed large amount of porosity as compared to the shocked samples, which remained nearly dense following reaction. In addition the shocked and reacted samples had a finer grain size (~6 μm) and microhardness values ~ 20% higher than the statically pressed reacted samples. In this paper, the results of this shock-assisted (post shock) reaction behavior of the monolithic powder mixtures forming Ti_5Si_3 intermetallic compound will be presented.

INTRODUCTION

Titanium-Silicon represents a highly exothermic intermetallic forming system, with associated heat of reaction of -577.4 kJ/mol (-138 kcal/mol), volume change of -27.8%, and a calculated reaction temperature of 2500 K, for the formation of Ti_5Si_3 [1]. Shock-compression experiments carried out on this system have shown shock induced chemical reaction initiation at pressures just above 1 GPa, and bulk temperatures below the melting point of Si [2, 3]. These studies also observed that initial particle size affects the propensity of shock-induced reaction, with fine (< 10μm) and coarse (> 100μm) powders reacting at shock pressures significantly higher than those for medium morphology powders (10 - 45 μm).

The high exothermicity and large volume change accompanying the shock induced reactions, produce a highly porous (< 60% dense) reaction product. Powder compacts shock-compressed at conditions below the reaction threshold, show an intimately mixed, densely-packed configuration of reactants. The reactant powders are in a highly activated state,

as revealed by XRD line broadening analysis of residual microstrain and particle size reduction [4].

The objective of the present work is to investigate the post-shock reaction behavior of such shock densified stoichiometric Ti-Si powder mixtures, and exploit the highly activated state of reactants for fabrication of net-shaped single phase Ti_5Si_3 intermetallic alloys with ultra-fine grained microstructures.

EXPERIMENTAL PROCEDURE

Ti and Si powders with particle size ~ 20-40 μm, mixed in a stoichiometry corresponding to Ti_5Si_3 were shock compressed to obtain dense compacts of intimately mixed reactants. The Sandia Poppa Bear fixture [5] with Baratol explosive (PB-B) was used to produce compacts with two-dimensional loading, with a calculated peak pressure of ~ 5 GPa. The single-stage light gas gun at Georgia Tech, was also used to produce compacts with planar wave loading (peak pressure ~ 1 GPa) using a fixture similar to the CalTech design [3]. Simulations performed using AUTODYN-2D [6] showed minimal effects of radial wave focussing. Compacts were also made by static pressing, to compare the reaction behavior of these compacts with those made by shock densification. The required static pressure for plastic flow (P_y) was calculated using Arzt and Fischmeister equation [7]:

$$P_y = 2.97 \; \rho^2 \; \frac{\rho - \rho_o}{1 - \rho_o} \cdot \sigma_y \qquad (1)$$

where s_y is the yield strength, r_o is the initial density and r is the final pressed density. The yield pressure was calculated to be 100 MPa for Ti, and 66 MPa for Si for densification from 20% tap density to 65% TMD. Reaction synthesis experiments were carried out on the densified compacts under inert Ar atmosphere using the Lindberg horizontal 1800 furnace.

RESULTS AND DISCUSSION

The shock-densified Ti-Si compacts made using the Poppa Bear Baratol and the gas gun fixture had densities measured to be in the range of 85% - 95% TMD. The statically densified compacts pressed at 780 MPa, had a density of ~ 60-65% TMD. Figures 1(a) - (b) show SEM micrographs of typical cross sections observed in the as shock-densified and statically-pressed state. Both compacts show a homogenous microstructure with an even distribution of Ti and Si particles. The shocked compacts were nearly fully dense while the statically pressed compacts showed significant porosity, consistent with density measurements. It is apparent that with static pressing from a tap density of ~ 45%, a significant amount of internal porosity is retained even with an applied pressure (780 MPa) exceeding the calculated yield pressure of Ti. X-ray diffraction line broadening analysis revealed that the microstrain in the Ti powders of both shock-densified and statically-pressed compacts was the same. However, the Ti crystallite size in the shock-densified compacts was reduced in contrast to that in the statically-pressed compacts. The reduction in crystallite size of Ti in the shocked compacts may be due to dynamic recovery and recrystallization. The recrystallized grain size is expected to influence the post-shock reaction behavior.

Reaction synthesis of the shock densified and statically pressed samples was carried out at 1000, 1050, 1100, 1150, 1200, 1250, and 1300°C for 3 hours, using an inert Ar atmosphere. In general it was observed that the shock densified Ti-Si samples, upon complete reaction, show an almost dense and fine grain (< 5-10 μm) microstructure, while the pressed samples show a large amount of porosity even upon reaction. Vickers microhardness measurements carried out on these samples after reaction synthesis, showed an average value of 791 VHN for the shock densified sample and 517 VHN for the pressed sample.

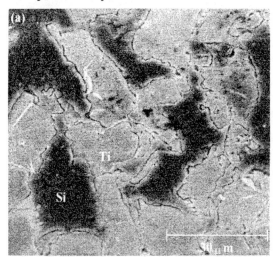

FIGURE 1. SEM Micrographs of a typical cross section observed in (a) shock-densified state, (b) statically pressed state (fracture surface).

REACTION BEHAVIOR OF POWDERS

Mixtures of Ti and Si powders undergo a self-sustaining combustion reaction upon heating at temperatures above the melting point of Si. In the case where powders are heated at low rates (≤10°C/min) a solid state diffusion reaction can be initiated at lower temperatures. However, at any given reaction initiation temperature, if the heat released due to the initial solid-state diffusion reaction is sufficient to raise the temperature of the reactants above the melting point of Si, then the reaction becomes dominated by the combustion process. Figure 2 shows a comparison of X-ray diffraction patterns, obtained from the shocked and pressed samples after the above mentioned heat treatments.

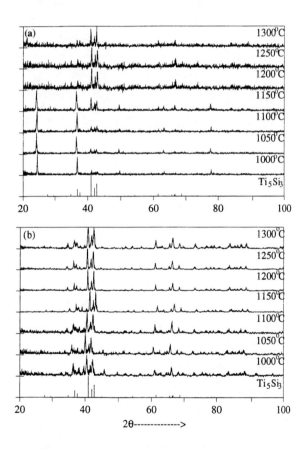

FIGURE 2. X-ray diffraction patterns obtained after synthesis at different temperatures from : (a) the shock-densified and (b) statically-pressed samples.

It can be seen from Fig. 2(a), that the pressed sample shows significant amount of reaction at a temperature as low as 1000°C, while the shock densified sample, Fig. 2(b), shows reaction initiation at 1000°C but no significant reaction until 1250°C. The dissimilar reaction behavior observed between the statically pressed and shock-densified powder compacts is attributed to different dominant reaction mechanisms. In both cases, the reaction initiates by defect-enhanced solid-state diffusion, an influence of the residual strain and crystallite size reduction. With continued reaction, the low porosity retained in the shocked compacts rapidly dissipates the heat generated due to the exothermic reaction, while the lack of heat dissipation in the statically pressed (60-65% dense) compact, allows build up of the reaction heat causing a local increase in temperature.

The local rise in temperature was calculated using a model developed based on transient heat flow. Ti and Si reactant particles forming Ti_5Si_3 product, the initial void volume, and the final porosity due to volume change, were considered as part of the system. Figure 3 shows the calculated temperature differentials at various furnace temperatures for the shock densified and statically pressed (30% initial porosity) samples of average particle size ~ 30 μm.

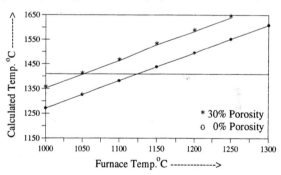

FIGURE 3. The calculated local temperatures as a function of the furnace temperatures.

It can be seen that for the statically pressed sample, the local temperature rise due to initial solid state reaction would exceed the melting point of silicon at a furnace temperature of ~ 1050°C, while for the shocked compact, the furnace temperature would have to be about 1125°C for the local temperature to exceed the m.p. of Si. Hence, the pressed compacts, react via a combustion mechanism at 1050°C following an initial solid state reaction. In contrast, as seen from Fig. 2(a), the shocked sample shows insignificant reaction at 1050°C, but shows significant amount of reaction at a furnace temperature of 1125°C. However, almost complete reaction in the shocked sample is not observed until the furnace temperature is 1250°C. Thus it appears that in the shocked compact, by the time the combustion onset temperature of 1125°C is reached, a significant extent of the reaction has already occurred via defect-enhanced solid-state diffusion. This leaves an insufficient mass of unreacted constituents to be able to generate the heat necessary to raise the bulk temperature above the m.p. of Si and thereby trigger the combustion reaction.

This can be clearly seen from Fig. 4, which compares the fracture surfaces of the shocked and pressed samples after furnace treatment at 1300°C. The shock densified sample, Fig. 4(a), shows a dense microstructure with sharp edged grains indicating a reaction largely dominated by solid-state diffusion. The pressed sample, Fig. 4(b), shows a microstructure revealing extensive liquid phase formation, indicating evidence of a reaction dominated by the combustion process. The grain size of the shock-densified samples was found to be ~ 6 µm. The lack of porosity in the shock-densified and reacted sample is attributed to the accommodation of voids via their migration to the surface during the solid-state diffusion process. In addition the Ti_5Si_3 compound formed can have a much higher maximum solid solubility range (25-28 wt% Si), which may not generate the large volume decrease similar to that in the stoichiometric compound.

CONCLUSIONS

Shock-densified Ti-Si powder mixture compacts react via defect-enhanced solid-state diffusion producing a Ti_5Si_3 compound with a grain size ~6 µm, and a hardness of ~ 791 VHN. The heat released due to the solid-state reaction is rapidly dissipated in the dense compacts which minimizes the local temperature increase. In statically-pressed powders, the increase in local temperature, causes the reaction to be taken over via a combustion process which results in significant retained porosity.

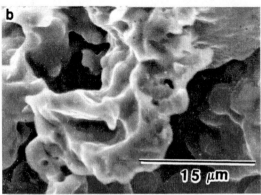

FIGURE 4. SEM Micrographs of fracture surfaces of the Ti-Si samples after synthesis at 1300°C in the furnace : (a) shocked and (b) pressed.

ACKNOWLEDGEMENTS

Work Supported by ARO Grant DAAH04-93-0062.

REFERENCES

1. Kubaschewski O., Alcock C. B., and Spencer P. J., *Materials Thermo-Chemistry*, 6th edition, Pergamon Press, 1993.
2. Thadhani N. N., Graham R. A. et al, *J. Appl. Phys.*, to be published Aug, 1997.
3. Vreeland T. Jr., and Mutz A. H., California Institute of Technology, Unpublished Results.
4. Royal T. E., Thadhani N. N., and Graham R. A., *Metallurgical and Materials Applications of Shock-Wave and High-Strain-Rate Phenomenon*, ed. Murr L. E., et al, Elsevier Science, B.V. 1995, pp 629-636.
5. Graham R. A., *High-Pressure Explosive Processing of Ceramics*, eds., Graham R. A. and Sawaoka A. B., Trans Tech Publications, pp 29-64
6. AUTODYN-2D, Century Dynamics Inc., California, USA
7. Fischmeister H. F. and Arzt E., *Powder Metall.*, 26(1983), 82

CP429, *Shock Compression of Condensed Matter – 1997*
edited by Schmidt/Dandekar/Forbes
© 1998 The American Institute of Physics 1-56396-738-3/98/$15.00

INERT HUGONIOT FOR A POROUS TITANIUM - TEFLON MIXTURE: EXPERIMENT AND CALCULATIONS

J.J. Davis[a,b]
A.J. Lindfors[b]

[a]*Department of Physics, American University, Washington DC 20016*
[b]*Research & Technology Group, Naval Air Warfare Center, China Lake CA 93555*

A study of shock-induced reactions in metal-polymer mixtures via a series of gas gun experiments have been performed. Tests on a mixture of titanium and Teflon which is by weight 80% Ti and 20% Teflon and approximately 92.5% of theoretical maximum density showed reactivity at 3.48 GPa and above. PVDF gauges were used to determine the pressure at the sample/driver interface. Some of the tests were performed with a second PVDF gauge embedded further in the sample to measure reaction. Impedance matching technique was used to determine the inert Hugoniot. A series of calculations were performed to determine the calculation model which best mimics the experimentally determined Hugoniot. The results from the calculations show that an isotherm mixture model matches well the experimental results over the low and middle pressure range studied.

INTRODUCTION

Shock-induced chemical reactions has received considerable attention recently. Most of the reported work has been on metal/metal and metal/metal oxide reactions. Our present work is done using metal - polymer systems that have been scaled up from prior drop-weight research in which these materials were shown to react under rapid shear and plastic deformation (1,2). Two goals were achieved during this research. First, Ti - Teflon has been shown to chemically react under shock.   Second, the inert Hugoniot for this material was developed. The inert Hugoniot is being used in hydrocode modeling to calculate the deformation of this material under rapid plastic flow. The experiment and modeling results are being used in the development of an ignition and growth model for these materials.

EXPERIMENTS AND RESULTS

Seven gas gun tests were performed on samples of titanium - Teflon at a mixture of 80% Ti + 20% Teflon by weight and pressed to 92.5% TMD. The samples were 1 inch diameter and 0.125 inch thick. The pellets could be sandwiched together with a gauge between them and/or have a very thin Al flyer plate behind the last pellet to give a particle velocity measurement via a velocity interferometer system for any reflector (VISAR) as shown in Fig. 1. The

impact flyer plate velocity ranged from 0.775 to 1.543 km/s which resulted in an input pressure range of 1.95 to 13.35 GPa into the samples.   The experimental arrangement eliminated the possibility of a reflected shock thus the input pressure was simply released by the rarefaction waves.

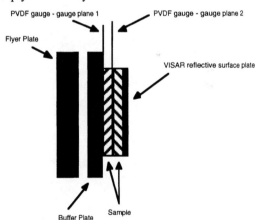

FIGURE 1. Experimental arrangement of samples and gauges for gas gun tests.

The experimental results are given in Table 1. Pressure time history for three of the tests are given in Figs. 2 through 4. The gas gun results show that chemical reaction is being initiated between 1.95 and 3.48 GPa.

TABLE 1. List of experimental results for gas gun experiments.

Test Number	Flyer/ Driver	Flyer plate velocity [km/s]	P [GPa]	ρ_0 [g/cc]	U_s [mm/μs]	u_p [mm/μs]
TITEF1	304 SS	1.369	13.35	3.469	3.563	1.08
TITEF2	PMMA	1.474	4.93	3.400	2.604	0.56
TITEF3	PMMA	1.187	3.48	3.480	2.412	0.41
TITEF4	PMMA	0.775	1.95	3.445	2.032	0.28
TITEF5	2024-T4 Al	1.543	10.65	3.392	3.620	0.87
TITEF6	304 SS	1.102	11.00	3.325	3.291	1.00
TITEF7	304 SS	1.135	11.25	3.463	3.270	1.135

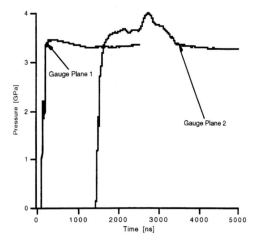

FIGURE 2. Pressure time histories from gas gun test TITEF3.

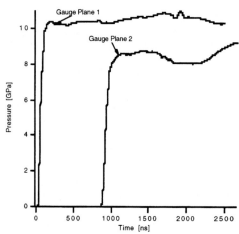

FIGURE 4. Pressure time histories from gas gun test TITEF5.

using the impedance matching technique by examining gauge plane 1 and knowing the Hugoniots for the flyer and buffer plates. In U_s-u_p space, the Hugoniot was found to be

$$U_s = 1.61\,mm/\mu s + 1.83\,u_p \qquad (1)$$

This is found by fitting a straight line through the data. In P-u_p space, the quadratic equation is found.

$$P = 6.37\,u_p + 5.41\,u_p^2 \qquad (2)$$

In P-V space, the Hugoniot is given in Fig. 5. The error bars are a standard ± 5% of specific volume.

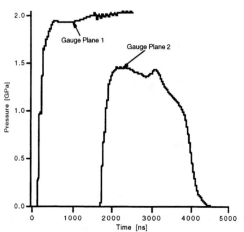

FIGURE 3. Pressure time histories from gas gun test TITEF4.

Information obtained in these experiments allowed for determination of a measured inert Hugoniot for this material at this % TMD. This was accomplished

CALCULATIONS

Two standard methods were used to calculate the Hugoniot: component averaging (wt % and vol %) and isotherm mixing. The calculations were then compared with the experimental data.

Component averaging is a standard method for determining the Hugoniot for mixtures (3). The parameters used in the calculations are given in Table 2. For weight %, the components are averaged on a weight bases.

$$C_0 = \sum_i m_i (C_0)_i = \left[0.8(5.22) + 0.2(1.84)\right]$$

$$(3)$$

$$= 4.54 \text{ mm/}\mu s$$

Likewise, S is 0.95. The calculated Hugoniot in U_s-u_p space for this mixture at TMD is

$$U_s = 4.54 \text{ mm/}\mu s + 0.95 u_p \qquad (4)$$

One can also perform the component mixing via an averaging based upon volume (4). From a fundamental approach, this method is more accurate because the sound speed, C_0, is dependent upon the amount of each type of material the wave passes in the sample, i.e., a volume consideration. C_0 is calculated as

$$C_0 = \sum_i v_i (C_0)_i = \left[0.6556(5.22) + 0.3444(1.84)\right]$$

$$(5)$$

$$= 4.06 \text{ mm/}\mu s$$

Likewise, S is 1.09. The calculated Hugoniot in U_s-u_p space by percent volume at TMD is

$$U_s = 4.06 \text{ mm/}\mu s + 1.09 u_p \qquad (6)$$

One can then calculate the Hugoniot in P-V space for a porous material based upon the Mie-Grüneisen equation of state given and the assumption $\gamma/V = \gamma_0/V_0$. Thus, the EOS for the powder can be written as

$$P = \frac{\left[2V - \left(\gamma_0/V_0\right)\left(V_0 V - V^2\right)\right] C_0{}^2 \left(V_0 - V\right)}{\left[2V - \left(\gamma_0/V_0\right)\left(V_{00} V - V^2\right)\right]\left[V_0 - S\left(V_0 - V\right)\right]^2} \qquad (7)$$

The results from the calculations are shown in Fig. 5.

A more precise way to perform this calculation is using the isotherms of the materials. This procedure

has been written in a FORTRAN program developed at the New Mexico Institute of Mining and Technology's Energetic Materials Research and Testing Center (EMRTC) called MIXTURE. The values used in the code are given in Table 2 with ρ_0 = 3.70 g/cc and ρ_{00} = 3.425 g/cc. C_v values were obtained by assuming C_v is approximately C_p.

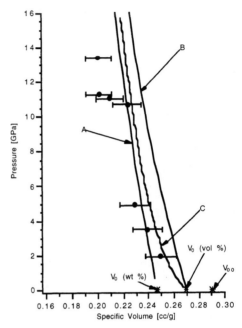

FIGURE 5. Comparison of pressure - volume curves calculated by component averaging using $\gamma/v = \gamma_0/v_0$ assumption: percent weight (A) and percent volume (B), MIXTURE code (C), and experimental data points (•). The initial specific volumes are also presented (✳).

The P-V curve computed by MIXTURE is compared to the experimental data and the component averaging by vol % and by wt % in Fig. 5. Component averaging by vol % did the poorest job in representing the experimental results. Component averaging by wt % did a good job of representing the data in the 3 to 11 GPa range. Since it calculates an incorrect TMD, this method does a poor job at the lowest pressures which is an important region in the hydrocode modeling work.

TABLE 2. Hugoniot parameters for titanium and Teflon.

Material	wt %	vol %	C_0 [mm/μs]	S	ρ_0 [g/cc]	γ	C_v [cal/gK]
Ti	80	65.56	5.22	0.761	4.52	1.09	0.126
Teflon	20	34.44	1.84	1.71	2.15	0.244	0.244

The MIXTURE code accurately fits the highest specific volumes and does good job at the middle region essentially equaling component averaging by percent weight calculation above 6 GPa. It also has the correct form as V goes to V_{00}. It is noticed that none of the calculations handle the highest pressure results. In Fig. 6, the MIXTURE code calculation is plotted against the experimental data and the P-V Hugoniot for pure Ti from <u>LASL Shock Hugoniot Data</u> (5). It is seen that at the higher pressures the experimental data lies on the pure Ti Hugoniot.

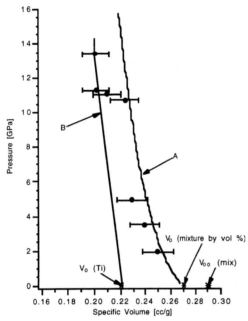

FIGURE 6. P-V curve calculated by the MIXTURE code (A) and the Hugoniot of pure titanium (B) are plotted with the experimental data points (●). The initial specific volumes are also presented (✳).

DISCUSSIONS AND CONCLUSIONS

To have an EOS that is satisfactory over the entire pressure range, a new way of calculating these materials is needed. This is due in part to the great differences in the mechanical and thermal properties of the individual components. Since the hydrocode modeling concentrates on the lower pressure region, the MIXTURE code can be used to calculate the inert Hugoniot for different porosities. Before the Hugoniot can be finalized, the crush-up path must be established. This has been done experimentally using a split-Hopkinson pressure bar (6). It was found that the MIXTURE code mates with the crush-up portion of the Hugoniot for this mixture.

The occurrence of chemical reaction in the metal/polymer materials under shock loading complements the prior work that has been done on these materials using shear initiation via drop-weight impacts. A great deal of research has been done on the Ti-Teflon system under various deformation conditions. Future work is planed to further study the observed transition of the mixture's Hugoniot to the pure Ti Hugoniot.

ACKNOWLEDGMENTS

The authors would like to express thanks to Dr. Diana Woody (NAWC) for providing the materials used in this study, Mr. Steve Finnegan (NAWC) for providing funding for the experimental work through his IR program, and Dr. Philip Miller (NAWC) for technical discussions. One of the authors (JJD) would like to thank Dr. Vasant Joshi of New Mexico Tech. for graciously providing the MIXTURE code. Mr. Scott Pockrandt is thanked for his technical support on the gas gun.

REFERENCES

1. Woody, D. L., Davis, J. J., and Deiter J. S., "Plastic Flow Generated Solid State Metal/Metal Reactions", in <u>Shock Compression of Condensed Matter - 1995</u>, Eds. Schmidt, S. C., and Tao, W. C., New York: AIP Press, 1996, pp. 717 - 720.

2. Davis, J. J., and Woody, D. L., "Reactions in Neat Porous Metal/Metal and Metal/Metal Oxide Compounds under Shear Induced Plastic Flow Conditions", in <u>Metallurgical and Material Applications of Shock-wave and High-Strain-Rate Phenomena</u> Eds. Murr, L. E., Staudhammer, K. P., and Meyers, M. A., Amsterdam: Elsevier, 1995 ch. 78 pp. 661-668.

3. Meyers, M. A. *Dynamic Behavior of Materials*. New York: John Whiley & Sons, 1994.

4. Bernecker, R.,. "The calculation of unreacted Hugoniots. I. TNT, RDX, and their mixtures", In *CPIA Pub. 582, Vol. 1: Proceedings of the 1992 JANNAF Propulsion Systems Hazards Subcommittee Meeting held in White Oak MD 27 April-1 May*, by the Chemical Propulsion Information Agency, Johns Hopkins University. Columbia MD: Chemical Propulsion Information Agency, 1992, pp. 285-302.

5. Marsh, S. P., *LASL Shock Hugoniot Data*, Berkeley: University of California Press, 1980.

6. Davis, J. J., *Characterization of plastic deformation and chemical reaction in titanium - polytetrafluoroethylene mixture*, Ph.D. diss. , The American University, 1997.

CP429, *Shock Compression of Condensed Matter – 1997*
edited by Schmidt/Dandekar/Forbes
© 1998 The American Institute of Physics 1-56396-738-3/98/$15.00

Recovery Studies of Impact-Induced Metal/Polymer Reactions in Titianium Based Composites

Diana L. Woody[a] and Jeffrey J. Davis[b]
J. Scott Deiter[c]

[a]*Airframe, Ordnance & Propulsion Division, Naval Air Warfare Center Weapons Division, China Lake CA 93555*
[b]*Research & Technology Group, Naval Air Warfare Center Weapons Division, China Lake CA 93555*
[c]*Naval Surface Warfare Center, Indian Head MD*

This paper will discuss the effect of rapid plastic flow upon a porous mixture containing titanium and Teflon. The unconfined samples were impacted at 13 m/s to induce a plastic flow. With the addition of Teflon, reactions have been observed. A two color infrared detector was used for real time emission measurements of the reacting materials. Micrographs show a region of decomposition surrounded by an unaltered area. X-ray diffraction results showed that the area of reaction had TiC whereas the surrounding area contained titanium. There was no solid Teflon recovered. The effect of the addition of aluminum and silicon to the mixture was also observed.

INTRODUCTION

The results of the small scale impact tests to be outlined in this paper have shown that reactions in porous metals can occur at much lower pressures if high plastic flow occurred. This paper will discuss one of the mixtures investigated containing titanium and Teflon in a 80/20 ratio by weight. Unconfined samples were impacted at 13 m/s to induce a plastic flow. The diagnostics consisted of a two-color infrared detector for real time emissivity measurements of the reacting materials and x-ray diffraction for recovery studies. The reactivity (formation of products) has been observed by post-shock analysis of the recovered samples. X-ray diffraction of the recovered sample revealed solid TiC and titanium but no recovered Teflon. Micrographs of the recovered sample revealed a melting pattern where the TiC was discovered differing from the surrounding area. CHEETAH calculations were performed on the Ti/Teflon mixture. The calculated products correlated with the information obtained from the x-ray diffraction.

This work represents a continuation of work previously done on the Ti -Si -Teflon system. Previous shock compression studies have been performed on Ti and Si mixtures at shock pressures between 0.80 and 7.5 ± 2.5 Gpa (1-3). In previous work by the authors the 5Ti + 3Si system was impacted to observe if it was possible to obtain the product, Ti_5Si_3 under lower impact velocities but high plastic flow conditions (4-5). The x-ray diffraction results on these samples showed that the formation of Ti_5Si_3 in the recovered sample was dependent upon the percentage of Teflon added to the system.

EXPERIMENT

The rapid plastic flow of the samples was obtained from a drop weight impact machine. The impact machine consisted of an anvil, accelerated guided drop weight, base, and release triggering device. This setup is illustrated in Fig. 1. The impact machine is described fully in another publication (6). Elastic shock cords were used to accelerate the drop weight to obtain impact velocities of 13 m/s. The impact of the drop weight on the anvil was planar to within 2 mrad.

The samples were 0.2 g and were in loose powder form prior to impact. Since the samples were loose powders, the porosity was not measured. The samples were opaque in both the visible and infrared ranges. Therefore, the light emanating from the impacted sample was from the edge of the sample. The Ti/Teflon mixture was formed by the mortar and pestle method. The Ti powder was 325 mesh from Cerac Inc. The Teflon used was 500 μm, 8A from Dupont. The two color infrared detector consists of a

HgCdTe element juxtaposed to an InSb element. Each element has dimensions of 0.101 cm by 0.101 cm with an active area of 0.010 cm². The elements were housed in a liquid nitrogen cooled Dewar and kept at an operating temperature of 77 K. The InSb element's spectral response was from 2 μm to 5 μm. The HgCdTe element was capable of detecting wavelengths from 5 μm to 12 μm. The signal from each infrared detector element was transmitted as a voltage through an initial voltage amplifier and then transferred to a LeCroy digital oscilloscope.

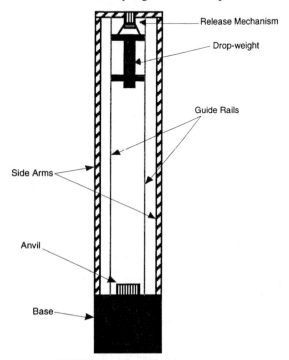

FIGURE 1. Impact Machine Apparatus.

RESULTS AND DISCUSSION

In order to further study the mechanism behind the chemical synthesis the system studied was reduced to Ti and Teflon. The micrographs and x-ray diffraction studies show that some reaction is taking place in the impacted Ti-Teflon mixture under the impact conditions. The micrographs show a definite smooth area in the center of the figure which appears to indicate decomposition occurring in the impacted sample. The smooth area in the middle of Fig. 2 indicates a chemical reaction such as a eutectic melt. On a macroscopic scale this corresponded to the blue-green area observed on the recovered sample. This is contrasted with the surrounding area shown in Fig. 3 which appears to show no decomposition.

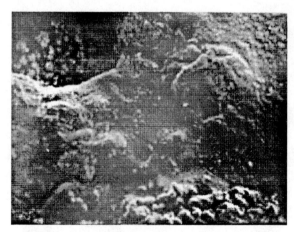

FIGURE 2. Micrograph of Impacted 80% titanium and 20% Teflon showing reaction.

FIGURE 2b. Micrograph of Impacted 80% titanium and 20% Teflon showing reaction.

FIGURE 3. Micrograph of Impacted 80% titanium and 20% Teflon surrounding area.

Significant changes in the chemistry of the starting products were observed by the x-ray diffraction results. X-ray diffraction of the recovered titanium-Teflon mixture revealed that TiC and Ti were the major solid products. CHEETAH calculations of the 80% Ti + 20% Teflon mixture were run at various densities. It was observed that the product concentrations varied with the density. The temperature on the other hand did not change. The mixture was not stoichiometric. Therefore, there was solid Ti at the end of the reaction. The calculations showed that the major solid product was TiC and that no solid Teflon remained. Therefore, the products obtained by the CHEETAH calculations correlated well with the x-ray diffraction results.

The effect of the addition of aluminum to the titanium Teflon mixture was observed via the x-ray diffraction. Aluminum was added in a small percentage to the titanium and Teflon mixture. The mixture was the following: 75% titanium, 5% aluminum, and 20% Teflon. X-ray diffraction was performed upon the recovered impacted mixture. It was observed that not all the Teflon was consumed in the impact of this mixture containing aluminum. The other recovered products were TiC and titanium.

A mixture of 75% Si, 5% Al, and 20% Teflon was impacted. The recovered sample was observed with x-ray diffraction. Si, Teflon, and aluminum were observed in the recovered sample. X-ray diffraction of the impacted sample containing 80% 5Ti + 3Si and 20% Teflon revealed TiC, TiSi, SiC, titanium, and silicon. All the Teflon was consumed for this recovered sample. It appears as though the addition of the aluminum inhibits full decomposition of the Teflon. Perhaps since the aluminum melts at such a comparatively lower temperature, 660°C, than the titanium, which melts at 1600°C, some of the heat during the rapid shear brought about by impact is being absorbed by the melting of the aluminum. The results of the impact of the various compositions are shown in Table 1.

TABLE 1. Results from X-ray Diffraction and Infrared Detectors of Impacted Mixtures.

Reactants	Products X-ray Diffraction	Emission (mV) Infrared Detectors
80% Ti, 20% Teflon	Ti, TiC	2950
75% Ti, 5% Al, 20% Teflon	TiC, Ti, Teflon	3200
80% 5Ti + 3Si, 20% Teflon	TiC, TiSi, SiC, Ti	3672
75% Si, 5% Al, 20% Teflon	Si, Al, Teflon	981

The micrographs and x-ray diffraction studies of the recovered samples of impacted titanium - Teflon mixture show that some reaction is taking place due to the impact. The micrographs show a definite area of decomposition occurring in the impacted sample. The x-ray diffraction showed that TiC was being produced but there was still some titanium left. It appears that the addition of titanium to the mixture augments the full decomposition of the Teflon upon impact. However, the addition of a small amount of aluminum to the titanium-Teflon inhibited the full decomposition of the Teflon that was observed in the recovered impacted sample. This could indicate that the lower melting point of aluminum could be affecting the decomposition of the Teflon and ultimately the chemical reactions occurring during the impact of the metal polymer mixtures.

ACKNOWLEDGMENT

This research was performed partly under the sponsorship of the 6.1 Independent Research Program at the Naval Surface Warfare Center, Silver Spring, Maryland, and partly under the Independent Research Program at the Naval Air Warfare Center, China Lake, California.

REFERENCES

1. Thadhani, N.N., Dunbar, E., and Graham, R.A., "Characteristics of Shock-Compressed Configuration of Ti and Si Powder Mixtures," Joint International Association for Research and Advancement of High Pressure Science and Technology and American Physical Society Topical Group on Shock Compression of Condensed Matter Conference, Colorado, June 1993.

2. Dunbar, E., Graham, R.A., Holman, G.T., Anderson, M.U., and Thadhani, N.N., "Time-Resolved Pressure Measurements In Chemically Reacting Powder Mixtures Joint International Association for Research and Advancement of High Pressure Science and Technology and American Physical Society Topical Group on Shock Compression of Condensed Matter Conference, Colorado, June 1993.

3. Krueger, B.R., Mutz, A.H., and Vreeland T., Metallurgical Transactions A, Vol. 23A, (1992).

4. Woody, D.L., Davis, J.J., and Miller, P.J., "Impact Induced Solid State Metal/Metal Reactions," in JANNAF Propulsion Systems Hazards Meeting, San Diego, California, 1994.

5. Woody, D.L., Davis, J.J, and Deiter, J.S., "Plastic Flow Generated Solid State Metal/Metal Reactions," in Shock Compression of Condensed Matter - 1995, Eds. S.C. Schmidt and W.C. Tao, AIP Press (1996) pp. 717-720.

6. Coffey, C.S., Devost, V.F., and Woody, D.L., "Towards Developing the Capability to Predict the Hazard Response of Energetic Materials Subjected to Impact," Ninth International Symposium on Detonation, Portland, OR, Sept. 1989.

CP429, *Shock Compression of Condensed Matter – 1997*
edited by Schmidt/Dandekar/Forbes
1998 The American Institute of Physics 1-56396-738-3/98/$15.00

SHOCK AND RECOVERY OF PTFE POWDER

W. Mock, Jr. and W. H. Holt

Naval Surface Warfare Center, 17320 Dahlgren Rd, Dahlgren, VA 22448-5100

G. I. Kerley

Kerley Publishing Services, P.O. Box 13835, Albuquerque, NM 87192-3835

Right-circular cylindrical containers with axial cavities were fabricated from 4340 steel (RC38) in two different cavity lengths (2.4mm and 17.7mm, respectively), with closure caps. Polytetrafluoroethylene (PTFE) powder specimens, 2.4mm long, were pressed into the base of each cavity, filling the cavity of the short container, and leaving a 15.3mm-length air-filled region inside the long container. The specimen porosities were all in the range 44-47%. The caps were secured to the containers by four high-strength screws. Momentum traps were provided for each container. In each experiment, a container was impacted by a gas-gun accelerated steel disk, and soft recovered in rags. For experiments at 0.9km/s impact speed, the PTFE in the short container showed melting and limited discoloration; in the long container, a jet of PTFE formed and damaged the inside of the cap, with significant discoloration of the residue. In an experiment at 0.3km/s impact speed with a long container, the PTFE showed no evidence of jetting or discoloration. Computational simulations using the CTH code and a newly-developed model for shock-induced dissociation of PTFE show results that are consistent with the experimental observations.

INTRODUCTION

Improved understanding of the response of materials to mechanical shock loading, and in particular the shock-induced decomposition of materials, can lead to extensions of existing models for more realistic simulations of material response to a wide range of shock loading conditions. Morris *et al.* (1) reported evidence of shock-induced dissociation of the initially-solid polymer, polytetrafluoroethylene (PTFE), leading to tetrafluoromethane gas and a residue of amorphous carbon. In order to develop equation-of-state models for the decomposition and/or oxidation reactions of polymer and other powder materials, it is important to determine the reaction products and the threshold stress conditions for initiating the

reactions, for given initial porosities. Initial porosity leads to the generation of much higher temperatures on shock compression than would be observed for material with no voids. Impact shock experiments at stress levels below and above the threshold stress for reaction will show differences in the post-shock residue composition.

The threshold conditions for the shock-induced decomposition of many of the polymeric solids are unknown, and the present work concerns measurements of the influence of shock stress amplitude on the dissociation of highly-porous PTFE powder, via the impact shock loading of closed steel containers (cells) using a gas gun (2).

EXPERIMENTAL TECHNIQUE

The specimen material was characterized with respect to initial particle morphology via optical and scanning electron microscopy, and was in the form of nearly-spherical white particles having obvious surface substructures. The average particle size was 534 microns (3).

The details of a cell and specimen configuration are shown in Fig. 1. Two different cell designs were used, each having the same cavity diameter, but with differing cavity lengths. The cap is secured to the cell by four high-strength steel screws (not shown in Figure 1). The momentum trap fits against the cap and has cavities for the screw heads. Momentum traps serve to mitigate shock damage to the cell and cap. The cell parts were fabricated from 4340 steel, heat treated to RC38 hardness.

at the base of the cell. For the shorter cell design, the cavity in the cell was only as long as the pressed specimen. The protrusion on the cap was in contact with the specimen. For these initial experiments the cell cavities and the pores in the powder contained ambient air. No attempt was made to contain or recover gases generated within the cells; planned future designs will incorporate gas seals and sampling ports to permit initial evacuation and post-shock mass spectrometric gas analyses. The cell, pressed powder specimen, cap, screws and momentum traps were assembled and placed in a holder for impact as shown in Fig. 2.

After impact, the impactor disk, holder parts, and cell were soft recovered in bales of rags. The cell was then opened to access the post-shock residue for chemical analyses. Experiments were performed for impactor disk speeds of 0.299km/s and 0.904km/s, as shown in Table 1. Figure 3 shows the recovered parts from Shot 565.

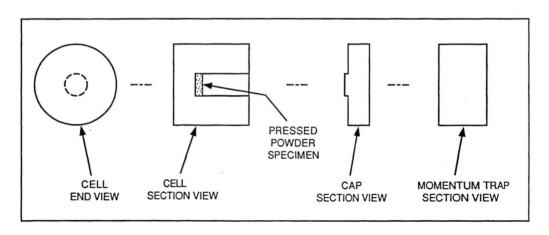

FIGURE 1. Schematic of cell configuration for impact shock loading and soft recovery of pressed porous material. The cell shown is the longer of two designs. The nominal specimen dimensions are 7.6mm diameter and 2.5mm thickness.

Pressed PTFE powder specimens having 44-47% initial porosity were prepared in the two types of cells. The powder was weighed, poured into the cell (with the cell axis vertical), leveled, and pressed with a flat-end rod to a measured depth to prepare *in situ* a specimen disk of prescribed initial density. After pressing, the material retained its disk shape

RESULTS AND DISCUSSION

At the lower impact speed, the recovered specimen material was essentially unchanged. At the higher impact speed, reaction products were evident in experiments for both the short and long cell designs. Material recovered from the short cell

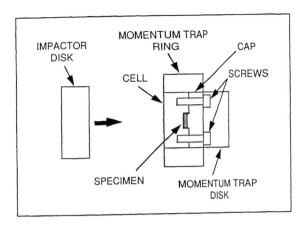

FIGURE 2. Schematic of impactor disk and assembled short cell, cap, screws, and momentum traps. The cell shown is the shorter of two designs. The cell assembly is placed in a precision holder and aligned on the muzzle of the gas gun. The impactor disk is carried on the end of a sabot. The barrel is evacuated ahead of the sabot so that impact occurs in vacuum to minimize gas cushion effects on impact.

FIGURE 3. Recovered parts from Shot 565. (a) Impactor disk. (b) Momentum trap disk. (c) Short cell showing black residue in cavity and on cell flat surface. (d) Cap showing black residue on surfaces that were against the cell.

TABLE 1. Parameters for Experiments.

Shot No.	Impact Speed (km/s)	Cavity Dimensions		Specimen Thickness (mm)	Specimen Mass (gm)	Initial Planar Shock Stress[a] (GPa)	Evidence Of Reaction[b]
		Diameter (mm)	Length (mm)				
565	0.904	7.57	2.49	2.49	0.131	1.5	Yes
566	0.904[c]	7.62	17.73	2.36	0.131	1.5	Yes
567	0.299	7.59	18.13	2.53	0.131	0.5	No

[a]Estimated initial planar stress in specimen material. Time-dependent radial stresses in the specimen due to deformation of steel container were 2-3 times these initial values.

[b]Visible black residue.

[c]Impact speed data for Shot 566 is not available. The value listed is the same as for Shot 565 since the gas gun firing parameters were the same for both shots.

showed both light and dark regions that were similar in composition to the unshocked material, but with small amounts of reaction products. For the long cell, the residue showed relatively large amounts of carbon and oxygen species (approximately an order of magnitude more than for the short cell). There was also evidence of inorganic fluorine compounds and oxidized iron. Very little residue was similar to the unshocked material. There was also evidence of jetting of the specimen material within the cell, with impact cratering damage to the inside of the cap.

Computational simulations using the CTH code

(4) were performed for the experiments, using a reactive equation of state model for PTFE that has been recently developed by Kerley (5). The PANDA code (6) was used to construct equations of state for the unreacted polymer and the dissociation products (carbon + tetrafluoromethane). An Arrhenius reaction rate was used to describe the time-dependence of the dissociation process. The simulations showed that the peak stress in the specimen was higher for the short cell experiments (probably due to the 5-6 GPa reinforcing compressive stress wave reflections within the short cavity). The specimen material temperature in the short cell exceeded 2000K. For the longer cell, the jetting of the specimen material and the resulting damage to the cap was predicted by the code (Fig. 4). Lateral shock compression waves produced a temperature increase of ~800K, with significantly higher temperatures occurring on impact of the jet with the cap. These calculated higher temperatures correlate with the observed significant increase in the quantities of reaction products, for the same initial impact stress.

ACKNOWLEDGMENTS

This work was supported by NSWC Independent Research Funds. F. J. Zerilli of NSWC is acknowledged for performing some additional non-reactive CTH simulations of the experiments. N. Turner of the Naval Research Laboratory is acknowledged for x-ray photoelectron spectroscopy analyses of the residues.

REFERENCES

1. C. E. Morris, J. N. Fritz, and R. G. McQueen, "The Equation of State of Polytetrafluoroethylene to 80 GPa," J. Chem. Phys., Vol. 80, pp. 5203-5217 (1984).
2. W. Mock, Jr. and W. H. Holt, Naval Surface Weapons Center Report NSWC TR-3473, Dahlgren, VA, 1976.
3. Average particle size information provided by DuPont Certification Office, P.O.Box 1217, Parkersburg, WV 26102-1217.
4. J. M. McGlaun, S. L. Thompson, and M. G. Elrick, "CTH: A Three-Dimensional Shock Wave Physics Code," Int. J. Impact. Engng., Vol. 10, pp. 351-360 (1990).
5. G. I. Kerley, Kerley Publishing Services, unpublished results.
6. G. I. Kerley, "User's Manual for PANDA II: A Computer Code for Calculating Equations of State," Sandia National Laboratories Report SAND88-2291, 1991.

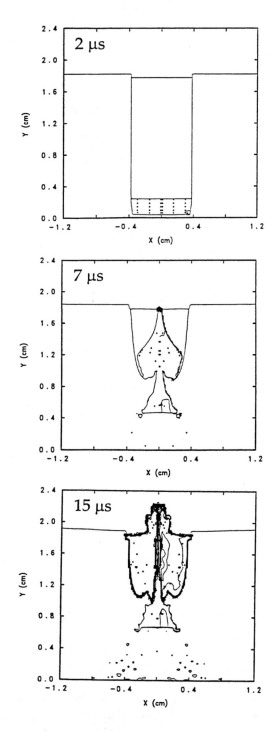

FIGURE 4. Computational simulations of the cell cavity region for Shot 566 at selected times after impact. The jetting of the specimen and impact damage to the cap correspond well with the experimental observations.

CP429, *Shock Compression of Condensed Matter – 1997*
edited by Schmidt/Dandekar/Forbes
© 1998 The American Institute of Physics 1-56396-738-3/98/$15.00

A WIDE VARIETY OF CARBON BEHAVIOR
- AMORPHOUS DIAMOND FABRICATED FROM C$_{60}$ FULLERENE
BY SHOCK COMPRESSION AND RAPID QUENCHING -

H. Hirai[1] and K. Kondo[2]

1: Institute of Geoscience, University of Tsukuba, Tsukuba 305, Japan
2: Materials and Structures Laboratory, Tokyo Institute of Technology
Nagatsuta, Midori, Yokohama 226, Japan

Shock compression can generate extreme conditions such as dynamic high pressure and high temperature, so it is effective technique to understand hidden nature of material and also to explore new advanced materials. Carbon is one of the most attractive and suitable material for such research, because of involving various potentials. This paper gives a brief overview of a wide variety of carbon behavior and a review of recent high pressure studies on C$_{60}$ fullerene. As a typical research applying shock compression effectively, our recent works of fabricating amorphous diamond from C$_{60}$ fullerene is mentioned. Potentials for developing new advanced materials are also described.

INTRODUCTION

Pressure is one of the most important parameters to understand physics and chemistry of materials. Especially shock compression, which generates extreme environment such as dynamic high pressure and high temperature, is useful technique for material science. Once a material is exposed to extreme environment, a certain transient state or nonequilibrium phase is likely to appear, which can allow us to find its hidden properties. We can, in this way, understand the essential behavior of the material, and also we can explore new advanced materials from the understandings. Besides, if an essential property of a material is established, utilizing this material as a sensor, we can diagnose shock compression state itself. As for materials treated for such study, it is more appropriate to use ones which are sensitive to changes in circumstance. Carbon possesses a wide potential for producing metastable, crystalline and noncrystalline phases due to various

bonding states. Then, we have investigated phase transition and its mechanism in carbon materials under shock compression.

In the course of this study, C$_{60}$ fullerene appeared as the third form of carbon, following diamond and graphite. C$_{60}$ fullerene exhibits a characteristic crystal structure and electronic state, it is thus expected to show unique behaviors different from those of graphite-based materials under high pressure. Comprehensive studies including physical and chemical properties of C$_{60}$ and structural changes especially under high pressure have been made since the discovery of method of mass production. Our studies on C$_{60}$ fullerene have concentrated especially on phase transitions in the pressure range higher than that where C$_{60}$ cages collapse. In this paper, we present a brief overview of a wide variety of carbon behaviors, summarize the recent high pressure studies on C$_{60}$ fullerene, and describe our works on fabricating amorphous diamond, along with the potential for developing new advanced materials.

A WIDE VARIETY OF CARBON BEHAVIOR

Although diamond and graphite are only phases which are established as thermodynamically stable phases (1), carbon possesses a wide variety of potentials for forming structures. This potential is especially promising under nonequilibrium conditions such as dynamic shock compression, chemical vapor deposition (CVD), and physical spattering (PS). Phases produced in such condition, involving amorphous as well as crystalline forms, are likely to show intermediate characteristics associated with the presence of sp^2 and sp^3 bonding. Intensive studies have focused on synthesis of carbon materials by various techniques.

Hexagonal diamond (lonsdaleite) is well-known and thermodynamically metastable phase consisting of exact sp^3 bond (2). Polytypes of diamond, consisting of cubic and hexagonal diamond units with different stacking sequence, have been thought out as analogous to ones of SiC (3). Recently, one of the polytypes was prepared by CVD method (4). A transparent graphite, which may be a transient sate to hexagonal diamond from graphite, was reported (5). Carbyne is also known metastable form consisting of one-dimensional chain with exact sp^1 bond (6). γ-carbon is one of forms often prepared by CVD, PS and shock compression method (7), and possible structure models, involving a modified sp^2 bond, were presented (8). In addition to these experimental works, hypothetical structures of carbon and their properties have been actively calculated using empirical potentials, atomic-orbital-based method, and ab initio local density approximation approaches. "H-6" and "bct-4" represent three-dimensional network consisting only of sp^2 bonds (3-d sp^2). The individual sp^2 bonds in the structures from flat plane (9). Another type of three-dimensional carbon network with sp^2 bond presented is usually known "schwarzites". Schwarzite is characterized by regions of negative curvature involving polygons larger than hexagons (10). bc-8, showing similar structure to Si-III, consists of modified sp^3 bond, and exhibit slightly higher density than diamond (11). Besides these structure, simple cubic, bcc, fcc structure were also presented. Most recently, fullerene group has come over as a new member of carbon (12). C_{60} represents highly modified sp^2 bond, near to sp^3 bond caused by containing pentagon ring (13).

Phases produced experimentally and those presented theoretically demonstrates existence of modified bonding state of sp^2 and sp^3 as well as intermediate bonding state of sp^2 and sp^3, in spite that carbon forms have been usually classified by exact hybridization state such as sp^1, sp^2 and sp^3. These forms are of particular interest in view of understanding the hidden nature of carbon and also for the usefulness of their physical and chemical properties.

C_{60} FULLERENE UNDER HIGH PRESSURE

C_{60} molecule is pseudospherical cage consisting of strained pentagonal rings adjacent to the benzenoid rings, which introduces highly chemical reactivity due to poor delocalization of the electron. And, the molecule itself is highly symmetric and extremely hard cage bonded each other by weak van der Waals interaction in solid C_{60} (C_{60} fullerene). Above room temperature the molecules are rotating almost freely, while below 260 K orientational ordering occurs. These features lead various potentials of structural change depending on surrounding temperature and pressure. High pressure studies of fullerenes focus on several aspects in different pressure ranges: i) A

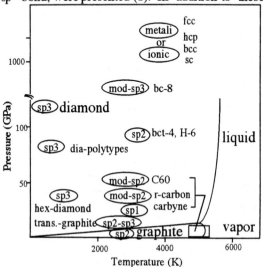

FUGURE 1. A wide variety of potentials of carbon.

pionir work was measurment of stability and compressibility of C_{60} fullerene at room temperature; Under hydrostatic pressure, C_{60} fullerene was stable up to 20 GPa, while non-hydrostatic loading caused a phase transformation to a lower symmetry structure (14). ii) Orientational ordering at low temperature gives rise to structure change from fcc to simple cubic lattice. Effect of pressure ($dT_c/dp=+104$ K/GPa) and pressure medium, N_2 and He, on orientational ordering were studied (15). The moleculs are arranged with the electron rich double bonds (6-6 bonds) facing electron depleted pentagons of their nearest neighbors rather than facing hexagon due to nergetic advantage at ambient pressure. While, the latter arrangement is favored due to its shorter inter-molecular distance at higher pressure. At ambient pressure C_{60} fullerene dose not develop chemical bonding between the individual molecules. iii) Effect of critical temperature of superconductivity, which develops when alkali or alkali earth atoms are introduced into the octahedral and tetrahedral sites, was examined ($dTc/dp=-0.78$K/GPa) (16). iv) Polymerization. With increasing pressure, inter-molecular distance decreases to lead chemical bonding originated by 2+2 cycloaddition reaction between the molecules. High pressure experiments and X-ray pattern simulations defined three pressure-induced phases; rhombohedral lattice, a tetragonal one (these two are two-dimentional polymers), and also an orthorhombic phase forming chains (17,18). With increasing temperature, continuous transition of fcc fullerene to an orthorhombic phase of polymerized C_{60} chains occurs, followed by polymerization into two-dimensional C_{60} polymer, above 800-900 °C the C_{60} cage collapse with formation of sp^2 amorphous phases. (19). v) Hard carbon materials one of which is harder than diamond were produced under higher pressure and temperature. The phases are partially characterized as that C_{60} molecules are retained and that they are covalently bonded to form a three-dimensional random network (20,21). Those are interesting in relation to 3-d sp^2 hypothetical structure. vi) C_{60} molecules collapse to form amorphous sp^3 carbon, amorphous diamond, and diamond polycrystallites under non-hydrostatic compression and shock compression at above $20\pm$

5GPa (22,23,24). Variety of the products is related to sample properties, pressure-temperature condition, and reaction kinetics. vii) Tentative P-T phase diagram of C_{60} fullerene was presented through these high pressure works.

AMORPHOUS DIAMOND FROM C_{60} FULLERENE BY SHOCK COMPRESSION

The existence of amorphous form diamond has been envisioned by many investigators especially those dealing with materials science and high pressure technology. Amorphous diamond is a transient state of carbon with great interest in carbon behavior and also with great potentialities for new material. Shock compression and rapid quenching (SCARQ) technique has a great advantage of freezing a transient state of material such as amorphous diamond, because this technique generates a adequately high-pressure at extremely short duration of nanosecond order, and also quenches a transformed phase with cooling rate of more than 10^8K/s (25). Making effective use of the potential of C_{60} fullerene to transformed to diamond, amorphous diamond was successfully fabricated by applying SCARQ technique (24).

Shock compression and rapid quenching

In order to quench a transient phase, C_{60} fullerene crystals were sandwiched by heat sink materials, gold disks. The C_{60} crystals were sprinkled down on a gold disk to prevent each crystals from superimposing or concentrating. Another disk was put on the disk, and this sandwich was pressed into a capsule so as to make the crystals contact intimately to the heat sink. The capsule was held by a protection system and impacted by a tungsten flyer accelerated by a powder gun. Peak pressures estimated by measured impact velocities were to be 50 to 55 GPa, and peak temperature was roughly estimated to be 2000 to 3000 K by using Hugoniot data of a vitreous graphite with density of 1.63 g/cm^3. The cooling rate was estimated to be 10^8-10^{10}K/s within a distance of 5 µm from the heat sink by one-dimensional thermal diffusion analysis.

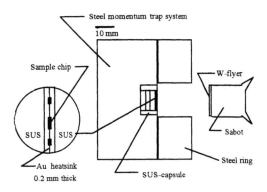

FIGURE 2. Cross section of SCARQ assembly

Characterization by spectroscopies

Transparent chips were recovered and characterized by optical microscopy, energy dispersive X-ray spectroscopy (EDX) detectable light element, X-ray diffractometry (XRD) for single crystal method, micro-Raman spectroscopy, transmission electron microscopy, and electron energy loss spectroscopy (EELS). The characterized properties are summarized below; The material was very stable under ambient condition. Elemental analyses by EDX and EELS exhibited that this material consists only of carbon. The XRD showed a halo, meaning amorphous. The electron diffraction pattern (EDP) represented diffuse bands similar to those common amorphous carbon. The Raman indicated absence of graphite and disordered graphite. The EELS gave a direct evidence that the phase has σ-electron state of sp^3 hybridization almost equivalent to that of typical diamond. XRD and EDP provide information of in long-range order, while Raman and EELS provide one of short-range order. All the characterizations revealed that the transparent material is amorphous in long-range order and, and is diamond in short-range order. Therefore the material can be interpreted as amorphous diamond.

Radial distribution function analysis

It became necessary to confirm the structure in short-range order by another technique, i.e., by establishing the radial distribution function (RDF) based on diffractometry, which gives the most probable distance between atoms and the number of pairs. In the present study the RDF was obtained from the electron diffraction pattern by using an imaging plate (IP), because each specimen chip had to be checked by EELS to determine whether it exhibited the sp^3 bond alone (26).

The first (0.152 nm), second (0.253 nm), and third (0.312 nm) nearest-neighbor distances obtained for the present material are essentially near those of crystalline diamond. No evidence of a characteristic peak at ~0.14 nm, associated with a graphite-like structure and the C_{60} fullerene-like structure, was observed in the correlation function. The first-neighbor distance (0.152 nm), the number of atomic pairs (4.5), and the second-neighbor distance (0.253 nm) indicate that a carbon atom in the material was accompanied by four carbon atoms as nearest neighbors. In addition, these four carbon atoms formed a nearly regular tetrahedron. The numbers of pairs of second and third peaks (13 and 10) showed that each apex atom of the tetrahedron was further surrounded by another tetrahedron. The configuration of these tetrahedra was almost the same as that of crystalline diamond. These topological analyses showed, in summary, that the carbon atoms in the present material were tetrahedrally coordinated, and that the tetrahedra were arranged in the same manner as those of crystalline diamond within a region corresponding to a unit-cell size of crystalline diamond, even though the distances among the tetrahedra were slightly larger. The present material is in itself amorphous but its degree of order is so high as to evaluate the third peak, indicating highly ordered amorphous. Therefore, this material can be called amorphous diamond.

Comparing the correlation function of the present amorphous diamond with those of other tetrahedrally coordinated amorphous materials, the present amorphous diamond is characterized by the third peak. In the other amorphous materials, this peak commonly is absent. This peak indicates that the configuration among the tetrahedra is similar to that of crystalline diamond. The amorphous diamond therefore displays a closer similarity to the crystalline form than do the a-Ge and ta-C. The present material is perfectly amorphous, according to conventional diffractometry, however, this can be distinguished clearly from an amorphous form with randomly distributed tetrahedral clusters. The nearest-neighbor data infer an amorphous nature in which highly ordered amorphous clusters, measuring less than 1 nm, are randomly packed.

Transition process from C_{60} fullerene

The amorphous diamond was synthesized in good reproducibility, but a small amount of altered C_{60} fullerene, less than 10%, remained in the product. Some larger chips recovered included a black core which consisted of variously altered C_{60} fullerene caused by temperature distribution during the shock compression. The variously altered fullerene quenched together with the amorphous diamond in these chips were the most suitable sample for examining the change from C_{60} fullerene to the amorphous diamond. As characterizing technique, the EDP gives information on the structure in the long-range order, while plasmon-loss spectra of EELS gives information on the electronic state and/or bonding states in the short-range order. By combining these analyses, the changes in the structure and the electronic state of the variously altered C_{60} fullerene and the coexisting amorphous diamond were studied to clarify the formation process of the amorphous diamond from C_{60} fullerene (27).

Three transient states were distinguished. The initial C_{60} fullerene exhibited a sharp and net EDP showing an fcc structure. Two bulk plasmon peaks observed in the EELS spectrum at approximately 6.4 and 26.7 eV were attributed to the collective excitation due to π electrons (π plasmon) and $\pi+\sigma$ electrons ($\pi+\sigma$ plasmon). At the first transient state, EDP was almost unchanged or a slightly broader, the π plasmon peak was extremely weakened or absent, and the $\pi+\sigma$ plasmon peak was shifted to lower energy (24.5 eV). According to the compression study (14), C_{60} fullerene is compressed without deformation of the cluster until the intercluster distance nears the intracluster atomic distance. The electron band structure and valence-electron density of fcc C_{60} fullerene calculated (13) represented that π bonding spreads outside the cluster more than the σ bond, showing intercluster overlap. Therefore, the first transient state can be regarded

as a compressed C_{60} fullerene in which the basic fcc structure is retained, and only the collective excitation of π electrons is prohibited owing to the decrease of the intercluster distance.

At the second transient sate, the π plasmon peak completely disappeared, and new shoulders appeared instead at approximately 12 and 17 eV. The new shoulders were attributed to single electron excitations of interband transition. The interaction between the molecules increases with decrease of the intercluster distance, which might result in a state where the interband transitions were enhanced and collective excitation of π electrons ceased. On the contrary, EDP became broader but net pattern was distinguished. Reducing of d-spacing and/or lowering symmetry, if any, was not evaluated because of low resolution. In any case, the basic structure

Photodiode Counts

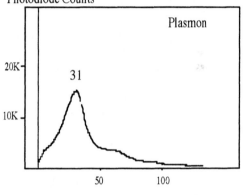

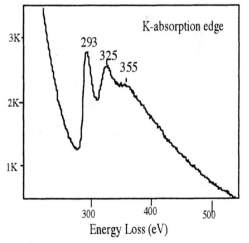

Energy Loss (eV)

FIGURE 3. EELS of amorphous diamond

679

composed of stacked C_{60} molecules was retained, although individual molecules might be somewhat deformed. Both the electronic state and the structure suggest that a certain intercluster bond was produced by a further decrease of the intercluster distance.

At the third state, EDP became a halo, indicating breakdown of the structure. An additional peak rose at approximately 35 eV besides the 14, ~20, and 24eV peaks in the EELS spectrum. The former peak can be assigned to the bulk plasmon peak of σ electrons for diamond, implying directly that the sp^3 bond was produced, while the long-range order structure was already broken.

The final state is amorphous diamond, where EDP showed two quite diffuse bands with maxima at approximately 0.22 and 0.12 nm. The EELS spectrum exhibited a shoulder (26.0 eV) and a peak (31 to 33 eV), which were almost equivalent to those of diamond. The slight shift of the σ-plasmon peak to lower energy indicates relatively lower density than that of diamond. Although the long-range order structure is not built up, the bonding state of sp^3 was completely reproduced. In this way, the transition was accomplished. If further thermal energy is supplied or some other factors facilitate diamond nucleation, diamond crystals might grow.

TOWARD NEW ADVANCED MATERIALS

The feature of amorphous diamond described above in itself demonstrates a remarkable evidence realizing a potential of C_{60} fullerene. And this also suggests that further new materials possessing various electronic states and structures can be introduced from amorphous diamond. Polymerization of fullerene is of the greatest interest for giving rise to new materials. The polymerized structures obtained at present are composed of sp^2-sp^3 complex hybridization. Possibility of bonding between polymerized layers producing a three-dimensional polymerized phase has been studied. In addition, clusrates, a new structure group, consisting of cages linked with three-dimensional network , has been a hot issue, concerning to polymerized fullerene structure. C_{46}, for instance, consist of C_{20} cages arranged in a manner of bcc, forming a three-dimensional network of which individual carbon atoms are sp^3 bond. Actually, same structure in which silicon substitutes carbon and Na and Ba are introduced was synthesized(28).

Carbon materials have been understood in term of a concept of sp hybridization, however, new materials derived from fullerene both experimentally and theoretically exhibit modified or complex hybridization sate. Those carbon materials demonstrate that for hybridization state of carbon not only discrete and highly symmetrical one, e.g. sp^1 and sp^2, but also continuous and asymmetrical one exist. If degree of freedom for hybridization state increases through polymerization and/or collapse, variety of carbon behavior will extremely expand and a lot of new materials will be produced.

REFERENCES AND NOTES

1. Bundy,F.P., Physica A **156**, 169 (1989).
2. Bandy,F.P. and Kasper,J.S., J. Chem. Phys. **46**, 3437 (1967).
3. Spear, K.E., et al., J. Mater. Res. **5**, 2277 (1990).
4. Frenklach, M.,et al., J. Appl. Phys. **66**,395 (1989).
5. Utsumi,W. and Yagi, T., Science **252**,1542 (1991)
6. Heimann, R.B. et al., Carbon **22**,147 (1984).
7. Palatnik, L.S. et al., Sov. Phys. JETP **60**, 520(1984).
8. Hirai, H.et al., Appl. Phys. Lett. **61**,414 (1992).
9. Liu, A.Y. et al., Phys. Rev. B **43**, 6742 (1991).
10. Mackay, A.L. and Terrones, H., Nature **352**, 762 (1991).
11. Fahy, S.and Louie, G., Phys. Rev. B **36**, 3373 (1987).
12. Kroto, H.W.et al., Nature **318**, 162 (1985).
13. Saito, S. and Oshiyama,A., Phys. Rev. Lett. **66**, 2637 (1991).
14. Duclos,S. J. et al., Nature **351**, 380 (1991).
15. Samara, G.A. et al., Phys. Rev. B **47**, 4756 (1993).
16. Sparn, G. et al., Science **252** 1829 (1991).
17. Y. Iwasa et al., Science **264**, 1570 (1994).
18. Nunez-Regueiro, M. et al., Phys. Rev. Lett. **74**, 278 (1995).
19. Nunez-Regueiro, M. et al., Phys. Rev. **54**, 12633 (1996)
20. Kozlov, M.E. et al., Appl. Phys. Lett. **66**, 1199 (1995).
21. Blank, V.D. et al., Phys. Lett. A **205**, 208 (1995).
22. Nunez-Regueiro, M. et al.,, Nature **355**, 237 (1992).
23. Yoo, C.S. et al., Appl. Phys. Lett. **61**, 273 (1992).
24. Hirai,H. et al., Appl. Phys. Lett. **64**,1797 (1994).
25. Hirai, H. and Kondo, K., Science **253**, 772 (1991).
26. Hirai, H. et al., Phys. Rev. B **52**, 6162 (1995).
27. Hirai, H. and Kondo,K., Phys. Rev. B **51**, 15555 (1995)
28. Kawaji, H. et al., Phys. Rev. Lett. **74**, 1427 (1995).

CP429, *Shock Compression of Condensed Matter – 1997*
edited by Schmidt/Dandekar/Forbes

MORE ON THE POSSIBILITY OF IMPACT ORIGIN OF CARBONADO

P.S. DeCarli

SRI International, Menlo Park, CA 94025

It has been suggested that carbonado, a type of polycrystalline cubic diamond, was formed by an ancient impact. Calculations based on experimental data predict a maximum size of about a cm diameter for impact formed carbonado. The largest known carbonado is 10 cm diameter.

INTRODUCTION

Carbonados are polycrystalline diamonds commercially mined from placer deposits in Brazil, Venezuela, and Central Africa. The carbon isotopic ratio is considered evidence that the source carbon was organic, consequently of crustal origin. Until the recent discovery of rocks that have been metamorphosed at very high pressures, it was considered unlikely that crustal carbon could be subducted to depths of hundreds of kilometers and then returned to the surface.

Smith and Dawson suggested that carbonados were formed by early impacts of crustal rocks.[1] Their main line of argument is that the inclusions in carbonado are crustal minerals that would not have survived deep sub-duction. They do not address the question of whether those mineral would have survived shock com-pression in the range above 20 GPa. The answer is no, but that argument is moot. The inclusions cited are found in pores connected to the surface; the inclusions appear to be contaminants that formed long after the organic carbon transformed to diamond.[2]

The question I wish to address is whether impact synthesis of carbonado is possible in a plausible geological setting, on the basis of laboratory experience in shock synthesis of diamond. This paper reports progress since the last report.[3]

BACKGROUND

Laboratory studies of shock synthesis and of uncatalyzed static high pressure synthesis have shown that there are two distinct diamond synthesis regimes.[4-10]

If the source carbon is dense, well-crystallized graphite, a mixture of cubic diamond and hexagonal diamond (lonsdaleite) can be made by a quasi-diffusionless mechanism at pressures above about 15 GPa and simultaneous temperatures in the range of about 1300 K to 2000 K.

If the source carbon is a less well-ordered graphitic carbon, cubic diamond can be formed by an inferred nucleation-and-growth mechanism at pressures above about 15 Gpa and temperatures above 3000 K. In this latter case, the newly formed diamond must be quenched to a lower temperature (to avoid graphitization) before the pressure is decreased below the diamond stability line.

Carbonado is purely cubic diamond, with no evidence of hexagonal stacking order. There is no question whether the requsite high pressures and temperatures could be produced in a large impact. The question is whether there exists a geologic setting in which carbonado could have been made and quenched.

GEOLOGIC CONSTRAINTS

The first task is to pick the optimal geologic setting for shock synthesis and quenching of diamond. One may presume that, at the time of diamond formation, the source carbon was enclosed in a rock matrix. Carbonado is found in alluvial deposits; there is no evidence bearing on the nature of the original rock matrix. We are therfore free to pick the optimal matrix, provided that it is geologically reasonable.

As noted above, high carbon shock temperatures, above 3000 K, are required for synthesis of cubic diamond. However, the temperature of the surrounding matrix must be below 2000 K on release of pressure, if rapid graphitization is to be avoided.(11)

These requirements imply that the optimal geologic setting would be porous carbon in a non-porous and relatively incompressible rock. At least some occurences of carbonado have been dated as Precambrian. Although common forms of natural porous carbon, such as coal and charcoal, do not appear until hundreds of millions of years later, coke-like porous carbon has been found in 2 billion year old formations.(12) The major constraint on the matrix rock is that it be of sedimentary origin. Sandstones and shales are much too compressible over the pressures of interest; carbonate rocks are the only option.

Finally, one must estimate the thickness of the overburden at the time of impact. The effective length of the pressure pulse, the time available for quenching the hot diamond, will be related to the distance to the nearest free surface. The overburden must not be so thick that the porosity is squeezed out of the carbon. 50 km is unreasonably thick; 5-10 km seems reasonable. The effective quenching time would be about a second.

SIMPLE   CALCULATIONS

The optimal geologic setting comprises porous carbon in a fully dense calcite matrix. For the best possible shock geometry, one can imagine that the carbon is in rod-shaped form, with the long axis parallel to the direction of shock propagation. Thanks to the impedance mismatch, a converging shock would be driven into the carbon. A Mach disc would form near the axis of the carbon rod.

A 31 GPa shock in the calcite would produce about a 40 GPa Mach region in the carbon. The calcite temperature would be low, an estimated 1200 K.

There would be a large temperature gradient in the carbon, from an estimated 3500 K in the Mach region to about 1500 K near the periphery.

The greatest advantage of this geometry is that the hot diamond would be quenched by heat flow to cooler carbon; carbon has a much higher thermal conductivity than calcite.

Let us assume that the Mach region can be as large as the largest carbonado. We can then estimate an upper limit to the size of carbonado that might be recovered on the basis of heat flow estimates.

We do not have very good thermal conductivity data for either diamond or compressed graphitic carbon at high temperatures and pressures. An elaborate heat flow calculation would be a futile exercise. However, we can make simple extrapolations from the results of numerous laboratory experiments on the shock wave synthesis of diamond.

In these laboratory experiments, diamond is made by shocking porous carbon; the diamond forms in hot spots and is quenched by conduction to the surrounding cooler carbon. The maximum size of diamond recovered is correlated with the shock duration. We recover 10 micron polycrystalline diamond from experiments having a duration of about a microsecond.

Thus, a one second quench time would imply a maximum carbonado diameter of about one centimeter. However Sergio, the largest known carbonado, is 10 cm diameter, implying either a quench time of 100 seconds or two order of magnitude higher cooling rates than we have estimated.

If the 10 micron diamonds in our lab experiments are quenched in one thermal time constant, the thermal conductivity of the diamond and compressed carbon would be in the range of 50-100 W/mK. This range is consistent with available data; it would be unreasonable to suggest that the conductivity could be even one order of magnitude higher.

CONCLUSION

The only viable mechanism for carbonado formation is deep subduction of crustal material. The shock hypothesis is not supported by any evidence; it doesn't seem to be possible on the basis of the arguments presented here.

ACKNOWLEDGEMENTS

I thank those colleagues who have helped me learn about carbonado, including Judith Milledge of University College London, Ron Girdler and David Collinson of the University of Newcastle-upon-Tyne, Colin Pillinger and Rob Hough of the Open University, and Rob Hargraves, Peter Heaney, and Barna De of Princeton. Jay Melosh first put me on the track of Precambrian carbon. I mourn the recent death of Gene Shoemaker, who shared his insight on cratering and general geological matters and who encouraged me to begin and continue this work.

REFERENCES

1. Smith, J.V., and Dawson, J. B., Geology 13, 342-343 (1985)
2. Dismukes, J.P., Gaines, P.R., Witzke, H., Leta, D.P., Kear, B.H., Behal, S.K., and Rice, S.B.,

Material Science and Engineering, A105/106, pp555-563 (1988)

3. DeCarli, P.S.,*Shock Compression of Condensed Matter-1995*, ed Schmidt, S.C. and Tao, W.C. pp 757-760, AIP Press, New York, 1996

4. Bundy, F.P., and Kasper, J. S., J. Chem. Phys. 46, 3437-3446 (1967)

5. DeCarli, P. S., and Jamieson, J. C., Science 133, 1821-1822

6. DeCarli, P.S., "Shock Wave Synthesis of Diamond and Other Phases" in Proceedings of MRC Symposium on Diamond, April 1995, in press, Materials Research Society, Sept. 1995

7. Trueb, L.F., J. Appl. Phys. 19, 4707-4716 (1968)

8. Trueb, L.F., J. Appl. Phys. 42, 503-510 (1971)

9. DeCarli, P. S., U.S. Patent 3,238,019 (1966)

10. Cowan, G.R., Dunnington, B. W., and Holtzman, A. H., U. S. Patent 3,401,019 (1968)

11. Davies, G., and Evans, T., Proc. Roy. Soc. London A238, 413-427 (1972)

12. Hargraves, R.B., Private Communication (1997)

CHAPTER IX

EXPLOSIVE AND INITIATION STUDIES

CP429, *Shock Compression of Condensed Matter – 1997*
edited by Schmidt/Dandekar/Forbes
© 1998 The American Institute of Physics 1-56396-738-3/98/$15.00

SHOCK INITIATION OF DETONATION IN NITROMETHANE

B.Leal*, H.N.Presles**, G.Baudin*

DGA/Centre d'Etudes de Gramat, 46500 GRAMAT
**ENSMA/Laboratoire de Combustion et de Détonique, 86960 FUTUROSCOPE*

The processes involved in the initiation of nitromethane (NM) have been the subject of many experiments and theoretical studies. These studies generally support the classical homogeneous model though some details of the buildup process are still controversial. In order to clarify these points, we have performed plate impact experiments to study the initiation of NM under conditions of steady one dimensional strain, for shock pressures ranging from 8.5 to 12 GPa. A six wavelength optical pyrometer, with 3 ns rise-time and a temperature range of 1500-6000 K, was used to determine the temperature during shock-to-detonation transition. A Fabry-Perot interferometer with a capacitor transducer and piezoelectric pins were also used to analyse the temperature profiles and to determine the sequence of events during the initiation process. According to our experimental results, it seems that, unlike Campbell et al assumptions, the superdetonation does not start at the plate/NM interface, but at a run distance inside the NM depending on the shock level.

INTRODUCTION

The mechanism by which detonation of a homogeneous high explosive (HE), especially nitromethane (NM), is initiated has for 30 years been the object of numerous studies both theoretical and experimental. The classical homogeneous model proposed by Chaiken in 1957 (1), is based on the phenomenon of an adiabatic thermal explosion and a chemical decomposition following a rate law of the Arrhenius type. A shock wave propagating in the nitromethane compresses and heats the latter. After an induction period, a detonation is initiated by thermal explosion at the explosive/driver interface (through which the initial shock wave was transmitted and which consequently underwent the longest heating). This detonation, propagating in a medium previously compressed (therefore at a greater speed than that attained in the same non-shocked medium), is known as "superdetonation". If the thickness of the explosive is sufficient, the detonation overtakes the leading shock wave. A highly non-stationary detonation now propagates

through the non-compressed NM and decays to the steady quasi-CJ detonation wave. Campbell *et al* (2) followed by Travis *et al* (3) and Dremin *et al* (4) produced considerable evidence which supported this homogeneous model. The situation took a new turn in 1970, after Walker and Wasley (5) proposed a second model in which the exothermic reactions take place in or just downstream of the shock front. Unfortunately, their work was limited to shock waves with amplitudes lower than 7 GPa, and because later studies performed at high shock pressures (Hardesty (6), Presles *et al* (7)) could not invalidate the thermal explosion model, the hypothesis of a pressure dependant initiation process has thus been carried forward. More recently, new studies have yielded refinements to the classical model. Thus, the analytical and numerical results of the theoretical study performed by Kapila *et al* (8) in gaseous systems highlight the mechanisms involved in the homogeneous initiation being more complex. The detonation initiated by the thermal explosion at the input boundary would initially be a weak detonation, slowing down and evolving into a superdetonation as it runs forward into the precompressed explosive (the rest of the process remaining unchanged). Sheffield (9) also takes up the idea of progressive build-up of the superdetonation in NM, attributing this time its formation to the accumulation of sonic compression waves. Thus, while studies concerning the initiation of homogeneous explosives, and in particular NM, generally agree as to the formation of a superdetonation in compressed media, a number of obscure points persist as to its build-up (how, where and when does it occur ?). In order to clarify these points, a series of plate impact experiments has been performed using NM, for shock pressures ranging from 8.5 to 12 GPa. The experimental set-up included a multi-wavelength pyrometer, a Doppler laser interferometer, a capacitor and piezo-electric pins which, jointly used, allowed the identification of the processes involved in the initiation phase, and more particularly, the determination of the point at which the superdetonation is formed. This work presents the experimental configuration employed, the measurements performed, and the conclusions drawn from the results.

EXPERIMENTAL CONFIGURATION

The explosive used is NM, with a 1.134 g/cm^3 density and a spectroscopic purity (>99%). The experiments are performed using an 11-m long, 98-mm diameter smooth bore single stage powder gun capable of projectile velocities up to 2400 m/s. In order to guarantee a simultaneous shock build-up process at the NM entry, the experimental set-up is adjusted so that tilts are better than 2 mrad (Fig. 1). Evacuation of the system to a pressure of less than 10^{-2} mbar prior to each experiment prevents the formation of an air cushion at impact. The impactors used in this study are copper disks with a 80 mm diameter and a thickness ranging from 6 mm to 15 mm, according to the required speeds. The target arrangement shown in Figure 1 is achieved as follows: the NM is confined in a polyethylene chamber (70 mm inside diameter, depth ranging from 15 to 25 mm, following the experiments), enclosed by a 4-mm-thick copper transfer plate. Polyethylene was chosen here to minimize the impedance mismatch between the NM and its cartridge. Both facings of the copper transfer plate are machined to nominal flatness and parallelism of

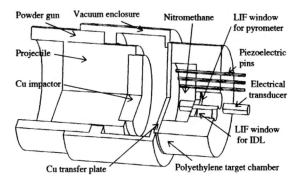

FIGURE 1. Experimental configuration

less than 3 and 20 µm respectively. In addition, the front surface of this plate is polished up to 0.01 µm rms roughness to prevent possible local hot spots at the interface from disturbing the initiation process. The impactor and barrier thicknesses were defined to ensure a constant stress state of one dimensional strain for sufficiently long test times (about 6 µs). Moreover, an external heating system provides a temperature regulation between 15 °C and 20 °C prior to each experiment.

A six-wavelength optical pyrometer (500, 650, 850, 1100, 1270 and 1510 nm) is used to determine the explosive temperature history during the detonation process. The destructive power of the involved phenomena requires a remote measurement means. Thus the thermal radiation is collected by an optical probe and directed towards the optoelectronic elements of the apparatus using an optical fiber. The remarkable characteristics of the pyrometer enable to cover a wide range of temperatures (1500 - 6000 K) with a 3 ns rise-time. Complementary information, required to interpret the temperature signals and get a better understanding of the phenomena governing the initiation of NM, is obtained from three other sources. First, a Fabry-Perot interferometer provides a record of the transfer plate velocity history during initiation. The plate reflectivity is locally enhanced near the target spot by addition of a 1-µm-thick aluminum deposit. An electrical transducer technique is added to these non-intrusive measurements. The experimental device is a 1-cm diameter copper cylinder, one end of which is located 5 to 8 mm from the transfer plate. Both plate and cylinder form the two electrodes of a parallel-plate capacitor used to observe the polarization phenomena inherent in any polar material associated with the shock or detonation wave. Finally, the shock and detonation velocities are measured using piezo-electric pins located at measured depths in the explosive. The pins and capacitor placements were chosen in order not to disturb the region of one-dimensional strain monitored by the pyrometer and the Fabry-Perot Interferometer.

RESULTS

A representative sample of the luminance temperature signals (at 500 nm and 1510 nm) and the corresponding capacitor record are displayed in Figure 2. The origin of the abscissas coincides with the instant the shock enters the NM. The polarization signals, similar to those previously obtained by Travis (10), are consistent with the classical homogeneous initiation model. Indeed, the shock entrance (time 0) as well as the superdetonation formation (time t_1) can be seen on the oscillogram as two positive electrical pulses. The third pulse, having the opposite sign, is due to the overtaking of the initial shock by the superdetonation (time t_2). Unlike the polarization signals, the pyrometric ones give no information about the shock entrance into the NM since, at that instant, the temperatures are too low to be detected. However, the two temperature jumps synchronous with the last two electrical pulses allow the various formation phases of the detonation to be identified. Moreover, assuming that the temperatures reached prior to the superdetonation are due to chemical reactions, pyrometric signals can provide additional information about the kinetics of the involved reactions. In particular, they may question the use of a single Arrhenius kinetic law with the classical activation energy value of 53 kcal/mol (2). Indeed, the latter expects an induction time (corresponding to the beginning of the exothermic reactions leading to the superdetonation formation) during which the explosive remains at a constant temperature, about 1000 to 1100 K (6), resulting from the shock-wave compression. Now, the temperatures measured by the pyrometer exceeding 1500 K far before the superdetonation formation, increase regularly to reach 2500 K at t_1. It is therefore delicate, in these conditions, to define an induction time.

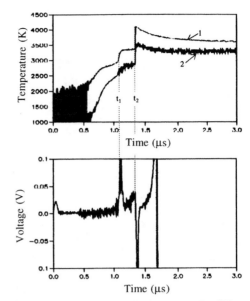

FIGURE 2. Sample of luminance temperatures (1=>500 nm and 2=>1510 nm) and polarization signal.

On the other hand, this gradual heating of the NM can be expressed in terms of two Arrhenius kinetic laws (11) : the first one, with a low activation energy (about a few kcal/mol), describing the first reactions rate, while the other, with a high activation energy (50 kcal/mol), would become dominant just prior to the superdetonation formation. We can also make the assumption that the temperature increase in the first part of the pyrometric signals has a physical origin such as gas compression inside micro-bubbles. The collapse of the bubbles, because of their small size, would not lead to NM chemical reaction.

Three plane shock experiments have been performed on NM at several pressure levels. The characteristics of these shots, and the results obtained, are displayed in Table 1. The pressure P, the particle velocity u and the density ρ induced in shocked NM were calculated from the projectile velocity v, using Rankine-Hugoniot equations, and Hugoniot for copper (D=3.94+1.489u (12)) and NM (D=1.647+1.637u (13)). Times t_1 and t_2, corresponding respectively to the superdetonation formation and to this superdetonation overtaking the initial shock, were read on the pyrometric signals. The depth x_2 at which the superdetonation overlaps the initial shock was inferred from the results provided by the piezo-electric pins.

TABLE 1. Characteristics of shots, and main results obtained

Shot No.	v m/s	P GPa	u m/s	ρ kg/m^3	t_1 μs	t_2 μs	x_2 mm	D* m/s	x_1 mm
1	1934	8.61	1708 1686[a]	1842	1.08	1.35	6.0	7185	2.2
2	2129	9.99	1870	1881	0.43	0.63	3.3	7114	1.1
3	2374	11.84	2072	1926	0.13	0.17	1.0	6967	0.5

[a] sole value measured by Fabry-Perot interferometer

The Cheetah code (14), using a BKW type equation of state and data from the BKWC database, allowed the superdetonation velocity (D*) estimation from the P and ρ characteristics of shocked NM. All these values finally lead to the computing of the superdetonation build-up distance x_1, given by the formula:

$$x_1 = x_2 - D^*(t_2 - t_1) - u.t_1 \qquad (1)$$

According to these calculations and measurements, the superdetonation would not form, as proposed by Campbell and al, exactly at the barrier/NM interface, but at a run distance x_1 inside the explosive. x_1 ranges from 0.5 to 2.2 mm (Table 1), depending on the shock pressure. Because of the uncertainty on each term of the formula (1), this conclusion is mainly supported by the 8.61 GPa and 9.99 GPa shots results.

CONCLUSION

Joint use of various measurement means, including an ultra-fast multi-wavelength pyrometer, allowed us to clarify the various phases in the shock initiation of NM. The profile and the high levels of the temperatures measured before the superdetonation formation exclude any interpretation by the Arrhenius kinetic law widely used up to now. Furthermore, in contrast with the model proposed by Chaiken (1), our results also show that the superdetonation forms inside the explosive at a depth which depends on the shock pressure. These new results support the Sheffield hypothesis according to which the superdetonation formation is not located at the explosive-driver interface. A few more experiments are planned to determine the origin of the temperature increase prior to the superdetonation.

REFERENCES

1. Chaiken, R.F., *"The kinetic theory of detonation of high explosives"*, M.S.Thesis, Polytechnic Inst. of Brooklyn, 1957.
2. Campbell, A.W., Davis, W.C. and Travis, J.R., *Physics of Fluids* 4, 498-510, (1961).
3. Travis, J.R., Campbell, A.W., Davis, W.C. and Ramsay, J.B., "Shock initiation of explosives-III. Liquid explosives", presented at *the Coll. Int. C.N.R.S. No.109. Les Ondes de Détonation*, p45, Paris, 1962.
4. Dremin, A.N., Savrov, S.D. and Andrievskii, A.N., *Comb. Expl. and Shock Waves* 1, 1 (1965).
5. Walker, F.E. and Wasley, R.J., *Comb. and Flame* 15, 233-246 (1970).
6. Hardesty, D.R., *Comb. and Flame* 27, 229-251 (1976).
7. Presles, H.N., Fisson, F. and Brochet, C., *Acta Astronautica* 7, 1361-1371, 1980.
8. Kapila, A.K. and Dold, J.W.,"A theoretical picture of shock-to-detonation transition in a homogeneous explosive", presented at the *9th Symp. Int. on Detonation*, Portland, Oregon, Aug.28-Sep.1, 1989.
9. Sheffield, S.A., Engelke, R. and Alcon, R.R., "In-situ study of the chemically driven flow fields in initiating homogeneous and heterogeneous nitromethane explosives", presented at the *9th Symp. Int. on Detonation*, Portland, Oregon, Aug.28-Sep.1, 1989.
10. Travis, J.R., "Electrical transducer studies of initiation of liquid explosives", presented at the *4th Symp. Int. on Detonation*,White Oak, Maryland, Oct. 12-15,1965.
11. Kipp, M.E. and Nunziato, J.W.,"Numerical simulation of detonation failure in nitromethane", presented at the *7th Symp. Int. on Detonation*, Anapolis, Maryland, June 16-19, 1981.
12. Steinberg, D.J., *"Equation of state and strength properties of selected materials"*, LLNL, 1991.
13. Mader, *Numerical modeling of detonations*, University of California Press, 1979, p.157.
14. Fried, L.E., CHEETAH 1.39 User's Manual, Laurence Livermore National Lab., Energetic Materials Center, 1996.

CP429, *Shock Compression of Condensed Matter – 1997*
edited by Schmidt/Dandekar/Forbes
© 1998 The American Institute of Physics 1-56396-738-3/98/$15.00

FAILURE AND RE-INITIATION DETONATION PHENOMENA IN NM/PMMA-GMB MIXTURES

J. C. Gois, J. Campos, I. Plaksin and R. Mendes

Lab. of Energetics and Detonics, Mechanical Eng. Dept., Fac. of Sciences and Technology
University of Coimbra, 3030 Coimbra, Portugal

The addition of a small amount of glass microballoons (GMB) on nitromethane/polymethylmethacrylate (NM/PMMA) mixture reduces strongly the failure thickness and increases the detonation sensitivity. Based in an explosive mixture of NM/PMMA (96/4 by weight) with 1% of GMB (QCel 520 FPS, mean particle diameter of 45 μm), the failure and re-initiation phenomena are investigated experimentally using the corner turning experiments. The printed erosion figure on a polished copper plate, is used as a witness surface. This witness plate not only shows the shock oblique waves but also, around the corner, the re-initiation points and curves. In the dark zone two re-initiation fronts are observed. Two detonation curve fronts can be observed, corresponding to behaviours of homogeneous and heterogeneous explosives. In order to evaluate the influence of shear and normal stress waves, the original set-up has been modified, fixing a thin sheet of a kapton barrier (50 μm thickness) at different angles from the corner. It has been observed fundamental changes in the original pattern of printed flow lines. Near the corner, in spite of the influence of GMB, the energy release is not high enough to ensure a stable detonation regime. This regime is reached clearly after the two fronts of re-initiation are joined.

INTRODUCTION

The addition of small amount of inert particles to liquid nitromethane (NM) has long been known to drastically change its detonation failure thickness and shock sensitivity (1). Glass microballoons (GMB) were used with success to sensitize emulsion explosives (2). The same effect was observed by Presles *et al.* (3,4) and Gois *et al.* (5-7) for NM-polymethylmethacrylate (PMMA) mixture. The shock sensitivity of NM/PMMA-GMB was found to increase with GMB mass fraction increased and particle diameter decreased (6,7). This increased shock sensitivity is followed by the decreasing of critical diameter and failure thickness (7,8).

The reduction of critical diameter can generally be explained by an increasing of the energy release rate, caused by particle collapse inducing hot-spots. As the particle concentration increases, the density of re-initiating sites grows, compensating the effect of the rarefaction waves avoiding the break-down of the reaction zone in the detonation front.

In previous work Gois *et al.* (8), have presented the influence of mass fraction of GMB on failure thickness of NM/PMMA-GMB mixtures. The failure thickness determines the charge size at which there exists a threshold limit for the propagation of detonation under steady conditions. Under this value the detonation fails. At the critical value, the pressure pulse of the initiation shock front and the CJ plane are just critical for an initiation detonation, after a run length equal to the reaction zone length and a total delay time equal to the reaction time.

Dremin theory (9) explains the conical failure wave generated when the detonation front reaches the section, corresponding to an abrupt increasing of cylinder charge diameter. If the main diameter is less than the critical diameter, the conical wave is carried out to extinction.

To a better understanding of these phenomena and the contribution of a heterogeneous particle, inside a homogeneous explosive (NM/PMMA), corner turning tests have been performed. A copper witness plate and several strips of optical fibres are used to measure respectively the grid of perturbations, the curvature at the conical failure zone and the velocity of shock waves, at different angles from the corner.

EXPERIMENTS AND RESULTS

Mixtures

Four per cent by weight of PMMA was added to NM, to increase its viscosity to avoid buoyancy effects of GMB (6). GMB (QCel 520 FPS, supplied by Asko Inc.), have a wall thickness of about 1 μm and a diameter range from 16 to 79 μm. The mean diameter obtained by laser diffraction spectrometry is 45±1 μm. The effective density measured by helium picnometry is 220±3 kg.m^{-3}. Adding 1% by weight of GMB to NM/PMMA matrix an initial density of 1090±5 kg.m^{-3} was obtained. The experiments were performed at an initial temperature of 15±3°C. The initiation of detonation of NM/PMMA-GMB mixture was performed with a PBX booster, based in RDX, with a D_∞=6200 m.s^{-1}.

Corner turning propagation

Corner turning propagation experiments have been performed for NM/PMMA-GMB mixture using copper and PMMA as confinement. A channel of 6 mm thickness was used with copper confinement. A 12 mm thickness was used with PMMA. Both configurations of the channel had 10 mm height and 80 mm length. The Fig. 1 shows the

typical experimental apparatus for two different set-up, using optical fibres. The first configuration (a) was adopted to measure the conical failure of detonation. The second set-up (b) was used to measure the extension of the dark zone and evaluate the radial shock wave propagation. The measurement is based on the optical signal, by 64 optical fibres strip. The strip was connected to a fast electronic streak camera (THOMSON TSN 506N).

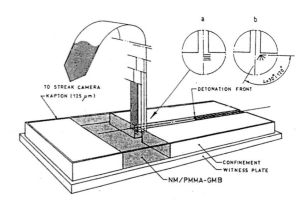

FIGURE 1   Experimental corner turning set-up.

Fig. 2 and 3 show typical results. Fig. 2 shows the change of curvature of the detonation front at the conical failure. An increasing of the conical angle, formed by printed flow lines from the corner to the vertex of the cone, is observed. This behaviour is in agreement with the decreasing of detonation front velocity, in the same region. After the vertex of the failure cone, the detonation velocity increases again.

The Fig. 3 shows the signal records of shock wave velocities at 0°, 45° and 90° from the corner, taking as reference the direction of the flow in the channel. This picture shows the shock wave velocity changing with the angle. When the angle increases, the shock wave velocity decreases. At perpendicular direction, immediately at the corner, no light was observed. A typical signal of curvature is observed at the last part of this record. A dark zone exists around the corner for an angle less than 45°.

692

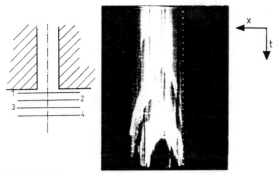

FIGURE 2. Experimental signal record obtained from the set-up of Fig. 1. using four strips of optical fibres spaced approximately 2.15±0.1 mm.

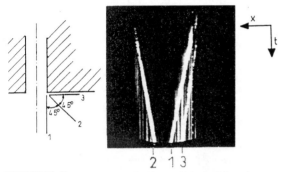

FIGURE 3. Experimental signal record obtained from the set-up of Fig.1. at the angles 0, 45 and 90° from the corner.

When a sheet of kapton of 50 μm thickness is fixed, from the corner, at different angles, the radial velocity of shock wave changes. For an angle of 30°, the radial velocities of shock wave are almost the same at 0° and 45°. No light was observed at 90° immediately at the corner. When the angle of the kapton sheet is 60°, the shock wave velocity at 45° is higher than at 0°. At 90° no signal was observed around the corner.

The erosion picture of the witness copper plate allows to observe the angle of conical failure, the radius of dark zone and the re-initiation trajectories of detonation products. Fig. 4 shows a typical erosion picture, around the corner, after being scanned and optimized. This picture allows to evaluate the extension of dark zone and its radius

(R), the angle of conical failure (α), the shape of re-initiation, as it is describe by Dremin theory and experimentally observed in NM/PMMA mixtures (7), and the shape of a semi-circular re-initiation front, characteristics of a spherical diverging detonation.

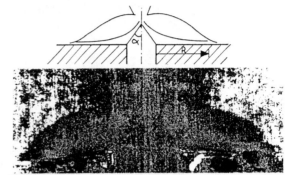

FIGURE 4. Erosion picture on witness cooper plate around the corner.

Without kapton sheet at the corner, an angle between 34° and 43° was measured in the conical failure, (mean value was 36°). The radius of dark zone, around the corner, was 16±0.5 mm. The two fronts of re-initiation, described above, start at the vertex of the conical failure, following the shape of an ellipse, until being joined at the radius of the dark zone. The interaction of the two fronts of re-initiation, allows to identify clearly the space of cellular structure of detonation front, when the detonation regime is reached.

By the orientation of the crater induced on the witness copper plate, by the collapse of GMB, the trajectory of the shock front around the corner was identified. Fig. 5 shows a photo, with magnification x55, obtained by optical microscope, of the region where the interaction of the two fronts of re-initiation is joined. In the dark zone, at the corner (*vd*. Fig. 6), small craters were observed, following an orientation with an angle near 45°. A complementary experiment was made using two opposite corner configurations. The result of this experiment is shown in Fig. 7, where it is identified clearly the angle of the interaction between the two waves, induced from the two opposite corners, in a channel with 10 mm thickness and 10 mm height.

The measured angle γ, referred to the direction of the flow in the channel, is 63.5°.

The orientation of the induced craters turns from an angle near 45°, at the corner, to one higher than 90°, immediately after the radius of dark zone.

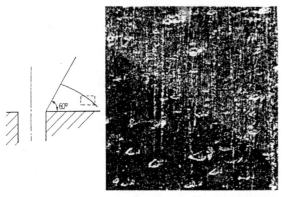

FIGURE 5. Trajectories of the craters, induced by the collapse of GMB, at the interaction of the two fronts of re-initiation.

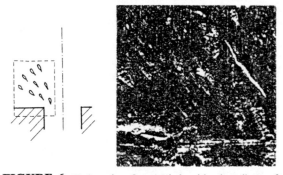

FIGURE 6. Trajectories of craters induced by the collapse of GMB at the corner.

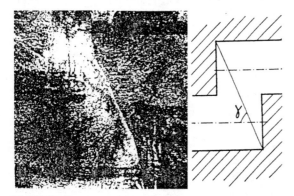

FIGURE 7. Angle induced by the interaction of two detonation fronts from two opposite corners.

DISCUSSIONS AND CONCLUSIONS

The structural features of heterogeneous explosives are difficult to identify experimentally. In the present work the angle of conical failure, the extension of dark zone and two re-initiations fronts starts at the vertex of the conical failure were evaluated. The presence of GMB induced hot spots into the reaction zone, just ahead of the front, is observed. This phenomenon initiates a hemispherical detonation front, moving into the induction zone and eventually overtaking the front. The cellular structure results from the collisions of such microexplosions. The interparticle distance of GMB determines the grid of regular pattern.

REFERENCES

1. Engelke R (1979) Effect of a physical inhomogeneity on steady-state detonation velocity. In Phys. Fluids, vol. 22, (9), pp. 1623-1630.
2. Yoshida M, Ilda M, Tanaka K, Fujiwara S, Kusakabe M, Shiino K (1985) Detonation behaviour of emulsion explosives containing microballoons. In Proceedings of the 8th Symposium (International) on Detonation, NSWC MP 86-194, Albuquerque, USA, pp. 993-1000.
3. Presles HN, Campos J, Heuzé O, Bauer P (1989) Effects of microballoons concentration on the detonation characteristics of nitromethane-PMMA mixtures. In Proceedings of the 9th Symposium (International) on Detonation, OCNR 113291-7, Portland, Oregon, USA, pp. 362-365.
4. Presles HN, Vidal P, Gois JC, Khasainov BA, Ermolaev BS (1995) Influence of glass microballoons size on the detonation of nitromethane based mixtures. In Shock Waves Journal, vol. 4, (6), pp. 325-329.
5. Gois JC, Presles HN, Vidal P (1993a) Effect of hollow heterogeneities on nitromethane detonation. In: Progress of Astronautics and Aeronautics ed, vol. 153 - Dynamics Aspects of Detonations, pp. 462-470
6. Gois JC, Campos J, Mendes, R (1993b) Shock initiation of nitromethane-PMMA mixtures with glass microballoons. In the Proceedings of the 10th Detonation Symposium, Boston, pp. 758-765.
7. Gois, J.C. (1995) Influência das micro esferas ocas de vidro na detonação da mistura nitrometano-polimetilmetacrilato. Ph.D. Thesis, University of Coimbra, Portugal.
8. Gois JC, Campos J, Mendes, R (1995) On extinction detonation behavior of NM-PMMA-GMB mixtures. In Proceedings of the Conference of the American Physical Society on Shock Compression of Condensed Matter, Seattle, Washington, Part 2, pp. 827-830.
9. Dremin, A.; Rozanov, S.D. and Trofimov, V.S. (1963) On the detonation of nitromethane. In Combustion and Flame, vol. 7, pp. 153-162.

CP429, *Shock Compression of Condensed Matter – 1997*
edited by Schmidt/Dandekar/Forbes
1998 The American Institute of Physics 1-56396-738-3/98/$15.00

Shock Reactivity of the Liquid Monopropellant Otto Fuel II

H. W. Sandusky and G. P. Chambers

Naval Surface Warfare Center, Indian Head Division, Indian Head, MD 20640-5035

When Otto Fuel was formulated in the 1960s, it would sustain detonation in a schedule 40 pipe with an inner diameter of 34.9 mm. Shock loading experiments with 95.2 and 50.8 mm diameter, pressed (1.56 g/cc) pentolite donors were recently conducted on Otto Fuel while heavily confined by thick-wall steel tubes with an inner diameter of 50.6 mm. When the donor shock was attenuated to the detonation pressure of Otto Fuel, the larger donor produced a high-velocity detonation. The smaller donor produced a low-velocity detonation, indicating that Otto Fuel is less detonable than it was in the 1960s. Other shock loading experiments using the smaller pentolite donor were conducted with less confinement from a thick-wall Lexan tube surrounding an aluminum tube with an inner diameter of 25.4 mm. Reaction had failed within 100 mm for a shock loading near the detonation pressure of Otto Fuel. When the aluminum tube was lined with bubble pack, which introduced an average but nonuniform porosity of 25%, violent reactions were attained for shock inputs ranging from 112 to 22 kbar.

INTRODUCTION

The potential hazards of the liquid propellant Otto Fuel II were evaluated after its formulation by Adams et al. (1). Detonation propagated over a filled 0.76 m length of steel pipe with a 34.9 mm inner diameter and 3.6 mm wall thickness (1 3/8" schedule 40 pipe) and failed to propagate detonation in a steel tube with a 25.4 mm inner diameter but a thicker wall of 9.6 mm. In these tests, the Otto Fuel was self-boosted by a larger diameter column (50.8 mm diameter by 127 mm long) to avoid over-driven detonation; this section of Otto Fuel was initiated by a 57.2 mm diameter by 171.5 mm long column of C-4 explosive. The large explosive booster appears to be necessary to initiate detonation in this size of sample. Otto Fuel failed to detonate when directly shocked with the 50.8 mm diameter by 50.8 mm long explosive column of 1.56 g/cc pressed pentolite in the large scale gap test (LSGT), which has a thicker (5.6 mm) and slightly larger inner diameter (36.5 mm) steel tube for confining the sample (2). Increasing the tube size to 50.5 mm inner diameter with a 6.35 mm wall thickness, while maintaining the same donor, did

result in detonation at ambient conditions for a gap thickness of 3.4 mm. Substantial confinement is also necessary to initiate detonation. When confined by only a cardboard tube, 152.4 and 203.2 mm diameter columns of Otto Fuel would not detonate.

Since hazards seldom involve non-ideal conditions, Adams et al. (1) bubbled carbon dioxide through Otto Fuel in a LSGT arrangement and attained detonation for a gap thickness of 6.6 mm. This is still shock insensitive behavior, relative to initiation at a gap of 17.8 mm as required for labelling an energetic material as a mass detonating explosive. In the various tests by Adams et al., high-velocity detonation (HVD) was observed. At the time of the Otto Fuel evaluation, low-velocity detonation (LVD) in liquid explosives was being studied by others (e.g. reference (3)). LVD can be a significant hazard because it is initiated with much less stimulus. Propagation occurs by reaction at cavitation sites created by a precursor wave in the confinement. This is similar to the increased sensitivity from bubbles in the Otto Fuel, but different in that no precursor wave was necessary to generate those bubbles.

695

The present study briefly revisited the sensitivity of Otto Fuel to HVD, in part to determine if the sensitivity has changed during the three decades after the original study. Most of the study, however, was devoted to determining conditions at which LVD could be observed.

EXPERIMENTAL ARRANGEMENT AND RESULTS FOR EXAMINING HVD

The arrangement shown in Figure 1 and a variation in which only the LSGT donor shock loaded the Otto Fuel were used in two experiments. The 690 g composite donor shown in Figure 1 combines a 160 g LSGT donor and one pellet from the expanded large scale gap test (ELSGT) donor. The longer shock pulse duration from this composite donor, relative to a LSGT donor, is probably more characteristic of the reaction time of Otto Fuel in a detonation wave. Over-driven detonation was avoided by attenuating the donor pulse with a PMMA gap to the ~12.1 GPa detonation pressure of Otto Fuel. About half of the

PMMA gap (T-Plug) was inserted into the tube past a shallow slot its top. By tilting the slotted side higher, the gap could be inserted while pushing out any entrapped air on top of the Otto Fuel. This technique was also used in the arrangement for the LVD experiments.

The distance-time data from ionization probes that projected 2mm into the Otto Fuel are shown in Figure 2 for both experiments. The larger donor initiated a steady HVD wave with an average front velocity of 5.8 mm/μs, which produced a 12.7 mm depression in the witness block. By comparison, the depression from this donor when the tube was filled with water was 2.9 mm. When the Otto Fuel was shock loaded by the smaller LSGT donor, the front velocity quickly declined to an average 4.5 mm/μs. The 6.3 mm depression in the witness block and a manganin gage measurement of ~6 GPa at the bottom of the Otto Fuel column verify a lower order event.

EXPERIMENTAL ARRANGEMENT AND RESULTS FOR EXAMINING LVD

The arrangement shown in Figure 3 was used in most experiments. The aluminum tube simulated a transport line and the Lexan tube simulated the confinement provided by immersing the transport line in liquid. In each experiment, except the first one, the inside of the aluminum tube was lined with bubble pack to introduce a nonuniform but repeatable porosity to encourage LVD. The first experiment had no bubble pack to verify that HVD or LVD would not

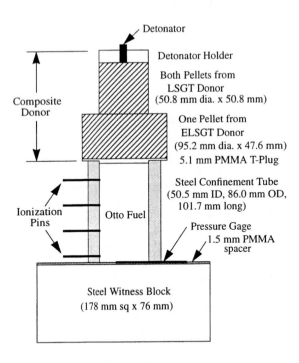

FIGURE 1. Arrangement for examining HVD with shock loading by a composite pentolite donor.

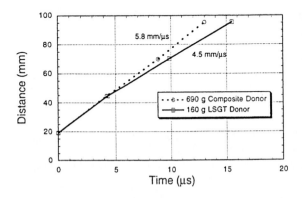

FIGURE 2. Reaction front progress in the HVD experiments.

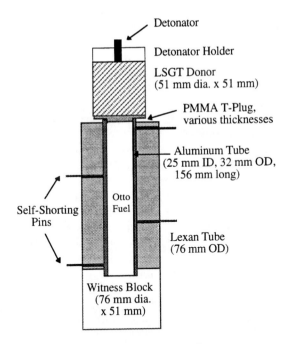

FIGURE 3. Arrangement for examining LVD.

TABLE 1. Summary of LVD Experiments with Arrangement in Figure 3.

Shock Input (GPa)	Porosity (%)	Final Vel. (mm/µs)	Recovered Al Tube
11.2	0	-	Bottom intact
11.2	25	2.4	Fragmented
4.7	25	2.5	Fragmented
2.2	25	2.8	Fragmented
1.9	25	2.2	Fragmented
1.9	12	-	Intact
1.7	25	-	Intact
1.0	25	-	Intact

propagate at half the diameter and with much less confinement than for the arrangement in Figure 1. In contrast to that arrangement, the passage of a shock/reaction front in the Otto Fuel was determined by self-shorting probes placed in contact with the outside of the aluminum tube. For the summary of data in Table 1, the reported velocity is that based on the response of the final probes.

In the initial experiment without any bubble pack, the 11.2 GPa shock input was near the detonation pressure of Otto Fuel. The front velocity in the first section of the tube was only 1.6 mm/µs, which would be near the sound velocity. There was no propagation of reaction based on the weak or missing response of probes near the bottom of the tube and the intact recovery of the last 50 mm of the tube.

A series of experiments were conducted with the inner wall of the tube fully lined with bubble pack, providing an average 25% porosity (9.5 mm dia. by 2.8 mm high bubbles spaced 10.9 mm apart in a triangular pattern). As shown in Table 1, propagating reaction occurred for shock loadings ranging from 11.2 to 1.9 GPa, but failed for loadings of 1.7 and 1.0 GPa. The propagating reactions are designated as LVD because 1) the front velocity was >2 mm/µs near the

end of the sample, 2) the entire aluminum tube fragmented, and 3) there was no remaining trace of Otto Fuel after the experiments. There was, however, no depression in the steel witness blocks. The two experiments with input pressures of 1.7 and 1.0 GPa had aluminum witness blocks in an attempt to obtain a depression, but LVD was not initiated.

Distance-time data from the self-shorting probes in those experiments exhibiting LVD are shown in Figure 4. The probe responses for the three highest input pressures (11.2, 4.7, and 2.2 GPa) were similar, with a somewhat increasing velocity for decreasing shock input. This effect may occur because a stronger shock input causes more damage to the apparatus, thereby reducing the end confinement. The probe responses for

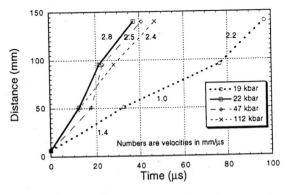

FIGURE 4. Progress of fronts in experiments exhibiting LVD.

697

the 1.9 GPa input indicates that the onset of LVD nearly failed. This experiment was also instrumented with a carbon resistor gage at the bottom of the Otto Fuel column for measuring pressure. The gage recorded a precursor wave with a peak pressure of 0.17 GPa followed 2 μs later by a wave that reached a peak of 3.0 GPa.

When reducing the porosity to 12% within the same arrangement by placing just four rows of bubbles lengthwise along the inner wall, LVD was not initiated at an input pressure of 1.9 GPa. In another modification of the arrangement, the inner wall was again lined with bubble pack but the explosive donor was exchanged for a detonator (0.2 g of explosive) inserted into the end of the aluminum tube in direct contact with the Otto Fuel. There was little reaction and most of the fuel was recovered. A third modification had a stainless steel tube (22.1 mm inner diameter and a 1.7 mm wall thickness) without the surrounding Lexan tube to simulate another type of fuel transport line. The inner wall was fully lined with bubble pack that had a smaller thickness for maintaining about the same porosity as in the larger diameter aluminum tube. For a porosity of 22% and a shock input of 1.9 GPa, the bottom end of the tube was recovered intact, indicating that reaction did not propagate.

DISCUSSION AND CONCLUSIONS

The column length of Otto Fuel in the HVD experiments was too short to determine if steady-state conditions were achieved, but sufficient to determine if detonation could be initiated. Even with high confinement and a shock input at the detonation pressure of Otto Fuel, a large donor is required to initiate HVD at a velocity of 5.8 mm/μs; therefore, detonation would fail in the LSGT, as reported by Adams et al. (1). It was also previously reported (1) that detonation propagated in a modified version of a LSGT with a larger, 50.8 mm diameter sample, whereas a higher confinement version of the same arrangement did not achieve HVD in the present study. There is the possibility that a near-HVD event in the previous study was mistaken for HVD because of punching a standard witness plate. The 4.5 mm/μs wave in the present study may have cleanly punched a witness plate, whereas the thicker witness block, as well as ionization pin and pressure gage

measurements, in the present study distinguished a lower-order event. Assuming the modified LSGT results of the previous study were correctly interpreted, it is concluded that Otto Fuel is somewhat less sensitive to HVD than when originally formulated.

Relative to HVD, LVD can propagate with much less shock input (a factor of six) in a smaller diameter sample that has considerably less confinement, if there is extensive porosity. For a 1.9 GPa shock input, which was near the threshold in the shown arrangement when there was 25% porosity in the sample, either reducing the porosity by a half or reducing the confinement did not permit LVD. The ~2.5 mm/μs propagation velocity for LVD was relatively consistent over a range of shock inputs from 11.2 to 1.9 GPa and was about half of the velocity observed for HVD.

REFERENCES

1. Adams, C. L. et al., "Hazard Evaluation of Otto Fuel II," Minutes of the Eight Explosives Safety Seminar on High-Energy Propellants, Huntsville, AL, Aug 9 -11, 1966.
2. Price, D. et al., "The NOL Large Scale Gap Test. III. Compilation of Unclassified Data and Supplementary Information for Interpretation of Results," NOLTR 74-40, Naval Surface Warfare Center, Mar 1974.
3. Watson, R. W. et al., "Detonations in Liquid Explosives - The Low-Velocity Regime," Proceedings of Fourth Symposium (International) on Detonation, ACR-126, Office of Naval Research, Oct 1965, pp. 373-380.

CP429, *Shock Compression of Condensed Matter – 1997*
edited by Schmidt/Dandekar/Forbes
© 1998 The American Institute of Physics 1-56396-738-3/98/$15.00

SHOCK-INITIATION CHEMISTRY OF NITROARENES

Lloyd L. Davis[†] and Kay R. Brower[‡]

[†]*DX-1, Mail Stop P-952, Los Alamos National Laboratory, Los Alamos, NM 87545*

[‡]*Department of Chemistry, New Mexico Institute of Mining and Technology, Socorro, NM 87801*

We present evidence that the shock-initiation chemistry of nitroarenes is dominated by the intermolecular hydrogen transfer mechanism discussed previously. The acceleration by pressure, kinetic isotope effect, and product distribution are consistent with the bimolecular transition state rather than rate-determining C-N homolysis. GC-MS analysis of samples which were subjected to a shock wave generated by detonation of nitromethane shows that nitrobenzene produces aniline and biphenyl, and *o*-nitrotoluene forms aniline, toluene, *o*-toluidine and *o*-cresol, but not anthranil, benzoxazinone, or cyanocyclopentadiene.   In isotopic labeling experiments *o*-nitrotoluene and TNT show extensive H-D exchange on their methyl groups, and C-N bond rupture is not consistent with the formation of aniline from nitrobenzene or nitrotoluene, nor the formation of *o*-toluidine from *o*-nitrotoluene.   Recent work incorporating fast TOF mass spectroscopy of samples shocked and quenched by adiabatic expansion indicates that the initial chemical reactions in shocked solid nitroaromatic explosives proceed along this path.

INTRODUCTION

The role of intermolecular reactions in the decomposition of nitroarenes in supercritical aromatic solvents was discussed previously by Minier et al.(1). It was subsequently asserted by Stevenson et al. (2) and Brill and James (3, 4) that C-NO_2 homolysis is the dominant reaction in the shock initiation of nitroarenes, based on the extrapolation of Tsang's (5) gas-phase shock tube Arrhenius plots to high temperatures. Brill and James (4) also state "Almost all studies employ only temperature as the intensive variable. The role of pressure on the rates and mechanisms of reactions of nitroaromatic explosives is essentially unknown."

When the reported pressure and kinetic isotope effects are considered, together with the observed product distributions, it is apparent that direct C-NO_2 homolysis is not the dominant mechanism in the shock-initiated reactions of nitroaromatic compounds.   Decomposition of nitroarenes is known to be strongly accelerated by pressure (1, 6). Minier et al. (1) reported large negative activation volumes, which is inconsistent with the mechanism that Brill proposed.

While rate-determining C-NO_2 homolysis of nitroaromatic compounds is known in the gas phase (5) it has never been documented in condensed-phase experiments. High shock or static pressure reduces the rate of this reaction by a factor of fifty, and accelerates the hydrogen abstraction mechanism by two to three orders of magnitude. We previously stated (6) that the shock-initiation chemistry of nitroarenes is dominated by intermolecular hydrogen atom transfer to the oxygen atom of the nitro group, as discussed previously.

NITROBENZENE

Minier et al. (1) showed that the decomposition of nitrobenzene between 590 and 650 K at 150 to 1360 bar, both neat and in supercritical aromatic solvents, proceeds via formation of a charge transfer complex and hydrogen atom transfer to the oxygen of the nitro group. H-atom transfer lowers the barrier to C-NO$_2$ homolysis significantly; the observed activation energy for the condensed-phase decomposition of nitrobenzene is less than that for the gas-phase Ar-NO$_2$ homolysis reaction by 126 kJ/mol. Expulsion of nitrous acid competes with homolysis of the HO-N bond yielding the hydroxyl radical and nitrosobenzene, which is directly reduced to aniline.

The large negative activation volume for nitrobenzene in benzene, $\Delta V^{\ddagger} = -46 \pm 6$ mL/mol., indicates that bond formation is accompanied by charge separation in the transition state. For nitrobenzene in perdeuterated benzene a primary kinetic isotope effect of 2.5 was found, indicating that abstraction of hydrogen from solvent is the rate-determining step. No isotope effect was found for perdeuterated nitrobenzene in benzene. Substituting toluene as the solvent increases the rate significantly due to the relative ease of abstracting benzylic hydrogen versus aryl.

In the shock recovery experiments discussed previously (6-8) nitrobenzene was decomposed by a shock generated by detonation of nitromethane. Decomposition of nitrobenzene in benzene, cyclohexane or toluene solvents produces aniline as well as benzene. Isotopic labeling shows that the original C-N bond is preserved in the product aniline. The observed rate is consistent with the Arrhenius parameters and pressure effect measured previously for the condensed-phase reduction. Aniline formation, discussed in detail by Minier, is inconsistent with decomposition via direct Ar-NO$_2$ homolysis.

o-NITROTOLUENE

Decomposition of o-nitrotoluene also proceeds via intermolecular hydrogen abstraction, although a competitive intramolecular cyclization reaction, leading to the formation of anthranil and its subsequent decomposition products, was also discussed by Minier et al. (1) and previous researchers noted therein. An analogous mechanism has been proposed by Sharma et al. for the thermal decomposition of TATB (9). For *o*-nitrotoluene in perdeuterated benzene a primary kinetic isotope effect of 1.7 was measured. Anthranilic acid, aniline, o-aminobenzaldehyde, benzoxazinone, and benzoylene indazole are minor products characteristic of the intramolecular cyclization reaction, although aniline may also result from oxidation of the methyl group of o-toluidine followed by decarboxylation. Stevenson et al. (2) raise the possibility that anthranil undergoes a ring-contraction reaction to cyanocyclopentadiene.

In shock recovery experiments using benzene solvent, the dominant product is o-toluidine; no other toluidine isomers are observed. Toluene and aniline are found as well, along with biphenyl. There are no detectable amounts of the products diagnostic of the intramolecular cyclization reaction; and no cyanocylopentadiene is observed. In our shock recovery experiments there is a primary deuterium kinetic isotope effect of 1.5, which compares favorably with the value of 1.7 for the static experiments.

Shock-induced decomposition of *o*-nitrotoluene in perdeuterated benzene solvent reveals that all of the products listed above and the starting material show H-D exchange, and the biphenyl is perdeuterated which indicates it is formed from coupling of phenyl radicals. Toluene formed in these experiments has incorporated one or two deuterium atoms, indicating that it is formed from o-nitrotoluene instead of methylation of benzene; and this is also true of the aniline. *O*-toluidine shows some H-D exchange but is predominantly protic, which indicates that it originated directly from the starting material by reduction of the nitro group. All of these observations support the mechanism proposed by Minier et al. (1).

Figure 1 shows some of the results obtained from GC-MS analysis of a shock-recovery experiment performed on *o*-nitrotoluene in perdeuterated benzene; a small quantity of β-pinene is also present as an internal thermometric standard. (8) This sample was confined in a steel capsule and shocked by detonation of nitromethane with 10% acetone and 2.5% DETA; the extent of reaction was 56%. Mass spectra are shown for the products with mechanistic significance.

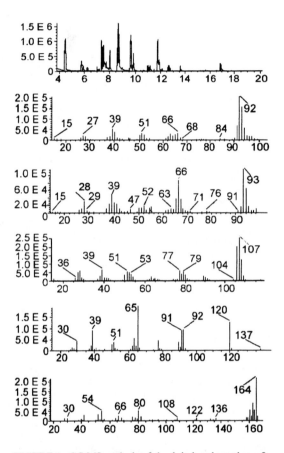

FIGURE 1. GC-MS analysis of shock-induced reaction of *o*-nitrotoluene in perdeuterated benzene, showing isotope distributions in the mass spectra of selected products. Top is total ion chromatogram, followed by mass spectra for toluene, aniline, o-toluidine, unreacted o-nitrotoluene, and perdeuterated biphenyl. All compounds have been positively identified and are consistent with Minier's mechanism.

TATB AND TNT

Time-of-flight mass spectral data for the shock-induced decomposition of TATB (1,3,5-triamino-2,4,6-trinitrobenzene) has been discussed previously. (10-12) This data was obtained with a unique collisionless mass spectrometer, designed and built specifically for studying the shock intitiation chemistry of explosives, which is described elsewhere. (13-15) Thorough calibration of the electric slappers using VISAR has been performed by Engelke and Blais. (16) The shock delivered to the samples is well-characterized; the reactive intermediates are expanded adiabatically into a hard vacuum and sampled via a molecular beam formed from a central region of the expansion fan, where compression and decompression is effectively uniaxial.

By comparing mass spectra recorded at different input stresses and examining the number of counts as a function of time for different peaks Östmark (12) showed that a mass peak at m/z = 242 is an authentic reaction product from the shock-induced decomposition of TATB, and not an artifact of the electron gun, nor present in the cracking pattern of unreacted TATB. Through deuterium and nitrogen-15 isotopic labeling, Greiner et al. (10) and Östmark (12) also demonstrated that this corresponds to loss of oxygen, and not NH_2, from the parent molecule. As in the simpler nitroaromatic compounds, direct reduction of nitro- to nitroso- is consistent with the intermolecular hydrogen atom transfer mechanism discussed previously. In the (unlabeled) TATB spectra reported previously (10) there is also a peak at m/z = 228, or 30 mass units lighter than the starting material. Greiner et al. identified this as loss of NO from the parent molecule. We attribute this to hydrogen atom transfer followed by C-N fission, rearrangement and recombination, and subsequently loss of nitroxyl as discussed previously. (1)

Shackelford (17-19) and Rogers, Janney, and Ebinger (20) have previously documented a primary deuterium kinetic isotope effect (DKIE) in the decomposition of TATB, TNT, and other solid explosives. They clearly state that homolytic C-H bond rupture must be the rate controlling step in the decomposition of these compounds, and Shackelford cites additional evidence obtained via electron paramagnetic resonance (EPR) spectroscopy.

Figure 2 shows a high-mass scan (near the mass of parent molecule) from the detonation mass spectrometer studies of TNT reported previously. (10) It is apparent that, as with TATB, loss of oxygen from the parent molecule dominates the process and any contribution from C-N homolysis (m/z=181) is negligible.

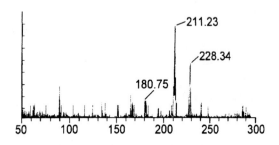

Figure 2. Scan 20 from detonation mass spectrometer study of TNT. Note that a mass peak at m/z=228 is recorded while the mass of TNT is 227. Ratio of relative intensities for peaks at m/z=211 to the starting material parent ion changes significantly with time, indicating that m/z=211 is due to formation of a reaction product and not just the cracking pattern of TNT.

CONCLUSION

There is a significant body of evidence supporting the contention that shock-induced reaction in nitroaromatic compounds proceeds through intermolecular hydrogen abstraction. In shock recovery experiments on nitrobenzene, *o*-nitrotoluene, and TNT we observe a primary DKIE and isotopic scrambling on methyl substituents. Product distribution is consistent with the hydrogen abstraction mechanism and not rate-determining C-N homolysis. Results of studies performed on solid nitroaromatic explosives with the LANL detonation mass spectrometer are consistent with this mechanism.

ACKNOWLEDGEMENTS

The authors thank Ray Engelke and Norm Blais for providing the data in Fig. 2, and for many hours of thoughtful discussions.

REFERENCES

(1) Minier, L.M.; Brower, K.R.; Oxley, J.C. *J. Org. Chem.* **56**, 3306-3314, 1991.

(2) Stevenson, C.D.; Garland, P.M.; Batz, M.L. *J. Org. Chem.* **61**, 5948-5952, 1996.

(3) Brill, T.B.; James, K.J. *J. Phys. Chem.* **97**, 8759-8763, 1993.

(4) Brill, T.B.; James, K.J. *Chem. Revs.* **93**, 2667-2692, 1993.

(5) Tsang, W.; Robaugh, D.; Mallard, G.D. *J. Phys. Chem.* **90**, 5968-5973, 1986.

(6) Davis, L.L.; Brower, K.R. *J. Phys. Chem.* **100**, 18775-18783, 1996.

(7) Davis, L.L.; Brower, K.R. *Rev. Sci. Instrum.* **66**, 3321-3326, 1995.

(8) Davis, L.L.; Brower, K.R. Shock Compression of Condensed Matter - 1995; AIP: New York, 1996; pp. 775-778.

(9) Sharma, J.; Forbes, J.W.; Coffey, C.S.; Liddiard, T.P. *J. Phys. Chem.* **91**, 5139-5144, 1987.

(10) Greiner, N.R.; Fry, H.A.; Blais, N.C.; Engelke, R. in Proceedings of the Tenth Symposium (International) on Detonation; ONR: Arlington, 1990; pp. 583-568.

(11) Blais, N.C.; Greiner, N.R.; Fernandez, W.J. in Chemistry and Physics of Energetic Materials, S.N. Bulusu, ed., Kluwer Academic; London, 1990; pp. 477-509.

(12) Östmark, H. in Shock Compression of Condensed Matter - 1995; AIP: New York, 1996; pp. 871-874.

(13) Blais, N.C.; Fry, H.A.; Greiner, N.R. *Rev. Sci. Instrum.* **64**, 174-183, 1993.

(14) Greiner, N.R.; Blais, N.C. in Proceedings of the Ninth Symposium (International) on Detonation; ONR: Arlington, 1990; pp. 953-961.

(15) Greiner, N.R. in Chemistry and Physics of Energetic Materials, S.N. Bulusu, ed., Kluwer Academic; London, 1990; pp. 457-475.

(16) Engelke, R.; Blais, N.C. *J. Chem. Phys.* **101**, 10961-10972, 1994.

(17) Shackelford, S.A.; Beckmann, J.W.; Wilkes, J.S. *J. Org. Chem.* **42**, 4201-4206, 1977.

(18) Shackelford, S.A. in Chemistry and Physics of Energetic Materials, S.N. Bulusu, ed., Kluwer Academic; London, 1990; pp. 413-432.

(19) Shackelford, S.A. in Chemistry and Physics of Energetic Materials, S.N. Bulusu, ed., Kluwer Academic; London, 1990; pp. 433-456.

(20) Rogers, R.N.; Janney, J.L.; Ebinger, M.H. *Thermochimica Acta*, **59**, 287-298, 1982.

CP429, *Shock Compression of Condensed Matter – 1997*
edited by Schmidt/Dandekar/Forbes
© 1998 The American Institute of Physics 1-56396-738-3/98/$15.00

SHOCK TO DETONATION TRANSITION IN HOMOGENEOUS EXPLOSIVES

P.KLEVERT

Commissariat à l'energie Atomique
BP 7-77181 Country (France)
pklebert@computserve.com

We present in this paper some experimental results about the initiation of homogeneous explosives. We show that the process generally adopted for this initiation is not checked at the used level of pressure. According to some previous studies, we propose here an initiation process by hot spots.

INTRODUCTION

The initiation process of explosives has been the subject of many experiments and theoretical speculations. All these investigations have contributed to our understanding of the problem. Despite the great number of experiments performed, many uncertainties subsist and the present paper aims at clarifying the SDT in liquid explosives.

We have undertaken experimental studies of shock initiation and detonation properties for homogeneous explosives. In all these studies, liquid nitromethane was used as a test explosive.

PREVIOUS INVESTIGATIONS ON SHOCK INITIATION IN HOMOGENEOUS EXPLOSIVES

Differences between the response of heterogeneous and homogenous explosive to the initiation of detonation have been reported by (1), (2). In heterogeneous explosives, it is shown that the shock which is loaded into the explosive initiates some chemical reactions localized around voids or defects (hot spots) ; energy coming from those reactions leads it to accelerate. This phenomenon gives a smooth transition between the initial shock loaded into the explosive, and the detonation.

In homogeneous explosives, like nitromethane, most of the studies support the classic thermal ignition model according to a one dimensional scheme. The model can be described using the conventional classical time-space diagram (3) (Fig.1).

It is assumed that an inert shock wave with velocity D_c enters the explosive at the driver-plate (x=0, t=0) and heats it. After an induction time τ_i, the explosive heated by shock reacts at the interface, where the high temperature prevailed the longest. Upon the initiation a detonation wave is generated and propagated at celerity D_{sd} into the unreacted, compressed liquid explosive. This D_{sd} velocity is greater than the Chapman-Jouguet detonation velocity D_{cj} for the unreacted material.

D_{sd} is referred to as superdetonation wave velocity. This superdetonation wave overtakes the initial shock wave ; the resulting overdriven detonation relative to the unshocked explosive ahead of it eventually decays to a steady CJ detonation wave.

This model describes the shock initiation in homogeneous explosives using a first order Arrhenius law for chemical decomposition.

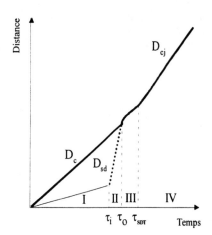

FIGURE 1. Shock to detonation transition in homogeneous explosive. I- initial 'inert' shock, II - Detonation in compressed NM, III-overshoot in initial NM, IV-Chapman-Jouguet Detonation.

Since 1958 however, a good many publications have begun to cast doubt on this view of things:
- Discrepancy appears for pressure level under 9 GPa,
- Initiation may come from hot spots (4),
All these studies point out that the shock initiation in homogeneous explosive is not necessarily completely described by an Arrhenius model.

PRESENT WORK

First, we have to determine the width of the SDT zone for a given shock pressure. Then, we record the associated particle velocity in this area. By those results, we are in a position to explain the shock to detonation transition in another way than by thermal explosion.

Experiments configuration and diagnostics

In most experiments, a plane wave lens was used to load a shock into the homogeneous explosive. It appears that this system can not allow us to know the exact level of pressure in the homogeneous explosive. Thus, we developed an experimental ∅100 mm diameter large system ; it allows to accelerate copper projectiles of 2200g to velocities in the range of 1.5mm/μs to 2mm/μs. We have chosen to work at a 8GPa pressure level.

First, using the vacuum pump the liquid nitromethane is lifted of bubbles after loading into the glass cup. The temperature of the explosive was carefully monitored using thermocouples placed in the neighborhood of the Cu-driver plate and in the upper part of the glass. The copper face of the liquid-cell nitromethane target was polished.

First step of the experiments

The goal is to determine the dimension of the shock to detonation transition zone. For that purpose, we used a streak camera. The glass tank is put in a background light produced by an argon flash ; the streak of the camera is placed in the middle of the tank. In that way, we could easily detect the transition zone though the darkness of the flow induced by the gas products of the detonation (Fig.3).

At the same time, laser doppler interferometry of the Cu/NM driver plate motion is performed by means of a 610 nm laser beam in order to check the imput pressure (5).

TABLE 1. Results from streak camera records

Shot	To (K)	Depth of catch	Overtake time : ι_0
1	307	10.7 mm	2.49 μs
2	301	12.8mm	3.01 μs
3	289	13.3mm	3.15 μs
4	299	12.3 mm	2.88 μs

The most important feature of these four shots is that we are able to record the evolution of the Cu/NM driver plate till the end of the transition zone when the signal vanishes abruptly (Fig.4). The corresponding time marks the end of the SDT (Zone III in Fig 2).

FIGURE 2. Smear camera record for shot n°1 (P=8 GPa,T=307 K).

TABLE 2. Summary of times obtained from the Streak Camera record and Velocity interferometer Camera Records

Shot	τ_0 from Streak Camera	τ_{SDT} from Streak Camera	End of LDI signal
1	2.49 µs	2.68 µsec	*
2	3.01 µs	3.60 µsec	3.66 µsec
3	3.15 µs	3.97µsec	4.01 µsec
4	2.88 µs	3.49 µsec	3.44 µsec

* Spike filter omitted from reflected beam optics 610nm

As we record the usual breaking of the slope in the time-space diagram, we believe that the initial shock is overtaken by a reactive wave which is coming from behind it. But concerning the nature of this wave, it's difficult to believe that this wave is a detonation (in a compressed material) because of the fact the laser beam is able to pass though this wave.

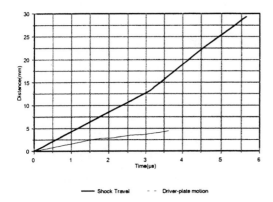

FIGURE 3. (t,y) diagram for Shot N°2

Second step of experiments

To continue our investigation, we record the particle velocity in the flow at different lagrangian position. For that purpose, we use the magnetic gauge measurement. The principle of it, was used by Shefflied (6).

The thickness of Al gauge which is made of Al is about 50 µm. It was put in the glass tank as shown by the Fig.4. The whole configuration is placed in a 600 Gauss magnetic field produced by a permanent magnet. The initial temperature is fixed at 34°C.

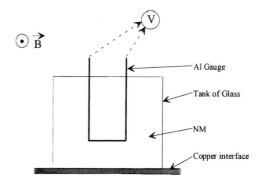

FIGURE 4. Experimental configuration for particle velocity measurements

We present three records in different lagrangian positions. The results of those different shots are shown in Fig.5.

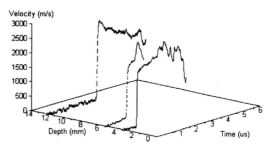

FIGURE 5. Velocity.vs.time evolution

DISCUSSIONS

All those experiments show some interessting features :

-First of all some remarks may be done concerning the LDI signals. We are able to follow the motion of the Cu/NM interface since the stable detonation takes place. This leads us to ask ourselves about the nature of the reactive wave which is supposed to arise inside the SDT zone. The composition of the detonation products in compressed and un-compressed NM are shown in the following table. Those results were obtained using the thermochemical code CARTE (7). It may be noticed, that the graphite proportion is greater in compressed than uncompressed medium.

TABLE 3. Detonation products composition (mole/kg)

Initial cond	P0, V0, E0	P=8GPa,V,E
nCO2	5,80	5,05
nGraph	6,66	11,32
nH2O	17,69	22,62
nCO	3,48	7,57E-5

- the particle velocity profile for this pressure level shows a progressive evolution and then leads to the detonation. It appears that the induction time of this reactive wave is lower than the one given by fisrt order Arrhenius law governing the thermal explosion model.

The process proposed in this paper takes into account the birth of a growing reactive wave in the compressed NM (growing of particle velocity) and the overtaking of the initial shock by this reactive wave (break of the slope in the (t,x) diagram). This wave is not a detonation wave since we record the LDI signals until the end of the SDT zone.

The initiation of this wave may be a 'chemical' process (with complex order of the reaction) or a 'mechanical' process as in solid explosives (hot spots). In the second one, the origin of hot spots may be localized around bubbles. The existence of these bubbles in degassed liquids has been show by (8). The difference with heterogeneous explosives is that an induction time exits for the hot spots formation. The process is going to evolve more and more and leads to a growing wave in the bulk of the explosive ; then this wave catches up with the initial shock and the resulting overdriven detonation decays into a CJ detonation. The results of (9) and (10) may be explained by this process of hot spots.

CONCLUSION

The experiments which have been performed, show that, at 8GPa pressure level, a new process of initiation takes place. LDI records and particle velocity measurements are at variance with the thermal explosion model which supports the birth of a detonation in the compressed explosive after an induction time at the interface. They suggest that reactive wavelets generated by hot spots lead to a reactive shock wave in the bulk of the compressed material. This new process takes place in a medium pressure range (8 to 9 GPa). We surmise that at high pressure the hot spots process will be too slow

compared to the chemical process. We consider submitting this presumption for further experiments.

ACKNOWLEDGMENTS

The author gratefully acknowledges members of the detonation physics group for helpful discussions. He directs special thanks to J.C. Protat, P. Manczur and D. Bouchu for their scrupulous experimental works

REFERENCES

1. CAMPBELL, A.W, DAVIS, W.C and TRAVIS, J.R, *Phys.Fluids*, **4**, 498-598 (1961).
2. CAMPBELL, A. W., DAVIS, W.C, RAMSAY, J. B. and TRAVIS, J.R., *Phys Fluids*, **4**, 511-517 (1960).
3. CHAIKEN, R. F," Kinetic theory of detonation of high explosives". Masters's thesis, Polytechnic Institute of Brooklyn (1958); also eight Symposium (international) on combustion, Baltimore Md., 759-764 (1962).
4. WALKER.F.E and WASLEY .R. J, *Combustion and Flame*, **15**, 233-334 (1970)
5. HARDESTY D. R, *Combustion and Flame*, **27**, 299-302 (1976).
6. SHEFFIELD, S. A, ENGELKE, R. and ALCON, R, "In-Situ Study of the Chemically Driven Flow Fields in Initiating Homogenous and Heterogeneous Nitromethane Explosives", presented at Ninth Symposium of Detonation, Portland, Oregon, August 27-Sept 1 1989.
7. TURKEL, M. L., CHANLET, F., KAZANDJAN, L. and DANEL J. F., HDP, Tours, FRANCE, May 1995.
8. KEDRINSKII, V., "Liquid Fracture at Explosive Loading", Shock Waves Symposium, Marseille, France (1993).
9. YAKUSHEVA, O. B., YAKUSHEV, V. V., DREMIN, A. N., *High Temperature - High Pressure*, **3**, 261-266 (1971).
10 . VOROBEV, A. A. and TROFIMOV, V. S., *Comb. Expl. Shock Waves*, **18**, 676-681 (1982).

CP429, *Shock Compression of Condensed Matter – 1997*
edited by Schmidt/Dandekar/Forbes
© 1998 The American Institute of Physics 1-56396-738-3/98/$15.00

IMPACT IGNITION OF NEW AND AGED SOLID EXPLOSIVES

Steven K. Chidester, Craig M. Tarver, and Chet G. Lee

*Lawrence Livermore National Laboratory,
P.O. Box 808, L-282, Livermore, CA 94551*

The critical impact velocities of 60.1 mm diameter steel projectiles required to produce ignition are measured for new and aged confined charges of the HMX-based solid explosives LX-10, LX-04, PBX- 9404, and PBX-9501. External blast overpressure gauges are employed to determine the relative violence of the explosive reactions. The experiment is modeled in DYNA2D using recently developed material strength models, and thermal energy deposition thresholds for impact ignition are found.

INTRODUCTION

Impact sensitivity of solid high explosives has long been an important concern in handling, storage, and shipping procedures. Several impact tests have been developed for specific accident scenarios, but these tests are generally neither reproducible nor amenable to computer modeling. The Steven impact test (1) was developed with these objectives in mind. One objective of this paper is to measure the relative violence of the explosive reactions of four HMX-based explosives: LX-10-1 (94.5% HMX, 5.5% Viton A binder); LX-04 (85% HMX, 15% Viton); PBX-9404 (94% HMX, 3% nitrocellulose, 3% CEF binder); and PBX-9501 (94.9% HMX, 2.5% BDNPA-F, 2.5% Estane binder, 0.1% DPA or Irganox). Blast wave overpressure gauges are used to measure the violence of the reactions in terms of grams of TNT equivalence, in analogy with the Susan test (2). The critical impact velocity and the violence of the reaction for 25 year old LX-04 are also measured. The Ignition and Growth reactive flow model is used to evaluate several empirical criteria for ignition under these impact conditions and to simulate the growth of explosive reaction following ignition as the confined explosive charge is producing gaseous reaction products.

EXPERIMENTAL

The experimental geometry for the Steven impact test is shown in Fig. 1. A 6.01 cm diameter steel

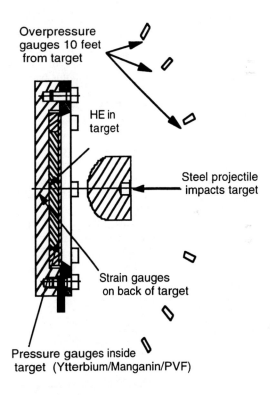

FIGURE 1. Geometry of the Steven impact test

projectile is accelerated by a gas gun into 11 cm diameter by 1.285 cm thick explosive charges confined by 0.3175 cm thick steel plates on the

707

impact face and 1.905 cm thick steel plates on the back side. The original Steven test used a 60.1 mm diameter tantalum rod or rounded projectile (1). DYNA2D calculations showed that the high explosive was driven to violent explosions by the frictional work done in the region where the tantalum projectile struck (1). Since this objective of this study was to determine thresholds for low order reactions and to measure relative reaction violence of these explosions, the projectiles were changed to steel to provide less frictional work on the explosive and to allow the 76.2 mm diameter gas gun to accelerate these projectiles to the higher velocities required to ignite LX-04. Four external blast overpressure gauges were placed ten feet from each target for direct comparison with Susan test data. As shown in Fig. 1, a variety of embedded pressure gauges are also being used to measure the internal pressure developed during ignition and growth of reaction. In addition, a microwave radar technique is being developed for use in measuring the velocities of the expanding steel confinement after explosion. Only the critical impact velocities and the blast overpressure gauge results are reported in this paper.

EXPERIMENTAL RESULTS

The results of 27 Steven tests are listed in Table 1. The type of explosive, the density of the charge, the projectile velocity, whether reaction was observed or not, and the average overpressure measured in pounds per square inch {Ave. O. P. (psi)} by the four gauges are included in Table 1. The critical impact velocities for the four HMX-based explosives are in the usual order of decreasing impact sensitivity: PBX-9404; LX-10-1; PBX-9501; and LX-04. In contrast to other impact tests, such as the Susan and Skid tests (2), the Steven test yields a distinct critical velocity below which no reactions are observed. With the possible exception of PBX-9404, the Steven test also yields relatively low order reactions which can be quantitatively measured by blast overpressure gauges and related to an equivalent amount of TNT. Figure 2 shows the average overpressures measured by the blast wave gauges for the four explosives as functions of projectile impact velocity. The use of steel projectiles allows low order reactions to be observed in LX-10, whereas the tantalum projectiles used by Chidester, et al. (1) produced only violent explosions in LX-10-1.

The abrupt increases in overpressure measured for PBX-9404 and LX-10-1 and the much slower

increases measured for PBX-9501 and LX-04 are similar to those obtained in the Susan test. A direct comparison of these Steven test results to Susan test results in terms of the TNT equivalent weight that would produce the same average overpressures at a distance of ten feet is shown in Fig. 3. Even though the Steven test uses approximately one half as much explosive as the Susan test, it produces larger overpressures at the same projectile impact velocity. This is due to the greater confinement in the Steven test, which uses two steel plates, compared to the thin aluminum cap that confines the explosive charge in the Susan test. This increased confinement allows the chemical reaction to grow further and produce more gas before the steel confinement of the Steven test is breached and the subsequent rarefaction waves slow the reaction. The confinement can easily be varied to test explosives under the conditions they will encounter during their lifetimes. Therefore, because of its reproducibility, two-dimensional and three-dimensional modelability, relative ease of instrumentation, and low cost, the Steven test has developed into the impact test of choice for hazard, surveillance, and vulnerability studies. Ignition and Growth reactive flow modeling results for LX-10-1 in the Steven test geometry are presented in the next section.

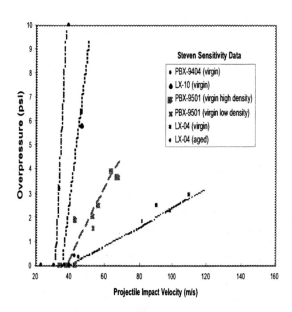

FIGURE 2. Blast overpressures from several HMX-based explosives in the Steven test as functions of impact velocity

TABLE 1. Summary of the Steven impact test experiments

Explosive	Density (g/cc)	Velocity (m/s)	Reaction	Ave. O.P. (psi)
PBX-9404	1.835	23	No	N.A.
PBX-9404	1.835	31	No	N.A.
PBX-9404	1.835	34	Yes	3.2
PBX-9404	1.835	39	Yes	10
LX-10-1	1.865	35	No	N.A.
LX-10-1	1.865	41.5	Yes	No records
LX-10-1	1.865	46.5	Yes	6.4
LX-10-1	1.865	47	Yes	5.8
PBX-9501	1.843	34	No	N.A.
PBX-9501	1.843	39	No	N.A.
PBX-9501	1.843	43	Yes	1.9
PBX-9501	1.843	53	Yes	2.05
PBX-9501	1.843	56	Yes	2.5
PBX-9501	1.843	63	Yes	3.93
PBX-9501	1.843	66.5	Yes	3.675
PBX-9501	1.843	68	Yes	No records
PBX-9501	1.829	38	No	N.A.
PBX-9501	1.830	43	No	N.A.
PBX-9501	1.829	53.4	Yes	1.55
LX-04	1.870	81	Yes	1.825
LX-04	1.871	90	Yes	2.525
LX-04	1.870	98	Yes	2.3
LX-04	1.870	110	Yes	2.925
LX-04 (aged)	1.868	37	No	N.A.
LX-04 (aged)	1.861	40	No	N.A.
LX-04 (aged)	1.862	43	Yes	0.4
LX-04 (aged)	1.862	45.5	Yes	0.325

REACTIVE FLOW MODELING

Previous DYNA2D modeling of the Steven test (1) concentrated on its mechanical aspects. The measured depths of dents in the targets that did not react were accurately calculated, and a constant frictional work criteria for LX-10-1 was developed. An analytical expression for ignition has since been developed by Browning (4). The Ignition and Growth reactive flow model uses two Jones-Wilkins-Lee (JWL) equations of state, one for the unreacted explosive and one for the reaction products:

$$p = A\, e^{-R_1 V} + B\, e^{-R_2 V} + \omega C_v T/V \qquad (1)$$

where p is pressure in Megabars, V is relative volume, T is temperature, ω is the Gruneisen coefficient, C_v is the average heat capacity, and A, B, R_1 and R_2 are constants. The reaction rate is:

$$dF/dt = I(1-F)^b(\rho/\rho_0-1-a)^x + G_1(1-F)^c F^d p^y$$
$$0 < F < F_{igmax} \qquad\qquad 0 < F < F_{G1max}$$
$$+ G_2(1-F)^e F^g p^z \qquad (2)$$
$$F_{G2min} < F < 1$$

where F is the fraction reacted, t is time in μs, ρ is

709

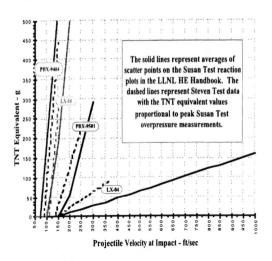

FIGURE 3. TNT equivalent output for the Steven and Susan tests as functions of projectile impact velocity

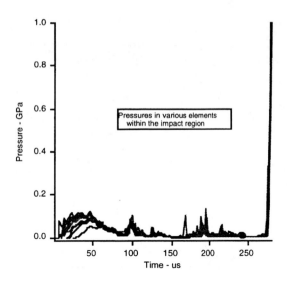

FIGURE 4. Pressure histories in LX-10 for a 40 m/s Impact

the current density in g/cm^3, ρ_0 is the initial density, p is pressure in Mbars, and I, G_1, G_2, a, b, c, d, e,g, x, y, and z are constants. This three term reaction rate law models the three stages of reaction observed during shock initiation of pressed solid explosives (3). For these low pressure (0.1 GPa), long time (several hundred μs) impacts, the first term in Eq. (2) uses x = 4 to simulate a constant input energy ignition criterion, which holds for PBX-9404 at both low and high pressures (5). The equations of state and growth of reaction rates are the standard ones for LX-10-1 shock initiation (3). Using I=1000 μs^{-1} in Eq. (2) yields an ignition rate similar to that predicted by the 0.37 cal/cm^2 frictional work criterion for LX-10-1 ignition used by Chidester et al. (1). Figure 4 shows the pressure histories for several LX-10-1 elements directly under the steel projectile for an impact velocity of 40 m/s. The maximum impact pressure is 0.12 GPa and lasts about 70 μs. Rapid reaction occurs about 280 μs after impact and at a fraction reacted of about 0.1%. Rapid reaction is not observed in a 1 ms long calculation at an impact velocity of 35 m/s. Other impact ignition criteria are being tested in the reactive flow model.

SUMMARY

The understanding of the effects of explosive aging on impact ignition and other hazards must improve

as systems are being deployed longer than their initial estimated lifetimes. Since many scenarios can not be tested, reliable reactive flow models based on data from well-instrumented, reproducible experiments are necessary. The Steven or Spigot Gun Test (6) is designed to provide such impact data.

ACKNOWLEDGMENTS

The authors would like to thank LeRoy Green, Roger Logan, and Cynthia Nitta for their support.

This work was performed under the auspices of the U.S. Department of Energy by Lawrence Livermore National Laboratory (contract no. W-7405-ENG-48).

REFERENCES

1. Chidester, S. K., Green, L. G., and Lee, C. G., *Tenth International Detonation Symposium*, ONR 33395-12, Boston, MA, 1993, pp. 785-792.
2. Dobratz, B. M. and Crawford, P. C., *LLNL Explosives Handbook*, Lawrence Livermore National Laboratory Report UCRL-52997 Change 2, 1985.
3. Tarver, C. M., Urtiew, P. A., Chidester, S. K., and Green, L. G., *Propellants, Explosives, Pyrotechnics* **18,** 117-127 (1993).
4. Browning, R. V., in *Shock Compression of Condensed Matter-1995,* Schmidt, S. C. and Tao, W. C., eds., AIP Press, New York, 1996, pp. 405-408.
5. Green, L. G., LLNL, private communication, 1997.
6. Idar, D. J., Lucht, R. A., Scammon, R., Straight, J., and Skidmore, C. B., *PBX 9501 High Explosive Violent Response/ Low Amplitude Insult Project*, Los Alamos National Laboratory Report LA-13164-MS, UC-741, 1997.

CP429, *Shock Compression of Condensed Matter – 1997*
edited by Schmidt/Dandekar/Forbes
© 1998 The American Institute of Physics 1-56396-738-3/98/$15.00

EFFECT OF MICROVOIDS ON THE SHOCK INITIATION OF PETN

J.L. Maienschein, P.A. Urtiew, F. Garcia, J.B. Chandler

Lawrence Livermore National Laboratory, Livermore, CA 94550

We demonstrate that the introduction of microvoids as glass microballoons sensitizes high-density solvent-pressed PETN to shock initiation. At input pressures ranging from 1.4-2.0 GPa, shock propagation velocities are higher and run distances to detonation are shorter for PETN sensitized by microballoons. By selecting the size and density of microballoons, we can therefore study the effect of void size and density on shock initiation by hot spots.

INTRODUCTION

It is accepted that hot spots, formed by shock interaction with voids or defects in explosives, control the onset of detonation; although many mechanisms have been postulated and modeled, the actual mechanism(s) are unknown. Our goal here is to quantify the effect on initiation of variations in the density and size of defects in an explosive. These data should prove useful in theoretical considerations of hot spot mechanisms.

We added a known quantity of voids to a nearly homogeneous explosive and determined their effect on shock initiation, using a one-dimensional impact with *in-situ* pressure gauges. We chose pentaerythritol tetranitrate (PETN), since it is nearly ideal (no reaction zone effects) and can be solvent-pressed to a low porosity. Glass micro-balloons introduced into the PETN provided a controlled size and density of defects. Results here show the feasibility of this approach.

PREPARATION OF SAMPLES

PETN samples 19 mm in diameter and 1.5-1.7 mm thick were solvent pressed with acetone. Densities of 92-95% of TMD (theoretical maximum density, 1.77 g/cc) were achieved by pressing 0.75 grams of powdered PETN (batch B-509) with 0.2 cc acetone in a uniaxial compaction die for 5 minutes at 200 MPa, after a 5 minutes soak at ambient pressure. A larger acetone volume of 2.0 cc, with three 5-minute pressing periods and 1 minute recovery times between pressure cycles, increased the density to 96-97% TMD.

Glass microballoons (Q Cel 650) were separated by sieving; we used the fraction that passed through a 25 micron sieve. From a photomicrograph, diameters ranged from 4-20 microns. Wall thickness appeared to be $\approx$ 10% of the diameter, although this was not accurately determined. A blend of 91 volume % PETN and 9 volume % microballoons was stirred by hand and pressed at the first pressing conditions to produce samples with densities of 86-88% TMD. We dissolved the PETN from a sample to verify that the glass microballoons remained intact, and confirmed that very few were broken.

SHOCK INITIATION EXPERIMENTS

We measured initiation and onset of detonation in samples subjected to impact from the 102-mm gun in the High Explosives Application Facility at LLNL. Targets had 4 or 6 layers of explosive, with a manganin pressure gauge between each layer and under the polycarbonate buffer layer on the impact surface. The manganin gauges were LLNL 50 milliohm style, 25 microns thick, laminated in 50 or 130 microns of Teflon® on each side. Each target was impacted by a polycarbonate sabot traveling 0.59-1.1 mm/μs, with impact pressures of 1.1-2.5 GPa. Data included pressure/time and shock propagation velocity. More details of this type of experiment are given in reference 1.

EXPERIMENTAL RESULTS

The shot conditions are shown in Table 1. Each of the three shots with PETN/microballoons has a matching shot with pure PETN at about the same

pressure. Comparison of these pairs of shots allows determination of the effect of the microballoons on shock initiation. Shock propagation velocities (Table 1 and Figs. 1-3) show that velocity acceleration is greater in samples with microballoons than in those without. At 1.2 GPa, the velocity increases ≈10% over a run distance of 5 mm with microballoons, but does not increase until a run distance of 6.3 mm without microballoons. At 1.4 GPa, both samples show a velocity increase, but it is more pronounced in the sample with microballoons. At 2.0 GPa, the microballoon sample accelerates earlier to detonation velocity. Thus all shots show increased sensitivity with microballoons.

TABLE 1. Sample specifications, impact conditions, and propagation velocities observed in shock impact experiments with PETN samples with and without microballoons.

Shot #	Mat'l	density g/cc	density TMD	impact vel mm/μs	impact pres. GPa	shock propagation velocity in each segment, mm/μs					total run distance to each segment, mm				
						1st	2nd	3rd	4th	5th	to 1st	to 2nd	to 3rd	to 4th	to 5th
4467	PETN	1.716	96.9%	0.587	1.15	2.85	2.90	2.90	2.87	2.79	1.46	2.92	4.35	5.79	7.23
4451	PETN/ μball	1.575	89.0%	0.689	1.16	1.81	1.96	2.03			1.72	3.44	5.17		
4465	PETN	1.640	92.7%	0.687	1.27	2.80	2.99	2.71	2.74	5.97	1.63	3.23	4.78	6.25	7.76
4466	PETN	1.653	93.4%	0.699	1.32	2.78	2.86	2.81	2.26	3.53	1.58	3.10	4.60	6.08	7.57
4447	PETN	1.678	94.8%	0.689	1.35	2.66	2.69	2.77			1.55	3.12	4.67		
4450	PETN/ μball	1.558	88.0%	0.840	1.45	2.02	2.33	3.13			1.71	3.40	5.11		
4448	PETN	1.685	95.2%	0.850	1.78	2.76	3.02	8.33			1.55	3.11	4.68		
4452	PETN/ μball	1.544	87.2%	1.090	1.98	2.45	4.59	8.97			1.69	3.40	5.13		
4468	PETN	1.720	97.2%	0.916	2.04	3.09	3.34	4.63	9.53	10.21	1.45	2.89	4.34	5.78	7.22
4449	PETN	1.688	95.4%	1.090	2.48	3.09	9.26	10.6			1.55	3.08	4.64		

Typical pressure data are shown in Fig. 4 for shot 4468 with pure PETN at 97.2% TMD and 2.0 GPa. Gauge depths are shown in Table 1 (first gauge is at buffer/PETN interface). We see slight reaction buildup at the first gauge, with more reaction at each successive gauge until the sample detonates at the 4th gauge, at a depth of 4.34 mm.

Comparison of individual gauge records shows that the presence of microballoons sensitizes the PETN to shock. Examples are shown in Fig. 5-7 for gauges at the third and fourth location in shots at 1.4 GPa; in all cases the gauge with micro-

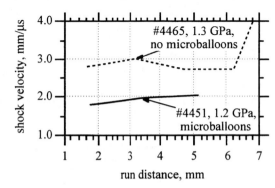

FIGURE 1. Shock propagation velocity, 1.2 GPa .

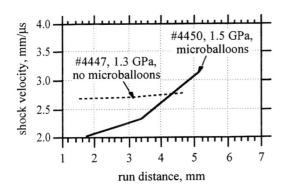

FIGURE 2. Shock propagation velocity, 1.4 GPa.

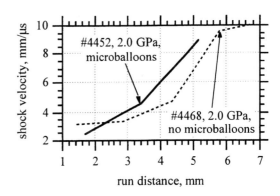

FIGURE 3. Shock propagation velocity, 2.0 GPa.

balloons is shifted in time to match onset times. In Fig. 5 the pressure at the third gauge build up faster following arrival of the shock for the microballoon sample; however, it is at a slightly longer run distance. In Fig. 6 the fourth pressure gauge shows detonation only in the sample with microballoons, but again the run distances are not strictly comparable. However, comparison of the third gauge at 3.4 mm run in the sample with microballoons with the fourth gauge at 4.67 mm run distance in the sample without microballoons shows almost identical profiles; since the gauge with microballoons is located at a shorter run distance than in the sample with no microballoons, this comparison clearly shows the increased sensitivity to initiation caused by the microballoons.

DISCUSSION OF RESULTS

In our samples, the density of samples was ≈ 88% TMD with microballoons, and ≈95% TMD without microballoons. Therefore, the void fraction was increased from 5% to 12%, or 2.5-times, by the inclusion of glass microballoons. This was sufficient to confer increased sensitivity to shock. To achieve our original goal of quantifying the effect of different sizes and densities of voids, a few improvements are needed. The size distribution of glass microballoons was wider than desired in this work, and was not accurately characterized. The homogeneity of microballoons in the pressed parts was unknown. The density of the pressed PETN/microballoon parts was also lower than desired, since we pressed those parts early on in the

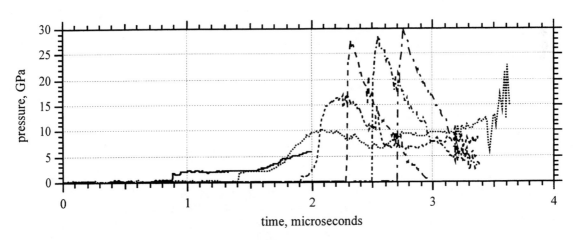

FIGURE 4. Manganin gauge pressure data in shot 4468, pure PETN at 97.2% TMD. Input pressure is 2.0 GPa.

713

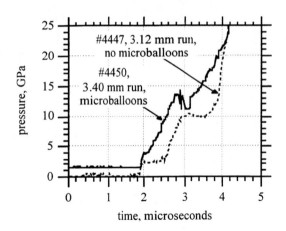

FIGURE 5. Data from the third pressure gauge, 1.4 GPa.

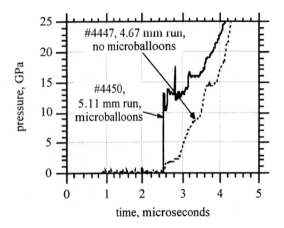

FIGURE 6. Data from the fourth pressure gauge, 14 GPa.

experimental sequence. A further limitation is the impedance mismatch between glass and explosive, which results in shock reflection in the sample which would not be present from a void. As an alternative to simply improving glass microballoon characterization and PETN pressing, the use of mono-size plastic microballoons offers good control of void size and eliminates the impedance mismatch. Plastic microballoons will probably not, however, withstand the pressure of solvent-pressing of PETN; a gelled liquid (to prevent balloon separation) or paste explosive would provide a good test matrix.

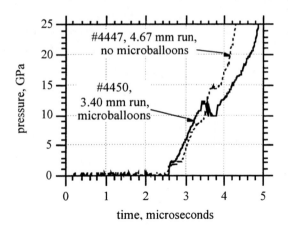

FIGURE 7. Data from the third pressure gauge with microballoons and fourth gauge without microballoons, 1.4 GPa.

CONCLUSIONS

We have demonstrated the practicality of embedding voids of known size and density into explosive, to characterize the effect of void density and size on shock initiation. The use of plastic microballoons and gelled or paste explosives offers improvements over the experimental system used here.

ACKNOWLEDGMENTS

We thank Jeffrey Wardell for his expert mechanical support of these experiments, and the HEAF gun crew for their role in firing the shots. This work was performed under the auspices of the U.S. Department of Energy by Lawrence Livermore National Laboratory under contract No. W-7405-ENG-48.

REFERENCES

1. P.A. Urtiew, T.M.Cook, J.L. Maienschein, C.M. Tarver, "Shock Sensitivity of IHE at Elevated Temperatures," **Proceedings Tenth International Detonation Symposium**, July 12-16, 1993, ONR 33395-12, p. 139.

CP429, *Shock Compression of Condensed Matter – 1997*
edited by Schmidt/Dandekar/Forbes
© 1998 The American Institute of Physics 1-56396-738-3/98/$15.00

SINGLE AND TWO INITIATION POINTS OF PBX

R. Mendes, I. Plaksin and J. Campos

Lab. of Energetics and Detonics, Mechanical Eng. Dep., Faculty of Science and Technology
University of Coimbra, 3030 Coimbra, Portugal

The initiation of PBX based on RDX (85% mass fraction-initial density >95% TMD) induced by single and double initiation points is studied experimentally. A thin 64 optical fibers strip, connected to a fast electronic streak camera allows to observe detonation wave profile and Mach stem zone evolution. The setup is formed by one or two initiation channels with 5 mm square cross section separated, between them, by 15 mm. The optical strips are placed along the wave collision zone and around the initiation point at several circles and different angles θ from channel axis. The results show explicitly the non monotonous (pulsing) behaviour of diverging detonation wave propagation with different velocities in central and periphery zones. At different run distances, the periphery zone ($40°<\theta<60°$) runs faster than the central one ($\theta>60°$). After this transition, the central zone accelerates and overtakes the periphery zone and the detonation front becomes almost spherical. Quantitative results clarify the models of colliding and diverging waves.

INTRODUCTION

Some papers have been presented about the complex problems of experimental tests and proposal characterization of the divergent detonation waves (DW) in high explosives (1-6). For (1-4) the authors considered a monotonous process for the diverging DW. When two divergent spherical DW collide, a regular reflection can be observed initially followed by an irregular reflection. A Mach stem can then be observed in the zone between the two spherical waves (1) and (5).

The initial stage of divergent (DW) formation as well as the evolution of Mach wave in the colliding of two spherical divergent DW is measured and presented. In order to characterize the divergent DW and colliding process between them a new high resolution method based on 64 optic fibers strip (OFS) has been used (7) and (8). The used explosive for this study was PBX based on RDX (85%RDX, 11.5% HTPB, 3.5% DOS). The RDX has a bimodal particle distribution (75% 96μm, 25% 22μm). The explosive density is 1.59 g/cm^3 that corresponds to 96.5% of its theoretical maximum density-TMD.

Detonation Velocity

The detonation velocity is measured using an optic fiber probe, the set of fibers is placed at normal direction relatively to the detonation wave direction. Other set of fibers is placed against the end of the detonation tube in order to measure the front DW curvature. The end of detonation tube was closed with two Kapton sheets with 125μm thickness.

The light collected by 64 OFS is recorded on a fast electronic streak camera (TSN-506-THOMPSON), typical result is shown in Fig. 1. Measured detonation velocity for 25 mm diameter is 7.82 mm/μs.

FIGURE 1. Electronic Streak camera record of detonation velocity and front curvature.

Detonation Pressure

The detonation pressure was measure taking the value of shock velocity inside a target induced by the shock of the front detonation wave. The obtained value is 23.8 ± 0.7 GPa.

Critical Diameter

The critical diameter of the explosive was tested for circular (diameter ϕ_c) and square (side ϕ_s) configurations. The experimental results based on the non propagation of detonation show $3.5 < \phi_c < 3.75$ mm and $\phi_s < 4$ mm respectively.

SINGLE POINT INITIATION

The single initiation divergent DW was characterized measuring the detonation velocity at 90° and 45° relative to the detonation wave propagation axis. The experimental setup used is presented in Fig. 2.

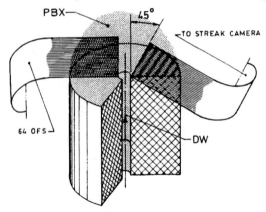

FIGURE 2. Experimental setup to measure the detonation velocity at 90° and 45°.

The initiation train is composed by a cylindrical channel with 5 mm diameter by 40 mm length of the same PBX explosive, initiated by a commercial n° 8 detonator.

The observation of the (x,t) diagram (Fig. 3) shows initially a low propagation value of the DW with an increasing value for X≈5 mm. Also the propagation of the diverging DW for 45° is higher than at 90°. The divergent DW curvature were measured with the experimental setup shown in Fig. 4. It shows a setup where the sets of 4 optic fibers each one are placed around the initiation point (θ=0° to θ=180°) at distance R, in a mushroom configuration.

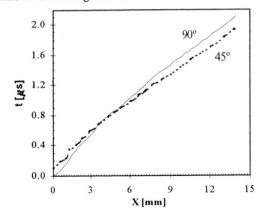

FIGURE 3. (X,t) diagram for divergent DW at 90° and 45°.

These fibers are enclosed in a non transparent film at both sides and epoxy, with 450 μm total thickness. Along these fibers 5 windows, from 12.5 to 22.5 mm radius, spaced of 2.5 mm were open and filled with epoxy and hollow glass micro balloons (HGMB) in order to generate a light pulse by shock compression.

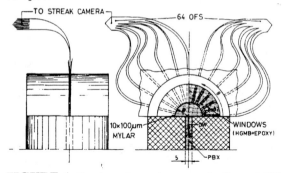

FIGURE 4. Experimental setup to measure the divergent DW shape as a function of radius and angle.

A streak camera record obtained by this method is in Fig. 5.

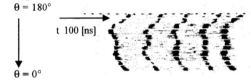

FIGURE 5. Streak camera record of detonation velocity and shape front as a function of radius and angle.

The record shows clearly the detonation wave arriving first at points that are 40°- 60° from normal direction to the DW propagation axis. Then the central part of DW accelerates and the detonation front is changing to spherical shape.

From the (R,t) diagram of the data in Fig. 6, the results show that the maximum deviation from ideal spherical shape are around 250 ns. and a pulsation behaviour of divergent DW front.

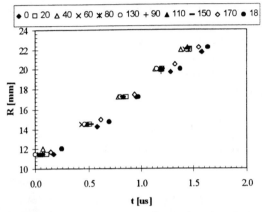

FIGURE 6. (R,t) diagram for divergent DW from 0° to 180°.

(R,t) diagram allows to draw the shape of the DW as a function of the angle for 5 time positions: $\Delta t = 0; 0.38; 0.70; 1.09$ and 1.37 µs, Fig. 7.

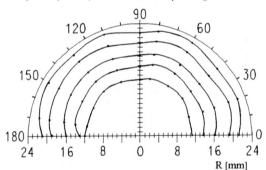

FIGURE 7. Profile of DW as a function of θ.

DOUBLE INITIATION AND MACH FORMATION

The interaction of two spherical DW was studied using the setup shown in Fig. 8. The probe is formed by 64 OFS. Also some windows were opened and filled by epoxy and HGMB for shock to light conversion. The distance between the cylindrical initiators (φ=5 mm) is 15 mm. The result of this experiment is shown in Fig. 9. Experimental results show clearly the regular

reflection, the transition for irregular reflection and the formation of Mach stem. The Mach stem is the bridge that connects the two detonation waves.

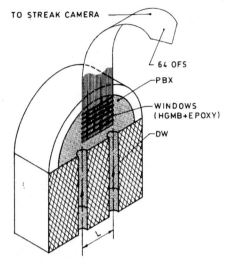

FIGURE 8. Experimental setup for double point initiation.

FIGURE 9. Electronic Streak camera record for two point initiation.

The evolution of the width of the Mach stem W_{Ms} as a function of the X distance is shown in Fig. 10. The angle where appear the first Mach stem, known by critical angle of formation of the Mach stem is between 59.7° and 63.9°.

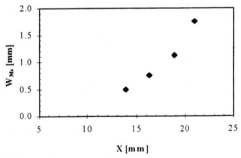

FIGURE 10. Mach stem width as a function of distance from the wall.

In order to characterise continuously the process of diverging DW and colliding process, one experiment of two points initiation of PBX in a slab sample was made. The top view of the experimental setup is shown in Fig. 11.

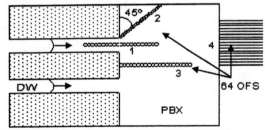

FIGURE 11. Experimental setup for flat double initiation.

The OFS 1 measure the evolution of DW in the propagation direction, the OFS2 measure the evolution the DW at 45°, the OFS 3 caracterize the evolution of the colliding point and the OFS 4 give us the shape of Mach wave. The obtained result is shown in Fig 12.

FIGURE 12. Electronic Streak camera record for two point initiation of a PBX in a slab configuration.

The (X,t) diagram of the OFS 1, 2 and 3 is presented in Fig. 13. The OFS 1 and 2 are show the same behaviour like for 1 point initiation.

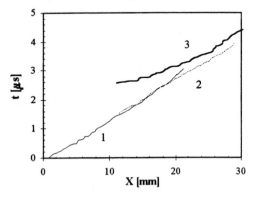

FIGURE 13. (x,t) for two point initiation in slab configuration.

The OFS 3 show that the propagation of Mach stem can be characterize by slowing down by non monotonous process.

DISCUSSION

Our results show that the process of formation the diverging DW is more complex than was presented by others authors. The results show the non monotonous (pulsing) behaviour of diverging DW. At different run distances, the periphery zone ($40° < \theta < 60°$) runs faster than the central one ($\theta > 60°$). After this transition, the central zone accelerates and overtakes the periphery zone and the DW front becomes almost spherical. The Mach wave formation by a collision process of two non monotonous diverging DW has a non monotonous behaviour also.

CONCLUSIONS

The application of the new high resolution method based on 64 OFS allowed us to find unknown detonation phenomena.

The results show explicitly the non monotonous (pulsing) behaviour of diverging detonation wave propagation with different velocities in central and periphery zones.

As the initial divergent detonation waves have a non monotonous behaviour the resulting colliding process is non monotonous also.

REFERENCES

1. Hull, L. M., Mach Reflection of Spherical Detonation Waves, in the *Tenth International Symposium* on *Detonation*, Boston, MA, July 12-16, 1993.
2. Aveillé, J., Baconin, J., Carion, N., and Zoé, J., Experimental Study of Spherically Diverging Detonation Waves, in the *Eighth International Symposium* on *Detonation*, Albuquerque, NM, July 15-19, 1985.
3. Chevalier, J. M., and Carion, N., Propagation Phenomena on the Detonation Wave Front, in the *Tenth International Symposium* on *Detonation*, Boston, MA, July 12-16, 1993.
4. Ferm, E. N., and Hull, L. M. Reflected-Shock Initiation of Explosives in the *Tenth International Symposium* on *Detonation*, Boston, MA, July 12-16, 1993.
5. Gerasimov, V. M., Gubachev, V. A., Vakin, V. A., Plaksin. I. E. and Shutov, V. I., *Sov. J. Chem. Phys.*, **12**(5), 1091-1097 (1994).
6. Quidot, M., and Groux, J., Reactive Modeling in Shock Initiation of Heterogeneous Explosives, in the *Ninth International Symposium* on *Detonation*, Portland, OR, Aug. 28- Sep. 1, 1989
7. Mendonça, M., Plaksin, I, Campos, J., and Gois, J., Emulsion Explosives with TNT, in the *Shock Waves in Condensed Matter*, Sep. 2-6, 1996.
8. Mendes. R., Plaksin, I., Ribeiro, J., and Campos, J., Electrical Explosion of the Aluminium Foil Behaviour of Explosions Products in the Process of Flyer Plate Launch, in the *Shock Waves in Condensed Matter*, Sep. 2-6, 1996.

CP429, *Shock Compression of Condensed Matter – 1997*
edited by Schmidt/Dandekar/Forbes
© 1998 The American Institute of Physics 1-56396-738-3/98/$15.00

FURTHER OBSERVATIONS ON HMX PARTICLE SIZE AND BUILDUP TO DETONATION

R. R. Bernecker[1] and R. L. Simpson[2]

1) Naval Surface Warfare Center, IHD, Indian Head, MD 20640
2) Lawrence Livermore National Laboratory, Livermore, CA 94550

Shock loading data for various particle sizes of HMX in unreactive (water) and reactive (FEFO) binder systems [1] have been re-analyzed. Traditional distance-to-detonation (x^*) values have been obtained for comparison to other wedge test data for systems using various particle sizes of HMX and RDX. In the log x^* - log P plane, the slope is nearly identical for 5 μm HMX/water and 5 μm HMX/FEFO samples, supporting the proposition that the slope in the log-log plane is constant for fine particle sizes of HMX. Analyses of predetonation distance-time (x,t) paths show similarities for a given particle size of HMX at various input pressures, suggestive of a common-curve buildup process.

INTRODUCTION

Development of pressure-dependent reaction kinetics associated with shock loading has become an integral part of the characterization of energetic materials. In large part, this is an ad hoc process with little insight coming from the fundamental aspects of the formulation, e.g., particle size distributions of the solids, solid loading fractions, etc. However, there are several data bases which can be used to gain more (mechanistic) insight pertaining to these reaction kinetics. For HMX systems, the pertinent data base is that of Simpson et al.[1]; one of the pertinent data bases for RDX systems is that of Moulard.[2] The Simpson et al. data are slightly more definitive since they basically used particles obtained from the same batch and which had not been recrystallized. (See the work of Baillou et al.[3] for some observations pertaining to the influence of recrystallization from various solvents on shock reactivity.)

In this paper the wedge test data of Simpson et al., for mixtures of HMX/water and HMX/FEFO, were re-analyzed to obtain a) distance-to-detonation (x^*) data as a function of input pressure (P) and b) spatial histories of the accelerating shock front. These two data sets complement the embedded gage data [1] for the validation of any selected reaction kinetics used to describe the buildup to detonation as a function of particle size.

DATA REDUCTION

Four different particle sizes of HMX were evaluated in water-filled samples: 5, 60, 110 and 1700 μm. For FEFO systems, the HMX particle sizes were 5 and 60 μm; see Ref. 1 for the complete compositions of the FEFO materials.

In contrast to the methodology of Ref. 1, the distance-to-detonation (x^*) was determined here by estimating the location where the steady state detonation wave began. Because of the spacing of the pins in these modified wedge experiments, it was necessary to use the trajectory of the accelerating shock, along with the path of the detonation wave. (The estimated uncertainties in x^* are of the magnitude of about 3%.) To accentuate the variation of predetonation shock velocities, relative distance ($\delta x = x - x^*$) and relative time ($\delta t = t - t^*$) data were also generated and are shown in some of the following figures.

The pertinent data for HMX/water mixtures are summarized in Table I while the data for mixtures of HMX/FEFO are summarized in Table II.

DISCUSSION

Traditionally x*-P data sets have been plotted in the log x* - log P plane and represented by Eq. 1.

$$\log x^* = A + B \log P \quad (1)$$

As discussed in Ref. 4, the slope in this plane (B or its absolute value, B*) provides an experimental measure of the <u>shock reactivity</u> of the energetic material (EM). One of the conclusion reached in Ref. 4 was that values for B* are quite similar for HMX/water samples with 5 μm HMX and for a cast plastic-bonded explosive (PBX) using 6 μm RDX in an HTPB binder.[2] It can be seen in Fig. 1 that this observation extends also to the FEFO mixture containing 5 μm HMX. The 5 μm HMX/water data (squares) are intermediary in <u>shock sensitivity</u> to that of the 5 μm HMX/FEFO samples and the cast PBX samples. The data in Fig. 1 suggest that one of the the wedge test parameters (B*) can be estimated *a priori* for EM's containing very fine particles of HMX (and perhaps for RDX), supposedly regardless of the nature of the binder system. (Another observation from Ref. 4 was that B* appears to be very similar for 5, 60, and 110 μm HMX/water samples.)

TABLE I. HMX/water Wedge Data

Shot	Density g/cc	P [1] GPa	X [1] mm	x* mm	t* μs
5 μm HMX (48 vol% HMX)					
FH-3	1.43	5.3	15	17.1	4.403
FH-1	1.43	6.0	11	12.7	3.063
FH-2	1.43	7.8	4.7	5.41	1.177
60 μm HMX (58 vol% HMX)					
FH-11	1.52	3.9	14	22.9	5.250
FH-10	1.53	6.1	4.8	6.44	1.376
FH-12	1.52	7.4	3.8	5.32	1.034
110 μm HMX (59 vol% HMX)					
FH-9	1.52	4.1	19	22.4	5.751
FH-7	1.54	6.0	5.8	7.53	1.639
1700 μm HMX (60 vol% HMX)					
FH-4	1.55	4.0	14	18.0	4.401
FH-6	1.54	5.6	9.2	13.75	2.883
FH-5	1.54	8.3	6.0	8.12	1.507

Table II. HMX/FEFO Systems Wedge Data

Shot	Density g/cc	P [1] GPa	X [1] mm	x* mm	t* μs
5 μm HMX (56 vol% HMX) (RX-08-GB)					
EVH-4	1.74	4.5	13	13.0	3.413
EVH-5	1.74	6.2	5.6	5.62	1.332
60 μm HMX (56 vol% HMX) (RX-08-GG)					
EVH-6	1.75	3.6	9.5	10.0	2.844
EVH-7	1.75	5.3	4.0	5.11	1.136
60 μm HMX (72 vol% HMX) (RX-08-EL)					
EVH-3	1.80	2.5	23	25.5	7.124
EVH-2	1.80	5.4	4.4	4.98	1.142
EVH-1	1.80	7.4	2.0	<2.0	--

While the nature of the binder system does not appear to influence values of B* for mixtures containing 5 μm HMX, this may not the case for the 60 μm HMX/FEFO samples. In Fig. 2 are plotted the data for the 60 μm HMX/water samples and the 60 μm HMX/FEFO samples (see Tables I and II, respectively). The two data (squares) for 60 μm HMX/FEFO (56% HMX) suggest that this material has a smaller B* value than for 60 μm HMX/FEFO (72% HMX) - the two x's in Fig. 2 . This is intuitively disturbing because the 56% HMX samples are then more <u>shock sensitive</u> than the 72% HMX samples. Hence, a line has been drawn to suggest

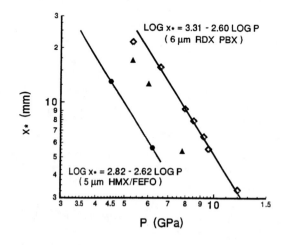

FIGURE 1. Comparison of Wedge Data for Formulations with Fine Nitramine Particle Sizes (▲ 5 μm HMX/water)

that these 56 and 72 vol% HMX samples have the same shock reactivity. Corollaries of this suggestion are a) the shock sensitivities of 60 μm HMX/FEFO samples are the same for solid loadings of 56 and 72 vol% HMX and b) the values of B* for 60 μm HMX/FEFO samples should be independent of solid loadings in the range of 56-72 vol% HMX.

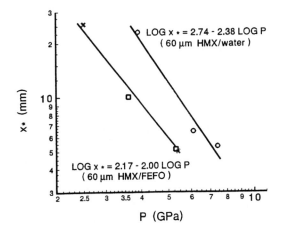

FIGURE 2. Comparison of 60 μm HMX in Inert and Reactive Binder Systems (□ 56 vol%, o 58 vol%, x 72 vol%)

To display the accelerative nature and "final" velocity of the predetonation shock wave, the x-t data were converted to relative values of x and t, as discussed above. For 5 μm HMX/water samples, there is a consistent pattern, as seen in Fig. 3, for input pressures of 5.3, 6.0 and 7.8 GPa. The "final" shock velocity is about 5.0 mm/μs. For the 60 μm HMX/water samples, the "final" shock velocity is about 6.5 mm/μs.

For the FEFO samples, there tended to be a little more scatter in the individual pin responses. Nevertheless, the same patterns appear to be present for the 5 μm HMX/FEFO samples, where the final velocities are 4.25 and 3.94 mm/μs for EVH-4 and EVH-5, respectively. For the 60 μm HMX/FEFO (72 vol% HMX) samples, the final velocities are 5.18 and 5.30, respectively, for EVH-2 and EVH-3. For the two samples (EVH-6 and EVH-7) of 60 μm HMX/FEFO (56 vol% HMX), the final velocities are not consistent with each other - in contrast to the patterns described above. Instead, the final velocity (5.34 mm/μs) for EVH-7 and the accelerative nature

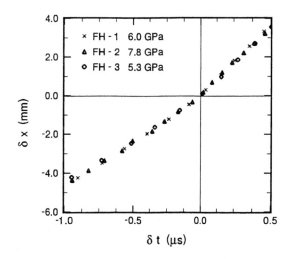

FIGURE 3. Common Predetonation Shock Paths for Samples of 5 μm HMX/water

of the δx-δt data are consistent with the 60 μm HMX/FEFO (72 vol% HMX) samples as seen in Fig. 4. (Note that the input pressures, 5.3 and 5.4 GPa, are nearly identical.) For EVH-6 the final velocity (3.62 mm/μs) and the accelerative nature of the relative x-t data are consistent with the 5 μm HMX/FEFO samples. The latter observations are taken as confirmation of the "non-consistent" nature of x* for EVH-6, relative to the other data for 60 μm HMX/FEFO samples in Fig. 2.

While the path to buildup to detonation may be

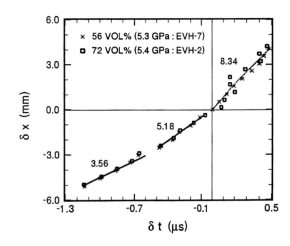

FIGURE 4. Common Predetonation Shock Paths for 60 μm HMX/FEFO Systems (numbers have units of mm/μs)

consistent ("equivalent"?) for a given particle size of HMX at a variety of input pressures (Figs. 3 and 4), contrasting features are observed for different particle sizes shock loaded at nearly identical pressures. This is illustrated in Fig. 5 for the shock loading of two HMX/water mixtures (FH-9 and FH-11) where both the input pressures and the values of x* are nearly identical. Fig. 5, in addition to the data for the 60 μm and the 110 μm samples, shows the extrapolated paths (solid lines) of the input shock and the detonation wave. (Dashed lines have been drawn to show regions of constant velocity in the late time predetonation regime.) It is readily seen that for the 60 μm HMX sample the input shock accelerates a little more quickly than for the 110 μm HMX sample. But more importantly, once the shock in the 60 μm HMX sample attains a velocity of 6.34 mm/μs, it takes over 6 mm more before a transition occurs. On the other hand, the shock in the 110 μm HMX sample attains a "final" velocity of only 5.41 mm/μs, basically 2 mm before a transition occurs. Hence, there is a distinct difference in shock-induced kinetics even though the shock sensitivity and experimental shock reactivity of 60 μm HMX/water and 110 μm HMX/water samples appear to be the same (within experimental error).[4] These are some of the aspects that challenge the successful modeling of reaction kinetics for the shock loading of samples containing different particle sizes of HMX (and in a variety of binder systems).

CONCLUSIONS

Wedge test data for samples of HMX/water and HMX/FEFO mixtures, with different particle sizes of HMX, have been re-analyzed to yield x*,t* data. In the log x* - log P plane, the slopes appear to be identical for 5 μm HMX samples, regardless of the nature of the "binder" (i.e., water vs. FEFO). In the case of HMX/FEFO systems, the x* - P data sets for 60 μm HMX/FEFO samples have been interpreted to suggest that both the shock reactivity and shock sensitivity are the same at solid loadings of 56 and 72 vol% HMX. This shock sensitivity aspect is essentially consistent with data for fine (Class 5) HMX in cast HMX/HTPB systems.[5]

Comparisons of the accelerative nature of the developing shock wave in both HMX/water and HMX/FEFO formulations show support for a common-curve buildup process for a given particle size. That is, the latter stage of the buildup appears to be independent of the input pressure. However, at a constant input pressure and for very similar values of x*, the developing shock wave definitely depends upon the contrasting reaction kinetics associated with the particle size of the HMX.

REFERENCES

1. Simpson, R. L., Helm, F. H., Crawford, P. C. and Kury, J. W., "Particle Size Effects in the Initiation of Explosives Containing Reactive and Non-Reactive Continuous Phases," Proc. Ninth Symposium (International) on Detonation, Arlington, VA, July 1990, p. 25.
2. Moulard, H., "Particular Aspect of the Explosive Particle Size Effect on Shock Sensitivity of Cast PBX Formulations," Proc. Ninth Symposium (International) on Detonation, OCNR 113291-7, Arlington, VA, July 1990 p. 18.
3. Baillou, F., Dartyge, J. M., Spyckerelle, C. and Mala, J., "Influence of Crystal Defects on Sensitivity of Explosives," Proc. Tenth International Detonation Symposium, Arlington, VA, July 1995, p. 816.
4. Bernecker, R. R., "Numerical Simulation of Shock Initiation in Simple Nitramine Systems," Proceedings 1996 JANNAF Propulsion Systems Hazards Subcommittee Meeting, CPIA Publication 645, November 1996.
5. Bernecker, R. R. and Anderson, E. W., "The Calculation of Unreacted Hugoniots. II. Cast Plastic-bonded Explosives and Cast Propellants," Proceedings 1996 JANNAF Propulsion Systems Hazards Subcommittee Meeting, CPIA Publication 645, November 1996.

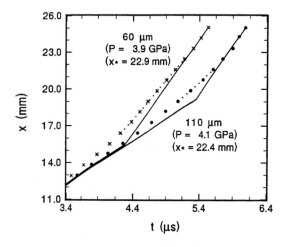

FIGURE 5. Contrasting Predetonation Paths for 60 and 110 μm HMX/water Samples

CP429, *Shock Compression of Condensed Matter – 1997*
edited by Schmidt/Dandekar/Forbes
© 1998 The American Institute of Physics 1-56396-738-3/98/$15.00

SENSITIVITY OF THE TATB-BASED EXPLOSIVE PBX-9502 AFTER THERMAL EXPANSION

Roberta N. Mulford and Joseph A. Romero

Los Alamos National Laboratory
Los Alamos, New Mexico 87545

The sensitivity of TATB-based explosive PBX-9502 is affected by non-reversible thermal expansion, or "ratchet growth." PBX-9502 is a plastic-bonded explosive consisting of 95 wt% TATB (2,4,6-trinitro-1,3,5-benzenetriamine) and 5 wt% Kel-F 800 binder (chlorotrifluoroethylene/vinylidine 3:1 copolymer). The magnitude of the increase in size and the corresponding increase in sensitivity is reported here for a particular pressing of PBX-9502, after repeated thermal cycling. The physical morphology of the expanded material is examined using scanning electron microscopy, in an effort to determine the increase in intergranular holes, intragranular cracks and fissures in the TATB crystals, and the change in the distribution of the Kel-F, all of which are suspected to affect the sensitivity of the material. These images support the proposed mechanism for ratchet growth.[1] Sensitivity, growth of the reactive wave behind the shock front, and Hugoniot data are obtained from in-material particle velocity gauge records of the shock initiation process. Increases in sensitivity with growth and with elevated temperature are summarized in Pop plots. Sensitivity increases commensurate with the increase in voids.

INTRODUCTION

TATB-based explosives show unexpected increases in sensitivity with temperature at temperatures between ambient and 150°C.[2] Non-Arrhenius behavior at these temperatures may arise because the influence of the changes in the physical structure of the material is stronger than is the thermal increase in chemical reactivity. TATB and its mixtures exhibit "ratchet growth," or hysteritic thermal expansion, producing a variety of microstructures in this temperature range as a function of both temperature and of thermal history. In particular, void content, void morphology, and density are governed by the thermal behavior of these materials, as well as by the well-studied factors of grain size and pressing protocol.

A detailed mechanism was developed by Cady,[2] from uniaxial thermal expansion data and previously published crystallographic studies.[3] This mechanism is supported by the SEM data presented here.

EXPERIMENTAL

Thermal cycling of the PBX-9502 was done by heating ten 2x2-inch machined cylinders of the explosive to 216 °C for one hour, in an explosion-proof oven. To prevent cracking of the sample, a heating rate of 2.5 °C/minute and cooling rate of ~0.5 °C/minute were used. Each cylinder was measured at 32 locations to ±0.0001 inch after each cycle, and flatness of ends checked to ±0.0001 inch.

The material used was 9502-17 from batch 86-03, with grainsize 99.3%<45μ, 67.3%<20μ. Samples were cut from a 9-inch cylindrical pressing of density ρ_0=1.886±0.003, with axes of sample cylinders parallel to the original pressing axis. Considerable variation in growth behavior may occur between different batches and pressings.[4]

Shock initiation experiments were done using a light-gas gun, in a configuration producing a one-dimensional and well-supported shock. Material response and run to detonation were measured using in-material electromagnetic (MIV) gauging. Expanded samples were remachined before testing, to provide flat impact and gauging surfaces. The gas gun and the data acquisition and analysis are described elsewhere,[5] as are shock initiation of heated PBX-9502.[6]

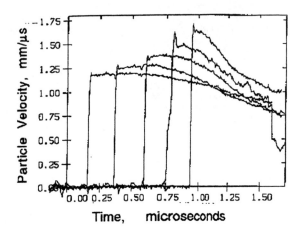

FIGURE 1. The development of the reactive wave inside the explosive sample. Each curve is particle velocity at a Lagrange position (material element) inside the explosive.

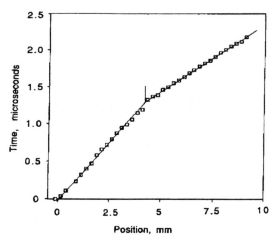

FIGURE 2. Acceleration of the shock front as the explosive reacts. Shock velocities are derived from linear fits to the data in the two regions.

Scanning electron microscopy (SEM) was done after expansion was complete. All SEM samples were cleaved from the cylinder to minimize distortion of surfaces. Samples were taken from both the outside and the center of the cylinder, and were examined both parallel and perpendicular to the pressing axis.

RESULTS

Ratchet growth produces an average volume expansion of 3.24% in these parts. Numbers as high as 10% have been reported,[4] with 4% being the expected value.[3] Expansion along the pressing axis exceeds that on the diameter by a factor of 1.7, because of alignment of the TATB crystallites during the pressing process. Densities of the parts were measured by immersion before and after cycling, and show an average increase of 3.24 %±0.10 % at 216°C, and 3.08 %±0.10 % at 211°C.

An asymptotic limit to growth is given by
$$g = g_{max} (1-e^{-f(\Delta T)n})$$
where g is growth in volume, ΔT is change in temperature, and n is the number of cycles.[7]

Shock initiation data from the in-material gauge experiments is summarized in Table 1. A gauge record is shown in Figure 1. The development of the shock wave into a detonation wave is also monitored using a shock-tracker gauge[8] inside the explosive. The obtained record of shock arrival times yields shock velocity, detonation velocity, and run to detonation.

The run to detonation is most conveniently expressed as a Pop plot, as shown in Figure 3. The Pop plot for the expanded material is displaced from that for ambient PBX-9502, but falls quite close to the curve for PBX-9502 at 75°C.

Hugoniot points for the expanded materials, given in Table 1, fall on the published Hugoniot for PBX-

TABLE 1. Shock initiation data for expanded and hot PBX-9502

sample number	temp °C	u proj	u_p	U_s	t*	P calculated	P impulse
1010	25	1.41(4)	1.209	5.259	1.25	10.9	12.71
1026	25	1.41(2)		5.553	1.20	10.9	12.2
1037	25	1.40(2)	1.144	5.25	1.5	10.9	12.98
1038	25	1.397(2)	1.191	4.9	1.9	10.9	9.41
elevated temperature data:							
1041	92	1.41(5)	1.03	3.5	1.2	6.81	
1044	100	1.41(2)	1.2	5.23	1	11.6	
1056	85	1.41(1)	1.01	4.81	1.3	9.18	
1057	100	1.41(1)	1.21	4.95	1.0	11.32	

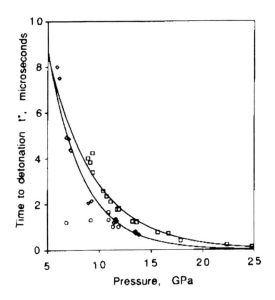

FIGURE 3. Pop plot, run to detonation t* as a function of input shock pressure. ☐ are from Ref. 9, ◊ are from reference 2a at temperature 75°C, and) are this work.
Outliers) are at 85°C (6Gpa) and 92°C (9GPa).

9502,[9] indicating that very little change in shock impedance has resulted from the expansion and alteration of the microstructure of the samples.

Micrographs of the expanded material, Figures 4 and 5, show the morphological changes associated with ratchet growth. Microcracking, the buckling of the TATB crystals under stress produced in the expansion, exfoliation (repeated splitting) of single TATB crystals, and localization of the Kel-F binder are evident in these images.

DISCUSSION

The mechanism of ratchet growth depends on synergistic distortion of the Kel-F binder and the TATB crystallites to minimize strain from thermal expansion at a given temperature.

Different processes occur in four temperature regimes. At low temperatures (to -139°C), the un-bending of the compressed TATB crystallites trans-fers strain to the Kel-F binder, which either breaks or stretches, (above or below T_g), permitting unbending and consequent expansion of the crystal network.

Above room temperature, the anisotropic expansion of each crystal leads to localized stresses in the matrix, requiring displacement of TATB crystallites. Between room temperature and 60°C, the elasticity of the Kel-F distorts the TATB, and the crystals are

maintained in their bent configuration. Thermal expansion within the bent crystals leads to exfoliation, or splitting between crystal planes, permanently increasing the volume of the TATB crystals. Above 60°C, the Kel-F flows, losing its adhesion and tendency to realign the crystals upon cooling, and filling intergranular voids. This rearrangement of the structure is preserved on cooling, leading to net expansion of volume. Above the pressing temperature of 90°C to 100°C, the crystals expand sufficiently to fill all intergranular volume.

The micrographs confirm the details of the mechanism in the high temperature region. Displacement of material by the expanding crystallites, exfoliation of individual crystallites, loss of restoring force on crystallites due to the relocation of the binder, and anisotropy of the thermal expansion of the TATB crystallites are evident.

Images of material broken parallel to the pressing axis (perpendicular to the crystal stacking axis) show the exfoliated crystals edge-on, (Figure 4) revealing the intragranular voids. The exfoliation produces long flat voids of maximum width 0.1μ, except where severe buckling opens a gap as large as 0.5μ.

Inter-granular voids can be identified on images (Figure 5) of surfaces cleaved perpendicular to the pressing axis, along the cleavage planes of the crystals. Intergranular voids have a maximum extent of almost uniformly 0.5μ, with occasional gaps aslarge as 1u. Uniformity appears to result from the maximum extension of crystal being consistent along the non-exfoliating axes.

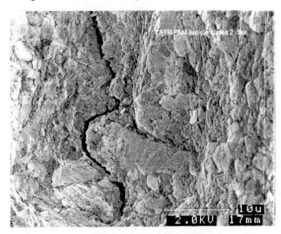

FIGURE 4. PBX-9502 after 5 cycles to 216°C. Magnification is 6000x. Note the crack at A, and hexagonal gap where crystal expanded along the non-exfoliating axis. Small round dots B are binder. Sample from center of part, viewed perpendicular to pressing axis, so that most crystals are cleaved.

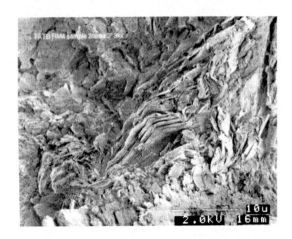

FIGURE 5. PBX-9502 after 5 cycles to 216°C. Magnification is 3000x. Note the bucking of the crystal at B, opening up multiple new voids inside the crystal by exfoliation C. Kel-F binder at D has melted and run away from its original location between crystals, opening up intergranular spaces at E.

The number of voids created during the thermal cycling is estimated from examination of 67 micrographs. Voids are segregated according to size to create the histogram shown in Fig. 6. Intragranular and intergranular voids are distinguished to estimate the size distribution typical of each. In conventional examinations of bulk density and sensitivity, intra- and intergranular hotspots are not distinguished.

The intragranular hotspots have increased in number and changed in character relative to the original PBX-9502, in which these voids are round holes of diameter 0.1μ to 0.2μ.

The large increase in the number of intergranular

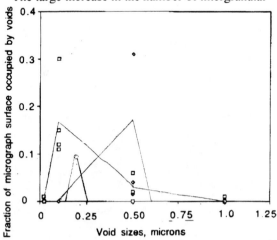

FIGURE 6. The distribution of created voids.
☐ are intragranular voids, ◊ are intergranular voids
◗ are values for unaltered PBX-9502.

voids results from one case where cracks opened up due to the lateral expansion of the crystal, and possibly due to the loss of binder in those spaces.

Comparisons of the growth of the reactive waves between expanded and unexpanded material shows a reactive wave close to the shock front in the expanded material, a profile typical of porous material.

SUMMARY

The majority of increase in numbers of voids falls among intragranular voids of $\sim 1\mu$ in size. Intergranular hotspots are also increased, with most falling about 0.5μ in size. If these additional hotspots cause the observed increase in sensitivity, then small internal voids in crystals may be effective as hotspots.

Because the reaction zone in TATB is long,[10] the additional intergranular voids can contribute effectively to heating in the reaction zone. The increase in the number of the intergranular hotspots is dependent on the binder. Thus the binder plays a critical role only if the relatively small increase in intergranular hotspots is governing the sensitivity.

1. H. Cady, 30th meeting of JOWOG-9, (1989).

2. J. C. Dallman and J. Wackerle, in the 10th Detonation Symposium, Boston, p. 130 (1993). P.A. Urtiew, T.M. Cook, J.L. Maienschein, and C.M. Tarver, ibid. p. 139 (1993).

3. John R. Kolb and H.F. Rizzo, Propellants and Explosives 4, 10-16, (1979).

4. Howard Cady, private communication.

5. R. Mulford, S. Sheffield, and R. Alcon, in "High Pressure Science and Technology," Colorado Springs, 1993, p. 1405-1409.

6. R.N. Mulford and R.R. Alcon, Proceedings of 1995 APS Topical Conference on "Shock Compression of Condensed Matter," p. 855, 1995.

7. Larry Hill, unpublished, 1994.

8. R.R. Alcon and R.N. Mulford, Proceedings of the APS Topical Conference on "Shock Compression of Condensed Matter," p. 1057, 1995.

9. J.J. Dick, C.A. Forest, J.B. Ramsay, and W.L. Seitz, J. of Appl. Phys., **63** (10), 4881 (1988).

10. W.L. Seitz, H.L. Stacey, Ray Engelke, P.K. Tang, and J. Wackerle, 9th Detonation Symposium, (Int'l), Portland, p.657, (1989)

CP429, *Shock Compression of Condensed Matter – 1997*
edited by Schmidt/Dandekar/Forbes
© 1998 The American Institute of Physics 1-56396-738-3/98/$15.00

SHOCK SENSITIVITY OF LX-04 AT ELEVATED TEMPERATURES

P. A. Urtiew, C. M. Tarver, J. W. Forbes and F. Garcia

Lawrence Livermore National Laboratory,
P.O. Box 808, L-282, Livermore, CA 94551

Hazard scenarios can involve multiple stimuli, such as heating followed by fragment impact (shock). The shock response of LX-04 (85 weight % HMX and 15 weight % Viton binder) preheated to temperatures near 170C is studied in a 10.2 cm bore diameter gas gun using embedded manganin pressure gauges. The pressure histories at various depths in the LX-04 targets and the run distances to detonation at several input shock pressures are measured and compared to those obtained in ambient temperature LX-04. The hot LX-04 is significantly more shock sensitive than ambient LX-04. Ignition and Growth reactive flow models are developed for ambient and hot LX-04 to allow predictions of impact scenarios that can not be tested directly.

INTRODUCTION

With safety issues playing a dominate role in present-day energetic materials technology, concern is increasing about the relative safety of solid high explosives exposed to extreme environmental conditions. High energy materials based on octahydro-1,3,5,7-tetranitro-1,3,5,7-tetrazocine (HMX) are particularly important. LX-04, which contains 85 weight % HMX and 15 weight % Viton binder, is a widely used HMX-based plastic bonded explosive. The sensitivity of LX-04 to single stimulus such as heat, impact, and shock has been studied. Hazard scenarios can involve multiple stimuli, such as heating to temperatures close to thermal explosion conditions followed by fragment impact, producing a shock in the hot explosive. This scenario has previously been studied for triaminotrinitrobenzene (TATB)-based insensitive solid explosives under various thermal and confinement conditions (1-3). The shock sensitivity of heated LX-04 is compared to that of ambient temperature LX-04 using embedded manganin pressure gauges (4) and reactive flow calculations (5).

EXPERIMENTAL

The experimental geometry for the heated LX-04 embedded gauge experiments is shown in Fig. 1. A 12.7 mm thick, 90 mm diameter aluminum flyer plate impacted a target consisting of a 6 mm thick, 90 mm diameter aluminum buffer plate, a 20 mm thick, 90 mm diameter LX-04 charge and a 6 mm thick, 90 mm diameter aluminum back plate. Heaters were placed within the aluminum plates and heated the LX-04 to approximately 170C at a rate of 10C/minute. When the nine thermocouples in the aluminum and LX-04 showed that the whole assembly was within a few degrees of 170C, the shot

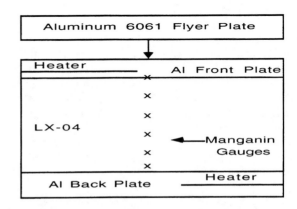

FIGURE 1. Geometry of the heated LX-04 embedded manganin pressure gauge experiments

was fired. Four shots with aluminum flyer velocities of 0.373, 0.515, 0.643, and 0.886 mm/μs produced shock pressures of 1.5, 2.2, 2.8, and 4.3 GPa, respectively. Six manganin pressure gauges placed along the LX-04 charge axis measured the pressure histories at 0, 5, 10, 13, 16 and 18 mm into LX-04 in the three lower pressure shots and at 0, 2, 4, 7, 10, and 15 mm into LX-04 in the highest pressure shot. One experiment was fired using ambient temperature LX-04 and Teflon flyer, buffer, and back plates with a velocity of 0.956 mm/μs to impart a shock pressure of 3.0 GPa for comparison.

REACTIVE FLOW MODELING

The Ignition and Growth reactive flow model uses two Jones-Wilkins-Lee (JWL) equations of state, one for the unreacted explosive and another one for the reaction products, in the temperature dependent form:

$$p = A\,e^{-R_1 V} + B\,e^{-R_2 V} + \omega C_v T/V \qquad (1)$$

where p is pressure in Megabars, V is relative volume, T is temperature, ω is the Gruneisen coefficient, C_v is the average heat capacity, and A, B, R_1 and R_2 are constants. The equations of state are fitted to the available shock Hugoniot data. The reaction rate law is:

$$dF/dt = I(1-F)^b(\rho/\rho_0-1-a)^x + G_1(1-F)^c F^d p^y$$
$$\quad 0<F<F_{igmax} \qquad\quad 0<F<F_{G1max}$$
$$+\, G_2(1-F)^e F^g p^z \qquad (2)$$
$$\quad F_{G2min}<F<1$$

where F is the fraction reacted, t is time in μs, ρ is the current density in g/cm^3, ρ_0 is the initial density, p is pressure in Mbars, and I, G_1, G_2, a, b, c, d, e, g, x, y, and z are constants. This three term reaction rate law models the three stages of reaction generally observed during shock initiation of pressed solid explosives (5). The equation of state parameters for ambient and hot LX-04, aluminum, and Teflon, and the Ignition and Growth rate law parameters are listed in Table 1. The reaction rates are similar to those used for LX-10 (6) and PAX2A (7). The hot spot growth rate parameter G_1 in Eq. (2) is varied to match the experimental data.

COMPARISON OF RESULTS

Table 2 contains the experimental flyer velocities,

impact pressures, and run distances to detonation for the ambient temperature LX-04 shot and the four hot LX-04 shots. Figure 2 shows the measured and calculated pressure histories at the six gauge locations in ambient temperature LX-04 impacted by a Teflon flyer plate at 0.956 mm/μs. The gauge records show rapid pressure increases, but detonation is not quite attained in the 20 mm thick charge. The agreement between the experiment and calculation is good, except the calculated shock front pressures increase slightly too rapidly. A growth reaction rate coefficient G_1 value of 100 is used for LX-04 in Fig. 2. This value is similar to the values of $G_1 = 120$ for LX-10 (94.5 % HMX, 5.5 % Viton) and $G_1 = 90$ for PAX2A, which is 85 % HMX in a different binder.

Two comparisons for hot (170C) LX-04 are shown in Figs. 3 and 4. Roth (8), in a shock initiation study of PBX 9404 at 150C, found that its density at that temperature had decreased from 1.84 to 1.77 g/cm^3. The density of LX-04 is assumed to decrease from 1.866 to 1.77 g/cm^3 at 170C (443K). The unreacted equation of state for hot LX-04 in Table 1 is adjusted for these initial conditions. Figure 3 contains the records for the 0.643 mm/μs aluminum flyer impact velocity experiment. The gauge records and calculations show that transition to detonation occurs very close to the 10 mm deep gauge position. Figure 4 shows the results for the 0.515 mm/μs

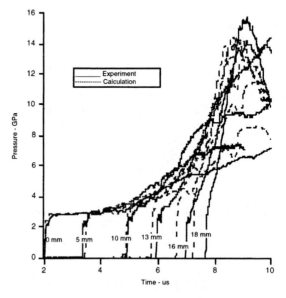

FIGURE 2. Pressure histories for ambient temperature LX-04 shock initiated by a Teflon flyer plate at 0.956 mm/μs

TABLE 1. Equation of State and Reaction Rate Parameters

1. Ignition and Growth Model Parameters for LX-04

Unreacted JWL	Product JWL	Reaction Rate Parameters	
A. Ambient temperature (T_o=298K; ρ_o =1.866 g/cm^3; Shear Modulus=0.05 Mbar; Yield Strength=0.002 Mbar)			
A=9522 Mbar	A=8.364 Mbar	I=7.43e+11	G_2=400
B=-0.05922 Mbar	B=0.1298 Mbar	a=0.0	e=0.333
R_1=14.1	R_1=4.62	b=0.667	g=1.0
R_2=1.41	R_2=1.25	x=20.0	z=2.0
ω=0.8867	ω=0.42	G_1=100	F_{igmax} = 0.3
C_V=2.7806e-5 Mbar/K	C_V=1.0e-5 Mbar/K	y=2.0	F_{G1max}=0.5
	E_0=0.095 Mbar	c=0.667, d=0.333	F_{G2min}=0.5

B. Hot LX-04 (T_o=443K; ρ_o=1.77 g/cm^3; Shear Modulus=0.0474 Mbar)
B=-0.0740834 Mbar G_1=210

2. Gruneisen Parameters for Inert Materials

$$p = \rho_0 c^2 \mu [1+(1-\gamma_0/2)\mu - a/2\mu^2] / [1-(S_1-1)\mu - S_2\mu^2/(\mu+1) - S_3\mu^3/(\mu+1)^2]^2 + (\gamma_0 + a\mu)E$$

where $\mu = (\rho/\rho_0)-1$ and E is thermal energy

Inert	ρ_0(g/cm^3)	c(mm/μs)	S_1	S_2	S_3	γ_0	a
6061-T6 Al	2.703	5.24	1.4	0.0	0.0	1.97	0.48
Teflon	2.15	1.68	1.123	3.98	-5.8	0.59	0.0

TABLE 2. Experimental flyer velocities, impact pressures, and run distances to detonation

Flyer Velocity	Impact Pressure	LX-04 Temperature	Experimental Run to Detonation Results	
(mm/μs)	(GPa)	(C)	Distance(mm)	Time(μs)
0.956 (Teflon)	3.0	25	20	7.8
0.373 (Al)	1.5	165-169	>20	>10
0.515 (Al)	2.2	167-171	14	5.4
0.643 (Al)	2.8	166-170	9	3.6
0.886 (Al)	4.3	166-170	4	1.3

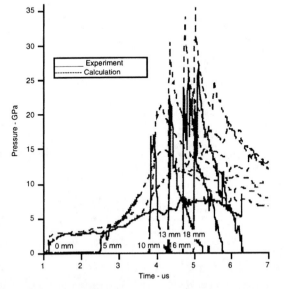

FIGURE 3. Pressure histories for hot (170C) LX-04 shock initiated by an aluminum flyer at 0.643 mm/μs

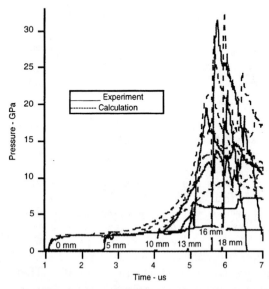

FIGURE 4. Pressure histories for hot LX-04 shock initiated by an aluminum flyer at 0.515 mm/μs

aluminum flyer experiment and the corresponding reactive calculations. The detonation transition takes place between the 13 mm gauge and the 16 mm gauge positions. The two sets of gauge records in Figs. 3 and 4 clearly demonstrate that LX-04 is significantly more shock sensitive at 170C than at ambient temperature. The highest aluminum impact velocity, 0.886 mm/μs, produces a 4.3 GPa shock in hot LX-04 and a 4 mm run distance to detonation.

For these three impact pressures, the Ignition and Growth model yields excellent results using a value of $G_1 = 210$, which is essentially double the ambient hot spot growth rate. The increases in shock front pressure measured by the embedded gauges in Figs. 3 and 4 for hot LX-04 are similar to those in Fig. 2 for ambient LX-04. This indicates that the amount of LX-04 ignited near the shock front does not increase substantially at 170C and that the increase in shock sensitivity is mainly due to the more rapid growth of reacting hot spots into the surrounding preheated explosive (9). The modeling results support this conclusion, since the only increase with temperature is in the G_1 coefficient. The lowest aluminum impact velocity, 0.373 mm/μs, in Table 2 imparts about 1.5 GPa into the hot LX-04. Little or no reaction is observed at this input shock pressure, and thus the run distance to detonation at 1.5 GPa is much greater than 20 mm.

The increased shock sensitivity of LX-04 at 170C is shown in terms of run distance to detonation versus input shock pressure in Fig. 5. Also shown in Fig. 5 are the "Pop Plots" for ambient LX-04 and PBX 9404 (10). Raising the temperature of LX-04 increases its shock sensitivity close to that of PBX 9404, which contains 94% HMX and an energetic binder based on nitrocellulose.

SUMMARY

The shock sensitivity of LX-04 heated to 170C has been quantitatively demonstrated to be greater than that of ambient temperature LX-04. Its run distance to detonation versus impact pressure curve is similar to that of PBX 9404. The Ignition and Growth reaction rates for LX-04 are similar to those previously developed for other HMX-based formulations. The only changes for hot LX-04 are a lower initial density and an increase in the hot spot growth coefficient G_1 from 100 to 210. This implies that the increased sensitivity of LX-04 at 170C is mainly due to the faster growth of hot spot

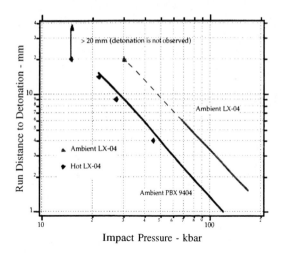

FIGURE 5. Run distance to detonation versus input shock pressure for ambient and heated LX-04 and PBX 9404

reactions following ignition. Additional experiments are planned in which the LX-04 is completely confined to limit thermal expansion during heating.

ACKNOWLEDGMENTS

This work was performed under the auspices of the U.S. Department of Energy by Lawrence Livermore National Laboratory (contract no. W-7405-ENG-48).

REFERENCES

1. Urtiew, P. A., Tarver, C. M., Maienschein, J. L., and Tao, W. C., *Combustion and Flame* **105,** 43-53 (1996).
2. Urtiew, P. A., Cook, T. M., Maienschein, J. L., and Tarver, C. M., *Tenth International Detonation Symposium,* ONR 33395-12, Boston, MA, 1993, pp. 139-147.
3. Dallman, J. C. and Wackerle, J., *Tenth International Detonation Symposium,* ONR 33395-12, Boston, MA, 1993, pp. 130-138.
4. Urtiew, P. A., Erickson, L. M., Hayes, B., and Parker, N. L., *Combustion, Explosion and Shock Waves* **22,** 597-614 (1986).
5. Tarver, C. M., Hallquist, J. O., and Erickson, L. M., *Eighth Symposium (International) on Detonation,* NSWC MP 86-194, Albuquerque, NM, 1985, pp. 951-961.
6. Tarver, C. M., Urtiew, P. A., Chidester, S. K., and Green, L. G., *Propellants, Explosives, Pyrotechnics* **18,** 117-127 (1993).
7. Baker, E. L., Schimel, B., and Grantham, W. J., in *Shock Compression of Condensed Matter-1995,* Schmidt, S. C. and Tao, W. C., eds., AIP Press, New York, 1995, pp. 409-412.
8. Roth, J., *Fifth Symposium (International) on Detonation,* ACR-184, Pasadena, CA, 1970, pp. 219-230.
9. Tarver, C. M., Chidester, S. K., and Nichols, A. L., *J. Phys. Chem.* **100,** 5794-5799 (1996).
10. Dobratz, B. M. and Crawford, P. C., *LLNL Explosives Handbook,* Lawrence Livermore National Laboratory Report UCRL-52997 Change 2, 1985.

CP429, *Shock Compression of Condensed Matter – 1997*
edited by Schmidt/Dandekar/Forbes
© 1998 The American Institute of Physics 1-56396-738-3/98/$15.00

SHOCK-INDUCED GRAIN BURNING IN MIXTURES OF AMMONIUM PERCHLORATE AND ALUMINUM

V. S. Joshi, P-A. Persson and K. Brower

Research Center for Energetic Materials, New Mexico Tech, Socorro, NM 87801, USA

Experiments have been performed to investigate the initiation of reaction in explosive-metal mixtures. Mixtures of ammonium perchlorate (AP) and aluminum (Al) were packed in capsules and exposed to a short duration pressure pulse, generated by the impact of an explosive driven flyer plate. This paper describes the development of a sealed fixture for complete recovery of the unreacted solids, gaseous and solid reaction products. The results of a series of experiments are presented, which provide new evidence of complete melting and densification of AP. Based on the critical information obtained from these experiments, we propose a mechanism for the initiation of reaction of AP, and that of Al in a strongly oxidizing environment.

INTRODUCTION

Large quantities of insensitive energetic materials are stored, handled and transported in many military and industrial applications. Safe processing of these materials requires knowledge of stimuli that can lead to initiation and detonation in different classes of these materials. These stimuli include accidental impact or large fire in the vicinity of the material. Improving safety in operation relies on the ability to predict hazards in such materials, and requires knowledge of material stability and reactivity under increased pressure and temperature during the progression of decomposition reaction. The current investigation is aimed at elucidating the mechanism of initiation of reaction via recovery and analysis of partially reacted samples in the initial stages of reaction.

BACKGROUND

Decomposition reactions of AP have been studied by isothermal cook-off experiments [1-2]. These experiments provided kinetics of reactions at low rates of heating. The mechanisms of decomposition are known to change at higher rates of heating [3], accompanied by an increase of reaction rates by several orders of magnitude. Initiation of reactions in dense AP crystals under high temperatures and rates

of heating have been studied by laser ignition [4], plasma arc discharge [5] and dynamic compression [6]. These studies emphasize the role of hot spots, defect generation and phase transformation during decomposition. Analysis of chemical species generated during decomposition of AP under different conditions has recently been carried out [7-9]. Despite these numerous investigations the understanding of the overall mechanism of decomposition of porous AP under shock-wave loading remains elusive.

The current research has its roots in the initial investigation by Polster [10], who compared the microstructure of pellets of AP before and after exposure to short duration shock pressure and determined that the grain size of the starting material reduced by a factor of two. He designed the initial fixture for the recovery experiments, known as the Polster fixture, in which an explosively driven copper driver plate compressed a pellet of AP at 90-95% density. This was followed by the work of O'Connor [11], who modified the fixture to capture the reaction product gases of Al + ammonium nitrate, Al + AP and Al +TNT mixtures. He found that the extent of reaction of Al increased with increasing proportion of Al in the mixture. He also concluded that Al particles were either completely reacted or completely unreacted.

The AP + Al powder mixture thus represents an interesting material system which has a fast and a slow reacting component. AP not only undergoes chemical reaction, but also can transform into to a cubic phase [12] upon exposure to high temperature, which if present, could indicate such exposure. Thus, the objective of the research presented here was to investigate the initiation of reaction of this system under shock-compression.

EXPERIMENTAL

Shock recovery experiments were performed over a range of shock pressures using the newly designed recovery fixture. We had become increasingly aware of the risk, present in both the Polster and modified Polster fixtures, of reaction products from the shock driver charge contaminating the reaction products from the test charge. The new fixture positively eliminates that risk by having the driver charge placed *outside* the sealed high-pressure stainless steel capsule. The driver charge accelerates a flyer plate, which transmits the shock to the sample charge inside through the end wall of the capsule. The capsule undergoes plastic deformation but stays sealed until it is opened for analysis after the experiment. A cylindrical momentum trap prevents the capsule from fracturing. Figure 1 shows the arrangement of this newly designed and improved fixture.

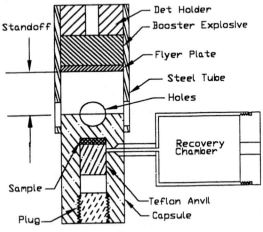

FIGURE 1. Schematic arrangement of the recovery fixture showing various features.

The cavity in the capsule has been designed to obtain a cylindrical sample with varying thickness.

This fixture can also be used for recovery work for other solid and liquid explosives since the Teflon anvil also can act as a seal. The gaseous reaction products from the sample pass through an orifice and side tube into the gas collection chamber. This chamber also serves as a trap for extremely fine, solid reacted product, which may leave the capsule with the evolving gas. The fine dust can be recovered by opening the collection chamber.

Sample mixtures consisting of AP with 90 μm grain size mixed with 20 weight percent Al powder with 44-87 μm grains size were packed at densities between 70 and 80% of the solid density. The powders were pressed in capsules at the desired packing densities and sealed. The capsules were then shock loaded by impact of an explosively driven flyer plate.

The shock-compressed samples were recovered by machining off the impact face of the capsules. The recovered samples were characterized by optical microscope at low magnification and scanning electron microscope at higher magnification to determine the changes in the morphology and other features. X-ray diffraction analysis were conducted on the transverse cross-sections of the recovered shocked compact. Recovered gases were analyzed by GC/FTIR to determine their proportions and the extent of the reaction.

RESULTS AND DISCUSSION

During the initial experiments, it was necessary to determine a reaction threshold condition. Most of the capsules either reacted fully or showed no signs of reaction. Once the optimum pressures were determined to obtain partially reacted samples, full analysis of the products was undertaken.

The recovered solids from the apparently fully reacted samples contained a variety of reaction products. Whereas most of the AP was consumed, Al particles were found in two distinct sizes: original particles which were loosely sticking to the wall of the capsule and which appeared to be unreacted, and very large chunks of Al which appeared to be partially reacted or melted and resolidified. X-ray diffraction results [13] confirmed the observations made by optical microscopy. In some of the large chunks, shiny Al particles seemed to be embedded in an essentially reacted matrix. This was new, since in previous experiments [12], we had not observed features similar to this sample. This material was taken up for detailed analysis.

The tiny particle was carefully mounted and ground to obtain a cross section. The Electron Microprobe was used to obtain the atomic fraction of the elements. A high magnification image, which was taken simultaneously, revealed extremely fine Al particles (much smaller than the starting powder) inside aluminum oxide particle, as shown in Figure 2.

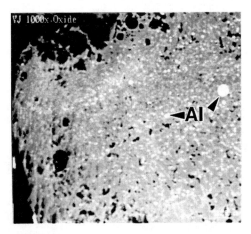

FIGURE 2. Cross section of a particle as observed under Electron Microprobe, revealing extremely fine aluminum (Al) particles embedded in an aluminum oxide matrix.

In other capsules where the reaction had progressed to a lesser extent, other interesting features were observed. A thin layer of compressed material, which originally was in contact with the capsule face (towards the impact side) was recovered. This material appeared to be melted and resolidified (glassy), although it was essentially unreacted. This layer was exposed to the highest incident pressure, and as such initiation of reaction could be expected here. However, the steel capsule appeared to have quenched this layer. The Al particles appeared deformed (flattened) and cracked on the surface. This indicates that the temperature at the surface was probably below the melting point of Al, although it could have been adequate to melt a substantial amount of AP to form a dense matrix. Recovery of this material has led us to believe that the reaction of AP probably did not start immediately.

In order to get some estimate of the lag, we decided to simulate the temperature-time excursions as seen by an individual particle of AP, at the surface and at the core. For this purpose, previously written FORTRAN program [14], which was verified for its accuracy, was modified to include the reactive energy term (heat of reaction) and the activation energy. The mesh was refined and the interface temperature was raised in successive runs to obtain a runaway condition in the sample. This is reflected as a sudden and extreme rise of temperature as a function of time. The program used the following data for AP:

Density= 1950 kg/m^3
Specific Heat= 1290 J/kg.K
Thermal conductivity = 0.5 W/m.K
Energy ($Ae^{(-Ea/RT)}$) = 1×10^{11} exp (-15193/T)

The data obtained was plotted as a temperature-time excursion as shown in Figure 3. The data shows that at about 1300K the heat generated within a particle exceeds the thermal diffusion, and runaway conditions occurs within a fraction of a microsecond thereafter.

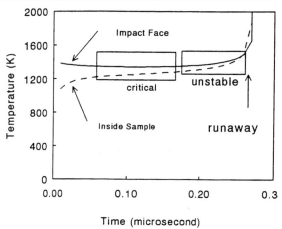

FIGURE 3. Temperature-time excursion of AP for two different locations- upper curve for the interface and lower curve for a point (interior) of the sample.

Transformation of granular AP to a transparent polycrystalline solid under shock compression suggests that extremely rapid heating and cooling can cause melting without decomposition. In order to resolve these observations, adiabatic inert gas compression [13] for short period of less than 100 μsecond has been applied to AP. In this method an empirical scale of melting point vs. compression ratio was created by exposing 50-μm particles of potassium salts of oxyanions to Ar gas initially at 2.2 atm and 25°C. The data points are: KNO$_3$- 334°C at 23:1, KBrO$_4$- 434°C at 29:1, KIO$_3$- 560°C at 40:1, KClO$_4$-

610°C at 50:1. At a compression ratio of 50:1 both AP and $KClO_4$ show signs of liquefaction, but AP is partially decomposed. It can be inferred that the melting point of AP is at least 600°C and probably not much higher.

The composition (mol %) of the gaseous products of pyrolysis of AP is N_2- 38%, N_2O- 20%, NO- 0.6 %, N_2O_4- 15%, O_2- 23% and CO_2- 3.7%. Water and HCl were not quantified. This mixture is exceedingly corrosive because it also contains NOCl and NO_2Cl in equilibrium with Cl_2 and NO_X. Capsule material made of stainless steel was severely corroded both in shock compression and adiabatic compression experiments, but Al is much more resistant. Complex formation by Fe and Cl may be an important factor. The products listed above appear as intermediates in most speculative mechanistic schemes for the pyrolysis of AP. In our experiments they remain in high concentration after the reaction is quenched.

The coexistence of Al and Al_2O_3, unreacted and fused AP, i.e. one each of the reactants and products, suggest that two different mechanisms are operative during and after the passage of the shock wave. One part of the chemical decomposition of AP occurs via heating leading to oxidation reaction of Al. Observation of cracks in the flattened Al grains suggests that Al particles may be quenching the heated AP particles. This would imply that all the heat is not available for reaction at an instant of time. Simulations also indicate that the runaway conditions could occur as a delayed event. While grain burning of AP may still be an earlier event than Al combustion, residual Al particles in both unreacted material and in reacted Al_2O_3 suggests that this oxidation reaction may be the slowest, occurring over a longer time frame than previously expected. In such a situation, reaction could have happened well past the duration of the high pressure shock state.

CONCLUSION

This is the first time that any melted and unreacted AP+Al mixture has been recovered under shock loading conditions. The exposure of AP and Al material to a highly non-equilibrium temperature excursion without any reaction is surprising. It is now clear that reaction is preceded by melting of AP, which could not be inferred earlier, since physical recovery of unreacted AP sample is a new evidence.

Recovery of glassy material in unreacted state is another important landmark in determining possible mechanisms for the reaction of AP.

ACKNOWLEDGMENTS

This work was supported by the Industry/University Cooperative Research Center for Energetic Materials (RCEM) at New Mexico Tech and by the US Navy Office of Naval Research (ONR).

REFERENCES

[1] Olson, D., Banks, M., and Roy, E., RCEM Technical Report A-01-92, New Mexico Tech, Socorro, NM, 1992.

[2] Kraeutle, K. J., NWC Technical Publication 7053, Ed. Boggs, T. L., Naval Weapons Center, China Lake, CA, May, 1990.

[3] Foltz, M.F., and Maienschein J.L., *Shock Compression of Condensed Matter-1995*, AIP Conf. Proc. **370**, pp. 239-42, 1995.

[4] Ramaswamy, A. L., Shin, H., Armstrong R.W., Lee, C. H., and Sharma J., *J. Mat. Sci.* **31**, pp. 6035-6042 (1996).

[5] Lee, R., Ph.D. Dissertation, University of Maryland, MD, 1996.

[6] Sandusky H. W., Beard B.C., Glancy B.C., Elban W.L. and Armstrong R.W., Mat. Res. Soc. Symp. Proc. **296**, pp. 93-98 (1993).

[7] Pace M.D., Mat. Res. Soc. Symp. Proc. **296** pp. 53-60 (1993).

[8] Korobeinichev O.P., Kuibida L. V., Paletsky A.A., and Shmakov A.G., Mat. Res. Soc. Symp. Proc. **418**, pp. 245-255 (1996)

[9] Brill T. R., Mat. Res. Soc. Symp. Proc. **296**, pp. 269-280 (1993).

[10] Polster, M., M.S. Thesis, New Mexico Tech, Socorro, NM, 1992.

[11] O'Connor E., M.S. Thesis, New Mexico Tech, Socorro, NM, 1995.

[12] Sharma J., Coffey C. S., Ramaswamy A. L., and Armstrong R.W., Mat. Res. Soc. Symp. Proc. **418**, pp. 257-264 (1996).

[13] Joshi, V. S., Persson, P-A., and Brower K., RCEM Technical Report A-01-97, New Mexico Tech, Socorro, NM, 1996.

[14] Joshi, V. S., Ph.D. Dissertation, New Mexico Tech, Socorro, NM, 1996.

CP429, *Shock Compression of Condensed Matter – 1997*
edited by Schmidt/Dandekar/Forbes
© 1998 The American Institute of Physics 1-56396-738-3/98/$15.00

ON DETONATION WAVE FRONT STRUCTURE
OF CONSENSED HIGH EXPLOSIVES

A.V.Fedorov, A.V.Menshikh, N.B.Yagodin

Russian Federal Nuclear Center, VNIIEF, 607190, Sarov, Russia

The present report describes detonation front particle velocity profile measurements for three high explosives (HE). Two were plasticized compositions based on PETN, HMX, and the third a TNT/RDX 50/50 mixture. Measurements were carried out using laser interferometry techniques with few nanosecond time resolution. Particle velocity profiles were recorded at HE–window (lithium fluoride [LiF]) interfaces. The detonation waves were diverging since they developed from a single initiation point. Recorded maximum particle velocities at the LiF window interfaces for PETN, HMX, and TNT/RDX mixture were 2.7 mm/µs, 3.15 mm/µs and 2.6 mm/µs, respectively. Duration of chemical reaction zone for HEs based on PETN and HMX were determined. The results of linear extrapolation of recorded particle velocity to zero time are presented, leading to estimates of von Neumann spike parameters for the PETN and HMX compositions.

INTRODUCTION

Detonation wave front structures are being examined more now than in the past. During the last 20 years many researchers have measured detonation wave parameters using different methods, including laser interferometry [1-13].

The present investigation measured detonation front particle velocity histories at HE–window interfaces, using either Fabry–Perot or ORVIS interferometer systems, for detonations emanating from a single initiation point. These methods provide time resolution of a few nanoseconds. Detonation fronts were recorded for the following HE compositions:

- composition 1: plasticized PETN of XTX 8003 and LX–13 type;
- composition 2: plasticized HMX of PBX 9404 type (HMX/polymeric binder 90/10) [12];
- composition 3: 50% TNT and 50% RDX.

Detonation product adiabats and Jouget state parameters for the compositions under examination were provided by A. A. Evstegneev.

EXPERIMENTAL

To reflect the incoming laser beam, an aluminum (Al) foil or sprayed Al layer was located at the HE–window interface. Single crystal lithium fluoride (LiF) was used as the window. Either a Fabry-Perot or an ORVIS interferometer transduced the Doppler shifted laser light from the HE–window interface so that the interface particle velocity was determined. The experimental setup is shown in Fig. 1.

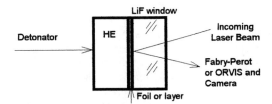

FIGURE 1. Experimental setup.

In the first experimental series, 6 - 20 µm thick Al foil was located at the interface; in later experiments, a 0.55 to 1.5 µm thick sprayed Al layer was used. The shock impedance of LiF and Al are quite close.

An electronic streak camera, Imacon-790, was used as the recording device. Camera space resolution was 10 line pairs/mm and time resolution was 0.2 - 3 ns when streak speeds of 2 - 30 ns/mm were used.

A 150 W ruby laser with a pulse duration of 100 - 200 µs and spectrum line width of $3 \cdot 10^{-3}$ cm^{-1} was used as the incoming laser beam. The beam was focused on the Al foil (or layer) with an ≈ 0.2 mm spot size.

Both HE cylinders and HE rods were used in various experiments. HE cylindrical samples were 13.6 to 30 mm dia. and 2 to 30 mm thick. HE rods were 8 to 14 mm dia. and 2 to 68 mm long.

Optical delay (etalon length) for the ORVIS setup was 60 to 200 mm, which produces velocity measuring errors of ±80 m/s to ±30 m/s. The length of the Fabry–Perot interferometer optic base was 10 to 100 mm, corresponding to velocity measuring errors of ± 200 m/s to ± 20 m/s. Time resolution for the various setups ranged from 1 to 15 ns, with the majority of the experiments done with setups producing 1 to 3 ns time resolution.

EXPERIMENTAL RESULTS

A set of plane-wave generator experiments was completed to determine if there was a substantial difference between detonation wave profiles produced by plane-wave initiation when compared to those from single point initiation. These experiments confirmed the two situations were not dramatically different. Within the error limits of the experiments, the particle velocity, U(t), for plane and diverging initiation schemes coincide. This report describes experiments in which the diverging detonation wave structure was recorded.

Experiments with composition 1

Over 10 experiments were done with Al foils; most had 6μm thick Al foils but some had 10 and 20 μm thick foils. HE samples of 25 mm dia. and 6 to 9 mm thick were used. Time resolution was 3 ns. Fig. 2 is a plot of the average particle velocity history (averaged over the entire experimental set) of the HE–window interface. The dotted lines correspond to the U(t) spread over the set of experiments; all experimental profiles were within the dotted lines.

Measured average particle velocities (for several different times) were U = 2.1 ± 0.1 mm/μs (for t = 3 ns), U = 1.8 ±0.1 mm/μs (for t = 5 ns), U = 1.65 ± 0.06 mm/μs (for t = 10 ns), and U = 1.57 ± 0.03 mm/μs (for t = 23 ns). The particle velocity spread in this time regime was not more than ± 100 m/s for this experimental set. (Most of the experiments for this composition were done with a 0.5 to 1.5 μm thick sprayed Al layer with a time resolution of about 1 ns.)

We used 22 mm dia. HE samples 2 to 12 mm thick and 10 mm dia. HE rods 12 to 68 mm long. Detonation velocity was measured in the 6 to 68 mm long HE samples and indicated a stationary

detonation. Particle velocity histories, U(t), for several of the experiments are shown in Fig. 3. U(t) profiles from the other experiments lie with a little spread in the same area as U(t) curves shown in Fig. 3.

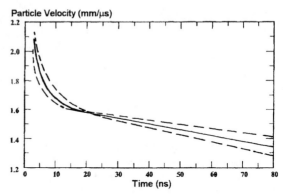

FIGURE 2. Particle velocity at HE–LiF interface for composition 1. Time resolution in the experiments was 3 ns. An Al foil 6 μm thick was located at the interface.

In this set of experiments the maximum HE-LiF interface particle velocity was measured at U = 2.7 ± 0.1 mm/μs (at t = 1 ns), which corresponds to P = 63 Gpa at the interface. Linear extrapolation of curves U(t) from t = 4 ns to t = 0 corresponds to a particle velocity spike at the HE-LiF interface of U = 2.95 ± 0.1 mm/μs.

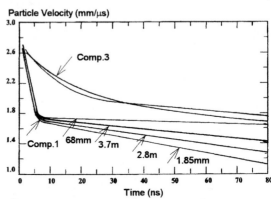

FIGURE 3. Particle velocity at HE–LiF interface for composition 1 and composition 3. An Al layer 0.5 to 1.5 μm thick was located at the interface.

From Fig. 3 it can be seen that all the curves converge at the point U = 1.66 ± 0.05 mm/μs, t = 10 ± 3 ns. It is a well known practice to determine chemical reaction zone dimensions and parameters of the Jouget state with high reliability by U(t) curves crossing for HE charges of different length [2]. Thus the duration of chemical reaction zone is 10 ns (corresponding to an 80 µm width). It is interesting to note that loading a LiF crystal along the detonation product adiabat from the Jouget state, U = 1.65 mm/µs must be realized (P = 32 GPa), which corresponds well with Fig. 3. The results of an experiment with a 68 mm long charge indicates that U(t) is approximately U = 1.65 mm/µs at t = 20 ns.

Experiments with composition 2

Experiments with composition 2 were made using an Al layer 1 - 1.5 µm thick and were done with about 1 ns time resolution. HE samples were 13.6 mm dia. and 10, 20 and 30 mm thick. Experimental results are shown in Fig. 4. Detonation velocity measurements indicated that stationary detonation was realized in the 30 mm long sample.

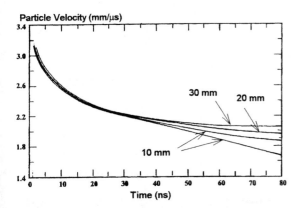

FIGURE 4. Particle velocity at HE–LiF interface for composition 2. An Al layer 1 - 1.5 µm thick was located at the interface.

The maximum particle velocity recorded was U = 3.15 ± 0.1 mm/µs (at t ≈ 1 ns), corresponding to P = 78 GPa at the HE–LiF interface. When determined as discussed above, the chemical reaction zone was about 36 to 44 ns long (corresponding to a length of about 350 µm). For more precise determination of chemical reaction zone parameters we intend to do experiments with less than 10 mm and over 30 mm thick samples.

Linear extrapolation of curves U(t) from t = 4 ns to t = 0 gives a spike value of U = 3.3 ±0.1 mm/µs. It is interesting to note that for composition 2, loading a LiF crystal along the detonation product adiabat from the Jouget state, U = 2.08 mm/µs must be realized (P = 44 Gpa), which corresponds well with Fig. 4.

Experiments with composition 3

Two experiments were done with composition 3 which were 25 mm dia. and 6 mm thick. These samples were initiated using an HE layer of composition 1, 4 mm thick. An Al layer 1.5 µm thick was used. U(t) profiles obtained from the experiments are shown in Fig. 3. A maximum interface particle velocity of U = 2.6 mm/µs (t = 2 ns) was measured corresponding to an HE-LiF interface pressure of 60 GPa. Using the process described earlier, the end of the reaction zone for this material would be expected at U = 1.62 mm/µs. More experiments will be required to determine a reaction zone for this material.

DISCUSSION OF RESULTS

Fig. 5 shows a P-U diagram which illustrates the states that must be realized in LiF at both the spike and the Jouget state for the composition 1 experiments. The detonation velocity, product adiabat and Jouget state parameters have been used to draw the figure, along with the LiF Hugoniot. Shock Hugoniots for KCl, NaCl, and water have been shown in the figure for reference.

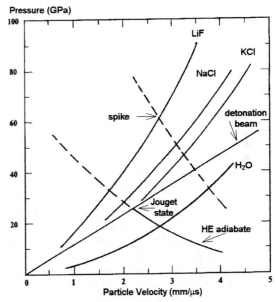

FIGURE 5. P – U diagram for composition 1.

To determine von Neumann spike parameters, it is necessary to follow a backward facing adiabat from spike state recorded at the HE–LiF interface down to the detonation beam. This is shown by the dotted line (for the U value at 1 ns). Since the unreacted Hugoniot for composition 1 is unknown,

737

we can only approximately estimate the von Neumann spike parameters in the HE. Assuming that the Zeldovich–von Neumann–Doering (ZND) theory applies, there is no chemical reaction in the detonation front. We chose to replace the HE with an inert simulant with nearly the same shock properties as the HE. Polytetrafluoroethylene (Teflon) [14] was chosen as the simulant to obtain spike state at t = 1 ns. The dotted line in Fig. 5, used to establish the spike point on the detonation beam, is the backward facing Teflon Hugoniot.

Using this method we determined the following von Neumann spike parameters: for composition 1– U = 3.45 mm/μs, P = 39.5 GPa, V = 0.355 cm^3/g; for composition 2 – U = 3.8 mm/μs, P = 61.5 GPa, V = 0.3 cm^3/g; and for composition 3 – U = 3.24 mm/μs, P = 41.5 GPa, V = 0.346 cm^3/g. The estimated von Neumann spike state at t = 1 ns exceeds the Jouget state in pressure and particle velocity for composition 1 by 61%, composition 2 by 54%, and composition 3 by 60%.

Linear extrapolation of the U(t) curves to t = 0 and estimation using the Teflon Hugoniot as described above, gives spike parameter values at t = 0 for composition 1 of P = 45.7 Gpa and U = 3.84 mm/μs and for composition 2, P = 64 Gpa and U = 3.91 mm/μs. The von Neumann spike state at t = 0 exceeds the Jouget state in pressure and particle velocity for composition 1 by 78% and composition 2 by 63%.

Similar von Neumann spike parameters were obtained in the report [9] for TNT, PETN and HMX, and in the report [15] for TNT and RDX.

CONCLUSIONS

The structure of a diverging detonation wave front was studied for compositions based on HMX, PETN and TNT/RDX 50/50 (compositions 1, 2, 3). Laser interferometry with few nanosecond time resolution was used to measure detonation wave particle velocity histories at an HE–LiF window interface. (Detonation wave velocity measurements on compositions 1 and 2 revealed that stationary detonation takes place.) Reaction zone duration for composition 1 was found to be 10 ns and for composition 2 about 36 - 44 ns.

Estimations of the von Neumann spike for the compositions revealed that at the t = 1 ns spike state, the Jouget state was exceeded by about 54 - 61% in pressure and particle velocity. To define the spike parameters more precisely, experiments using windows having various shock impedances are needed.

Plane wave lens experiments showed that within experimental error, measured particle velocity histories, U(t), coincide for plane and diverging wave experiments.

ACKNOWLEDGEMENTS

We express our gratitude to L. A. Gatilov, V. M. Bel'sky, B. L. Glushak, I. P. Khabarov, A. A. Evstigneev, A. B. Medvedev, G. S. Doronin, and A. N. Dremin for valuable scientific help.

We also thank O. A. Burtseva, V. S. Sergeev, I. A. Vidashov and V. L. Kotasonova for their assistance in experimental activities and report releasing.

REFERENCES

1. Дремин А.Н. Хим. Физика, 1995, т.14, № 12,с. 22-40.
2. Альтшулер Л.В., Доронин Г.С., Жученко В.С. ФГВ, 1989, т. 25, № 2, с. 84-103.
3. Sheffield S.A., Bloomquist D.D., Tarver C.M. J.Chem. Phys. 1984. v.80(8), p. 3831-3844.
4. Seitz W.L., Stacy H.L., Engelke R. Et al. Proc. 9th Symp. (Int.) on Detonation. Office of Naval Research, USA, 1989, p. 657.
5. Green L.E., Tarver C.M., Erskine D.Y. Proc. 9th Symp. (Int.) on Detonation. Office of Naval Research, USA, 1989, p. 670.
6. Lubyatinsky S.N., Loboyko B.G. Proc. of Russian-American symposium of energetic materials, Livermore, 1994.
7. Tarver C.M., Green L.G., Urtiew P.A., Tao W.C. Proc. of Russian-American symposium of energetic materials, Livermore, 1994.
8. Tang P.K. Proc.10 th Symp.(Int.) on Detonation, USA, 1993, p. 947- 953.
9. Альтшулер Л.В., Ашаев В.К., Доронин Г.С., и др. Материалы 4-го симпозиума по горению и взрыву (В сборнике «Детонация»), Черноголовка, 1980.
10. Erskine D.Y. ,Tarver C.M.,Green L.G. Proc.6 th Conference «Shock Compression of Condensed Matter» , Albuquerque,1989.
11. Steinberg D.,Chau H. Proc.8 th Symp.(Int.) on detonation, Albuquerque,1986.
12. Taibinov A.V., Alekceev A.V.,Filin V.P., at all .Proc. of Russian-American symposium of energetic materials, Livermore, 1994.
13. Fedorov A.V.,Mikhailov A.L.,Poklontsev B.A. Proc. 22 th (Int.) Congresse on high-speed photography and photonics, Santa Fe,1996.
14. Morris C.E.,Fritz J.N., McQueen R.G. J. Chem.Phys., 80 (10), 1984, p.5203-5218.
15. Уткин А.В., Канель Г.И., Материалы 8-го симпозиума по горению и взрыву, (В сборнике «Детонация и ударные волны»), Черноголовка, 1986.

CP429, *Shock Compression of Condensed Matter – 1997*
edited by Schmidt/Dandekar/Forbes
© 1998 The American Institute of Physics 1-56396-738-3/98/$15.00

DETONATION WAVE PROFILES IN HMX BASED EXPLOSIVES[†]

R. L. Gustavsen, S. A. Sheffield, and R. R. Alcon

Los Alamos National Laboratory, Los Alamos, NM 87545

Detonation wave profiles have been measured in several HMX based plastic bonded explosives including PBX9404, PBX9501, and EDC-37, as well as two HMX powders (coarse and fine) pressed to 65% of crystal density. The powders had 120 and 10 μm average grain sizes, respectively. Planar detonations were produced by impacting the explosive with projectiles launched in a 72-mm bore gas gun. Impactors, impact velocity, and explosive thickness were chosen so that the run distance to detonation was always less than half the explosive thickness. For the high density plastic bonded explosives, particle velocity wave profiles were measured at an explosive/window interface using two VISAR interferometers. PMMA windows with vapor deposited aluminum mirrors were used for all experiments. Wave profiles for the powdered explosives were measured using magnetic particle velocity gauges. Estimates of the reaction zone parameters were obtained from the profiles using Hugoniots of the explosive and window.

INTRODUCTION

In the early 1940's Zel'dovich, von Neumann, and Doering independently advanced the theory of steady one-dimensional (1-D) detonation beyond the earlier Chapman Jouguet (CJ) theory by finding a solution of the flow equations with a resolved chemical reaction zone(1). In this model, called the ZND model, the detonation process consists of a shock wave that takes the material from its initial state to a "von Neumann" spike point on the unreacted Hugoniot. The ZND reaction zone is traversed by proceeding down the detonation Rayleigh line from the spike point to the CJ condition, i.e., the fully reacted state. An interesting characteristic of ZND theory is that the pressure and particle velocity decrease from the spike point to the CJ state, even though energy is being released by the chemical reaction in this region. From the CJ state the explosive products expand in a Taylor wave. ZND theory (and detonation in general) is discussed in Ref. 2.

The reaction zone time (time to go from the spike point to the CJ state) is thought to be a sensitive function of the chemical composition of the explosive, as well as its density and grain size. However, since very little is known about the chemistry occurring or the explosive grain crushing and burning characteristics in the detonation environment, the effect of various parameters is unknown. It is in search of some understanding in this area that this study was made.

Reaction zone studies have been made in the past using a number of different techniques to directly or indirectly observe the reaction zone in a detonating explosive. These have included plate-push experiments, detonation front curvature measurements, emitted light measurements, and laser velocity interferometry measurements. In all measurements, the front is perturbed by the measurement.

Laser velocity interferometry has yielded the best reaction zone measurements. In this method, a window with a thin, diffusely reflecting mirror is placed in contact with the explosive which is detonated. Laser light reflected from the mirror is Doppler shifted when the velocity changes and an interferometer transduces this into particle velocity of the interface vs. time. Several studies have used this technique in various interferometer setups (ORVIS and Fabry-Perot) to estimate reaction zones in various explosives(3-5). Time resolution can range from 10 ns down to subnanosecond, depending on the interferometer and the recording technique.

This study is an application of the VISAR interferometer technique to the study of high-density plastic-bonded HMX-based explosive reaction zones. In addition, the electromagnetic gauging technique is used to study reaction zones in two low density pure HMX explosives.

[†] Work performed under the auspices of the U.S. Dept. of Energy.

EXPERIMENTAL DETAILS

VISAR Experiments on High Density Explosives

Materials – Three different plastic bonded HMX based high explosives were used in this part of the study: PBX9501, PBX9404, and EDC-37. PBX9501 has 95wt% HMX and 5% binders. The material used had a density of 1.83 g/cm³. PBX9404 has 94wt% HMX and 6% binders. The material used here was made in 1987 and had a density of 1.84 g/cm³. EDC-37 has 91wt% HMX and 9% binders and the material used had a density of 1.84 g/cm³. In all three materials, some of the binders were energetic, i.e., nitrocellulose or some other material with nitrate groups.

Experimental Setup – The experimental setup for the high density explosives is shown in Fig. 1. Gas gun driven projectiles faced with Vistal (a pressed alumina ceramic) produced planar, sustained shock input conditions to the explosives which were 50.8 mm diameter and 12.5 mm thick.

A polymethylmethacrylate (PMMA) window was used as the interferometer window in all experiments. (Two kinds were used; Rohm and Haas Type II UVA Plexiglas was used on the PBX9501 and Polycast PMMA made to Mil-Spec. P5425D was used on the PBX9404 and EDC-37.) The window was lapped flat, polished to a slightly diffuse finish, and then aluminum was vapor deposited on it. An 8 μm thick sheet of Kapton was epoxied on top of the aluminum to protect it. An explosive cylinder was then glued to the window. The combined thickness of all glue bonds for a typical experiment was a few microns.

Impact velocity for all experiments was about 0.92 mm/μs resulting in an input to the explosive of about 6 GPa. At this input, runs to detonation were, respectively, 3.2 mm for PBX9404, 3.8 mm for PBX9501, and 6.8 mm for EDC-37, so a planar detonation traveled 5.7 to 9.3 mm before colliding with the window.

VISAR Setup – Interface particle velocity measurements were made using two VISARs set at different velocity per fringe constants; 1.818 and 0.806 mm/μs/fringe respectively. The particular VISARs used were Valyn Model VLNV-04-C (6). These VISARs have photomultiplier tubes that convert the light to electrical signals which were recorded by a Tektronix TDS-684 digitizer at 0.4 ns/point.

Several experiments were done in an effort to determine the time resolution of the photomultiplier/ digitizer recording system. A ring-up experiment in a sapphire plate was used to produce a series of sharp shocks which were recorded by the VISAR system. Analysis of this and other experiments led us to estimate that the time resolution of the system is about 2.5 to 3 ns.

Magnetic Gauge Experiments on Porous HMX

Materials - Two different batches of HMX powder were used in this part of the study: *"Coarse"* HMX (Holston HMX Lot 920-32) has particles that look like granulated table sugar and a mean particle size of about 120 μm; *"Fine"* HMX (Holston HMX Lot HOL-83F-300-023) has particles that look like powdered sugar and a mean particle size of about 10 μm. Thus, these two batches have about an order of magnitude difference in the average particle size. Additionally, the coarse HMX has crystals with sharp corners and edges. The fine HMX has particles with a rounded appearance indicating that this material was probably prepared by milling.

Experimental Setup – The experimental setup for the low density HMX experiments has been described earlier(7,8). Briefly, gas gun driven projectiles were used to obtain planar sustained-shock inputs. HMX powder was confined in sample cells which had a polychlorotrifluoroethylene (Kel-F) front face and a polymethylmethacrylate (PMMA) cylindrical plug back. The front face was attached with screws to a Kel-F confining cylinder with an outside diameter of 68.6 mm and an inside diameter of 40.6 mm. The pressed HMX (between the Kel-F and PMMA) samples were nominally either 6 or 8 mm thick with a nominal density of 1.24 g/cm³. The back plug was pressed into the Kel-F confining cylinder to compact the HMX and was held in place with an interference fit. Magnetic particle velocity "stirrup" gauges (with 10 mm long active ends) were epoxied to the Kel-F front and the PMMA back so they would contact the HMX. Particle-velocity histories were measured at both the front and back of the HMX sample, although only

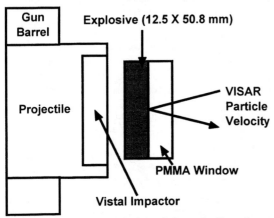

FIGURE 1. Cross section view of the projectile and target.

the back gauge records will be presented here. Wave profiles were recorded on fast digitizers.

Projectiles faced with Vistal impacted the Kel-F target face with a velocity of about 0.77 mm/μs producing inputs of about 2.5 GPa to the HMX. Full detonation is reached in runs of about 3 mm(9).

RESULTS

Five experiments were completed on the high density explosives; two each on PBX9501 and PBX9404 and one on EDC-37. Figure 2 shows the best particle velocity profiles obtained for each of the explosives. All have a spike followed by a reaction zone and then the Taylor wave. The spikes are at nearly the same particle velocity. This might be expected since all three explosives are similar in chemical makeup. However, we may not be able to resolve differences because of the 2.5-3 ns resolution of this system. The profile shapes after the spikes are somewhat different, with the PBX9404 remaining slightly higher than EDC37 and PBX9501. It is not yet known if these differences are significant; more identical experiments will have to be completed on each material to demonstrate a difference.

In all the experiments, the PMMA windows went opaque, but the PBX9501 experiments (using Rohm and Haas UVA II windows) were analyzable for about 1.3 μs while those for PBX9404 and EDC-37 (using Polycast windows) were only analyzable for the first 200 ns. This is almost certainly due to differences in the two types of PMMA, and will be looked into more carefully in the future. We note that Seitz et al. (4) used only Rohm and Haas UVA II windows in their reaction zone studies of PBX9502.

Figure 3 shows detonation wave profiles obtained in the 1.24 g/cm³ coarse and fine particle HMX. The records shown are for 8-mm thick samples initiated with about 2.5 GPa. Similar records were obtained for 6 mm thick samples. Estimates based on sample thickness and a detonation velocity of 6.55 mm/μs indicate that the fine particle HMX achieves detonation in about 100 ns, and the coarse particle HMX in about 600 ns. The time resolution is obviously much lower than the VISAR waveforms but the gauges are robust enough to record for several microseconds.

The profiles, particularly for the coarse HMX, are not characteristic of ZND type flow. This is probably because of the extreme heterogeneity of the processes that are occurring as the detonation moves through the porous bed of HMX. It is not known what causes the considerable (incredible) noise on the records but we surmise it is caused by crystal breakage, reaction, and detonation processes. The detonation front, particularly in the coarse HMX, is probably very rough. The fine particle profiles were reproducible from shot to shot but the coarse particle profiles were not.

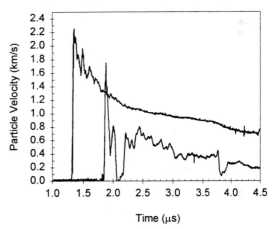

FIGURE 3. Detonation wave profiles obtained for fine particle and coarse particle HMX shown left to right respectively. The time difference is the difference in initiation times.

ANALYSIS

We have analyzed the waveforms in an attempt to estimate reaction zone parameters from the data. This has been done by finding Hugoniot and isentrope intersections in the pressure-particle velocity plane. The relevant Hugoniots and the product isentrope for PBX9501 are shown in Figure 4. As one might suppose, the parameters obtained are only as good as the assumptions made for the equations of state.

Briefly, the spike point is determined by the intersection of the Rayleigh line and the unreacted

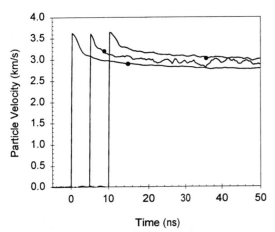

FIGURE 2. Wave Profile obtained for PBX9501, EDC-37, and PBX9404 shown left to right, respectively. The time origins for the EDC-37 and PBX9404 profiles have been shifted. The calculated ends of the reaction zones are shown as dots (see text).

PBX9501 Hugoniot. This is matched down to the PMMA Hugoniot. Similarly, the product isentrope is used to match from the estimated CJ point to the PMMA Hugoniot. This gives the points S:M and CJ:M in particle velocity. These data are then used to determine the nearness of the measurement to the spike point anticipated and to determine the reaction zone time based upon when the particle velocity goes below the CJ:M condition. Doing the analysis in this way leads to estimates of the von Neumann spike pressure of slightly over 50 GPa and reaction zone times (lengths) of 15 ns (130 μm) in PBX9501. A similar analysis for the other two materials leads to widely varying data because the Hugoniots (both unreacted and products) are not well known, particularly for EDC-37. Until further experiments are done, we will only estimate that the spike pressure for this class of materials is near 50 GPa and the reaction zone time (length) is about 15-25 ns (130 to 250 μm) long.

Reaction zone lengths and times could not be determined for the porous HMX samples. The fine particle HMX waveform appears to have reached a peak midway between CJ and the calculated spike point. The coarse particle HMX peak just barely reaches the estimated CJ point.

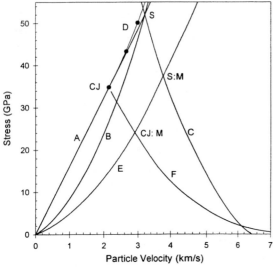

FIGURE 4. Hugoniots and isentropes relevant to calculating the reaction zone parameters. PBX 9501 is shown here. Explanations of the labels follow. (A)Detonation Rayleigh line. (B) Unreacted Hugoniot for the explosive. (C) Reflection of the unreacted Hugoniot for the explosive. (D)Explosive products Hugoniot. (E) PMMA window Hugoniot. (F)Explosives products isentrope calculated using the JWL EOS. (S)ZND spike point. (S:M) Spike point matched onto the PMMA Hugoniot. (CJ) Chapman-Jouguet or CJ state. (CJ:M) CJ state matched onto the PMMA Hugoniot. The matched spike and detonation points are those which would be observed.

ACKNOWLEDGMENTS

Robert Medina is thanked for his assistance operating the gas gun. Mel Baer of Sandia National Laboratory is thanked for useful discussion regarding the detonation profiles in porous HMX.

REFERENCES

1. Zeldovich, Ya. B., Sh. Eksp. Teor. Fiz. **10**, 542 (1940); von Neumann, J., "Theory of Detonation Waves," (1942), *in John von Neumann Collected Works*, Vol. 6, (ed. L. A. J. Taub): Macmillian, New York (1963); Doering, W., Ann. Phys. **43**, 421 (1943).

2. Engelke, R., and Sheffield, S.A. "Explosives," in *Encyclopedia of Applied Physics*, Vol. 6, VCH Publishers, Inc., 1993, p. 327.

3. Sheffield, S. A., Bloomquist, D. L., and Tarver, C. M., J. Chem. Phys. 80, 3831 (1984).

4. Seitz, W.L., Stacy, H.L., Engelke, R., Tang, P.K, and Wackerle, J. , "Detonation Reaction-Zone Structure of PBX9502" in *Proceedings of the Ninth Symposium (International) on Detonation*, Office of the Chief of Naval Research Report No. OCNR 113291-7, 1989, pp. 657–669.

5. Tarver, C. M., Breithaupt, D. R., and Kury, J. W., J. Appl. Phys. **81**, 7193 (1997).

6. Valyn International, Albuquerque, NM sells this VISAR system.

7. Sheffield, S. A., Gustavsen, R. L., Alcon, R. R., Graham, R. A., and Anderson, M. U., "Particle Velocity and Stress Measurments in Low Density HMX," in the *High Pressure Science and Technology – 1993*, Edited by S. C. Schmidt, J. W. Shaner, G. A. Samara, and M. Ross, American Institute of Physics (AIP) Conference Proceedings 309 (1994), pp. 1377-1380.

8. Gustavsen, R. L., Sheffield, S. A., and Alcon, R. R., "Low Pressure Shock Initiation of Porous HMX for Two Grain Size Distributions and Two Densities," in *Shock Compression of Condensed Matter – 1995*, American Institute of Physics (AIP) Conference Proceedings 370 (1995), pp. 851-854.

9. Dick, J. J., *Combustion and Flame* **54**, 121-129 (1983).

CP429, *Shock Compression of Condensed Matter – 1997*
edited by Schmidt/Dandekar/Forbes
© 1998 The American Institute of Physics 1-56396-738-3/98/$15.00

DENSITY EFFECT ON DETONATION REACTION ZONE LENGTH IN SOLID EXPLOSIVES

S. N. Lubyatinsky, B. G. Loboiko

Russian Federal Nuclear Center – Institute of Technical Physics
P. O. Box 245, Snezhinsk, Chelyabinsk region 456770 Russia

Density effect on detonation reaction zone length have been studied on RDX and PETN using a photoelectric technique to record the radiation intensity history of the shock front in chloroform placed on the charge face. Charge density was found to drastically affect the reaction zone length as well as the charge appearance. The charges pressed to 0.92 of crystal density were completely opaque and exhibited the von Neumann spike of about 0.3 mm in length, typical for high explosives. The charges solvent-pressed to 0.99 of crystal density were agatized or translucent and did not exhibited the von Neumann spike, which implies that its length did not exceed 0.03 mm. The following explanation is offered. In translucent, practically non-porous, charges the detonation front is a strong plane shock inducing almost instant reaction. In charges consisting of separate crystals the detonation front becomes three-dimensional. As a result some fraction of explosive is compressed almost isentropically and reacts relatively slowly, so that it can be measured.

INTRODUCTION

It was found earlier that RDX and PETN solvent-pressed to about 0.99 of TMD become agatized (resembling agates) or translucent and exhibit no von Neumann spike lasting longer than the experimental time resolution of 5 ns (1-3). Such a short reaction zone was assumed to be due to the high homogeneity of these almost voidless explosives. To test this hypotheses, detonation reaction zone measurements were performed on RDX and PETN charges pressed without solvent to lower densities at which explosive crystals are more separated and the charges are opaque.

EXPERIMENTS

The measurements were carried out by recording the radiation intensity history of the shock front in chloroform placed on the end of a detonating explosive using a photoelectric technique (see Fig.1). The technique has high sensitivity (about 0.2% in particle velocity), high time resolution (about 5 ns), and long recording time (>1 μs) (3, 4).

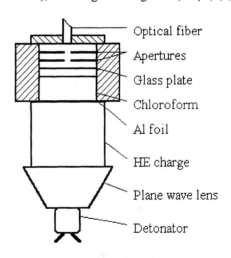

FIGURE 1. The experimental setup.

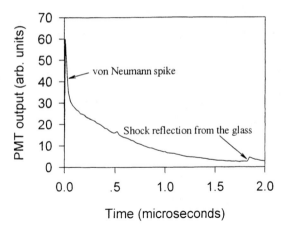

FIGURE 2. The oscilloscope record for a 1.67 g/cm³, 40 mm long, RDX charge.

The 10 μm aluminum foil, screening detonation radiation, is attached to the face of the 40 mm diameter charge with an epoxy. The two 1.5 mm apertures restrict the viewed portion of the shock front to 3 mm in diameter. The shock front radiation is transmitted by a 1.5 mm plastic optical fiber to a SNFT-3 photomultiplier, whose output is recorded on a SUR-1 oscilloscope. A representative oscilloscope record is shown in Fig. 2. The von Neumann spike and the shock reflection from the glass plate can clearly be seen. In transforming the oscilloscope record $I(t_f)$ into the interface velocity history $U_P(t_i)$ the problem of variability of the optical fiber properties is solved through calibration against the reference point with known parameters $U_{P0}(I_0)$ using the earlier derived relations

$$t_i = Z \cdot t_f, \qquad (1)$$

$$U_P = \left(U_{P0}^{-b} - A \cdot \ln\left(\frac{I}{I_0}\right) \right)^{-1/b}, \qquad (2)$$

where $Z=0.52\pm0.02$, $A=(0.039\pm0.001)$ (km/s)$^{-1.17}$, and $b=1.17$ (4). The particle velocity U_{P0} behind the shock wave produced in chloroform by the detonation products expanding from the CJ point was calculated using the earlier determined CJ parameters of RDX and PETN (5) and the chloroform Hugoniot (6). The photomultiplier output I_0, corresponding to the reference point U_{P0} was found through least-squares fitting the oscilloscope record to the equation

$$I = I_0 + I_1 \cdot t_f + I_2 \cdot t_f^2, \qquad (3)$$

in the region unaffected by the reaction zone. This approach is consistent with the standard practice of CJ parameters determination neglecting the reaction zone length. The parameters of the studied charges are given in Table 1, the corresponding interface velocity histories $U_P(t_i)$ are shown in Figs. 3 and 4.

In the experiment with a 40 mm long, 1.67 g/cm³ RDX charge a check of calibration was carried out against the time interval t_0 between the shock entry into the chloroform and its reflection from the glass plate. Using the U_S-U_P relationship for chloroform the particle velocity history at the shock front $U_P(t_f)$ was transformed into the shock velocity history $U_S(t_f)$ and the distance traveled by the shock front in chloroform was calculated

$$X_0 = \int_0^{t_0} U_S(t_f) dt_f = 9.35 \text{mm}, \qquad (4)$$

that was in turn compared with the distance from the charge end to the glass plate, equal to 10 mm. The 6.5% discrepancy between these two values is probably due to the experimental errors of both t_0 and CJ parameters used in the calculation. It should be noted that measuring t_0 requires a longer and slower oscilloscope sweep thus reducing the precision of the reaction zone measurements. In view of this, t_0 was not measured in the other experiments. At the same time, for explosives with unknown CJ parameters, t_0 can be directly used for calibration. In this case I_0 is chosen arbitrarily and U_{P0} is chosen so that the calculated and experimental values of X_0 coincide.

TABLE 1. Reaction Zone Parameters

Explosive	Charge Length (mm)	Charge Density (g/cm³)	T_J (μs)	X_J (mm)
Agatized RDX (3)	80	1.78	<0.005	<0.03
RDX	40	1.73	0.07	0.37
RDX	40	1.67	0.05	0.28
RDX	80	1.67	0.06	0.34
RDX/Inert 94/6 (4)	80	1.80	0.06	0.34
RDX/TNT 50/50 (3)	100	1.65	0.12	0.59
Agatized PETN (3)	80	1.75	<0.005	<0.03
PETN	40	1.74	0.07	0.37
PETN	80	1.63	0.06	0.32
PETN/Inert (6)	-	1.65	0.04	-

744

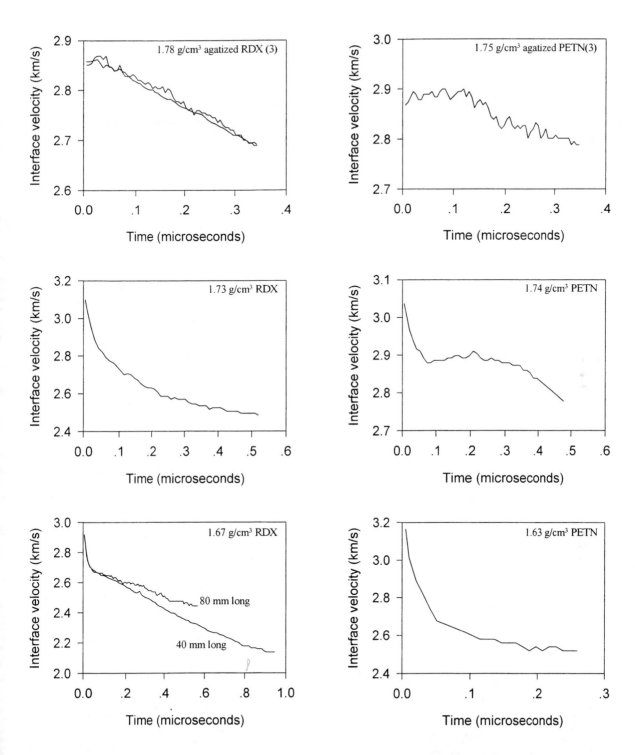

FIGURE 3. Interface velocity histories for RDX.

FIGURE 4. Interface velocity histories for PETN.

745

The nonagatized explosives all exhibited the von Neumann spike with the rise time of about 5 ns (the time resolution of the technique). The reaction started at the front without delay. The reaction rate being initially maximal decreased as the reaction proceeded. Within the reaction zone the interface velocity profiles were exponential, as previously observed for a variety of explosives (3, 4). The reaction zone parameters were found as follows (3, 4). The smoothed profile $U_P(t_i)$ was differentiated and the function $\ln(-dU_P/dt_i)$ was plotted against t_i. The plot consisted of two straight lines corresponding to different laws of shock attenuation. The point of their intersection T_J was associated with the CJ point. The reaction zone length was calculated with the formula

$$X_J = \int_0^{T_J} \left(D - U_P \right) dt_i, \qquad (4)$$

derived from analysis of the x-t diagram. D is the detonation velocity of the explosive. The reaction zone lengths are listed in Table 1 in comparison with those of some RDX and PETN based explosives.

RESULTS

The nonagatized RDX and PETN at densities of 0.92-0.98 of TMD have the reaction zone length of about 0.3 mm and are close in this parameter to the other RDX and PETN based explosives. A density change of about 5% has no noticeable effect on the reaction zone length.

The agatized RDX and PETN at a density of 0.98-0.99 of TMD have the reaction zone lengths that are at least an order shorter (<0.03 mm) (3). As they do not almost differ from the nonagatized in density the conclusion can be made that the parameters of the von Neumann spike depend strongly on explosive structure.

The following reaction mechanism can be assumed. Most solid explosives (excepting monocrystals and agatized) consist of explosive crystals separated either by air or by binder . This result in a three-dimensional structure of the detonation front. Different explosives particles reach the von Neumann spike pressure in different ways. A fraction of explosive is compressed to the von Neumann spike pressure in a single shock wave of about 50 GPa amplitude and reacts practically instantly. At the same time some fraction of explosive is compressed by a sequence of shock waves almost isentropically and reacts at relatively low rate, so that it can be measured.

Agatized charges are characterized by much closer and wider contact between separate explosive crystals. As a result, in this case the detonation front is plane and practically all explosive is compressed to the von Neumann spike pressure in a single shock wave reacting in a time shorter than the time resolution of the technique.

In other words, the heterogeneity of solid explosives results in both "hot spots" increasing the reaction rate in relatively weak shock waves (of the order of 5 GPa) and "cool spots" decreasing the reaction rate in relatively strong shock waves (of the order of 50 GPa). If the "hot spots" make themselves evident at shock initiation, the "cool spots" make themselves evident at detonation.

REFERENCES

1. Voskoboinikov, I. M.< Gogulya, M. F., *Khimicheskaya Fizika (Russian J. of Chemical Physics)* **3**, 1036-1039 (1984).
2. Ashaev, V. K., Doronin, G. S., Levin, A. D., *FGV (Russian J. of Physics of Combustion and Explosion)* **24**, 95-99 (1988).
3. Lubyatinsky, S. N., Loboiko, B. G., "Study of Chemical Reaction Zone Structure in Detonating High Explosive Using a Photoelectric Technique," in *Proceedings of the Symposium on Energetic Materials Technology*, Pleasanton, California, May 18-25, 1994.
4. Lubyatinsky, S. N., Loboiko, B. G., "Reaction Zone Measurements in Detonating Aluminized Explosives," in *Proceedings of the APS Conference on Shock Compression of Condensed Matter*, Hotel Sheraton, Seattle, Washington, August 13-18, 1995, pp. 779-782.
5. "LASL Explosive Property Data," Ed. T. R. Gibbs, A. Popolato. University of California Press. Berkely - Los-Angeles - London. 1980.
6. Dick, R. D., *J. of Chemical Physics* **74**, 4053-4061 (1981).
7. Ashaev, V. K., Levin, A. D., Mironov O. N., *Pisma v ZhTF (Russian Letters to the J. of Technical Physics)* **6**, 1005-1009 (1980).

CP429, *Shock Compression of Condensed Matter – 1997*
edited by Schmidt/Dandekar/Forbes
1998 The American Institute of Physics 1-56396-738-3/98/$15.00

SHOCK RESPONSE OF THE EXPLOSIVE PBXN-103

G.T. Sutherland, P.K. Gustavson, E.R. Lemar, J. O'Connor

Naval Surface Warfare Center, Indian Head Division, Indian Head, Maryland, 20640

For the explosive PBXN-103, multiple gauge gas gun experiments were performed to obtain Hugoniot and Lee-Tarver reactive rate law parameters. Simulations based on these parameters produced pressure histories in good agreement with measured pressure histories. These parameters were also used to simulate other tests such as failure diameter, the large scale gap test, and the modified gap test. Again, good agreement was found between the observed and calculated results.

INTRODUCTION

In earlier work[1], gas gun experiments were conducted with the explosive PBXN-103 to obtain experimental shock loading characteristics. In this paper, these pressure-time histories have been used to obtain an unreacted Hugoniot and parameters for the Lee-Tarver reactive rate law [2]. These computer simulations followed the approach of Miller and Sutherland [3], and Miller [4]. The successful development of these Lee-Tarver parameters will also allow the numerical simulation of other experimental data (observations), such as detonation velocity decrement, gap tests, wave curvature, and fragment impact.

EXPERIMENT

PBXN-103 contains [5,6] 40 wt.% Ammonium Perclorate (AP), 27 wt.% Aluminum and 33 wt.% energetic binder. The average number diameter for the AP and Al particles was 220 μm and 14 μm respectively. The measured density was 1.881 g cm^{-3}.

A light gas gun was used to impact PBXN-103 samples with PMMA, Al, or Cu flyers. Multiple manganin or PVDF gauges were placed at various depths inside the sample to measure pressures produced by the shock wave and any chemical reaction in the explosive. Experimental details [1] appeared in a previous paper.

HYDROCODE CALCULATIONS

The shock response of PBXN-103 was simulated using the CTH [7] code. The reactive EOS is a tabular EOS that was generated from a PBXN-103 [8] JWL EOS (Equation 1) using the PANDA [9] program.

$$P = A\left(1 - \frac{\omega}{R_1 V}\right)e^{(-R_1 V)}$$
$$+ B\left(1 - \frac{\omega}{R_2 V}\right)e^{(-R_2 V)} + CV^{-(\omega + 1)} \quad (1)$$

$$where \ \ V = \rho_o / \rho$$

A Lee-Tarver reactive rate equation determines the state between no and full reaction. To limit the number of adjustable parameters, we used the two term Lee-Tarver reactive rate law (Equation 2) with 3 adjustable parameters to give the fraction of the explosive reacted (F) as a function of pressure and density. The first term corresponds to "hot spot initiation" and the second term to growth leading to detonation. The parameters I, G, and Z were varied until the best agreement of four experimental records in Ref. 1 and calculated stress-time profiles was obtained. The values of I = 1.0; G = 14,000.0 and Z = 4.0 gave the best results. The amount of reaction for the ignition term was not limited.

$$\frac{\partial F}{\partial T} = I(1-F)^{(2/3)}\left(\frac{\rho}{\rho_o}-1\right)^4 \quad (2)$$
$$+ G(1-F)^{(2/3)}F^{2/9}P^Z$$

Table 1. JWL Parameters

Parameter	Value	Parameter	Value
A (Mbar)	18.83	w	0.4
B (Mbar)	1.375	Dcj (cm/µs)	0.59
C	0.03298	E_o (Mbar)	0.095
R_1	8.00	Pcj (Mbar)	0.15
R_2	4.00		

RESULTS

Hugoniot information (see Figure 1) [1] was obtained from measured projectile and shock velocities. A resulting shock velocity (U) - particle velocity (u) relationship of U = 0.24 cm/µsec + 1.95 u was obtained. This result is the same as was obtained by Roslund and Colburn [5,10] et al. The measured shock velocities were corrected for the time required for the shock wave to travel through the thick Teflon gauge packages used. Pressures measured by the front surface manganin gauges were not in agreement with pressures calculated from the above Hugoniot. We plan to look at the manganin gauge calibration for pressures below 10 GPa.

Reactive stress-time profiles appear [1] in Figures 2 and 3. The wave arrival times and the general structure of the calculated and observed profiles are in good agreement. Electrical noise was observed to be superimposed on the records; the resulting profiles are smoothed with a least squared fit applied over part of the experimental record. The noise may be due to the aluminum particles in the PBXN-103 moving in the magnetic field produced by the large current flowing through the manganin gauges.

MODELING OF OTHER TESTS

To test the applicability of simulation based on the Lee-Tarver model, other detonation physics experiments were modeled. As part of the work reported in Ref. 1, the large-scale gap test (LSGT) was performed with PBXN-103 samples from the same casting. Other experiments used explosives from other castings; these explosives may of different densities, be different in age, and have ingredients from a different batch or source. We want to determine how well the model parameters predict experimental results for explosives from the different castings.

For the NOL LSGT, a measured gap thickness of 2.413 cm was obtained; this corresponds (using the LSGT calibration [11]) to a pressure at the end of the gap of 5.94 GPa. The calculated gap was between 2.26 and 2.31 cm and corresponds to a pressure of about 6.1 GPa [11]. In our simulations, however, the calculated gap pressure was considerably higher.

To investigate the above difference in the pressures obtained from the LSGT calibration and the CTH simulation we modeled the LSGT donor system. We observed (Figure 4) that our simulations of the LSGT donor system and the reported calibration were not in agreement. This may be due to uncertainty [11] in the calibration or the parameters [12] used to simulate the cast Pentolite explosive in the donor system.

The failure diameter is the diameter of an explosive cylindrical charge in which a detonation wave will not propagate. The reported failure diameter is between 2.7 to 2.9 cm [5]. Our simulations showed a value of 3.7 and 3.6 cm. Other researchers [13] have reported that Lee-Tarver simulations predict larger than observed failure diameters.

The modified gap test (MGT) [14] consists of the LSGT donor that inputs a shock into a 5.08 cm diameter acceptor and 1.27 inch thick explosive acceptor [14]. The ejecta velocity of the ejecta cloud is measured. A free surface velocity and a peak pressure near this surface is calculated assuming no reaction. The difference between the measured ejecta velocity from Ref. 14 and calculated free surface velocity is plotted as a function of the above pressure. As can be seen by

748

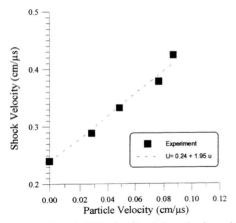

Figure 1. Shock Velocity-Particle velocity relationship obtained from gas gun experiments.

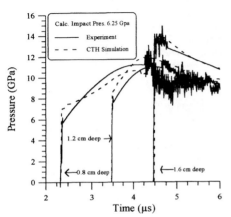

Figure 2. Experimental and calculated stress-time profiles for exp. 96-506.

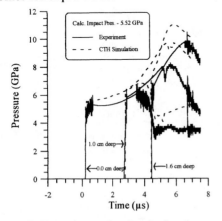

Figure 3. Experimental and calculated stress-time profiles for exp. 96-507.

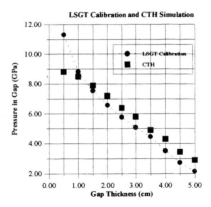

Figure 4. Nol LSGT Calibration and CTH Simulaiton

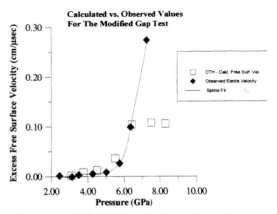

Figure 5. Calculated and observed results for the Modified Gap Test.

749

Figure 5 there is good agreement between simulated and measured velocities.

CONCLUSIONS & FUTURE WORK

Experiments were performed to allow unreacted Hugoniot and Lee-Tarver reactive rate parameters to be obtained for the explosive PBXN-103. Good agreement was obtained between the calculated and measured stress-time profiles. The Lee-Tarver model was used to simulate other detonation physics experiments for PBXN-103. Reasonable agreement was obtained between calculated and measure values for the MGT and the LSGT.

Future work includes simulating other detonation physics experiments such as wave curvature, fragment impact, and donation velocity decrement. Recent PBXN-103 experimental measurements [15] and generated Lee-Tarver parameters [15] from other researchers should be compared with this data base. A procedure [16] to automate the process of obtaining the Lee-Tarver parameters would be helpful.

We need to further investigate the difference observed between measurements and simulations for the large scale gap test.

ACKNOWLEDGMENTS

The authors thank Drs. Miller (NAWC, China Lake), Sandusky, Bernecker, Forbes (LLNL) for helpful discussions. Dr. Deiter, Mr. Volk, and Mr. Chang are thanked for assisting in the experimental work. Mr. Walker and Mr. Burns are thanked for providing a SESAME table for fully reactive PBXN-103. This work was performed for the Office of Naval Research Explosives and Undersea Warheads Program.

REFERENCES

1. Sutherland, G.T., Gustavson, P.K., Lemar, E.R. and O'Connor, J., Shock Response of the Explosive PBXN-103, 1996 JANNAF Propulsion Systems Hazards Subcommittee Meeting, Volume II, Chemical Propulsion Information Agency, Columbia, MD.
2. Lee, E.L., and Tarver, C.M., Phys. Fluids 23, 2362(1980).
3. Miller, P.J., and Sutherland, G.T., Reaction Rate Modeling of PBXN-110, Shock Waves in Condensed Matter-1995, editors Schmidt, S.C., and Tao, W.C., American Institute of Physics, Woodbury, NY.
4. Miller, P.J., Proceedings of the Material Research Society Fall 1995 Meeting, Boston Massachusetts, Vol. 418, p. 325, 1996.
5. Hall, T.N. and Holden, J.R., Navy Explosive Handbook, NSWC MP 88-116, NSWC MP 88-116, (Naval Surface Warfare Center, Dahlgren, VA, 1988), pp. 2-24.
6. Stosz, M., The Development of Nitrasol as a New Under Water Explosive, NOLTR 62-204, 1963, (Naval Surface Warfare Center, Dahlgren, VA, 1963)
7. Bell, R.L., Brannon, R.M., Elrick, M.G., Farnsworth, A.V., Hertel, E.S., Kerley, G.I, Petney, W.V., Silling, S.A., Taylor, P.A., CTH User's Manual and Input Instructions, version 2.00, Sandia National Laboratory, Albuquerque, N.M. 1995.
8. Breithaupt, D., Lawrence Livermore National Laboratory, unpublished data.
9. G.I. Kerley, User's Manual for Panda II: A Computer Code for Calculating Equations of State, SAND88-2291, Sandia National Laboratory, Albuquerque, NM 87185.
10. Roslund, L.A, and Colburn, N.L., The Isentrope of PBXN-103 Explosive Products Below the Chapman Jouguet State, NOLTR 74-112, 1974, NSWC, Dahlgren, VA.
11. Erkman, J.O., Edwards, D.J., Clairmont, A.R., and Price, D., Calibration of the NOL Large-Scale Gap Test; Hugoniot Data for Polymethyl Methacrylate, NOL TR 73-15, Apr. 1973, Naval Surface Warfare Center, Dahlgren, VA.
12. The values for the HEBURN option of CTH for cast 50/50 Pentolite are obtained form the Cheetah thermal-equilibrium code.
13. Miller, P.J., Private communication.
14. Lemar, E.R., Liddiard, Forbes, J.W., Sutherland, G.T., and Wilson, W.H., The Analysis of Modified Gap Test Data for Several Selected Insensitive Explosives, NSWC TR 89-290, Naval Surface Warfare Center, Dahlgren, VA, 1993.
15. Linfors, A., and Miller, P.J., NAWC, China Lake, CA, to be published.
16. Baker, E.L., Schimel, B., Grantham, W.J., Numerical Optimization of Ignition and Growth Reactive Flow Modeling For PAX2A, Shock Waves in Condensed Matter-1995, editors Schmidt, S.C., and Tao, W.C., American Institute of Physics, Woodbury, NY.

750

CP429, *Shock Compression of Condensed Matter – 1997*
edited by Schmidt/Dandekar/Forbes
© 1998 The American Institute of Physics 1-56396-738-3/98/$15.00

HIGH EXPLOSIVE CORNER TURNING PERFORMANCE AND THE LANL MUSHROOM TEST

L.G. Hill, W.L. Seitz, C.A. Forest, & H.H. Harry

Los Alamos National Laboratory, Los Alamos, New Mexico 87545 USA

The Mushroom test is designed to characterize the corner turning performance of a new generation of less sensitive booster explosives. The test is described in detail, and three corner turning figures-of-merit are examined using pure TATB (both Livermore's Ultrafine and a Los Alamos research blend) and PBX9504 as examples.

INTRODUCTION

The use of insensitive high explosives in main charges has inevitably shifted safety concerns toward detonators and boosters. In particular, one would like to replace conventional HMX-based boosters with a less sensitive material. In response to this need Livermore has developed Ultrafine TATB (UF-TATB) and Los Alamos has developed PBX9504 (70 wt.% TATB, 25 wt.% PETN, 5 wt.% binder). These materials are less shock sensitive than conventional boosters, but the increased safety comes at a price: less sensitive explosives generally have thicker reaction zones, and corner turning performance depends inversely on the product of the local wave curvature and the reaction zone width. Consequently, to maximize safety one must generally accept a booster material with marginal corner turning properties. Engineering design becomes critical, as one must decide how marginal the system can be given reasonable tolerances on factors like particle size and pressed density. To make such decisions one needs a corner turning test optimized for the sensitivity of the materials in question. The Mushroom test is designed for this purpose.

All corner turning tests use the same basic idea: one delivers a small (modestly exceeding the failure diameter) but strong pressure stimulus* to

the sample. The detonation then spreads with some difficulty, meaning that there is a large departure from ideal wave propagation (i.e., Huygens' construction) and a substantial "dead zone" in which detonation fails. The propagation of this marginal wave is quite sensitive to the material properties of the explosive, and the goal of a corner turning test is to characterize its departure from ideal propagation. Other than flash radiography (a useful but involved method), one must infer wavefront properties by observing detonation breakout from the test charge, and devise suitable figures-of-merit from these observations.

Besides being tuned for a particular class of explosives, the mushroom test has a few distinguishing features that set it apart from other corner turning tests (e.g., refs. 1→3). One is that it uses a small amount of test explosive (about 8 gm), a virtue when material is scarce. Another is that it uses a hemispherical sample. This has several benefits: 1) the configuration is similar to that of real boosters, 2) the test samples the entire wavefront, which (assuming a constant detonation velocity) allows one to reconstruct its shape, and 3) the observation surface closely matches the shape of the emerging wave. The last attribute allows a very sensitive breakout measurement but, even more importantly, the angle between the emerging wave and the observation surface is always less than the critical angle, above which the presence of the latter affects the wave's shape.

---

*This is opposite to an *initiation* test, which delivers a relatively large, weak, pressure stimulus to the sample.

751

EXPERIMENTAL DETAILS

Figure 1 shows a scale drawing of the Mushroom test shot assembly. The test is named for its distinctive geometry—a 25.4 mm diameter pressed hemispherical "cap" of test material, initiated by a 6 mm-diameter × 24 mm-long "stem" composed of four 1.69-g/cc PBX9407 (94 wt.% RDX, 6 wt.% binder) pellets. The stem provides a strong and repeatable pressure input irrespective of the reproducibility of the Reynolds RP-2 detonator. Its 6 mm diameter was chosen by trial and error, using PBX9504 as the test explosive.

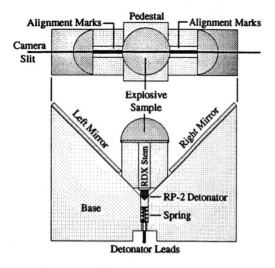

FIGURE 1. Scale drawing of the Mushroom test.

Samples are pressed directly to shape in a hemispherical die. Since the final volume is fixed, the sample density may be controlled simply by inserting the appropriate amount of mass. The bulk density of the finished pieces is also measured by immersion, as corner turning is quite sensitive to this parameter.

The sample is epoxied at four spots around the edges to a PMMA pedestal. The stem passes through a hole in the pedestal and contacts the equator of the test sample. The pedestal is glued to a PMMA base, which holds the detonator and two 45° mirrors. The detonator and pellets are spring-loaded against the test sample to ensure contact between the pieces. The assembly is designed so that all pieces are automatically and accurately located with respect to one another.

Detonation breakout is observed with an internal-slit rotating mirror streak camera. The experiment is viewed directly and from two sides via the mirrors. The image of the camera slit spans the center of the hemisphere and its mirror images to give a composite streak record. The slit is precisely aligned by centering it between two blackened scribe marks on the pedestal (Fig. 1), which are visible in the mirror views. The sample surface is painted with aluminum hexafluorosilicate to enhance the light output upon breakout.

Figure 2 shows a sample (negative) streak camera record. The two diagonal bands on top are the detonation wave traveling through the stem, as seen through the PMMA pedestal. Their termination indicates the entrance of the detonation wave into the pellet, and provides the time origin used for the breakout data. The three images on the bottom are detonation breakout and the subsequent product light from the three views.

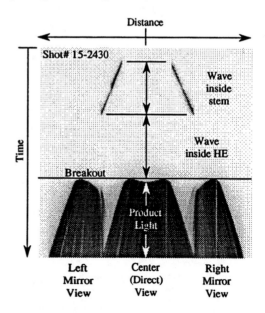

FIGURE 2. Mushroom test dynamic streak record.

The test is fired cold (-55 C) to observe worst case corner turning behavior. The shot is cooled at 1 C/min using dry nitrogen, and is soaked at temperature for 30 minutes. The styrofoam shot box has a double-glass observation window, and dry nitrogen is blown between the panes and also over the top pane to prevent condensation.

DATA REDUCTION

The streak records of Fig. 2 are digitized and combined to produce Fig. 3—a composite plot of breakout time vs. polar angle. For ideal propagation the wave would break out first at 90° because that path is the shortest. However, the real wave is retarded at larger angles because its strength is lower on the fringes. Consequently the wave breaks out at an intermediate angle first, giving the breakout plot its characteristic "w" shape.

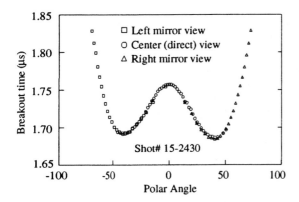

FIGURE 3. Composite breakout plot.

To achieve the level of consistency between the three views exhibited by Fig. 3 one must correct for certain experimental errors. The first is that one cannot distinguish in the dynamic record between the breakout locus and the edge of the detonation product cloud. Since breakout cannot extend beyond the undetonated charge, one may superimpose a still picture of the shot (not shown in Fig. 2) on the film above the dynamic record. The edges of the charge determined from it may then be used to "window out" spurious points.

Processing the data in this way one finds that the three views do not overlay to within the random scatter in the data. One reason is that the still and dynamic records are rarely in perfect alignment due, for example, to a slight amount of play in the mirror bearings. One may compensate for misalignment by laterally shifting the still picture (or rather, the "window" points it defines) relative to the dynamic record prior to data processing. By trial and error one may then find an offset for which any remaining discrepancy between views is symmetric about the ordinate.

One then finds that the side and direct views agree in the vicinity of first breakout but deviate slightly near the pole and equator. This problem is closely related to the first; specifically, for sufficiently oblique angles the direct view observes the edge of the expanding product cloud rather than the breakout locus—and likewise for the mirror views. Thus, for angles greater than first breakout one should discard points from the direct view that deviate from those of the mirror views. Likewise, for angles less than first breakout one should discard points from the mirror views that deviate from those of the direct view.

ANALYSIS

The objective is to quantify how well the detonation spreads in a sample, compared to other samples tested under nominally identical conditions. For easy comparison it is desirable to distill the breakout plot of Fig. 3 into a single figure-of-merit. Regardless of how one defines such a quantity, it is beneficial to perform a series of tests at various pressed densities, and to plot the chosen figure-of-merit versus density. One may then fit a curve to the data and evaluate it at the nominal density. This is easier than pressing samples to exactly the nominal density; moreover, the curve's slope indicates the sensitivity of corner turning performance to density variations.

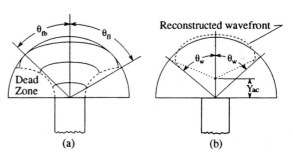

FIGURE 4. Definitions of First breakout angle and Failure locus angle (a), and Apparent center of initiation (b).

Three figures-of-merit are defined in Fig. 4. One is the first breakout angle θ_{fb} (Fig. 4a), corresponding to the two minima (averaged together) in Fig. 3. The better the wave spreads the less the lag at the edges, and the closer θ_{fb} is to the ideal 90° value. Fig. 5 shows data for PBX9504, UF-TATB, and a similar Los Alamos material,

753

Fine-particle TATB (FP-TATB). For each material θ_{fb} decreases linearly with pressed density. An inverse relationship is expected because less dense pressings have a larger internal void fraction, the collapse of which provides the mechanical work that initiates reaction.

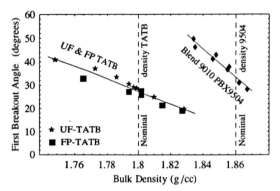

FIGURE 5. First breakout angle.

A second figure-of-merit (Fig. 4a) is the angle θ_{fl} at which the failure locus meets the observation surface. The failure locus separates the detonating region from the so-called "dead zone," and is indicated by an abrupt extinction of the breakout light. Fig. 6 shows θ_{fl} for PBX9504, UF-TATB, and FP-TATB. For sufficiently low density θ_{fl} decreases linearly for all materials. For PBX9504, only the three highest density samples had θ_{fl}'s less than the maximum value of 90°. For both TATB materials θ_{fl} rolls off exponentially starting at about the nominal density of 1.8 g/cc. For sufficiently high densities PBX9504 may do the same, but it differs from pure TATB in that it is sensitized by PETN rather than void space.

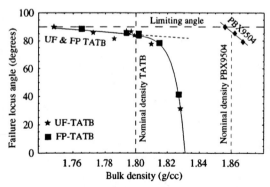

FIGURE 6. Failure locus angle.

The third figure-of-merit presented is derived from wavefront reconstruction. The emerging wave shape may be calculated (Fig. 4b) given the shape of the observation surface and assuming the wave speed to be constant (independent of curvature). A circle is fit to the reconstructed wave at the mean breakout time, in a chosen angular window θ_w. Its center relative to the geometric center of the sample, Y_{ac}, is the *apparent center of initiation*. If the wave expansion is close to spherical then $|Y_{ac}|$ is small, but if spreading is poor the apparent center moves up into the sample.

Figure 7 shows Y_{ac} for PBX9504, UF-TATB, and FP-TATB. The detonation velocity for each material lot was taken to be the ratio of the sample height to centerline transit time, averaged over all the samples in the lot. In each case θ_w is chosen to be equal to θ_{fl}. It is seen that Y_{ac} increases linearly with density except for the two highest TATB densities, for which it decreases. It is clear from Fig. 6 that this reversal does not indicate improved corner turning, but is a peculiarity of the failing wave.

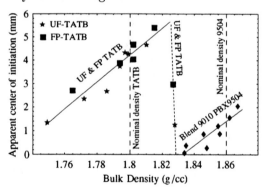

FIGURE 7. Apparent center of initiation.

ACKNOWLEDGEMENTS
We thank A. Anderson for contributing to the analysis, and G. Vasilik & D. Murk for firing support. This work was supported by the U.S. DOE.

REFERENCES
1. Cox, M., and Campbell, A.W., 7th Symp. (Int.) on Detonation (1981).
2. Forbes, J.W., Lemar, E.R., and Baker, R.N., 9th Symp. (Int.) on Detonation (1989)
3. Hutchinson, C.D., Foan, G.C.W., Lawn, H.R., and Jones, A.G., 9th Symp.(Int.) on Detonation (1989)

CP429, *Shock Compression of Condensed Matter – 1997*
edited by Schmidt/Dandekar/Forbes
© 1998 The American Institute of Physics 1-56396-738-3/98/$15.00

INTERACTION OF DOUBLE CORNER TURNING EFFECT IN PBX

I. Plaksin, J. Campos, M. Mendonca, R. Mendes and J. C. Gois

*Lab. of Energetics and Detonics, Mechanical Eng. Dept., Fac. of Sciences and Technology
University of Coimbra, 3030 Coimbra, Portugal*

The corner turning effect in PBX has been studied. The setup built with Cu or PMMA plates, has a channel with a square cross section of 5 x 5 mm and 10 x 10 mm, corresponding respectively to 1.33 and 2.67 of PBX failure diameter. A thin optical fibre (250 µm) strip, connected to a fast electronic streak camera, and the printed erosion figure on an witness plate (base plate and internal walls of setup) allows clear quantification of the detonation wave turning phenomena and the printed current lines of the products of detonation. The optical probes allow in real time, an original, direct front observations of multiple zones, by the registration of the interacting waves. Even in the cases of Cu-confinement, the corner turning effect shows the existence of non-detonated PBX ("dark zone"). The correlation between the non-monotonic (x-t) diagrams and light pulsations, recorded during the propagation of detonation wave inside the channels and after the corner section, proves the existence of pulse behaviour and cellular structure of detonation front propagation in PBX.

INTRODUCTION

In spite of more than 50 years history of investigations certain features of detonation wave (DW) corner turning phenomenon in heterogeneous HE, Fig.1, such as the process of "dark zone" (DZ) formation near the corner (1), as well as the predominate propagation of detonation front (DF) in some specific directions (2,3), etc., still remain unclear today. With the exception of a few contributions (2-4) the propagation of diverging DW in PBX has been studied experimentally beginning at the radius of DF curvature for r>5mm.

The main outcomes of the preceding investigations are: (i) in the DZ near the corner no high order reaction occurred, (ii) shape of the diverging DF is smooth, monotonically divergent, but with some deviation from ideal spherical.

The most justified mechanism of DZ formation was proposed in (5), where the DZ (the area of unreacted HE) is originates midway along the confinement PBX interface, as a result of pre-shocking by shock travelling in the confinement.

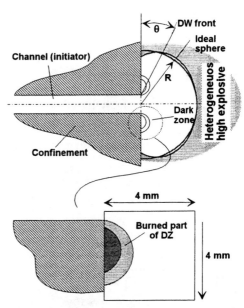

FIGURE 1. Corner turning phenomenon in heterogeneous HE.

Our recent experiments (6) show explicitly the non-monotonic (pulsing) behaviour of diverging

DW propagation in PBX, in different zones of DF. We are considering the obtained results as significant experimental evidence that the diverging DW in PBX, in the intial phase of formation (r < 20 mm), has a cellular structer, while the DZ is the first vixualised manifestation of the cellular front interaction with the shock-loaded confinement in the corner area.

This paper focuses attention on the study of DW propagation in channels and initial phase of DW corner turning, inside a small zone 4 x 4 mm. The used explosive was a PBX based on RDX (85 wt % RDX, 11.5 wt % HTPB, 3.5 wt % DOS) (6). The RDX has a bimodal particle size distrubtion (75 wt % - 96 µm, 25 wt % - 22 µm). The PBX density is 1.59 g/cm^3 that corresponds to 0.965 TMD. Detonation velocity is 7.81 mm/µs. The measured failure diameter is 3.75 mm.

A new high resolution optical methods based on a 64 optical fibre (ϕ 250 µm each) strip, connected directly to a fast electronic streak camera (Thompson - TSN 506 N), was developed for characterization of 2D shock and detonation waves within a few mm of the PBX surface. Spatial and temporal resolution for each fibre are not worse than respectively 250 µm and 3 ns.

The printed erosion figure on the Cu witness plates allows observation of the current lines of the detonated material in multiple zones.

EXPERIMENTAL RESULTS AND DISCUSSION

DW propagation inside of initiation channel

Experiments were performed to record the (x-t) diagram simultaneously with the DF curvature inside of the channels of PBX, confined by Cu and PMMA plates, Fig. 2.

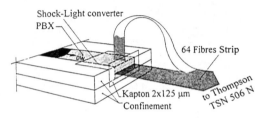

FIGURE 2. Arrangement for experiments on DW propagation inside the channels.

Direct fibre connection between PBX and electronic streak camera (without any intermediate optics) allows recording with minimum of losses, the light emitted by DF before touching the optical probe.

This non-disturbed layer of PBX (2~2.5 mm of thickness) is partially transparent for the DF light emission. Typical photochronograms are shown in Fig.3.

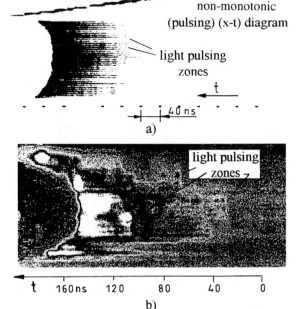

FIGURE 3. Typical photocronograms: a) copper confinement, channel 10x10 mm; b) PMMA confinement, channel 5x5 mm.

Analysis has been performed for records obtained in the experiments with channel sizes 5 x 5 mm and 10 x 10 mm, formed into both the Cu and PMMA matrixes, showing that: (i) DW propagation inside the channels is characterised by light pulsations, corresponding to the local front accelerating and slowing down, as observed in the (x-t) diagrams, while the mean detonation velocity remains constant; (ii) temporal-spatial characteristics of light pulsing zones are: period - 10-30 ns, width (recorded by used optical probes) is more than 0.5 mm and depended on the confinement and channel sizes (observed maximum size was 1.75 mm), (iii) DF is non-smooth. The dimensions of DF shape irregularities, as well as inherent plumage/delay intervals between the adjacent non-monotonic portions of DF, correspond to spatial-temporal characteristics of light pulsing zones observed before the front touched the optical probes.

Obtained results constitute the significant evidence that the DF in PBX inside the channel has a cellular structure. These results could be interpreted also as evidence of the break-down concept of DW structure developed by K. Shchelkin, Ya. Troshin and A. Dremin (7).

In accordance with this approach, the DF has the structure shown in Fig. 4.

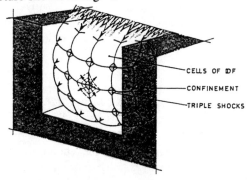

FIGURE 4. Interpretation of multifront structure of DW in PBX inside the channel.

Corner turning research tests

The corner turning tests have been conducted using the setup shown in Fig. 5. The optical probe, having a fan configuration, was placed on the PBX in the corner area, covering a square $\cong$ 4 x 4 mm.

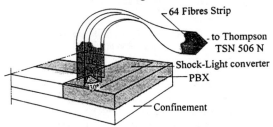

FIGURE 5. Arrangement for corner turning tests.

An example of a streak record obtained in the experiment, with a channel 5x5 mm confined by copper is shown in Fig. 6.

The digital analysis of photochronogram has been done and allows to build up the (r-t) diagrams of DF propagation along the fibre lines (Fig. 7) and then the DF isochrones (Fig. 8).

The printed erosion figures, obtained in the same experiment are shown in Fig. 9.

FIGURE 6. The photochronogram obtained in the experiment with copper confinement, channel 5x5 mm. 40 ns between two markers.

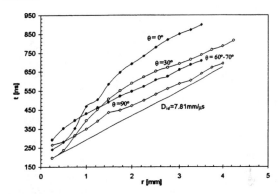

FIGURE 7. (r-t) diagrams of DF during the corner turning process obtained in the experiment for copper confinement with channel 5x5 mm.

The obtained results show the occurrence of: (i) the pre-shock effect inside a small area of PBX, near the corner, by the lateral shock wave travelling in confinement; (ii) the preferential DF development in the 50° to 90° of θ-range relative to PBX-confinement interface; (iii) the non-monotonic propagation of different parts of DF, alternating by zones of acceleration and slowing down.

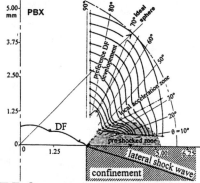

FIGURE 8. Isochrones of DF during the corner turning process obtained in the experiment for copper confinement with channel 5x5 mm. The time interval between isochrones is 40 ns.

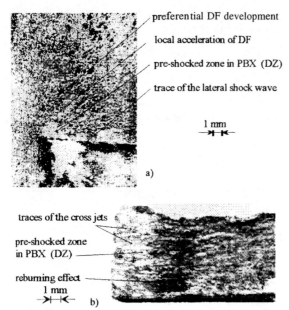

preferential DF development

local acceleration of DF

pre-shocked zone in PBX (DZ)

trace of the lateral shock wave

1 mm

a)

traces of the cross jets

pre-shocked zone in PBX (DZ)

reburning effect

1 mm

b)

FIGURE 9. Fragments of erosion figure printed on Cu plates: a) base plate, b) internal wall.

The effect of PBX weak pre-shocking in a small area adjacent to internal wall is characterized by shock propagation with mean velocity of 2 mm/μs, creating the desensitization in this region, defined above as a DZ. The lateral cells of DF (Fig. 4) had, however, to propagate through the pre-shocked region, which first part is reburned and the second one remained unreacted. The reburning process in part of DZ is indicated clear on the photochronogram (Fig. 6) for $\theta = 0°$. The process of DF propagation through the DZ is followed by a penetration of cross jets, caused by the triple shocks of the lateral DF cells inside of initiation channel, Fig. 4. The traces of these cross jets are printed clear on the internal walls of Cu-confinement (Fig.9 b). When the shock waves created by cross jets have overtaken the non-disturbed PBX, the detonation occurred. Both the effects, preferential development and cross jets, could be interpreted as the result of the initial cellular structure of DW in the channel. As a result of the interaction between the zone of DF preferential development and the stream along the DZ, the local acceleration of DF occurs in the range 30° to 45° of θ. The subsequent phase of DW formation (r>10 mm) is described in (6).

CONCLUSIONS

The corner turning effect of DW in PBX RDX/HTPB/DOS has been studied. The developed original high resolution optical method, coupled with the witness plates, allows the clarification of the initial phase of the corner turning phenomenon. The pulsating behaviour of the DW inside the initiation channel is being interpreted as a result of its cellular structure. The cellular structure of DW in the channel is responsible for both of the observed effects in the complex corner turning phenomena: DZ intersection by cross jets and DF preferential development in the range of 50° to 90° from the PBX-confinement interface. The subsequent DW development is governed by several complex interaction processes.

ACKNOWLEDGEMENTS

The authors acknowledge especially Eng\u00aa. Noémia Cabrita, for the help in PBX preparation, as well as Mr. João Moreira and Eng°. José Ribeiro for technical support in pursuance of research.

REFERENCES

1. Herzberg, G. and. Walker, G. R, Initiation of High Explosives, *Nature*, 161 647-8, April 24, 1948.
2. Bonthoux, F., Deneuville, P., de Longueville, Y., Diverging Detonations in RDX and PETN Based Cast-Cured PBX, in the *Senventh Symposium. (International) on Detonation*, Annapolis, Maryland, Jun. 16-19, 1981.
3. Quidot, M. and Groux, J.,"Reactive Modelling in Shock Initiation of Heterogeneous Explosives", in the *Ninth Symposium. (International) on Detonation*, Portland, Oregon, Aug. 28-Sep. 1, 1989.
4. Cox, M. and Campbell, A. W., Corner Turning in TATB, in the *Seventh Symposium. (International) on Detonation*, Annapolis, Maryland, June 16-19, 1981.
5. Graeme, A., Leiper and Kennedy, D. L., Reactive Flow Analysis and its Application in the *Ninth Symposium. (International) on Detonation*, Portland, Oregon, Aug. 28-Sep. 1, 1989.
6. Mendes, R., Plaksin I., and Campos, J., Single and Two Initiation Points of PBX. in *Topical Conference on Shock Compression of Condensed Matter*, Amherst, MA., Jul. 27-Aug.1, 1997.
7. Dremin, A. N., Novelties of Detonation Phenomenon Study, in the *Tenth Symposium. (International) on Detonation*, Boston, MA, Jul. 12-16, 1993.

CP429, *Shock Compression of Condensed Matter – 1997*
edited by Schmidt/Dandekar/Forbes
© 1998 The American Institute of Physics 1-56396-738-3/98/$15.00

AQUARIUM TEST EVALUATION OF A PYROTECHNIC'S ABILITY TO PERFORM WORK IN MICROSECOND TIME FRAMES

J. W. Forbes, B. C. Glancy, T. P. Liddiard, W. H. Wilson

Naval Surface Warfare Center, Indian Head Division, Indian Head, Maryland 20640

Pyrotechnic materials can release tremendous thermal energy upon reaction. A pyrotechnic's ability to do work, when mixed with other materials to produce a working fluid at high pressure and temperature, is studied in this work. An experimental technique is used to measure underwater expansion of cylinders containing porous pyrotechnic materials shock compressed by a surrounding annular explosive charge. Expansion velocity enhancement due to reaction in the pyrotechnic core is detected as an increase in outer wall velocity over that obtained with a solid inert core. The outer wall expansion is measured for 180 μs after wall motion begins, so that relatively late-time energy release from the core may be detected.

INTRODUCTION

Certain solid state reactions, such as those which occur in mixtures of metals with metal oxides, can release tremendous thermal energy. On a volume basis, the heat of reaction typical of these systems can be two to five times the heat of detonation typical of a high explosive. It is thus of interest to determine if energy released by these solid state systems can be used in applications requiring high power mechanical work, applications traditionally employing high explosives. To do so, at least two conditions need to be met. First, the solid state reaction kinetics must be sufficiently rapid, such that the energy release rate is comparable to that of a detonating explosive. Second, a condensed working fluid must be made available, to convert the thermal energy released to useful mechanical work, via. subsequent expansion.

In this study, experiments were designed and run to produce shock-induced, rapid reaction in cores of highly porous solid state materials. Reaction was driven by convergent shock from surrounding explosive. Various polymer powders were mixed with a metal-metal oxide powder system and pressed to form test cores. The polymer materials were used both to act as a binder in the pressing, and to provide

working fluid as the solid state reaction progressed. The ability of the polymer-solid state material combinations to perform work was evaluated by comparing their underwater cylinder expansion to that of an inert solid core surrounded by an explosive driver identical to that used for the porous reactive cores.

BACKGROUND

Enikolopian (1) has reported that, in Bridgman anvil experiments in which solid state materials were compressed at high stress combined with large shear strain, solid state reactions occurred at rates from 3 to 8 orders of magnitude greater than reactions between the same materials in liquid phase. Over the last two decades, Graham and co-workers at Sandia National Laboratory have studied shock-induced solid state reactions in porous materials. They have found that solid-state reaction kinetics are significantly impacted by a number of factors not generally held to be important in conventional thermochemical reactions. Graham (2) describes a number of effects which accelerate reaction kinetics when highly porous solids are compacted by shock waves. Under these conditions, even normally highly brittle solids experience large plastic deformation as the material flows to fill voids. High levels of material mixing and shear flow occur in this process, greatly enhancing solid state reaction rates. Defect density is greatly increased as the shock propagates through the solid grains, also enhancing reaction. Shock loading removes contaminants from particle surfaces, which otherwise would act as reaction barriers. As porous solids are shock compacted, an irreversible process, they become very hot; this also accelerates reaction.

EXPERIMENTAL DESCRIPTION

Figure 1. illustrates features of the ring-charge configuration used in the experiments. The powder materials were pressed into 14-mm inner-diameter stainless steel tubes. These tubes were encased by annular charges of the Navy explosive PBXN-110. The web thickness of the explosive annulus was 7.9-mm. Thus, the volume ratio of porous sample material to driver explosive was approximately 0.26, for the half-length of the charge surrounding the test sample. The first half-length of the annular explosive in each experiment surrounded a solid core of polymethylmethacrylate (PMMA). This provided direct comparison of the outer cylinder expansion for regions surrounding porous solid-state reactants to that for regions surrounding a solid inert material, for the same explosive driver. One control experiment was run in which the core along the entire length was solid PMMA. The charge and core assemblies were contained in 38.1-mm outer-diameter PMMA tubes of 3.1-mm wall thickness.

The PBXN-110 charges were initiated at the end containing the solid PMMA core by a booster comprised of two cast Pentolite pellets, each 25.4-mm thick and 50.8-mm in diameter; the booster was initiated by an RP-80 EBW detonator. The experiments were run in an aquarium, backlighted by argon bombs. Expansion of the cylinders was observed via. a Jacobs framing camera, recording at an interframe time of ~1.8 µs. Because the charges were detonated in water, expansion was slower than it would be in air. This was done to allow more time at high compression for the solid state reaction to occur and affect the rate of wall expansion observed.

The CTH hydrocode was used to model detonation in the configuration with a solid PMMA core. This simulation indicated that converging shock from the PBXN-110 ring forms a Mach disc in the core, with a peak pressure of ~370-kb.

A solid state system of aluminum and manganese oxide powders was used for each of the reactive metal porous core experiments. Two different polymer binder systems were used: Teflon (TFE) and polyvinylidene fluoride (PVF$_2$). Table 1 shows details of the starting powders. Three experiments with reactive powder porous cores were run. In two experiments, the proportion of TFE used was varied; one experiment used the PVF$_2$ binder. Table 2 shows mix details for the three experiments, including mass ratios and pressing densities of the samples. The powders were weighed out and hand mixed to give homogeneous mixtures of the desired proportions. These mixtures were then pressed by stages into the stainless steel tubes. The final density of the pressed sample mixtures for each experiment was approximately 60% of theoretical maximum density (TMD).

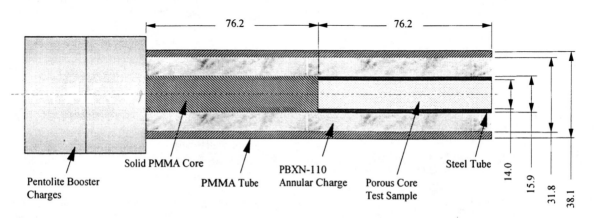

FIGURE 1. Ring charge configuration used in the experiments, with Pentolite booster charge. All dimensions in mm.

760

TABLE 1. Starting powders used in the experiments.

Powder	Source	Mean Size μm	Distribution
MnO_2	Cerac M-1054 Special Lot # 63230	1.8	99% $1-5$ μm
Al	Alcoa 1401 Special	6.7	100% < 16 μm
TFE	ICI Americas Fluon G	3.3	94% $1-10$ μm
PVF_2	Scientific Polymer Products #102	3.4	93% $1-6$ μm

TABLE 2. Porous mixes used in the experiments.

Experiment #	Core Materials (Mass Ratios)	TMD g/cc	Sample Density g/cc
JK-152	$Al/MnO_2/TFE$ (0.41 : 1 : 0.43)	3.387	2.03
JK-154	$Al/MnO_2/PVF_2$ (0.41 : 1 : 0.5)	3.018	1.81
JK-155	$Al/MnO_2/TFE$ (0.41 : 1 : 0.85)	3.076	1.85

EXPERIMENTAL RESULTS

In Figure 2, frames from the high-speed camera records from two experiments, JK-151 and JK-152, show the expanding cylinders at relatively late time. JK-151 was a control test in which the entire core was solid PMMA. Note the difference in the character of expansion of the cylinders. In the test with the inert core, the wall shape is nearly cylindrical (linear sides in cross-section), indicating that expansion is smooth and uniform along the full length. In test JK-152, the expanded cylinder is curved outward slightly at the down-stream end, which contained the core of solid-state reactive material. This occurred because this portion of the expanding cylinder, though exposed to the detonation front later in the test, has caught up to and passed the expansion of the first half-length of the cylinder, which surrounded an inert PMMA core. The same general character of expansion, in which the part of the cylinder with a porous reactive material eventually expanded at higher velocity than that with an inert solid PMMA core, was observed in all three experiments.

The cylinder expansion history for each experiment was obtained from a frame by frame analysis of the camera records. For each experiment, expansions at the midpoints of the inert and reactive core regions were determined. A plot of the expansion history from JK-152 is shown in Figure 3. The curves show outer-wall radial-position vs. time, starting with first motion at the measurement location. For approximately the first 30 μs of expansion, the motions of the two regions are essentially identical. However, expansion of the region surrounding reactive material suddenly jumps ahead at about 30 μs, and from that time on it is increasingly greater

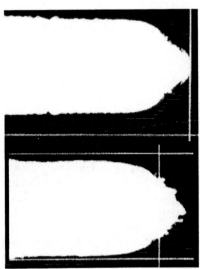

FIGURE 2. High-speed camera frames from control test JK-151 (upper) and porous core test JK-152 (lower).

761

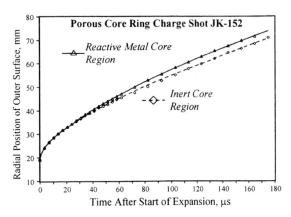

FIGURE 3. Plot of cylinder position vs. time after start of motion, for two measurement locations, test JK-152.

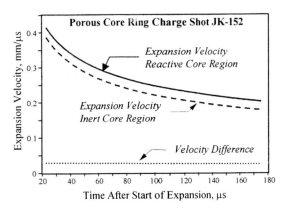

FIGURE 4. Expansion velocities derived from position-time data, test JK-152.

than that of the inert region. This expansion character was typical for all three experiments run with the various reactive materials. In each case, the initial expansion was the same as that for the cylinder surrounding an inert solid core for the first 30 μs or so. In each case there was then a sudden boost in expansion velocity beyond that of the inert region. Figure 4 shows a plot of wall velocities derived from the position vs. time data for test JK-152. It is of interest to note that, while the wall velocities from both regions fall off with time (as expected in a cylinder expansion test), the velocity advantage of the region with the reactive core is maintained for at least as long as the camera coverage lasted. The velocity difference for the two regions remained about constant. This general behavior was also typical of all three experiments with the reactive materials.

DISCUSSION

These experiments provide some insight about the kinetics of reaction in the porous pyrotechnic materials used in the test cores. Excess cylinder expansion velocities were observed in each case, but only after a delay of a few tens of microseconds. This time delay should be viewed as an upper bound on shock-driven reaction kinetics in these pyrotechnic materials. I.e., an observed time delay, from start of motion until sudden increase in expansion velocity at the outer surface, included not only the time of pyrotechnic material reaction and energy release, but also formation of a working fluid, and propagation of this fluid's expansion effect from the compressed core region to the expanding outer sur-

face of the PMMA cylinders. Nevertheless, these experiments demonstrate that reaction in these pyrotechnic materials can be shock driven to release energy and perform expansion work in microsecond time scales comparable to that of work done by non-ideal explosives. The ratio of driver explosive to pyrotechnic material in the core used in these experiments was not optimized. In order to better model and design similar ring-charge experiments in the future, there is a need for the mechanisms of ignition and growth of reaction in shock-driven pyrotecnic materials to be determined and understood.

ACKNOWLEDGEMENTS

This work was sponsored by the Independent Research program of NSWC, Indian Head Division. The authors wish to thank Mr. John Volk for pressing the core samples and fabricating the test assemblies, and Mr. Robert Baker and Mr. Dale Ashwell for performing the experiments.

REFERENCES

1. N. S. Enikolopyan, "Super-fast chemical reactions in solids," *Russian Journal of Physical Chemistry*, 63 (9), 1261-1265, 1989.
2. R.A. Graham, "Issues in Shock-Induced Solid State Chemistry," *Third Symposium International sur le Comportment des Milieux Denses sous Haute Pressions Dynamiques*, La Grande-Motte, France, June 1989.

CP429, *Shock Compression of Condensed Matter – 1997*
edited by Schmidt/Dandekar/Forbes
© 1998 The American Institute of Physics 1-56396-738-3/98/$15.00

BLAST WAVES FROM NON-IDEAL EXPLOSIVES

Dr. Van Romero and Pharis E. Williams

Energetic Materials Research and Testing Center, New Mexico Tech, Socorro, NM 87801

The non-ideal behavior of explosives comes from different ways which retard the energy release from the explosive. These include a lack of oxygen balance which results in energy being released after the shock wave from the detonation has gone into the surrounding air and the detonation products react with this fresh source of oxygen as it is included within the shock wave. Also included are slow and/or multiple reactions which cause energy to be released late in the reaction zone of the detonation when the pressure of the detonations has dropped until the local sound speed has fallen below the detonation velocity. All energy released after this point cannot keep up with the detonation shock wave and must wait to catch up with the blast wave that propagates into the air. Both of these non-ideal characteristics of the explosive reduce the irreversible losses to the air close to the explosive charge by reducing the peak pressure, and therefore, the temperature compressive heating of the air. Less irreversible losses to the air means more energy propagates to greater distance. This presentation covers the research conducted into the influence of these non-ideal effects upon the propagation of peak pressures and the positive and negative impulses from non-ideal explosives.

INTRODUCTION

The blast waves from ideal explosives have been related to each other using TNT as the standard[1]. This method of comparing the blast waves works very good for explosives which release their energy very rapidly and which are oxygen balanced. Some explosives release their energy more slowly and this method of comparing the blast waves to TNT gives incorrect results[2,3]. The underlying reason for the error comes from the non-ideality of these explosives. With the increase of terrorism worldwide, the issue of ensuring structural integrity of buildings increases the need for accuracy in the prediction of blast waves from the explosives that the terrorist will likely use.

TECHNICAL DISCUSSION

Suppose we consider explosive energy released from three sources: 1) the energy released from a nuclear explosion, 2) the detonation of ideal high explosives, and 3) the detonation of non-ideal explosives.

Now let us examine these differences one at a time. First, compare the blast from a nuclear bomb to one from ideal high explosives. The pressure from the nuclear bomb is higher than that from the ideal high explosives. The concentrated energy release from more nearly a point source means that higher temperatures and pressures will be generated. On the other hand, compared to the ideal high explosives, the pressures at long range from the nuclear blast will be lower. The higher temperatures in air means more irreversible energy lost from the nuclear blast than from the ideal high explosives; therefore, the dissipation is greater for the higher pressures. This means, at long range, less energy left in the blast from the nuclear explosion and, hence, less pressure than from the ideal high explosives blast.

An additional phenomena associated with a nuclear blast is the radiation, which heats the air ahead of the shock wave. This leads to more dissipation of energy irreversibly from the nuclear than from the ideal high explosives. This adds to the difference in the long range pressures between these two type of blasts.

A phenomena, referred to as *after burning*, is associated with explosives, both ideal and non-ideal, that have reaction products that are not balanced with respect to oxygen content. This means that if an explosive is not oxygen balanced, then, as air is included within the shock wave going out into the air, the reaction products will react with the oxygen in this included air to release more energy. Any energy released in *after burning* does not contribute to higher initial pressures and, therefore, these explosives reactions (which are not oxygen balanced) do not generate the same amount of irreversible losses that an ideal high explosive would generate for equal total energy.

Comparing the blast from an ideal high explosive to that from a non-ideal explosive, one finds similar discussion. That is, for the ideal explosive with the higher detonation (or CJ) pressure, the short range temperatures and pressures will be higher. The long range pressures will be lower, because the higher initial temperatures and pressures in the air lead to greater irreversible losses, which dissipate greater amounts of energy and lead to lower pressures for the ideal explosive.

Thus far, one may see that if the energy is released slowly, the tendency is for there to be less detonation pressure, which leads to less dissipation and, therefore, more long range pressures.

PAST ANALYSIS
In 1947, John von Neumann considered energy released instantaneously from a point source. He found that if one assumes that the entropy change in the air can be ignored, then a closed form analytical solution may be obtained[4]. Bethe followed with a study of the release of energy from a high explosive blast[4]. In his study, he considered different assumptions. First, he considered the case where $\gamma-1$ is small, all the material is near the shock front, the pressure is virtually a constant at radii less than 0.8 R, and that his analytical solution should have von Neumann's point source solution as an asymptote. In a second, different approach to the characteristics of a blast at very large distances, Bethe assumed that near the origin, the shock was of very short duration. He then showed that the impulse was given by

$$p\theta = \frac{const}{R}, \qquad (1)$$

the pulse shape remains unchanged, and the positive impulse approximates the negative impulse. By again considering short duration shocks near the origin, Bethe also considered the character of blast at intermediate distances and arrived at a similar result as above except that the shape was allowed to change. For the purpose of discussing the blast waves from non-ideal explosives at distance, it is important to note that all these attempts basically assume instantaneous energy release into the air and then assume certain conditions of the air. In particular, it should be noted that the assumption of no change in the entropy of the air is always made.

MODELING THE REACTION RATES OF EXPLOSIVES
Almost all numerical models in reactive hydrocodes use a single reaction rate with various terms in it. The various terms in the single reaction rate are supposed to be able to model different reactions or reaction influences. These models include models named Forrest Fire, programmed burn, and initiation and growth models. All produce good results when used with ideal high explosives, which release their energy very rapidly. All do very poorly on non-ideal explosives, which release their energy through multiple reaction rates. This is to be expected, since they were not developed to address the non-ideal explosives. Non-ideal explosives are better modeled using multiple reaction rate codes, developed specifically for explosives, which have multiple independent and dependent reactions[5].

The expected result of better modeling of the energy released from non-ideal explosives is the release of

764

energy as a function of time and the local conditions, such as, the local sound speed. This would give a better prediction of the amount of energy released quickly, under conditions where the energy can catch up to and support the detonation shock wave, and the amount of energy that is released more slowly and under conditions where the energy cannot catch up to the detonation shock wave in the explosive, and must wait to catch up after the shock has gone into the air and slowed down. This late arriving energy violates all of the assumptions made in the early analytical models trying to determine the character of the blast from explosives.

INFLUENCE OF REACTION RATES ON THE BLAST

Given the expectation of late energy being released into the air due to either *after burning,* or the arrival of the late energy, which is catching up with the shock wave after it reaches the air, how might one investigate the influence of this late energy on the character of the blast at a distance from the source? This may be investigated by considering what happens when two of the assumptions made in Bethe's analysis are changed. First, consider changing the assumption of instantaneous energy release to the assumption that some energy will be released into the air after the shock has proceeded into it. This violates the assumption of constant entropy. If one still assumes that the air may be described by an ideal gas, then it may be shown that the entropy may be written as:

$$S = \frac{R_g}{(\gamma-1)}\left[\ln\left(p\rho^{-\gamma}\right) - \ln\left(p_0\rho_0^{-\gamma}\right)\right] + S_{NI}\left[1-e^{(-bn)}\right]. \quad (2)$$

The last term in this expression for the entropy is an assumed function of time, which might represent late energy arriving and increasing the entropy. Notice that this increase in entropy reaches an asymptote beyond which there is no further increase. This would be the case when all energy has been released and catches up with the shock wave.

If one now uses this entropy to calculate the pressure for an ideal gas they find that

$$\begin{aligned} P &= P_0\left(\frac{V}{V_0}\right)^{-\gamma} exp\left[\left(\frac{(\gamma-1)S_{NI}}{R_g}\right)\left(1-e^{-bt}\right)\right] . \\ &= P_0\left(\frac{V}{V_0}\right)^{-\gamma} f(t) \end{aligned} \quad (3)$$

From this, one may see that the late energy adds to the entropy and, in turn, adds to the pressure.

The wave propagation velocity is given by

$$c + u \approx c_0\left(1 + \frac{(\gamma+1)}{2}f(t)\left[\frac{V_0 - V}{V_0}\right]\right) \quad (4)$$

where the function of time, f(t), is defined above.

When looking at the wave impulse due to the late energy addition, one finds

$$p_f\theta = \frac{constant \quad f(t)}{R} \quad (5)$$

so that one sees that both the pressure and the impulse are increased by the late energy from the non-ideal explosive.

If the above predictions concerning the propagation of pressure and impulse are correct then the use of a single TNT equivalence number for a non-ideal explosive cannot work. Rather, we must use a varying equivalence depending upon the distance. However, in the absence of having a complete solution we must turn to experimentation to see the affect of the non-ideality.

EXPERIMENTS

There is a need in the counter terrorist explosive research to develop improvements to the various codes assisting the structural design engineers in determining the blast loads on the structures they are designing. This includes the codes that obtain their results by looking up a TNT equivalence as well as the reactive hydrocodes. There also is a need to provide a method of verifying the ability of codes to properly calculate the blast waves from explosive shapes as well as types. In order to provide experimental data for this verifying effort a number of experiments were conducted which measured the air blast from cylinders of explosives of different length to diameter ratios and explosive types. The data from these experiments are being processed now.

An excellent example of the data that presents the best experimental confirmation of the variation of the TNT equivalence of a non-ideal explosive is the comparison of the data from the side of a TNT cylindrical charge and a cylinder of terrorist explosive. First, we may compare the incident pressures from the side of the TNT cylinder with past data collected from TNT hemispheres in order to verify the data collection system. Figure 1 displays the comparison of the data from the TNT cylinder with the past hemisphere data[6].

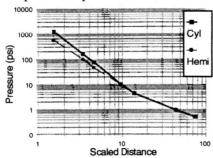

FIGURE 1. Cylinder vs Hemisphere

Figure 2 shows the incident pressures from the side of the TNT cylinder and the cylinder of terrorist explosive called Alpha.

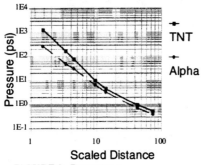

FIGURE 2. Pressures

Figure 3 shows the ratio of the incident pressure from the Alpha to those of the TNT cylinders. This is the TNT pressure equivalence as a function of the scaled distance from the cylinders.

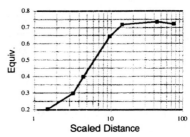

FIGURE 3. Pressure Equivalence

An additional data point is that the detonation velocity of the explosive Alpha in the diameter of the cylinder experiment was 3.18 km/sec. When compared with the 6.95 km/sec detonation velocity of the TNT, this gives a shock, or detonation, equivalence of 0.209, or 21%. It may be noted that the first recorded pressure equivalence was 0.205, again 21%.

CONCLUSION

From the above discussion, it may be argued that late energy from non-ideal explosives not only adds to the pressure and impulse at distance from the blast, but does so in an exponential manner. The data from the air blast from cylinders of TNT and a terrorist explosive shows just this kind of behavior. It is very evident from the data that the pressures show a detonation equivalence of 21% very close to the charge and that this equivalence increases to about 72% at a scaled distance of 14. The late energy comes from both slow reaction rate of the terrorist explosive and its lack of oxygen balance. It is evident that the assumption of a single number for the TNT equivalence for non-ideal explosives does not give a good approximation of the blast wave from non-ideal explosives.

REFERENCES

1. Cooper, Paul W., "Explosives Engineering," VCH, 1996.
2. Williams, P.E., "The Influence of the Reaction Rate of Explosives on Blast Effects," 5th International Symposium on the Analysis and Detection of Explosives," 4-8 Dec., 1995.
3. Romero, V. and Williams, P. E., "The Characterization of Non-Ideal Explosives," Specialty Symposium on Structures Response to Impact and Blast, Tel Aviv, 6-10 October, 1996.
4. Bethe, H. A., "Blast Wave," Los Alamos Scientific Laboratory report, LA-2000, 27 March, 1958.
5. Williams, P.E., 1994, "The Importance of the Reaction Rate Laws in Modeling DDT", proceeding of the JANNAF Propulsion Systems Hazards Subcommittee meeting, 1-4 August, 1994, San Diego, CA.
6. Kingery, Charles N., "Airblast Parameters from TNT Spherical Air Burst and Hemispherical Surface Burst," US Army Research and Development Center, Ballistic Research Laboratory, Technical Report ARBRL-TR-02555, April, 1984.

CHAPTER X

ELECTRICAL, MAGNETIC, AND OPTICAL STUDIES

CP429, *Shock Compression of Condensed Matter – 1997*
edited by Schmidt/Dandekar/Forbes
© 1998 The American Institute of Physics 1-56396-738-3/98/$15.00

THE OPPORTUNITY OF THE USE OF SAPPHIRE AT MULTIPLE SHOCK-WAVE COMPRESSION OF HYDROGEN

**V.I.Postnov, D.N.Nikolaev, V.J.Ternovoi,
A.S.Filimonov, V.E.Fortov and V.V.Yakushev**

Institute of Chemical Physics in Chernogolovka, Moscow region, Russia, 142432

The conductivity of cooled gaseous hydrogen with initial density 0,022g/cc was measured inside the single-crystal sapphire cell. The transition of hydrogen to a high electroconducting state during the fourth circulation of the shock wave (40-50 GPa) was registered. It has been concluded that the behaviour of brittle substances, (e.g. sapphire) used as isolators, should be taken into account to interpret the dielectric-metal transition in low initial density substances.

INTRODUCTION

One of the main problems of dynamic research on the insulator - metal transition is choice of a measuring cell material. As such materials, Teflon and sapphire are most frequently used, because they preserve insulating properties up to pressures in excess of 1 Mbar [1-6].

At present the insulator - metal transition is investigated by means of technique previously used to study this transition in elementary sulfur [7].The main idea is to minimize the parasitic inductance of the measuring circuit by means of accommodating a shunting resistor together with the research sample directly into the assembly being loaded. The shunt material was manganin, and Teflon an insulator. The value change of the shunting resistor at compression was determined from the indications of a manganin gauge, located in the same measuring cell. This method was used to record the resistivity at the level 10^{-2} - 10^{-3} Ohm cm behind the transition of sulfur and iodine to a high conducting state under the multiple compression in the dynamic pressure range 30 - 110 GPa [8].

In this work we have applied the above technique to record the transition of gaseous hydrogen to a high conducting state.

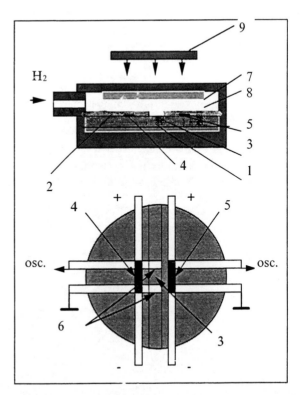

FIGURE 1. The assembly for measuring hydrogen gas conductivity.

U.S. researchers [1] have already measured the resistivity of liquid hydrogen inside a single-crystal sapphire cell under shock wave compression over the pressure range 93-180 GPa. It was found that the resistivity decreased by four orders as the pressure increased from 93 GPa up to 140 GPa and remained constant at 10^{-3} Ohm cm in the range (140-180) GPa. The received data were explained by transition of liquid H_2 to a two-nuclear liquid metal at p=140 GPa.

EXPERIMENT AND RESULTS

In our experiments we have also used synthetic single-crystal sapphire as a insulator. The experimental assembly (Fig. 1) was previously pumped out and filled gaseous hydrogen up to pressure 100 atm. Two sapphire 40 mm disks 4 mm (1) and 0.25 mm (2) thick were placed inside of the assembly. The thinner disk was cut so that there was a 3 mm gap (3) between its parts. Two manganin gauges were placed between the thick and thin disks. The initial resistance of gauges was 0.05-0.07 Ohm. The first gauge (4) was used for measuring the pressure history, the second one(5), whose measuring electrodes (6) were placed in a 3 mm gap (3), was a shunting resistor at measurement of the conductivity of the shock-compressed hydrogen. A sapphire disk (7) of 1 mm thickness and 30 mm diameter was mounted on the upper internal surface of the assembly. The distance between disks (7) and (2) was varied from 3 up to 6 mm. The electrical contact wires were passed through the lateral surface of the assembly using ceramic insulation. Immediately before the start of experiments, the assembly was cooled to temperature of liquid nitrogen in order to increase the initial density of hydrogen (8). This state was controlled by a thermocouple.

Assembly was loaded by a steel impactor (9) (the thickness 1.5 mm and diameter 30 mm) at a speed of 5 - 6 km/s.

It should be to note, that the structure of the described assembly (sapphire-hydrogen-sapphire) is similar to one described in [1], with the difference that in our case H_2 had the initial gas density of 0.022 g/cc, whereas the liquid H_2 density was 0.071 g/cc.

The oscillogram of one of experiments is shown in fig. 2. It is seen, that the resistance first rises over the time $t_1 - t_2$, then during $t_2 - t_3$ becomes equal to the initial value 0,05 Ohm of the shunt and finally smoothly decreases during the time $t_3 - t_4$. Based on the analysis of the multiple compression of hydrogen between sapphire plates it was determined, that the time $t_1 - t_2$ of 240 ns, corresponded to the second and third waves of compression, and the time $t_2 - t_3$ corresponded to the fourth wave of compression and the maximum of the pressure. Now, using of the manganin gauge (4) indication, it would be necessary to correct the measured resistance, which represented the parallel connection of resistance of the manganin shunt (5) and the hydrogen layer to determine the resistivity of the shock wave compressed hydrogen sample. But the indications of the second gauge, which recorded the pressure history, appeared unexpected. The pressure maximum p_{max} of the multiple compression was at a level 30 GPa, i.e. essentially below the pressure (130 GPa) which could be obtained in the system: sapphire-gas-sapphire under loading with our steel impactor. The increase in thickness of the top sapphire disk (7) up to 1.8 mm has only resulted in increasing the pressure (p_{max}) existence time almost without changing its amplitude.

We assumed, that the this result has been probably due to: 1) the resistance of the manganin gauge was shunted by the sapphire conductivity, 2) sapphire was not the best choice for such experiments, because of its peculiar behaviour when released from $p_{max} = 130$ GPa to the pressure of the first wave of hydrogen shock compression, $p_1 = 1.5$ GPa.

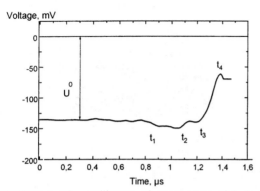

FIGURE 2. The oscillogram of measurement of hydrogen conductivity. U_0 is the initial voltage on the manganin shunt.

To check the first assumption we measured the conductivity of a sapphire sample with resistance measured perpendicularly to the plane of the shock wave front under dynamic pressure of 130 GPa. The main optic axis of sapphire was in the plane of the sample. The scheme is shown in fig. 3. The sapphire sample (1) 1.8 mm thick was placed on a steel plate (2), which was the first electrode. A second electrode (3) 10 mm in diameter was a copper foil placed between the sample (1) and other sapphire disk (4) to realize the uniaxial shock wave compression. The oscillogram of the experiment is shown on Fig. 4. The deviation of the beam U_0 corresponds to shunt resistance R_{sh}=8.64 Ohm. Here we can note some the shock wave effects on the sapphire sample. It is known, that sapphire has the property of shock polarization. According to this phenomenon the moment t_1 is the arrival time of the shock wave in the sample, and t_2 is the time when it approaches the electrode (3). The time t_2 - t_3, when the sample was loaded, has been calculated from the impactor speed and the sapphire Hugoniot.

The moment t_3 is the time of the unloading wave arrival. By the time t_4 the sample resistance is again at the level R_{sh}. As a result, the sapphire resistivity ρ_{load} and $\rho_{release}$ was determined at p=130 GPa and subsequent release using the data of Al_2O_3 shock compression. ρ_{load} was 170 Ohm cm and $\rho_{release}$ was 10 Ohm cm.

DISCUSSION AND CONCLUSION

Consequently, the shunting of resistance of manganin gauges under shock wave compression and possibly in a unloading wave did not occur. R_{load} measured in our experiments appeared close to the value in [2], obtained in measurements in the plane parallel to the shock wave.

As to the second assumption, the following should be noted. It is known, that the shock compression of brittle substances, to which also sapphire belongs , is accompanied by processes of non-uniform plastic currents, adiabatic shear, local heating. These phenomena results in heterogeneity of shock compressed substances. The most information on these phenomena has received in research of crystal and melted quartz. In optical experiments [9], in conductivity measurements [10], in experiments with recovery ampoules [11] it was found, that SiO_2 was a heterogeneous material under shock wave and under release, alike consisting of firm microblocks, surrounded by molten layers. According to the analysis [2] sapphire also displays of heterogeneity under shock compression [12,13]. In our experiments (Fig. 4) it was found , that the specific resistance of Al_2O_3 was changing with time under unloading and was 10

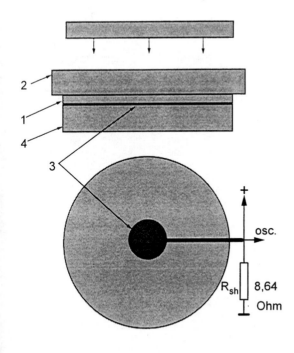

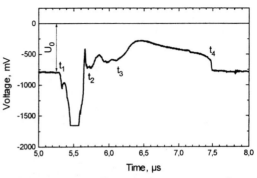

FIGURE 3. The assemble of measurement of sapphire conductivity.

FIGURE 4. The oscillogram of measurement of sapphire conductivity. U_0 is the initial voltage on R_{sh}.

771

Ohm cm. Extrapolation of the resistivity - temperature dependence [14] to the fusion area of Al_2O_3 under normal pressure gave the resistivity which was close to this value of specific resistance. It would fair to assume, that sapphire, even if locally, melted in conditions of unloading. This can be the reason for surface instability of the sapphire plate (7) adjacent to gas, which account for low values of the dynamic pressure, registered in our experiments. It is known that the conditions initiating these instabilities can even appear the interface of brittle and condensed substances (Teflon) under dynamic compression [15].

In the light of the results obtained above it should be considered that the registration under multiple dynamic compression of the transition of the hydrogen with initial density 0,022 g/cc at temperature a liquid nitrogen and initial pressure 100 atm to the high conducting state with specific resistance at a level 0,005 Ohm cm takes place in the region of 4 circulation of the shock wave. This would correspond to 40-50 GPa for the steel impostor velocity and Hugoniots of steel and sapphire. But the cause of the initiation of high conductivity in this case appears to be related not only to shock wave compression of hydrogen but also to the interaction between H_2 and melted Al_2O_3.

ACKNOWLEDGMENTS

This work was supported by Russian Foundation for Basic Research, grant number 96-05-65742

REFERENCES

1. Weir S.T., Mitchel A.C., Nellis W.J., Phys. Rev.Lett **76**, 1860-1863 (1996).
2. Weir S.T., Mitchel A.C., Nellis W.J., J.Appl. Phys. **80**, 1522-1525 (1996).
3. Hauke R.S., Duerre D.E., Huebel J.G., Keeler R.N., Wallace W.C., J.Appl. Phys. **49**, 3298-3303 (1978).
4. Kalashnikov N.G., Kuleshova L.V., Pavlovskij M.N., J.Prikl. Mech. Teh. Fiz. (Rus.), 187-191 (1982).
5. Champion A.R., J.Appl. Phys. **43**, 2216-2220 (1972).
6. Karahanov S.M., Bordzilovskij S.A., "The Conductivity of Teflon under multiple compression up to 150 GPa", Proc. of IV Vsesouz. sov. "Detonation", Telavi, USSR, November, 1984.
7. Nabatov S.S., Dremin A.N., Postnov V.I., Yakushev V.V., Pisma v JETF (Rus.) **29**, 407-410 (1979).
8. Nabatov S.S., Dremin A.N., Postnov V.I., Yakushev V.V., "Electroconductivity measurement of condensed matter under multiple shock wave compression up to 1 Mbar", Proc of the VI Vsesouz. simp. Goren. Vzr. (Rus.), Alma-Ata, USSR, September 23-26, 1980.
9. Kondo K., AhrensT.J., Sawaoka A., J.Appl. Phys. **54**, 4382-4385 (1983).
10. Postnov V.I.,Nabatov S.S., Yakushev V.V., " Investigation of fused quartz behaviour behind shock wave front by electroconductivity measurement method ", Proc. the HERF Conference, Novosibirsk, USSR, August 18-22, 1986.
11. Ananin A.V., Breusov O.N., Dremin A.N., Pershin S.V., Tatsij V.F., Journ. Fiz. Goren. Vzr. (Rus), 578-584 (1974).
12. Urtiev P.A., J.Appl. Phys. **45**, 3490 (1974).
13. Yoo C.S., Holmes N.S., Ross M., Webb D.J., Pike C., Phys. Rev. Lett. **70**, 3931- 3935 (1993).
14. Cherepanov A.M., Tresvjatskii V.T., Highly refractory materials and oxide products(Rus), Moscow: Metallurgy,1964, p.75.
15. Kanel G.I., Molodets A.M., Journ. Teh. Fiz. **XLVI**, 398-407 (1976).

CP429, *Shock Compression of Condensed Matter – 1997*
edited by Schmidt/Dandekar/Forbes
© 1998 The American Institute of Physics 1-56396-738-3/98/$15.00

SHOCK INDUCED POLARIZATION IN SOME LIQUIDS

Y.Hironaka, M. Nicol, and K.Kondo

Materials and structures laboratory, Tokyo Institute of Technology,
4259 Nagastuta Midori Yokohama 226 Japan.

Shock induced polarization is one model used to explain the electromotive forces originating from shock compressed materials. This model supposes that dipoles are induced at the shock front and decay behind the front with a characteristic relaxation time. The resulting polarization is sensed with an oscilloscope which measures the current flowing between two electrodes surrounding the sample being shocked. The polarization model provides a good explanation of the variation of the current from this capacitor with time. Here, we report the shock induced electromotive forces for two classes of materials : several phases of 4-methoxybenzylidene-4-n-butylaniline (MBBA) and dilute solutions of electrolytes in water. These cases shows a strong dependence of the signals on material composition and phase.

INTRODUCTION

Shock induced electromotive forces (EMF's) have been detected in many materials by sensing currents in external circuits containing no other EMF sources connected to pairs of electrodes which surrounded the dielectric being shocked[1]. Here, we described measurements of some aspects of the shock induced polarization for two types of materials: (1) the solid, nematic, and isotropic liquid phases of 4-methoxybenzylidene-4-n-butylaniline(MBBA) and (2) dilute aqueous solutions of the electrolyte, potassium fluoride. For MBBA at ambient pressure, the nematic phase is stable from 295K to 320K. We studied the dependencies of the magnitudes of the induced dipoles on pressure and on the thickness of the samples and the initial temperature of the MBBA. We also discuss the propagation of the induced dipoles in the shocked sample and what these results imply about microscopic models of the polarization.

Several models explain these EMF's in terms electric polarization appearing at the compression front in a medium between the plates of a parallel capacitor and subsequent relaxation of the polarization behind the shock front. We find Allison phenomenological model especially useful for correlating experimental data. The microscopic basis of this effect is understood less.

The magnitudes of the shock induced polarization might be estimated by considering how shocks accelerate charges in molecules and relate this to molecular polarization. Because electric polarization generally relates to an applied electric field, the relative acceleration of the charges might be related to the external electric field by:

$$m\frac{d\dot{x}}{dt} = 2qE = \frac{2q^2}{\alpha}x .$$

(1)

where m, q and α are the mass, charge and polarizability of the molecule, respectively. E is the electric field and x is the displacement between the centers of negative and positive charge. Equation (1) and energy conservation give,

$$\frac{1}{2}m\dot{x}^2 = \frac{p^2}{\alpha}$$

(2)

where p is the molecular polarization. If the particle velocity, v, relates to the velocity in the equation (2) with proportional coefficient, b, the polarization density will be,

$$P = b\sqrt{\frac{1}{2}\rho\chi} \cdot v$$

$$(3)$$

where ρ is the density of the material, and χ is the polarizability density. Qualitatively, the equation (3) explains the relationship between particle velocity and polarization.

Implicit in this approach is the idea that the lighter electrons respond more rapidly to the shock than the nuclei or atomic cores. If this was the case, the induced polarization should align opposite to the direction in which the shock propagates. Our experiments show, however, that the polarization aligns along the propagation direction. The relaxation times we measured also are longer than those for electron polarization. At least in these respects, the molecular basis of shock induced polarization remains unknown.

EXPERIMENTAL RESULTS

The details of the sample cell, shock wave generation, and detection electronics are described elsewhere[1]. The initial phase of the MBBA was controlled by a heater surrounding the sample cell. MBBA (Tokyo KASEI) and KF(Wako chemical) were used as received.

Figure 1 shows the typical voltage signal obtained for the isotropic liquid phase of MBBA. At time T_1, the shock wave enters in MBBA and induced polarization rise rapidly, in less than 30ns. We interpret the rapid fall of the polarization signal

at time T_2 as the arrival of shock wave at the rear electrode. In contrast with the behavior of more conventional liquids[1], the EMF for liquid MBBA rises again at later times, as suggested in Figure 1. This rise is reproducible. We estimated the shock velocity from initial sample thickness and the interval, T_2-T_1, and calculated the particle velocity in the sample using the shock impedance matching method. The dotted line in Figure 1 shows the result of a calculation fitting these data to Allison's model[2]. From this calculation, we conclude that the shock induced polarization is 1.16×10^{-5} C/m^2 at 6.79GPa. The maximum EMF is around 50mV.

For nematic MBBA, the EMF begins to rise when the shock enters the sample but increases more slowly than for the isotropic phase, as Figure 2 shows.

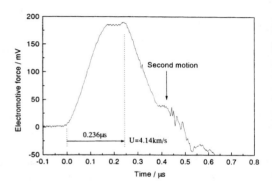

FIGURE 2. The EMF signal obtained for compression of the nematic phase of MBBA. The initial temperature and thickness of this sample were 300K and 1mm, respectively.

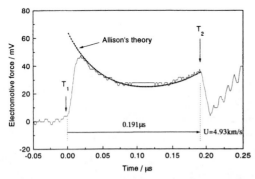

FIGURE 1. Typical voltage signal obtained during shock compression of the isotropic liquid phase of MBBA. The pressure was estimated to be 6.8GPa. The initial temperature and thickness of the sample were 300K and 1mm, respectively.

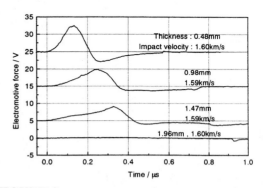

FIGURE 3. The EMF signal obtained for compression of solid MBBA initially at 271K.

After about 0.12 μ s, the EMF saturates at about 180mV and remains constant while the shock crosses the sample. Then, the EMF falls until a "second motion" signature appears.

When solid MBBA is shocked, the induced EMF also rises slowly; see Figure 3. For this phase, both the gradient of this EMF rise and the peak magnitude decrease with increasing sample thickness. No "second motion" EMF is detected for this phase. We associate the initial rise of the EMF for all three phases with the compressed region of the sample. That the rise times for the nematic and solid phases are slower than for the isotropic liquid suggest that the reorientational inertia is greater in these phases. The "second motion" response for the fluid phases of the liquid crystal-forming material may be a consequence of the larger rotational inertia of individual or groups of MBBA molecules, possibly in the uncompressed area in font of the shock, or of the phase transition at high pressues.

Measurements for pure water and aqueous solutions of KF reveal another influence on the shock induced EMF; see Figure 4 and 5. All of the water data were collected for pressures near 7GPa. The EMF signal for pure water is very small, and the maximum EMF increases to some extent with electrolyte concentration. For mole fractions greater than 2.0×10^{-3}, the initial EMF is negative at these pressures.

Although Allison's theory may not be appropriate for these solutions because of their large electrical conductivity, models for conductive samples are both

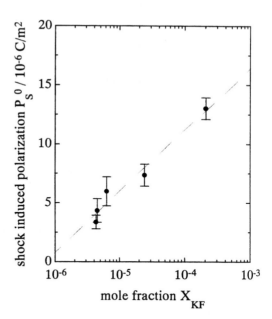

FIGURE 5. The relationship between the magnitude of the shock induced polarization and mole fraction of KF at pressures near 7GPa.

more complicated and hard to apply to extra meaningful numbers. Therefore, we used Allison's model to estimate the polarization shown in Figure 5.

If we consider the KF solutions in terms of an ionic atmosphere model, that is, each ion surrounded by nearby ions having the opposite charge, each region of the unshocked solution is uncharged and nonpolar. However, the passing shock may upset the local distribution of ions, possibly along lines used to develop Equation (3). The polarization observed for dilute solutions would result if the fluoride ions move more rapidly in response to the external fields. This also suggests that intermolecular as well as intramolecular interactions affect the induced EMF's.

However, more experiments are needed to develop a complete understanding of the mechanisms of the shock induced polarization.

ACKNOWLEDGMENT

The authors are grateful to T. Ogura who assist in performing the experiment.

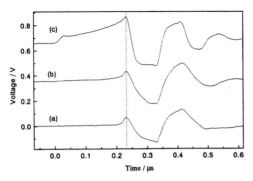

FIGURE 4. The shock induced EMF's for aqueous solutions of potassium fluoride: a, pure water; b, $X_{KF}=2.0 \times 10^{-5}$; c, $X_{KF}=2.0 \times 10^{-4}$.

REFERENCES

1. Kanchan Saxena, Yoichiro Hironaka, Hisako Hirai, and Ken-ichi Kondo., Appl. Phys. Lett. 68, 920-922 (1996)
2. F. E. Allison, J. Appl. Phys. 36. 7, 2111-2112 (1965)

CP429, *Shock Compression of Condensed Matter – 1997*
edited by Schmidt/Dandekar/Forbes
© 1998 The American Institute of Physics 1-56396-738-3/98/$15.00

A STUDY OF SEMICONDUCTOR-METAL TRANSITION IN SHOCKED MONOCRYSTAL SILICON

S. D. Gilev, and A. M. Trubachev

Lavrentyev Institute of Hydrodynamics, Lavrentyev Prosp. 15, Novosibirsk 630090, Russia

An experimental investigation of the electrical behavior of monocrystal silicon in the shock stress range up to $20\,GPa$ is performed. The conductivity measurement technique of a condensed matter is elaborated for the dielectric-metal transition in a shock wave. The technique permits one to decrease the current relaxation time to its limit, to improve the measurement accuracy and temporal resolution by one order of magnitude as compared to the known scheme. As the normal stress P_x rises, the silicon conductivity σ increases monotonously by over five orders of magnitude and reaches the value about $4 \cdot 10^4\,Ohm^{-1}cm^{-1}$. The $\sigma(P_x)$ dependence comprises two parts: the sharp increase and the "plateau". The dependence break is fixed at $P_x \approx 12\,GPa$. The metallic silicon conductivity is of an intrinsic nature, its state is equilibrium and highly defective.

INTRODUCTION

Shock compression behavior of silicon has been the subject of investigations over three decades. *Pavlovskii* [1], then *Gust and Royce* [2] researched the shock wave structure and the compressibility of monocrystal silicon. *Coleburn et al.* [3] performed the first electrical measurements in silicon. Based on a sharp decrease of the silicon resistance, the fact of silicon transition to the metallic state was established under stresses exceeding the Hugoniot Elastic Limit (HEL). Measurements of the silicon resistance in the elastic region were taken by *Rosenberg* [4] (stress P_x up to $3.4\,GPa$). In observing a multiwave structure *Goto et al.* [5] revised the monocrystal silicon Hugoniot. The phase transition stress was determined as $P_x = 13.4 \pm 0.2\,GPa$ independently of the crystal direction. It was shown that, with exceeded HEL, the monocrystal state was close to the isotropic one.

At present there is a violent discrepancy between the shock wave and static compression experiments concerning the metallic state of silicon. On the one hand, according to [3,4] silicon is metallized in a shock wave under stresses exceeding HEL ($5.6, 8.4\,GPa$ for different crystal directions). With further increasing a shock stress the silicon conductivity decreases considerably [3]. On the other hand, under static compression the transition to the metallic state is registered with confidence for pressure about $12.5\,GPa$ [6-8], the conductivity rises monotonously with increasing the pressure and reaches at $15\,GPa$ the conductivity for such metals as Ni, Zn [9,10]. Such apparent disagreement between the dynamic and static investigations gives rise to questions of the electrical measurements in a shock wave.

The object of the paper is the experimental investigation of the monocrystal silicon transition to the metallic state under shock compression. We are engaged in the electrical properties of the silicon metallic phase, the transition characteristics, the nature and mechanism of the shock-induced conductivity.

The problem is attracted considerable interest, on the one hand, for its fundamental important investigations of the matter metallization under high compression, on the other hand, for some applications of the shock-induced conductivity for governing the electromagnetic energy flows in the high-power energetic systems (see our review [11]).

MEASURING SCHEME

The problem of measuring the condensed matter conductivity under dielectric-metal transition in a shock wave is known in the early (19)60s. The problem is caused by the electromagnetic transients in the measurement circuit comprising a shunt and a specimen connected parallel to one another [12,13]. The current relaxation time of the measuring cell is about $\tau \approx L/R$, where L is the inductance of the shunt-specimen circuit, R is the circuit resistance. If the specimen conductivity is high, then the relaxation time τ is high too. For the metallic conductivity of matter, the time τ exceeds the lifetime of the high stress region in a shock wave. To investigate the matter metallization there is a need to decrease the inductance of the shunt-specimen circuit. In [14-16] the shunt was placed outside the shock wave zone and the measurement loop was rather large. The relaxation time in the circuit was considerably reduced by *Nabatov et al.* [17] by locating a shunt immediately adjacent to the specimen, in the shock wave zone. However this scheme prevents as before the recording of the conductivity conforming to classic metals.

A radical improvement of the time resolution of the shunt measuring cell can be attained by extremely approaching the shunt and the specimen: the shunt (made of thin metallic foil) is put on the specimen, so that the shunt will be contacted by the specimen over the entire surface (Fig. 1).

A plane shock wave propagates into the measuring cell from the top down. Electrodes are placed in the same plane as the foil, they are connected to a measuring cable outside the shock effect zone. The cell current is produced by an outer source and remains constant for a measuring time.

The distinctive features of the scheme are the following: 1) the scheme has the lowest inductance of the shunt-specimen circuit, 2) the electrical

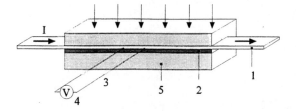

Figure 1. The measuring scheme of the matter conductivity in the dielectric-metal transition in a shock wave. The numbers denote: 1 = metallic foil, 2 = test matter (specimen), 3 = electrodes, 4 = measuring unit (oscilloscope), 5 = the dielectric.

contact is supported over the interface shunt-specimen, hence one can expect that the contact resistance will be as low as possible, 3) the shock wave direction is perpendicular to the current flow, therefore no polarization signals are available.

Specimen metallization causes redistribution of the current between the shunt and the specimen and decreases the voltage registered. To obtain the matter conductivity two cell states are used: the initial state (just before a shock action) and the final one (on compressing the specimen and on completing the transients). The time resolution of the cell is determined by the hydrodynamic and electromagnetic relaxations. The hydrodynamic relaxation is caused by stress smoothing through the shunt, the specimen, and the surrounding dielectric. It takes a characteristic time about $t_h \approx 2x_0/D$ (x_0 is the layer thickness, D is the shock velocity). The electromagnetic relaxation time is about $t_e \approx \mu_0 \sigma x_0^2$ (σ is the electroconductivity). To perform exact measurements the relaxation time has to be much smaller than that observed $t_h, t_e \ll t$. Therefore a limitation on the specimen thickness can be obtained as $x_0 \ll min \left\{ \sqrt{t/\mu_0\sigma}, \; tD/2 \right\}$. Considering as typical values $t \approx 1 \, \mu s$, $\sigma \approx 5 \cdot 10^5 \, Ohm^{-1}cm^{-1}$ (copper conductivity), $D \approx 5 \, Km/s$, one can derive that the layer thickness has to be much smaller than $100 \, \mu m$. Thus the high conductivity measurements in the scheme can be supplied for rather thin specimens. In this case there arises an uncertainty of the matter state caused both by multiple compression and by the boundary effect. The same restrictions take place in investigating the metal-metal transition in a shock wave by the standard thin foil technique [12].

The current in the measuring cell was excited by a capacity discharge in the LC-circuit controlling by a thyristor. The measurements were conducted at the current maximum (the current rise time was about $70\,\mu s$). The current (up to $700\,A$) was measured by a inductive gage. The constantan foil (100–$200\,\mu m$ thick, $10\,mm$ width) was used as a shunt. The shunt choice is due to small change of the constantan resistance in a shock wave. The silicon specimens were the plates $0.23 \div 0.41\,mm$ thick, $10\,mm$ $10\,mm$ width, and about $35\,mm$ in length. The specimen resistivity was $4 \div 7\,Ohm\,cm$ at the room temperature.

The experiments revealed a high effect of the current circuit and the loading system on the voltage recorded. The motion of a metallic flyer in a magnetic field of the shunt current produces the large eddy currents and the considerable stray are recorded by the oscilloscope. The stray magnitude depends on the mass velocity of the metallic plate relative to the shunt, system geometry and the conductivity of the shunt material.

EXPERIMENTAL RESULTS

Two voltage records taken in experiments with monocrystal silicon are shown on Fig. 2.

The conductivity is calculated by the formula

$$\sigma = \frac{h_s k}{h}\frac{1}{\rho_c}\left(\frac{V_0}{V}-1\right).$$

Here h_s, h is shunt and specimen thickness, respectively, k is the coefficient of silicon compressibility, ρ_c is the constantan resistivity, V_0 is the initial voltage, V is the voltage reached on completing the transients. The conductivity error in the experiments is about 10%, the time resolution is about $200\,ns$ at the maximum shock stress.

The experiments allow us to make some conclusions regarding the character and properties of the silicon metallic phase. The occurrence of the conductivity corresponds to the arrival of a shock wave into the specimen with a precision of about $100\,ns$. The impactor experiments show that after a sharp change the conductivity of compressed silicon

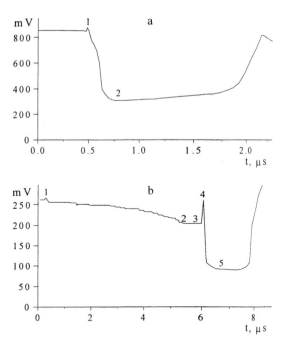

FIGURE 2. The voltage records of experiments on measuring the conductivity of monocrystal silicon. a - contact loading by the high explosive ($P_x = 14.8\,GPa$, $\sigma = (3.0 \pm 0.3)\cdot 10^4\,Ohm^{-1}cm^{-1}$). The numerical symbols denote: 1 = the shock wave arrival at the shunt, 2 = the silicon metallic state. b - loading by the metallic impactor ($P_x = 17\,GPa$, $\sigma = (3.5 \pm 0.2)\cdot 10^4\,Ohm^{-1}cm^{-1}$). 1 = the impactor begins to move, 2 = the impactor strikes on the dielectric, 3 = the shock wave arrival at the shunt, 4 = an occurrence of the silicon conductivity, 5 = the silicon metallic state.

remains constant for the time. This testifies that the state of shock compressed silicon is equilibrium one. A comparison of the experiments performed with the specimens of different impurity shows that the conductivity of the metallic silicon is of an intrinsic nature.

The stress dependence of the silicon conductivity $\sigma(P_x)$ is shown on Fig. 3. As P_x rises, the silicon conductivity σ increases monotonously by more than five orders of magnitude. One can note two parts on the $\sigma(P_x)$ dependence: the sharp growth and the "plateau". The break in the $\sigma(P_x)$ dependence is fixed at $P_x \approx 12\,GPa$. The first part of the dependence connects the initial conductivity

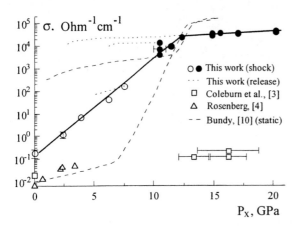

FIGURE 3. The dependence of silicon conductivity σ on shock stress P_x. Other available results are also presented. Data [3,4] were recalculated to the conductivity based on the specimen dimensions and the resistance.

value defined by silicon purity and the "plateau" conductivity corresponding to the metallic state of silicon. The maximum conductivity of silicon is about $4 \cdot 10^4 \ Ohm^{-1}cm^{-1}$ and appropriates to that of "poor" metals such as lead.

The general trend of the conductivity versus the stress in dynamic measurements is in remarkable agreement with the static measurements by *Bundy and Kasper* [10] (the monotonous rise with increasing the stress, the break at $12-13 \ GPa$).

Moreover, the metallic silicon conductivity for dynamic and static compression differs greatly (by 5 times for $P_x = 15 \ GPa$). The analysis shows that the temperature effect is small and the conductivity difference is mainly caused by the generation of the crystal structure defects in a shock wave. It turns out that the change of the silicon resistance due to the crystal structure defects exceeds analogous values for the classic metals [12,18] by more than order of magnitude. This suggests the highly defective state of the metallic silicon in a shock wave.

The investigation of the silicon electrical behavior in release wave points out the severe asymmetry of the forward and reverse transitions. Unloading silicon brings into the temporal conservation of its metallic state. Our observations testify to the metastability of the high stress phase for typical times of a shock wave experiment.

CONCLUSIONS

The technique presented let us solve the problem of recording the dielectric-metal transition in a shock wave, which was known even in the early (19)60-s. The present experiments are the first successful attempt to study a shock metallization process of silicon by directly measuring a matter conductivity. The metallic conductivity of shocked silicon turns out by order less than the copper conductivity, which does not confirm the results by *Coleburn et al.* [3]. The tendency for the conductivity versus the stress in dynamic conditions agrees with the static data. The direct measurements show the metallic state of shocked silicon is highly defective. This is indicative of some essential distinctions between the deformation mechanism of silicon and metals in a shock wave.

REFERENCES

1. Pavlovskii, M. N., *Sov. Phys.-Solid State* **9**, 2514 (1968).
2. Gust, W. H., and Royce, E. B., *J. Appl. Phys.* **42**, 1897-1905 (1971).
3. Coleburn, N. L., Forbes, J. W., and Jones, H. D. *J. Appl. Phys.* **43**, 5007-5012 (1972).
4. Rosenberg, G., *J. Phys. Chem. Solids* **41**, 561-567 (1980).
5. Goto, T., Sato, T., and Syono, Y., *Jap. J. Appl. Phys.* **21**, L369-L371 (1982).
6. Weinstein, B. A., and Piermarini, G. J., *Phys. Rev. B.* **12**, 1172-1186 (1975).
7. Werner, A., Sanjurjo, J. A., and Cardona, M., *Sol. State Commun.* **44**, 155-158 (1982).
8. Hu, J. Z., and Spain, I. L., *Sol. State Commun.* **51**, 263-266 (1984).
9. Bundy, F. P., *J Chem. Phys.* **41**, 3809-3814 (1964).
10. Bundy, F. P., and Kasper, J. S., *High Temp.-High Press.* **2**, 429-436 (1970).
11. Gilev, S. D., and Trubachev, A. M., *Shock Compression of Condensed Matter - 1995*, Woodbury, New York: AIP Press. 1996, pp. 933-936.
12. Keeler, R. N., and Royce, E. B., *Physics of High Energy Density*, New York, London: Academic Press, 1971, pp. 106-125.
13. Yakushev, V. V., *Combustion, Explos. Shock Waves* **14**, 131-146 (1978).
14. Brish, A. A., Tarasov, M. S., and Tsukerman, V. A., *Sov. Phys. JETP* **11**, 15 (1960).
15. Kuleshova, L. V., *Sov. Phys.-Solid State* **11**, 886 (1969).
16. Mitchell, A. C., and Nellis, W. J., *J. Chem. Phys.* **76**, 6273-6281 (1982).
17. Nabatov, S. S., Dremin, A. N., Postnov, V. I., and Yakushev, V. V., *Sov. Tech. Phys. Lett.* **5** (1979).
18. Dick, J. J., and Styris, D. L. *J. Appl. Phys.* **46**, 1602-1617 (1975).

CP429, *Shock Compression of Condensed Matter – 1997*
edited by Schmidt/Dandekar/Forbes
© 1998 The American Institute of Physics 1-56396-738-3/98/$15.00

DYNAMIC ELECTROMECHANICAL CHARACTERIZATION OF THE FERROELECTRIC CERAMIC PZT 95/5

R. E. Setchell, L. C. Chhabildas, M. D. Furnish, S. T. Montgomery, and G. T. Holman

Sandia National Laboratories, Albuquerque, NM, 87185

Shock-induced depoling of the ferroelectric ceramic PZT 95/5 has been utilized in pulsed power applications for many years. Recently, new design and certification requirements have generated a strong interest in numerically simulating the operation of pulsed power devices. Because of a scarcity of relevant experimental data obtained within the past twenty years, we have initiated an extensive experimental study of the dynamic behavior of this material in support of simulation efforts. The experiments performed to date have been limited to examining the behavior of unpoled material. Samples of PZT 95/5 have been shocked to axial stresses from 0.5 to 5.0 GPa in planar impact experiments. Impact face conditions have been recorded using PVDF stress gauges, and transmitted wave profiles have been recorded either at window interfaces or at a free surface using laser interferometry (VISAR). The results significantly extend the stresses examined in prior studies of unpoled material, and ensure that a comprehensive experimental characterization of the mechanical behavior under shock loading is available for continuing development of PZT 95/5 material models.

INTRODUCTION

Many shock-activated pulsed power supplies have utilized a lead zirconate titanate ceramic having a Zr:Ti ratio of 95:5 and modified with 2% niobium, subsequently referred to as PZT 95/5. The nominal state of this material is ferroelectric (FE), but it is near an antiferroelectric (AFE) phase boundary. A remanent polarization can be produced by electrical poling, and the bound charge can be liberated into an external circuit by shock compression into the AFE phase. The poled ceramic has a complex dynamic behavior, with nonlinear coupling between mechanical and electrical variables. The electrical response of this material under shock loading was examined in some detail nearly twenty years ago (1) when pulsed power sources were under development. Interest in understanding the complex behavior of PZT 95/5 has been renewed recently due to new design and certification requirements. In particular, sufficient under-standing of this behavior must be established so that these requirements can be addressed through numerical simulations. Because few relevant studies have been performed during the past twenty years, we have initiated an extensive experimental study to improve our understanding of PZT 95/5 and to provide well-characterized data for assessing material models under development.

The first phase of this study addresses the mechanical behavior of unpoled PZT 95/5. In an early study, Doran (2) examined a similar material using explosively driven shock waves. Using material at densities of 7.67-7.89 g/cm^3, he found an obvious cusp in the Hugoniot curve at a pressure near 4 GPa, and a weaker cusp at a pressure near 0.2 GPa. He suggested that the strong cusp could be the Hugoniot elastic limit, and the weaker cusp the onset of the FE to AFE phase transition. Using planar-impact techniques, Dick and Vorthman (3) measured impact-face conditions and recorded transmitted wave profiles in unpoled material at densities from 7.29 to 7.37 g/cm³. They also measured electrical response and transmitted wave profiles in poled material. At a fixed peak stress of approximately 1.6 GPa, the transmitted wave profiles were strongly ramped with a weak two-wave structure. A more extensive study of shock wave propagation in unpoled material was performed some years later by Chhabildas (4). Using PZT 95/5 material at a density of 7.27-7.34 g/cm³, he recorded transmitted wave profiles under both uniaxial-strain and pressure-shear loading conditions at pressures from 0.9 to 4.6 GPa. Above 2.6 GPa a three-wave structure was observed due to the onset of pore compaction in the lower density material.

In the present study we have used unpoled PZT 95/5 samples having densities between 7.28 and 7.32 g/cm³. Planar impact conditions have generated axial stresses from 0.5 to 5.0 GPa, with particular conditions chosen primarily to cover ranges not addressed in previous studies.

EXPERIMENTAL CONFIGURATION

The general experimental configuration used in the present study is shown in Fig. 1. A 63.5-mm diameter, compressed-gas gun is used to conduct the planar-impact experiments. In most experiments a 3.2-mm thick, fused-silica impactor is

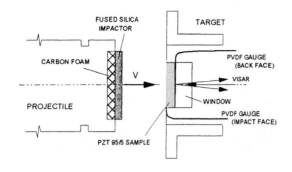

Figure 1. Experimental configuration used in the present study.

mounted on a layer of low-density (0.2 g/cm³) carbon foam to provide loading to a steady shock

condition followed after 1.1 μs by a release wave. A PVDF gauge package consisting of the 0.025-mm PVDF film and a 0.025-mm film of insulating Teflon is bonded to the front surface of the PZT 95/5 sample at the impact surface. The rear surface of the PZT 95/5 sample is bonded to either a fused silica or sapphire window with a diffusively reflecting surface at the interface. Laser interferometry (VISAR) is used to obtain particle velocity histories at this interface. On one experiment a second PVDF gauge package was included at the window interface.

CHARACTERISTIC LOADING RECORDS

In one experiment a thick fused-silica impactor was used to produce sustained shock loading without release, and a second PVDF gauge was positioned at the window interface to obtain a transmitted wave profile for comparison with the VISAR records. The resulting profiles are shown in Fig. 2. The transmitted wave profiles show the characteristic three-wave structure observed by Chhabildas (4). The PVDF gauge at the impact surface showed a transient rise during the first two microseconds. As will be seen in a subsequent

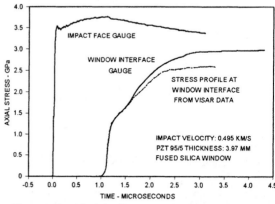

Figure 2. Combined data from a sustained-shock experiment.

figure, all of the impact gauge records showed a slow rise following impact. In the only previous study that examined impact conditions (3), a steady stress state was reported at 1.6 GPa levels. A possible explanation for the present observations is an anomalous gauge response that has been observed in impact-face gauges having the same configuration of electrical leads as in the present experiments (5). In the VISAR profile shown in Fig. 2, particle velocity has been converted to stress using the fused silica Hugoniot curve (6). This profile shows good agreement with the PVDF gauge at the window interface for the first 0.7 μs after wave arrival, then deviates progressively.

Figure 3 shows the charge histories obtained from PVDF gauges at the impact surface for three experiments with identical target and projectile assemblies but different impact velocities. The curves represent the time integral of the recorded current histories. In addition to the slow rise after impact, all of these records show a release to non-zero levels. Since the impact face stress releases to zero under these experimental conditions, the remanent charge levels indicate that charge has been added or lost. This suggests that the 0.025-mm layer of insulating Teflon between the gauge and the piezoelectric PZT 95/5 sample may not have been sufficient.

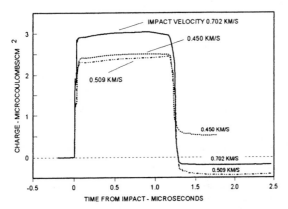

Figure 3. Charge histories from PVDF gauges at the impact face.

Figure 4 shows transmitted wave profiles recorded with VISAR at the window interface for the same three experiments as in Fig. 3. A small ramp at the start of each profile represents the behavior of the FE phase. The ramping behavior corresponds to compressibility increasing with increasing pressure. The front of this feature propagates at the longitudinal acoustic speed, which was found to be

4.16 ± 0.01 km/s in these samples. The corresponding transit time determined from the PVDF gauges shown in Fig. 2 was also 4.16 km/s. The low-amplitude ramp is followed by a fairly steep rise reflecting the transition to the AFE phase. The final wave feature represents relatively slow pore-compaction processes.

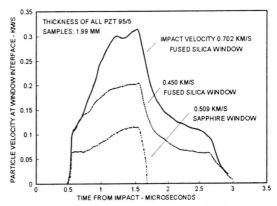

Figure 4. Particle velocity histories at the window interface.

Figure 5 shows the results of an approximate method that uses VISAR records to calculate the stress-strain paths followed by the PZT 95/5 samples during loading and unloading (7).

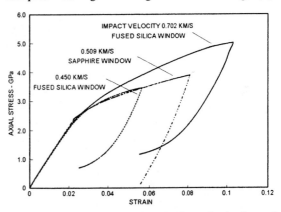

Figure 5. Approximate stress-strain loading and unloading paths calculated from the profiles shown in Fig. 4.

Although the assumptions required for this analysis (8) are not well met in this material, the curves in Fig. 5 are very consistent up to the onset of pore compaction at approximately 2.5 GPa. Well above this level the stress-strain path depends upon the stress at impact. The strong deviation of the release paths from the loading paths results from several factors, including the irreversible pore compaction and differences in the reverse phase-transition kinetics.

HUGONIOT STATES

The approximate final states reached during loading (Fig. 5) can be plotted and compared with similar end states calculated by Chhabildas (4) and impact-face conditions measured by Dick and Vorthman (3). The comparison is quite good, as shown in Fig. 6. Also shown in this figure are the

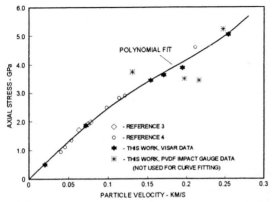

Figure 6. Hugoniot measurements and end states calculated from transmitted wave profiles.

initial impact states indicated by the PVDF gauges, which appear biased as could be expected (Fig. 3). The combined points (except for the PVDF gauge results) were fitted with a fourth-order polynomial using a least-squares routine. The coefficient values are: $C_1 = 26.86$, $C_2 = 5.271$, $C_3 = -363.8$, and $C_4 = 930.6$, where velocity is in km/s and axial stress is in GPa.

SUMMARY

The present work represents the first phase of extensive experimental characterizations of the dynamic electromechanical response of the ferroelectric ceramic PZT 95/5. The use of PVDF

gauges to record impact face conditions was limited by an apparent gain or loss in gauge charge, probably due to inadequate insulation. VISAR measurements of transmitted wave profiles were made for various impact conditions and sample thicknesses, showing the multi-wave loading structure and the release behavior. Approximate end states were calculated from these profiles. The current results confirm the material behavior described in previous studies and provide a continuous characterization of this behavior from 0.5 to 5.0 GPa.

ACKNOWLEDGMENTS

The authors would like to thank David E. Cox for his skillful efforts in preparing and conducting the experiments, and Mark U. Anderson for his advice and assistance in those efforts. Sandia is a multiprogram laboratory operated by Sandia Corporation, a Lockheed Martin Company, for the United States Department of Energy under Contract DE-AC04-94AL85000.

REFERENCES

1. For example, see: Lysne, P. C. J. Appl. Phys. **48**, 1020-1023 (1977).

2. Doran, D. G., J. Appl. Phys. **39**, 40-47 (1968).

3. Dick, J. J., and Vorthman, J. E., J. Appl. Phys. **49**, 2494-2498 (1978).

4. Chhabildas, L. C., "Dynamic Shock Studies of PZT 95/5 Ferroelectric Ceramic," SAND84-1729, Sandia National Laboratories, Albuquerque, NM (Dec 1984).

5. Bauer, F., Moulard, H., and Graham, R. A., "Piezoelectric Response of Ferroelectric Polymers Under Shock Loading: Nanosecond Piezoelectric PVDF Gauge," in Shock Compression of Condensed Matter - 1995, eds. S. C. Schmidt and W. C. Tao, New York: AIP Press, AIP Conference Proceedings 370, 1996, pp. 1073-1076.

6. Barker, L. M., and Hollenbach, R. E., J. Appl. Phys. **41**, 4208-4226 (1970).

7. Grady, D. E., and Young, E. G., "Evaluation of Constitutive Properties from Velocity Interferometer Data," SAND75-0650, Sandia National Laboratories, Albuquerque, NM (Aug 1976).

8. Fowles, R., and Williams, R. F., J. Appl. Phys. **41**, 360-363 (1970).

CP429, *Shock Compression of Condensed Matter – 1997*
edited by Schmidt/Dandekar/Forbes
© 1998 The American Institute of Physics 1-56396-738-3/98/$15.00

ELECTRICAL CONDUCTIVITY OF SHOCK-COMPRESSED PVDF FILMS.

V.V.Yakushev, T.I.Yakusheva

Institute of Chemical Physics in Chernogolovka RAS, Moscow Region, 142432, Russia

Time-dependent conductivity of unpolarized PVDF films has been studied over a wide range of shock pressures. Two types of PVDF films were investigated. They were a 30-micron uni-axially-stretched modified PVDF film (4% tetrafluoroethylene) and a 25-micron bi-axially-stretched pure PVDF film. Shock wave experiments were performed on two conductivity cell configurations. The first represented four layers of the film clamped between two metal electrodes in the geometry of a parallel-plate capacitor. The second was a gauge arrangement with a pattern similar to that used by Bauer of ISL, France, mounted in Teflon for the tests. Observations were made for 1 to 2 microseconds after shock arrival. The onset of electrical conductivity was detected at 10 GPa for the modified film, whereas, this event was detected at approximately 28 GPa for the pure film.

INTRODUCTION

Ferroelectric poly (vinylidene fluoride) (PVDF) and vinylidene fluoride copolymers are promising in dynamic pressure gauges applications [1-7]. The development of the gauges is generating an interest in the measurement of the shock-induced conductivity of these materials.

As far as we know there is only indirect evidence [6] that below 40 GPa the resistance of the shock loaded PVDF gauge (the 23 μm-thick Kureha film providing a 4 mm^2 active area) remains very high compared to the 50 Ω level. Only at 60 GPa the resistance of the gauge drops to about 100 Ω.

This paper describes the experiments aimed at a better understanding of the electrical phenomena taking place in the gauges during the process of shock compression.

The objective of this investigation was: First, to study electrical conductivity of the polymers under shock compression; Second, to measure the time dependent electrical resistance between electrodes of unpolarized sensitive elements of the pressure gauges; Third, to study shock-induced electrical signals in unpolarized samples of the polymers.

EXPERIMENTS

Sample materials

Two types of polymer films were investigated. They were a 30-micron uni-axially-stretched modified PVDF (m-PVDF) film made of co-polymer vinylidene fluoride with 4% tetrafluoroethylene supplied by Plastpolymer Okhta Research and Production Association of Russia [8] and a 25-micron bi-axially-stretched pure PVDF film (p-PVDF) from Solvay, Belgium.

The experiment design

Uniaxial shock compression of the samples in the 5 - 30 GPa pressure range was achieved by impact of explosively accelerated planar aluminum or stainless-steel flyers 50 or 90 mm diameter and 3 to 7 mm thick. The velocities of the impactors which can be obtained have been tabulated for various explosive systems. In specific cases, a 5Ω manganin gauge was included in the experimental set-up to record the loading history. The pressure profiles were also computed with a 1-D Lagrangian code.

The resistance was measured with two longitudinal conductivity cell configurations shown schematically in Fig. 1. The cells were mounted on an electrically grounded metal driver plate 120 mm diameter and 2 mm thick.

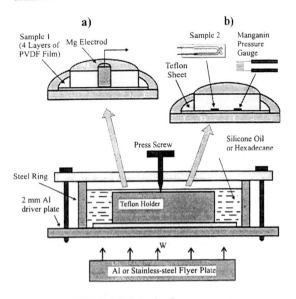

FIGURE 1. Schematic of measurements.

In the first configuration (a), four layers of the film under investigation were clamped between the driver plate and a magnesium backing electrode (6 mm in diameter) in the geometry of a parallel-plate capacitor. The face of the driver plate toward the film was mirror polished.

In the second configuration (b) sample arrangements with 5-mm² active areas and patterns similar to those used by Bauer of ISL, France, were used. The non-polarized films were electroded by thermal vacuum deposition of Cr/Cu combination. A chromium layer (0.02 - 0.03 μm thick) yielded a good electrode adhesion with the films. A copper layer (1.0 - 1.2 μm) provided sufficiently low resistance of the electrodes (< 0.5 Ω). The electrical insulation of the samples from the driver plate was provided by a Teflon sheet 0.28 mm thick.

When assembling the samples were clamped between the sheet and a rectangular-shaped Teflon

holder of about 40x40 mm² area and 6 - 8 mm thick with a screw. The assembly was immersed in a bath with a silicone oil or hexadecane for an electrical insulation of leads and eliminating of air gaps. It has been shown that the liquids remain an insulator under the experimental conditions.

The most obvious advantage of the second configuration is that the results can be directly extended to the pressure gauges. However there are two limitations in this approach. The shock wave reflections from the copper electrodes, probably, are not negligible and can contribute to the resistance measured. We should not exclude also the influence of the over-film defects.

The first cell configuration is more suitable for measuring volume conductivity of the films. On the one hand magnesium is a relatively good shock impedance match to Teflon and PVDF. On the other hand, if the conductivity results from the film defects or inclusions and their concentration is not too high, a multilayer sample allows to localize their influence within one film layer.

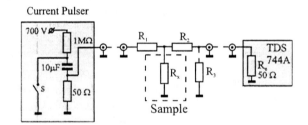

FIGURE 2. Schematic of the measurement curquit.

The circuitry covered a large resistance range is shown schematically in Fig. 2. A 700 V current pulser energized the samples several tens μsec before the arrival of the pressure pulse. Resistors R_1, R_2 and R_3 were mounted on the experimental assembly and destroyed at each test.

An electronic recording system consisted of a 1 GHz bandwidth Tektronix TDS 744A digitizer connected to coaxial transmission cables 8 meters in length.

The shunt resistor R_3 was used only to measure the resistance when it was less than 50 Ω.

The sample resistance was calculated from the equation: $R_x = \dfrac{U_x}{U_0 - U_x}\left(\dfrac{1}{R_1} + \dfrac{1}{R_{eq}}\right)^{-1}$ where U_x is the voltage across R_e; U_0 is the same voltage if $R_x = \infty$; $R_{eq} = R_2 + R_3 R_e / (R_3 + R_e)$. The electrical conductivity, Σ, at a given time was then calculated using measured R_x, the electrode area of the sample, and the sample thickness.

RESULTS AND DISCUSSIONS

p-PVDF in the first cell configuration

Figure 3a shows the shock-induced electrical signal recorded for the p-PVDF using the cell (a) at the zero applied voltage. The first positive pulse on the left side of the trace is the usual shock-induced polarization (SIP) current which occurs when a strong shock enters a polar polymer [9].

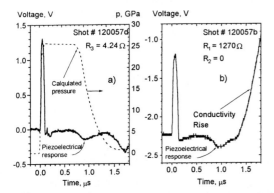

FIGURE 3. (a) Shock-induced electrical signal observed for p-PVDF with no applied voltage. (b) Resistance measurement under the same conditions. Time = 0 is the moment at which the shock front enters sample.

The integral of the current over the pulse duration is the remanent polarization accumulated by the sample. In this shot, the polarization was 0.075 $\mu C/cm^2$. As the result the sample becomes a piezoelectric and generates the negative voltage pulse at unloading which occurs on the time moment $t \cong 0.7$ μsec.

Figure 3b demonstrates a drop in resistivity associated with the ensuing relief wave. In this connection it should be noted that the drop is similar to some observations of Kuleshova [10] and Champion [11] on Teflon. Therefore we can not exclude the possibility that the drop is a result of the Teflon conductivity. We are planning additional experiments to clear up the question.

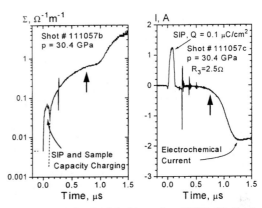

FIGURE 4. (a) Conductivity history for p-PVDF. (b) Shock-induced electrical signal observed with no applied voltage. The arrows correspond to entrance of the relief wave into the samples.

The results of measurements at 30.4 GPa shock pressure presented in Fig. 4a shows that the conductivity of the p-PVDF grows at constant pressure as well as at unloading. An interesting phenomenon can be seen in Fig. 4b. In the experiment a galvanic cell with shock-compressed conducting PVDF or its destruction products as an electrolyte was formed. The cathode for this cell was the stainless-steel driver plate (Fig. 1a) and the anode was the magnesium electrode. The negative current on the right side of the trace is the output signal of the cell. This phenomenon has been studied in liquid dielectrics [12] and testifies that the shock-induced conductivity of a substance under investigation (PVDF in this context) is of an ionic nature.

A total of six tests were conducted for the p-PVDF with the circuit of Fig. 1a between 24 and 30.5 GPa. From the data obtained it might be inferred that the conductivity shows a marked and reproducible increase for shock pressure more than approximately 28 GPa.

PVDF films in the second cell configuration

Unfortunately a major part of the experiments for the p-PVDF in this configuration were not reproducible, especially at the pressures more than 10 - 15 GPa. It is evident, for example, from a comparison of the voltage histories shown in Fig. 5.

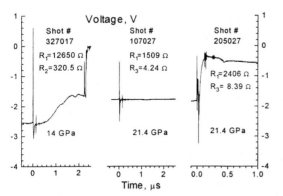

FIGURE 5. Test results

The experimental results for the m-PVDF are summarized in Fig. 6 as the conductivity histories at different shock pressures. In contrast to the p-PVDF, in this case both experimental arrangements of Fig. 1 provide nearly the same results.

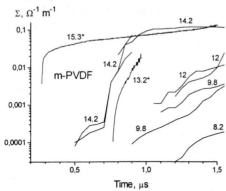

FIGURE 6. Summary of conductivity histories for the m-PVDF at different pressures. The numbers near the curves indicate the pressures. The experiments of Fig 1a are indicated by the asterisks.

Figure 6 shows in particular that the conductivity increases with time at a constant pressure. It seems likely that it is accounted for by development of a chemical reaction in the compressed and heated sample.

To get additional information, the microstucture of the films has been investigated using both an optical and a scanning electron microscopes. The study demonstrated that there are many non-identified inclusions in the form of grains in the p-PVDF film in contrast with the m-PVDF film which looks more homogeneous. The grains manifest themselves as closed near-circular lines on the film surface. Their characteristic size ranges up to a few tens of microns. Most likely the grains have a different phase constitution than the basic component of the film. Probably the complicated structure of the p-PVDF films is responsible for the low reproducibility of the conductivity experiments with the single film samples.

ACKNOWLEDGMENTS

The authors appreciate the continuous interest and support of the work by Dr. P.A.Urtiew of LLNL and Dr. J.A. Charest, president of DYNASEN Inc. We would like to thank Dr. J.A. Charest also for the p-PVDF samples. This work was supported by the Russian Foundation for Basic Research under grant №. 97-02-17575

REFERENCES

1. Bauer F., in: *Shock Waves in Condensed Matter - 1981*, 1982, pp. 251-266.
2. Urtiew P.A. and Erickson L.M., in: Techniques and Theory of Stress Measurements in Shock Wave Applications, *Proc. of the summer ASME Meeting,* **AMD- 83**, N.Y., 1987, pp. 29-35.
3. Fogelson D.J., Lee L.M., Gilbert D.W., Conley W.R., Graham R.A., Reed R.P., ibid, pp. 615-618.
4. Charest J.A., and Lynch C.S. *in Shock Compression of Condensed Matter - 1991*, 1992, p.897.
5. Bauer F., Graham R.A., Anderson M.U., Lefebvre H., Lee L.M.and Reed R.P., ibid, pp. .887-890.
6. Chartagnac P., Decaso P., Jimenez B., Bouchu M., Cavailler C. and Delaval J., ibid, pp. 893-896.
7. Urtiew P.A. *Himicheskaya Fizika,* **12**, 579-601 (1993).
8. Sherman M.Ya., Lesnykh O.D., Vlader N.B., Artem'ev B.A., Myasnikov G.D., Lobanov A.M., Zolotova V.I. Plastmassi, № 10, 46-48 (1990).
9. Hauver G.E., *J.Appl. Phys.*, **36**, 2113-2118 (1965).
10. Kuleshova L.V. , *Sov. Phys. Solid State,* **11**, 886 (1969).
11. Champion A.R. , *J.Appl. Phys.* **43**, 2216-2220 (1972).
12. Dremin A.N. and Yakushev V.V. in *Proc. Fifth Symp. (Internat.) on Detonation - 1970*, 1972, pp. 399-402.

CP429, *Shock Compression of Condensed Matter – 1997*
edited by Schmidt/Dandekar/Forbes
© 1998 The American Institute of Physics 1-56396-738-3/98/$15.00

ADVANCES IN FERROELECTRIC POLYMERS FOR SHOCK COMPRESSION SENSORS

F. Bauer*, H. Moulard*, G. Samara**

* *Institut Franco-Allemand de Recherches, (ISL), Saint-Louis, France*
** *Sandia National Laboratories, Albuquerque, NM 87185 USA*

Our studies of the shock compression response of PVDF polymer are continuing in order to understand the physical properties under shock loading and to develop high fidelity, reproducible, time-resolved dynamic stress gauges. New PVDF technology, new electrode configurations and piezoelectric analysis have resulted in enhanced precision gauges. Our new standard gauges have a precision of better than 1% in electrical charge release under shock up to 15 GPa. The piezoelectric response of shock compressed PVDF gauges 1 mm^2 in active area has been studied and yielded well-behaved reproducible data up to 20 GPa. Analysis of the response of these gauges in the " thin mode regime " using a Lagrangian hydrocode will be presented. P(VDF-TrFE) copolymers exhibit unique piezoelectric properties over a wide range of temperature depending on the composition. Their properties and phase transitions are being investigated. Emphasis of the presentation will be on key results and implications.

INTRODUCTION

Although ferroelectricity and piezoelectricity in Polyvinylidene Fluoride were discovered by Kawai (1) in 1969 and subsequently confirmed by Kepler (2), the materials commercially available did not exhibit reproducible properties due to the critical importance of mechanical and electrical processing history. Subsequently, the mechanical and electrical processes needed to achieve precisely known and reproducible electrical properties were developed (3).

The availability of reproducible samples provides an opportunity to study the materials and to describe the piezoelectric response of PVDF under the destructive, very high pressure conditions achieved in controlled shock loading (4). Previous work (3) has shown that selected and

precisely poled standard 9 mm^2 PVDF gauges could respond precisely to pressures of 25 GPa. For 1 mm^2 gauges, at pressures up to 12 GPa, well-defined signals are observed. But the usefulness of 1 mm^2 PVDF gauges was hampered by observations of differences in responses even at low shock pressure.

The need to develop high fidelity, reproducible time-resolved dynamic stress gauges especially with small active areas, has led us to continue these studies of shock compression response of PVDF polymers, and to understand the physical properties under shock loading. The present paper will briefly summarize the advances in new poling technology, as well as the development of high fidelity, reproducible, time-resolved dynamic stress gauges with small areas. The shock induced polarization for 1 and 9 mm^2 precisely poled gauges are compared. The analysis of the response of these gauges and of standard PVDF gauges in the " thin mode regime " using a Lagrangian hydrocode has been revisited. Effects of pressure on the dielectric properties of P(VDF-TrFE) copolymers have been also investigated and will be summarized.

PROGRESS IN TECHNOLOGY

In order to enhance the precision of the gauges and to avoid some deviations in the gauge response as reported before (3), we have developed new poling equipment. This equipment allows us to adjust, in real time via a high voltage and data acquisition computer controlled system, the predetermined remanent polarization as well as the maximum displacement current measured at the coercive field to an individual sample. Further, with appropriate attention to the history of each sample, space charge in the samples can be eliminated (3), (4). A higher degree of reproducibility is achieved when the maximum displacement current at the coercive field is stabilized. Each sample fabricated in this process is characterized with an individual poling history with well-defined electrical properties, better than 2 %, which can be reproduced at will.

HIGH PRESSURE APPARATUS: PRECISE IMPACT LOADING

In the impact experiment the symmetry conditions for identical impactor and sample materials require that precisely one-half the velocity at impact be imparted to the sample. In the electrical measurement circuit the expensive CVR is replaced by a CMS resistance. Sample response is determined by recording the short-circuited current during the time the shock waves are reverberating within the samples until mechanical equilibrium is achieved corresponding to the longitudinal stress in the standard material. The electrical charge is determined by numerical integration of the recorded current. A range of impact velocities from 0.25 km/s to 1.8 km/s are achieved with a powder gun which accelerates the

projectile to a preselected velocity. PVDF samples are placed on the impact surface of either z-cut quartz, sapphire crystals or selected copper which serve as the standard materials to define the stress. Typical times to achieve equilibrium in the 25 micron films are 50 to 150 nanoseconds

PIEZOELECTRIC POLARIZATION AT PRESSURE

Figure 1 shows the results of experiments to peak pressures of about 12 GPa for both 1mm^2 and 9 mm^2 PVDF gauges. Data are shown for standard materials as well as those with the copper, the fused quartz ramp loading or KelF impactors. Data for 1mm^2 PVDF including " Barker pillow " isentropic loading (where the stress is measured simultaneously with a 9 mm^2 PVDF) are also shown on the same Figure 1.

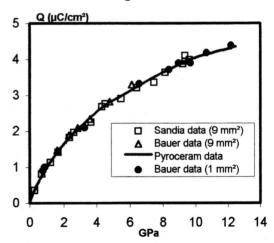

FIGURE 1. Observed charge vs stress data for the various loading paths show that the final charge is independent of path. At low stress the continuous response data is compared to shock response (3), (4).

The deviation of experimental points about their representative values is thought to be within the experimental uncertainty of the recording instruments and less than 1% in electrical charge release. The behavior indicates a strongly non-linear character. The shock induced polarization for both precisely poled gauges under shock loading is observed to be identical. The experimental results are in excellent agreement with the Graham-Sandia data (4), and recalled here.

LAGRANGIAN ANALYSIS

Even though the piezoelectric polarization is observed to be continuously nonlinear with stress (3), the observed polarization with the computed true strain of PVDF is non linear in the true strain values region (0 - 0.03), and appears to be approximately linear in the range (0.03 - 0.45). Figure 2 gives the electrical charge versus true strain for both ISL and Sandia data (3)

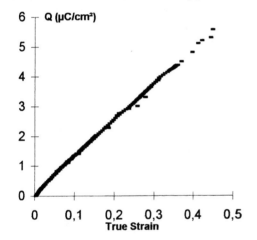

FIGURE 2. Electrical charge vs the computed true strain ε.

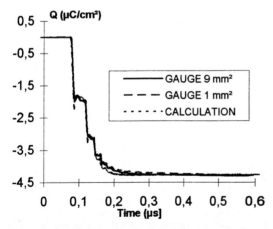

FIGURE 3. Comparison between experimental and computed data.

In introducing (5) the experimental (Q-ε) relation into the hydrocode DYNA or SHYLAC, we can compute the theoretical current or charge profile for a given test. Figure 3 shows the computation-experiment comparisons for both 1 mm^2 and 9 mm^2 PVDF gauges in the same symmetric copper impact test (shock pressure : 12 GPa). The calculation allows us to determine the stress too. We observe that the experimental and computed charge versus stress are in very good agreement with the published data (0 - 10 GPa) of R. A. Graham (4) on our PVDF gauges, Figure 4.

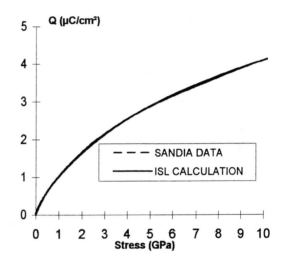

FIGURE 4. Experimental and computed charge vs stress.

P (VDF$_{1-X}$-TRFE$_X$) COPOLYMERS

P(VDF-TrFE) copolymers (6) exhibit tailorable ferroelectric, piezoelectric and structural properties that may be superior to those of PVDF for some shock gauge applications. Consequently, we have been investigating these properties as functions of static and dynamic pressure.

The phase diagram for P(VDF$_{0.77}$ TrFE$_{0.23}$) was determined from dielectric spectroscopy measurements as functions of temperature, hydrostatic pressure and frequency following established procedures (6). The important features in the phase space for PVDF and its copolymers are (with increasing temperature) a prominent molecular relaxation process centered around

T_β=270K and the melting transition (T_m). In addition the copolymers exhibit a ferroelectric transition (T_C) below T_m. All of these transitions have strong influences on the electrical and mechanical responses of these polymers. The transition at T_β, T_C, and T_m are well-defined features in the real (ε') and imaginary (ε'') parts of the dielectric response, Figure 5. All of these features shift to higher temperatures with increasing pressure, as shown.

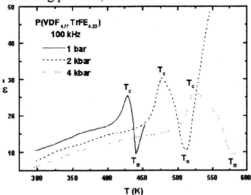

FIGURE 5. Real part of the dielectric constant vs temperature showing the large influence of pressure on the various phase transitions in the copolymer.

In Figure 6, the initial slopes of the phase boundaries dT_x/dP for T_β, T_C and T_m are 11±1 (independent of frequency), 24±1 and 41±1 K/kbar, respectively. The slope dT_β /dP is the same as for PVDF. The slopes dT_C/dP and dT_m/dP, exhibit strong dependence on composition (6). Specifically and for comparison, dT_C/dP = 30±2K/kbar for P(VDF$_{0.70}$ TrFE$_{0.30}$) and dT_m/dP=29±2K/kbar for PVDF and 53.4K/kbar for P(VDF$_{0.70}$ TrFE$_{0.30}$). The melting curve of PVDF is the dashed curve, Fig. 6.

CONCLUSION

Advances in poling process of 1 mm^2 and 9 mm^2 PVDF gauges have been achieved. The shock induced polarization for both precisely poled gauges sustained to shock loading greater than 12 GPa is observed to be identical. Computed data via numerical analysis are in good agreement with the experimental and published data. It is clear (6) that pressure strongly stabilizes the ferroelectric phase.

The electrical output of the gauge is determined solely by the piezoelectric response, i.e., there is essentially no domain switching. The much larger $T_m(P)$ slope suggests that the copolymer may have advantages over PVDF for high pressure (>100kbar) applications. Shock experiments should be designed to avoid or take the strong relaxational response into consideration.

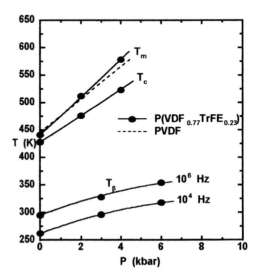

FIGURE 6. The Temperature Pressure phase diagram for P(VDF$_{0.77}$TrFE$_{0.23}$).

ACKNOWLEDGMENTS

This work in ISL was supported by the DRET under Contract and in Sandia by the United States Department of Energy under Contract DE-AC04-94AL85000. We express our appreciation to I.S.L. and Sandia staffs.

REFERENCES

1. H. Kawai, *J. Appl. Phys.* **8**, pp 975, (1969).
2. R.G. Kepler, *Ann. Rev. Phys. Chem.* **29**, 497 (1978).
3. F. Bauer and R.A. Graham, *Ferroelectrics*, **171**, 95 (1995) and references therein.
4. R.A. Graham, *Solids under High Pressure Shock Compression*, New York: Springer Verlag, 1993, pp 103-113.
5. H.Moulard, F.Bauer, *Proceedings of Shock Waves in Condensed Matter*,North-Holland, pp 1065-1068, 1995
6. G. A. Samara and F. Bauer, *Ferroelectrics*, **135**, 385 (1992) and references therein.

CP429, *Shock Compression of Condensed Matter – 1997*
edited by Schmidt/Dandekar/Forbes
© 1998 The American Institute of Physics 1-56396-738-3/98/$15.00

INVESTIGATION OF SHOCK COMPRESSED PLASMA PARAMETERS BY INTERACTION WITH MAGNETIC FIELD

**S.V.Dudin, V.E.Fortov, V.K.Gryaznov,
V.B.Mintsev, N.S.Shilkin, A.E.Ushnurtsev**

Institute of Chemical Physics in Chernogolovka RAS. 142432 Russia

The Hall effect parameters in shock compressed air, helium and xenon have been estimated and results of experiments with air and helium plasma are presented. Explosively driven shock tubes were used for the generation of strong shock waves. To obtain magnetic field a solenoid was winded over the shock tube. Calculations of dense shock compressed plasma parameters were carried out to plan the experiments. In the experiments with the magnetic field of ~5 T it was found, that air plasma slug was significantly heated by the whirlwind electrical field. The reflected shock waves technique was used in the experiments with helium. Results on measurements of electrical conductivity and electron concentration of helium are presented.

Direct measurements of the electron concentration behind powerful shock waves are of great importance for the understanding of the physical processes at high energy density. In the present work Hall effect parameters of shock-compressed air, helium and xenon has been estimated for clearing up the possibility to determine the electron concentration by its interaction with magnetic field and preliminary experiments with air and helium are described.

For investigation of Hall effect the well-known method advanced by Van der Pauw [1] was applied, with coefficient R_H being obtained from change in resistance R at impressing homogeneous magnetic field B perpendicularly to a flat-parallel pattern of d thickness: $R_H = Rd//B/$. Knowing R_H one can determine the concentration of carries as: $N_e = r_H/R_H e$, where r_H is a Hall-factor equal for the case of scattering on Coulomb potential to 1.93 [1].

To realize such a method in dynamic experiments it is necessary to ensure the uniform of magnetic field in a moving plasma homogeneous slug. In such a case it seems to be attractive the use of explosively driven shock tubes [2]. This facility permits one to obtain the homogeneous plasma slug behind incident and reflected shock waves with a characteristic sizes of some centimeters and life-time of some microsecond, enough to carry out the necessary measurements. To generate strong magnetic field a solenoid must be applied on the body of shock tube, with the direction of magnetic field coinciding with the direction of moving plasma. On one hand, for magnetic field to freely penetrate into the plasma behind incident shock wave it is needed to form a flow with Reynolds number $Re_m = \mu_o \sigma U D < 1$. Here σ is the plasma electric conductivity, U is the particle velocity of its movement, D is the characteristic length scale. On the other hand, the conductivity of plasma must be higher than that of detonation products so that interaction with magnetic field be governed by plasma existence. Moreover, to plan the experiments is necessary to know expected values of Hall resistance which must be of enough great values.

With a goal of optimal planning such a test the plasma parameters have been calculated at modeling these experiment conditions. Calculations of incident and reflected shock wave adiabats were made within the incident shock wave velocity range 3-12 km/s. A mixture of oxygen, nitrogen, and argon was taken for the air which changed its state within the given velocity range from the state of molecular gas to a partially ionized plasma. Therefore, in the calculations existence of both molecules, including multiatomic molecules and molecular ions, and atomic ions were taken into account, with initial air pressure P_o being constant and equaled to 1 bar. At the same time, as for helium and xenon, the shock wave adiabats were calculated at initial pressures 1, 3, 10, and 30 bar. In the given region of thermodynamical parameters important is only the Coulomb interaction which was taken into account in the frame of Debye approach for a grand canonical ensemble [3]. Fig.1 shows the air plasma composition in dependence on shock wave velocity. It should be noted that in the lower part of the velocity range the level of air

background of significant Coulomb interaction of particles. This is obvious because the effect of decreasing the ionization potential on plasma composition was omitted in [4]. Xenon plasma reached a significant amount of ionization already at velocities higher than 4 km/s, and in the upper part of the velocity range the ionization was completed for all initial pressures with attending the second ion in the case of initial pressure P_o = 1 bar. In the case of air the parameters of Coulomb nonideality reached 1 for high velocities and of xenon these values exceeded 1 at P_o = 1 bar and exceeded 4 at P_o = 30 bar.

Besides the compositions and thermodynamical functions it was calculated the electrical conductivity and Reynolds magnetic number Re_m. The former was calculated in the approach [5], when scattering of electrons on ions and neutrals was taken into account, and at calculating the effective cross-sections of electron scattering on charged particles it was taken into consideration Debye screening and inter-region correlations in the frame of Ziman modified theory [5, 6]. Fig.2 represents the dependence of

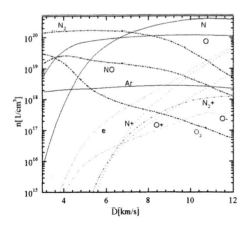

Fig.1. Composition of shock compressed Air.

ionization is ignorably low, reaching values about 1% at velocities higher than 7 km/s and about 10% only in the upper part of the range. The calculations of plasma compositions and thermodynamic functions well agree at low velocities with the available data [4], differing from them only at a marked ionization against a

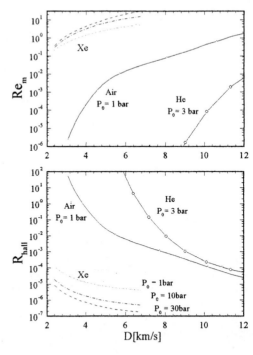

Fig.2. Re_m and R_{hall} vs shock wave velocity.

magnetic Reynolds numbers Re_m and Hall resistance R (at B=5 T and d=5 mm) for air and xenon on shock wave velocity for the calculated range. Also the values of Re_m of incident shock wave and values of R behind reflected shock wave in helium at P_0= 3 bar are placed on the figure.

To realize the needed velocity range linear explosively driven shock tubes were used, with shock waves formed at expansion of high explosive detonation products into the gas investigated (Fig.3). A solenoid, inside of which the magnetic field ~5 T was generated, was winded on the body of the shock tube. Using high speed streak and frame cameras VFU the experiments on visualization of flux were carried out. It was shown that at the distance 5-10 cm from the cut of charge the one-dimensional and enough stable

Fig.3: Explosively driven shock tube for the experiments with magnetic field

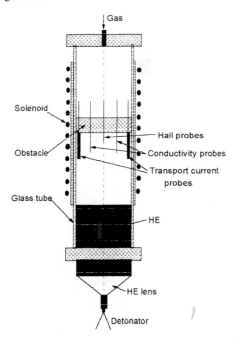

flow was formed. In the air it has been found out from the experimentation that at moving the plasma slug through a region of highly nonuniform magnetic field the electric breakdown in plasma and its warming by induced whirlwind electric fields occurred. As a result the plasma

became of high conduction and forced out the magnetic field to the channel walls. Therefore, the regime of flow has been produced with high magnetic Reynolds numbers.

To avoid this phenomenon experiments have been done with the helium. At the typical shock wave velocity of 10 km/s this gas have very low electrical conductivity comparable with that of expanded detonation products. Significant ionization starts when the flow run into the obstacle. This circumstance allow to diminish whirlwind currents behind incident shock wave and to realize regime of "frozen" magnetic field behind reflected shock wave.

To measure electrical conductivity of helium behind reflected shock wave we have used well known four probes method [6]. Additional two electrodes for Hall voltage measurements was inserted perpendicular to that of for conductivity measurements. The probes have different lengths for fixing the moments of the shock wave arrival. The measured values of electrical conductivity (16 $Om^{-1}c^{-1}$) and electron concentration ($3 \cdot 10^{18}$ cc^{-1}) are in a good agreement with the calculated values.

This method looks perspective for measurements of electron concentration of weakly ionized xenon with strong interparticle interaction.

References
1. K.Zeeger. "Physics of Semiconductors". Moscow, Mir, 1977.
2. V.B.Mintsev, V.E.Fortov, "Explosively driven shock tubes", *Teplofizika Visokih Temperatur*, v.20, N4, 1982, p.745, Russian.
3. A.A.Likalter, *Zh. Eksperim. Teor. Fiz.*, v.56, N1, 1969, p.240, Russian.
4. N.M.Kuznetsov. "Themodynamical Functions and Shock Adiabats of Air at High Temperatures". Moscow, Mashinostr., 1965, Russian.
5. V.K.Gryaznov, Yu.V.Ivanov, A.N.Starostin, V.E.Fortov, "Thermophysical properties of nonideal Ar and Xe plasma",*Teplofiz. Visokih Temperatur*, v.14, N3, 1976, p.643, Russian.
6. V.B.Mintsev, V.K.Gryaznov, V.E.Fortov, "Electrical conductivity of high temperature nonideal plasma", *Zh. Eksperim. Teor. Fiz.*, v.79, No1, 1980, p.116, Russian.

795

CP429, *Shock Compression of Condensed Matter – 1997*
edited by Schmidt/Dandekar/Forbes
© 1998 The American Institute of Physics 1-56396-738-3/98/$15.00

SHOCK TEMPERATURES OF SODA-LIME GLASS MEASURED BY AN OPTICAL PYROMETER

T. Kobayashi, T. Sekine, O.V. Fat'yanov*, E.Takazawa, and Q.Y. Zhu**

National Institute for Research in Inorganic Materials, Namiki 1-1, Tsukuba, Ibaraki 305, Japan

Shock temperatures of soda-lime glass in a pressure range 52-110 GPa have been determined by using a radiation pyrometer in conjunction with a two-stage light gas gun. This pyrometer consists of two parts, i.e., an OMA which provides a radiation spectrum over a visible range(~450 nm window) and a 4-channel PMT system which provides a time-varying behavior of shock temperatures. Obtained spectra are well fit by the Planck function with moderate emissivities, indicating that relatively homogeneous thermal radiation is the main component of radiation. Obtained shock temperatures range from 2800 K to 5700 K and they seem to represent glass temperatures in the liquid state. The Hugoniot is well described by a linear shock velocity-particle velocity relation, $u_s = 0.16 + 1.92\ u_p$ km/s. The results suggest that radiations from shocked materials in the liquid state, in contrast to radiations from shocked solids, are more thermal and fit the Planck function well, even at lower temperatures, because emission characteristics of compressed solids such as those due to heterogeneous deformation cannot be generated in shocked materials in the liquid state.

INTRODUCTION

In order to specify the states of shock compressed materials satisfactorily, it is necessary to know shock temperatures as well as shock pressures and volumes. Shock pressure and volume can be determined experimentally within an experimental error of a few %. However, shock temperature is more difficult to determine experimentally or theoretically.[1] Direct measurements of shock temperatures of solids and liquids have been performed mainly by observing thermal radiation spectra of shock compressed materials.[2-6] The radiation spectra are compared with the Planck function to obtain emissivities and shock

temperatures as color temperatures. Radiation pyrometry is useful not only to determine shock temperatures but also to study certain phase transitions with a large enthalpy change, e.g. melting of solids[7-11] and molecular dissociation in liquids.[12-14] It can also be used to estimate melting curves. Shock-induced melting can be difficult to identify with conventional techniques such as the inclined mirror method because the density change is often very small.

Over the last two decades, the technique of radiation pyrometry has been improved.[9,15-17] Mainly two methods have been used; (1) measurements of radiation spectra with an OMA (Optical Multichannel Analyzer) which cover the whole spectra in and near the visible region and (2) measurements of radiation intensities at several wavelengths using photomultipliers or photodiodes.

---

* Present address: Tokyo Institute of Technology, Midokiku, Yokohama 227, Japan.
** Present address: University of Sussex Briton BN19 QT, UK.

This gives time-varying shock temperatures. If the number of channels is sufficient, (2) may be the better method but if not, it seems that both methods should be employed simultaneously. The method (1) is more reliable to tell whether the observed spectrum is a thermal one or it involves other kind(s) of luminescence [4,6] because a detailed comparison between the Planck function and the OMA spectrum is possible. The method (2) is more suited for non-transparent materials because the radiation intensity may change with time in a complicated manner during the passage of shock front through the material.

In the present study, we have constructed an optical pyrometer as a part of the in situ diagnostic system of shock compressed materials generated by the two-stage light gas gun at NIRIM. As the first target, soda-lime glass (the most commonly used glass) was chosen because it is transparent and can be compared with other SiO_2 compounds.[8]

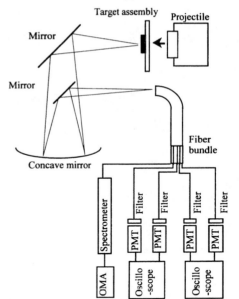

FIGURE 1. Schematic diagram of fiber-coupled optical pyrometer.

EXPERIMENTAL

The layout of the optical pyrometer is shown in Fig.1. Four fibers (fused silica with a 200 mm diameter and a 0.2 numerical aperture) were used for each channel. The photomultipliers used were R1894 (rise time<0.8 ns, Hamamatsu) with interference filters (20 nm FWHM) centered at 400, 500, 600, 800 nm. PMT outputs were recorded on 500 MHz oscilloscopes (HP54522A, Hewlett-Packard) with 50 ohm input impedance. The optical multichannel analyzer (IRY-512 / par, Prinston Instruments) with a 512-channel diode array and a spectrometer (Monosped 18, Jarrell-Ash) were used to record radiation spectra with a spectral resolution ~ 3 nm. The gated radiation signal was integrated for a period of 100 ns. The gate pulse for the OMA was given by a pulse generator (FG-200, Princeton Instruments). Recorded radiation intensity was calibrated into absolute scale (radiance in $W / m^3 \cdot str$) using a quartz tungsten halogen lamp (63355, Oriel Instruments) as a standard lamp. Calibration of the wavelength was done using atomic lines of a mercury lamp. Obtained OMA spectra and PMT outputs were compared with the Planck function to determine shock temperatures and emissivities by the least squares fit. The uncertainties in shock temperatures and emissivities are estimated to be about ±5% and ±30%, respectively. Shock experiments were performed using a two-stage light gas gun at NIRIM (2ST-1) with Helium or Hydrogen as the driving gas.

3.0 mm thick commercial soda-lime glass with a density 2.52 g/cm^3 was used as the target. The composition of this glass is roughly 71 wt% SiO_2, 13 wt% Na_2O, 9 wt% CaO and 4 wt% MgO. The typical area of the target was 1.5 x 1.5 cm^2 and the edges were masked by black-painted aluminum foil to reduce edge effects, thus the radiating area of about 1.0 cm^2 on the glass sample was observed by the pyrometer. The timing trigger pulse for the OMA and the PMT system was generated by a trigger pin which was placed on the driver plate.

RESULTS AND DISCUSSION
Radiation Spectra and PMT Measurements

Typical radiation spectra taken by the OMA are shown in Fig. 2. Solid curves represent the Planck function with appropriate temperatures and emissivities (the graybody distribution function) chosen to fit the experimental spectra. Observed spectra are all fit by the graybody distribution function very well and no sign of emissions of other sources such as electronic transitions is seen. Moderate values of emissivities obtained from all the spectra indicate that relatively homogeneous thermal radiation was realized at the shock front. It was observed that the peak position smoothly moves towards blue as the temperature increases.

In Fig.2, PMT results (circles with error bar) are also plotted with the OMA spectra, showing reasonable agreement between the two results. For these runs, the OMA signal was integrated for 100 ns and the PMT outputs plotted in the figure were those measured at 50 ns from the beginning of each OMA signal accumulation. The time-varying PMT outputs are shown in Fig. 3 where almost a flat feature of the radiation intensities is seen. It is known that the transmitance of SiO_2 is reduced markedly by shock compression so that only the radiation from the vicinity of shock front is observed through the remaining transparent part of the sample. This also means that radiations from other parts behind the shock front such as driver plate are not reflected into the observed radiation. These facts are responsible for the flat feature of the PMT outputs. These time-varying PMT signals allowed us to determine the time intervals for shock waves to travel through samples and thus shock velocities. When the flyer surface is inclined with respect to the target surface upon collision, a slow response of the PMT outputs were observed. However, We have confirmed that the shock temperature itself is not influenced significantly by the inclination of the flyer.

Shock Temperature

Fig. 4 shows a shock temperature vs. shock pressure plot for soda-lime glass. Shock temperatures from OMA spectra and those from PMT measurements coincide well with each other. Almost a linear relationship between shock temperatures and shock pressures is seen. It is reported that shock-induced liquefaction of fused quartz occurs at ~70 GPa with the shock temperature of liquefied fused quartz ~4500 K.[8] Since softening of soda-lime glass under ambient pressure

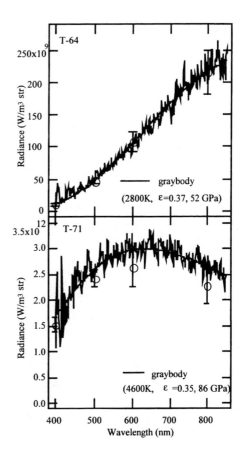

FIGURE 2. Radiation spectra recorded by OMA and radiation intensities measured by 4-channel PMT system. Solid curves are the Planck functions with appropriate temperatures and emissivities.

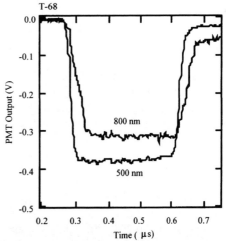

FIGURE 3. Time-varying PMT outputs. Results of only two channels are shown to avoid congestion.

799

takes place at much lower temperature compared to fused quartz, measured shock temperatures of soda-lime glass should also be considered as those of liquefied soda-lime glass.

Theories[18] to estimate melting temperatures of solids (mainly metals) gives a very rough estimate for the shock temperature of liquefied soda-lime glass to begin at ~2400 K which is lower than the lowest temperature observed in this study. The fact that the radiation of soda-lime glass at temperatures as low as 2800 K is still thermal and is fit by the graybody function well may be considered to provide another support for the assumption that the shocked soda-lime glass in this study was liquefied. Radiation spectra of fused quartz at temperatures as high as ~4000 K (before liquefaction) are not fit well by the graybody distribution function due to heterogeneous deformations.[4,6] In this case, the main contribution of the emission comes from localized spots of the sample and the emissivity is very low, which is explained by heterogeneous deformation mentioned above. Thus the reported shock temperatures of fused quartz probably do not represent thermal temperatures. For shocked fused quartz, it is in the solid state (superheated solid) up to ~4000 K and an emission caused by a heterogeneous deformation is observed superimposed on the thermal radiation. However, shocked soda-lime glass at 2800 K is already in the liquid state and thus no such emissions originating from the solid nature can be generated. This suggests that radiations from liquids can be thermal at much lower temperatures than those from solids and the radiation pyrometry can be used reliably to determine relatively low shock temperatures of liquids as well as high shock temperatures.

It is interesting to investigate the behavior of shock temperatures of glasses in the lower temperature-pressure region to see how the effect of shock-induced liquefaction is accommodated in the behavior of shock temperature of glasses.

ACKNOWLEDGMENTS

Authors wish to thank Mr. Ohsawa for his technical assistance and Ms. Ohtsuka for preparing the manuscript.

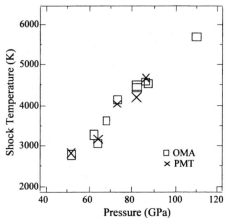

FIGURE 4. Shock temperature vs. shock pressure for soda-lime glass. Temperatures shown in the figure are believed to represent liquefied glass temperatures.

REFERENCES

1. Sugiura, H, Kondo, K., and Sawaoka, A., *Rev. Sci. Instrum.* **51** (6) 750 (1980).
2. Kormer, S.B., Sinitsyn, M.V., Kirilof, G.A., and Urlin, V.D., *Sov. Phys. JETP* **21** 689 (1965).
3. Lyzenga, G.A., and Ahrens, T.J., *Rev. Sci. Instrum.* **50** (11) 1421 (1979).
4. Kondo, K, Ahrens, T.J., and Sawaoka, A., *J. Appl. Phys.*, **54** (8) 4382 (1983).
5. Boslough, M.B., Ahrens, T.J., and Mitchell, A.C., *J. Geophys. J.R. astr. Soc.* **84** 475 (1986).
6. Schmitt, D.R., and Ahrens, T.J., *J. Geophys. Res.* 94, **B5** 5851 (1989).
7. Uetiew, P.A., and Grover, R., *J. Appl. Phys.* **48** (3) 1122 (1977).
8. Lyzenga, G.A., Ahrens, T.J., and Mitchell, A.C., *J. Geophys. Res.* 88 **B3** 2431 (1983).
9. Yoo, C.S., Holmes, N.C., Ross, M., Webb, D.J., and Pile, C., *Phys. Rev. Lett.* **70** (25) 3931 (1993).
10. Boness, D.A., and Brown, J.M., *Phys. Rev. Lett.* **71** (18) 2931 (1993).
11. Holland, K.G., and Ahrens, T.J., *Science*, **275** 1623 (1997).
12. Radousky, H.B., Nellis, W.J., Ross, M., Hamilton, D.C., and Mitchell, A.C., *Phys. Rev. Lett.* **57** (19) 2419 (1986).
13. Nellis, W.J., Radousky, H.B., Hamilton, D.C., Mitchell, A.C., Holmes, N.C., Christian, K.B., and Van Thiel, M., *J. Chem. Phys.*, **94** (3) 2244 (1991).
14. Nellis, W.J., Ross, M., and Holmes, N.C., *Science* **269** 1249 (1995).
15. Sugiura, H., Kondo, K., and Sawaoka, A., *Rev. Sci. Instrum.* **51** 750 (1980).
16. Boslough, M.B., and Ahrens, T.J., *Rev. Sci. Instrum.* **60** (12) 3711 (1989).
17. Holmes, N.C., *Rev. Sci. Instrum.* **66** (3) 2615 (1995).
18. Schlosser, H., Vinet, P., and Ferrante, J., *Phys. Rev.* **B40** (9) 5929 (1989).

CP429, *Shock Compression of Condensed Matter – 1997*
edited by Schmidt/Dandekar/Forbes
© 1998 The American Institute of Physics 1-56396-738-3/98/$15.00

SHOCK-INDUCED LUMINESCENCE IN POLYMETHYLMETHACRYLATE

W.G. PROUD, N.K. BOURNE AND J.E. FIELD

Shock Physics, PCS, Cavendish Laboratory, Madingley Road, Cambridge, CB3 0HE, UK.

Light emission during impact events is common and may be produced by various processes. These include the compression of gas between impactor and target, fractoemission and bulk luminescence. In this paper the results of a series of experiments on polymethyl methacrylate (PMMA) are presented. The relative intensity of emissions in the near infrared and visible regions have been measured and spectra obtained. Bulk luminescence is shown to be relatively weak at pressures of up to 22.5 GPa compared to fractoemission from the PMMA during release. At the maximum pressure used, 22.5 GPa, the relative intensity of the outputs of the photodiodes suggests a polymer temperature over 2000 K. The integration of the light emitted during the shock compression can be fitted to a black body giving an estimated temperature of 3700 ±400 K.

INTRODUCTION

Light emission from shocked polymers was reported in the 1950's when PMMA was shocked up to several hundred GPa (1). Previous research has suggested that light emission occurs at levels as low as 5.0 GPa (2). Shocking PMMA up to 200 GPa gives a spectral output corresponding to a black body temperature of 8500 K(1). More recent work on organic liquids has given temperatures of the same magnitude for similar shock conditions (3). In these cases the output is fitted to a black or grey body radiator. The motivation for such experiments includes the design of adequate vessels for shock spectroscopy experiments and high-speed photography. Light emission from trapping of residual gas between impactor and target has been shown to be significant at pressures of a few Pascal. Even low strain-rate tensile tests have shown broad band release of photons from polymers.

Recent experiments in the Cavendish have focused upon light emission from shock-compressed gases, linked to sonoluminescence, using PMMA vessels (4). Commonly in VISAR studies (5) PMMA is used as the window and manganin gauges are often mounted using PMMA blocks (6). In order to account for the possible bulk luminescence in such experiments a series of tests have been performed to measure the light output from PMMA in the pressure range up to 22.5 GPa.

It was decided to use a cell consisting of a disc of PMMA between anvils and ring the pressure up to levels of interest.

EXPERIMENTAL

Plate impact experiments were carried out on the 50 mm bore gun at the University of Cambridge (7). Impact velocity was measured to an accuracy of 0.5% using a sequential pin-shorting method and tilt was fixed to be less than 1 mrad by means of an adjustable specimen mount. Impactor plates were made from lapped tungsten alloy and were mounted onto a polycarbonate sabot with a relieved front surface in order that the rear of the flyer plate remained unconfined. The targets consisted of a thin disc of PMMA (ICI), 0.5 mm thick and 47 mm diameter fitted into a recessed tungsten disc. This assembly was backed by a sapphire window 50 mm diameter, 5 mm thick. Targets were flat to within 10 μm across the surface.

The rear of the PMMA was viewed by three fibres (3M FG-200-LCR), fixed at 10 mm from the sapphire rear surface, which were fed into two photodiodes and a UV/visible spectrometer. One photodiode (Electro-optics Technology ETC2010) was sensitive over the wavelength range to 300-1000 nm and the other to near infra-red 1.0 - 1.6 μm (ETC 4000). The rise and decay times of both diodes was less than 1 ns. The output of the photodiodes was fed into a Tektronics TDS 744a oscilloscope and sampled at 1 GS s^{-1}.

The spectrometer used was an EGG 1235 spectrometer with a 1455 intensified gated photodiode array. This was arranged to monitor wavelengths from 400 - 800 nm and gated for 2 μs integration time. The gate was timed to coincide with the entry of the shock wave into the sample and the compression process. This gating ensured that the release part of the shock process was *not* seen by the spectrometer.

An Ultranac FS501 high-speed camera capable of framing rates of up to 20 million frames per second was used in some experiments to monitor impact flashes.

RESULTS

The sample chamber in the impact facility has a rough vacuum of 1 mbar present at the point of firing. Direct impact on PMMA results in large, sustained light output due to the compression of the residual gas betwen the impactor and the target. This is illustrated in the high speed photograph (figure 1) in which a tungsten impactor travelling at 350 m s^{-1} impacts a PMMA block.

In frame 1, taken at impact, the initial flash across the whole of the impactor face is visible. By frame 2 this has decayed to a ragged area in the centre of the target. From frames 3 to 5 the gas in the recessed holes containing the screws holding the plate to the sabot luminesces due to the ongoing compression. In frames 3 and 4, but most strongly in frame 6, ink marks made to label the sample light up. The photodiode output both visible and near infrared reached several volts in magnitude and lasted up to20 μs. This is to be compared with the mV level of

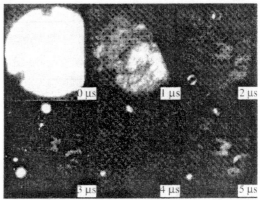

Figure 1. High speed photograph of impact by a tunsten impactor travelling at 350 ms-1 striking a PMMA block. Exposure time 500 ns, interframe time 500 ns.

output found later for the bulk luminescence. In order to eliminate this effect the PMMA was placed in a recess at the rear of a tungsten plate.

Figure 2 shows the prediction of a 1-D hydrocode applied to the cell. The pressure in the PMMA rises over 1 μs to a value of 18.0 GPa when the cell is struck by a tungsten projectile travelling at 750 m s^{-1}. On average each step in the ringing up of the plate takes 100 ns and the PMMA reaches its maximum pressure, under these conditions, before the longitudinal release wave comes in from the sapphire window. Repeat simulations with the projectiles travelling at 910 m s^{-1} indicate a maximum pressure of 22.5 GPa in the PMMA.

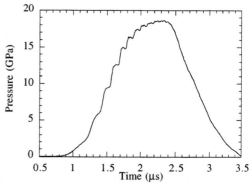

Figure 2. Simulation of pressure in PMMA in the sample configurations used.

The output of the photodiodes (figure 3) shows that for a cell loaded to 18.0 GPa there is very little output from the infra-red photodiode, whilst the output of the visible photodiode shows a small (reproducible) rise over half a microsecond. Applying the results to the simulation above it is possible to say that light is emitted when the PMMA has reached a pressure of *ca.* 7 GPa. The output when the cell is struck to 22.5 GPa shows a higher level of both visible and infra-red output commensurate with a higher shock-induced temperature.

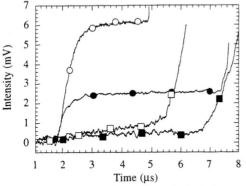

Figure 3. The output of the photodiodes. Circles indicate visible photodiode output, squares correspond to infrared. Open markers show the output for a cell struck to 22.5 GPa, filled for a sample shocked to 18 GPa.

The sensitivity of the infra-red photodiode is a thousand times greater than that of the visible one. Taking this, the wavelength sensitivity of the photodiodes and other system factors such as fibre transmission factors into account and comparing this with the integration of black body power output over appropriate wavelength ranges suggests a body above 2000 K. A simple estimate, comparing the output registered here with other known light sources, gives a total power output of a few watts per square centimetre.

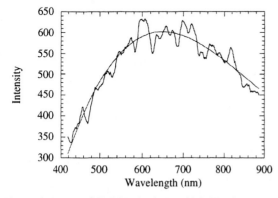

Figure 4. Spectra of PMMA shocked to 22.5 GPa. The smooth line shows a black body curve corresponding to a temperature of 3700 K.

The spectral output for the cell loaded to 18.0 GPa was very weak, displayed no clear lines and gave a temperature of *ca.* 3000 K when fitted to a black body curve. At a higher pressure the spectrum shown in figure 4 resulted. Again there are no clear lines and a black body fit of 3700 ±400 K gave good agreement. It is recognised that other possible analyses (such as grey bodies) exist and that the spectrum represents the output of a sample being rung-up. Some variation in temperature can be expected in the PMMA, due to pressure variations, though cooling will be negligible on the time scales of the experiments. However, the fit used represents the simplest situation and the temperatures associated with the spectra are thus to be taken as approximate.

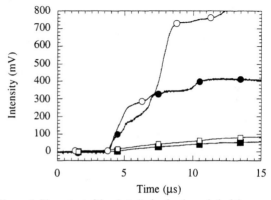

Figure 5. The output of the photodiodes for the period of the experiment up to the point at which the fibre probes are contacted by the sample. Open markers indicate samples struck at 910 m s^{-1}, filled struck at 750 m s^{-1}, circles correspond to the visible output and squares to the near infrared.

The output of the photodiodes reached a plateau upon maximum compression and maintained it over the 3 μs or so corresponding to the compression part of the shock process. When the release waves enter the PMMA and the sample starts to break up, the output increases a hundred-fold (figure 5). At this point the tungsten, sapphire and PMMA can separate and the 1 mbar of residual gas in the sample chamber can enter into the cracks, contributing to the emission process. The bulk of this light is probably due to fractoemission in the PMMA. Again the more violent impact results in greater fragmentation of the sample and so greater light intensity.

The ratio of the outputs of the photodiodes changes since the infra-red diodes now register significant light output. This reduction in the ratio between the two photodiode outputs indicates a broader spectral distribution in the light.

DISCUSSION AND CONCLUSIONS

The results indicate the care that must be taken in making such measurements. The light output due to shock compression is small compared to that due to gas compression and sample fracture. This may lead to emission due to the fracture processes on release being erroneously assigned to the compression phase.

A pressure of around 7 GPa is required to start light output, this is in good agreement with the previously reported level of 5 GPa. The photodiode output is predominantly in the visible region indicating a hot body.

The spectral characteristics suggest a temperature of 3700 ±200 K in the polymer shock compressed to 22.5 GPa. This is consistent with a temperature of 8500 K found with PMMA shocked to 200 GPa. All these temperatures are estimates based on a black body spectrum though it is realised that more sophisticated analyses could be applied.

Figure 6 shows the temperature estimated from a numerical integration based on the method of Walsh and Christian (8). This calculation assumes a system involving a single pass of a shock wave and so differs from the multi-pass experiment reported here. The system in this experiment have a behaviour lying between that of the isentrope and the hugoniot. As can be seen, a difference of a few GPa at the higher pressures can make several hundred degrees of difference in the temperatures. It must be remembered that the spectra are time integrated over several microseconds, the whole of the compression phase.

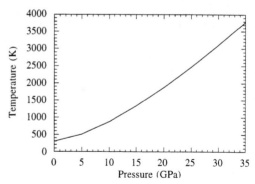

Figure 6. Temperature of the PMMA calculated following ref (8). Parameters used in analysis from Steinberg (9).

Shorter gating and greater spectrometer sensitivity is required to gather information during the compression process.

ACKNOWLEDGEMENTS
The authors acknowledge funding from SERC. We thank Mr D.L.A. Cross and Mr R.P. Flaxman for technical support.

REFERENCES
1. Zeldovich, I., Kormer, S.B. and Sinitsyn, M.V., Sov. Phys. Dokl., **3**, 938-939 (1958).
2. Cremers, D.A., Marston, P.L. and Duvall, G.E., High Temp - High Press., **12**, 109-112 (1980).
3. Nellis, W.J., Ree, F.H., Trainor, R.J., Mitchell, A.C. and Boslough, M.B., J. Chem. Phys., **80**, 2789-2799 (1984).
4. Bourne, N.K., Proud, W.G. and Field, J.E., J. Acoust. Soc. Am., **100**, 2716 (1996).
5. Barker, L.M. and Hollenbach, R.E., J. App. Phy., **41**, 4208-4226 (1970).
6. Bourne, N.K., (1994), *Plate impact experiments on PMMA*, University of Cambridge PCS report, Prepared for DRA agreement no. EMR/2029/268, PCS/SP1005.
7. Bourne, N.K., Rosenberg, Z., Johnson, D.J., Field, J.E., Timbs, A.E. and Flaxman, R.P., Meas. Sci. Technol., **6**, 1462-1470 (1995).
8. Walsh, J.M. and Christian, R.H., Phys. Rev., **97**, 1544-1556 (1955).
9. Steinberg, D., (1996), *Equation of State and Strength Properties of Selected Materials*, Lawrence Livermore National Laboratory, UCRL-MA-106439.

CP429, *Shock Compression of Condensed Matter – 1997*
edited by Schmidt/Dandekar/Forbes
© 1998 The American Institute of Physics 1-56396-738-3/98/$15.00

SURFACE TEMPERATURE MEASUREMENTS OF HETEROGENEOUS EXPLOSIVES BY IR EMISSION

B. F. Henson, D. J. Funk, P. M. Dickson, C. S. Fugard and B. W. Asay[1]

Los Alamos National Laboratory, Los Alamos, NM 87545

We present measurements of the integrated IR emission (1-5 μm) from both the heterogeneous explosive PBX 9501 and pure HMX at calibrated temperatures from 30°C to 250°C. The IR power emitted as a function of temperature is that expected of a black body, attenuated by a unique temperature independent constant which we report as the thermal emissivity. We have utilized this calibration of IR emission in measurements of the surface temperature from PBX 9501 subject to 1 GPa, two dimensional impact, and spontaneous ignition in unconfined cookoff. We demonstrate that the measurement of IR emission in this spectral region provides a temperature probe of sufficient sensitivity to resolve the thermal response from the solid explosive throughout the range of weak mechanical perturbation, prolonged heating to ignition, and combustion.

INTRODUCTION

It has long been recognized that solid phase temperature is a key observable for understanding the behavior of energetic materials. Experiments designed to measure temperatures associated with shock initiation in plastic bonded explosives have included both fast thermocouple (1) and IR radiometric (2) techniques.

Current problems of interest in these explosives, including the violence of reaction after prolonged heating and the thermal response when subject to mechanical deformation, require the development of techniques to measure surface temperature which allow both high sensitivity and high spatial and temporal resolution. We have recently demonstrated the ability to obtain IR emission from the surface of unconfined explosives subject to weak impact (3).

In this report we present quantitative measurements of the integrated IR emission (1-5 μm) from both the heterogeneous explosive PBX 9501, and the pure components at calibrated temperatures from 30°C to 250°C. The IR power emitted as a function of temperature from these materials is that expected of a black body, attenuated by a unique temperature independent constant we report as the thermal emissivity. We also report preliminary measurements of surface temperature, based on this calibration, in both mechanical and thermal ignition experiments spanning a temperature range of 30 °C to 400°C and microsecond to millisecond timescales.

EXPERIMENTS AND RESULTS

We apply IR Radiometric techniques based on InSb (Indium Antimonide) detection to determine the surface temperature from measurements of IR radiant power (4). Determination of surface temperature from radiant power over a frequency

---

[1] This work sponsored by the U.S. Department of Energy under contract number W-7405-ENG-36.

interval v_i to v_f is accomplished via the Plank black-body distribution

$$dP = \frac{4\pi h}{c^2}\frac{A_d}{2}\,\varepsilon\sin^2(\theta)\int_{v_i}^{v_f}\frac{v^3\eta(v)\,\sigma(v)\,dv}{\exp(hv/kT)-1} \quad (1)$$

where P is the radiant power, θ is the acceptance half angle, A_d the surface area element, ε the temperature independent emissivity, T the temperature, v the frequency, and h, c and k the usual fundamental constants. The frequency distribution is convoluted with the detector quantum efficiency, $\eta(v)$, and responsivity, $\sigma(v)$. We apply a single lens imaging geometry in which the surface area element from the sample is equal to the known detection element area and θ is determined by the focal distance and the lens radius. The convolution functions $\eta(v)$ and $\sigma(v)$ for InSb semiconductor materials are well known. The radiant power as a function of temperature is thus a function of the single unknown constant, the emissivity ε.

Calibration Experiments

We have calibrated the detectors used in this work by imaging the surfaces of both MIKRON and Electro-Optics standard black-body ovens. We used the single lens geometry described previously. Assuming unit emissivity from the standard surface, the measured voltage as a function of temperature directly yields the detector responsivity upon numerical integration of Eq. (1).

The measurements of surface emissivity from the various explosive components were then performed in an analogous manner. Explosives and component materials were held at known temperatures and the InSb detector voltage measured to generate a curve of voltage as a function of temperature. These curves have been plotted for several independent measurements of PBX 9501 and pure HMX surfaces in Fig. (1). Plotted as the logarithm of voltage against $1/T$ a black-body response is linear with a constant slope, hv/k, independent of material and with an intercept which is a temperature independent function of the emissivity, ε. This is the observed shape in $\log(V)$

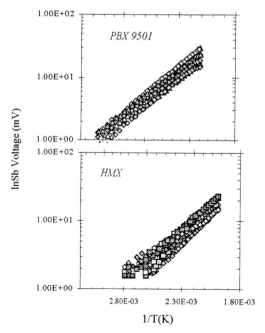

FIGURE 1. Detector voltage as a function of inverse temperature in Kelvin for PBX 9501 and pure HMX. Both samples were pressed. The data are from several different samples of each material.

vs. $1/T$ for the samples of Fig. (1) and all other materials reported here. The varying intercept in these data reflect a variable emissivity from the surfaces of different samples of the same material. This variance is considerably larger than the statistical uncertainty in the individual determination of ε from a single curve and reflects an inherent, morphological variation. This will directly affect the accuracy of the inversion of voltage data to generate surface temperatures. The measured emissivity from a number of components of PBX 9501 are summarized in Table 1. We have

Table 1. Summary of measured emissivity from components of PBX 9501 from 50 to 200°C.

Sample material		emissivity, ε	σ
PBX 9501	(pressed)	0.75	0.14
HMX	(pressed)	0.64	0.14
Estane	(polyurathane rubber)	0.25	0.09
Binder[a]		0.73	0.09

[a]The binder is a 50/50 blend of Estane and bis dinitropropyl acetal/formal.

utilized this calibrated radiometric system in two preliminary dynamic experiments.

Mechanical Impact

We first present the measured temporal and spatial temperature field from the surface of a PBX 9501 sample during deformation resulting from a 1 GPa impact. The details of the experiment, including measurement and calculation of the surface displacement field as a function of time during deformation, are described in companion papers in this symposium (5,6). Briefly, a rectangular sample of PBX 9501 (25 mm x 10 mm x 5 mm) is confined in a steel assembly at the end of a light gas gun. One inch diameter sapphire windows confine the large area sides of the sample. The sample is subject to impact by a brass bullet (190 m/s) impinging on a rounded steel plunger of 19 mm radius which couples the weak shock into the narrow end of the sample. The two dimensional weak shock propagates through the sample, and the spatial and temporal profiles of the resulting temperature field are recorded from the surface through the IR transparent windows.

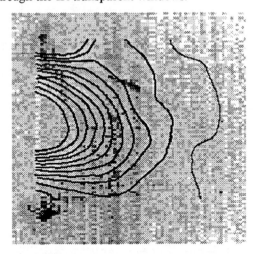

FIGURE 2. Imaged surface of PBX9501 subject to 1 GPa impact on the left side by a rounded plunger (19 mm radius). Darker regions indicate higher temperatures.

The spatial profile of the surface temperature ~100 μs after impact is shown in Fig. 2. The detector, from Santa Barbara Inc., consists of 40 μm elements in a 256x300 array. The darker

regions indicate higher temperatures, and reflect the symmetry of the plunger from the rounded impact surface at left, tapering to cold material ~1 cm ahead. The surface temperature varies from ~50°C in the heated region to sharp temperature increases near ~100°C at the boundaries. Superimposed on this image is the displacement vector field, which clearly demonstrates the correlation between particle motion, boundaries and heating.

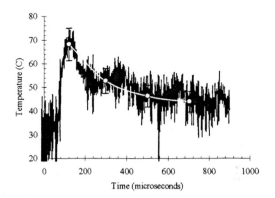

FIGURE 3. Surface temperature recorded subsequent to 1GPa impact (Time=0). The error bars correspond to uncertainty in calculated temperature due to emissivity uncertainties reported in Table 1.

The temporal profile of the temperature field over approximately a millisecond is shown in Fig. 3. This detector, from Cincinnati Electronics, consists of two parallel linear arrays of eight 80 μm elements, with 10 μs risetime amplification. The error bars reflect the uncertainty in temperature which results from the uncertainty in surface emissivity reported in Table 1.

Unconfined Cookoff

We have also conducted preliminary experiments to measure the surface temperature of PBX 9501 during spontaneous thermal ignition. In these experiments the Cincinnati Electronics detector was utilized with 1 μs rise time electronics. The samples were right circular cylinders with both length and diameter of 1 cm. The samples, which were placed in an open oven and heated over approximately three hours, appeared unchanged to external observation until the temperature reached approximately 235°C, when ignition occurred and the sample completely

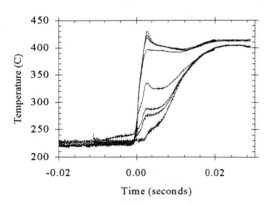

FIGURE 4. Temperature recorded from eight adjacent surface elements 80x80µm. Recorded from unconfined PBX 9501 during ignition subsequent to prolonged heating.

decomposed in approximately a second. The IR signal from the surface was measured, with single lens imaging as above, both during the heating trajectory and during the first 40 ms of rapid decomposition. The ignition data for one experiment, showing the signal from eight detectors imaging different areas of the surface, are shown in Fig. (4).

The data exhibit several reproducible features. These include the small 'pre-ignition' rise in temperature from the surface prior to t=0 and the rapid temperature rise defined as ignition. The temperature rises to a constant from some areas of the surface quickly, with much slower rise to the same constant temperature from other areas of the surface.

CONCLUSIONS

The calibration experiments, ranging from ambient through critical temperatures of PBX 9501, indicate a reliable measurement methodology for these materials.

The powerful physical description enabled by such detailed temperature measurement is clear from the preliminary results reported here. The spatial and temporal temperature fields of Figs. (2)

and (3), and the correlation of these fields with both the impact symmetry expected of the rounded plunger and the measured surface displacement (5) will place a considerable constraint on the mechanism of heating in prospective constitutive models. Data such as the detailed temporal behavior of the surface temperature during ignition in the unconfined cookoff experiment are revealing many of the physical and chemical mechanisms involved in coupling thermal decomposition with ignition and even combustion.

These experimental examples, and many other problems of current interest concerning explosive safety illuminate the need for precise measurement of the thermal fields generated by a number of coupled thermal and mechanical perturbations. These examples represent the beginning of a program to obtain such measurements in our laboratory.

ACKNOWLEDGEMENTS

The authors acknowledge the support of Phillip Howe and Joseph Repa throughout this effort.

REFERENCES

1. Bloomquist, D. D. and Sheffield S. A., Thermocouple Temperature Measurements in Shock Initiated PBX-9404, *Seventh Symposium on Detonation*, Annapolis, MD, 1004-1009, 1981.
2. Von Holle, W. G. and Tarver, C. M., Temperature Measurements of Shocked Explosives by Time Resolved Infrared Radiometry - A New Technique to Measure Shock-Induced Reaction, *Seventh Symposium on Detonation*, Annapolis, MD, 993-1003, 1981.
3. Asay, B. W., D. J. Funk, G. W. Laabs, and P. J. Howe. *Measurement of Strain and Temperature Fields During Dynamic Shear of Explosives.* in *Shock Compression of Condensed Matter, American Physical Society Topical Conference.* 1995. Seattle, WA.: AIP Press.
4. McCluney, R., *Introduction to Radiometry and Photometry*, Artech House, Boston, 1994.
5. B. W Asay, B. F. Henson, P. M. Dickson,C. S. Fugard, and D. J. Funk. *Direct Measurement of Strain Field Evolution During Dynamic Deformation of an Energetic Material.* in *Shock Compression of Condensed Matter.* 1997. Amherst MA: AIP Press.
6. K. S. Haberman, J. G. Bennett, B. W. Asay, B. F. Henson and D. J. Funk. *Modeling, Simulation and Experimental Verification of Constitutive Models for Energetic Materials.* in *Shock Compression of Condensed Matter.* 1997. Amherst MA: AIP Press.

CP429, *Shock Compression of Condensed Matter – 1997*
edited by Schmidt/Dandekar/Forbes
© 1998 The American Institute of Physics 1-56396-738-3/98/$15.00

REAL-TIME CHANGES INDUCED BY PULSED LASER HEATING IN AMMONIUM PERCHLORATE AT STATIC HIGH PRESSURES

G.I. Pangilinan* and T.P. Russell

Chemistry Division, Code 6110, Naval Research Laboratory, Washington DC 20375
* under contract with Nova Research Inc.

Time-resolved absorption measurements of ammonium perchlorate (AP) decomposition under static high pressures are presented. Chemical changes induced by pulsed laser heating (8 μs at 514 nm) are inferred from the absorption spectra. Absorption measurements of the decomposition process with 1 μs temporal resolution were collected from 400 to 630 nm, during and up to 20 μs after the heating pulse. AP decomposition as a function of initial pressure and laser fluence was monitored to 2 GPa and to 11 J/cm², respectively. The role of pressure and laser fluence in the fast decomposition of AP will be discussed. Global reaction rates and possible reaction mechanisms will be addressed.

INTRODUCTION

Understanding the physics and chemistry of the rapid decomposition of condensed matter at high pressures and temperatures is one of the outstanding challenges in energetic material research.(1) Real-time experimental studies are currently being pursued.(2,3) Recently, spectroscopic measurements of fast laser-induced decomposition at elevated pressures was demonstrated(2,3) on trinitroethylamine tetrazine. There is a need to apply this technique to other energetic materials, preferably with those that are widely-used and better characterized.

Ammonium perchlorate (NH_4ClO_4, AP) is a widely used material. It is a common ingredient in propellants and explosives. Various models have been proposed to describe its slow thermal decomposition and combustion,(4) but to our knowledge, no experimental work has been done to address its fast (<20 μs) decomposition at elevated pressures. Because of the number of different elements present in AP, and their different oxidation states, over 1000 reaction steps are possible. The goal, therefore, of this research is to probe the fast laser-induced reactions of AP and (a) find evidence for reaction intermediates that will reduce the subset of important plausible reactions, and (b) describe the global kinetics of the reaction.

In this work, the real-time processes in AP samples at different pressure and laser fluence values (P > 0.5 GPa, and laser fluence > 3 J/cm²) in which decomposition takes place, are examined.

EXPERIMENTAL CONFIGURATION

Details of the method to obtain time-resolved absorption from statically compressed samples can be found elsewhere, (3-5); therefore, only a brief description is provided. The AP samples were obtained from Dr. T. Boggs at China Lake. The crystals were loaded in Merrill-Bassett cells with cubic zirconia anvils.(5) A 200-250 μm diameter hole was drilled in a 0.2 mm thick tantalum (Ta) sheet to make a gasket. The gaskets were pre-indented prior to drilling for

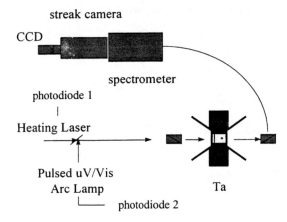

Figure 1. Experimental Configuration

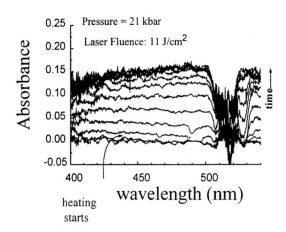

Figure 2. Typical time-resolved (1 μsec separation) absorbance spectra from decomposing AP showing absorbance increase in time. There is no vertical offset in the spectra; the dip centered at 514 nm is due to attenuating filters and is ignored.

experiments in which the initial pressure is above 1.5 GPa. Ta gaskets were used because unlike other metals, Ta does not catalyze AP reactions.(6) The AP samples were pressed to fill the gasket. A small ruby sphere 10 μm in diameter was loaded to provide a pressure gauge.(7) AP was studied in the pressure range 0.6 GPa ≤ P ≤ 2.1 GPa.

The experimental arrangement to measure absorbance from a heated sample is shown in Figure 1. The heating source is a dye laser (514 nm, 8 μs full width at half maximum pulse duration) with a measured laser fluence of 3, or 11 J/cm². This laser fluence is sufficient to initiate a reaction, at all pressures studied. Simultaneously, a pulsed uv-Vis source that provides light from 350-650 nm is introduced. The pulse duration of the uv-Vis lamp is longer than 25 μs. The light transmitted through the cell is collected and sent to a monochromator (SPEX 270M), which disperses the light according to wavelength. The light is further dispersed in time by a streak camera (Hamamatsu 1487) and is recorded on a CCD detector.

An initial uv-Vis transmission is recorded through the pressurized AP sample without the heating source, to give a reference intensity $I_o(t,\lambda)$. The uv-Vis light transmitted to the sample when the heating pulse is turned on is similarly obtained to yield $I(t,\lambda)$. The absorbance is calculated from $A(t,\lambda) = \log_{10}(I_o/I)$. Photodiodes that monitor the light outputs of the heating laser and the uv-Vis

source are used to determine the start of the heating pulse, assigned to be t = 0.

RESULTS

Figure 2 shows typical time-resolved absorbance curves of laser heated AP from 400 to 560 nm. At times prior to laser heating, the absorbance is flat and has zero value as expected from unreacted material. At the onset of the heating pulse, a broad flat absorbance is seen to increase continuously and reaches a maximum (at 8 μs in Figure 2). A feature slowly grows in the absorbance that becomes more evident at late times when the absorbance reaches a fairly constant spectral shape in time.

To verify the existence of the broad absorption feature, an experiment was performed to measure absorbance at longer wavelengths, centered at 550 nm. The absorbance again increased at the onset of the heating pulse, and plateaus to a constant shape at later times. Three absorbance curves at 15-18 μs are shown in Figure 3B. Together with the absorption curves centered at shorter wavelengths (redrawn from Figure 2) shown in Figure 3A, an absorption feature centered at about 480-490 nm is discerned.

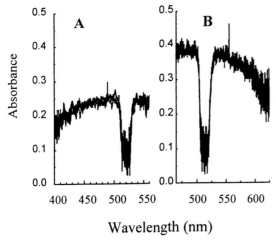

Wavelength (nm)

Figure 3. Three consecutive one μsec absorbance spectra of AP from 400 to 630 nm, showing the existence of a broad peak. The initial pressure is 0.6 GPa, and the laser fluence used is A) 3 J/cm², and B) 6 J/cm² respectively.

To investigate the effects of pressure, time-resolved absorbance measurements of reacting AP at initial pressures of 0.6, 1.2, 1.6, and 2.0 GPa and a constant laser fluence (3 J/cm²), were obtained. In addition, experiments at 0.6, 0.9, 1.5, and 2.1 GPa, with laser fluence of 11 J/cm², were studied to understand the effects of laser fluence on the decomposition. In all these measurements, the broad absorbance feature shown in Figure 3 is observed at later times. The rate at which the broad absorbance reaches a maximum value showed a dependence on the initial pressure, and laser fluence.

DISCUSSION

Absorbance from Transient Species

The broad absorbance feature was observed in all experiments performed. It arises from transient species from decomposing AP. ClO_2, ClO_3, $HOCl$, $NOCl$, and NO_2Cl are the only intermediate species currently identified in AP combustion which have absorption in the region 400 to 600 nm. Identifying the particular specie(s) at the present time, based on comparison with absorbance data at ambient pressures and temperatures is not possible. All five molecules have broad vibro-elctronic absorption at

ambient pressure between 400-600 nm, and at these high pressure and temperature regimes, the absorbance may be modified from their ambient values. However, the observed absorption feature does reduce the number of important intermediates from the numerous possibilities. Two methods to help distinguish the five molecules are time-resolved emission and fluorescence measurements. These experiments will be carried out in the future.

Global Reaction Kinetics

The absorbance increase is used to investigate global kinetics of AP decomposition. It is shown in Figure 3 that the absorbance peaks around 480 to 490 nm. The change of the average absorbance between 480 and 490 nm was determined to measure the temporal dependence of the absorbance. The absorbance values are then normalized such that the maximum change in absorbance is equal to 1. The change in normalized absorbance at different initial pressures during pulsed laser heating (3 J/cm²) is shown in Figure 4A. Similar plots are shown in figure 4B, using a higher laser fluence of 11 J/cm².

At low initial pressures, the absorbance increases after some finite delay, or induction time. As the initial pressure is increased, the induction time is reduced. The curves of the normalized absorbance rapidly converge above 0.9 GPa, at 11 J/cm².

The reaction growth can be modeled assuming that the normalized absorbance is directly related to the molar fraction (α) of AP decomposed. There are numerous expressions that model reaction rates in the literature. Two expressions that appropriately describe the decomposition of nitromethane(8) and β-RDX(9) were attempted. These are a diffusion limited reaction, and an autocatalytic, branching chain reaction, where the rate is determined from the slope of $\ln(1-\alpha)$, vs. time, and $\ln(\alpha/(1-\alpha))$ vs. time, respectively.

Figure 5 shows plots of $\ln(1-\alpha)$ vs. time, for AP decomposition at 0.6, 1.2, 1.6, and 2.0 GPa during pulsed laser heating (3 J/cm²). Similar plots are obtained for the higher laser fluence (11 J/cm²). The data points reasonably fit lines. In contrast, plots of $\ln(\alpha/(1-\alpha))$ vs. time show a distinct

811

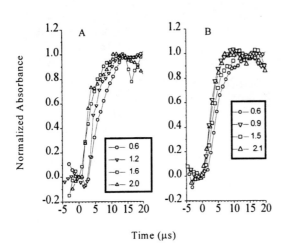

Figure 4. Normalized absorbance for eight different experiments: A)with 3 J/cm², and B)11 J/cm² fluence. Initial pressures are given in GPa.

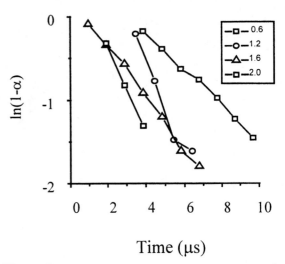

Time (μs)

Figure 5. Plots of ln (1-α) for different pressures with 3J/cm² fluence. The data points fit lines, indicating that the diffusion limited model gives a good description of the reaction growth.

concavity. This implies that the reaction growth in the laser induced decomposition of AP is simulated better with the diffusion-limited reaction model.

Other factors need to be considered, to place more physical meaning to the decomposition kinetics. Heating pulse duration, sample pressure, and sample morphology will be evaluated to give more physical significance to the kinetic expressions. In addition, the method of energy transfer from the laser pulse to the AP sample, and how the reaction propagates to completion need to be considered. These will be addressed in future work.

ACKNOWLEDGMENTS

The research was conducted under funding from the Office of Naval Research and the Naval Research Laboratory.

REFERENCES

1. See for example Schmidtt, S.C., and Tao, W.C., eds., *Proceedings Shock Compression of Condensed Matter - 1995,* New York: AIP Press, 1996
2. Russell, T.P., Allen, T.M., Rice, J.K., and Gupta, Y.M., *J. de Physique IV,* **5**, C4:553-559 (1995).
3. Russell, T.P., Allen, T.M., and Gupta, Y.M., *Chem. Phys. Lett,* **267**, 351-358 (1997).
4. Ermolin, N.E., Korobeinichev, O.P., Tereschenko, A.G., and Fomin, V.M., *Comb. Explos. Shock Waves,* **18**, 180-189 (1992); Galwey, A.K., Herley, P.J., and Mohamed, M.A., *Thermochimica Acta,* **132**, 205-215 (1988); Hackman, E.E., Hesser, H.H., and Beachell, H.C., *J. Phys. Chem.,* 76, 3545-3554 (1972).
5. Russell, T.P., and Piermarini, G.J., *Rev. Sci. Instrum,* **68**, 1835-1840 (1997).
6. Foltz, M.F., and Maienschein, J.L., *Mater. Letters,* **24**, 407-414 (1995).
7. Piermarini, G.J., Block, S., Barnett, J.D., and Forman, R.S., *J. Appl. Phys.* **46**, 2774 (1975); Block, S., and Piermarini, G.J., *Phys. Today,* **29**, 44 (1976).
8. Piermarini, G.J., Block, S., and Miller, P.J., *J. Phys. Chem.,* **93**, 457-462, (1989).
9. Piermarini, G.J., Block, S., and Miller, P.J., *J. Phys. Chem.,* **91**, 3872-3878 (1987).

CP429, *Shock Compression of Condensed Matter – 1997*
edited by Schmidt/Dandekar/Forbes
© 1998 The American Institute of Physics 1-56396-738-3/98/$15.00

MECHANISM OF CHEMICAL DECOMPOSITION IN A SHOCKED CONDENSED EXPLOSIVE

Y. A. Gruzdkov and Y. M. Gupta

Shock Dynamics Center and Department of Physics, Washington State University, Pullman, WA 99164-2814

A detailed understanding of shock induced chemical decomposition in explosives at the molecular level remains an outstanding problem. Time resolved optical spectroscopic techniques were used in our laboratory to address this problem. Nitromethane (NM), an insensitive high explosive, was investigated in this work as a model material. In particular, our emphasis was on understanding the chemical mechanism governing amine sensitization of shocked NM. The previously suggested reaction mechanisms are summarized briefly. We acquired the necessary data using optical absorption, fluorescence, and emission spectroscopies. Spectra of neat NM and mixtures of NM with amines shocked to within 12 - 17 GPa using step-wise loading were obtained. A transient intermediate was detected in sensitized NM with the optical absorption and fluorescence techniques. In a series of experiments with various amines, the intermediate was identified as a radical anion of NM. Several possible radical anion mechanisms were considered and evaluated. The base catalysis by amines is favored as the most plausible mechanism.

I. INTRODUCTION

A shock induced chemical process of considerable practical importance is the decomposition of energetic materials [1]. A variety of methods has been developed to examine this process. It has been recognized that only a combination of continuum measurements (pressure, particle velocity) and time-resolved spectroscopic techniques can provide the necessary macroscopic and microscopic insight into this challenging problem [2]. While continuum measurements on a number of shocked energetic materials have been carried out for a long time [1], real time optical spectroscopic methods are a relatively recent development [2]. Our emphasis in this work was on achieving an understanding of the molecular mechanisms governing shock induced chemical decomposition of liquid nitromethane through time-resolved optical spectroscopy. These results are

used with previous continuum studies to obtain a comprehensive description of the chemical process.

Nitromethane (NM) is an insensitive high explosive that serves as a good prototypical energetic material [3,4]. The simplicity of its chemical structure makes it attractive for mechanistic studies. It is also the simplest member of the family of nitrocompounds. For these reasons, NM is a very well-studied material; good reviews regarding earlier work on NM may be seen in Refs. 3 and 4. Therefore, information can be drawn from a large body of scientific literature including spectroscopic data at ambient pressure, static high pressure data, and continuum data under shock loading.

It has been known since the late 1940s that NM can be sensitized by the addition of small amounts of amines [5]. The mechanism of shock sensitization, although widely believed to be chemical in nature, is not well understood due to the lack of data at the microscopic level. The objective of this work was to obtain such data by examining in real time the molecular changes in shocked sensitized NM. A brief summary of the previous work on the sensitization effect is given in the next section. Section III describes briefly the experimental methods used; in Section IV we present and discuss our results. The main findings are summarized in the last section.

II. BACKGROUND

Considerable evidence exists for the sensitization of NM by amines. It has been demonstrated, using the gap test, that the strength of the shock wave required to initiate a detonation in NM is lowered considerably by the presence of a number of amines [6,7]. Essentially the same observation has been made for failure diameters [8,9]; the addition of 0.03 wt.% of diethylenetriamine results in a 43 % decrease in the failure diameter of NM [8]. In slow thermal decomposition experiments [3,10], it has been shown that the activation energy of reaction decreases from 130 kJ/mol in neat NM to around 85 kJ/mol in the NM - amine mixtures.

Several molecular mechanisms have been put forward to explain the sensitization. Chronologically, the first explanation was proposed by Engelke et al. [11]. They suggested that the aci-anion of NM was somehow involved in the rate determining step. To support this idea, they argued that other factors known to sensitize NM, such as the addition of organic bases, UV irradiation, and static high pressure, provide higher levels of the aci-anion as well [12-16]. Although, no exact mechanism was offered.

Politzer et al. have pursued this idea further [17]. Using density functional calculation, they found that, of the various possibilities examined, the most energetically favorable reaction is the one involving the aci-anion and amine. They emphasized that the activation energy for this reaction was less than the CN bond dissociation energy in NM.

Cook and Haskins questioned the importance of the aci-anion for sensitization [18]. Based on their own calculations, they proposed that NM formed hydrogen bonded complexes with amines. Depending on the geometry of the complex, two reactions occurring via CN bond rupture were proposed.

Based on slow thermal decomposition measurements, Constantinou et al. [3,10] proposed that sensitization is due to a weakening of the C-N bond of NM, but not through hydrogen bonding. They suggested that a charge transfer complex is responsible for it. The decomposition of charge transfer complexes was thought to give rise to the initial stage in the thermal decomposition observed for the NM-amine mixtures. This extra stage was interpreted as an indication of a new reaction pathway. It was shown also that the mixtures decompose through a process that follows first order kinetics while neat NM decomposes through a cubic autocatalytic process.

As can be seen from the reviewed publications, there is no agreement among different authors on the molecular mechanism of NM sensitization. Clearly, experimental data at the microscopic level are necessary to resolve the matter. Ideally, the detection and identification of transient intermediates would help to determine the mechanism of sensitization. We, therefore, undertook time-resolved optical spectroscopic experiments under shock loading to address this need.

III. EXPERIMENTAL METHODS

Shock waves were generated by impact between a sapphire impactor, mounted on a projectile, and the sapphire front window of the sample cell. The projectile could be accelerated to any velocity up to 1.2 km/s using a single stage gas gun. The description of the cell design can be found elsewhere [19]. Typically, the sample was 0.2 mm thick. The shock wave reverberated between the cell windows causing pressure ring-up in several steps; the final pressure was maintained in the sample for about 1 μs. In this work, NM was shocked typically to 12 - 14 GPa and 700 - 800 K.

The configuration for time-resolved absorption spectroscopy used in this work was similar to that used in earlier work [19]. Briefly, a pulse of light from a xenon flashlamp was directed through the sample into an optical fiber; it was then delivered to a spectrometer. The spectrometer dispersed it in wavelength, and the streak camera dispersed it in time. As a result, a two-dimensional image with

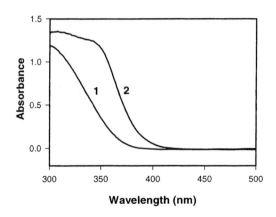

FIGURE 1. Absorption spectra of nitromethane at ambient pressure (1) and 14.5 GPa (2). Adapted from Ref. [4].

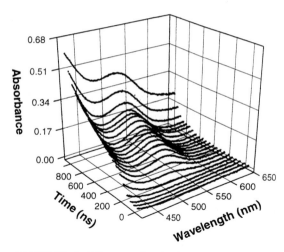

FIGURE 2. Time-resolved absorption spectra of nitromethane-triethylamine (37.5 mM) mixture shocked to 11.8 GPa. At 0 ns shock enters the sample; by 210 ns 95 % of the final pressure is reached. Spectra were taken with 49 ns resolution.

wavelength on the horizontal axis and time on the vertical axis was produced at the output of the streak camera. This image was digitally recorded by a CCD detector as a series of transmission spectra separated by ca. 50 ns time intervals. There was sufficient transmitted light intensity for measurements within 420 - 650 nm spectral and - 0.3 - +1.2 optical density range. More detailed description of this and some other experimental techniques used in this work can be found elsewhere [19-21].

IV. RESULTS AND DISCUSSION

Intermediate Detection and Identification

Time-resolved UV-vis absorption and Raman spectroscopies were used previously to examine chemical decomposition in neat NM. Comprehensive summary of these results is given in Ref. [4]. Briefly, no evidence for chemical reaction was found for pressures up to 14 GPa and temperatures up to 800 K. Typical pressure induced changes in UV absorption spectra of neat NM are shown in Fig. 1. They consisted of a broadening of the absorption band (λ_{max} = 270 nm) and an increase in the peak absorbance of the band. No absorbance was found at wavelengths longer than ca. 400 nm.

Unlike neat NM, amine sensitized NM shows signs of chemical reaction at pressures as low as 10 GPa [19]. Absorption spectra of the triethylamine - NM mixture shocked to 11.8 GPa are shown in Fig. 2. As can be seen, after the shock wave enters the sample an absorption band at 525 nm begins to grow. It continues to grow after the final pressure is reached at 210 ns. Other changes to the spectra include: the appearance of another band's edge at ca. 450 nm and a growth of flat absorbance seen as vertical translation of the spectra after ca. 800 ns. Certainly the most striking feature in the spectra is that a new absorption band developed. We assign it to an intermediate formed in the mixture during early stages of decomposition. In a series of experiments we identified the intermediate as a radical anion of NM, $CH_3NO_2^{\bullet-}$. Below we provide more experimental evidence to support the identification and discuss it in more detail.

The electronic structure of the radical anion is known from quantum chemical calculations, ESR spectroscopy, and energy loss spectroscopy [22-24]. Three electronic states of $CH_3NO_2^{\bullet-}$ have been identified. The equilibrium C-N distance for the $^2A_1$ ground state is 2.0 Å which is longer than in a

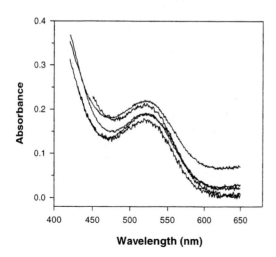

[chart: Absorbance (0.0 to 0.4) vs Wavelength (nm) (400 to 650)]

FIGURE 3. Absorption spectra of nitromethane - amine mixtures taken at 600 ns after shock entered the sample. The samples were shocked to ca. 12 GPa. Five different amines were used: n-butylamine, ethylenediamine, diethylamine, triethylamine, and 1,4-diazabicyclo(2.2.2)octane.

neutral molecule (1.475 Å) and the potential energy well is much shallower (0.56 eV as opposed to 2.52 eV). There are two $^2B_1$ electronically excited states. Because of symmetry, the lower $^2B_1$ state is not coupled (electric dipole interaction) to the ground state; the transitions between these two states are restricted. Transition to the upper $^2B_1$ state is dipole allowed and has an estimated energy of 2.3 eV [24]. This value matches well the energy of the absorption band detected in our experiments. The latter provides the first supporting point for the identification.

Furthermore, as follows from the electronic structure of the radical anion it should fluoresce from the upper $^2B_1$ excited state. We used this property to verify our identification independently in a separate set of experiments. We did observe the expected fluorescence when the intermediate was excited with the 514 nm laser pulse. Time evolution of the intensity of fluorescence coincided with the kinetics of the absorption peak height.

Finally, if the above identification is correct, the same intermediate is expected to form in NM mixtures with any amine regardless of its chemical structure. Fig. 3 shows the intermediates spectra

detected in five different NM mixtures with primary, secondary, tertiary, and di- amines. As evidenced by the spectra, exactly the same intermediate was produced in all of the mixtures in full agreement with the prediction.

Thus, in these experiments, we have established that the shock induced decomposition in NM - amine mixtures occurs via formation of the radical anion of NM, $CH_3NO_2^{\bullet-}$.

Reaction Mechanism and Kinetics

It is generally accepted (see also Section II) that cleavage of the C-N bond is a key stage in the decomposition of NM [10,17,18]. Sensitization is likely to include a process that lowers the C-N bond dissociation energy in comparison to neat NM. From this point of view, the radical anion is an attractive pathway. As mentioned above, capture of an electron by NM elongates the C-N bond and lowers the dissociation energy from 245 kJ/mol ($CH_3 + NO_2$) to around 50 kJ/mol ($CH_3 + NO_2^-$) [17,24]. Therefore, the unimolecular decomposition of radical anions is expected to occur quite easily. In addition, NM has a positive adiabatic electron affinity which is around 50 kJ/mol. The latter means that radical anion is thermodynamically more favorable than the neutral molecule. This fact further strengthens the arguments in favor of the radical anion pathway. Below we analyze several possible radical anion mechanisms.

The simplest mechanisms to form radical anions would be either a direct electron transfer from amine or heterolytic dissociation of charge transfer complexes to produce a pair of radical ions. The counter ions formed there would be aminium radicals. They are well-established transient intermediates; they form in a great number of reactions including electron transfer from alkylamines [25].

A significant body of spectral data is available for aminium radicals. To permit detection in our experiments, a radical should possess a strong absorption band in the visible and be long-lived. The aminium radicals of 1,4-diazabicyclo(2.2.2) octane (DABCO) and N-methylaniline seemed to meet these requirements. They have absorption bands at around 460 nm [26]. The aminium radical of DABCO is also exceptionally long-lived among other aminium radicals because of the significant delocalization of the unpaired electron. If the mentioned above reactions were responsible for generation of radical anions, we should have detected the aminium radical counter ions along with the radical anions. As can be seen in Fig. 3 this did not happen in the case of DABCO. N-methylaniline was found not to sensitize NM at all. These results indicated that simple mechanisms such as electron transfer from amines or decomposition of charge transfer complexes are not operative in NM - amine mixtures.

The analysis of experiments with different amines suggested that the amine basicity rather than electron donor properties is responsible for sensitization. A plausible mechanism in this case would be the base catalysis by amines:

$$CH_3NO_2 + RNH_2 \rightleftharpoons [CH_2NO_2]^- + RNH_3^+$$
$$[CH_2NO_2]^- + CH_3NO_2 \rightarrow CH_3NO_2^{\bullet-} + CH_2NO_2^{\bullet}$$
$$CH_3NO_2^{\bullet-} \rightarrow CH_3^{\bullet} + NO_2^-$$

The first stage here is an acid-base equilibrium between NM and amine to produce aci-anions and the protonated amine. In the second rate determining stage, aci-anions react with NM to produce radical anions and nitromethyl radicals. In the third stage, radical anions decompose unimolecularly via C-N bond scission.

There are two predictions of this mechanism that can be experimentally verified. First, the steady state kinetic regime should be observed for radical anions at sufficiently low pressures/temperatures. Second, the steady state concentration of radical anions will be proportional to the square root of initial amine concentration. Both of the predictions were confirmed experimentally. We did observe the steady state kinetics at ca. 12 GPa/730 K. Fig. 4 shows that the steady state concentration was indeed proportional to the square root of amine concentration.

Combining together all the experimental evidence presented we conclude that the base catalysis by amines is the most plausible operative mechanism in NM - amine mixtures.

V. SUMMARY

Mechanism of amine sensitization of nitromethane was investigated with time-resolved optical spectroscopy. NM - amine mixtures react

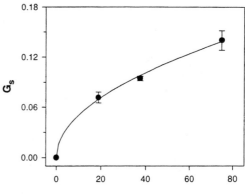

FIGURE 4. Steady-state concentration of the radical anion of nitromethane (G_s) as a function of ethylenediamine concentration in nitromethane - ethylenediamine mixtures shocked to 12 GPa. Solid line is the least squares fit with the square root function.

via formation of an intermediate that gives rise to a transient absorption peak at 525 nm. Based on our analysis and the data available from the literature, we identify the intermediate as the radical anion of nitromethane, $CH_3NO_2^{\bullet-}$. Possible reaction mechanisms are discussed and evaluated. Among the several considered, the base catalysis by amines is favored. Among the mechanisms proposed earlier the aci-anion hypothesis of Engelke [8] is consistent with this mechanism. The sensitization of NM is brought about by the significantly lower C-N bond dissociation energy in the radical anion than in the neutral molecule of NM.

ACKNOWLEDGEMENT

D. Savage and K. Zimmerman are thanked for their assistance in the experimental effort. Discussions with Dr. J. M. Winey and Dr. G. I. Pangilinan are gratefully acknowledged. This work was supported by ONR grant N00014-93-1-0369 (Program Manager: Dr. R. S. Miller).

REFERENCES

1. Cheret, R., *Detonation of Condensed Explosives*, New York: Springer-Verlag, 1993.

2. Gupta, Y.M., *J. de Physique IV*, Colloque C4, **5**, C4-345 (1995).
3. Constantinou, C.P. *The nitromethane-amine interaction.* Ph.D. Dissertation, Cambridge University, 1992.
4. Winey, J.M. *Time-resolved optical spectroscopy to examine shock-induced decomposition in liquid nitromethane.* Ph. D. Dissertation, Washington State University, 1995.
5. Eyster, E.H.; Smith L.C.; Walton S.R., *US Naval Ordnance Lab. Report No. NOLM 10336*, 1949
6. Cook, M.D.; Haskins, P.J. In: *Proc. 19th Int. Annual Conf. of ICT on Combustion and Detonation Phenomena*; Karlsruhe, Germany, 1988, 85-1.
7. Forshey, D.R.; Cooper, J.C.; Doyak, W.J., *Explosivstoffe Nr.*, **6**, 125 (1969).
8. Engelke, R., *Phys. Fluids*, **23**, 875 (1980).
9. Kondrikov, B.N.; Kozak, G.D.; Raikova, V.M.; Starshinov, A.V., *Dokl. Akad. Nauk SSSR*, **233**, 402 (1977).
10. Constantinou, C.P.; Mukundan, T.; Chaudhri, M.M., *Phil. Trans. R. Soc. Lond. A*, **339**, 403 (1992).
11. Engelke, R.; Earl, W.L.; Rohlfing, C.M., *Int. J. Chem. Kin.*, **18**, 1205 (1986).
12. Piermarini, G.J.; Block, S.; Miller P.J., *J. Phys. Chem.*, **93**, 457 (1989).
13. Miller P.J.; Block, S.; Piermarini, G.J., *J. Phys. Chem.*, **93**, 462 (1989).
14. Engelke, R.; Schiferl, D.; Storm, C.B.; Earl, W.L., *J. Phys. Chem.*, **92**, 6815 (1988).
15. Engelke, R.; Earl, W.L.; Rohlfing, C.M., *J. Phys. Chem.*, **90**, 545 (1986).
16. Engelke, R.; Earl, W.L.; Rohlfing, C.M., *J. Chem. Phys.*, **84**, 142 (1986).
17. Politzer, P.; Seminario, J.M.; Zacarias, A.G., *Mol. Phys.*, **89**, 1511 (1996).
18. Cook, M.D.; Haskins, P.J., *Proc. 9th Symp. (Int.) Detonation*, 1989, 1027; *Proc. 10th Symp. (Int.) Detonation*, 1993, 870.
19. Constantinou, C.P.; Winey, J.M.; Gupta, Y.M., *J. Phys. Chem.*, **98**, 7767 (1994).
20. Gupta, Y.M., *High Pressure Research*, **10**, 717 (1992).
21. Pangilinan, G.I.; Gupta, Y.M., *J. Phys. Chem.*, **98**, 4522 (1994).
22. Ramondo, F., *Can. J. Chem.*, **70**, 314 (1992).
23. Compton, R.N.; Reinhardt, P.W.; Cooper, C.D., *J. Chem. Phys.*, **68**, 4360 (1978).
24. Lobo, R.F.M.; Moutinho, A.M.C.; Lacmann, K.; Los, J., *J. Chem. Phys.*, **95**, 166 (1991).
25. Chow, Y.L.; Danen, W.C.; Nelsen, S.F.; Rosenblatt, D.H., *Chem. Rev.*, **78**, 243 (1978).
26. Shida, T. *Electronic absorption spectra of radical ions.* Elsevier, 1988.

CP429, *Shock Compression of Condensed Matter – 1997*
edited by Schmidt/Dandekar/Forbes
© 1998 The American Institute of Physics 1-56396-738-3/98/$15.00

PICOSECOND VIBRATIONAL SPECTROSCOPY OF SHOCKED ENERGETIC MATERIALS

Jens Franken, Selezion A. Hambir and Dana D. Dlott

School of Chemical Sciences, University of Illinois at Urbana-Champaign, Urbana, IL 61801

The dynamic response of a thin film of the insensitive high explosive 5-nitro-2,4-dihydro-3H-1,2,4-triazol-3-one (NTO) to ultrafast shock compression has been investigated by picosecond coherent anti-Stokes Raman spectroscopy (CARS). Vibrational spectra were obtained in the 1200 cm^{-1} to 1450 cm^{-1} region with a time resolution on the order of 100 ps. The frequency shifts and widths of the two vibrational transitions in this region show an entirely different behavior when subjected to a shock load of about 5 GPa. An additional weak band at 1293 cm^{-1} appears temporarily while the shock front is within the NTO layer.

INTRODUCTION

In this paper a study on the shock compressed high explosive NTO (5-nitro-2,4-dihydro-3H-1,2,4-triazol-3-one) (figure 1) (1-4) is presented using a new technique, which has been recently developed in our group (5-10). This technique

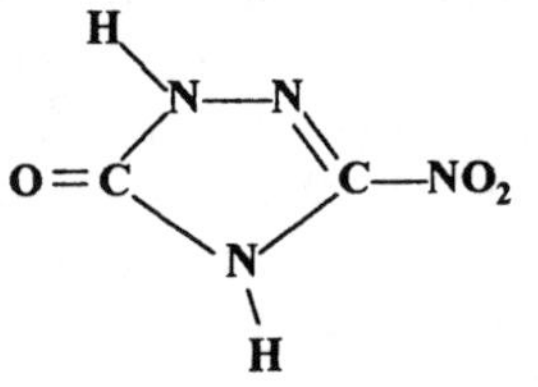

FIGURE 1. Molecular structure of NTO.

combines picosecond coherent anti-Stokes Raman spectroscopy (CARS) with a microfabricated multilayer shock target array. This multilayer sample can easily be customized to allow shock experiments on virtually any solid material by simply changing the composition and/or the order of one or more layers in the sample, while keeping the shock generating layer the same. This combination has already been shown in previous work to be very useful in determining instantaneous pressures and temperatures right behind the shock front as well as the entropy increase across the shock front, due to the irreversibility of the shock compression (5-10). The time resolution can be as high as 25 ps (8,9), where the achievable time resolution depends on the probe laser pulse widths as well as on the sample layer thickness.

In the search for powerful but safe, i.e. insensitive, high explosives NTO has been identified as a potential candidate. It shows similar explosive performance characteristics to RDX (1,3,5-trinitro-1,3,5-triazocyclohexane), but is less sensitive to shock and impact (3).

EXPERIMENTAL

The shock waves in our experiments have been generated by irradiating a thin poly-(methylmethacrylate) (PMMA) film with 1064 nm IR light (fluence ≈ 0.4 J/cm^2) from a Nd:YAG laser having a 150 ps pulse duration (figure 2) (7,8). This film has been doped with the near-IR absorbing dye IR165 as well as with RDX. The dye IR165 converts the optical energy within a few ps into heat,

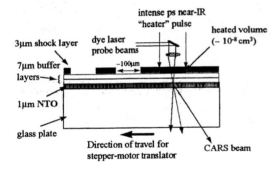

FIGURE 2. Schematic drawing of the multilayer sample array.

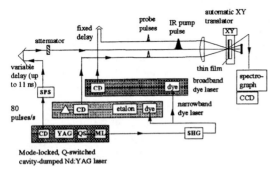

FIGURE 3. CARS spectrometer. A Nd:YAG laser generates the near-IR pump pulse used to generate shock waves. It also pumps a broad band and a narrow band dye laser to generate the CARS probe pulses. (CD = cavity dumper; QS = Q-switch; ML = mode locker; SPS = single pulse selector; SHG = second-harmonic generator; CCD = charge-coupled device).

leading to ultrafast, isochoric heating of the entire shock layer, thus generating a pressure jump. The addition of RDX has been found to increase the maximum shock pressure by more than 30% (6). This shock layer (d ≈ 3 μm) has been spin-coated on top of two buffer layers of poly(vinyl alcohol) (PVA) and PMMA. The purpose of the buffer layers is to avoid dissolving the sample in the shock layer during the preparation of the sample array, as well as to give the shock layer time to ablate before the shock wave reaches the sample layer. The dye IR165 in the shock generating layer exhibits a very large background in the CARS spectra, which might even entirely cover the signal from the sample layer (6,7). Therefore to obtain high-quality, low-background vibrational spectra of the sample layer it is essential to delay the arrival of the shock wave at the sample layer long enough

so that the by the IR pulse irradiated part of the shock layer has time to ablate completely.

Both buffer layers (d_{total} ≈ 7 μm) were spin-coated on top of a thin NTO layer (d ≈ 1 μm), which was sprayed from acetone solution onto a glass plate using an airbrush. Maximum shock pressures in the sample layer of about 5 GPa were generated at a repetition rate of 80 Hz. The shock effects were probed using broadband multiplex coherent anti-Stokes Raman spectroscopy (CARS). One dye laser is operated in broadband, the other one in narrow band configuration (figure 3). The dye laser beams are focused at the center of the IR beam. The CARS spectra are captured by a CCD camera (1100x330 pixels) attached to a 3/4m spectrograph. After each laser shot a translation stage moves the sample to a new spot. Each CARS spectrum corresponds to an accumulation over about 5,000 laser shots.

The NTO was purified prior to its use by re-crystallization from hot aqueous solution. The IR spectrum was consistent with that of the α polymorph of NTO (1).

RESULTS

Figure 4 shows experimental CARS spectra of the NTO sample at various delay times. The delay time t_d = 0 indicates the arrival of the IR pump pulse at the sample. The shock is generated by this pulse. After the shock builds up and propagates through the buffer layers, the front reaches the NTO sample at about 1600 ps. The two vertical dotted lines mark the frequencies of the two bands at ambient pressure and temperature. The spectrum in figure 4a, taken long after the shock layer has disappeared but before the shock front reached the NTO layer, shows two bands at about 1355 cm^{-1} and 1325 cm^{-1}. The band at 1355 cm^{-1} has been assigned to the symmetric stretching vibration of the nitro group (2,11); the assignment of the 1325 cm^{-1} band is not known. Figures 4b-e show spectra while the shock front is inside the NTO layer. The band at 1355 cm^{-1} shows a large blueshift, which is maximum at about 2000 ps delay time (figure 4d). The 1325 cm^{-1} band shows a small redshift which peaks at the same time. Figures 4f-h show spectra

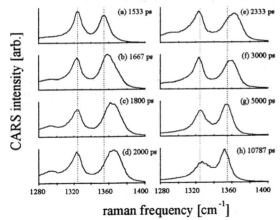

CARS intensity [arb.]

(a) 1533 ps
(b) 1667 ps
(c) 1800 ps
(d) 2000 ps
(e) 2333 ps
(f) 3000 ps
(g) 5000 ps
(h) 10787 ps

raman frequency [cm^{-1}]

FIGURE 4a-h. Experimental CARS spectra of a thin film of NTO at various delay times. (a) The shock front has not reached the NTO layer, (b) - (e) the shock front is inside the sample layer and (f) - (h) during pressure relaxation after the shock front has left . The vertical dotted lines indicate zero frequency shift.

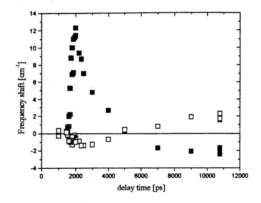

FIGURE 5. Time dependence of the frequency shifts of the NTO 1355 cm^{-1} (■) and 1325 cm^{-1} (□) bands.

after the shock front has left the sample layer. During the shock unloading the blueshift of the 1355 cm^{-1} band decreases and turns into a small redshift, while the small redshift of the 1325 cm^{-1} band turns into a small blueshift. Figure 5 quantifies the time dependence of the frequency shifts. Using CARS spectroscopy, it is possible to determine the instantaneous temperature and pressure of the material being studied, if at least two vibrational transitions are monitored simultaneously, provided each responds differently to T and P. That is accomplished by measuring the vibrational frequencies as a function of pressure and temperature under static conditions (5,6,10). At long delay times, after the sample has undergone a cycle of irreversible shock compression and reversible unloading, there will be a residual frequency shift due to the heat remaining in the sample. Measuring this shift permits the calculation of the entropy increase across the shock front (6,10).

Comparison of figure 4h with, for instance, the spectrum in figure 4a shows a decrease in the peak height of the 1325 cm^{-1} band relative to the 1355 cm^{-1} band during the 3 to 11 ns period. This effect was studied quantitatively as follows. The CARS intensity is proportional to the square of the number density (12). The square root of the spectra are

computed, and each spectrum is corrected for contributions from the buffer layers and the glass prior to their analysis (7). Then the spectra are fit to a pair of Voigt lineshape functions by varying the peak location, peak width (FWHM) and peak area. Using this procedure, it was found that the ratio of the areas of the two bands was the same before and after the shock front was in the NTO layer. The change in relative peak heights can be explained as a time-dependent increase of the width of the 1325 cm^{-1} band. The time dependence of the peak widths are shown in figure 6. The width of the 1355 cm^{-1} band behaves as expected. The FWHM

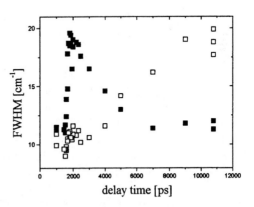

FIGURE 6. Time dependence of the FWHM of the NTO 1355 cm^{-1} (■) and 1325 cm^{-1} (□) bands. The band widths were obtained by fitting the experimental CARS spectra, which were corrected for contributions from the buffer layers and the glass prior to their analysis, to Voigt functions.

increases sharply when the shock front enters the NTO layer and it is a maximum during the period of peak shock loading. The increase in FWHM is attributed to two sources. First, when the front moves through NTO, the CARS spectra are a superposition of shocked and ambient spectra with different peak shifts (6,8). Second, the FWHM of the shocked NTO should increase with the increased temperature characteristic of shock loading. During shock unloading, the width decreases and returns to its original ambient value after about 6 ns. The width of the 1325 cm^{-1} band shows completely different behavior. It does not seem to change at all under shock loading but it increases steadily after about 3 ns. The FWHM even increases during the 5-10 ns period, after the NTO layer has already unloaded to ambient pressure. At about 11 ns delay time the width increase appears to level off. Thus this broadening process has a time constant of a few ns. The origin of this process is now being studied. It may possibly be due to the formation of defects in the NTO crystals during the material relaxation process which occurs during shock loading.

Closer inspection of the spectra in figure 4 reveal the temporary appearance of an additional weak band at about 1293 cm^{-1}. This band is only visible while the shock front is inside the NTO layer (figures 4b-e). Fitting the CARS spectra to Voigt functions as described above shows that the area of this band peaks at a value between 5 and 8% of the main peaks. This additional band might be due to some reversible crystal deformation, thereby temporarily lowering its symmetry. Alternatively it is a hot band (v=1$\rightarrow$2) transition of one of the larger peaks. In that case, one can estimate the temperature at peak shock load to be several hundred degrees, which seems reasonable. We have not so far seen evidence for any shock-induced chemical reaction in NTO, which is consistent with the idea of NTO being a quite insensitive explosive (3). However the use of larger shock pressures, or the use of more sensitive explosives should allow us to make real-time observations of shock-induced energetic materials chemistry.

SUMMARY

The use of ps CARS with our microfabricated multilayer shock target array is very useful for investigating shock effects in polycrystalline materials such as NTO. The time resolution is very good, presently on the order of 100 ps for a 1 μm sample layer. Some very interesting effects have been observed, namely the appearance of a third, smaller intensity transition during shock unloading, which may be a hot band, and the nanosecond time scale broadening of one of the peaks, which might be caused by material relaxation processes occurring after the shock unloads.

ACKNOWLEDGEMENT

This work was supported by Air Force Office of Scientific Research contracts F49620-94-1-0108 and F49620-97-1-0056, Army Research Office contract DAAH04-96-1-0038, and National Science Foundation grant N00014-95-1-0259. We would like to thank Prof. Wight, University of Utah, for sending us a sample of NTO. Furthermore we would like to thank Dr. D. E. Hare and Dr. G. Tas for their support.

REFERENCES

1. Lee K.-Y., Gilardi R., *Mat.Res.Soc.Proc.* **296**, 237 (1993).
2. Sorecsu D.C., Sutton T.R.L., Thompson D.L., Beardall D., Wight C.A., *J. Molecular Structure* **384**, 87 (1996).
3. Lee K.-Y., Chapman L.B., Coburn M.D., *J.Energ.Mater.* **5**, 27 (1987).
4. Ritchie J.P., *J.Org.Chem.* **54**, 3553 (1989).
5. Hare D.E., Franken J., Dlott D.D., *Chem.Phys.Lett.* **244**, 224 (1995).
6. Franken J., Hambir S.A., Hare D.E., Dlott D.D., *Shock Waves* **7**, 135 (1997).
7. Hambir S.A., Franken J., Hare D.E., Chronister E.L., Baer B.J., Dlott D.D., *J.Appl.Phys.* **81**, 2157 (1997)
8. Tas G., Franken J., Hambir S.A., Hare D.E., Dlott D.D., *Phys.Rev.Lett.* **78**, 4585 (1997).
9. Tas G., Hambir S.A., Franken J., Hare D.E., Dlott D.D., *J.Appl.Phys.* **82**, 1 (1997).
10. Hare D.E., Lee I.-Y.S., Hill J.R., Franken J., Suzuki, H., Baer B.J., Chronister E.L., Dlott D.D., *Mater.Res.Soc.Proc.* **418**, 337 (1996).
11. Prabhakaran K.V., Naidu S.R., Kurian E.M., *Thermochimica Acta* **241**, 199 (1994).
12. Eesley G.L., *Coherent Raman Spectroscopy*, Oxford: Pergamon (1981).

CP429, *Shock Compression of Condensed Matter – 1997*
edited by Schmidt/Dandekar/Forbes
© 1998 The American Institute of Physics 1-56396-738-3/98/$15.00

ULTRAFAST VIBRATIONAL SPECTROSCOPY OF SHOCKS IN MOLECULAR MATERIALS: THE FIRST 100 PS

Selezion A. Hambir, Jens Franken, Jeffrey R. Hill and Dana D. Dlott

School of Chemical Sciences, University of Illinois at Urbana Champaign, Urbana, IL 61801

A system which uses a picosecond laser of moderate energy (150 µJ) to generate high repetition rate (100/s) shocks in a monolithic microfabricated shock target array is described. The sample is a molecular material incorporated as a thin layer of the array. Initial results used to characterize the shocks are obtained on polycrystalline anthracene. Coherent Raman spectroscopy is used to determine the shock front risetime (< 25 ps), the shock pressure (4.2 GPa), the shock-induced temperature jump (~350 deg), the shock falltime (1.5 ns), and the shock velocity (~4 km/s). The application of this nanoshock technique to biological materials, specifically the heme protein myoglobin, is discussed briefly.

INTRODUCTION

The goal of our research is to use short pulses from a convenient tabletop laser to generate shock waves in solids at a high repetition rate. We wish to generate reproducible, planar, constant velocity shocks with a controllable rise and fall time, which provide single-stage compression to at least a few GPa. In laboratories across the world, researchers are using ultrafast spectroscopy to study solid-state dynamics. Exciting new techniques have been developed, including time-resolved absorbance, fluorescence, Raman and infrared spectroscopies, and nonlinear spectroscopies such as photon echoes, four-wave mixing, second harmonic generation, transient gratings, etc. The techniques we are developing are intended to make it easy for laser researchers to study shock-induced phenomena, thereby bringing the benefits of shock wave research to many more laboratories.

Here we describe experiments in shock generation and detection using polycrystalline anthracene (1-4). Anthracene is regarded as a model system for the study of molecular materials, and an excellent database of anthracene properties now exists. The experiments on anthracene are intended to probe the details of our shock wave generation and detection system. Subsequently we intend to extend the work to materials of greater technological significance. Secondary high explosives are an obvious choice, and preliminary results on NTO are described in another paper in these Proceedings. Shock waves in biological materials is an area which is well suited to our methods. Some preliminary results using a biocompatible array and a heme protein are discussed briefly at the end of this paper.

SHOCK GENERATION AND DETECTION

Our experiments use a monolithic microfabricated shock target array, which consists of a series of thin layers on a substrate whose area is large enough (10 cm x 10 cm) to accommodate about 10 million individual shock target elements. An individual target element is ~100 µm in diameter, and it contains about 1 ng of material. The target array used for anthracene experiments, containing a 700 nm thick layer of polycrystalline anthracene, termed the "optical nanogauge", is diagrammed in Fig. 1.

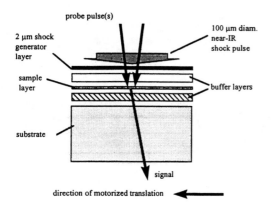

FIGURE 1. Schematic diagram of monolithic microfabricated shock target array. The anthracene sample layer is 700 nm thick. The shock transit time is 180 ps. From ref. 4.

In the experiments, shock dynamics in anthracene are monitored with broadband multiplex picosecond coherent anti-Stokes Raman spectroscopy (CARS) (5). We focus on the most intense anthracene Raman transition, denoted ν_4, at a nominal frequency of 1404 cm^{-1}. Raman spectra as a function of temperature and pressure indicate the blueshift of ν_4 is sensitive to density and insensitive to temperature, and the width of the ν_4 transition increases with temperature but is insensitive to density (1-4). Raman spectra of ν_4 can thus be used to estimate the instantaneous temperature and pressure. The details of this method, and a critical evaluation of its accuracy are given in refs. 1 and 4.

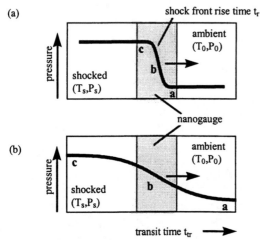

FIGURE 2. A shock moving through a thin layer. (Top) risetime is faster than transit time. (Bottom) risetime is slower than transit time. From ref. 4.

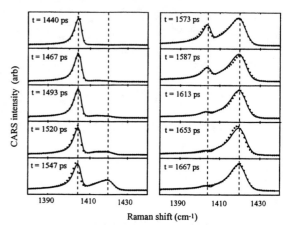

FIGURE 3. CARS spectra of anthracene with shock moving through 700 nm layer. The shock arrives at ~1500 ps. The two-peak structure shows the risetime is very much less than 180 ps. Detailed analysis shows the risetime < 25 ps. From ref. 4.

Figure 2 is a schematic of a shock front moving through a thin layer of material, from ref. 3. In our experiments, the shock front transit time through this 700 nm layer is $t_{tr} \sim 180$ ps. When the CARS spectrum is obtained, the spectrum is obtained from the entire layer at once. If the shock front risetime t_r is short compared to t_{tr}, then a CARS spectrum with the front in the layer will be seen with two peaks, one from ambient (T_0, P_0) material ahead of the shock front and one from shocked material (T_s, P_s) behind the front. When the risetime t_r is long compared to t_{tr}, the pressure rises uniformly in the thin layer and only a single peak which shifts between the locations of the ambient and the shock peaks will be seen.

Figure 3 shows data obtained while a shock produced by a near-IR 150 ps duration, 150 μJ pulse moves through the anthracene. Each displayed spectrum is the average of ~5000 reproducible shocks obtained over a ~60s time period. The shock front reaches the anthracene at about 1.5 ns (the arrival of the near-IR pulse at the target is at time denoted t = 0). Then a characteristic double peak spectrum is seen while the front moves through the anthracene. When the front has passed through the anthracene, a spectrum is observed with a peak blueshift of 16 cm^{-1} and a FWHM of 6 cm^{-1}, about 1.5 times greater than the width at 300K. These values indicate a shock pressure of 4.2 GPa and a shock pressure of ~350°C (3).

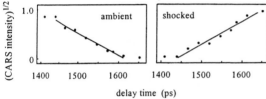

FIGURE 4. Relative values of the square root of the CARS intensities of the ambient and shocked spectra in Fig. 3. Knowing the layer thickness, the velocity can be determined to be 4.3 km/s, which is consistent with the shock pressure estimated from CARS and the shock Hugoniot. From ref. 3.

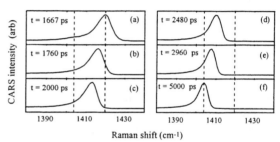

FIGURE 5. CARS spectra of anthracene during shock unloading, from ref. 3.

Even a brief glance at Fig. 3 shows the shock front risetime is very much less than the 180 ps transit time. In refs. 2 and 3, we show how a mathematical method termed (6) singular value decomposition (SVD) can be applied to these spectra. SVD resolves the data in Fig. 3 into a minimal set of eigenspectra and time-dependent coefficients. With SVD, we can determine the shock front risetime and shock front velocity (3). If the risetime is zero, the spectra in Fig. 3 will be solely a superposition of the ambient and shocked spectra. If the risetime is finite, there will be an additional contribution from material at the shock front, where the pressure rises from P_0 to P_s over a finite distance. SVD analysis shows the risetime $t_r < 25$ ps. Consequently the shock front is < 100 nm wide. Since an anthracene molecule is about 1 nm, the shock front is less than 100 molecules wide. Using SVD to look at the time dependent intensities of the two peaks, we can tell at any instant how much of the 700 nm thick layer is ahead of and behind the shock front. That allows us to measure the shock velocity in real time as the front moves just a few tens of nm. Results of the velocity measurements are shown in Fig. 4. Keep in mind the intensity of a CARS peak is proportional to the square of the amount of material (5).

Figure 5 shows data obtained during shock unloading. In this data we see a single peak with a spectral peak which smoothly moves from the shocked position to close to the ambient location. That is the expected behavior for an unloading process which occurs more slowly than the 180 ps shock transit time. The time constant for unloading is 1.5 ns.

Figure 6 shows the waveform of the shock pulse deduced from the spectra in Figs. 3 and 5. We term this a "nanoshock" pulse because the duration is a

few ns, and the shock compresses a few ng of material. By varying the thickness and composition of the shock generation and buffer layer, we exert some control over the nanoshock pulse. We have made the risetime range from <25 ps to 800 ps, and the falltime range (1,3) from 1.5 ns to 5 ns.

The nanoshock pulse in Fig. 6 has two features, a picosecond risetime and a nanosecond falltime. Each feature is remarkable in its own way. The <25 ps risetime of the shock is a bit of a surprise considering the shock front is moving through a polycrystalline layer of a rather complicated molecular material. This risetime is faster than the characteristic time for relaxation of vibrational excited states in anthracene. When this occurs, the passage of the shock front produces highly non-equilibrium vibrationally excitations in the material (3).

During the 1.5 ns unloading process, the temperature of anthracene falls at a tremendous rate-- several hundred degrees per ns. There are many ways to rapidly heat materials with a laser, but to our knowledge there is no other way to cool a material at such a tremendous rate--a hundred billion deg/s. As far as we know, the fastest cooling rates

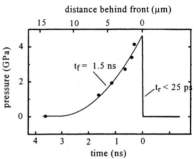

FIGURE 6. Waveform of the nanoshock deduced from data in Figs. 3-5. From ref. 3.

otherwise obtained in materials are perhaps a million degrees/sec. Ultrarapid cooling processes can be exploited to freeze materials into metastable configurations and to rapidly quench chemical reactions. Photochemical reactions can be controlled by turning the light on and off. By analogy, a thermochemical reaction could be turned on by ultrafast shock compression heating and turned off by the cooling from ultrafast shock unloading.

SHOCK IN BIOLOGICAL MATERIALS

A novel application for shock waves being pursued in our laboratory involves shocks in biological materials. Our first experiment involves the dynamics of the heme protein myoglobin (Mb). Mb is a molecular nanomachine (see Fig. 7) whose dimensions are about 4 nm, whose function is oxygen storage (7). Mb consists of a complicated set of interlocking protein helices which fold around a colored molecule called heme. Oxygen binds at an active site at the heme. Like all large biomolecules, Mb will have quite complicated mechanical properties. The folding and unfolding dynamics of proteins is one of the hottest topics in biochemistry and biophysics today.

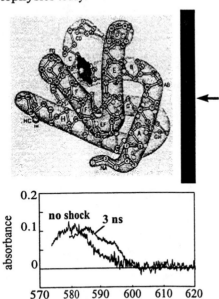

FIGURE 7. (top) Schematic illustration of shock front hitting myoglobin heme protein in water. (bottom) visible absorption spectrum and shifted spectrum, 3 ns after shock.

Shocking Mb is expected to result in a large structural distortion, since Mb undergoes denaturation (8) (unfolding) at ~80°C or 8 kB. After the fast cooling process caused by shock unloading, Mb will be frozen into a metastable configurational state which will undergo quite interesting relaxation processes which can be optically monitored.

In preliminary experiments, we have fabricated biocompatible shock target arrays, where the Mb protein is present in its native folded configuration. The sample layer is a thin layer of Mb solution in aqueous buffer, with glycerol added to stabilize the layer. The effects of shock are detected via transient visible absorption of the strongly colored heme group. A portion of the visible absorption spectrum is shown in Fig. 7 without shock, and 3 ns after shock. These very early experimental results illustrate the significance of the nanoshock technique for non-conventional materials, or precious materials available only in tiny quantities.

ACKNOWLEDGMENT

We acknowledge the contributions of Dr. Guray Tas now at Department of Physics, Brown University, and Dr. David E. Hare, now at Lawrence Livermore National Laboratory. This research was supported by Air Force Office of Scientific Research F49620-94-1-0108 and F49620-97-1-0056, National Science Foundation DMR 94-04806, and US Army Research Office DAAH04-96-1-0038. The work on proteins was supported by Office of Naval Research N00014-95-1-0259.

REFERENCES

1. S. A. Hambir, J. Franken, D. E. Hare, E. L. Chronister, B. J. Baer and D. D. Dlott, J. Appl. Phys. 81, pp. 2157-2166 (1997).
2. J. Franken, S. Hambir, D. E. Hare and D. D. Dlott, Shock Waves 7, 135 (1997).
3. G. Tas, J. Franken, S. A. Hambir and D. D. Dlott, Phys. Rev. Lett. 78, pp. 4585 (1997).
4. G. Tas, S. A. Hambir, J. Franken, D. E. Hare and D. D. Dlott, J. Appl. Phys. 82, 1 (1997).
5. G. L. Eesley, *Coherent Raman Spectroscopy*, Pergamon, Oxford, 1981.
6. R. I. Shrager and R. W. Hendler, Anal. Chem. 54, 1152 (1982).
7. E. Antonini, M. Brunori, *Hemoglobin and myoglobin in their reactions with ligands*, North Holland, Amsterdam, 1971.
8. A. Zipp and W. Kauzmann, Biochemistry 12, 4217 (1973).

826

CP429, *Shock Compression of Condensed Matter – 1997*
edited by Schmidt/Dandekar/Forbes
© 1998 The American Institute of Physics 1-56396-738-3/98/$15.00

DETERMINATION OF THE RESPONSE OF PENTAERYTHRITOL TETRANITRATE TO STATIC HIGH PRESSURE UP TO 4.28 GPa BY NEUTRON DIFFRACTION. *

J. J. Dick[a] and R. B. von Dreele[b]

a Group DX-1, MS P952, b MLNSC, MS H805,
Los Alamos National Laboratory, Los Alamos, New Mexico 87545

Neutron powder diffraction experiments were performed on pentaerythritol tetranitrate explosive up to 4.28 GPa. For deuterated samples the changes in lattice parameters, intramolecular torsional angles and molecular rotation were measured. The lattice parameter changes were different from those observed in protonated samples. However, there is no evidence of a phase transition or change in molecular symmetry.

INTRODUCTION

Some orientations of pentaerythritol tetranitrate (PETN) crystals have anomalously high shock initiation sensitivity around 4 to 5 GPa. Anomalous luminescent emission and initiation of detonation have been observed for two orientations of single crystals of (PETN) in shock experiments near 4 GPa.(1) The crystals were more sensitive at 4.2 GPa than at 8.5 GPa. From the data available it was not clear what was responsible for this anomaly.

There have been observations of this anomaly by other workers. An unexpected, rapid rise in longitudinal stress after an induction time of about 0.3 μs was noted in an impact-face, quartz-gauge experiment at 4.1 GPa in a [110] crystal by P. M. Halleck and J. Wackerle.(2) This was the first published work on this subject. Recently observations of the induction time for onset of luminescent emission in [110] crystals was obtained over the stress range of 3.5 to 12 GPa by D. Spitzer.(3) The minimum induction time was observed at about 5 GPa; induction times are longer for shock stresses both smaller and larger than 5

GPa. Their results are consistent with our observations in Ref.1.

There is a report of reaction propagation rates in polycrystalline PETN in a diamond anvil cell that may be related to the anomaly.(4) The reaction propagation rate was measured from 2 to 20 GPa. A local peak in rate was observed at about 5 GPa, about the same location as the peak shock sensitivity for [110] crystals at low shock stresses. It is not clear what the connection is between the two kinds of measurements. In the diamond anvil cell the propagation rates are more than an order of magnitude slower than shock speeds and apparently represent deflagrations at high pressure. In addition, they performed Fourier transform infrared (FTIR) spectral measurements that they interpreted as caused by an asymmetric deformation of the molecule to a lower symmetry conformation on compression.

In order to answer the question if any phase transition or change in molecular conformation is associated with the anomalous sensitivity at 4 to 5 GPa, neutron powder diffraction data under pressure was obtained using the neutron beam at LANSCE in Los Alamos.

*Work performed under the auspices of the U. S. Department of Energy

EXPERIMENTAL TECHNIQUE

Static high pressure was obtained using "Paris-Edinburgh" cell(5) that is capable of hydrostatic pressures up to 10 GPa. Previously data has been obtained on nitromethane to 5.5 GPa.(6) Nitromethane is an explosive molecule CH_3NO_2. It is much smaller than PETN $C_5H_8N_4O_{12}$ currently under study. In PETN there is a central carbon with a tetrahedron of 4 carbons around it. To each of the 4 carbons is bonded 2 hydrogens and an ONO_2 group. In order to reduce background counts from incoherent scattering, PETN was initially studied in fully deuterated form d_8-PETN, more than 99 mol% deuterium. A mixture of 47 mg of d_8-PETN and 23 mg of NaCl was pressed into a pellet 4 mm in diameter and 3.5 mm thick. The pellet had a void content of about 15%. This pellet was placed in a slurry of crushed aerogel and Fluorinert-70 (3-M Company) in the cavity of the pressure cell. The cavity between the gasketed tungsten carbide anvils is 8.5 mm in diameter. Upon compression the Fluorinert-70 infiltrated the pellet providing a hydrostatic medium. The NaCl provides the pressure standard.

Diffraction patterns were obtained in about 12 hours at each of 5 pressures from 0.154 to 4.28 GPa. Structure refinements were performed via the Rietveld method using the GSAS suite of computer programs.(7) "Soft" constraints were included for all near-neighbor interatomic distances; the torsion angles were not affected by these constraints. Final residuals, Rwp, of ca. 2% were obtained for all refinements.

EXPERIMENTAL RESULTS AND DISCUSSION

A neutron powder diffraction for d_8-PETN at 0.154 GPa is shown in Fig. 1. There is no change in symmetry from ambient. It is still in space group $P4\bar{2}_1c$ and remains there to 4.28 GPa. Therefore, the molecule has not undergone any asymmetric deformation. The FTIR experiments published in Ref.4 showed apparent asymmetric deformation of the molecule. In that case a thin film of PETN was compressed against a KBr pellet in a diamond anvil cell. The resultant nonhydrostatic stresses may have caused the spectrum changes that were interpreted as due to asymmet-

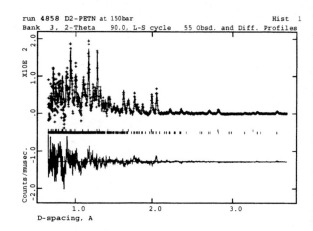

FIGURE 1. Time-of-flight neutron powder diffraction data for d_8-PETN at 0.154 GPa. The observed data have been normalized by the incident intensity; the computed background from the Rietveld refinement has been subtracted. The upper curve is the fit to the data. The lower curve is the difference between the fit and the measured counts. Between the curves there are two sets of bars. The upper set marks the position of NaCl peaks, and the lower set marks the position of PETN peaks.

ric deformation of the molecule.

The pressure vs volume compression data for PETN are shown in Fig. 2. The compression for deuterated PETN appears to be the same as for x-ray data for protonated PETN(8) in this pressure range within experimental precision. While the apparent volume decrease of the unit cell is the same for both, the unit cell deformation is different for protonated and deuterated PETN. This is shown in the lattice parameter changes with pressure for the two types in Fig. 3. The values at zero pressure are from the x-ray and neutron data refinement reported by Conant *et al.*(9) From x-ray spectra they found that the lattice parameters for protonated and deuterated PETN were the same. However, with pressure the lattice parameters vary differently. For deuterated PETN the a-axis shortens more and the c-axis shortens less than protonated PETN . It does so in such a way that the unit cell volumes at a given pressure are the same within experimental precision.

There are significant changes in the molecular conformation with compression. Torsional angle changes require the least energy in accommo-

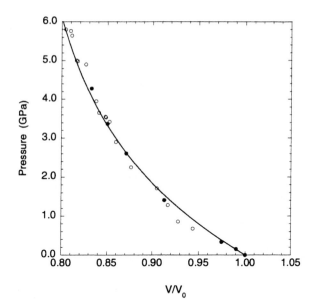

FIGURE 2. Pressure vs volume data for PETN. The open circles are x-ray data from Olinger *et al.* for protonated PETN, and the closed circles are from neutron diffraction of d_8-PETN. The solid line is a fit to the protonated data given in the article by Olinger *et al.*.

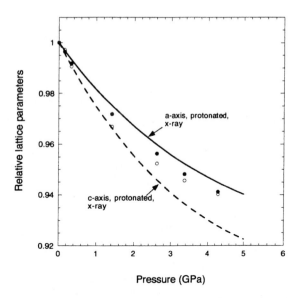

FIGURE 3. Lattice parameter changes for protonated and deuterated PETN vs pressure. The solid circles are the d_8-PETN a-axis data, and the open circles are the d_8-PETN c-axis data from neutron experiments. The solid line is the fit to protonated, a-axis, x-ray data and the dashed line is the fit to protonated c-axis data of Olinger *et al.* The values at zero pressure are from Conant *et al.*

dating the closer approaches between neighboring molecules. The angle changes are as large as 8°. The changes for the C-C, C-O, and O-N torsions are displayed in Fig. 4. The angle changes are referenced to zero at the first measured point 0.154 GPa since the conformation of d_8-PETN at zero pressure is not known with certainty. It is interesting to note that both the C-C and C-O torsional angle changes go through a minimum near 1.4 GPa. The changes from the angle at 0.154 GPa are 8° and 4°, respectively. As noted earlier, these changes are symmetric within the molecule. Precision of the angle measurement is about 0.25°.

In addition to the intramolecular conformation changes, the entire molecule performs a rotation about the c axis. These rotations are shown in Fig. 5. It changes by a total of 2°. It is interesting to note that this angle change goes through a local maximum at 1.4 GPa where the C-C and C-O torsional angle changes go through a minimum. As the lattice parameters are reduced under compression, the intermolecular forces get stronger. The intramolecular conformation changes and molecular rotation act in concert so as to minimize the energy of interaction.

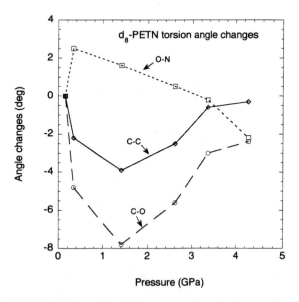

FIGURE 4. Intramolecular torsional angle changes for d_8-PETN vs pressure. The data is referenced to zero at the lowest pressure from the neutron diffraction experiments on d_8-PETN, 0.154 GPa.

Initially work was performed on d_8-PETN to reduce the background signal due to incoherent scattering. This scattering is much stronger from protons than deuterons. However, since small but significant changes occurred in molecular conformation as well as molecular rotation under compression in the deuterated case, we decided to attempt to obtain similar information for protonated PETN. All the shock experiments on PETN single crystals were performed on protonated samples. Powder neutron diffraction data was obtained. There was more background, so the runs were for 24 hours instead of 12. The data has not been fully analyzed yet.

Since this work began an article has been published showing that the anomalous shock initiation sensitivity at 4 to 5 GPa is connected with a 2-wave, elastic-plastic wave structure with large elastic precursor waves.(10) These observations are consistent with the model of initiation by sterically hindered shear during the high-strain-rate, shock compression in the plastic wave. In this case the PETN is uniaxially precompressed to about 3 GPa by the elastic wave before the sterically hindered shear begins. In addition to reduction of lattice parameters, some amount of molecular rotation and conformation changes must occur during this precompression. Perhaps this has the effect of enhancing the sterically hindered shear and causing more molecular excitation followed by decomposition.

ACKNOWLEDGMENTS

The deuterated PETN was prepared by Mike Coburn and Don Ott at Los Alamos a number of years ago. Elaine Hickman pressed up the pellet samples.

REFERENCES

1. J. J. Dick, R. N. Mulford, W. J. Spencer, D. R. Pettit, E. Garcia, and D. C. Shaw, J. Appl. Phys. **70**, 3572 (1991).

2. P. M. Halleck and Jerry Wackerle, J. Appl. Phys. **47**, 976 (1976).

3. D. Spitzer, Ph. D. thesis, Universite Louis Pasteur de Strasbourg, p. 153, 1993.

4. M. F. Foltz, in *Tenth International Detonation Symposium*, (Office of Naval Research, Arlington, VA, 1993) p. 579.

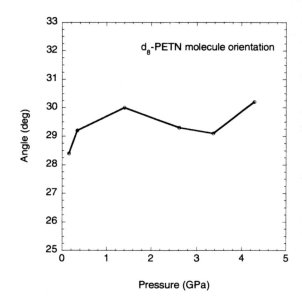

FIGURE 5. Molecular rotation for d_8-PETN about the c axis. The angle is referenced to zero with the C-C bond of the asymmetric unit in the a-c plane.

5. R. J. Nelmes, J. S. Loveday, R. M. Wilson, J. M. Besson, P. Pruzan, S. Klotz, and S. Hull, Phys. Rev. Letters **71**, 1192 (1993).

6. R. B. von Dreele, High Pressure Res. **14**, 13 (1995).

7. A. C. Larson and R. B. von Dreele, Los Alamos Report LAUR-86-748 (1994).

8. B. Olinger, P. M. Halleck and H. H. Cady, J. Chem. Phys. **62**, 4480 (1975).

9. J. W. Conant, H. H. Cady, R. R. Ryan, J. L. Yarnell, and J. M. Newsam, Los Alamos Report LA-7756-MS, 1979 (unpublished).

10. J. J. Dick, J. Appl. Phys. **81**, 601 (1997).

CHAPTER XI

INTERFEROMETRY
AND GAUGES

CP429, *Shock Compression of Condensed Matter – 1997*
edited by Schmidt/Dandekar/Forbes
© 1998 The American Institute of Physics 1-56396-738-3/98/$15.00

THE ACCURACY OF VISAR INSTRUMENTATION

L. M. Barker

Valyn International, 12514 Menaul Blvd. N. E., Albuquerque, New Mexico 87112

VISAR instrumentation accuracy depends on the accuracy of determining the VPF constant and the Fringe Count at any time. Sources of uncertainty in these two factors are discussed. Experimental calibrations of VISAR accuracy are reviewed. VISAR accuracies of 0.2% to 1% of the peak velocity are common.

INTRODUCTION

A VISAR (1) (Velocity Interferometer System for Any Reflector) makes use of the Doppler shift in the wavelength of laser light which is reflected from a moving surface. The VISAR interferometer analyzes the light, producing interference fringes in proportion to the amount of the Doppler shift. The surface velocity is then calculated by multiplying the interferometer's Velocity Per Fringe (VPF) constant by the number of fringes produced. Essential to the operation of a VISAR is a delay time which is introduced into one of the light paths of the inter-ferometer, the magnitude of which determines the measurement sensitivity of the VISAR.

In order to simplify the following discussions of VISAR accuracy, only VISARs which use etalon delay elements, and only those which use photomultiplier detectors are considered, unless otherwise stated.

Sources of uncertainty in the VPF constant are considered first, after which the accuracy of measuring the number of fringes (the fringe count) is discussed. Finally, two experimental calibrations which confirm calculations of VISAR accuracy are reviewed.

VELOCITY PER FRINGE CONSTANT

An accepted equation for the VPF constant is (2)

$$\text{VPF} = \lambda_o/[2\tau(1+\delta)(1+\Delta v/v_o)] , \qquad (1)$$

where λ_o is the laser light wavelength, τ is the delay time in the interferometer, δ accounts for the extremely slight change in refractive index of the delay elements because of the Doppler shift in wavelength, and $\Delta v/v_o$ corrects for changing index of refraction effects in a window material when a transparent window is in contact with the surface whose velocity is to be measured. Although Eq. 1 generally provides a very good approximation of the VPF, it assumes that all of the transparent elements in the interferometer have the same value of δ, a material-dependent parameter. A more exact development of the VPF equation, obtained by generalizing the derivation in Ref. 2, defines an effective delay time, τ', which replaces the product $\tau(1+\delta)$ in Eq. 1. The value of τ' is the sum of the effective $\Delta\tau'$s of the various transparent elements in the interferometer ($\tau' = \Sigma\Delta\tau'_i$), where

$$\Delta\tau'_i = (2h_i/c)(n_i - 1/n_i)(1+\delta_i). \qquad (2)$$

Here $2h_i$ is the total distance traveled by the interferometer beam through the ith element, n_i is its refractive index, and c is velocity of light. Note that the appropriate dispersion factor, $(1+\delta_i)$, appears for each element in its value of $\Delta\tau'_i$. The use of τ' instead of $\tau(1+\delta)$ in Eq. 1 allows accurate VPFs to be calculated for mixed delay element material types in VISARs.

Another change which increases accuracy is the addition of the factor $[\frac{1}{2}(1+\cos\theta)]$ in the denominator of Eq. 1. In this term, θ is the average angle of reflection of the light collected for return to the VISAR interferometer (See Fig. 1). The assumption in Eq. 1 is that $\theta = 0$, i.e., the collected reflected light travels normal to the surface, just like

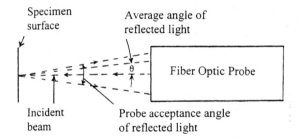

Specimen surface

Average angle of reflected light

Fiber Optic Probe

θ

Incident beam

Probe acceptance angle of reflected light

Figure 1. Geometry of incident beam and reflected light. The angle θ is the weighted average of the reflection angle of the light collected by the probe.

the incident light, in which case the above factor is unity. However, when the average angle of the reflected light is more than a few degrees, the Doppler shift is decreased slightly by the above factor, thus affecting the VPF. In Valyn International's standard fiber optic probe, for example, inclusion of the factor increases the VPF by 0.25%. The improved equation for VPF, then, is

$$VPF = \lambda_0 / [2\tau'(\tfrac{1}{2}(1+\cos\theta))(1+\Delta v/v_0)] . \quad (3)$$

It is still assumed in Eq. 3 that the light onto the specimen surface has a zero degree angle of incidence, of course. However, the error is less than 0.1% for angles less than 3.6°.

An analysis of the factors in Eq. 3 which determine the VPF gives the following approximate percent error attributable to each:

n_i = index of refraction uncertainties	~0
λ_0 = laser light wavelength	<0.02%
$(1+\delta_i)$ = dispersion effect in elements	<0.1%
θ = average reflection angle	<0.1%
$2h_i$ = dist. light travels in elements	<0.1%
Δx = back-off setting of delay leg mirror	<0.1%
$(1+\Delta v/v_0)$ = calibrated window factor	½ to 1%

The last three items above warrant some discussion. The value of $2h_i$ depends on micrometer measurements of element thicknesses, which are quite accurate for thicker elements (smaller VPFs). However, the ± 5 μm accuracy of these measurements becomes a larger percent uncertainty when element thicknesses are quite small, and especially when an element in the delay leg must be partially offset by a compensating element in the reference leg to attain a higher VPF.

The uncertainty of setting the back-off, Δx, of the delay leg mirror of the interferometer is a function of the VISAR's design and/or the care used by the maker of the VISAR. Although Δx does not appear in Eq. 2 for τ', it is important because it adds or subtracts from the calculated τ' because of the resulting difference in distance that light travels in the delay leg. The Δx accuracy of the Valyn VISAR is about ±10 μm, which is felt to be considerably smaller than most. Nevertheless, when Δx is small, i.e., for higher VPFs, the percent error can increase. The percent uncertainties given above attributable to $2h_i$ and Δx are valid for VPFs up to about 2 km/s/fr, above which they increase approximately in proportion to the VPF.

When a window is used on the VISAR specimen, the accuracy of the VPF depends also on the accuracy with which the term $\Delta v/v_0$ has been experimentally measured. The most common VISAR window materials include PMMA, fused silica, sapphire (3,4), and lithium fluoride (4). The uncertainty in VPF due to these window materials ranges from about ½% to 1%, depending on the material and the velocity range measured. When no window is used on the specimen, there is no error from $\Delta v/v_0$, of course.

Considering the several sources of error in calculating the VPF, the RMS uncertainty in VPF is about ± 0.1 to ± 0.2% for VPFs up to about 2 km/s/fringe, if no window is used. When a window is used, it dominates the other uncertainties, making the VPF accuracy about ± ½% to ± 1%. Since the equation for the velocity, v, at any given time, t, is

$$v(t) = VPF \times F(t), \quad (4)$$

where F(t) is the fringe count, the percent uncertainty in v(t) attributable to the uncertainty in the VPF is the same as the percent uncertainty in the VPF constant.

DETERMINING THE FRINGE COUNT

In the VISAR interferometer, physics dictates that the relation between the output signal light amplitude and the fringe count is a sine wave. VISAR fringe data seldom look like sine waves, however, because they are recorded as a function of time, not fringe count. The fringe count is the number of *complete oscillations* of the fringe record since the velocity was zero, an integer we shall call $N_F(t)$, plus any remaining *fraction of a fringe*, $f_F(t)$, up to the time in question. The value of $f_F(t)$ is determined from the knowledge that the light amplitude vs. the fringe count is a sine wave. The equation for the fringe count is then

$$F(t) = N_F(t) + f_F(t). \qquad (5)$$

N_F normally has zero error. The only exception is when a large jump in velocity produces a burst of fringes whose frequency is too high to be recorded. Even in these cases, however, N_F can almost always be determined exactly by a knowledge of the boundary conditions of the experiment. Evaluating f_F, on the other hand, depends on the recording system linearity and noise on the data trace.

The nonlinearities in the electronic elements, such as the photomultipliers and the digitizing oscilloscopes, taken together, should not exceed about 3% of the oscilloscope signal amplitude. Shot noise, which arises mainly from the random fluctuation in the number of photoelectrons generated at the PMT photocathode, is the primary noise source. Based on experience, the standard deviation of the shot noise is generally less than 5% of any given VISAR PMT signal. Combining with the 3% uncertainty due to nonlinearities leads to under 6% uncertainty in the signal amplitude. However, a 6% uncertainty in amplitude leads to less than ± .02 fringe uncertainty in f_F, because one fringe consists of four excursions of the signal amplitudes, two for each of the two 90° out-of-phase VISAR signals. Several other minor factors may raise the uncertainty in f_F to about ± .02 fringe.

Since there is zero uncertainty in $N_F(t)$, the total uncertainty in the fringe count, $F(t)$, is just the uncertainty in $f_F(t)$, or ± .02 fringe. Although this agrees with the uncertainty estimated in the original VISAR paper of 1972 (1), intervening improvements have further improved accuracy. The most important of these is the push-pull modification of Hemsing (5), which increases light efficiency and reduces shot noise by about half. Another improvement is the use of an optical fiber to transport the Doppler-shifted laser light from the specimen to the VISAR interferometer. Without the fiber, any specimen tilt and/or displacement causes small variations in the optical paths through the interferometer, producing false fringe shifts. Recent improvements at Valyn International allow easy, precise optimizing of the phase difference and interferometer alignment just prior to the shot, further contributing to accuracy.

Although the above improvements have reduced the calculated fringe count uncertainty to less than ± .015 fringe, only a ± .02 fringe uncertainty is often claimed for reasons of conservatism and because extensive experimental verifications of accuracy have been lacking. The percent error in the fringe count, $F(t)$, is therefore stated as $[.02/F(t)] \times 100\%$, and the corresponding error in the velocity is the same percent, of course. Note that unlike the error attributable to the VPF, which is a constant up to 2 km/s/fr, the error in $F(t)$ is inversely proportional to $F(t)$. The overall uncertainty in $v(t)$ of a VISAR measurement is the RMS value of the VPF and $F(t)$ uncertainties. Figure 2 shows the approximate accuracies which can be expected of Valyn VISAR measurements with 1 to 2 ns time resolution.

EXPERIMENTAL ACCURACY CALIBRATION

Perhaps the most accurate calibration of etalon-delay VISAR measurements, and possibly the only published one, was done by Barker and Schuler (2) in 1974. In that study, two experiments were performed in which very careful electrical probe measurements of projectile velocity, accurate to within ± 0.1%, were compared to VISAR measurements of the same projectiles, with a calculated accuracy of ± 0.2%. The final result was that the VISAR measurement agreed with the probe measurement to within ± 0.1% in each experiment.

Unpublished calibration experiments have been done by R. E. Setchell, G. T. Holeman, and O. B. Crump, Jr. of Sandia National Laboratories as a consistency check on their instrumentation. The experiments involve an impact of fused silica onto fused silica, with a fused silica window, such that the peak velocity must be exactly half the impact velocity (assuming uniform fused silica material properties). The latest two experiments used Valyn fiber optic probes and an etalon delay VISAR. The projectile velocity measurements were accurate to ± 0.1%. The VISAR produced only about 1.2 fringes in the first experiment, and agreed with half

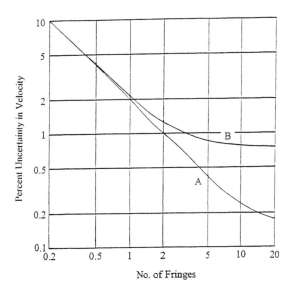

FIGURE 2. Percent uncertainties in velocity vs. number of fringes produced by the VISAR interferometer. Interferometer sensitivity is adjustable for more or fewer fringes. Curve A is for measurements without a window. Curve B is representative when a calibrated window material is used on the specimen surface.

the projectile velocity to within 0.2%, which compares with a calculated uncertainty of 1.7% from curve B of Fig. 2. The second experiment produced 1.7 fringes, and agreed with the expected velocity to within 0.2%, compared with a predicted uncertainty of about 1.3% from curve B of Fig. 2.

One might expect the better agreement than predicted by curve B in these experiments because the peak velocity was obtained by averaging a large number of data points, effectively removing the shot noise as an inaccuracy factor. Nevertheless, the agreements are still better than might be expected, which adds to confidence that the calculated uncertainties are at least not overly optimistic. According to Setchell, these experiments typically show only ½% or less measurement discrepancies, which is well within expectations from Fig. 2.

DISCUSSION

Many details must be left out in treating the complicated subject of accuracy in a paper of this length. For instance, the VISAR's intrinsic time resolution is equal to its delay time, which can be from a small fraction of a nanosecond to 10 ns or more in lens delay VISARs. However, the final time resolution cannot be better than the rise-time of the system used to detect and record the interferometer's light fringes, of course.

An additional accuracy attribute of VISARs is the absence of any perturbation of the velocity by the VISAR measurement, at least in the case of free surface measurements. However, some aids to VISAR measurements <u>can</u> affect the measured velocity. These include reflective materials applied to specimen surfaces to enhance reflectivity, buffers used to preserve surface integrity during the shock event, and of course windows attached to specimen surfaces, together with any adhesive layers associated with these items.

Finally, it is important to remember that VISAR accuracies of ½ to 1% have been easily and routinely attained without going to *heroic* efforts to optimize certain parameters, such as balancing photomultiplier outputs, obtaining perfect interferometer alignment, or setting the phase angle to exactly 90°. VISAR accuracy is reasonably tolerant of small deviations from optimum alignment.

ACKNOWLEDGEMENTS

The author gratefully acknowledges the support of Robert E. Setchell, G. T. Holeman, and O. B. Crump, Jr. of Sandia National Laboratories in sharing their data on calibration experiments to evaluate VISAR accuracy.

REFERENCES

1. Barker, L. M., and Hollenbach, R. E., *J. Appl. Phys.* **43,** 4669-4675 (1972).
2. Barker, L. M., and Schuler, K. W., *J. Appl. Phys.* **45, 3692-3693 (1974).**
3. Barker, L. M., and Hollenbach, R. E., *J. Appl. Phys.* **41,** 4208-4226 (1970).
4. Wise, J. L., and Chhabildas, L. C., in *Shock Waves in Condensed Matter*, ed. by Y. M. Gupta, Plenum Press, New York, 1986, 441-454.
5. Hemsing, W. F., *Rev. Sci. Inst.,* **50,** 73-78, (1979).

CP429, *Shock Compression of Condensed Matter – 1997*
edited by Schmidt/Dandekar/Forbes
© 1998 The American Institute of Physics 1-56396-738-3/98/$15.00

INVESTIGATION OF MICROSCALE SHOCK PHENOMENA USING A LINE-IMAGING OPTICALLY RECORDING VELOCITY INTERFEROMETER SYSTEM[*]

Wayne M. Trott and James R. Asay

Sandia National Laboratories, Albuquerque, NM 87185-0834

An optically recording velocity interferometer system (ORVIS) can be operated in a line-imaging configuration that effectively combines subnanosecond temporal resolution with high spatial resolution (length scales < 10 μm). This technique easily captures very small temporal variations in the onset of motion across the face of a small-scale (400-μm diameter) laser-driven flyer. In another application, line-imaging ORVIS has been used to obtain spatially resolved particle velocity vs. time information at flyer impact on a lithium fluoride witness plate. Subnanosecond differences in flyer arrival time are clearly resolved and the results also show subtle amplitude variations in the pulse delivered at different locations of the acceptor. Observed velocity field variations in laser acceleration of a patterned flyer target demonstrate the feasibility of using line ORVIS in studies of instability formation and growth. These results indicate that this diagnostic can be applied to a wide variety of shock phenomena.

INTRODUCTION

Velocity interferometry is a well-established and frequently utilized technique for measuring the velocity-time history of many types of samples, including explosively driven or laser-driven flyer plates and a wide variety of shock loaded materials. In numerous applications, time-resolved data from different interferometer designs (such as VISAR, ORVIS and Fabry-Perot) have led to improved understanding of physical and chemical processes associated with high-rate deformation. The most commonly employed approach involves "single point" measurements wherein an average velocity is determined (as a function of time) over a spot diameter typically ranging from 50 to 200 μm. In this configuration, neither macroscopic scale spatial variations in the response of samples larger than the

spot size nor small scale variations within this diameter can be determined.

The wealth of information potentially available in spatially resolved measurements provides a compelling motivation to develop more complex approaches to velocity interferometry. Interesting applications for advanced diagnostics include examination of the physical mechanisms of high-rate deformation in heterogeneous materials (e.g., granular explosives and polycrystalline piezoelectric materials) and studies of the formation and growth of hydrodynamic instabilities in thin, ablatively accelerated samples. In 1986, Gidon and Behar (1) reported a Fabry-Perot design that captured the velocity field of an entire surface at a single time. Subsequent developments included line-imaging Fabry-Perot and VISAR interferometers (2,3) for continuous measurement of the velocity of a line segment on a moving surface and an extended full-field Fabry-Perot method (4) for determining velocity as a function of both position and time (using a framing camera to record data in a time sequence).

[*]This work was supported by the United States Department of Energy under Contract DE-AC04-94AL85000. Sandia is a multiprogram laboratory operated by Sandia Corporation, a Lockheed Martin Company, for the United States Department of Energy.

Recently, Baumung et al. (5) presented a simplified design for a high-resolution, line-imaging interferometer that can be operated with a continuous wave laser source of modest power (~1 W). In this paper, we explore the feasibility of using a line-imaging ORVIS assembly of similar design to examine microscale phenomena in several experiments involving laser-driven flyers. Data revealing fine-scale spatial and temporal variations at both flyer launch and flyer impact have been obtained. In addition, we report preliminary tests on flyer targets prepared with periodic thickness variations in order to evaluate line-imaging ORVIS as a diagnostic for hydrodynamic instability formation and growth.

EXPERIMENTAL

Many elements of the experimental design for measuring the velocity of laser-driven flyers with ORVIS have been described previously (6). In modifying ORVIS, the method of Baumung et al. (5) was used to achieve a line focus at the target assembly; i.e., a cylindrical lens was introduced in the incoming laser path to expand the beam horizontally ahead of the usual achromat doublet which focuses the laser and collects the diffusely reflected light from the target. The resulting illuminated area at the achromat lens focus was significantly longer than the flyer target diameter (400 μm) and ~50 μm wide. In effect, the light collection optics in ORVIS act as a long-working-distance microscope coupled to the interferometer assembly. This arrangement allows the image size (determined by the light collection optics) to be varied essentially independently of the fringe spacing (determined by the angle and spacing of the interferometer mirrors), providing substantial latitude in adjusting both field of view and spatial resolution. Different configurations of image size and fringe spacing were used, depending on the spatial resolution requirements of each test. The interferometer fringe displacement (linearly proportional to velocity) was viewed by a electronic image-converter streak camera and recorded on a 576 x 385 element intensified CCD detector.

The driving laser used in this work was a Q-switched Nd:Glass oscillator (λ = 1.054 μm, ~18 ns pulse duration). Flyer targets were prepared by physical vapor deposition on polished output ends of

FIGURE 1. Surface profile image of patterned flyer target (first coating--0.8-μm-thick aluminum).

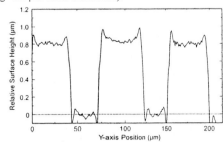

FIGURE 2. Expanded view of surface profile of patterned flyer target (along y-axis) after final coating of aluminum applied.

400-μm-diameter multimode optical fibers. Tests focusing on the flyer launch and impact processes utilized a composite material consisting of a 0.25-μm-thick layer of Al_2O_3 imbedded in different thicknesses of aluminum. In the flyer impact (witness plate) experiments, particle velocity measurements were performed on lithium fluoride acceptor windows. These windows were equipped with a thin (50 nm) coating of sputtered Al that provided both adequate fringe brightness in the interferometer and sufficient opacity to background light generated in the flyer acceleration process. Accurate spacing between the coated fiber face (flyer target) and the Al-coated windows was provided by shim stock of known thickness. The patterned flyers used to induce hydrodynamic instability growth were prepared in two stages. First, a fine mesh screen was placed in contact with the output ends of the fibers, providing a crude mask. A 0.8-μm coating of Al was then applied to the accessible regions of the fiber faces. The resulting pattern was examined using a WYKO RST Plus optical surface profiler; one profile image is shown in Fig.1. In the second step, an 8-μm "overcoat" of Al was applied. The original pattern of steep-walled steps was preserved, as seen in Fig. 2. Hence, the targets featured 10% thickness variations with periodic spacing ("wavelength") of 80 μm.

RESULTS AND DISCUSSION

To examine the flyer launch process, we adjusted the light collection optics to compress the image size so that a line segment spanning the entire 400-μm flyer diameter could be viewed by the streak camera and CCD detector. The interferometer mirrors were set to produce 16-17 fringes across the image, resulting in a spatial resolution of ~25 μm along the line segment. The fringe displacement record for one test is shown in Fig. 3. Despite an early loss of fringe intensity, the record clearly reveals a systematic center-to-edge variation in the onset of flyer motion. The maximum difference in this case was 3.5 ns. A more uniform launch was observed at higher incident energies. This target response was undoubtedly related to a nonuniform driving laser intensity distribution at the fiber output face. Indeed, measurements of the spatial profile exiting an uncoated fiber (using a beam profiling analysis system) showed a center-to-edge distribution (falloff) consistent with the observed flyer behavior. With improvements in optical coupling and alignment, line-imaging ORVIS should be able to provide a detailed picture of the flyer shape (along one segment) vs. time.

Using ORVIS in the conventional "single point" mode, velocity-time records were obtained for the 18-μm-thick composite flyer material employed in the witness plate experiments. The flyer response at 30 mJ incident energy is shown in Fig. 4. Comparison of the velocity and calculated displacement curves indicates that an impact velocity near 1.7 km-s^{-1} should be obtained with a 0.003" spacing between the flyer target and the acceptor. From analysis of the Hugoniot relationships for Al and LiF (7), the resulting acceptor particle velocity is expected to be approximately 0.86 km-s^{-1}. Accounting for the index of refraction effects in LiF (7), the apparent peak velocity, u_{app}, should be ~1.09 km-s^{-1}.

A line-imaging witness plate record obtained with this impact condition is shown in Fig. 5. For this shot, a high-magnification image of the impact plane was used (~10 μm per fringe cycle). The delay leg in the interferometer was set to produce a velocity per fringe constant of 1.093 km-s^{-1}. Prior to flyer impact, the backreflection from the acceptor surface was highly specular. Only a small portion of this light was accepted by the coupling mirror for the interfer-

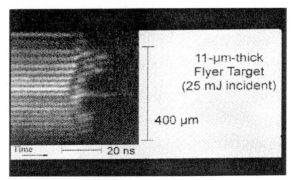

FIGURE 3. Line ORVIS velocity-time record of the initial acceleration of an 11-μm-thick composite flyer (25 mJ incident energy). Camera streak speed: 5 ns-mm^{-1}.

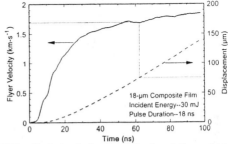

FIGURE 4. Typical velocity-time behavior of 18-μm-thick composite flyer used in LiF witness plate experiment.

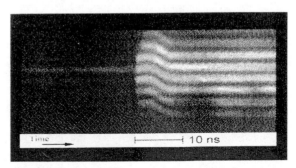

FIGURE 5. Line ORVIS record of flyer impact on LiF witness plate (30 mJ incident energy). Camera streak speed: 2ns-mm^{-1}.

ometer, accounting for the dim pre-impact fringes. Immediate brightening occurred on impact. Consistent with the simple analysis above, the peak fringe displacement was almost exactly one cycle; i.e., u_{app} ~ 1.09 km-s^{-1}. The most significant result of this test, however, is the fine resolution of different impact arrival times across this portion of the acceptor. The shock jump at each fringe location was recorded in 1-2 pixels, within the 0.26-ns interferometer resolution for this shot. Across the line seg-

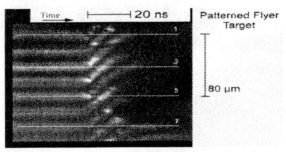

FIGURE 6. Line ORVIS record showing velocity-time behavior of different segments of patterned flyer (30 mJ incident energy).

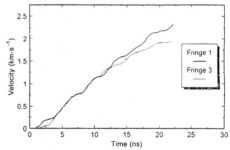

FIGURE 7. Reduced line ORVIS data for two fringes in patterned flyer target experiment.

ment imaged in Fig. 5, flyer curvature resulted in >1 ns spread in arrival time. In comparing adjacent fringes, differences of ~0.2-0.3 ns are clearly evident. The record also captures subtle variations in the pulse shape at each fringe location. The demonstrated combination of high spatial and temporal resolution available in this technique should find application in a variety of microscale studies in addition to flyer performance tests.

A high-magnification image was also used in the experiments on patterned flyers. The length of the line segment viewed by the streak camera and detector was ~160 μm, corresponding to two step cycles near the center of the target (cf. Figs 1 and 2). The interferometer mirrors were set to produce four fringes per 80 μm "period." Figure 6 illustrates the results obtained in one test. As in the flyer launch data described above (cf. Fig. 3), early loss of fringe intensity was a problem. This may have resulted from vignetting due to flyer curvature or from a loss of flyer integrity related to the sharp features on the target surface. Careful inspection of the individual fringe displacements, however, reveals differences in slope (acceleration) that are significant and that also appear to "track" the four-fringe cycle. Analysis of two individual fringes (labelled "1" and "3" in Fig. 6) confirms this visual appearance, as illustrated in Fig. 7. The velocity-time curves at these two positions are very similar in the first 10-15 ns after onset of motion. After this time, however, the local velocities diverge with the difference approaching 0.25 km-s⁻¹ at the end of the useful record. A similar trend occurs in the other step cycle (fringes "5" and "7" in Fig. 6).

These preliminary tests indicate that the resolution capabilities of line-imaging ORVIS provide a promising new approach to instability studies. Much addi-

tional testing is needed to identify with confidence the source of observed velocity variations in the flyer studies. In particular, an optimized (e.g., sinusoidal) perturbation geometry would be desirable.

ACKNOWLEDGMENTS

The excellent technical assistance of Jaime N. Castañeda is gratefully acknowledged. Flyer targets were skillfully prepared by Catharine Sifford and Juan Romero.

REFERENCES

1. Gidon, S., and Behar, G., "Instantaneous Velocity Field Measurements: Application to Shock Wave Studies," *Applied Optics* **25**, 1429-1433 (1986).
2. Mathews, A. R., Warnes, R. H., Hemsing, W. F., and Whittemore, G. R., "Line-imaging Fabry-Perot Interferometer," in *SPIE Proc. No. 1346*, San Diego, CA, 1990, pp. 122-132.
3. Hemsing, W. F., Mathews, A. R., Warnes, R. H., and Whittemore, G. R., "VISAR: Line-imaging Interferometer," in *SPIE Proc. No. 1346*, San Diego, CA, 1990, pp. 133-140.
4. Mathews, A. R., Boat, R. M., Hemsing, W. F., Warnes, R. H., and Whittemore, G. R., "Full-field Fabry-Perot Interferometer," in *Shock Compression of Condensed Matter--1991*, eds. S. C. Schmidt, et al., New York: Elsevier Science Publishers, 1992, pp. 759-762.
5. Baumung, K., Singer, J., Razorenov, S. V., and Utkin, A. V., "Hydrodynamic Proton Beam-Target Interaction Experiments Using an Improved Line-imaging Velocimeter," in *Shock Compression of Condensed Matter--1995*, eds. S. C. Schmidt and W. C. Tao, Woodbury, NY: AIP Press, 1996, pp. 1015-1018.
6. Trott, W. M., and Meeks, K. D., "Acceleration of Thin Foil Targets Using Fiber-Coupled Optical Pulses," in *Shock Compression of Condensed Matter--1989*, eds. S. C. Schmidt, et al., New York: Elsevier Science Publishers, 1990, pp. 997-1000.
7. Wise, J. L., and Chhabildas, L. C., "Laser Interferometer Measurements of Refractive Index in Shock Compressed Materials," in *Shock Waves in Condensed Matter--1985*, ed. Y. M. Gupta, New York: Plenum Press, 1986, pp. 441-454.

CP429, *Shock Compression of Condensed Matter – 1997*
edited by Schmidt/Dandekar/Forbes
© 1998 The American Institute of Physics 1-56396-738-3/98/$15.00

SIMULTANEOUS PVDF/VISAR MEASUREMENT TECHNIQUE FOR ISENTROPIC LOADING WITH GRADED DENSITY IMPACTORS

M. U. Anderson, L. C. Chhabildas and W. D. Reinhart[*]

Sandia National Laboratories, Albuquerque, NM 87185-1421[1]
[*]Ktech Corporation, Albuquerque, NM, 87110

A simultaneous PVDF/VISAR measurement technique was used for isentropic-loading experiments with a polymethyl methacrylate (PMMA) specimen. The experiments used a graded density impactor accelerated onto a tantalum driver backed with PMMA and then lithium fluoride windows for each experiment. Simultaneous measurements made at each window interface provided precise transit time and particle velocity measurements which can be used to determine the stress-vs-strain loading path using Lagrangian analysis techniques. The experimental technique provides access to 40 GPa stress levels in PMMA under isentropic-loading conditions.

INTRODUCTION

Shock loading techniques allow the capability for characterization of material properties at high pressures and strain rates[1]. During shock loading, a substantial temperature increase can accompany the pressure increase. The use of finite-rise-time loading produces quasi-isentropic compression of materials by introducing a series of small shocks that provide access to high pressure equilibrium states with lower temperatures than those obtained at Hugoniot shock states (quasi-isentropic loading will be referred to as isentropic loading in this paper for brevity, although isentropic loading is achieved only in the limit as the amplitude of each shock jump approaches zero).

A variety of isentropic-loading techniques have been developed that make use of the unique material properties of fused silica and glass ceramics to transform a shock input wave into an acceleration wave when used as buffer materials.

(2,3,4) The stress limitations are approximately 3 and 20 GPa, respectively, due to their stress-strain behavior. (5) Graded density impactors have also been used for isentropic loading using both powder sedimentation techniques (6) and multiple layer impactors. (7,8)

The present technique development is intended to extend the level of stress achievable under isentropic loading by use of a multiply-layered, graded-density impactor accelerated onto a high impedance buffer material backed by the specimen and optical window. The material impedances and thicknesses in this technique were optimized with numerical simulations. There are two primary motivations for this technique development: material characterization under isentropic loading, and exploration of the response of the piezoelectric polymer PVDF under isentropic-loading conditions beyond the present limit of 10 GPa. (9) The present report describes the details of this isentropic-loading technique, and presents

---

[1] Sandia is a multi-program laboratory operated by Sandia Corporation, a Lockheed Martin Company for the USDOE under contract DE-AC04-94AL85000.

preliminary results that are encouraging in regard to the optical use of PMMA at high stress levels.

EXPERIMENTAL TECHNIQUE

Isentropic loading over the velocity range of 0.6 to 2.2 km/sec was achieved by impacting a graded density impactor onto the tantalum driver-plate/specimen/window assembly to provide an isentropic loading to the specimen as seen in Fig. 1. The lithium fluoride window provides optical access to the specimen. The present study uses a polymethyl methacrylate (PMMA) specimen which provides optical access to the driver/specimen interface. PVDF sensors are mounted at each VISAR reflecting surface.

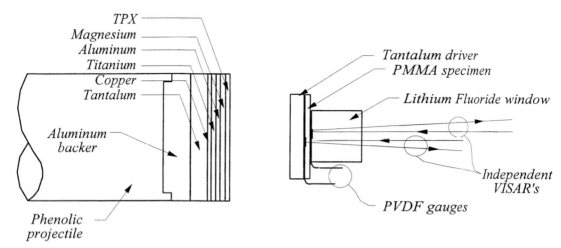

Figure 1. Isentropic-loading technique with multiple-layer laminated impactor accelerated onto tantalum driver in target. PMMA specimen is backed with lithium fluoride to allow optical access to front and rear surfaces of PMMA. Impactor layers increase in density and decrease in thickness from the TPX impact layer to the tantalum rear surface.

The simultaneous PVDF/VISAR measurement technique used in these measurements is based on the technique of Setchell (10) and prior material response characterization of the high explosive HMX (11), employing both laser interferometer (VISAR) (12) and piezoelectric polymer (PVDF) (9) diagnostics in simultaneous measurements at two independent locations. The two independent VISAR measurements, when used in conjunction with the precise measurement of shock propagation time through the sample from the PVDF sensors, allow determination of the stress-vs-strain loading path of the specimen from the Lagrangian wave analysis method of Fowles and Williams. (13) This isentropic loading technique will also provide a method for examination of the loading-path effects on PVDF response at stress levels up to 40 GPa.

An 88-mm diameter, powder-driven, single-stage gun was used for these experiments. Projectile velocity was measured at impact with ±0.5% accuracy. Measured tilt angle at impact is typically 1 milliradian or less. The graded density impactor was made from multiple layers laminated together, each layer having an increasing density and decreasing thickness from front to back. The aluminum backer is an integral part of the phenolic projectile. The experimental details of the impactor and target materials are listed in Table 1.

The mechanical shock impedance (Z) of each material is defined as the product of the unshocked density (ρ_o) and the Y intercept from each material's tabulated U_s-vs-u_p fit at u_p=0. The impedance of the impactor varies from 0.1326 gm/cm^2 μsec for TPX to 5.713 gm/cm^2 μsec for tantalum.

The tantalum driver thickness was chosen from numerical simulations to allow smoothing of the individual shock reflections at each interface layer within the graded density impactor. The thickness

TABLE 1. Experimental configuration and dimensions for isentropic-loading technique.

Exp #	Impact vel.	Impactor material	Impactor thickness	Target material	Target thickness
	(mm/usec)		(mm)		(mm)
51105	0.606	Ta/Cu/Ti/Al/Mg/TPX	4.579/.388/.314/.530/.589/1.056	Ta/PMMA/LiF	5.999/.988/25.451
51123	0.911	Ta/Cu/Ti/Al/Mg/TPX	4.506/.391/.312/.528/.587/1.064	Ta/PMMA/LiF	6.337/.967/25.350
51125	1.77	Ta/Cu/Ti/Al/Mg/TPX	4.443/.380/.319/.526/.5911.019	Ta/PMMA/LiF	3.177/.996/25.277
51124	2.12	Ta/Cu/Ti/Al/Mg/TPX	4.443/.379/.317/.518/.597/1.022	Ta/PMMA/LiF	1.983/.997/25.209
51106	2.21	Ta/Cu/Ti/Al/Mg/TPX	4.516/.394/.315/.536/.592/1.067	Ta/PMMA/LiF	1.996/.971/25.458

was constrained to prevent the high velocity tantalum reflected shock from the rear of the impactor to overtake the low velocity TPX shock, thus creating a shock jump input condition to the specimen. The tantalum thickness at the rear of the impactor was chosen to delay the shock release from the rear surface from overtaking the loading wave into the target.

VISAR measurements were made at the driver/specimen interface and the specimen/backer interface by depositing 2000Å-thick reflecting surfaces over opposite halves of the two interfaces.

PVDF measurements were made in current-mode operation (14) with the recording instrumentation arranged to provide a common time-base between the two sensor locations.. The amplitude of piezoelectric current generated by PVDF in current-mode operation is proportional to the stress difference between the front and rear surfaces of the 25 μm thick film. Shock arrival at the front surface of

PVDF generates a current pulse with a few nanosecond rise time to peak current, thus allowing a precise measurement of loading wave arrival times. Typical accuracy in transit time measurement is ±2 nanoseconds based on typical recording intervals of 1 nanosecond. The specimen transit time is calculated by subtracting shock transit times through the PVDF and insulation layers using shock properties for each film.

RESULTS

The measured particle velocity profile at the front surface of the PMMA specimen shows initial shock jumps to 11% of the peak amplitude, followed by an acceleration loading that approaches isentropic compression of the specimen as seen in Fig. 2. The measured front-surface particle velocity reaches a maximum of 1.94 mm/μsec, corresponding to 40

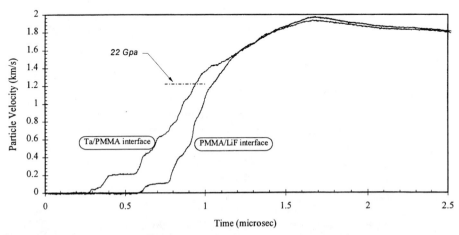

FIGURE 2. Particle velocity profiles from front and rear surfaces of PMMA specimen under isentropic loading. Mechanical equilibrium is achieved within specimen before release arrives. 22 GPa transparency limit for shock loading is shown.

GPa, well beyond the 22 GPa limit for transparency under shock-loading. (15)

The overlaid measurement profiles from front and rear specimen surfaces show that equilibrium is achieved within the specimen before release arrives.

The desired result from these isentropic-loading experiments is stress-vs-strain loading paths for PMMA up to 40 GPa. The Langrangian analysis used to calculate stress-vs-strain paths (12) makes the assumption that the wave velocity for any given level of particle velocity is constant. Multiple thicknesses of specimens need to be studied at the same input conditions in order to satisfy this assumption. Analysis of the present data is in progress.

The PVDF measurements are not shown due to an unanticipated complication. The PVDF response was obscured at high stress levels by the shock-induced polarization of the uninsulated PMMA specimen. The onset of shock-induced polarization from the PMMA specimen was observed at 1.0 GPa superposed onto the piezoelectric current generated from the PVDF sensor, thereby preventing the observation of quantitative stress histories. Accurate arrival times were measured at low stresses only. This observation demonstrates the need for electrical insulation and shielding of the PMMA specimen. Teflon insulation is typically used due to the low level of piezoelectric activity (10^{-7} C/m^2), in contrast to PMMA (10^{-4} C/m^2)which is larger by 3 orders of magnitude. (16)

CONCLUSIONS

This simultaneous PVDF/VISAR measurement technique combined with the isentropic-loading technique provides the capability for exploring the isentropic response of materials to higher stresses. PMMA appears to remain transparent during isentropic-loading to 40 GPa. Care must be taken to properly insulate the PVDF gauges when using samples that may show high-stress piezoelectric behavior.

ACKNOWLEDGMENTS

The authors thank Heidi Anderson for her talented preparation of the experimental assemblies and R. E. Setchell for the manuscript review.

REFERENCES

[1] Davison, L. W., R. A. Graham, *Phys. Rep.*, **55**, pp. 257-379 (1979).

[2] Barker, L. M., R. E. Hollenbach, *J. Appl. Phys.*, **41**, pp. 4208-4226 (1970).

[3] Setchell, R. E., *Combustion and Flame*, **54**, pp. 171-182 (1983).

[4] Germain-Lacour, M., M. de Gliniasty *Shock Waves in Condensed Matter-1981*, AIP, pp. 481-485. (1981).

[5] Asay, J. R., L. C. Chhabildas, *High Pressure Science and Technology*, Pergamon Press, pp. 958-964 (1980).

[6] Barker, L. M., *Shock Waves in Condensed Matter-1983*, North-Holland, pp. 217-224 1982).

[7] Adadurov, G. A., and V. I. Gol'danskii, *C. M. Backman, T. Johannisson and L. Tegner*, **V1**, Arkitektkopia Sweden, pp. 18-32. (1982).

[8] Chhabildas, L. C., J. R. Asay, and L. M. Barker, SAND88-0306, Sandia Nat. Lab.

[9] Graham, R. A., M. U. Anderson, F. Bauer, and R. E. Setchell, *Shock Compression of Condensed Matter-1991*, North-Holland, pp. 883-886 (1992).

[10] Setchell, R. E., *Shock Waves in Condensed Matter-1987*, pp. 623-626 (1988).

[11] Anderson, M. U., and R. A. Graham, *Shock Compression of Condensed Matter-1995*, pp. 1101-1104 (1996).

[12] Barker, L. M. and R. E. Hollenbach, *Rev. Sci. Instrum.*, **36**, pp. 1617 (1965).

[13] Fowles, G. R., and R. F. Williams, *J. Appl. Phys.* **41**, pp. 360 (1970).

[14] Anderson, M. U., D. E. Wackerbarth, R. A. Graham, *Shock Compression of Condensed Matter-1989*, pp. 805-808 (1990).

[15] Chhabildas, L. C., H. J. Sutherland, and J. R. Asay, *J. Appl. Phys.*, **50**, pp. 5196-5201, (1979).

[16] Graham, R. A., *J. of Phys Chem.* **83**, pp. 3048-3056, (1979).

CP429, *Shock Compression of Condensed Matter – 1997*
edited by Schmidt/Dandekar/Forbes
© 1998 The American Institute of Physics 1-56396-738-3/98/$15.00

MAGNETIC GAUGE INSTRUMENTATION ON THE LANL GAS-DRIVEN TWO-STAGE GUN[†]

R. R. Alcon, S. A. Sheffield, A. R. Martinez, and R. L. Gustavsen

Los Alamos National Laboratory, Los Alamos, NM 87545

The LANL gas-driven two-stage gun was designed and built to do initiation studies on insensitive high explosives as well as equation of state and reaction experiments on other materials. The preferred method of measuring reaction phenomena involves the use of in-situ magnetic particle velocity gauges. In order to accommodate this type of gauging in our two-stage gun, it has a 50-mm-diameter launch tube. We have used magnetic gauging on our 72-mm bore diameter single-stage gun for over 15 years and it has proven a very effective technique for all types of shock wave experiments, including those on high explosives. This technique has now been installed on our gas-driven two-stage gun. We describe the method used, as well as some of the difficulties that arose during the installation. Several magnetic gauge experiments have been completed on plastic materials. Waveforms obtained in some of the experiments will be discussed. Up to 10 in-situ particle velocity measurements can be made in a single experiment. This new technique is now working quite well, as is evidenced by the data. To our knowledge, this is the first time magnetic gauging has been used on a two-stage gun.

INTRODUCTION

The two-stage gun is a compressed-helium driven, two-stage light gas gun (based on a design from Ernst Mach Institute[1]) designed to perform shock initiation studies on insensitive high explosives (see Fig. 1). It has a 100-mm diameter by 7.6-m long pump tube and a 50-mm diameter by 7.6-m long launch tube. The relatively large launch tube diameter of 50 mm was chosen to provide an experimental area large enough to allow one-dimensional multiple magnetic gauge experiments to be done. A gas breech, capable of operating at 15,000 psi, is the driver for the pump piston. Three large hydraulic clamps are used to clamp the breech to the pump tube, the pump tube to the transition section, and the transition section to the launch tube. Helium is used as the driver gas for both the launch projectile and the pump piston. The 1-m diameter target chamber provides room for the electro-magnet which produces the magnetic field.

Projectile velocities in excess of 3 km/s have been achieved with the breech charged to only 8000 psi. The gun design and performance have been described previously(2,3,4)

We have used magnetic particle velocity gauging for 15 years to measure the details of initiation in solid and liquid explosives initiated by projectile impact in a single-stage gun. This technique allows us to make up to 10 in-situ particle velocity measurements in a single experiment so that the shape of the reactive initiation shock can be monitored as it grows from the input shock to a detonation. With this type of experimental information available, functions can be developed to describe the global reaction process occurring. In this paper we briefly describe the gauge technique and how it is has been implemented on the two-stage gun. We believe this is a new technique as far as two-stage guns are concerned, and a number of changes to the gun design have been required to implement it.

[†] This work supported by the United States Department of Energy.

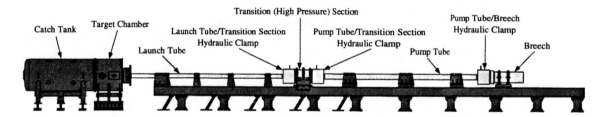

FIGURE 1. Schematic of the LANL gas-driven two-stage gun.

MAGNETIC GAUGING TECHNIQUE

Magnetic gauging was first used in Russia in 1957 (5) and later tried out by several groups. It has been used at a number of shock laboratories outside Russia over the years, but has achieved a prominent position in shock measurements over the years only in Russia and at Washington State University (WSU), SRI International, and Los Alamos National Laboratory. People at Physics International and WSU,(6,7) along with several students who learned this technique, have been responsible for its implementation in the U.S.

Principle of Operation – The gauge function is quite simple. When a conductor in a closed loop moves in a magnetic field, a voltage is induced in the circuit because part of the loop cuts magnetic field lines as it moves. Output voltage depends on the magnetic field strength (B), the length of the conductor (l) cutting the field lines, and the velocity it is moving (v). This can be written as $E = Blv$ where E is the voltage. B and l are measured before the experiment and E is measured as a function of time during the experiment. From this the mass or particle velocity (at the particular Lagrangian position of the gauge) as a function of time can be obtained if the assumption is made that it moves with the material. In solid samples this is the case; liquids are more complicated (8).

Gauge Design – The LANL magnetic gauge technique development work was started by Vorthman and Wackerle in about 1980 (9). This technique involves the use of a thin gauge membrane (60-μm thick) that is embedded in the sample material so that in-situ particle velocity measurements are made. The gauge package is a sandwich of 25-μm-thick FEP Teflon, a 5-μm-thick piece of aluminum foil (etched to the gauge

configuration after gluing) glued to the Teflon, and another piece 25-μm-thick FEP Teflon glued on top as insulation. One gauge pattern is shown in Fig. 2.

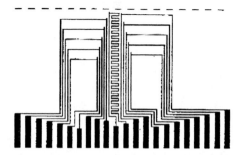

FIGURE 2. Details of the gauge pattern for a 10 gauge membrane with a shock tracker to track the shock velocity. Gauges are staggered so that 10 different positions can be measured in one experiment.

Target Design – In solid targets the sample is machined with a bottom and top designed so the gauge membrane can be glued in at an angle as shown in Fig. 3. Generally the angle is 30 degrees so that the gauge elements are ≈1/2 mm apart on the sample axis. The active gauge elements do not shadow each other. Typically the two-stage gun samples are 43-mm diameter by about 23-mm thick.

The target assembly is secured to a target plate with the position of the gauge ends carefully noted. The target plate is placed in the target chamber so that it is between the pole pieces of the electromagnet and the gauge ends are perpendicular to the magnetic field lines (and the gauge leads are parallel to the lines).

Magnet -- The magnetic field is developed by a large electromagnet located in the gun target chamber that is turned on just before the experiment. The two-stage gun magnet is capable

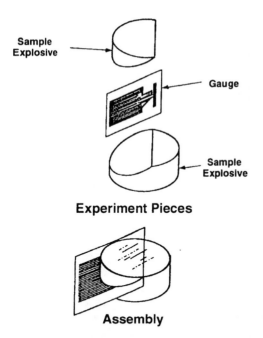

Experiment Pieces

Assembly

FIGURE 3. Schematic of the sample with the gauge membrane. Angle is usually 30 degrees. The assembly top is lightly machined after the gluing operation to make sure the top is flat.

of producing a field up to 2000 gauss but is normally operated at about 1000 gauss. The field is mapped before each experiment, just prior to installing the target assembly. A picture of the target chamber with the magnet installed is shown in Fig. 4. A 10-inch diam aluminum tube with a ½-inch thick wall, centered between the magnet pole pieces and around the target, is used to protect the magnet and direct shot debris into the catch tank.

FIGURE 4. Picture of the target chamber with the electromagnet system installed. In the center is the plate with 25 holes used for mapping the magnetic field. The center hole of the plate is on the axis of the gun barrel. The target chamber is about 1-m diameter.

Two-Stage Gun Implementation – Initially the magnetic field in the two-stage gun was not as uniform as desired. This was due to two things: 1) the launch tube is 4340 steel, a magnetizable material, so it perturbs the field lines, and 2) it was necessary to put the target close to the end of the launch tube because the projectiles are short (2 to 3 inches long) and we want the projectile to impact the target before it is completely out of the launch tube to minimize tilt at impact. A mapping of the field in the region of the gauge allowed us to determine the field was uniform to about 2 percent. In addition, the most uniform region was about 2 inches past the center of the magnet due to lalunch tube distortion of the field. To help correct this problem, a 4-inch long stainless steel launch tube extension was installed and the launch tube was repositioned to accommodate it.

After this change was made, a new mapping of the magnetic field yielded the contour lines shown in Fig. 5. This figure shows that the uniformity of the field in the region of the gauge to be better than one percent. This is as good as the single-stage gun magnet which was what we were hoping for. The contour lines indicate the magnet is still positioned about 0.2 inch high but this is not considered a large amount considering the size of the magnet.

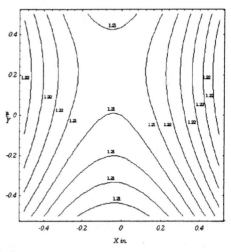

FIGURE 5. Contour plots of field calibration data from the two-stage gun after modifications to the launch tube. Measurements were taken on the centerline of the magnet. The field varies from 1210 to 1220 gauss, indicating that the errors due to the magnetic field measurement are less than 1 percent.

MAGNETIC GAUGE EXPERIMENTS ON
THE TWO-STAGE GUN

Several multiple magnetic gauge experiments have been conducted on the two-stage gun with varying degrees of success, principally because we were working the bugs out of the system. These experiments involved either with polychlorotrifluoroethylene (Kel-F) or polymethylmethacrylate (PMMA) targets that were impacted by polycarbonate (Lexan) projectiles. The targets were as shown in Fig. 3 with 10 gauge elements as shown in Fig. 2. The finished target was a cylinder 43 mm diam by 23 mm high.

Shot 2s-11 was a PMMA target impacted by a Lexan projectile at a velocity of 2.8 km/s. A gauge package with ten gauges was used but only nine gauge voltages were successfully recorded. The particle velocity data from the shot are shown in Fig. 6. The particle velocity measured was about 1.35 mm/μs. Data from the shock tracker gave a shock velocity of 4.76 mm/μs. This translates into a pressure in the PMMA of 7.6 GPa. This data point agrees reasonably well with other PMMA data (10). We take this to mean that the magnetic gauging technique is working properly on the two-stage gun and will provide good data. The waveforms have some rounding at the top which is evidence that there is still viscoelastic behavior even at 7.6 GPa in PMMA.

In Fig. 6, five particle velocity waveforms came from one side of the experiment and 4 came from the other side. These are not evenly spaced because there were problems with projectile tilt at impact on this experiment. If the tilt had been very small (less than 0.1 mradian) the waveforms would have been equally spaced in time. The offset from this equal spacing indicates that there was a few mradians of tilt in the experiment. There are several aspects of the gun setup that have a bearing on this: 1) the straightness of the launch tube, 2) the mating of the heavy catch tank to the target chamber, and 3) the alignment of the target with respect to the launch tube muzzle. These are being looked into at the present time to find the problem and eliminated it. Tilt of less than a mradian is desired.

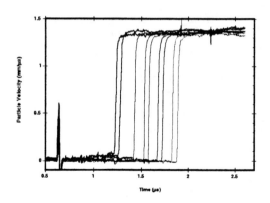

FIGURE 6. Particle velocity waveforms from Shot 2s-11. There are only 9 gauge measurements because 1 gauge was not recorded properly.

REFERENCES

1. Stilp, A. J., "The New EMI Terminal Ballistics and Hyper-velocity Impact Range," unpublished report, Ernst Mach Institut, Freiburg, Germany, August 1987.
2. Martinez, A. R., Sheffield, S. A., Whitehead, M. C., Olivas, H. D., and Dick, J. J., "New LANL Gas Driven Two-Stage Gun," in *High-Pressure Science and Technology -- 1993*, Eds. S. C. Schmidt, J. W. Shaner, G. A. Samara, and M. Ross, American Institute of Physics (AIP) Conference Proceedings No. 309 (1994) p. 1643.
3. Sheffield, S. A., and Martinez, A. R., "New LANL Group M-7 Two-Stage Gun: Double Diaphragm and Wrap-Around Gas Breech," presented at the 43rd Aeroballistic Range Association (ARA) Meeting, Columbus, Ohio, Sept. 28 - Oct. 2, 1992.
4. Sheffield, S. A., and Martinez, A. R., "Testing of New LANL Gas-Driven Two-Stage Gun," presented at the 45th Aeroballistic Range Association (ARA) Meeting, Huntsville, Alabama, Oct. 10-14, 1994.
5. Pohkil & Dremin
6. Koller, L. R., "Generation and Measurement of Simultaneous Compression-Shear Waves in Arkansas Novaculite," Ph.D. Thesis at Washington State University under R.Fowles, 1978.
7. Young, C., Fowles, R., and Swift, R. P., in *Shock Waves and the Mechanical Properties of Solids*, Eds. J. J. Burke and V. Weiss, Syracuse University Press, 1971, p. 203.
8. Gustavsen, R. L., and Sheffield, S. A., "Response of Inclined Electromagnetic Particle Velocity Gauges in Shocked Liquids," in *High Pressure Science and Technology -- 1993*, Eds. S. C. Schmidt, J. W. Shaner, G. A. Samara, and M. Ross, American Institute of Physics (AIP) Conference Proceedings 309 (1994), p. 1703.
9. Vorthman, J. E., "Facilities for the Study of Shock In-duced Decomposition of High Explosives," in *Shock Waves in Condensed Matter -- 1981*, Eds. W. J. Nellis, L. Seaman, and R. A. Graham, American Institute of Physics (AIP) Conference Proceedings No. 78 (1982) p. 680.
10. Carter, W. J., and Marsh, S. P., "Hugoniot Equation of State of Polymers," Los Alamos National Laboratory Report, LA-13006-MS, July 1995.

CP429, *Shock Compression of Condensed Matter – 1997*
edited by Schmidt/Dandekar/Forbes
© 1998 The American Institute of Physics 1-56396-738-3/98/$15.00

SIMULTANEOUS MANGANIN GAUGE AND VISAR MEASUREMENTS OF SHOCK WAVE PROFILES

N.K. Bourne and Z. Rosenberg*

Shock Physics, PCS, Cavendish Labs, Madingley Road, Cambridge, CB3 0HE, UK.
*RAFAEL, PO Box 2250, Haifa, Israel.

Shock wave profiles have been measured by many techniques probing the stress or particle velocity histories of shocked materials. Amongst the most commonly used methods are velocity interferometry and embedded foil stress gauges. It is relatively unusual to employ both these techniques simultaneously in an experiment to compare the responses of the two. Examples will be given of the relative responses of the two techniques in metals, ceramics and glasses to illustrate the reproducibility of the gauge.

INTRODUCTION

Over the past 50 years a range of sensors has been developed to follow the state of a material which has been dynamically loaded. The aim of an impact experiment is usually to characterise the variables that describe the state of the impactor or target, to define the pressure, volume and temperature as a function of time. Invariably these parameters are not measured directly. Instead related quantities such as stress, strain or particle velocity are.

One of the most commonly used of coupled mechanical/electrical properties is that of piezo–resistivity which is the change of the electrical resistivity of a material with applied external stresses. One such material is manganin (an alloy of 84% copper, 12% manganese, and 4% nickel) which has been calibrated over a wide pressure range (1).

The principle recent advances in dynamic instrumentation have been in the development of non-invasive interferometric techniques for the monitoring of the velocity of the external surfaces of impact targets. A major step forward came with the realisation that the Doppler shift of the light could be used to produce interference patterns from rough surfaces and which varied much more slowly so that

they could be followed with relative ease. This system called VISAR (*v*elocity *i*nterferometry *s*ystem for *any* *r*eflector) was originally developed by Barker at Sandia (2, 3) and has now reached a position of dominance in the measurement of normal velocity at a surface.

In a previous paper we have presented an analysis of the ringing that is present on the traces obtained by manganin gauges placed in the so-called backsurface configuration in which a polymethyl-methacrylate (PMMA) backing is used to match the impedance of the epoxy used to surround the gauge (4, 5). In this work we use this backing as a window to make a simultaneous VISAR measurement with the spot placed adjacent to the gauge. This simultaneous measurement allows direct comparison of the two techniques and verifies the accuracy of previously obtained gauge results.

EXPERIMENTAL

Plate impact experiments were carried out on the 50 mm bore gun at the University of Cambridge. Stress profiles were measured with commercial manganin stress gauges placed on the rear face of the specimens and supported with PMMA blocks. These gauges (Micromeasurements type LM-SS-125CH-

048) were calibrated by Rosenberg *et al* (1)). The signals were recorded using a fast (1 GS s^{-1}) digital storage oscilloscope and transferred onto a micro-computer for data reduction. Impact velocity was measured to an accuracy of 0.5% using a sequential pin-shorting method and tilt was fixed to be less than 1 mrad by means of an adjustable specimen mount. Impactor plates were made from lapped tungsten alloy, copper and aluminium discs and were mounted onto a polycarbonate sabot with a relieved front surface in order that the rear of the flyer plate remained unconfined.

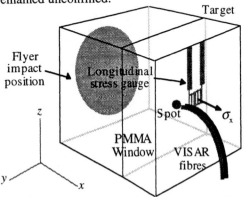

FIGURE 1. Experimental arrangement for simultaneous gauge/VISAR experiments

The VISAR used was an adapted VALYN system (6) with polymer fibres illuminating the target and feeding into the enclosure and with the output of the four fringe photomultipliers recorded directly onto a 2 GS s^{-1} storage oscilloscope. The design is of standard push-pull form and subtraction of the signals was carried out in software on a computer which downloaded the recorded data from the scope. The delay etalons limit the resolution of the system used in this configuration to *ca.* 2 ns. The phase data was unwrapped using the method of Hemsing (7). Where necessary a flash coating was evaporated onto the rear of the target to aid reflection. A schematic of the experimental arrangement is shown in Fig. 1.

The following data represent typical traces selected from ongoing research in the shock loading of metals, glasses, and ceramics. Each series of traces are shown in pairs and are recorded simultaneously from the same experiment. The VISAR traces are presented as dotted traces whilst the manganin histories are solid lines. A range of

stresses is used to illustrate features of each response and the series of experiments presented is ordered according to the stress induced. In all cases remarkable agreement is seen. However, some features appear on one trace and not the other and it is these that we discuss below.

RESULTS AND DISCUSSION

Figure 2 shows two shots conducted on the lead glass DEDF. This has been the subject of extensive investigation in our group over the last few years (8, 9). Each trace is conducted at close to the Hugoniot elastic limit (HEL) of the material which has been determined to be 4.5 GPa. Each DEDF tile was of the same thickness (8 mm) but the aluminium flyer was varied in order that the releases from the rear of flyer and the PMMA backing material superposed at different positions within the tile. In neither case was any spall seen. The DEDF contains *ca.* 30% silica and 70% PbO by weight so that it is poorly conducting in ambient conditions. There is no significance to the separation in rise time of the two pairs of traces. They have been offset for ease of presentation. It will be noted that velocity and stress histories follow one another very closely upon loading and release. It will be seen that the maximum stress in the gauge is *ca.* 1.5 GPa which is the HEL of manganin. Thus the gauge material is behaving elastically throughout the loading. There are slight deviations between the velocity and stress at the very end of the pulses but these may be attributed to the arrival of lateral releases in the gauge which gives rise to an extra strain contribution to the signal resulting in an apparent rise in stress.

The most significant differences are ringing on the gauge traces at the bottom and top of the shock wave where the rate of rise of the pulse is at a maximum. This ringing is superposed upon a steady response which follows the VISAR signal. This ringing has been noticed before and has been the subject of further analysis (4, 5). In those works we showed that conducting materials showed an initial dip at the arrival of the shock wave which corresponded to a capacitive linkage between the gauge and the conducting target. On the top of the pulse, a ringing appeared which has a frequency dependent upon the inductance of the gauge, the capacitance of the gauge and target and of the connecting cabling. Some authors have attempted to explain such ringing by mechanical arguments. The figures presented here show conclusively that the ringing has an electrical origin and that once deconvolved from such effects, the gauges follow precisely the observed particle velocity history at the PMMA/target interface.

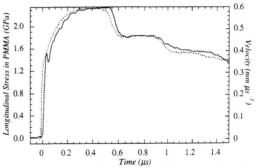

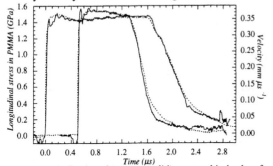

FIGURE 2. VISAR (dotted), gauge (solid) mounted in backsurface configuration. Lead glass DEDF impact with a 3 and 6 mm thick Al flyer at 518 m s⁻¹.

FIGURE 3. VISAR (dotted), gauge (solid) mounted in back-surface configuration. Symmetrical impact of a 97.5% alumina flyer of thickness 3 mm and target 6 mm at 688 m s⁻¹.

Figure 3 shows simultaneous VISAR and gauge traces from impact upon a 97.5% alumina. This material and others of differing alumina content have been investigated in depth by Murray (10, 11) and the figure can be regarded as a typical trace. Again the agreement between VISAR and gauge is very good. The velocity trace is displaced 20 ns before the gauge. The stress range now extends to 2.4 GPa which results in plastic deformation of the gauge.

There are several features to be noted. Again there is ringing on the gauge trace of the type discussed above which has the same period and source. The initial dip has however disappeared. This is as a result of the insulating nature of the ceramic under ambient conditions. We have recently shown that the break in slope after the rapid elastic rise is related to a surface fracture zone near the impact plane (12). It will be seen that the gauge shows the start of ringing at this time which may also indicate that the material is conducting at this time as a result of free charge produced on fracture.

The dip has been used in the past as a measure of the HEL of the ceramic but as Grady has discussed this may simply be a lower strain-rate dependent yield (13). In particular, several authors have shown that this threshold decays with distance (10). The comparison with the VISAR signal shows that the

break occurs later and at a slightly higher point in the trace in the stress history. This is to be expected since the gauge rise is limited to 30 ns by its inductance so that it slightly overestimates the threshold. The other feature of the trace is the appearance of a small reloading signal after release in the gauge trace which is not apparent in the VISAR trace. This might be erroneously interpreted at a small spall strength but may result from the plastic response of the gauge.

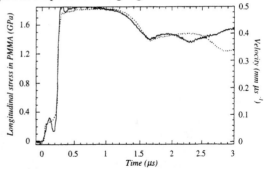

FIGURE 4. VISAR (dotted), gauge (solid) mounted in backsurface configuration. Impact of a copper flyer of thickness 3 mm onto a 12 mm mild steel target at 536 m s⁻¹.

Figures 4 and 5 show the response of metals as seen by gauges and VISAR in the backsurface configuration. In both figs mild steel (see 14 for material properties) is impacted with copper and tungsten alloy flyers respectively. In figure 5 the 13 GPa phase transition stress is exceeded in the steel pushing the gauge to a stress of *ca.* 3.5 GPa. In both figures the dip is seen before the arrival of the elastic precursor which has been discussed above for conducting materials.

It will be seen that there is a second dip in the gauge trace when the target and gauge are further pushed close together by the arrival of the main plastic wave. The effect is larger in Fig. 5 at this point since the acceleration is greater. In the top of the shock in Fig. 4 and on the plateau in Fig. 5 there is a Gibbsian overshoot and some damped ringing as expected. Again this may be analysed using previous work (5).

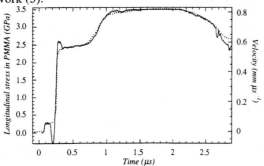

FIGURE 5. VISAR (dotted), gauge (solid) mounted in backsurface configuration. Impact of a tungsten alloy flyer of thickness 5 mm on a 12 mm mild steel target at 707 m s⁻¹.

There is no ringing in Fig. 5 when the reflected compression pulse from the phase transformed material arrives at the gauge. This is because the wave is slow rising (as it results from a reflected release fan). This illustrates that a fast rising pulse (<100 ns) is necessary to observe ringing on gauge histories. Traces for materials such as soda-lime or borosilicate glasses whose open structure results in mechanical ramping of the shock front show no deviation of the stress and particle velocity histories and no ringing and have thus not been included here. It is to be emphasised that the results presented here are representative of the *worst* cases of deviation of gauges from particle velocity histories. When an embedded gauge is used to measure the phase transition in this material directly, the wave rises more rapidly and this is sufficient to result in an overshoot at the bottom and the top of the history. This is illustrated in 14 in these proceedings.

There is deviation of the velocity and stress histories evident in both figures in the release phase after *ca.* 2.5 μs. This may be due to permanent plastic deformation in the gauge but it is more likely (particularly since the gauges were slightly off-axis to accommodate the VISAR spot) that lateral releases have reached either measuring location by this stage resulting in deviations.

CONCLUSIONS

We have confirmed that manganin gauges follow the stress history of uniaxial strain impact experiments and have presented simultaneous VISAR traces that are immune to electrical interference as a comparison. We have noted several features of gauge behaviour that must be borne in mind when using them. Firstly, for slow (>100 ns) rising pulses the gauges follow precisely the VISAR signal. For fast rising signals a ringing is observed due to the reactance of the gauge and its associated cabling and environment. We have shown that this ringing is apparent particularly at the threshold for elastic behaviour in ceramics and that this point slightly overestimates that same threshold seen in the VISAR signal.

ACKNOWLEDGMENTS

The authors gratefully acknowledge the support of DERA, in particular Dr B. Goldthorpe, for making available funds to purchase the VISAR. We thank Prof. J.E. Field, Dr W.G. Proud, Dr I.G. Cullis and Dr P. Church for useful discussions, and D.L.A. Cross for technical support. NKB acknowledges EPSRC for a fellowship.

REFERENCES

1. Rosenberg, Z., Yaziv, D. and Partom, Y., J. Appl. Phys., **51**, 3702-3705 (1980).
2. Barker, L.M. and Hollenbach, R.E., J. Appl. Phys., **43**, 4669-4675 (1972).
3. Barker, L.M. and Schuler, K.W., J. Appl. Phys., **45**, 3692-3693 (1974).
4. Bourne, N.K. and Rosenberg, Z., in *Shock Compression of Condensed Matter 1995,* (AIP, Woodbury, New York, 1996), pp. 1053-1056.
5. Bourne, N.K. and Rosenberg, Z., Meas. Sci. Technol., **8**, 570-573 (1997).
6. Barker, L.M., User manual, Valyn VISAR. (1996).
7. Hemsing, W.F., Rev. Sci. Instrum., **50**, 73-78 (1979).
8. Bourne, N.K., Rosenberg, Z. and Ginzburg, A., Proc. R. Soc. Lond. A, **452**, 1491-1496 (1996).
9. Bourne, N.K., Millett, J.C.F. and Rosenberg, Z., Proc. R. Soc. Lond. A, **452**, 1945-1951 (1996).
10. Murray, N.H., Bourne, N.K. and Rosenberg, Z., in *Shock Compression of Condensed Matter 1995,* (AIP, Woodbury, New York, 1996), pp. 491-494.
11. Murray, N.H., *PhD Thesis,* Cambridge University, 1997.
12. Bourne, N.K., Millett, J.C.F., Rosenberg, Z. and Murray, N.H., J. Mech. Phys. Solids, in press (1997).
13. Grady, D.E., in *Shock Compression of Condensed Matter 1995,* (AIP, Woodbury, New York, 1996), pp. 9-20.
14. Millett, J.C.F., Bourne, N.K., and Rosenberg, Z. in these proceedings (1997).

CHAPTER XII

EXPERIMENTAL TECHNIQUES

CP429, *Shock Compression of Condensed Matter – 1997*
edited by Schmidt/Dandekar/Forbes
© 1998 The American Institute of Physics 1-56396-738-3/98/$15.00

HIGH STRAIN RATE CHARACTERIZATION OF LOW-DENSITY LOW-STRENGTH MATERIALS

O. Sawas, N. S. Brar, and R. A. Brockman

University of Dayton Research Institute, Dayton OH 45469-0182.

The Conventional Split Hopkinson Bar (CSHB) is a reliable experimental technique for measuring high strain rate properties of high-strength materials. Attempts to use the CSHB for similar measurements in more compliant materials, such as plastics and foams, are limited by the maximum achievable strain and high noise-to-signal ratios. This work introduces an all-polymeric split Hopkinson bar (APSHB) experiment, which overcomes these limitations. The proposed method uses polymeric pressure bars to achieve a closer impedance match between the pressure bars and the specimen materials, thus providing both low noise-to-signal ratio data and a longer input pulse for higher maximum strain. Data reduction procedures for APSHB that account for the viscoelastic behavior of the pressure bars are presented. Comparing the high strain rate response of 1100 Al obtained from CSHB and APSHB validates these procedures. Stress-strain data at strain rates of 500-2000/s for polycarbonate, polyurethane foam, and styrofoam are presented.

INTRODUCTION

The split Hopkinson bar (SHB) is the most widely-used method for investigating the dynamic behavior of high-strength materials in the range of strain rate of 100/s-10,000/s [1]. Much interest has developed recently for providing similar measurements for much more compliant materials such as plastics, rubbers, and foams. The mechanical impedance and the wave propagation velocity of conventional pressure bar materials (steel, aluminum alloy) are extremely large compared to those of most plastics, and foams. Consequently, serious problems, such as unacceptably high noise-to- signal ratios and short loading time which limits the maximum achievable strain, arise when conventional SHB (CSHB) techniques are used to investigate the high strain rate properties of such low-density, low-strength materials. This problem is shown in Figure 1, which presents a typical stress-strain curve at a strain rate of 500/s for low-density polyurethane foam. Note that the noise-to-signal ratio is high enough to render the stress-strain curve suspect, and that the test was discontinued after the specimen was strained to only 12 % due to limitations on the length of the loading pulse.

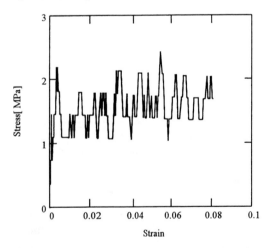

Figure 1. Stress-strain curve for polyurethane foam obtained using the CSHB

ALL-POLYMERIC SPLIT HOPKINSON BAR

An all-polymeric split Hopkinson bar (APSHB) can be used to overcome both the difficulties of bar/specimen impedance mismatch and short loading pulse duration. Polymeric bar materials have better mechanical impedance matches than metals with most plastics and foams, and their lower moduli enable them to produce strain gage signals with substantial amplitude that curtails the noise-to-signal ratio problem. In addition, the lower wave propagation velocities in polymeric bars produce longer loading pulse duration that allows sufficient time for specimens to achieve large strain levels. Polymers, however, are viscoelastic in their dynamic stress-strain behavior, which invalidates conventional analyses of CSHB data. Viscoelastic constitutive equations-of-state of the bar materials govern wave propagation and relate strain to stress, and strain-to-particle velocities. These are needed to properly analyze the experimental strain gage data in order to determine the specimen stress, strain rate, and strain.

Assuming that the specimen deforms uniformly and neglecting slight differences in the properties of the transmitter and the incident bars in the general SHB configuration the specimen stress, strain rate, and strain are related to stress (σ_t) and particle velocity (v_r) in the bar through the following equations (1):

$$\sigma_s(t) = \frac{A_b}{A_s} \sigma_t(x_2, t) \qquad (1)$$

$$\dot{\varepsilon}_s = \frac{-2}{l_s} v_r(x_1, t) \qquad (2)$$

$$\varepsilon_s = \frac{-2}{l_s} \int_0^t v_r(x_1, t) dt \qquad (3)$$

Where A_b, and A_s are the bar and the specimen cross sectional area; l_s is the specimen gage length. Due to the dispersive nature of wave propagation in APSHB, the pulses measured at the strain gage do not represent those at the bar-specimen interface. Furthermore, because of the viscoelastic nature of the bar material, stress and particle velocity can not be determined from the measured strain on the bar through simple multiplication with a constant; instead, linear viscoelastic equations are required.

SHORT TIME VISCOELASTIC PROPERTIES OF THE POLYMERIC PRESSURE BARS

The linear viscoelastic properties of the pressure bars are very important in the interpretation of the APSHB data. A wave propagation method was adapted to determine the very early time linear viscoelastic properties of the polymeric bar materials. This method is based on the work of Sackman and Kaya [2] and can be used to either determine the material viscoelastic properties

(identification problem) or to solve viscoelastic wave propagation in material with known viscoelastic properties (prediction problem). For the identification problem, the method requires measurements of a single quantity such as strain at two different locations along a bar that is overrun at different times by the propagating wave. These measurements can then be related analytically to the linear viscoelastic material properties. The development of the method procedure requires the mathematical formulation of the linear viscoelastic wave propagation in semi-infinite circular rods. This formulation and a detailed derivation of the procedures are presented in [3]. The major steps of this method as applied to the identification problem are shown in the form of a flow chart in Figure 2.

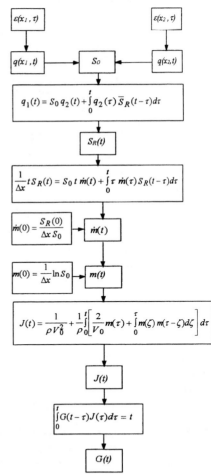

Figure 2. Flow chart showing the major steps of the wave propagation method as applied for the identification problem

APSHB EXPERIMENTAL SETUP

The APSHB shown in Figure 3 was constructed from cast acrylic bars 25.4 *mm* in diameter. Incident and transmitter bars were nearly 2.5 *m* long, with strain gage pairs mounted near the center of each bar. The two pressure bars were mounted and aligned longitudinally in Teflon bushings that supported them rigidly along a single horizontal axis, while permitting free axial movement. Test specimens were placed between the two bars. Anvils 1.0 *mm* thick made from high-strength titanium alloy were placed between the specimen and the two bars to prevent damage to the bars during specimen compression and minimize localized elastic deformation at the specimen/bar interfaces. The specimen/anvil interfaces are lubricated before each test with petroleum jelly to reduce friction and allow radial expansion of the specimen. The 1.07 *m* long striker bar was accelerated to the desired impact velocity by a "slingshot" mechanism where the driving force was supplied by a torsion bar spring.

Striker bar   Incident bar   Specimen   Transmitter bar

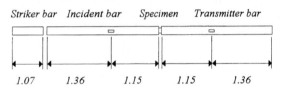

1.07   1.36   1.15   1.15   1.36

Figure 3. Schematic sketch of the APSHB setup.

APSHB VISCOELASTIC DATA ANALYSIS

The following steps were taken to determine the specimen stress, strain rate, and strain from the strain measured at the strain gage location on the pressure bars:

1. Determine the viscoelastic properties of the polymeric pressure bar material employed in the APSHB.
2. Using the predetermined material properties, solve the resulting linear viscoelastic wave propagation equations in terms of strain.
3. Using the solution of the linear viscoelastic wave propagation in terms of strain, determine the strain at the bar/specimen interfaces from the strain measured at the strain gage locations.
4. Based on the predetermined material properties of the pressure bars, establish a linear viscoelastic strain-stress and strain-particle velocity relationship

5. Using the stress-strain relationship, determine the stress at the bar/specimen interfaces from the predetermined strain.
6. Using the strain-particle velocity relationship, determine the particle velocity at the bar/specimen interfaces from the predetermine strain .
7. Substitute the values of the predetermined stress and particle velocity at the bar/specimen interfaces into Eqs. (1)-(3) to obtain the specimen stress, strain rate, and strain.

VALIDATION OF THE APSHB DATA

A total of 16 test were preformed on 1100 Al to validate the APSHB viscoelastic data analysis. The first eight tests were conducted with the CSHB at strain rates of 400, 1000, and 4000/s. The results show conclusively that this material is strain rate insensitive over the range investigated. Identical specimens were then tested with the APSHB at nominal strain rates of 500, 1000, and 1500/s. Stress-strain results for one typical test at a strain rate of 1000/s are presented in Figure 4. The solid curve represents our best estimate of 1100 Al stress-strain from the CSHB. The lower curve represents the raw strain gage data obtained from the APSHB analyzed assuming that the polymeric bar beehive elastically. Note that this curve under-predicts the stress by about 15% and over-predicts the strain by more than 26%. The last curve represents the APSHB results that were analyzed by the viscoelastic analysis described earlier. This comparison confirms that the APSHB data and the viscoelastic data reduction model are very accurate.

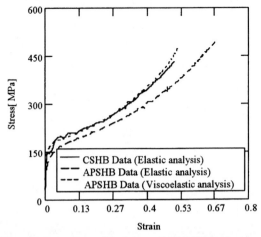

Figure 4. Comparison of the data on Al-1100 Obtained using CSHB and APSHB with elastic and viscoelastic analysis

EXPERIMENTAL RESULTS

High strain rate response of three compliant materials (polycarbonate, polyurethane foam, and styrofoam) was investigated using the APSHB. These materials are important for many engineering applications and were chosen to cover wide ranges of densities (50-1200 kg/m^3) and strengths (1-300 MPa). Polycarbonate can be characterized to some extent using both the CSHB and the new APSHB, yet the data quality and maximum strain levels achievable with the APSHB far exceed the capabilities of any CSHB yet constructed. High strain rate data on materials such as styrofoam, on the other hand, can only be obtained using the APSHB technique. High strain rate responses of the three materials are presented in Figures 5-7.

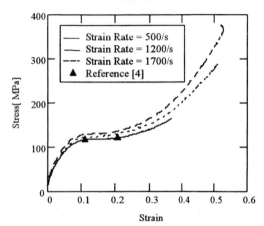

Figure 5. Stress-strain curves for polycarbonate at three rates

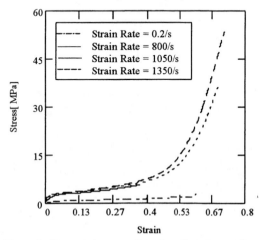

Figure 6. Stress-strain curves for polyurethane foam at four rates

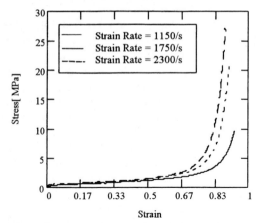

Figure 7. Stress-strain curves for styrofoam at three rates

Polycarbonate data obtained in the present study using APSHB agree well with those reported by Walley and Field at strain rate of 1980/s obtained using the drop weight technique [4].

CONCLUSIONS

The present work demonstrates that an all-polymeric split Hopkinson bar (APSHB) can be used to measure high strain rate characteristics of low-density low-strength materials such as plastics, and foams. A novel experimental technique and associated data reduction procedures have been developed. The validity of the test method and the viscoelastic data reduction procedures has been demonstrated on Al 1100. They also have been shown to produce high quality accurate data on polycarbonate, polyurethane, and stryofoam foam. These data demonstrate that the APSHB is capable of effective testing over a range of material properties that are far beyond the capabilities of earlier SHBs, and is of engineering interest for a variety of modern applications.

Acknowledgment: This work was partially supported by ASSEERT grant No. DAAG55-97-1-0150 from Army Res. Off.

REFERENCES

1. Nicholas, T., "Tensile testing of materials at high rates of strain," *Exp. Mech.***21**, 117-185, 1980.
2. Sackman, J.L. and Kaya, I. "Very early time characterization of linear viscoelastic material," Ph.D. Thesis, University of California, Berkeley, 1968
3. Sawas. O, "High strain rate characterization of low-density low-strength materials," Ph.D. Thesis, University of Dayton, 1997.
4. Walley S. M and J. E Field, Strain rate sensitivity of polymers, DYMAT Journal, Vol. 1, No. 3, p. 211, 1994.

CP429, *Shock Compression of Condensed Matter – 1997*
edited by Schmidt/Dandekar/Forbes
© 1998 The American Institute of Physics 1-56396-738-3/98/$15.00

WAVE GENERATIONS FROM CONFINED EXPLOSIONS IN ROCKS

C. L. Liu and Thomas J. Ahrens

Seismological Laboratory, California Institute of Technology, Pasadena, CA 91125

In order to record P- and S-waves generated from confined explosions in rocks in the laboratory, a method is developed based on the interactions between incident P- and SV-waves and free-surfaces of rocks. The relations between particle displacements of incident P- and SV-waves, and the strains measured using strain gauges attached on free-surfaces of rocks are analytically derived. P- and SV-waves generated from confined explosions in Bedford limestone are recorded.

INTRODUCTION

Virtually all the methods that have been proposed for discrimination of underground explosions ($m_b \leq 4$) from earthquakes and mining explosions(1 and 2) are based on various P-to-S amplitude ratios. Although there are many previous studies of seismic radiation patterns from decoupled explosions(3 - 6), it is still unclear what controls the radiation pattern of S-waves in tamped and decoupled explosions (5). Therefore, study of P- and S-wave generation from confined explosions is important for discrimination purposes. In order to investigate waves generated from small-scale laboratory explosions in rocks, a measurement method is required to monitor both P- and S-waves. Conventional seismic recording systems and methods for laboratory scale high strain rate experiments(7) can not be utilized. Based on the analysis of the interactions between P- and SV-waves and free surfaces, we have developed a method to monitor P- and SV-wave profiles using two perpendicular strain gauges attached to the free-surfaces of samples. The method and some initial experimental data are presented below.

MEASUREMENT METHOD

When elastic P- and S- waves generated from explosions in rocks reflect at free surfaces, these generate different displacement-time histories.

We determine P- and SV-wave displacements using the strains measured along two perpendicular directions at a series of stations on free-surfaces of the rock samples.

Data reduction method

The strain gauges are attached at positions along the intersection of the plane containing the axis of the spherical wave front and the sample free-surface as shown in Fig.1. The strains recorded by the gauges include the contributions from incident P- or S-waves and reflected P- and S-waves. The relation between the measured strains and incident P- and S-wave particle displacements are derived as follows:

P-wave reflections at free surfaces

The displacement reflection coefficients for incident planar P-waves at free surfaces (8) are $PP = (B - A)/(B + A)$, $PS = 2\frac{\beta}{\alpha}\sin(2\theta)\cos(2j)/(A + B)$, where PP and PS are reflection coefficients of P- and SV-wave displacements due to incident P-waves, and α and β are P- and S-wave velocities, respectively. $A = \cos^2(2j)$ and $B = (\frac{\beta}{\alpha})^2\sin(2j)\sin(2\theta)$, where θ and j are P-wave incident angle and S-wave reflection angle, respectively (Fig.1).

The displacements of particles on free surfaces after P-wave reflections are

$$u_{par}^p = u_p^I[(1 + PP)\sin\theta + PS\cos j] = H_{par}u_p^I, \quad (1)$$

$$u_{per}^p = u_p^I[(1-PP)\cos\theta + PS\sin j] = H_{per}u_p^I, \quad (2)$$

where u_p^I is the particle displacement of incident P-waves; u_{par}^p and u_{per}^p are the particle displacements along the directions shown in Fig.1, respectively. Having substituted PP and PS into Eqs.(1) and (2), the two coefficients are $H_{par} = 2\cos\theta\sin(2j)/(A+B)$ and $H_{per} = 2\cos\theta\cos(2j)/(A+B)$.

SV-wave reflections at free surfaces

For incident SV-waves, the reflection coefficients for P-waves (SP) and SV-waves(SS)(8) are $SP = \frac{\beta}{\alpha}\sin(4\theta)/(A_s + B_s)$, $SS = (A_s - B_s)/(A_s + B_s)$, where $A_s = \cos^2(2\theta)$, $B_s = (\frac{\beta}{\alpha})^2\sin(2\theta)\sin(2j)$. θ and j are SV-wave incident angle and P-wave reflected angle, respectively.

The displacements of particles on free surfaces after SV-wave reflections are

$$u_{par}^{sv} = u_{sv}^I[(1+SS)\cos\theta + SP\sin j] = G_{par}u_{sv}^I, \quad (3)$$

$$u_{per}^{sv} = u_{sv}^I[(SS-1)\sin\theta - SP\cos j] = G_{per}u_{sv}^I, \quad (4)$$

where u_{par}^{sv} and u_{per}^{sv} are particle displacements after reflection along the directions shown in Fig.1, and u_{sv}^I is the particle displacement of incident SV-waves. The two coefficients are determined to be $G_{par} = 2\cos(2\theta)\cos\theta/(A_s + B_s)$, $G_{per} = -2\frac{\beta}{\alpha}\cos j\sin(2\theta)/(A_s + B_s)$.

Particle displacements of incident P-waves

Because u_{per}^p is perpendicular to free surfaces, and incident waves are assumed to be spherical, the strain due to u_{per}^p along direction 1, ε_1^{per}, is simply expressed as

$$\varepsilon_1^{per} = H_{per}u_p^I/r_0, \quad (5)$$

where r_0 is the distance between the center of explosive source and the free surface of samples at $\theta = 0$.

u_{par}^p does not result in any strains along direction 1, so the total strain induced by the incident P-waves is determined to be $\varepsilon_1^p = H_1 u_p^I/r_0$, where $H_1 = H_{per}$. Therefore, the strain along direction 1 yields the particle displacement of incident P-waves as

$$u_p^I = r_0\varepsilon_1^p/H_1. \quad (6)$$

Since both u_{per}^p and u_{par}^p have contributions to strains along direction 2, we need to consider the resultant displacements. The length of the gauge after reflection, Δs, is equal to $(r_n^2 + (\frac{\partial r_n}{\partial\theta})^2)^{\frac{1}{2}}\delta\theta$, where $\delta\theta \approx \frac{l_s\cos\theta}{r_n}$, l_s is the initial length of the strain gauge, r_n is the distance between the explosive source center to the position of a gauge upon P-wave reflection. Here r_n can be expressed as $r_n \approx r + u\cos(\eta-\theta)$, where r is the distance between the explosive source center and the gauge before P-wave reflection, u is the resultant displacement of the point at θ on the free surface and η is the angle between u and u_{per}^p. u and η are given by $u = u_p^I 2\cos(\theta)/(A+B)$ and $\eta = 2j$.

From the expressions above, Δs is given as

$$\Delta s \approx (r_n + \frac{1}{2r_n}(\frac{\partial r_n}{\partial\theta})^2)\delta\theta. \quad (7)$$

Because $\frac{u_p^I}{r} \ll 1$, Δs is rewritten as

$$\Delta s = (r(1+\frac{\tan^2(\theta)}{2})+u_p^I(W(\theta)(1-\frac{\tan^2(\theta)}{2})+\tan(\theta)\frac{dW}{d\theta}))\delta\theta, \quad (8)$$

where $W(\theta) = 2\cos(\theta)\cos(\eta-\theta)/(A+B)$.

From the definition of strains,

$$\varepsilon_2^p = \frac{\Delta s - l_s}{l_s} = \frac{H_2 u_p^I}{r_0}, \quad (9)$$

where $H_2 = \cos(\theta)(W(1 - \tan^2(\theta)/2) + \tan\theta\frac{dW}{d\theta})/(1 + \tan^2(\theta)/2)$.

The particle displacement of an incident P-wave determined from a gauge along direction 2 is

$$u_p^I = r_0\varepsilon_2^p/H_2. \quad (10)$$

Particle displacements of incident SV-waves

Using the same above formulation, the relations between incident SV-wave particle displacements and strains along the two directions are obtained.

The particle displacements of incident SV-waves from the strains along direction 1 and 2 are

$$u_{sv}^I = r_0\varepsilon_1^{sv}/G_1, \quad (11)$$

and

$$u_{sv}^I = r_0\varepsilon_2^{sv}/G_2, \quad (12)$$

where $G_1 = G_{per}$,

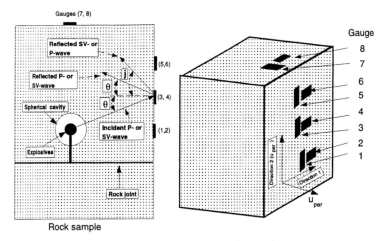

Gauges (7, 8)

Reflected SV- or P-wave

Reflected P- or SV-wave

Spherical cavity

Incident P- or SV-wave

Explosives

Rock joint

Rock sample

Gauge

Direction 2 (u per)

Direction 1

U_{per}

Figure 1.   Schematic arrangement of experiments

$$G_2 = \cos(\theta)(W_s(1 - \frac{\tan^2(\theta)}{2}) + \tan\theta\frac{dW_s}{d\theta})/(1 + \frac{\tan^2(\theta)}{2}),$$

$$W_s = \frac{2\cos(\eta - \theta)\cos(\theta)}{A_s + B_s}(\cos^2(2\theta) + 4(\frac{\beta}{\alpha})^2\cos^2 j\sin^2(\theta))^{\frac{1}{2}},$$

and $\tan\eta = -\alpha\cos\theta/(\beta\cos j)/\tan(2\theta)$.

Characteristics of strains in different directions

Figure 2 shows the dependence of H_1, H_2, G_1 and G_2 on incident angle that were calculated using the equations derived above. For incident P-waves, the constant, H_1, is relatively insensitive to θ, and H_1 changes from 2.0 at $\theta = 0^o$ to 1.4 at $\theta = 60^o$ for Bedford limestone. The constant, H_2, is very sensitive to θ, and it varies from 2 to -0.4 when θ varies from 0 to 60°. It can be seen that the strains induced by compressional P-waves along direction 1 are always positive, but the strains along direction 2 are positive when θ is less than 47° and negative when θ is larger than 47°. This change in polarity is controlled by the ratio of the projection of P-wave displacements along direction 1 to that along the direction that is perpendicular to the free surface. If the strain induced by the displacement along direction 2 is less than that due to the displacement along the perpendicular direction, the strain is positive, otherwise, the strain is negative.

From the above calculation, the gauges along direction 1 are not sensitive to an incident SV-wave, however, the gauges along direction 2 are very sensitive to an incident SV-wave. The polarities of the strains along direction 2 are always negative, and the polarities of the strains along direction 1 are determined by the direction of SV-wave particle motion .

Eqs.(6), (10), (11) and (12) give the relations between strains along the two directions and incident P- and SV-wave particle displacements. If strains along the two directions can be recorded, the P- and SV-wave amplitudes can be determined experimentally.

EXPERIMENTS AND RESULTS

The method described above was used to monitor P- and SV-waves generated from confined explosions in rocks. The rock sample (Bedford limestone) was assembled with two blocks as shown in Fig.1. The rock sample with strain gauges was placed inside a tank that was pressurized to 10 bars.

The recorded strains for one of the experiments are shown in Figs.3 and 4. The characteristics of the strains recorded by the gauges are the same as predicted from our derived equations. The strains along direction 1 induced by incident P-waves are always positive, while the strains along direction 2 change polarities as P-wave incident angle increases (Figs.3 and 4). The strains induced by incident SV-waves along direction 2 are negative and are much larger than that along direction 1.

From the records, it is straightforward to determine P and SV-wave amplitudes for the experiment using the expression given above. From the P- and SV-wave velocities of Bedford limestone, the expected S-wave arrivals are labeled on the records. The time difference between the expected and the recorded is less than $2\mu s$ for the experiment.

CONCLUSIONS

In this work, a method has been developed for measuring P- and SV-wave amplitudes generated from explosions in rocks. The relations between the strains given by gauges placed on the free surfaces of rocks and incident particle displacements are derived analytically. The experimental results showed that the characteristics of recorded strains along the two directions are in good agreement with the predictions.

ACKNOWLEDGMENTS

Research was sponsored by Air Force Technical Applications Center. Contribution 6212, Division of Geological and Planetary Science, California Institute of Technology.

REFERENCE

1. Blandford, R. R., AFTAC-TR-95-002, 1995.

2. Helmberger, D. V. and Woods, B., *Monitoring a Comprehensive Test Ban Treaty*, edited by Husebye, E.S. and Dainty, A.M., Kluwer Academic Publishers, the Netherlands, 1996, 365-383.

3. Glenn L. and Goldstein P., *J. Geophys. Res.*, **99**, 11,732 - 11,730(1994).

4. Glenn L., Ladd A., Moran B., and Wilson K., *Geophys. J. R. astr. Soc.*, **81**, 231- 241(1985).

5. Murphy J. , op. cit in Ref.(2), pp 247-293.

6. Sykes L., op. cit in Ref.(2), pp 225-245.

7. Kim S., Clifton R. and Kumar P., *J. Appl. Phys.*, **48**, 4132-4139(1977).

8. Aki, K. and Richards, P., *Quantitative Seismology Theory and Methods,* W. H. Freeman and Company, 1980, ch.5, 320.

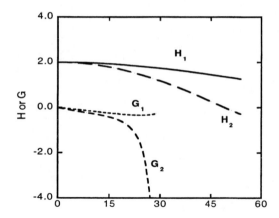

Figure 2.   Incident angle (degree)

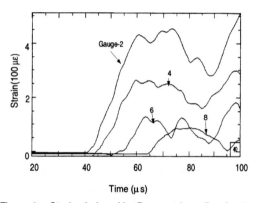

Figure 3.   Strains induced by P-wave along direction 1

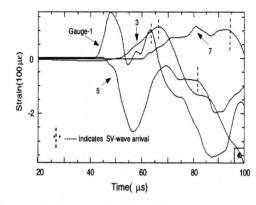

Figure 4.   Strains induced by incident SV-waves along direction 2

CP429, *Shock Compression of Condensed Matter – 1997*
edited by Schmidt/Dandekar/Forbes
© 1998 The American Institute of Physics 1-56396-738-3/98/$15.00

ELECTROMAGNETIC CYLINDRICAL COMPRESSION TEST UNDER LARGE STRAIN AT HIGH STRAIN RATE

J. PETIT

DGA/DCE/Centre d'études de Gramat, 46500 Gramat, France.

M. KAZEEV, P. LEVIT, Yu. TOLSTOV

Kurchatov Institute/NFI, Kurchatov Sq., Moscow 123182, Russia.

The behavior of materials under high strain rates ($> 10^4$ s^{-1}) and large strains (> 50 %) is unattainable with conventional tests. Their behavior at these strain rates and strains is then deduced by extrapolating results from tensile, compression, torsion, plate impact or pressure shear experiments. Results from an electromagnetic cylindrical compression test can be used to validate this extrapolation. The principle as well as the limits of this exploitation through experimentation-1D calculation comparisons are illustrated with the initial tests performed on aluminum, copper and tantalum materials.

INTRODUCTION

Numerical simulation is widely used to assist experimental studies but also to study involved physical processes. It is the only method to obtain non measurable quantities such as stresses or thermo-mechanical state of materials. Numerical results are only reliable if the selected materials models have been validated under corresponding field of use. This paper presents a validation method of behavior models for high strain and high strain rate through the analysis of electromagnetic cylindrical compression experiments.

EXPERIMENTAL SEQUENCE AND PRINCIPLE OF ANALYSIS

A capacitor bank is discharged in a massive single turn solenoid. The sample tube is placed in the middle of the solenoid. The magnetic field and the current induced in the tube result in its cylindrical compression. The experiments on copper which are reported in this paper were performed with the TROB-100 apparatus of the Kurchatov Institute. Aluminum and tantalum experiments were performed on the CYCLOPE apparatus designed by the Kurchatov Institute for the CEG. The maximum stored energies are respectively 100 and 150 kJ for these apparatuses and the maximum pulsed magnetic fields are respectively 40 and 53 MA/m. The deformation of the tube and its compression symmetry are observed respectively with streak and frame cameras while the magnetic field outside the tube is recorded. Details on the apparatus, the measurements and their accuracy were given in a previous paper (1).

The magnetic field is applied during the whole tube motion. Therefore, there is no 'free flight' phase during which an energy balance could allow determination of an average value of the yield strength as it is performed in the ring expansion tests (2,3).

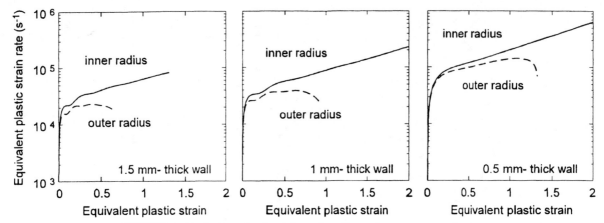

FIGURE 1 : Equivalent plastic strain and strain rate in copper experiments.

The electromagnetic cylindrical compression experiment can be used as an appropriate tool to validate behavior models by experiment-simulation comparisons in domains that are inaccessible to conventional experiments. The measured magnetic field is used as an input in the code. The numerical closure of the tube is then compared to the experimental one to validate constitutive model.

EXPERIMENT-CALCULATION COMPARISON

The numerical simulations were performed with UNIDIM (4), a one-dimensional code inspired from WONDY (5) and used in its cylindrical geometry. This code was adapted to take into account electromagnetic stress : Laplace volume forces were

added in the momentum conservation equation and a Joule heating term was introduced in the energy conservation equation (6,1).

Evaluation of copper models

The experiments performed with copper samples enabled us to test a series of models and to study the limits and accuracy of the validation method. The sample tubes were manufactured from an OFE-OK grade copper rod. The average initial grain size of the copper was 50 μm. The inner diameter of the tubes is equal to 18.6 mm. Tubes of different wall thickness were used to investigate three different domains of the strain rate regime (see fig. 1)

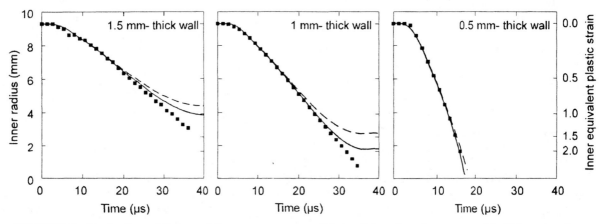

FIGURE 2 : Evaluation of Z-A modeling, ▪ exp. , —— without dislocation drag, – – with dislocation drag.

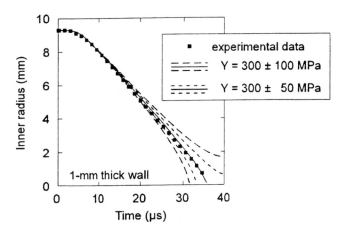

FIGURE 3 : Evaluation of perfectly plastic models of copper.

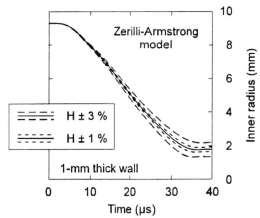

FIGURE 4 : Effect of the magnetic field input.

Two constitutive models of copper were used to evaluate the consistency of the presented analysis : the model proposed for an OFHC copper by Zerilli and Armstrong in its first version (7) and their next one including a term for dislocation drag that gives a sharp increase of the stress at high strain rate (8). The copper used in this study was not characterized and the primary coefficients published with these models were used as default values. Only the measured average grain size was taken into account.

When the wall thickness is 1.5 mm, a difference between simulations and experiment appears when the inner radius reaches about 6.5 mm (fig. 2). When the wall thickness is 1 mm, this difference appears later. When the wall thickness is 0.5 mm, the difference becomes very small. So, it seems that the constitutive model has no effect on the tube motion when the closure is too fast. It is a limit of this kind of analysis based on the motion.

Simulations were performed with different values of the yield strength Y in a model of perfectly plastic behavior in the case of 1 mm-thick wall tube. They allow to assess the overestimation of the Zerilli-Armstrong models and consequently the accuracy of the models evaluation (fig. 3). They show that, in this case, it is possible to clearly observe the effect of an error of 50 MPa on the mean threshold stress.

The measured value of the magnetic field H is employed as the input data in the numerical simulations. The accuracy of this measurement can be evaluated to 3 % for this series of experiments achieved with only one probe. Numerical simulations were implemented to evaluate its influence on a numerical result (fig. 4). These last results compared with the preceding ones (fig. 3) show that the effect of a 3 % inaccuracy on the field is, for this experiment, equivalent to an inaccuracy of about 20 % on the yield strength evaluation. Enhancing the accuracy of the magnetic field measurement is accordingly essential to a more reliable validation of the behavior models. To reach an accuracy of 1 % is possible and seems enough (fig. 4) for the future series of experiments.

Evaluation of aluminum and tantalum models

Tubes of aluminum and tantalum were also tested. The characteristics of these tubes are :
- 99.9% Al ; mean grain size of 1 mm ; inner diameter, 18,6 mm ; wall thickness, 1 mm,
- 99.5% Ta ; mean grain size of 35 µm ; inner diameter, 16 mm ; wall thickness, 1 mm.

The strain rates reached are close to that reached with the copper tube of 1 mm thick wall. In the case of aluminum the Zerilli-Armstrong model for a 99.99% aluminum (9) and the rate independent Steinberg- Cochran-Guinan model (10) for a 1100-0 aluminum (11) were used in the simulation. For the tantalum, the Zerilli-Armstrong model (12,9) for

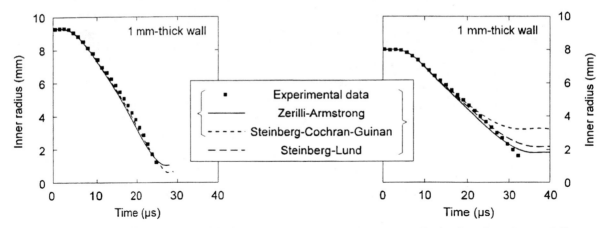

FIGURE 5 : Evaluation of aluminum modeling.

FIGURE 6 : Evaluation of tantalum modeling.

which σ_G and k were supposed to be respectively equal to 125 MPa and 14 MPa mm$^{1/2}$, and the rate independent (S-C-G version, 10) and rate dependent (S-L version, 13) Steinberg models (11) were tested. The results are presented on the figures 5 and 6. Simulations of experiment with aluminum produced results close to the experimental observation but tantalum results deviated from experimental measurements.

CONCLUSION

The results obtained in this initial study with three materials (Cu, Al, Ta) demonstrate the feasibility of using electromagnetic cylindrical compression experimental results coupled with simulation to evaluate the constitutive models in the regime of strains greater than 50 % and strain rates between 10^4 and 10^5 s^{-1}. At strain rate greater than 10^5 s^{-1} , this method based on the analysis of the tube motion becomes unsuited. The continuation of the evaluation work of this method requires the testing of a perfectly known material whose behavior model coefficients have been previously determined from conventional experimental results.

REFERENCES

1. Petit J., Alexeev Yu. A., Ananiev S.P., Kazeev M.N., "The electromagnetic compression : a tool to test behavior modeling under large strain at high strain rate", to be published in the Proceedings of the Eurodyamat 97 conference, Toledo, Spain, September 1997.

2. Warnes R.H., Karpp R.R., Follansbee P.S., *Journal de Physique*, Colloque C5, sup. au n°8, Tome 46, pp C5-583 C5-590, 1985.

3. Gourdin W.H., *J. Appl. Phys.*, Vol 65, N°2, pp 411-422, 1989.

4. Hereil P.L., "UNIDIM : un code monodimensionnel aux différences finies", CEG Technical Report, to be published.

5. Kipp M.E., Kipp R.J., "WONDY V : A One-Dimensional Finite-Difference Wave-Propagation Code", Sandia National Laboratories Albuquerque, SAND--81-0930 Report, 1982.

6. Knoepfel H., *Pulsed High Magnetic Fields*, North Holland Publishing Company, Amsterdam, 1970.

7. Zerilli F.J., Armstrong R.W., *J. Appl. Phys.*, Vol 61, N°5, pp 1816-1825, 1987.

8. Armstrong R.W., Zerilli F.J., *Journal de Physique*, Colloque C3, sup. au n°9, Tome 49, pp C3-529 C5-534, 1988.

9. Zerilli F.J., Armstrong R.W., "Constitutive relations for the plastic deformation of metals", High-Pressure Science and Technology--1993, AIP Conference Proceedings 309, AIP Press, 1994.

10. Steinberg D.J., Cochran S.G., Guinan M.W., *J. Appl. Phys.*, Vol 51, N°3, pp 1498-1504, 1980.

11. Steinberg D.J., "Equation of State and Strength Properties of Selected Materials", LLNL UCRL-MA-106439 Report, 1996 updated.

12. Zerilli F.J., Armstrong R.W., *J. Appl. Phys.*, Vol 68, N°4, pp 1580-1591, 1987.

13. Steinberg D.J., Lund C.M., *J. Appl. Phys.*, Vol 65, N°4, pp 1528-1533, 1989.

CP429, *Shock Compression of Condensed Matter – 1997*
edited by Schmidt/Dandekar/Forbes
© 1998 The American Institute of Physics 1-56396-738-3/98/$15.00

DIFFERENTIAL HUGONIOT AND EXPERIMENTAL ESTIMATE OF THE GRÜNEISEN PARAMETER FOR SiO_2

Dennis Grady

Applied Research Associates, 4300 San Mateo Blvd., A-220, Albuquerque, New Mexico 87110

An experimental technique is explored for measuring the difference in two nearby Hugoniots with high accuracy. The method is used to provide accurate estimates of the local Grüneisen properties of the test material. Silicon dioxide (SiO_2) in the initial α-quartz and fused silica states is tested in the 40-70 GPa range.

INTRODUCTION

The Hugoniot provides a measure of the compressibility of a material when subjected to high compressive pressures. A more complete equation of state requires knowledge of the thermal contribution to pressure – usually characterized by the material Grüneisen property $\gamma = \upsilon \, \partial p / \partial E|_{\upsilon}$.

Various methods have been pursued to measure the magnitude and functional dependence of this property at high pressure. Several common methods have included the use of distended solids to vary the thermal state on the Hugoniot, and the measurement of sound velocities at the Hugoniot state to assess the local isentrope properties. In both methods calculations of Grüneisen parameters involve differencing the data usually leading to quite large uncertainties in γ.

Here we have explored an experimental method which directly measures the difference between similar Hugoniots with high accuracy. Assuming differences at the high pressure Hugoniot states achieved are due to thermal characteristics of the material, Grüneisen properties are calculated The material chosen for study was SiO_2 in initial α-quartz and fused silica states.

EXPERIMENTAL METHOD

Impact experiments were performed on a two-stage light gas launcher. Projectile and target configuration are shown in Figure 1. The SiO_2 test materials were mounted on the projectile. The higher shock impedance material at the front, was y-cut α-quartz with a reported density of 2653-2658 kg/m^3. The second was glass (fused-silica Dynasil 1000) with a reported density of 2201 kg/m^3. The samples were backed with an aluminum support plate and carried on a Lexan (polymeric) sabot. SiO_2 sample diameters were approximately 25 mm. Critical experimental thickness properties are given in Table 1.

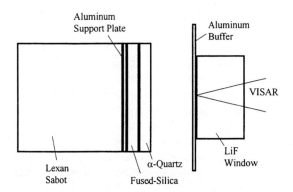

FIGURE 1. Experimental configuration

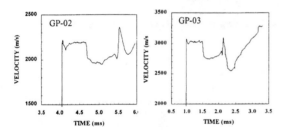

FIGURE 2. VISAR velocity profiles.

The stationary target consisted of a 19-mm diameter by 19-mm length cylindrical lithium fluoride (LiF) window preceded by a thin aluminum plate. Aluminum matches the shock impedance of LiF closely and was included in the experiment because of past experience of maintaining laser light reflected off of a surface subjected to direct shock impact. Motion characteristics imparted by the impact were recorded at the aluminum-LiF interface with VISAR diagnostics (1). The velocity-per-fringe constant was set to 197.44 m/s per fringe for all tests.

TABLE 1. Experimental Parameters

Test #	Glass (mm)	Quartz (mm)	Al Bfr (mm)	Al Bkr (mm)	Vel. (km/s)
GP-02	3.258	3.171	.490	.499	4.98
GP-03	3.239	3.239	.487	.236	6.61

EXPERIMENTAL RESULTS

The measured velocity histories for the two successful experiments are shown in Figure 2. The velocity amplitude was determined through the addition of an integral number of fringe jumps (eleven for GP-02 and fifteen for GP-03) until agreement with the Hugoniot for quartz reported by McQueen (2) was achieved. The initial velocity amplitude persisting for 0.5-0.6 µs corresponds to the α-quartz impact on LiF. Irregularities in the first 0.05-0.1 µs corresponds to equilibration of the thin aluminum buffer and glue bond preceding the LiF. The decreasing step in velocity (decrement wave) of about 200-300 m/s which persists for 0.6-0.7 µs corresponds to arrival of the decrease in shock amplitude from the glass-quartz interface. Finally the spike in velocity towards the end of the history is a signature from the aluminum plate backing the glass sample.

EXPERIMENTAL ANALYSIS

Hugoniot States

Hugoniot states for both α-quartz and fused silica are determined from the experiment. Those for α-quartz are determined directly while Hugoniot states for fused silica are determined as a difference from the α-quartz Hugoniot. As noted earlier, existing Hugoniot data (2) are necessary to adjust the number of integral fringes to the α-quartz Hugoniot states. The velocity-per-fringe constant is sufficiently insensitive, however, that there is no ambiguity in selecting the correct integral fringe jump. Hugoniot states measured for α-quartz are compared with the linear shock-velocity-versus-particle-velocity representation of existing α-quartz Hugoniot data in the high pressure region provided by Trunin et al. (3) and by McQueen (2), in Figure 3.

Hugoniot states in material two (fused silica), given the Hugoniot states in material one (α-quartz) for particle velocity, pressure, specific volume and specific internal energy are, respectively,

$$u_2 = u_1 + \Delta u, \qquad (1)$$

$$p_2 = p_1 - \rho_{10} c \Delta u, \qquad (2)$$

$$\upsilon_2 = \upsilon_{20} - u_2^2 / p_2, \qquad (3)$$

$$E_2 = \tfrac{1}{2} u_1^2 + u_1 \Delta u. \qquad (4)$$

Both Δu and c are determined from the velocity histories in Figure 2. The velocity decrement Δu used to calculate Hugoniot states in Equations 1-4 is related to measured velocity decrement Δu_{meas} in Figure 2 through a factor $R = (Z_1 + Z_2)/2Z_1$, such that $\Delta u = R \Delta u_{meas}$, to account for the acoustic impedance mismatch during transmission of the velocity step through the quartz-LiF interface. The measured property c is the velocity of the decrement wave through the quartz and is determined through a subtraction of the calculated shock transit through quartz from the total time between shock arrival and decrement wave arrival in Figure 2. (Some minor corrections for the aluminum buffer must be accounted for.) Relevant

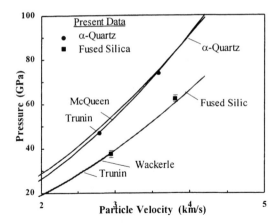

FIGURE 3. Hugoniot states for quartz and fused silica.

properties are provided in Table 2. Decrement Hugoniot states for fused silica are provided in Figure 3 and compared with fused silica Hugoniots reported by Trunin *et al.* (3) and by Wackerle (4) which essentially overlay. Note the uncertainties (error bars) are in the decrement between the α-Quartz and fused-silica Hugoniots.

Sound Velocities

Lagrangian acoustic wave velocities determined from the transmission time of the decrement wave through α-quartz are shown in Figure 4. (We take the liberty of calling this decrement wave an acoustic or sound velocity even though its amplitude is of order 10 GPa as seen in Table 2.) Comparisons

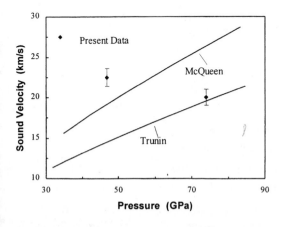

FIGURE 4. Comparison of present sound velocity data with the derivative of quartz Hugoniot data (Lagrangian velocities).

are made with the local slope of the Hugoniot, $c^2 = dp/d\rho|_H$. Hugoniot parameters for quartz provided by both McQueen (2) and Trunin *et al.* (3) are used and indicated the scatter in this calculated property. Decrement wave speeds in excess of the Hugoniot slope value would suggest crystal strength in the 50-70 GPa shock pressure range as reported by Morgan and Fritz (5) for quartz. The fact that the higher pressure acoustic wave has a lower velocity is disturbing. Uncertainties in the measurement is a possibility: The calculated shock wave transit time which is subtracted from the measured round trip transit is about 75% of the total. The round trip transit time was measured at the midpoint of the dispersing decrement wave. On the other hand decreasing sound velocity with increasing Hugoniot pressure have been reported elsewhere (*e.g.*, McQueen (2), Chhabildas and Grady (6)) for quartz although pressures at turn-over differ. This behavior of Hugoniot sound velocities has been attributed to increase in Poisson's ratio as melt is approached.

TABLE 2. Experimental Results

Test #	Δu_{meas} (m/s)	c (km/s)	Δp (GPa)	γ
GP-02	212	22.5	9.4	1.64-1.80
GP-03	262	20.0	11.6	1.64-1.87

Grüneisen Properties

Assuming that α-quartz and fused-silica are in the same thermodynamic state on the high-pressure Hugoniot and that the measured difference in Hugoniot curves is due strictly to differences in thermal pressure (very dubious assumptions) the present data will be used to estimate the thermodynamic Grüneisen parameter for SiO_2. There are several ways that this can be done. Here we calculate from the measured velocity data the differences in specific volume and internal energy at the coexistence pressure p_2 (See Equation 2) after reflection of the principal shock at the quartz-glass interface. The relation,

$$\frac{\gamma}{\upsilon} \cong \frac{p_h'}{E_h' - \Delta E/\Delta \upsilon|_p} , \qquad (5)$$

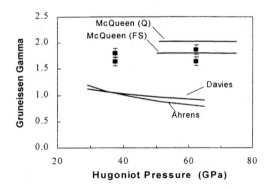

FIGURE 5. Grüneisen parameter for silicon dioxide.

is then used to estimate the Grüneisen parameter where p_h' and E_h' are calculated from the Hugoniot relation for SiO_2. Specific volume and internal energy at the pressure p_2 for quartz are calculated from,

$$\bar{\upsilon}_1 = \upsilon_{10} - \frac{u_1^{\,2}}{p_1} + \frac{\Delta u}{\rho_{10}c} \ , \qquad (6)$$

$$\bar{E}_1 = \tfrac{1}{2}u_1^{\,2} - \frac{p_1 \Delta u}{\rho_{10}c} \ . \qquad (7)$$

For purposed of thermal energy estimates E_{10} and E_{20} are assumed equal. When combined with Equations 3 and 4 the volume and energy differences used in Equation 5 are obtained. Values for γ , assumed to apply on the quartz Hugoniot are plotted in Figure 5 and compared with earlier results.

Volume dependent Grüneisen parameters for SiO_2 developed from extensive Hugoniot data by Ahrens et al. (7) and by Davies (8) provide the two lower curves in Figure 5. Grüneisen properties estimated by McQueen (2) from Hugoniot sound velocity data at Hugoniot pressures in excess of 50 GPa on both quartz and fused silica are indicated by respective lines in Figure 5. For a given silicate material calculated Grüneisen values from McQueen's data were constant over the range investigated. Symbols indicate results from the present technique where upper points assume Hugoniot properties of McQueen while lower points that of Trunin et al. (3) The present technique suggest local Grüneisen

properties in excess of the earlier results of Ahrens et al. (7) and Davies (8). Results are in reasonable agreement with the values of McQueen (2).

DISCUSSION

The present technique illustrates a method for estimating local Grüneisen properties of matter with reasonably high accuracy. This follows since the volume and energy differences needed to calculate γ scale directly with the measured velocity differential. It is also important that initial densities of the two materials be known with high accuracy. The assumptions that the initial energies of α-quartz and fused silica are the same (or negligibly different), and that energy differences at the high-pressure states achieved are strictly thermal in nature should be critically examined and this was not done here. The substantial difference between the present measurements of the Grüneisen parameter and those of McQueen (2) ($\gamma = 1.7\text{-}2.0$) with the earlier estimates of Ahrens et al., (7) and Davies (8) ($\gamma = 0.8\text{-}1.0$) need not imply that one or the other is necessarily wrong. The latter study represents a more global description of the equation-of-state surface. The present technique and that of McQueen (2) can be expected to provide a more local measure of the Grüneisen parameter.

REFERENCES

1. Barker, L. M. and Hollenbach, *J. Appl. Phys.*, **43**, 4669-4675 (1972).
2. McQueen, R. G., *Shock Compression of Condensed Matter-1991*, Schmidt, S. C., Dick, R. D., Forbes, J. W., Tasker, D. G.,eds., Elsevier, 1992, pp.7578-7581.
3. Trunin, R. F., *et al.*, Izv. Acad. Sci. USSR Phys. Solid Earth, 1, 13-20 (1970).
4. Wackerle, J., *J. Appl. Phys.*, **33**, 922-937 (1962).
5. Morgan, J. A. and Fritz, J. N., *High Pressure Science and Technology*, K. D. Timmerhaus and M. S. Barger, Eds., Plenum, 1979, pp. 109.
6. Chhabildas, L. C. and Grady, D. E., *Shock Compression of Condensed Matter-1983*, Asay, J. R., R. A. Graham, G. K. Straub, eds., Elsevier, 1984, pp.175-178.
7. Ahrens, T. J., Takahashi, T., Davies, G. F., *J. Geophys. Res.*, 77, 310-316 (1970).
8. Davies, G. F., *J. Geophys. Res.*, 77, 4920-4933 (1972).

CP429, *Shock Compression of Condensed Matter – 1997*
edited by Schmidt/Dandekar/Forbes
© 1998 The American Institute of Physics 1-56396-738-3/98/$15.00

A CLOSED WATER-FILLED CYLINDER TEST FOR CHARACTERIZING NON-IDEAL EXPLOSIVES

R. Guirguis, R. McKeown, and J. Kelley

Naval Surface Warfare Center, Indian Head, MD 20640

A small-scale test in which the detonation products of an explosive are confined at high pressure in a closed cylinder completely filled with water is introduced. Numerical simulations of the test show that although the dynamic effects of the shock induced into the water reduce the pressure below the theoretical maximum achievable if all processes were quasi-static, the residual equilibrium pressure is still high enough for the slow reactions of non-ideal explosives to proceed at a finite measurable rate.

INTRODUCTION

Non-ideal underwater explosives usually contain a significant portion of slow reacting components - for example, a mixture of aluminum and ammonium perchlorate particles. Only a fraction of the explosive's energy is released early enough to contribute to sustaining the detonation front which, therefore, propagates at lower velocity and pressure than in ideal explosives. The slow-reacting components release the remaining energy late, after the Chapman-Jouguet (CJ) surface, often even after the bubble has expanded several times the volume of the charge [1].

The slow release of energy poses a challenge for conventional testing methods. The detonation products have to remain longer at high temperature and pressure in order to allow the non-ideal components enough time to react. This is accomplished by either using a large explosive charge or by confining the detonation products. In underwater tests, the inertial resistance of the water helps slow the expansion of the products, but because the bubble expands in all directions, the resulting spherical divergence quickly reduces the pressure and temperature inside.

In the Moby-Dick test, the bubble expansion is restricted to one-direction only in order to reduce the divergence and slow the rate of pressure decay [2]. In this paper, numerical simulations are presented that show the feasibility of confining the detonation products at high pressure in a closed cylinder full of water. The purpose of the paper is to explain the problems involved in designing a closed cylinder that

can withstand the dynamic load resulting from detonating a small PETN charge inside, and to estimate the residual pressure in a small aluminum cylinder of reasonable dimensions, after all pressure waves induced by the detonation are dissipated.

CLOSED-CYLINDER DESIGN

The test presented here is a closed version of the Guirguis Hydro-Bulged Cylinder (GHBC) test in which a small (grams) explosive (PETN) charge was detonated at the center of an open-ended water-filled seamless metal (aluminum) cylinder. A streak camera recorded the radial expansion of the outer wall, while a LASER interferometer measured its velocity. The final deformation of the cylinder proved to be a good measure of the total energy of the explosive [3]. The pressure at the interface between water and metal was measured using a tourmaline crystal. The hydrocode DYSMAS used in this paper was validated by comparing its predictions to the measured pressures [4]. DYSMAS predictions of the final deformation were also compared to the test results [5].

To characterize non-ideal explosives, the cylinder is closed such as to confine the detonation products indefinitely at high pressure. Thick, rigid aluminum plugs, slightly larger than the inner (nominal) diameter, are interference-fitted into the cylinder. Aluminum rings, as thick as the end plugs, but slightly smaller than the outer (nominal) diameter, are also interference-fitted around the cylinder ends, as illustrated in Fig. 1. Firmly clamping the ends of the cylinder into the covers in such a fashion is needed in order to contain the force of the explosion. It also makes modeling the mechanical joint easier, in comparison to other fastening methods such as threading or welding. The pressure is measured off the axis at the interface between the water and the thick, relatively rigid bottom cover. The rigidity of the cover makes using a piezo-electric transducer possible. The cylinder expansion is measured using strain gauges.

The Second Law of Thermodyamics indicates that if all the processes involved are quasi-static, the pressure will monotonically build up to the maximum static pressure that can be contained inside the cylinder. The detonation products are held in a fixed-volume bubble equal to the volume of the explosive charge, until the pressure and all other thermodynamic properties equilibrate. The corresponding final state is the point on the reacted Hugoniot having the same density as the solid explosive. Then the bubble is allowed to isentropically expand, i.e., slowly, such as to remain uniform, and adiabatically, without any heat loss to the water. The pressure inside the bubble decays as it expands. The water is compressed, its pressure increases, which in turn, stretches the cylinder walls. Eventually, the pressure inside the bubble becomes equal to the pressure of the water, whence both bubble and cylinder cease to expand.

For simplicity, les us assume that the cylinder is manufactured out of an elastic-perfectly plastic metal.

At any point during the quasi-static expansion of the bubble, the increase in internal volume due to the cylinder radial expansion has to balance the difference between the increase in the volume of bubble beyond the initial charge size, and the reduction in volume of the water due to bulk compression. Before the elastic limit is exceeded, this balance results in a water pressure that depends on the volume of the bubble, the amount and bulk compressibility of the water, as well as the cylinder dimensions, and Young's modulus and Poisson's ratio of the metal. However, because metals yield, the maximum static pressure p_e that can be contained inside the cylinder is independent of the properties of the explosive or the water. It only depends on the yield behavior of the metal and the cylinder dimensions.

As the water pressure rises, the hoop stress σ_θ at the inner wall increases until it reaches the yield stress σ_Y. When σ_θ becomes equal to σ_Y throughout the whole thickness t of the cylinder, the pressure ceases to increase. With further expansion of the bubble, the cylinder continues to expand, but the pressure remains almost constant at p_e. Because the inner radius r_i of the cylinder is much smaller than the radius of curvature of the deformed cylinder wall in the azimuthal plane, the maximum pressure

$$p_e = \sigma_Y \frac{t}{r_i} \ . \qquad (1)$$

If the equation of state (EOS) of the detonation products and the (adiabatic) bulk compressibility of the water are known, we can calculate the cylinder's final deformation at p_e by balancing the volumes, as explained above. For simplicity, let us assume that the axial profile of the radial expansion of the inner wall is parabolic, vanishing at the edges where the cylinder walls are rigidly clamped into the covers, and reaching its maximum value δr_i at the mid plane. The radial expansion increases the internal volume of the cylinder by $4\pi r_i \, \delta r_i \, L_i /3$, where L_i denotes the inner height of the cylinder.

Equation 1 indicates that for a given metal, we can increase p_e by increasing the thickness of the cylinder and/or reducing its diameter. However, as will demonstrated below, the detonation introduces into the water strong pressure waves that deform the cylinder, even if it is open, i.e., even if the corresponding static pressure is zero [3]. This dynamic deformation exceeds the deformation resulting from a static pressure p_e. The cylinder overexpands and as a result, the residual equilibrium pressure inside the cylinder after

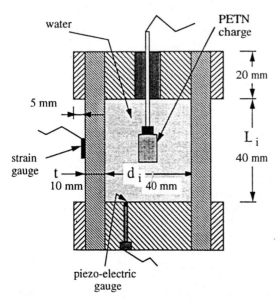

FIGURE 1. Closed-cylinder configuration.

all pressure waves are dissipated, is smaller than p_e. If we reduce the radius r_i, the pressure waves that interact with the wall will be stronger and the cylinder will overexpand even more. On the other hand, if we attempt to reduce the dynamic deformation by increasing r_i, the cylinder will contain more water for the same weight of explosive. Water being compressible, this might drop the final pressure below p_e, in addition to reducing p_e itself, as indicated by Eq. 1.

Here, we attempt to confine at 1 kbar the detonation products of a small PETN charge, 8 mm in diameter and 10 mm long, weighing 890 milligrams, in a cylinder made of aluminum 5086, for which σ_Y = 2.08 kbar, the same material used in the GHBC test [3]. Selecting r_i = 20 mm and t = 10 mm, Eq. 1 results in a maximum value for the pressure p_e = 1.04 kbar. Although the length of the cylinder does not appear in Eq. 1, if L_i is too large, the larger volume of water inside the cylinder will reduce the final pressure, as explained above. Here, we select L_i = 40 mm, the same as the inner diameter of the cylinder. Using the JWL EOS to describe the detonation products of PETN and the Tait EOS for water, we find δr_i = 0.232 mm, $\varepsilon_{\theta i} \equiv \delta r_i / r_i$ = 0.0116 = 1.16%. For comparison, the strain at the elastic limit for aluminum 5086 is approximately 0.2%, indicating that 890 milligrams PETN are enough to exceed the elastic limit in this cylinder. Hence, the final pressure is indeed equal to p_e.

DYNAMIC EFFECTS

The experiment was modeled using the coupled Eulerian-Lagrangian hydrocode DYSMAS. The results are illustrated in Figs. 2, 3, and 4. The final deformation of the cylinder, after all the pressure waves have died out, is illustrated in Fig. 2. Due to shock focussing, the largest plastic strains, between 17 and 23%, are concentrated at the inner corners, where the cylinder is clamped into the covers. Most of the plastic strain is between 6 and 12%, and is limited to the cylinder walls. Except for a narrow region on the axis where the strain is approximately 5%, the thick aluminum plugs deform little. That is why it is recommended to mount the piezo-electric transducer through the bottom plug flush with the inner surface but off the axis, as illustrated in Fig. 1.

The pressure history on the axis, at the inner surface of (any of) the aluminum plug(s) is shown in Figs. 3a and 3b on two different scales. Figure 3a il-

lustrates the short-term pressure history. The large oscillations in pressure are caused by the different reflections, when the shock interacts with the walls as well as the bubble. Due to fast deformation of the metal, the pressure vanishes on the axis causing cavitation of the water. Figure 3b shows that the pressure inside the cylinder eventually decays to a residual equilibrium pressure only 565 bars high. This pressure is only half as high as the maximum static pressure p_e (= 1,040 bars).

The history of the radial expansion at the mid plane of the inner and outer walls of the cylinder is shown in Fig. 4. At the inner wall, δr_i = 2.40 mm, which is much larger than the 0.232 mm predicted for static loading. At the outer wall, δr_o = 1.60 mm.

CONCLUSIONS

Due to dynamic effects, the cylinder overexpands, resulting in a smaller residual equilibrium pressure than the theoretical maximum pressure achievable if all processes were quasi-static. Still, this residual pressure is high enough for the slow reactions in a non-ideal underwater explosive to proceed at a finite measurable rate because the rate is usually a weak function of the pressure, typically $\sim p^{1/6}$ [1, 2]. The rate at 565 bars is 90% of the rate at 1.04 kbar.

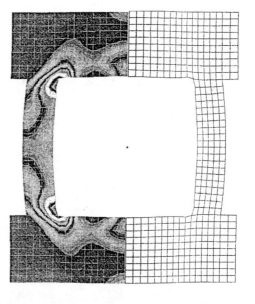

FIGURE 2. Final deformation of the cylinder and equivalent plastic strain contours.

By definition, the late energy release starts at CJ [1], but it is still considered adequate to measure the reaction rate at pressures as low as 1 kbar because the reaction only slows down by a factor of 2.15, when the pressure decreases from 100 to 1 kbar.

It was shown that it is possible to confine the detonation products of 890 milligrams of PETN at 565 bars using a 10 mm thick aluminum cylinder. Using copper or steel, higher equilibrium pressures can be confined because the inertial resistance of the heavier metals reduces the dynamic effects of the pressure waves induced into the water, and because these metals have higher elastic limits (for copper alloys, σ_Y = 2-6 kbars; for steel, σ_Y = 3-10 kbars).

An important advantage of the experiment is that the detonation products and the water are completely confined within the cylinder. The products are preserved for further sampling, if for example, we want to determine their chemical composition, or for proper disposal, if some of the resulting species should not be released in the atmosphere.

REFERENCES

1. Guirguis, R. H., "Relation Between Early and Late Energy Release in Non-Ideal Explosives," *Proceedings of the 1994 JANNAF PSHS*, pp. 383-396, 1995.
2. Miller, P. J., and Guirguis, R. H., "Experimental Study and Model Calculations of Metal Combustion in Al/AP Underwater Explosives," *Proceedings of the 1992 MRS Symposium*, vol. 296, pp. 299-304, 1993.
3. Sandusky, H., Chambers, P., Zerilli, F., Fabini, L., and Gottwald, W., "Dynamic Measurements of Plastic Deformation in a Water-Filled Aluminum Tube in Response to Detonation of a Small Explosive Charge," *Proceedings of the 67th Shock and Vibration Symposium*, 1996.
4. Chambers, G., Sandusky, H., Zerilli, F., Rye, K., and Tussing, R., "Pressure Measurements on a Deforming Surface in Response to an Underwater Explosion," will appear in the *Proceedings of this APS Topical Group Meeting*.
5. Zerilli, F., Naval Surface Warfare Center, Indian Head Division, *unpublished work*, 1997.

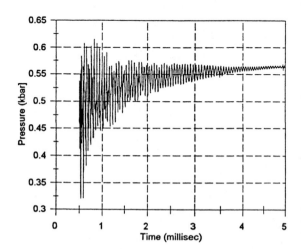

FIGURE 3b. Pressure history on the axis between 500 µs and 5 ms.

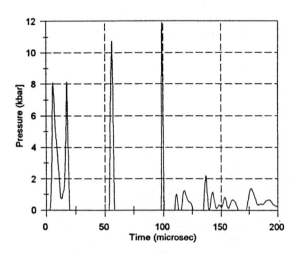

FIGURE 3a. Pressure history on the axis between 0 and 200 µs.

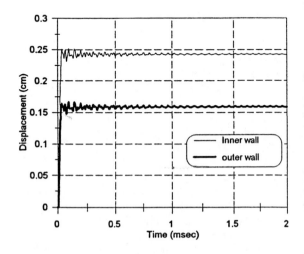

FIGURE 4. Radial expansion of the wall.

CP429, *Shock Compression of Condensed Matter – 1997*
edited by Schmidt/Dandekar/Forbes
© 1998 The American Institute of Physics 1-56396-738-3/98/$15.00

SENSITIVE DETECTION OF SHOCK FRONT AND FREE SURFACE VELOCITY HISTORY FOR POLYMERIC MATERIALS BY NEW INCLINED-PRISM METHOD IN 1 GPa PRESSURE REGION

Yasuhito Mori and Kunihito Nagayama

Department of Applied Physics, Faculty of Engineering, Kyushu University, Fukuoka 812 Japan

Based on the total internal reflection of light at the bottom surface of the prism, very sensitive detection of the shock arrival at free surface of a target has been demonstrated for polymeric materials in 1 GPa stress region. New experimental technique for measuring the shock Hugoniot in this stress region has been established especially for exploring various compressible materials.

INTRODUCTION

Strong viscoelasticity of polymeric materials may induce some drastic and interesting shock properties like non-linear Hugoniot of shock velocity vs particle velocity, stress and velocity relaxation after shock front. Since the polymeric materials have very large compressibility even for the lower stress region of around 0.5 GPa, these interesting properties are enhanced especially in this stress region. Most of the mechanisms of these properties, however, still have not been explained physically.

Shock Hugoniot curve has been one of the most fundamental characteristics of condensed media obtained by impact shock experiments. In case of polymers exhibiting relaxation process, not only Hugoniot data but also velocity or stress history should be measured simultaneously. The most suitable and popular method of fulfilling this requirement is the inclined-mirror method [1] established for higher stress region more than 10 GPa. For the lower region, however, the method is inadequate because of the ambiguous shock detection [2]. Although the VISAR technique [3] is the most reliable and established method for the velocity history measurement, it is not easy to measure the Hugoniot data simultaneously.

To realize more sensitive shock detection, that is, to obtain the precision shock Hugoniot of polymeric materials for around 1 GPa stress region, we have applied the principle of the inclined-prism method [4,5] proposed by Eden and Wright [6]. The principle for the shock detection in this method was based on the light extinction of *total internal reflection* from prisms placed on the free surface of a target assembly. This method had been successfully applied to higher stress region by using high-explosives and prisms with special polygon shape by Gust et al. [7]. However, no serious attempts have been made for the lower stress region, although the method seems effective for the region.

The present experiments are the first experimental verification of the feasibility of using total internal reflection to obtain the precision shock Hugoniot and velocity history for the relatively low shock stress region. Typical streak photographs for polymethyl-methacrylate (PMMA) [4,5] and polyethylene (PE) are shown. It is found from other experiments that the plane shock wave propagating into the polymeric materials is stationary with regard to shock velocity. The obtained Hugoniot data for PMMA are good agreement with the previous data [8,9]. The u_s-u_p Hugoniot for PE exhibits non-linear form like that for PMMA.

SHOCK REGISTRATION SYSTEM

We have established a shock wave registration system consisting of a single-stage compressed gas gun, a long-pulsed dye laser as a light source and a very compact high-speed streak camera [2,10,11]. Combination of these equipments serves a reliable possibility of shock study of condensed media. Since the gas gun can accelerate the projectile only up to 500 m/s, the attainable stress region is limited to about 1 GPa in case the polymeric materials are used as a target assembly.

The launch tube of the gun has 40 mm in diameter and 2 m long. This gun has several features effective for the precision optical observation of the shock wave phenomena. Projectile velocity is measured by beam cut method with the accuracy of about 0.5 %. The projectile measurement system plays also an important part for the reliable synchronization between the μs shock events and the long-pulsed dye laser radiation of 20-30 μs duration. The streak camera of rotating mirror type is used. Since the writing arm length of the camera is only 150 mm, we can use a low-sensitivity but high-resolution film of ISO 32 - 100, although the attainable streak velocity is limited up to 3 km/s. A double slit of 10 μm width and 2 mm spacing is used to realize the time resolution of less than 10 ns and to measure the shock wave tilt angle precisely.

DETECTION OF SHOCK ARRIVAL BY TOTAL INTERNAL REFLECTION

To realize precision optical measurement of shock Hugoniot curves of polymeric materials, at first we had tested a conventional inclined-mirror method [2]. However, the detection of shock arrival at the free surface of a target assembly was insensitive in 1 GPa stress region. Our experience shows that the weak shock induces insufficient destruction of mirrors.

For sensitive detection of shock arrival even for 1 GPa stress region, we have constructed a new experimental technique using *total internal reflection* (TIR) by prisms. Figure 1 shows the principle of the sensitive detection of shock arrival based on this phenomenon. Incident laser beam is illuminated in a small angle to free surface of a target so that the TIR occurs at the bottom face of the triangular prism. The phenomenon of the TIR occurs at the boundary of two materials whose index of refraction is different. Here, their materials are the prism and narrow gap region between the bottom surface of the prism and the free surface of the target.

The necessary condition of the TIR is the existence of evanescent wave penetrated into the gap region of smaller index of refraction. To realize the TIR, therefore, the gap spacing has to be greater than the penetration depth of the evanescent wave. For the sensitive and precise shock detection, however, the gap spacing must be as narrow as possible. The calculation shows that the penetration depth of the evanescent wave is around

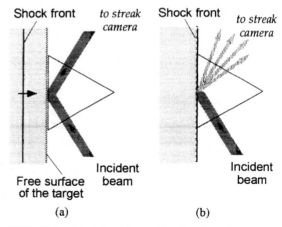

FIGURE 1. Principle of the sensitive detection of shock arrival at the free surface of a target by utilizing total internal reflection at the bottom surface of a triangular prism; (a) before and (b) after the shock arrival at the free surface.

0.15 μm for visible light [5]. Sub-micron spacing may be expected by pressing the bottom face of the prism onto the free surface.

Before the shock arrival at the free surface, as shown in Fig. 1(a), the gap spacing remains the sub-micron order. Therefore, the streak camera records the intense light of the TIR from the prism. Since the narrow gap closes by the movement of the free surface by the shock arrival at the surface, the condition of the TIR changes to that of the reflection depending on the roughness of the free surface as shown in Fig. 1(b). The principle for the detection of shock arrival of the present method is based on the sudden extinction of the intense light reflected from the prism by shock arrival.

TARGET ASSEMBLY

Inclined-Prism Method

Figure 2 shows a schematic of typical target assembly for inclined-prism method using triangular prisms. Among the prisms, three prisms are used for shock front detection. While, an inclined-prism is used for continuous recording of free surface velocity history. The inclined angle is determined according to the expected free surface velocity. As shown in Fig. 2, the target assembly is illuminated from two different directions. Because the light reflected from the inclined-prism should be parallel to the lights reflected from the other prisms. In the present method, an exact tilt angle of shock wave is measured by using a double slit necessary for the accurate evaluation of shock velocity.

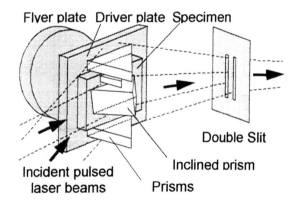

FIGURE 2. A schematic of a target assembly for inclined-prism method using triangular prisms.

FIGURE 3. A typical streak photograph for PMMA target obtained by the present inclined-prism method. The projectile of 56.53 g with PMMA flyer was accelerated up to 255.3 m/s by helium gas of 20 bar.

Shock Impedance Mismatching

We have used the same material for flyer, driver plate and specimen in most of the experiments to assure the particle velocity at the impact surface to be exactly half of the flyer velocity. In some of the experiments, however, we have used copper flyer to obtain higher shock stress inaccessible by plastic flyers. This can be justified only when shock impedance is constant during shock propagation. Necessary and sufficient condition for this is the constancy of shock velocity with propagation distance. We are also studying the shock velocity history by using another assembly. Until now, we have noticed no indication of unsteadiness of the velocity. We have designed a target assembly without inclined-prism to obtain Hugoniot point much more easily based on the shock impedance mismatching principle. It is found that both assemblies with or without inclined-prism gives essentially equivalent results.

RESULTS AND DISCUSSION

Polymethyl-methacrylate (PMMA)

Figure 3 shows a typical streak photograph for PMMA target. The lights reflected from three prisms for shock wave detection extinguish suddenly by the shock arrival at the free surface of the target. According to the analysis of this photograph, the values of shock velocity and shock stress are 3.13 km/s and 0.47 GPa, respectively. Even for around 0.5 GPa, therefore, very sensitive detection of the shock arrival is realized. The tilt angle of the shock wave can be estimated to be 2.20 degrees

from the streak image recorded by the double slit. From these analyzed values, the tilt angle of the moving free surface can be calculated to be 0.18 degrees. This angle automatically corresponds to the impact angle between the flyer plate and the target assembly, since the same material is used as the flyer and target.

The free surface velocity history was recorded by an inclined-prism whose inclined angle is 1.82 degrees. In the obtained velocity history, a relaxation structure can be recognized. The free surface velocity jumps up to 221.3 m/s at the shock front and then approaches to 253.1 m/s gradually within a relaxation time of about 1.5 μs. This velocity history corresponds to the velocity relaxation of shock wave front, which propagated into PMMA up to about 7.88 mm from the impact surface of the target. Since the maximum velocity of the free surface (253.1 m/s) agrees almost with the impact velocity of the projectile (255.3 m/s), it can be said that the analysis of the present method is very reliable.

Polyethylene (PE)

Figure 4 shows a typical streak photograph for PE target obtained by using the assembly without inclined-prism. The shock velocity and the tilt angle of the shock wave can be estimated to be 2.95

FIGURE 4. A typical streak photograph for PE target obtained by using three triangular prisms without inclined-prism. The projectile of 26.92 g with copper flyer was accelerated up to 440.2 m/s by helium gas of 20 bar.

km/s and 2.93 degrees, respectively. From the impedance mismatch technique, on the other hand, the particle velocity behind the shock wave and the shock stress can be estimated to be 408.8 m/s and 1.11 GPa, respectively. Therefore, the free surface velocity for the final state can be estimated to be 817.6 m/s.

Hugoniot Curves

We have made a series of experiments for PMMA and PE to measure the precise Hugoniot data. The u_s-u_p Hugoniot data are plotted in Fig. 5. The previous data for PMMA are also plotted in this

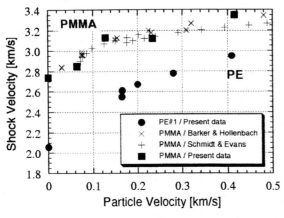

FIGURE 5. The u_s-u_p Hugoniot curves for PMMA and PE.

figure [8,9]. Since the Hugoniot data obtained for PMMA coincide very well with the previous data, the present method seems promising even for the lower shock stress region.

As shown in Fig.5, the Hugoniot curve for PE exhibits non-linear form like that of PMMA. The shock Hugoniot for polymeric materials, however, might be affected by the molecular weight distribution. In addition to this, the shock Hugoniot for the crystalline polymers like PE seems to depend also on the degree of crystallinity. Therefore, the systematic shock experiments are necessary for the solution of the shock characteristics of polymeric materials.

SUMMARY

Even in the lower shock stress region of around 0.5 GPa, very sensitive detection of shock wave in condensed media has been realized by the present inclined-prism method in which the total internal reflection from prisms is used. The reliable Hugoniot curves and/or the unknown shock characteristics of polymeric materials will be measured for this shock stress region by the present method.

REFERENCES

1. T. J. Ahrens, W. H. Gust and E. B. Royce, *J. Appl. Phys.*, **39**, 4610-16 (1968).
2. Y. Mori and K. Nagayama, *Proc. 2nd Symposium on High Speed Photography and Photonics, Tohoku University, Sendai, Japan*, 159-193 (1996).
3. L. M. Barker and R. E. Hollenbach, *J. Appl. Phys.*, **43**, 4669-75 (1972).
4. Y. Mori and K. Nagayama, *Proc. 21st Intl. Symposium on Shock Waves, Australia*, 1997 in press.
5. Y. Mori, T. Tamura and K. Nagayama, *Rev. Sci. Instrum.*, to be submmiting.
6. G. Eden and P. W. Wright, *Proc. 4th Int. Symposium on Detonation*, 573 (1966).
7. W. H, Gust and E. B. Royce, *J. Appl. Phys.*, **42**, 1897-1905 (1971).
8. D. N. Schmidt and M. W. Evans, *Nature*, **206**, 1348-49 (1965).
9. L. M. Barker and R. E. Hollenbach, *J. Appl. Phys.*, **41**, 4208-26 (1970).
10. K. Nagayama and Y. Mori, *Proc. 20th Int. Cong. High-Speed Photography and Photonics*, SPIE **1801**, 154-159 (1993).
11. K. Nagayama, *Shock Waves in Material Science*, Springer-Verlag, Tokyo, 1993, Chap. 9, pp.195-224.

CP429, *Shock Compression of Condensed Matter – 1997*
edited by Schmidt/Dandekar/Forbes
© 1998 The American Institute of Physics 1-56396-738-3/98/$15.00

ABOUT MEASUREMENTS OF STOPPING POWER BEHIND INTENSE SHOCK WAVES

V.Gryaznov, M.Kulish, V.Mintsev, V Fortov

Institute of Chemical Physics in Chernogolovka, Chernogolovka, Moscow reg., 142432, Russia

B. Sharkov, A. Golubev, A.Fertman, N.Mescheryakov

Institute for Theoretical and Experimental Physics, B. Cheremushkinskaya 25,117259, Moscow, Russia.

D.H.H.Hoffmann, M.Stetter, C.Stöckl

Geselschaft f, r Schwerionenforschung, D-64220, Darmstadt, Germany

D.Gardes

Institute National de Physique Nucleare, 91406, Orsay, France

Method of generation of plasma targets with electron densities $n_e \geq 10^{21}$ cm^{-3} behind strong shock wave for study of energy losses of protons and heavy ions is discussed. The problems of matching of large scale accelerator facility and explosive technique are considered. It is suggested to use small (<150 g TNT) vacuum pumped explosive metallic chambers with fast valves in such experiments. Construction of small-sized explosively driven generators of strong shock waves is described. Estimations of stopping power in strong shock waves in hydrogen, xenon and argon were carried out. It is shown that to get main contribution of free electrons it is necessary to have velocities of shock wave in xenon and argon more than 20 km/s and in hydrogen more than 60 km/s.

INTRODUCTION

For the investigations of heating of matter induced by particle beams the energy losses in a dense plasma at high pressures and temperatures are crucial. Experiments carried out at GSI with the discharge plasma up to the electron densities ~10^{19} cm^{-3} show on a considerable contribution of free electrons to the process of beam-plasma interaction. With the increasing of plasma density influence of the effects of Coulomb interparticle interaction are expected to be of great importance. Shock wave technique make it possible to produce plasma with electron densities up to 10^{22} 1/cc [1]. For this goal explosively driven plasma generators have been developed [1]. In such devices plasma slug with uniform parameters [1] is created behind a plane front of intense shock wave generated by detonation of chemical high explosive. Standard shock wave plasma generators [1] contain more then 500 g of high explosive, that allows to produce shock compressed strongly coupled plasma with temperatures of 1-10 eV, pressures of 1-200 kbar and

Coulomb coupling parameter of 1-5 [2]. Explosively driven plasma also looks very attractive from the point of view of absence of strong electromagnetic fields like in discharges, which significantly effect the beam transport.

To use standard explosive devices in beam areas of accelerator facilities it is necessary to build large-scaled explosive chambers and to solve the problems of matching deep vacuum beam lines with the explosive technique. As the first step we suggest to use small (<150 g TNT) vacuum pumped explosive metallic chambers with fast valves in such experiments. To minimize the explosive plasma generators numerical simulations of plasma shock compression and special series of shock wave experiments were carried out. They show possibility to construct small-sized linear and cumulative explosively driven generators which yields velocities of shock front about 6-20 km/s whereas amount of high explosive does not exceed 30-150 g. Using these devices we assume to investigate experimentally (i) effect of strong interparticle interactions in plasma on energy losses of fast ions experimentally (i) effect of strong interparticle interactions in plasma on energy losses of fast ions and (ii) stopping power of plasma at high ionization degrees (due to free electrons).

ESTIMATIONS OF STOPPING POWER BEHIND POWERFUL SHOCK WAVES

To calculate shock compressed plasma parameters computer code SAHA-4 [2] was used. Computing procedure is based on the chemical picture of plasma [3], in which particle densities are determined from conditions of interactions are taken into account. This code allows to calculate composition, equation of state and thermodynamic functions of plasmas in wide range of densities and temperatures. Using this code Hugoniots of hydrogen and noble gases were calculated in a range of shock velocities, typical for explosively driven plasma generators under consideration.

Computer simulations of explosively driven plasmas show that using of simplest (linear) scheme of shock tube make it possible to obtain ionization degree about several units and high electron densities for gases with rather high molecular weight. For example shock wave with velocity 6 km/s in xenon at initial pressure 1 bar produces plasma electron density more than 10^{20} per cubic centimeter and ionization degree about 1 (complete first ionization). Higher plasma densities can be reached without any problem by increasing of initial pressure of investigated gas.

Shock compression of more light gases gives lower electron densities at the same shock velocities and one needs more powerful explosive charge to obtain the same ionization degree. More high electron densities can be reached if technique of reflected shock waves is used. It is one of possible ways to achieve more high plasma parameters. In hydrogen due to its low molecular weight noticeable ionization degree after incident shock can be obtained only at front velocities more then 40 km/s. These shock velocities are too high for linear plasma generators. Reflected shock

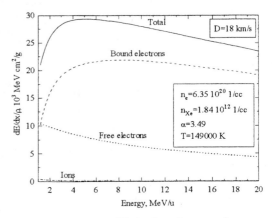

FIGURE 1. Energy losses of Xe ion beam in xenon plasma.

only slightly increases ionization degree. These parameters of shock cannot be obtained with simple linear plasma generator and it has to be modified. It is known that use of cumulative version of plasma generator allows to produce shock waves of very high velocities up to 100 km/s [4].

Stopping power of explosively produced plasma was calculated in frames of approximation [2]. All calculations were carried out for Xe probe ions. Results of calculations for xenon plasma are represented in Fig.1. These parameters of shock (D=18km/s) are typical for using of cumulative plasma generators and they are allow to investigate experimentally effects of Coulomb interparticle interaction on stopping power.

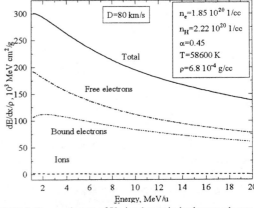

FIGURE 2. Energy losses of Xe ion beam in hydrogen plasma.

Contribution of free electrons to energy losses is comparable to bound ones at low energy limit of probe ion but estimated range (about few millimeters) makes it technically difficult to investigate contribution of free electrons for these kind of plasma targets but these parameters can be increased. This effect can be studied in hydrogen plasma (Fig.2). At shock velocities about 80km/s that is reachable by using of compact cumulative explosive device, ionization degree is about 0.5. It is seen that is enough for study of free electron effects because stopping power of free electrons is higher then energy losses due to bound electrons.

EXPERIMENTS

Scheme of linear small-sized plasma explosively driven generator is demonstrated in Fig.3.

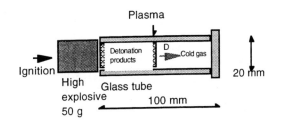

FIGURE 3. Explosive shock tube.

Detonation products in this device push metal impactor that shapes flat shock front in investigated gas. Glass tube remains immobile during the time of about 1 μs in passage of plasma through the channel. The flow of plasma is practically one dimensional. Shock velocities of 6-8 km/s were obtained in this generator. By applying of conical section the shock velocities up to ~15 km/s were reached. Plasma slug of cylindrical form with diameter from 0.5 to 2 cm and thickness of 1 cm is created. This plasma object exists about several microseconds that is enough to carry out measurements of with ion beams.

To protect all the equipment from the explosion plasma generator is placed into special compact vacuum pumped (up to 10^{-2} torr) steel chamber with the diameter of 80 cm, which allows to apply charges up to 150g of TNT. This chamber is specially designed for the ion beam experiments and

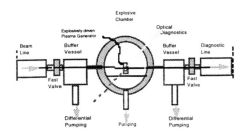

FIGURE 4. Experimental setup.

provides complete matching with beam line and other experimental facility. Scheme of experimental setup with explosive chamber and plasma generator inside is presented in Fig.4. 3 MeV, 60 μs proton beam delivered by 148.5 Mhz ISTRA-36 RFQ linaq of ITEP were used for energy losses experiments. Differential pumping was proved to be sufficient for insulation of the deep vacuum beam line from low vacuum explosive chamber. To protect beam line from the moving detonation products fast valves were used. They were closed before the moment when explosive flow run into the chamber walls. A position of the proton beam on the scintillator after analyzing magnet was fixed by PCO-camera. Plasma temperature and shock velocity in every experiment were measured as well.

First shots with weakly nonideal xenon plasma shows good performance of scheme described above. Measured values of energy losses are in a range of several hundred KeV and are in agreement with the theoretical estimations. Experiments with the strongly coupled explosively driven plasma are in a way now.

This work was supported by INTAS grant N94-1638.

REFERENCES

1. V.Fortov, I.Yakubov. *Physics of Nonideal Plasma* (Hemisphere, 1990).
2. W. Ebeling, A.Foerster, V. Fortov, V.Gryaznov, A. Polishchuk, *Thermophysical Properties of Hot Dense Plasmas* (Teubner, Stuttgart-Leipzig 1991).
3. W. Ebeling, Physica **43** (1969) 293.
4. V.Mintsev, V.Fortov. Explosively driven shock tubes. High Temperatures, **20** (1982) 584.

CP429, *Shock Compression of Condensed Matter – 1997*
edited by Schmidt/Dandekar/Forbes

X-RAY DIFFRACTION TECHNIQUE WITH VISAR SUPPORT FOR STUDY OF SHOCK-COMPRESSED SINGLE CRYSTALS

E.Zaretsky

Pearlstone Center for Aeronautical Engineering Studies,
Department of Mechanical Engineering, Ben-Gurion University of the Negev
P.O.Box 653, Beer-Sheva 84105, Israel

A new geometry of X-ray diffraction survey of single crystal in the monochromatic X-ray radiation was used to obtain a diffraction patterns of shock-compressed NaCl single crystals. Shock compression of the 2-mm [100]-oriented single crystals backed with 0.7-mm graphite window was produced by 6-mm Al impactors accelerated up to velocities 320 - 330 m/sec in 25-mm pneumatic gun. The velocity of the free surface of the graphite window was monitored by VISAR. In two shots the 30-nsec pulse X-ray source was triggered after the arrival of the elastic precursor and the plastic wave front at the NaCl-window interface. No evidence of the transversal strain were found in the diffraction patterns of the shock-compressed NaCl. The material seems to be able to maintain the state of 1-D strain under intensive plastic deformation.

INTRODUCTION

Attempts to obtain a diffraction pattern containing a microscopic information about the strain state of the material with shock were done by a variety of the research groups during two decades [1] - [9]. In most of these works the pulse X-ray diffraction patterns were obtained from shock-compressed single crystals. The intensity of the existing pulse X-ray sources is insufficient for obtaining the diffraction pattern from polycrystalline sample. This, together with the conditions of one-dimensional shock loading of the single crystal sample, forced the researchers to use the survey scheme which is suitable for obtaining the diffraction pattern of monochromatic X-ray radiation from plane polycrystalline sample, the Bragg-Brentano scheme. In this case the vector of the direction of the shock propagation $\mathbf{x}$ belongs to the plane containing the incident and the reflected X-ray beams together with the normal $\mathbf{n}$ of the crystalline planes normal to $\mathbf{x}$. The lattice compression may be found by differentiating the Bragg's formula

$$2d_{hkl} \sin \theta_{hkl} = \lambda \qquad (1)$$

where d_{hkl} is the (hkl)-planes spacing, and λ is the wavelength of the characteristic radiation of the X-ray source, θ is the glide angle of the incident X-ray beam. The Eq. (1) gives for the lattice strain (the angles θ are small)

$$\varepsilon_x = \frac{\delta a}{a} = \frac{\delta d_{hkl}}{d_{hkl}} = -\frac{\delta \theta_{hkl}}{\theta_{hkl}}, \qquad (2)$$

where a is the lattice parameter. It is clear, that use of the Bragg-Brentano scheme in the study of shock-compressed single crystals allows to obtain the information about the lattice strain in the direction of the wave propagation only. It is very attractive to try to obtain the complete information about the strain state of the compressed material.

The purpose of the present work is to develop and to apply a new scheme of the X-ray diffraction survey to study of shock-compressed NaCl single crystal whose transient states should be continuously monitored by VISAR [10].

MULTIPLE-PEAK X-RAY DIFFRACTION IN SHOCK WAVE EXPERIMENTS

The reflection from (200) planes (the strongest one in the case of the crystals with rock-salt structure) remains unchangeable when the crystal is rotated around its axis parallel the [200] direction.

Taking into account that for crystals of cubic symmetry the plane spacing d_{hkl} is related to the lattice period a, $d_{hkl} = a/\sqrt{h^2 + k^2 + l^2}$, the equation (1) may be rewritten

$$\sin\theta_{hkl} = \sin\theta_o \sqrt{h^2 + k^2 + l^2}/2, \qquad (3)$$

where $\sin\theta_{200} = \lambda/a = \sin\theta_o$. Since the plane (200) is normal to the vector [200], the rotation of the crystal on angle φ around axis [200] should result in arrival of the plane (hkl) to the position when Bragg's conditions (3) are satisfied. When the sample surface is parallel to the crystal (200) plane the angle θ between the plane and the incident beam vector $\mathbf{q}$ remains unchanged for any rotation angle φ, as well as the projection $\mathbf{q_o}$ of $\mathbf{q}$ on the (200) plane. In such case $\mathbf{q} = q_o[-\bar{i}(\cos\theta_o) - \bar{j}(\sin\theta_o) + \bar{k} \cdot 0]$.

When the crystal is rotated around its [200] axis (the same as [h00]) the Miller indexes of the plane (hkl) will be changed according the equations:

$$k' = k\sin\varphi + l\cos\varphi, \quad l' = l\cos\varphi - k\sin\varphi \qquad (4)$$

This, with the condition $[hk'l'] \cdot \mathbf{q} = -[hk'l'] \cdot \mathbf{q}'$ of the change of the sign of the component q'_{200} of the reflected beam $\mathbf{q}'$ parallel to the [200] direction and the condition (3) yields for the crystal rotation on the angle φ the Bragg's condition of the appearance of the reflection from the (hkl) plane

$$\frac{\tan\theta_o}{2} = \frac{h\tan\theta_o - (l\sin\varphi + k\cos\varphi)}{h^2 + k^2 + l^2}. \qquad (5)$$

For the film which is located at the distance L_o from the sample axis and whose plane is normal to the plane of primary beam $\mathbf{q}$ and the crystal sample axis [200] the position of the reflection from the (hkl) plane may be found from the condition of the invariability of the sign of the glide, with respect to the (hk'l') plane, component of the vector $\mathbf{q}$, $[hkl] \times \mathbf{q} = [hk'l'] \times \mathbf{q}'$, and from the equation of film plane, $y = L_o$:

$$x = -L_o \tan\theta_o \frac{(1-h)}{1 + \tan\theta_o(l\sin\varphi + k\cos\varphi)}$$
$$z = L_o \tan\theta_o \frac{(l\cos\varphi - k\sin\varphi)}{1 + \tan\theta_o(l\sin\varphi + k\cos\varphi)}, \qquad (6)$$

where angle φ should be obtained from the Bragg's condition (5) of the reflection from the given plane family (hkl).

The idea of the above scheme is close to the idea of the rotation method (e.g., Warren [11]). The scheme allows to obtain multiple-peak diffraction pattern from the single crystal radiated by the monochromatic X-rays. The rotation angle φ for which at least two reflections exist, the reflection from the (200) plane plus a reflection from any other plane (hkl), always may be found. When the single crystal undergoes a shock deformation according to the scheme shown in Fig. 1 the X-ray diffraction pattern changes. These changes may occur as a diffraction spot shift, as a broadening of the diffraction peaks (including asymmetrical broadening of the peaks), as the change of the peaks intensity.

For a crystal lattice having the rhombohedral symmetry the lattice strains ε_{xx}, ε_{yy} and ε_{zz} may be related to the variation of the indexes h, k, l :

$$\varepsilon_{xx} = \frac{\delta d_{h00}}{d_{h00}} = -\frac{\delta h}{h}$$
$$\varepsilon_{yy} = \frac{\delta d_{0k0}}{d_{0k0}} = -\frac{\delta k}{k} \qquad (7)$$
$$\varepsilon_{zz} = \frac{\delta d_{00l}}{d_{00l}} = -\frac{\delta l}{l}$$

The Eqs. (7) are useful for the application of the suggested diffraction scheme for a study of the strain state of the crystal, containing shock. The angles θ_o and φ stay unchanged during the shock experiment, while the spatial grid of the points (hkl) experiences the deformation, Eqs. (7). The Eqs. (6) and (7) allow to relate the initial position of the diffraction peak (x_i, z_i) with its position $(x_i + \delta x, z_i + \delta z)$ in the pattern corresponding to the shocked material .

for the reflection 200 :
$$\frac{\delta x_{200}}{x_{200}} = -\varepsilon_{xx}; \quad \delta z_{200} = 0$$
for the reflection 220 : $\qquad (8)$
$$\frac{\delta x_{220}}{x_{220}} = -\varepsilon_{xx}; \quad \frac{\delta z_{220}}{z_{220}} = -\varepsilon_{yy}$$

The terms with $\tan^2\theta_o$ are neglected in (8). The use of Eqs. (8) makes it possible the complete determination of the crystal strain state on the basis of the multiple-peak X-ray diffraction pattern.

EXPERIMENTAL

Two NaCl single crystals, I and II, of 2.1-mm thickness backed with polycrystalline pyrolitic graphite 0.7-mm window were shocked by impactors, the disks of aluminum alloy 6061-T6, accelerated in the 3-m barrel of 25-mm pneumatic gun up to velocities about 328 (I) and 322(II) m/sec. The scheme of the experimental setup with the diagnostic tools is shown in Fig. 1. Velocity and trigger (lapped together with the sample front surface) electrical pins were mounted in the sample body.

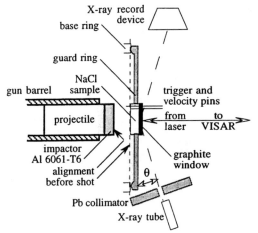

FIGURE 1: Scheme of the experimental setup for pulse X-ray diffraction measurements accompanied by continuous monitoring of the velocity of the free surface of the graphite window.

In different experiments the 30-nsec pulse X-ray source (based on Scandiflash 300 X-ray pulse system with 1-mm focus Mo anode) was triggered at the different instances after arrival of the shock wave at the NaCl-graphite interface. The X-ray diffraction pattern was captured by the record system containing the 75-mm scintillating screen, fiber optic cone, Hamamatsu V1366P Image Intensifier and Hamamatsu C4346-01 CCD camera connected with IBM486 computer via Data Translation Frame Grabber EZ-55. The instance of the X-ray appearance was detected by 14-m optic fiber (10-nsec rise time scintillator) coiled around the X-ray tube and coupled with 2.5-nsec R647 Hamamatsu photomultiplier tube. Both the shots were accompanied with continuous monitoring of the velocity of the graphite free surface by the VISAR. To enhance the graphite surface reflectivity the 10-μ aluminum foil was glued (glue layer of about 4 μ) on the surface. The impactor-sample misalignment does not exceed 0.2 milliradian in both the shots.

RESULTS AND DISCUSSION

The results of the VISAR measurements in one of two shots are shown in Fig. 2. Since the impact conditions were close in both the shots the instants of the appearance of the X-ray radiation in both the experiments are shown with the same velocity profile. As is clear from Fig. 2 the diffraction pattern was captured in the shot I between the arrivals of the elastic precursor and the plastic front at the NaCl-graphite interface. In shot II the pattern was captured after arrival of the plastic front.

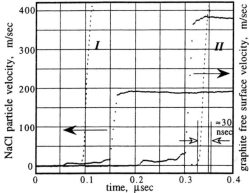

FIGURE 2: VISAR measurements of the free surface velocity of graphite window and the particle velocity at the NaCl-graphite interface (circles). Impact velocity is 322 m/sec (shot II). The dashed curves I and II show the time instants of the X-ray flash appearance in the corresponding shot..

The diffraction pattern obtained at the shot II is shown in Fig.3. Analysis of the change of the intensity distribution of the patterns, Figs. 4 and 5, allows to obtain information about the peaks shifts and the strain state of the crystal. As is clear from Fig. 4 both the shots result in the shift of the (200) and (220) peaks in the horizontal direction. The strain values corresponding to the shifts of the peaks (200) and (220) in Fig. 4 are $\varepsilon_{xx} = 0.004 \pm 0.002$ (shot I) and $\varepsilon_{xx} = 0.050 \pm 0.002$ (shot II). These values agree well with stain values calculated from hydrodynamic considerations. In NaCl the elastic precursor velocity is $C_l = 4700$ m/sec, the shock front velocity D depends on the particle velocity U as $D = 3440 + 1.45U$ [4]. Calculated (Fig. 2) particle velocities are: elastic precursor amplitude $U_{HEL} = 9 \pm 2$ m/sec, plastic wave amplitude $U_{pl} = 189 \pm 5$ m/sec. This corresponds to the values of 1-D compression: $\varepsilon = 0.002 \pm 0.0004$ behind the elastic precursor and $\varepsilon = 0.050 \pm 0.001$ behind the plastic front.

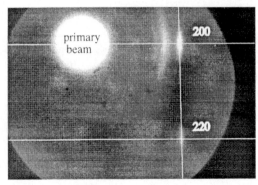

FIGURE 3: Typical X-ray diffraction pattern: shock-compressed NaCl single crystal in shot II. Ark-like reflections are the reflections from polycrystalline graphite. The horizontal and vertical intensity distributions (see. Figs. 4, 5) was measured along the straight lines.

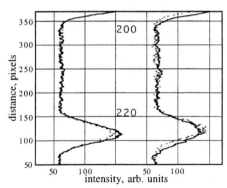

FIGURE 5: Vertical (see Fig. 3) sections of the diffraction patterns of the shots I(left) and II (right). Dashed line -initial state, solid line - shocked state.

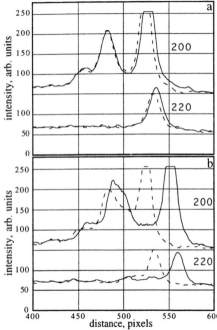

FIGURE 4: Shots I (a) and II(b). Horizontal sections of the diffraction patterns via 200 (top) and 220 (bottom) spots, see Fig. 3. Dashed line - initial state, solid line - shocked state.

No shift of (220) peaks in vertical direction was found, see Fig. 5. The crystal strain in the direction normal to the direction of the wave propagation seems to be zero (see Eqs 8). The coincidence of the strain value obtained from the diffraction pattern with the strain obtained from the hydrodynamic data with assumption of unidimensional compression confirms that the material is able to maintain the state of 1-D strain under intensive plastic deformation. The plastic

deformation of the rocksalt lattice should be necssarily accompanied by the dislocation glide in the {110} planes, the glide planes of both the primary and the secondary glide systems of NaCl lattice. This glide, seems, does not result in the change of the spacing of the glide planes in the direction normal to the direction of the wave propagation in the sample.

REFERENCES

1. Johnson Q., A.Mitchell, R.N.Keeler, and L.Evans, *Phys. Rev. Lett*, **25**, (1970) 1099

2. Egorov L.A., E.V.Nitochkina, and Yu.K.Orekin, *JETF Lett.*, **16**, (1972), 4

3. Kondo K., T.Mashimoto, A.Sawaoka, and S.Saito, in. High Pressure Science and Technology, Proceedings of the 6th AIRAPT Conference, (1977) p.883

4. Jamet F., and F.Bauer, in: Actes du Symposium International sur le Comportment des Milleux Denses sous Hauter Pressions Dynamiques, Editions du Comissariat a l'Energie Atomique, (1978) p.409

5. Miller F., and E.Schulte, *Z. fur Naturforsch.* **33a**, (1978) 918

6. Zaretsky E.B., G.I.Kanel', P.A.Mogilevsky, V.E.Fortov, *High Temperature Physics*, **29**, (1991) 1002

7. Wark J.S., G.Klein, R.R.Whitlock, D.Riley, and N.C.Woolsey, *J. Appl. Phys.*, **68**, (1990) 4531

8. Whitlock R.R., J.S.Wark, *Phys. Rev. B*, **52**, (1995) 8

9. Podurets, A.M., A.I.Barenboim and R.F.Trunin, in Shoc Compression of Condensed Matter-1995, Eds. S.C.Schmidt and W.C.Tao (AIP, 1996) p.231

10. Barker L.M. and Hollenbach R.E., *Journ.Appl.Phys.*, **43**, (1972) 4669

11. Warren B.E., X-Ray Diffraction (Dower Publications Inc., New-York, 1990) p.85

CP429, *Shock Compression of Condensed Matter – 1997*
edited by Schmidt/Dandekar/Forbes
© 1998 The American Institute of Physics 1-56396-738-3/98/$15.00

DYNAMIC MEASUREMENT OF TEMPERATURE USING NEUTRON RESONANCE SPECTROSCOPY (NRS)

**D.J. Funk, B.W Asay, B.I. Bennett, J.D. Bowman,
R.M. Boat, P.M. Dickson, B.F. Henson, L.M. Hull,
D.J. Idar, G.W. Laabs, R.K. London, J.L. Mace, G.L. Morgan,
D.M. Murk, R.L. Rabie, C.E. Ragan, H.L. Stacy, V.W. Yuan**[*]

Los Alamos National Laboratory, Los Alamos, New Mexico 87545 USA

Accurate temperature measurements in dynamic systems have been pursued for decades and have usually relied on optical techniques. These approaches are generally hampered by insufficient information regarding the emissivity of the system under study. We are developing NRS techniques to measure temperature in dynamic systems and overcome these limitations. Many neutron resonances have narrow intrinsic Breit-Wigner widths such that the resonance is substantially broadened by the atomic motion even at room temperature. Thus, accurate measurement of the Doppler contribution allows one to infer the material temperature, and for the conditions achieved using standard high explosives, the probe itself is not perturbed by the high temperature and pressure. Experiments are conducted using a pulsed spallation source at LANSCE with time-of-flight measurement of the neutron spectra. In initial experiments, we have demonstrated that measurements with ten percent accuracy are possible. We have fielded dynamic tests, most of which were neutron-flux limited. An overview of the approach and the status of our experimental campaign are discussed.

INTRODUCTION

Accurate temperature measurements in dynamic systems have been pursued for decades and have usually relied on optical techniques. These approaches are generally hampered by insufficient information regarding the emissivity of the system under study, particularly when looking at detonating energetic materials. This results from the complex chemical change that the detonating system is undergoing, from neat organic crystals with a discrete set of electronic states and molecular vibrations embedded in a binder, to gas phase products, with a different set of discrete states. Typically, when making an optical measurement, one makes the gray body approximation and measures the photon flux at a few discrete bands to extract a temperature from the light emission. In shocked "opaque" systems, one is also hindered by the short duration of the experiment; release waves and conduction can occur faster than the photon integration time scale necessary to accurately measure the flux. We are developing NRS techniques to measure temperature in dynamic systems and overcome some of the limitations of optical systems. Many elements have neutron resonances with narrow enough intrinsic Breit-Wigner (Lorentzian) widths such that the resonance is substantially broadened by atomic motion even at room temperature. Thus, accurate measurement of the Doppler contribution allows one to infer the material temperature. Moreover, since the opacity

[*] This work sponsored by the U.S. Department of Energy under contract number W-7405-ENG-36.

of many interesting (energetics, metals) materials to neutrons is much less than it is to infrared, visible, or ultraviolet light, we probe the internal temperature, in contrast to measuring the surface temperature through an emission technique. Also, NRS is much less affected by the opacity of soot or other particulate than are optical methods.

EXPERIMENTAL DETAILS

The energy dependence of the cross section of an isolated neutron resonance can be expressed (using a single Breit-Wigner form) as:

$$\sigma \propto \frac{\Gamma^2}{(E - E_r)^2 + (\frac{\Gamma}{2})^2} \tag{1}$$

where Γ is defined as:

$$\Gamma = \Gamma_{(n)} + \sum_r \Gamma_{(r)} \tag{2}$$

Here, σ is the cross section, $\Gamma_{(n)}$ is the neutron width and the $\Gamma_{(r)}$ are the partial widths for reactions r, E is the center-of-mass collision energy of the neutron-nucleus system, and E_r is the energy of the resonance. Due to the fact that the resonance depends on the center-of-mass collision energy and the fact that the target nucleus is never at rest (i.e. average motion due to it's oscillation in the potential well of electronic energy due to both zero-point and additional average thermal energy), the zero-order cross-section must be convolved with the velocity distribution of the target nucleus. This is determined by the phonon distribution of the target atom (in its surroundings) or in the case of a gas, a Maxwell-Boltzman distribution of velocities, and the convolution written as:

$$\sigma(v, T) = \int_{-\infty}^{+\infty} \sigma(v)G([v - v'], T)dv' \tag{3}$$

where T is the temperature, G(v,T) the distribution of velocities (as a function of temperature) and we have

written the cross section in terms of velocity, v, as the convolution can occur in velocity space (or energy space; there is a one-to-one correspondence). In the simplest system, i.e., that of an ideal gas, the temperature dependence of the Gaussian or Doppler contribution to the resonance can be expressed as:

$$\Delta \approx \frac{A}{A+1}[\frac{0.05 \bullet E \bullet T}{A \bullet 290}]^{1/2} \tag{4}$$

where Δ is the Doppler width, E is the resonance energy in eV, T is the temperature in Kelvins, and A is the atomic number of the target nucleus. Thus, we see that the width changes as the square root of temperature for a fixed collision energy. Therefore, accurate measurement of the width of a given resonance can be related directly to the temperature of the target atom provided sufficient detail is known about the phonon distribution that the target atom carries or if the target atom can be treated as an ideal gas (well above the Debye limit). Alternatively, one can use a calibration curve built up from width vs. temperature data as a source of temperature correlation. Moreover, the shift in resonance energy due to the target atom's relative velocity (say, behind a shock front) can be used as a particle velocity probe. Shown in Fig. 1 are simulations of a resonance ([182]W) under static (T = 300 K and v = 0.0 m/s) and dynamic conditions (T = 3000 K and v = 2.0 km/s). Measurement of the width and centroid position can be accomplished (in principal), through the simple measurement of the transmission of a white neutron flux (delta-function pulse) through the sample of interest, recording the intensity of the flux as a function of time (as Time-of-Flight [TOF] is related to the neutron energy), and abstracting the width and centroid from the spectroscopic data.

Finally, we are interested in the timescale over which measurements can be made. The width in time of a resonance lineshape depends on the time spread introduced by the moderator, on the intrinsic energy width of the resonance, and on the energy width contributed by Doppler broadening. In traversing a thin sample, neutrons which are spread out in energy will also end up spread out in the time they take to pass through the sample. The quantitative size of the time spread (transit time, Δt) depends in turn on the mean energy of the neutrons and on the distance of the sample from the neutron source:

$$\Delta t = \frac{t}{2} \frac{\Delta E}{E} \cong \frac{36.15 \cdot L}{E^{1.5}} \qquad (5)$$

where E is the resonance energy in eV, and L is the distance in meters from sample to source. The total time width of the resonance sets a lower limit on the achievable transit time for the considered experimental configuration. With a sample at 7 meters and neutrons of energies less than 10 eV, the transit times are on the scale of several microseconds near 10 eV - to tens of microseconds near 1 eV.

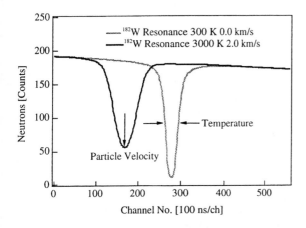

FIGURE 1. Simulated Time-of-Flight (TOF) neutron spectrum of the ^{182}W resonance at 21.10 eV. Note the shift in the centroid, the increased width, and decreased peak cross section at high temperature and particle velocity.

Thus, moving the sample closer to the moderator and using higher energy intrinsically "narrow" resonances minimizes the time (we are then "moderator limited") it takes to make a measurement, and thus the time required to keep a shocked state steady.

All experiments are carried out using spallation sources located at LANSCE (Los Alamos Neutron Science CEnter). Neutron generation and detection is accomplished as follows: H$^-$ is accelerated to 800 MeV in a linac, stripped to protons with a carbon foil and inserted into the Proton Storage Ring (PSR) which has a circumference such that the round-trip time for a single 800 MeV proton is 360 ns. The protons are accumulated and a 250 ns pulse (the ring

is filled to 3/4 of its circumference) is kicked out and carried to either a water moderated W target or a polyethylene moderated ^{238}U target. Fast neutrons born in the targets undergo multiple collisions in the moderators to slow them to epithermal energies. The materials with large cross section and small intrinsic widths lay in this energy regime, and are useful as dopants for these studies. Time zero is detected and all timing signals derived from a pulse generated by a current monitor located on the path from the PSR to the target. The "white" neutrons are collimated using steel and polyethylene rings, and the resulting collimated neutron beam passes through the sample prior to TOF detection by a ^{6}Li glass scintillator mounted on a blue sensitive phototube. The ^{6}Li captures the neutrons, resulting in the generation of an excited ^{7}Li, which subsequently decays to a triton and an alpha particle with excess energy. The excess energy results in the excitation of a Ce fluor, with the subsequent fluorescence detected by phototube. The timing of the shot is set so that the neutrons of the proper energy (i.e. those in the resonance) traverse the sample during the time in which the state is steady.

RESULTS

In order to test the feasibility of recording single shot spectra for the measurement of temperature, we set up an oven in which we heated thin samples of gold and indium obtained by electroplating them onto 1/8" steel substrates. The oven was resistively heated using 2" long 1/4" diameter cartridge heaters from Omega Engineering. Temperature was controlled to typically only +/- 5 C through the use of an Omega 3000N process controller, fedback through a type K thermocouple located in the oven at the same location as the target. We found that measurements with 10% accuracy were possible with single-shot statistics.

Shown in Fig. 2 is one of our first examples of dynamic neutron spectroscopy. In this experiment, a thin Ta foil (25 microns) was mounted onto a 1.2 cm diameter, 1.2 cm thick PBX 9501 pellet. The pellet was initiated with an RP-3 (an EBW coupled to a 29 mg pellet of PETN, RISI) and a change in particle velocity observed. The pellet was initiated towards the neutron flux, thereby decreasing the energy of the resonance in the laboratory frame. Due to the location of the sample (7 m from the moderator) and the energy of the resonance (Ta 4.28 eV), the resonance is observed statically prior to detonation, and then again at lower energy after detonation.

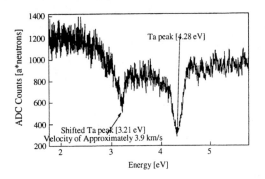

FIGURE 2. Particle velocity measurement of a thin Ta foil (25 microns) off of a pellet of PBX 9501.

In a separate experiment, NRS measurements were made of a linear silver metal jet. To model and/or design a metal jet requires knowledge of it's constitutive properties from ambient conditions to those at high strain and strain rate. Thus, the

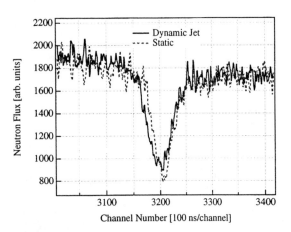

FIGURE 3. Comparison of dynamic (——) and static (- - -) spectra of a silver metal jet. The jet's extracted temperature is a function of its thickness and the spectral background.

measurement of temperature and its comparison to model calculations is a sensitive measure of computational accuracy. Silver provides a nice tool

for spectroscopic application in this case, as it jets well, and a one mm jet has the appropriate thickness for a spectroscopic depth of approximately two, (i.e. $\rho\sigma l = 2$), with resonances at high enough energy to help minimize the time that the state needs to remain steady. Approximately 340 gm of PBX 9501 is line initiated, a silver liner collapsed, a jet formed, and contained in a vessel rated for 3.5 lb. TNT equivalent. X-radiographic records were taken separately at PHERMEX to determine the thickness of the jet as it is important to know for accurate spectroscopic modeling. Shown in Fig. 3 is a comparison of static and dynamic data.

FUTURE WORK

One of the fundamental problems that remains in shock and detonation physics is the development of a "complete" equation of state. Theoretical efforts to construct realistic and accurate equations of state extend back over fifty years. Variables amenable to direct measurement in dynamic explosive experiments have been pressure, particle velocity, and shock velocity. A missing critical piece of information is a precise determination of temperature. We are presently conducting experiments on two shocked systems to measure temperature using NRS; shocked molybdenum and shocked PBX 9502 (triamino-trinitro-benzene with Kel-F 800 as a binder). We are using a new plane wave lens system that is capable of launching 2.5" diameter, 0.25" thick Al flyers between 3.6 and 4.1 km/s, depending on the amount of HE used in the system. For molybdenum, we have doped a one mm section with [182]W using a ball-milling CIP/HIP process developed at Los Alamos. Present studies are using the 3.6 km/s flyer, which yields pressures of 634 kbar and temperatures near 700 K. In our studies of PBX 9502, we have deuterated it to minimize the attenuation of the neutron flux by hydrogen, and we have added [182]WO$_3$ to the binder in a 2 mm section to start. Initial studies with this system are using the 3.6 km/s flyer which nearly overdrives the detonation at 270 kbar.

CP429, *Shock Compression of Condensed Matter – 1997*
edited by Schmidt/Dandekar/Forbes
© 1998 The American Institute of Physics 1-56396-738-3/98/$15.00

SOUNDING EXPERIMENTS OF
HIGH PRESSURE GAS DISCHARGE

Joachim K. Biele

Wehrtechnische Dienststelle für Waffen und Munition, Box 17 64, 49707 Meppen, Germany

A high pressure discharge experiment (200 MPa, $5 \cdot 10^{21}$ molecules/cm³, 3000 K) has been set up to study electrically induced shock waves. The apparatus consists of the combustion chamber (4.2 cm³) to produce high pressure gas by burning solid propellant grains to fill the electrical pump chamber (2.5 cm³) containing an insulated coaxial electrode. Electrical pump energy up to 7.8 kJ at 10 kV, which is roughly three times of the gas energy in the pump chamber, was delivered by a capacitor bank. From the current-voltage relationship the discharge develops at rapidly decreasing voltage. Pressure at the combustion chamber indicating significant underpressure as well as overpressure peaks is followed by an increase of static pressure level. These data are not yet completely understood. However, Lorentz forces are believed to generate pinching with subsequent pinch heating, resulting in fast pressure variations to be propagated as rarefaction and shock waves, respectively. Utilizing pure axisymmetric electrode initiation rather than often used exploding wire technology in the pump chamber, repeatable experiments were achieved.

INTRODUCTION

Although low pressure discharge is well known in physics, Lorentz force effects were not found to be of significant influence. In high energy electrical accelerator research (1), however, plasma armatures are utilized to transfer propulsive Lorentz forces to push a rigid body specimen (often referred to as plasma armature accelerator) (2).

As described here, for the creation of shock waves a high density highly pressurized plasma source was designed. It is made up of a combination of two features: First, closed vessel technology as routinely utilized in solid propellant testing. Second, electrical discharge technology as utilized in pulse power engineering (2).

PRINCIPLES OF APPROACH

The goal of the experiments is focused on high power fast gas heating. Gun propellant gas is chosen as working fluid, initially produced by combustion under closed vessel conditions. The advantage of this approach is both the limitation of flame temperature to about 3000 K and the capability of controlling gas density and energy by loading density of the vessel, deposited prior to the ignition as unburnt propellant mass/volume ratio.

The requirements of the vessel are that it be constructed ruggedly enough to withstand pressure of several 100 MPa and at the same time high electric voltage of about 10 kV. This especially is a severe task to the insulating material which has to withstand both the mechanical and electrical rigors.

Gas density as well as pressure in closed vessel burning are closely related to the molecular weight of the propellant material: Loading density of 0.2 g/cm³ of conventional propellant with average molecular weight of 25 g/mole is gasified by combustion to give $5 \cdot 10^{21}$ molecules/cm³. This gas density at flame temperature, taking into account covolume of 1 cm³/g, generates a pressure of about 250 MPa (assumption of no heat losses).

A propellant gas at or close to flame temperature appears to be a combustion plasma with only very little electric conductivity. Therefore, discharge initiation is needed. This is often achieved by exploding a metal wire between both electrodes using a large current surge. Energetic relationships as well as behavior and interaction of metal particles with the propellant gas (a mixture of mainly carbon monoxide, water, molecular hydrogen as well as nitrogen, and carbon dioxide) are usually not easily understood. As learnt from first observations, the effects to be described here must be checked under more refined conditions. Consequently, in order to achieve reliable results a novel test setup had to be designed.

TEST SETUP

The closed steel vessel (Fig. 1) is comprised of two chambers connected to each other by a duct. The combustion chamber (volume 4.2 cm³) contains the propellant grains to be gasified. The smaller electrical pump chamber (volume 2.5 cm³) is pressurized from the combustion chamber via the duct (4 mm diameter) prior to electrical pumping.

The pump chamber is internally equipped with a coaxial anode insulated against the chamber wall cathode using a poly-acetal plastic (trademark Sustarin). Spacing between the electrodes is 7 mm at the base. A capacitor bank is designed to be switched to the anode in the pump chamber using common spark switching technique. Trigger event is to be positioned right after that point at which all the propellant is gasified. This is predetermined by time delay started at ignition of the propellant grains.

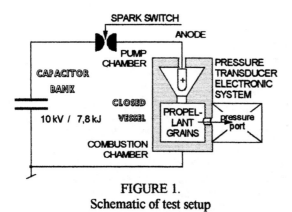

FIGURE 1.
Schematic of test setup

EXPERIMENTS AND RESULTS

Measurements of pressure as well as voltage and current were taken all vs. time.

Pressure was measured piezoelectrically (3) (Kistler transducer type 6211, natural frequency 140 kHz, rise time 2 µs, sensitivity 20 pC/MPa). Based on the potential difference between faces of the quartz sensor element as the result of applied pressure, the pressure transducer has to be positioned to prevent grounding interferences. Disturbances like these may originate when discharging the capacitor bank. Therefore, the pressure measurement system (including transducer, charge amplifer, transient recorder with sampling time interval of 20 µs, and power supply batteries) as a whole was directly attached to the outer wall of the combustion chamber. Consequently, no conjunction of trigger events of the capacitor bank discharge and the pressure/time relationship was available.

From the location of the pressure port in the combustion chamber the recorded pressure data of pressure variations caused by the pump cycle within the pump chamber represent those mechanically filtered by the duct between chambers and transferred to the pressure port. Therefore, pressure variations as originating in the pump chamber have to be expected here steeper and higher in amplitude than detected. More detailed information about the pump action driving the pressure variations can be seen from the voltage-current relationship.

Using commercially available equipment voltage and current raw data all vs. time were measured and stored on transient recorder (sampling time interval 0.5 µs). Considerable effort was needed in order to eliminate reactances of closed vessel and cables to enable measurement within an accuracy of about 10 %. Data reduction was done next to give absorbed power vs. time and voltage vs. current dependence.

A sequence of results from a typical test is presented in Figures 2 and 3.

Pressure vs. time trace is separated (Fig. 2, left and right) into the overall history within the time frame of 50 ms showing the point of all propellant gasified at the first pressure peak (Fig. 2, left). This is followed by a cooling-down phase of pressure decrease until interrupted by a rapid pressure variation 10 ms later. Focusing on this specialty Fig. 2, right, shows the details of pressure variation

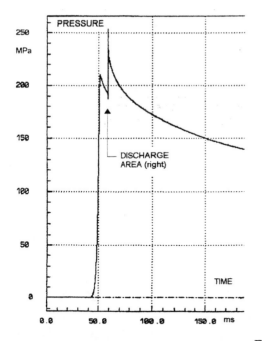
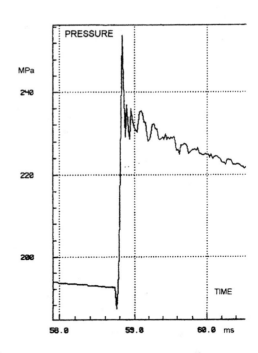

FIGURE 2.

Typical test results of overall pressure time history (left) and discharge driven pressure variation (right)

originating from discharge driven pump action: Starting with a significant underpressure peak down from the declining slope, a rapid rise by 65 MPa within less than 100 μs is observed, precursing an elevated cooling-down slope.

The electrical pump chamber is considered to act as nonlinear resistance in the circuit with the charged capacitance as power source (see Fig. 1). Power absorption vs. time of a typical test is depicted in Fig. 3, left: The amount of 7,8 kJ was electrically available from the charge on the capacitor. This is more than twice the energy chemically deposited within the pump chamber volume at the all-propellant gasified peak pressure event.

The power absorption history reveals two major spikes: The first one starting at 125 μs rises up to 200 MW indicating ignition of discharge, the second one occurring at about 140 μs exceeds the value of that first peak considerably.

Further characteristic details of the discharge process may be taken from the current vs. voltage dependence shown in fig. 3, right: Ignition of discharge was started at 10 kV by random bursts leading to a constant current of 25 kA at decreasing voltage down to 4 kV. On the low voltage level a

current spike is developed exceeding 100 kA. Unfortunately, the peak current value was not observable. This spike is mapped into the absorption history (Fig. 3, left) emanating there from increasing power and coming down again to reach the state of almost no power flow.

The remaining process is interpreted from Fig. 3 to be the end of discharge turning over to ohmic heating by oscillation (including unidentified voltage interferences).

Consequently, it is argued that the spiking process coincides with shock wave generation. From the current and power absorption spike the shock wave generation is confirmed to be Lorentz force driven, bringing about the magnetic pinch phenomenon: If the gas is confined within the pump chamber so that the pressure falls rapidly outside that region, a rarefaction wave must travel into the combustion chamber to be detected with the piezoelectric transducer. Related to this, the magnetic pressure must rise equally rapid in order to energize the gas mechanically and to launch the shock wave following the rarefaction wave at that same location of pressure measurement.

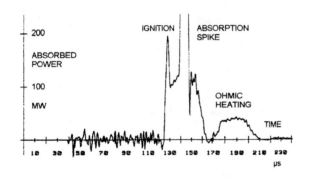 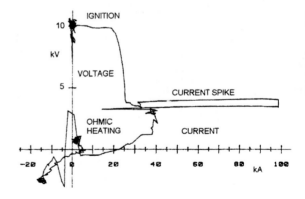

FIGURE 3.
Typical test result of absorbed electrical power vs. time (left)
as well as of current and voltage dependence (right).

CONCLUSIONS

Having first noted underpressure peaks under different conditions (one single chamber for combustion and pumping with about two orders of magnitude more volume and electrical energy, using exploding wire technology for ignition) its origin could not be identified at that time. As shown in this paper, Lorentz forces are very likely to pressurize the propellant gas in addition to ohmic heating. From this fact it is concluded that Lorentz forces may be exploited to enhance gas-operated accelerators.

ACKNOWLEDGEMENTS

The author appreciates the expert experimental work of L. Mohring and B. Trame as well as A. Husteden. H. F. Heuer's managerial effort is gratefully acknowledged. Special thanks are extended to Annelene Gebbeken for carefully preparing the manuscript.

REFERENCES

1. Francis, G., *Handbuch der Physik*, Berlin/Göttingen/ Heidelberg: Springer Verlag, 1956, pp. 53 - 208.
2. Thio, Y.C., *Proceedings of the 6th International Symposium on Ballistics*, American Defense Preparedness Association, Orlando Fl., Oct. 27 - 29, (1981), pp. 532 - 545.
3. Oliver, F. J., *Practical Instrumentation Transducers*, New York, 1971, p. 154.

CHAPTER XIII

IMPACT PHENOMENA AND HYPERVELOCITY STUDIES

CP429, *Shock Compression of Condensed Matter – 1997*
edited by Schmidt/Dandekar/Forbes

AN ANALYTIC PENETRATION MODEL FOR A DRUCKER-PRAGER YIELD SURFACE WITH CUTOFF

James D. Walker and Charles E. Anderson, Jr.

Southwest Research Institute, Engineering Dynamics Department
San Antonio, TX 78228

An analytic model has been developed for the penetration of materials for which the constitutive response can be characterized by a Drucker-Prager yield surface with a maximum flow stress cutoff. This constitutive model represents extensively fractured glasses and ceramics, where large pressures lead to large resistance to shear deformation. The analytic penetration model using this yield surface is based on the ideas of Walker and Anderson (1) where the momentum balance is explicitly solved along the centerline. Target response occurs in a hemispherical region containing an interior boundary (one of the unknowns) demarcating the two domains of the yield surface. Model results are compared to hydrocode calculations and experimental penetration data for glass.

INTRODUCTION

Analytic models have been developed that do an excellent job of predicting depth of penetration into metal targets. This paper presents a model for the penetration into thick glass and ceramic materials by a projectile. Initial work on a pressure dependent yield surface using this approach was performed in (2). It is assumed that the ceramic or glass fractures early in the penetration process (3) and therefore the projectile is considered to penetrate failed material. The strength of the failed material is assumed to increase with increasing pressure up to a maximum value. This model is usually referred to as the Drucker-Prager constitutive model. This constitutive model will lead to a penetration model based on a centerline momentum balance, an approach that has proved very successful in metal target penetration problems (1).

THE MODEL

The projectile and target, considered axisymmetric, will be assumed to lie along the z axis. The location of the interface between the projectile and the target is denoted $z_i(t)$, with $z_i(0) = 0$. The rear of the projectile is denoted $z_p(t)$, and $z_p(0) = -L_o$, where L_o is the initial length of the projectile. The velocity along the centerline in the projectile and target is written $u_z(z)$. The interface velocity is $u = u_z(z_i)$ and the velocity of the back end of the projectile is

$v = u_z(z_p)$. The target is assumed to be semi-infinite.

A central theme of the model is the use of the momentum balance along the z axis. On the axis itself, where $x = y = 0$, $u_x = u_y = 0$ by symmetry. Since the x and y directions are equivalent, the momentum balance along the centerline simplifies to:

$$\rho \frac{\partial u_z}{\partial t} + \frac{1}{2}\rho \frac{\partial (u_z)^2}{\partial z} - \frac{\partial \sigma_{zz}}{\partial z} - 2\frac{\partial \sigma_{xz}}{\partial x} = 0 \quad (1)$$

The following assumptions are made based on the examination of numerical (hydrocode) simulations of long-rod impacts that we have performed for numerous impact simulations at different impact velocities and with various materials:

1) A velocity profile along the centerline in both the projectile and the target is specified. This is the same profile assumed in (1) for penetration into metal targets.
2) The back end of the projectile is decelerated by elastic waves, with a magnitude proportional to the yield strength of the projectile. Again, this is the same as in (1).
3) A shear behavior in the target material is specified. This is where the pressure dependent yield surface appears.

With suitable expressions resulting from assumptions 1 and 3, the axial momentum equation can then be integrated to obtain an equation of motion for the location of the interface between the target and the projectile. Assumption 2 provides an equation for the deceleration of the rear of the projectile.

Initial conditions for the model include an initial interface velocity and an expression that relates the crater radius to the impact velocity. The first is obtained from the equations for one-dimensional plate impact (the particle velocity in the Rankine-Hugoniot relations). Currently we are using the crater radius versus impact velocity expression from (1), although this is based on metal data. A cavity expansion solution used to estimate the extent of plastic flow within the target. The interior boundary will be found by assuming stress deviator continuity along the centerline.

Velocity Profiles

The velocity in the projectile is constant over most of the projectile length, except for a small region near the target-projectile interface. This velocity profile is approximated by a bilinear expression, with s as the extent of the plastic zone along the axis:

$$u_z(z) = \begin{cases} u - \dfrac{v-u}{s}(z-z_i) & (z_i-s) \le z < z_i \\ v & z_p \le z < (z_i-s) \end{cases} \quad (2)$$

For the target, velocity fields in numerical simulations have a hemispherical behavior, and due to this, a flow field written in spherical coordinates is used. In particular, the velocity is based on the curl of a vector potential since this is automatically divergence free, thus a volume preserving flow, a feature of rigid plasticity. Letting $r(z) = z - z_i(t) + R$ where R is the *crater* radius, the following form of the velocity along the centerline in the target is assumed

$$u_z(z) = \begin{cases} \dfrac{u}{\alpha^2-1}\left[\left(\dfrac{\alpha R}{r(z)}\right)^2 - 1\right] & R \le r(z) < \alpha R \\ 0 & r(z) \ge \alpha R \end{cases} \quad (3)$$

αR is viewed as the extent of the plastic zone in the target. The crater radius R is assumed constant.

Pressure Dependent Yield

The assumptions of rigid plasticity applied to the three dimensional flow field give the normal derivative of the shear stress as (see (1)):

$$\left.\frac{\partial s_{xz}}{\partial x}\right|_{x=0} = -\frac{7Y_t}{6z} \quad (4)$$

The shear term within the projectile will be considered negligible.

The pressure dependent yield is given by a zero pressure flow stress Y_0 and a slope b (Fig. 1):

$$Y_t = Y_0 + bp \quad (5)$$

This form is usually referred to as a Drucker-Prager constitutive model. Due to symmetry, the following state of stress along the centerline is derived:

$$\sigma_{zz} = -p - \frac{2Y}{3} \quad (6)$$

Inserting Eqn. (6) into the yield expression gives

$$Y_t = \frac{Y_0}{1+2b/3} - \frac{b}{1+2b/3}\sigma_{zz} \quad (7)$$

Combination of Eqns. (5) and (6) allows the normal derivative of the shear stress to be written in terms of the stress. With

$$A \equiv \frac{2Y_0}{1+2b/3}, \qquad B \equiv \frac{2b}{1+2b/3} \quad (8)$$

within the region with the pressure dependent yield, Eqn. (1) gives

$$\rho\frac{\partial u_z}{\partial t} + \rho\, u_z\frac{\partial u_z}{\partial z} - \frac{\partial \sigma_{zz}}{\partial z} + \frac{7}{6z}(A - B\sigma_{zz}) = 0 \quad (9)$$

where σ_{zz} is a function of pressure, and hence of z. Using the integrating factor

$$z^\delta, \qquad \text{where} \qquad \delta \equiv \frac{7B}{6} \quad (10)$$

the solution to Eqn. (9) is

$$\sigma_{zz}(\tilde\alpha R) = \left(\frac{\alpha}{\tilde\alpha}\right)^\delta \sigma_{zz}(\alpha R) - \frac{\rho_t \dot u R}{\tilde\alpha^\delta(\alpha^2-1)}\left\{\alpha^2\frac{\alpha^{\delta-1}-\tilde\alpha^{\delta-1}}{\delta-1}\right.$$
$$\left. - \frac{\alpha^{\delta+1}-\tilde\alpha^{\delta+1}}{\delta+1}\right\} - \frac{Y_0}{b}\left(\left(\frac{\alpha}{\tilde\alpha}\right)^\delta - 1\right)$$
$$- \frac{2\rho_t\alpha^4 u^2}{\tilde\alpha^\delta(\alpha^2-1)^2}\left\{\frac{\alpha^{\delta-2}-\tilde\alpha^{\delta-2}}{\delta-2} - \frac{\alpha^{\delta-4}-\tilde\alpha^{\delta-4}}{\delta-4}\right\}$$
$$- \frac{2\rho_t u\alpha R\dot\alpha}{\tilde\alpha^\delta(\alpha^2-1)^2}\left\{\frac{\alpha^{\delta+1}-\tilde\alpha^{\delta+1}}{\delta+1} - \frac{\alpha^{\delta-1}-\tilde\alpha^{\delta-1}}{\delta-1}\right\}$$

$$(11)$$

where we have explicitly integrated from $z = \tilde\alpha R$ to $z = \alpha R$. The density has been assumed to be approximately constant so that ρ could be pulled out of the integrals.

The region where the flow stress equals the cutoff $\overline Y$ ranges from the crater radius R to an intermediate radius $\tilde\alpha R$ (Fig. 2), since the stresses are higher near the projectile-target interface. The pressure dependent portion of the flow surface then ranges from this intermediate radius to the outer edge αR. The intermediate radius $\tilde{\alpha}R$ is an interior boundary and is one of the unknowns.

The integral of the momentum balance within the region between R and $\tilde{\alpha}R$ where the flow stress is the constant $\overline{Y}$ and the integral of the momentum balance over the length of the projectile are found in a similar fashion. Solution to Eqn. (1) from each region, with $\sigma_{zz}(z_p) = 0$ and $\sigma_{zz}(\alpha R) = 0$, combine to form the integrated momentum balance along the centerline:

$$
\begin{aligned}
&\rho_p \dot{v}(L-s) + u\Bigg[\rho_p s + \frac{\rho_p \alpha^2 s^2}{(\alpha^2-1)R} + \frac{\rho_t R}{\alpha^2-1}\Bigg\{\alpha^2 - \frac{\alpha^2}{\tilde{\alpha}} - \tilde{\alpha} + 1 \\
&\quad + \frac{1}{\tilde{\alpha}^\delta}\left(\alpha^2 \frac{\alpha^{\delta-1}-\tilde{\alpha}^{\delta-1}}{\delta-1} - \frac{\alpha^{\delta+1}-\tilde{\alpha}^{\delta+1}}{\delta+1}\right)\Bigg\}\Bigg] \\
&= \frac{1}{2}\rho_p(v-u)^2 - \frac{7}{3}\overline{Y}\ln(\tilde{\alpha}) - \frac{Y_0}{b}\left\{\left(\frac{\alpha}{\tilde{\alpha}}\right)^\delta - 1\right\} \\
&\quad - \frac{1}{2}\rho_t u^2\left(\frac{\alpha}{\tilde{\alpha}}\right)^4\left(\frac{\tilde{\alpha}^2-1}{\alpha^2-1}\right)^2 \\
&\quad - \frac{2\rho_t \alpha^4 u^2}{\tilde{\alpha}^\delta(\alpha^2-1)^2}\left\{\frac{\alpha^{\delta-2}-\tilde{\alpha}^{\delta-2}}{\delta-2} - \frac{\alpha^{\delta-4}-\tilde{\alpha}^{\delta-4}}{\delta-4}\right\} \\
&\quad - \frac{2u\alpha\dot{\alpha}}{(\alpha^2-1)^2}\Bigg[\frac{\rho_t R}{\tilde{\alpha}}(\tilde{\alpha}-1)^2 - \frac{\rho_p s^2}{R} \\
&\quad + \frac{\rho_t R}{\tilde{\alpha}^\delta}\left\{\frac{\alpha^{\delta+1}-\tilde{\alpha}^{\delta+1}}{\delta+1} - \frac{\alpha^{\delta-1}-\tilde{\alpha}^{\delta-1}}{\delta-1}\right\}\Bigg]
\end{aligned}
\tag{12}
$$

The deceleration of the tail of the projectile is from (1), and the time rate of change of the length of the projectile is the difference between the penetration speed and the speed of the tail:

$$
\dot{v} = -\frac{\sigma_p}{\rho_p(L-s)}\left\{1 + \frac{v-u}{c} + \frac{s}{c}\right\}
\tag{13}
$$

$$
\dot{L} = -(v-u)
\tag{14}
$$

Next comes the step of solving for the interior boundary $\tilde{\alpha}$. At this boundary, the axial stress deviator is assumed continuous. Thus, at the interior interface we have (see Eqns. (5) and (6))

$$
-\frac{2}{3}\overline{Y} = -\frac{2}{3}\{Y_0 + bp(\tilde{\alpha}R)\}
\tag{15}
$$

Using Eqns. (5) and (6), this gives

$$
\sigma_{zz}(\tilde{\alpha}R) = -\frac{2\overline{Y}}{3} - \frac{\overline{Y}-Y_0}{b}
\tag{16}
$$

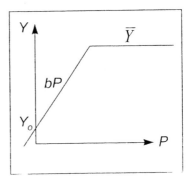

FIGURE 1. Drucker-Prager Constitutive Model

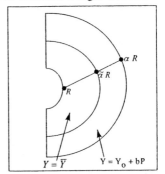

FIGURE 2. Target Geometry

This is then combined with Eqn. (11) to produce a nonlinear equation for $\tilde{\alpha}$, which must be solved simultaneously with Eqns. (12) and (13) for u and v. We do this with an iterative quasi-Newton procedure, Eqn. (16) is solved with a one-dimensional Newton step. A linear system for u and $\dot{v}$ from Eqns. (12) and (13) for the given $\tilde{\alpha}$ is then solved. Next, the new u and $\dot{v}$ are placed into Eqn. (16) for the calculation of a new $\tilde{\alpha}$, and the cycle repeats itself until convergence is achieved. As a first step, however, one solves the problem without cutoff ($\tilde{\alpha} \equiv 1$) to see if the stresses are such that the cutoff is even reached.

The calculation of α is based on the *intact* material's bulk modulus and low pressure flow stress (1):

$$
\left(1 + \frac{\rho_t u^2}{Y_t}\right)\sqrt{K_t - \rho_t \alpha^2 u^2} = \left(1 + \frac{\rho_t \alpha^2 u^2}{2G_t}\right)\sqrt{K_t - \rho_t u^2}
\tag{17}
$$

The intact material determines the size of the zone. (Note the intact material strength is not Y_0, which is the crushed materials low pressure strength). s is found by matching the slopes of the assumed velocity profile in the target and in the projectile at the target-projectile interface:

$$\frac{u - v}{s} = -2 \frac{u}{\alpha^2 - 1} \frac{\alpha^2}{R} \qquad (18)$$

Examples

To examine the model, it is compared against both experimental data for glass and hydrocode calculations with an identical constitutive model. The experiments into glass involved tungsten projectiles impacting glass targets at two different impact velocities, and flash x-ray radiographs of the penetrating projectiles were taken to give penetration depth into the glass versus time (4). These data were collected for two different impact velocities: 1.25 km/s and 1.70 km/s.

Figures 3 and 4 shows the best fit found for both velocities (dot-dashed line). The constitutive parameters are $Y_0 = 100$ MPa, $b = 2.01$, $\overline{Y} = 1.5$ GPa, and $Y_{cav} = 1.0$ GPa, where Y_{cav} is the strength of the intact material. The match with the data is very good. The CTH (5) results (solid line) is for a slightly different constitutive model, with $Y_0 = 0$ MPa and $b = 2$. These are close. When the model is run with $Y_0 = 0$, as in the CTH constitutive model, the agreement is very good, showing that the analytic model agrees well with the hydrocode calculations.

SUMMARY

A penetration model has been presented for penetration into a target material that has a pressure dependent yield stress. The model includes transient effects. The model results compare well with hydrocode calculations (with the same constitutive model) and experimental penetration data into glass. The model requires the calculation of an internal boundary in the target, demarcating the domains of the target corresponding to the cutoff flow stress and the pressure dependent flow stress. The results are important in that they demonstrate that a complicated constitutive model can be solved analytically within the framework of a fundamentally-based penetration model.

REFERENCES

1. J. D. Walker and C. E. Anderson, Jr., *Int. J. Impact Engng*, **16**(1), 19-48 (1995).
2. D. L. Littlefield, C. E. Anderson, Jr., and S. R. Skaggs, "Analysis of Penetration of Steel and Al$_2$O$_3$ Targets," High Pressure Science and Technology - 1993, **2**, 1793-1796 (1993).

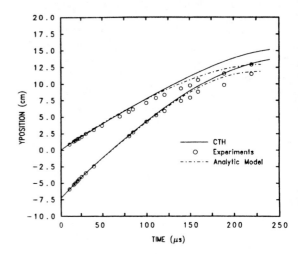

FIGURE 3. Model compared to data and CTH for 1.25 km/s impact.

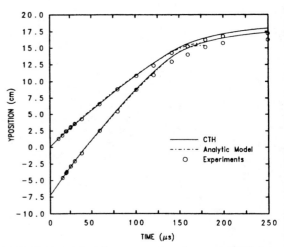

FIGURE 4. Model compared to data and CTH for 1.70 km/s impact.

3. J. D. Walker and C. E. Anderson, Jr., "An Analytical Model for Ceramic Faced Light Armors," Proceedings, 16th International Symposium on Ballistics, San Francisco, 23-28 September (1996)
4. C. E. Anderson, Jr., V. Hohler, J. D. Walker, and A. J. Stilp, "Penetration of Long Rods into Steel and Glass Targets," Proc. 14th Int. Symp. on Ballistics, **1**, 145-154 (1993).
5. J. M. McGlaun, S. L. Thompson and M. G. Elrick, *Int. J. Impact Engng*, **10**, 351-360 (1990).

CP429, *Shock Compression of Condensed Matter – 1997*
edited by Schmidt/Dandekar/Forbes
© 1998 The American Institute of Physics 1-56396-738-3/98/$15.00

NUMERICAL STUDY OF PENETRATION IN CERAMIC TARGETS WITH A MULTIPLE-PLANE MODEL

H.D. Espinosa, G. Yuan, S. Dwivedi, and P.D. Zavattieri

School of Aeronautics and Astronautics, Purdue University,
West Lafayette, IN 47907

The penetration mechanics in different material/structure systems has been investigated by numerical simulations with the finite element code EPIC95. A multi-plane microcracking model was implemented to simulate ceramic fragmentation and comminution. Two kinds of confined structures, depth-of-penetration (DOP) and interface-defeat (ID) configurations, were examined in the simulations. The results revealed that the penetration process is found to be less dependent on the ceramic material than usually assumed by most investigators. By contrast, the penetration process is highly dependent on the multi-layered configuration and the target structural design (geometry, and boundary conditions). From a simulation standpoint, we found that the selection of the erosion parameter plays an important role in predicting the deformation history and interaction of the penetrator with the target. These findings show that meaningful light weight armor design can only be accomplished through a combined experimental/numerical study in which relevant ballistic materials and structures are *simultaneously* investigated.

INTRODUCTION

Design of light weight armor depends on fundamental understanding of material performance, constitutive/failure models, and structural performance. Ceramics have high hardnesses and low densities and have been considered good candidates materials for defeat of long-rod tungsten projectiles.

During the past ten years, several constitutive models for brittle materials have been developed to describe impact behavior. A multiple-plane model (MPM) that can predict damage induced anisotropy was developed by Espinosa et al. (1). It appears that the MPM can provide relevant insight into the design of ceramic armor material/structure systems. Hauver et al. (2) found that the performance of ceramic targets during penetration depends on the nature of ceramic confinement. They proposed a new structure known as interface defeat (ID) configuration. The objective is to keep the confinement on the ceramic plate so that the penetrator is consumed by lateral flow at the ceramic-cover plate interface.

In this paper, a parametric study is carried out for the penetration of two different multi-layered confined ceramic target plates. Basically, the DOP and ID configurations are simulated with a particular emphasis in understanding the variation of measurable quantities as a function of target configuration, ceramic type and confinement. In all calculations, a penetrator striking velocity of 1.5 Km/s is used.

MODEL

The multiple-plane microcracking model, used in this study, is based on the assumption

901

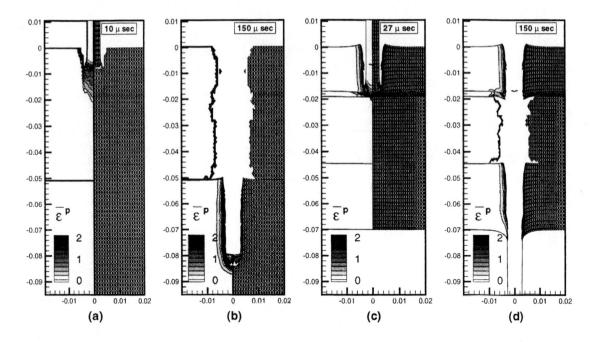

FIGURE 1: Contours of effective plastic/cracking strain for DOP (a-b) and ID (c-d) configurations with alumina, d/L=0.05, and erosion=1.5. Dimensions are given in meters.

that microcracking and/or slip can occur on a discrete number of orientations. Slip plane properties, friction, size, density, etc. and their evolution are *independently* computed on each plane. The macroscopic response of the material is computed by additive decomposition of the strain tensor into elastic and inelastic parts. In contrast to scalar representations of damage, the present model is broad enough to allow the examination of *damage induced anisotropy* and damage localization in the interpretation of impact experiments. In particular, the effective behavior of the solid is predicted to be rate dependent due to crack kinetics effects. Model parameters were identified independently through plate and rod impact simulation of experiments.

ANALYSES AND RESULTS

Effect of ceramic material: The DOP target plate has been analyzed with both Al_2O_3 and SiC ceramics for a penetrator with d/L of 0.05.

The erosion parameter has been chosen as 1.5 and analyses carried out till the elapsed time of 150 μs. The penetration of alumina at two time cycles together with a plot of effective plastic/cracking strain is shown in Figure 1. It was found that the response of Al_2O_3 and SiC to penetration are similar without much appreciable difference. However, the crater shape in the case of the SiC target is larger than that in Al_2O_3, specially at the ceramic-steel interface. This may be due to the higher wave speed in SiC which results into more time for wave interactions at the interface.

Effect of target structure: As mentioned earlier, Hauver proposed the ID configuration in which the ceramic is maintained under constant confinement with the help of a cover plate and an interface layer made of softer graphite material. Analyses have been carried out for a d/L of 0.05 for both the DOP and the ID configurations. The ceramic used in the simulations was

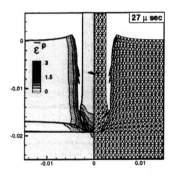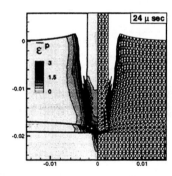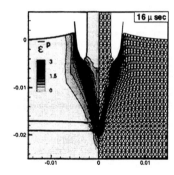

FIGURE 2: Contours of effective plastic/cracking strain for erosion values of 1.5, 3, and 9, respectively.

Al_2O_3 and an erosion effective cracking strain of 1.5 was chosen. The target width and layer thickness were selected such that a meaningful comparison can be made. Figure 1 shows the penetration event together with the plot of effective plastic/cracking strain at various time cycles. Significant wave damage in the ceramic is observed in the the DOP configuration. see Fig. 1a. In the case of ID, the nose of the penetrator reaches the surface of the ceramic at 27 μs after defeating the cover plate. At this time, the tail velocity of the penetrator is identical for both structures. Subsequently, it reduces significantly more in the ID configuration than in the DOP configuration. Moreover, wave damage in the ceramic, for the ID configuration (see Fig. 1c), is much less than it is for the DOP configuration. This is expected from variations in ceramic confinement. Hence, the model predicts a superior behavior in the case of multi-layered ceramic target structures in confirmation with experimental evidence. However, our experimental observations (3) indicate that the penetrator nose can be partially spread in the lateral direction during interface defeat. This feature can be seen as lateral flow of graphite in the present simulation, see Fig. 1c. Also, the penetrator perforates the target in agreement with our experimental observations (3), see Fig. 1d.

Effect of erosion parameter: EPIC95 employs a critical effective plastic/cracking strain as an

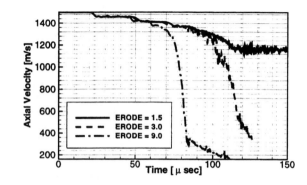

FIGURE 3: Penetrator tail velocity for different values of the erosion parameter.

element erosion parameter. A higher value of the erosion parameter means that it is more difficult to erode the material. In all the above analyses, its value has been kept constant as 1.5. In order to study its effect on penetration, the ID configuration has been analyzed for two additional values of the eorsion parameter, namely, 3.0 and 9.0. The results show that for increasing values of the erosion parameter, the penetrator material flows upwards and develops a triangular nose, see Fig. 2. The sharpness of the nose increases as the penetration continues which facilitates further penetration by focusing the inelastic deformation right ahead of the penetrator. As a result, the penetrator reaches the graphite-

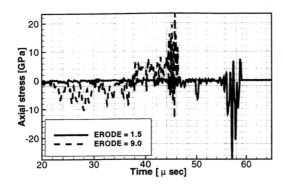

FIGURE 4: Axial stress at ceramic-back plate interface for two values of the erosion parameter.

ceramic interface at 27, 24 and 16 μs for erosion values of 1.5, 3.0 and 9.0, respectively. The difference in the nose shape provides different initial conditions when the penetrator hits the ceramic-graphite interface. Despite the faster rate of penetration, with increasing erosion value, overall penetration of the projectile is not achieved when the erosion value is set to 9. The erosion value affects not only the deformed shape of the penetrator but also the whole history of stresses and particle velocities. Penetrator tail velocity histories and in-material stress histories, for various values of the erosion parameter, are given in Figs. 3 and 4 respectively.

CONCLUSIONS

The above analyses show that the response of multi-layered ceramics targets is relatively independent of the ceramic material for the two types of ceramics considered, i.e., Al_2O_3 and SiC. The interface defeat (ID) configuration proposed by Hauver has distinct advantages over the depth of penetration (DOP) configuration. Ceramic wave damage is less in the case of ID and its resistance to penetration dramatically improved by the confinement provided by the cover plate.

The value of the erosion parameter is an important input for the simulation. The above analyses show that the penetrator nose deforms and gains a conical shape as the value of the

erosion parameter is increased. This facilitates penetration initially. However, the presence of hardened material reduces the velocity of the penetrator and the peak stress more efficiently. Therefore, the selection of an erosion parameter needs detailed experimental measurements of in-material stresses, free surface velocities and recovered penetrator shapes.

It can be concluded that the integration of a multiple-plane microcracking model into the finite element code EPIC95 has been successful. It is able to predict the response of different types of the multi-layered ceramic target in confirmation with experimental evidence. Limitations have been encountered in the use of EPIC95. Only one erosion parameter can be selected for the penetrator and various target plates. Interfaces need to be modeled as perfect or frictional interfaces without the possibility of modeling progressive decohesion. Mesh adaptivity is not available to avoid excessive mesh distortion Erosion is needed to advance the calculation with a reasonable time step. All these features have been addressed by Espinosa et al. (1). Software capable of simulating the interface defeat configuration without erosion is currently under development.

ACKNOWLEDGMENTS

The research reported in this paper was supported by the Army Research Office through award Nos. DAAH04-96-1-0142 and DAAH04-96-1-0331.

REFERENCES

1. Espinosa, H.D., Zavattieri, P.D., and Emore, G.L., "Adaptive FEM Computation of Geometric and Material Nonlinearities with Application to Brittle Failure," to appear in *Special Issue of Mechanics of Materials*, 1997.

2. Hauver, G., Netherwood, P., Benck, R, and Keoskes, L., "Enhanced Ballistic Performance of Ceramics," 19th Army Science Conference, Orlando, FL, 20-24 June, 1994.

3. Brar, N.S., Espinosa, H.D., Yuan, G. and Zavattieri, P.D., "Experimental Study of Interface Defeat in Confined Ceramic Targets," this volume.

CP429, *Shock Compression of Condensed Matter – 1997*
edited by Schmidt/Dandekar/Forbes
© 1998 The American Institute of Physics 1-56396-738-3/98/$15.00

OPTIMIZATION OF THE PERFORMANCE OF SEGMENTED-TELESCOPIC PENETRATORS

D. L. Littlefield

Institute for Advanced Technology, The University of Texas at Austin, Austin Texas 78759

The performance of segmented-telescopic penetrators is investigated in this numerical study. It is shown that the penetration depth of single segmented-telescopic penetrators impacting a semi-infinite target is exceptional. However, the penetration depth of multiple segment trains can be marginal; the degradation in performance is a result of segment-to-segment interference effects. Techniques for improving the performance of multiple segment trains are suggested.

INTRODUCTION

It is well known that the penetration performance of a long rod can be increased if the rod is subdivided into a series of segments with small length-to-diameter (L/D) ratios, spaced apart by an appropriate amount[1-3]. However, there are several practical difficulties in applying this principle to increase the performance of a penetrator; for instance, if tubes or low-density spacers are used to space the segments then there is increased parasitic mass and can be launch difficulties. An alternative is to consider a package that can be launched in a collapsed configuration, then extended in flight to achieve a separated, segmented grouping. This arrangement has been referred to by previous researchers as a *segmented-telescopic,* or *seg-tel* penetrator. Previous experimental and numerical studies of this arrangement have shown that the performance of segmented-telescopic penetrators can be very sensitive to the specifics of their geometry[4]. As such, it becomes very important to understand how variations in segment shape influence the performance in order to make the most effective use of the arrangement. In this paper, a numerical study is performed to examine the performance of generic segmented-telescopic penetrator geometries. Techniques for improving their performance are suggested.

ORIGINAL SEG-TEL CONFIGURATION

Figure 1 illustrates an original seg-tel configuration that was studied. The all-tungsten alloy device is comprised of a small L/D rod with a tube connected to the back section of the rod. Using this arrangement, the seg-tels can be stacked together to form a long rod.

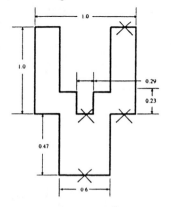

FIGURE 1. Original seg-tel configuration.

CALCULATIONS

Simulations of the seg-tel penetrator impacting a semi-infinite armor steel target were carried out using the Eulerian wave propagation code CTH[5]. All the calculations were two-dimensional axisymmetric; approximately 30 zones were used across the diameter of the penetrator. Square zoning was maintained in strong interaction regions between the target and penetrator; beyond these regions the zone dimensions were increased at a constant rate.

The performance of a single seg-tel penetrator is illustrated in Figure 2, where the positions of different points on the penetrator are shown as a function of time. The impact velocity V of the penetrator was 2.6 km/s. Here the rod interface and tail refer to points located in the rod portion of the seg-tel and positioned along the centerline, and the tube interface and tail to points in the tube portion (see the X's in Figure 1). The original length of the tube portion was 0.76 cm. The deepest point of penetration is 1.78 cm and is achieved by the tube portion; the penetration efficiency (the penetration depth P normalized by the length of the tube L) is 2.34. This value represents a significant improvement over the penetration efficiency that can be achieved with a long rod (typically about 1.5 at this impact velocity), or a $L/D \sim 1$ penetrator (typically about 2.0 at this velocity).

The results in Figure 2 also suggest that the tube portion of the seg-tel penetrates with a higher interface velocity than the rod portion; the tube interface surpasses the rod interface about 5 μs after impact. This is illustrated more clearly in Figs. 3 - 4, where pressure and equivalent plastic strain contours are shown at 5 and 15 μs, respectively. At 5 μs, Figure 3 shows the tube portion of the penetrator just beginning to surpass the rod portion. The lower interface velocity in the rod is a result of the additional lateral confinement provided by the tube during penetration; the additional confinement increases the pressure and axial extent of the plastic zone in the rod, causing a more rapid deceleration of the rod than would occur without the tube. By 15 μs, penetration by the tube ceases; it has completely eroded. However, a significant portion of the rod still

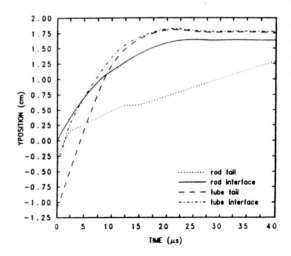

FIGURE 2. Penetration depth versus time.

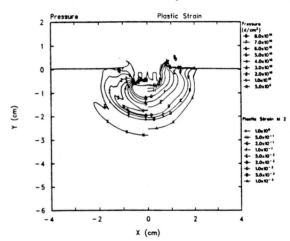

FIGURE 3. Pressure and plastic strain contours at 5 μs.

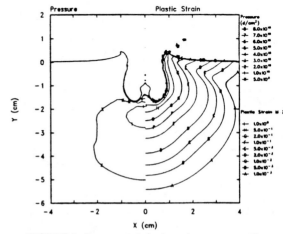

FIGURE 4. Pressure and plastic strain contours at 15 μs.

remains, and continues to penetrate for several microseconds until it completely erodes.

Since contributions to the penetration depth occur for both the rod and tube portion of the seg-tel, it is easy to see why a seg-tel penetrator outperforms a $L/D \sim 1$ cylindrical penetrator. However, the overall penetration time (when normalized by L/V) required for a seg-tel penetrator is significantly longer than the time required for a $L/D \sim 1$ cylindrical penetrator; this is due to the additional deceleration that occurs for the rod portion of the seg-tel. The increase in penetration time can result in substantial inter-segment interference when multiple segment trains are impacted into semi-infinite targets. Shown in Figs. 5 - 7 are pressure and plastic strain contours for the penetration of a train of four seg-tels, initially spaced $2.5D$ apart, at 20, 40 and 90 μs, respectively. At 20 μs, Figure 5 shows the impact of the second seg-tel occurring at the base of the base of the penetration channel; however, a portion of the first seg-tel still remains at the base of the crater and interferes with penetration. The interference becomes accentuated with the impact of subsequent seg-tels, as is suggested in Figure 6, where the impact of the third seg-tel is shown. The final penetration channel is shown in Figure 7. The depth of penetration is 5.21 cm (excluding the last few millimeters of penetration, which can be regarded as an artifact of the perfect symmetry imposed on the problem). When normalized by the *collapsed length* L_c (that is, the length the penetrator would have if the seg-tels were stacked together), the penetration efficiency is 1.71, representing only a modest improvement over the value that can be obtained with a long rod.

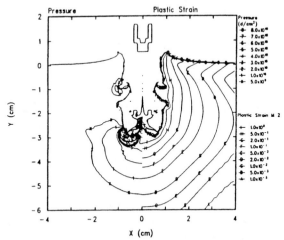

FIGURE 6. Pressure and plastic strain contours at 40 μs.

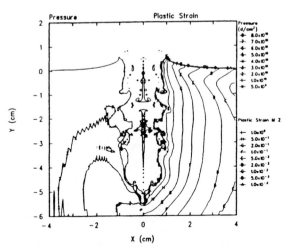

FIGURE 7. Pressure and plastic strain contours at 90 μs.

IMPROVED SEG-TEL CONFIGURATION

The multiple-segment performance of the seg-tel configuration shown in Figure 1 can be improved if alternative geometries are considered. Results from the numerical simulations suggest the changes that should be made. For example, in

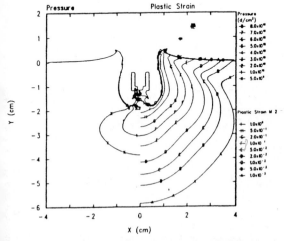

FIGURE 5. Pressure and plastic strain contours at 20 μs.

order to decrease the penetration time for the seg-tel and decrease the inter-segment interference, the tube portion should be minimized in size and mass to the extent possible. The calculations also suggested that interference may occur with the sidewalls of the penetration channel; this can be somewhat alleviated by increasing the impact velocity. Shown in Figure 8 is an improved seg-tel configuration guided by these observations. The tube portion of the seg-tel has a thinner wall and is assumed to be comprised of a low-density material; for example a graphite-epoxy composite. The rod portion is somewhat longer and has a pointed nose. The performance of this penetrator is illustrated in Figure 9, where a plot of position versus time for four points on the seg-tel (refer to the X's in Figure 8) is shown. The impact velocity was 3 km/s. Here the tube portion of the penetrator does not surpass the position of the rod portion. The majority of penetration has occurred by 15 μs, and the final depth of penetration is 2.20 cm. The original length of the tube was 0.90 cm, yielding a penetration efficiency of 2.44.

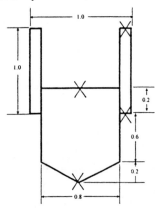

FIGURE 8. Improved seg-tel configuration.

The improved seg-tel configuration shown in Figure 8 results in a considerable increase in multiple segment performance. Using this geometry, a calculation was performed for the impact of four seg-tels, spaced 2.5D apart, at a velocity of 3 km/s. The final depth of penetration was 7.69 cm; the penetration efficiency based on the collapsed length was 2.13. This represents an improvement of about 42% over what is achievable

with a long rod, and 25% over the original seg-tel configuration.

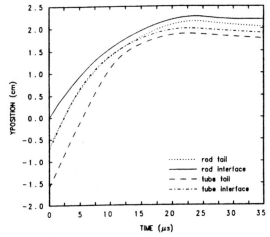

FIGURE 9. Penetration depth versus time.

CONCLUSIONS

The performance of segmented-telescopic penetrators was investigated in this study. Results from the simulations were used to suggest ways to improve the performance of multiple segment trains.

ACKNOWLEDGEMENTS

This work was supported by the Army Research Laboratory under contract No. DAAA21-93-C-0101.

REFERENCES

1. Orphal and R. R. Franzen, "Penetration mechanics and performance of segmented rods against metal targets", *Int. J. Impact Engng.*, **10**, 427-438 (1990).
2. Naz and H. F. Lehr, "The crater formation due to segmented rod penetrators", *Int. J. Impact Engng.*, **10**, 413-425 (1990).
3. Hohler and A. J. Stilp, "Penetration performance of segmented rods at different spacing – comparison with homogeneous rods at 2.5 - 3.5 km/s", *Proc. 12th Int. Symp. On Ballistics*, **3**, 178-187 (1990).
4. Anderson, Jr., J. D. Walker, and T. R. Sharron, "Investigations of seg-tel penetrators", *Int. J. Impact Engng.*, to appear.
5. McGlaun, S. L. Thompson, and M. G. Elrick, "CTH – a three-dimensional shock wave physics code", *Int. J. Impact Engng.*, **10**, 351-360 (1990).

CP429, *Shock Compression of Condensed Matter – 1997*
edited by Schmidt/Dandekar/Forbes
© 1998 The American Institute of Physics 1-56396-738-3/98/$15.00

TENSILE DAMAGE EFFECTS IN STEEL PLATE PERFORATION BY A TUNGSTEN ROD

Martin N. Raftenberg

U.S. Army Research Laboratory, Aberdeen Proving Ground, Maryland 21005-5066

The EPIC Lagrangian wavecode was used to study sensitivities of rolled homogeneous armor (RHA) target hole size and the length and speed of a tungsten heavy alloy (WHA) residual rod to RHA spall pressure. The target had a 50.8-mm thickness, and the rod had a 112.0-mm length, a 20.9-mm diameter, and a striking speed of 1.52 km/s. Corresponding to a spall pressure of −0.5 GPa, good agreement with experiment was obtained in terms of target hole size and residual length for two different RHA fits to the Johnson-Cook strength model. Adiabatic shear banding in RHA is proposed to be the mechanism by which spall pressure is locally changed from its measured value of −3.0 GPa.

INTRODUCTION

Raftenberg and Kennedy (1) presents results from an experiment in which a 50.8-mm-thick RHA plate was perforated by a WHA rod. The plate had a square face with a 457-mm edge length. Brinell hardness of the face was 302. The rod was a right circular cylinder of 112.0-mm length and 20.9-mm diameter. The WHA had a composition of 91W-6Ni-3Co and a Rockwell C hardness of 41.6. The impact speed was 1.52 km/s. A 7075-T651 aluminum drag flare was attached to the rear of the rod. In (2), Raftenberg used the EPIC wavecode (3) to model this experiment, with the focus on representations of the drag flare, the RHA yield function, the WHA shock Hugoniot, and the Grüneisen parameter for both RHA and WHA. All results were reasonably successful in predicting residual length, but they underpredicted the final perforation-hole diameter. The focus now is on damage modeling in RHA.

MODELING PROCEDURE

The plate was represented as a circular disc with a 500-mm radius to reduce the problem to axial symmetry. Slideline erosion (4) was based on an

equivalent plastic strain of 1.5. Meshes were composed of three-node triangular elements. RHA and WHA meshes were composed entirely of elements arranged in sets of four-element quadrilaterals (rectangles prior to deformation). The "original" meshes contained 5 such quadrilaterals across the rod's radius and 25 across the plate's thickness. The "refined" meshes had 10 and 50, respectively.

Undamaged dilatational behavior of RHA, WHA, and Al were governed by the Mie-Grüneisen equation of state. The shock Hugoniot curve served as the reference curve. Hugoniot pressure, p_H, was related to compression, μ, by the cubic

$$p_H(\mu) = K_1\mu + K_2\mu^2 + K_3\mu^3. \tag{1}$$

Grüneisen parameter, Γ, was related to compression by

$$\Gamma(\mu) = \frac{\Gamma_o}{1+\mu}. \tag{2}$$

Undamaged deviatoric behavior of RHA, WHA, and Al were governed by von Mises plasticity. Elastic shear modulus, G, was treated as a material constant. Flow stress, Y, was computed from the Johnson-Cook strength model (5),

$$Y = \left[C_1 + C_2 \left(\varepsilon^p \right)^{C_4} \right] \left[1 + C_3 \ln \left(\frac{\dot{\varepsilon}^p}{1.0 \, \text{s}^{-1}} \right) \right] \times$$

$$\left[1 - \left(\frac{\Theta - \Theta_r}{\Theta_m - \Theta_r} \right)^{C_5} \right]. \tag{3}$$

ε^p is equivalent plastic strain, $\dot{\varepsilon}^p$ is equivalent plastic strain rate, and Θ, Θ_r, Θ_m are current, room, and melting temperatures, respectively. Hardening was assumed to be isotropic. Damage behavior of WHA and Al was represented with the Johnson-Cook fracture model (6). All material constants were assigned values based on the literature. These values are listed in Table 4 of (2).

Two sets of values for C_1, C_2, ... C_5 were assigned to RHA. One set, called "JC83," first appeared in (5) for 4340 steel. The second set, called "GCWL94," appeared in (7) for RHA. Both sets are listed in Table 1. The JC83 database included strains up to 1.0 and strain rates up to 650 s^{-1}. The GCWL94 database included strains up to 0.2 and strain rates as high as 7,000 s^{-1}.

TENSILE DAMAGE MODEL FOR RHA

Figure 14 in (8) shows an RHA specimen cut from a plate perforated by a shaped charge jet. The figure shows voids that have coalesced to form a fracture surface. The spheroidal nature of the voids suggests that their growth under tension had been in a ductile mode. A simple model to introduce into a finite element effects of such voids consists of two parts: a failure criterion and an algorithm for post-failure stress reduction.

TABLE 1. RHA Strength Constants

Material Parameter	JC83	GCWL94
C_1 (GPa)	0.7922	0.900
C_2 (GPa)	0.5095	1.305
C_3	0.014	0.0575
C_4	0.26	0.90
C_5	1.03	1.075

The tensile failure criterion is based on a negative-valued material constant, the "spall pressure," p_{fail}. Material in a given finite element is deemed to fail instantaneously at the first time step in which the condition

$$p \leq p_{fail} < 0 \tag{4}$$

is met. Here, p is current pressure within the element. Considering only current pressure neglects the influence of load duration on damage.

In the literature, axial spall stress, σ_{fail}, has been determined for various materials by means of plate impact experiments. Bless reported an axial spall stress of -6.0 GPa for RHA specimens (9). Since in plate impact tests the condition of uniaxial strain is closely satisfied, spall pressure, p_{fail}, and σ_{fail} are related by

$$\sigma_{fail} + p_{fail} = \frac{4Y}{3}, \tag{5}$$

as shown in (2).

Table 2 evaluates p_{fail} of RHA for several reasonable values of Y.

The post-failure algorithm, applied in the element at the time step of failure detection and at every time step thereafter, is that the element is not allowed to support any hydrostatic tension or deviatoric stress. The element's ability to support hydrostatic compression is undiminished. This algorithm treats the effects of tensile damage as occurring instantaneously. This is a good approximation if the time scales associated with these effects are much smaller than other time scales in the problem.

TABLE 2. p_{fail} vs. Y for RHA

Y (GPa)	p_{fail} (GPa)
1.0	-4.7
1.5	-4.0
2.0	-3.3
2.5	-2.7

RESULTS AND DISCUSSION

The problem was run with the original and the refined meshes until 1 ms after initial impact.

Figures 1, 2, and 3 plot as functions of p_{fail} results at 1 ms for the perforation-hole diameter averaged through the thickness, $\bar{D}$, the residual length, L_r, and the residual speed, v_r, respectively. For each value of p_{fail}, four solutions were obtained: JC83 strength constants and the original mesh, JC83 and refined mesh, GCWL94 and original mesh, and GCWL94 and refined mesh. Also included are experimental results reported in (1) for $\bar{D}$ and L_r.

Throughout the range $-4.0 \leq p_{fail} \leq -1.5$ GPa, the solutions show little dependence on p_{fail}. With either the original or the refined mesh, JC83 constants led to larger values of $\bar{D}, L_r$, and v_r than did the GCWL94 constants. This was to be expected, since the GCWL94 constants consistently lead to a larger flow stress for given levels of ε^p, $\dot{\varepsilon}^p, \Theta$. It is also true that throughout this range of p_{fail}, with either the JC83 or the GCWL94 constants, the refined mesh consistently led to larger values of $\bar{D}$, and smaller values of L_r than did the original mesh; no consistent trend was observed in v_r.

All computational results for $\bar{D}$ in this p_{fail} range are smaller than the experimental measurement. With

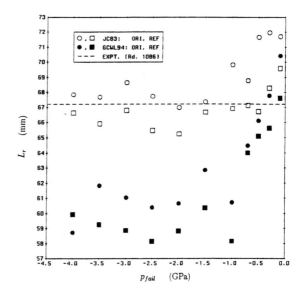

FIGURE 2. Residual length vs. spall pressure.

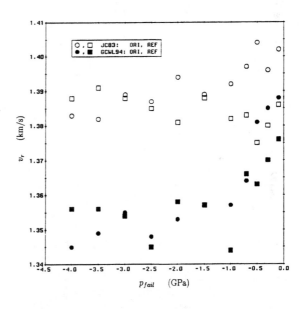

FIGURE 3. Residual speed vs. spall pressure.

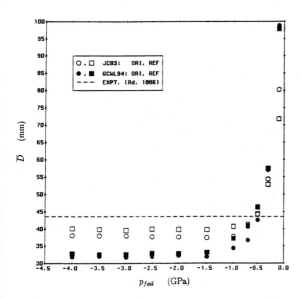

FIGURE 1. Through-thickness-averaged perforation hole diameter vs. spall pressure.

JC83, the original mesh results of about 38 mm are smaller by about 6 mm, and the original mesh results of about 40 mm are smaller by roughly 4 mm. It is not clear how much of the remaining discrepancy can be made up by further mesh refinement. Each refined-mesh problem required at least 48 CPU hours on a Silicon Graphics R8000 processor, and no further mesh refinement has been attempted to date. With JC83, original and refined mesh results for L_r mainly bounded the experimental measurement in this p_{fail} range. GCWL94 results for $\bar{D}$ and L_r with both sets of meshes were substantially smaller than experiment.

At a p_{fail} of about -1.0 GPa, the tensile failure model began to exert a strong influence on the solutions. $\bar{D}$ began its dramatic increase for both sets of strength constants. With both sets of meshes and both sets of RHA strength constants, $\bar{D}$ crossed the experimental result at a p_{fail} of about -0.5 GPa. Moreover, with GCWL94 constants, L_r and v_r also increased beginning at a p_{fail} of about -1.0 GPa. L_r achieved close agreement with experiment at a p_{fail} of -0.5 GPa.

These results raise the question of whether a physical mechanism exists to locally change spall pressure by an order of magnitude from -3.0 to -0.5 GPa. Ref. (8) contains evidence to implicate adiabatic shear banding. Figures 7 and 16 show that cracks emanating from the perforation hole boundary run along shear bands. Material surrounding the shear bands is not fractured. Figures 17 and 20 show that recovered RHA fragments are bounded by shear bands, suggesting that shear banding is key to the fragmentation process.

The following scenario is proposed: In both the target and the rod, a shock wave travels from the site of initial impact, leaving compressed material in its wake. Shear banding occurs in the compressed RHA in the vicinity of the hole boundary, where deviatoric stresses are largest. When the shocks reach the rear surface of the target and rod, they reflect to create tensile waves. The tensile wave in RHA travels back toward the entrance face. That in WHA travels toward the target interface and enters the RHA. As regions of the hole boundary become unloaded, release waves form. Interactions between the reflected tensile waves and these release waves lead to regions of large hydrostatic tension. This tensile stress, while insufficient to fail the original RHA material (i.e., less negative than -3.0 GPa), encounters weaker material within shear bands. That material within transformed shear bands is weaker in tension than the original material is a reasonable hypothesis. If the shear band is still hot, the material

within is presumably more ductile and more susceptible to void growth. If the material within has had time to quench, it is harder than the original RHA (8) and is more susceptible to brittle fracture.

REFERENCES

1. Raftenberg, M. N., and Kennedy, E. W., *Proceedings of the 15th International Symposium on Ballistics*, edited by Mayseless, M. and Bodner, S. R., vol. 1, pp. 315–321, 1995.
2. Raftenberg, M. N., *Structures Under Extreme Loading Conditions - 1996*, New York: ASME, 1996, pp. 205–220.
3. Johnson, G. R., *Journal of Applied Mechanics* **98**, 439–444 (1976).
4. Johnson, G. R., and Stryk, R. A., *Communications in Numerical Methods in Engineering* **12**, 885–896 (1996).
5. Johnson, G. R., and Cook, W. H., *Proceedings of the 7th International Symposium on Ballistics*, The Hague, pp. 541–547, 1983.
6. Johnson, G. R., and Cook, W. H., *Engineering Fracture Mechanics* **21**, 31–48 (1995).
7. Gray, G. T., III, Chen, S. R., Wright, W., and Lopez, M. F., Constitutive Equations for Annealed Metals Under Compression at High Strain Rates and High Temperatures, LA-12669-MS, Los Alamos National Laboratory, 1994.
8. Krause, C. D., and Raftenberg, M. N., Metallographic Observations of Rolled Homogeneous Armor Specimens from Plates Perforated by Shaped Charge Jets, ARL-MR-68, U.S. Army Research Laboratory, Aberdeen Proving Ground, Maryland, 1993.
9. Bless, S. J., Spall Criteria for Several Metals, AFWAL-TR-81-4040, Air Force Wright Aeronautical Laboratories, Wright-Patterson Air Force Base, Ohio, 1981.

CP429, *Shock Compression of Condensed Matter – 1997*
edited by Schmidt/Dandekar/Forbes
1998 The American Institute of Physics 1-56396-738-3/98/$15.00

MODELING OF IN-SITU BALLISTIC MEASUREMENTS USING THE RAJENDRAN-GROVE AND JOHNSON-HOLMQUIST CERAMIC MODELS

A.M. Rajendran and K.P. Walsh

Weapons and Materials Research Directorate, Army Research Laboratory, Aberdeen Proving Ground MD 21005

This paper presents results from numerical simulations of a ballistic experiment in which a long-rod tungsten projectile strikes and penetrates a layered target consisting of two silicon carbide (SiC) ceramic tiles backed by an RHA steel block. In the experiment, the penetrating rod's tail-end velocity history was recorded by a specially designed Doppler radar system, the stress-time histories at the tile/tile and tile/steel interfaces were recorded using embedded stress gauges, and the rod's residual depth of penetration (DOP) into the steel block was measured. The 1995 version of the EPIC finite element code was employed to simulate this experiment. The results obtained from the numerical simulations using the Rajendran-Grove (RG) and Johnson-Holmquist (JH2) ceramic models were compared with the experimental data. This paper further discusses the abilities of the RG and JH2 models to reproduce the measured data from ballistic as well as plate impact experiments.

INTRODUCTION

To validate ceramic constitutive models, it is necessary to obtain data from both plate impact and ballistic tests. Plate impact tests are performed to investigate material response to uniaxial (strain) loading. In ballistic tests, however, materials experience complex multiaxial loading conditions, thus posing a serious challenge to damage model validation.

This paper presents results from numerical simulations of a long tungsten rod striking a layered target consisting of two ceramic tiles backed by a steel block. Chang [1] reported measuring the rod's tail-end velocity history during penetration using a specially designed Doppler radar with high resolution in both velocity and time. In addition, stress-time histories at the tile/tile and tile/steel interfaces were measured using embedded stress gauges. The 1995 version of the EPIC finite element code [2] was employed in the simulations.

In the simulations, we used two distinctly different models to describe the high strain rate and shock response of the ceramic material: 1) the Rajendran-Grove (RG) model [3], and 2) the Johnson-Holmquist (JH2) model [4]. Each model has about seven to eight parameters. Using

previously determined RG and JH2 model constants, we simulated Chang's ballistic experiment and compared the model predictions with the recorded data.

RAJENDRAN-GROVE MODEL

In the RG model, the total strain is decomposed into elastic strain (e_{ij}^e) and plastic strain (e_{ij}^p). The elastic strain consists of the elastic strain of the intact matrix material and the strain due to crack opening/sliding. Plastic flow is assumed to occur in the ceramic only under compressive loading when the applied pressure exceeds the pressure at the Hugoniot elastic limit (HEL). The stress-strain equations for the microcracked material are given by, $s_{ij} = M_{ijkl}\, e_{kl}^e$. The components of the stiffness tensor M are described by Rajendran [5]. The pressure is calculated through the Mie-Gruneisen equation of state. In the ceramic model, microcrack damage is measured in terms of a dimensionless microcrack density γ which is defined as, $\gamma = N_o^*\, a^3$. N_o^* is the average number of microflaws per unit volume and a, the maximum microcrack size, is treated as an internal state variable. The initial values of these two parameters are material model constants. Microcracks are assumed to extend when the stress state satisfies a generalized Griffith criterion. This criterion requires the fracture toughness K_{IC} as well as a dynamic friction coefficient μ as model constants.

During microcrack extension, the crack density (γ) increases and stress relaxation occurs. The crack extension (damage evolution) law is derived from a fracture mechanics based relationship for a single crack propagating under dynamic loading conditions: $\dot{a} = n_1 C_R [1 - (G_c / G_1)^{n_2}]$, where C_R is the Rayleigh wave speed, G_c is the critical strain energy release rate for microcrack growth, and G_1 is the applied strain energy release rate. The n_1 is assumed to be one for tension and n_2 is set equal to one for both tension and compression. The ceramic is assumed to pulverize under compression when γ reaches a critical value of 0.75. The strength of the post-fractured ceramic is described by: $Y = \alpha + \beta P$, where α and β are model constants. In summary, there are five constants that have to be determined to describe the microcracking of the intact ceramic and two to describe the strength of the pulverized ceramic. The corresponding model constants for silicon carbide that were determined from planar plate impact data [6,7] and DOP tests [8] are: $a = 10$ μm, $N_o^* = 10^9$ /m^3, $K_{IC} = 4$ MPa $\sqrt{m}$, $\mu = 0.7$, $n_1 = 0.1$ (compression), $\alpha = 0$, and $\beta = 0.6$.

JOHNSON-HOLMQUIST MODEL

The JH2 model is a phenomenological model based on an elastic-viscoplastic approach. The strength of the ceramic is assumed to vary with pressure, strain rate and tensile strength. Basically there are two surfaces: one corresponds to $D = 0$, and the other to $D = 1$.

Once damage initiates, the flow surface reduces to an intermediate state as indicated by the dashed line in Fig. 1. As in the Johnson-Cook fracture model for metals, the damage (D) increases with effective plastic strain and is defined as,

$$D = \sum \Delta \varepsilon^p / \varepsilon_f^p \qquad (1)$$

where $\Delta \varepsilon^p$ is the incremental effective plastic strain and ε_f^p is the effective plastic strain at fracture, given by,

$$\varepsilon_f^p = D_1 (P^* + T^*)^{D_2} \qquad (2)$$

D_1 and D_2 are damage model parameters, and P^* and T^* are dimensionless pressure and tensile strength, respectively. The dimensionless intact strength (σ_i^*) and fracture strength (σ_f^*) relationships are given by,

$$\sigma_i^* = A(P^* + T^*)^N (1 + C \ln \dot{\varepsilon}^*) \qquad (3)$$

$$\sigma_f^* = B P^{*M} (1 + C \ln \dot{\varepsilon}^*) \qquad (4)$$

where $\dot{\varepsilon}^*$ is the dimensionless strain rate and A, B, M, N, and C are model constants. Johnson and Holmquist described their model in detail in Ref. 4. The JH2 model constants for silicon carbide are: $A = 0.93$, $N = 0.64$, $B = 0.17$, $M = 0.7$, $C = 0.004$, $D_1 = 0.003$, $D_2 = 2.8$, and $\sigma_{HEL} = 15$ GPa.

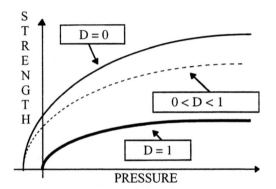

FIGURE 1. Strength variation with respect to pressure for JH2 model.

PLATE IMPACT SIMULATIONS

In the RG model constant calibration scheme, we considered two plate impact tests; one was conducted at an impact velocity below the Hugoniot elastic limit (HEL) [6] and the other was performed at an impact velocity above the HEL [7]. The model constant calibration scheme used the particle velocity vs. time data (VISAR signals) measured at the back surface of the target. The initial crack size was adjusted to match the low velocity spall signal.

As can be seen from Fig. 2, both the JH2 and RG models reasonably matched the low velocity experimental (below HEL) data. The JH2 model missed the spall signal arrival time, but predicted the rebound velocity level very well.

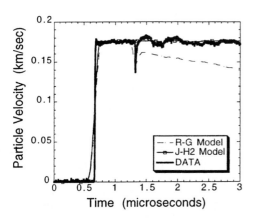

FIGURE 2. Comparisons of RG and JH2 model simulation results with below-HEL plate impact data.

The RG model matched the spall time very well; however the peak amplitude of the spall signal was much lower than the level in the experiment. Further adjustment of the model parameters did not increase this level. The best fit was obtained with $a_0 = 10$ microns, and $N_o^* = 10^9 / m^3$.

Using the calibrated model constants, the high velocity (above HEL) experiment was simulated. Figure 3 shows the corresponding comparison between the models and data. Again, both models matched the data very well. However, the RG model reproduced the elastic unloading and the spall portions of the data extremely well. In the EPIC simulations, we used the Mie-Gruneisen equation of state (EOS). The same EOS parameters for silicon carbide were used in all the simulations.

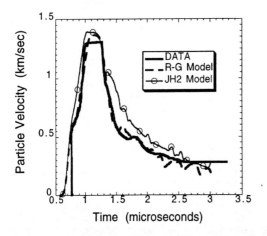

FIGURE 3. Comparisons of RG and JH2 model simulation results with above-HEL plate impact data.

SIMULATIONS OF AN IN-SITU BALLISTIC EXPERIMENT

The experimental technique for tail-end velocity history measurement involved a 94-GHz Doppler Radar Hyper-Velocimeter system. The configuration of the target is shown in Fig. 4. The first manganin stress gauge, G1, was embedded between the first and the second SiC ceramic tiles and the second gauge, G2, was between the second tile and the RHA block. Lateral confinement of the ceramic tiles was provided by a 25.4 mm wide steel frame with the same thickness as the two ceramic tiles. The gap between the tiles and the frame was ~1 mm and it was filled with epoxy. The first gauge had premature failure. Therefore, the only stress measurement data available was from the bottom gauge (G2). The depth of penetration of the projectile into the steel back plate was also measured from the recovered target.

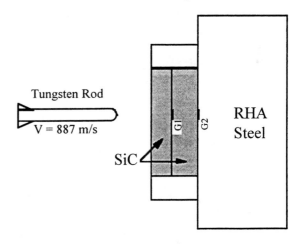

FIGURE 4. Ballistic experiment configuration.

We simulated the experiment reported by Chang [1] with both the RG and JH2 models, using the previously calibrated model constants for silicon carbide. The computed tail velocity history was obtained for each model. Figure 5 compares the smoothed RG model prediction with the experimental measurement; as the figure shows, the model matched the data extremely well. Since the data was very noisy beyond 110 microseconds, it was not possible to validate the model-predicted tail velocity profile beyond this time. However, the simulated depth of penetration (DOP) matched the experimental measurement.

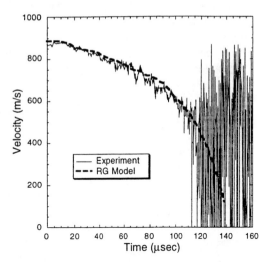

FIGURE 5. Comparison of RG model prediction with measured tail velocity history.

In Fig. 6, the smoothed computed tail velocity histories using the RG and JH2 models are compared. Interestingly, these two models with different approaches (microcracking damage versus effective-plastic-strain-based damage) produced very similar results.

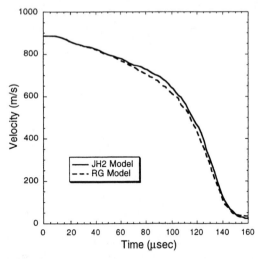

FIGURE 6. Comparison of the calculated tail velocity histories from the RG and JH2 model simulations.

In general, the velocity history of the tail end of the long rod is insensitive to the penetration rate of the front end of the long rod. Consequently, matching the tail velocity measurement with the model prediction is not a "stringent test" to verify a

model's capabilities. Neither model accurately predicted the measured stress histories at the gauge locations; both models predicted lower stress levels than the measurements. However, the RG model matched the data slightly better than the JH2 model.

SUMMARY

Though the RG and JH2 ceramic models are based on two different approaches, both models reproduced the measured velocity vs. time data from two one-dimensional plate impact experiments and a multiaxial ballistic experiment reasonably well. Both models assume plastic flow above the HEL in compression, and employ the same Mohr-Coulomb type strength model for the pulverized ceramic. The main difference between the models is the absence of a microcracking-based damage description in the JH2 model. Therefore, the effect of this difference on the computational results may be ascertained from simulations of uniaxial stress tests in which the deformation process is dominated by tensile cracking rather than plastic flow under compression. In summary, further validation is required to evaluate the capabilities of these two models.

ACKNOWLEDGMENTS

The authors greatly appreciate the funding support of Dr. James Thompson of TARDEC, Warren, MI.

REFERENCES

1. Chang, A.L., ARL-TR-1187, Army Research Laboratory, August 1996.
2. Johnson, G.R., Stryk, R.A., Petersen, E.H., Holmquist, T.J., Schonhardt, J.A., and Burns, C.R., Alliant Techsystems Inc., Minnesota ,1994.
3. Rajendran, A.M., and Grove, D.J., International Journal of Impact Engineering, Vol. 18, No. 6, 1996, pp. 611-631.
4. Johnson, G.R., and Holmquist, T.J., *Shock Compression of Condensed Matter-1993*, edited by, Schmidt, et al., AIP Press, NY, 1993, pp. 733-736.
5. Rajendran, A.M., International Journal of Impact Engineering, Vol. 15, No. 16, 1994, pp. 749-768.
6. Bartkowski, P. and Dandekar D.P., *Shock Compression of Condensed Matter-1995*, edited by S.C. Schmidt, and W.C. Tao, AIP Press, New York, 1993, pp. 535-538.
7. Grady, D.E., and Moody, R.L., SAND96-0551, Sandia National Laboratory, NM 87185, 1996.
8. Rajendran, A.M., and Grove, D.J., *Shock Compression of Condensed Matter-1995*, edited by S.C. Schmidt, and W.C. Tao, AIP Press, New York, 1993, pp. 539-542.

CP429, *Shock Compression of Condensed Matter – 1997*
edited by Schmidt/Dandekar/Forbes
© 1998 The American Institute of Physics 1-56396-738-3/98/$15.00

PENETRATION OF TUNGSTEN-ALLOY RODS INTO COMPOSITE CERAMIC TARGETS: EXPERIMENTS AND 2-D SIMULATIONS

Z. Rosenberg[†], E. Dekel[†], V. Hohler[‡], A.J. Stilp[‡] and K. Weber[‡]

[†]*RAFAEL, P.O. Box 2250, Haifa, Israel*
[‡]*EMI, D-79104, Freiburg, Germany*

A series of terminal ballistics experiments, with scaled tungsten-alloy penetrators, was performed on composite targets consisting of ceramic tiles glued to thick steel backing plates. Tiles of silicon-carbide, aluminum nitride, titanium-diboride and boron-carbide were 20-80 mm thick, and impact velocity was 1.7 km/s. 2-D numerical simulations, using the PISCES code, were performed in order to simulate these shots. It is shown that a simplified version of the Johnson-Holmquist failure model can account for the penetration depths of the rods but is not enough to capture the effect of lateral release waves on these penetrations.

INTRODUCTION

The ballistic resistance of strong ceramics, against long-rod penetrators, has been the subject of intense research in the past decade ((1)-(3), for example). Several ceramics, notably alumina, boron and silicon carbides, titanium diboride and aluminum nitride, have ballistic efficiencies of 2-3 relative to armor steels, against long-rod penetrators. Some of the issues concerning this efficiency are still open, such as its velocity dependence or its changes with tile thickness. Also, optimal tile dimensions are, apparently, configuration dependent as is the thickness and type of cover plate which is used in such experiments.

The purpose of the present paper is to highlight the role of the lateral dimensions of ceramic tiles on their performance. We do this by presenting experimental depth of penetration (DOP) measurements on several ceramic/steel composite targets, for which tile thickness varied between 20-80 mm. The lateral dimensions of these tiles ranged between 75-150 mm (square tiles). 2-D simulations were performed in order to account for the measured DOP values and, in particular, to simulate the lateral effects, as observed in these shots.

EXPERIMENTS

All experiments were performed with scaled tungsten alloy rods 5.8 mm in diameter, 72.5 mm long. The tiles were bonded to hard steel backing (compressive strength of 1.5 GPa) into which residual penetration was measured. Static properties as obtained from the manufacturers of these tiles are given in Table 1 below. The penetrator's density was 17.6 g/cc and its yield strength 1.2 GPa.

The tiles were laterally supported with a soft steel frame (20 mm thick) which, although not entirely rigid, did supply some support to the tiles. More details on the experiments can be found in (4).

All the experimental results, in terms of penetration depth versus tile thickness (multiplied by tile density, in order to present aerial densities), are given in Figure 1. As is clearly seen, all data are bounded by the two straight line which represent ballistic efficiencies of 1.7 and 2.7, as shown in the figure. Moreover, a close examination reveals that

TABLE 1. Ceramic Properties

Material	Density (g/cc)	Bending Strength (MPa)	Compressive Strength (MPa)	Young Modulus (GPa)
SiC	3.15	430	2500	430
AlN	3.23	350	2100	310
TiB$_2$	4.45	380	3000	570
B4C	2.5	400	2800	450

most of the data points show a gradual decrease in efficiency with increasing tile thickness, which is clearly a lateral release effect. In an earlier study (5), we have seen that for alumina tiles (AD85 manufactured by Coors) a ratio of at least 5, between tile diameter and thickness, is needed to avoid lateral effects on tile performance. This is what we also see in Figure 1 since only tiles which are 20 mm thick achieve the high efficiency expected of these ceramics (2.7). Thicker tiles, having diameter to thickness ratio of less than 5, have lower efficiencies (down to 1.7). Note that for the thickset tiles (60 mm) the diameter-to-thickness ratio is close to 1. An interesting phenomenon concerns TiB$_2$ tiles with lateral dimensions of 100×150 mm. These tiles show higher ballistic efficiency than the square ones, although they were cut from the same source. One can speculate that lateral release waves are less severe in these tiles because they do not interact simultaneously at the tile's center area, as do release waves in square tiles.

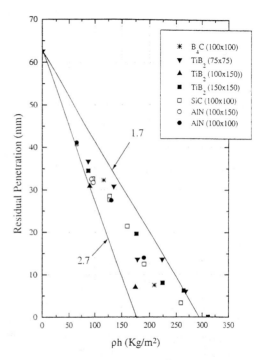

FIGURE 1. Summary of the DOP results.

SIMULATIONS

Two-dimensional simulations were performed with the Eulerian processor of the PISCES 2DELK code described in (6). The failure model we used for the ceramic tiles is also described in (6). Essentially, this is a simplified version of the Johnson-Holmquist model (7) which includes a single free parameter. This parameter (f) controls the strength properties of the comminuted material with the aid of the strength of the intact material. A single experiment is chosen to calibrate f (<1) which is then used to simulate the results for the rest of the experiments. Figure 2 shows schematically the basic features of our simplified model with the single parameter f.

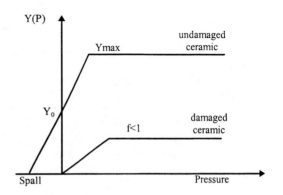

FIGURE 2. Our Simplified Version of the Johnson-Holmquist Model.

Since the only lateral-free experiments were those with the 20 mm tiles, we chose these experiments to obtain the values of f for each ceramic tile. The dynamic strength parameters of the intact materials are obtained from the spall strength, compressive strength and the Hugoniot elastic limit. It is interesting to note that the spall strengths are very close to the static bending stresses 0.3-0.5 GPa, while the static compressive strengths (uniaxial stress) are lower than those inferred from the HEL of these materials. We tried various values for the dynamic yield strengths in the range of 4-8 GPa in order to obtain the closest agreement between experiment and simulation, as far as DOP for the thicker tiles is concerned. The fact that lateral release waves influenced the residual penetration depths for these tiles put strong constraints on the values of the three strength parameters: spall strength, compressive strength and maximum strength (from HEL), for the intact material. Table 2 lists several runs for SiC where we found f values for the thin tile (20.6 mm) and checked it for the thicker ones (40.4 and 60.4 mm). For these tiles the experiments gave residual penetrations of 28.5 mm (40.4 mm tile) and 12.5 mm (60.4 mm tile).

As is clearly seen, none of the runs in this series reproduced the experimental results for the thicker tiles. Moreover, a simple calculation shows that the ballistic efficiencies of the simulated thicker tiles are the same as those of the thin (20.4 mm) tile. Thus, the lateral effects did not become apparent with these simulations, although we varied each of the three strength parameters by a factor of about 2. This means that our simple version of the Johnson-Holmquist model cannot account for lateral effects and one needs to introduce at least one other parameter in order to achieve this goal. The original model (see (5)) includes 13 parameters covering all the aspects of strength (strain rate sensitivity, hardening, etc.). The nature of this additional parameter is not self-evident and more work is needed to find out which parameter will produce the lateral release effects.

As a compromise, for the time being, we decided to approach this problem in a somewhat different way. Instead of calibrating the damage parameter (f) through the thinnest tile, we chose the 40.4 mm thick SiC tiles for this task. Using σ_{spall} = 0.5 GPa, Y_0 = 4 GPa and Y_{max} = 8 GPa we obtain a value of f = 0.35, which resulted in the following residual depths: 44.4 mm for the 20.6 mm tile and 12.6 mm for the 60.4 mm tile. These values are in fair agreement with the experimental ones (40.8 and 12.6 mm for the 20.6 and 60.4 mm tiles, respectively). Thus, a reasonable agreement can be obtained if one chooses a medium thickness tile for the calibration process. Moreover, the computed ballistic efficiencies of the tiles in these simulations decrease with tile thickness from 1.95 to 1.7, showing some effect of lateral release. Hence, to a first order this procedure results in quite reasonable values for the ballistic efficiency of these tiles.

A similar situation is obtained for the TiB_2 ceramic for which we chose the 39.6 mm thick tile as the calibration experiment. Table 3 shows the comparison between experimental results and two sets of simulations with σ_{spall} = 0.35 GPa and Y_0 = 3 GPa.

The agreement is not as good as with the SiC tiles, and, in particular, the lateral release effects are not manifested in the simulations, as they are in the experiments. Thus, the calculated ballistic efficiency is constant also for the TiB_2 ceramic.

We would like to note here that recent experiments (8), (9), on similar ceramics, with scaled rods impacting at 1.5-4.6 km/s, also result in ballistic efficiencies in the range of 2.7-1.5 (decreasing with impact velocity). It may turn out

TABLE 2. Several Attempts to Simulate SiC Results

Run #	Strength (GPa)				Simulation Results for DOP of 40.4 and 60.4 mm Tiles	
	Spall	Compressive	Maximum	f		
1	0.375	2.5	5.0	0.62	22.3	4.6
2	0.750	2.5	7.4	0.61	20.8	1.0
3	0.750	2.5	5.0	0.76	23.5	5.0
4	0.500	4.0	8.0	0.45	21.6	3.6

TABLE 3. Simulation Results for TiB$_2$ Tiles

Test #	Tile Thickness	Experimental DOP	Simulation DOP (mm)	
	(mm)	(mm)	Y_{max} = 5 GPa, f = 0.45	Y_{max} = 8 GPa, f = 0.31
2005	19.8	34.5	42.2	42.2
2006	39.6	19.6	20.4	19.4
2008	50.3	8.0	9.0	7.7
2009	59.4	6.2	1.6	0.0

that this is also a lateral release phenomenon rather than a real velocity dependent effect, as stated by the authors of (8) and (9).

SUMMARY

A series of terminal ballistic experiments with scaled tungsten-alloy rods was performed in order to evaluate the ballistic efficiency of various ceramics against long-rod penetrators. The tiles included silicon and boron carbide, titanium diboride and aluminum nitride. Impact velocity was 1.7 km/s and tile thicknesses ranged between 20-80 mm. Using depth of penetration measurements, it was found that the ballistic efficiency of the various tiles decreases with increasing tile thickness from 2.7 to 1.7. This is very likely a lateral release effect resulting from the relatively small lateral dimensions of the tiles. 2-D simulations using our simplified version of the Johnson-Holmquist model (with a single free parameter) were not able to simulate this effect in a satisfactory manner. It is evident that at least one more parameter has to be added to our model in order to fully account for the results in the present study.

REFERENCES

1. Mescall, J.G., and Tracey, C., in Proc. 1986 Army Science Conf., West Point, N.Y., 17-19 June, 1986, Vol. III, p. 41.
2. Rosenberg, Z., and Tsaliah, J., *Int. J. Impact Eng.*, 9, 247, (1990).
3. Anderson, C.E., and Morris, B.L., *Int. J. Impact Eng.*, 12, 167, (1993).
4. Hohler, V., Stilp, A.J., and Weber, K., *Int. J. Impact Eng.*, 17, 409, (1995).
5. Rosenberg, Z., and Yeshurun, Y., (unpublished results).
6. Rosenberg, Z., Dekel, E., Yeshurun, Y., and Bar-On, E., *Int. J. Impact Eng.*, 17, 697 (1995).
7. Johnson, G.R., and Holmquist, T., in "Shock waves and high strain-rate phenomena in materials", M.A. Meyer, L.E. Murr and K.P. Staudhammer (eds.), Marcel Dekker, (1995), p. 1075.
8. Orphal, D.L. and Franzen, R.R., *Int. J. Impact Eng.*, 19, 1 (1997).
9. Orphal, D.L. et al., *Int. J. Impact Eng.*, 19, 15 (1997).

CP429, *Shock Compression of Condensed Matter – 1997*
edited by Schmidt/Dandekar/Forbes
1998 The American Institute of Physics 1-56396-738-3/98/$15.00

RESPONSE OF NONLINEAR ELASTIC SOLIDS TO OBLIQUE PLATE IMPACT

Mike Scheidler

U.S. Army Research Laboratory, APG, Maryland 21005-5066

We give a theoretical analysis of the nonlinear elastic response in the interior of the target in an oblique plate impact test, prior to interactions with any reflected waves. Approximate relations are derived between the changes in stress, strain, particle velocity, and wave speeds across the shear wave by neglecting shear strain terms of order six in the internal energy function. The results are valid for any finite (elastic) longitudinal strain ahead of the shear wave. They apply to isotropic materials and to appropriately aligned transversely isotropic and orthotropic materials.

INTRODUCTION

In an oblique plate impact test a flyer plate, inclined relative to the axis of the projectile, impacts a parallel target plate, which we assume is at rest and stress free. Let (X, Y, Z) denote the Cartesian coordinates of a material point in this natural reference state, with the X-axis normal to the target face. The particle velocity $\mathbf{v}_F$ imparted to the face of the target has nonzero components both normal and parallel to the target face. The coordinate axes are oriented so that $\mathbf{v}_F = (u_F, v_F, 0)$, with the normal (X) component $u_F > 0$ and the transverse (Y) component $v_F > 0$. The impact is assumed weak enough that the target response is elastic. We assume the target is homogeneous, and restrict attention to points in the interior and times prior to the arrival of waves reflected from free surfaces or material interfaces. Then the motion is independent of the Y and Z coordinates. However, for a general anisotropic material, the motion may have nonzero components in the Z-direction at points not on the impact face. For isotropic materials as well as some anisotropic materials (appropriately aligned with the coordinate axes), there is no motion in the Z-direction. This is the case considered here.

PRELIMINARIES

If (x, y, z) is the position at time t of the material point initially at (X, Y, Z), then $x = X + d_1(X, t)$, $y = Y + d_2(X, t)$, $z = Z$. The normal and transverse components of the particle velocity are $u = \partial x/\partial t$ and $v = \partial y/\partial t$. Let $\boldsymbol{P}$ and $\boldsymbol{T}$ denote the 1st Piola-Kirchhoff and Cauchy stress tensors. Then $\sigma \equiv -P_{XX} = -T_{XX}$ and $\varepsilon \equiv 1 - \partial x/\partial X$ are the normal or longitudinal components of stress and strain, taken positive in compression; and $\tau \equiv -P_{YX} = -T_{YX}$ and $\gamma \equiv -\partial y/\partial X$ are the shear stress and shear strain. The signs have been chosen so that ε, γ, σ, τ, as well as u and v, should be nonnegative for the impact problem considered here. Since ε and γ are the only nonzero strain components, the internal energy e per unit mass depends only on ε, γ, and the entropy per unit mass s. Then

$$e = \hat{e}(\varepsilon, \gamma, s), \quad \sigma = \rho_0 \frac{\partial e}{\partial \varepsilon}, \quad \tau = \rho_0 \frac{\partial e}{\partial \gamma}, \quad (1)$$

where ρ_0 is the density in the natural state. Heat conduction is neglected in the analysis, so that response is isentropic except for entropy jumps across shocks. The equations of motion are

$$\rho_0 \frac{\partial u}{\partial t} + \frac{\partial \sigma}{\partial X} = 0, \quad \rho_0 \frac{\partial v}{\partial t} + \frac{\partial \tau}{\partial X} = 0. \quad (2)$$

The strain rate and velocity gradient satisfy

$$\frac{\partial \varepsilon}{\partial t} + \frac{\partial u}{\partial X} = 0, \qquad \frac{\partial \gamma}{\partial t} + \frac{\partial v}{\partial X} = 0. \qquad (3)$$

Smooth solutions are governed by (2) and (3), with σ and τ given by (1) and s constant; this is a quasilinear system in the four unknowns u, v, ε, γ. The (Lagrangian) characteristic wavespeeds are the roots $\pm U$ and $\pm V$ of

$$2\rho_0 \begin{Bmatrix} U^2 \\ V^2 \end{Bmatrix} = \frac{\partial \sigma}{\partial \varepsilon} + \frac{\partial \tau}{\partial \gamma} \pm \sqrt{\left(\frac{\partial \sigma}{\partial \varepsilon} - \frac{\partial \tau}{\partial \gamma}\right)^2 + 4\left(\frac{\partial \tau}{\partial \varepsilon}\right)^2}, \qquad (4)$$

where the fast wave speeds $\pm U$ correspond to the + sign and the slow wave speeds $\pm V$ to the − sign in (4). The constitutive inequalities

$$\frac{\partial \sigma}{\partial \varepsilon} > \frac{\partial \tau}{\partial \gamma} > 0, \qquad \frac{\partial \sigma}{\partial \varepsilon}\frac{\partial \tau}{\partial \gamma} > \left(\frac{\partial \tau}{\partial \varepsilon}\right)^2 \qquad (5)$$

guarantee that there are four real distinct characteristic wave speeds (i.e., $U > V > 0$), and that the fast waves are primarily longitudinal. Waves with negative speed would be generated by reflections and hence are not considered here. The right inequality in (5) is equivalent to

$$\frac{\partial^2 e}{\partial \varepsilon^2}\frac{\partial^2 e}{\partial \gamma^2} > \left(\frac{\partial^2 e}{\partial \varepsilon \partial \gamma}\right)^2, \qquad (6)$$

which implies that e is a strictly convex function of ε and γ.

The one-dimensional plane waves for this oblique impact problem are centered simple waves and/or centered shocks. For nonlinear isotropic elastic solids, some of the earlier papers on this problem are Bland (1), Davison (2), and Abou-Sayed & Clifton (3). Isotropy implies that e is an even function of γ for fixed ε and s, and that the fast wave is purely longitudinal, i.e., there are no changes in v, γ, or τ. The fast wave brings the material to an intermediate state of uniform uniaxial strain. The slow wave propagates into this uniaxially strained material. The slow wave is primarily a transverse or shear wave. However, nonlinear elastic effects cause 2nd order changes in u, ε, and σ across the shear wave. The case where both waves are simple waves is illustrated in Fig. 1.

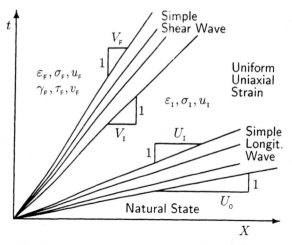

FIGURE 1. Simple wave solutions for oblique plate impact. Quanties in th4e intermediate state (ahead of the shear wave) and final state (behind the shear wave) were denoted by I and F subscriptes, respectively.

For anisotropic materials there may be motion in the Z direction. Even if there is no motion in the Z direction, the fast wave need not be purely longitudinal. However, if the material is orthotropic relative to the X, Y, Z axes, or transversely isotropic about the X or Z axis, then it can be shown that there is no motion in the Z direction, and that e is an even function of γ. Furthermore, it can be shown that this last property implies all of the results discussed previously for the isotropic case. With these restrictions, most of the analysis for isotropic materials in references (1)–(3) can be extended to the anisotropic case. Since e is an even function of γ for fixed ε and s,

$$e \overset{6}{\approx} \hat{f}(\varepsilon, s) + \tfrac{1}{2}\hat{g}(\varepsilon, s)\gamma^2 + \tfrac{1}{4!}\hat{h}(\varepsilon, s)\gamma^4. \qquad (7)$$

Here and below, $a \overset{n}{\approx} b$ means $a = b$ to within an error of order γ^n. Note that no approximations are made for the dependence of e on ε or s. In particular, the results in this paper are valid for arbitrary finite (elastic) longitudinal strains ahead of the shear wave. We study only the shear wave in this paper, and consider only the case where it is a simple wave (as opposed to a shock). s_0 and s_I denote the entropy in the natural and intermediate states . If the longitudinal wave is simple, $s_I = s_0$; if it is a shock, $s_I > s_0$.

922

SIMPLE SHEAR WAVES

Set $f(\varepsilon) = \hat{f}(\varepsilon, s_I)$, $g(\varepsilon) = \hat{g}(\varepsilon, s_I)$, and $h(\varepsilon) = \hat{h}(\varepsilon, s_I)$. Since the deformation in the simple shear wave is isentropic, (7), (1), and (4) imply

$$e \stackrel{6}{\approx} f(\varepsilon) + \tfrac{1}{2} g(\varepsilon) \gamma^2 + \tfrac{1}{4!} h(\varepsilon) \gamma^4, \qquad (8)$$

$$\frac{\sigma}{\rho_0} = \frac{\partial e}{\partial \varepsilon} \stackrel{4}{\approx} f'(\varepsilon) + \tfrac{1}{2} g'(\varepsilon) \gamma^2, \qquad (9)$$

$$\frac{\tau}{\rho_0} = \frac{\partial e}{\partial \gamma} \stackrel{5}{\approx} g(\varepsilon) \gamma + \tfrac{1}{6} h(\varepsilon) \gamma^3, \qquad (10)$$

$$V^2 \stackrel{4}{\approx} g(\varepsilon) - \left(\frac{(g')^2}{f'' - g} - \frac{h}{2} \right)(\varepsilon) \gamma^2, \qquad (11)$$

where a prime denotes differentiation with respect to ε. Note that σ and V are even functions of γ for fixed ε, while τ is an odd function of γ.

In the intermediate uniaxially strained state ahead of the shear wave, $\gamma = \tau = v = 0$, and by (9)–(11), (4), and (5)$_1$, $\sigma_I = \rho_0 f'(\varepsilon_I)$ and

$$U_I^2 = f''(\varepsilon_I) > g(\varepsilon_I) = V_I^2. \qquad (12)$$

The rate of change of the speed V_I of the shear wave front (which is a transverse acceleration wave) with respect to the uniaxial strain ahead of the wave is

$$V_I' \equiv \frac{dV_I}{d\varepsilon_I} = \frac{g'(\varepsilon_I)}{2\sqrt{g(\varepsilon_I)}}. \qquad (13)$$

In the centered simple shear wave, $V = X/t$; and ε, σ, u, γ, τ & v are constant along the straight-line characteristics (i.e., functions of X/t only). The characteristic form of the quasilinear system yields the differential relations

$$\frac{du}{dv} = \frac{d\sigma}{d\tau} = \frac{d\varepsilon}{d\gamma} = \frac{-\dfrac{\partial \tau}{\partial \varepsilon}}{\dfrac{\partial \sigma}{\partial \varepsilon} - \rho_0 V^2}, \qquad (14)$$

$$\frac{dv}{d\gamma} = V, \quad \frac{d\tau}{dv} = \rho_0 V, \quad \frac{d\tau}{d\gamma} = \rho_0 V^2. \qquad (15)$$

γ, τ & v are strictly increasing with passage of the wave, and any two of these variables is an odd function of the other. ε, σ, u & V are even functions of any one of the variables γ, τ or v; hence the total derivatives in (14) are 0 at the wavefront,

where $\gamma = \tau = v = 0$. Indeed, $\partial \sigma / \partial \varepsilon - \rho_0 V^2 > 0$ by (5)$_1$, while by (10), $\partial \tau / \partial \varepsilon = 0$ at the wavefront. If $\partial \tau / \partial \varepsilon \neq 0$ for $\gamma \neq 0$, then (15) holds with v, γ & τ replaced by u, ε & σ, respectively. Inequality (5)$_3$ places no restriction on the sign of $\partial \tau / \partial \varepsilon$; if it changes sign within the shear wave, then by (14), u, σ & ε do not vary monotonically through the wave.

Next, we quantify the above statements by deriving approximate relations which hold throughout the shear wave. On using (9)–(13) in (14)$_3$ and noting that $\varepsilon = \varepsilon_I$ when $\gamma = 0$, we obtain

$$\varepsilon \stackrel{4}{\approx} \varepsilon_I - \tfrac{1}{2} Q(\varepsilon_I) \gamma^2, \qquad (16)$$

$$Q(\varepsilon_I) \equiv \frac{g'}{f'' - g}(\varepsilon_I) = \frac{2 V_I V_I'}{U_I^2 - V_I^2}. \qquad (17)$$

On substituting (16) into (11) and using (12), (13), (17), and $V = V_I$ when $\gamma = 0$, we obtain

$$V^2 \stackrel{4}{\approx} V_I^2 - R(\varepsilon_I) \gamma^2, \quad V \stackrel{4}{\approx} V_I - \frac{R(\varepsilon_I)}{2 V_I} \gamma^2, \quad (18)$$

$$\begin{aligned} 2 R(\varepsilon_I) &\equiv \frac{3 (g')^2}{f'' - g}(\varepsilon_I) - h(\varepsilon_I) \\ &= \frac{12 V_I^2}{U_I^2 - V_I^2} (V_I')^2 - h(\varepsilon_I). \end{aligned} \qquad (19)$$

From (15)$_1$, (18)$_2$, and $v = 0$ when $\gamma = 0$, we get

$$v \stackrel{5}{\approx} V_I \gamma - \frac{R(\varepsilon_I)}{6 V_I} \gamma^3, \qquad (20)$$

which may be inverted to give

$$\gamma \stackrel{5}{\approx} \frac{v}{V_I} + \frac{R(\varepsilon_I)}{6 V_I^2} \left(\frac{v}{V_I} \right)^3. \qquad (21)$$

Then (18)$_2$, (21), and (15)$_2$ imply

$$V \stackrel{4}{\approx} V_I - \frac{R(\varepsilon_I)}{2 V_I} \left(\frac{v}{V_I} \right)^2, \qquad (22)$$

$$\frac{\tau}{\rho_0} \stackrel{5}{\approx} V_I v - \frac{R(\varepsilon_I)}{6 V_I^3} v^3. \qquad (23)$$

Similar arguments yield the relations

$$\varepsilon \stackrel{4}{\approx} \varepsilon_I - \tfrac{1}{2} Q(\varepsilon_I) (v/V_I)^2, \qquad (24)$$

$$\sigma \stackrel{4}{\approx} \sigma_I - \tfrac{1}{2} Q(\varepsilon_I) \rho_0 v^2, \qquad (25)$$

$$u \stackrel{4}{\approx} u_I - \tfrac{1}{2} Q(\varepsilon_I) v^2 / V_I. \qquad (26)$$

923

Dropping the v^3 terms in (21) and (23) and eliminating $Q(\varepsilon_\mathrm{I})$ from (24)–(26) yields

$$\gamma \overset{3}{\approx} v/V_\mathrm{I}, \qquad \varepsilon - \varepsilon_\mathrm{I} \overset{4}{\approx} (u - u_\mathrm{I})/V_\mathrm{I}, \quad (27)$$

$$\tau \overset{3}{\approx} \rho_0 V_\mathrm{I} v, \qquad \sigma - \sigma_\mathrm{I} \overset{4}{\approx} \rho_0 V_\mathrm{I}(u - u_\mathrm{I}). \quad (28)$$

More accurate material-independent approximations for γ and τ can be obtained by solving (22) for $R(\varepsilon_\mathrm{I})$ and substituting into (21) and (23):

$$\gamma \overset{5}{\approx} \left[1 + \tfrac{1}{3}(V_\mathrm{I}/V_\mathrm{F} - 1)(v/v_\mathrm{F})^2\right] v/V_\mathrm{I}, \quad (29)$$

$$\tau \overset{5}{\approx} \rho_0\left[V_\mathrm{I} - \tfrac{1}{3}(V_\mathrm{I} - V_\mathrm{F})(v/v_\mathrm{F})^2\right] v. \quad (30)$$

Since V must decrease with passage of the wave, (18) implies that $R(\varepsilon_\mathrm{I}) \geq 0$ is necessary for the centered shear wave to be simple (as opposed to a shock) for sufficiently weak impacts, while $R(\varepsilon_\mathrm{I}) > 0$ is sufficient. It can be shown that $R(\varepsilon_\mathrm{I}) < 0$ is sufficient for a shear shock. There are no restrictions on $Q(\varepsilon_\mathrm{I})$, however. By (17) and (12), we see that $Q(\varepsilon_\mathrm{I})$ and V_I' have the same sign. Hence, (24)–(26) imply that if $Q(\varepsilon_\mathrm{I})$ or V_I' is negative [resp. positive], then ε, σ, u increase [resp. decrease] across the shear wave for sufficiently weak impacts; if $Q(\varepsilon_\mathrm{I})$ or V_I' is zero, then ε, σ, u are constant to within an error of order γ_F^4.

Solving $(18)_2$ for γ in terms of V shows that γ has a square root singularity at $V = V_\mathrm{I}$, i.e., at the wavefront; likewise so do τ and v. Since $V = X/t$, the strain rate, stress rate, and particle acceleration are infinite at the wavefront, as noted by Abou-Sayed & Clifton (3) for isotropic materials.

At the rear of the shear wave (i.e., at the final state), (29) and (30) reduce to

$$\gamma_\mathrm{F} \overset{5}{\approx} \left(\frac{2}{3} + \frac{1}{3}\frac{V_\mathrm{I}}{V_\mathrm{F}}\right)\frac{v_\mathrm{F}}{V_\mathrm{I}} \overset{5}{\approx} (1 + \tfrac{1}{3}\Delta_\mathrm{s})\frac{v_\mathrm{F}}{V_\mathrm{I}}, \quad (31)$$

$$\tau_\mathrm{F} \overset{5}{\approx} \rho_0(\tfrac{2}{3}V_\mathrm{I} + \tfrac{1}{3}V_\mathrm{F})v_\mathrm{F} \overset{5}{\approx} (1 - \tfrac{1}{3}\Delta_\mathrm{s})\rho_0 V_\mathrm{I} v_\mathrm{F}, \quad (32)$$

where $\Delta_\mathrm{s} \equiv V_\mathrm{I}/V_\mathrm{F} - 1 = V_\mathrm{I} t_\mathrm{R}(X)/X$, and $t_\mathrm{R}(X) = X/V_\mathrm{F} - X/V_\mathrm{I}$ is the rise time of the shear wave at X. By (22),

$$\frac{t_\mathrm{R}(X)}{X} = \frac{1}{V_\mathrm{F}} - \frac{1}{V_\mathrm{I}} \overset{4}{\approx} \frac{R(\varepsilon_\mathrm{I})}{2V_\mathrm{I}^3}\left(\frac{v_\mathrm{F}}{V_\mathrm{I}}\right)^2$$

$$\approx \frac{1}{V_0}\left[\frac{6(V_0')^2}{U_0^2 - V_0^2} - \frac{h(0)}{2V_0^2}\right]\left(\frac{v_\mathrm{F}}{V_0}\right)^2. \quad (33)$$

The top formula in (33) includes dependence of the rise time on the uniaxial strain ε_I ahead of the shear wave. The bottom formula neglects this dependence; it is obtained by setting $\varepsilon_\mathrm{I} = 0$ and using (19), and contains errors of order γ_F^4 and $\varepsilon_\mathrm{I}\gamma_\mathrm{F}^2$. Here U_0 and V_0 are the longitudinal and shear wave speeds in the natural state, and V_0' denotes the value of V_I' at $\varepsilon_\mathrm{I} = 0$. Thus $U_0 \& V_0$, V_0', and $h(0)$ are 2nd, 3rd, and 4th order elastic constants, respectively. An analogous formula in Abou-Sayed & Clifton (3), eqn. (46), is off by a factor of 3/2 (the result of neglecting changes in ε across the shear wave), and missing the $h(0)$ term (the result of neglecting 4th order strain terms in the internal energy).

From (22) and (19) we obtain an approximation for the 4th order coefficient h in (8) evaluated at the uniaxial strain ε_I ahead of the shear wave:

$$h(\varepsilon_\mathrm{I}) \overset{2}{\approx} \frac{12V_\mathrm{I}^2}{U_\mathrm{I}^2 - V_\mathrm{I}^2}(V_\mathrm{I}')^2 - 4V_\mathrm{I}^3\frac{V_\mathrm{I} - V_\mathrm{F}}{v_\mathrm{F}^2}. \quad (34)$$

V_I' can be estimated by differentiating the $V_\mathrm{I}(\varepsilon_\mathrm{I})$ curve obtained from a series of oblique plate impact tests, or by solving (26) and (17):

$$V_\mathrm{I}' \overset{2}{\approx} -\left(\frac{u_\mathrm{F} - u_\mathrm{I}}{v_\mathrm{F}^2}\right)(U_\mathrm{I}^2 - V_\mathrm{I}^2). \quad (35)$$

When this is substituted into (34), we obtain

$$h(\varepsilon_\mathrm{I}) \overset{2}{\approx} 12\left(\frac{u_\mathrm{F} - u_\mathrm{I}}{v_\mathrm{F}^2}\right)^2(U_\mathrm{I}^2 - V_\mathrm{I}^2)V_\mathrm{I}^2$$

$$- 4V_\mathrm{I}^3\frac{V_\mathrm{I} - V_\mathrm{F}}{v_\mathrm{F}^2}. \quad (36)$$

The right-hand sides of (35) and (36) involve particle velocities and wave speeds only. These could be measured in a single test, at least if the longitudinal wave is simple. If the longitudinal wave is a shock with speed $\mathbb{U}$, then U_I^2 can be approximated by $U_\mathrm{I}^2 \approx 2\mathbb{U}^2 - U_0^2$ to within an error of order ε_I^2, or calculated from $(12)_1$ if f is known.

REFERENCES

1. Bland, D. R., *Z. Angew. Math. Phys.* **16**, 752–769 (1965).
2. Davison, L., *J. Mech. Phys. Solids* **14**, 249–270 (1966).
3. Abou-Sayed, A. S., and Clifton, R. J., *J. Appl. Phys.* **47**, 1762–1770 (1976).

CP429, *Shock Compression of Condensed Matter – 1997*
edited by Schmidt/Dandekar/Forbes
© 1998 The American Institute of Physics 1-56396-738-3/98/$15.00

INVESTIGATION OF YAWED IMPACT INTO A FINITE TARGET

C. E. Anderson, Jr.[1], S. J. Bless[2], T. R. Sharron[1], S. Satapathy[2], M. J. Normandia[2]

[1]*Southwest Research Institute, Materials and Structures Division, P. O. Drawer 28510, San Antonio, TX 78228-0510 (USA);* [2]*Institute for Advanced Technology, The University of Texas at Austin, 4030-2 W. Braker Lane, Austin, TX 78759-5329 (USA)*

It is well known that impact inclination is detrimental to penetration performance for long-rod projectiles into semi-infinite targets. The drop-off in performance, measured as the depth of penetration normalized by the initial projectile length, is generally attributed to interference of the tail of the projectile with the side walls of the penetration cavity. However, little has been done to study the effect of impact inclination on the performance of long rods penetrating finite targets. A reverse ballistic experiment and three-dimensional numerical simulations are used to examine the effect of impact inclination against an oblique plate. In particular, the interaction of the projectile with the plate, to include asymmetric enlargement of the penetration channel and side loading of the projectile, are examined in the simulations, and compared to experimental results.

INTRODUCTION

Penetration mechanics of yawed long rod projectiles is a subject of great importance in terminal ballistics. In this paper we treat the coupling of yaw and obliquity effects. The subject has received recent attention in low velocity rigid body penetration [1,2]. However, we examine the case of high velocity eroding rod projectiles, which to our knowledge, has received little attention. The projectiles under consideration are tungsten heavy alloy. Targets are RHA (armor steel) plates which are easily perforated by the rods. We wish to understand the effect of pitch (defined as angle of attack in the plane normal to the plates) on penetration erosion and deflection.

Experiments are conducted in the reverse impact mode. The experimental technique is described in Ref. [3]. Simulations are conducted with CTH 3-D [4], using the material properties described in Ref. [5]. Define the impact obliquity by

$$\theta = \sin^{-1}\left(sgn\,(\boldsymbol{V} \times \hat{\boldsymbol{n}}) \frac{|\boldsymbol{V} \times \hat{\boldsymbol{n}}|}{|\boldsymbol{V}|} \right) \qquad (1)$$

where $\hat{\boldsymbol{n}}$ is the unit outward target normal, and the pitch by

$$\alpha = \sin^{-1}\left[sgn\,(\boldsymbol{V} \times \boldsymbol{L}) \frac{|\boldsymbol{V} \times \boldsymbol{L}|}{|\boldsymbol{V}||\boldsymbol{L}|} \right]. \qquad (2)$$

where $\boldsymbol{L}$ is the rod length expressed as a vector, pointing from tail to tip. Using this notation, positive pitch implies a rod pitched away from the target, while negative pitch means a rod pitched into the target. Clearly for $\theta = 0$, impact results will not depend on the sign of α. We consider cases of fixed large obliquity, and we seek the effect of changing the sign of pitch, as depicted in Fig. 1.

Cagliostro, *et al.* [6], and Johnson and Cook [7] performed computational studies to examine the effects of pitch up and pitch down conditions on the penetration of a finite-thick oblique plate. In Refs. [6-7], they examined residual velocity and residual length of the rod, the trajectory deflection in degrees, and the angular rotation rate of the center of mass for the initial impact pitch conditions of +10°, 0°, and -10°,

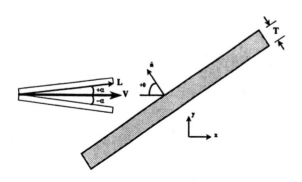

FIGURE 1. Schematic of Geometry Showing Plate Obliquity and Impact Conditions of Positive and Negative Pitch.

for a plate obliquity of 65°. However, they conducted their computations in a transformed coordinate system, where the pitch angles were 10.3°, 0°, and -9.3°, with the respective obliquity angles of 54.9°, 65°, and 73.5°. Neither set of investigators transformed their results back to the original coordinate system.

EXPERIMENTS

We conducted two experiments that address the subject of pitch-obliquity coupling. The condition of the tests are given in Table 1. The experiments were essentially identical except for the pitch reversal and a slight change in penetrator dimensions. It can be seen that there is a profound coupling of yaw and obliquity effects, even though the total angle is the same.

The case of positive pitch has about 1.4 times the erosion of negative pitch case. The transverse velocity (the velocity that is normal to the original velocity vector) profiles after plate perforation were measured from X-ray shadowgraphs, and are plotted in Fig. 2; the eroded nose position is defined as 0 mm. In both experiments, the rod was broken into four pieces.

NUMERICAL SIMULATIONS

Numerical simulations provide a way to conduct a systematic parametric study. Test No. 265—the results are given in Table 2—was selected to be one point of the parametric study. The case of fixed obliquity of 53.1°, impact inclinations of +6.9°, 0°. and -6.9°, is studied. The impact velocity was held constant at 2.38 km/s. For the initial computational study, the length-to-diameter ratio (L/D) of the projectile was chosen as 20 to minimize computer requirements.

The penetration velocities versus time for the L/D 20 projectiles are shown in Fig. 3. It is observed that all rods achieve the "semi-infinite" steady-state penetration velocity of approximately 1.3 km/s, but that the duration at this velocity depends upon the pitch angle. The positive pitch rod takes significantly longer to perforate the plate, and as a consequence, the rod erodes more. This is reflected in the ΔL values in Table 2. The plate is very "ductile" in the computations (failure, such as plugging, is not accurately modeled) so it is expected that considerably more projectile erosion will occur in the simulation than the experiment, as shown in Table 2. Severe deformation of the projectile for the cases with nonzero pitch leads to rather large uncertainties in the length of eroded rod.

TABLE 1. Positive/Negative Pitch Tests

Test	Pitch (deg)	Obliquity (deg)	Velocity (km/s)	D (mm)	T (mm)	L_o (mm)	ΔL (mm)
259	-9.7	70.4	2.46	2.19	4.75	79	9
238	+10.2	70.4	2.39	2.00	4.75	72	13

926

TABLE 2. CTH Computations and Experiment

Run No.	Pitch (deg)	Obliquity (deg)	Velocity (km/s)	D (mm)	T (mm)	L_0 (mm)	ΔL (mm)
Test-265	+ 6.9	53.1	2.37	2.49	4.75	88.6	7.0 ± 0.5
R22	+ 6.9	53.1	2.38	2.49	4.75	49.8	17 ± 4
R21	0	53.1	2.38	2.49	4.75	49.8	8.4 ± 0.2
R24	- 6.9	53.1	2.38	2.49	4.75	49.8	10.6 ± 2
R23	- 6.9	53.1	2.38	2.49	4.75	88.6	10.7 ± 2

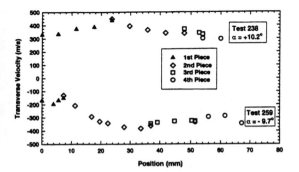

FIGURE 2. Measured Transverse Velocities

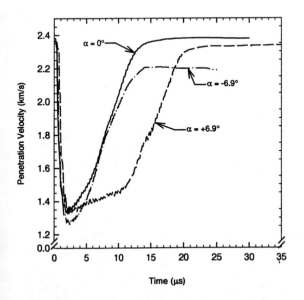

FIGURE 3. Penetration Velocities

MODEL FOR TRANSVERSE VELOCITY

A simple model has been developed to estimate the transverse velocity imparted to the rod as a result of side loading from the plate. The transverse velocity histories of 5 tracer points are shown in Fig. 4. It is observed that the transverse velocity imparted to the rod occurs over a very short time interval, associated with the time that the element of the rod is in contact with the plate. The model assumes a nominally steady-state condition in which the front of the rod has perforated the plate and the remainder of the rod is dragging the side of the plate due to projectile pitch. At the relative velocities (for the side load), the weaker of the two materials deforms, which in the cases here is the RHA plate. Thus, a slot is cut into the plate, and a reaction force is imparted to the rod. Newton's second law gives:

$$\rho_p \frac{\pi D^2}{4} \frac{T}{\cos(\theta+\alpha)} \frac{|\Delta v|}{\Delta t} = \sigma_t \frac{DT}{\cos(\theta+\alpha)} \quad (3)$$

where σ_t is the flow stress of the armor plate, and Δv is the change in the transverse velocity. (It is assumed that the rod is "pushing" against the plate and cutting a slot in the plate; if $\alpha = 0$, then the lateral force on the rod is zero.) The section of the rod is in contact with the plate for approximately

$$\Delta t = \frac{T}{V\cos(\theta+\alpha)} \quad (4)$$

927

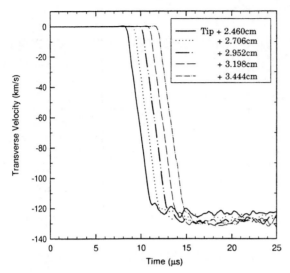

FIGURE 4. Transverse Velocity Histories for Selected Tracer Points ($\alpha = -6.9°$)

The Δt calculated from Eqn. (4) for simulation R24 is 2.88 μs, which is within 4% of the intervals of rapid acceleration shown in Fig. 4. Similarly, for simulation R22, Δt is calculated to be 3.99 μs. This value is approximately 25% less than calculated from the computations, but the 3.99 μs time interval captures 90% of the change in the transverse velocity (there is a decrease in the acceleration, particularly as the rod approaches a steady-state transverse velocity). Therefore, solving for $|\Delta v|$ gives:

$$|\Delta v| = \frac{4\,\sigma_t T}{\pi\,\rho_p\,DV\cos(\theta+\alpha)} \qquad (5)$$

Inserting the appropriate material and geometric values (using 1.7 GPa for σ_t) gives 142 m/s for $\alpha = -6.9°$, and 197 m/s for $\alpha = +6.9°$. These values are in excellent agreement with the "steady-state" values from the computations (Fig. 5). The model does not do as well for the two experiments in Table 1; whether this is due to rod breakage or other deficiencies in the model is unclear at this time.

It was decided to confirm that the transverse velocity does indeed reach and maintain a steady-state value. Simulation R24 was repeated, but with L/D=36. The results are shown in Fig. 5, confirming the assumption that the rod achieves a steady-state transverse velocity. From Eqn. (5), it is seen that

thinner plates or a faster impact velocity will result in a lower "steady-state" transverse velocity.

The dependence of the transverse velocity as either (or both) plate obliquity and pitch angle is given by the derivative of Eqn. (5):

$$\frac{d(\Delta v)}{d(\theta+\alpha)} = \Delta v \tan(\theta+\alpha) \qquad (6)$$

Equation (6) suggests that the transverse velocity changes more rapidly (fixed θ) for increases in positive pitch than for a comparable increase in negative pitch. Thus, for fixed plate obliquity, positive pitch is a more adverse impact condition for the projectile in terms of erosion and transverse deflection of the rod. Similar derivatives show the sensitivity to plate thickness, and the product of rod diameter and impact velocity.

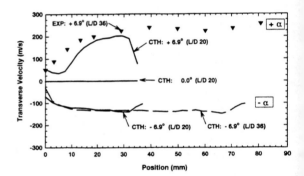

FIGURE 5. Transverse Velocities

REFERENCES

1. A. J. Piekutowski, M. J. Forrestal, K. L. Poorman, T. L. Warrent, *Int. J. Impact Engng.*, **18**, 877-887 (1996).
2. W. Goldsmith, E. Tam, D. Tomer, *Int. J. Impact Engng.*, **16**, 479-498 (1995).
3. M. Guillot, R. Subramanian, D. Berry, "A Method for Launching Oblique Plates from a Light Gas Gun," *47th Aeroballistic Range Assoc. Meeting*, Zurich, Switzerland, Oct. 14-17, 1996.
4. J. M. McGlaun, S. L. Thompson, and M. G. Elrick, *Int. J. Impact Engng*, **10**, 351-360 (1990).
5. C. E. Anderson, Jr., V. Hohler, J. D. Walker, and A. J. Stilp, *Int. J. Impact Engng.*, **16** (1), 1-18 (1995).
6. D. J. Cagliostro, D. A. Mandell, L. A. Schwalbe, T. F. Adams, and E. J. Chapyak, *Int. J. Impact Engng.*, **10**, 81-92 (1990).
7. G. R. Johnson and W. H. Cook, *Int. J. Impact Engng.*, **14**, 373-383 (1993).

CP429, *Shock Compression of Condensed Matter – 1997*
edited by Schmidt/Dandekar/Forbes
© 1998 The American Institute of Physics 1-56396-738-3/98/$15.00

A DISCREET IMPACT MODEL FOR EFFECT OF YAW ANGLE ON PENETRATION BY ROD PROJECTILES

Minhyung Lee and Stephan J. Bless

Institute for Advanced Technology, The University of Texas at Austin, Austin, TX 78759

When a long-rod penetrator impacts a target at a high yaw angle, interference of the penetrator with the sidewall of a crater is the mechanism for the degraded penetration performance. The yawed penetrator is described as a series of cylindrical elements. Using this assumption, the model considers the effective diameters of the penetrator whether or not it interacts with the crater wall and a revised crater profile is calculated with the effective diameters.

INTRODUCTION

The conventional interpretation of yaw effects is in terms of critical angle, the yaw angle at which the rear of the penetrator will just strike the edge of the entrance hole in the target. An approximate formula for critical yaw angle was given by Bjerke et al. (1),

$$\gamma_c = sin^{-1}(D_c - D_p / 2L), \qquad (1)$$

where D_c is the hole diameter and D_p is the penetrator diameter.

Figure 1, for example, shows data for yawed penetrators from our laboratory and other sources. It appears that the higher the L/D_p ratio, the more penetration may be degraded. Of particular importance in Figure 1 is the line labeled "Bjerke et al." because this is a widely used empirical relation (1). It is obvious that the empirical fit is not valid for all values of L/D_p and all yaw angles.

There have been several attempts to model yawed rod penetration. Bless et al. (2) presented an empirical formula for relatively short steel rods penetrating steel targets. Jamet et al. (3) adopted conventional rod penetration models by using an effective length in place of the actual length of the yawed rods, where the effective length is the projection of the projectile length onto the axis normal to the target.

Yaziv et al. (4) presented an analytical model for predicting the final penetration and compared results with experiments. They assumed that for small yaw angles, rod erosion controls penetration, but for larger yaw angles, the rod penetrates like a rigid body. Later, they presented an analytical model which predicts the penetration of yawed long-rods in the 0 to 90 range, with some assumptions (5). Bukharev and Zhukov (6) consider the forces exerted by the target on the penetrator directed antiparallel to the impact velocity and also arising from the sidewall interaction, directed normal to the rod axis. All of these models are only accurate over the specific ranges of material, L/D ratios and velocities for which they are calibrated.

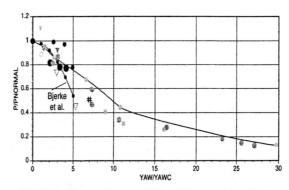

FIGURE 1. Decrease in penetration with yaw angle. Abscissa is ratio of yaw to critical yaw; ordinate is ratio of penetration with yaw to penetration without yaw. Data from references (4, 6, 7) for L/D_p between 10 and 30.

YAW MODEL

Penetration/Diameter

Consider a yawed long-rod of density ρ_p, diameter D_p, and length L impacting at normal incidence an infinite target of density ρ_t. The impact velocity of the rod is V and yaw angle is θ. The model of the cratering mechanism of yawed long-rods is displayed in Fig. 2. Hence, the yawed penetrator is assumed to be comprised of a series of continuous cylindrical disk elements (n). Each element of the penetrator hits a crater profile which is created by the impact of preceding elements. The incremental penetration depth achieved by the impact of each element can be determined by,

$$P_e = s(V, \rho_p, \rho_t, R_t, Y_p) L_e, \qquad (2)$$

where R_t is the target resistance, Y_p dynamic strength of projectile material, and L_e the length of the disk element given by,

$$L_e = L\cos\theta/n. \qquad (3)$$

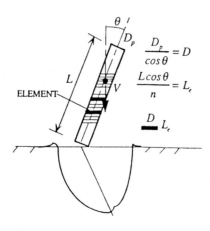

FIGURE 2. Impact Geometry.

From the modified Bernoulli equation (8) the constant penetration rate can be obtained by,

$$s = -\frac{\Delta P}{\Delta L} = \frac{U}{V - U} = \frac{-\varphi + \sqrt{\varphi + \alpha^2(1 - \varphi)}}{\varphi + \alpha^2}, \qquad (4)$$

where U is the penetration velocity, and

$$\varphi = \frac{R_t - Y_p}{\frac{1}{2}\rho_p V^2}, \quad \alpha = \sqrt{\rho_t/\rho_p}. \qquad (5)$$

Equation (3) is valid until a sidewall collision occurs after which some degradation in penetration may occur.

The diameter of the crater achieved by an impact of each element can be determined by,

$$D_e = m(V, \rho_p, \rho_t, R_t, Y_p)D, \qquad (6)$$

where

$$D = D_p/\cos\theta. \qquad (7)$$

From the momentum balance principle (9), m is given by,

$$m = \sqrt{Y_p/R_t + 2\rho_p(V - U)^2/R_t}. \qquad (8)$$

Equations (2) to (8) are sufficient to establish an algorithm to model the penetration. With no yaw they reduce to Tate's solution (8). When a sidewall collision occurs, there is a grazing impact on the penetration crater wall. Physically, the grazing effect cannot be explained by the same relation between the crater and projectile diameter as a normal impact. By replacing the element diameter with effective diameter, however, Eqn. (6) can still be used in the analysis. In other words, we assume that the extent of cratering is calculated with the effective diameter which is determined in the next section.

Interaction Algorithm

The geometry of a disk element impacting the crater wall and the way the crater is formed is displayed in Fig. 3. From the geometry, a condition that the element will collide with a sidewall is given by,

$$x_s + D/2 > a_e, \qquad (9)$$

where a_e is the lateral coordinate of the penetration crater profile, and x_s is the lateral coordinate of the center of the disk element.

930

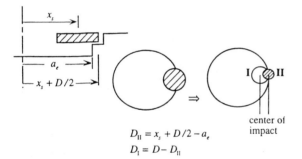

$$D_{II} = x_s + D/2 - a_e$$
$$D_I = D - D_{II}$$

FIGURE 3. Geometry of the interaction.

With no sidewall interaction, a whole disk element impacts a crater bottom and incremental penetration and diameter are calculated using Eqns. (4) and (8). With a sidewall collision, it is assumed that only the portion of the diameter of the element which overlaps the crater profile edge contributes to an enlargement of the crater. The assigned effective diameter, D_{II}, is given by,

$$D_{II} = x_s + D/2 - a_e \qquad (10)$$

and $D_I = D - D_{II}$. This reduced diameter is used in Eqn. (6) to calculate the increment of hole diameter throughout the depth of penetration.

Penetration Degradation

There are three possible cases:

CASE I
The first approximation is that there is no degradation in penetration in spite of the interaction with the crater wall.

CASE II
A second approximation is that once a collision has occurred then no further contribution to penetration takes place. This assumption may be valid for shaped charge jet analysis.

CASE III
For yawed long-rods at normal incidence into semi-infinite targets, the most reasonable approximation is that the degradation in penetration of the disk element is proportional to the portion of the diameter of the element which does not overlap the crater profile edge:

$$P_e = sL_eD_I/D. \qquad (11)$$

RESULTS

We consider the case of $L/D = 20$ tungsten rod ($Y_p = 2$ GPa, $\rho = 17.4$ g/cm^3) penetrating steel ($R_t = 5$ GPa) at $V = 2.6$ km/s. We find that the shapes of the craters change with yaw angle. The results are shown in Fig. 4. The craters are roughly triangular shaped, as discussed in (5). In Fig. 5, the penetration results (normalized by penetration with no yaw) are shown as a function of yaw angle normalized by a critical yaw, and the results for the three different cases are compared with the data (1). Implementation of the third assumption for the degradation in penetration brings the calculated results into good agreement with the empirical equation.

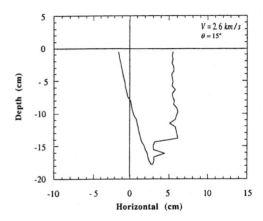

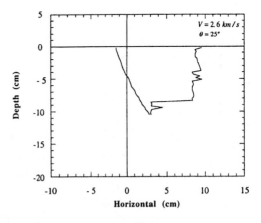

FIGURE 4. Sample crater profiles. Rod $L/D = 20$.

931

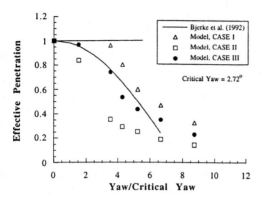

FIGURE 5. Normalized penetration versus normalized yaw. Rod *L/D* = 20.

Figure 6 shows the normalized penetration versus normalized yaw angle for *L/D* = 10 tungsten rods against RHA targets at 1.4 km/s. The experimental data from (2) are also shown in the figure. The present model overpredicts the normalized penetration at small yaw angle and underpredicts at large yaw angle. However, the observation evident in Fig. 1 that yaw degradation is less for shorter *L/D* rods is accounted for. In Fig. 7, the normalized penetration results are shown as a function of yaw angle normalized by a critical yaw for the impact of *L/D* = 10 at 2.6 km/s. It appears that the degradation in penetration falls off more slowly, which is also consistent with the data in Fig. 1. The accuracy of the model in the important range of $2 < \gamma/\gamma_c < 5$ is hard to assess because of considerable data scatter in that region.

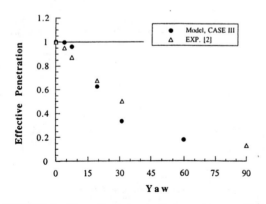

FIGURE 6. Normalized penetration versus yaw. Rod *L/D* = 10 at 1.4 km/s.

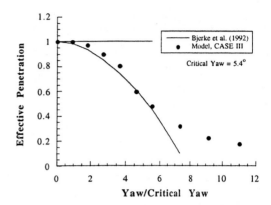

FIGURE 7. Normalized penetration versus normalized yaw. Rod *L/D* = 10 at 2.6 km/s.

ACKNOWLEDGMENT

This work was supported by the US Army Research Laboratory (ARL) under contract DAAA21-93-C-0101.

REFERENCES

[1] T. W. Bjerke, G. F. Silsby, D. R. Scheffler and R. M. Mudd, "Yawed Long-Rod Armor Penetration," *Int. J. Impact Eng.*, vol. 12, no. 2, pp. 281-292, 1992.

[2] S. J. Bless, J. P. Barber, R. S. Bertke and H. F. Swift, "Penetration Mechanics of Yawed Rods," *Int. J. Engng Sci.*, vol. 16, pp. 829-834, 1978.

[3] F. Jamet, M. Giraud and G. Weihrauch, "Experimental Simulations of the Terminal Ballistics of EFP," 10th International Symposium on Ballistics, San Diego, CA, 1987.

[4] D. Yaziv, Z. Rosenberg and J. P. Riegel, "Penetration Capability of Yawed Long-Rod Penetrators," 12th International Symposium on Ballistics, San Antonio, TX, 1990.

[5] D. Yaziv, J. D. Walker and J. P. Riegel, "Analytical Model of Yawed Penetration in the 0 to 90 Degrees Range," 13th International Symposium on Ballistics, Stockholm, Sweden, 1992.

[6] Y. I. Bukharev and V. I. Zhukov, "Model of the Penetration of a Metal Barrier by a Rod Projectile with an Angle of Attack," *Fizika Goreniya I Vzryva* 31, 104-109, 1995.

[7] Data for circular and non-circular rods with 10<L/D<30 available at the Institute for Advanced Technology, 1997.

[8] A. Tate, "Long-Rod Penetration Models - Part II. Extensions to the Hydrodynamic Theory of Penetration," *Int. J. Engng Sci.*, vol. 28, no. 9, pp. 599-612, 1986.

[9] M. Lee and S. Bless, "Cavity Dynamics for Long-Rod Penetration," 16th International Symposium on Ballistics, San Francisco, CA, September 23-28, 1996.

CP429, *Shock Compression of Condensed Matter – 1997*
edited by Schmidt/Dandekar/Forbes
© 1998 The American Institute of Physics 1-56396-738-3/98/$15.00

PENETRATION OF THICK TARGETS BY YAWED LONG RODS

Stephan Bless and Sikhanda Satapathy

Institute for Advanced Technology, The University of Texas at Austin
4030-2 W. Braker Lane, Austin, TX 78759

We present a model for the steady-state penetration phase of yawed long rods. The projectile-target interaction is comprised of axial deceleration and erosion due to frontal pressure, and lateral deceleration due to cavity expansion pressure. By solving the resulting ODEs through time-stepping, the evolution of the projectile/ crater shape is obtained.

BACKGROUND: STAGES OF YAWED PENETRATION

Impact of a yaw on eroding rod projectiles is a complex event that may consist of three distinct phases. In the *cavitation phase*, the impact crater grows laterally due to inertia. Penetration proceeds in the direction of the impact velocity. If the crater size is large enough, the entire rod can pass through without interacting with the side-wall of the crater, and the effect of initial yaw is minimal. Bjerke, et al. (1) presented a formula to calculate the maximum yaw (critical yaw) for this condition. The critical yaw is a function of maximum crater diameter, which in turn depends on the impact velocity and the projectile diameter. Bjerke, et al. assumed that the crater grows rapidly enough that the final size is attained almost instantaneously. Shinar, et al. (2) presented a time dependent crater evolution solution which may lead to more accurate results.

If the initial yaw is larger than the critical yaw, a *transient phase* starts when the rod comes in contact with the crater wall. In such a situation, penetration due to the frontal portion of the rod, which is already inside the crater at the instant of initial lateral contact, is not affected. The rear portion of the rod then continues to engage the crater wall. The penetration of the part of the penetrator that experiences lateral de-

flection continues as an extension of the *cratering phase*. In a companion paper we present a model for this stage (3).

If the rod is long enough, or diameter small enough, the effect of the initial crater will become insignificant and there will be a *steady phase*. The slot-cutting in the steady state phase is self similar. In this paper, we analyze this third phase, which yields the crater shape for long rod cases.

STEADY-STATE SLOT CUTTING

Consider a rod of length, L, diameter, d, impacting a semi-infinite target at an yaw angle of γ (see Fig.1). We distinguish between the instantaneous rod shape and the path followed by a point on the rod (particle path). Instantaneous particle paths define the shape of the rod. Two angles define the geometry of penetration at any instance of time: θ, the angle that the instantaneous velocity vector of the particle makes with the X-axis, and ϕ, the angle between the tangent to the instantaneous rod shape and X-axis at the point considered. Resolution of the velocity vector into normal (V_n) and parallel (V_p) components, and along the X-axis (V_x) and Y-axis (V_y) are shown in Figure 1.

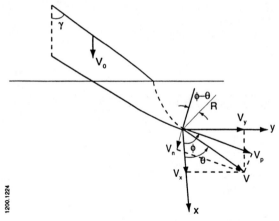

FIGURE 1. Slot cutting geometry.

The target-penetrator interaction is assumed to be due to: i) frontal resistance to penetration along the rod axis, and ii) lateral resistance to slot-cutting. We assume that Tate's model (4) applies in the axial direction, by virtue of which erosion and deceleration in the axial direction can be calculated. The lateral resistance is given by cavity expansion pressure, which acts in the normal direction to the rod. No erosion is envisioned in the lateral direction, which is true for cases with low lateral velocity.

From Tate's model, we obtain the following equations in the rod's axial direction:

$$\left.\begin{aligned}
\tfrac{1}{2}\rho_p(V_p - u)^2 + Y &= \tfrac{1}{2}\rho_t u^2 + R_t \\[4pt]
\dot{L} &= -(V_p - u) \\[4pt]
\dot{V}_p &= -\frac{Y}{\rho_p L}
\end{aligned}\right\} \quad (1)$$

where ρ_p is the projectile density, ρ_t is the target density, u is the tip velocity of the rod, Y is the projectile strength, R_t is the target resistance.

In the rod-normal direction, the cavity expansion speed is V_n. Forrestal, et al. (5) have derived a formula relating the pressure required to open a cylindrical cavity at a speed V_n as follows.

$$\sigma_c = \sigma_0 + b\rho_t V_n \quad (2)$$

where σ_0 is the static cylindrical cavity expansion solution. For RHA steel, with a flow stress of 1.2 GPa, σ_0 is 3.3 GPa (5). As an interim approximation, we use B = 1.837, which is the value Forrestal derives for aluminum (5).

We use this cavity expansion pressure as an approximation to the lateral stress present at the projectile-target interface. Considering an element of length ΔL, we express Newton's 2nd law for the element as follows.

$$d\Delta L\left[-\left(\sigma_0 + bV^2\right)\right] = \frac{\pi d^2 \rho_p}{4}\Delta L\left[V_n^2 - \frac{V^2}{R}\cos(\phi - \theta)\right] \quad (3)$$

The first term in the square bracket of the RHS represents the acceleration of the point in the target in contact with the rod element. The second term in the square bracket of the RHS is the normal component of the acceleration of the rod element with respect to the target point that is required to maintain the rod in a curvilinear motion. The angle, θ is given by,

$$\tan(\theta) = V_y/V_x \quad (4)$$

Differentiating eq. (4) with respect to time we obtain

$$\dot{\theta} = \frac{V_x \dot{V}_y - V_y \dot{V}_x}{V^2} \quad (5)$$

From geometry, V_x and V_y are related to V_n and V_p through

$$\left.\begin{aligned}
V_x &= V_p \cos\phi + V_n \sin\phi \\
V_y &= V_p \sin\phi - V_n \cos\phi
\end{aligned}\right\} \quad (6)$$

Inserting eq. (6) into eq. (5) we obtain

$$\dot{\theta} = \dot{\phi} - \cos(\phi - \theta)\frac{\dot{V}_n}{V} + \sin(\phi - \theta)\frac{\dot{V}_p}{V} \quad (7)$$

934

where we have used $V_p = V\cos(\phi-\theta)$ and $V_n = V\sin(\phi-\theta)$.

Solving eqs. (3) and (7) along with the relation, $R = V/\dot\theta$ we obtain

$$\dot\theta = \frac{\dot\phi + \left(\dfrac{4}{\pi\rho_p d}\right)\left(\dfrac{a+bV_n^2}{V}\right)\cos(\phi-\theta) - \dfrac{Y}{V\rho_p L}\sin(\phi-\theta)}{1+\cos^2(\phi-\theta)} \quad (8)$$

and

$$\dot V_n = \frac{V\dot\phi\cos(\phi-\theta) - \left(\dfrac{4}{\pi\rho_p d}\right)\left(a+bV_n^2\right) - \dfrac{Y}{2\rho_p L}\sin 2(\phi-\theta)}{1+\cos^2(\phi-\theta)} \quad (9)$$

At any given time step, the angle, ϕ and its derivative are given by,

$$\tan\phi = \frac{dy}{dx} \quad \text{and} \quad \dot\phi = \cos^2\phi \frac{dV_y}{dV_x} \quad (10)$$

Finally, to conserve mass, incompressibility in the rod material dictates that axial derivative of V_p be zero. Thus V_p is taken to be a constant for the whole rod at any given instance.

Eqs. (1), (8) and (9) form a set of ODEs which can be time integrated along with eq. (10) and the initial conditions to obtain a time dependent evolution of the rod shape. It should be noted that when the normal component of the particle velocity, V_n becomes zero, $\dot\theta$ is given by

$$\dot\theta = \left(\frac{4a}{\pi\rho_p d}\right)\frac{1}{V\cos(\phi-\theta)} \quad (11)$$

instead of eq. (8).

RESULTS

We considered a tungsten projectile of $L/D = 30$ interacting with an RHA target in steady-state. The other material parameters and the constants used are: $\rho_p = 17000$ Kg/m^3, $\rho_t = 7850$ Kg/m^3, $\rho_t = 5$ Gpa. Figure 2 shows the cases for $\phi_0 = 5^0$ and 20^0, with $V_0 = 2.5$ Km/s. Figure 3 shows the cases for $V_0 = 2$ Km/s and 3 Km/s, with $\phi_0 = 5^0$. It is important to note that since the first two phases, namely, the cavitation phase and the transient phase are ignored in this solution, the depth of penetration obtained is misleading. However, the solution should be used to obtain the evolution of the crater shape with time, in the steady-state slot-cutting process.

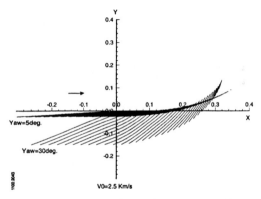

FIGURE 2. Effect of yaw angle.

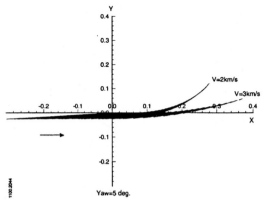

FIGURE 3. Effect of velocity.

935

DISCUSSION

The results show that the rod tends to become horizontal. Although at first this may be surprising, upon reflection it seems reasonable that an infinitely long rod (e.g. a traversing jet) cannot penetrate forever, but must asymptote to a finite depth.

Our solution frame-work is promising since other cases of practical importance can be modeled. An initial obliquity can be modeled by simply changing the initial condition for θ. For finite plates, the exit yaw and yaw rate of the rod may be obtained once suitable values for a and b are chosen and a break-out criteria is modeled. We plan to conduct penetration experiments with yawed long rods to obtain necessary experimental data for comparison with our analytical model.

ACKNOWLEDGMENT

This work was supported by the U.S. Army Research Laboratory (ARL) under contract DAAA21-93-C-0101.

REFERENCES

1. Bjerke, T. W., Silsby, G. F., Schefler, D. R., and Mudd, R. M., 1992, "Yawed Long-Rod Armor Penetration," *Int. J. Impact Engng.*, Vol. 12, pp. 281.
2. Shinar, G. I., Barnea, N., Ravid, M., and Hirsch, E., 1995, "An Analytical Model for the Cratering of Metallic Targets by Hypervelocity Long Rods," 15th Int. Symp. on Ballistics, Jerusalem, Israel.
3. Lee, M., S. Bless, "A Discreet Impact Model for Effect of Yaw Angle on Penetration by Rod Projectiles," presented at the Topical Group Conference on Shock Compression of Condensed Matter, Amherst, MA, 27 July - 1 August 1997.
4. Tate, A, 1967, "A Theory for the Deceleration of Long Rods After Impact," *J. Mech. Phys. Solids*, Vol. 15, 387-399.
5. Forrestal, M. J., Okajima, K., and Luk, V. K., 1988, "Penetration of 6061-T651 Aluminum Targets with Rigid Long Rods," Vol. 55, pp. 755.

936

CP429, *Shock Compression of Condensed Matter – 1997*
edited by Schmidt/Dandekar/Forbes
© 1998 The American Institute of Physics 1-56396-738-3/98/$15.00

ANALYSIS OF TRANSVERSE LOADING IN LONG-ROD PENETRATORS BY OBLIQUE PLATES

G. C. Bessette and D. L. Littlefield

Institute for Advanced Technology, University of Texas at Austin, Austin, TX 78759

The loading history for the hypervelocity impact of a long-rod striking a moving flat plate at an oblique angle was analyzed using the EPIC code. The loads were reduced to their axial and lateral components to provide insight into the penetration problem as well as indicate temporal regimes where loading must be considered in the design and analysis of long rod penetrators. Furthermore, the effect of an EPIC input parameter, the erosion strain, used for the numerical treatment of eroding interfaces was assessed with attention focused on the predictions of peak loads, mass loss, and final configuration of the rod.

INTRODUCTION

The focus of this paper will be to study the loading history on a long-rod impacting a moving oblique plate for hypervelocity impact. Numerical experiments were run using the EPIC code to calculate loading history, mass loss, and final configuration of the rod. The loads were reduced to their axial and lateral components to provide insight into the penetration problem as well as indicate temporal regimes where loading must be considered in the design and analysis of long rod penetrators. Particular attention was focused on the contribution of lateral loading.

Hypervelocity impact problems involving ductile materials are characterized by significant plastic flow and erosion at the projectile/target interface. The severe deformation at the interface is difficult to treat numerically. The approach used by EPIC for handling erosion is to input an erosion strain, whereby elements are deleted when the equivalent plastic strain exceeds this value. Parametric studies were performed to assess the effect of erosion strain on the outcome of the results.

PROBLEM FORMULATION

The problem to be analyzed is a cylindrical tungsten alloy penetrator impacting a flat RHA steel plate. The problem setup is illustrated in Fig 1. The length and diameter of the projectile are 480 mm and 16 mm, respectively. The projectile strikes the plate with an impact velocity and obliquity of 2.6 km/s and 30 degrees, respectively. The target plate is 350 mm long and has a depth of 150 mm measured from the centerline. The plate is 38.1 mm thick and has a vertical velocity of 330 m/s.

The Elastic-Plastic Impact Code, Version 96 (EPIC96) was used to model the problem. EPIC uses an explicit Lagrangian finite element formulation (1-3). Element masses are lumped at the nodes and a linear displacement field within an element is assumed. During a cycle, the statically equivalent forces at the nodes are determined from the element stresses. The nodal accelerations are derived from the nodal forces and masses, and subsequently integrated to obtain nodal velocities and displacements. Strain rates and volume changes are derived from the velocity field, which, in turn, are necessary parameters for deriving the stresses in the element for a rate-dependent material. Stress is decomposed into its deviatoric and dilatational

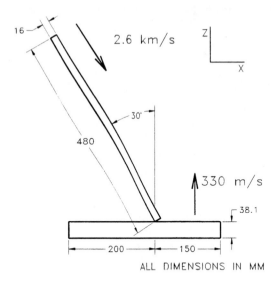

2.6 km/s

16

30°

480

330 m/s

38.1

200

150

ALL DIMENSIONS IN MM

FIGURE 1. Problem setup for a cylindrical tungsten alloy penetrator impacting a moving flat RHA plate.

components. A constitutive model is used to relate the deviatoric behavior to strains, strain rates, history, etc. An equation of state relates pressure to internal energy and density. Once the element stresses are obtained, the problem is integrated forward in time and the cycle repeated.

The eroding interface algorithm in EPIC was developed to treat excessive mesh distortion that occurs at the contact surface (4, 5). It allows for a calculation to proceed and, in most cases, avoids difficulties involved with remeshing of the problem. EPIC handles erosion via a strain-based criteria where an element is eroded away when the equivalent plastic strain exceeds some threshold. The threshold is termed the erosion strain and is given as the input parameter ERODE to the code. When the erosion criteria is met, the element essentially disappears. However, mass and momentum are retained at the nodes associated with the eroded element. The slidelines are then redefined based upon the new configuration of the problem. ·

EPIC was used to generate the finite element mesh. The problem was modeled using symmetric constant-strain tetrahedral elements. Elements were

uniformly spaced in the projectile. For the target, a graded mesh was used in the x-direction and a uniform mesh spacing in the z-direction. Due to the problem setup, a plane of symmetry exists at the origin of the y-axis. The library material constants for both tungsten and 4340 steel based on a Johnson-Cook material model were used with cumulative damage and fracture allowed. A double pass approach was used for the contact surface to achieve greater accuracy (6).

Three analyses were conducted. The first was a mesh sensitivity analysis to determine an acceptable level of mesh refinement. Coarse, medium, and fine meshes were considered where element widths were 8 mm, 4 mm, and 2.6 mm, respectively. The erosion strain was kept constant at 1.5.

The second analysis considered the loading history on a penetrator and subsequent response of both projectile and target. EPIC generates instantaneous values for the global linear and angular momentum. These can be used to determine the instantaneous global forces and moments as follows

$$\mathbf{f} = d(m\mathbf{v})/dt \qquad (1)$$

$$\mathbf{M} = d(\mathbf{r} \times m\mathbf{v})/dt \qquad (2)$$

The forces can be rotated into a coordinate system coincident with the longitudinal axis of the projectile to provide the axial and lateral components of loading. The axial and lateral axes were chosen to coincide with the original orientation of the projectile. Clearly, our penetrator deforms, so the definitions of axial and lateral axes become ambiguous; but, it will be shown that this orientation is reasonable for the impact conditions and geometry under consideration and can provide valuable insight into the physics of the problem.

The third and final analysis assessed the effect of an EPIC input parameter ERODE, the erosion strain, on the outcome. Specifically, the effect on residual velocity, mass, rod length, and loading history were of interest. The total eroded mass was taken as the sum of all masses associated with eroded elements.

MESH SENSITIVITY

The adequacy of the mesh was tested based on convergence of the normalized residual mass, velocity, final rod length, and target hole diameter (see Table 1 for definitions). All of the normalized parameters appeared insensitive to the degree of mesh refinement for the problem under consideration. Similar observations were noted in (6). Intuitively, the degree of mesh refinement does affect the quality of the solution. It appears that the criteria chosen are not a suitable indicator of the quality of the mesh. At this time, no suggestion can be offered as to an appropriate indicator for mesh resolution. Given the size of the problem, the baseline mesh was chosen for the remaining analyses and is thought to be a reasonable choice for balancing the quality of the solution and CPU time.

LOADING HISTORY

A typical load-time trace is provided in Fig 2. Initial penetration is characterized by high axial and lateral loads. These loads ramp up to a peak value. The ramping corresponds to the blunt nose of the projectile embedding itself into the target. A brief period of steady state penetration is noted shortly thereafter. This period is short-lived (5 to 20 μs) as the target is not very thick. At approximately 40 μs, the projectile nose perforates the target. By this time, there is contact between projectile and the backside crater lip. The contact area increases with time. At 120 μs, approximately 10 element widths are in contact. At 160 μs, full contact along the entire edge of the target is noted. This time corresponds to a peak in the late-time lateral loading. Axial loading increases at late-times due to the upward motion of the target. It is important to note that the duration of lateral loading is significant (approximately 60 μs).

After complete perforation, the projectile is slightly bent. However, the longitudinal axis does not vary by more than a few degrees from that of the original impact obliquity, thereby allowing for a reasonable calculation of axial and lateral loadings in a manner consistent with the original penetrator orientation. This would not be possible if either excessive deformation occurred or large amounts of rotation were noted.

EFFECT OF EROSION STRAIN

The effect of varying the erosion strain on the loading and various normalized parameters can be observed in Table 1. Loads are averaged over the indicated range. The magnitude of loading increased with erosion strain. This is not surprising as increasing erosion strain has the effect of increasing the energy absorption of the material. The effect on the normalized parameters was relatively insignificant in the 1.5 to 2.5 range. However, lower erosion strains appeared to greatly affect the degree of mass loss and final length. No data was obtained for higher erosion strains due to a numerically unstable energy balance likely arising from excessive mesh distortion. The effect on the peak loads was also relatively insignificant for erosion strains in the 1.5 to 2.5 range.

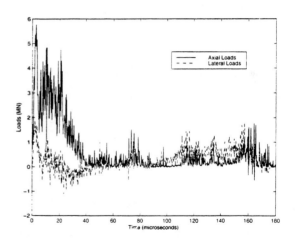

FIGURE 2. Typical results for loads on a projectile.

The eroded mass loss as a function of erosion strain for the baseline projectile is presented in Fig 3. There is little difference in the rate of mass loss until 20 μs. At 20 μs, the projectile nose is approximately at two-thirds of the target thickness and bulging of the back target plate is observed. Once the back surface breaks away and the projectile nose perforates the target, the effect of erosion strain becomes significant. As shown, the smaller the erosion strain, the greater the mass loss.

939

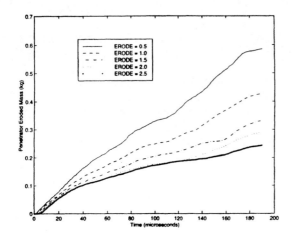

FIGURE 3. Eroded mass as a function of the erosion strain.

an artifact of the eroding interface algorithm. Furthermore, there was separation along the projectile/target interface resulting from element removal where a "sliding" behavior should have been exhibited. Thus one concludes that the contact boundary conditions are not satisfied by the eroding interface algorithm. Care must be exercised when employing this algorithm. This is especially true when considering multiple targets. In this situation, accurate modeling of the deformed configuration is critical as strength of the projectile is calculated on the element level.

ACKNOWLEDGMENTS

This work was supported by the U.S. Army Research Laboratory (ARL) under contract DAAA21-93-C-0101.

CONCLUSIONS

An analysis was performed to assess the axial and lateral loads on a long rod impacting an oblique plate. Both lateral and axial loads were shown to be significant at late times. These loads increase due to the combined action of increasing contact area and the upward moving plate. Both are on the same order of magnitude with lateral loads usually slightly greater than the corresponding axial loads. However, the duration of lateral loading is longer.

The erosion strain in the 1.5 to 2.5 range does not significantly affect either the normalized parameters or the peak loads. However, qualitatively, it is difficult to reconcile the amount of eroded material with that observed in testing. Postprocessing of the EPIC results displayed an excessive amount of lost material (eroded elements) on the underside of the projectile. This degree of mass loss is generally not observed in testing and is

REFERENCES

1. Johnson, G.R., "Analysis of Elastic-Plastic Impact Involving Severe Distortions", *J. Appl. Mech.*, **98**, 3, pp. 439-444 (1976).

2. Johnson, G.R., "High Velocity Impact Calculations in Three Dimensions", *J. Appl. Mech.*, **99**, 1, pp. 95-100 (1977).

3. Johnson, G.R., Stryk, R.A., Holmquist, T.J., and Beissel, S.R., *User Instructions for the 1996 Version of the EPIC Code*, Alliant Techsystems, Inc. (1996).

4. Johnson, G.R. and Stryk, R.A., "Eroding Interface and Improved Tetrahedral Element Algorithms for High-Velocity Impact Computations in Three Dimensions", *Int. J. Impact Eng.*, **5**, pp. 411-421 (1987).

5. Belytschko, T. and Lin, J.R., "A Three-Dimensional Impact-Penetration Algorithm with Erosion", *Int. J. Impact Eng.*, **5**, pp. 111-127 (1987).

6. Johnson, G.R. and Schonhardt, J.A., "Some parametric sensitivity analyses for high velocity impact computations", *Nuclear Eng. and Design*, **138**, pp. 75-91 (1992).

TABLE 1. Comparison of Results for Baseline Projectile

Erosion Strain	M^*	V^*	L^*	D^*	Initial Loading (Average Peak Values)			Averaged Late-time Loading	
					f_{ax} (MN)	f_{lat}(MN)	M_y (MN)	f_{ax} (MN)	f_{lat}(MN)
0.5	0.64	0.93	0.79	4.31	3.50	1.00	-0.20	0.6	0.5
1.0	0.73	0.94	0.84	4.13	5.00	1.50	-0.30	0.5	0.5
1.5	0.80	0.94	0.89	4.13	5.50	1.75	-0.40	0.5	0.7
2.0	0.82	0.95	0.92	4.06	5.75	2.00	-0.50	0.4	0.8
2.5	0.85	0.95	0.91	4.13	6.00	2.25	-0.55	0.4	0.9
3.0	No data due to numerical instability								

$M^* = M_{residual}/M_o$ $V^* = V_{residual}/V_o$ $L^* = L_{final}/L_o$ $D^* = D_{crater}/d_o$ f_{ax} - axial force f_{lat} - lateral force

CP429, *Shock Compression of Condensed Matter – 1997*
edited by Schmidt/Dandekar/Forbes
© 1998 The American Institute of Physics 1-56396-738-3/98/$15.00

SHOCK WAVE REFLECTION BEHAVIOR IN DOUBLE-LAYER METEOROID BUMPER SYSTEMS

H. Nahme, A.J. Stilp and K. Weber

Fraunhofer-Institut für Kurzzeitdynamik, Ernst-Mach-Institut (EMI), Eckerstrasse 4, 79104 Freiburg, Germany

In laminated meteor bumpers the shock wave transfer is affected by the shock impedance mismatch and the order of the layers. For titanium-tungsten targets it has been found that the higher pressure changes, caused by loading and unloading waves, over a longer time are responsible for the higher degree of fragmentation compared to the tungsten-titanium target layer sequence. Evidence for this is given by planar impact tests combined with the VISAR measurement technique. The velocity-time diagrams show different wave propagations inside the target components, dependent on the order of the layers.

INTRODUCTION

The conventional shielding system for spacecraft protection is the dual-plate Whipple shield. During the last decades the Whipple shield has been improved by both variation of geometrical configuration and type of shield materials to achieve a higher protection efficiency as well as weight savings of the structure. The three alternative concepts are the multi-shock shield (1), the mesh double-bumper shield (2) and the "Stuffed" Whipple shield (3). Still another concept is the laminated meteor bumper shield (4). Double-layer shielding systems have been investigated experimentally at EMI in the velocity range 3-8 km/s (5).

TEST PARAMETERS AND SET-UP

For investigation of the shock wave transfer and reflection behaviors in double-layer meteoroid bumper systems with layers of distinctly different shock impedances planar plate impact tests (6) in combination with a VISAR (7) were carried out.

Some test parameters are summarized in Table 1. The 10 mm thick aluminum (Al) disc projectile was accelerated by a powder gun with a bore diameter of 70 mm to an impact velocity of around 600m/s. The double-layer targets consist of 4 mm titanium (Ti) and 1 mm tungsten (W) plates with shock impedances given in Table 1. The plates are arranged in the Ti-W and the reversed W-Ti layer sequences. The layers are glued together with a two-component epoxy adhesive. With the VISAR the velocity history of the target rear side has been recorded with a time resolution of 2 ns.

TABLE 1. Test parameters and material properties of projectile and bumper shield layers

Material	Al disc	Ti layer	W layer
Material density [g/cm³]	2.7	4.51	19.3
Bulk sound velocity [m/s]	5300	5020	4030
Shock velocity (u_p=300m/s) [m/s]	6300	5700	5250
Plate thickness [mm]	10	4	1
Shock impedance at test conditions / 10^6 [kg/m^2 s]	17.0	25.7	101

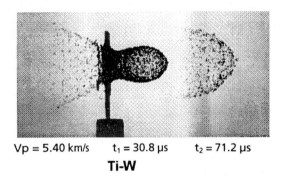 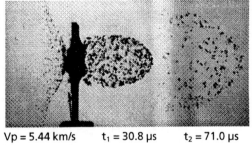

| Vp = 5.40 km/s | t₁ = 30.8 µs | t₂ = 71.2 µs | | Vp = 5.44 km/s | t₁ = 30.8 µs | t₂ = 71.0 µs |

Ti-W   **W-Ti**

FIGURE 1. Hypervelocity impact tests of 4 mm **titanium**-1 mm **tungsten** and 1 mm **tungsten**-4 mm **titanium** double-layer meteoroid bumpers struck by a 10 mm diameter aluminum sphere at $v_P \approx 5.4$ km/s

EXPERIMENTAL RESULTS

Phenomenology of Debris Cloud and Backsplash of Double-Layer Bumpers

The effectiveness of meteoroid shielding systems against hypervelocity impact of projectiles depends on their ability to break up, liquify and vaporize the projectile material and disperse it over an area behind the shield as large as possible. This capability is determined by geometrical configuration and material properties of the shield.

The flash X-ray photographs of Fig. 1 depict the time dependent formation process of debris cloud (debris behind the target) and back splash (debris in front of the target) for the hypervelocity impact of a 10 mm diameter Al sphere on double- layer shields with 4 mm Ti - 1 mm W and the reversed 1 mm W - 4 mm Ti, respectively, at times $t_1 \approx 31$ µs and $t_2 = 71$ µs at an impact velocity of $v_P \approx 5.4$ km/s. In the case of the Ti-W target a relatively large crater hole is formed in the Ti front layer, distinctly larger than the perforation hole in the W front layer of the W-Ti target. Furthermore, shapes of debris cloud and back splash are strongly dependent on the order of the two layers. With Ti as front layer material a vase-shaped debris cloud and funnel-shaped back splash is formed. However, if the front layer consists of W the debris cloud looks like a truncated ellipsoid and the funnel-shaped back splash becomes distinctly flatter. The debris cloud caused by the Ti-W target consists of smaller fragments than in the case of the W-Ti target, i.e.,

the order of the target layers also determines the degree of fragmentation. Furthermore, it can be seen that for the Ti-W target the debris cloud fragments are spread over a smaller area than for the W-Ti target. The X-ray shadowgraphs show that in the case of the W-Ti-configuration the fragment sizes decrease with increasing impact velocity and approaches those of the Ti-W shield. On the other hand the lateral velocity of the debris cloud increases stronger for W-Ti- than for Ti-W plate order with increasing impact velocity.

VISAR Measurements

In order to support the explanations of the different shielding probabilities of Ti-W- and W-Ti configurations given previously (5), planar-plate-impact experiments have been performed with aluminum projectile plates of 10 mm thickness. From the X-t diagrams of Fig. 2 together with the corresponding σ-u_p-diagrams the velocity-time histories and thus the wave propagation inside the sample plates can be explained to fairly high degree of accuracy. The velocity time histories confirm the great influence of the sequence of the target plates. Because of the small plate thickness especially of the W-plate the wave pattern inside the plate quickly becomes quite confusing. So the wave propagation and the material stress states were traced by means of the X-t- and σ-u_p-diagrams as far as differences were clearly recognizable. This includes all waves before

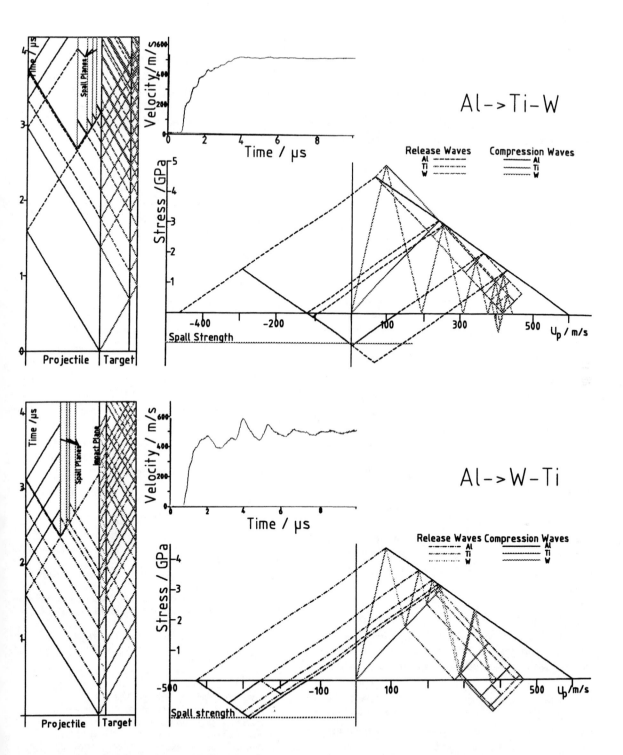

Figure 2. Velocity – time – histories, X-T-diagrams and σ - u_p – diagrams for W-Ti- and Ti-W shield configuration

spallation inside the projectile plate occurred. It can be seen from the diagrams that for the impact conditions used here spallation only occurs inside the projectile plate at locations, where release waves superpose causing tension stress. Because the adjacent spall signals can not be detected with the VISAR the samples and projectiles have been recovered softly, cut and polished. In all but one case the spall planes have been found near the location forecasted by the X-t- diagrams.

No spall planes have been found inside the Ti-plates, indicating that the tension stress level did not exceed the spall strength. The very brittle W-plates were completely fragmented during the tests so that no post loading examination was possible. The relatively long time periods over which the velocity histories have been recorded show, that the plate setup remained undamaged and no high tension stress occured across the W-Ti interface for at least 4µs after impact. While for both the W-Ti- and Ti-W configuration a stepwise increase of the free surface velocity was observed, definite differences for later stages of the loading and unloading process have been found. For the W-Ti-configuration the wave responsible for the first prominent free surface velocity reduction has been clearly identified as a consequence of the shock impedance mismatch between Ti and W. The next prominent velocity reduction around 3.3 µs is due to the release wave from the projectile rear side The subsequent velocity variations can be explained to a good level of confidence by compression and release waves propagating inside the Ti-plate with the W-Ti-interface having been broken.

For the Ti-W configuration much less velocity variations have been observed. The compression stress levels inside both the W- and Ti-plates show large amplitudes at the beginning but then approach low stress levels. Due to the more regular alternation of compression and release waves the X-t-diagram looks much more regular than in the case of the W-Ti-arrangement. This finds its expression in a much smoother velocity-time-history with no prominent features.

The wave propagation diagrams show, that the occurrance of tension stress inside the Ti- and W-plates during the first few microseconds of the

loading process depends on the combination of the two plates. At impact velocities exceeding those 600m/s used for these tests, spallation should occur also inside the shield plates. This remains to be proven by further experiments.

CONCLUSIONS

While for single sheet bumpers the protection efficiency against debris/meteoroid impact depends on their mass per unit area and their material properties, the effectiveness is additionally influenced by the shock impedance ratio and the order of the layers in the case of laminated targets. The comparison of the fragmentation behavior of the two configurations with the results of the wave propagation study indicate that the higher degree of fragmentation of projectile and shield material but smaller spread of the fragment cloud for the Ti-W layer sequence than for the reversed W-Ti target is a direct result of the wave propagation properties. The observation, that different arrangements of the same materials influence both the fragment sizes and the spread of the debris cloud shows, that with the impedance mismatch of the shield components an additional degree of freedom for debris shield design is available.

REFERENCES

1. Cour-Palais, B.G., and Crews, J.L., *Int. J. Impact Engng.*, **10**, Nos. 1-4, pp. 135-146 (1990).
2. Christiansen, E.L., and Kerr, J.H., *Int. J. Impact Engng.*, **14**, pp. 169-180 (1993).
3. Christiansen, E.L., Crews, J.L., Williamsen, J.E., Robinson, J.H., and Nolen, A.M., *Int. J. Impact Engng.*, **17**, Nos. 1-6, pp. 217-218 (1995).
4. Riney, T.D., *High-Velocity Impact Phenomena* (edited by Kinslow), New York and London: *Academic Press*, 1970, pp. 166-177.
5. Stilp, A.J., and Weber, K., „Debris clouds behind double-layer targets", presented at the 1996 Hypervelocity Impact Symposium, Freiburg, Germany, Oct. 7-9, 1996.
6. Barker, L.M., Shahinpoor, M., and Chhabildas, L.C., *High-pressure shock compression of solids* (edited by Asay, J.M., and Shahinpoor, M.), New York: Springer, 1993, pp. 43-73.
7. Barker, L.M., and Hollenbach, R.E., *J. Appl. Phys.*, **43**, pp. 4669-4675 (1972).

CP429, *Shock Compression of Condensed Matter – 1997*
edited by Schmidt/Dandekar/Forbes
© 1998 The American Institute of Physics 1-56396-738-3/98/$15.00

THE IMPORTANCE OF MATERIAL PROPERTIES FOR CRATERING AND PENETRATION IN GEOLOGIC MATERIALS

D. R. Curran

Poulter Laboratory, SRI International, Menlo Park, California 94025

Accurate cratering and penetration predictions in many solids are difficult because the yield strength, particularly for wet or jointed rock, depends in complicated ways on strain rate and confining pressure. Computational and experimental results are presented that illustrate the evolution of crater morphology from bowl-shaped to saucer-shaped as the shear strength increases, and the evolution of penetration depth from shallow to deep as the block mobility increases.

BACKGROUND AND APPROACH

Cratering and penetration have important applications in several shock physics areas. These include micrometeorite and space debris impacts on space assets, shrapnel damage in the National Ignition Facility chamber, armor penetration, earth penetrators, nuclear cratering, and asteroid impacts. The crater sizes and penetration depths in these applications range from micrometers to kilometers, covering nine orders of magnitude!

In this paper we will first note that for all of the above applications the yield strength often plays the dominant role in determining crater morphology and depth of penetration. Second, we will show that the specific physical processes underlying yielding can sometimes produce counterintuitive cratering and penetration behavior. Finally, we draw some conclusions regarding experiments required to develop respectable yield models.

PROCESSES GOVERNING CRATERING AND PENETRATION

Two processes govern cratering and penetration, (1) compaction of empty pores to form a compaction crater or penetration tunnel, and (2) nonelastic "plastic" flow of material out of the crater or around the penetrator. The "compaction strength", or pressure required to compact the pores, is proportional to the yield strength of the matrix material. A simple formula derived by Carroll and Holt (1) that has been found to agree well with experimental data is:

$$P = 2Y\ln(\phi^{-1/3}) \qquad (1)$$

where P is the compaction pressure, Y is the yield strength of the matrix material, and ϕ is the relative volume porosity. Eq(1) states that the compaction pressure is on the order of the yield strength over a wide range of porosity.

When the yield strength is large compared to inertial or gravitational stresses, the crater and penetration depths scale as follows (2):

Crater diameter, $D \propto (E/Y)^{1/3}$

Penetration depth, $X \propto E/Y$ (2)

where E is the energy coupled to the target.

In short, for both porous and nonporous solids, if Y is large enough it can dominate cratering and penetration. Therefore, to accurately predict cratering and penetration in a specific solid, one must first have a realistic yield model.

PHYSICAL "MICROMECHANISMS" THAT GOVERN YIELDING

Yielding is defined as the occurrence of nonelastic slip, i.e. the strain does not return to zero when the stresses are relaxed to zero. In granular or fractured geologic materials, or in fragmented ceramics, such slip can occur via sliding of fragments or granules past each other. The resistance to frictional slip depends on confinement by adjacent material, friction between the granules, and granule inertia. The frictional stress depends in turn on the coefficient of friction, the contact area, and lubrication and "effective stress" effects due to the presence of water or soft material between the granules or fragments.

Realistic models of frictional materials can give counterintuitive results for cratering and penetration. We next show two such examples, one involving the effect of water-saturation, and the other involving the effect of granule mobility.

EXAMPLE 1: EXPLOSIVE CRATERING IN WATER-SATURATED CALCITE GROUT

An experiment was performed in which a small explosive charge was detonated tangent to a sample of 40% porous, water-saturated calcite grout (3). The sample was in two pieces which could be separated after the experiment to observe the crater morphology and sub crater damage. Figure 1 shows the post test observations compared with a calculation using a yield model that gave good agreement with the observations. Figure 2 compares two calculations, one using the "correct" yield model, and one using an "intuitive" yield model.

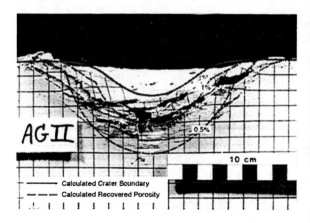

FIGURE 1. Experimental result and comparison with calculation.

As discussed in (3), the "intuitive" model used was the common Terzaghi "effective stress" model. This model produces low yield strengths due to the "unloading" and lubrication effects of water in the cracks that connect the pores in the material. However, as discussed in (3) and (4), confining pressures that are a significant fraction of the bulk modulus of water (about 2 GPa) will squeeze the water out of thin cracks, resulting in a relatively stiff, strong material. An effective stress model that accounts for this pressure hardening effect was used for the "correct" model calculation shown.

Similar calculations for large scale (megaton TNT equivalent) charges showed similar results because the calculated stresses at the crater boundary were higher than the gravitational stresses, and the crater morphology was strength-dominated.

In short, when pressure hardening is included in the yield model for water-saturated material, the expected crater morphology may change dramatically from that expected from the usual Terzaghi model.

EXAMPLE 2: CRATERING AND PENETRATION IN BRITTLE, FRAGMENTED MATERIAL

Cratering and penetration of ceramic armor or hard rock with low initial porosities occurs by "plastic" flow of the fragment granules or blocks, where the flow occurs via slip between the blocks. As discussed in references (5) - (7), the geometry of such granular flow

946

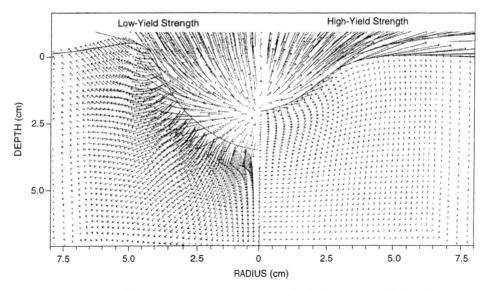

FIGURE 2. Predictions of "inituitive" and "correct" models. (Calculated velocity vectors at 200 ms for experiment AG).

can be treated with a "macrodislocation" model, which focuses on the motion of lines of vacancies between the granules, where the vacancies must be large enough to allow motion of granules into them.

Figures 3 and 4 show calculations with this model of the penetration of a 8 cm diameter, 10 cm long steel rod impacting a confined alumina ceramic target at 250m/s. The model allows comminution of the granules from their initial specified size (joint spacing) to smaller sizes determined by the stress histories,

assumed initial flaw sizes, and fracture toughness. In the calculations shown, granular slip was allowed on thirteen specified planes. Cylindrical symmetry was assumed.

Figures 3 and 4 show the result of varying the initial granule size from small (100 μm) to larger (0.5 cm).

One might intuitively expect that the shear stress would relax quickly to the static yield surface, and the

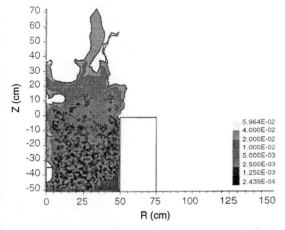

FIGURE 3. Final penetration depth of steel penetrator in a confined granular alumina target. Initial granule size = 0.01 cm. (Contours are levels of porosity).

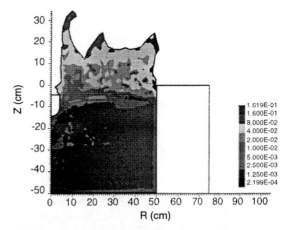

FIGURE 4. Final penetration depth of steel penetrator in a confined granular alumina target. Initial granule size = 0.5 cm. (Contours are levels of porosity).

coefficient of friction alone would determine the penetration depth. However, Figures 3 and 4 show that initial granule size may govern penetration. The reasons are outlined below.

As described in (7), on a given slip plane the Orowan equation for nonelastic strain rate can be written approximately:

$$\partial\gamma_p/\partial t \approx M[(\tau - \mu\sigma_n)/\rho]1/2 \qquad (3)$$

where M is a macrodislocation mobility function (or inverse viscosity) given by:

$$M = (f\sqrt{b})/B \qquad (4)$$

[Eq(3) is an update of the linear stress-dependence of the plastic stain rate equation of references (5)and (6).] In the above equations, γ_p is the nonelastic shear strain on the slip plane, τ is the shear stress on the plane, μ is the coefficient of friction, σ_n is the normal compressive stress on the plane, ρ is the target material density, "f" is the fraction of granules that have an adjacent macrodislocation (a space large enough for the granule to slide into), B is the granule size, and bB is the dislocation size (the macro Burger's vector).

In the comminution model of (6) and (7), "f" is determined by the free boundary condition. Thus, the strain rate is proportional to $\sqrt{b}/B$, and the initial value of the granule size B can govern penetration, as shown in Figures 3 and 4. In the calculation of Figure 4, the initial granule size of 0.5 cm was reduced by comminution to below 0.1 cm. Thus, the comminution process is also important in determining the depth of penetration.

CONCLUSIONS

Cratering and penetration in solids are often governed by the material yield strength which, in porous, fragmented, frictional materials, depends in complex ways on water saturation and the fragment configurations. To identify, construct, and calibrate realistic yield models, one must be able to perform experiments that exercise the relevant physics. Spherical explosive cavity experiments such as those reported in Refs [6] and [8] have been shown to be useful for these purposes. An advantage of such experiments is that the material is forced through strain paths relevant for cratering or penetration. Pre and post test examination of the specimens can quantify the amount of block comminution and flow. Furthermore, dynamic stress and strain diagnostics relatively close to the cavity boundary can provide further information needed for development of realistic yield models.

ACKNOWLEDGMENTS

This work was partially supported by the Army Research Office under the supervision of Dr. John Bailey, and partially supported by SRI International internal research. Thanks are due to T. Cooper for implementing the models and performing the hydrocode calculations in this presentation, and to R. W. Klopp and D. A. Shockey for helpful discussions.

REFERENCES

1. Carroll, M. M., and Holt, A. C., *J. Appl. Phys.*, **43**, 1626 - 1636, 1972.
2. Holsapple, K. A., and Schmidt, R. M., *J. Geophys. Res.*, **85(B12)**, 1849 - 1870, 1982.
3. Curran, D. R., Aidun, J. A., Cooper, T., and Tokheim, R.E., *Journal de Physique*, **Colloque C3, Supplement No. 9, Tome 49**, 449 - 455, 1988.
4. Curran, D. R., *Journal de Physique*, **Colloque C8, Supplement III, 4**, 243 - 247, 1994.
5. Curran, D. R., Seaman, L., Cooper, T., and Shockey, D.A., *Int. J. Impact Engng*, **13, no. 1**, 53 - 83, 1993.
6. Curran, D. R., Seaman, L., Klopp, R. W., De Resseguier, and Kanazawa, C., in *Fracture and Damage in Quasibrittle Structures*, eds. Bazant, Z. P., Bittnar, Z., Jirasek, M., and Mazars, J., London, E & FN Spon, 1994, pp. 245 - 257.
7. Curran, Klopp, R. W., and Cooper, T., *in preparation.*
8. Florence, A. L., Cizek, J. C., and Keller, C. E., in *Shock Waves in Condensed Matter - 1983*, eds. Asay, J. R., Graham, R. A., and Straub, G. K., North-Holland, 1984, pp. 521 - 524.

CP429, *Shock Compression of Condensed Matter – 1997*
edited by Schmidt/Dandekar/Forbes
© 1998 The American Institute of Physics 1-56396-738-3/98/$15.00

COMPARISON OF H_2O AND CO_2 ICES UNDER HYPERVELOCITY IMPACT

M.J.Burchell[1], J.Leliwa-Kopystynki[2], B.Vaughan[1] and J.Zarnecki[1]

[1] *Unit for Astrophysics and Space Sciences, Physics Dept., Univ. of Kent, Canterbury, Kent, CT2 7NR, UK.*
[2] *University of Warsaw, Institute of Geophysics, ul. Pasteura 7, 02-093 Warszawa, and Space research Centre of the Polish Academy of Sciences, ul. Bartycka 18A, 00-716, Warszawa, Poland.*

We have studied hypervelocity impacts at 5 km s^{-1} of 1 mm diameter projectiles on two types of ice: H_2O and CO_2. We compare crater morphology for the resulting impact craters and find that the craters are smaller in the CO_2 ice than in H_2O ice. However the crater shape, as expressed by the ratio depth/diameter, is indistinguishable between the two types of ice. Further, we have investigated the dependence of crater size and shape on projectile density. We find that for densities of 2750 to 7850 kg m^{-3}, the ratio depth/diameter does not have the strong dependence on density suggested elsewhere in the literature. This is true for both types of ice. We thus suggest that whilst the crater size may depend strongly on ice type, the shape does not and is also almost independent of projectile density.

INTRODUCTION

Impact cratering on ices has been the subject of study for many years. In the field of Solar System sciences it is particularly important as there are many ice covered surfaces in the Solar System, and they have been subject to periods of bombardment in the past. The resulting craters are observable by space probes, and understanding their size and shape is an on-going task.

Unfortunately, the scales of impacts in the Solar System are beyond the range of laboratory experiments. The main problems are in projectile size and impact speed. In the Solar System sizes of 10's to 1000's of metres can be expected for impactors, with impact speeds of 10's of km s^{-1}. However, these are unobtainable in the laboratory. Most laboratory work has understandably been carried out on the micron to cm size scale. However it is unfortunate that it has also been carried out mostly at $\leq$ 1 km s^{-1}. Thus not only is it below the natural speed for impacts in the Solar System, but the work is also below what is usually

defined as the hypervelocity range (typically greater than a few km s^{-1}). Some work (1,2) has been at hypervelocities, but represents the smaller part of published data concerning impacts on ices, e.g. (3) and references therein. A further complication is that all the work has been carried out on water ice or water ice-silicate mixtures.

When trying to generalise the behaviour of ices under hypervelocity impacts, one thus has a very restricted data set. In this paper, we present data for two types of ice and for a range of projectile densities. We can thus for the first time look at the way hypervelocity cratering depends on both the projectile and the ice. No attempt is made to extrapolate to Solar System scales as there are still many unanswered questions (e.g. dependence of cratering on ice temperature etc.)

METHOD

The experiments were carried out using the two-stage light gas gun of the Unit for Space Sciences and Astrophysics at the University of Kent at Canterbury (UK). The gun fires projectiles carried in a sabot, which is then discarded in flight. The velocity of a projectile is measured by the interruption of two laser light curtains through which it passes. Projectiles used here were 1.0 mm diameter spheres of aluminium, titanium and stainless steel (densities 2710, 4540 and 7850 kg m^{-3} respectively). The target chamber was held at a vacuum of 0.2 to 0.3 mbar during each shot. In this chamber was placed a cold plate (cooled by flow of liquid nitrogen) with an ice target in a metal holder above it. The target temperature was monitored by a thermocouple. For H_2O ice targets, the typical range of target temperatures was 255 ± 4 K. For CO_2 targets the cooling was used to lower the temperature of the target holder to 240 K. This was above the sublimation point of CO_2, hence the target temperature of the target itself was at its sublimation point.

The H_2O ice targets were made in two fashions. Clear ice targets were made by boiling distilled water (to reduce the dissolved gas content) and then freezing in cylindrical moulds 20 cm across and 12 cm deep. Cloudy ice targets were made by packing crushed ice into the mould and then adding distilled water, taking care to free any trapped air bubbles during filling. The resulting ice was mostly clear with slight clouding, indicating a small degree of porosity. The quasi-static compressive strength of the ice was measured in the laboratory as between 2 and 10 MPa for both types of H_2O ice. This is compatible with the 10 MPa value given in (4). Several samples of both types of H_2O ice were measured and although the clear ice was on average weaker, the two data sets overlapped considerably.

The CO_2 targets were purchased from BOC (UK), they were uniform in appearance with crystal sizes of approximately 1mm. By measuring the density we found a possible porosity of (8 ± 4) %. We measured the compressive strength to be some 2.5 times that of the water ice. There is some uncertainty on this value as, due to the CO_2 ice's opaque nature, propagation of internal cracks could not be observed. Instead large cracks visible on the external face viewed by the observer were

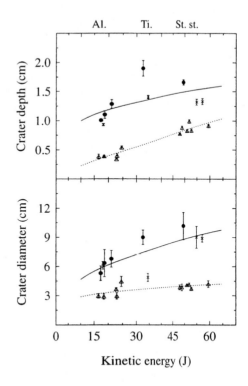

FIGURE 1. Crater depth and diameter vs. impact kinetic energy for • pure water ice, × cloudy water ice and △ solid CO_2

taken to indicate brittle failure (indeed one block being tested at strain rates of order 10^{-5} s^{-1} exploded before any cracks were observed).

In all cases the target surfaces were smooth to better than 50 microns. After each shot of the gun the target chamber was opened and the crater measured. A depth gauge was used to obtain depth profiles of the crater along several (usually 4) diameters equally spaced around the crater. The original surface was fit with a straight line, and the total excavated crater diameter in the original surface plane was found for each profile and averaged (allowing for non axial symmetry). The depth in the crater centre was found relative to the original target surface. Crater volume was found by taking the excavated area in each profile, rotating around the central point to obtain a

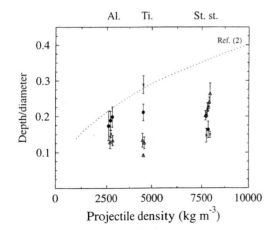

FIGURE 2. Crater depth/diameter ratio vs. projectile density for ● pure water ice, × cloudy water ice and △ solid CO_2

volume, and again averaging to allow for non-axial symmetry.

RESULTS

Nine shots were carried out on H_2O ice with a mean velocity of 5.18 ± 0.18 km s^{-1} (5 on clear blocks and 4 on cloudy blocks). Eleven shots were carried out CO_2 ice with a mean velocity of 4.86 ± 0.30 km s^{-1}. No systematic variation of velocity with projectile density was observed, although the three shots of titanium projectiles on CO_2 targets were slightly low in velocity (and are hence displaced leftward on plots of the data vs. kinetic energy). In Figure 1 we show the crater depth and diameter vs. impact kinetic energy. No systematic difference is observed between the two types of H_2O ice. We find that the crater diameter (cm) and depth (cm) are related to impact energy E (J) power laws of the form $y=aE^b$, where the coef.'s a and b are given in table 1 (the fits are shown on all figures as solid/dashed lines for H_2O and CO_2 respectively). The crater depth in CO_2 appears to be increasing faster with energy than in the H_2O ice, with the situation reversed for the diameter. Previous results for impacts on H_2O ice (1,3,5) of similar energy as here (for (3,5) velocities range from 100 to 1000 m s^{-1}), show that crater diameter can be related to impact energy by a power law

whose exponent is in the range 0.29 to 0.40. The value found here (0.39) is at the high end of the range, but is not incompatible. The absolute magnitudes of the previous results span those reported here.

In Figure 2 we show the depth/diameter vs. projectile density (note that at each fixed density the data is shown slightly distributed in density to avoid data points laying on top of each other). Although there is a scatter on the data, no difference is apparent between the data for any target type. Indeed the scatter is such to prevent a sensible fit of either a straight line or a power law for depth/diameter vs. density. Instead we average the data and find that depth/diameter for H_2O ice (both types combined) = 0.16 ± 0.01 with standard deviation $\sigma = 0.05$, and for the CO_2 ice = 0.18 ± 0.01, $\sigma = 0.06$. This indicates that the shape is not only relatively insensitive to projectile density, but is also indistinguishable between the H_2O and CO_2 ice.

The lack of dependence of shape on projectile and target densities contradicts the predictions of (1), where for H_2O ice, data for impacts by low density projectiles at approximately 6 km s^{-1} is combined with that of (5) for lead projectiles at a few hundred m s^{-1}. The result was that depth/diameter was related to the ratio of (projectile/target) densities by a power law with exponent 0.37 (although an exact normalization is not given). In (2) (again for impacts on H_2O ice) a similar result is found, namely that depth/diameter = $0.0058\rho^{0.46}$ (ρ projectile density in kg m^{-3}) shown in Figure 2 as a dotted line. The discrepancy between (1,2) and the results here lies in the behaviour at projectile densities greater than 3000 kg m^{-3}. In (2) (impacts of glass beads on H_2O ice at 5 km s^{-1}) the depth/diameter = 0.16 ± 0.01, as here. This was then combined with the only previously available data at larger projectile

TABLE 1. Coef. of fits to the data y = aEnergy^b.

	H_2O		CO_2	
	a	b	a	b
Depth	0.56	0.25	0.033	0.82
Diameter	1.88	0.39	1.79	0.20
Volume	0.14	1.34	0.022	1.17

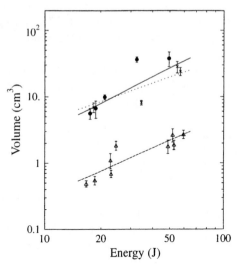

FIGURE 3. Crater volume vs. impact kinetic energy for • pure water ice, × cloudy water ice and △ solid CO_2.

method of ice preparation or is an effect of the different velocity ranges.

CONCLUSIONS

We find that the size of craters in H_2O and CO_2 ice differ substantially but that the shape does not. We also find that the shape is not dependent on projectile density, despite previous predictions. The previous work relies solely on one (non-hypervelocity) data set for projectile densities greater than 3000 kg m^{-3}, but the results presented here clearly rule out the expected dependence.

We also find that there are some differences between craters in H_2O ice in the present work at 5 km s^{-1} and previous work at ≤ 1 km s^{-1}. At similar energy the same crater volume is found, but the value of the power of the energy differs by 40%. One of the aims of ice impact studies is to be able to extrapolate over many orders of magnitude to Solar System scales. Simple scaling with energy in itself is well known not to be appropriate e.g. (6), but most published work has been for impacts at $\leq$ 1 km s^{-1} with the implicit assumption that velocity effects can be ignored. It may be necessary to test this in more detail.

ACKNOWLEDGEMENTS

This work is financed by a grant from PPARC (UK). JLK acknowledges support from the Royal Society (UK) and the Polish National Academy.

densities (5), for the impacts on H_2O ice of lead bullets at a few hundred m s^{-1}. There the craters are relatively deep, although the average depth/diameter ratio in (5) does not use 2 impacts (there were 8 good impacts) on the grounds of being oddly shallow. It may also be significant that the impacts in (5) were not hypervelocity. It is this data for lead which, when combined with the data at lower densities, gives the dependence on density found in (1) and (2). It seems that no other data exists for projectile densities above 3000 kg m^{-3}, except for that reported here.

We finally look at crater volume V (shown in Figure 3 vs. impact kinetic energy). A good fit of V is found in terms of E (Table 1). The craters in CO_2 ice are smaller than in H_2O ice (factor of ≈ 11 in volume), but the power of the dependence on impact energy is similar. We compare the results to those of (3), who for impacts on H_2O ice by aluminium (with similar energies as here but velocities of 300 to 900 m s^{-1}) found a relation $V = 0.52E^{0.93}$ (V in cm^3, E in J, shown on Figure 3 as a dotted line). This agrees well in absolute normalization, but has a smaller power dependence. It is not clear if this difference is due to a difference in the range of densities used, the

REFERENCES

1. Croft, S.K., "Hypervelocity Impact Craters in Icy Media", Abstracts of 12th Lunar & Planetary Science Conference (Houston), 190-192, 1981.
2. Frisch, W., "Hypervelocity Impact Experiments with Water Ice Targets", *Hypervelocity Impacts In Space*, UK: Univ. of Kent at Canterbury, ISBN: 0-904938-32-8, 1992, pp. 7-14.
3. Kato, M. *et al.*, *Icarus* **113**, 423-441 (1995).
4. Arakawa, M. and Maeno, N., *Physics and Chemistry of Ice*, Japan: Hokkaido Univ. Press, 1992, pp 464-469.
5. Croft, S.K., Keiffer, S.W. and Ahrens T.J., *J. Geophys. Res.* **84** (B14), 8023-8032 (1979).
6. Holsapple, K.A., *J. Impact Engng* **5**, 343-355 (1987).

CP429, *Shock Compression of Condensed Matter – 1997*
edited by Schmidt/Dandekar/Forbes
© 1998 The American Institute of Physics 1-56396-738-3/98/$15.00

DISTRIBUTION OF MASS IN DEBRIS CLOUDS

Andrew J. Piekutowski

University of Dayton Research Institute, 300 College Park Ave., Dayton, OH 45469-0182

Results of a study which quantitatively determined the distribution of material in debris clouds are presented. The determinations relied on an interpretation of the dimensions of fragments taken from the radiographs of debris clouds produced by the impact of 2017-T4 aluminum spheres with various thicknesses of 6061-T6 aluminum bumpers. Bumper and projectile fragments formed three major features in well-developed debris clouds—an ejecta veil, an external bubble of debris, and an internal structure which contained a front element, a center element, a large central fragment, and a rear element. The ejecta veil and the external bubble of debris consisted almost entirely of bumper material. The internal structure consisted of the fragmented projectile and a small amount of bumper material.

INTRODUCTION

The use of a thin sacrificial outer sheet or bumper to disintegrate a fragment which threatens the integrity of the wall of a spacecraft was first proposed by Whipple (1). Knowledge of the mass, velocity, and spatial distribution of the fragments in the debris cloud formed by the impact of the original fragment with the bumper, facilitates the selection of materials used in the construction of the rear or main wall of the spacecraft. However, only very limited debris-cloud mass distribution are available. The test results presented in this paper significantly add to the body of debris-cloud mass-distribution data published in an earlier work which described the formation of debris clouds produced by the hypervelocity impact of aluminum spheres with thin aluminum sheets (2). In Ref. 2, mass-distribution data for tests using 9.53-mm-diameter spheres were presented. This paper compares the results of these tests with the results of tests using larger and smaller spheres. The model for the distribution of mass in the debris cloud was developed using fragment-size data obtained from analyses of the radiographs of the debris clouds.

EXPERIMENTAL PROCEDURES

Twelve impact tests were performed at the University of Dayton Research Institute (UDRI) Impact Physics Laboratory using a 50/20 mm, two-stage, light-gas gun. The tests were performed using 6.35-, 9.53-, 12.70-, and 15.88-mm-diameter, 2017-T4 aluminum spheres with nominal masses of 0.373, 1.275, 3.000, and 5.860 g, respectively. Various thicknesses of 6061-T6 aluminum sheets were used as bumpers. All bumpers were installed and impacted normal to the range center line.

Four pairs of fine-source, soft, flash x-rays were accurately positioned on the target chamber to provide a simultaneous orthogonal view of the projectile before impact and three orthogonal views of the debris clouds. The fourth pair of x-ray heads was fired when the debris cloud was about 30 cm downrange of the bumper. The delay in firing the fourth pair of heads allowed the debris cloud to expand and permitted a more detailed examination of the cloud structure.

In previous works describing the formation of debris clouds (2,3), the debris clouds were shown to have three major structural features—an ejecta veil on the impact or front side of the bumper, an expanding bubble of bumper debris on the rear side of the bumper, and a significant internal structure composed of projectile debris located inside and at the front of the external bubble of bumper debris. The internal structure, shown in Fig. 1, was the

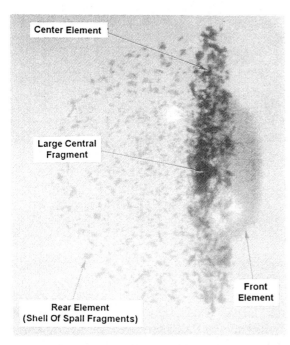

FIGURE 1. Radiograph showing the various elements which are formed in the internal structure of a debris cloud produced by the impact of an aluminum sphere with a thin aluminum sheet.

most significant feature of the debris cloud in terms of potential for damage to assets located downrange of the bumper. The front element consisted of finely divided fragments and/or molten droplets of bumper and projectile material. The disk-like center element was composed of a large number of solid slivers, comma-shaped, and/or chunky pieces of fragmented projectile and also contained a single large chunky projectile fragment which was located at the center of the disk and on the debris-cloud center line. The rear element of the structure was a hemispherical shell of fragments spalled from the rear surface of the sphere.

The determination of the distribution of the mass in a debris cloud relied on an interpretation of the dimensions of fragments which were taken from the radiographs of the debris clouds, since the

recovery and identification of thousands of individual fragments using other techniques was not practical. While use of two-dimensional objects (i.e., shadows of fragments on film) to estimate the volume and mass of three dimensional objects (i.e., the fragment) is a subjective process at best, the procedures which were used provided consistent data.

The fragments in the center element were closely spaced and overlapped, making an accurate measurement of the size and number of these fragments nearly impossible. However, the large central fragment and fragments in the spall shell were reasonably clear in most of the radiographs. The mass of material in the center element was determined to be the difference between the pre-impact mass of the sphere and the combined mass of the front element, the spall shell, and the large central fragment.

The mass of the sphere involved in the formation of the front element was assumed to be the same as the mass of the plug of bumper material that experienced quasi one-dimensional loading during the impact (see Ref. 2). Computation of the mass of the bumper plug was accomplished with use of the following assumptions and by applying the principle of conservation of momentum to the impact process. Work by Nysmith and Denardo (4) had shown that a negligible amount of projectile momentum was transmitted to the bumper sheet during impact. Consequently, the post-impact momentum of the disintegrated sphere and the bumper plug was assumed to be equal to the pre-impact momentum of the sphere. The velocity of the leading edge of the center element was assumed to be the velocity of the center of mass of the debris cloud. Accordingly, the mass of the bumper plug, m_b, was determined with use of the following relationship:

$$m_b = m(V_0 - V_C) / V_C$$

where m is the mass of the sphere, V_0 is the impact velocity, and V_C is the velocity of leading edge of the center element.

The Martin's statistical diameters of fragments making up approximately 20 percent of the spall shell were determined using procedures described in Ref. 5 and illustrated in Fig. 2. The median

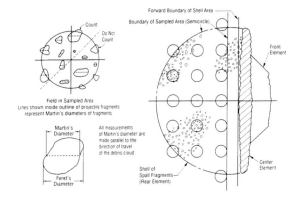

FIGURE 2. Illustration of the sampling procedure used to estimate the number and the size of particles in the shell of spall fragments.

Martin's diameter of fragments in the spall shell was determined by assuming a log-normal distribution of the measured Martin's diameters of the fragments. The mass of the median fragment was computed by assuming the median Martin's diameter to be the diameter of a spherical fragment with the density of the unshocked projectile. The total number of fragments in the spall shell was estimated by multiplying the number of fragments counted in the sampled area by the ratio of the total shell area (in the radiographs) to the sampled area. The mass of material in the spall shell, M_S, was estimated for each test by multiplying the mass of the median fragment in the spall shell by the total number of fragments in the shell.

As shown in Fig. 3, the large central-fragment height, H, width, W, and thickness, T, were

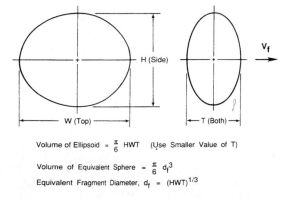

Volume of Ellipsoid $= \frac{\pi}{6}$ HWT   (Use Smaller Value of T)

Volume of Equivalent Sphere $= \frac{\pi}{6} d_f^3$

Equivalent Fragment Diameter, $d_f = (HWT)^{1/3}$

FIGURE 3. Dimensions of the ellipsoid and the computational procedures used to determine the equivalent diameter of the large central fragment.

obtained from late-time-view radiographs of the debris clouds using an optical comparator at a magnification of 10X. The measured dimensions were used to compute the equivalent diameter, d_f, of a sphere having the same volume as an ellipsoid with the dimensions of the large central fragment. Every precaution was taken to insure the measurements were of the large fragment and not several overlapping fragments. The mass of the large central fragment, m_f, was computed by assuming the density of the equivalent-diameter sphere was the same as that of the original projectile.

RESULTS AND DISCUSSION

The results of the fragment-size analyses are presented in Table 1 in order of increasing bumper-thickness-to-projectile-diameter ratio, t/D, and increasing impact velocity for each set of tests with the same t/D ratio. In Table 1, the results of the computations of m_b are presented as m_p, the mass of the sphere that was part of the front element, since these masses were assumed to be equal.

The mass of the center element, M_C, is shown as a percentage of the mass of the sphere in the extreme right-hand column of Table 1. Given the nature of the procedures used to estimate the masses of the various debris-cloud elements, this percentage was surprisingly consistent for the tests. In general, the mass of the center element (exclusive of the large central fragment) was shown to be 65 to 70 percent of the mass of the sphere.

An interpretation of the numerous observations made regarding the distribution of projectile and bumper material in the debris cloud is portrayed in the illustration presented in Fig. 4. This illustration identified the source of the material which eventually formed the ejecta veil, the external bubble of debris, and the three elements of the internal structure of a fully-developed debris cloud. The figure was drawn to scale for a test with a t/D ratio of 0.1. Debris clouds formed by impacts with other t/D ratios would have the same general distribution of material although the partitioning, on an element-by-element basis, would vary.

TABLE 1. Distribution of Projectile Mass in Fully-Developed Debris Clouds
(For 9.53-mm-diameter spheres, unless noted otherwise)

Shot Number	$\dfrac{t}{D}$	Impact Velocity, (km/s)	Front Element, m_p (mg)	Large Fragment, m_f (mg)	Spall Shell, M_S (mg)	Center Element M_C (mg)	Center Element Percent of Original Wt.
4-1395	0.026	6.70	11.6	242.2	189	832	65
4-1394	0.049	5.45	38.1	144.6	190	902	71
4-1762	0.049[a]	6.15	186.8	324.0	825	4524	77
4-1790	0.049	6.24	33.6	69.1	346	827	65
4-1358	0.047[b]	6.26	58.6	240.9	431	2270	76
4-1789	0.048[c]	6.40	7.7	17.7	93	255	68
4-1360	0.049	6.62	26.0	37.5	337	874	68
4-1744	0.049	7.38	44.9	10.8	254	965	76
4-1359	0.062	6.78	42.1	38.5	451	743	58
4-1621	0.084	4.62	67.1	162.8	163	882	69
4-1289	0.084	6.68	53.1	12.5	378	831	65
4-1716	0.135	4.71	96.0	49.5	291	838	66

[a] 15.88-mm-diameter sphere.   [b] 12.70-mm-diameter sphere.   [c] 6.35-mm-diameter sphere.

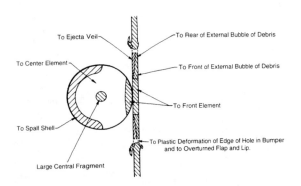

FIGURE 4. Estimated distribution of mass in a fully-developed debris cloud produced by an impact with a t/D ratio of 0.1.

The front element was composed of the lens-shaped region of the sphere and bumper sheet that were subjected to quasi one-dimensional shock loading during the impact. The spall shell was formed from a thin layer of material at the rear of the sphere. The center element and the large central fragment were derived from the remainder of the sphere material. The ejecta veil consisted almost exclusively of bumper material from a thin, tapered flat ring on the impact side of the sheet. The external bubble of debris was formed from a flat ring of bumper material from the downstream side of the sheet. The small gap shown between the outer edges of flat rings and the inside of the hole represents the material which was not actually separated from the sheet, but was plastically deformed and displaced during the later stages of hole growth.

SUMMARY AND CONCLUSIONS

Two thirds of the post-impact projectile mass appeared to be concentrated in the center element for all combinations of impact velocity, sphere diameter, and t/D ratio used in the tests. The mass of the large central fragment decreased as impact velocity or t/D ratio increased. However, the decrease in the large-central-fragment mass was usually offset by an increase in the mass of the front element and the spall shell

REFERENCES

1. Whipple, F.L., *The Astronomical Journal*, **1161**, p. 131,1947.
2. Piekutowski, A.J., *Formation and description of debris clouds produced by hypervelocity impact*, NASA CR-4707, 1996.
3. Piekutowski, A.J., *J. Impact Engng.* **14**, pp. 573-586, (1993).
4. Nysmith, C.R. and B.P. Denardo, Experimental investigation of the momentum transfer associated with impact into thin aluminum targets, NASA TN D-5492, 1969.
5. Piekutowski, A.J., *High-Pressure Shock Compression of Solids II*, Springer-Verlag, 1995, ch.6, pp. 150-175.

CP429, *Shock Compression of Condensed Matter – 1997*
edited by Schmidt/Dandekar/Forbes
© 1998 The American Institute of Physics 1-56396-738-3/98/$15.00

LIGHT FLASHES CAUSED BY IMPACTS
AGAINST THE MOON

I.V.Nemtchinov, V.V.Shuvalov, N.A.Artemieva,
B.A.Ivanov, I.B.Kosarev, I.A.Trubetskaya

Institute for Dynamics of Geospheres, 38 Leninskiy prospect, Moscow 117939, Russia

Impacts of meteoroids on airless cosmic bodies (asteroids, comets, Moon) result not only in cratering and fragmentation of a target, but cause a formation of radiating vapor plumes. Though the duration of the flash is rather short and the coefficient of the kinetic energy conversion into the radiation impulse is rather small (in comparison to the impacts onto the planets having atmospheres), it may be sufficient for detection of such impacts.

Radiation efficiency of the impact is determined by many factors: mechanical and chemical composition of the target and projectile, optical properties of the eject material, impact parameters (velocity, trajectory inclination, event scale, etc.). An estimate of the light impulse created by a 1m-in-radius silica body impacting the lunar surface with the velocity of 50km/s has been made by Melosh et al. (1). In view of some impending experimental work (2) it appeared important to review and refine the estimates of luminous efficiencies and other characteristics of impact originated light flashes, e.g., shape, duration, spectra for various velocities, sizes, compositions, etc.

THERMODYNAMICAL PROPERTIES

Equation of state is critical for temperature determination, especially at the late stage of vapor release. We considered the impactor and target to consist of similar materials. We tried ANEOS for quartz (3), Russian equation of state for granite (4) (which will be called below as RUGEOS) and gas equation of state for H-chondrite vapor (5). Analysis of shock and release adiabats show that RUGEOS is more appropriate for description of the initial stage of impact (shock compression and plume formation) for real meteoroids and lunar soil. The most difficult is a consideration of condensation process. The theory of condensation is not developed enough yet, and different assumptions give conflicting results. If equilibrium is established, the plume temperature remains constant for a long time due to release of evaporation en-

ergy. This leads to prolongation of radiative impulse. If there is no condensation at all, vapor temperature drops very fast due to adiabatic cooling. So we may argue that RUGEOS gives the upper limit of radiation, and gas equation of state for vapor gives the lower limit. In our numerical simulations we used both EOS.

OPTICAL PROPERTIES

Opacities of the vapor are much more sensitive to chemical composition than thermodynamical properties. Even small impurities can significantly change the absorption coefficients. For this reason we have used calculated optical properties of H-chondrite, which is more representative for meteoroids composition and lunar soil than pure silica (6).

Fig.1 demonstrates absorption coefficients for typical parameters of a vapor cloud : T=3000K and

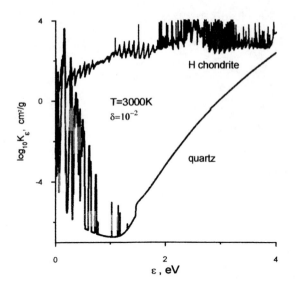

FIGURE 1. Mass absorption coefficients versus photon energy for H-chondrite (thin black line) and pure quartz (thick gray line).

δ=0.01, where T is the temperature and δ is the relative density (with respect to normal density 0.00109g/cm^3). The difference between pure SiO$_2$ and H-chondrite is clearly seen, for the visible range of the wavelengths this difference reaches 5-8 orders of magnitude.

NUMERICAL SIMULATIONS

To determine luminous efficiencies for impacts against the Moon we have performed 2D and 3D numerical simulations of these impacts, using a multimaterial Eulerian code similar to the one widely used in the USA, i.e., the CTH code (7). Radiation fluxes and spectra at different moments of time were determined by direct solution of transfer equation for various photon energies along a great number of rays. Simple estimates and the results of numerical simulations show that radiated energy is much less than the thermal energy of the plume for the entire period of interest. So radiative cooling does not change the plume temperature, and the influence of radiation on gasdynamic flow was not taken into account.

Fig.2 demonstrates typical temperature and density distributions for a vertical impact. The plume has a conical shape with rarefied inner core and dense envelope formed due to ejection

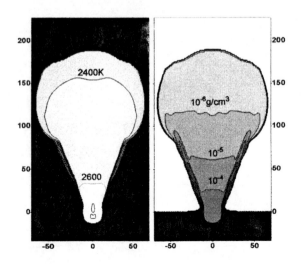

FIGURE 2. Density (left-hand image) and temperature (right-hand image) distributions at 250µs after 1cm-in-radius body impact with the velocity of 20km/s. All distances are measured in cm.

of partially evaporated matter. In the case of equilibrium condensation it is this envelope that radiates the main portion of energy.

Fig.3-4 present radiative fluxes and luminous efficiencies (the ratio of emitted energy to kinetic energy of a projectile) for different velocities in assumption of equilibrium condensation. Most of

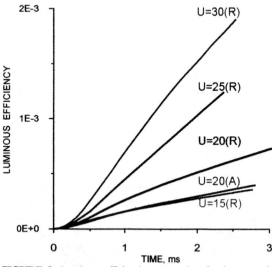

FIGURE 3. Luminous efficiencies versus time for the runs with equilibrium condensation. The values of impact velocity are indicated near the curves. Letters A and R correspond to simulations using ANEOS and RUGEOS respectively. Projectile radius equals 1cm.

958

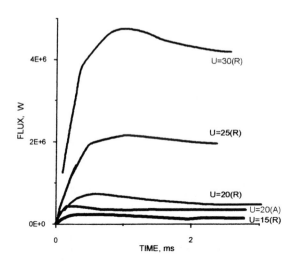

FIGURE 4. Radiative fluxes versus time for the runs with equilibrium condensation (all designations are the same as in Fig.3).

the energy is radiated at the late stage of the plume evolution, when vapor temperatures are almost constant due to energy release in the process of condensation. The usage of RUGEOS gives total radiation fluxes, which are twice as great as those obtained with ANEOS. It follows from our simulations that a difference in luminous efficiency between runs with condensation (using ANEOS or RUGEOS) and without it (with gas equation of state) reaches an order of magnitude. And this difference appears about 100μs after the collision. As it was expected, luminous efficiency considerably increases with the increase of velocity.

Analysis of spectra obtained in the course of simulations shows that in the case of equilibrium condensation only 10-20% of radiation is emitted in the visible. So for better detection IR sensors should be used.

Luminous efficiency depends on the impactor size as well. An increase of projectile radius from 1 to 10 cm leads to increase in luminous efficiency by a factor of 1.5-3.0.

Fig. 5 demonstrates the density and temperature distributions for an oblique impact with an angle of 30° to the horizon and a velocity of 25 km/s. Density and temperature under shock loading are defined by the vertical component of the impact

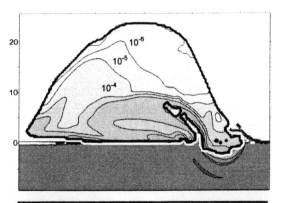
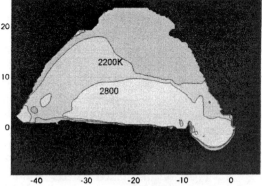

FIGURE 5. Density (upper image) and temperature (lower image) distributions at 10μs after oblique (30° to horizon) 1cm-in-body impact with the velocity of 25km/s. All distances are measured in cm.

velocity (and are smaller to those in the case of vertical impact). The projectile does not penetrate deep into the target, it slides over the surface, forming a high- velocity jet. As in the case of the vertical impact the temperature of the vapor cloud is equal to the temperature of phase transition. The target material also arises above the surface, but occupies a rather small region near the crater. It follows from our simulations that at the initial stage of impact (first 30-50μs) luminous efficiency for oblique impact is higher than that for a vertical collision. This nonobvious effect is explained by a difference in the depth of penetration. In the vertical impact meteoroid penetrates the target for a distance of several characteristic sizes, and the radiating surface is defined for a long time by the crater radius. Almost all the high-temperature mass is screened by cold dense surface material. In the oblique impact the vapor cloud just after the collision is much bigger which results in increase of luminous efficiency.

SURFACE GEOMETRY

Luminous efficiency of impact generated explosions depends strongly on availability or lack of the atmosphere. The thermal energy of the plasma cloud formed in the process of collision is quickly transformed into kinetic energy. Only a small fraction of impact energy (<0.1%) may be reradiated within this short period of time. If the density of surrounding air is high enough, the air decelerates the jet and its kinetic energy is transformed into the thermal energy, which dramatically increases luminous efficiency. There is no such effect for a target in vacuum conditions. But if the angle of trajectory inclination (to the horizon) is small enough the jet expands along the surface. Some part of vapor may be decelerated by different natural obstacles (crater rims, boulders, walls of the ramparts etc.). This effect may substantially increase luminous efficiency. But even if there are no obstacles, trajectory inclination affects emitted radiation.

PROJECTILE AND TARGET STRUCTURE

In reality the target (especially in the case of impacts onto the Moon covered by a thick lunar regolith) and even the impactor itself may be porous ones, which makes the problem much more difficult. Moreover the size of grains may be comparable with impactor radius. In this case temperature and even mechanical equilibrium is not established, and very high temperature (and radiation) may be achieved even with a rather small velocity of about 3-7 km/s.

Laboratory experiments (8) demonstrate that rather high temperature of about 7,000-8,000K may be obtained with very small velocities of 1.5-3.0 km/s. This problem is now being studied theoretically.

It should be noted here that the direct extrapolation of the laboratory experiment results (for a example similar to those of Schultz (9)) for interpretation of real impacts may be wrong because of unrealistic ratio of porous target grains and projectile sizes in comparison to those for a real impactor and lunar soil grains (6).

DISCUSSION AND CONCLUSIONS

The results of numerical simulations show that a lower limit of luminous efficiencies is 10^{-6}-10^{-5} (depending on impactor velocity). The results depend on equation of state in use and especially on equilibrium or nonequilibrium of condensation process. The lower limit is determined by radiation of hot noncondenced material. The energy emitted during this stage is small because of fast adiabatic cooling. The assumption of equilibrium condensation gives an upper limit of luminous efficiency (up to 10^{-3}). But in many cases the real value may be considerably greater due to the specific geometry and target structure. The value of luminous efficiency of about 10^{-5}-10^{-4} is believed to be rather feasible and may be used to forecast a probability of lunar impacts. We are looking forward to continue these investigations on the base of real observations.

ACKNOWLEDGMENTS

We are very grateful to Prof. Peter Franken and Prof. Igor Sobelman for valuable remarks and discussions. The work is partially supported by the USA Department of Defense.

REFERENCES

1. Melosh, H.J., Artemjeva, N.A., Golub', A.P., Nemtchinov,I.V., Shuvalov,V.V., and Trubetskaya I.A. *Lunar and Planet. Sci. Conf. XXIV.(abstr)*, 975-976 (1993)
2. Franken, P., Crowe, D., Sobelman, .I, personal communications
3. Thompson, S.L., and Lauson, H.S, *Sandia National Laboratory Report SC-RR-71 07-14* (1972)
4. Zamyshliaev, B.V., and Yevterev, L.S., *Models of dynamic deforming and failure for Ground Media*, Moscow: Nauka, 1990, 215 p. (in Russian)
5. Kosarev I.B., Loseva T.V., Nemtchinov, I.V., *Solar System Research* **30**, 40, 265-278 (1996)
6. Papike, J.J., and Vaniman, D.T., *Mare Crisium. The view from Luna 24.* (Merrill, R.R., and Papike, J.J., eds.) Lunar and Planet.Inst, Houston, Tex., 1978, pp.281-289.
7. McGlaun, J.M., Thompson, S.L., and Elrick, M.G., *Int.J.Impact.Engng.***10,** 351-360 (1990).
8. Belyakov, G.V., Rodionov, V.N., Samosadny, V.M. *Fizika gorenia i vzryva (Physics of combustion and explosion)* **13,** 614-619 (1977) (in Russian).
9. Schultz P., *JGR* **101,**21,117-21,136 (1996).

CHAPTER XIV

HIGH VELOCITY LAUNCHERS AND SHAPED CHARGES

CP429, *Shock Compression of Condensed Matter – 1997*
edited by Schmidt/Dandekar/Forbes
© 1998 The American Institute of Physics 1-56396-738-3/98/$15.00

OPTIMIZATION STUDIES OF A THREE-STAGE LIGHT GAS GUN

Lewis A. Glenn

Lawrence Livermore National Laboratory, Livermore, California 94550

ABSTRACT

A new gasdynamic launcher is described, in which intact projectiles weighing at least 1 gram can be accelerated to mass velocities of 15-20 km/s. The system employs a conventional 2--stage light gas gun, with the barrel modified and filled with helium to act as a pump tube for a third stage. The key feature of the launcher is that the peak pressure in the third stage can be maintained below 2.5 GPa, thus assuring high efficiency and the integrity of the projectile.

INTRODUCTION

The maximum velocity attainable in a gasdynamic launcher is limited by the maximum sound speed in the driver gas. For a conventional 2–stage light gas gun, the limit is ~ 10 km/s. To achieve higher velocities requires adding additional stages. Staging methods have been reported whereby a pusher of high shock impedance impacts a stack of target plates in which the impedance decreases from plate to plate in the desired direction of motion (1, 2), and velocities up to 14 km/s have been obtained in this manner. Chhabildas and his colleagues at Sandia National Laboratory have reversed this strategy with a graded density pusher, and they have demonstrated velocities up to 15.8 km/s with a 6 mm-diameter × 0.56 mm thick titanium projectile (3). A difficulty with this method is that the loading pulse is applied in a very short period of time and to achieve such a high velocity in this period requires a very high pressure, up to 100 GPa or even more. This means that the projectile thickness must be kept very thin to prevent spall fracture, and the impactor design is highly constrained to avoid shock-melting or even vaporizing the projectile. Also, such high pressures

effect large energy losses via *pdV* work on the walls of the launcher; in the 15.8 km/s experiment cited above, the kinetic energy efficiency of the third stage (ratio of the kinetic energy of the flyer to that of the impactor) was only 0.6%.

It may be possible to avoid many of these problems by slightly modifying the configuration of the conventional 2–stage system, as shown in Fig. 1. Normally, a vacuum is drawn in the 2nd stage launch tube of the gun, so that after the burst diaphragm is broken by the compressed hydrogen, the 2nd stage projectile is accelerated with no downstream resistance. If instead, this launch tube is closed at the downstream end with a second burst diaphragm, and filled with an appropriate mass of light gas, the second-stage projectile will act to compress the gas in much the same manner as does the piston in the first stage. The third stage projectile is placed downstream of the second burst diaphragm in a smaller diameter barrel, which now becomes the final launch tube.

We are interested in the solution of the following problem: Given the fixed characteristics of a two-stage light gas gun (pusher mass and velocity, m_2

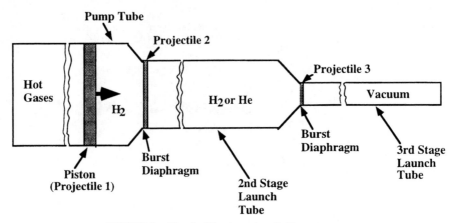

FIGURE 1. Sketch of the three-stage light gas gun.

and v_2, cross sectional area and length of barrel, A_2 and L_2), find the largest projectile mass, m_3, that can be accelerated to a final velocity, v_3, subject to the constraint that the maximum pressure on the projectile, $\hat{p}$ is less than some prescribed value. It is required to also find the corresponding third stage characteristics, A_3 and L_3, and the interstage gas mass, m_g. Under certain simplifying assumptions, it can be shown that an exact solution to the above-posed problem exists (4). Our approach in what follows will be to use the analytical results to derive the approximate solution space and then to perform hydrodynamic code calculations to assess the effects of removing the simplifying assumptions.

ANALYTICAL MODEL

If it is assumed that 1) the walls of the gun tubes remain rigid, 2) there is a step change in cross section between stages and the pusher mass, m_2, is brought to rest precisely at this location, 3) pressure gradients in the interstage gas can be ignored, 4) a monatomic γ-law equation of state describes the gas, and 5) after the first shock, the gas is effectively isentropically compressed, then the following solution for the third stage mass is derived in (4):

$$m_3 = \frac{4k}{3}\frac{m_g}{g^2}\left[\left(1+\frac{1}{\alpha}\right)-\frac{1}{k}\left(\frac{\Phi}{1+\Phi}\right)\right] \quad (1)$$

where g is the velocity gain, v_3/v_2, k is the fraction of the peak internal energy of the gas that goes into increasing the kinetic energy of the projectile from its value at the time of peak pressure to its final value; k is essentially a free parameter, however, choosing a value of k much larger than 0.75 leads to diminishing returns, the energy saved being at the expense of greatly increased barrel length. The quantity, α, derives from solution of the equation:

$$\left(\frac{1+\alpha}{\alpha}\right)^{5/2}\frac{\alpha(1-\alpha)}{1-\alpha+[\alpha(1+\alpha)]^{1/2}}=\frac{3A_2L_2}{4}\frac{\hat{p}}{m_2v_2^2} \quad (2)$$

All the quantities on the right-hand side of equation (2) are given, so that α is readily evaluated. The quantity, Φ in equation (1) is then found from the relation:

$$\Phi=\frac{[\alpha(1+\alpha)]^{1/2}}{1-\alpha} \quad (3)$$

and the gas mass, m_g, is determined by:

$$m_g=\frac{3}{4}m_2\left(\frac{\alpha}{1+\Phi}\right) \quad (4)$$

The cross sectional area of the third stage launch tube, A_3, is then found from the relation:

$$A_3=\left(A_2^2\Phi\frac{m_3}{m_2}\right)^{1/2} \quad (5)$$

and finally, the length of the third stage launch tube, L_3, is given by:

964

CD-Rom Installation, Instructions and
System Requirements for:

AIP
CONFERENCE
PROCEEDINGS
429

TOOLS & BUTTONS

The tool bar contains "button" icons that help you view the articles. To use a tool click on a button with your mouse.

 Page Only closes the overview area of the window.

 Bookmarks and Page opens the overview area to display book marks. Click on the bookmark to move to that location.

 Thumbnails and Page uses overview to display thumbnail images of article pages.

 Hand scrolls text in the direction you indicate.

  Zoom magnifies or reduces page display.

 Select text highlights text for copying on to the clip board with the copy command.

   Browse moves within a docu ment one page at a time or jumps to the first or lat page.

  Go Back/Forward retraces your viewing steps.

 Actual Size displays a page at 100%.

 Fit Page scales a page to fit within the window.

 Find locates text in the document.

$$L_3 = \frac{\left(\dfrac{1+\alpha}{1+\Phi}\right)\dfrac{m_2 v_2^2}{\hat{p}}}{3A_3(1-k)^{3/2}} \qquad (6)$$

To illustrate, the intermediate LLNL two-stage light gas gun has been shown to fire a projectile, $m_2 = 37$ g at a velocity $v_2 = 6.2$ km/s. The launch tube bore diameter is $D_2 = 28$ mm (so that $A_2 = 6.16$ cm^2) and the length is $L_2 = 9$ m. Assuming it is desired that the projectile velocity from the added third stage be $v_3 = 20$ km/s, $g = 3.23$. If the third stage projectile is made of titanium alloy ($\rho_3 = 4.51$ g/cm^3), we choose $\hat{p}$ to be 2.5 GPa, less than half the ~ 5.5 GPa spall strength. Equations (1)-(6) then yield $\alpha = 0.267$, $\Phi = 0.793$, $m_3 = 1.65$ g, and $A_3 = 1.158$ cm^2, so that the bore diameter of the third stage barrel is 12.1 mm and the projectile length is $m_3/A_3\rho_3 = 3.16$mm. Also, $m_g = 4.13$ g, which leads to a fill pressure of 0.46 MPa assuming helium is employed, and $L_3 = 9.3$ m. Although this represents a quite acceptable parameter set, it might be desirable to reduce the barrel length. This could be accomplished by reducing the projectile mass (length) somewhat. Another alternative is to accept a slightly lower final projectile velocity. Code calculations show that the 1.65 g projectile attains 0.95 v_3 in the first 3.9 m, and is already at 0.9 v_3 in 2.7 m.

We have not accounted for losses thus far. These are from three principal sources, gasdynamic drag, heat loss to— and pdV work done by the gas on the walls. We intend to choose $\hat{p}$ well below the spall strength of the projectile material, so as to minimimize pdV work on the walls. Heat loss by radiation transport can be shown to be insignificant as long as the gas temperature remains below ~ 10 eV. Gasdynamic drag and convective heat transport from the turbulent boundary layer adjacent to the wall are discussed below.

HYDROCODE CALCULATIONS

GGUN2 is an extension of earlier codes (GGUN(5) and IGUN(6,7)) employed to study the performance of the two-stage light gas gun and various implosion and multistage launchers. It is an arbitrary Lagrange-Eulerian (ALE) code that solves the equations of motion in one dimension, but takes into account arbitrary flow cross section (including discontinuities), convective heat transport to the walls, and gasdynamic wall drag. Tabular equations-of-state can be employed, as well as Grüneisen, polynomial, and γ-law models, and JWL-explosive and Noble-Abel propellant burn packages have recently been added. There is also a coupled wall motion algorithm that allows the gun-barrel walls to move in response to the developing internal pressure. The equations of motion and other details have been described earlier (7).

The GGUN2 code was validated by comparing the computational results with experimental data from the first two stages of the three-stage light gas gun discussed in the previous section; it was shown in (4) that excellent agreement was obtained. Excellent agreement was also obtained by comparing third stage calculations with 2D simulations performed with the CALE code (8).

Helium is the preferred working fluid for all stages beyond the second because a higher final sound speed can be produced for given initial and final pressures. This is true also for the second stage. For example, when we recalculated the two-stage launcher discussed above, using helium in place of hydrogen (all other parameters remained the same, including the initial gas density and temperature), the final velocity increased by about 3%, notwithstanding the lower atomic mass of the hydrogen. However, practical considerations preclude the use of helium in the second stage. Although the velocity was slightly higher, the peak gas temperature in the AR (transition) section with helium was 6360 K compared with 1880 K with hydrogen. The slight increase in performance is more than countered by the potentially severe erosion that would be engendered by repeated use of the gun with helium. In the third stage, however, this is no longer a problem since the barrel in this stage cannot be salvaged and must be replaced after each shot.

Figure 2 shows the performance when the results of the analytical model are used to design the third stage. Both the velocity of, and the peak pressure on, the third stage projectile are shown for 3 cases. All 3 calculations employed tabular equations of

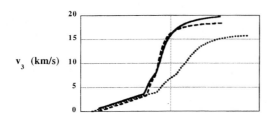

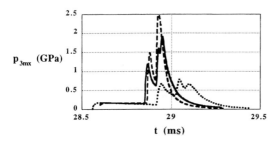

FIGURE 2. Calculated velocity of and peak pressure on third stage projectile. Solid curve includes radial wall motion, dashed curve adds convective heat transfer, and dotted curve adds gasdynamic drag.

state for both the hydrogen in the second stage and the helium in the third. The solid curves depict the results when wall motion is included, but not heat transfer or drag. Assuming fully developed turbulent flow, the dashed curves include convective heat transfer from the helium plasma to the walls, and the dotted curves include both heat transfer and gasdynamic drag.

DISCUSSIONS & CONCLUSIONS

The calculated results exhibit the effects of all important loss mechanisms, pdV work on the walls (kept small by maintaining relatively low pressures in the system), heat transfer, and turbulent drag. The latter effect appears most important, and reduces the maximum velocity from 20 to 16 km/s. At the same time, the peak pressure on the projectile drops from 2.5 GPa to approximately 0.8 GPa. Subsequent calculations have shown that shaving the length of the projectile to decrease the mass to 1 g (all other parameters fixed, and all loss mechanisms activated) increases the predicted

velocity to 18 km/s and decreases the peak pressure to less than 0.5 Gpa.

REFERENCES

1. Laptev, V. I., and Trishin, Yu. A., "Increase of initial velocity and pressure upon impact on an inhomogeneous target", *Zh. Prikladnoi Mekhaniki i Tekhnickeskoi Fiziki*, **6**, 128-132 (1974); Eng. transl. Sov. J. Appl. Mech. and Tech. Phys., p. 837-841 (1976).
2. Fowles, G. R., Leung, C., Rabie, R., and Shaner, J., "Acceleration of flat plates by multiple staging", *High--Pressure Science and Technology, Sixth AIRAPT Conference -1977*, 911-919, Plenum Press, New York (1979).
3. Chhabildas, L. C., Kmetyk, L. N., Reinhart, W. D., and Hall, C. A., *Int. J. Impact Engrg.*, **17**, 183-194 (1995).
4. Glenn, L. A., "On how to make the fastest gun in the West", *International Workshop on New Models and Numerical Codes for Shock Wave Processes in Condensed Media - 1997*, Oxford, UK (see also Lawrence Livermore National Laboratory Report UCRL-JC-126057, January 1977).
5. Glenn, L. A. "Performance analysis of the two-stage light gas gun", *Shock Waves in Condensed Matter - 1987*, 653-656, Elsevier Science Publishers, The Netherlands (1988). (see also Lawrence Livermore National Laboratory Report UCRL-96021, May 1987).
6. Glenn, L. A., Latter, A. L., and Martinelli, E. A., "Multistage gasdynamic launchers", *Shock Compression of Condensed Matter - 1989*, 977-984, Elsevier Science Publishers, The Netherlands (1990).
7. Glenn, L. A., *Int. J. Impact Engrg.*, **10**, 185-196 (1990).
8. Tipton, R., "CALE - a *C*-Language, arbitrary Lagrange Eulerian hydrocode for UNIX systems", Lawrence Livermore National Laboratory Report, to be published.

CP429, *Shock Compression of Condensed Matter – 1997*
edited by Schmidt/Dandekar/Forbes
© 1998 The American Institute of Physics 1-56396-738-3/98/$15.00

THREE DIMENSIONAL NUMERICAL SIMULATIONS
OF THE EFFECT OF CYCLIC PERTURBATIONS
ON LINER ACCELERATION

Gabi Luttwak[1] and Meir Mayseless

RAFAEL, P.O. Box 2250, Haifa 31021, ISRAEL

We investigate in numerical simulations the effect of cyclic angular disturbances on the liner motion and deformation. The calculations were carried out in the three-dimensional codes MSC/Dytran and Autodyn-3D. Inert inserts inside the charge caused the disturbances. The results show the dependence of the amplitude of the perturbation on the distance of the inserts from the liner. We have also considered the case of cyclic variations in the casing width. We discuss the symmetry of the problem and the numerical techniques utilized in the simulations.

INTRODUCTION

A typical shaped charge or self forging projectile has axial symmetry. In the presence of most perturbations the axial symmetry is broken and a three dimensional calculation has to be carried out to model the configuration. In this work we consider the case of cyclic perturbations over the charge circumference.

In this case the computational domain can be limited to an angular slice depending on the wave length of the perturbation. In the case of four perturbations as shown in Fig.1 there are 4 planes of symmetry and a slice of 45^0 could be taken.

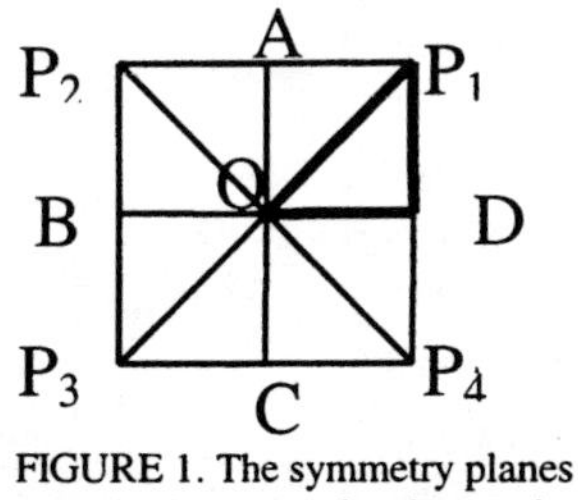

FIGURE 1. The symmetry planes for four perturbations over the charge circumference

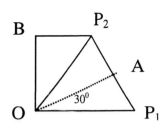

FIGURE 2. The case of hexagonal symmetry.

---

[1] On sabbatical at Century Dynamics, 2333 San Ramon Valley Blvd., San Ramon CA 94583

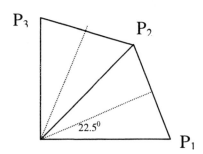

FIGURE 3. The case of octagonal symmetry.

Similarly for hexagonal symmetry (Fig. 2) a slice of 30^0 and for octagonal symmetry (Fig. 3) an angular slice of 22.5^0 should be taken.

In all of the above cases, there are two symmetry planes which divide the computational domain into four quadrants. Thus in Figs.2-3 only one quadrant is shown with the additional symmetry planes shown for each case. The finite difference schemes used in the simulations work best for an orthogonal mesh with zones of hexahedrons. Therefore, most of the calculations have been carried out for a full quadrant of the computational model.

Such perturbations are employed to achieve aerodynamically stabilized self forging projectiles (Weimann (1)). The cyclic disturbances form a projectile with a star shaped tail that has good flight stability. This kind of disturbances can be generated by varying the width of the casing or by inserting inert materials inside the explosive.

In the present work, we have carried out three-dimensional numerical simulations of the liner motion in the presence of cyclic perturbations. For simplicity, a 100mm long cylindrical charge with a radius of 50mm was chosen. The cyclic perturbations were caused either by an inert cylindrical insert inside the charge, or by varying the casing width. In the first case we have also checked the sensitivity of the perturbation to the distance of the inserts from the liner.

THE DYTRAN SIMULATIONS

In the Dytran (2) simulations, the charge was inclosed in a steel casing. Both the casing and the liner were 1mm thick. They were both modeled as a thin shell and they were coupled with the ALE (Arbitrary Lagrangian Eulerian) technique (4) to the charge. Perspex cylinders with a length of 40mm, and a radius of 10mm were inserted inside the charge at a radius of 35mm, while the distance from the liner to the beginning of the Perspex cylinder was between 2 and 10mm in the different calculations. When the distance was higher than 7.5mm the perturbations in the shape of the shell were quite small. In this case we have plotted the isocontours of the axial position z on the surface of the shell to in order be able to better follow the amplitude of the perturbations.

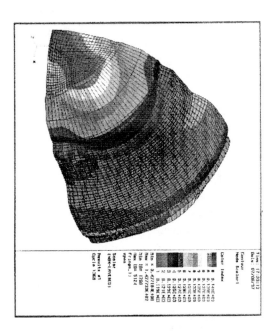

FIGURE 4. The perspex cylinders are inserted 7.5mm from the liner.Isocontours of z are shown at T=100μsec

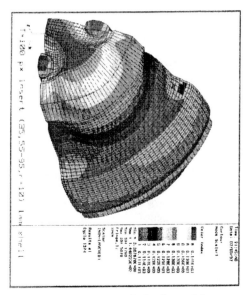

FIGURE 5. The Perspex cylinders inserted 5mm
from the liner. The isocontours of z are shown at
T=100μsec

In other calculations, we tried to generate the
perturbations by varying the thickness of the

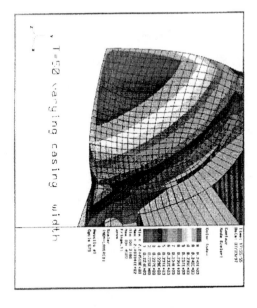

FIGURE 6. The liner at T=50μsec for the case
of varying casing width.

casing. In Fig.6 we can see the charge and the liner.
The contours of constant axial z coordinates at
T=100μsec do not show significant perturbation in
the shape of the liner.

THE AUTODYN SIMULATIONS

In the AUTODYN-3D (3) the main charge was
modeled in Lagrangian coordinates and it was
coupled to the shell by a sliding-impacting interface.
The cylindrical charge had the same dimensions, but
it was unconfined. The perturbations were generated
by inserting Teflon into a block of 4x4x3 zones in
the charge as seen in Fig.7.The charge was meshed
with 14x14x40 zones.

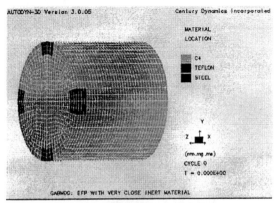

FIGURE 7. Setup of the Autodyn simulations.

FIGURE 8. The charge at T=10 μsec.

969

In Fig.8 we can see the charge at T=10μsec. In Fig.9 we can see the projection of the liner into the z-plane at T=150μsec. In Fig.10 we can see the liner as seen from the side.

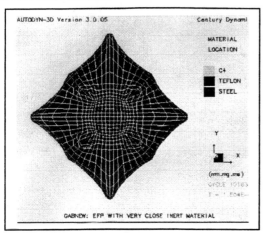

FIGURE 9. The liner projection into the z-plane at T=150μsec.

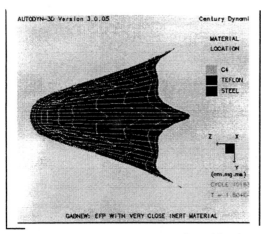

FIGURE 10. The liner at 150μsec in a side view.

The results again show that the amplitude of the disturbances is very sensitive and grows as the distance of the inserts to the liner decreases.

CONCLUSIONS

The amplitude of the perturbations increased as the disturbances were inserted closer to the liner. Variations in the thickness of the casing did produce weaker perturbations than the plastic inserts in the charge near the liner. In our calculations, we took a relatively high strength steel with a yield of 0.5GPa. This may have caused the damping of small amplitude disturbances in these calculations. For the cylindrical charge considered, the liner remains thin, and this allows us to model it as a thin shell. For a self forging projectile, the liner collapses toward the axis, and the liner should be modeled as a material region.

AKNOWLEDGEMENTS

The first author wants to thank Naury Birnbaum for his help and also for enabling him to complete this paper while starting his sabbatical at Century Dynamics.

REFERENCES

1. Weimann K., "Flight Stability of EFP with Star Shaped Tail", in the Proc. of the 14th Int. Symposium on Ballistics,Quebec,Canada, pp. 755-763 (1993)
2. MSC/Dytran, User Manual, The MacNeal Schwendler Corp. (1996)
3. AUTODYN-3D, User Documentation, Century Dynamics
4. G. Luttwak , C. Florie, A.Venis, "Numerical Simulation of Soft Body Impact", Shock Compression of Condensed Matter 1991,S.C.Schmidt et al. editors, pp999-1002,North Holland,(1992)

CP429, *Shock Compression of Condensed Matter – 1997*
edited by Schmidt/Dandekar/Forbes
© 1998 The American Institute of Physics 1-56396-738-3/98/$15.00

HIGH VELOCITY FLYERS ACCELERATED BY MULTI STAGE EXPLOSIVE SLABS

Sun Chengwei, Zhao Feng, Wen Shanggang, Li Qingzhong, Liu Cangli

Laboratory for Shock Wave and Detonation Physics Research,
Southwest Institute of Fluid Physics, P.O.Box 523, Chengdu Sichuan 610003, P.R.China

The problem of accelerating metallic flyers to ultra high speed with strong detonating explosive slabs has been analyzed and numerically simulated in this paper, where the next stage slab is impacted by the flyer of previous stage and accelerates the next stage flyer to a higher speed. There is a high plateau in the detonation products pressure profile of the slab, to which the effective acceleration is attributed. A combination of impedance matched flyers of the final stage is impacted by the strong detonating explosive driven flyer at speed 6-7 km/s, and could be sped up over 10 km/s. This kind of high speed impactors have the advantages of simple structure, lower cost, smart design and promising in many applications of high dynamic pressure loading and high velocity impact.

INTRODUCTION

In order to study dynamic behaviors of materials under very high pressure ($10^2 \sim 10^3$ GPa), and high velocity impact phenomena, many dynamic loading techniques, such as light gas gun, high power laser and underground nuclear explosion, have been developed. But these devices are expensive, bulky, and with many limitations in use. By means of several stages of the explosive slab-flyer combinations the final stage flyer could be sped to a very high velocity over 10 km/s. This test set-up is simple and useful as a dynamic loading method (1-3).

In this paper, the strong detonation propagation in the explosive slab impacted by the previous stage flyer and the acceleration of this stage flyer by the strong detonation products have been analyzed with one dimensional gas dynamics theory and the Gurney model. Furthermore, the complete dynamic process in a three stage device has been numerically simulated with the one dimensional reactive hydrodynamic elastic and plastic code SSS. The calculation indicates that the velocity of the last molybdenum flyer can be over 14 km/s. This prediction should be verified and improved by the experiments to be conducted.

STRONG DETONATION PROPAGATION IN EXPLOSIVE SLAB

Steady Strong Detonation

Let D_J denote the normal steady detonation speed and γ the isentropic index of detonation products of the explosive slab which is impacted by a planar rigid piston at a constant velocity u_0. A strong detonation will form in the slab if u_0 is greater than the CJ particle velocity $u_J = D_J/(\gamma+1)$. Behind the detonation front the products flow field is uniform and with a particle velocity u_0. If the initial pressure p_0 ahead of the front is neglected, the steady strong detonation speed D can be deduced as (4)

$$D = \frac{D_J^2 + (\gamma+1)^2 u_0{}^2}{2(\gamma+1)u_0} \tag{1}$$

The strength of a strong detonation can be characterized by the parameter z defined as

$$z = [1 - (D_J/D)^2]^{\frac{1}{2}} \tag{2}$$

971

Obviously, $0 \le z \le 1$, $z = 0$ or 1 denotes CJ detonation or the infinite strong detonation respectively. With the parameter z, the flow variables just behind the strong detonation front are

$$u = \frac{D(1+z)}{\gamma+1}, \quad v = \frac{v_0(\gamma - z)}{\gamma+1}, \quad p = \frac{\rho_0 D^2(1+z)}{\gamma+1},$$

$$c^2 = \frac{\gamma D_J^2}{(\gamma+1)^2}(1+z)(\gamma - z) \qquad (3)$$

where the reaction zone thickness is neglected, u, v, p and c are particle velocity, specific volume, pressure and sound speed of detonation products respectively, $\rho_0 = 1/v_0$ is explosive density. In the following, we use above symbols to denote the corresponding dimensionless variables whose dimension factors are ρ_0, D_J and explosive slab thickness l. Then the Riemann invariants can be expressed as

$$\alpha = u + \frac{2c}{\gamma-1} = \frac{3\gamma-1}{\gamma^2-1} + \frac{2z}{\gamma+1} - \frac{(\gamma+1)z^2}{4\gamma(\gamma-1)} + O(z^3)$$

$$\beta = u - \frac{2c}{\gamma-1} = -\frac{1}{\gamma-1} + \frac{(\gamma+1)z^2}{4\gamma(\gamma-1)} + O(z^3) \qquad (4)$$

The Quasi-Steady Approximation of Strong Detonation Propagation

After the impactor flyer M_1 with initial velocity u_0 hits a explosive slab, a strong detonation is induced in the slab. Then the detonation products push the flyer backwards, therefore, the strong detonation front is sped down. The interaction between M_1 and the strong detonation front proceeds until the detonation front becomes a normal one. In order to uncouple this interaction, the quasi-steady approximation is assumed that the rigid flyer M_1 acts as a piston and moves under the pressure of uniform detonation products described by Eqs. (1) and (3), where u_0 should be replaced by the flyer's instant velocity U_1. The flyer acceleration can be described by

$$M_1 dU_1 / dt = -p = -DU_1 \qquad (5)$$

where M_1, U_1 and t are flyer's dimensionless mass, velocity and time respectively. Denote x_D as the detonation front position, $dx_D/dt = D$ yields

$$dx_D / dU_1 = -M_1 / U_1 \qquad (6)$$

under the initial condition $t = 0$: $x_D = 0$, $U_1 = u_0$ Integrate Eq. (6) and get

$$U_1 / u_J = tg[arctg(u_0 / u_J) - t / 2M_1]$$

$$x_D = M_1 \ln(u_0 / U_1) \qquad (7)$$

U_1 decreases as t increases. At $t = t_J$, U_1 becomes u_J, then the strong detonation becomes the normal detonation or CJ detonation. The transition time is

$$t_J = 2M_2\left[arctg(u_0 / u_J) - \pi/4\right] \qquad (8)$$

The Linear Approximation of Strong Detonation Propagation

If z is small and the terms of z^2 and higher order are neglected, the Riemann invariant β is approximately equal to β_J, and the detonation front is controlled by the invariant α depending on the impactor flyer. The α family of characteristics will be a truncated fan with the center coordinates as (5):

$$x_* = -\frac{1+4u_0}{\upsilon(1+2u_0)^2}, \quad t_* = -\frac{2}{\upsilon(1+2u_0)^2} \qquad (9)$$

where $\upsilon = 16/27M_1$, The point for flyer M_1 to impact the explosive is taken to be the origin of (x,t) coordinate. The locus of strong detonation front is described by the following equations

$$dx_D / dt = D(z),$$
$$(x_D - x_*)/(t - t_*) = u + c = \alpha(z) \qquad (10)$$

and the initial condition $t = 0$: $x_D = 0$
If we take the linear approximation

$$D(z) \approx 1 + z^2 / 2, \alpha(z) \approx 1 + z / 2$$

then obtain

$$x_D - x_* = \frac{(t-t_*)[3+C(t-t_*)]}{2+C(t-t_*)},$$

$$D = 1 + \frac{2}{[2+C(t-t_*)]^2} \qquad (11)$$

where $\quad C = 2\upsilon(1-2u_0)(1+2u_0)^2/(4u_0-1)$.
The approximate solution (11) is only valid for $1/4 \le u_0 \le 1/2$, where $u_0 \ge 1/4$ is the necessary condition for strong detonation, and for $u_0 = 1/2$, Eq. (11) yields z=1, i.e., the infinite strong detonation. Obviously, this limitation is unreal due to the linear approximation, and hence Eq. (11) can be only employed near $u_0 = 1/4$.

ACCELERATION OF FLYER BY STRONG DETONATION PRODUCTS

1-D Unsteady Gas Dynamics Theory

Because of the interaction between the impactor flyer and the detonation front, there is not any exact solution to the strong detonation propagation and the resulting movement of the secondary flyer driven by it. But it is still possible to obtain some approximate solutions for the secondary flyer acceleration on the basis of the quasi-steady or the linear approximation mentioned above. It is easy to understand that the thickness of explosive slab is quite important. Under the action of the same impactor flyer, the detonation speed and pressure plateau in a thin explosive slab are higher, but the products flow is with less mass and affected by serious rarefactions. On the other hand, for thicker explosives, the detonation may attenuate to a normal one and the products flow may become the Taylor wave before reaching the secondary flyer so that the acceleration will degenerate.

For the quasi-steady approximation, the selection of explosive slab thickness is based on the principle that the detonation front just runs through the explosive slab at t_J. The secondary flyer M_2 moves under the truncated center rarefaction wave, and can be calculated by the one dimensional gas dynamics theory.

Improved Gurney Model

In order to estimate the final velocities of $U_{1\infty}$ and $U_{2\infty}$ for M_1 and M_2, the Gurney model is a convenient approach. Assume that flyers M_1 and M_2 are both rigid, and the product particle velocity between them is of linear distribution. We improve the Gurney model with the modification of M_1 having an initial velocity $U_0 > 0$ and get

$$\begin{cases} U_{1\infty} = [2M_1U_0 - (1+2M_2)U_{2\infty}]/(1+2M_1) \\ U_{2\infty} = [-H + \sqrt{H^2+4FI}]/2F \end{cases}$$

$$(12)$$

where

$$F = (1/3 + M_2) + G/3 + (1/3 + M_1)G^2$$
$$G = -(1+2M_2)/(1+2M_1)$$
$$H = 2M_1U_0[1+6(1/3+M_1)G]/3(1+2M_1)$$
$$I = u_g^2 + M_1U_0^2(1+8M_1/3)/(1+2M_1)^2$$

When $U_0 = 0, H = 0, I = u_g^2$, Eq. (12) becomes the standard Gurney model. For explosive JO-9159, the Gurney velocity is $u_gD_J = 0.286\,\text{cm/}\mu\text{s}$. Eq.(12) gives $U_{1\infty}D_J = 0.0335\,\text{cm/}\mu\text{s}$, $U_{2\infty}D_J = 0.6762\,\text{cm/}\mu\text{s}$ for $U_0D_J = 0.47\,\text{cm/}\mu\text{s}$, they are close to the numerical simulation and related test results. Therefore Eq.(12) can be used as a basic tool to optimize the test set-up design.

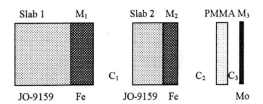

FIGURE 1. A illustration of the numerical simulation model

NUMERICAL SIMULATION

The numerical simulation model is illustrated in Figure 1, where C_1, C_2 and C_3 are cavities, and their widths are 10, 8 and 0.005 mm respectively. The thickness of slab 1, M_1, slab 2, M_2 and M_3 are 50,2.5,5, 1 and 0.1 mm respectively. The third stage of this system also adopts an inert flyer combination of impedance matching.

The calculation shows that the strong detonation speed reaches 11.1 km/s initially, but soon attenuates to the normal detonation state. The impact of flyer M_1 contributes a part of energy to the explosive slab 2 and results in a slowly increasing pressure profile in its products (see Fig.2). Therefore flyer M_2 can reach a marked high velocity range close to the detonation speed. Fig. 3 shows the free surface velocity u_2 of flyer M_2. Its final velocity is about 6.7~7.0 km/s. If we use a single stage explosive set-up, the acceleration of a 2.5 mm thick steel flyer up to 7 km/s requires a 25 cm thick slab of JO-9159, whose diameter should be over 50 cm in order to maintain a planar area of detonation wave at its end. This comparison indicates the potential advantage of the multi-stage explosive device.

As shown in Fig.4 the free surface velocity u_3 of the last molybdenum flyer M_3 can be up to 14.4 km/s within a acceleration time 0.28μs..

The precision of the numerical simulation depends not only on the calculation method but also on the thermodynamic constants of flyer materials as well as detonation products. Unfortunately there is no precise experimental data available yet for the strong detonation products. Hence, the last flyer's velocity calculated may be too high. But the possibility to obtain a very high velocity flyer using the strong detonation device is encouraging. It should be pointed out that the flatness of flyers at the impact moments, their stability and reliability in acceleration are quite important to the related experiments.

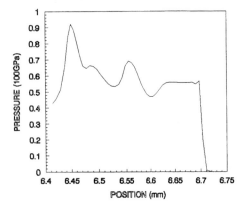

FIGURE 2. Pressure in detonation products of explosive slab 2

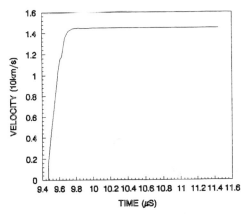

FIGURE 4. Free surface velocity history of flyer M_3

REFERENCES

1. R.Trebinski, W.Trzcinski et al, J. Tech. Phys., Vol. 21, No. 1, 1-62, 1980
2. R.Swierczynski, J.Tyl et al, J. Tech. Phys., Vol. 25, No. 2, 207-224, 1985
3. L.C.Chhadildas, L.N.Kmetyk et al, Int. J. Impact Eng., Vol. 17, 183-194, 1995
4. SUN Chengwei, WEI Yuzhang et al, Applied Detonation Physics, to be published
5. SUN Chengwei, An Estimation of the Inverse Driving Problem by Explosive, in Collected Papers on Detonation, Vol. 2. 108-114, Southwest Institute of Fluid Physics, 1994

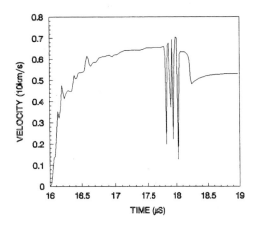

FIGURE 3. Free surface velocity of flyer M_2

CP429, *Shock Compression of Condensed Matter – 1997*
edited by Schmidt/Dandekar/Forbes
1998 The American Institute of Physics 1-56396-738-3/98/$15.00

FUZZY LOGIC APPLIED TO SHAPED CHARGE JET PENETRATION OF GLASS COMPOSITE

Dennis S. Pritchard

U.S. Army Research Laboratory, Aberdeen Proving Ground, MD 21005

A shaped charge warhead, designed to produce an ideal jet, always has at least some small manufacturing flaws that degrade jet formation. This degradation is reflected in a vague or fuzzy sense when its effect in turn degrades penetration performance. Fuzzy logic was used to relate warhead performance to manufacturing errors. Fuzzy penetration versus standoff points are predicted to lie within a confidence band with boundaries perpendicular to a linear least-squares fit (LLSQ) to 13 well-distributed experimental data points for a shaped charge penetrating semi-infinite S2 Glass composite. On a penetration versus standoff plot, the LLSQ line at zero standoff intercepts the penetration axis at 8.01 charge diameters (CD), and at 20.00 CD standoff (the outer experimental limit) the penetration is 2.89 CD. For any standoff over the experimental range, predicted penetration is within $\pm 11\%$ of the least-squares value at that standoff. This percentage interval forms the upper and lower boundary for the confidence band.

INTRODUCTION

Fuzzy logic is a formalization of reasoning for dealing with uncertainties in an approximate manner. This formalization may be addressed by considering classic set theory and its relation to fuzzy sets (1). Let X be the universe with elements denoted by x. Membership in a classical subset A of X is viewed as a characteristic function μ_A from X to the valuation set $\{0,1\}$, such that:

$$\mu_A(x) = 0 \ \textit{iff} \ x \notin A \ \vee \ 1 \ \textit{iff} \ x \in A. \qquad (1)$$

If the valuation set is allowed to be the real interval [0,1], A is called the "fuzzy set" by Zadeh (2), where μ_A is the grade of membership of x in A. The closer the value of μ_A is to 1, the more x belongs to A. Fuzzy set membership is therefore subject to degree evaluation and cannot be viewed in a black or white sense as with classic set theory. Based on thorough fuzzy set analysis, a fuzzy logic penetration algorithm was formulated for a typical laboratory warhead penetrating semi-infinite S2 Glass.

EXPERIMENTS

Thirteen tests were conducted with the warheads detonated at various standoffs to obtain penetration versus standoff data that were also compared to results found for rolled homogeneous armor (RHA). All targets were fabricated as shown in Fig. 1. The penetration versus standoff results in Fig. 2 are unique. Starting from the zero standoff point, the penetration prediction curve for the S2 Glass can be modeled as the solid straight line that, until it reaches a point that shows some correlation with the bendover point on the RHA penetration curve, is above the RHA performance, but goes below it (but not by much) past that point as standoff increases. The inverse S2 Glass penetration

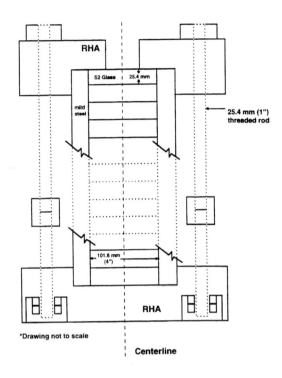

Figure 1. Target Schematic.

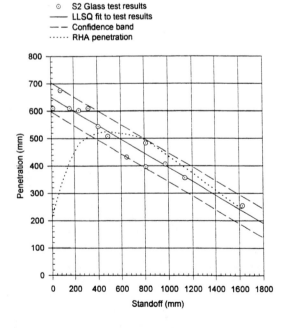

Figure 2. Penetration versus Standoff.

curve is indicating that the normal effect of the elongating jet per the DSM penetration model is not being allowed to take place in some manner by the composite. The jet is elongating to breakup regardless of the glass composite penetration reaction, which should logically lead to the penetration behavior exhibited to some degree as shown by the RHA; however, the failure mode of the composite is modifying itself as standoff distance increases in, as yet, some unknown manner to keep its penetration curve linear at all standoffs, regardless of the shaped charge jet formation behavior.

FUZZY LOGIC APPLICATION

An ideal jet is normally encountered in a computer simulation. Real jet performance deviates from ideal behavior in a degrading manner. Ideal jet behavior can never be ultimately obtained because machining and other errors come into the picture to degrade jet performance. The worst jets would come from shaped charge warheads of the poorest manufacture. The worst constructed target, say of semi-infinte RHA, would include nonhomogeneous material problems, hard and soft spots, etc. Although metallurgists design ideal materials, real-world manufacturing techniques can only come close to this ideal. Other nonmetallic target materials suffer similar problems.

There is no such condition as perfect alignment in any manufacture device. All warheads are manufactured with machine tolerances that allow mechanical fits within limits. For instance, the tolerance of the cone to fit the confinement body for the charge here is +0.00000 CD to -0.00031 CD for the cone base and +0.00125 CD to -0.00000 CD for the body cavity that it fits. Clearly, a zero difference (especially given one wants to avoid a distortional squeeze fit) would be a statistical fluke. Prior work by DiPersio, Simon, and Merendino (3) and DiPersio, et al. (4) indicated that the penetration depth achieved from a particular nonprecison (loose machined part tolerance) warhead is far less than that for a precison (tight machined part tolerance) warhead at equal standoffs.

Misalignment may not be quite so critical as the geometric irregularites in the machined parts that, once in place, cannot be corrected by physically moving, for example, a nonconcentric liner around, but misalignment is always present to some (hopefully tolerable) degree. It is difficult, if not impossible, to separate the two as they affect penetration.

Geometric irregularities in liner, explosive, and confinement components are also unavoidable, but very small irregularities must be accepted. These qualities irreparably ruin shaped charge jet formation symmetry.

Variations in liner metallic material properties occur. This includes variations from theory and variations over liner geometries, such as length and thickness. Majerus, Golaski, and Merendino (5) found that the uniformity of copper liner grain size, for example, may contribute as much as at least 0.25 CD difference in semi-infinite RHA penetration. Uniform grain size is a critical metallurgical variable in the making of a liner. Variations in the average grain size from shaped charge lot to lot may generate dispersed penetration results. Duffy and Golaski (6) improved the penetration performance of a shaped charge design by making a smaller average grain size than before. It is not known how small variations in this grain size from shape charge lot to lot would affect variations in penetration. Also, It is not know how variations in grain size over a longitudial distance along a shaped charge liner affects jet formation.

The inconsistency of explosive mixtures plays a part in the summation of deviatoric shaped charge jet formation behavior. Explosive variation toward the center of the charge leads to round-to-round variation in the pressure pulse along the forming jet, in turn leading to round-to-round variations in the stretching of the jet, for example.

Random inconsistencies may compound. The liner could be out of alignment in one direction and the detonator out of alignment 180^0 in the opposite direction to create opposing misalignments about the centerline of a warhead. Of course, this is a worse condition than just having either out of alignment, but this is only one way manufacturing errors compound. The desire is to have a tight tolerance on any misaligned parts so their misalignment-produced errors in shaped charge formation cannot significantly superpose.

These manufacturing errors were taken into consideration when the confidence bands were determined for the shaped charge penetration results.

A linear least-squares fit (LLSQ) and not some other fit was made to the penetration versus standoff numbers for the semi-infinite S2 Glass because of the S2 Glass penetration behavior before and after its progression through the vicinity of the maximum RHA penetration point as shown for comparison in Fig. 2. The composite penetration before the maximum RHA penetration point should behave like the first part of the RHA curve, except the mechanical failure of the glass composite compensates for the stretching jet, restricting penetration depth as standoff increases instead of letting the penetration increase with standoff as in the baseline RHA penetration case. When the crossover point is reached at the maximum RHA penetration and the S2 Glass curve goes beyond it, then behavior occurs that is more associated with the normal particulated jet penetration behavior as seen in RHA. As there is a tendency to make the penetration hole smaller at the penetration-target interface, more fiber gets ground up into smaller fragments and thus absorbs more energy. This is an offset to the stretching jet becoming more efficient. As compared to RHA, the glass composite energy absorption varys greatly as a function of the changing mechanical loading conditions associated with localized S2 Glass failure. To support this statement, it is seen on post-test evaluation that fibers have been pulverized in ranges from dust size particles to fibers a few millimeters long. A ragged edge penetration hole is seen with the S2 Glass as opposed to the smooth melted surface apparent with RHA penetration. The ragged edges restrict material flow out of the penetrated hole differently than that failure mode found along the melted edges of holes in metallic material penetrated by jets. Comparison to many other's work on RHA penetration shows good agreement as far as data scatter goes, except their data are not scattered in as tight of a pattern as standoff increases as the S2 Glass numbers show. Perhaps S2 Glass is more self-correcting penetration-wise in some unkown manner as compared to RHA.

The $\pm 11\%$ penetration difference for the laboratory warhead performance into the S2 Glass appears to fit comfortably within acceptable performance limits. This is an acceptable baseline determined by analyzing and fitting test data. The lower limit corresponds to the "worst found" experimental data point. If more tests are conducted, a new worst found point might replace this one, but even so, it should not be radically different valued from the now existing one. The upper bound is merely a logically mirrored image of the lower bound as shown in Fig. 2. The fluctuations of the experimental data points about the LLSQ-determined line are probably, for the most part, due to the misalignment of the shaped charge components about the centerlines of the respective shaped charges creating the data. However, the total root causes for the fluctuations may be explained in general only in a qualitative sense of the errors contributing to jet degradation and its related drop in penetration performance as compared to an ideal jet. No quantitative solution leading to a precise summation of effects is possible now.

Fuzzy set membership in this paper is determined to be of two kinds, performance and linear, based on test results in this report. The first is the performance membership (μ_p) as defined by comparing the test penetration to that of the theoretically obtainable penetration at the same standoff as predicted on the upper dashed confidence line shown in Fig. 2. The other membership is the linear membership (μ_l), which is also performance related, but is, in addition, forced to lie along a straight line. For prediction purposes, the three linear memberships are useful and are given as: the dashed confidence line above the LLSQ line with penetrations at +11% of the LLSQ values with correspondingly equal standoffs (μ_l set equal to 1.00), the solid LLSQ line ($\mu_l = 0.89$), and the -11% dashed line ($\mu_l = 0.78$) in Figure 2. Any penetration versus standoff point predicted along either the -11% or LLSQ line has that associated membership number as compared to any found along the +11% or 1.00 membership line. All linear membership concepts are new concepts proposed by the author. Supermembership would entail finding a line with a value greater than the current maximum μ_l (1.00), which would determine the renormalized value of 1.00 for this data set. An optimum supermembership line could be determined from testing the best assembled warhead and target configuration within the most stringent assembly parameters as discussed previously. A proposed supreme supermembership line consisting

of values lying along a line determined by firing a perfect jet into a perfect target is of course not possible in an experimental sense. Calculations to find this line can be attempted theoretically, but the line can never be found experimentally. Performance membership defines for each test how well the warhead performed compared to a possible reasonably high performance standard as defined by the +11% confidence line (target value) from Fig. 2 with the μ_p simply the ratio of the actual value over the target value. The 0.89 linear curve can be considered as the "performance curve" or "average performance curve." Points along this line would be used for warhead design purposes, with the confidence bands being kept in mind to give an overall feel for accuracy.

CONCLUSION

It is recommended that more tests using varied shaped charge warhead designs be conducted. Different glass composites, other than S2 Glass, could be examined also. Some penetration modeling and code runs could be performed with their results compared to those given here. The fuzzy logic treatment here can be used in other terminal ballistics problems.

REFERENCES

1. Dubois, D. and H. Prade, *Fuzzy Sets and Systems: Theory and Applications*, New York, Academic Press,1980, pp.9-10.
2. Zadeh, L. A., *Information and Control*, **vol. 8,** pp. 338-53, 1965.
3. DiPersio, R., J. Simon, and A. B. Merendino. "Penetration of Shaped-Charge Jets into Metallic Targets." BRL Report Number 1296, U.S. Army Ballistic Research Laboratory, Aberdeen Proving Ground, MD, September 1965.
4. DiPersio, R., W. H. Jones, A. B. Merendino, and J. Simon. "Characteristics of Jets from Small Caliber Shaped Charges with Copper and Aluminum Liners." BRL Memorandum Report No. 1866, U.S. Army Ballistic Research Laboratories, Aberdeen Proving Ground, MD, September 1967.
5. Majerus, J. N., S. K. Golaski, A. B. Merendino. "Influence of Liner Metallurgy, Apex Configuration and Explosive/Metal Bond Strength upon Performance of Precision Shaped-Charges." BRL Technical Report ARBRL-TR-02451, U.S. Army Ballistic Research Laboratory, Aberdeen Proving Ground, MD,December 1982.
6. Duffy, M. L. and S. K. Golaski. "Effect of Liner Grain Size on Shaped Charge Jet Performance and Characteristics." Technical Report BRL-TR-2800, U.S. Army Ballistic Research Laboratories,Aberdeen Proving Ground, MD, April 1987.

CHAPTER XV

LASER AND PARTICLE BEAM-MATTER INTERACTION

CP429, *Shock Compression of Condensed Matter – 1997*
edited by Schmidt/Dandekar/Forbes
© 1998 The American Institute of Physics 1-56396-738-3/98/$15.00

ATHERMAL ANNEALING
OF
NEUTRON-TRANSMUTATION-DOPED SILICON

J. Grun[a], C.K. Manka[b], C. A. Hoffman[c], J. R. Meyer[c], O. J. Glembocki[d], S. B. Qadri[e], and E. F. Skelton[e]

Naval Research Laboratory Washington, DC 20375

D. Donnelly and B. Covington

Sam Houston State University, Huntsville, Texas 77341

We demonstrate a new mechanism for annealing silicon that does not involve the direct application of heat as in conventional thermal annealing or pulsed laser annealing. A laser pulse focused to high power on a small surface spot of a neutron-transmutation-doped silicon slab is shown to anneal regions *far outside* the illuminated spot where no heat was directly deposited. Electrical activation of donors throughout the slab was uniform and comparable to that of thermally annealed control samples. We conjecture that the annealing was caused by mechanical energy introduced by the laser pulse.

INTRODUCTION

Annealing is critical to the production of semiconductor devices. During this process dopants within a semiconductor crystal lattice can move to locations where the dopants become electrically active i.e., contribute to electrical conduction in the semiconductor. Annealing is also needed to remove structural damage in crystals. Current semiconductor annealing methods are based on thermal processes that are accompanied by diffusion which degrades the definition of device features or cause new problems, such as increased junction leakage. This will be a serious obstacle for the production of next-generation ultra-high-density, low-power semiconductor devices.

We report here (1) the experimental demonstration of an entirely new semiconductor annealing method which is much *faster* than thermal annealing and does not involve the direct application of thermal energy. We show that a laser pulse focused to high power on a small surface spot of a neutron-transmutation-doped silicon slab can initiate a process which electrically activates dopants and also removes structural damage induced by the transmutation process in regions far *outside* the laser illuminated spot where no heat was directly deposited. The electrical characteristics of successfully athermally annealed slabs in our

[a] Plasma Physics Division
[b] Research Support Instruments, Hunt Valley, MD
[c] Optical Sciences Division
[d] Electronic Science & Technology Division
[e] Condensed Matter & Radiation Sciences Division

experiment are comparable to what is attainable with commercial furnace annealing.

It is important to realize that we are describing a technique different from conventional pulsed laser annealing (PLA). In PLA a laser pulse focused to relatively low fluence (of order joule/cm^2) is used to heat and anneal areas of the wafer directly beneath the laser spot. In our work, a laser pulse focused to a much higher fluence (of order kjoule/cm^2) was found to anneal areas far removed from the laser spot.

EXPERIMENT

To demonstrate athermal annealing, ~25mm x ~25mm x 2 mm thick neutron-transmutation-doped (NTD) Si slabs were doped to a concentration of 10^{15}cm^{-3}. (In the NTD process neutrons from a reactor irradiate a Si sample for 110 hours, transmuting ^{30}Si to ^{31}P. The neutron flux was measured to be 1E13 n/cm^{2}sec at 0.25-0.5 eV, 6E11 n/cm^{2}sec at 0.5-10 eV, and modeled to be ~ 3E12 n/cm^{2}sec at 10 eV to 1 MeV. The process creates a uniform distribution of phosphorus through the entire slab along with point defect densities several orders of magnitude higher than the phosphorus concentration.) Seven slabs were placed inside a vacuum chamber and irradiated by one or two pulses from a 1.06-μm wavelength, 5-ns FWHM duration, ~10 joule laser pulse focused to a 1-mm diameter spot. After laser irradiation, the samples were analyzed for changes in activation, carrier density, mobility, resistivity, and crystal structure as a function of position across the entire slab and particularly in areas far from the illuminated spot. Electrical measurements were made using a 4-point probe, an advanced Hall method, and far-infrared spectroscopy. Structural changes in the crystal lattice were measured with x-ray diffraction, x-ray topography, and Raman spectroscopy. Unannealed and thermally annealed slabs were used as controls. In 2 out of 7 samples we observed excellent annealing, the evidence for which is described below.

RESULTS

Far-infrared spectroscopy provides clear evidence for the activation of donor species. In this technique, the absorbance spectrum from 150 to 500 cm^{-1} of a sample cooled to 5.5K is measured by a Fourier transform infrared spectrometer (FTIR). Electrically active donors are known to exhibit Lyman absorption lines at <800 cm^{-1}, whose strengths are proportional to the concentration of electrically active donors. Figure 1 shows the absorbance spectra of unannealed, thermally annealed, and two athermally-processed slabs. As expected, the spectrum of the unannealed sample shows no Lyman lines because the phosphorus is not electrically active. In contrast, both athermally annealed slabs show distinctive phosphorus Lyman lines up to the 5p line. The widths of the lines are on the order of 0.55 cm^{-1} (0.07 meV), which is consistent with published measurements on thermally annealed NTD-Si with similar donor concentrations. The integrated area of a Lorentzian fit to the 2p$^\pm$ line at 316 cm^{-1} is 15.4 ± 1.0 cm^{-2} for the two athermally-treated samples, and 18.5 ± 0.5 cm^{-2} for the thermally annealed sample. Thus, we estimate activated donor concentrations of 6.6 ± 0.4 x 10^{14} cm^{-3} and 7.9 ± 0.2 x10^{14} cm^{-3} for the athermally and thermally annealed cases respectively.

Figure 2 shows the temperature dependence of the mobilities and carrier densities in the same two samples. After annealing, a low temperature process was used to attach electrical leads to the corners of the samples for Hall characterization using the Van der Pauw method. Measurements were performed at temperatures between 20K and 300K and at magnetic fields from 0 to 7 tesla. Analysis of the results, performed using the Quantitative Mobility Spectrum Analysis (QMSA) method, shows the presence of a single electron species whose concentration corresponds to 1.1x10^{15}cm^{-3} activated donors. A fit of the standard freeze-out relation to the electron-concentration vs. temperature data implied a donor binding energy of 43 meV, which agreed well with published results for Si:P. The mobility agreed with theoretical predictions for thermally annealed, uncompensated n-type silicon, as well as with previous experimental results for comparably-doped melt-grown Si:P. These results also demonstrated that lattice damage was removed to an extent that it has no detectable effect on the mobility.

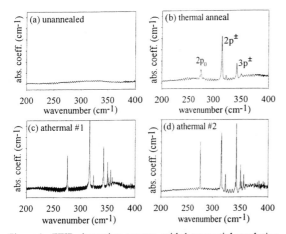

Figure 1. FTIR absorption spectra, with 1-cm spatial resolution, showing activation of P donors by annealing. Shown are spectra from (a) an unannealed NTD Si:P slab, (b) a thermally annealed sample(1 hour at 900 °C in a nitrogen atmosphere), and (c) and (d) athermally annealed, with a 10-joule laser pulse, NTD samples. Integrated areas under a Lorentzian fit to the $2p^{\pm}$ line indicate comparable levels of activation in (b)-(d). The apparent split in the $2p^{\pm}$ line in (d) is noise. Other measurements on these same samples are shown in Figures 2 and 3 below and described in the text.

Four-point probe measurements, with 1-mm spatial resolution, on the athermally annealed samples showed that electrical activation was uniform with no systematic position-dependent variation across either the front or back surfaces of the sample. In particular, resistivity near the slab edges and corners, where shock wave reflections are expected to occur, was not measurably different from resistivities closer to the center of the slab. The n-type sheet resistivity of 56 ± 1 ohms/square for the athermally annealed samples compared with an n-type sheet resistivity of 130 ohms/square for a thermally annealed NTD slab, p-type 1000 ohms/square for an undoped wafer, and an unmeasurably high sheet resistivity for an unannealed NTD sample.

X-ray topographs and rocking curve measurements on unannealed NTD samples showed no differences from bulk Si, indicating that damage in the form of polycrystalline islands was not formed during the NTD process. The measurement does not rule out damage in the form of small regions of amorphous silicon or clusters of vacancies. No residual strain or excess dislocations were introduced by the laser pulse in annealed regions far away from the focal spot. The unit-cell length parameter at 2-mm or more from the focal spot center was 5.430 ± 0.002 angstroms, a value consistent with that of undamaged silicon. Closer to the focal spot the unit cell parameter increased to ~5.445 angstroms, indicating that the lattice is under a residual tensile strain of 0.2-0.3%.

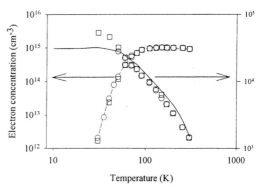

Figure 2. Hall measurements confirming electrical activation of athermally annealed NTD Si:P slabs. The symbols O and □ refer to two different samples. Electron concentration is shown on the left axis. A fit of the standard freeze-out relation to the dashed data curve implies $1.1 \times 10^{15} cm^{-3}$ activated donors with a binding energy of 43 meV. Electron mobility is shown on the right axis. The solid curve is the mobility for a melt-grown Si:P wafer with donor density of $9.6 \times 10^{14} cm^{-3}$ and acceptor density of $2.0 \times 10^{14} cm^{-3}$. Higher mobilities of athermally annealed slabs at <40°K imply lower compensation.

Raman spectroscopy can be a sensitive probe of material crystallinity, particularly in the case of polycrystalline or amorphous silicon. Raman spectra show that the thermally annealed sample had a sharp longitudinal optical (LO) phonon mode at 521.0 ± 0.2 cm^{-1}. A most noticeable effect of NTD on this mode was the reduction in its peak intensity by about a factor of two everywhere on the front and back of the sample. Correlated with this was a slight red shift of 0.5 ± 1 cm^{-1}. Athermal annealing blue-shifted the line back to 521 cm^{-1} and its intensity recovered to within 10% of the thermally annealed sample. X-ray measurements showed no polycrystalline islands or strain that could account for the reduction in Raman intensity and the red shift.

DISCUSSION

The mechanisms for the absorption of high-intensity laser energy and for the creation and transport of mechanical and thermal energy beyond the illuminated spot are very well known. At an irradiance of 3 x 10^{11} watts/cm^2 ~100% of the laser energy is absorbed by a plasma above the slab surface. This plasma, which is very hot (~200 eV), expands rapidly (~200 km/sec) generating a back-pressure (~ 0.2 Mbar) which drives a shock wave into the sample's interior. After traveling a short distance (~ 0.5mm) the shock is weakened by rarefaction waves and by geometrical expansion so that quickly its pressure is drastically reduced (e.g., ~ 1/1000 of original strength by 1 cm). Further decrease in pressure is gradual. Before they decay away, the pressure waves and accompanying rarefaction waves reverberate within the sample. In contrast, temperature falls with distance very quickly so than no significant heating occurs outside the illuminated spot.

We conjecture that the athermal annealing in our samples is caused by mechanical energy since that is the only form of energy that could have traveled so far (~ 1cm) beyond the spot where the laser energy was deposited. The mechanism cannot involve bulk heating because the laser does not have sufficient energy to heat the entire sample significantly. Temperatures capable of annealing i.e., 900 °C, exist only near the edge of the illuminated spot. This is consistent with our observation that melting, which occurs at temperatures greater than 1414°C , exists only within the immediate neighborhood of the focal spot. Heating by visible or x-ray radiation from the laser-heated plasma can not play a role since the back of the slab, which was shielded from this radiation, was annealed as effectively as the front. Similarly, the laser's 1.06-μm radiation cannot be responsible because almost all of the laser energy is absorbed by the plasma (formed at ~ 10^8 watts/cm^2) above the slab surface. Thus, mechanical energy is the most plausible candidate to explain the observed activation.

In summary, we have annealed NTD silicon slabs without the direct application of heat. Details of the process such as thresholds, etc. need to be determined and a comprehensive theoretical understanding is yet to be developed. It remains to be shown whether the process can anneal semiconductors other than NTD silicon, whether its reproducibility can be improved, and whether it will be effective on industrial-scale wafers.

ACKNOWLEDGEMENTS

We thank Mr. Kirk Evans, Mr. Rayvon Burris, Mr. Levi Daniels, and Mr. Nicholas Nocerino for their able technical assistance, and Dr. Richard Singer, Dr. Larry Larson, Dr. Mike Bell, Dr. Richard Hubbard, and Dr. Benjamin V. Shanabrook for their sound advice and assistance. Shock characteristics at distances larger than 1 mm are based on simulations performed by Dr. B. Stellingwerf and Dr. C. Wingate at LANL. Using the SPHINX SPH code they modeled shocks and rarefaction waves inside a 1-mm thick, 10-cm diameter aluminum disk irradiated at its center by a 4-joule pulse focused to a 1-mm spot. This project was supported by The Defense Advanced Research Projects Agency and The Office of Naval Research.

1. For more complete versions of this work see: J. Grun, et al, Phys. Rev. Lett. **78**, pg. 1584 (1997); D. Donnelly, et al, to be published in Appl. Phys. Lett. **71**, (5), 1997.

CP429, *Shock Compression of Condensed Matter – 1997*
edited by Schmidt/Dandekar/Forbes
© 1998 The American Institute of Physics 1-56396-738-3/98/$15.00

COATING DEBONDING INDUCED BY CONFINED LASER SHOCK INTERPRETED IN TERMS OF SHOCK WAVE PROPAGATION.

M. Boustie, C. Seymarc, E. Auroux, T. de Rességuier, J.P. Romain

Laboratoire de Combustion et de Détonique (U.P.R. au C.N.R.S. n° 9028)
E.N.S.M.A. - B.P. 109 - 86960 Futuroscope Cedex (France)

Debonding of coatings on a substrate is currently achieved by the so-called laser spallation technique. Here, we give an interpretation of the debonding by laser shock based on the impedance mismatch between the substrate and the coating rather than on the spallation process. An analysis in the (pressure-particle velocity) plane coupled with (space-time) diagrams provides the stress history at the interface and shows the possible traction history, depending on the applied pressure loading and the nature of the target. From this study, the different configurations that can lead to a separation of the layers under laser confined irradiation are evidenced according to the impedance mismatch between both components. A case of this analytical survey is experimentally studied. Free surface velocity measurements by the electro-magnetic technique are performed on a 75 μm-thick aluminum coating stuck on a 1.5 mm-thick copper plate irradiated by various laser intensities in water confined geometry. A debonding threshold is observed on the records.

INTRODUCTION

High power laser shocks have been widely used to study the spallation behavior of various materials under extreme loading conditions (1). This technique generating tensile stress within the material when applying a shock onto a surface has been used in the case of two layer materials to achieve the debonding at the interface (2, 3). Two configurations are of particular interest for direct or indirect cleaning operations. The shock is respectively applied either on the substrate (cf fig. 1-a) or on the coating (cf fig. 1-b); according to the geometrical and the loading parameters, the tensile stress history can be modified and the bond strength can be overpassed, leading to the removal of the film. We studied how shock waves propagation generates a possible traction at the interface in both configurations in the case of a water confined laser interaction with a target composed of a 75 μm coating over a 1.5 mm

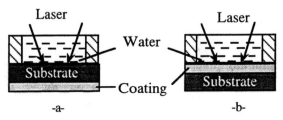

FIGURE 1. Sketch of the studied configurations : irradiation of the substrate (a) or of the coating (b)

substrate. Various impedance mismatches were studied on space-time diagrams coupled with pressure-particle velocity analysis to compare the interface stress history evolutions. The case of a copper substrate with an aluminum coating was experimentally studied : free surface velocity measurements were performed and evidence of a threshold for the removal of the coating was found.

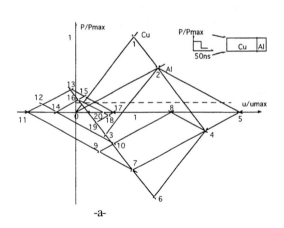
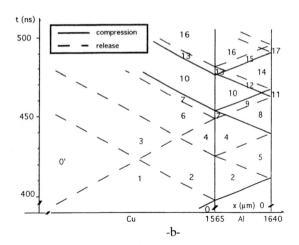

FIGURE 2. Pressure-particle velocity states (a) and space-time diagram (b) modeling the propagation of a water confined laser irradiation of a copper substrate with an aluminum coating

ANALYTICAL STUDY

In order to understand the mechanical behavior of the bond of a two layer target in both cases of irradiations shown on figure 1, a simplified modeling is proposed. To remain close to the experimental context, at first, we have considered a 1565µm copper substrate with a 75µm aluminum coating. Without changing the thicknesses, a relative impedances mismatch commuting was studied. The applied pressure corresponding to a water confined laser irradiation of 7GW/cm^2 in infra-

red wavelength with an initial pulse duration of 20ns has been simplified as a main square pulse with an amplitude $P_M=4GPa$ and a duration of 50ns due to the broadening induced by the confinement, followed by a tail at 2GPa (4). In order to extrapolate the results for other intensities, the pressure has been normalized relative to P_M and the material velocity to the particle velocity $u_M=P_M/\rho_0 c_0$. Concerning the shock propagation into the target, we have assumed a 1D propagation with a unique speed for all the waves (shocks and release) equal to the sound

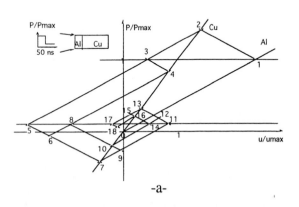
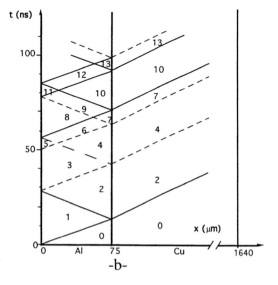

FIGURE 3. Pressure-paticule velocity states (a) and space-time diagram (b) modeling the propagation of a water confined laser irradiation of an aluminum coating on a copper substrate

986

speed of the considered material. The material velocity of free surfaces and interfaces has been neglected.

Shock applied to the substrate (cf fig. 1-a)

Under the above assumptions, the space time diagram of the different states reached inside the target can be built according to the corresponding (p-u) plane (cf fig. 2-a). A zoom of the space time diagram near the interface (cf fig. 2-b) allows to draw the analytical stress history of the bond (cf fig. 4-a) by reading the corresponding states on the (p-u) plane. On this particular case, we can see that the interface is first submitted to a shock and that the traction of the bond begins only a bit later with a rather low level but with a duration equivalent to the main pulse duration.

Shock applied to the coating (cf fig. 1-b)

For this configuration, a few microns of the coating were burnt into plasma and expelled from the coating. However, for thick coatings such as the ones considered here (some tens of microns), the ablation of the whole coating is not only due to the plasma formation and it can be explained by the

mechanisms of shock waves propagation. With the same assumptions as in the previous part, we get the stress history at the interface (cf fig. 4-b) thanks to the space-time diagram analysis of the shock propagation into the aluminum coating on a substrate considered as infinite as regards to the respective thicknesses of the layers (cf fig. 3). The stress profile of the interface exhibits a traction in that case that could contribute to the removal of the coating. Compared to the case of the irradiation of the substrate, the traction of the interface is shorter, corresponding only to the transit time of the back and forth of the wave into the coating.

When switching the relative impedances of the substrate (Z_s) and of the coating (Z_c), a similar space-time analysis for both cases considered provides the stress histories of the interface in the case of a water confined loading (cf fig. 4-c and d.). From that, we can conclude that there is no possibility to remove a coating of higher impedance than the substrate by irradiating the coating since the wave propagation yields no traction in that case (cf fig. 4-d). On the other hand, all other cases lead to a traction at the interface, making possible the debonding.

EXPERIMENTAL STUDY

In order to check experimentally the achievement of the debonding according to the analytical study, experiments involving irradiated copper coated with adhesive aluminum were performed.

The targets were irradiated with a water confinement by an infra-red laser pulse of 20ns with a maximum intensity of 7GW/cm^2 (below the breakdown threshold in the water) (4). Free surface velocity measurements of the aluminum coating were performed with an electromagnetic velocity gage (5). Several experiments with decreasing intensities were carried out until no debonding is observed. The corresponding velocity records are reported in figure 5. For intensities above 0.5GW/cm^2, the separation of the coating has been obtained, as shown on the cross sectional picture of figure 6. The velocity records show also the spallation of the coating with a first peak of velocity corresponding to the shock break-out of the target, and then an almost uniform velocity corresponding to the coating velocity when separated. However, for the record at 0.4GW/cm^2 where no decohesion is observed, we can notice the rebounds of the shock

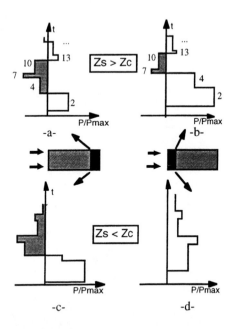

FIGURE 4. Evolution of the stress history of the interface of a two layer target under confined laser irradiation. Comparison of the cases of irradiation of the substrate (a-c) or of the coating (b-d) when the respective impedances of both components are inverted.

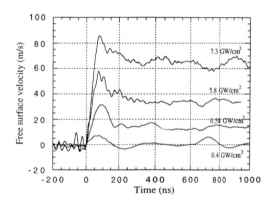

FIGURE 5. Free surface velocity records at the rear face of a 75μm adhesive aluminum coating over a 1565μm copper substrate irradiated with an infra-red 20ns water confined laser pulse at various intensities.

FIGURE 6. Cross section micrography of the debonding of an adhesive 75μm aluminum over a 1565μm copper substrate irradiated by a water confined laser pulse at 5.8GW/cm^2

into the whole target, approximately 600ns after it has first reflected from the back. Such records provide an experimental basis to detect the debonding threshold and could help in determining a criterion for separation by comparison with numerical simulations.

CONCLUSION

By this study, we showed how water confined laser irradiation of a substrate with a coating can generate a traction at the interface, irrespective of the impedance mismatch. In the cases of the irradiation of the coating, we observed that such an irradiation could not generate any tensile stress if the substrate has a lower impedance than the coating. In the other case, we showed that traction induced by the transmission of shock wave can be at the origin of the removal of the coating, such as it occurs in laser cleaning surfaces process. Some experiments showed the possibility to detect the debonding threshold by free surface velocity measurements, providing an experimental basis for developping debonding criterions with the help of numerical simulation.

REFERENCES

1. Curran D.R., Seaman L., Shockey D.A., *Phys. Today*, **47** (46), 1977
2. Nutt G.L., *J. Mater. Res.*, **7** (1), 203-213, 1992
3. Yuan J., Gupta V., *J. Appl. Phys.*, **74** (4), 2388-2410, 1993
4. Peyre P., Fabbro R., Berthe L., Dubouchet C., *Journal of Laser Applications*, **8**, 135-141, 1996
5. Romain J.P., Zagouri D., "Laser shock studies using an electromagnetic gauge for particle velocity measurements", in *Shock Waves in Condensed Matter at Williamsburg*, 801-804, 1991

CP429, *Shock Compression of Condensed Matter – 1997*
edited by Schmidt/Dandekar/Forbes
© 1998 The American Institute of Physics 1-56396-738-3/98/$15.00

Pressure Dependent Laser Induced Decomposition of RDX

Tod R. Botcher, H. D. Ladouceur, T. P. Russell

Chemistry Division, Code 6110, Naval Research Laboratory, Washington DC 20375

The pressure dependence of laser induced decomposition of RDX (hexahydro-1,3,5-trinitro-1,3,5-triazine) and RDX-d6 has been studied from ambient pressure to 5.2 GPa. RDX samples were loaded in Merrill-Bassett anvil cells and cooled to a minimum of 23 K at the desired starting pressure. Individual samples were each exposed to a single 8 ns pulse of 532 nm light from a Nd:YAG laser. Thermal transport from the decomposing RDX and RDX-d6 into the anvils arrested the decomposition, enabling the identification of intermediates and products formed under high pressure-high heating rate conditions. A two dimensional finite elements model was employed to estimate the quenching rates and heat flow within the samples. Single crystals, pressed powders, and solutions were studied to determine the effect of mechanical stress on the sensitivity towards laser initiation. The threshold energy for reaction initiation and product distribution is given as a function of pressure.

INTRODUCTION

The reaction of an energetic material propagating at 1 m/sec - 8 km/sec (deflagration to detonation regime) is an extremely complex mechanical, physical and chemical process. Very little information of the actual reaction sequence or kinetics in high temperature, high pressure conditions like those found during deflagrations or detonations has been determined. Advanced development of the fundamental reaction sequences will provide a more complete understanding of the energy release and help improve the fundamental understanding of issues such as initiation and sensitivity.

Recently, advanced experimental techniques have begun to mimic deflagration conditions. Among these techniques are simultaneous thermogravimetry modulated beam mass spectrometry, simultaneous mass and temperature change/FTIR method, and laser assisted decomposition studies, in both the gas and condensed phases.[1]

The field of light induce decomposition of energetic materials has also been heavily studied. Initially investigated using intense discharge lamps and incorporating lasers during the 1960's, this method has been used to conduct many studies of rapidly heated materials, but none into the high pressure regime.[2]

The experiment described here probes chemistry under heating rates, pressures, and time scales representative of the detonation/deflagration regime. A thin film RDX is loaded inside the sample chamber of a gem anvil cell. The sample is pressurized, up to as much as 5.2 GPa, and rapidly heated using a single pulse (532nm) from a Nd:YAG laser. After initiation, thermal transport into the anvils rapidly quenches the reaction, trapping products and intermediates, which are analyzed by FTIR.

EXPERIMENTAL PROCEDURE

Quenching Estimates

The times required for the samples of RDX to quench after pulsed laser heating are investigated by solving a time-dependent axis-symmetric heat conduction equation[3]. Since the sample configuration involves materials of diverse thermal properties and geometric shapes, a time-dependent finite element code (PDEASE2) is utilized to model the anvil cell. The materials properties data required for the model calculations include thermal conductivities, densities, and heat capacities as a function of temperature. These values are tabulated in *The Thermophysical Properties of Matter*[4]. Since the thermal properties vary with temperature, the heat conduction problem is non-linear.

The cell is cylindrically symmetric and, therefore, a 2-D code can be used to accurately estimate the thermal transport in 3 dimensions.

Calculations are performed for samples in 3 varying configurations. The first has the sample of RDX imbedded within the salt matrix, the others have the RDX at the salt/anvil interface, using the thermal transport properties for either cubic zirconia or diamond to estimate the quenching. The temperature profile of the cell is obtained as a function of time, and the time required for quenching is also monitored.

Matrix Isolation

The matrix isolation apparatus consists of a closed-cycle helium refrigerator (APD Cryogenics CSW 202A) onto which a Merrill-Bassett anvil cell is mounted and is similar that described elsewhere.[5] The cell is loaded with NaCl, a small amount of RDX, and a ruby for measuring pressure.[6] The RDX was obtained from Bill Koppes and Al Stern of the Naval Surface Warfare Center Indian Head, Md. in a purified state to ensure the absence of HMX. No further purification of the samples was performed. An inconel 600 gasket is used to confine the sample. The sample is cylindrically shaped, with a diameter of 250 mm and a thickness of ~200 mm, of which ~3 mm are RDX and the remainder NaCl. The layer of RDX is either imbedded between salt layers, or between the salt and anvil, depending on the desired quenching rate

The cell is attached to the cold finger of the refrigeration unit, and the system is evacuated. For non room temperature experiments, the cell is cooled to 27 K. A single 8 ns pulse from a frequency doubled Nd:YAG laser (Continuum SLI-10) is used to initiate reaction. The energy required to initiate decomposition is investigated as a function of pressure. A sample is subjected to laser pulses of successively higher energy until decomposition is observed. In this manner the pressure dependence of the threshold to

decomposition is determined for samples at 298 K and 27 K, in cubic zirconia anvil cells and at 27 K in diamond anvil cells.

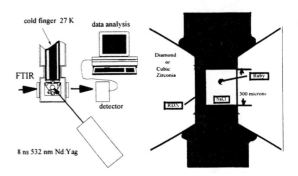

Figure 1. Schematic of experimental setup (left) and of sample confined in anvil cell (right).

The reaction products and intermediates are also investigated as a function of pressure. Thermal transport into the anvils rapidly quenches the reaction, trapping products and intermediates. FTIR spectra are obtained both prior to and after laser heating, at both 298 K and 27 K. Spectra are taken of the region from 4000 cm^{-1} to 400 cm^{-1} at 2 cm^{-1} resolution collecting 5000 scans. Samples are studied in the pressure region from ambient pressure up to 5.2 GPa.

RESULTS AND ANALYSIS

Quenching Rates

The two dimensional finite element code provides two important pieces of information. First, most of the energy is lost to the anvils, and very little is transferred into the gasket. Without even considering material differences, the surface area in contact with the anvils is so large compared to that in contact with the gasket, that the edge effects can all but be ignored. The problem largely becomes a one dimensional calculation. Secondly, the configuration in the anvil cell has a large effect on quenching times. When imbedded into the salt matrix, the sample quenches to 500 K on the order of 50 msec, when in contact with a cubic zirconia anvil, this value drops to ~25 msec, and when in contact with a diamond anvil, the quenching time is reduced to 10 msec.

Pressure Dependence of Initiation

The energy required to initiate decomposition was investigated as a function of pressure. An individual sample was subjected to laser pulses of successively higher energy until decomposition was observed. In this manner the pressure dependence of the threshold to decomposition is determined for samples at 298 K and 27 K. The threshold for RDX decomposition shows a strong dependence on initial pressure (figure 3). In samples heated at pressures < 0.2 GPa, initiation requires a fluence in excess of 40 J/cm². At these energies, anvil damage occurs, so determination of actual initiation fluence is not possible. Going to higher pressures, the energy required decreases, reaching a minimum of 3 J/cm² at a pressure of 1.3 GPa.

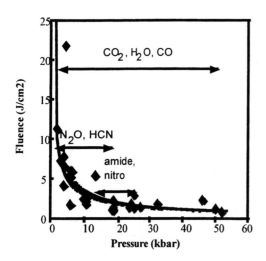

Figure 2. Pressure dependence of laser induced decomposition of RDX

However, two plausible explanations may be postulated to discuss this phenomenon. We currently believe the strong dependence on pressure is most likely due to physical characteristics of the solids. First, at low pressures, the voids and defects are relatively large and are randomly scattered throughout the energetic material. As the pressure is increased, the volume decreases. This will reduce the void and defect sizes through crushing

and will effectively increase the number density of defect sites per unit volume in the RDX. Therefore, as the pressure is increased the scattering efficiency is increased as the pressure increases. Second, as the pressure is increased, it can also be assumed that at the defect sights will provide an increase in the total stress in the sample. This increase in stress will thereby cause an improved absorption of the laser energy as a function of pressure increasing the sensitivity to laser initiation. The curve of fluence versus pressure goes nearly linear in the 15-10 Kbar region, which corresponds to the elastic plastic transition of RDX.

Product Isolation

Post analysis by infrared spectroscopy of the threshold initiation studies provides information on the products produced. In the room temperature pyrolysis studies the only products detected are H_2O, CO_2, and CO, indicating complete reaction of RDX at room temperature under high pressure and pulsed laser heating conditions. However, when the initial cell and sample temperature is 27 K, the chemical reactions are arrested and intermediate products are trapped. A series of reactions were carried out in cubic zirconia and diamond to determine the influence of the quenching rate on products observed.

At low pressure, RDX absorption bands are observed in the post pyrolysis spectrum, indicating that the reaction was quenched prior to complete consumption. At higher pressures, intermediates are trapped. In the cold post pyrolysis spectrum, there are broad absorptions in the region of 1700-1200 cm⁻¹, in addition to the CO_2, H_2O, and the doublet at 3730 and 3610 cm⁻¹. Upon warming to room temperature, the broad features between 1700-1200 cm⁻¹ have diminished, and absorptions at 1400 and 815 cm⁻¹ are now prevalent, as are the bands at 3130 and 3045 cm⁻¹. These features appear as a pair of doublets in the pyrolysis of RDX-d6, indicating that only part of the absorption is due to vibrations involving H atoms. Due to the location of the band, and products observed in previous RDX studies, we feel that these absorptions are due to two types of reactive intermediates; one with an amide like functionality,

991

the other being some nitro- species. Upon warming, these intermediates are no longer trapped in the matrix, and react to the final products observed

intermediates were quenched from the reaction scheme as a function of pressure. Observed products include CO_2, H_2O, CO, N_2O, HCN, a species absorbing in the NO_2/N_2O_4 region, and an unidentified doublet absorbing at 3730 and 3610 cm^{-1} (OH stretch) Trapped intermediates absorb in the region from 1700 to 1200 cm^{-1}, with a feature at 1310 cm^{-1}, most likely amide and nitro like species. Upon warming, these convert to a product or products (yet to be identified) which absorb at 1400 cm^{-1}, 815 cm^-, and a doublet at 3130 and 3045 cm^{-1}.

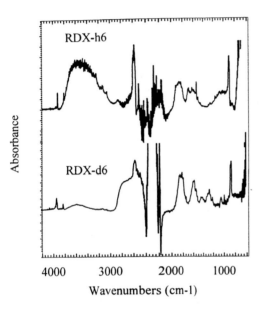

Figure 3. Spectra post pyrolysis samples of RDX and RDX-d6. Note that the doublet in the 1700cm^{-1} region splits into a pair of doublets, indicating absorbing species which contain and do not contain hydrogen are both present

CONCLUSION

The high pressure matrix isolation technique described in this paper has been used to investigate the laser induced decomposition of RDX. RDX sensitivity towards laser induced initiation at 532 nm is greatly affected by the physical state of the material (as modified by increased pressure). We feel this pressure dependence of the initiation is demonstrative of the efficiency of energy input into the system. At higher pressures, specifically above the elastic/ plastic transition, the energy input required to initiate decomposition is minimized. After initiation, final products as well as reaction

ACKNOWLEDGMENTS

The research was conducted under funding from the Office of Naval Research and the Naval Research Laboratory. The National Research Council provided funding for Tod R. Botcher as a post-doctoral fellow.

REFERENCES

[1]. Botcher, T. R.; Ladoucer, H. D.; and Russell, T. P., *Rus. Journ. Of Chem. Phys.* 1996,(Submitted). And references therein

[2]. Ramaswamy, A. L. "Laser Ignition of Secondary Explosives", *Thesis*, June 1993, Cambridge England. And references therein.

[3]. Ladouceur, H. D. ; Russell, T. P. *J. Appl. Phys.* Submitted.

[4]. *Thermophysical Properties of Matter* (IFI/Plenum: New York, 1970).

[5]. Rice, J. K. ;Russell, T. P. *Chem Phys. Lett.*, **1995**, *234*, 195.

[6]. Eremets, M. I. *High Pressure Experimental Methods*, (Oxford University Press: New York, 1996).

CP429, *Shock Compression of Condensed Matter – 1997*
edited by Schmidt/Dandekar/Forbes
© 1998 The American Institute of Physics 1-56396-738-3/98/$15.00

ION-BEAM DRIVEN SHOCK DEVICE USING ACCELERATED HIGH DENSITY PLASMOID BY PHASED Z-PINCH

Kazuhiko Horioka

Department of Energy Science, Tokyo Institute of Technology, 4259 Nagatsuda, Yokohama 464, Japan

Tatsuhiko Aizawa and Minoru Tsuchida

Department of Metallurgy, University of Tokyo, 7-3-1 Hongo, Tokyo 113, Japan

A high density Z-discharge plasma in a tapered capillary tube was electro-magnetically accelerated by a phased implosion of a capillary Z-pinch. The implosion was just about timed to the plasmoid drift by the shaped capillary wall. The feasibility of high energy acceleration was experimentally demonstrated using a 100 mm long thin capillary and a fast pulse power generator. For filling gas of 100 Pa of Ar, the axial drift velocity of the plasma was 7×10^7 cm/s, which corresponds to 70 keV argon atoms.

INTRODUCTION

Three methods are well-known to generate high shock pressure by acceleration of high density plasma or particles: intense ion beams [1], plasma gun [2] and rail gun [3]. The first candidate has a space charge limit so that sufficient current density can never be obtained. The second method utilizes only low density plasma with the order of 10^{13} cm^{-3} so that a sufficient number of particles can never be accelerated. The final approach has its speed limit to be realized and often suffers from severe erosion. Hence, a new frontier awaits to propel shock physics and chemistry by using the high density plasma. The phased implosion method [4] is preferable to acceleration of plasma and particles since both the implosion or pinch time and its scheduling can be precisely controlled. In the present paper, a new timed Z-pinch method is proposed as a new device to generate high shock pressure [5]. In the present

method, the plasma density can be compressed to the order of 10^{18} to 10^{19} cm^{-3}, and high density plasma can be accelerated with axial shock pressure, resulting in high-velocity launching of a flyer. In the present paper, systematic experiments are performed to demonstrate that high energy plasma flow can be electro-magnetically driven by the timed capillary Z-pinch, and to characterize the ion velocity and its current density. The estimated value of ion speed from the plasma measurement reaches 7×10^7 cm/s corresponding to 70 to 100 KeV for Ar. Furthermore, laser doppler velocimetry is used to measure the flyer velocity and to investigate the relationship between the accelerated plasma and the generated shock pressure.

PHASED Z-PINCH IMPLOSION IN CAPILLARY

To overcome the intrinsic limits of plasma density

and velocity to the conventional implosion methods, a new concept for high density plasma acceleration is proposed on the basis of a simple calculation of pinching time of Z-pinched plasma. Axially phased implosion of the current sheet acts as a virtual piston to compress the plasma and drive a high energy directional plasma flow along the total length of a narrow discharge channel at very high current level. This type of acceleration has no intrinsic speed limit and can continuously create impulse on the plasma along the total length of a tube. Using the snowplow model to time response of the Z-pinched plasma, the pinch time τ_p of coaxial discharge is estimated by

$$\tau_p = (\pi r_0^2/I_0)\,(N_i m_i/\mu_0) \quad \sim \quad (r_0^2/I_0), \qquad (1)$$

where $N_i m_i$ is the initial gas density, μ_0 the permeability in vacuum, I_0 the discharge current, and r_0 the initial radius. Since the rapid compaction drives a plasma of small radius with high current level in pinching and the pinch time strongly depends on r_0 in the above equation, the control of pinch time by slight shaping of the discharge wall enables us to accelerate the plasmoid by the timed pinching. Its fundamental mechanism is schematically depicted in Fig. 1. A high current fast Z-discharge with the order of 10^5 A leads to generation of an azimuthal magnetic field B_θ, enough to accelerate the plasma radially inward.

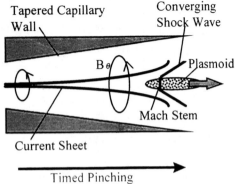

Tapered Capillary Wall

Converging Shock Wave

B_θ

Plasmoid

Mach Stem

Current Sheet

Timed Pinching

Fig. 1 Schematic diagram of phased Z-pinch implosion.

Due to accumulation and compression of the plasma up to GPa level by the snowplow effect, a high density plasmoid is formed at the discharge axis. The time delay for arrival of this plasmoid is controlled to the pinching process along the discharge axis; in addition to the inward pinching by B_θ the magnetic flux gradient B_z in the driving direction is also generated by the phased pinching. Then, the generated plasma can be radially compressed by the self B_θ and axially driven by B_z, simultaneously. When the plasma implosion time is exactly phase-matched to the plasmoid transit time, a significant part of the input energy can be converted to the kinetic energy to drive the plasma in the axial direction. This capillary plasma is expected to have an internal structure, where the current sheet drives a converging cylindrical shock wave ahead of itself. If this shock wave grows into the Mach stem, strong axial shock wave may be generated in the inside of the plasma column.

EXPERIMENTAL APPARATUS

A fast pulse generator LIMAY-I was used to drive the capillary discharge. A high voltage pulse was switched on the electrode after a weak pre-ionization discharge. Under this condition, the load current of 80 kA with a rise time of 20 ns can be driven from the 3 ohm - 70 ns pulse forming line through a pre-pulse suppression SF_6 switch. A schematic diagram of the present experimental set-up is depicted in Fig. 2. The capillary of 100 mm length has a thin conical wall, the inlet and exit diameters of which were 4 mm and 8 mm, respectively. This capillary was slightly tapered to implode sequentially and to compress the exit plasma within the pulse width (70 ns) of the discharge in the order of 10^5 A. Under this situation, the plasma is thought to move axially in a leaky-stream-tube mode without occurrence of discharge instability; hence, significant plasma accumulation can never take place. As shown in Fig. 2, the load current was monitored by a Rogowskii coil of self-integral type.

For differential pumping, the discharge section and the diagnosis chamber was separated by a pinhole of about 0. 2 mm diameter. In order to avoid the damage by the plasma, the pinhole was located at 30 mm from the exit of the capillary tube. The initial static densities of the inert gas were in the order of 10^{17} cm^{-3}, and the background pressure of the time-of-flight chamber was kept below 1.33×10^{-1} Pa (10^{-3} mmHg) by differential pumping.

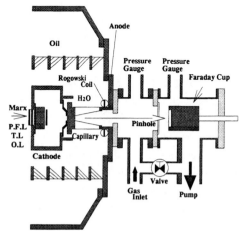

Fig. 2 Experimental arrangement for proof-of-principle experiments.

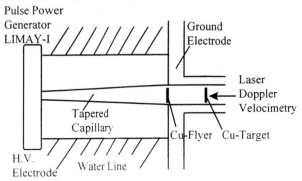

Fig. 3 Schematic diagram of preliminary experimental set-up for flyer acceleration.

In the powder compaction experiments, the set-up in the right-hand side from the exit of plasma was replaced with a recovery section. Figure 3 depicts a schematic view of a preliminary experimental set-up. To be noted here are: [1] Cu flyer is located

just at the exit of capillary to make full use of plasma acceleration, [2] Flyer velocity history is monitored by using the laser doppler velocimetry.

EXPERIMETAL RESULTS

The peak of the plasma signal was found to be shifted as a function of distance and amount of gas loading. When the initial filling pressure was set to be 100 Pa in Ar, the drift velocity of the plasma became about 7×10^7 cm/s. In other words, Ar atoms can travel with the kinetic energy of about 70 keV. The measured current density through a pinhole, which was located at 30 mm from the exit, was 200 A/cm^2. Since the plasma expands rapidly from the exit of the capillary, actual current density of the plasma flux at the capillary tube exit must be more than 200 A/cm^2. In order to check whether the high energy plasma might be caused by the Z-

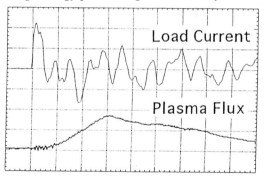

(a) Case when a positive load current was applied.

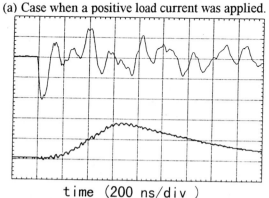

time (200 ns/div)

(b) Case when a negative load current was applied.

Fig. 4 Polarity effect on the ion flux signal.

discharge instability, the polarity of the pulse power generator was inverted. As shown in Fig. 4, the plasma signal had the same form in spite of the polarity inversion. This assures that the high energy plasmoid should be purely electro-magnetically accelerated but not by the anormalous electric fields.

A thin Cu metal plate with 0. 1 mm thickness, 5 mm diameter and 99. 9 % purity was placed at the exit of the capillary tube. The charging energy E_g in the condenser bank or the pulse power generator was varied to control the flyer velocity V_f. Figure 5 shows the variation of the Cu flyer velocity with E_g. Monotonic increase of V_f with E_g assures successful controllability of this shock loading device. When $E_g = 1. 1kJ$, the flyer was accelerated up to 2. 5 km/s. This implies that about 5 % of E_g ($\sim$ 50 J) was converted to the kinetic energy of the small projectile.

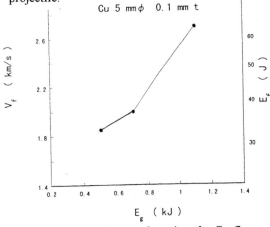

Fig. 5   Acceleration test by using the Cu-flyer.

DISCUSSION

High density plasma can be electromagnetically accelerated in a tapered capillary discharge. Hence, the drift velocity of high energy plasma by 7×10^7 m/s is just corresponding to the kinetic energy of Ar particles with 70 keV. The present idea may be basically related to a pinch engine [6] or a particle acceleration by zippering of plasma focus discharges [7].

The point to be noted is that the plasma compression and acceleration should be controlled by the tapered capillary discharge.   In other words, the plasma implosion is exactly synchronized to the plasmoid drift by a properly shaped wall including the effect of plasma accumulation along the axis. Since the plasma acceleration has close relationship with the sweep velocity of pinching, both the initial background gas pressure and wall shape of capillary tube have significant influence on the implosion instability.

CONCLUSION

The present ion beam shock loading device by phased implosion of capillary Z-pinch is capable to accelerate relatively dense plasma in the order of 10^{19} cm^{-3} up to sufficiently higher velocity than 10^8 cm/s.   If the implode configuration is constructed to strengthen the Mach stem, it can also drive the axial plane shock wave. Both cold and hot alloy-powder compaction experiments are taking place to investigate the effect of pulse intensity and duration time on the shock compaction behavior.

ACKNOWLEDGEMENTS

This study is financially supported in part by   the Pioneer-Research Fund from JAERI (Japan Atomic Energy Research Institute).

REFERENCES

[1] K. Baumung et al., *Shock compaction of condensed matter.*   AIP press 1015 (1995).

[2] S. Humphries et al., Rev. Sci. Instrum. **52** 162 (1981).

[3] D.A. Tidman et al., J. Appl. Phys., **51** 1975 (1980).

[4] T. Hosokai et al., Jap. J. Appl. Phys. (in press).

[5] K. Horioka et al., Proc. 4th Int. Conf. Dense Z-pinches (1997, Vancouver) (in press).

[6] P.J. Hurt, J. Appl. Phys., **27** 436 (1960).

[7] T,W, Hussey et al., J. Appl. Phys., **59** 2677 (1986).

CP429, *Shock Compression of Condensed Matter – 1997*
edited by Schmidt/Dandekar/Forbes
© 1998 The American Institute of Physics 1-56396-738-3/98/$15.00

USE OF Z-PINCH SOURCES FOR HIGH-PRESSURE SHOCK WAVE EXPERIMENTS

C.H. Konrad, W.M. Trott, C.A. Hall, J.S. Lash, R.J. Dukart, B. Clark, D.L. Hanson, G.A. Chandler, K.J. Fleming, T.G. Trucano, L.C. Chhabildas and J.R. Asay

*Sandia National Laboratories,
Albuquerque, New Mexico 87185-1181*

Recent developments have demonstrated the use of pulsed power for producing intense radiation sources (z-pinches) that can drive planar shock waves in samples with spatial dimensions significantly larger than possible with other radiation sources. In this paper, we will discuss the use of z-pinch sources for shock wave studies at multi-Mbar pressures. Experimental plans to use the technique for absolute shock Hugoniot measurements and with accuracies comparable to that obtained with gun launchers are discussed.

Introduction

Shock wave techniques have long been a principal tool for determining the high pressure (EOS) of materials in regimes inaccessible by other methods (1). A variety of shock wave techniques have been developed for producing well-controlled shock planar shock waves to study dynamic material response. For ultra-high EOS measurements, underground nuclear tests (2) have been used to produce high shock wave pressures. Recently, there has been considerable interest in using energy deposition techniques for shock wave measurements. Evans et al. (3) have developed direct deposition laser techniques using impedance matching to produce shock waves in copper to pressures of about 20 Mbar. Baumung et al. (4) are developing proton beam techniques to launch thin flyer plates to high velocities. Most recently, Cauble et al. (5) have developed a technique for making absolute measurements of the equations of state of low atomic number materials in regimes of extremely high pressure. Fortov et al.

(6) have used x-radiation from a Z pinch source to develop uniform shock waves up to 3 Mbars for EOS applications. We are developing methods for making precise time-resolved measurements of planar shock waves produced by Z pinch techniques. This technique uses imploding metal plasma produced by self-magnetic fields to produce high temperature x-ray environments in hohlraum enclosures to drive shock waves through an ablation process. Previous experiments have demonstrated that planar shock waves can be produced with this approach. (7). Stepped aluminum samples, ranging in thickness from 100 μm to 300 μm thickness and with lateral dimensions of about 9 mm in diameter, are used as samples. VISAR interferometry is used to measure particle velocity behind the shock wave and fiber optic breakout "pins" are used to determine shock velocity between prescribed steps in the target. The combined diagnostics provide an absolute measurement of the shock Hugoniot and also allow EOS information on the isentropic unloading response.

Sandia is a multiprogram laboratory operated by Sandia Corporation, a Lockheed Martin Company, for the United States Department of Energy under Contract DE-AC04-94AL85000.

Experimental Technique

The target, Figure 1a, consists of an aluminum disk top hat design. A Lithium Fluoride (LiF) window, 3 mm thick with a diffused, vacuum deposited aluminum mirror is bonded to the top hat. The target is mounted to a secondary hohlraum, as shown in Figure 1b.

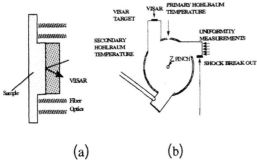

(a)   (b)

Figure 1. Experimental technique for shock wave studies on Z. (a) target configuration, (b) hohlraum design.

Additional secondary hohlraums are used to detect shock planarity and breakout. Fiber optic probes are used on the first step to measure shock arrival. The shock breakout probes consist of an active and return fiber set approximately 0.5 mm off the surface. The reduction of light due to shock arrival is coupled through the return fiber to a streak camera. A narrow bandpass (1 nm FWHM) filter centered on the laser line frequency was inserted in the return fiber prior to the streak camera to reduce the effect of fiber fluoresce. The alignment of the fibers to a spectral surface is very difficult and not dependable. Because of this, a diffuse surface is used to reflect light into the return fiber. Additional techniques such as a multifiber bundle are also being considered.

A miniature fiber optics probe has been developed to focus the incident VISAR beam on the LiF target interface and to collect the reflected diffuse light for use in the VISAR. The probe has an efficiency of 10% - 12% and has been used successfully for VISAR measurements.

All VISAR diagnostics were tested in the radiation environment of the Z accelerator. The experiments consisted of an irradiating 200 micron core, radiation hardened fiber within 100 mm of the z-pinch load. This resulted in a fluorescence signal that routinely saturated the detectors. Fiber darkening was tested by passing an argon ion laser beam down the fiber and observing the decrease in transmission during and after the radiation dose. Both effects were mitigated by installing a 1.2 nm FWHM spectral filter centered at 514.5 nm, which reduced the fluorescence signal by factors of 100-1000.

Other tests examined the use of LiF shock windows and miniature glass lenses as components of VISAR probes. The LiF windows darkened in about 20 ns to a transmissivity of 60% and recovered in ~ 400 ns, while the lenses darkened at about 100 ns after z pinch.

Optical fibers provide a convenient means to transmit diagnostic signals in experiments utilizing a high-energy pulsed power source such as PBFA-Z. A concern is the temporal dispersion that occurs in optical waveguides. This phenomenon must be considered and characterized in any experimental seeking < 1ns resolution.

Neglecting material dispersion or absorption, the maximum pulse spread t_d is given by (8):

$$t_d = \frac{z}{c} n_{co} \left\{ \frac{n_{co}}{n_{cl}} - 1 \right\}$$

where z is fiber length, c is the speed of light, and n_{co} and n_{cl} represent the refractive indices of the fiber core and cladding, respectively. For a multimode fused silica fiber this relation yields a maximum pulse spread of 56 ps/m. Observed values are likely to fall closer to the r.m.s. width (8) which is ~ $t_d / \sqrt{12}$; a spread of 16 ps/m. The ray dispersion equations suggest that intensity-based measurements can be made at temporal resolutions approaching 100 ps provided that fiber lengths are limited to ~10 meters. More stringent constraints exist in applications involving velocity interferometry, where the fiber dispersion must be considered in relation to the desired optical time delay. In tests with an ORVIS, we found that accurate profiles can be recorded when the signal is coupled through short fiber sections; coupling through a long fiber significantly degrades performance.

The fringe records in Figure 2 were generated by laser-accelerated flyer plates launched under nearly identical driving conditions. Fig. 2a exhibits the typical response of a conventional, open-beam ORVIS using, in this case, an optical time delay of 228 ps (corresponding to a velocity-per-fringe constant of 1.09 km-s^{-1}). In Fig. 2b, light reflected from the accelerating target was coupled through a 41.5-m-long, 200-μm-diameter fiber. This element introduces a nominal r.m.s. pulse spread of 0.67 ns. With this relatively large dispersion (three times longer than the interferometer delay time), ORVIS was clearly unable to record the correct fringe motion, completely "losing" the appropriate fringe positions at onset of flyer motion and in a later time period (as well as apparently generating anomalously low slopes at other times). The deleterious dispersion effects were found to be even more pronounced in witness plate tests producing particle velocity pulses <6 ns in duration. Experiments to define the limits on fiber length relative to the fringe constant parameter of the interferometer are in progress.

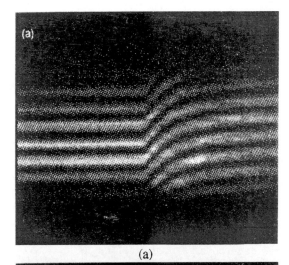

(a)

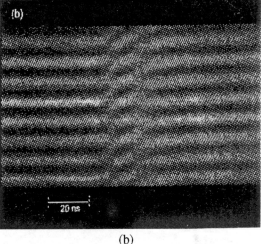

(b)

Fig. 2. ORVIS velocity-time records of laser-driven flyers. Time proceeds from left to right. Fringe Constant is 1.09 km-s^{-1}.(a) open-beam ORVIS; (b) fiber optic coupled signal.

Numerical Simulations

We simulated the ablation pressures in various materials using a 15 ns wide trapezoidal radiation source (Planckian) in the ALEGRA (9) code. This code utilizes an implementation of the SPARTAN SPN package developed by Morel and Hall (10,11). The calculations were run in 1-D Cartesian geometry. Sixty energy groups over the range to 10 keV were used. Materials studied included CH, Al, Ti, and W, using SESAME EOS tables and opacity determined by XSN. Peak temperatures for the assumed radiation pulse were chosen over the interval of 50 to 200 eV, which is characteristic of the secondary radiation hohlraums.

The purpose of this study was to investigate ablation pressure scaling with atomic number of the target. Two types of calculations were undertaken. First, direct illumination of the targets was performed. Second, calculations were done with a CH ablator..

For direct illumination, we found that ablation pressures increased more rapidly with peak source temperature than expected for tungsten and titanium For the lower Z materials, our results were in reasonable agreement with both Dukart's calculations and analytic approximations to the scaling behavior (12). Numerical calculations of pressure wave profiles expected in aluminum for a 100 eV x-ray temperature are shown in Figure 3.

Summary

In summary, we are developing a new diagnostic using z pinch sources for producing high pressure shock waves. VISAR measurements of particle velocity and fiber optic measurements of particle velocity and shock speed are used for absolute EOS measurements. Preliminary data obtained with the technique are encouraging.

References

1. Asay, J.R., and G.I. Kerley, J., *Impact Engng,.* **5**, 69 (1986).
2. Ragan, C.E., III, *Phys. Rev.* **A25**, 3360 (1982).
3. Evans, A.M., et al, *Laser and Particle Beams* **14**, 113 (1996).
4. Baumung, K., et al.,ibid, 181.
5. Cauble, R., et al., preprint, submitted to Physical Review Letters (1997).
6. Fortov, V., et al., *Shock Comprssion of Condensed Matter 1995,* Proceedings 370,Part 2,1255-1258,1995.
7. Olson, R.E., et al. Submitted to Phys. Plasmas, Volume 4, May1997
8. Snyder, A.W., and Love, J. D., *Optical Waveguide Theory*, NY, Chapman and Hall, 51-88.(1983)
9. Budge, K.G. and Peery, J.S. *J. Impact Engng..* **14**, 107-120 (1993).
10. Morel, J.E., Dendy, Jr., M.L., Hall, and S.W. White, Differencing Scheme, *Computational Physics* **103**, 286-299(19xx).
11. Morel, J.E., preprint (1997).
12. Dukart, R.J., Private communication (1997).

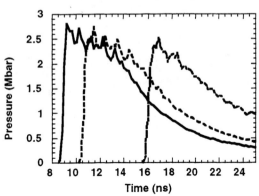

FIGURE 3. Pressure profiles in aluminum for a 100eV drive at distances of 75, 100 and 175 μm.

AUTHOR INDEX

A

Abeln, S. P., 411
Ahrens, T. J., 115, 133, 859
Aizawa, T., 651, 993
Alcon, R. R., 575, 739, 845
Anderson, Jr., C. E., 897, 925
Anderson, M. U., 553, 841
Anderson, W. W., 115
Arad, B., 459
Armstrong, R. W., 215, 471, 563
Artemieva, N. A., 957
Asay, B. W., 273, 289, 567, 571, 579, 805, 887
Asay, J. R., 837, 997
Atou, T., 151
Auroux, E., 985

B

Baker, E. L., 357
Barbee III, T. W., 35
Bardenhagen, S. G., 281
Barker, A. K., 513
Barker, L. M., 833
Barrett, J. J. C., 293, 297, 329
Baudin, G., 687
Bauer, F., 789
Baumung, K., 155
Bedford, A., 231
Belak, J., 211
Bennett, B. I., 887
Bennett, J. G., 273
Bernecker, R. R., 719
Bessette, G. C., 937
Biele, J. K., 891
Billingsley, J. P., 199
Bless, S. J., 231, 489, 925, 929, 933
Blumenthal, W. R., 411, 583
Boat, R. M., 887
Boettger, J. C., 129
Bogach, A. A., 223, 443
Bogatch, A. A., 447
Bordzilovsky, S. A., 545
Borschevsky, A. O., 95
Botcher, T. R., 989

Boteler, J. M., 537
Bourne, N. K., 137, 493, 533, 801, 849
Boustie, M., 219, 985
Bowman, J. D., 887
Brar, N. S., 451, 497, 521, 855
Brockman, R. A., 855
Brower, K. R., 699, 731
Brown, J., 333
Browning, R. V., 277
Budil, K. S., 55
Bukiet, B., 321
Burchell, M. J., 949
Burnside, N. J., 571

C

Cady, C. M., 583
Campbell, E. M., 3
Campos, J., 341, 691, 715, 755
Cannon, D. D., 411
Carpenter, W. R., 411
Carton, E. P., 549, 619
Cauble, R., 3, 55
Celliers, P., 55
Chambers, G. P., 397, 695
Chandler, G. A., 997
Chandler, J. B., 711
Chang, S. N., 415
Che, R., 99
Chen, G. Q., 133
Chen, S. R., 435
Chhabildas, L. C., 119, 501, 505, 557, 781, 841, 997
Chidester, S. K., 707
Cho, Y. S., 401
Choi, J. H., 415
Choi, K. Y., 369
Christopher, F. R., 389
Clark, B., 997
Clifton, R. J., 463, 475, 517, 521
Coffey, C. S., 563
Collins, G. W., 55
Cook, M. D., 305, 337
Corey, E. M., 43
Cottenot, C. E., 427
Counihan, P. J., 419

Covington, B., 981
Cowperthwaite, M., 385
Crawford, A., 419
Curran, D. R., 945
Curtis, J. P., 333

D

Dandekar, D. P., 525
Da Silva, L. B., 55
Davis, J. J., 663, 667
Davis, L. L., 699
Davis, R. F., 643
DeCarli, P. S., 681
Deiter, J. S., 667
Dekel, E., 459, 917
Demol, G., 353
Denoual, C., 251, 427
de Resseguier, T., 219, 985
Dey, T. N., 285
Dick, J. J., 827
Dick, R. D., 471, 591
Dickson, P. M., 567, 805, 887
Dionne, J. P., 317
Dlott, D. D., 819, 823
Dobler, E. A., 635
Donnelly, D., 981
Dragon, A., 179
Droughton, J., 31
Dudin, S. N., 489
Dudin, S. V., 793
Dukart, R. J., 997
Duprey, K. E., 475
Durães, L., 341
Dwivedi, S., 901

E

Elban, W. L., 563
Elert, M. L., 293, 297, 329
Eliaz, N., 459
Eliezer, D., 459
Eliezer, S., 459
Embury, J. D., 423
Espinosa, H. D., 431, 497, 901
Evans, A. M., 79

A1

SUBJECT INDEX